T0142196

Lecture Notes in Networks and Systems

Volume 87

The series "Lecture Notes in Networks and Systems" publishes the latest developments in Networks and Systems—quickly, informally and with high quality. Original research reported in proceedings and post-proceedings represents the core of LNNS.

Volumes published in LNNS embrace all aspects and subfields of, as well as new challenges in, Networks and Systems.

The series contains proceedings and edited volumes in systems and networks, spanning the areas of Cyber-Physical Systems, Autonomous Systems, Sensor Networks, Control Systems, Energy Systems, Automotive Systems, Biological Systems, Vehicular Networking and Connected Vehicles, Aerospace Systems, Automation, Manufacturing, Smart Grids, Nonlinear Systems, Power Systems, Robotics, Social Systems, Economic Systems and other. Of particular value to both the contributors and the readership are the short publication timeframe and the world-wide distribution and exposure which enable both a wide and rapid dissemination of research output.

The series covers the theory, applications, and perspectives on the state of the art and future developments relevant to systems and networks, decision making, control, complex processes and related areas, as embedded in the fields of interdisciplinary and applied sciences, engineering, computer science, physics, economics, social, and life sciences, as well as the paradigms and methodologies behind them.

**** Indexing: The books of this series are submitted to ISI Proceedings, SCOPUS, Google Scholar and Springerlink ****

More information about this series at http://www.springer.com/series/15179

Elena G. Popkova · Bruno S. Sergi
Editors

Digital Economy: Complexity and Variety vs. Rationality

Editors
Elena G. Popkova
Plekhanov Russian University of Economics
Moscow, Russia

Bruno S. Sergi
Harvard University
Cambridge, USA

University of Messina
Messina, Italy

ISSN 2367-3370 ISSN 2367-3389 (electronic)
Lecture Notes in Networks and Systems
ISBN 978-3-030-29585-1 ISBN 978-3-030-29586-8 (eBook)
https://doi.org/10.1007/978-3-030-29586-8

This Springer imprint is published by the registered company Springer Nature Switzerland AG
The registered company address is: Gewerbestrasse 11, 6330 Cham, Switzerland

Introduction

Formation of the digital economy is one of the most contradicting tendencies of modern times. On the one hand, automatization allows increasing effectiveness and developing the production of complex goods, ensuring high precision of spending the resources and minimizing production and consumption waste. Execution of routine processes by machines opens a wide field for creative activities of a modern human: from production management to design of new technologies, processes, and systems.

On the other hand, a new type of thinking—digital—is formed. The sociocultural factors, which become less significant under the influence of globalization, go to the background during decision-making. Thus, the digital economy challenges the idea of humanism and returns human society to "natural selection," which result is distorted treatment of social justice as victory of the strong over the weak.

One of the most obvious problems of the digital society is the reduction of its inclusiveness, which is one of the key achievements of civilization and socioeconomic progress. Growth of social differentiation and increase of differentiation of countries as to the level of economic development could increase the imbalance of the modern global economic system and cause its crisis, which would cover the economic, ecological, and social spheres.

The signs of formation of the digital economy are observed in all modern economic systems of any level and geographic position. That is why the scientific and practical problem of search for the methods of solving the contradiction of the digital economy and provision of its balance becomes very important. In the presented volume, complexity and manysidedness of the digital economy are opposed to its rationality and unidirectionality, and a strategic scientific and methodological approach to state and corporate management of the digital economy for provision of its functioning and development in the interests of the society is developed.

The volume contains the most progressive and science-driven research works that were presented at the 9th National Scientific and Practical Conference "Digital Economy: Complexity and Variety vs. Rationality," which took place on April 17–18 in Vladimir, Russia. The target audience of the volume include representatives of the academic society, researchers, and experts who are interested in the issues of building the digital economy and the digital society; undergraduates and postgraduates of various specialties; and companies that use or are interested in using digital technologies in their economic practice; public officers and government workers, who conduct monitoring, control, and regulation of the digital economy.

The volume consists of seven parts:

- The Digital Economy as a Perspective Direction of Modernization of Economic System;
- The Tools for Building the Digital Economy in Russia;
- Specific Features and Perspectives of Digital Modernization of the Regional Economy;
- Social Consequences of Transition to the Digital Economy;
- The Financial Component of Formation of the Digital Economy;
- The Legal Challenges of Digital Reality for Society and State;
- The Main Challenges and Threats to the Profession of Jurisprudent in the Conditions of Economy's Digitization.

The conference was organized by the Institute of Scientific Communications (Volgograd, Russia) and Vladimir State University named after Alexander and Nikolay Stoletovs (Vladimir, Russia). The conference organizers received more than 700 applications, and more than 1,000 scholars from various universities and research institutes of Russia participated. The best works, which contribute into the development of the theory and practice of the digital economy, were selected.

It is necessary to note the work "Global Competitiveness of the Digital Economy: the Problem of Measuring and Management" (Irina V. Mukhomorova, Elena S. Akopova, L. K. Pavlova, and V. V. Sheveleva), which formulates and successfully solves one of the important problems of the modern national economy, region, and company—determining the competitiveness of the forming and developing digital economy and increasing its level.

Special gratitude should be expressed to Elena V. Belokurova, Sergey V. Pizikov, Elena S. Petrenko, and Gaukhar K. Koshebayeva for their work "The Institutional Model of Building the Digital Economy in Modern Russia." The scholars consider the institutional processes that take place in the economic system in the conditions of formation and development of the digital economy and offer a complex of recommendations for effective management of these processes.

In the paper "Contradiction of the Digital Economy: Public Well-Being vs. Cyber Threats," Elena G. Popkova and Kantoro Gulzat oppose social and technological manifestations of the digital economy and develop a method of solving their contradiction for simultaneous increase of effectiveness of economic activities and population's quality of life in the conditions of the digital economy, as well as support for its sustainability, stability, and security in the long term.

In the work "Specific Features of Strategic Planning of the Activities of Entrepreneurial Structures in the Conditions of Digital Transformation of the Modern Economy," Alexandr V. Tekin and Olga V. Konina show the causal connections of digital modernization of a modern company and offer recommendations for managing this process for protecting the interests of all interested parties.

Each chapter of this volume deserves attention and contains results that contribute to scientific knowledge in the sphere of the digital economy.

Contents

The Legal Challenges of Digital Reality for Society and State

The Digital Economy as a Perspective Direction of Modernization of Economic System

Modernization of the Russian Agro-Industrial Complex in the Conditions of Increase of Food Security

Alexandra V. Glushchenko[1(✉)], Gilyan V. Fedotova[2,3], Natalya V. Gryzunova[4], Shakhnoza S. Sultanova[5], and Viktoria M. Ksenda[6]

[1] Volgograd State University, Institute of Economics and Finance, Volgograd, Russian Federation
audit@volsu.ru
[2] Volga Region Research Institute of Production and Processing of Meat and Dairy Products, Volgograd, Russian Federation
g_evgeeva@mail.ru
[3] Volgograd State Technical University, Volgograd, Russian Federation
[4] Plekhanov Russian University of Economics, Moscow, Russian Federation
nat-nnn@yandex.ru
[5] RUDN University, Moscow, Russian Federation
shahnozia@mail.ru
[6] Volgograd State University, Volgograd, Russian Federation
KsendaVM@volsu.ru

Abstract. Subject/Topic. The issues of increase of food security of the state have become very important due to increase of the planet's population and the requirements of increase of living standards. The necessity for increasing the intensity of production of food products in the conditions of stable production capacities of agriculture grows. In this situation, agrarians look for new innovational solutions to the problem of increase of the production volumes without large growth of final products' price. Another problem is growth of the cost of food products, which also requires the search for new technological solutions for increasing the efficiency in agriculture. Certain states strive to reduce their export of food – due to growth of prices – satisfying their own need for food and food reserves. However, such measures only increase the problem in the global scale – so there's a necessity for the global modernization of the sphere.

Thus, application of innovational technologies in agriculture could become a new vector for modernization of the traditional agricultural methods of production and activation of hidden potentials of the sphere.

Goal/Tasks. The purpose of this research is theoretical substantiation of the current processes, search for practical methods of stimulation and increase of efficiency of the Russian agriculture, and development of the directions of further modernization of the sphere. According to the set goal, the authors formulate and solve the following tasks: analysis of the situation with the dynamics of the world population, analysis of the consumer sector and its profitability level, evaluation of dynamics of population's real incomes, conclusion on the number of the poor, and substantiation of the necessity for search for new solutions for the problems of food security.

E. G. Popkova and B. S. Sergi (Eds.): ISC 2019, LNNS 87, pp. 3–12, 2020.
https://doi.org/10.1007/978-3-030-29586-8_1

Methodology. The research is conducted with application of the methods of graphic presentation of information, statistical analysis of data, financial analysis, trend analysis, and the methods of comparison, analogy, and systematization.

Results. To increase the level of the Russian food security, it is necessary to provide economic access to high-quality food for the Russian population, which is impossible without government support for agricultural manufacturers and new solutions in this sphere.

Conclusions/Significance. The value of the work consists in emphasis on external and internal threats to national food security, which increase despite the growth of external import.

Keywords: Food · Modernization · Population · Food security · Food reserves · Incomes

JEL Code: I31 · I38 · P23 · P36

1 Introduction

The performed evaluation of the level of modern development of the Russian agricultural sphere in view of the tendency for building the information society and information economy allows showing the importance of modernization of the traditional sphere. The current digital transformation of agriculture is a perspective direction of development of the sphere, which allows solving a lot of problems connected to ineffectiveness and low profitability of production. Absence of the normative and legal regulation complicates transition to digital technologies, but informatization tools are already implemented in practice.

Implementation of IT technologies at all stages of the agricultural cycle allows for rationalization of the whole process – from preparation to sale of final products. Separate attempts of the Russian farmers to implement new tool in work do not provide a large-scale effect, as there's a necessity for systemic planning and state support for digitization. The work in this direction is already conducted by the Ministry of Agriculture of the RF, but there's a project of the national program "Digital agriculture" and a road map FoodNet (Smart agriculture).

The proclaimed wide transition to the digital economy led to the necessity for development and transformation of the existing systems of provision of society's life. The spheres of the AIC – as one of the strategic complexes of the national economy – are no exception. In these conditions, there's a task of building a new model of the ecosystem of digital transformation of agriculture, which will stimulate building "smart", "rational", and partnership relations in the processes of exploitation of information resources in favor of society. For this, a systemic analysis was performed, which took into account the influence of challenges and threats to the level of digitization of agriculture, for further development of new tools and mechanisms of its implementation.

At present, there are three main variants of development of the digital economy: #DigitizeEU (international program of the EU for modernization of industry, 2011 – until now), Made in China 2025 (created on the basis of INDUSTRY 4.0, 2013 – until now), Digital economy (Russian national technological initiative, 2017 – until now).

The program, "Digital economy" was adopted by the Government of the RF in July 2017, and it will last until 2024. The sphere of agriculture is not among the top-priority spheres for digitization. In later 2017, the Ministry of Agriculture of the RF offered creating a national sub-program "Digital agriculture". The Ministry created the Analytical center, which dealt with monitoring of the state of agricultural lands and conducted negotiations with Roskosmos and Rosgidromet regarding creation of a common data base of space images and climate data. The largest agrarian universities open departments of digitization of agriculture, which are to train personnel in this sphere. The Ministry of Agriculture is going to adopt – at the government level – the program "Digital agriculture", for its inclusion into the program "Digital economy" [1]. The landmarks are certain target indicators that are set in this program for the whole period of its implementation (Table 1). The indicators reflect the implementation of the main measures for digitization of agriculture. According to their content, we see that digitization is to cover the companies of the sphere, the specialists in this sphere, cargo transportation in Russia and abroad, and the volumes of digitized agricultural lands.

Table 1. Target indicators of the project of the program "Digital agriculture".

Indicator	2018	2021	2024
Share of the AIC companies that use the ICT[a]	Less than 1%	20%	60%
Share of coverage of the agricultural lands with the ICT	Less than 10%	30%	70%
Share of the AIC companies that are equipped with the means of objective control and that transfer data for receipt of subsidies in the electronic form	Less than 10%	50%	100%
Products sold on electronic platforms	Less than 10%	50%	100%
Number of private meteorological stations on the agricultural production lands	Less than 1 million	3 million	7 million
Number of the AIC cargos that are transported within the EAEU with connection to the platform of transport and logistics	Less than 10%	50%	80%
Export	USD 25 billion	USD 30 billion	USD 50 billion
% of jobs that are connected to the ICT	<1%	8%	20%

[a]ICT – information and communication technologies.

Apart from the project of the national program, there is a road map FoodNet (Smart agriculture), presented by the Agency of Strategic Initiatives and the business society in 2017. The road map FoodNet (Smart agriculture) is a part of the National Technological Initiative [2]. The document is brought down to creation of an intellectual market of production and distribution of food and products with individual logistics. According to the plan, Russian companies are to account for 5% of the global market in five top-priority segments by 2030. They include "smart" agriculture (production uses authomatization, artificial intelligence, and Big Data), accelerated selection, accessible organics, and "new sources of raw materials" (e.g., processing of the biomass of sea grass and insects, implementation of pseudo grain crops, etc.), and personalized nutrition.

According to the Federal State Statistics Service, expenditures for implementation of the ICT into agriculture constituted 0.2% of the total volume of expenditures for the ICT in Russia (2017). This indicator is the lowest for the spheres, but the sphere of ICT has large potential, which is shown by the global practice and the experience of such leaders of agricultural startups as the USA, China, India, Canada, and Israel.

2 Materials and Methods

Theoretical and applied issues of evaluation of effectiveness of state management of economy and its spheres are studied in multiple works: (Aristovnik and Obadić [12]), (Bondarenko [14]), (Comerio and Batini [16]), (Lacko et al. [17]), (Clara et al. 2017), and (Fedotova et al. [6]).

Scientific and methodological isues of managing the process of informatization and modernization of socio-economic systems are studied in the works (Bogoviz et al. [13]), (Clara et al. 2017), (Nasir et al. [18]), (Sukhodolov et al. [19]), and (Fedotova et al. [21]).

The problems of increase of food security and food independence of the state are studied in (Gorlov et al. [6]) and (Fedotova et al. [7]).

However, despite the large number of publications on adjacent topics, the issues of search for economic solutions of the problem of food security and development of the directions of modernization of agriculture and the role of state management in this model are poorly studied. In view of search for solutions to the problem of food security and the indicators that are regulated at the national level, it is necessary to evaluate the situation with the modern position of the sphere of agriculture. These issues are studied in the article.

3 Results

Agriculture is a sphere of the agro-industrial sector of economy that is a complex system which includes the interconnected elements that create an integrated system. The sign of the openness of this system is a certain influence of external and internal environment on its work.

The concept of development of agriculture should be developed and implemented in view of the policy of formation of integrated information space of the RF, as digitization – as a regularity of development of the spheres of the national economy – is inevitable for Russia. In this case, it is necessary to envisage the following directions: digitization of the production process, digitization of social infrastructure of rural territories, and digitization of the process of training of skilled personnel for agriculture.

The imperatives for increasing the volumes of production in agriculture are the following arguments:

(1) The world population grows. In thirty years, the humankind will need 1.7 times more food than it produces now. This requires serious modernization of agriculture (Fig. 1). According to the UN forecasts, world population will each 9.8 billion by 2050; to feed it, it is necessary to increase food production by 70%.
(2) Growth of population requires larger crop and cattle production by intensification and increase of effectiveness with the same agricultural areas.

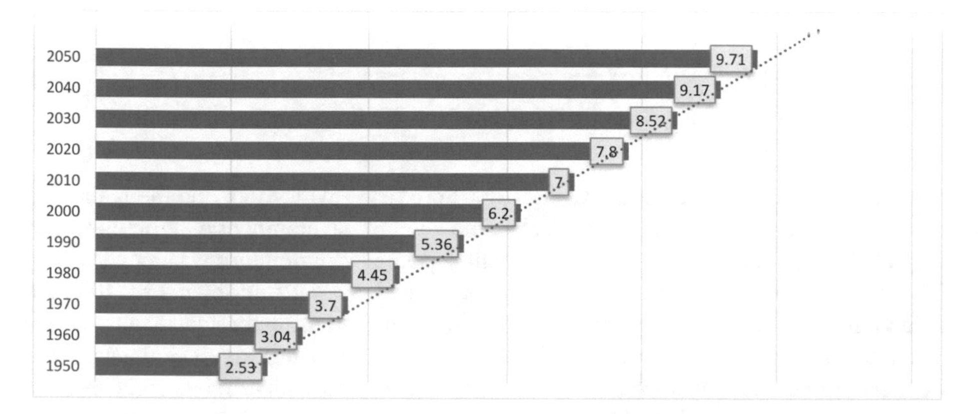

Fig. 1. Dynamics of population of the earth, according to the UN forecasts [3]

The above dynamics of the number of population of the Earth shows the necessity for intensification of agricultural production, which leads to implementation of a complex program of modernization of the whole sphere [4]. Assessment of economic accessibility of high-quality food products for the Russian population showed that the Russians have deficit of food, try to save money on food products, and try to buy cheaper food [5]. The existing situation should be amended, as high-quality food directly influences the nation's health, life span, and capability for reproduction (Fig. 2).

The data of the official statistics on the cost of grocery basket show that the main food products become more expensive, while incomes of the Russians do not grow. That's why any program of modernization of agriculture should be aimed at solving this problem – feeding people [7, 9].

The current world policy of economic sanctions against Russia clearly showed the existing problems of the Russian agro-industrial sector and high connection of the food sector to import of food products and the international food products. This position was further complicated by Russia's accession to the WTO.

Russia's accession to the WTO was not supported by the representatives of the AIC, as this led to reduction of weak peasant farms, unprofitable companies of food industry, growth of unemployment, absence of growth points in rural territories, and increase of social tension in society [8]. The Russian economy was not prepared for the

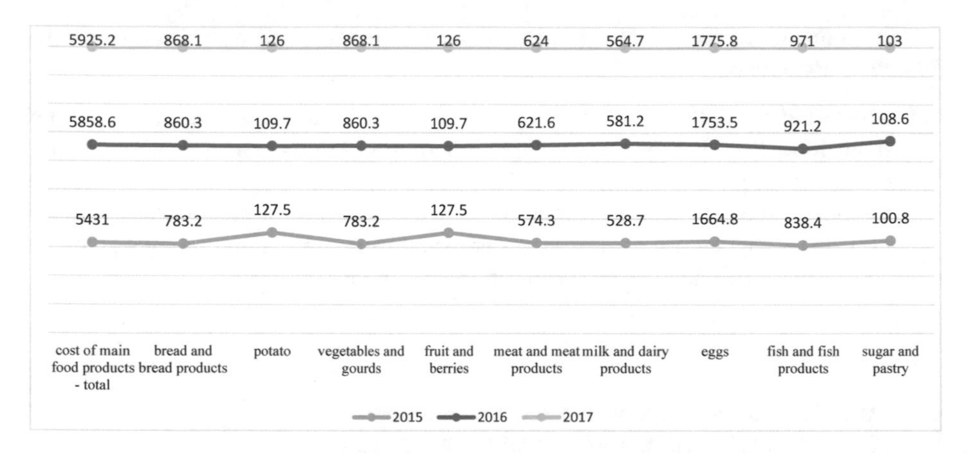

Fig. 2. Cost of grocery basket of Russians, RUB per month [6, p. 67].

conditions of the WTO, as there was no 100% provision with food, and the remaining part was compensated by food import; also, the level of development of agriculture was very low. As per the WTO rules, government support for agriculture was reduced – by 2018 the volume of government support dropped to USD 4.4 billion, and by 2020 the reduction will constitute 44%. Opening the Russian market to foreign manufacturers and reduction of tariff protection from 13–14% to 10.8% negatively influenced the Russian agricultural manufacturers, who cannot compete with foreign companies in prices, assortment, or technical equipment. These factors weaken food security – which is not just provision with food products but also their economic accessibility.

The only solution in this situation is to reduce retail prices – by more than 10–15% - and preserve or even increase marginality of business of agricultural manufacturers, preserving the quality of products [10, 11]. This will be possible in case of cardinal transformation of the whole process of production and sales of agricultural products – which is actually called digital transformation. Digital transformation should influence all spheres and sectors of agriculture – from preparation to production to selling final products to consumers.

The main obstacles on the path of successful informatization of the sphere are low level of authomatization and mechanization of peasant farms, absence of reserve funds for purchase of agricultural machinery, high risk of production due to climatic conditions, and absence of the legal base of regulation of the practice of technologies application.

Nevertheless, as of year-end 2017, Russia was ranked 15th in the world as to the level of digitization of agriculture, and the market of information and computer technologies was estimated at RUB 360 billion – according to the data of the Ministry of Agriculture of the RF, announced in February 2018 at the conference “Precise farming 2018” [10].

Unmanned flying vehicles are being developed in the sphere of agriculture even despite the unfavorable legal regulation. The most active participants of the market are *Bespilotniye Tekhnologii* (*Drone Technologies*) (Novosibirsk), *Geoskan* (St. Petersburg), *Avtonomniye Aerokosmicheskiye Sistemy - Geoservis* (*Autonomous Aerospace Systems – Geoservice*) (Krasnoyarsk), and *ZALA AERO* (Izhevsk) [11, 12, 18].

The above examples are scarce, so there's a need for complex modernization of the sphere through its large-scale digitization. In the digital environment, the main task of IT is maximum authomatization of all stages of the agricultural cycle for the purpose of reducing the losses, increasing efficiency of business, and distributing resources in the optimal manner. At the next stages, authomatization has to take the sphere to a higher level of digital integration. Thus, integration of the obtained data with various intellectual IT applications, which conduct the processing in real time, leads to a revolutionary shift in decision making for a farmer, providing the results of analysis of multiple factors and substantiation for further actions. The more sensors and field controllers are connected to the network and exchange data, the smarter the information system becomes and the more valuable information for user it can bring [17, 19, 20].

For example, in the course of the season a farmer has to make more than forty different decisions: which seeds to plant, how to treat them, how to treat plants with diseases, etc. Lack of information for decision making leads to the fact that up to 40% of the crop are lost in the process. Up to 40% of the crop are lost during crop harvesting, storage, and transportation. According to the scholars, 2/3 of the factors of losses could be controlled with the help of automatized systems of management (Hi-Tech Management).

Based on scientific calculations, the information system can create recommendations for treatment of plants or instructions for their automatic execution by robots.

For example, predictive analytical model helps to determine that increase of temperature by two degrees stimulates hatching of insects and increase of humidity above the optimal level might lead to an outbreak of disease. Management of these factors creates a real value of modeling the micro-climatic conditions: in case of a greenhouse it is possible to avoid increase of temperature; in case of a field it is possible to observe the plot and use chemicals against parasites. It is for the first time in the history of agriculture that a farmer is able to control natural factors, design precise business processes, and forecast the result with mathematical precision [13, 15].

The sphere of cattle breeding will also see changes in the aspect of transition from incident management to proactive management of the whole production cycle.

Thus, the problem of food security is expressed in economic inaccessibility of high-quality food products for 19.3 million Russians. This indicator has been growing for several years. For reducing the social tension, the Government of the RF adopted the "Concept of development of domestic food assistance", which defines the notion of domestic food assistance as direct supplies of food products for interested parties and transfer of finances for purchasing food products in view of food rations.

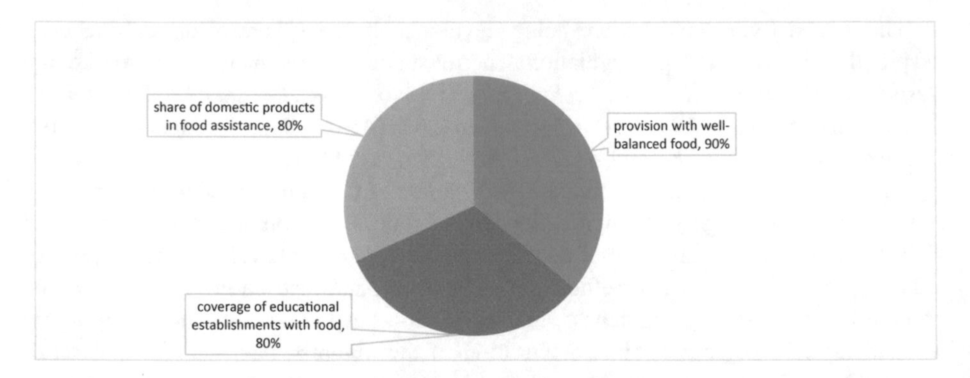

Fig. 3. The expected results of implementation of the Concept of development of domestic food assistance, % [20].

Figure 3 shows that the main directions of food assistance are provision of population with high-quality food, supply of hot food to educational establishments, and growth of the share of domestic products in food assistance.

The tools for implementing this policy are as follows:

- shipment of food products directly to recipients;
- issue of food products through the networks of social stores;
- organization of free dinners in social meal stations;
- issue of free food gift baskets;
- financing of purchase of certain food products (financial compensation).

The above mechanisms are able to provide population with high-quality food products for some time, but they are only temporary measures that depends on additional financing in the budget. The problem has to be solved in complex. It is necessary to increase domestic food base, implement the latest achievements of technologies in agriculture, and finance the development of new food supplements for enrichment of food products.

Growth of the volumes of production and export of domestic food resources should be reduced, as it is unprofitable. It is necessary to build and develop own processing base and to process raw materials, thus obtaining final products. Thus, it would be possible to satisfy domestic need for food and then enter the international markets of food products.

Implementation and further development of domestic food assistance to citizens allows solving the problem of economic accessibility of high-quality food products for the poor – but only for some time. However, the large-scale transformation of the system of product provision of population regardless of the social group will require complex modernization of the whole sphere of the Russian AIC.

4 Conclusion

It is necessary to bear in mind that the scales of implementation of digital technologies in agriculture directly depend on the general level of development of market relations in the agrarian sphere and on the state of information economy in the country on the whole. In Russia, this level does not conform to the modern requirements. Formation of digital resources is slowed down by lack of financial assets, in particular absence of conditions of stimulation of investments, absence of the practice of such work and experience, absence of knowledge on modern digital technologies, and incompetence of agrarians in this issue. That's why within this project it is necessary to use the accumulated experience [21]. Thus, transition to the digital economy, announced by the Government of the RF in July 2017 is an important vector of development of the modern state policy, which is to ensure the necessary level of development of all spheres of national economy of Russia. The process of transition and transformation of the existing systems of organization of the activities of agricultural companies requires transformation of the whole economic model according to the requirements of the Fourth industrial revolution and the concept "Industry 4.0".

References

1. Explanatory note to the offer on implementation of a new direction of the program "Digital economy of the RF". https://iotas.ru/files/documents/Пояснит.записка%20eAGRO%20fin%20000.pdf. Accessed 28 Dec 2018
2. The Agency for strategic initiatives to deal with "smart" agriculture. https://www.rbc.ru/business/07/09/2017/59afd0429a79473485bdb58c. Accessed 29 Dec 2018
3. Human population. Materials of the official web-site of the UN. http://www.un.org/ru/sections/issues-depth/population/. Accessed 28 Dec 2018
4. Decree of the President of the RF "Regarding adoption of the Doctrine of food security of the RF" No. 120 dated January 30 (2010). The information system "ConsultantPlus". http://www.consultant.ru/document/cons_doc_LAW_96953/. Accessed 22 Dec 2018
5. The National program of development of agriculture and regulation of the markets of agricultural products, resources, and food for 2013–2020. Decree of the Government of the RF dated June 14, 2012, No. 717. http://www.mcx.ru/. Accessed 25 Dec 2018
6. Gorlov, I.F., Fedotova, G.V., Sazonov, S.P., Sergeev, V.N., Yuldashbaev, Y.A.: The cognitive approach to studying the problems of food security: monograph, 168 p. Volgograd Institute of Management Publ., Volgograd (2018)
7. Fedotova, G.V., Chugumbaev, R.R., Chugumbaeva, N.N., Sukhinin, A.V., Kuzmina, E.V.: Increase of Economic Security of Internet Systems of Credit Organizations. The Future of the Global Financial System: Downfall or Harmony, vol. 57, pp. 922–931. Springer Nature Switzerland AG (2018)
8. Nidziy, E.N., Chugumbaev, R.R.: The issues of application of network integration of companies of textile industry in the conditions of dynamics of external environment. Bulletin of higher educational establishments, vol. 2, no. 374, pp. 55–58.The technology of textile industry (2018)

9. Decree of the Government of the RF dated July 3, 2014, No. 1215-r "The concept of development of domestic food assistance in the RF". Consultant Plus. http://www.consultant.ru/document/cons_doc_LAW_165323/eebc5130ddb6b6f16e6b900d50630c2b7b871b86/. Accessed 29 Dec 2018
10. Revenues, expenditures, and conditions of households. The data of the web-site of the Federal State Statistics Service. http://www.gks.ru/wps/wcm/connect/rosstat_main/rosstat/ru/statistics/population/level/. Accessed 9 Nov 2018
11. Drones are up. Regions call for more active implementation of digital technologies into agriculture. Materials of the web-site of the Russian newspaper. https://rg.ru/2018/09/18/reg-cfo/selskoe-hoziajstvo-perejdet-na-cifru.html. Accessed 09 Nov 2018
12. Aristovnik, A., Obadić, A.: The impact and efficiency of public administration excellence on fostering SMEs in EU countries. Amfiteatru Econ. **17**(39), 761–774 (2015)
13. Bogoviz, Aleksei V., Sergi, Bruno S.: Will the circular economy be the future of Russia's growth model? In: Sergi, B.S. (ed.) Exploring the Future of Russia's Economy and Markets: Towards Sustainable Economic Development, pp. 125–141. Emerald Publishing, Bingley (2018)
14. Bondarenko, Y.G.: Public administration efficiency increase in investment management. Actual Prob. Econ. **172**(10), 89–90 (2015)
15. Clara, A.M.C., Canedo, E.D., De Sousa Júnior, R.T.: Elements that orient the regulatory compliance verification audits on ICT Governance. In: ACM International Conference Proceeding Series, Part F128275, pp. 177–184 (2017)
16. Comerio, M., Batini, C.: Efficiency vs efficacy driven service portfolio management in a public administration (Invited Paper). In: Proceedings - 2015 IEEE 8th International Conference on Service-Oriented Computing and Applications, SOCA 2015, 7399103, pp. 139–146 (2016)
17. Lacko, R., Hurný, F., Rozkošová, A.: Conceptual framework for evaluating the efficiency of public administration. In: IDIMT 2017: Digitalization in Management, Society and Economy - 25th Interdisciplinary Information Management Talks, pp. 253–260 (2017)
18. Nasir, A., Shahzad, M., Anwar, S., Rashid, S.: Digital governance: improving solid waste management through ICT reform in Punjab. In: ACM International Conference Proceeding Series, Part F132087, 3136600 (2017)
19. Sukhodolov, A.P., Popkova, E.G., Kuzlaeva, I.M.: Methodological aspects of study of internet economy. Stud. Comput. Intell. **714**, 53–61 (2018)
20. Fedotova, G.V., Kulikova, N.N., Kurbanov, A.K., Gontar, A.A.: Threats to food security of the Russia's population in the conditions of transition to digital economy. In: Popkova, E.G. (ed.) The Impact of Information on Modern Humans, vol. 622, pp. 542–548. Springer International Publishing AG (part of Springer Nature), Switzerland (2018). Proceedings of HOSMC (Russian Federation), 23–24 November 2017. Kozma Minin Nizhny Novgorod State Pedagogical University, Institute of Scientific Communications. https://link.springer.com/content/pdf/bfm%3A978-3-319-75383-6%2F1.pdf
21. Fedotova, G.V., Ilyasov, R.K., Gontar, A.A., Ksenda, V.M.: The strategy of provision of tax security of the state in the conditions of information economy. In: Gashenko, I.V., Zima, Yu. S., Davidyan, A.V. (eds.) Optimization of the Taxation System: Preconditions, Tendencies and Perspectives. Series Studies in Systems, Decision and Control, pp. 217–228. Springer (2019). https://www.springer.com/us/book/9783030015138

Economic and Legal Problems of State and Municipal Procurement in the Russian Federation

Yuri A. Kozenko[1], Lyubov V. Perekrestova[1], Diana A. Kurazova[2](✉), Olga S. Tereshkina[3], and Olga A. Golodova[4]

[1] Volgograd State University, Institute of Economics and Finance, Volgograd, Russian Federation
{kozenkoja, PerekrestovaLV}@volsu.ru
[2] Chechen State University, Grozny, Russian Federation
diana.kurazova.89@list.ru
[3] MIREA - Russian Technological University, Moscow, Russian Federation
olyteresh@mail.ru
[4] Volgograd State University, Volgograd, Russian Federation
ogolodova@volsu.ru

Abstract. **Purpose/Tasks:** The purpose of the article is to evaluate the economic and legal problems of implementing state and municipal procurement in the modern conditions. According to the OECD, procurement for state and municipal needs (Public procurement) account for 12% of GDP in the OECD members. According to the Deputy Minister of Economic Development of the RF, procurement for state and municipal needs, together with procurement of government corporations, account for 25% of GDP, reaching RUB 30 trillion.

According to the set purpose, the authors formulate and solve the following tasks: general evaluation of the state of the system of organization of state and municipal procurement, analysis of the legal basis of implementation of procurement, determination of the problems of transparency of procurement for state and municipal purposes, imperfection of evaluation of effectiveness of state and municipal procurement, and transfer of the contractual system into the electronic form.

Methodology: The authors use the methods of normative and legal analysis and evaluation of the modern situation in the sphere of state and municipal procurement, as well as comparison, analogy, and logical analysis.

Results: In the modern world of hi-tech, organization of the process of supplies for state and municipal purposes acquires high importance. State is one of the largest customers in the market of commodities and services, so contracts for supplies are very attractive for the private sector. However, the existing problems that accompany implementation of the process of procurement do now allow ensuring equal access to entrepreneurs to this market. The authors systematize the main problems in this sphere.

Conclusions/Significance: Peculiarity of this work consists in systematization and complex substantiation of the current processes of informatization – primarily, in the sphere of organization of state and municipal procurement.

E. G. Popkova and B. S. Sergi (Eds.): ISC 2019, LNNS 87, pp. 13–22, 2020.
https://doi.org/10.1007/978-3-030-29586-8_2

Keywords: State · Municipality · Procurement · Information systems · Auction

JEL Code: H57 · H60 · H61 · H72

1 Introduction

The contractual system in the sphere of procurement of goods, works, and services for state and municipal needs (hereinafter – the Contractual system) functions in the modern form on the basis of the Federal Law dated April 5, 2013, No. 44-FZ. This Federal Law replaced the Federal Law dated July 21, 2005 No. 94-FZ “Regarding placement of orders for supply of goods, execution of works, and provision of services for state and municipal needs” (hereinafter - Federal Law No. 94-FZ).

The preconditions for changing Federal Law No. 94-FZ included insufficient elaboration of this law in the part of the procedure of procurement and in the part of absence of the systemic character in approaches to the procurement. This led to a lot of amendments to Federal Law No. 94-FZ: 37 changes were introduced in the course of 8 years.

It would seem that Federal Law No. 44-FZ has no drawbacks of this kind, as the concept of this normative legal act envisages the systemic approach to state and municipal procurement and detailed elaboration of all elements of the institute of procurement. However, Federal Law No. 44-FZ was amended even before its entry into legal force in January 2014. During four years of its action (2014–2018), 57 changes were implemented into Federal Law No. 44-FZ.

This cannot but influence the quality of this law, which complicates its application by the subject of the contractual system. It should be noted that the number of sub-legislative normative legal acts adopted in execution of Federal Law No. 44-FZ exceeded the number of sublegislative normative legal acts adopted in execution of Federal Law No. 94-FZ (108 normative legal acts).

However, it should be noted that on the whole the contractual system in the sphere of procurement for state and municipal purposes, as an independent complex legal institute, is already formed, and any further changes will only correct, but not change, the sense of this economic and legal phenomenon.

Procurement of government corporations, state-funded establishments, state and municipal unitary companies, and companies with public or municipal participation is a separate issue. This procurement is regulated by Federal Law dated July 18, 2011, No. 223-FZ “On procurement of goods, works and services of certain kinds of legal entities” (hereinafter - Federal Law No. 223-FZ).

It should be noted that at the beginning of its action Federal Law No. 22-FZ determined only the general goals and principles and the main provisions of corporate procurement. However, by now Federal Law No. 223-FZ has acquired the features of Federal Law No. 44-FZ as to detalization of legal regulation of corporate procurement. This could be explained by the fact according to the data of the Ministry of Economic Development of the RF, in the first half of 2016 96% of corporate procurement was conducted on the non-competitive basis: 51% - purchase from the sole supplier, 45% - “other methods”, and only 4/7% of corporate procurement were conducted on the competitive basis – via actions and competitions. The Ministry of Economic Development mentions 3,800 methods of determining the suppliers, including "Procurement on the basis of sustainable economic ties", "The sole source", and even "The favorite method of procurement" – these are all euphemisms for "Procurement from the sole supplier". This urged the legislator to adopt Federal Law dated December 31, 2017, No. 505-FZ "On amendments in certain legislative acts of the Russian Federation", which changed Federal Law No. 223-FZ cardinally and made it similar according to the content to Federal Law No. 44-FZ. The legislator tries to limit corporate procurement from the sole supplier, envisaging the requirement on a closed list of the cases of procurement from the sole supplier (Article 3.6. Federal Law No. 223-FZ).

Certain specifics in the contractual system belong to Federal Law dated December 29, 2012 No. 275-FZ "About the state defensive order", which regulates the relations on formation and peculiarities of placement and execution of the state defensive order and control in this sphere. The contents of this Federal Law show that the state defensive order is a very specific manifestation of the contractual system, with a lot of exceptions from the general requirements of the contractual system, for the purpose of provision of state tasks in the sphere of national security. The state defensive order exists in the contractual system as an element that uses the existing system of administration of public procurement but has a large autonomy.

2 Materials and Method

The theoretical and applied issues of evaluation of effectiveness of state management of economy are studied in the works (Aristovnik and Obadić [10]), (Bondarenko [12]), (Comerio and Batini [14]), (Vertakova et al. [23]), (Plotnikov et al. [20]), (Smotrickay [7]), (Lacko et al. [16]), (Méndez Reátegui et al. [17]), (Mesquita and dos Santos 2015), (Pessoa et al. [19]), (Kojuhov 2015), (Marinov and Angelov [3]), and (Ulyanov [9]).

The scientific and methodological issues of managing the process of informatization of socio-economic systems are considered in the works (Bogoviz et al. [11]), (Clara et al. [13]), (Dai [15]), (Hawes and Li 2017), (Zuev and Bolbakov [1]), (Sakil 2017), (Sukhodolov et al. [22]), (Nidziy and Chugumbaev [5]), (Rogova [6]), and (Popkova et al. [21]).

However, despite the large number of publications on adjacent topics, the issues of assessment of effectiveness of organization of procurement and the role of state management in this model are poorly studied. In the aspect of development of the optimization model of information economy, it is necessary to reconsider the indicators – which are regulated at the state level – and supplement them for increasing the efficiency of implementation of the measures of state and municipal procurement.

3 Results

Transparency of public procurement is one of the key principles of the Contractual system. This principle is established in Article 7 of the Federal Law No. 44-FZ, according to which the Russian Federation provides free access to the information on the contractual system in the procurement sphere. Openness and transparency of information are provided also by placing it in the integrated information system.

Establishment of the principle of openness and transparency of information on the contractual system in Federal Law No. 44-FZ conforms to the corresponding international practice, which emphasizes the basic character of the principle of transparency of procurement for state and municipal purposes.

The Institute of Public Procurement of the USA defines transparency as a timely and clear access to information. Transparency allows guaranteeing determination of any deviations from fair and equal relations – which makes such deviations improbable. Transparency protects the integrity of the process and the interests of the organization, the interested parties, and the society [1, p. 7–8]. The deviations from fair and equal relations include also corruption.

This definition of transparency is common for any process or phenomenon. Thus, transparency in the contractual system could be viewed as one of the aspects of transparency of state management that is implemented through the mechanisms of the contractual system.

A lot of researchers define transparency as the basic characteristic of the environment of political management, which ensures development of democracy and civil society through openness of the political system, the procedure of making of political decisions, and public control over the activities of government bodies [2, p. 114].

A very important aspect in this definition is characteristics of transparency as the environment of political management – i.e., totality of the conditions in which certain political processes take place. As to the contractual system, characteristics of transparency as the environment of management is opened through management of information and information flows, which contain the data on the needs of public and legal entities, public finances and their distribution, quality of management, satisfaction of public and legal entities, and other important data. Moreover, a modern human can imagine this environment on the Internet, where the above information is produced and disseminated [3, p. 79].

According to Part 1 Article 4 of the Federal Law No. 44-FZ, for the purpose of information provision of the contractual system in the sphere of procurement, the unified information system (UIS) is created, which interaction with other information systems ensures:

(1) formation, processing, storing, and provision of data (including automatized) to participants of the contractual system in the sphere of procurement;
(2) control over the correspondence:
    (a) information on the volume of financial provision, included into the procurement plans, information on the volume of financial provision for procurement, which is established and which the customer is aware of;

(b) information that is included in the scheduled plans of procurement and information containing in the procurement plans;
(c) information containing in notifications on procurement and documents on procurement and information containing in schedules plans;
(d) information containing in protocols of selection of suppliers (contractors) and information containing in the procurement documents;
(e) conditions of the project of the contract sent in the form of an electronic document to a participant of the procurement which whom the contact is concluded and information containing in the protocol of selection of supplier (contractor);
(f) information on the contract containing in the list of contracts concluded by the customers;

(3) usage of improved qualified electronic signature for signing electronic documents, envisaged by Federal Law No. 44-FZ.

According to Part 3 Article 4 of Federal Law No. 44-FZ, UIS contains procurement plans, scheduled plans, information on implementation of procurement plans and scheduled plans; information on the conditions, prohibitions, and limitations of merchandise admissibility from other countries or group of other countries and admissibility of works and services that are performed and provided by foreign parties, list of foreign countries and groups of foreign countries with which the RF has international agreements on mutual application of the national regime during procurement, and conditions of such national regime; information on procurement envisaged by Federal Law No. 44-FZ and on execution of contracts; register of contracts concluded by the customers; unified register of the procurement participants; register of unfair suppliers (contractors); library of typical contracts and typical contract terms; register of bank guarantees' register of complaints, scheduled and unscheduled inspections, their results, and issued instructions; register of sole suppliers of the goods which production is created or modernized and (or) mastered on the territory of the RF; register of international financial organizations created according to the international agreements which participants include the RF and of international financial organizations with which the RF has international agreements; results of monitoring of procurement and audit and control in the sphere of procurement; reports of customers, envisaged by Federal Law No. 44-FZ; register of goods, works, and services for provision of state and municipal needs; normative and legal acts that regulate the relations in the contractual system; information on the commodity markets prices for goods, works, and services that are purchased for provision of state and municipal needs and on the customers' requests for quotations for goods, works, and services; other information and documents which placement in the UIS is envisaged by Federal Law No. 44-FZ, Federal Law No. 223-FZ and the corresponding normative and legal acts.

The UIS has the following web address: zakupki.gov.ru. As could be seen, the array of the information in the UIS is very large, which causes several problems.

The first problem of the technical character is a complex system, so failures are very frequent. Since the start of its work, the UIS has been showing insufficient reliability. In early 2017, the UIS did not function due (according to Deputy Minister of Economic Development of the RF E. Elin) to introduction of new control functions in the UIS with absence of proper automatization of the control process and conflict of the UIS interface, electronic budget, and regional systems. Systemic failures of the UIS influence the interaction between participants of the contractual system and information flows and distort the perception of information on procurement for state and municipal needs.

The volume of information uploaded into the UIS constantly grows. The Federal Treasury that is responsible in the Ministry of Finance of the RF for development of the UIS offers transferring all documents into the electronic form, including the documents on execution of contracts.

The second problem has the political character. The UIS is an important information resource, the struggle for which among the government bodies is very urgent. In particular, the Ministry of Economic Development of the RF, which, at the initial stage of formation of the USI was responsible for control over its development and functioning, was ousted in 2017 by the Ministry of Finance, which according to the Government's Decree dated April 14, 2017 "Regarding the changes in certain Decrees of the Government of the RF" received authorities for developing policy and normative and legal regulation in the contractual system in the sphere of procurement for state and municipal needs [7]. It should be also noted that the Ministry of Finance of Russia has not received control over procurements of the Ministry of Defense of Russia.

It should be also noted that interaction between the UIS and the online trade platforms – Sberbank-AST, Roseltorg, RTS-Tender, MMVB – National online platform, Zakaz RF, and the Russian Auction House – is not flawless. Apart from the problems of the technical character that appear in the process of integration of the UIS systems and trade platforms, the latter oppose the intensification of data exchange between the UIS and the online trade platforms, which, in their opinion, will lead to the UIS's monopolizing the functions of trade platforms and elimination of the existing participants of the market [4].

The third problem consists in the closed character of the data on procurements for defense and national security. In particular, the Ministry of Defense of Russia, The Foreign Intelligence Service of the Russian Federation, Rosatom State Corporation, and certain other establishments are not obliged to place the data on classified procurement in the UIS, which distorts the information picture of procurements for the public needs on the whole.

Effective spending of budget assets in the process of procurement for state and municipal purposes is one of the main tasks of the contractual system, according to Part 1 Article 1 of Federal Law No. 44-FZ. It is obvious that in order so solve socially important issues public and legal entities have to use budget assets with maximum effectiveness, as these budget assets are mostly tax payments from individuals and companies.

From the point of view of economic science, effectiveness is studied through the prism of efficiency and efficient performance. According to I.I. Smotritskaya, effectiveness characterizes the level of achievability of the planned result, while efficiency is translated as efficient performance or minimization of costs of a product item or results of works or provided services [5–7].

The Swedish National Audit Office uses the concept of triple "E" for inspection and evaluation of the work of establishments and organizations and the state of execution of budget programs:

(1) effectiveness – results that are compared to the goal and resources that are used for achievement of this goal.
(2) efficiency – ratio of product in the form of goods, services, etc. and resources that are used for their production.
(3) economy – minimization of the cost of resources used for activities, in view of the corresponding quality.

V.L. Kozhukhov notes that the following components of the procurement process are to be assessed:

- spent resources (inputs), which include financial, labor, and material resources;
- outputs – volume of goods, works, and services that are received as a result of inputs;
- results (outcome) – the desired state or goal at which the process is aimed [2, 11, 12].

The main criterion of evaluation of effectiveness in Russia is economy during placement of the order. In particular, the Methodology of evaluation of effectiveness of public procurement in Kostroma Oblast determines this indicator as the most important one. Thus, minimization of expenditures for purchase of the necessary goods, works, and services in the decisive criterion of effectiveness of the contractual system in the RF.

This imbalance in evaluation of effectiveness of state and municipal procurement in favor of economy and not in favor of efficiency led to a strange situation when the main method of competitive procurement – auction – caused the excessive popularity of such method of non-competitive procurement as procurement from the sole supplier. Government corporations and economic entities with public participation that work according to Federal Law No. 223-FZ, being commercial companies, are aimed at result, not economy, so they prefer to place orders with the sole supplier or prefer other methods of procurement in which the price is not the main or decisive character.

According to A. Ulyanov, auctions are an obvious leader among the procurement methods – their share of 60% (in the sum of contracts) in 2015–17 was even called the "auction imbalance" in the report by the Center of Strategic Developments of A. Kudrin. The share of auctions in public procurement of the EU countries is striving towards zero, and in South Korea and the UK constitutes 5% [8, 9].

The auction in which the only criterion of victory is reduction of price for the offered goods, works, or services creates a problem of dumping and aggravation of quality of goods and result of execution of works and provision of services, as participants of procurement with a real offer, which conforms to the quality requirements, have to leave the market, giving way to those offering lower price and neglecting quality. The mechanisms of anti-dumping, envisaged by Article 37 of Federal Law No. 44-FZ, do not prevent the low-quality result of procurement.

According to the Accounts Chamber of the RF, the volume of violations during public procurement within 44-FZ constituted RUB 104.6 billion in 2017, which is twice as much as in previous year [22]. According to the Ministry of Finance of the RF, the total volume of economy constituted RUB 398.6 billion [23] as of year-end 2017.

Growth of the volume of violations in absolute numbers shows that "outcome" – result of procurement procedures – is achieved less frequently than in previous year, due to overestimation of "output" – ratio of price and quality of goods, works, and services, which is received as a result of reduction of prices in the course of the auction. Book debts and fines for public contracts with the customers does not show that customers receive the planned result from the procurement – on the contrary, it is a proof of bad interrelations between the customer and the supplier and the customer's dissatisfaction with the result of execution of the contract by the supplier.

It seems necessary to change the methodology of evaluating the effectiveness of state and municipal procurement, in which the price for goods, works, and services is the main criterion. It is more important to evaluate the effectiveness of procurement from the point of view of achievement of the planned result of procurement, satisfaction of customer (beneficiary) with the result of procurement: quality of product, speed of supply or execution of works, and provision of proper service. This criterion – not price – should be decisive for evaluating the effectiveness of procurement.

This change will also lead to change of the ratio of action and methods of procurement: auction imbalance in the contractual system could be overcome with the help of change of the criteria of evaluation of procurement effectiveness.

The Decree of the Government of the RF dated April 28, 2018 No. 824-r "On the sole aggregator of trade, all customers of the federal level since November 1, 2018 have been conducting procurement of small volumes via the sole aggregator of trade – i.e., online store. This form of online procurement has been using by corporate customers since 2015, including those working according to Federal Law No. 223-FZ.

It is recommended for regional and local authorities to joint this form of procurement, and certain regions have already implemented online stores for small procurement on their territories [13–15].

However, transfer of the contractual system into the electronic form creates a lot of problems, solving which will require a lot of resources.

In particular, an important problem is access of local authorities to the optical fibre network, which allows working with the UIS. Not all cities and villages have access to the optimal fibre network, so certain local authorities will not be able to conduct any procurement – due to prohibition of the paper form of procurement. This is especially important for towns and villages of Siberia and the Far East, which do not have the corresponding communications.

Besides, as was mentioned above, the UIS is not stable, has a lot of failures during work, and sometimes can lose information that is important for the participants of the contractual system. Without stable work of the UIS, the electronic form of procurement will complicate – not simplify – the position of participants of the contractual system. The participants of the contractual system depend on the UIS not only in the period of informing on the procurement but also in the moment of its implementation, execution, and closure of the deal.

4 Conclusions

It should be noted that the register of the above problems and their elimination will help reducing the openness and transparency of the contractual system and, therefore, leveling the state and society's efforts on fighting corruption and increase of effectiveness of using the budget resources. However, transparency of functioning of the contractual system in Russia is an important factor that allows assigning Russia to the leading countries in the sphere of opening data on public procurement.

It is possible that Russia should consider the experience of participation of countries in application of the Open Contracting Data Standard – OCDS, developed in 2015 by Open Contracting Partnership with the World Bank and the World Wide Web Foundation [16–19].

According to the creators of the OCDS, their initiative consists in publication and usage of open, accessible, and timely information on government contracts for attraction of the public and business for determining and solving the problems (corruption). At present, the Committee of civil initiatives *GosZakupki* (Public Procurement) works according to the standards of open data on the contracts [20].

The user interface of the UIS is not adapted for intuitive usage by participants of the procurement and does not allow correcting any mistakes. In view of large fines for violation of the laws on the contractual system, including during exploitation of the UIS, unintentional mistakes or even guiltless action will provoke large fines.

Thus, it sees necessary to ensure the high level of technical support for the UIS users and to organize the permanent teaching process of using all UIS functions for all interested parties. Also, it is necessary to reconsider the norms of The Code of the Russian Federation on Administrative Offenses for violations of the UIS requirements, including reduction or exclusion of responsibility for unintentional violations of requirements to the electronic form of procurement and usage of program and technical complexes in this sphere.

References

1. Zuev, A.S., Bolbakov, R.G.: Regarding the telecommunication services on the basis of the technologies of virtual reality. Russ. Technol. J. **5**(6(20)), 3–10 (2017)
2. Kozhukhov, V.L.: Formation of the system of indicators of evaluation of effectiveness for audit of public procurement. Bull. Plekhanov RUE, 6(84), 107–115 (2015)
3. Marinov, M., Angelov, G.: Monitoring of management of debit debt of a company. Russ. Technol. J. **6**(2(22)), 67–81 (2018)
4. Nadtochiy, Y.B., Budovich, L.S.: Intellectual capital of organization: essence, structure, and approaches to evaluation. Russ. Technol. J. **6**(2(22)), 82–95 (2018)
5. Nidziy, E.N., Chugumbaev, R.R.: The issues of application of the network integration of companies of the textile industry in the conditions of dynamics of the external environment. Bull. High. Educational Establishments, (2(374)), 55–58 (2018). The technology of the textile industry
6. Rogova, V.A.: Personnel problems of development of hi-tech in Russia in the mirror of the global innovations index. Russ. Technol. J. **6**(4(24)), 105–116 (2018)

7. Smotritskaya, I.I.: The macro-economic approaches to evaluation of integral effectiveness of the system of public procurement. ETAP. Economic theory, analysis, and practice, no. 5 (2014). https://cyberleninka.ru/article/v/makroekonomicheskie-podhody-k-otsenke-integralnoy-effektivnosti-sistemy-gosudarstvennyh-zakupok
8. Sorokin, S.A., Benenson, M.Z., Sorokin, A.P.: Methodology of evaluation of efficiency of heterogeneous calculation systems. Russ. Technol. J. **5**(6(20)), 11–19 (2017)
9. Ulyanov, A.: Repairing the auction imbalance in public procurement. News. https://www.vedomosti.ru/opinionarticles/2018/04/18/767018-auktsionnii-kren-goszakupkah. Accessed 23 Dec 2018
10. Aristovnik, A., Obadić, A.: The impact and efficiency of public administration excellence on fostering SMEs in EU countries. Amfiteatru Econ. **17**(39), 761–774 (2015)
11. Bogoviz, A.V., Ragulina, Y.V., Morozova, I.A., Litvinova, T.N.: Studies in systems. Decis. Control **135**, 147–154 (2018)
12. Bondarenko, Y.G.: Public administration efficiency increase in investment management. Actual Probl. Econ. **172**(10), 89–94 (2015)
13. Clara, A.M.C., Canedo, E.D., De Sousa Júnior, R.T.: Elements that orient the regulatory compliance verification audits on ICT Governance. In: ACM International Conference Proceeding Series, Part F128275, pp. 177–184 (2017)
14. Comerio, M., Batini, C.: Efficiency vs efficacy driven service portfolio management in a public administration (Invited Paper). In: Proceedings - 2015 IEEE 8th International Conference on Service-Oriented Computing and Applications, SOCA 2015, 7399103, pp. 139–146 (2016)
15. Dai, X.: Politics of digital development: informatization and governance in China. In: Digital World: Connectivity, Creativity and Rights, pp. 34–51. Taylor and Francis (2013)
16. Lacko, R., Hurný, F., Rozkošová, A.: Conceptual framework for evaluating the efficiency of public administration. In: IDIMT 2017: Digitalization in Management, Society and Economy - 25th Interdisciplinary Information Management Talks, pp. 253–260 (2017)
17. Méndez Reátegui, R., Coca Chanalata, D.G., Alosilla Díaz, R.: The efficiency of public administration: A compartive analysis of the peruvian and ecuadorian cases from a neo institutional approach" | [La eficiencia de la administración: Un análisis comparado desde el enfoque neo institucional de los casos peruano y ecuatoriano]. Revista General de Derecho Administrativo (43), 12–19 (2016)
18. Nasir, A., Shahzad, M., Anwar, S., Rashid, S.: Digital governance: Improving solid waste management through ICT reform in Punjab. In: ACM International Conference Proceeding Series, Part F132087, 3136600 (2017)
19. Pessoa, A.A.M., Justino, A.N.P., De Farias, F.H.C., Da Silva, J.M.D., Lima, P.T.D., De Sousa, V.R.M.: Analysis of the efficiency, efficacy and effectiveness in public administration: The case of IDEMA/RN | [Análise da eficiência, eficácia e efetividade na administração pública: O caso do IDEMA/RN]. Espacios, **37**(8), 8 (2016)
20. Plotnokov, V., Fedotova, G.V., Popkova, E.G., Kastyrina, A.A.: Harmonization of Strategic planning indicators of territories' socioeconomic growth. Reg. Sectoral Econ. Stud. **15–2** (July–December), 105–114 (2015)
21. Popkova, E.G., Popova, E.V., Sergi, B.S.: Clusters and innovational networks toward sustainable growth. In: Sergi, B.S. (ed.) Exploring the Future of Russia's Economy and Markets: Towards Sustainable Economic Development, pp. 107–124. Emerald Publishing, Bingley (2018)
22. Sukhodolov, A.P., Popkova, E.G., Kuzlaeva, I.M.: Methodological aspects of study of internet economy. Stud. Comput. Intell. **714**, 53–61 (2018)
23. Vertakova, Y., Plotnikov, V., Fedotova, G.: The system of indicators for indicative management of a region and its clusters. Procedia Econ. Finan. **39**, 184–191 (2016)

Global Competitiveness of the Digital Economy: The Problem of Measuring and Management

Irina V. Mukhomorova[1(✉)], Elena S. Akopova[2], L. K. Pavlova[2], and V. V. Sheveleva[2]

[1] Russian State Social University, Moscow, Russia
mukhomorova@mail.ru

[2] Rostov State University of Economics, Rostov-on-Don, Russia
kafedra_kil@mail.ru, popova_plk@mail.ru, beloveronika@yandex.ru

Abstract. Purpose: The purpose of the paper is to improve the scientific and methodological provision of the digital economy, which is aimed at measuring and managing its global competitiveness.

Design/Methodology/Approach: The authors develop a new approach to measuring global competitiveness of the digital economy, which reflects its advantages for all interested parties. The methodological tools of the research include structural & functional and logical analysis, synthesis, induction, deduction, and formalization. The information and empirical basis of the research includes the materials of IMD World Digital Competitiveness Ranking for 2018.

Findings: It is substantiated that the existing rating of IMD contains detailed and useful statistical information, which could be recommended for measuring the global competitiveness of the digital economy of the modern countries of the world – but not in the initial ("raw") form but in the form that is processed with the help of systematization and analysis through the prism of achieving various goals of formation of the digital economy. The IMD ranking reflects the level of achievement of only one (and not the most important) goal of formation of the digital economy – strengthening of the country's reputation at the world arena (prestige and place marketing). A new scientific and methodological approach to measuring and managing the competitiveness of the digital economy is developed; it takes into account the level of achievement of other five goals of formation of the digital economy: development of business, increase of export, increase of population's living standards, innovative development, and growth of economy's effectiveness. They correspond to the five offered treatments of competitiveness of the digital economy, of each of which separate methods of measuring and management are offered.

Originality/Value: Approbation of the developed scientific and methodological approach to measuring and managing the competitiveness of the digital economy by the example of Russia in 2018 showed that its effectiveness is very high, attractiveness for living and export activity are medium, and innovativeness and favorability for digital business are low. This allowed offering practical recommendations for managing the competitiveness of the digital economy of modern Russia.

E. G. Popkova and B. S. Sergi (Eds.): ISC 2019, LNNS 87, pp. 23–29, 2020.
https://doi.org/10.1007/978-3-030-29586-8_3

Keywords: Global competitiveness · Digital economy · Measuring · Management · Modern Russia

JEL Code: O31 · O32 · O33 · O38

1 Introduction

Most of the modern countries of the world joined the Fourth industrial revolution and started the programs of formation of the digital economy. This led to the global competition of digital economies, which grows annually and becomes more complicated with invention and distribution of new digital technologies. In the new global economic landscape, an important role belongs to scientific and methodological provision of the digital economy, which, firstly, should provide the opportunity for objective, precise, and authentic measuring of its global competitiveness on the whole and for determining its pros and cons, and, secondly, reflect (in the generalized formulation) the perspectives and logic of managing the digital economy for increasing its global competitiveness.

At present, the scientific and methodological provision of the digital economy is based on the annual index and IMD World Digital Competitiveness Ranking. One-time statistical accounting (e.g., "The Global Information Technology Report", presented by the specialists of the World Economic Forum in 2016, which lost its topicality 2019) and regional practices of statistical accounting (e.g., Digital Economy and Society Index, also known as DESI, calculated only for the EU) cannot be alternatives to the IMD ranking, despite its certain drawbacks.

Firstly, the IMD ranking covers only 63 countries (2018) – i.e., less than half of the countries of the world – which does not allow for evaluation of the general state of global competition of the digital economy and for full-scale international comparisons. Secondly, the IMD ranking is poorly informative from the positions of measuring of the digital economy, as its open official version does not open the initial values of the indicators, and provides only the countries' positions for each indicator. The ranking's results are predictable and coincide with the results of other rankings of countries of the world (according to the criterion of innovative activity, criterion of competitiveness of economy, etc.), which complicates their usage in the process of managing the competitiveness of the digital economies of countries of the world.

Thirdly, the structure of the indicators and their systematization in the IMD ranking are contradictory from the scientific point of view and do not fully conform to the interests of the modern practice of managing the competitiveness of the digital economy. For example, expenditures for R&D and the indicators of efficiency of R&D are assigned to the same category – Scientific concentration – though they reflect completely different categories – expenditures (which are to be reduced) and results (which are to be increased). Also, the logic of assigning the indicator "female researchers" to the list of the indicators of digital competitiveness of economy is doubtful, as this indicator should be used during evaluation of sustainability of development of the economic system, and it's unclear how it characterizes the digital economy.

Thus, there's a problem of improvement of the scientific and methodological provision of the digital economy, which is aimed at measuring and management of its global competitiveness – which is the purpose of this paper.

2 Materials and Method

There are a lot of works on the digital economy and measuring and management of its global competitiveness: Dyatlov et al. (2018), Lichtenthaler (2017), Mitrović (2015), Petrenko et al. (2018), Polyakova et al. (2019), Popkova (2019), Popkova et al. (2019), Ross and Liechtenstein (2018), Torres Taborda (2018), and Trappey et al. (2016).

The performed content analysis of these publications showed the approach to studying the digital economy as an element of prestige of the economic system and the tool of its place marketing. This is a very treatment of the digital economy, which slows down its formation and development. The interested parties – population and business – do not see practical advantages from formation of the digital economy and express a protest against redistribution of national resources to the damage of social projects and projects for supporting business in favor of the projects in the sphere of the digital economy.

That's why there's a need for a new approach that would reflect the advantages of the digital economy for all interested parties. It is developed in this paper. For this, the authors use the structural & functional analysis, synthesis, induction, deduction, and formalization. The information and empirical basis of the research includes the materials of the IMD World Digital Competitiveness Ranking.

3 Results

The developed new scientific & methodological approach to measuring and management of competitiveness of the digital economy is presented in Table 1.

Table 1. The new scientific & methodological approach to measuring and management of competitiveness of the digital economy.

Purpose of formation of the digital economy	Treatment of global competitiveness of the digital economy	Indicators for measuring global competitiveness of the digital economy	Tools of managing global competitiveness of the digital economy
Development of business	Competitiveness of economy as the territory for doing digital business	– risk of investing; – development of digital infrastructure (in view of components)	Development of digital infrastructure and reduction of investing risks

(continued)

Table 1. (*continued*)

Purpose of formation of the digital economy	Treatment of global competitiveness of the digital economy	Indicators for measuring global competitiveness of the digital economy	Tools of managing global competitiveness of the digital economy
Increase of export	Competitiveness of digital products that are manufactured in the economy in the global markets	– volume of hi-tech export; – volume of export of hi-tech products	Increase of export of high technologies and hi-tech export
Increase of population's living standards	Competitiveness of economy as the territory of residence of digital human	– level of development of e-government; – level of digital security	Development of e-government and increase of the level of digital security
Innovative development of economy	Competitiveness as the level of innovativeness of the digital economy	– share of economic subjects that use various traditional (Internet) and breakthrough (Big Data, blockchain) digital technologies	Provision of development, distribution, and usage of breakthrough digital technologies (AI, quantum technologies, the Internet of Things)
Growth of economy's effectiveness	Competitiveness as effectiveness of the digital economy	– ratio of the volume of production of hi-tech products to expenditures for R&D	Provision of growth of efficiency of R&D and manufacture of hi-tech products

Source: compiled by the authors.

Table 1 shows that we distinguished five goals of formation of the digital economy, for each of which the corresponding treatment of competitiveness of the digital economy and is offered and the methods of measuring and management are recommended.

The 1st goal: development of business (for accelerating economic growth). For determining the level of achievement of this goal, it is suggested to determine competitiveness of economy as the territory for doing digital business. The indicators for measuring competitiveness should be risk of investing and development of digital infrastructure in view of the components (e.g., digital personnel and education, accessibility of digital technologies, and general conditions for doing business). The tools of managing global competitiveness of the digital economy include the development of digital infrastructure and reduction of risks of investing.

2nd goal: increase of export on the basis of digital (hi-tech) products, which are in high demand in the world markets. For determining the level of achievement of this goal, it is offered to determine competitiveness of the digital products that are manufactured in this economy in the world markets. The indicators for measuring competitiveness should be the volume of export of high technologies and the volume of export hi-tech products. The tools of managing global competitiveness of the digital economy include increase of export of high technologies and hi-tech products.

3rd goal: increasing the population's living standards. For determining the level of achievement of this goal, it is offered to determine competitiveness of economy as the territory of residence of digital human. The indicators for measuring competitiveness should be the level of development of e-government and the level of digital security. The tools of managing global competitiveness of the digital economy includes development of e-government and increase of the level of digital security.

4th goal: innovative development of economy. For determining the level of achievement of this goal, it is offered to determine competitiveness as the level of innovativeness of the digital economy. The indicators of measuring competitiveness should be the share of economic subjects that use various traditional (Internet) and breakthrough (Big Data, blockchain) digital technologies. The tools of managing global competitiveness of the digital economy include provision of development, distribution, and usage of breakthrough digital technologies (AI, quantum technologies, and the Internet of Things).

5th goal: growth of effectiveness of economy. For determining the level of achievement of this goal, it is offered to determine competitiveness as effectiveness of the digital economy. The indicators for measuring competitiveness should be the ratio of the volume of manufacture of hi-tech products to expenditures for R&D. The tools of managing global competitiveness of the digital economy include provision of growth of efficiency of R&D and manufacture of hi-tech products.

It is offered to measure the level of digitization of economy separately (without connection to competitiveness) with the help of such indicator as the share of the sphere of information and communication technologies in the structure of GDP, the share of business and population that use digital technologies, etc. It is also recommended to compile rankings of countries as to the level of competitiveness of the digital economy for different regions of the world and different categories of countries. As an approbation of the offered scientific and methodological approach we measured various components of competitiveness of the digital economy of Russia in 2018 based on the data of IMD (2019). The following results were obtained:

- competitiveness of economy as the territory for doing digital business is very low: high risk of investing (57th position), lack of protection of intellectual property (52nd position);
- competitiveness of digital products that are manufactured in this economy in the world markets is medium: medium level of hi-tech export (36th position);
- competitiveness of economy as the territory for residence of digital human is medium: medium level of development of e-government (30th position) and digital security (37th position);
- competitiveness as the level of innovativeness of the digital economy is low: foundation in the traditional digital technologies, low level of usage of Big Data (58th position);
- competitiveness as effectiveness of the digital economy is high: efficiency of R&D as to the number of publications is very high (6th position), and expenditures for R&D are medium (34th position).

Based on the result of measuring competitiveness of the digital economy of Russia in 2018, we offer the following recommendations for managing this competitiveness (for increasing it):

- top-priority: paying attention to increasing the economy's competitiveness as the territory for doing digital business by reducing the risk of investing (and protecting intellectual property);
- top-priority: increasing the level of innovativeness of the digital economy through distribution of breakthrough digital technologies;
- additionally: increasing competitiveness of digital products (increase of hi-tech export) and competitiveness of economy as the territory of residence of digital human (by developing e-government and increasing the level of digital security).

4 Conclusion

Thus, it is possible to conclude that the existing ranking of IMD contains detailed and useful statistical information, which could and should be used for measuring global competitiveness of the digital economy of the modern countries of the world - but not in the initial ("raw") form but in the form that is processed with the help of systematization and analysis through the prism of achieving various goals of formation of the digital economy. The IMD ranking reflects the level of achievement of only one (and not the most important) goal of formation of the digital economy – strengthening of the country's reputation at the world arena (prestige and place marketing.

The paper presents a new scientific & methodological approach to measuring and management of competitiveness of the digital economy, which takes into account the level of achievement of other five goals of formation of the digital economy: development of business, increase of export, increase of population's living standards, innovative development, and growth of economy's effectiveness. They correspond to the five offered treatments of competitiveness of the digital economy: (1) competitiveness of economy as the territory for doing digital business, (2) competitiveness of digital products that are manufactured in this economy in the world markets, (3) competitiveness of economy as a territory of residence of digital human, (4) competitiveness as the level of innovativeness of the digital economy and (5) competitiveness as effectiveness of the digital economy.

Each treatment corresponds to the methods of its measuring and management. Approbation of the developed scientific and methodological approach to measuring and managing the competitiveness of the digital economy by the example of Russia in 2018 showed that its effectiveness is very high, attractiveness for living and export activity are medium, and innovativeness and favorability for digital business are low. This allowed offering practical recommendations for managing competitiveness of the digital economy of modern Russia.

References

Dyatlov, S.A., Lobanov, O.S., Zhou, W.B.: The management of regional information space in the conditions of digital economy. Econ. Region **14**(4), 1194–1206 (2018)

IMD. World Digital Competitiveness Ranking 2018 (2019). https://www.imd.org/wcc/world-competitiveness-center-rankings/world-digital-competitiveness-rankings-2018/. Accessed 22 Apr 2019

Lichtenthaler, U.: Shared value innovation: Linking competitiveness and societal goals in the context of digital transformation. Int. J. Innov. Technol. Manag. **14**(4), 1750018 (2017)

Mitrović, D.: Broadband adoption, digital divide, and the global economic competitiveness of Western Balkan countries. Econ. Ann. **60**(207), 95–115 (2015)

Petrenko, E., Pizikov, S., Mukaliev, N., Mukazhan, A.: Impact of production and transaction costs on companies' performance according assessments of experts. Entrep. Sustain. Issues **6** (1), 398–410 (2018). https://doi.org/10.9770/jesi.2018.6.1(24)

Polyakova, A.G., Loginov, M.P., Serebrennikova, A.I., Thalassinos, E.I.: Design of a socio-economic processes monitoring system based on network analysis and big data. Int. J. Econ. Bus. Admin. **7**(1), 130–139 (2019)

Popkova, E.G.: Preconditions of formation and development of industry 4.0 in the conditions of knowledge economy. Stud. Syst. Decis. Control **169**, 65–72 (2019)

Popkova, E.G., Ragulina, Y.V., Bogoviz, A.V.: Fundamental differences of transition to industry 4.0 from previous industrial revolutions. Stud. Syst. Decis. Control **169**, 21–29 (2019)

Ross, G., Liechtenstein, V.: Management of financial bubbles as control technology of digital economy. Adv. Intell. Syst. Comput. **724**, 96–103 (2018)

Torres Taborda, S.L.: Professional competitiveness strategy based on the appropriation of digital tools. Case of the American University Corporation, Medellin, Colombia | [Estrategia de competitividad profesional basada en la apropiación de herramientas digitales. Caso Corporación Universitaria Americana, sede Medellín - Colombia]. Espacios, **39**(50) (2018)

Trappey, C.V., Trappey, A.J.C., Mulaomerovic, E.: Improving the global competitiveness of retailers using a cultural analysis of in-store digital innovations. Int. J. Technol. Manag. **70**(1), 25–43 (2016)

The Perspectives of Provision of New Quality of Growth of Economic Systems in the Digital Economy

Rustam A. Yalmaev[1(✉)], Vladislav A. Shalaev[2], Aydarbek T. Giyazov[3], and Gulzat K. Tashkulova[4]

[1] Chechen State University, Grozny, Russia
r.yalmaev@chesu.ru
[2] Tyumen Industrial University (Nizhnevartovsk Branch), Nizhnevartovsk, Russia
shhel77.77@mail.ru
[3] Kizil-Kiya Institute of Technology, Economics and Law, Kizil-Kiya, Kyrgyzstan
aziret-81@mail.ru
[4] Kyrgyz National University Named After J. Balasagin, Bishkek, Kyrgyzstan

Abstract. Purpose: The purpose of the research is to substantiate the perspectives of provision of new quality of growth of economic systems in the digital economy and to develop the scientific recommendations for their practical implementation.

Design/Methodology/Approach: The authors use the methods of economic statistics (econometrics) – in particular, trend analysis, calculation of direct average, and building the regression curves. The research objects are developed countries from Major advanced economies (G7) and developing countries from BRICS. The information and analytical basis of the research includes statistical materials of Cornell University, INSEAD, WIPO, Sustainable Development Solutions Network, the United Nations Development Programme, the IMD, and the International Monetary Fund for 2013 and 2018.

Findings: The perspectives of provision of new quality of economic systems' growth in the digital economy are determined. This allows expanding the concept of the digital economy, which is to consider not only as a goal in itself (this approach could be applied only for developed countries) but also as a tool of provision of new quality of economic systems' growth and a source of social (human development: opening human potential, implementation of socially important projects), ecological (sustainable development), and economic (innovative development) advantages.

Originality/Value: It is shown that at present the potential of the digital economy in provision of new quality of economic systems' growth is not realized in full, due to existence of "markets gaps" (domination of private interests over public interests, which hinders the implementation of the corresponding initiatives in entrepreneurship). The authors' recommendations are offered for overcoming them. These recommendations will allow reducing the disproportions in development of the modern economic systems not only at the macro-level but also at the micro-level, creating significant advantages for each human.

E. G. Popkova and B. S. Sergi (Eds.): ISC 2019, LNNS 87, pp. 30–38, 2020.
https://doi.org/10.1007/978-3-030-29586-8_4

Keywords: New quality of growth · Economic growth · Innovative development · Human development · Sustainable development · Economic systems · Digital economy

JEL Code: Q01 · M14 · O31 · O32 · O33 · O38 · O47

1 Introduction

Formation of the digital economy is one of the most vivid and popular tendencies in the modern global economic system and, at the same time, one of its most contradictory manifestations. External purposes of formation of the digital economy, which are connected to provision of global competitiveness of the modern economic systems, increase of effectiveness, acceleration of the rate of economic growth, and increase of import, often contradict the internal goals of these systems, which consist in provision of new quality of economic growth by increasing the population's living standards (human development) and innovative and sustainable (stable, well-balanced, with low ecological costs) development.

Considering digital modernization of economy as a goal in itself, governments redistribute public resources in its favor, refusing from implementation of socially important projects. This is especially characteristic of developing countries, which strive – by means of increasing the level of global digital competitiveness – to strengthen their positions in the global markets. However, improvement of quantitative indicators (e.g., acceleration of the rate of economic growth) in these cases is not usually accompanied by improvement of qualitative indicators (e.g., increase of quality of economic growth). Due to this, the advantages from overcoming the underrun from developed countries could be visible only at the macro-economic level and could be very limited – though the key priority of reduction of disproportions in development of the global economic system is leveling the quality of life of people and creating the advantages for each human.

Thus, an important task of economics is solving the contradiction of the digital economy. The working hypothesis of the research is the supposition that the digital economy possesses the potential of not only quantitative acceleration of the rate of economic growth but also increase of its quality. The purpose of the research is to substantiate the perspectives of provision of new quality of growth of economic systems in the digital economy and to develop the scientific recommendations for their practical implementation.

2 Materials and Method

The conceptual foundations and practical experience of formation of the digital economy in the modern economic systems and the obtained advantages, which are connected to developing hi-tech productions and their becoming a growth pole of the modern economy, are studied in the works Bogoviz et al. (2019), Bogoviz et al. (2019), González and Nuchera (2019), and Negrea et al. (2019). The scientific and methodological issues of evaluating the quality of growth of economic systems in the modern economic conditions are studied in the works Long and Ji (2019), Popkova (2018), Popkova et al. (2018), Sun et al. (2018), and Xu et al. (2019).

As a result of content analysis of the existing literature on the selected topic, we made a conclusion on the insufficient elaboration of the topic of interconnection between the digital economy and quality of growth of the modern economic systems. For studying it and verifying the offered hypothesis, we use the methods of economic statistics (econometrics) – in particular, trend analysis (for assessing the growth of the indicators' values in 2018 as compared to 2013), calculation of direct average (for determining annual growth), and building the regression curves (for determining the dependence of quality of growth of economic systems on the level of their digital competitiveness in 2018).

The research objects are developed countries from Major advanced economies (G7) and developing countries from BRICS. The basic year is 2013, as it is the period of the start of programs of digital modernization of the modern countries' economies, and final year in 2018, as it is the year with the most statistical information available.

The indicators of the quality of economic growth are the index of innovations, calculated by Cornell University, INSEAD, and WIPO; the index of sustainable development, calculated by the Sustainable Development Solutions Network; and the index of human development, calculated by the United Nations Development Programme. Also, the authors analyze the value of the global digital competitiveness index, which is calculated by the IMD, and the rate of economic growth (growth of GDP in constant prices), which is calculated by the International Monetary Fund.

3 Results

The initial statistical data, selected for the research, and the results of their analysis are presented in Tables 1 and 2 and in Fig. 1.

Table 1. Dynamics, assessment of growth and direct average of the index of innovations and the index of sustainable development.

Category	Country	Index of innovations			Index of sustainable development		
		Value in 2013, points 1–100	Value in 2018 m points 1–100	Growth in 2018, %	Value in 2013, points 1–100	Value in 2018, points 1–100	Growth in 2018, %
Major advanced economies (G7)	Canada	57.6	52.98	−8.02	76.8	76.8	0.00
	France	52.83	54.36	2.90	77.9	81.2	4.24
	Germany	55.83	58.03	3.94	80.5	82.3	2.24
	Italy	47.85	46.32	−3.20	72.7	74.2	2.06
	Japan	52.23	54.95	5.21	75	78.5	4.67
	UK	61.25	60.13	−1.83	78.1	78.7	0.77
	USA	60.31	59.81	−0.83	72.7	73	0.41
BRICS	Brazil	36.33	33.44	−7.95	64.4	69.7	8.23
	China	44.66	53.06	18.81	59.1	70.1	18.61
	India	36.17	42.53	17.58	48.4	59.1	22.11
	Russia	37.2	37.9	1.88	66.4	68.9	3.77
	South Africa	37.6	35.13	−6.57	53.8	60.8	13.01
Direct average for 12 countries		–	–	1.83	–	–	6.68

Source: compiled by the authors based on the materials of Cornell University, INSEAD, WIPO (2019), and Sustainable Development Solutions Network (2019).

Table 1 shows that average growth of the index of innovations in 12 countries of the selection constituted 1.83% in 2018 as compared to 2013, and growth of the index of sustainable development constituted 6.68%.

Table 2 shows that average growth of the index of human development in 12 countries of the selection constituted 1.85% in 2018, as compared to 2013; and growth of the rate of economic growth constituted 5.30%.

Table 2. Dynamics, evaluation of growth and direct average of the index of human development, rate of economic growth and the index of global digital competitiveness.

Category	Country	Index of human development, 0–1			Rate of economic growth (annual growth of GDP in constant prices), %			Global digital competitiveness
		Value in 2013, points 1–100	Value in 2018, points 1–100	Growth in 2018, %	Value in 2013, points 1–100	Value in 2018, points 1–100	Growth in 2018, %	2018, points 1–100
Major advanced economies (G7)	Canada	0.911	0.926	1.65	2.475	1.956	−20.97	95.201
	France	0.889	0.901	1.35	0.576	1.65	186.46	80.753
	Germany	0.928	0.936	0.86	0.592	1,532	158.78	85.405
	Italy	0.876	0.88	0.46	−1.728	0.815	−147.16	64.958
	Japan	0.899	0.909	1.11	2	0.586	−70.70	82.170
	UK	0.915	0.922	0.77	1.911	1.457	−23.76	93.239
	USA	0.916	0.924	0.87	1.677	2.519	50.21	100.000
BRICS	Brazil	0.748	0.759	1.47	3.005	1.748	−41.83	51.693
	China	0.729	0.752	3.16	7.8	6.168	−20.92	74.796
	India	0.607	0.64	5.44	6.54	7.685	17.51	57.066
	Russia	0.804	0.816	1.49	1.279	1.444	12.90	65.207
	South Africa	0.675	0.699	3.56	2.489	1.569	−36.96	56.876
Direct average for 12 countries		–	–	1.85	–	–	5.30	–

Source: compiled by the authors based on IMD (2019), International Monetary Fund (2019), and United Nations Development Programme (2019).

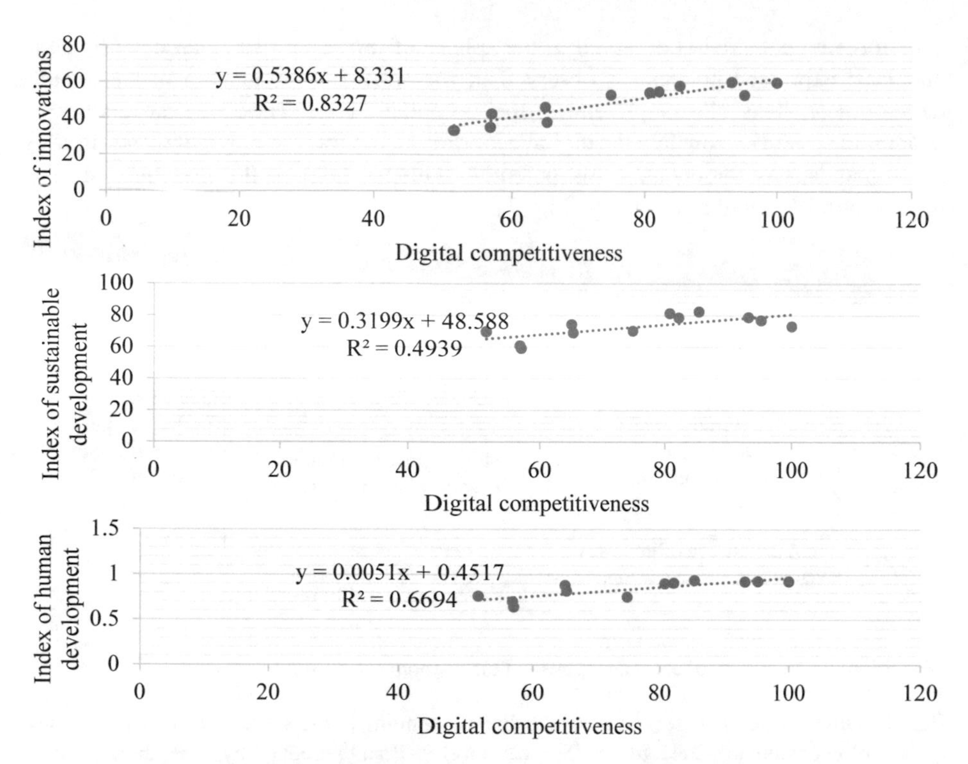

Fig. 1. Regression curves that show dependence of the quality of economic systems' growth on the level of their digital competitiveness in 2018. Source: compiled by the authors.

Figure 1 shows that growth of the digital competitiveness index by 1 point in 2018 led to growth of the index of innovations by 0.5386 points (correlation – 83.27%), growth of the sustainable development index by 0.3199 points (correlation – 49.39%), and growth of the human development index by 0.0051 points (correlation – 66.94%). In this work we refused from deep regression analysis, as in order to obtain precise and authentic results it is necessary to conduct comprehensive research (study all countries of the world) – and even in this case the results could be distorted due to the influence of social (random) and regulatory (absence of special measures of state regulation) factors.

However, from the logical point of view, the determined regression and correlation dependencies are enough for scientific confirmation of the existence of potential of the digital economy in provision of positive influence on the innovative, sustainable, and human development of the modern economic systems. As the influence of digital competitiveness on innovative development of the modern economic systems is most expressed, it is the basis of the potential of the digital economy in provision of new quality of economic systems' growth.

Moderate potential is observed in the sphere of provision of human development, and least expressed potential is observed in the sphere of stimulation of sustainable development. This allows presenting the potential of provision of new quality of economic systems' growth in the digital economy in the form of a pyramid and offering recommendations for opening this potential with the help of the measures of state regulation of the digital economy (Fig. 2).

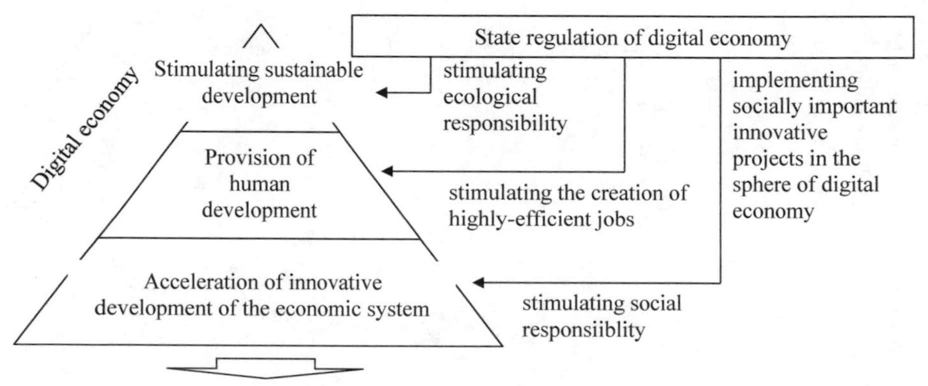

Fig. 2. Directions of state regulation of the digital economy in the interests of provision of new quality of economic systems' growth. Source: developed and compiled by the authors.

Figure 2 shows that for the digital economy's stimulating the sustainable development of the modern economic systems we offer state stimulation of ecological responsibility of manufacturers and consumers. Digital technologies allow increasing resource effectiveness of production and consumption, but they are not used in this sphere – which requires state regulation.

For the digital economy to stimulate human development, state regulation of the labor market through creation of highly-efficient jobs is recommended. Due to this, the digital economy will turn from the risk of unemployment into the source of increase and opening of human potential and maximization of effectiveness of usage of intellectual resources in the modern economic systems.

For the digital economy to stimulate the acceleration of innovative development of the modern systems, state stimulation of social responsibility and implementation of socially important innovative projects in the sphere of the digital economy, including on the basis of the mechanism of public-private partnership, is offered. This will allow building a socially oriented market economy, and innovations will turn from the source of risk into the source of social advantages.

4 Conclusion

Thus, the collected proofs conform the offered hypothesis on the existence of the perspectives of provision of new quality of growth of economic systems in the digital economy. This allows expanding the concept of the digital economy, which is to consider not only as a goal in itself (this approach could be applied only to developed countries) but also as a tool of provision of new quality of economic systems' growth and a source of social (human development: opening the human potential, realization of socially important projects), ecological (sustainable development), and economic (innovative development) advantages.

At present, the potential of the digital economy in provision of new quality of economic systems' growth is not realized in full, due to the existence of "market gaps" (domination of private interests over public interests, which hinders the implementation of the corresponding initiatives in entrepreneurship). For overcoming them we offer the recommendations that will allow reducing the disproportions in development of the modern economic systems at the macro-level and micro-level, creating significant advantages for each human.

References

Bogoviz, A.V., Alekseev, A.N., Ragulina, J.V.: Budget limitations in the process of formation of the digital economy. In: Lecture Notes in Networks and Systems, vol. 57, pp. 578–585 (2019)

Bogoviz, A.V., Lobova, S.V., Ragulina, J.V.: Distortions in the theory of costs in the conditions of digital economy. In: Lecture Notes in Networks and Systems, vol. 57, pp. 1231–1237 (2019)

Cornell University, INSEAD, WIPO. Global Innovation Index 2018 (2019). https://www.wipo.int/publications/ru/details.jsp?id=4330. Accessed 23 Apr 2019

González, R.L.H., Nuchera, A.H.: Dynamics of service innovation management and co-creation in firms in the digital economy sector. Contaduria y Administracion **64**(1), 1802 (2019)

IMD: World Digital Competitiveness Ranking 2018 (2019). https://www.imd.org/wcc/world-competitiveness-center-rankings/world-digital-competitiveness-rankings-2018/. Accessed 23 Apr 2019

International Monetary Fund. World Economic Outlook Database (2019). https://www.imf.org/external/pubs/ft/weo/2017/01/weodata/weoselgr.aspx. Accessed 23 Apr 2019

Long, X., Ji, X.: Economic growth quality, environmental sustainability, and social welfare in China - provincial assessment based on genuine progress indicator (GPI). Ecol. Econ. **159**, 157–176 (2019)

Negrea, A., Ciobanu, G., Dobrea, C., Burcea, S.: Priority aspects in the evolution of the digital economy for building new development policies. Qual. - Access Success **20**(S2), 416–421 (2019)

Popkova, E.G.: Contradiction of economic growth in today's global economy: economic systems competition and mutual support. Espacios **39**(1), 20 (2018)

Popkova, E.G., Bogoviz, A.V., Pozdnyakova, U.A., Przhedetskaya, N.V.: Specifics of economic growth of developing countries. Stud. Syst. Decis. Control **135**, 139–146 (2018)

Sun, Y., Rahman, S., Dai, C.: Measuring quality of economic growth incorporating environment and social welfare: a total productivity approach with an application on China. J. Donghua Univ. (English Edition) **35**(5), 411–417 (2018)

Sustainable Development Solutions Network. Sustainable development goals index (2019). http://sdgindex.org/reports/. Accessed 23 Apr 2019

United nations development programme. Human development reports (2019). http://hdr.undp.org/en/2018-update/download. Accessed 23 Apr 2019

Xu, H., Jilenga, M.T., Deng, Y.: Institutional quality, resource endowment, and economic growth: evidence from cross-country data. Emerging Markets Finan. Trade **55**(8), 1754–1775 (2019)

Specifics of Building the Digital Economy in Developed and Developing Countries

Stanislav Benčič[1(✉)], Yuliana A. Kitsay[2], Aziza B. Karbekova[3], and Aidarbek Giyazov[4]

[1] Pan-European University, Bratislava, Slovakia
bencic7777@gmail.com
[2] Immanuel Kant Baltic Federal University, Kaliningrad, Russia
Juliana_kn666@mail.ru
[3] Jalal-Abad State University, Jalal-Abad, Kyrgyzstan
aziza-karbekova@mail.ru
[4] Kyzyl-Kiya Institute of Technology, Economics and Law, Kyzyl-Kiya, Kyrgyzstan

Abstract. Purpose: The purpose of the paper is to determine the specifics of building the digital economy in developed and developing countries, to determine the optimal scenario of further digital modernization of the global economic system, and to develop the recommendations for its practical implementation.

Design/Methodology/Approach: The research is performed at the level of the global economy on the whole by the example of developed and developing countries, which are leaders in their categories and which occupy the medium and peripheral positions. All integral indicators of digital competitiveness of their economies, which are distinguished and calculated by IMD as of 2018, are studied. The methodology of the research includes analysis of variation (calculation of direct average, standard deviation, and coefficient of variation), forecasting, and scenario analysis.

Findings: It is determined that in developed countries the basis of digital competitiveness of economy is the high level of integration of information and communication technologies and devices, and the barrier on the path of its increase is low interest of business in digital modernization. Developing countries have an opposite situation – low level of integration of information and communication technologies and devices with high interest to digital modernization from business.

Originality/Value: It is shown that the optimal scenario of development of the modern global digital economy in the short-term (2019–2024) is integration of developed and developing countries, which envisages their close interaction and cooperation in the sphere of creation and development of digital infrastructure, as well as implementation of digital technologies into the activities of business structures on the basis of the clustering mechanism. This will allow reducing the gap in the level of global digital competitiveness of economies of developed and developing countries to 20.592 points (by 28.88%) by 2024.

Keywords: Digital economy · Digital modernization · Clusters · Integration · Developed countries · Developing countries

E. G. Popkova and B. S. Sergi (Eds.): ISC 2019, LNNS 87, pp. 39–48, 2020.
https://doi.org/10.1007/978-3-030-29586-8_5

JEL Code: F02 · F15 · O31 · O32 · O33 · O38

1 Introduction

The programs of digital modernization, which are actively implemented by various countries of the world, are aimed at simultaneous completion of two significant and contradictory functions. The first function consists in implementation of the own potential of each economic system that forms the digital economy and in increase of its global competitiveness. Thus, developed countries, which were the first to start the processes of digital modernization of their economies, are interested in preserving their leading positions in the world markets of high technologies and hi-tech products. In their turn, developing countries adopt the national programs of digital modernization for entering these markets.

The second function is connected to acceleration of growth rate of the global economy, formation of the global digital economy, which would cover all countries of the world, and overcoming of disproportions in its development. This function is based on the global goals in the sphere of sustainable development. When studying this function, it is necessary to pay attention to the fact that the digital economy is not only digital production but also digital society (consumption). That's why even developed countries are interested in quick formation of digital economy in developing countries for expanding the sales markets for their hi-tech products, which consumption requires the presence of digital competencies. Reduction of the level of differentiation of developed and developing countries in the global economy is a guarantee of social stability and thus conforms to the interests of all its participants.

The modern economic science has to solve a complex task of well-balanced achievement of these two functions. The purpose of this paper is to determine the specifics of building the digital economy in developed and developing countries, to determine the optimal scenario of further digital modernization of the global economic system, and to develop the recommendations for its practical implementation.

2 Materials and Method

The topic of the digital economy is studied in detail in the existing research literature. The universal character of the digital economy, which forms in the process of the Fourth industrial revolution (also called the transition to Industry 4.0), and its influence on all participants on the modern international economic relations are noted on the works Popkova (2019) and Popkova et al. (2019). The works Balnaves (2019), Bogoviz et al. (2019a), Bogoviz et al. (2019b), Lazović and Durickorić (2014), Mueller and Grindal (2019), Nathan et al. (2019), Negrea et al. (2019), Pritvorova et al. (2018), Sako (2019), Tsai et al. (2019), Turdubekov et al. (2018) note the necessity for considering the specifics of developed and developing countries during management of the processes of digital modernization of their economy.

Table 1. The indicators of digital competitiveness in the selection of developed countries in 2018.

Country	Position in ranking	Digital competitiveness index, points	Talent, position	Training & education, position	Scientific concentration, position	Regulatory framework, position	Capital, position	Technology framework, position	Adaptive attitudes, position	Business agility, position	IT integration, position
USA	1	100.000	11	21	1	16	1	9	1	9	8
Singapore	2	99.422	1	1	19	2	8	1	20	18	3
Sweden	3	97.453	10	5	3	12	10	7	9	10	11
Australia	13	90.226	8	32	11	6	18	19	2	28	6
South Korea	14	87.983	26	8	7	27	44	2	3	47	20
Austria	15	86.770	12	7	18	24	38	21	25	5	10
Luxembourg	24	81.490	33	26	44	9	4	35	29	17	13
Estonia	25	80.845	34	17	39	25	21	15	24	29	22
France	26	80.753	21	33	17	5	25	28	32	36	19
Direct average	–	89.438	17	17	18	14	19	15	16	22	12
Standard deviation from the average, position	–	7.888	11.66	12.03	15.01	9.43	14.91	11.65	12.37	13.91	6.62
Coefficient of variation, %	–	8.82	67.28	72.19	84.95	67.39	79.38	76.52	76.80	62.93	53.17

Source: compiled by the authors based on IMD (2019).

Table 2. The indicators of digital competitiveness in the selection of developing countries in 2018.

Country	Position in ranking	Digital competitiveness index, points	Talent, position	Training & education, position	Scientific concentration, position	Regulatory framework, position	Capital, position	Technology framework, position	Adaptive attitudes, position	Business agility, position	IT integration, position
Qatar	28	78.873	15	38	59	32	24	30	16	8	26
Lithuania	29	76.059	27	16	31	28	35	22	41	24	31
China	30	74.796	18	46	21	26	30	40	23	19	41
Kazakhstan	38	65.504	44	6	55	22	59	42	47	43	44
Thailand	39	65.272	42	44	45	34	28	23	55	34	55
Russia	40	65.207	40	12	23	38	58	38	39	62	43
Mongolia	61	48.056	60	24	60	58	55	61	31	61	62
Indonesia	62	45.776	51	61	58	57	34	60	61	46	60
Venezuela	63	24.795	63	60	22	63	63	63	63	51	63
Direct average	–	60.482	40	34	42	40	43	42	42	39	47
Standard deviation from the average, position	–	17.679	17.13	20.48	17.21	15.45	15.51	16.01	16.45	18.81	13.56
Coefficient of variation, %	–	29.23	42.83	60.05	41.40	38.84	36.17	38.02	39.36	48.66	28.72

Source: compiled by the authors based on IMD (2019).

However, there's a deficit of scientific research of the processes of the digital economy at the level of the global economy, as most publications are concentrated on the experience of separate countries of the world. This does not allow for systemic evaluation of global consequences of formation of the digital economy and scenarios of its development in the short-term. The determined gap in the existing scientific knowledge is to be filled by this paper.

This research is performed at the level of the global economy on the whole by the example of developed and developing countries, which are leaders in their categories and have medium and peripheral positions. All integral indicators of digital competitiveness of their economies, which are distinguished and calculated by IMD as of 2018, are studied. The methodology of the research includes analysis of variation (calculation of direct average, standard deviation, and coefficient of variation), forecasting, and scenario analysis. The initial data for the selection of developed countries are systematized and presented in Table 1, and the data for the selection of developing countries are presented in Table 2.

As is seen from Table 1, direct average of the digital competitiveness index in the selection of developed countries constitutes 89.438 points. Coefficient of variation is moderate (8.82%), which reflects homogeneity of the selection and guarantees the receipt of precise results of its analysis. High (more than 10%) variation of countries' positions in the ranking is explained by selecting the countries with various levels of the economy's digital competitiveness for provision of representativeness of the selection (possibilities of dissemination of the research results for all developed countries of the world).

As is seen from Table 2, direct average of the digital competitiveness index in the selection of developing countries constitutes 60.482 points. Coefficient of variation is moderate (29.23%), which reflects homogeneity of the selection and guarantees the receipt of the results of its analysis. High (more than 10%) variation of positions of the counties in the ranking is explained by selecting countries with various levels of digital competitiveness of economy for provision of representativeness of the selection (possibility of distribution of the research results to all developing countries of the world).

3 Results

Figure 1 shows that the highest level of digital competitiveness of economies of developed countries is observed for the indicator of integration of information and communication technologies and devices (12th position in the world on average). Also, they have favorable normative and legal conditions for development of the digital economy (14th position in the world) and high level of development of digital infrastructure (15th position in the world). At the same time, the investment support for digital modernization of economy is weak (19th position in the world), as well as interest to it from business (22nd position in the world) (Fig. 2).

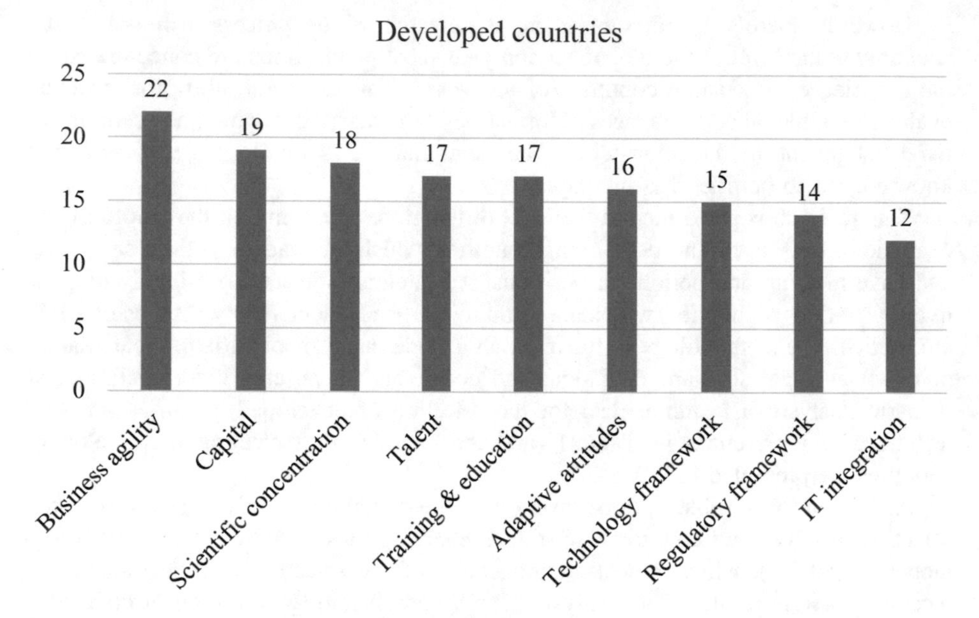

Fig. 1. Direct average of the indicators of digital competitiveness of developed countries in 2019, position. Source: calculated and compiled by the authors.

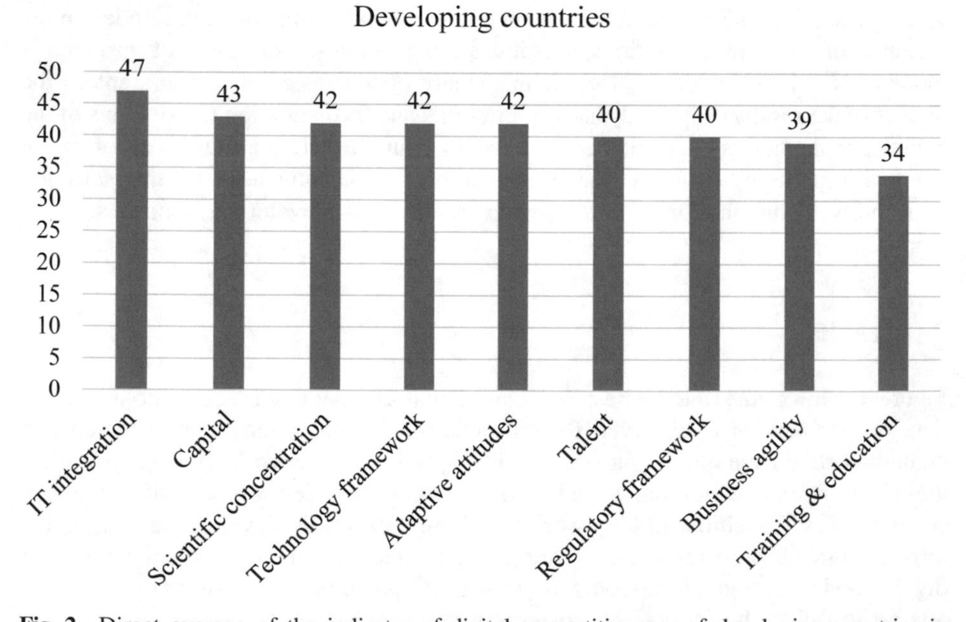

Fig. 2. Direct average of the indicator of digital competitiveness of developing countries in 2019, position. Source: calculated and compiled by the authors.

Figure 1 shows that the highest level of digital competitiveness of economies of developing countries is observed for the indicator of training of digital personnel (34th position in the world on average). Interest of business in opportunities of the digital economy is also very high (39th position in the world). The most serious barrier on the path of increasing the digital competitiveness of economies of developing countries is low level of integration of information and communication technologies and devices (47th position in the world).

As is seen, pros and cons of developed and developing countries are opposite to each other. This allows envisaging that unification of their efforts for joint implementation of digital modernization of economy will allow for the fullest realization of their potential but reduction of the gap between them. According to this, three forecast scenarios of digital modernization of the modern global economy for 2019–2024 are compiled (Figs. 3, 4 and 5).

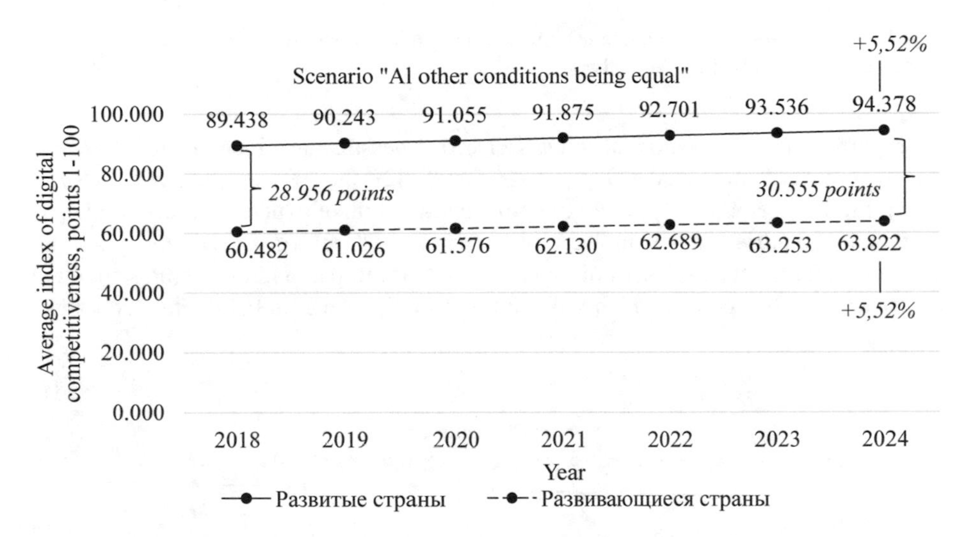

Fig. 3. Scenario of digital modernization of the global economy “All other conditions being equal”, points. Source: calculated and compiled by the authors.

Figure 3 shows that as of now (2018) the gap in the average level of digital competitiveness of developed and developing countries constitutes 28.956 points. All other conditions being equal – i.e., without serious changes of the set tendency of digital modernization of the modern global economy – this will lead to equal increase of the level of digital competitiveness of economies of developed and developing countries and its growth by 5.52% by 2024. As a result, the gap in the level of digital competitiveness of economy of developed and developing countries will reach 30.555 points.

Fig. 4. Scenario of digital modernization of the global economy "Crisis", points. Source: calculated and compiled by the authors.

Figure 4 show that a more negative scenario is possible, at which the growth rate of digital competitiveness of developed countries (7.42% in 2024 as compared to 2018) much more exceeds the growth rate of digital competitiveness of economies of developing countries (0.54% in 2024 as compared to 2018). As a result, the gap in the level of digital competitiveness of economies of developed and developing countries will reach 305.265 points, which will lead to a crisis of the global digital economy.

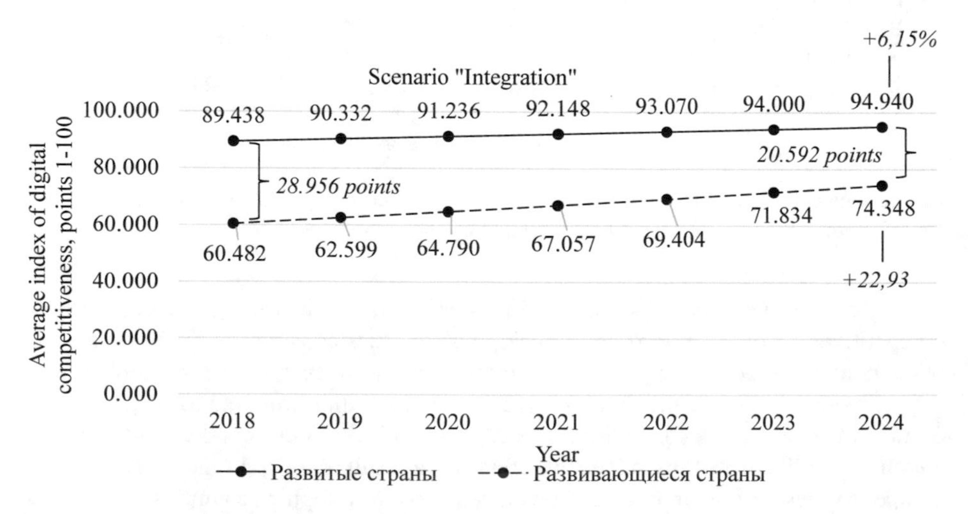

Fig. 5. Scenario of digital modernization of the global economy "Integration", points. Source: calculated and compiled by the authors.

Figure 5 shows that the optimal scenario is the scenario of integration of developed and developing countries, which allows for joint overcoming of their weaknesses and increasing advantages, maximizing the aggregate synergetic effect – reduction of the gap in the level of digital competitiveness of economies of developed and developing countries to 20.592 points by 2024. Growth of the level of digital competitiveness of developed countries in 2024, as compared to 2018, will constitute 6.15%, of developing countries – 22.93% (as their potential is not developed). A perspective mechanism of integration of developed and developing countries in the sphere of formation of the digital economy is formation of transnational digital clusters (in the sphere of infrastructure and entrepreneurship).

4 Conclusions

Thus, as a result of the research it was determined that the basis of economy's digital competitiveness in developed countries is high level of integration of information and communication technologies and devices, and a barrier on the path of its increase is low interest of business in digital modernization. Developing countries have an opposite situation, caused by the low level of integration of information and communication technologies and devices with high interest in digital modernization from business.

The optimal scenario of development of the modern global digital economy in the short-term (2019–2024) is integration of developed and developing countries, which envisages their close interaction and cooperation in the sphere of creation and development of digital infrastructure, and implementation of digital technologies into activities of business structures on the basis of the mechanism of clustering. This will allow reducing the gap in the level of global digital competitiveness of economies of developed and developing countries to 20.592 points (by 28.88%) by 2024.

References

Balnaves, M.: Digital economy planning in Kuwait. In: 11th CMI International Conference, 2018: Prospects and Challenges Towards Developing a Digital Economy within the EU, PCTDDE 2018, 8624795, pp. 32–37 (2019)

Bogoviz, A.V., Alekseev, A.N., Ragulina, J.V.: Budget limitations in the process of formation of the digital economy. In: Lecture Notes in Networks and Systems, vol. 57, pp. 578–585 (2019a)

Bogoviz, A.V., Lobova, S.V., Ragulina, J.V.: Distortions in the theory of costs in the conditions of digital economy. In: Lecture Notes in Networks and Systems, vol. 57, pp. 1231–1237 (2019b)

IMD: World Digital Competitiveness Ranking 2018 (2019). https://www.imd.org/wcc/world-competitiveness-center-rankings/world-digital-competitiveness-rankings-2018/. Accessed 26 Apr 2019

Lazović, V., Duricković, T.: The digital economy in developing countries-challenges and opportunities. In: 2014 37th International Convention on Information and Communication Technology, Electronics and Microelectronics, MIPRO 2014 – Proceedings, 6859817, pp. 1580–1585 (2014)

Mueller, M., Grindal, K.: Data flows and the digital economy: information as a mobile factor of production. Digit. Policy, Regul. Gov. **21**(1), 71–87 (2019)

Nathan, M., Vandore, E., Voss, G.: Spatial imaginaries and tech cities: place-branding East London's digital economy. J. Econ. Geogr. **19**(2), 409–432 (2019)

Negrea, A., Ciobanu, G., Dobrea, C., Burcea, S.: Priority aspects in the evolution of the digital economy for building new development policies. Qual. – Access Success **20**(S2), 416–421 (2019)

Popkova, E.G.: Preconditions of formation and development of industry 4.0 in the conditions of knowledge economy. Stud. Syst. Decis. Control **169**, 65–72 (2019)

Popkova, E.G., Ragulina, Y.V., Bogoviz, A.V.: Fundamental differences of transition to industry 4.0 from previous industrial revolutions. Stud. Syst. Decis. Control **169**, 21–29 (2019)

Pritvorova, T., Tasbulatova, B., Petrenko, E.: Possibilities of blitz-psychograms as a tool for human resource management in the supporting system of hardiness of company. Entrep. Sustain. Issues **6**(2), 840–853 (2018). https://doi.org/10.9770/jesi.2018.6.2(25)

Sako, M.: Technology strategy and management free trade in a digital world: Considering the possible implications for free trade, traditionally based on non-digital goods, for a modern global economy that is increasingly based on intangible products and services enabled by digital technologies. Commun. ACM **62**(4), 18–21 (2019)

Tsai, I.-C., Wu, H.-J., Liao, C.-H., Yeh, C.-H.: An innovative hybrid model for developing cross domain ICT talent in digital economy. In: Proceedings of 2018 IEEE International Conference on Teaching, Assessment, and Learning for Engineering, TALE 2018, 8615150, pp. 745–750 (2019)

Turdubekov, B.M., Karbekova, A.B., Makhmudova, G.U.: Formation of the economic strategy of development of civilized entrepreneurship. Bull. Jalal-Abad State Univ. **2**(37), 104–108 (2018)

Sustainable Development of the Digital Economy on the Basis of Managing Social and Technological Threats

Maynat M. Chazhaeva[1(✉)], Alla A. Serebryakova[2], Gulzat K. Tashkulova[3], and Nurgul K. Atabekova[4]

[1] Chechen State University, Grozny, Russia
mchm-1976@mail.ru
[2] Immanuel Kant Baltic Federal University, Kaliningrad, Russia
a.serebryakova@inbox.ru, ASerebryakova@kantiana.ru
[3] Kyrgyz National University named after J. Balasagin, Bishkek, Kyrgyzstan
[4] Kyrgyz State Law Academy, Bishkek, Kyrgyzstan

Abstract. Purpose: The purpose of the paper is to determine the perspectives of bringing the model of the digital economy in accordance with the current requirements to its sustainability and to develop the recommendations for managing its threats, the key of which – in the case of the digital economy – are social and technological threats.

Design/Methodology/Approach: The authors use scenario modeling for studying the accumulated experience of formation of the digital economy (2013–2018) and determining the scenarios of development of the digital economy in the long-term from the positions of its sustainability. The analyzed indicator is growth rate of GDP in constant prices. The research objects are two countries that occupy the leading positions in the global rating of countries as to the level of competitiveness of the digital economy (IMD) – the USA and Singapore; two countries with peripheral positions in this rating – China and Russia; and two countries at the bottom of the rating – Indonesia and Venezuela.

Findings: It is determined that practical implementation of the model of the digital economy does not guarantee its sustainable development – its growth could take place according to one of the three scenarios: stable development (absence of fluctuations of growth rate of GDP in stable prices – e.g., in Indonesia and China); unstable development and crisis (vivid fluctuations of growth rate of GDP in constant prices – e.g., in Venezuela and Russia); sustainable growth (the most preferable scenario, which envisages increase of growth rate of GDP in constant prices – e.g., in Singapore and the USA). Social and technological factors largely determine the scenario according to which the digital economy develops.

Originality/Value: The most probable threats to the digital economy are determined. For managing them, the authors offer recommendations, which practical implementation will stimulate the neutralization of social and technological threats, will allow preventing crises, and will ensure stability or sustainable growth of the digital economy in the long-term.

Keywords: Sustainable development · Digital economy · Management · Social threats · Technological threats

E. G. Popkova and B. S. Sergi (Eds.): ISC 2019, LNNS 87, pp. 49–56, 2020.
https://doi.org/10.1007/978-3-030-29586-8_6

JEL Code: D81 · G01 · H12 · Q01 · O31 · O32 · O33 · O38

1 Introduction

The digital economy is a new method of organization of economic systems, at which digital technologies are widely implemented and actively used. This is a new model, a landmark, to which the modern countries of the world strive. That's why high requirements from interested parties are set to it – this model has to provide opportunities for increasing the economy's competitiveness and to create the potential for growth of effectiveness of economic activities, as well as to ensure increase of population's living standards.

During studying the digital economy attention is usually paid to the possibilities of obtaining socio-economic advantages from it. Though, as the results of multiple scientific studies confirm, the digital economy fully conforms to the requirements of creation of advantages for the interested parties, this is not enough for acknowledging its preference as compared to other existing models (e.g., the preceding models of the post-industrial economy), as uncertainty as to potential threats to the digital economy and perspectives of their management remains.

As the world experience of practical implementation of the model of the post-industrial economy showed, the other side of quick economic growth could be formation of economic bubbles and, as a result, emergence of the crisis of an economic system. That's why among the requirements to a new model of economy the important role belongs to the criterion of its sustainability – i.e., ability to ensure expected advantages over a long period of time, without causing economic crises. The purpose of the paper is to determine the perspectives of bringing the model of the digital economy in accordance with the current requirements to its sustainability and to develop the recommendations for managing its threats, the key of which – in the case of the digital economy – are social and technological threats.

2 Materials and Method

Development of the conceptual model of the digital economy and overview and critical consideration of the accumulated experience of its practical implementation are studied in multiple works: Ansong and Boateng (2019), Bogoviz et al. (2019a), Bogoviz et al. (2019b), Sako (2019), etc. Sustainable development of the modern economy is studied in the works Kurniawan and Managi (2018), Morozova et al. (2018), Niță (2019), Popkova et al. (2019), Shakhovskaya et al. (2018), Solarin et al. (2019), etc.

Despite this, the problem of sustainable development of the digital economy, its potential social and technological threats, and possibilities and perspectives of their management are poorly studied and require further research. The authors use the method of scenario modeling for studying the accumulated experience of formation of the digital economy (2013–2018) and determining the scenarios of development of the digital economy in the long-term from the positions of its sustainability.

The analyzed indicator is growth rate of GDP in constant prices (measured in %). The research objects are two countries that occupy the leading positions in the global rating of countries as to the level of competitiveness of the digital economy (IMD 2019) – the USA (1st position, 100.00 points) and Singapore (2nd position, 99.422 points); two countries that occupy peripheral positions in this rating – China (30th position, 74.796 points) and Russia (40th position, 65.207 points); and two countries that are the last in the rating – Indonesia (62nd position, 45.776 points) and Venezuela (63rd position, 24.795 points).

The source of statistical data is the materials of the International Monetary Fund. For precise and correct compilation of the scenarios of development of the digital economy in the long-term period from the positions of its sustainability, the authors perform the research within the time period of formation of the digital economy (2013–2018) and before (2000–2012) and after its (forecast of the experts of the International Monetary Fund). The results of the performed scenario modeling are presented in Figs. 1, 2 and 3, each of which reflects one of the possible scenarios of development of the digital economy in the long-term from the positions of its sustainability.

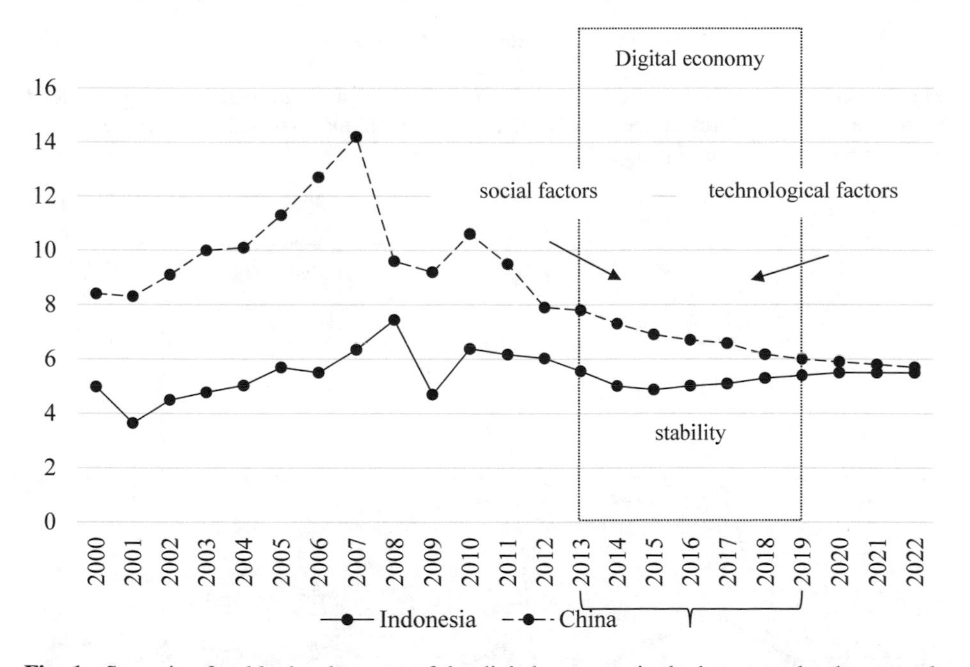

Fig. 1. Scenario of stable development of the digital economy in the long-term by the example of Indonesia and China (growth rate of GDP, %). Source: built and analyzed by the authors based on International Monetary Fund (2019).

Figure 1 shows that economies of Indonesia and China are stable on the whole – they showed growth with a peak in 2007–2008 and slight decline in 2009 under the influence of the global recession. In 2013–2018, growth rate of their GDP in constant prices was stable (unchanged) due to the neutral influence of the social and technological factors.

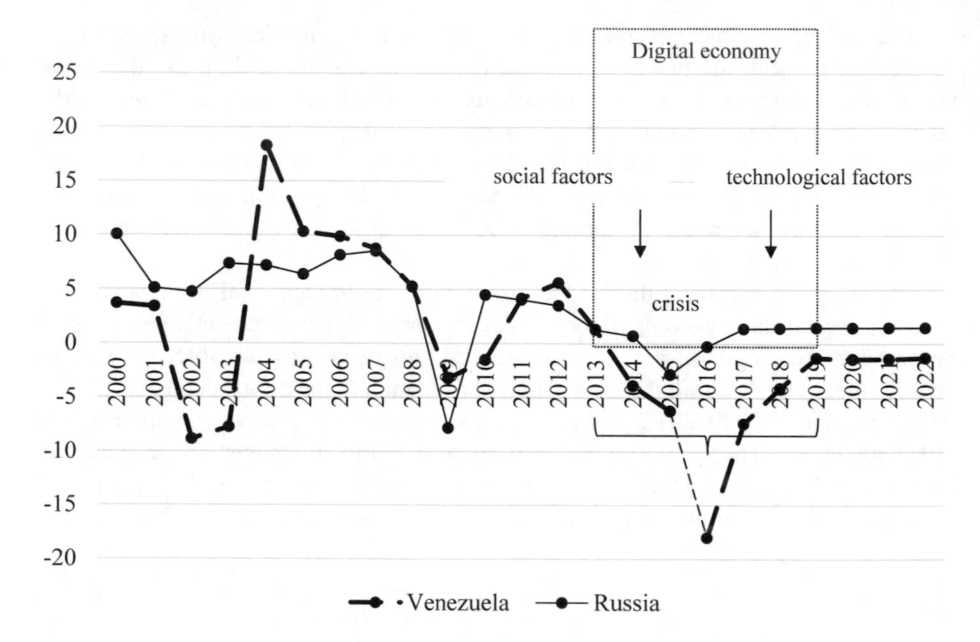

Fig. 2. Scenario of the crisis of the digital economy in the long-term period by the example of Indonesia and China (growth rate of GDP, %). Source: built and analyzed by the authors based on International Monetary Fund (2019).

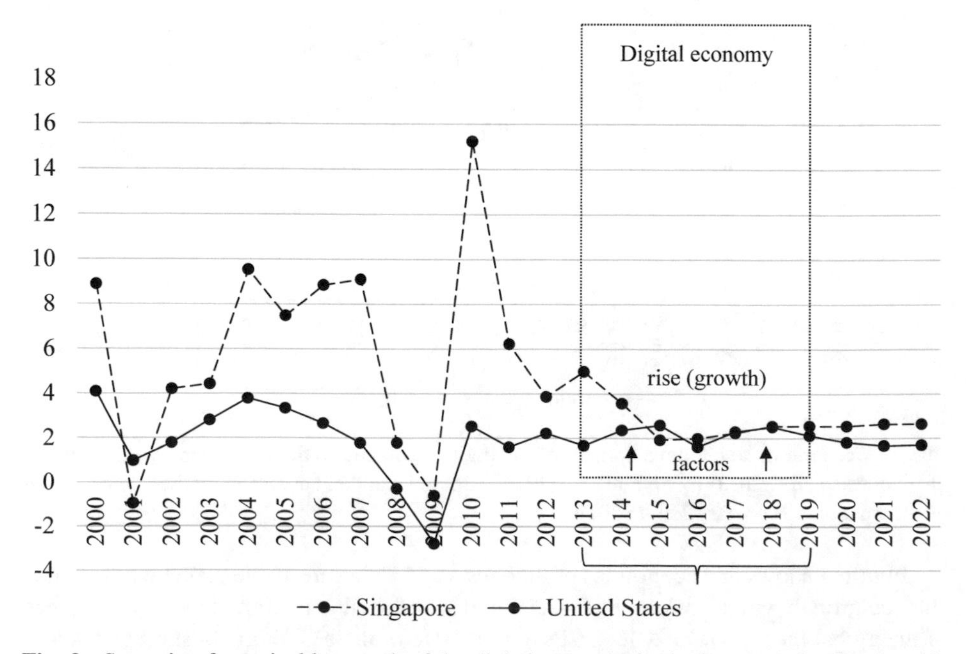

Fig. 3. Scenario of sustainable growth of the digital economy in the long-term by the example of Indonesia and China (growth rate of GDP, %). Source: built and analyzed by the authors based on International Monetary Fund (2019).

Figure 2 shows that economies of Indonesia and China are peculiar for constant fluctuations of growth rate of GDP in constant prices. The period of 2015–2016 was another crisis of these economies, and its reason cannot be precisely determined. However, the social and technological factors have a vivid negative influence on the digital economy.

Figure 3 shows that in Singapore and the USA growth rate of GDP in constant prices is also unstable, with its peaks in 2000 and 2010, and declines in 2001 and 2009. The period of 2016–2018 is peculiar for poorly expressed growth of economies of these countries, though it cannot be directly connected to the digital economy. However, the social and technological factors have favorable influence on the digital economy, supporting its sustainable growth and stability.

3 Results

The determined scenarios of development of the digital economy in the long-term period showed that the social and technological factors require management, as they could create the threats and cause crises. The most probable social and technological threats to the digital economy are determined, and measures for managing them are offered (Table 1).

Table 1. Social and technological threats to the digital economy and the offered measures for managing them.

Threats to the digital economy		Offered measures of managing the threats
Social	Increase of social differentiation according to the criterion of possession of digital competencies	– provision of mass accessibility of educational services on mastering of digital competencies; – social marketing of digital competencies by the state, aimed at actualization of their necessity and advantages for a modern human (employee and consumer)
	Reduction of accessibility of goods and services for the population with low level of digital literacy and inaccessibility of digital devices	– development of infrastructure of the digital economy; – creation of the centers of mass access to digital technologies, support for obtaining digital goods and services, and increase of digital literacy

(*continued*)

Table 1. (*continued*)

Threats to the digital economy		Offered measures of managing the threats
Technological	Instability of the work of digital devices	– development of infrastructure of the digital economy; – establishment of responsibility of suppliers of the infrastructure for the work of digital devices; – creation of the centers of reserve support of functioning of digital devices; – insurance of risks of termination of work of digital devices
	Lack of protection of digital information and data	– simplified and quick consideration of requests on the issues of protection of digital information and data; – development of the systems of digital security and provision of their wide accessibility

Source: compiled by the authors.

Table 1 shows that one of the social threats is increase of social differentiation according to the criterion of possession of digital competencies (e.g., various opportunities for employment and career building, different level of income). For managing this threat, it is offered to ensure wide accessibility of educational services on mastering of digital competencies and social marketing of digital competencies by the state, which is aimed at actualization of their necessity and advantages for a modern human (employee and consumer).

The second social threat is reduction of accessibility of goods and service for population with the low level of digital literacy and inaccessibility of digital devices. For managing this threat, it is recommended to develop the infrastructure of the digital economy (increase of accessibility of digital devices and technologies), to create the centers of mass access to digital technologies, to support obtaining digital goods and services, and to increase digital literacy.

The first determined technological threat to the digital economy is instability of the work of digital devices, for which management it is offered to develop the infrastructure of the digital economy, establish responsibility of suppliers of the infrastructure for the work of digital devices, create the centers of reserve support for functioning of digital devices, and insure the risk of termination of work of digital devices.

The second technological threat to the digital economy is connected to lack of protection of digital information and data. For managing it, quick consideration of requests and applications on the issues of protection of digital information and data is recommended, as well as development of the systems of digital security and provision of their wide accessibility.

Due to implementation of the offered recommendations, the social and technological threats will be neutralized and there could be achieved positive influence of the social and technological factors on the digital economy, which is especially important for the countries with a high risk of crisis (e.g., Venezuela and Russia). As a result, the digital economy will be showing stability and sustainable development in the long-term.

4 Conclusion

Thus, practical implementation of the model of the digital economy does not guarantee its sustainable development – its growth could take place according to one of the three scenarios: stable development (absence of fluctuations of growth rate of GDP in constant prices – e.g., in Indonesia and China), unstable development and crisis (vivid fluctuations of growth rate of GDP in constant prices – e.g., in Venezuela and Russia), and sustainable growth (the most preferable scenario, which envisages increase of growth rate of GDP in constant prices – e.g., in Singapore and the USA).

Social and technological factors largely determine the scenario according to which the digital economy develops. The most probably connected threats are increase of social differentiation according to the criterion of possession of digital competencies, reduction of accessibility of goods and services for the population with low level of digital literacy and inaccessibility of digital devices, instability of the work of digital devices, and lack of protection of digital information and data. For managing them, the authors' recommendations are offered. Their practical implementation will stimulate neutralization of the social and technological threats, prevent the crises, and ensure stability or sustainable growth of the digital economy in the long-term.

References

Ansong, E., Boateng, R.: Surviving in the digital era – business models of digital enterprises in a developing economy. Digit. Policy Regul. Gov. **21**(2), 164–178 (2019)

Bogoviz, A.V., Alekseev, A.N., Ragulina, J.V.: Budget limitations in the process of formation of the digital economy. In: Lecture Notes in Networks and Systems, vol. 57, p. 578–585 (2019a)

Bogoviz, A.V., Lobova, S.V., Ragulina, J.V.: Distortions in the theory of costs in the conditions of digital economy. In: Lecture Notes in Networks and Systems, vol. 57, pp. 1231–1237 (2019b)

IMD: World Digital Competitiveness Ranking 2018 (2019). https://www.imd.org/wcc/world-competitiveness-center-rankings/world-digital-competitiveness-rankings-2018/. Accessed 23 Apr 2019

International Monetary Fund: World Economic Outlook Database (2019). https://www.imf.org/external/pubs/ft/weo/2017/01/weodata/weoselgr.aspx. Accessed 23 Apr 2019

Kurniawan, R., Managi, S.: Economic growth and sustainable development in Indonesia: an assessment. Bull. Indones. Econ. Stud. **54**(3), 339–361 (2018)

Morozova, I.A., Popkova, E.G., Litvinova, T.N.: Sustainable development of global entrepreneurship: infrastructure and perspectives. Int. Entrep. Manag. J. 1–9 (2018)

Niță, S.C.: Sustainable development and economic growth. Qual. - Access Success **20**(S2), 428–432 (2019)

Popkova, E.G., Inshakov, O.V., Bogoviz, A.V.: Regulatory mechanisms of energy conservation in sustainable economic development. In: Lecture Notes in Networks and Systems, vol. 44, pp. 107–118 (2019)

Sako, M.: Technology strategy and management free trade in a digital world: Considering the possible implications for free trade, traditionally based on non-digital goods, for a modern global economy that is increasingly based on intangible products and services enabled by digital technologies. Commun. ACM **62**(4), 18–21 (2019)

Shakhovskaya, L., Petrenko, E., Dzhindzholia, A., Timonina, V.: Market peculiarities of natural gas: case of the Pacific Region. Entrep. Sustain. Issues **5**(3), 555–564 (2018). https://jssidoi.org/jesi/article/167

Solarin, S.A., Shahbaz, M., Hammoudeh, S.: Sustainable economic development in China: modelling the role of hydroelectricity consumption in a multivariate framework. Energy **168**, 516–531 (2019)

The Algorithm of Modern Russia's Transition to the Digital Economy

Aleksei A. Shulus[1(✉)], Anna Zarudneva[2], Stanislav Yatsechko[3], and Olga Fetisova[4]

[1] G.V. Plekhanov, Russian University of Economics, Moscow, Russia
shulus@bk.ru
[2] Volgograd State Technical University, Volgograd, Russia
volgoeconA@yandex.ru
[3] Financial University under the Government of the Russian Federation, Moscow, Russia
[4] Institute of Management and Regional Economics, Volgograd State University, Volgograd, Russia
fetissova@volsu.ru

Abstract. Purpose: The purpose of the paper is to solve the problem of implementing the national program "Digital economy of the Russian Federation" and to develop the algorithm of modern Russia's transition to the digital economy.

Design/Methodology/Approach: The authors determine the general (universal) algorithm of transition of a modern economic system to the digital economy. Evaluation of efficiency of modern Russia's transition to the digital economy is performed. Collection, systematization, and analysis of the existing statistical information are conducted with the help of the method of comparative analysis, which allows comparing the level of digitization of the Russian economy and economies of other countries of the world and determining the stage of transition to the digital economy in Russia. Then, the barriers and perspectives of finishing this transition are determined, and the authors' recommendations are offered – based on which the algorithm of transition to the digital economy of modern Russia is compiled.

Findings: It is proved that modern Russia successfully performs transition to the digital economy, together with the most progressive developed and developing countries of the world. As of the 2019 data, Russia is finishing the third stage of this transition (out of four stages of the universal algorithm), which is connected to provision of mass usage of the Internet. The next (final) stage should be mass distribution of breakthrough digital technologies.

Originality/Value: The algorithm of modern Russia's transition to the digital economy is offered. It will allow overcoming the barriers on the path of finishing this process with joint efforts of the state, business, and society. Main attention is paid to the issues of formation of effective demand for breakthrough digital technologies. It includes three consecutive steps: development of information society, formation of technological reserves, and implementation of breakthrough digital technologies; thus, Russia's transition to the digital economy will be finished by 2024.

E. G. Popkova and B. S. Sergi (Eds.): ISC 2019, LNNS 87, pp. 57–63, 2020.
https://doi.org/10.1007/978-3-030-29586-8_7

Keywords: Digital economy · Global competitiveness · Innovative development · Digital modernization · Digitization · Modern Russia

JEL Code: O31 · O32 · O33 · O38

1 Introduction

Implementation of digital technologies into the economic practice of modern Russia initially has the character of modernization of the economy for increasing the level of its global competitiveness. This envisaged limited digitization – selected application of the most accessible and popular technologies in certain spheres of industry by the companies that are most inclined for manifestation of innovative activity. Different specialized projects were used for this. They were aimed at acceleration of usage of digital technologies in society (formation of information society), business (development of online trade and hi-tech industry) and the state (formation of e-government).

Under the influence of the global digital competition, the scale of the described process was increased, and the national program "Digital economy of the Russian Federation" was adopted by the decree of the Government of the Russian Federation (2019) No. 1632-r. This program seeks the goal of modern Russia's transition to the digital economy as a new type of economic system, in which digital technologies dominate. Thus, the problem of scientific search for the methods of practical implementation of this transition has become every important.

This paper is to solve this problem and to develop the algorithm of modern Russia's transition to the digital economy. For achieving this goal, a general (universal) algorithm of transition of a modern economic system to the digital economy is compiled at first. Then, evaluation of efficiency of modern Russia's transition to the digital economy is performed. After that, the barriers and perspectives of finishing this process are determined, and authors' recommendations are offered. Based on this, the algorithm of modern Russia's transition to the digital economy, which takes into account its specifics, is compiled.

2 Materials and Method

The fundamental and applied issues of transition to the digital economy, which consist in mass authomatization of production and consumption of goods and services in the economic system on the basis of digital technologies in the interests of growth of effectiveness of usage of production factors (primarily, growth of labor efficiency) and increase of global competitiveness, are studied in detail in the works Ansong and Boateng (2019), Bezrukova et al. (2019), Meier and Manzerolle (2019), Nathan et al. (2019), Negrea et al. (2019), Petrenko et al. (2019), Popkova (2019), Popkova et al. (2019), Sako (2019), and Yu et al. (2019).

Though the specifics of this process in modern Russia is not studied sufficiently, the accumulated knowledge and practical experience (of developed and developing countries) allow compiling a universal algorithm of the economic system's transition to the digital economy (Fig. 1).

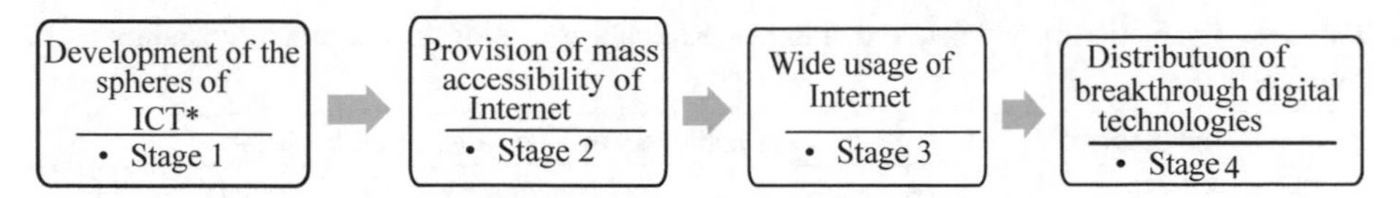

*ICT – information and communication technologies.

Fig. 1. The universal algorithm of the economic system's transition to the digital economy. Source: compiled by the authors.

As is seen from Fig. 1, at the first stage the development of the spheres of information and communication technologies is performed. For example, according to the data of the National Research University "Higher School of Economics" (2019), the highest level of development of the spheres of ICT is observed in South Korea, where their share in the structure of gross added value constitutes 10.3%. In the USA, Estonia, and Japan, the share of the spheres of ICT in the structure of gross added value constitutes 6.0%, in the UK – 5.4%, in Germany – 5.0%, in France – 4.6%, and in Italy – 3.6%. At the second stage, wide accessibility of the Internet is provided. In the countries with the highest level of global digital competitiveness (as of early 2019) – e.g., the USA, Singapore, Canada, the UK, etc. – accessibility of the Internet strives towards 100%.

At the third stage, mass usage of the Internet is provided. For example, 31% of the population of Finland use the Internet for search for a job (22% in the USA and 15% in South Korea); 20% of Americans use the Internet for remote education (14% in the UK and 6% in Germany), 82% of the population of the UK use the Internet for ordering goods and services (57% in the USA and 32% in Japan).

At the fourth stage, distribution of breakthrough digital technologies – e.g., the Internet of Things, blockchain, etc. – takes place. The state could use them for improving the practice of state management of economy on the basis of artificial intelligence for protecting online data with the help of quantum technologies. In business, breakthrough digital technologies envisage primarily robotization, which covers primarily the processing industry. For example, in the UK there are 74 robots per 10,000 employees, in China – 74 robots per 10,000 employees, and in Brazil – 10 robots per 10,000 employees (International Federation of Robotics 2019). "Smart technologies" could be used in society.

For evaluating the efficiency of modern Russia's transition to the digital economy, the authors perform collection, systematization, and analysis of the existing statistical information and use the method of comparative analysis for comparing the level of digitization of the Russian economy and other economies and precise determination of the stage of transition to the digital economy in Russia.

3 Results

The results of the performed evaluation of the efficiency of modern Russia's transition to the digital economy are given in Table 1.

Table 1. Evaluation of efficiency of modern Russia's transition to the digital economy as of early 2019.

Stage	Indicator of efficiency	Results that are achieved in Russia as of early 2019		
		State	Business	Society
1	Share of the sphere of ICT[a] in economy	2.7% of GDP[b]	3.4% of GAV[c]	1.7% of employment
2	Access to the Internet	100%	86% of organizations	76.3% of households
3	Usage of the Internet	94.5% of government bodies; 42.3% of state services are provided in the electronic form	81.6% of organizations, 18.1% for purchases, 12.1% for sales	73% of population use the Internet, 32% for ordering goods and services
4	Usage of breakthrough digital technologies	Artificial intelligence, quantum and other technologies are not used	3 industrial robots per 10,000 employees in the processing industry	"smart technologies" are used by less than 1% of population

[a]ICT – information and communication technologies;
[b]GDP – gross domestic product;
[c]GAV – gross added value.
Source: compiled by the authors based on the materials of the National Research University "Higher School of Economics" (2019), International Federation of Robotics (2019).

Table 1 shows that according to the data of the National Research University "Higher School of Economics", as of early 2019 the share of the sphere of ICT constitutes 2.7% of Russia's GDP; 3.4% of gross added value is created in it and 1.7% of employment is provided – the values of these indicators are at the level of the most progressive developed and developing countries of the world. This shows formation of the spheres of ICT and their important role in functioning and development of the modern Russia's economy, which, in its turn, shows successful finish of the first stage of transition to the digital economy.

Access of public authorities and local administration to the Internet is fully provided; 86% of organizations and 76.3% of households of Russia also have access to the Internet. The level of Internet coverage of all economic subjects in modern Russia is very high and similar to the level of the most progressive countries of the world – which shows successful finish of the second stage of transition to the digital economy.

The Internet is regularly used by 94.5% of public authorities and local administration, but only 42.3% of state services are provided online. The Internet is used by 81.6% of modern Russian organizations, but only 18.1% - for purchases and 12.1% - for sales. 73% of the Russian population use the Internet, but only 32% - for online orders of goods and services. Thus, there are large perspectives of increasing the activity of usage of the Internet in Russia in the directions that are preferable for the digital economy – e-government and online trade. Successful finish of this stage of transition to the digital economy in modern Russia is hindered by the low level of trust online trade due to insufficient protection of its security and low reliability of the Internet; thus, the advantages of Internet trade (e.g., low price with the same quality) are rarely obtained in practice.

Breakthrough digital technologies are poorly used in modern Russia. According to the calculations of the International Federation of Robotics (2019), there are 3 industrial robots per 10,000 employees in Russia. "Smart technologies" are used by less than 1% of population, and AI, quantum technologies, and other technologies are not used by the state. Ubiquitous and mass usage of breakthrough digital technologies in modern Russia is hindered by their deficit and high price. For finishing modern Russia's transition to the digital economy we suggest using the following algorithm (Fig. 2).

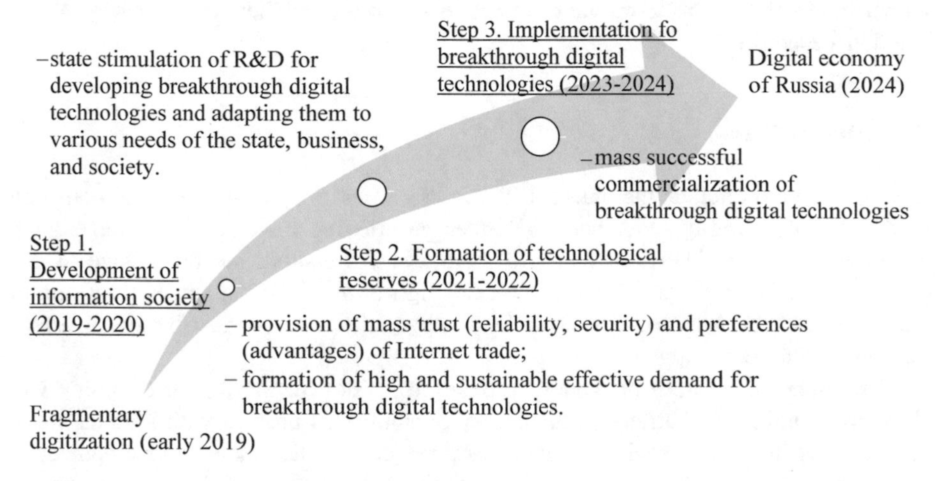

Fig. 2. The algorithm of modern Russia's transition to the digital economy. Source: compiled by the authors.

As of early 2019, Russia is already at the finishing stages (between 3rd and 4th) of transition to the digital economy (according to the universal algorithm), so the offered algorithm has steps within finishing the 3rd stage and successful implementation of the 4th stage. The first step is development of information society in 2019–2020. For this, it is recommended to provide mass trust (reliability and security) and preferences (advantages) of Internet trade, which envisages development of the normative & legal and infrastructural (including security systems) of provision of the digital economy. It is also suggested to form high and sustainable effective demand for breakthrough digital technologies by increasing the level of population's digital literacy, training of digital personnel, and increasing the level of disposable incomes of the population.

The second step is formation of technological reserves in 2021–2022. State stimulation for R&D for developing breakthrough digital technologies and their adaptation to various needs of the state, business, and society is recommended. This could envisage the development of the system of grants to specialized research institutes and the system of tax stimulation of the activities of innovations-active business structures, as well as other accessible measures. The key criterion of provision of state stimuli should be simultaneous combination of creation of additional value for interested parties and mass accessibility (price) of the technologies.

The third step is implementation of breakthrough digital technologies in 2023–2024. This requires mass successful commercialization of breakthrough digital technologies, which allows making them widely accessible as to quantity (overcoming the deficit and formation of manufacturers' competition) and as to price. As a result, it is possible to finish the process of Russia's transition to the digital economy in 2024, which conforms to the terms that are set in the program "Digital economy of the Russian Federation".

4 Conclusions

It is possible to conclude that modern Russia us successfully conducting the transition to the digital economy, together with other progressive developed and developing countries of the world. As of early 2019, Russia is finishing the third stage of this transition (out of four stages of the universal algorithm), which includes provision of mass usage of the Internet. The next (final) stage will be mass distribution of breakthrough digital technologies.

The offered algorithm of modern Russia's transition to the digital economy will allow overcoming the barriers on the path of finishing this process with joint efforts of the state, business, and society. It includes three consecutive stages: development of information society, formation of technological reserves, and implementation of breakthrough digital technologies – thus providing the completion of Russia's transition to the digital economy by 2024. The main attention in the offered algorithm is paid to the issues of formation of effective demand for breakthrough digital technologies.

This requires reduction of their prices and increase of disposable incomes of the population. Both these tasks could be performed in modern Russia, but they the state has to be interested in solving them – as they require financing from the federal budget. The issues of financial provision of completion of modern Russia's transition to the digital economy according to the offered algorithm require further thorough research.

References

Ansong, E., Boateng, R.: Surviving in the digital era – business models of digital enterprises in a developing economy. Digit. Policy Regul. Gov. **21**(2), 164–178 (2019)

Bezrukova, T.L., Kuksova, I.V., Kirillova, S.S., Gyiazov, A.T.: Forecasting development of forest complex in the formation of digital economy. In: IOP Conf. Ser.: Earth Environ. Sci. **226**(1), 012063 (2019)

Meier, L.M., Manzerolle, V.R.: Rising tides? Data capture, platform accumulation, and new monopolies in the digital music economy. New Media Soc. **21**(3), 543–561 (2019)

Nathan, M., Vandore, E., Voss, G.: Spatial imaginaries and tech cities: place-branding East London's digital economy. J. Econ. Geogr. **19**(2), 409–432 (2019)

Negrea, A., Ciobanu, G., Dobrea, C., Burcea, S.: Priority aspects in the evolution of the digital economy for building new development policies. Qual. - Access Success **20**(S2), 416–421 (2019)

Petrenko, E., Pritvorova, T., Dzhazykbaeva, B.: Sustainable development processes: service sector in post-industrial economy. J. Secur. Sustain. Issues **7**(4), 781–791 (2018). https://doi.org/10.9770/jssi.2018.7.4(14)

Popkova, E.G.: Preconditions of formation and development of industry 4.0 in the conditions of knowledge economy. Stud. Syst. Decis. Control **169**, 65–72 (2019)

Popkova, E.G., Ragulina, Y.V., Bogoviz, A.V.: Fundamental differences of transition to industry 4.0 from previous industrial revolutions. Stud. Syst. Decis. Control **169**, 21–29 (2019)

Sako, M.: Technology strategy and management free trade in a digital world: Considering the possible implications for free trade, traditionally based on non-digital goods, for a modern global economy that is increasingly based on intangible products and services enabled by digital technologies. Commun. ACM **62**(4), 18–21 (2019)

Yu, H., Goggin, G., Fisher, K., Li, B.: Introduction: disability participation in the digital economy. Inf. Commun. Soc. **22**(4), 467–473 (2019)

International Federation of Robotics: Robotization of the processing industry (2019). https://rb.ru/story/countries-with-greatest-density-of-robots/. Accessed 26 Apr 2019

National Research University "Higher School of Economics": Digital economy – 2019: a short statistical collection (2019). https://www.hse.ru/data/2018/12/26/1143130930/ice2019kr.pdf. Accessed 26 Apr 2019

Government of the Russian Federation: Program "Digital economy of the Russian Federation", adopted by the decree dated July 28, 2017, No. 1632-r (2019). http://static.government.ru/media/files/9gFM4FHj4PsB79I5v7yLVuPgu4bvR7M0.pdf. Accessed 26 Apr 2019

The Institutional Model of Building the Digital Economy in Modern Russia

Elena V. Belokurova[1(✉)], Sergey V. Pizikov[2], Elena S. Petrenko[3], and Gaukhar K. Koshebayeva[4]

[1] Tyumen Industrial University (Nizhnevartovsk branch), Nizhnevartovsk, Russia
e.belokurowa@yandex.ru

[2] Higher School of Economics and Management, South Ural State University, Chelyabinsk, Russia
psv_uk@mail.ru

[3] G.V. Plekhanov, Russian University of Economics, Moscow, Russia
petrenko_yelena@bk.ru

[4] Karaganda State Technical University, Karaganda, Kazakhstan
gauhark@bk.ru

Abstract. Purpose: The purpose of the research is to develop the institutional model of building the digital economy in modern Russia.

Design/Methodology/Approach: The authors use the methodology of the institutional economic theory – scientific induction, namely, the method of similarity. This method is used for determining the common causal connections of the preconditions, normative & legal regulation, and the economic practice at different stages of the process of institutionalization of the digital economy in Russia. The method of structuring is used for distinguishing the stages of this process. The method of modeling of socio-economic processes and systems is used for systemic presentation of this process.

Findings: It is determined that the process of institutionalization of the digital economy is implemented in three stages: formation of the system of e-government, information society, and the digital economy. This process has its specifics (it is predetermined by external factors, has a reverse, and faces the absence of opportunities of practical implementation of the government plan) – i.e., it is created artificially (by the government), as opposed to the existing economic practice.

Originality/Value: It is substantiated that Russia is peculiar for contradiction of interests of the state, which are established in the legal norms, and interests of the society, which are reflected in the economic practice, in the process of building the digital economy. It is proved that this contradiction hinders the formation of the digital economy in Russia and reduces its global competitiveness. The perspectives of further institutionalization of the digital economy in Russia are outlined, and recommendations for overcoming the contradiction of the digital economy in Russia, its quick institutionalization, and growth of its global competitiveness are offered.

Keywords: Institutionalization · Digital modernization · Digital competitiveness · Digital economy · Modern Russia

JEL Code: B52 · E02 · E14 · M15 · O14 · O31 · O32 · O33 · O38

E. G. Popkova and B. S. Sergi (Eds.): ISC 2019, LNNS 87, pp. 64–70, 2020.
https://doi.org/10.1007/978-3-030-29586-8_8

1 Introduction

The modern global economy is characterized by competition between the countries for building the digital economy; according to the IMD Business School (2019), 63 developed and developing countries already joined it. Despite the almost simultaneous start of the national programs of formation of the digital economy, developed countries are the leaders in the global rating of digital competitiveness. For example, the USA occupies the first position (100 points), followed by Singapore (99.422 points) and Sweden (97.453 points). Among the developing countries, the Malaysia (80.631 points) has the highest position. Russia is ranked 40th (65.207 points).

Thus, it is possible to offer a hypothesis that formation of the digital economy in developing countries has the institutional specifics, which consist in the fact that unlike developed countries, where this process is harmoniously involved in the process of their evolution, in developing countries this process is artificial and contradicts the existing socio-economic reality, thus requiring more flexible state regulation. Studying the process of institutionalization of the digital economy in different countries is an important task of the modern economic science, as it allows determining their specific problems, suggesting possible solutions, and outlining a perspective trajectory of further development of the digital economy in the interests of increase of its global competitiveness.

Russia's experience deserves more attention, as Russia occupies a position between developed and developing countries, showing high level of development as to certain indicators (e.g., GDP and level of innovative activity) and moderate level of development as to other indicators (e.g., living standards). Formation of the digital economy could be a tool of finishing the market transformation of the Russian economy, and thus its experience could be useful for other countries with transitional economy (e.g., the CIS). This predetermined the purpose of the research – development of an institutional model of building the digital economy in modern Russia.

2 Materials and Method

The concept of the digital economy is studied in a lot of works of the modern authors. As a result of their systematization, we distinguished two conceptual approaches to treatment of the essence of the digital economy. According to the first approach, it is a new type of socio-economic systems. This means that the digital economy is to oust the preceding (i.e., pre-digital) economic practice. This approach is presented in the works Ansong and Boateng (2019), Bogoviz et al. (2019), Bogoviz et al. (2019), Petrenko et al. (2018), and Popkova (2019).

Within the second approach, it is defined as a new method of organization of economic activities and the socio-economic relations that emerge in this process. This means that the digital economy is to supplement the existing economic practice and, though the digital economy could dominate in the socio-economic system, a parallel (alternative) economic practice is realized. This approach is described in the works Galushkin et al. (2019), Ünal (2018), and Vanberg (2018).

Within the both distinguished approaches the digital economy is treated as economic practice that envisages the usage of the modern information and communication – digital – technologies. The alternative (or preceding – depending on the approach) practice is not connected to their usage. Here we use the second approach, as it provides the most correct description of the modern economic reality in the global economic system, in which the digital economy co-exists with the alternative economic practice.

As a result of the performed overview of the research literature on the selected topic, we determined that though the conceptual foundations of the digital economy are determined, the problem of its institutionalization is not elaborated sufficiently in the modern socio-economic systems and, in particular, in Russia.

At the same time, there is a large layer of scientific research in the sphere of the institutional economic theory. Though the object of the research is not the digital economy, they describe the common logic of the process of institutionalization of economic practices, according to which this process is conducted in three generalized (certain authors distinguish more stages) consecutive stages:

1. Emergence of an objective need to solve the current socio-economic problems. By the example of the digital economy this could be problems of increasing the effectiveness of information management, growth of accessibility of goods and services in the economy, transnationalization of business, and increase of global competitiveness;
2. Spontaneous formation of scattered practices of solving the existing problems by representatives of the society, state, and business; selection and establishment (regular repeating) of the successful practices. By the example of the digital economy this could be the practices of using the digital technologies in various types of economic activities (e.g., online trade and online banking);
3. Normative (legal) establishment of the legal foundations of implementing the successful practices for their sustainable reproduction in the economy. By the example of the digital economy this means regulation of requirements, conditions, and principles of usage of digital technologies in various types of economic activities, as well as introduction of the system of sanctions for their unlawful usage (e.g., violation of intellectual property rights and rights for protection of personal information).

The deficit of scientific research on the topic of institutionalization of the digital economy predetermines the necessity for its further research by analyzing the process of formation of the digital economy through the prism of the logic of the institutional theory. For this we use the methodology of the institutional economic theory – scientific induction – for determining the common causal connections of the preconditions, the normative & legal regulation, and economic practice at different stages of the process of institutionalization of the digital economy in Russia. The method of structuring is used for distinguishing the stages of this process. The method of modeling of socio-economic processes and systems is used for systemic presentation of this process.

3 Results

As a result of studying the experience of formation of the digital economy in modern Russia we compiled an institutional model of this process (Fig. 1).

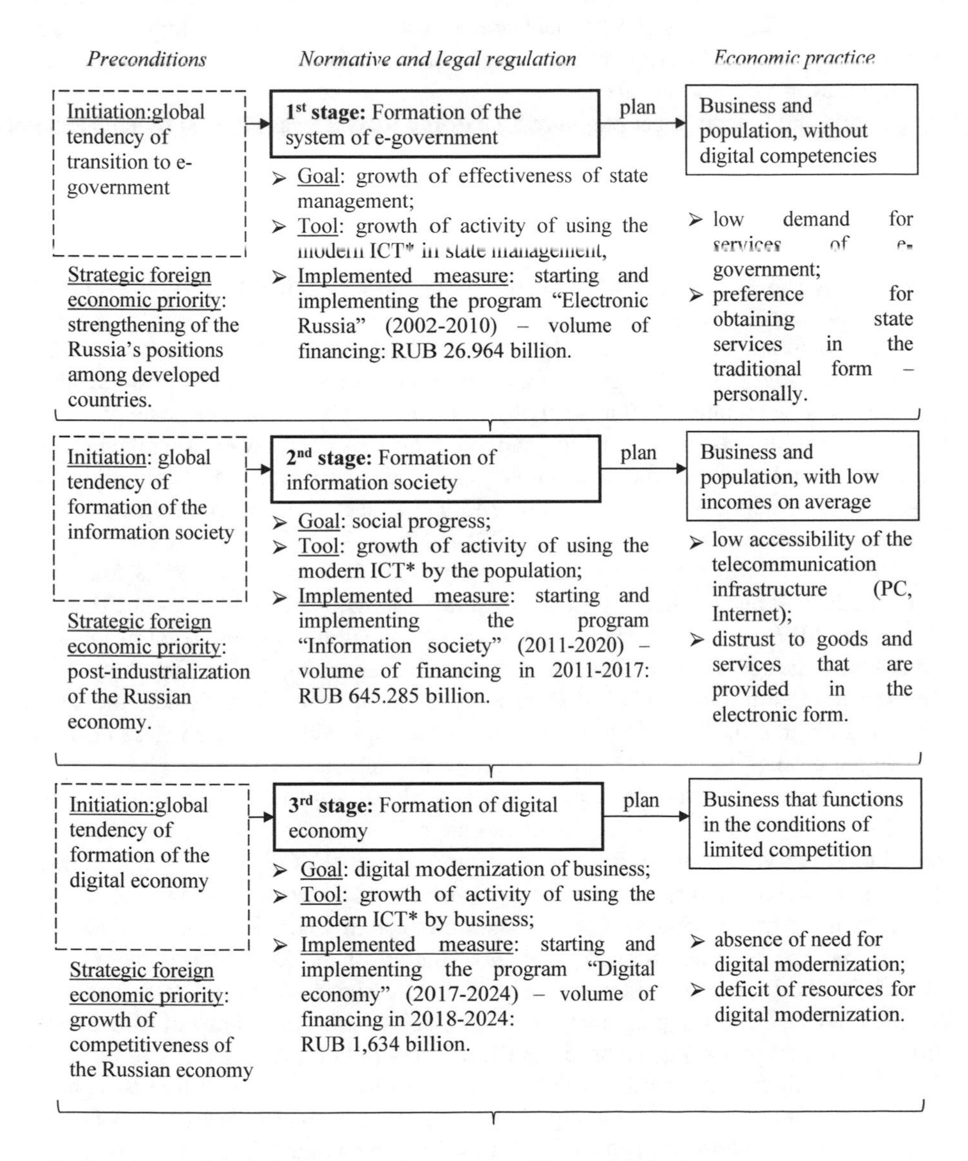

Fig. 1. The institutional model of formation of the digital economy in modern Russia. Source: compiled by the authors.

Figure 1 shows that the process of institutionalization of the digital economy in modern Russia took place in three stages. This process was initially started by the global tendency of transition to e-government. The strategic foreign economic priority then was strengthening of the Russia's positions among developed countries. Thus, the first stage of "Formation of the system of e-government" began. At this stage, the goal of increase of effectiveness of state management was set. The tool of its implementation had to be growth of activity of using the modern information and communication technologies in state management.

For this, the federal target program "Electronic Russia" was adopted by the Decree of the Government of the Russian Federation (2019b) on January 28, 2002, No. 65. The program was set for the period of 2002–2010. The volume of its federal financing constituted RUB 26.964 billion – i.e., 0.04% of GDP for all nine years. Business and population, which did not have mass digital competencies, set a low demand for the services of e-government, preferring state services in the traditional form (obtained in state structures personally).

A precondition to the second stage was the global tendency of formation of the information society. The strategic foreign economic priority was post-industrialization of the Russian economy. At that stage, the purpose of acceleration of social progress was set. The tool of its implementation had to be growth of activity of using the modern information and communication technologies by the population. For this, the state program "Information society" was adopted by the Decree of the Government of the Russian Federation (2019a) on April 15, 2014, No. 313.

The program was set for the period of 2011–2020. The volume of its financing (from the federal budget) in 2011–2017 constituted RUB 645.285 billion – i.e., 0.387% of GDP in 2017, for seven years. Business and population, with low level of income, did not have the access to the telecommunication infrastructure (PC and Internet). They also showed distrust to goods and services that are provided in the electronic form, preferring purchasing goods in traditional retail stores, for bringing their risks down to the minimum (e.g., the risk of supply, return, and exchange of products).

A precondition for the third stage was the global tendency of formation of the digital economy. The strategic foreign economic priority in that moment was growth of global competitiveness of the Russian economy. The goal of digital modernization of business was set at the time. The tool of its implementation has to be growth of activity of usage of the modern information and communication technologies by business.

For that, the national program "Digital economy of the Russian Federation" was adopted by the Decree of the Government of the Russian Federation (2019c) on June 28, 2017, No. 1632-r. The program was set for 2018–2024. The volume of its financing (from the federal budget) is to constitute RUB 1,634 billion – i.e., 0.979% of the 2018 GDP, for all eight years. Business that functions in the conditions of limited competition does not have a need for digital modernization. If also faces the deficit of resources (own, investment, and credit) for its implementation.

The inductive analysis of the process of institutionalization of the digital economy in modern Russia allowed determining the following features that are common for all three distinguished processes. Firstly, influence of the external factors. The domestic demand for application of the modern information and communication technologies was and remains low due to their high cost, quick moral wear, and the necessity for possession of digital competencies, which mastering is not fully ensured in the process of standard education (e.g., higher education).

Secondly, reverse direction of the process of institutionalization. The state compiled a plan not on the basis of the accumulated experience (forecast) but on the basis of priorities. Due to this, the plan is poorly connected to the modern economic practice and cannot be implemented. Instead of direct institutionalization, which envisages demand for legal norms from society and business, Russia is peculiar for reverse institutionalization, which is connected to creation of legal norms and requirement of observing them by society and business, which inevitably leads to opportunism.

Thirdly, absence of the possibilities for practical implementation of the government plan. The normative and legal provision at each stage is fragmentary. The government describes the priorities of the national programs in detail and compiles the plans of their implementation, focusing on the target results. However, the legislative basis for implementing the economic practices that are necessary for achieving these results is incomplete.

For example, stimulation of development of e-commerce at the second stage was ineffective due to uncertainty of the legal field in which it should be conducted. Lack of clarity of rights and responsibilities of sellers and buyers and absence of government guarantees of their observation/execution caused distrust to e-commerce. High cost of computer equipment – due to underdevelopment of the domestic production and high custom fees – and high cost of Internet services made them inaccessible for mass consumers.

4 Conclusions

Thus, in the course of the research the offered hypothesis has been proved. It is shown that the process of institutionalization of the digital economy has its specifics (predetermined by the external factors, has a reverse direction, and faces the absence of possibilities of practical implementation of the government plan) – i.e., it is caused artificially (by the government), contrary to the existing economic practice. Therefore, Russia is peculiar for contradiction of interests of the state, which are established in legal norms, and interests of the society, which are reflected in the economic practice, in the process of formation of the digital economy.

The perspectives of further institutionalization of the digital economy in Russia are connected to overcoming of the determined contradiction. At the modern stage (2019), it is necessary to stimulate the demand for digital technologies from society and business and to ensure their mass accessibility. The following measures of regulation of this process are offered: development of competitiveness of domestic production of digital equipment and provision of high level of competition in the commodity markets, services markets, and markets of telecommunication services (Internet services).

It is also recommended to strengthen the normative and legal provision of application of digital technologies (e.g., in the sphere of e-commerce), guarantee of its observation, modernization of the educational standards, and control over their observation for mass mastering of digital competencies by the population. Due to implementation of the offered recommendations it would be possible to overcome the contradiction of the digital economy in Russia, to ensure its quick institutionalization, and to provide growth of global competitiveness of the Russia's digital economy.

References

Ansong, E., Boateng, R.: Surviving in the digital era – business models of digital enterprises in a developing economy. Digit. Policy Regul. Gov. **21**(2), 164–178 (2019)

Bogoviz, A.V., Alekseev, A.N., Ragulina, J.V.: Budget limitations in the process of formation of the digital economy. In: Lecture Notes in Networks and Systems, vol. 57, p. 578–585 (2019)

Bogoviz, A.V., Lobova, S.V., Ragulina, J.V.: Distortions in the theory of costs in the conditions of digital economy. In: Lecture Notes in Networks and Systems, vol. 57, pp. 1231–1237 (2019)

Galushkin, A.A., Nazarov, A.G., Sabyna, E.N., Skryl, T.V.: The institutional model of formation and development of industry 4.0 in the conditions of knowledge economy's formation. Stud. Syst. Decis. Control **169**, 219–226 (2019)

IMD business school: World Digital Competitiveness Ranking 2018 (2019). https://www.imd.org/wcc/world-competitiveness-center-rankings/world-digital-competitiveness-rankings-2018/. Accessed 01 Apr 2019

Petrenko, E., Pritvorova, T., Dzhazykbaeva, B.: Sustainable development processes: service sector in post-industrial economy. J. Secur. Sustain. Issues **7**(4), 781–791 (2018). https://doi.org/10.9770/jssi.2018.7.4(14)

Popkova, E.G.: Preconditions of formation and development of industry 4.0 in the conditions of knowledge economy. Stud. Syst. Decis. Control **169**, 65–72 (2019)

Ünal, E.: An institutional approach and input–output analysis for explaining the transformation of the Turkish economy. J. Econ. Struct. **7**(1), 3 (2018)

Vanberg, G.: Constitutional political economy, democratic theory and institutional design. Pub. Choice **177**(3–4), 199–216 (2018)

Government of the Russian Federation: Decree dated April 15, 2014, No. 313 "Regarding the adoption of the state program of the Russian Federation" Information society (2011–2020) (2019a). http://www.consultant.ru/document/cons_doc_LAW_162184/. Accessed 02 Apr 2019

Government of the Russian Federation: Decree dated January 28, 2002, No. 65 "Regarding the federal target program" Electronic Russia (2002–2010) (2019b). http://base.garant.ru/184120/. Accessed 02 Apr 2019

Government of the Russian Federation: Decree dated June 28, 2017, No. 1632-r regarding the adoption of the program "Digital economy of the Russian Federation" (2019c). http://static.government.ru/media/files/9gFM4FHj4PsB79I5v7yLVuPgu4bvR7M0.pdf. Accessed 02 Apr 2019

Specific Features of Strategic Planning of the Activities of Entrepreneurial Structures in the Conditions of Digital Transformation of the Modern Economy

Alexandr V. Tekin[1(✉)] and Olga V. Konina[2]

[1] Volgograd State Technical University, Volgograd, Russia
alexander.green.tekin@gmail.com
[2] Moscow State Pedagogical University, Moscow, Russia
koninaov@mail.ru

Abstract. The authors make an attempt of consolidation of the main directions and specific features of digital transformation of the process of strategic planning of entrepreneurial activities and budgeting technology (as a complex technology of planning and managing the entrepreneurial activities) in the modern conditions of the digital economy. The following issues are considered: evaluation of the role of the modern digital technologies in modification of business processes of entrepreneurial activities; evaluation of interconnection of entrepreneurial activities and the digital economy as the basic drivers of development of the modern economic systems; the basic technological, systemic, and complex effects (proved potential of digital transformation) and specific features of the process of strategic planning (on the whole and separate functional directions) of the modern entrepreneurial activities in the conditions of economy's digital transformation; the main directions and enlarged algorithm of digital transformation of budgeting as a complex technology of strategic planning of entrepreneurial activities. The economic category of information is treated as the main type of resources of entrepreneurial activities and production factor in the modern digital economy. Also, the authors provide a definition of "digitization" of business and determine the perspective technologies and services of the digital economy that are important for digital transformation of budget planning of entrepreneurial activities.

Keywords: Information and communication technologies · Entrepreneurial activities · Entrepreneurial structure · Strategic planning · Budgeting · Specific features · Digital transformation · Digital economy

JEL Code: M150 · L26 · G300

1 Introduction

Economic activities of the modern corporate or entrepreneurial structures are impossible without usage of the achievements in the sphere of information and communication technologies. These technologies, which were considered very complex in implementation in the past, have established themselves in the current business practice.

E. G. Popkova and B. S. Sergi (Eds.): ISC 2019, LNNS 87, pp. 71–83, 2020.
https://doi.org/10.1007/978-3-030-29586-8_9

The preconditions for that were active research of the place and role of information in business and quick development of digital technologies and systems.

Information, as an important component of business, was actively studied by theoreticians and practitioners of economics, though the formalized idea of its role and place in economy was obtained only in the 20th century. In particular, P. Drucker in late 20th century confirmed importance of information as one of the main resources of business (P. Drucker "Management Challenges for the 21st Century", 1999) [1].

Information is not treated as an inseparable component of the economic environment (especially the knowledge economy), which could be viewed as an important production factor (together with entrepreneurial abilities and labor and material and financial resources) and a valuable source of development and making of decisions.

The second precondition is directly connected to the information and technological breakthrough in late 20th – early 21st century: appearance of mobile communication, the Internet, development of computer systems and technologies, remote transfer and expanded access to information, formation of data bases; creation of social networks and communities, etc. – all this is reflected not only in specific features and characteristics of the modern economy but also in the activities of the components of its economic agents.

The role of modern digital technologies in modification of business processes is very high: digital transformation influences the whole chain "supply-production-sales" with the corresponding digitization of the movement of financial flows, accounting activities, etc.

Thus, digitization (digital transformation) has been actively forming a top-priority vector of development of economic systems and economic subjects for more than twenty years. As a matter of fact, it means the process of appearance, implementation, and scaling of information systems and technologies in the practice of business for the purpose of minimization of risks, increase of competitiveness, formation of new sales markets, development of the existing sales markets, increase of customer orientation, reduction of cists, and increase of speed of work and business activity. Digitization of business and rationalization of management on the basis of large arrays of various data take place. On the other hand, digital transformation of business is the product of business and entrepreneurial activities. From this point of view, for determining the specific features of planning and implementation of modern entrepreneurial activities in the conditions of the digital economy it is necessary to determine the interconnection between the studied categories.

2 Interconnection Between Entrepreneurship and the Digital Economy

The problems of development of entrepreneurship and entrepreneurial activities within the modern Russian economy are very topical.

The key triggers of emergence of this topicality are a lot of reasons: from necessity of provision of competitiveness to the necessity of restructuring of the Russian economy, including on the basis of import substitution.

In its turn, growth of competitiveness and implementation of the processes of diversification and restructuring in the national economy are impossible without foundation on development of entrepreneurship and involvement of the subjects of the entrepreneurial sector into this process.

Such subjects (as compared to corporate structures) are more susceptible to transformation changes and thus are able of quick market, technological, and managerial adaptation and reorientation; they "consciously" function in the conditions of high risks (internal and external) and aim at "moving to the market" the innovations of various character, which are necessary for import substitution and restructuring of the country's economic model in favor of non-resource economy of "knowledge and information" (in the long-term – export economy of "knowledge and information").

The possibility of quick and more effective generation of innovations in the conditions of high risks within the entrepreneurial activities confirms the importance of the latter for the global economic development.

On the other hand, the above implies that the role of digital transformation of business in the modern economic environment is high to the extent to which the entrepreneurial activities predetermine the emergence and development of certain digital technologies and digital transformation of business.

Regarding the history of the issue of interconnection of digitization of economy and entrepreneurial activities, it should be noted that the modern digital economy appeared as a result of entrepreneurial initiatives, which are implemented at the level of entrepreneurial agents and at the level of corporate entrepreneurship. An example could be the activities of such prominent Western entrepreneurs of late 20th – early 21st century as Steve Jobs, Bill Gates, Paul Allen, Mark Zuckerberg, etc. and the activities of large IT and other corporations (IBM, Google, Apple, etc.).

From this point of view, digital transformation was initially an innovative and high-risk product of entrepreneurial activities, obtained in the course of implementing individual entrepreneurial initiatives and in the course of implementation of corporate entrepreneurial projects.

As a result, creation of certain information and communication technologies and systems, as well as attempts of their implementation into the corporate or entrepreneurial practices (entrepreneurial management), was conducted for minimizing the risks of economic activity and optimizing the innovative activities – i.e., approaching to permanent generation of innovations.

The interconnection between entrepreneurial activities and digital transformation is shown in Fig. 1.

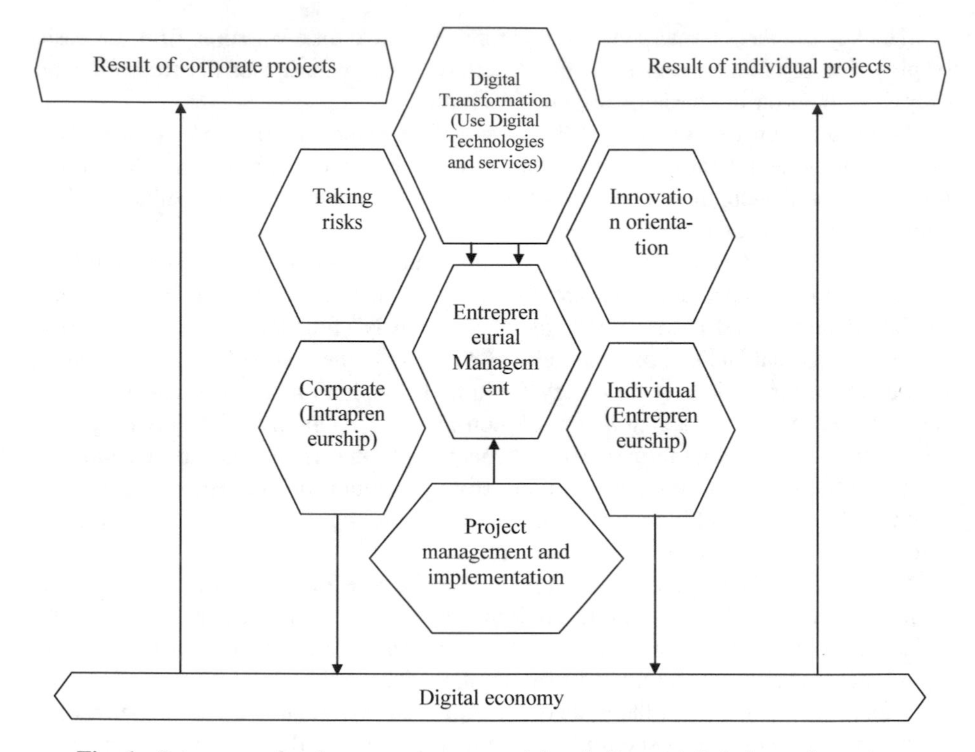

Fig. 1. Interconnection between entrepreneurial activities and digital transformation.

The studied approach to understanding the essence and specifics of digital transformation allows supposing that the digital economy is a product of entrepreneurial activities, which, however, moved beyond the limits of entrepreneurship. Like entrepreneurial abilities, information (as the main type of resources within the digital economy) has acquired a status of a separate production factor [3]. Like the development of entrepreneurship, development of the digital economy is paid a lot of attention in Russia – as they are the basic drivers of economic growth.

The current practice and scientific research show that digital transformation develops actively in the countries with developed entrepreneurial sector (within which the means of the digital economy are developed, implemented, transformed, modified, and scaled) [4]. The countries without developed concepts of development of entrepreneurship are usually outsiders in the sphere of digitization as well.

Characteristics of interconnection of entrepreneurship and digitization of economy in the context of determination and evaluation of specific features of strategic planning of entrepreneurial activities in the conditions of digital transformation would have been incomplete without determining the directions of mutual transformation influences of the two studied categories.

The directions of such mutual transformation influences cannot be studied without consideration of the above interconnection (Fig. 2).

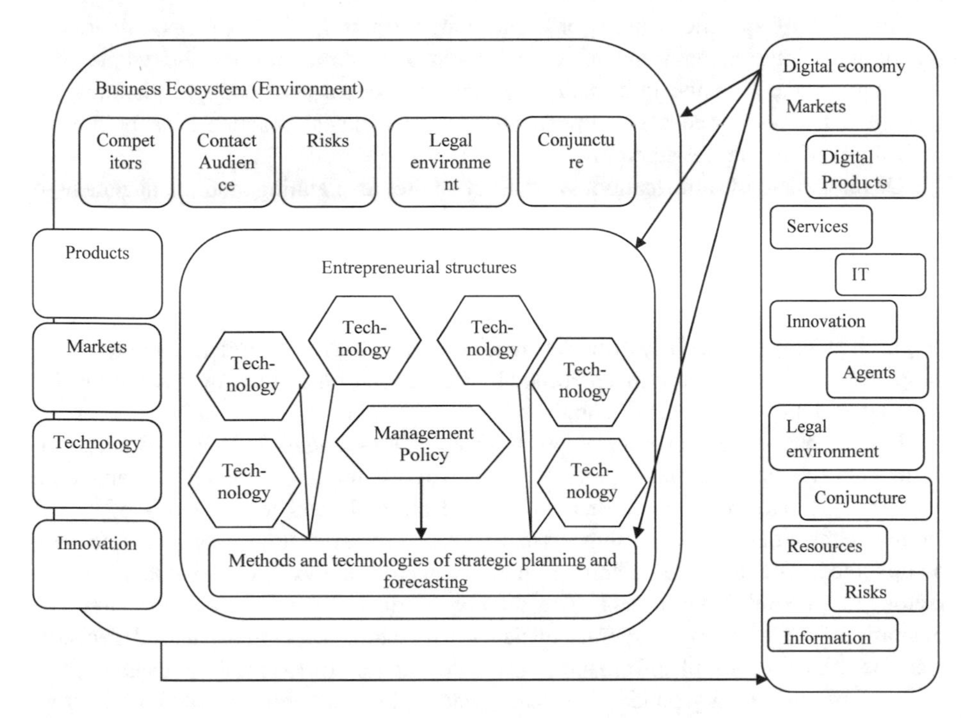

Fig. 2. Directions of mutual transformation influences of entrepreneurial activities and their digital transformation.

It is possible to note that entrepreneurial innovations, technologies, and markets on the whole determine the infrastructure and eco-system of the modern digital economy.

In its turn, the digital economy and its factors influence the general entrepreneurial environment (eco-system of entrepreneurship) and its separate elements, including the specific subjects of entrepreneurship. In the latter case, digital transformation could influence separate internal functional components (supply, sales, personnel, finances, etc.) and supra-functional business processes, of which the basic one is strategic planning and forecasting of entrepreneurial activities and their development.

3 Strategic Planning of Entrepreneurial Activities in the Conditions of Digital Transformation

The practice of the a modern digital transformation of business processes shows that if the entrepreneurial environment is not susceptible to it, the transformation does not take place. Quite on the contrary, a lot of modern entrepreneurial initiatives and projects are aimed at modification and development, search for innovations within the information technologies and systems that are used primarily for the purposes of strategic planning and management. This thesis confirms that the transformation is not a unilateral process.

Speaking of specific features of digital transformation of the process of strategic planning or internal environment of entrepreneurial structures on the whole, it is necessary to mention the information systems of tactical and strategic planning and management, which had been implemented in the business practice even before formalization of the digital economy.

Development of information systems of strategic planning and management of business is shown in Fig. 3.

The basis of digital transformation of the process of strategic planning of entrepreneurial activities has been the process of automatization of management, which started in the middle of the 20th century with emergence of MRP-systems. The main emphasis in usage of such systems was made on optimization of logistics and resources supply chains. This process was preceded by formation of the initial massive of the data on logistical and time costs of supply.

Passing the evolutional development from MPR-systems to ERP-systems, economic agents received more perfect systems of planning of operational and other activities [5]. The require information and the methods and technologies of its collection, consolidation, processing, transformation, transfer, and usage became more complicated. The focus of planning shifted towards analysis and evaluation of the factors of internal environment. The totality of these systems was often used for planning of activities by corporate management, while the entrepreneurial decisions were peculiar for lack of information on the key factors of external environment.

Against the background appearance and wide distribution of the Internet, informatization, development of telecommunication technologies and systems, and development and implementation of CRM-, SCM- and CSPR-systems, entrepreneurs received more acceptable (from the point of view of the character of entrepreneurial solutions) tools of planning of their activities. Focus of planning shifted towards comprehensive analysis and evaluation of the factors of external business environment. The following "market" reorientation of business with the continuing development of the digital systems and services allowed forming the eco-system of the modern digital economy. Within such structure of economic interrelations, the modern systems of strategic planning of entrepreneurial activities are open for usage of such services and technologies the digital economy as remote services, blockchain, cloud technologies, Big Data, artificial intelligence, etc. ERPII-systems are becoming very popular in the segment of corporate entrepreneurship.

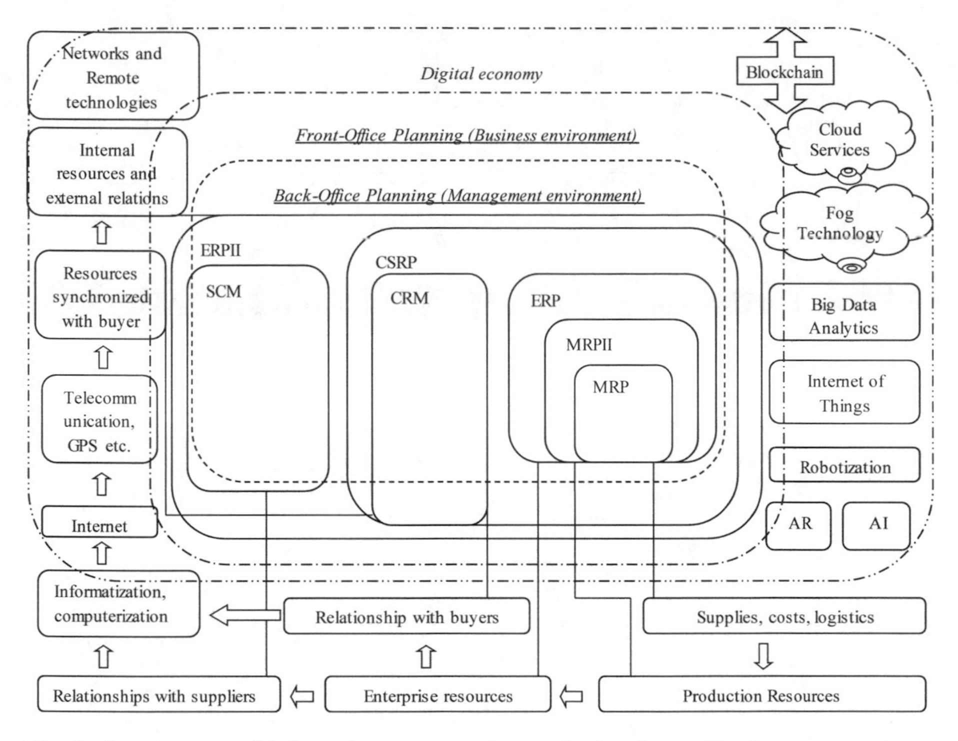

Fig. 3. Development of information systems of strategic planning and business management.

Which innovative information technologies and services will influence the process of strategic planning of entrepreneurial activities more is difficult to say now. Moreover, usage of the leading information systems is usually accessible only for subjects of large business. Remote services and financial data bases (including on regional specifics of business) form the current entrepreneurial situation within the micro-, small and medium (i.e., mainly regional) entrepreneurship. An example of such service could be the specialized portal of entrepreneurial activities [5].

One way or another, in the modern conditions the process of digital transformation and modification of strategic planning influences almost all subjects of entrepreneurship: even modern micro-business and self-employment can rarely work without the data obtained from the Internet and/or social networks.

Strategic planning of entrepreneurial activities, as a supra-functional business process within an entrepreneurial structure, also has its functional spheres (components), in which all stages of the planning cycle are realized: from strategic planning of operational activities to marketing planning. They are digitalized with certain effects (Fig. 4).

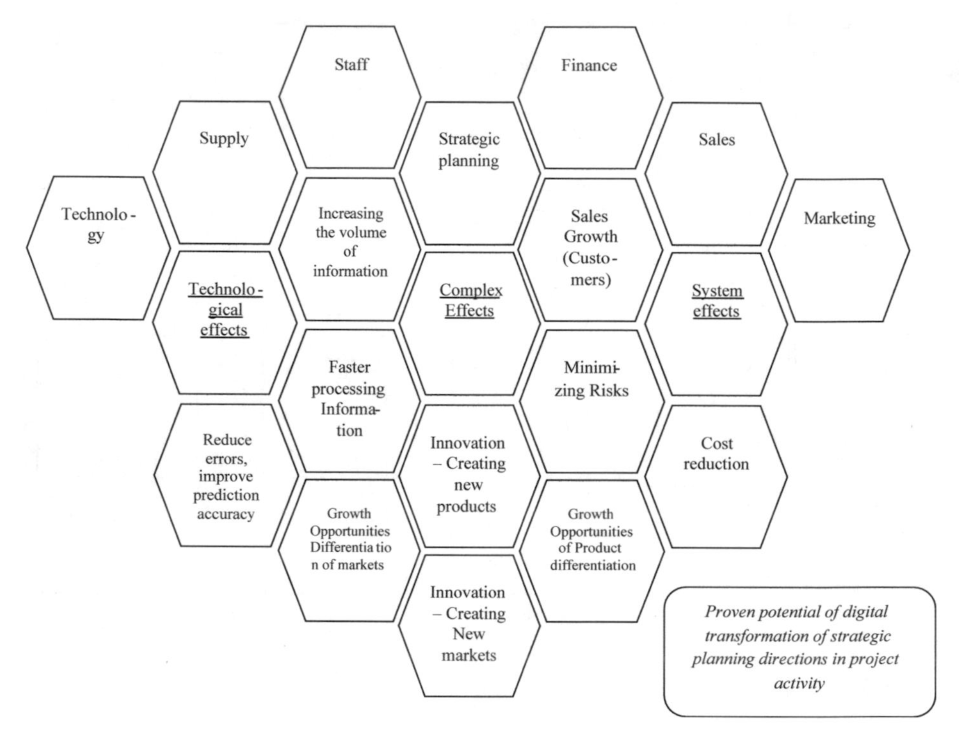

Fig. 4. The key effects and directions of digital transformation of the functional spheres of strategic planning of entrepreneurial activities.

Thus, digital transformation of strategic planning of entrepreneurial activities leads to certain effects, which include:

- technological effects (related to digital transformation of forms, methods, and technologies of planning). They might include increase of the volumes of analyzed information, growth of speed of its processing, and reduction of errors in such processes;
- systemic effects (related to positive influence of digital transformation planning on economic effectiveness of entrepreneurial activities). They include primarily minimization of costs (especially, transaction costs), growth of sales (revenues), more precise forecasting of risks, and more effective determination of possible ways of their elimination, compensation, and hedging;
- complex effects (related to increase of effectiveness of entrepreneurial activities on the whole, with determination of new ways of their diversification and development), of which the most significant are determination and increase of capabilities of the market and product diversification, including by means of product innovations (as a result – creation of new markets). We think that these effects are of top-priority in the process of decision making as to the necessity for digital transformation of strategic planning of entrepreneurial activities.

Obtaining the above effects has been proved by the existing practice of entrepreneurial activities – especially when the product is new digital technologies or services.

The above effects are largely determined by peculiarities of implementation of the modern strategic planning of entrepreneurial activities in the conditions of digital transformation of economy. These include as follows:

- more substantiate accounting of the factors and risks of the external environment as compared to planning of "internal" factors; strategic planning includes a lot of "incoming" data, with improvement of the mechanism of their validation and increase of precision of plans and forecasts;
- acceleration of processing of external data – "quicker planning";
- usage of complex predicative models (models of factor interaction), which also allows using someone else's positive and negative experience and finding benchmarks (models) of entrepreneurial tactical and strategic decisions;
- expansion and deepening of the horizon of strategic planning, better determination of the specific features of its scenarios.

In the conditions of digital transformation, strategic planning of entrepreneurial activities acquires more proactive, predicative, and market-oriented character for all functional spheres of the process of planning.

It should be noted that the lists of the above effects and specific features of digital transformation of strategic planning of entrepreneurial activities are not complete. Moreover, some of the above effects and specific features could have negative character (e.g., processing of external incorrect data, their replacement, hackers' attacks, failures in the work of information and remote systems, complexity of resetting of predicative models, absence of personnel with the required qualification; social tension in the process of dismissal of personnel due to digitization, etc.). However, presence of these and other negative aspects has not yet reached the "critical mass" for refusal from digital transformation, including the processes of strategic planning and managing the entrepreneurial activities.

It should be noted that the whole totality of these specific features and effects of digital transformation of the process of strategic planning is manifested not only in its functional spheres but it is reflected on specific technologies, through which this planning is implemented.

Thus, one of the most current and important – due to high adaptability and integration – as well as perspective (from the point of view of possibilities of digital transformation) technologies is budgeting.

4 The Directions of Digital Transformation of Budgeting as a Technology of Strategic Planning of Business

From the methodological point of view, budgeting is a complex technology of strategic planning, which is used primarily in corporate management.

Due to high adaptability, flexibility, integration, and absence of regulation of managerial accounting, as well as due to the totality of the implemented functions, it is possible to use it for strategic planning of entrepreneurial activities. As was mentioned earlier, digital transformation of strategic planning is expressed by digital transformation of the technology of this planning – i.e., it is manifested in digital transformation of budgeting.

Modern budgeting, implemented on the basis of the modern digital platforms, systems, and services or with the usage of the information systems of strategic planning and management of business, is completely integrated into the corporate or entrepreneurial environment, as well as project activities, performing its tasks. Packages of specialized software and sets of digital solutions could be developed for this integration.

Digital transformation could be aimed at the technology of budgeting (i.e., the process of budget planning – the budget process, its "external" manifestation) and the tools and mechanisms of budget planning (i.e., the process of budgeting of certain resources, expenditures, and benefits – the process of compilation and implementation of a certain system of budgets, the "internal" component of the process of budgeting).

The directions of digital transformation of budgeting are presented in Fig. 5.

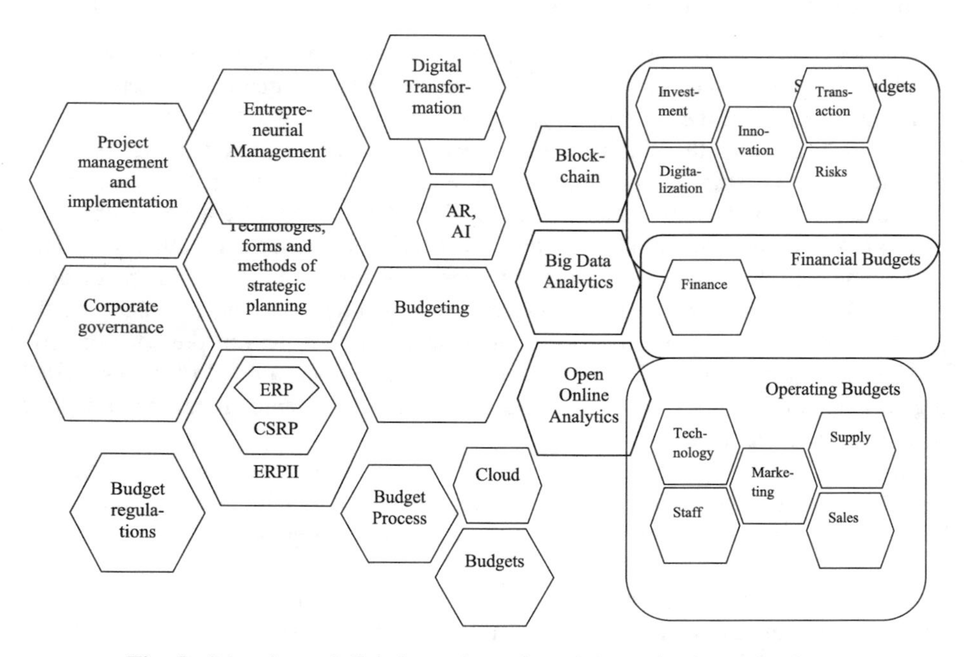

Fig. 5. Directions of digital transformation of the budgeting technology.

It should be noted that the perspective technologies of digital transformation of budget planning of entrepreneurial activities could be the main products and services of the digital economy. Thus, the top-priority direction of digital transformation of budgeting within strategic planning of entrepreneurial activities could be digitization of so called specific budgets (in terms of source [7]), as these tools of the modern budgeting form its importance and necessity for entrepreneurial activities. These budgets could include budget of innovations, investments, transaction costs, budget of compensation (hedging of risks), digital transformation, etc.

Regarding the directions of digital transformation of budgeting, in the first case transformation influences the regulation and sub-processes of functioning of the budgeting system – from its implementation and setting to potentially possible (required) modification.

In the second case, transformation influences specific types of budgets, methods of their formation, provision of interconnection, modification (when necessary – formation, exclusion, balancing of articles, revenues, expenditures, and inflows and outflows of finances) and performance (including the methods of performance control).

The algorithm (main stages and directions) of digital transformation of the budget process and the process of budgeting is shown in Fig. 6.

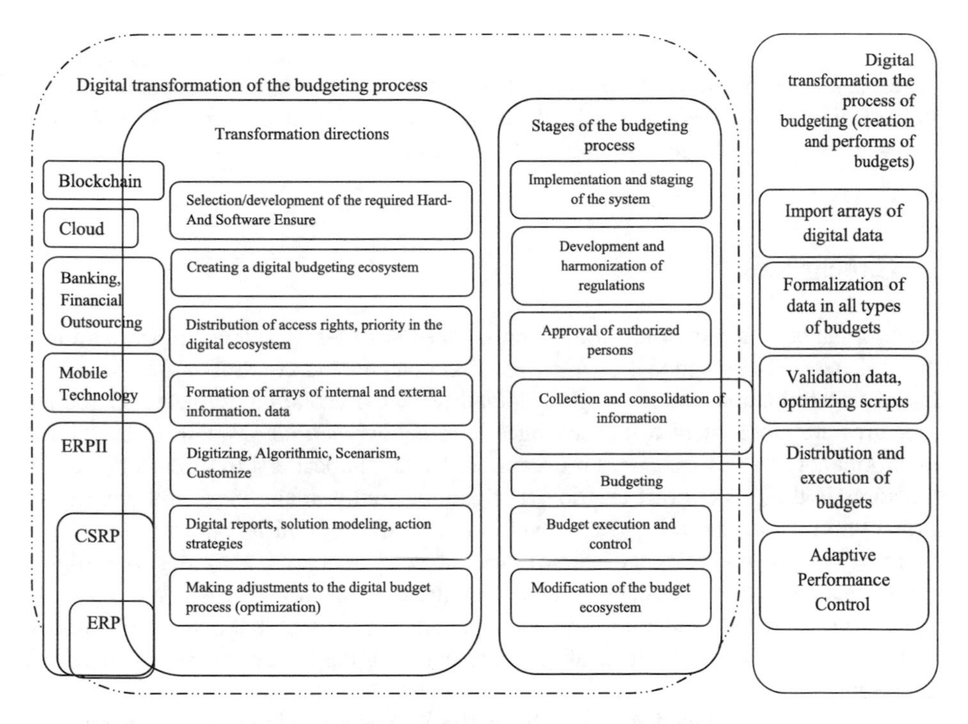

Fig. 6. Digital transformation of budget planning.

As was mentioned earlier, for implementation of these transformation of budgeting, the existing and perspective digital systems of financial planning and services and products of the digital economy could be used. Among these, special attention should be paid to remote and cloud technologies and blockchain technologies.

Specific features and effects of digital transformation of budgeting, apart from the specific features that constitute the modification of the business process of strategic planning as such are as follows:

- increase of flexibility of budgeting by means of quicker reaction to the changing conditions (i.e., quicker modification of scenarios that are set into certain budgets);
- potential possibility of simultaneous implementation of two methodologies of budgeting ("top-down" and "bottom-up") – for provision of larger correspondence of operative and tactical tasks to the strategic goals.

Based on such algorithm, it is possible to perform modernization of the practice of budgeting of financial resources within strategic planning of entrepreneurial activities.

Of course, such digital transformation of budgeting requires the corresponding digital, methodological, technical, and financial provision, which only the subjects of large business can afford. Also, wide dissemination and relative cheapening of the modern digital solutions could stimulate the development of digitization of the processes of budget strategic planning at the level of the subjects of small and medium business.

5 Conclusions

Transformation and modernization of the modern business and development and creation of new business models, which are necessary in the conditions of economy's digital transformation, require high adaptability of the economic subjects. Corporate structures are mote latent to any changes than entrepreneurial structures, and digital transformations in this case are no exception. Hence the leadership of entrepreneurial structures in the process of transition to digital economy, including in the process of formation of the digital eco-system "within itself" and beyond its structure.

Initially, the basic stage of digitization of entrepreneurial activities was digital transformation of the used technologies of its strategic planning. This is a top-priority aspect, to which special attention should be paid due to the fact that quality, timeliness, flexibility, and other parameters of strategic planning determine success of potential entrepreneur.

It should be concluded that the above requirements to digital transformation, as well as potential advantages and "narrow spots" of digitization and usage of budgeting as a complex and current technology of strategic planning of entrepreneurial activities (built on the modern digital solutions), form the importance of continuation of scientific research in this direction of modern economics.

References

1. Drucker, P.: Management Challenges for the 21st Century. http://library.globalchalet.net/Authors/Startup%20Collection/%5BDrucker,%201999%5D%20Management%20Challenges%20for%20the%2021st%20Century.pdf. Accessed 10 Apr 2019
2. Moskovtsev A.F, Yurova, O.V.: Entrepreneurial management, 132 p. Volgograd State Technical University, Volgograd (2013)
3. Official web-site of the Government of the Russian Federation. Passport of the national program "Digital economy". http://government.ru/info/35568/. Accessed 14 Apr 2019
4. Official web-site of the National Research University "Higher School of Economics". Statistical collections of Higher School of Economics. Indicators of the digital economy. https://www.hse.ru/primarydata/iio. Accessed 15 Apr 2019
5. Tekin, A.V., Konina, O.V.: The role of information and communication technologies in the process of strategic management of entrepreneurial structures activities: the budget and financial aspect. In: Popkova, E.G., Ostrovskaya, V.N. (eds.) Perspectives on the use of New Information and Communication Technology (ICT) in the Modern Economy, Conference papers, Pyatigorsk, Russia 1 February 2018. Advances in Intelligent Systems and Computing book series (AISC), vol. 726, pp. 269–278. Institute of Scientific Communications, Center for Marketing Initiatives, Pyatigorsk State University, Springer International Publishing AG (Part of Springer Nature), Switzerland (2019). https://www.springer.com/gp/book/9783319908342. Accessed 13 Apr 2019
6. Official web-site of the business navigator of small and medium entrepreneurship. https://smbn.ru. Accessed 13 Apr 2019
7. Tekin, A.V., Konina, O.V.: The role of specific budgets in provision of strategic development of the modern entrepreneurial structures. Audit and financial analysis, no. 6, pp. 95–105 (2018)

World Currencies: Analysis of the Conditions for Global Demand Generation and Internationalization Level

Inna V. Kudryashova(✉)

Volgograd State University, Volgograd, Russia
kudryashovaiv@volsu.ru

Abstract. The paper focuses on the changes in the economic, institutional, and organizational conditions for the world currency internationalization. The changes revealed are compared to the dynamics of the global demand for the US dollar, the euro, the pound sterling, the Japanese yen, and the Chinese yuan as a unit of account, medium of exchange and store of value, both in public and private sectors. The analysis concludes that there is no expected response of the world market participants to the choice of the world currency that corresponds to the conditions created in the issuer countries/group of countries. It is shown that increased use of the euro and the Chinese yuan as the world currencies is possible only after the issuers expand their contribution to the creation of the world product, provide full convertibility of the Chinese currency, and develop their internal financial markets.

Keywords: World currency · Currency internationalization · Functions of world currencies · International reserves · Exchange rates · Currency market

JEL Code: F31 · F33

1 Introduction

Internationalization of a currency is possible only under certain conditions which characterize the issuer as a reliable participant of the international monetary and financial system. However, after the acquisition of the world currency status these conditions can evolve. Besides, the principles of the current monetary system allow functioning of several world currencies which acquired this status at different times. Such circumstances can change the international demand for these currencies.

The present paper focuses on the comparison of the changes in the conditions ensuring the functioning of the US dollar, the euro, the pound sterling, the Japanese yen, and the Chinese yuan as the world currencies, and also the dynamics of their use as a unit of account, medium of exchange and store of value.

E. G. Popkova and B. S. Sergi (Eds.): ISC 2019, LNNS 87, pp. 84–96, 2020.
https://doi.org/10.1007/978-3-030-29586-8_10

2 Literature Review

The 1976 Jamaica Currency Agreement marked the beginning of the modern monetary system. The notion of a reserve currency was changed to a freely convertible currency. In economic literature, such a currency is referred to as the world currency. It is viewed as a kind of international monetary units used as money all over the world. Currently, the International Monetary Fund (IMF) defines the US dollar, the euro, the pound sterling, the Japanese yen and the Chinese yuan as the world currencies.

The studies on the theory and practice of the world currency formation and functioning are divided into three groups. The first ones focus on the conditions for enhancing the international status of a national currency, and the functions of the world currencies.

Kenen (1983) characterizes the peculiarities of the world currencies functioning as a unit of account, medium of exchange and store of value, both in the public and private sectors.

Frankel (1995) describes the main conditions for internationalization of a national currency. Such conditions are history, magnitudes of output and trade, the country's financial markets, and confidence in the value of the currency.

Recent studies expanded the list of these conditions. Genberg (2009) shows that a necessary condition for the acquisition of the world currency status is full capital account liberalization. Krasavina (2008) includes in these conditions the need to build up international reserves by the issuer.

Kudryasova (2017) combines all the conditions for currency internationalization into two groups. The first group includes the economic conditions that generate the global demand for the currency, and also the conditions that provide stability of the currency and the ability of the issuer to maintain this stability. The second group includes the institutional and organizational conditions.

Other studies analyze the experience of internationalization of the currencies. Chitu et al. (2012) in their joint research focused on the US dollar internationalization. Tavlas and Ozeki (1992) examine the development of the Japanese yen as a world currency. Ito (2011) characterizes the stages of enhancing the status of the Chinese yuan.

Finally, the third group of studies analyzes the role of the world currencies in the current world and the level of their internationalization. For instance, Pollard (2001) analyzes the use of the US dollar and the euro as the world currencies. Eiji and Makoto show that the role of the Japanese yen is decreasing in the modern monetary system.

However, a detailed research on the changes in the conditions necessary for the acquisition of the world currency status and as a result the changes in the international use of a world currency has not been conducted yet.

3 Methodology and Materials

The analysis of the conditions for internationalization and maintaining the world currency status is conducted on the basis of several indicators. First of all, we analyze the dynamics of the shares of the issuers in the world production and export. Then, we examine the development level of the financial markets of the issuers. To assess the stability of the US dollar, the euro, the pound sterling, the Japanese yen and the Chinese yuan we examine the inflation dynamics and exchange rates in the countries/group of countries under analysis. The ability of the USA, the Euro area, the United Kingdom, Japan and China to maintain stable prices for their currencies relative to other currencies are analyzed on the basis of the world reserves dynamics. The institutional and organizational conditions for the internationalization of the world currencies are viewed from the convertibility perspective.

An assessment of the internationalization level of the modern world currencies is conducted on the basis of the analysis of the international use of the US dollar, the euro, the pound sterling, the Japanese yen and the Chinese yuan as a unit of account, medium of exchange and store of value.

The period of 1981–2018 is the period under analysis as the modern approach of the IMF to the recognition of a currency as a world currency entered into force in 1981.

The statistical information on the GDP dynamics and export volume provided by UNCTAD has been used as the database for the present research.

The qualitative indicators characterizing the degree of the development of separate segments of financial markets of the countries/group of countries under analysis and the data on the inflation have been provided by the World Bank.

The information on the volume and composition of the international reserves of the countries/group of countries is available at the Data Template on International Reserves and Foreign Currency Liquidity of the IMF.

The statistical data on the volume of the international bonds and deposits at foreign banks are available at the Statistics section at the official website of the Bank for International Settlements (BIS).

The present study involves the comparative analysis method, historical method, method of absolute and relative indicators, table and graphical methods.

4 Results

4.1 The Conditions for the Acquisition and Maintaining the World Currency Status: Changes in 1981–2017

The analysis has shown considerable changes in the conditions needed for the acquisition and maintaining the status of the world currency for the period under review.

The main macroeconomic condition for a national currency to be in great international demand is the growth of the issuer's share in the global GDP. This share must be considerable and much greater than those of other countries, and the issuer's economy must be open and play a key role in the international trade.

Since 1980, when the list of the world currencies was defined, the USA have been the largest world producer (23.4% in 2017) (Fig. 1). However, in recent decades, the Euro area (15.7% in 2017), and China (15.2% in 2017) have been competing with the USA in terms of the share in the world production. The share of the United Kingdom and Japan decreased by 21% and 37% and amounted to 3.4% and 6.1% in 2017, respectively.

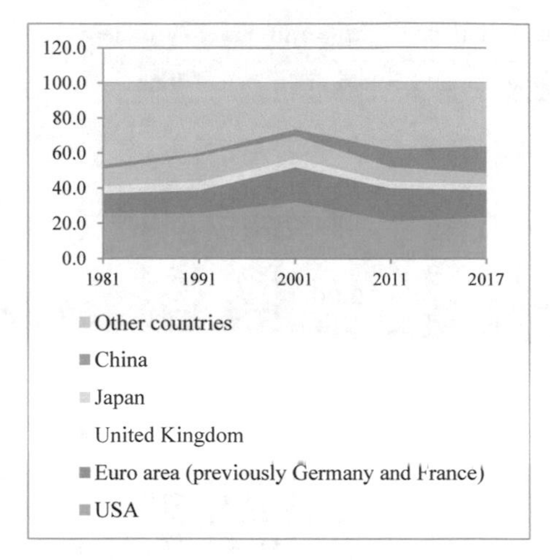

Fig. 1. The country composition of the global GDP in 1981–2017, %

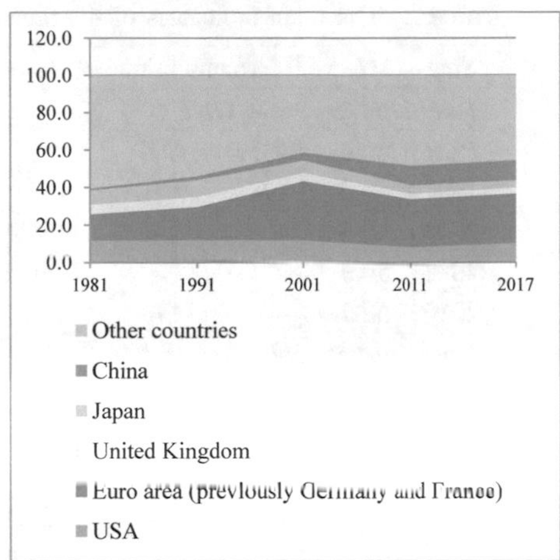

Fig. 2. The share of the modern international currencies in the world export in 1981–2017, %

For the period under analysis, the total share of the countries issuing the world currencies in the world export increased by 15% (Fig. 2). However, the USA share in the world export remained almost the same. The Euro area's share increased considerably. In 2017, the Euro area exported 2.5 times more compared to the USA (26.4% of the total world export). Moreover, China strengthened its position as a world exporter. Currently, its share in the world export is similar to that of the USA and amounted to 10.7% in 2017. The shares of the United Kingdom and Japan decreased significantly.

Another key condition for a national currency to acquire the world currency status and to internationalize its monetary functions is an open and free of tight state regulation developed financial market in the issuer/group of issuers.

An open national financial market of the issuer where non-residents can effectively place their deposits and official reserves in the national currency and a financial market that provides alternative funding sources even in times of international liquidity crises can generate a strong global demand for such a currency.

For the period under analysis, the USA lost their leading position in terms of financial market development (Table 1). Today, only the US internal corporate bonds market provides foreign investors with more opportunities compared to other issuers.

Meanwhile, China's financial market increased considerably its development level. For instance, in 2017, the correlation of the bank assets and GDP was higher than that of other countries under analysis and amounted to 168%. China's stock market capitalization was similar to that of the Euro area – 71.2% and 77.6%, respectively. China's internal corporate bonds market compared to GDP was slightly less than that of Japan – 52.5% and 55.2%, respectively.

The United Kingdom's banking sector of the financial market showed a significant growth, and so did the stock markets of Japan and the United Kingdom.

Table 1. The main indicators of the financial markets of the world currency issuers

Year	USA	Germany	France	Euro area	United Kingdom	Japan	China
Total bank assets to GDP (%)							
1981	60.4	103.0	79.3	–	27.3	169.9	No data
1991	58.2	108.2	103.0	–	102.0	226.0	77.2
2001	55.3	–	–	83.6	123.4	195.6	113.7
2011	59.3	–	–	110.6	179.9	189.9	128.8
2016	62.7	–	–	102.7[a]	130.5	158.3	168.0
Stock market capitalization to GDP (%)							
1981	41.4	7.5	6.6	–	20.4	18.5	No data
1991	58.8	19.4	25.9	–	80.9	38.3	No data
2001	138.3	–	–	59.1	149.8	46.9	41.6
2011	107.4	–	–	30.4	118.7	76.4	52.2
2017	164.8	–	–	77.6	118.1	112.8	71.2
Outstanding domestic private debt securities to GDP (%)							
1991	70.4	No data	58.0	–	13.7	40.9	2.9
2001	98.8	–	–	24.6	17.3	53.8	7.9
2011	91.9	–	–	37.8	12.3	72.5	29.5
2016	No data	–	–		No data	55.2	52.5
Outstanding domestic public debt securities to GDP (%)							
1991	55.9	No data	24.1	–	24.4	44.6	2.3
2001	41.1	–	–	34.9	27.8	95.3	9.9
2011	82.6	–	–	46.6	58.6	190.8	15.6
2016	No data	–	–		No data	181.9	28.8

[a]Data for 2014

Stability of a currency is also an important economic condition for the acquisition and maintaining of the world currency status. Actually, it means that the currency is characterized by low inflation rate and does not experience sharp fluctuations of the price compared to other currencies.

Since the moment of the acquisition of the world currency status, the inflation rate in the issuers/group of issuers was low (Fig. 3). For instance, in the period of 1981–2017, the US inflation rate based on consumer price index ranged from (−0.4) to 10.3%, the UK – from 0.4 to 11.9%, and Japan's – from (−1.4) to 4.9. The index values of 10.3% in the USA and 11.9% in the United Kingdom were registered only once – in 1981. In other years, the price dynamics in these countries was much lower. Since the euro introduction in 1999, the inflation rate in the Euro area has not exceeded 2.6%; since 2016 and during the period when the Chinese yuan functions as a world currency, China's inflation rate has been lower than 2% a year.

The exchange rate regimes of the US dollar, the euro, the pound sterling and the Japanese yen are floating and do not exclude the interference of the monetary authorities in the price formation process at the currency market. The issuers possessing the largest reserve assets in the world are able to smooth sharp fluctuations of the exchange rates of their currencies: in 2000–2002 – in the Euro area, and in 2009–2012 – in Japan. In general, the exchange rate fluctuations of these currencies were relatively small (Fig. 4).

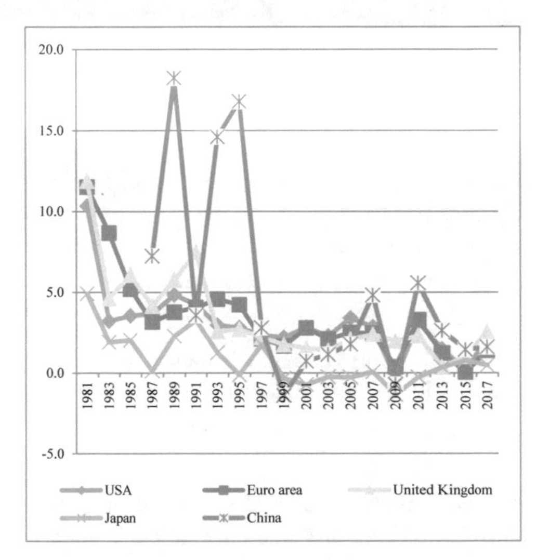

Fig. 3. The inflation dynamics in the world currency issuers/group of issuers in 1981–2017, % Source: World Bank. Open Data. Available at: http://data.worldbank.org/indicator/FP.CPI.TOTL.ZG

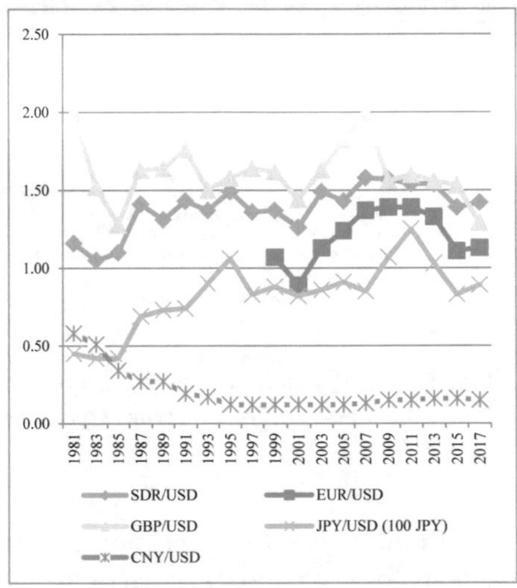

Fig. 4. The dynamics of the world currency exchange rates in 1981–2017 Source: World Bank. Open Data. Available at: http://data.worldbank.org/indicator/FP.CPI.TOTL.ZG

Despite a gradual liberalization of the Chinese yuan exchange rate regime, the price for China's national currency is determined not by the market, but by the People's Bank of China thus making it relatively stable.

Finally, building-up international reserves by the issuer is also a key condition for the acquisition of the world currency status.

The size of the international reserves of the world currency issuers changed considerably (Fig. 5). Until 1995, the USA had the largest international reserves. However, in the years that followed, Japan, and later the Euro area and China began to get the USA ahead in terms of the international reserves size. It should be noted that by 2017, the US total international reserves were 1.9 times less that those of the Euro area, and 2.8 times less compared to Japan, and 7.2 times less compared to China.

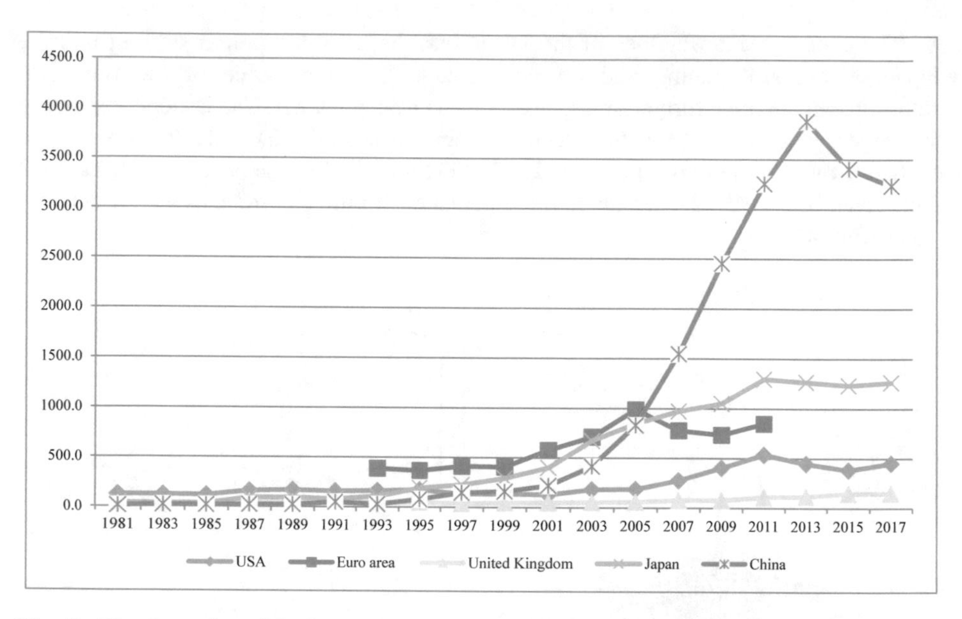

Fig. 5. The dynamics of the international reserves of the world currency issuers in 1981–2017, bln. dollars Source: World Bank. Open Data. Available at: http://data.worldbank.org/indicator/FP.CPI.TOTL.ZG

Besides economic conditions, there are also institutional and organizational conditions for internationalization of the monetary functions of some currencies. They are, first of all, absence of foreign exchange restrictions for current account and capital operations of the balance of payments.

The increased use of the US dollar in the international settlements (after the end of the World War II) is mainly explained by the absence of foreign exchange restrictions. The convertibility of the currencies of West European countries, later the Eurozone with a single currency – the euro, was achieved by the beginning of the 1960s. The Japanese yen also became a freely convertible currency after the lifting of foreign exchange restrictions in the late 1970s. In 2016, when the Chinese yuan acquired the world currency status, the convertibility of the yuan was restricted for capital operations. Though, China has lifted capital operation restrictions selectively, the yuan convertibility is planned to be increased only by 2020.

4.2 The Analysis of the Internationalization Level of the World Currencies

An assessment of the world currency internationalization has been conducted on the basis of the analysis of the dynamics of the use of the world currencies by non-residents as a unit of account, medium of exchange and store of value from 1981 up to the present.

The world currencies are used as the units of account, first of all, in the public sector; they are used as the anchor for pegging local currencies. Second, they are used in the private sector for denominating trade and financial transactions.

In 1981–2017, the US dollar was used most commonly for pegging local currencies (the share of the US dollar in the total amount of the currencies with pegged exchange rate changed from 29% in 1993 to 48% in 2015) (Fig. 6).

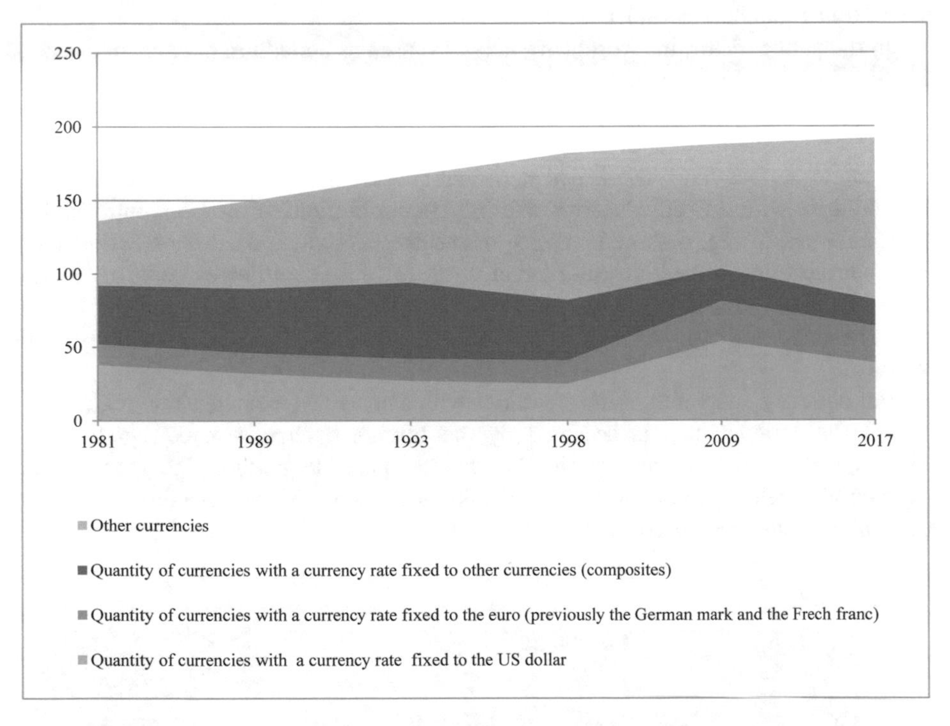

Fig. 6. The quantity of currencies with a currency rate fixed to the world currencies in 1981–2017 Source: International Monetary Fund. IMF Annual Report, 1981, 1989, 1993, 1998, 2009. Available at: http://www.imf.org/en/publications/areb; Annual Report on Exchange Arrangements and Exchange Restrictions, 2018. Available at: https://www.imf.org/en/Publications/Annual-Report-on-Exchange-Arrangements-and-Exchange-Restrictions/Issues/2018/08/10/Annual-Report-on-Exchange-Arrangements-and-Exchange-Restrictions-2017-44930

The share of the euro (previously the combined share of the German mark and the French franc) in the total amount of the currencies with pegged exchange rate did not exceed 28% (in 2015).

The only time when the pound sterling was used as the anchor currency during the period under analysis was registered by the IMF in Gambia in 1981. The use of the Japanese yen and the Chinese yuan as the anchor currencies has not been reported yet.

Consolidated statistical data on the currency composition of invoices in the international trade and financial operations are not available. However, the studies conducted show that the US dollar remains the main currency used in the private sector as a unit of account. According to the data available, in 1980, the share of the US dollar in foreign trade invoices was 56.1%, and in 1992 – 47.8% (European Economic Commission, 1994). G. Gopinath (2015) revealed that by 2015 the share of the dollar as a currency of invoices had been 3.1 times more than the US dollar share in the world export. It means that most non-American exporters invoice in dollars. The euro share in the foreign trade invoices is 1.2 times more than the export share of the Euro area which means that the euro is less used as a unit of account by non-residents in the international purchase contracts.

In the public sector, the world currencies are used as a medium of exchange and as intervention currencies beyond the issuer country/region; in the private sector, they are used as the currencies of payment in the international trade and financial contracts, and also as the vehicle currencies in exchange operations.

Despite the fact that consolidated statistical data are not available, it is obvious that the countries with a fixed exchange rate use anchor currencies for interventions. The data presented in Fig. 6 show that 28% of the IMF countries used the US dollar for their interventions at the currency market. In 2017, this share decreased to 10%. The share of the countries that use the euro (previously the German mark and the French franc) as an intervention currency remained the same during this period – slightly more than 10%.

For currency interventions, the countries with a non-fixed exchange rate regime buy or sell the currency that is accepted by the foreign exchange market participants without any limitations and within the shortest possible period. The data on the interventions held by the largest central banks of the world show a leading role of the US dollar (Kudryashova 2017).

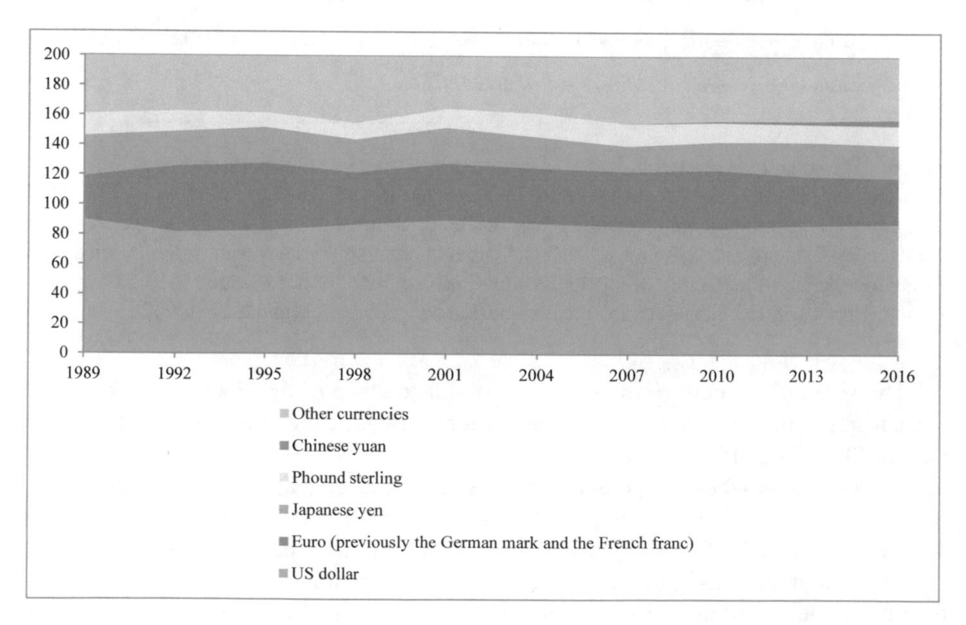

Fig. 7. The shares of some currencies at the world currency market, % Source: Bank for International Settlements. Statistics. Available at: https://www.bis.org/

Mostly the US dollar, the euro, the pound sterling, and yen are used as vehicle currencies at the world market for almost thirty years (Fig. 7). In the late 1980s, 90% of the world currency transactions were carried out in the US dollars. The US dollar retained its leading position even after the euro introduction. In 2016, the US dollar was used as a vehicle currency in 88% of the foreign exchange operations, and the euro – in 31%, the yen – in 22%, the pound sterling – in 13%, the yuan – in 4%.

The available statistical data do not allow us to assess the changes in the currency composition of the international payments for the period under review. However, we can speak about leading positions of the US dollar and the euro in such operations in 2018 – 41.5% and 33%, respectively. The pound sterling and the yen were used less for the same transactions (6.8% and 3.4%). In this period, the role of the yuan as a payment currency in the international transactions enhanced considerably. In January 2012, the yuan was ranked the 20th, and by December 2018, the yuan moved to the 5th place in the rating (the yuan accounted for 2.1% of the total international payments).

In the public sector, the world currencies are used as a store of value to form currency reserves; in the private sector – for investment in the form of deposits, credits, bonds, etc.

In recent decades, the share of the US dollar in the international currency reserves slightly decreased (almost by 10% points). However, the US dollar still dominates as a store of value in the public sector (Fig. 8). It is followed by the euro with a 20–25% share in the world currency reserves, which is higher than the combined share of the mark and the franc.

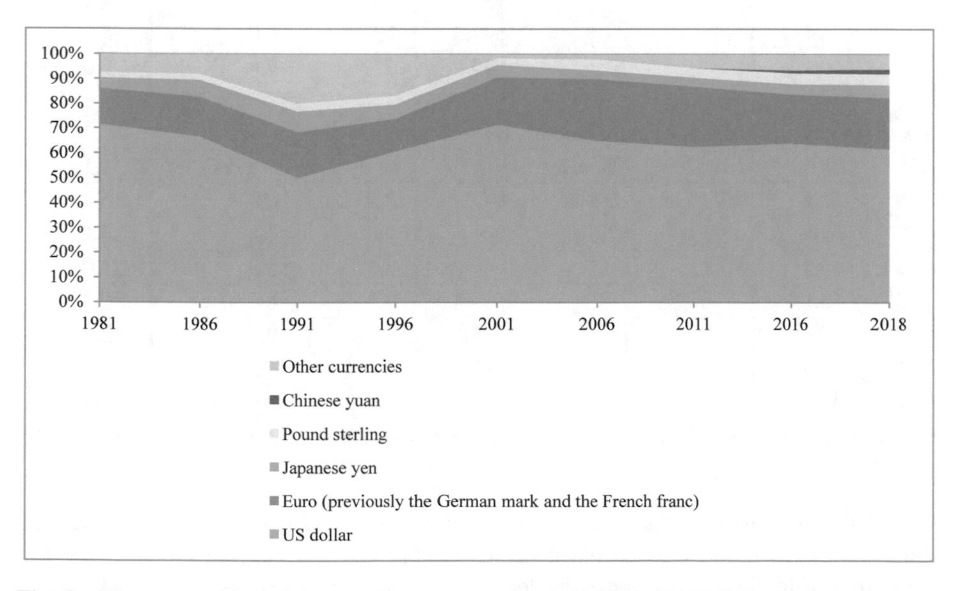

Fig. 8. The composition of the world currency reserves in 1981–2018, % Source: International Monetary Fund. Data Template on International Reserves and Foreign Currency Liquidity. Available at: http://data.imf.org/?sk=2DFB3380-3603-4D2C-90BE-A04D8BBCE237&ss=1481575224638

The share of the pound sterling in the world currency reserves increased by 50% in 1981–2018, but still is rather small (4.4% in 2018). The yen share was not characterized by a stable trend and was 3.1% in 2018. After the inclusion of the yuan in the SDR currencies in October 2016, the IMF began to calculate its share in the currency reserves composition, and at the end of 2018 it was 1.9%.

The combined share of the US dollar, the euro (up to 1999 the German mark and the French franc), the Japanese yen and the pound sterling in the currency composition of the outstanding international bonds was characterized by a stable growth and reached 95% in 2018 (Table 2). However, the share of investment in debt securities nominated in dollars decreased by more than 10% points. The share of the international bonds nominated in the euro exceeded not only the combined share of the mark and franc but in some years the share of the US dollar.

Table 2. The currency composition of the world private portfolio in 1981–2018, %

Year	US dollar	German mark	French franc	Euro	Japanese yen	Pound sterling	Other currencies
The currency composition of the outstanding international bonds							
1981	.	No data	No data	–	6.9	No data	40,5
1985	56.7	9.1	0.6	–	7.6	3.4	22.6
1991	39.1	9.6	2.8	–	12.2	8.3	28.0
1996	37.7	12.0	6.1		14.0	8.3	22.0
2001	50.7	–	–	31.7	6.0	7.4	4.2
2006	36.4	–	–	47.3	2.8	8.2	5.2
2011	40.7	–	–	42.2	2.7	7.4	7.0
2018	44.5	–	–	39.7	1.8	8.4	5.6
The currency composition of non-banking sector deposits in foreign banks							
1981	68.9	No data	No data	–	No data	No data	31.1
1985	69.2	9.3	1.4	–	3.8	0.9	15.4
1991	57.8	14.2	2.5	–	4.0	1.4	20.1
1996	46.5	16.2	1.2	–	3.3	5.8	27.0
2001	45.1	–	–	25.4	3.3	6.0	20.2
2006	47.8	–	–	27.6	2.8	9.0	12.8
2011	50.0	–	–	27.1	2.8	6.3	13.8
2018	52.2	–	–	24.6	2.6	5.4	15.2

Source: Bank for International Settlements. Statistics. Available at: https://www.bis.org/

As for the currency composition of the deposits placed with foreign banks, the share of the US dollar also decreased considerably and the share of the euro did not reach that of the US dollar but exceeded the combined share of the mark and franc.

5 Conclusion

In recent decades, the economic conditions necessary for the acquisition of the world currency status by the US dollar, the euro, the pound sterling, and the Japanese yen have changed considerably. Moreover, the conditions for the recognition of the Chinese yuan as a world currency by the IMF have been created. In fact, these changes are as follows: loss of the leading position in the world production and export by the USA, and a growing share of the Euro area and China in the world production and international trade.

Besides, now the USA are not able to provide better opportunities for funds allocation in all segments of the financial market compared to other world currency issuers.

The changes that occurred could lead to considerable enhancing the role of the euro and the Chinese yuan as the international unit of account, medium of exchange and store of value for non-residents.

However, the American dollar is primarily used as a world currency globally. It is only the demand for the US dollar as a store of value that decreased, and conversely, for the euro it increased still with a leading position of the US dollar.

This situation can be explained by the following circumstances: first, despite the establishment of the Euro zone during the period analyzed the euro has not become the absolute world leader. The same is true for China, where there are still restrictions on the use of the national currency for capital operation and its financial market is rather closed for non-residents. And finally, there is one more factor that explains the US dollar leading position as a world currency; it is inertial behavior of the world economy entities (Frankel (1995)). As a result of inertia, the participants of the world market can choose the world currencies used previously, even if the issuers lost their leading positions in the world economy. Overcoming current restrictions on the euro and yuan internationalization will lead to increased use of the single European currency and the Chinese currency as a unit of account, medium of exchange and store of value.

References

Bank for International Settlements. Statistics. https://www.bis.org/

Chitu, L., Eichengreen, B., Mehl, A.J.: When did the Dollar overtake Sterling as the Leading International Currency? Evidence from the bond Markets. National Bureau of Economic Research, 18097 (2012). http://www.nber.org/papers/w18097

Genberg, H.: Currency Internationalization: analytical and Policy Issues. HKIMR Working Paper 31/2009 (2009). http://dx.doi.org/10.2139/ssrn.1628004

International Monetary Fund. IMF Annual Report, 1981, 1989, 1993, 1998, 2009. http://www.imf.org/en/publications/areb;

International Monetary Fund. Annual Report on Exchange Arrangements and Exchange Restrictions (2018). https://www.imf.org/en/Publications/Annual-Report-on-Exchange-Arrangements-and-Exchange-Restrictions/Issues/2018/08/10/Annual-Report-on-Exchange-Arrangements-and-Exchange-Restrictions-2017-44930

International Monetary Fund. Data Template on International Reserves and Foreign Currency Liquidity. http://data.imf.org/?sk=2DFB3380-3603-4D2C-90BE-A04D8BBCE237&ss=1481575224638

Gopinath, G.: The international price system. In: Jackson Hole Symposium, vol. 27. Federal Reserve Bank at Kansas City (2015). https://www.bls.gov/mxp/internationalpricesystem.pdf

Eichengreen, B.: Sterling's Past, Dollar's Future: Historical Perspectives on Reserve Currency Competition. NBER Working Papers 11336 (2005). https://econpapers.repec.org/paper/nbrnberwo/11336.htm

Eiji, O., Makoto, M.: The Japanese yen as an international currency. East Asian Econ. Rev. **21** (4), 317–342 (2017). https://doi.org/10.11644/KIEP.EAER.2017.21.4.333

Frankel, J.: Reports of the Dollar's Demise are Greatly Exaggerated (1995). https://www.hks.harvard.edu/fs/jfrankel/RESCURR$.FA5.PDF

Ito, T.: The internationalization of the RMB: Opportunities and pitfalls. In: Manuscript for CDRF–CFR Symposium (2011). https://www.cfr.org/report/internationalization-rmb

Kenen, P.B.: The Role of the Dollar as an International Currency, p. 40. Group of Thirty, New York (1983)

Krasavina, L.N.: The Russian ruble as the world currency: a strategic challenge. Money Credit **5**, 11–18 (2008)

Kudryashova, I.V.: Chinese yuan as a world currency: the imperatives and realities. World Econ. Inter. Relat. **9**, 36–44 (2017)

Pollard, P.: The creation of the euro and the role of the dollar in international markets. Federal Reserve Bank (2001). https://doi.org/10.3886/ICPSR01247.v1

Society for Worldwide Interbank Financial Telecommunications, SWIFT RMB Monthly Tracker. https://www.swift.com/our-solutions/compliance-and-shared-services/business-intelligence/renminbi/rmb-tracker/document-centre#topic-tabs-menu

Tavlas, G.S., Ozeki, Y.: The internationalization of currencies: an appraisal of the Japanese Yen. International Monetary Fund, Occasional Paper 90 (1992). http://web.pdx.edu/~ito/Tavlas_Ozeki.pdf

van de Koolwijk, P.J.A.: The international use of main currencies. Statistical Overview of Recent Developments. European Economic Commission (1994). http://aei.pitt.edu/41563/1/A5661.pdf

UNCTAD. Data center. http://unctadstat.unctad.org/wds/TableViewer/tableView.aspx?ReportId=96

World Bank. Open Data. http://data.worldbank.org/indicator/FP.CPI.TOTL.ZG

The Tendencies of Healthcare and Sports in the Conditions of the Digital Economy

Elena S. Berezhnaya(✉), Vladimir A. Bondarev, Yevgeniya V. Zazulina, Natalya V. Koloskova, and Anna V. Strichko

Institute of Service and Business (Branch), Don State Technical University, Shakhty, Russian Federation
doesb@mail.ru, helenbondareva@yandex.ru, zazulina78@list.ru, natasha-fks62@mail.ru, annastrichko@gmail.com

Abstract. Purpose: The purpose of the paper is to determine the tendencies of healthcare and sports in the conditions of the Russian digital economy.

Design/Methodology/Approach: The authors use the method of regression analysis for studying the dependence of emergence of new diseases and popularity of sports activities among Russians on activity of usage of Internet and mobile communications. The information and empirical basis of the research is statistical materials of the Federal State Statistics Service and the results of a sociological survey that was conducted by RiaRating in 2018. The timeframe of the research is 2013–2018 and the forecast period of 2019–2022.

Findings: It is determines that in the conditions of the digital economy the disease rate increases – but not for all diseases. For example, there is no statistical connection between endocrine diseases, digestion disorders, and dysmetabolism with the usage of digital technologies. Usage of Internet stimulates the growth of nervous diseases rate and does not influence other diseases' rates. Usage of mobile communications influences the population's health in Russia to a larger extent, causing growth of disease rate of morbid growths and nervous system diseases and leading to reduction of disease rate of certain germ diseases, parasitical diseases, eye diseases, congenital anomalies (birth defects), deformations, and chromosome breakages. There's also a sustainable and vivid connection between increase of activity of usage of mobile communications and growth of the share of Russian who play sports.

Originality/Value: It is substantiated that the tendencies of healthcare and sports in the conditions of the digital economy in Russia differ from the global tendencies that are noted by the World Health Organization.

Keywords: Health · Sports · Digital economy · Modern Russia

JEL Code: I15 · L83 · O31 · O32 · O33 · O38

E. G. Popkova and B. S. Sergi (Eds.): ISC 2019, LNNS 87, pp. 97–103, 2020.
https://doi.org/10.1007/978-3-030-29586-8_11

1 Introduction

Studies of the technological progress traditionally put emphasis on the economic advantages that are connected to growth of labor efficiency and increase of the living standards due to improved satisfaction of public needs. Thus, the digital economy is treated as an economic system that ensures growth of economic effectiveness as compared to the preceding post-industrial economy. At the same time, the foundation for building the digital economy is the information society – population (consumers and workers) that actively uses digital technologies. This determines the topicality of studying the social consequences of transition to the digital economy.

One of the most obvious and generally acknowledged social consequences of automatization of the production and distribution processes is reduction of moving activities of the representatives of most professions in the conditions of the digital economy and the connected problem of obesity. According to the experts of the World Health Organization (2019a), over the last forty years the number of people with obesity grew by three times; in 2016, the share of people above 18 with obesity constituted 13%. One of the reasons of this problem is decrease of the popularity of sports activities. Another negative social consequence of the digital economy is establishment of psychological dependence on social networks and replacement of personal communication with communication in social networks (World Health Organization (2019b)).

The working hypothesis of the research consists in the idea that wide distribution of digital technologies causes more serious social consequences than is traditionally noted – these consequences are connected with systemic aggravation of the state of health. The purpose of the paper is to determine the tendencies of healthcare and sports in the conditions of the Russian digital economy. The research is performed by the example of Russia due to significant differences in the level and directions of usage of the digital technologies in countries of the world and the specifics of national social consequences of the digital economy.

2 Materials and Method

Literature overview on the selected topic showed that certain social consequences of establishment of the digital economy are considered in the works (Bogoviz et al. 2019), (Eime et al. 2015), (Johnson 2019), (Kang 2015), (Popkova 2017), (Popkova et al. 2019), (Popkova et al. 2018), and (Sergi 2019). However, the tendencies of healthcare and sports in the conditions of the digital economy have been studies fragmentarily, and most existing publications use the methods of expert evaluations, which predetermine large subjectivity of the research results, reducing their precision and authenticity and perform selective studies on the basis of certain social categories of population or companies. Thus, it is impossible to compile a comprehensive picture of social consequences of the digital economy.

In this paper, we use the method of regression analysis for determining the dependence of emergence of new diseases according to the Federal State Statistics Service (2019a) and popularity of sports activities among Russians based on the results of a sociological survey by RiaRating (2019) in 2018 on the activity of usage of Internet and mobile communications according to the Federal State Statistics Service (2019b). For simplification of calculations, we use the following legend:

- x_1: usage of Internet by the population during last three months, % of the total population;
- x_2: number of portable devices of mobile communications per 1,000 people, units;
- y_1: certain germ and parasitical diseases, number of patients with the first diagnosis in life, thousand people;
- y_2: growths, number of patients with the first diagnosis in life, thousand people;
- y_3: blood and bloodmaking organs diseases and certain diseases of the immune mechanism, number of patients with the first diagnosis in life, thousand people;
- y_4: endocrine diseases, digestion disorders, and dysmetabolism, number of patients with the first diagnosis in life, thousand people;
- y_5: nervous system diseases, number of patients with the first diagnosis in life, thousand people;
- y_6: eye diseases, number of patients with the first diagnosis in life, thousand people;
  - y_7: ear and mammillary tubercle diseases, number of patients with the first diagnosis in life, thousand people;
  - y_8: diseases of the circulatory system, number of patients with the first diagnosis in life, thousand people;
  - y_9: diseases of the respiratory system, number of patients with the first diagnosis in life, thousand people;
  - y_{10}: diseases of the digestive system, number of patients with the first diagnosis in life, thousand people;
  - y_{11}: skin and subcutaneous tissue diseases, number of patients with the first diagnosis in life, thousand people;
  - y_{12}: musculoskeletal system and conjunctive tissue diseases, number of patients with the first diagnosis in life, thousand people;
  - y_{13}: diseases of the genitourinary system, number of patients with the first diagnosis in life, thousand people;
  - y_{14}: pregnancy, delivery, and postpartum period, number of patients with the first diagnosis in life, thousand people;
  - y_{15}: congenital abnormalities (birth defects), deformations, and chromosome breakages, number of patients with the first diagnosis in life, thousand people;
- y_{16}: traumas, intoxications, and certain other consequences of the influence of external reasons, number of patients with the first diagnosis in life, thousand people;
- y_{17}: share of Russians that play sports, % of population.

The initial data for the regression analysis are given in Tables 1 and 2. As transition to the digital economy started in 2013 (e.g., evaluation of the global digital competitiveness of economy started in 2013), we use the data since 2013 for the analysis. For bringing the number of observations to the minimum required level (for correct results of the regression analysis) we supplemented the selection with the forecast data for 2019–2024.

Table 1. Dynamics of emergence of new diseases and activity of usage of Internet and mobile communications in Russia in 2013–2018 and the forecast for 2019–2024.

Year	x_1	x_2	y_1	y_2	y_3	y_4	y_5	y_6	y_7
2013	69.1	1,933.3	4,434	1,629	668	1,527	2,364	5,023	4,014
2014	73.5	1,908.4	4,504	1,693	668	1,636	2,370	5,067	4,050
2015	70.1	1,937.8	4,116	1,672	692	1,953	2,257	4,878	3,893
2016	73.1	1,977.9	4,086	1,668	688	2,038	2,231	4,787	3,863
2017	76.0	2,002.6	3,765	1,708	709	2,720	2,105	4,562	3,718
2018	79.0	2,027.6	3,470	1,749	730	3,630	1,987	4,348	3,578
2019	82.1	2,052.9	3,197	1,791	752	4,845	1,875	4,143	3,443
2020	85.4	2,078.6	2,947	1,834	774	6,466	1,770	3,949	3,314
2021	88.8	2,104.5	2,715	1,877	797	8,630	1,670	3,763	3,189
2022	92.3	2,130.8	2,502	1,922	821	11,518	1,576	3,586	3,069
2023	96.0	2,157.4	2,306	1,968	846	15,373	1,488	3,418	2,954
2024	99.8	2,184.4	2,125	2,016	871	20,518	1,404	3,257	2,843

Source: compiled by the authors based on Federal State Statistics Service (2019a, b).

Table 2. Dynamics of emergence of new diseases and popularity of sports activities in Russia in 2013–2018 and the forecast for 2019–2024.

Year	y_8	y_9	y_{10}	y_{11}	y_{12}	y_{13}	y_{14}	y_{15}	y_{16}	y_{17}
2013	4,285	48,568	5,055	6,740	4,634	7,147	2,778	298	13,285	45
2014	4,205	48,708	5,342	6,767	4,647	7,164	2,801	307	13,183	48
2015	4,563	49,464	5,163	6,437	4,410	6,793	2,618	297	13,235	51
2016	4,649	51,573	5,229	6,241	4,332	6,689	2,450	302	13,063	54
2017	5,044	54,764	5,409	5,779	4,050	6,260	2,161	306	12,845	57
2018	5,472	58,152	5,595	5,351	3,786	5,859	1,906	310	12,630	60
2019	5,937	61,750	5,788	4,955	3,539	5,484	1,681	314	12,419	63
2020	6,442	65,571	5,987	4,588	3,308	5,132	1,482	319	12,211	66
2021	6,989	69,628	6,193	4,248	3,093	4,803	1,307	323	12,007	69
2022	7,582	73,936	6,406	3,934	2,891	4,496	1,153	327	11,807	72
2023	8,227	78,510	6,627	3,643	2,703	4,207	1,017	332	11,609	75
2024	8,925	83,368	6,855	3,373	2,527	3,938	897	336	11,415	78

Source: compiled by the authors based on Federal State Statistics Service (2019b) and RiaRating (2019).

3 Results

The main results of the performed regression analysis are shown in Tables 3 and 4. They present the obtained values of the determination coefficients (R^2, it should be above 0.9), which show close connection of the variables and estimate coefficients that reflect the direction and value of the change of y with the change of x by 1, and p-values (should not exceed 0.05), which reflect correctness of the regression dependencies.

Table 3. Regression dependence of emergence of new diseases on activity of usage of Internet and mobile communications in Russia.

–	Characteristics of dependence	y1	y2	y3	y4	y5	y6	y7	y8	y9
x1	R^2	−0.98	0.99	0.99	0.95	−0.98	−0.98	−0.98	0.99	0.99
	Estimated coefficient	−0.52	0.85	−0.11	0.00	2.77	0.68	20.26	−2.02	0.51
	p-value	0.02	0.01	0.02	0.06	0.02	0.02	0.19	0.19	0.19
x2	R^2	−0.99	0.97	0.99	0.92	−0.99	−0.99	−0.99	0.98	0.99
	Estimated coefficient	−8.40	12.12	−0.11	−0.02	44.46	−10.87	339.81	−33.89	8.60
	p-value	0.02	0.02	0.85	0.06	0.02	0.02	0.19	0.19	0.19

Source: calculated by the authors.

Table 4. Regression dependence of emergence of new diseases and popularity of sports activities on activity of usage of Internet and mobile communications in Russia.

–	Characteristics of dependence	y10	y11	y12	y13	y14	y15	y16	y17
x1	R^2	0.99	−0.98	−0.98	−0.98	−0.97	0.99	−0.99	0.98
	Estimated coefficient	−3.15	12.12	−29.31	→0	0.01	0.24	−0.01	0.25
	p-value	0.19	0.19	0.19	0.28	→0	→0	→0	→0
x2	R^2	0.97	−0.99	−0.99	−0.99	−0.99	0.95	−0.99	0.99
	Estimated coefficient	−53.21	203.39	−492.00	0.17	−0.05	−10.02	−0.52	3.52
	p-value	0.19	0.19	0.19	→0	→0	→0	→0	→0

Source: calculated by the authors.

The data of Tables 3 and 4 show that coefficients of determination for all indicators exceed 0.9 – which shows close connection between the studied indicators. However, statistically significant (p-values do not exceed 0.05) at the significance level $\alpha = 0.05$ are dependencies $y_1(x_1)$, $y_2(x_1)$, $y_3(x_1)$, $y_4(x_1)$, $y_5(x_1)$, $y_6(x_1)$, $y_{13}(x_1)$, $y_{14}(x_1)$, $y_{15}(x_1)$, $y_{16}(x_1)$, $y_{17}(x_1)$, $y_1(x_2)$, $y_2(x_2)$, $y_5(x_2)$, $y_6(x_2)$, $y_{13}(x_2)$, $y_{14}(x_2)$, $y_{15}(x_2)$, $y_{16}(x_2)$, $y_{17}(x_2)$. The most significance ones (estimated coefficients exceed 1) are dependencies $y_5(x_1)$, $y_1(x_2)$, $y_2(x_2)$, $y_5(x_2)$, $y_6(x_2)$, $y_{15}(x_2)$ and $y_{17}(x_2)$. This allows for the following conclusions:

- increase of the number of devices of portable mobile communications in Russia by 1 unit per 1,000 people leads to reduction of the disease rate of certain germ diseases and parasitical diseases (y_1) by 8,400 people;
- increase of the number of devices of portable mobile communications in Russia by 1 unit per 1,000 people leads to growth of the disease rate of growths (y_2) by 12,120;

- increase of usage of Internet by Russians in the course of last three months by 1% leads to increase of the disease rate of nervous system diseases (y_5) by 2,770 people;
- increase of the number of devices of portable mobile communications in Russia by 1 unit per 1,000 people leads to increase of the disease rate of the nervous system diseases (y_5) by 44,460;
- increase of the number of devices of portable mobile communications in Russia by 1 unit per 1,000 people leads to reduction of the disease rate of eye diseases (y_6) by 10,870 people;
- increase of the number of devices of portable mobile communications in Russia by 1 unit per 1,000 people leads to reduction of the disease rate of congenital anomalies (birth defects), deformations, and chromosome breakages (y_{15}) by 10,020;
- increase of the number of devices of portable mobile communications in Russia by 1 unit per 1,000 people leads to increase of the share of Russians that play sports (y_{15}) by 3.52%.

4 Conclusion

Thus, as a result of the research it is determined that contrary to the offered hypothesis not all diseases' rates grow in the conditions of the digital economy. For example, there is no statistical connection between endocrine diseases, digestion disorders, and dysmetabolism with the usage of digital technologies. Usage of Internet stimulates growth of the disease rate of only the nervous system diseases and does not influence the disease rate of other types of diseases.

Usage of mobile communications influences the population's health in Russia to a larger extent, causing growth of disease rate of morbid growths and nervous system diseases and leading to reduction of disease rate of certain germ diseases, parasitical diseases, eye diseases, congenital anomalies (birth defects), deformations, and chromosome breakages.

We also determined a sustainable and vivid positive connection between increase of activity of usage of mobile communications and growth of the share of Russians that play sports. Therefore, the tendencies of healthcare and sports in the conditions the digital economy in Russia differ from the global tendencies that are noted by the World Health Organization. Applied research by the example of other countries will be a perspective direction of further scientific studies.

References

Bogoviz, A.V., Lobova, S.V., Ragulina, J.V.: Distortions in the theory of costs in the conditions of digital economy. In: Lecture Notes in Networks and Systems, vol. 57, pp. 1231–1237 (2019)

Eime, R.M., Sawyer, N., Harvey, J.T., Casey, M.M., Westerbeek, H., Payne, W.R.: Integrating public health and sport management: SPORT participation trends 2001–2010. Sport Manag. Rev. **18**(2), 207–217 (2015)

Johnson, M.R.: Inclusion and exclusion in the digital economy: disability and mental health as a live streamer on Twitch.tv. Inf. Commun. Soc. **22**(4), 506–520 (2019)

Kang, S.: Trend of the sports convergence technology for health promotion. In: 2015 5th International Conference on IT Convergence and Security, ICITCS 2015 – Proceedings, 7293030 (2015)

Popkova, E.G.: Economic and Legal Foundations of Modern Russian Society: A New Institutional Theory. Advances in Research on Russian Business and Management. Information Age Publishing, Charlotte (2017)

Popkova, E.G., Sergi, B.S.: Will Industry 4.0 and other innovations impact Russia's development? In: Exploring the Future of Russia's Economy and Markets, 34–42, pp. 51–68. Emerald Publishing (2019)

Popkova, E.G., Litvinova, T., Mitina, M.A., French, J.: Social advertising: a Russian perspective. Espacios **39**(1), 17 (2018)

Sergi, B.S. (ed.): Exploring the Future of Russia's Economy and Markets: Towards Sustainable Economic Development. Emerald Publishing, Bingley (2018)

World Health Organization. Obesity and overweight (2019a). https://www.who.int/ru/news-room/fact-sheets/detail/obesity-and-overweight. Accessed 12 Apr 2019

World Health Organization. Public health implications of excessive use of the Internet and other communication and gaming platforms (2019b). https://www.who.int/substance_abuse/activities/addictive_behaviours/en/. Accessed 12 Apr 2019

RiaRating. Overview showed the number of Russians that play sports (2019). https://ria.ru/20180828/1527330741.html. Accessed 11 Apr 2019

Federal State Statistics Service. Information society statistics (2019a). http://www.gks.ru/wps/wcm/connect/rosstat_main/rosstat/ru/statistics/science_and_innovations/it_technology/

Federal State Statistics Service. Russia in numbers (2019b). http://www.gks.ru/wps/wcm/connect/rosstat_main/rosstat/ru/statistics/publications/catalog/doc_1135075100641. Accessed 12 Apr 2019

The Key Directions of Management of a Company that Conducts Digital Modernization

Olga V. Konina(✉)

Moscow State Pedagogical University, Moscow, Russia
koninaov@mail.ru

Abstract. Purpose: The purpose of the article is to determine the key directions of management of a company that conducts digital modernization, to determine the level of global competitiveness of management of Russian entrepreneurship that conducts digital modernization, and to develop the recommendations for its increase.

Design/Methodology/Approach: Based on statistical and analytical data of the IMD and the World Economic Forum, the author performs evaluation of digital competitiveness of entrepreneurship in Russia according to the distinguished directions of management, with the help of the international organizations' indicators and the author's formula. The results of the evaluation are presented graphically with the help of the polygon of competitiveness – which allowed determining the low level of global competitiveness for all directions of management of a company that conducts digital modernization in Russia in 2018.

Findings: The author determines the key directions of management of a company that conducts digital modernization: risk management, marketing management, personnel management, financial management, management of innovations, management of information and communication technologies, and integration management.

Originality/Value: A conceptual model of management of a company that conducts digital modernization is developed – it ensures systemic implementation of the distinguished directions of management and offers recommendations for its practical implementation, which create such advantages as controllability and manageability of the process of company's digital modernization, acceleration of the process of company's digital modernization, and obtaining of corporate benefits from the process of company's digital modernization (by means of state support and increase of competitiveness).

Keywords: Management · Company · Digital modernization · Modern Russia · Competitiveness

JEL Code: F12 · G34 · L26 · O31 · O32 · O33 · O38

E. G. Popkova and B. S. Sergi (Eds.): ISC 2019, LNNS 87, pp. 104–111, 2020.
https://doi.org/10.1007/978-3-030-29586-8_12

1 Introduction

The modern entrepreneurship functions and develops under the influence of three factors. 1st factor: technological progress. Prominent achievements in science and technology – digital revolution – create new opportunities for optimization of the entrepreneurial activities on the basis of automatization of certain processes and the whole production and distribution systems. 2nd factor: global competition. Reduction or full elimination of customs barriers leads to expansion of the presence of foreign rivals in domestic sectorial markets, increasing the level of their concentration and stimulating the market players (companies) to look for new sources of competitive advantages.

3rd factor: state regulation. One of the latest tendencies of state regulation of economy on the whole and entrepreneurship in particular in most countries of the world is the course at formation of the digital economy. Thus, active state stimulation (in the form of toughening of norms and standards and in the form of tax and infrastructural support) of implementing digital technologies into entrepreneurship takes place. Systemic influence of these mutually reinforcing factors leads to digital modernization of modern entrepreneurship, which covers all sectorial markets and all entrepreneurial processes.

In the countries with developed market economy (OECD) this process is largely determined by the market factors (technological progress and global competition), due to which digital modernization is a natural manifestation of entrepreneurial activities and is conducted progressively, and digital technologies are gradually built into the entrepreneurial processes and are adapted to the specifics of companies.

Contrary to this, digital modernization is a lot of countries with forming market economy is caused primarily by the factor of state regulation. It has been started artificially – which often contradicts own strategies of companies' development; digital technologies are implemented in a quick regime without the necessary adaptation to the specifics of companies, which complicates their practical application and leads to contradictory results (e.g., improvement of statistical indicators of digital modernization with simultaneous reduction or stability of the level of digital competitiveness).

Thus, the current object of studying the modern economic science is management of a company that conducts digital modernization in the countries with forming market economy – which include modern Russia. The purpose of the article is to determine the key directions of management of a company that conducts digital modernization, to determine the level of global competitiveness of management of Russian entrepreneurship that conducts digital modernization, and to develop recommendations for increasing it.

2 Materials and Method

The issues of management of a company that conducts digital modernization are not paid sufficient attention in the existing economic literature. However, some works study certain issues of management of a company that conducts digital modernization, which does not allow for systemic presentation of this process. Examples could be the works Raschke and Mann (2017), Yamamoto (2017), Quattrociocchi et al. (2018), and Polyakova et al. (2019).

In other studies the research is performed at the fundamental level: Raschke and Mann (2017), Bogoviz et al. (2019), Bogoviz et al. (2019), and Popkova (2019). In other works the authors use the leading experience of countries with developed markets: Julia et al. (2018) and Schmidt et al. (2018). This does not allow evaluating the competitiveness of the existing practice of management of companies that conduct digital modernization in the countries with forming market economy.

Thus, there's deficit of scientific research in the sphere of management of companies that conduct digital modernization in the countries with forming market economy – which is to be overcome by this article. Studying the transformation processes that take place in activities of the company that conducts digital modernization allowed determining the key directions of management of a company that conducts digital modernization: risk management, marketing management, personnel management, financial management, management of innovations, management of information and communication technologies, and integration management.

Based on statistical and analytical data of the IMD and the World Economic Forum, evaluation of digital competitiveness of entrepreneurship in Russia for the distinguished directions of management was performed with the help of the international organizations' indicators. Evaluation of digital competitiveness was performed with on the following formula:

$$DC = 100 - (VI * 100\%/NP) \tag{1}$$

where DC – это digital competitiveness of entrepreneurship for the direction of management, %;
VI – value of the indicator, position in the global rating of countries;
NP – number of countries that participate in the rating.

The results of the performed evaluation based on the 2018 data are given in Table 1.

The results of the performed evaluation are generalized and presented graphically with the help of the polygon of competitiveness (Fig. 1).

Table 1. The key directions of management of a company that conducts digital modernization.

Direction of company's management	Indicator of implementing the direction of management in entrepreneurship	Value of the indicator in Russia in 2018, position	Digital competitiveness of entrepreneurship for the direction of management, %
Risk management	Investment risk	57 out of 63	9.52
	Cyber security	37 out of 63	41.27
	Software piracy	54 out of 63	14.29
	Direct average		21.69

(*continued*)

Table 1. (*continued*)

Direction of company's management	Indicator of implementing the direction of management in entrepreneurship	Value of the indicator in Russia in 2018, position	Digital competitiveness of entrepreneurship for the direction of management, %
Marketing management	E-Participation	28 out of 63	55.56
	Agility of companies	61 out of 63	3.17
	11.04 Nature of competitive advantage	72 out of 137	47.45
	11.08 Extent of marketing	59 out of 137	56.93
	Direct average		40.78
Personnel management	Employee training	41 out of 63	34.92
	Digital/Technological skills	29 out of 63	53.97
	Direct average		44.44
Financial management	Venture capital	55 out of 63	12.70
	Internet retailing	37 out of 63	41.27
	Direct average		26.98
Management of innovations	Knowledge transfer	54 out of 63	14.29
	Innovative firms	46 out of 63	26.98
	High-tech exports (%)	36 out of 63	42.86
	12.04 University-industry collaboration in R&D	42 out of 137	69.34
	Direct average		38.37
Management of information and communication technologies	Use of big data and analytics	58 out of 63	7.94
	Communications technology	30 out of 63	52.38
	Direct average		30.16
Integration management	Public-private partnerships	50 out of 63	20.63
	11.03 State of cluster development	88 out of 137	35.77
	Direct average		28.20

Source: compiled and calculated by the author based on IMD (2019), World Economic Forum (2019).

As is seen from Fig. 1, for all directions of management of a company that conducts digital modernization in Russia in 2018 there is low level of global competitiveness: risk management (21.69%), marketing management (40.78%), personnel management (44.44%), financial management (26.98%), management of innovations (38.37%), management of information and communication technologies (30.16%), and integration management (28.20%). This shows the necessity for improving the above practices of management.

Fig. 1. Global competitiveness of management of the Russian entrepreneurship that conducts digital modernization, in 2018, %. Source: calculated and compiled by the author.

3 Results

The key reason of low global competitiveness of management of the Russian entrepreneurship that conducts digital modernization is low initiative of entrepreneurship in the sphere of management, caused by expectation of support from the state. For overcoming this reason, the author developed a conceptual model of management of a company that conducts digital modernization, which reflects the logic of systemic implementation of all distinguished key directions of management (Fig. 2).

According to Fig. 2, in process of digital modernization the company is recommended to conduct digital marketing management that allows for highly-effective (with expanded possibilities of conducting marketing research, promotion of products, search or optimal suppliers, and suppliers of continuous contact with them) management of its relations with suppliers of resources, spare parts, intermediary products, and consumers. It is also offered to conduct integration management during management of relations with rivals (through formation of clusters of digital companies) and the state (through the projects of public-private partnership) for attracting additional resources into digital modernization and distributing (reducing) investment resources.

It is necessary to conduct management of information and communication technologies through provision of companies' access to the new information and communication technologies (e.g., Big Data processing technologies), which allow for automatization of its production and/or distribution processes and increase of their effectiveness. It is important to conduct risk management through acquisition of the modern technologies of provision of digital security and protection of corporate data and information.

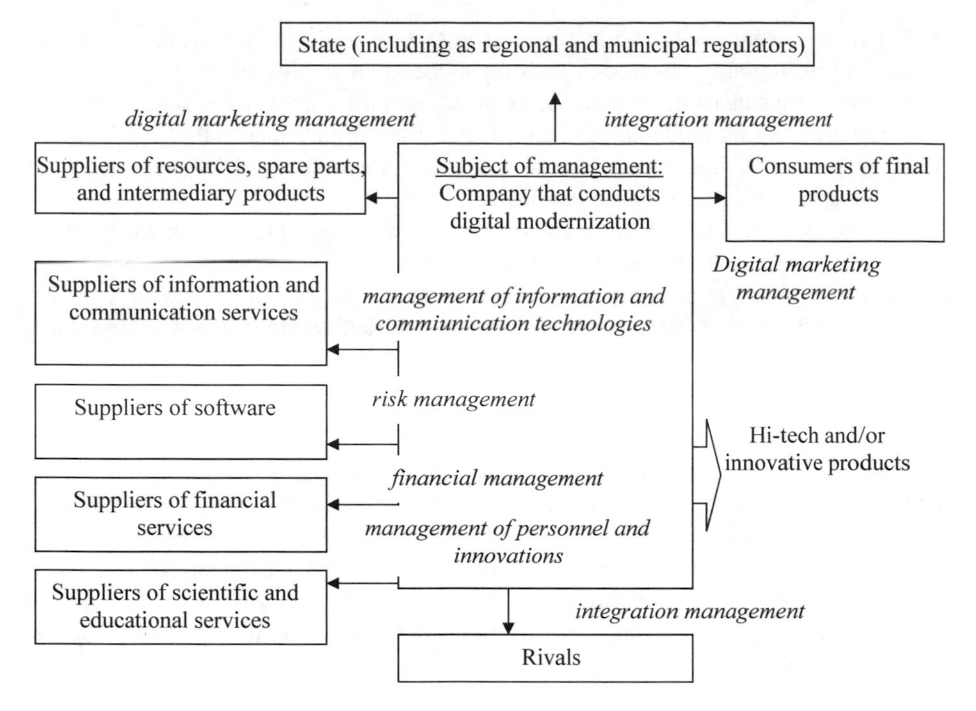

Fig. 2. The conceptual model of management of a company that conducts digital modernization. Source: compiled and developed by the author.

It is recommended to conduct financial management through attraction of credit and venture investment resources into the company's digital modernization and to obtain access to new financial tools. It is offered to conduct personnel management through attraction of digital personnel and receipt of corporate services in the sphere of development of digital competencies with the company's employees and management of innovations through organization of usage of digital technologies for manufacture of hi-tech and/or innovative products, which are peculiar for high level of global competitiveness.

4 Conclusion

The performed research allowed determining the following key directions of management of a company that conducts digital modernization:

1. Risk management that is aimed at provision of preserving corporate digital data and reliability and security of usage of digital technologies by the company;
2. Marketing management that is aimed at promotion of digital modernization of the company with the usage of digital technologies (e.g., Internet marketing);

3. Personnel management that is aimed at attraction of digital personnel and development of digital competencies with the company's employees;
4. Financial management that is aimed at attraction of financial resources for the company's digital modernization and its access to digital financial technologies;
5. Management of innovations that is aimed at usage of digital technologies for manufacture of hi-tech and/or innovative products;
6. Management of information and communication technologies that is aimed at the company's access to digital infrastructure (e.g., Internet);
7. Integration management that is aimed at strengthening of the company's market positions by means of cooperation with rivals and partnership with the state, attraction of additional resources into digital modernization, and distribution (reduction) of its investment resources.

As a result of using the developed conceptual model of management of a company that conducts digital modernization (the model ensures systemic implementation of the distinguished directions of management) and observing the offered recommendations for its practical implementation, the following advantages will be achieved: controllability and manageability of the process of company's digital modernization, acceleration of the process of company's digital modernization, and obtaining of corporate benefit from the process of company's digital modernization (by means of state support and increase of competitiveness).

References

Bogoviz, A.V., Alekseev, A.N., Ragulina, J.V.: Budget limitations in the process of formation of the digital economy. In: Lecture Notes in Networks and Systems, vol. 57, pp. 578–585 (2019)

Bogoviz, A.V., Lobova, S.V., Ragulina, J.V.: Distortions in the theory of costs in the conditions of digital economy. In: Lecture Notes in Networks and Systems, vol. 57, pp. 1231–1237 (2019)

IMD: World Digital Competitiveness Ranking 2018 (2019). https://www.imd.org/wcc/world-competitiveness-center-rankings/world-digital-competitiveness-rankings-2018/. Accessed 22 Apr 2019

Julia, K., Kurt, S., Ulf, S.: How digital transformation affects enterprise architecture management – a case study. Int. J. Inf. Syst. Project Manag. **6**(3), 5–18 (2018)

Polyakova, A.G., Loginov, M.P., Strelnikov, E.V., Usova, N.V.: Managerial decision support algorithm based on network analysis and big data. Int. J. Civ. Eng. Technol. **10**(2), 291–300 (2019)

Popkova, E.G.: Preconditions of formation and development of industry 4.0 in the conditions of knowledge economy. Stud. Syst. Decis. Control **169**, 65–72 (2019)

Quattrociocchi, B., Mercuri, F., D'Arcangelo, D., Cristini, V.: Knowledge management to compete in the digital era: skills evolution of enterprise systems. In: Proceedings of the European Conference on Knowledge Management, ECKM, vol. 2, pp. 733–740 (2018)

Raschke, R.L., Mann, A.: Enterprise content risk management: a conceptual framework for digital asset risk management. J. Emerg. Technol. Account. **14**(1), 57–62 (2017)

Schmidt, R., Möhring, M., Keller, B., Zimmermann, A., Toni, M., Di Pietro, L.: Digital enterprise architecture management in tourism – state of the art and future directions. In: Smart Innovation, Systems and Technologies, vol. 73, pp. 93–102 (2018)

World Economic Forum. The Global Competitiveness Report 2017–2018 (2019). http://www3.weforum.org/docs/GCR2017-2018/05FullReport/TheGlobalCompetitivenessReport2017–2018.pdf. Accessed 22 Apr 2019

Yamamoto, S.: Enterprise requirements management knowledge towards digital transformation. In: Lecture Notes in Electrical Engineering, vol. 449, pp. 309–317 (2017)

Contradiction of the Digital Economy: Public Well-Being vs. Cyber Threats

Elena G. Popkova[1(✉)] and Kantoro Gulzat[2]

[1] Plekhanov Russian University of Economics, Moscow, Russia
210471@mail.ru
[2] Kyrgyz National University named after J. Baiasagyn, Bishkek, Kyrgyzstan
gulzat.tashkulova@mail.ru

Abstract. Purpose: The purpose of the paper is to substantiate the logical interconnection of social consequences of the digital economy and its consequences for cyber security and to develop – based on this interconnection - the scientific and methodological provision of systemic monitoring of the digital economy and the model of managing the digital economy for maximizing public well-being and overcoming the cyber threats.

Design/Methodology/Approach: For determining the interdependence (cross-correlation) between human development index (indicator of public well-being) and cyber security index (indicator of cyber threats), the authors use the method of correlation analysis; for determining their dependence on the index of digital competitiveness (indicator of the digital economy) the authors use the method of regression analysis. As a result, the authors determine low information content of the statistical indicators that reflect the integral characteristics of public well-being and cyber security and come to the conclusion on inapplicability of the standard (e.g., econometric) methods in the work. That's why the method of logical analysis is used for analyzing the qualitative essence of social consequences of cyber threats of the digital economy.

Findings: It is proved that social consequences of the digital economy and its consequences for the cyber security are interconnected and could and should be studied and managed as a whole. It is substantiates that when treating the cyber threats it is necessary to focus not on their technical components (causes of threats that are determined by the specifics of digital technologies) but on social components – i.e., objects of threats – various categories of the population. In this case, the influence on public well-being becomes a common basis of the studied consequences.

Originality/Value: Based on the determined interconnection, the authors develop the scientific and methodological provision of systemic monitoring of effectiveness of the digital economy, which envisages comparison of its social advantages and allows determining the top-priority objects of the digital economy management. Approbation of the offered methodological provision by the example of the digital economy of Russia in 2018 showed its low effectiveness. In order to increase it, it is recommended to use the developed model of managing the digital economy for maximizing public well-being and overcoming the cyber threats.

Keywords: Public well-being · Cyber threats · Digital economy · Effectiveness · Management · Modern Russia

E. G. Popkova and B. S. Sergi (Eds.): ISC 2019, LNNS 87, pp. 112–124, 2020.
https://doi.org/10.1007/978-3-030-29586-8_13

JEL Code: F52 · H56 · L86 · L97 · M14 · M15 · O31 · O32 · O33 · O35 · O38

1 Introduction

The digital economy has contradictory influence on the modern economic systems. On the one hand, it opens new opportunities for opening human potential and solving the social problems (e.g., the problems of formation of inclusive social environment and satisfaction of population's needs), which ensures growth of public well-being. On the other hand, the digital economy causes new threats to functioning and development of economic systems, which are of the cybernetic nature. The most typical example of these threats is the necessity and high complexity of provision of information security (security of creation, storing, processing, and exchange of digital information).

These manifestations of the digital economy are studied separately. Due to this, the supporters of the digital economy focus on its social consequences, and its opponents focus on its consequences for the cyber security. This does not allow compiling a full picture of consequences of the digital economy and determining the expedience of its development or, on the contrary, the selection of the alternative model of development of the modern economic systems. In addition to this, state management of the digital economy is complicated, as management of social consequences and management of consequences for the cyber security are separate processes. Delimitation of these directions predetermines large pressure on the federal budget, as it leads to implementation of the independent regulation measures and financing of two separate controlling bodies.

Thus, the problem of scientific study and the search for the means of overcoming the contradiction of the digital economy, connected to opposition of its social consequences and consequences for cyber security, is very topical. The working hypothesis of the research is that these consequences are interconnected and could and should be studied as a whole. The purpose of the paper is to substantiate the logical interconnection of social consequences of the digital economy and its consequences for cyber security and to develop – based on this interconnection - the scientific and methodological provision of systemic monitoring of the digital economy and the model of managing the digital economy for maximizing public well-being and overcoming the cyber threats.

2 Materials and Method

The existing scientific literature studies the expected and the existing social advantages of the digital economy – one of which is growth of population's living standards and social risks that emerge in the conditions of the digital economy, which include the risk of growth of unemployment. These issues are discussed in the works Holford (2019), Janchai et al. (2019), Komarova (2019), Konkolewsky (2017), Malakhova et al. (2018), Martynenko and Vershinina (2018), Muntaner (2018), Petrenko et al. (2018), Popkova et al. (2019), and Qerimi and Sergi (2017).

Despite this, quantitative evaluation of social consequences the digital economy is complicated due to incompatibility of the data – impossibility to compare the qualitative indicator of living standards, which is measured in points (or monetary units in connection to GDP per capita) and quantitative indicator of unemployment rate, which is measured in number of population or per cent.

The cybernetic (technical) advantages of the digital economy, which consists in more convenient creation, storing, processing, and exchange of digital information and possibilities of automatization of these processes, and cybernetic threats to the digital economy, which are connected to provision of security of these processes, are studied in detail in the existing studies and publications. These include Ansong and Boateng (2019), Boban (2014), Filyak (2018), Frolova et al. (2018), MacAk et al. (2017), Mueller and Grindal (2019), Popkova (2019), and Teoh and Mahmood (2017).

Quantitative evaluation of the consequences of the digital economy for cyber security is complicated due to incompatibility of the data – impossibility to compare the share of automatized productions and the number of attacks of cyber criminals. However, despite the high level of elaboration of the social consequences of the digital economy and its consequences for cyber security separately, their contradiction is poorly studied and requires further research.

The global statistics uses separate indicators for measuring public well-being (e.g., human development index) and cyber threats (e.g., cyber security index). For determining their interdependence (cross-correlation) the authors use the method of correlation analysis; for determining their dependence on the digital economy (measured with the help of the digital competitiveness index) the authors use the method of regression analysis.

In order to ensure representativeness of selection of the data, it contains the data for the countries of every region of the world, according to the International Telecommunication Union. The objects are the countries that are peculiar for the highest level of cyber security – Top 3 countries from each region of the world, except for Africa (the data only for South Africa are available).

Table 1. Cyber security index, human development index, and index of digital competitiveness in the selection of countries in 2018.

Region of the world	Country	Cyber security index, points 0–1	Human development index, points 0–1	Digital competitiveness index, points 1–100
		y_1	y_2	x
Africa	South Africa	0.652	0.699	56.876
America	USA	0.926	0.624	100.00
	Canada	0.892	0.926	95.201
	Mexico	0.629	0.774	56.385
Middle East	Saudi Arabia	0.881	0.853	61.869
	Qatar	0.860	0.856	78.873
	Jordan	0.556	0.735	57.195

(*continued*)

Table 1. (*continued*)

Region of the world	Country	Cyber security index, points 0–1	Human development index, points 0–1	Digital competitiveness index, points 1–100
		y_1	y_2	x
Asia-Pacific	Singapore	0.898	0.932	99.422
	Malaysia	0.893	0.802	8,.631
	Australia	0.890	0.939	90.226
Europe	UK	0.931	0.922	93.239
	France	0.918	0.901	80.753
	Lithuania	0.908	0.858	76.059
	Russia	0.836	0.816	65.207

Source: compiled by the authors based on IMD Business School (2019), International Telecommunication Union (2019), United Nations Development Programme (2019).

Based on the data of Table 1 we calculated cross correlation of y_1 and y_2, which constituted 0.5045. Therefore, the cyber security index and human development index are interconnected by 50.45% (weak connection). The results of the regression analysis did not show significant dependence of y_1 on x and y_2 on x. That's why we shall restrain to building regression curves (Fig. 1).

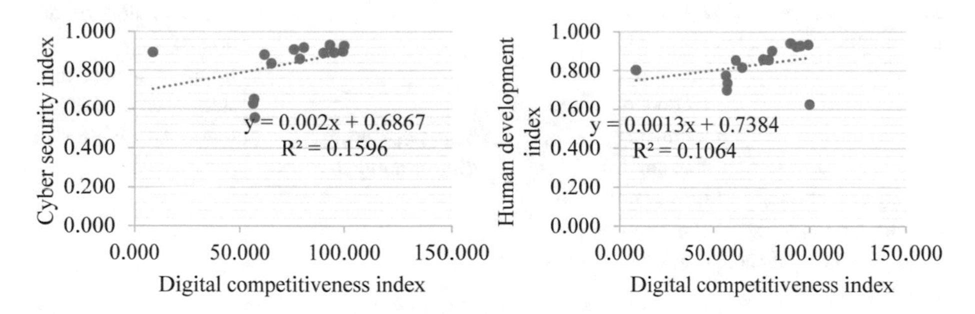

Fig. 1. Regression curves that reflect dependence of the cyber security index and human development index in the digital competitiveness index in the selection of countries in 2018. Source: calculated and built by the author.

Figure 1 shows very low values of estimate coefficients and coefficients of determination – due to which thorough regression analysis is inexpedient. The obtained results show that the existing statistical data are not applicable to the standard analysis of social consequences of the digital economy and its consequences for cyber security, as there is no statistically significant connection between the indicators. Thus, there's a necessity for developing a special scientific and methodological provision, which allows for analysis of these consequences of the digital economy on the basis of the existing statistical data.

3 Results

Due to low information content of the generalized statistical indicators, which reflect the integral characteristics of public well-being and cyber security, let us use the method of logical analysis and analyze qualitative essence of social advantages and cyber threats to the digital economy (Tables 2 and 3).

Table 2. The logic of influence of the digital economy on public well-being.

Consequences of the digital economy	Social advantages	Conditions of gaining advantages	Influence on public well-being
Transformation of the labor market on the basis of automatization	Creation of highly-efficient jobs	Mass automatization of production and distribution processes in entrepreneurship	Growth of living standards (on the basis of increase of incomes)
	Increase of the intellectual component of labor		Expansion of opportunities for self-realization, increase of social justice
	Growth of social mobility, expansion of opportunities for global labor migration		
Modernization of production, distribution, and consumption	Optimization of spending of natural resources		Growth of living standards (on the basis of reduction of prices)
	Minimization of production waste		Improvement of the state of environment
	Rationalization of consumer decisions	Mass mastering of digital competencies	Fuller satisfaction of needs
Emergence of new benefits and increase of accessibility of the existing benefits	Increase of effectiveness of provision of state services		
	Development of production according to individual orders		
	Increase of inclusiveness of the social environment	Mass distribution of “smart” technologies	Leveling of social disproportions

Source: compiled by the authors.

Table 2 shows that the digital economy has three important consequences, which lead to social advantages. 1st consequence: transformation of the labor market on the basis of automatization of production and distribution processes in entrepreneurship. Due to automatization of routine operations, highly-efficient jobs are created – which leads to growth of living standards (on the bass of increase of incomes) and increase of the intellectual component of labor, which opens expanded opportunities for self-realization.

In the conditions of the digital economy, the most important criterion of personnel selection and career building is digital competencies. Due to this, employees with different age and sex characteristics and countries of residence become equal in their opportunities, which ensures growth of social mobility and expansion of opportunities for global labor migration.

Table 3. The logic of emergence of cyber-attacks in the conditions of the digital economy.

Consequences of the digital economy	Cyber threats	Conditions of threats emergence	Influence on public well-being
Transformation of the labor market on the basis of automatization	Growth of unemployment rate and increase of tension in the labor market due to reduction of its capacity and toughening of criteria of employment and career building	Mass automatization of production and distribution processes in entrepreneurship	Reduction of living standards (on the basis of reduction of incomes and emergence of additional expenditures)
	Increased danger of labor due to necessity to be among machines		Increase of the number of traumas and accidents at production
Modernization of production, distribution, and consumption	Growth of energy capacity of production		Total deficit of energy
	Growth of probability of anthropogenic disasters		Aggravation of the state of environment
	Inaccessibility of goods and services for consumers who do not have digital competencies	Fragmentary mastering of digital competencies	Reduction of accessibility of goods and services, complication of the process of obtaining them, lower satisfaction of the needs
	Threat of unstable accessibility and loss of digital data	Domination of the digital form of provision of data	
	Threat of distortion and theft of digital data by cyber criminals		
Emergence of new benefits	Threat to privacy (violation of personal space)	Mass distribution of "smart" technologies	Psychological tension of people

Source: compiled by the authors.

2nd consequence: modernization of production, distribution, and consumption on the basis of high-precision digital technologies. This stimulates optimization of spending of natural resources (i.e., reduction of cost) and growth of living standards (based on reduction of prices) and leads to minimization of production waste and, therefore, to improvement of the state of environment. Mass distribution of technologies of intellectual decision support stimulates rationalization of consumer decisions – i.e., selection of the most optimal products in the market according to the set criteria.

3rd consequences: emergence of new (created with the help of digital technologies) and increase of accessibility of the existing benefits. This ensures increase of effectiveness of provision of state services (e.g., possibility of online registration) and development of production according to individual orders, which stimulates fuller satisfaction of the needs. Mass distribution of "smart" technologies leads to increase of inclusiveness of the social environment (equal opportunities and quality of life for all people, including the handicapped persons), due to which social disproportions are leveled.

Table 3 shows that cyber threats to the digital economy go beyond the limits of information security and cover all technical manifestations of usage of digital technologies, which, as is seen, influence public well-being. They are predetermined by the same consequences the digital economy as social advantages and influence public well-being.

1st consequence: transformation of the labor market on the basis of automatization. It causes growth of unemployment rate and increase of tension in the labor market due to reduction of its capacity and increase of criteria of employment and career building. For a lot of employees this will mean reduction of living standards – at least for the period of unemployment on the basis of reduction of incomes and emergence of additional expenditures due to the necessity for payment for educational services for mastering the digital competencies. Also, increase danger of labor emerges due to necessity for being among machines – which leads to increase of the number of traumas and accidents at production.

2nd consequence: modernization of production, distribution, and consumption. Manual labor, which requires minimum energy resources (e.g., for lighting of buildings), will be replaced by digital technical devices that require large energy resources. This will cause growth of energy capacity of production and total deficit of energy, which will lead to growth of prices for it. Also, the probability of industrial disaster, which aggravate the state of the environment, will grow.

A serious problem will be inaccessibility of digital goods and services for consumers who do not possess digital competencies, as well as the threat of unstable accessibility and losses of digital data – e.g., impossibility of performing payment operations during update of the online banking systems or impossibility to identify the person by an electronic passport without Internet access. This will lead to reduction of accessibility of goods and services, complication of the process of their receipt, and lower satisfaction of the needs. Also, a threat of distortion and theft of digital data by cyber criminals arises.

3rd consequence: emergence of new benefits on the basis of "smart" technologies – e.g., automatic determination of violations of law with the help of video surveillance and data recognition by AI. The causes a threat of privacy (violation of personal space) and psychological tension of people (including reduction of labor efficiency).

As is seen, social advantages and cyber threats in the conditions of the digital economy depend on the same conditions: mass automatization of the production and distribution processes in entrepreneurship, mastering of digital competencies, usage of the digital form of provision of data, and mass distribution of "smart" technologies, which are a logical substantiation of their interconnection. This allows for and predetermines the necessity for complex manifestation of social consequences of the digital economy and its consequences for information security on the basis of regulation of the determined conditions – for which the following model is developed (Fig. 2).

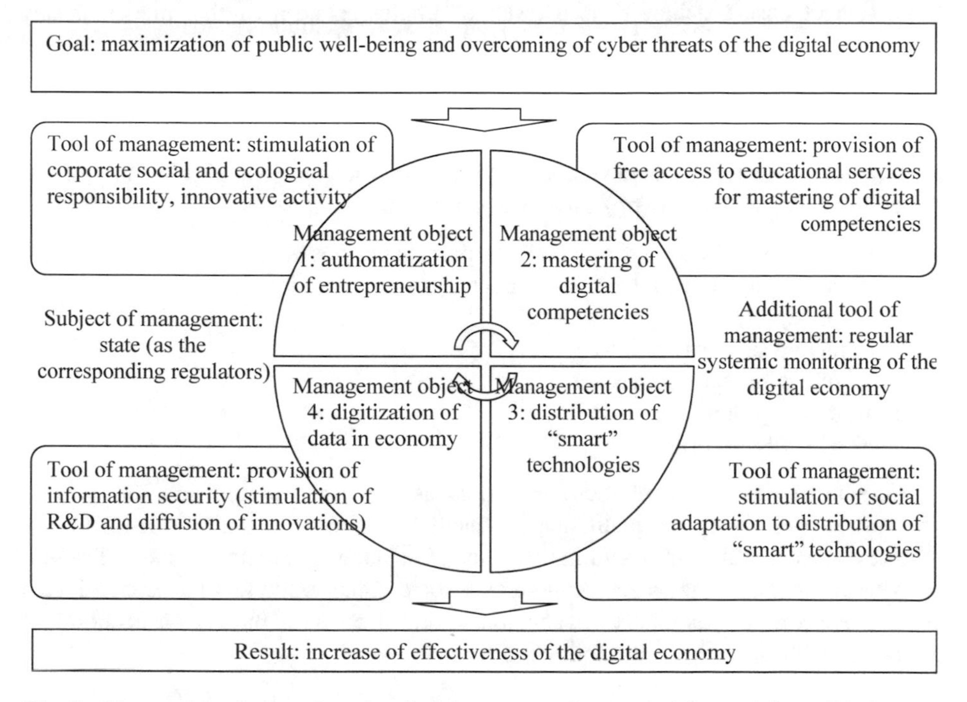

Fig. 2. The model of managing the digital economy for maximizing public well-being and overcoming the cyber threats. Source: compiled by the authors.

Figure 2 shows that the developed model seeks the goal of maximizing public well-being and overcoming cyber threats to the digital economy. It is based on the mechanism of management, which subject is the state as the corresponding regulators (e.g., the Ministry of Science and Education) and objects are the determined conditions of obtaining social advantages and appearance of cyber threats in the conditions of the digital economy, namely:

1. Automatization of entrepreneurship, which management tool should be stimulation of corporate social and ecological responsibility and innovative activity;
2. Mastering of digital competencies, which management tool is provision of free access of all interested parties (consumers, entrepreneurs, employees, and the unemployed) to educational services for mastering of digital competencies by the conditions of minimization of their expenditures (by means of state co-financing);

3. Distribution of “smart” technologies, which management tool should be stimulating the social adaptation to distribution of “smart” technologies;
4. Digitization of data in economy, which management should be performed with the help of such tool as provision of information security by stimulation of R&D and diffusion of innovations in this sphere.

Social advantages are positive results of the digital economy, and cyber threats are its costs, which are borne by the society. That’s why maximization of public well-being (increase of positive results) and overcoming of cyber threats to the digital economy (reduction of costs), achieved as a result of implementation of this model, means increase of effectiveness of the digital economy.

An additional tool of management in the presented model, which is not connected to any conditions in the digital economy, is regular systemic (combining social advantages and cyber threats) monitoring of its effectiveness. Special scientific and methodological provision is developed for this. The indicators of effectiveness of the digital economy within this provision are as follows:

- GDP per capita in view of population’s purchasing power;
- coefficient of tension in the labor market;
- index of happiness;
- consumption of electric energy;
- index of ecological effectiveness;
- death rate at production;
- index of cyber security.

Growth of the values of these indicators as compared to the previous year is determined. Depending on qualitative treatment (positive or negative) of growth, the category of each indicator is determined – social advantages (achieved due to leveling of cyber threats) or social costs (borne by the society due to the fact that cyber threats are not overcome). Quantitative evaluation of social costs of the digital economy is performed with the help of the following formula:

$$A_{de}/C_{de} = (I_{max} - I_{min}) * 100/I_{min}, \quad (1)$$

where A_{de} – social advantage of the digital economy;
C_{de} – social cost of the digital economy;
I_{max} – higher value of the indicator of two studied time periods (current and preceding);
I_{min} – lower value of the indicator of two studied time periods (current and preceding).

Formula (1) envisages evaluation of per cent change of the indicator's value. After quantitative evaluation of all social advantages and costs of the digital economy, the index of effectiveness of the digital economy is calculated with the following formula:

$$I_{dee} = \left(\sum A_{de}\right)/\left(\sum C_{de}\right), \quad (2)$$

where I_{dee} – index of effectiveness of the digital economy.

As is seen from Formula (2), the offered index is calculated with the classical formula of effectiveness by finding a sum of all advantages as compared to sum of all expenditures. The higher the value of the index, the higher the effectiveness of the digital economy. Let us perform approbation of the offered scientific and methodological provision of systemic monitoring of effectiveness of the digital economy by the example of Russia in 2018 (Table 4).

Table 4. The initial values of the indicators of effectiveness of the digital economy in Russia in 2018 and quantitative evaluation of its social advantages and costs.

Indicator	Value in 2017	Value in 2018	Treatment of the tendency	Category	Quantitative evaluation of advantages and costs
GDP per capita in view of purchasing power of population, USD	24,130	24,890	+	Social advantages	(24,890 − 24,130) * 100/24,130 = 3.15
Coefficient of tension in the labor market[a]	3.3	2.7	+	Social advantages	(3.3 − 2.7) * 100/2.7 = 22.22
Index of happiness, points	5.963	5.810	–	Social costs	(5.963 − 5.810) * 100/5810 = 2.63
Consumption of electric energy (kWh per capita)	6,539.211	6,602.658	–	Social costs	(6602.658 − 6539.211) * 100/6602.658 = 0.97
Index of ecological effectiveness, points	83.52	63.79	–	Social costs	(83.52 − 63.79) * 100/63.79 = 30.93
Death rate at production, number of cases	82,195	82,226	–	Social costs	(82226 − 82195) * *100/82226 = 0.04
Index of cyber security, points	0.788	0.836	+	Social advantages	(0.836 − 0.788) * 100/0.788 = 6.09

[a]Ratio of average annual number of the unemployed (according to the methodology of the International Labour Organisation) to average annual number of vacancies applied by employers to the employment services. Calculation is performed based on the data of Federal State Statistics Service and Federal Service of Labor and Employment.

Source: compiled and calculated by the authors based on the materials of Federal State Statistics Service (2019), Helliwell et al. (2019), International Labour Organization (2019), Global socio-economic indicators 2019 (2019), Yale Center for Environmental Law & Policy (Yale University), Center for International Earth Science Information Network (Columbia University), World Economic Forum (2019).

Let is consider an example of calculations. GDP per capita (PPP) of Russia in 2018 showed growth (as compared to 2017). This is a positive tendency (marked with “+”), so it is assigned to the category of social advantages. Quantitative evaluation of this advantage is performed in the following way: $A_{de} = (24{,}890 - 24{,}130) * 100/24{,}130 = 3.15$. From the higher value of the indicator (24,890, in 2018) of the two studied time periods (current and preceding) we subtract the lower value of the indicator (24,130, in 2017), multiply by 100, and divide by the lower value of the indicator (24,130, in 2017). Result: 3.15.

Index of happiness in Russia in 2018 showed decline (reduced as compared to 2017). This is a negative tendency (marked with “−”), so it is assigned to the category of social costs. Quantitative evaluation of this advantage is performed in the following way: $C_{de} = (5.963 - 5.810) * 100/5{,}810 = 2.63$. From the higher value of the indicator (5.963, in 2017) of the two studied time periods (current and preceding) we distract the lower value of the indicator (5,810, in 2017), multiply by 100, and divide by the lower value of the indicator (5,963, in 2017). Result: 2.63.

Based on the data of Table 4 we calculate the index of effectiveness of the digital economy in Russia in 2018: $I_{dee} = (3.15 + 22.22 + 6.09)/(2.63 + 0.97 + 30.93 + 0.04) = 31.46/34.57 = 0.91$. Therefore, aggregate social costs (34.57) exceed social advantages (31.46), so the value of the index is below 1 (0.91), which shows low effectiveness of the digital economy in Russia in 2018 and the necessity for paying attention to its management with emphasis on the determined social costs – in particular, automatization of entrepreneurship by stimulation of corporate social and ecological responsibility and innovative activity.

The recommended object of management has been selected with the exclusion method. Thus, growth of GDP per capita in view of purchasing power of population together with reduction of tension in the labor market shows successful mastering of digital competencies and overcoming of unemployment. “Smart technologies” have not become very popular and are not yet taken into account. Growth of the cyber security index shows successful digitization of the data in the Russian economy and overcoming of the problem of information security.

Therefore, all the determined costs – reduction of the happiness index, increase of death rate at production, growth of consumption of electric energy, and reduction of the index of ecological effectiveness – are either caused by mass automatization of the production and distribution processes in entrepreneurship or are not connected to the digital economy (sociological survey is necessary for specifying this issue).

4 Conclusion

As a result of the research it is possible to conclude that the offered hypothesis is correct – social consequences of the digital economy and its consequences for cyber security are mutually connected and thus could and should be studied and managed as a whole. It is substantiated that when treating cyber threats it is necessary to focus not on their technical component (causes of threats that are predetermined by the specifics of digital technologies) but on the social component – i.e., objects of threats (various groups of the population). In this case, the influence on public well-being becomes a common basis for the studied consequences.

Based on the determined interconnection, we developed a scientific and methodological provision of systemic monitoring of effectiveness of the digital economy, which envisages comparing its social advantages and costs and allows determining the top-priority objects of management of the digital economy. Approbation of the offered methodological provision by the example of the digital economy of Russia in 2018 is performed, which showed its low effectiveness and the necessity to pay attention to automatization of entrepreneurship during state regulation of the digital economy by stimulating corporate social and ecological responsibility and innovative activity. When conducting this management, it is recommended to use the developed model of managing the digital economy for maximizing public well-being and overcoming the cyber threats.

Acknowledgments. The research was performed with financial support from the Russian Fund of Fundamental Research within the scientific project No. 18-010-00103 A.

References

Ansong, E., Boateng, R.: Surviving in the digital era – business models of digital enterprises in a developing economy. Digit. Policy Regul. Gov. **21**(2), 164–178 (2019)

Boban, M.: Information security and the right to privacy in digital economy - the case of Republic of Croatia. In: 2014 37th International Convention on Information and Communication Technology, Electronics and Microelectronics, MIPRO 2014 – Proceedings, 6859804, pp. 1503–1508 (2014)

Filyak, P.Yu.: Information security in the context of the digital economy. In: CEUR Workshop Proceedings, 2109, pp. 25–30 (2018)

Frolova, E.E., Polyakova, T.A., Dudin, M.N., Rusakova, E.P., Kucherenko, P.A.: Information security of Russia in the digital economy: the economic and legal aspects. J. Adv. Res. Law Econ. **9**(1), 89–95 (2018)

Helliwell, J., Layard, R. Sachs, J.: World happiness report (2019). http://worldhappiness.report/ed/2018/. Accessed 01 Apr 2019

Holford, W.D.: The future of human creative knowledge work within the digital economy. Futures **105**, 143–154 (2019)

IMD business school: World Digital Competitiveness Ranking 2018 (2019). https://www.imd.org/wcc/world-competitiveness-center-rankings/world-digital-competitiveness-rankings-2018/. Accessed 01 Apr 2019

International Labour Organization. Summary of work-related mortality. World Bank division: work-related mortality (2019). http://www.ilo.org/moscow/areas-of-work/occupational-safety-and-health/WCMS_249278/lang–en/index.htm. Accessed 01 Apr 2019

International Telecommunication Union. Global Cybersecurity Index 2018 (2019). https://www.itu.int/en/ITU-D/Cybersecurity/Pages/global-cybersecurity-index.aspx. Accessed 01 Apr 2019

Janchai, W., Siddoo, V., Sawattawee, J.: A systematic review of work integrated learning for the digital economy. Int. J. Inf. Commun. Technol. Educ. **15**(1), 67–78 (2019)

Komarova, O.: Social issues of the transition to a digital economy. In: Proceedings of the 32nd International Business Information Management Association Conference, IBIMA 2018 - Vision 2020: Sustainable Economic Development and Application of Innovation Management from Regional expansion to Global Growth, pp. 7024–7032 (2019)

Konkolewsky, H.-H.: Digital economy and the future of social security. Administration **65**(4), 21–30 (2017)

MacAk, T., Maitah, M., Sergi, B.S.: Assessment of similarities between model behaviour and actual product from economic and technical perspectives. Int. J. Manag. Decis. Making **16**(2), 172–185 (2017)

Malakhova, E.V., Garnov, A.P., Kornilova, I.M.: Digital economy, information society and social challenges in the near future. Eur. Res. Stud. J. **21**, 576–586 (2018)

Martynenko, T.S., Vershinina, I.A.: Digital economy: the possibility of sustainable development and overcoming social and environmental inequality in Russia | [Economía digital: La posibilidad del desarrollo sostenible y la superación de la desigualdad social y ambiental en Russia]. Espacios **39**(44), 18–25 (2018)

Mueller, M., Grindal, K.: Data flows and the digital economy: information as a mobile factor of production. Digit. Policy Regul. Gov. **21**(1), 71–87 (2019)

Muntaner, C.: Digital platforms, gig economy, precarious employment, and the invisible hand of social class. Int. J. Health Serv. **48**(4), 597–600 (2018)

Petrenko, E., Pritvorova, T., Dzhazykbaeva, B.: Sustainable development processes: service sector in post-industrial economy. J. Secur. Sustain. Issues **7**(4), 781–791 (2018)

Popkova, E.G.: Preconditions of formation and development of industry 4.0 in the conditions of knowledge economy. Stud. Syst. Decis. Control **169**, 65–72 (2019)

Popkova, E.G., Fetisova, O.V., Zabaznova, T.A., Alferova, T.V.: The modern global financial system: social risks vs. technological risks. In: Lecture Notes in Networks and Systems, vol. 57, pp. 1013–1019 (2019)

Qerimi, Q., Sergi, B.S.: The nature and the scope of the global economic crisis' impact on employment trends and policies in South East Europe. J. Int. Stud. **10**(4), 143–153 (2017)

Teoh, C.S., Mahmood, A.K.: National cyber security strategies for digital economy. J. Theor. Appl. Inf. Technol. **95**(23), 6510–6522 (2017)

United Nations Development Programme. Human Development Reports (2019). http://hdr.undp.org/en/2018-update/download. Accessed 01 Apr 2019

World Bank. Global socio-economic indicators (2019). https://data.worldbank.org/indicator. Accessed 01 Apr 2019

Yale Center for Environmental Law & Policy (Yale University), Center for International Earth Science Information Network (Columbia University), World Economic Forum. Environmental performance index (2019). https://epi.envirocenter.yale.edu/downloads/epi2018policymakerssummaryv01.pdf. Accessed 01 Apr 2019

Federal State Statistics Service of the Russian Federation (Rosstat). Regions of Russia. Socio-economic indicators (2019). http://www.gks.ru/wps/wcm/connect/rosstat_main/rosstat/ru/statistics/publications/catalog/doc_1138623506156. Accessed 01 Apr 2019

Musical Mega-event as an Instrument of Area Branding in the Digitization Era

Alexandra M. Ponomareva[1(✉)] and Anna A. Tkachenko[2]

[1] Rostov State Economic University (RINH) Rostov-on-Don, Russian Federation and Southern Federal University, Rostov-on-Don, Russian Federation
alexandra22003@rambler.ru
[2] Rostov State Economic University (RINH), Rostov-on-Don, Russian Federation
anatkach@yandex.ru

Abstract. The paper presents approaches to the problem of the use of musical events and mega-events as instruments for the formation and promotion of the city/area brand, development of their infrastructure and socio-economic indicators. The author's conclusions are based on descriptive-analytical, semantic, comparative and structural analysis of existing projects of music festivals. The research was based on information about more than 100 music festivals – both Russian and foreign, international, some of which are musical mega-events. The goal of research is to describe and categorize the phenomenon of a musical mega-event, to present the directions and principles of strategic planning of a musical mega-event as an instrument of development of the city/area brand using digital-technology. Planning of a musical mega-event as an international music and festival event that has the potential to form and promote the city/area brand and to develop their infrastructure must be based on the principles of harmonization of the concept of brands and development strategies of the city/area and event, complexity of the impact on all segments of the city/area marketing, consistency in the organization and control of investment flows, openness to the demands of civic institutions, people, entrepreneurs and investors in making decisions based on the creation of social and professional councils using digital technology, permanence, continuity and multidimensionality of monitoring and evaluation of results of a musical mega-event.

Keywords: City branding · Digital-technology · Event-related marketing · Musical event · Musical mega-event · Area branding

JEL Classification Codes: M310 · M370 · O350

1 Introduction

The timeliness of the topic presented in our research is predetermined, on the one hand, by the need to form and develop new modes of interaction of a musical discourse with its recipients, and on the other hand – by the active development of brands of cities/areas and the instruments of their promotion. Musical event-related marketing is at the intersection of interests of the music management, area branding and urban

E. G. Popkova and B. S. Sergi (Eds.): ISC 2019, LNNS 87, pp. 125–132, 2020.
https://doi.org/10.1007/978-3-030-29586-8_14

planning. The issues of interaction between urban and music space, the sounding of a city, its potential for becoming the platform for the development of various musical directions and cultures, as well as potential for using music for the formation of a comfortable urban environment, and musical marketing as an instrument of formation of the city/area have become the topic of our research.

The processes of branding of Russian cities and the entire country have been actively promoted by financing at the level of federal, regional and municipal budgets in recent years. At the same time, the correlation between strategies for development of urban areas, their identity and image of the city brand is often poorly traceable, and decisions in this area do not take into account the interests of all target segments – tourists, residents, investors, and entrepreneurs. Our hypothesis is based on the research of Krasnova, E.M., Tkachenko, A.A., Sedova N.N., Baskakova, Y.M., Turbina, E., Yatsyk, A. and consists in the fact that planning and development of the city in accordance with changes in society, economics, technology investing with a view to forming urban infrastructure and urban environment must correlate with the resolution of issues associated with the area branding, and event-related marketing, including musical events, is one of the instruments of such correlation. The sa-called musical mega-events - international events which do not only promote the city/area brand, but develop its infrastructure at the same time - are of particular note. The theoretical and methodological background of our research consists of research papers in the field of the music management, project management, event-related marketing, urban planning and branding.

2 Methodology

The research is targeted at music and festival events; the subject of research consists in the use of these events as an instrument which influences brand formation and promotion as well as the infrastructure of the area/city. Descriptive-analytical method, method of semantic analysis, comparative and structural research methods were used as research methods. The research presented in the paper is based on secondary data – information that is posted on websites of music festivals and cities in which such music festivals are held. The secondary research stage is a preparatory stage in relation to the planned primary research - expert surveys and interviews with the consumers of brands of music festivals and cities. The information about those music festivals which perform the function of promotion of the city/area was used as the data for study. A special place here is held by festivals, which according to their scale and significance have the status of mega-events, develop the infrastructure of the area, improve its socio-economic indicators, for example, Donauinselfest (Vienna, Austria, 3.1 million visitors), Mawazine (Rabat), Morocco, 2.6 million visitors), Summerfest (Milwaukee), Wisconsin, 850,000 visitors), Rock In Rio (Rio de Janeiro, 700,000 visitors), Coachella Valley Music and Art Festival (California), Indio, California, 675,000 visitors), Essence (New Orleans, Louisiana, 550,000 visitors), New Orleans Jazz and Heritage Festival, 435,000 visitors), Electric Daisy Carnival (Puerto Rico, Mexico, New York, Las Vegas, 375,000 visitors), Tomorrowland (Boom, Belgium, 360,000 visitors), Ultra (Miami, Florida, 330,000 visitors) ("TOP-10: The largest music festivals in the world" (2014), available at: http://funpress.ru/events/1756-top-10-krupneyshie-muzykalnye-festivali-v-mire.html#ixzz5m6KiWaEb). Festivals that have a high potential to consolidate mass audience include, inter alia, the Grammy Award, which is held in Los Angeles, a Nashestviye Rock Festival (Bolshoye Zavidovo), PinkPop Festival (Landgraaf, Netherlands), Roskidle Festival (Roskidle, Denmark), Open'er (Gdynia, Poland), Rock Werchter (Leuven, Belgium), Nos Alive (Algés, Portugal), Festival Internacional de Benicassim (FIB) (Benicassim, Spain), Melt! (Gräfenhainichen, Germany), Off-Off-festival (Katowice, Poland), Wild Mint (different venues, Russia), Retro FM Legends (Moscow, St. Petersburg, Russia), A-Zov (Krasnodar Territory, Russia) etc.; a separate group consists of academic music festivals as part of the concentrated marketing of city/area brands, e.g. the Abu Dhabi Festival, the Newport classical and jazz music festivals, the English Bach Festival (Oxford, London, Athens), the Bayreuth Festival (Germany), Mad Day (Festival) (Nantes, France), Beethoven Festival (Bonn, Germany), Brandenburg Summer Concerts (Germany), World Music Days (held every year in another city), Glyndebourne Festival Opera (Glyndebourne estate, Great Britain), the Donaueschingen Festival (Germany), Salzburg Festival (Germany), Festival de Lucerne (Switzerland), W LIVE - Marathon in memory of Vasyl Slipak (Lviv), Valletta International Baroque Festival (Malta), Royan Festival (France), Ohrid Summer Festival (Republic of Macedonia), Prague Spring International Music Festival (Czech Republic), The Proms (London, Great Britain), Sobinov's Festival (Saratov, Russia), Stockholm Early Music Festival (Sweden), Hector Berlioz Festival (La Côte-Saint-André, France), Stavanger International Chamber Music Festival (Norway), Florence Musical May (Italy), Odessa Classics (Ukraine), AlterMedium (Moscow, Yekaterinburg, Russia), White Lilac International Festival (Kazan, Russia), Voronezh Festival of Cello Performers (Russia), Classics over Volga (Tolyatti, Russia), A. Sakharov International Art Festival (Nizhny Novgorod, Russia), etc.

3 Results

The object of promotion of the communication instrument "musical event" may consist in an educational institution, e.g. a conservatory, a mass medium, for example Afisha publication, Retro FM radio station, or a commercial product, e.g. a beer brand. A musical project can also be a standalone commercial product.

Our research has been focused on those musical events which are an element of the city/area marketing. They can be conveniently classified into two large groups depending on the marketing function of the music festival. A musical event most often exercises the function of promotion of the city/area, formation and development of its positive image, increase in tourist traffic. Another type of music festivals has a considerable impact on the infrastructure and socioeconomic development of the city/area. We shall refer to such music festivals as musical mega-events.

Mega-event is an instrument of event-related marketing, which, according to the study of tender documents and current branding processes, is actively engaged in the formation and development of area brands. Mega-events include major sports events (e.g. the Olympics, Universiade, World Football Championship) (Yatzyk, 2014), international exhibitions, Economic Forums, coronations, anniversaries of persons or events, music festivals, other cultural events (Krasnova 2013). Similar mega-events have recently been held in such Russian cities as Sochi, Vladivostok, St. Petersburg, Kazan, Saransk, Rostov-on-Don, Yekaterinburg, Kaliningrad, Krasnoyarsk, etc. Hundreds of thousands of people become direct participants of the mega-event, and they are watched by tens of millions of people thanks to the use of television, mobile applications, Internet networks, and social media. Digital instruments of promotion of the mega-event significantly accelerate the processes of formation of the city/area brand, too.

Event-related marketing implies designing and implementation of an event within the scope of which target groups contact the brand and become emotionally involved in a cultural, sports, entertaining, educational, etc. event. We shall single out an individual direction within the scope of event-related marketing – art-marketing, which includes musical events as an integral part of it.

Musical mega-event is an international music and festival event that has the potential to form and promote the city/area brand and to develop their infrastructure.

The main functions of musical mega-events with regard to the city/area consist in the promotion of their investment attractiveness, infrastructure development and branding. The essence of a mega-event as an instrument of formation and development of the city brand on the basis of urban planning principles, which implies the inclusion of marketing instruments of infrastructure development and socio-economic indicators of the region in the branding scope, has been presented in the work authored by Ponomareva, A. and Tkachenko, A. (Ponomareva and Tkachenko, 2017). We shall present our own vision of investment, communication and infrastructure flows within the scope of the use of a musical mega-event as a branding instrument (Fig. 1).

A special type of management is applied with regard to musical mega-events - creative and innovational management with the use of marketing approaches. The ambivalence principle is implemented within the scope of this approach: musical mega-event is an instrument of branding and development of city brand/area and their infrastructure, and, at the same time – the object of promotion, and the core principle in development and promotion of the mega-event is the adherence to the correlation between the concept and the content of a mega-event and the system of its communication management. The analysis of events on the basis of collected data has shown that musical mega-events can be conveniently classified into periodic events (e.g. annual or quadrennial events) and one-time events, as well as into stationary mega-events the venue of which is permanent, e.g. Coachella Valley Music and Art Festival, and mobile mega-events, e.g. “Eurovision” in which the venue changes at regular times.

We shall provide several examples of musical mega-events of different kinds and describe the specificity of their impact on the city/area brand.

The annual Grammy Awards ceremony held by The Recording Academy (Grammy Awards (2019) is one of the best known musical mega-events; music awards portal available at: https://www.grammy.com/?fbclid=IwAR2wodSfK0hemczBfWVyV_Z1zwHmZJqiXcnMP-4xEzUxtPbNQHvGVKicc6Y); it is held in Los Angeles and attracts to celebrity musicians from all over the world. Its function with regard to the brand of Los Angeles consists in support and development of the city image as a music capital, the center of the music scene. Direct coverage of this event is not large-scale, but this mega-event receives global coverage through the use of digital instruments.

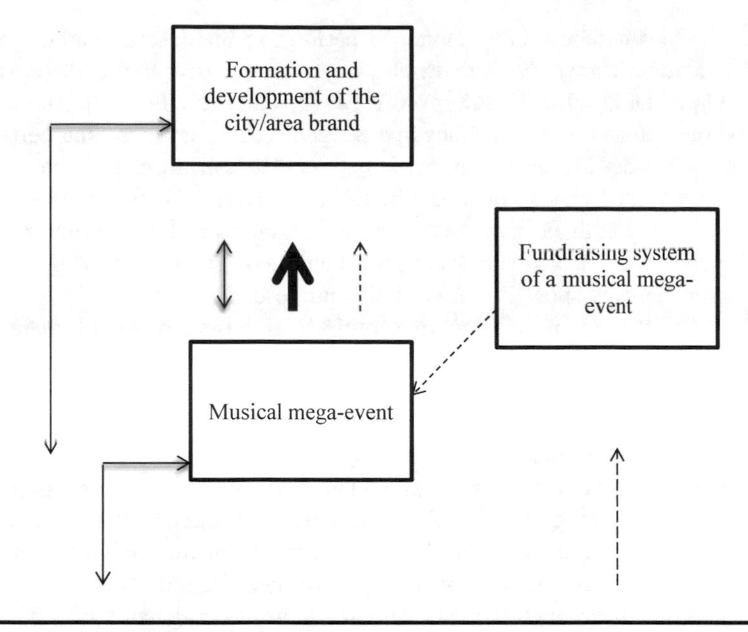

Infrastructural changes in the area

Investment flows providing the opportunity for the implementation of a musical mega-event, as well as formation and development of the city/area brand

Communication flows

Fig. 1. Investment, communication and infrastructure flows within the framework of a musical mega-event

An American town called Newport, which is home to a slightly more than 24 thousand residents, became internationally renowned for its international music festivals - Newport Music Festival and Newport Jazz Festival, music festival portal available at: https://www.newportmusic.org/discover-newport/). These musical mega-events build up an internationally acclaimed image of the city brand among the admirers of music, and increase both domestic and international tourist traffic.

The festival of academic music in the desert called "Abu Dhabi Festival" which has been held since 2004 has become so significant that today it is not only an instrument of promotion of Abu Dhabi brand, but also an instrument of promotion of the entire country

UAE (Makarytchev (2008 Honor to perform in the desert, available at: https://www.classicalmusicnews.ru/interview/chest-vystupit-v-pustyne/?fbclid=IwAR1thSxgSGuI-VW0pbXDttkLwLnzlbNfG6ipWGWGETu_V2uuIjImqhftpE0). The choraguses make no bones about the fact that they invest significant means in the star performers of classic music in order to make them participate in this festival, to promote new values in the country, with a view to making Abu Dhabi a Capital of Culture in the XXI century.

The Bayreuth Festival has been an integral part of infrastructure of the Bavarian city for over 100 years; it has been established by Richard Wagner and is held in a theater that was specially built for this purpose.

The analysis of existing megaprojects and the history of their implementation shows that they can have both positive and negative impact on the city/area brand and their citizens, investors, and other stakeholders. Turbina, E., analyzing the international negative experience in the implementation of mega-events, regards disregard of interests of the population and shifting financial expenses associated with the organization of a global event on the shoulders of the taxpayers, corruption, understatement of budgets of mega-events when planning with their further bloating, lack of systematicity, fragmentary nature of the implementation of projects preventing the infrastructure development of the city (Turbina, E. (2012), "Polis and mega event", available at: http://www.strana-oz.ru/2012/3/polis-i-megasobytiya). That being said, for many cities, the holding of such events became a starting point in the game-changing infrastructure development, brought the city to a new level of tourist attraction and allowed the state and private investors to invest vast amounts of money in enterprises, transportation network, and in such spheres of living environment of the policy as safety and accessibility of the urban environment. Mega-event creates a new economic activity which increases the budget of the area at the expense of new jobs, taxes, increase in profitability in some industries and, as a result, the purchasing power of the population.

The examples provided above carry inference that musical mega-events contribute to the modernization of the current infrastructure of the city, creation of special urban facilities based on the raised investment, create opportunities for qualitative shifts in the perception of significance of the city for its citizens and its socio-economic development, transform and strengthen the existing area brand.

A topical academic cross-disciplinary problem that is currently facing the researchers consists in the development of a scientific model of a mega-event based on available positive and negative experience in the implementation of mega-events, which would provide the opportunity to further develop such events with a focus on the positive socio-economic and infrastructural effect. This path has been taken by the researchers of the All-Russian Public Opinion Research Center within the framework of the project "Mega-Events and Civic Society" (Sedova, N., Baskakova, J. (2014), "Expeditionary project ACSPO", available at: https://wciom.ru/nauka_i_obrazovanie/ehkspedicii_vciom/), aimed at studying the effects of megaprojects in various cities of Russia at all stages of development of a mega-event.

4 Conclusions/Recommendations

Strategic planning of a musical mega-event with a view to using it for the development of the city/area brand through the use of digital technology must include the following directions:

- development of investment proposal for domestic and international investors of a musical mega-event;
- provision of legal coverage for the status of a musical event as a mega-event aimed at promotion of the city/area brand;
- creation of the talent pool of a musical mega-event which would consolidate musicians, experts in urban planning, city/area branding, event-related marketing, and volunteers;
- development of the concept of a musical outward-looking mega-event, which, on the one hand, would be able to attract interest of users at international scale, and on the other hand – to offer potential for development of the city/area brand, its infrastructure, significance, attractiveness both for outside and inside target audiences;
- development of the digital-marketing system of a musical mega-event which would be capable of performing the function of the simultaneous promotion of the city and the music festival;
- formation of the system of interaction between civic society, entrepreneurs and authorities for the resolution of issues related to planning and organization of a musical mega-event in the interests of all target segments of the city/area brand;
- development of the basic documents of a musical mega-event based on the principles of social and environmental marketing, which is important for achieving the socio-economic efficiency of a planned musical mega-event and preventing its adverse effects.

According to practical examples of national and foreign festivals, musical mega-event can act as a fillip to solving another major problem of big cities – the excessive concentration of urban infrastructure in the city center.

In order to make the mega-event an effective city branding instrument, it is important to comply with the following principles:

- mega-event must correlate with the identity of the area (city, region); only an accurate hit in the "code of the city" will provide a positive result in the development of the megaproject in years to come; harmonization of the concept of a mega-event with the "code of the city", the essence, the identity of the city brand - is the first and, perhaps, one of the most important stages predetermining the socio-economic efficiency of the megaproject in the long term;
- harmonization of strategic planning of the city branding and promotion of the mega-event based on the identified unique features of the area;
- implementation of the principle of integrated impact on all target audiences (investors, entrepreneurs, international tourists, immigrants, citizens, who are both residents of the city and domestic tourists) in the process of harmonization of strategies of the city branding and promotion of the mega-event;

- the principle of consistency and organization of public control over the movement of investment flows;
- the principle of openness and creation of mechanisms to respond the demands of civic institutions, residents, entrepreneurs and investors in making decisions based on the creation of social and professional councils using digital technology;
- the principle of constancy and continuity in the assessment of the implementation of a mega-event means that during the development phase, criteria will be set for the assessment of efficiency of the mega-event for every stage, and the results of this assessment will be publicly available to all participants of the mega-event, investors, entrepreneurs, authorities, and city residents.

References

Krasnova, E.: Cultural mega-events and their role in the development of cultural tourism in cities (on the example of the project "European Capital of Culture"). Knowledge of the countries of the world: history, culture, achievements, no 3, pp. 26–33 (2013)

Makarytchev, M.: Honor to perform in the desert (2008). https://www.classicalmusicnews.ru/interview/chest-vystupit-v-pustyne/?fbclid=IwAR1thSxgSGuI-VW0pbXDttkLwLnzlbNfG6i-pWGWGETu_V2uuIjImqhftpE0. Accessed on 24 April 2019

Ponomareva, A., Tkachenko, A.: Mega-event as a marketing and investment tool for territorial branding, in Staffing investment development and the competitiveness of municipalities improving (best domestic and foreign practices) 2017. In: Proceeding of the Round Table with International Participation in Rostov on Don, Russia 2017, RANEPA, Rostov on Don, pp. 152–162 (2017)

Sedova, N., Baskakova, J.: Expeditionary project ACSPO (2014). https://wciom.ru/nauka_i_obrazovanie/ehkspedicii_vciom/. Accessed on 24 April 2019

TOP-10: The largest music festivals in the world (2014). http://funpress.ru/events/1756-top-10-krupneyshie-muzykalnye-festivali-v-mire.html#ixzz5m6KiWaEb. Accessed on 24 April 2019

Turbina, E.: Polis and mega event (2012). http://www.strana-oz.ru/2012/3/polis-i-megasobytiya. Accessed on 24 April 2019

Yatzyk, A.: Sports mega events and popular culture: global and local in the landscapes of the 2013 Universiade in Kazan. Topography of popular culture: a collective monograph (2014). https://public.wikireading.ru/162685. Accessed on 24 April 2019

Coachella, music festival portal (2019). https://live.coachella.com. Accessed on 14 April 2019

Grammy Awards, music awards portal (2019). https://www.grammy.com/?fbclid=IwAR2wodSfK0hemczBfWVyV_Z1zwHmZJqiXcnMP-4xEzUxtPbNQHvGVKicc6Y. Accessed on 14 April 2019

Newport music festival, music festival portal (2019). https://www.newportmusic.org/discover-newport/. Accessed on 14 April 2019

Improvement of the Development of the Social Infrastructure of the Regions on the Basis of Public-Private Partnership

Anipa M. Zulpueva, Almash A. Kutmanbekova(✉), Asel S. Kayipkulova, Ainura M. Khamzaeva(✉), and Salamat U. Astanova(✉)

Osh Technological University named by M.M. Adyshev, Osh, Kyrgyzstan
almash-k@mail.ru, ainura.hamzaeva@mail.ru, s-astanova@mail.ru

Abstract. This article deals with the development of the social infrastructure of the regions on the basis of public-private partnership.

Keywords: Public-private partnership · Health · Education · Culture · Sport

1 Importance of the Topic

The paper studies the problem of development of the spheres of the social sphere in the conditions of economic instability and the necessity for implementing the leading forms and methods of cooperation of the state and business on implementing the joint projects, the most popular of which is developed countries is public-private partnership (PPP).

At present, in the conditions of economic instability and increase of problems in the sphere of international relations, one of the top-priority directions of state policy in Kyrgyzstan is development of domestic business and increase of cooperation between the state and the entrepreneurial community. In this sense, development of the leading and modern forms of interaction of the state and private sector of economy acquires a special role. The leading experience of development and application of the institute of public-private partnership could be used for solving important socio-economic tasks on implementation of socially important projects in the country's regions with joint efforts of the state and business. It could be stated that public-private partnership is at the stage of formation yet; there's no state law that would regulate all issues of interaction of the state and business within public-private partnership. Public-private partnership in the conditions of budget limitation is an effective tool of implementing socially important and strategic investment projects in regions. This form of cooperation of the state and business is expressed primarily in such spheres as healthcare, education, electric energy, and transport. Projects in these spheres are very important in the social aspect, but state assets are not always sufficient for implementation of these projects, so investments of private business are attracted for joint financing and management.

Apart from this, a positive aspect of the mechanism of PPP is private partner's transferring his professional experience, which includes design, construction, and

E. G. Popkova and B. S. Sergi (Eds.): ISC 2019, LNNS 87, pp. 133–136, 2020.
https://doi.org/10.1007/978-3-030-29586-8_15

exploitation of large infrastructural object. As of now, the sphere of PPP passes from the stage of formation to the stage of establishment, but, its further development requires solution of a range of tasks. Apart from the issues that are connected to improvement of the normative and legal basis, financial mechanisms, and creation of an optimal model of distribution of risks between participants, there's insufficient clarity of state priorities in the direction of development of infrastructure [2]. Therefore, planning of further application of the mechanisms of PPP should be conducted in view of priorities of the state as to directions of development of the social, transport, energy, and communal infrastructure and the spatial development on the whole.

It is possible to note that the level of development of social infrastructure for any subject of Kyrgyzstan plays one of the key roles. Influencing the population's quality of life, social infrastructure includes a range of spheres: healthcare, education, fitness and sports, and art. In order to determine the perspective directions in this sphere, it is expedient to determine the current state of infrastructure in the territorial aspect.

Public-private partnership is contractual relations between state structures and private business, which envisage mutually profitable cooperation between the parties during implementation of projects with socially important significance, and the financial risks, expenditures, and results are distributed in the proportions, according to the agreement with a legal power.

There are various opinions on the necessity for adoption of the law that would regulate public-private partnership. Some specialists think that this law is necessary, while others do not agree with them. We share an opinion that such normative document is necessary.

Public-private partnership develops independently in different regions, based on the laws in the subjects of Kyrgyzstan.

Four key forms of PPP are envisaged:

(1) implementing the complex investment projects;
(2) concluding concession agreements;
(3) participating in registered capitals of open joint-stock companies by the terms of agreements and contracts that are envisaged by the laws of Kyrgyzstan;
(4) other forms that do not contradict the current laws.

The schemes of PPP are frequently used in the world due to construction of the objects of healthcare, education, in the sphere of transport and road infrastructure, and during reconstruction and construction of sports objects. Popularity of PPP in these spheres could be explained by a range of advantages for infrastructural projects on the whole: minimization of the project's budget, the best ratio of price and quality that is attractive for private investors, application of the best managerial and construction technologies, and reduction of commercial risks by means of state's participating in the project.

Implementation of projects with application of the mechanisms of interaction of state and private business abroad has been taking place for quite some time, and this mechanism is constantly improved.

In the global practice, the best experience of successful and wide application of the forms of PPP belongs to developed countries of Western Europe and the USA. It should be noted that the rates of development of PPP are different in various countries. The highest results in this direction of cooperation of the state and business were achieved by such countries as the UK, France, Germany, Spain, Singapore, the USA, and Canada.

For development of public-private partnership and implementation of the projects in the regions of Kyrgyzstan, it is necessary to create favorable conditions for formation of this institute. Among the main factors that hinder wide application of the mechanisms of PPP in the Kyrgyz economy the following could be distinguished:

(1) insufficiently developed normative and legal basis, including the absence of the state law on public-private partnership;
(2) absence of the mechanisms of effective financing and insurance of projects for all participants of the market, a significant political component for implementing long-term projects;
(3) ineffectively working legal institutes, which would guarantee the rights of private owners that participate in the projects;
(4) limited access of potential participants of PPP to large projects due to high requirements that are set by the state.
(5) low level of diversity and insufficient usage of the potential of perspective forms and tools of public-private partnership, which limits the possibilities of its usage.

For providing the development of public-private partnership in cities and regions of Kyrgyzstan, it is necessary to develop and implement a complex of measures of the systemic character:

(1) Development of the conceptual foundations, theoretical principles, and practical standards of public-private partnership and the strategic plan of development of projects PPP in Kyrgyzstan, which envisage formation of the systemic interaction of the state and business structures on the legislative basis with detailed tools and the organizational and economic mechanism of cooperation, which is implemented according to the common algorithm.
(2) Creation of the special state and regional centers that are necessary for coordination, control, monitoring, and evaluation of the implemented projects within public-private partnership with the corresponding authorities for solving the legal, organizational, financial, and economic problems.
(3) Development of legal, social, and government institutes and independent expert communities which effective activities would ensure transparency, trust, and effective cooperation during interaction between the state and business.
(4) Systemic training of managerial and expert personnel who are competent in the sphere of public-private partnership with formation of favorable public opinion and public control during transferring the management and usage of the objects of state and municipal property for solving the socio-economic tasks to representatives of business.

Thus, it is possible to make to following conclusions.

(1) It is necessary to improve the scientific and methodological foundations of public-private partnership and develop and create conditions for wider practical application of the mechanisms and models of PPP in the cities, borrowing the leading experience of developed countries, during implementation of socially important projects.
(2) An important priority in regulation of development of economy of large cities is not only development of the normative and legal basis of public-private partnership with adoption of the state law on PPP, improvement of the institutional environment for improvement of interaction of the state and business, but also expansion of rights of municipal authorities in the financial and economic regulation of the territories' development.

References

1. Tsarev, D., Ivanyuk, A.: Public-private partnership: a legal aspect. Financial gazette, no. 17, p. 12 (2009)
2. Ilyin, I.E.: State and business: in the course of mutually profitable partnership. Organization of sales of bank products, no. 1, pp. 87–96 (2009)
3. Varnavsky, V.G.: Management of public-private partnership abroad. The issues of state and municipal management, no. 2, pp. 134–147 (2012)
4. Public-private partnership in the social sphere [E-source]. http://studme.org/1151051319786/. Accessed 7 June 2015

Macro-strategic Planning of Innovational Development and Studying the Three-Component Approach to Modernization Growth

Bahadyr J. Matrizaev[1(✉)], Allakhverdieva Leyla Madat kyzy[2], and Muslima K. Sultanova[2]

[1] Financial University under the Government of the RF, Moscow, Russian Federation
matrizaev@mail.ru
[2] Moscow State University of Humanities and Economics, Moscow, Russian Federation
leila26uz@mail.ru, darmuz@mail.ru

Abstract. The article presents a comparative analysis of the experience of macro-planning of innovative strategies and technological modernization of economies in the countries with emerging economies. The analysis was performed on the basis of the authors' three-component approach, which takes into account the differences between the intensity of technological modernization, structural changes, and global interaction. The authors confirm that there are no common mechanisms of technological modernization within economies of the countries with emerging economies. Instead, we see the proofs of several unique contours of technological modernization with various compromises between intensity, structural changes, and the character of interaction with the global economy. The authors develop the theoretical basis and build a general hypothesis on the characteristics of the processes of technological modernization.

Keywords: Macro-strategic planning · Innovative strategy · Innovative potential · Technological modernization · Models of innovative growth · Human capital

JEL Code: O21 · E61 · F62

1 Introduction

Certain studies (Lee and Kim 2009) offered an opinion that technological development is the mandatory factor of sustainable growth – in particular, for the countries with average level of income. However, the Neo-Schumpeterian theory states that the moving forces of growth are different for countries with various levels of income and technologies (Aghion and Howitt 1992). Similarly, it is possible to conclude that there are no universal indicators that would measure growth (including technological growth).

E. G. Popkova and B. S. Sergi (Eds.): ISC 2019, LNNS 87, pp. 137–144, 2020.
https://doi.org/10.1007/978-3-030-29586-8_16

The new theory of growth shows that technology is an important factor of economic growth - however, it cannot be brought down to the narrow variable – e.g., R&D or general efficiency of the factors that were obtained by the exogenous method. Technology, as a driver of growth, is a multi-dimensional phenomenon. This is seen in authoritative overviews – the Global Competitiveness Index and the Global Innovation Index – which shows the necessity for conceptualization of technological growth as a multi-dimensional phenomenon. Against this background, the new theory of growth is peculiar for calls for application of new methodologies of measuring technological growth for understanding modernization of technologies, emphasizing on the problem of countries with the average level of income (Radosevic and Yoruk 2016). An effective reaction to these calls might be development of the models that differentiate intensity of technological modernization, which is reflected in various types of technological possibilities; width of coverage of technological modernization; and necessity for global interaction for technological modernization (Martizaev 2018a, b). Thus, the authors conceptualized the technological modernization as a result of interaction of these three dimensions and substantiated a range of general hypotheses on technological modernization. Approbation of the developed multi-dimensional statistical structure on the basis of patent indicators has been performed as to measuring of technological modernization of economies of BRICS, as compared to certain countries with developed economies (EU-15, the USA, and Japan) for 1980-2015.

2 Methodology

The practical significance of this research consists in determining the technological strategies of various countries of BRICS and the necessity for demand for our approach in other countries with quickly developing economies. The authors' approach is empirically substantiated and it has high theoretical significance. In this context we consider the applied approach as an important theoretical basis that is aimed at amending the generally accepted drawbacks of the indicators, which, according to T. Koopmans, are often "measuring without theory" (Koopmans 1947). This work also shows that there are no common mechanisms of technological modernization within economies of the countries with emerging economies. Instead of this we see the proofs of several unique contours of technological modernization with various compromises between intensity, structural changes, and character of interaction with the global economy.

3 Results

Let us consider the intensity of modernization of technologies within the performed research. As was already noted, the intensity of technological modernization is connected to accumulation of various forms of potential, which also reflect various technological levels of economy. Our thesis could be supported by the studies of Bell and Pavitt (1997), in which they emphasize two forms of processes of accumulation in the latest industrial companies and countries. One of these processes of accumulation is accumulation of technologies that are embodied in physical capital and the connected human capital, which is necessary for exploitation of the objects at this level of effectiveness. This accumulation was defined as production possibilities. Such ability to accumulate requires good operational effectiveness and skilled technical and labor force. Another process, which has not yet become generally recognized in the traditional theory of growth, is accumulation of innovative potential.

Besides, Bell (2009) states that the first process of accumulation is connected to the companies' abilities to use the existing technologies in production. This "overtaking" could be reflected in the indicator of labor efficiency and reduction of difference in efficiency between "delayed" companies and the companies that are at the stage of the leading technologies. The second process of accumulation is connected to companies' abilities to create new technologies and change the technologies that they already use. The sense of such "overtaking" consists in reduction of the gap between copying and implementation of the existing technology, on the one hand, and its improvement or creation, on the other hand. In this process the "overtaking" companies reduce the gap between themselves and the companies of innovative leadership, and "overtaking" the gaps is difficult to measure – however, it could be assesses based on the growing difference between the levels of innovative potential and speed with which the companies move along these levels (Lall 1992). The empirical literature has a range of works that are devoted to successful examples of transition from production potential to innovative potential of "overtaking companies" in the countries of East Asia (Hobday 1995, Ernst 2013), Latin America (Dutrenit 2000), and Central and Eastern Europe (Radosevic and Yoruk 2018). However, in the course of technological modernization of economy the production potential is still very important. It should be noted that the production and innovative potential, as well as intensity of R&D/knowledge are present in each economy. Like R&D, which are valuable as a generator of knowledge but also as the tool of mastering of knowledge (Cohen and Levinthal 1990), the production and innovative potentials strengthen each other. This does not mean that there's a certain fixed optimal proportion between various types of opportunities and/or R&D. Also, technological modernization might not be fully reflected by increase of the share of certain types of activities and reduction of other types of activities. According to certain researchers, individual value of production potential, potential of R&D, and innovative potential as drivers of growth vary depending on the achieved income, level of technologies, and structural peculiarities of the economy (Radosevic and Yoruk 2016). Their interaction and complementarity are very important. High share of the global leading technologies in the economy with weak production potential (or in economies where companies have weak potential for mastering of R&D) will lead to the enclave growth with limited dissemination of technologies and side effects of efficiency.

The following hypothesis is offered as an original methodology:

Hypothesis 1. Countries with different levels of income conduct various types of production, innovative, and R&D activities. In this context, their technological modernization is presented in the form of mutually reinforcing interconnections between production, innovations, and R&D, which lead to increase of technological activity.

This means that technological modernization is not a linear and autonomous process of growth of independent production, technological, and research opportunities, but a non-linear process, which includes several threshold levels (Radosevic and Yoruk 2016). Transition from one stage to another is not guaranteed and requires a new set of mutually supplementing technical, financial, and organizational preconditions. Our factual data, which are based only on the patent data, do not allow verifying all three aspects of increasing the technological intensity (production, innovations, and R&D). However, using international and national patents, we can show the stage of transition to technological modernization – i.e., from the stage "before" the leading technologies to the stage "in" the leading technologies.

Let us describe the width of technological modernization. Technological modernization is something more than intensity and scales of technological activities that are observed in the process of overtaking. The recent research show the important of scales and coverage of the structural actors. It is possible to see that earlier studies development is described as an evolutional process that passes through several stages (Rostow 1960). In particular, Walt Whitman Rostow in the work "The Stages of Economic Growth: A Non-Communist Manifesto" notes that this was based on the idea of industrial life cycles and "leading sectors" which stimulate economic growth at certain stages. In addition to the Rostow's thesis it is possible to present the arguments of Tunzelmann (1995), who states that a common feature of these models is the idea that "all economic systems (markets) pass through the same stages in the same order, though not necessarily in the same time". Despite certain similarity of the above arguments, it should be noted that there is no common theory of structural changes but a range of theoretical approaches to various methodological characters, aimed at explanation of structural shifts between the wide sectors and spheres in these sectors (Krüger 2008). There's a common understanding that technological changes influence the structural changes in the way that the spheres with relatively low rates of growth of efficiency reduce from the point of view of their shares, while the spheres with higher growth rates of efficiency expand. However, the empirical studies on the role of structural changes show that they have positive and negative contributions into aggregate growth of efficiency. As a lot of these consequences are average, the structural changes provide weak influence. Thus, we think that instead of focusing on the structural changes at the level of spheres, it is expedient to track changes in the structure of technological knowledge.

Also, we found in a range of empirical studied the theses that do not support the idea that economic growth correlates with the share of hi-tech sectors in economy (Sandven et al. 2005). We also found clear suppositions on implementation of hi-tech activities in low-tech spheres, as well as low-tech activities in the spheres that are classified as hi-tech, i.e., science-driven (Tunzelmann et al. 2005). In several cases, instead of reflecting the structural changes in shares at the sectorial level we observe the change of the character of spheres and services and their convergence. These changes are peculiar for the growing role of knowledge intensive business services (KIBS) and the growing importance of knowledge-intensive activities (KIA) in all sectors of economy. It is possible to state that accumulation of production and innovative potential within the overtaking model of economic development is connected to the changes in basic implementation of knowledge. As is known, these changes reflect the structural changes in creation and mastering of knowledge in favor of the high share of hi-tech knowledge and higher intensity of economic activities.

Based on the performed analysis we offer the following hypothesis:

Hypothesis 2. Countries with low level of income are often associated with low share of knowledge intensive types of activities, whole countries with medium and high level of income increase their share of knowledge intensive economic activities. Sectorial concentration of countries seems to have the U structure as to income per capita. In particular, Imbs and Wacziarg (2003) show that economy grows in two stages of diversification. First sectorial diversification grows – but there's a level of income per capita beyond which sectorial distribution of economic activity starts to shrink. The best basis of successful overtaking economy seems to be a non-linear trend. Lee (2013) shows that technological diversification, not specialization, is one of the main factors in achievement of high levels if income. While the supporters of New Structural Economics[1] show that the path to modernization of technologies is based on "industrial copying", which uses hidden comparative advantages in the process of transition from low to medium level of income, Lee (2013) shows that countries with medium level income use the "bypass" or temporarily specialize on so called "short cycles" of technology. He shows that South Korea and Taiwan entered the lower number of spheres of knowledge with higher technological capabilities, but with higher number of sectors. However, in the course of growth of South Korea and Taiwan they entered the group with high level of income as a result of the process of significant technological diversification.

Against this background, it is possible to offer the following hypothesis.

Hypothesis 3. Countries with low level of income mainly copy foreign technologies and are peculiar for narrow specialization of domestic technological knowledge. The prospering countries with medium level of income could temporarily specialize in the narrow spheres with high technological capabilities, but the path of technological modernization (it could be non-linear) is peculiar for the growing diversification of knowledge.

4 Conclusions

Growth and modernization of technologies are not fully independent processes – they are connected to the global interaction. For example, Akamatsu (Akamatsu 1962) describes modernization technologies as an interactive process between "leaders" and "followers". This argument could be connected to various directions of studies that are oriented at development, which concern direct foreign investments and modernization of global production and sales chains. It is possible to state that all channels of global interaction potentially influence the intensity of technological modernization within the overtaking process. The inflow of direct foreign investments has been traditionally associated with accumulated technological advantages that come from the founding

[1] New Structural Economics is a new part of the modern economics that was founded by the researchers of the World Bank and is devoted to new theoretical and practical views on economic development after the 1997-1998 crisis.

country, which are moved to the accepting country, where they are distributed in the domestic economy. As a matter of fact, Findlay (1978) stated that the potential of distribution of technologies through direct foreign investments is positively connected to relative technological gap between the economy of the founding country and the receiving country. He referred to the "contagion effect", at which technical innovations are copied effectively during personal contact between those who already know the innovation and those who implement it. This approach, at which the production effectiveness of the accepting country is modeled only as a growing function of foreign capital, was criticized by Wang and Blomstrom (1992). Their arguments acknowledge the expenditures for transfer of technologies in transnational corporations and expenditures for training of domestic companies. Thus, the external factors of direct foreign investments positively depend on technical and managerial competency of a foreign subsidiary company and on the decision of the domestic company on investing into training. Certain other studies (Makino et al. 2002) showed the presence of strategies of direct foreign investments that are based on knowledge. There is also an existing direction of research (Grossman and Helpman 1991), which shows technological training with importers/exporters. In view of the fact that foreign branches often show higher levels of import and/or export as compared to domestic companies, accumulation of the technology through trade and direct foreign investments could be considered to the mutually reinforcing effects. International licensing and flows of knowledge in the hidden form are also the important channels of technology transfer. However, they are closely connected and cannot be separated from trade or flows of direct foreign investments. In the theory of global production and sales chains (Sturgeon and Gereffi 2009), modernization is expressed in various forms: increase of effectiveness by means of reorganization of the production system or implementation of the leading technologies; increase of the quality of products, when the company implements more complex production lines; functional modernization, when the company acquires new functions (or refuses from the existing ones) for raising the general qualification of activities. That's companies from the countries with forming market economy entering the global production and sales chains creates opportunities for technological modernization on the basis of training and interaction.

Therefore, we think that the final recommendation could be the last hypothesis.

Hypothesis 4. In the countries with low level of income global interaction is very important for obtaining access to the leading technologies. However, countries with low level of income have weak organizational opportunities, and their patent knowledge is often commercialized by foreign claimers. In the course of growth of incomes of countries and modernization of technological potential they could enter the process of joint generation of knowledge. In the countries with high level of income creation of the leading technologies is based on domestic subjects, which could actively receive and commercialize the technological knowledge from abroad. We state that countries with low level of income win from transferring the technology by means of imported direct foreign investments and by means of export/import. At this stage, countries with low level of income have weak organizational opportunities for commercialization of their patent knowledge. At the later stages, countries with medium level of income start participating in the processes of modernization – mainly modernization of processes and products - and could participate in the activities on joint generation of knowledge with foreign partners. Developed countries with medium level of income also start conducting the functional modernization of knowledge intensive business functions and in the cycle "modernization and creation of national leading companies". Also, it is possible to predict the increase of exported direct foreign investments, which is partially predetermined by striving for mastering of technologies and reverse transfer of technologies for compensation for unfavorable conditions in the countries of origin.

References

Solow, R.M.: Technical change and the aggregate production function. Rev. Econ. Stat. **39**(3), 312–320 (1957)

Matrizaev, B.D.: Macro-strategies of innovative development and the global economic growth: Macro-economic analysis, trends, and forecasts. – M.: URSS, 256 p. (2018a)

Matrizaev, B.D.: Global innovative leadership: macro-contours and modeling of its conceptual basis. Municipal academy, no. 1, pp. 85–91 (2018b)

Aghion, P., Howitt, P.: A model of growth through creative destruction. Econometrica **60**(2), 323–351 (1992)

Akamatsu, K.: A historical pattern of economic growth in developing countries. Dev. Economies **1**(1), 3–25 (1962)

Bell, M., Pavitt, K.: Technological accumulation and industrial growth: contrasts between developed and developing countries. In: Archibugi, D., Michie, J. (eds.) Technology, Globalisation and Economic Performance, pp. 25–45. Cambridge University Press, Cambridge (1997)

Bell, M.: Innovation capabilities and directions of development. STEPS Working Paper (Brighton: STEPS Centre) Working Paper no. 33, p. 18–25 (2009)

Cohen, W.M., Levinthal, D.A.: Absorptive capacity: a new perspective on learning and innovation. Adm. Sci. Q. **35**(1), 128–152 (1990)

Ernst, D.: Industrial Upgrading Through Low-Cost and Fast Innovation—Taiwan's Experience. East-West Center Working Paper, Honolulu, p. 133 (2013)

Dutrenit, G.: Learning and Knowledge Management in the Firm: From Knowledge Accumulation to Strategic Capabilities, pp. 57–78. Edward Elgar, Aldershot (2000)

Fagerberg, J., Godinho, M.: Innovation and catching-up. In: Mowery, D.C., Fagerberg, J., Nelson, R. (eds.) The Oxford Handbook of Innovation, pp. 514–543. Oxford University Press, New York (2005)

Grossman, G., Helpman, E.: Innovation and Growth in the Global Economy. MIT Press, Massachusetts (1991)

Hobday, M.: East Asian latecomer firms: learning the technology of electronics. World Dev. **23**(7), 1171–1193 (1995)

Imbs, J., Wacziarg, R.: Stages of diversification. Am. Econ. Rev. **93**(1), 63–86 (2003)

Koopmans, T.C.: Measurement without theory. Rev. Econ. Stat. **29**(3), 161–172 (1947)

Krüger, J.J.: Productivity and structural change: a review of the literature. J. Econ. Surv. **22**(2), 330–363 (2008)

Lall, S.: Technological Capabilities and Industrialization, pp. 34–42 (1992)

Lee, K.: Schumpeterian Analysis of Economic Catch-Up: Knowledge, Path-Creation, and the Middle-Income Trap, pp. 35–49. Cambridge University Press, Cambridge (2013)

Lee, K., Kim, B.: Both institutions and policies matter but differently for different income groups of countries: determinants of long-run economic growth revisited. World Dev. **37**(3), 533–549 (2009)

Lin, J.Y., Rosenblatt, D.: Shifting patterns of economic growth and rethinking development. J. Econ. Policy Reform. **15**(3), 171–194 (2012)

Makino, S., Lau, C., Yeh, R.: Asset-exploitation versus asset-seeking: implications for location choice of foreign direct investment from newly industrialized economies. J. Int. Bus. Stud. **33**(3), 403–421 (2002)

Radosevic, S., Yoruk, E.: Why do we need a theory and metrics of technology upgrading? Asian J. Technol. Innovation **24**, 8–32 (2016)

Romer, P.M.: Endogenous technological change. J. Polit. Econ. **98**(5), S71–S102 (1990)

Verspagen, B.: A new empirical approach to catching up or falling behind. Struct. Change Econ. Dyn. **2**(2), 359–380 (1995)

Nelson, R.R.: Recent evolutionary theorizing about economic change. J. Econ. Lit. **33**(1), 48–90 (1991)

Radosevic, S., Yoruk, E.: Technology upgrading of middle-income economies: a new approach and results. Technol. Forecast Soc. Change **129**, 56–75 (2018)

Rostow, W.W.: The Stages of Economic Growth: A Non-Communist Manifest, Second Enlarged Edition, 1971st edn, pp. 35–42. Cambridge University Press, Cambridge (1960)

Sandven, T., Smith, K., Kaloudis, A.: Structural change, growth and innovation: the roles of medium and low tech industries, 1980-2000. In: Hirsch-Kreinsen, H., Jacobson, D., Laestadius, S., Smith, K., Lang, P. (eds.) Low-Tech Innovation in the Knowledge Economy, pp. 31–63. Peter Lang Publishing, Frankfurt (2005)

Sturgeon, T.J., Gereffi, G.: Measuring Success in the global economy: International trade, industrial upgrading, and business function outsourcing in global value chains. Transnatl. Corporations **18**(2), 1–35 (2009)

Von Tunzelmann, G.N.: Technology and Industrial Progress: The Foundations of Economic Growth, pp. 10–15. Edward Elgar Publishing, Aldershot (1995)

Von Tunzelmann, N., Acha, V.: Innovationin'low-tech'industries. In: Fagerberg, J., Mowery, D., Nelson, R. (eds.) The Oxford Handbook of Innovation, pp. 407–432. Oxford University Press, Oxford (2005)

Wang, J., Blomström, M.: Foreign investment and technology transfer. Eur. Econ. Rev. **36**(1), 137–155 (1992)

Findlay, R.: An "Austrian" model of international trade and interest equalization. J. Polit. Econ. **86**(6), 989–1007 (1978)

Trends in the Sharing Economy: Bibliometric Analysis

Natalia M. Filimonova[1](✉), Nadezhda V. Kapustina[2], Vyacheslav V. Bezdenezhnykh[3], and Nana A. Kobiashvili[4]

[1] Vladimir State University, Vladimir, Russian Federation
natal_f@mail.ru
[2] Russian University of Transport, NRU "Higher School of Economics", Moscow, Russian Federation
kuzminova_n@mail.ru
[3] Financial University under the Government of the Russian Federation, Moscow, Russian Federation
savrula@gmail.com
[4] Moscow Polytechnic University, Moscow, Russian Federation
nanisa@yandex.ru

Abstract. The digital economy development has a great impact on existing economic systems and predefines the necessity of identification of the new business-consumers-state interaction models and patterns. Today's information flows in all areas emphasize the need to find appropriate methods and approaches to the structure optimization and research field determination. This article aims to identify prospective sharing economy development areas based on a bibliometric analysis. In the course of our research, we applied the CiteSpaceV software that enabled us to determine promising research areas based on the co-citation analysis as well as to visualize scientific vectors. The Web of Science information base was selected for picking up publications, reflecting various sharing economy development features in the period from 2010 to 2018. As a result of the study, key sharing economy development vectors have been defined: web, to-peer renting, business model friend, future research, to-peer economy system, transforming consumption, collaborative network, and authentic economy system.

Research areas, structured based on the Web of Science publications, can be a backbone for predicting promising research areas.

Keywords: Sharing economy · Publications · Fields of knowledge · Web of Science · Gig economy · Collaborative economy · P2p economy · Pear-to-pear economy · Collaborative consumption

JEL Code: 057

E. G. Popkova and B. S. Sergi (Eds.): ISC 2019, LNNS 87, pp. 145–154, 2020.
https://doi.org/10.1007/978-3-030-29586-8_17

1 Introduction

Digitalization of economic and management processes expedites the process of coordinated activities between service providers and consumers and systematizes information flows, thus accelerating data transfer and retrieval processes. Also, the digital economy development results in the emergence of new business models and interaction forms between business, consumers and the state.

The application of the sharing economy concept with the "use, rather than own" rule as a basic principle is one of available advanced management forms and methods. The sharing economy is a completely new business process organization model that radically differs from traditional approaches.

A collaborative consumption model was first described in detail by Botsman and Rogers (2010). The collaborative consumption is an economic model that is based on the collective use of goods and services as well as rents and barters instead of ownership. Shared use of various resources such as houses, cars, parking lots, equipment, tools, knowledge, and skills is practiced. The development of technologies and a wide spread of the Internet and social networks has accelerated its development, while environmental problems and the economic crisis have motivated people to participate in this process (Maykova 2015).

Researchers from the University of Utrecht (the Netherlands) identified two groups of reasons for participation in the shared use process: external reasons (economic benefits, practical feasibility, receiving praises) and social reasons (helping another person, meeting new people). It's also worth noting that the participation interest grows if the participation does not imply any monetary contributions, especially in case if it concerns the exchange of inexpensive items such as screwdrivers or scythes. As the cost grows, the people's desire to share items tends to fall; at the same time, a higher educational level of persons involved in the exchange increases the likelihood of their participation. Also, it was noted that women are more inclined to collaborative consumption, than men (Maykova 2015).

The fundamental principle of the sharing economy concept is that it is more convenient and profitable for consumers to pay for getting temporary access to a product, rather than to own it. In the modern world, the sole owner of both expensive property (yachts, airplanes, country houses, foreign residences) and rather inexpensive property (cars, sports and construction equipment) becomes costly and often unprofitable. In general, buying and maintaining a car, which is typically used only over 3% of its lifetime and thus deems rather irrational, is not cheap also. Furthermore, it's worth mentioning a changed consumer behavior model in society. The market for rental cars, houses, equipment, inventory, etc., is developing quite intensively simply because people don't want to buy or have no place to store things, or because it's not obvious that this particular item may be needed ever later, rather than they are not able to buy desired items. Since consumers want freedom, new impressions and trips around the world, the property becomes a real ballast that requires continuous care and considerable maintenance costs (Hartmann 2016).

Thus, sharing becomes an important trend and, possibly, a new economy and society development vector (Kelly 2017). First of all, it is applicable to such resources like information, environmental assets and public relations, which need fair distribution, cooperation and horizontal interaction for providing more efficient usage (Shor et al. 2015).

It also should be noted that two key factors, i.e. consumer and service provider confidence and changed consumer thinking, affect the sharing economy development (Rinne 2019, Popkova 2017, Popkova et al. 2018, Sergi et al. 2019, Sergi 2003, Sukhodolov et al. 2018).

Bursting growth is one of the main features of entrepreneurs operating based on the sharing economy principles. For instance, Uber Technologies Inc., a mobile taxi operator established in 2009, plans an initial public offering (IPO) for 2019, and the IPO costs, according to preliminary estimates, are about USD120 billion (Rinne 2019). However, it's worth noting that some of them suffer fairly large losses. For example, Ofo, a Chinese bicycle rental company, is on the verge of bankruptcy, since 11 million users require returning the deposit (Yang 2018).

At the same time, as the collective resource use techniques acquire a paradigm nature, a certain adaptation of traditional economic activities, in which the use of resources is usually based on private property and profit-making, to these techniques it required. At the same time, a broader understanding of the shared use may arise, which is not limited to the idea of equal co-production of profit-making goods but also encompasses the interaction of various-scale economic entities, allowing each entity to receive additional benefits and increase social welfare.

2 Methodology

The promising research areas for the sharing economy development have been defined through a bibliometric analysis, using research area identification techniques. This method is widely used in various fields of knowledge to identify promising research areas (Ekanayake et al. 2019).

In our study, the below-mentioned stages were implemented to determine promising research areas. At the first stage of the research, keywords for promising research area determination were defined. The keywords to be used were as follows: "sharing economy", "gig economy", "collaborative economy", "p2p economy", "pear-to-pear economy", "collaborative consumption", taking into account logical conditions and combinations. At the second stage of the research, a scientific and information base for the research was selected. Currently, lots of peer-reviewed literature databases, such as Scopus, Web of Science, Agris, PubMed, Google are available. For our study, we selected the Web of Science database because it represents scientific metric fundamentals for assessing the performance of scientists' activities in Russia. At the third stage, an original pool of scientific publications was created, using the keywords. From the Web of Science, 1579 publications were retrieved. A preliminary analysis of selected publications showed that some of them are inconsistent with the subject of the research, and deal with various aspects related to the knowledge areas such as environmental science, green economics, law, etc.

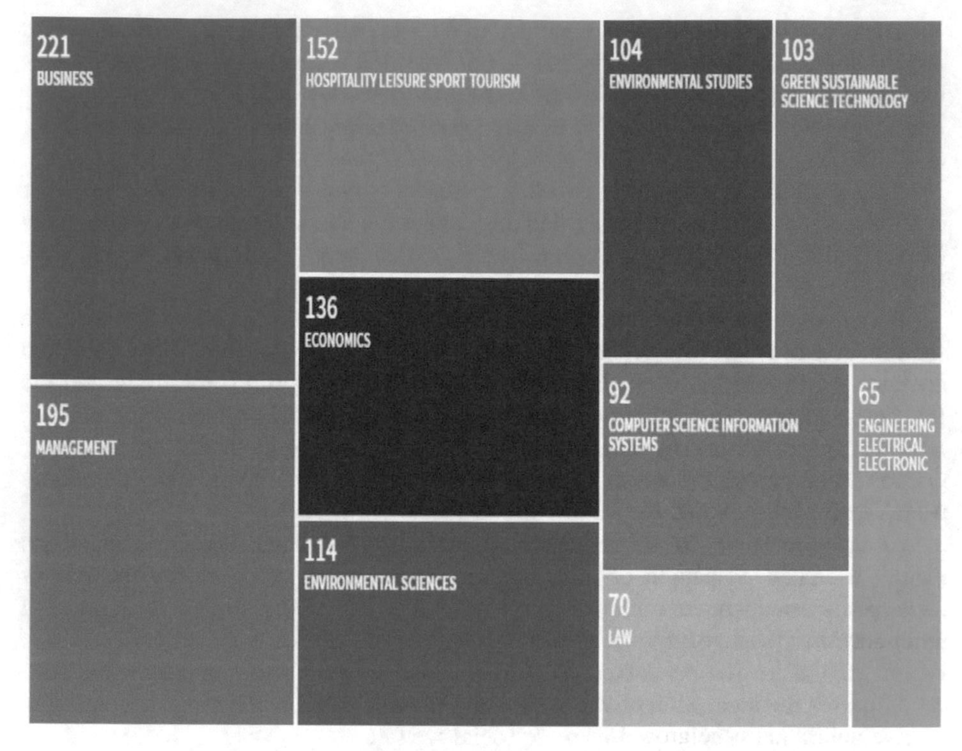

Fig. 1. Distribution of articles by fields of knowledge

Scientists study various topics and specific features of these companies as well as the company growth and failure factors, etc. Based on the data presented in Fig. 4, the research structure on the topic "sharing economy" was analyzed. As a result of this study, it was revealed that more than 50% of all studies fall on business (17.65%), management (15.58%), tourism (12.24%) and economy (10.86%).

This fact required us to refine the articles reflecting the subject under study. As a result, following the research topic clarification, 1311 articles were picked up. At the next stage, during the review of publications retrieved from the Web of Science using the freely distributed product CiteSpaceV (Chen et al. 2010), we constructed research areas allowing promising research vectors to be determined.

3 Results

As a result of the investigation into Google's search queries for Uber (Fig. 1), a notably growing interest to the companies, operating in the sharing economy, was observed.

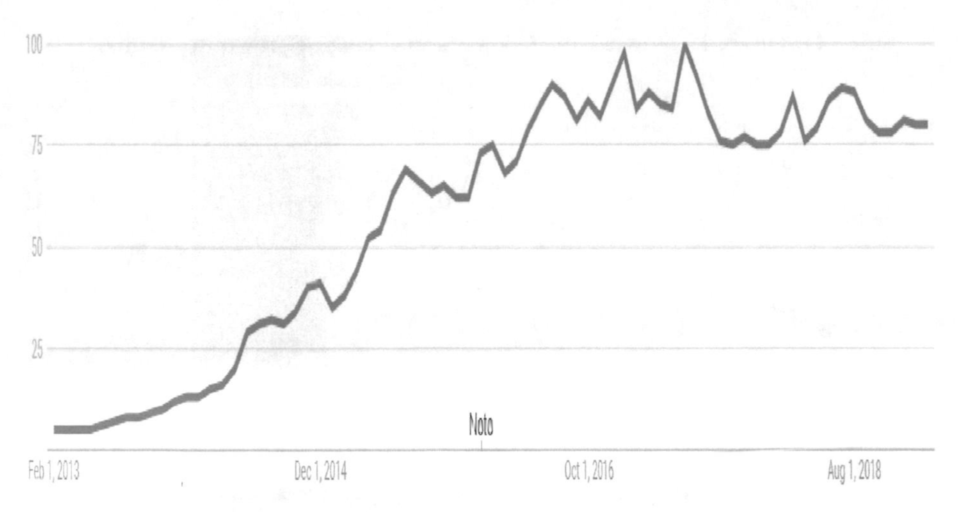

Fig. 2. Google users' interest growth to Uber

It's interesting to note that the Google users expressed virtually no interest to Uber in 2013; however, by 2014 the interest jumped almost 10-fold, is clear evidence of a sharp increase in the company's popularity. The Google users' peak interest to Uber falls on 2016-2017, but after that, in 2018, it drops slightly, but still remains at a consistently high level so far.

Figure 2 shows the growing interest of Google users to AIRBNB.

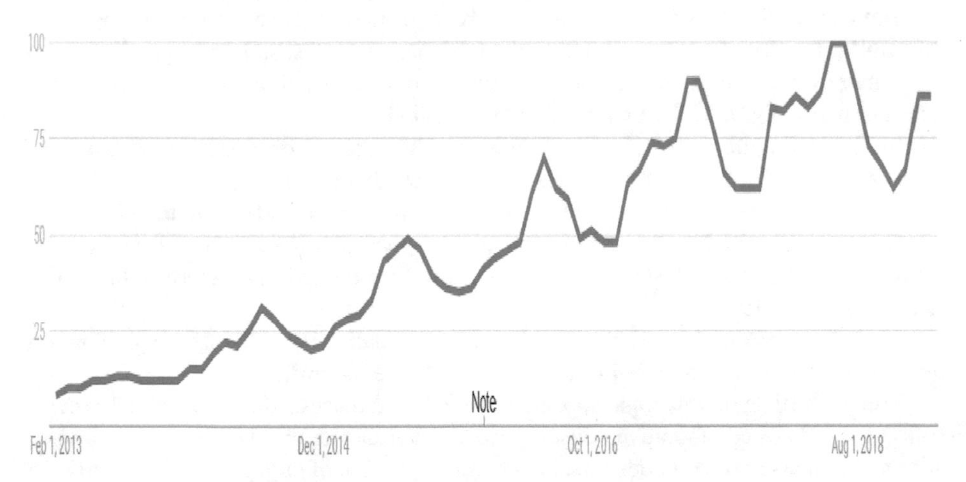

Fig. 3. Google users' interest growth to AIRBNB

As follows from Fig. 2, the growth in the Google users' interest to AIRBNB in 2013 was on par with that for Uber, i.e., users expressed virtually no interest to this company. In 2014, it grew up, but not so strongly as for Uber. The peak growth was registered in 2017-2018, during which the fall to the 2015 level was also observed, followed by a sharp rise (almost 2-fold).

Based on Fig. 3, let's review the Google users' interest growth to Netflix.

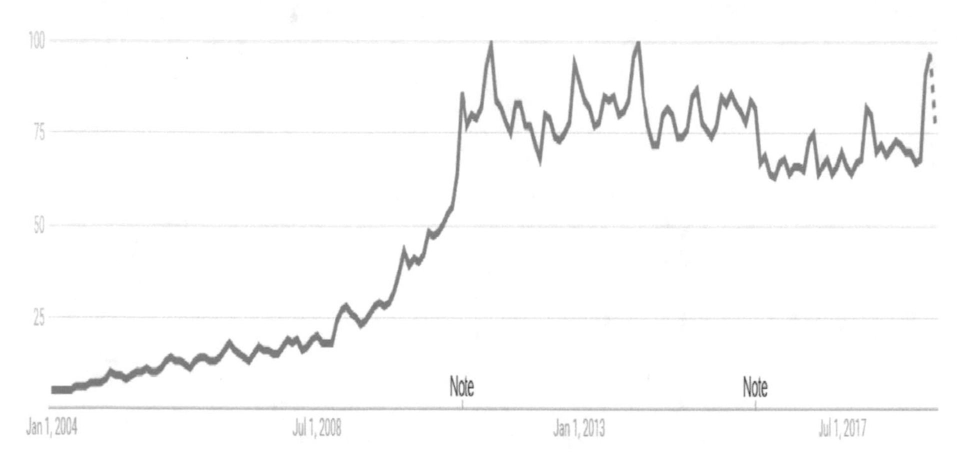

Fig. 4. Google users' interest growth to Netflix

Google users' interest to Netflix was reviewed starting with the year 2004. It was growing only slightly during the first five years, and only after 2008 it began to grow at a fast pace, with the peak interest falling during 2010-2011. However, ever since and to the present time, Google users' interest to Netflix remained at the 2010 level, with occasional slight falls and follow-up upraises to the same level.

Having studied the Google users' interest growth to such companies as Uber, AIRBNB, Netflix, a current general trend of active and sustainable growth of the consumer interest to organizations, operating on the sharing economy basis and actively using digital technologies, should be noted.

In line with the growing user interest to the sharing economy companies, the scientific community's interest to these companies also tends to grow.

At present, the world scientific community pays great attention to the sharing economy issues, which is confirmed by a year-by-year growing number of publications on this topic. Figure 5 illustrates the authors' publication activity pattern in the period from 1978 to 2019.

As follows from Fig. 5, the number of publications on the sharing economy development issues, available in the Web of Science, is on the rise.

The first article on this topic appeared in 1978 and discussed the concerned issue as follows: "Acts of collaborative consumption are the events in which one or more persons consume economic goods or services in the process of engaging in joint activities with one or more others". This concept was known for quite a long time and assumed a shared use of consumer goods. For example, such organizations as rental centers, libraries, cinemas, etc., allowing users to share goods and services without acquiring them for personal use, existed. At present, an approach, involving a shared use of goods without purchasing them for personal use and without purchasing personal services, is used. However, digital technologies are widely used in the current context, enabling the market of potential customers to be extended through digital platforms.

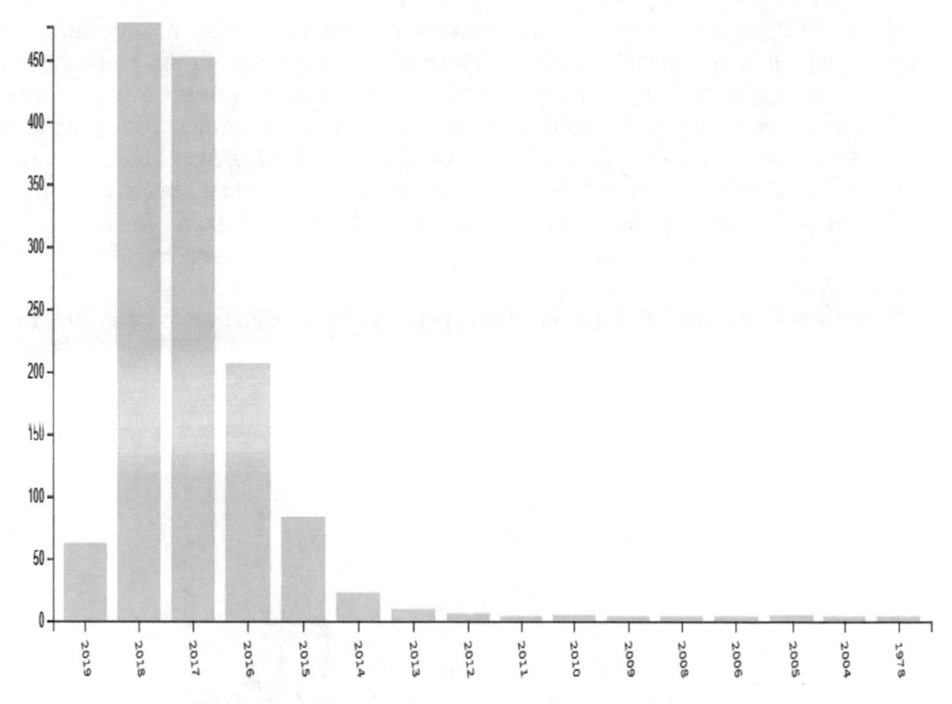

Fig. 5. The dynamics of the authors' publication activity

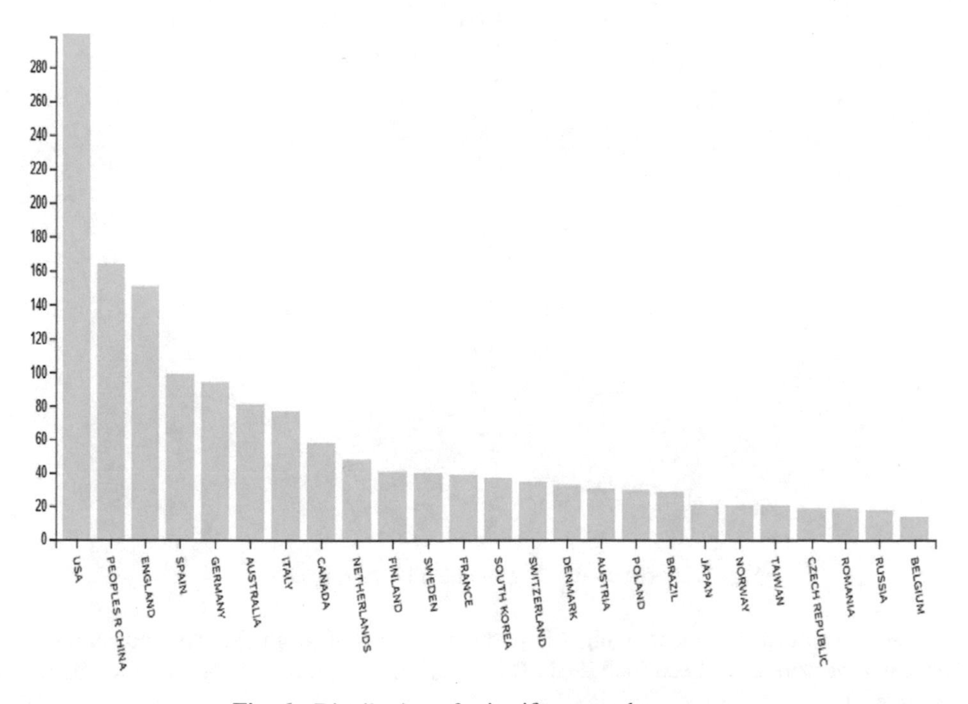

Fig. 6. Distribution of scientific papers by country

Since 2000, studies on the sharing economy development under new conditions began to appear, but a notable growth is observed since 2014, with the most active interest being registered only recently, since 2016 and up to the present time.

Having studied the distribution of scientific interest to the sharing economy issues across the countries (Fig. 6), it's worth mentioned that the United States ranks first in terms of the number of publications on the topic studied, exceeding almost twice the number of publications in China (Rank 2) and in England (Rank 3). As for Russia, it ranks only 24th in the said list. Although recently a growing interest of Russian researchers to this topic is observed.

Using the Citespace, we have identified eight major research areas presented in Fig. 7.

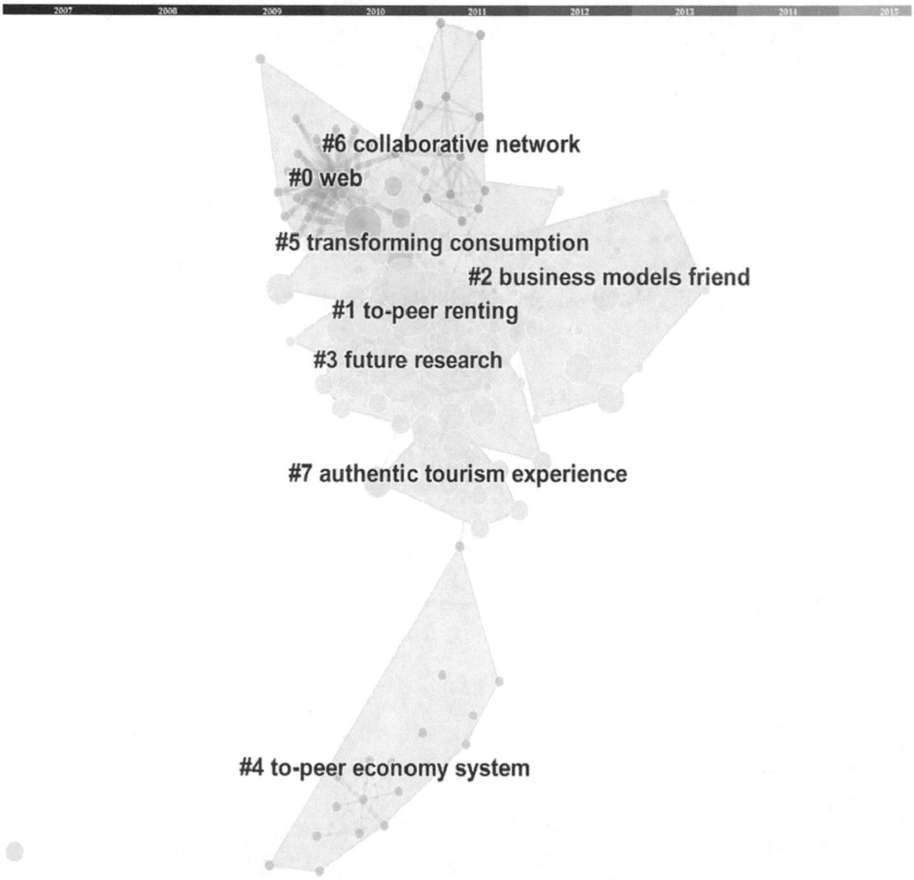

Fig. 7. Research fronts in the field of sharing economics

As a result of the research, eight key research areas of promising sharing economy research vectors have been identified. The description thereof will be provided below (Table 1).

Table 1. Main research fronts in the field of sharing economics

ClusterID	Size	Silhouette	Mean(Year)	Label (LSI)
0	15	0.879	2010	Web
1	21	0.647	2012	To-peer renting
2	13	0.739	2014	Business model friend
3	16	0.646	2014	Future research
4	10	0.955	2011	To-peer economy system
5	12	0.83	2012	Transforming consumption
6	7	0.727	2009	Collaborative network
7	9	0.905	2015	Authentic economy system

The largest cluster (#0) has 15 members and a silhouette value of 0.879. It is labeled as *web*. The most active citer to the cluster is 0.57931 Belk (2014).

The second largest cluster (#1) has 21 members and a silhouette value of 0.647. It is labeled as *to-peer renting*. The most active citer to the cluster is 0.23191 Heather et al. (2015).

The third largest cluster (#2) has 13 members and a silhouette value of 0.739. It is labeled as *business models friend*. The most active citer to the cluster is 0.3211 Aloni, E Pluralizing (Erez 2016).

4 Conclusions

Using bibliometric methods is a good and convenient tool for planning and evaluating scientific works. As a result of the research, three major clusters in the sharing economy field have been identified. The study of the clusters helped us to identify significant research areas to facilitate further study of various sharing economy development issues. The cluster that emerged in 2015 and includes articles studying user requirements such as authenticity, allowing the review of a daily lifestyle of citizens living in one or another places, is of particular interest, because travelers require authentic, experience-oriented opportunities with a deeper interaction with the locals. And the sharing economy has arisen in part as a reaction to these needs.

References

Belk, R.: Sharing versus pseudo-sharing in Web 2.0. Anthropologist **18**(1), 7–23 (2014)

Botsman, R., Rogers, R.: What's mine is yours: the rise of collaboration consumption. Harper Business (2010)

Chen, C., et al.: The structure and dynamics of cocitation clusters: a multiple-perspective cocitation analysis. J. Am. Soc. Inform. Sci. Technol. **61**(7), 1386–1409 (2010)

Ekanayake, E.M., et al.: Mapping the knowledge domains of value management: a bibliometric approach. Eng. Constr. Architectural Manag. **26**(3), 499–514 (2019)

Erez, A.: Pluralizing the Sharing Economy. Washington Law Review, vol. 91 (2016). https://heinonline.org/HOL/Page?handle=hein.journals/washlr91andid=1429anddiv=andcollection=. Accessed 24 Jan 2019

Hartmann, O.: How shared using changes our life and how can we make money on it (2016). http://www.forbes.ru/karera-i-svoy-biznes/347961-kak-sovmestnoe-potreblenie-menyaet-nashu-zhizn-i-kak-na-etom-zarabotat. Accessed 11 Jan 2019

Heather, P., et al.: Examining temporary disposition and acquisition in peer-to-peer renting. J. Mark. Manag. **31**(11–12), 1310–1332 (2015)

Kelly, K.: The Inevitable. Understanding the 12 technological forces that will shape our future. Penguin Books, New York (2017)

Maykova, A.: How can sharing economy improve a city's functioning (2015). https://strelka.com/ru/magazine/2015/07/02/share-economy. Accessed 12 Jan 2019

Popkova, E.G.: Economic and legal foundations of modern russian society: a new institutional theory. In: Advances in Research on Russian Business and Management. Information Age Publishing, Charlotte (2017)

Popkova, E.G., Bogoviz, A.V., Pozdnyakova, U.A., Przhedetskaya, N.V.: Specifics of economic growth of developing countries. Studies in Systems, Decision and Control, vol. 135, pp. 139–146 (2018)

Rinne, A.: 4 big trends for the sharing economy in 2019 (2019). https://www.weforum.org/agenda/2019/01/sharing-economy/. Accessed 15 Jan 2019

Sergi, B.S., Popkova, E.G., Bogoviz, A.V., Ragulina J.V.: Entrepreneurship and economic growth: the experience of developed and developing countries. Entrepreneurship and development in the 21st century, pp. 3–32. Emerald publishing limited, Bingley (2019)

Sergi, B.S.: Economic Dynamics in Transitional Economies: The Four-P Governments, the EU Enlargement, and the Bruxelles Consensus. International Business Press, New York (2003)

Shor, J.B., Walker, E.T., Lee, C.W., et al.: On the sharing economy: sharing, caring and prosit. Contexts **14**(1), 12–19 (2015)

Sukhodolov, A.P., Popkova, E.G., Litvinova, T.N.: Models of Modern Information Economy: Conceptual Contradictions and Practical Examples, pp. 1–38. Emerald Publishing Limited, Bingley (2018)

Yang, Y.: Chinese Ofo biking is on the verge of bankruptcy (2018). http://ekd.me/2018/12/kitajskij-bajkshering-ofo-na-grani-bankrotstva-11-mln-polzovatelej-potrebovali-vernut-zalog/. Accessed 14 Jan 2019

Automation of Financial Information Exchange: Implementation into the Russian Legislation

Inna N. Kolkareva(✉), Anna V. Dudchenko, Alena A. Eremeeva, and Ruslan N. Frolov

Krasnodar branch of G.V. Plekhanov Russian University of Economics, Krasnodar, Russian Federation
innakolkareva@yandex.ru, A-dyd4enko@yandex.ru, allo_14@mail.ru, docent-1976@mail.ru

Abstract. This article discusses the process of implementation of the international unified act on the automatic exchange of financial information in Russia. The authors consider in detail the issue of its influence on the modern national law of the Russian Federation. The purpose of the study is the legal basis for the process of automating the exchange of information between fiscal and control and supervisory authorities of the subjects of public law. Methodology. The methodological basis of the study combines the use of various methods of scientific knowledge: dialectical, formal-logical, synthesis, synthesis, analogy and comparative law. Results: 1. The significance of the legal regimes of the Russian Federation: tax and banking secrets and the state of their modern legal regulation in the context of international rules and standards. 2. Developed criteria for the optimal integration of international legal norms on the automatic exchange of information into the national legal system in order to change the structure of the rule of law of this state. 3. It justifies the need, at the level of the Government of the Russian Federation, to develop a system for automatic identification and risk-profiling of data arising in the course of the activity of each entity.

Consideration of the issues of practical significance and implementation of the provisions of the unification norms on digitalization and the binding nature of information flows is aimed at minimizing the negative effects of the shadow economy and countering capital outflow from Russia.

The conducted research contributes to the theory of financial and tax law, as it specifies the legal consequences that arise after the transformation of the content of banking and tax secrets.

Recommendations: The results can be useful for practitioners in the financial sector and all legal entities acting as a party to tax legal relations.

Keywords: Information · Globalization · International law · National law · Banking secrecy · Credit organizations

JEL Code: K10

E. G. Popkova and B. S. Sergi (Eds.): ISC 2019, LNNS 87, pp. 155–160, 2020.
https://doi.org/10.1007/978-3-030-29586-8_18

1 Introduction

The urgent need for capital controls of individuals and legal entities has led states to begin international cooperation in the area of financial information exchange. This need is related to the fact that tax residents of a state conduct their activities in the jurisdictions of different states and in offshore zones. Offshore zones are territories in which special regimes are in place related to registration, licensing, and taxation. "Offshores" are attractive tax and customs benefits. And these instruments contribute mainly to the export of capital outside the state, which leads to the inability to answer for obligations to counterparties and financial instability.

The result of the work of states in the field of economics and the exchange of financial information is a series of international legal acts regulating the exchange of such information. The Organization for Economic Cooperation and Development (hereinafter - the OECD) plays a significant role in the process of formation and development of this area. The work of the OECD is to advise and stimulate states aimed at accomplishing the tasks of the organization. As part of the OECD activities, in 2014, a fundamental document was developed that establishes clear boundaries for the exchange of financial information in an automatic mode - the Standard on the automatic exchange of financial information (hereinafter - the Standard). For the purposes of the Standard, economic information refers to a report of a credit institution on the accounts of individuals and legal entities, as well as their income and expenses for the reporting period. Taxpayer details to be transferred include: passport, TIN, place of residence, etc.

OECD standards, having an international character, must be implemented in national legislation. This process implies not just the inclusion of international legal norms, but the adaptation of the rule of law under the norms of state law for their more effective application.

The adoption of a number of international instruments in the field of exchange of financial information indicates the interest of states in obtaining sufficient funds to the budgets of all levels.

In our opinion, the practical implementation of unified international standards governing the exchange of information seems appropriate, since these norms have a high degree of structure and universality.

The processes of globalization dictate to modern society and world states the need to jointly regulate competition between public entities, monitor and track financial resources and receive reliable and up-to-date information on capital flows. Problems of fiscal dumping or unfair tax competition are becoming more common in a given state, creating the conditions for international tax planning of unused liquidity. Official statistics indicate that the volume of the shadow economy in the member states of the European Union (hereinafter - the EU) is at least 20% of GDP - about $ 1 trillion. Euro per year. It is these factors that have clearly manifested themselves in recent decades that have led states to the formation of universal rules of conduct for resolving this problem.

2 Background and Methodology

Prior to the development and implementation of the Standard, the main international acts in the field of digitalization of financial information were: the Convention on Mutual Administrative Assistance in Tax Matters 1988, the European Union Directive on Savings Taxation 2005, the US Foreign Account Tax Compliance Act (FATCA) 2010.

The development of new international legal norms was necessary both for the collection and joint exchange of reliable tax information, and the possibility of applying adequate measures to counter illegal operations in the financial sphere.

Consider a concrete example of the effectiveness of the implementation of the provisions of the Standard. For example, if Russian credit organizations do not participate in the implementation of FATCA provisions and do not provide information on the accounts of non-residents of the United States, such a refusal to cooperate will have the following consequences:

- compulsory withdrawal of 30% of the amount of the bank transfer of the client's international account;
- violation of legal regimes of banking and tax secrets, personal data;
- reputational risks, etc.

3 Discussion and Results

For the Russian Federation, the implementation of the norms regulating information exchange is associated with certain conflicts, contradictions, since the transfer of financial information violates banking and tax secrets, and also entails the responsibility of credit institutions for the transfer of data with restricted access. The latter problem is solved by the legislator, since the transfer of information containing banking secrets is governed by the rules of law relating to tax control, i.e. after the transfer of such data, the tax authority is responsible.

In the context of the provisions of the Standard, the issue of preserving the legal regimes of banking and tax secrecy is of genuine interest and very significant discussions in the professional and scientific environment.

At present, the provision of information to tax authorities, which allows countries to control funds of legal entities and individuals, directly depends on information received by tax authorities from credit organizations. According to the legislation of the Russian Federation, this information relates to the legal regime of bank secrecy and commercial activity in general.

Based on the content of the provisions of the Standard, it can be concluded that the regime of bank secrecy is preserved, and tax secrecy in respect of non-residents is violated.

For the optimal integration of international legal norms into the national legal system, certain changes in the structure of the rule of law of this state are also required. Therefore, the implementation of the requirements is carried out not only with the consent of the state, but requires compliance with certain criteria:

- availability of effective public administration; the activities of state bodies in the implementation of accepted international legal obligations;
- the existence of an institutional and organizational mechanism, that is, the rule of law must be fulfilled or applied by the competent authority.

Based on this, it should be understood that the refusal or late integration of a legal act can be considered as a violation of international relations and entail certain consequences (in particular, sanctions).

Recently, the following acts were adopted for the implementation of the main provisions of the Standard for the mutual exchange of information in the Russian Federation:

- Federal law on ratification of the Convention on mutual administrative assistance in tax matters;
- Federal Law on Amendments to the Tax Code of the Russian Federation;
- Decree of the Government of the Russian Federation on the conclusion of agreements on the exchange of information on tax disputes;
- Order on the distribution of responsibilities between the federal bodies of state power and organizations on the interaction of the Russian Federation and the Organization for Economic Cooperation and Development;
- Model agreement on the exchange of information on tax matters between the Government of the Russian Federation and the Government of a foreign state.

In 2014, after conducting a procedure for reviewing the compliance of Russian legislation with OECD international criteria, it was decided to join Russia in the organization in 2015. At that time, the status of Russia assumed the value of the "interacting" participant, however, the accession was suspended due to a number of specific events that occurred in Russia at that time, and the status was changed to the "joining" state. The adoption of this decision entailed the termination of joint projects, negotiations, etc.

However, even despite the organization's decision, in 2016 Russia managed to bring the national legal system in accordance with the Standard and international provisions on automatic exchange, and the Russian Federation joined the rules for the exchange of financial information for tax purposes.

In September 2018, Russia submitted the first report on non-residents to the OECD and in the same year received information about the assets of residents abroad. The report was provided by the competent authorities in the field of tax policy: the Ministry of Finance of the Russian Federation, the Ministry of Economic Development of Russia and the Federal Tax Service (based on the order of the Government of the Russian Federation No. 409-P).

According to the Federal Tax Service for 2017, Russia automatically received and processed information about the financial activities of Russian tax residents from the competent authorities of 58 states (territories). The information was also provided by such "non-transparent" jurisdictions as the Cayman Islands, the British Virgin Islands, Belize and others. Russia provided relevant information on the financial accounts of citizens and legal entities of 42 states (territories).

By the end of 2019, a number of jurisdictions are expected to join the automatic exchange of financial information with the Russian Federation - Austria, Switzerland, the Bahamas, Hong Kong, Grenada, Monaco, Macau, etc.

Consider several articles of the Standard that have been successfully integrated into the national law of the Russian Federation:

- The standard contains the main provisions of informational relations: on the composition of the information to be exchanged; types of competent authorities (tax authorities, credit organizations, non-state pension funds, investment funds, insurance companies, etc.) that accumulate and report to foreign tax authorities; types of accounts and tax residents to be recorded. The tax resident status of the Russian Federation is determined in accordance with Articles 207 and 246.2 of the Tax Code of the Russian Federation, which is confirmed by issuing the Federal Tax Service of Russia or its territorial tax authority, a document confirming the tax resident status of the Russian Federation;
- All financial statements listed in the Standard include information on account balances, interest, shares, sale opportunity and release from financial assets held in different accounts registered to taxpayers (Section B, Article I).

The standard includes general requirements for the amount of information provided and implies the presentation of data on account holders of individuals or legal entities. It should be noted that the list of financial information is an exhaustive list. At the same time, in 2019 it was significantly expanded and must necessarily include information on financial accounts for all contracts concluded after the date of its application (after July 20, 2018). Thus, the information should cover both the implemented relationship, and assumed to be executed.

All newly concluded contracts should be analyzed in accordance with stricter requirements with regard to the financial situation of non-residents, their identification, sum criteria for the value of contracts.

In conclusion, it should be stated that the development of law always lags behind the actual relations between the subjects of legal relations and the main task of state authorities is to find a compromise between the interests of the individual and the state as a whole. Therefore, the legislator should fill in the gaps in the legislation of the Russian Federation, immediately responding to world trends and the needs of modern realities.

At the moment, it is impossible not to agree with the point of view of most experts that digitalization of the Russian economy without implementing standards for financial transactions and submitting automatic reports on the state of accounts is impossible.

In our opinion, it is automatic exchange that should become one of the most important institutions of tax law and ways of obtaining information. The use of automatic exchange of financial information allows the tax authorities of Russia to receive data on undeclared assets in foreign jurisdictions by pressing the button in a matter of seconds.

Thus, the implementation of automatic exchange standards by Russia will be one of the key ways to increase the efficiency of tax administration. At present, the mechanism of automatic exchange of financial information is all-encompassing and extends even to those territories that for decades were outside the legal influence of most countries of the world. The current tendency to adhere to the provisions of the Standard suggests that outside their action may remain those areas that are not suitable for maintaining effective international business and storage of assets. Placing your money in credit institutions of "financial marginal" will be meaningless. On the contrary, participation in the automatic exchange will be one of the signs that positively affect the business image of a country.

4 Conclusions

We believe that the practical implementation of international treaty standards and the inclusion of the institution of automatic exchange of financial information in the national legislation of Russia are logical and should be distinguished by a clearer structuredness and forethought. In the near future, at the legislative level, it is imperative to develop a system for automatic identification and risk-profiling of data arising in the process of activity.

References

Beznoschenko, N.V.: The ratio of banking and tax secrets. Taxes, no. 9, pp. 13–14 (2017)

Lauts, E.B.: Anticrisis legislation and legislation in the field of state regulation of business activities. Lawyer, no. 15, p. 4 (2017)

Lukashuk, I.I.: International law. A common part. - M.: BEK, p. 224 (1997)

Mashkova, E.V.: Issues of efficiency and features of the activities of the bodies of the European Free Trade Association. Moscow University Herald, Series 11, Right. no. 2, p. 110–111

Suvorov, V.Y.: The Implementation of International Law, 65c. Publishing House of the Sverdlovsk Law Institute, Yekaterinburg (1992)

Communication from the commission to the European parliament and the Council on concrete ways to reinforce the fight against tax fraud and tax evasion including in relation to third countries. In: COM (2012) final, Brussels, p. 2–3, 27 June 2012

Council Directive 2014/107/EU of 9 December 2014 amending Directive 2011/16/EU as regards mandatory automatic exchange of information in the field of taxation, Official Journal of the European Union, L 359, p. 1 – 29, 16 December 2014

HARMFUL TAX COMPETITION An Emerging Global Issue. http://www.uniset.ca/microstates/oecd_44430243.pdf. Accessed 30 March 2019)

OECD, Standard for Automatic Exchange of Financial Account Information in Tax Matters. https://read.oecd-ilibrary.org/taxation/standard-for-automatic-exchange-of-financial-account-information-for-tax-matters_9789264216525-en#page1. Accessed 30 March 2019

Digital Human: Principles of Behavior in the Market and Internal Contradictions

Viktor G. Antonov[1], Nurgul K. Atabekova[2], Ulyana A. Pozdnyakova[3](✉), and Andrey I. Novikov[4]

[1] State University of Management, Moscow, Russia
v.antonov1949@yandex.ru
[2] Kyrgyz State Law Academy, Bishkek, Kyrgyzstan
[3] Institute of Scientific Communications, Volgograd, Russia
wua@list.ru
[4] Institute of the Service Sphere and Entrepreneurship (branch) of Don State Technical University, Shakhty, Russia
novikov-1962@bk.ru

Abstract. Purpose: The purpose of the paper is to determine the principles of behavior of digital human in the market and his internal contradictions and to determine the conditions of solving these contradictions for stimulating his support for formation of the digital economy.

Design/Methodology/Approach: For determining the contribution of a digital human into creation of the digital economy the authors use the methods of correlation and regression analysis. The source of statistical data is materials of the rating of the global digital competitiveness of countries according to the IMD for 2018. The research objects are leaders in the global rating of the United Nations Development Programme (2019) as to the level of human development and Russia.

Findings: It is determined that in "pure" market conditions digital human perceives the digital economy as a source of risk, uncertainty, and threats to his security. That's why he voluntarily refuses from using digital technologies, shows low flexibility of behavior in the market, and low innovative activity.

Originality/Value: It is substantiated that digital human does not always performs the functions on development of the digital economy, and in unfavorable conditions and absence of external stimulation (from the state) he could show the behavior in the market that could slow down the development of the digital economy. That's why an important task of state stimulation of regulating the process of development of the digital economy is creation of favorable conditions (primarily, mass accessibility of digital technologies and devices, provision of digital security, and protection of intellectual property and stimuli (in cooperation with entrepreneurial structures) for practical application of digital competencies of digital human.

Keywords: Digital human · Behavior in the market · Internal contradictions · Digital economy

JEL Code: E24 · J24 · O15 · O31 · O32 · O33 · O38

E. G. Popkova and B. S. Sergi (Eds.): ISC 2019, LNNS 87, pp. 161–167, 2020.
https://doi.org/10.1007/978-3-030-29586-8_19

1 Introduction

A new subject of economic activities and economic relations that emerged in the conditions of the digital economy is digital human. This individual, which is a retail consumer in the commodity and service markets, is also a seller in the labor market. A specific feature of digital human is high level of digital literacy and possession of digital competencies, due to which he's able to use digital technologies. Formation of society, which consists of digital humans, and maximization of their share in the structure of the national social system, are envisaged in the strategies of formation of the digital economy as a key direction of their implementation due to the two following reasons.

Firstly, only digital human can use the possibilities and obtain advantages from the digital economy. Consumption of hi-tech and innovative products in the modern economic conditions usually requires digital skills. This is also true for online trade, participation of consumer in which requires digital devices and means of digital payment.

That's why demand for products of the digital economy could be set only by digital human. Participation in the system of e-government as a consumer is possible only for digital human. Creation of digital humans is to reduce social opposition to transition to the digital economy due to overcoming the population's deprivation from the digital economy. Digital human also has to stimulate the globalization of the economic system, as he actively uses the Internet.

Secondly, digital modernization of entrepreneurship requires attraction of digital personnel (social role of digital human), which could implement and apply digital technologies in the production and distribution processes. Attraction of digital personnel enables growth of innovative activity of modern entrepreneurship and increase of flexibility and effectiveness (on the basis of usage of breakthrough technologies – e.g., Big Data).

However, the share of the digital economy (as a totality of hi-tech production, online trade, and e-government) in the modern economic systems remains low (below 10% in the structure of GDP) and increases very quickly for finishing the Fourth industrial revolution in the planned period (until 2022–2024). This determines actuality of studying the contribution of a digital human into creation of the digital economy. The purpose of the paper is to determine the principles of behavior of digital human in the market and his internal contradictions and to determine the conditions of solving these contradictions for ensuring his stimulation for formation of the digital economy.

2 Materials and Method

Two social role, performed by a digital human, are usually considered separately in the existing economic literature. The role of digital human in the commodity markets in the conditions of the digital economy is outlined in the works Frunză (2019), Popkova (2017), Popkova et al. (2017a), and Popkova et al. (2017b). Demand for digital personnel in the labor market in the conditions of the digital economy and their usage in digital entrepreneurship as a production factor is considered in publications Polyakova et al. (2019), Pritvorova et al. (2018), Sibirskaya et al. (2019), and Vanchukhina et al. (2018).

However, the principles of behavior of digital human in both markets are not determined, and his internal contradictions are not studied sufficiently.

For determining of contribution of a digital human into creation of the digital economy the authors use the methods of correlation and regression analysis. The authors evaluate the influence of the level of digital literacy (Digital/Technological skills) on activity of population's participation in the system of e-government (E-Participation) and online trade (Internet retailing) and inclination to globalization (Attitudes toward globalization), as well as influence of the level of training of digital personnel (Employee training) on activity of usage of Big Data technologies and analytics in entrepreneurship (Use of big data and analytics), knowledge exchange (Knowledge transfer), flexibility (Agility of companies), and innovative activity (Innovative firms).

The source of statistical data is materials of the rating of the global digital competitiveness of countries according to the IMD for 2018. The research objects are leaders in the global rating of the United Nations Development Programme (2019) as to the level of human development and Russia. The selection of statistical data is shown in Table 1.

Table 1. Selection of data for analysis of behavior of digital human in the market and his internal contradictions, position among 63 countries of the world.

Country, position in the rating of human development points	Human as consumer				Human as an employee				
	Digital/ Technological skills	E-Participation	Internet retailing	Attitudes toward globalization	Employee training	Use of big data and analytics	Knowledge transfer	Agility of companies	Innovative firms
	x_1	y_1	y_2	y_3	x_2	y_4	y_5	y_6	y_7
Norway (1st position, 0.953 points)	12	23	6	15	5	14	17	15	16
Switzerland (2nd position, 0.944 position)	20	51	10	25	4	24	1	16	4
Australia (3rd position, 0.939 points)	42	2	9	27	38	23	26	42	21
Ireland (4th position, 0.938 points)	22	34	8	5	29	28	12	5	6
Germany (5th position, 0.936 points)	54	23	16	28	3	41	9	28	9
Iceland (6th position, 0.935 points)	4	40	–	15	37	25	22	12	2
Hong Kong (7th position, 0.933 points)	10	–	24	3	26	18	19	1	48
Sweden (8th position, 0.933 points)	13	23	15	2	10	9	11	4	14
Singapore (9th position, 0.932 points)	16	8	26	8	20	21	8	26	–

(*continued*)

Table 1. (*continued*)

Country, position in the rating of human development points	Human as consumer				Human as an employee				
	Digital/ Technological skills	E-Participation	Internet retailing	Attitudes toward globalization	Employee training	Use of big data and analytics	Knowledge transfer	Agility of companies	Innovative firms
	x_1	y_1	y_2	y_3	x_2	y_4	y_5	y_6	y_7
Netherlands (10th position, 0.931 points	6	5	7	7	9	17	5	24	5
Russia (49th position, 0.816 points)	29	28	37	61	41	58	54	61	46

Source: compiled by the authors based on IMD (2019), United Nations Development Programme (2019).

3 Results

For determining the influence of independent variables (x) on dependent variables (y), evaluation of their autocorrelation was performed (Table 2).

Table 2. Autocorrelation of development of the commodity and service market depending on behavior of digital human.

	y_1	y_2	y_3	y_4	y_5	y_6	y_7
x_1	−0.164211	0.083479	0.535228	–	–	–	–
x_2	–	–	–	0.397754	0.742287	0.375233	0.440253

Source: calculated and compiled by the authors.

Table 2 shows that all determined dependencies are very weak. The results of deep analysis showed that the only statistically significant dependence is $y_5(x_2)$ – its detailed characteristics are shown in Table 3.

The data from Table 3 show that increase of the level of training of digital personnel by 1 position leads to increase of knowledge transfer by 0.7289 position in the global rating. Determination coefficient constitutes $R^2 = 0.7423$ – therefore, the change of knowledge transfer by 74.235 is explained by the change of the level of digital personnel training. Significance F does not exceed 0.05 and constitutes 0.0089 – therefore, regression dependence is correct at the level of significance $\alpha = 0.05$.

The obtained results (absence of regression dependencies) show that digital human during behavior in the market uses the following principles:

- principle of maximization of obtained advantages: increase of digital literacy and mastering of digital competencies are caused not by internal motives of a modern human but by external stimuli from the state and employers in the interests of obtaining additional revenues, successful employment, and career building;
- principle of critical reconsideration of the digital economy: digital human mistrusts digital technologies and devices, as he clearly realizes their advantages and drawbacks;

Table 3. Characteristics of regression dependence of knowledge transfer depending on the level of training of digital personnel.

Regression statistics	
Multiple R	0.7423
R-square	0.5510
Adjusted R-square	0.5011
Standard error	10.1971
Observations	11

Dispersion analysis

	df	*SS*	*MS*	*F*	*Significance F*
Regression	1	1148.3631	1148.3631	11.0441	0.0089
Leftover	9	935.8187	103.9799		
Total	10	2084.1818			

	Coefficient	*Standard error*	*t-Stat*	*P-Value*	*Lower 95%*	*Upper 95%*
Y-intercept	2.0174	5.3893	0.3743	0.7168	-10.1741	14.2090
x_2	0.7289	0.2193	3.3233	0.0089	0.2327	1.2250

Source: calculated and compiled by the authors.

– principle of minimization of risk: digital human perceives digital technologies and devices as sources of the high risk level and consciously avoids their usage, especially breakthrough digital technologies, with insufficient development of technologies of provision of digital security.

These principles contradict the internal nature of digital human. Firstly, the mastered digital technologies are not used in practice, despite the obvious interest to their approbation and establishment. Secondly, instead of high susceptibility to innovations, digital human protests against them. Therefore, mastering of digital competencies is a self-foal of digital human. The existing model of behavior of digital human in the market hinders the development of the digital economy. Its correction requires external (state) influence on digital human (Fig. 1).

Figure 1 shows that in "pure" economic conditions, digital human perceives the digital economy as a source of risk, uncertainty, and threats to his security. That's why he voluntarily refuses from usage of digital technologies and shows low flexibility of behavior in the market and low innovative activity. External state influence on digital human should be connected to creation and support for favorable conditions for using digital technologies and manifesting high flexibility and innovative activity (e.g., mass accessibility, reliability, and safety of digital technologies and devices, protection of intellectual property, etc.).

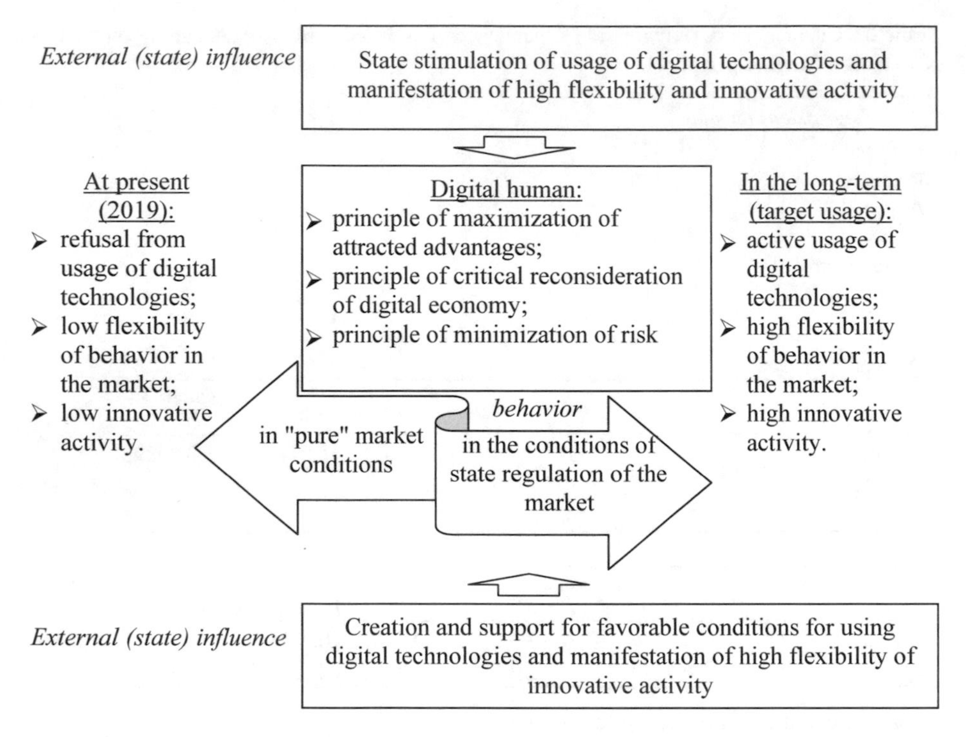

Fig. 1. Digital human: principle of behavior in the market and internal contradictions. Source: developed and built by the authors.

External state influence on digital human should also include state stimulation of usage of digital technologies and manifestation of high flexibility and innovative activity (e.g., legislative establishment of additional labor payment for usage of digital technologies and manifestation of innovative activity). Due to this, it is possible to achieve target behavior of digital human, connected to active usage of digital technologies, manifestation of high flexibility of behavior in the market, and high innovative activity.

4 Conclusions

Thus, the results of the research showed that digital human does not necessarily performs the functions on development of the digital economy, and, in case of unfavorable conditions and absence of external stimulation (from the state) he might even show the behavior that would slow down the development of the digital economy. Unlike pre-digital human, who does not possess digital competencies and protests against them, digital human better understands the risks of usage of digital technologies and strives to avoid them.

This is confirmed by the tendencies of mass cash-out of electronic money just after their receipt, the tendency of online stores' customers' outflow after accumulation of unsuccessful experience of purchase of their products, and a lot of other tendencies in the modern digital economy. Despite the high interest to digital technologies and the skills of their practical usage, digital human could show low innovative activity as a measure of risk management.

That's why an important task of state regulation of the process of development of the digital economy is creation of favorable conditions (primarily, mass accessibility of digital technologies and devices), provision of digital security and protection of intellectual property and stimuli (during cooperation with entrepreneurial structures) for practical application of digital competencies by digital human. It should be concluded that the performed research determined the economic nature of digital human. For determining his social nature, it is necessary to perform sociological and marketing research in future works on this topic.

Acknowledgments. The reported study was funded by RFBR according to the research project No. 18-010-00103 A.

References

Frunză, S.: Human condition and the sacred in the digital era. J. Study Relig. Ideol. **18**(52), 156–169 (2019)

IMD: World Digital Competitiveness Ranking 2018 (2019). https://www.imd.org/wcc/world-competitiveness-center-rankings/world-digital-competitiveness-rankings-2018/. Accessed 22 Apr 2019

Polyakova, A.G., Loginov, M.P., Serebrennikova, A.I., Thalassinos, E.I.: Design of a socio-economic processes monitoring system based on network analysis and big data. Int. J. Econ. Bus. Adm. **7**(1), 130–139 (2019)

Popkova, E.G.: Economic and Legal Foundations of Modern Russian Society: A New Institutional Theory. Advances in Research on Russian Business and Management. Information Age Publishing, Charlotte (2017)

Popkova, E.G., Bogoviz, A.V., Lobova, S.V.: Vacuum in the structure of human capital: a view from the position of the theory of vacuum. In: Human Capital: Perspectives, Challenges and Future Directions, pp. 163–181 (2017b)

Popkova, E.G., Morozova, I.A., Litvinova, T.N.: New challenges for human capital from the positions of its infrastructural role in the system of entrepreneurship. In: Human Capital: Perspectives, Challenges and Future Directions, pp. 257–275 (2017a)

Pritvorova, T., Tasbulatova, B., Petrenko, E.: Possibilities of blitz-psychograms as a tool for human resource management in the supporting system of hardiness of company. Entrepreneurship Sustain. Issues **6**(2), 840–853 (2018). https://doi.org/10.9770/jesi.2018.6.2(25)

Sibirskaya, E., Popkova, E., Oveshnikova, L., Tarasova, I.: Remote education vs traditional education based on effectiveness at the micro level and its connection to the level of development of macro-economic systems. Int. J. Educ. Manage. **33**(3), 533–543 (2019). https://doi.org/10.1108/IJEM-08-2018-0248

United Nations Development Programme: Latest Human Development Index (HDI) Ranking (2019). http://hdr.undp.org/en/2018-update. Accessed 29 Apr 2019

Vanchukhina, L.I., Leybert, T.B., Khalikova, E.A., Khalmetov, A.R.: New approaches to formation of innovational human capital as an element of institutional environment. Espacios **39**, 22–32 (2018)

Managing the Digital Economy: Directions, Technologies, and Tools

Larisa V. Shabaltina[1(✉)], Elena N. Egorova[2], Igor A. Agaphonov[3], and Lilia V. Ermolina[3]

[1] Plekhanov Russian University of Economics, Moscow, Russia
lvs-28@mail.ru
[2] Russian State Social University, Moscow, Russia
elegni@yandex.ru
[3] Samara State Technical University, Samara, Russia
yuhan@mail.ru, Ermolina@mail.ru

Abstract. Purpose: The purpose of the article is to develop a concept of management of the digital economy, which allows conducting the management in the current directions with application of new tools and modern technologies.

Design/Methodology/Approach: Two practices of management of the digital economy are distinguished. The first practice (which is peculiar for developing countries) envisages management on the basis of traditional technologies, including in new directions. The second practice (peculiar for developed countries) is connected to development of the system of e-government, which envisages application of new (digital) technologies, but within the traditional directions of state management, the key of which is provision of state services (in the electronic form). The authors compare (with the help of the method of analysis of variation) the effectiveness of implementation of the above directions of state management of the digital economy within the first (by the example of Russia) and second (by the example of the UK) practice on the basis of statistical data of IMD for 2018.

Findings: Analysis of the official statistics of the IMD showed that effectiveness of implementing various directions of state management of the digital economy is too differentiated in Russia and in the UK. Therefore, both existing practices do not guarantee high effectiveness of management.

Originality/Value: A new concept of state management of the digital economy, which conforms to all criteria that are set to this management, is offered: implemented in new directions with the help of new tools and modern digital technologies. The developed concept is based on digital technologies of the future, which are still at the stage in development and which by 2024 will be commercialized and distributed. This ensures the possibility of application of the offered concept in the long-term.

Keywords: State management · Digital economy · Directions · Technologies · Tools · UK · Russia

JEL Code: O31 · O32 · O33 · O38

E. G. Popkova and B. S. Sergi (Eds.): ISC 2019, LNNS 87, pp. 168–174, 2020.
https://doi.org/10.1007/978-3-030-29586-8_20

1 Introduction

Digital economy is a global investment and innovative project of modern times, which is realized by joint efforts of the state, business, and society. This project goes beyond the limits of creation of a new (digital) sphere and envisages formation of a new type of economic system, within which specific socio-economic relations appear during development, dissemination, and usage of digital technologies. Thus, a need for a new practice of state management appears, which should take into account the specifics of the digital economy and conform to the three following criteria.

1st criterion: state management of the digital economy should be conducted in new directions, which did not exist earlier, as new objects of management appear in it. 2nd criterion: new tools, which were not available earlier, should be applied for state management of the digital economy in the new directions. 3rd criterion: the managerial process should involve the modern digital technologies, as the preceding (traditional) technologies do not allow achieving high level of effectiveness of state management of the digital economy and lead to its uncontrollability and development of the shadow economy.

At present, there are two managerial practices. The first one envisages management of the digital economy on the basis of traditional technologies, which is conducted also in the new directions. It is applied in Russia (30th position in the world as to e-government development, according to the IMD, 2019), China (48th position) and other developing countries. The second practice is connected to development of the system of e-government, which envisages application of new (digital) technologies, but within the traditional directions of state management, the key of which is provision of state service (in the electronic form). It is used in the UK (1st position), Australia (2nd position), South Korea (3rd position) and other developed countries.

The working hypothesis of the research is that both existing practices do not fully conform to the needs of the digital economy, so there's a need for new practice that overcomes their drawbacks and embodies their advantages. The purpose of the article is to develop a concept of management of the digital economy that allows conducting this management in the current directions and with application of new tools and modern technologies.

2 Materials and Method

The existing practice of state management of the digital economy in new directions, but with application of traditional technologies, is studied in the works Polyakova et al. (2019), Pritvorova et al. (2018), Sibirskaya et al. (2019), and Vanchukhina et al. (2018). The practice of management of the digital economy with the help of the system of e-government with application of new technologies, but in the traditional directions, is studied in the works Kassen (2019), Khatib et al. (2019), Moreno-Enguix et al. (2019), Popkova et al. (2019a), and Popkova et al. (2019b). The new objects of state management that appear in the conditions of the digital economy and predetermine the necessity for implementation of new directions of this management are as follows:

- digital business: business structures that use digital technologies and are a driver of growth of the digital economy, due to which the new directions of its state management are stimulation of creation (for measuring the effectiveness of implementation of this direction by the IMD experts the indicator "regulatory framework" is used), innovative activity, and development (indicator "capital") of digital business;
- digital society: consumers of goods and services in the digital form and personnel for the digital economy, due to which the new directions of its state management are provision of social adaptation to the digital economy (indicator "talent") and training of personnel for the digital economy (indicator "training & education");
- digital infrastructure: the existing digital technologies (e.g., the Internet, cloud technologies, etc.) and devices (e.g., computer and mobile devices), which are applied in the conditions of the digital economy, due to which a new direction of its state management is provision of mass access to the digital technologies and devices (indicator "technological framework");
- digital technologies: innovative digital technologies and devices, which are used in the conditions of the digital economy, due to which a new direction of its state management is stimulation of R&D in the sphere of digital technologies (indicator "scientific concentration");
- digital security: protection of digital data from failures in the work of equipment and technologies, and from actions of digital criminals that appear in the conditions of the digital economy – due to which a new direction of its state management is provision of general digital security (indicator "IT integration").

For scientific and practical purpose of the research and verification of the offered hypothesis, let us compare the effectiveness of implementation of the above directions of state management of the digital economy within the first (by the example of Russia) and second (by the example of the UK) practice on the basis of the statistical data of the IMD for 2018 (Fig. 1).

According to Fig. 1, variation of effectiveness of implementing new directions of state management that appear in the conditions of the digital economy is very high in Russia and the UK in 2018 (exceeds 10%), constituting 40.92% and 57.79%, accordingly. This shows low effectiveness of both practices of implementing this management and confirms the necessity for developing a new, more effective, concept.

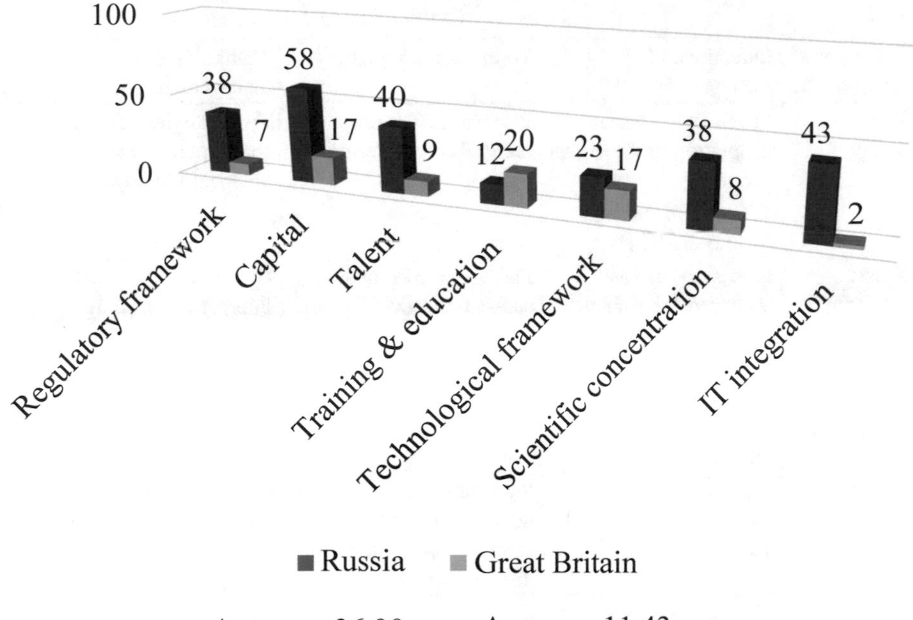

Average: 36.00;
Deviation from average: 14.73;
Variation: 40.92%

Average: 11.43;
Deviation from average: 6.60;
Variation: 57.79%

Fig. 1. Effectiveness of implementing the new directions of state management that appear in the conditions of the digital economy, in Russia and the UK in 2018 (positions 1-63). Source: compiled by the authors based on IMD (2019).

3 Results

For increasing the effectiveness of state management of the digital economy in the distinguished directions we determined the perspective tools and technologies (Table 1).

Table 1. Directions and perspective technologies and tools of management of the digital economy.

Objects of management	Directions of management	Tools of management	Technologies of management
Digital business	Stimulation of creation of digital business	Formation of favorable business climate	Ubiquitous computing, Internet of Things (for selecting the objects for support)
	Stimulation of innovative activity and development of digital business	Formation of favorable investment and innovative climate	

(*continued*)

Table 1. (*continued*)

Objects of management	Directions of management	Tools of management	Technologies of management
Digital society	Provision of social adaptation to the digital economy	Provision of mass accessibility of remote training	Technologies of virtual and alternate reality (for practical training)
	Training of personnel for the digital economy		
Digital infrastructure	Provision of mass access to the digital technologies and devices	Formation of global high speed Internet	Global system of satellite Internet connection
Digital technologies	Stimulation of R&D* in the sphere digital technologies	Provision of growth of efficiency of labor in R&D	Roots (for authomatization of R&D*)
Digital security	Provision of comprehensive digital security	Stimulation of development of new means of provision of digital security	Quantum, cloud, and blockchain technologies

Source: compiled by the authors.

As is seen from Table 1, stimulation of creation of digital business requires formation of favorable business climate. Stimulation of innovative activity and development of digital business requires formation of favorable investment and innovative climate. It is recommended to use ubiquitous computing and the Internet of Things for selecting business objects for state support. Provision of social adaptation to the digital economy and training of personnel for the digital economy requires mass accessibility of remote education. Here it is recommended to use the technologies of virtual and alternate reality (for practical training).

Provision of mass access to digital technologies and devices envisages formation of global hi-speed Internet, for which it is recommended to form a global system of satellite Internet connection. Stimulation of R&D in the sphere of digital technologies envisages provision of growth of labor efficiency in R&D, which is possible on the basis of automatization with the help of robots. Provision of comprehensive digital security envisages stimulation of development of new means of provision of digital security on the basis of quantum, cloud, and blockchain technologies. Based on the above, we developed a concept of management of the digital economy (Fig. 2).

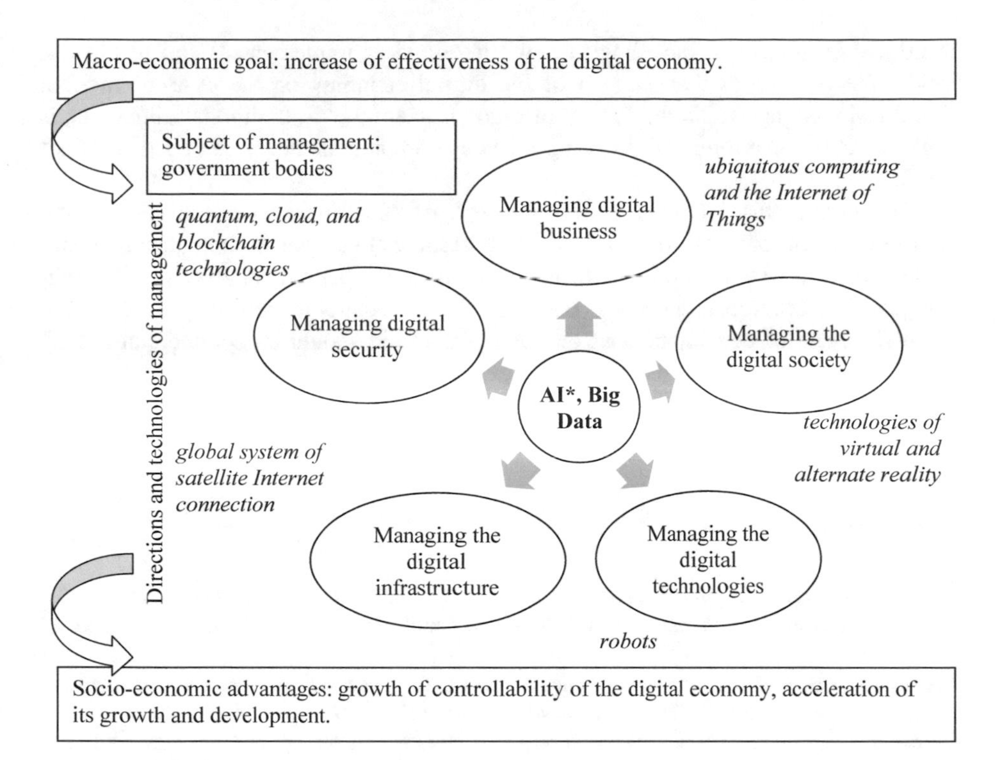

Fig. 2. The concept of management of the digital economy. Source: compiled by the authors.

Figure 2 shows that the offered concept seeks the macro-economic goal of increasing the effectiveness of the digital economy. The systemic character of state management of the digital economy is achieved on the basis of artificial intelligence and the technologies of Big Data processing. This allows for successful coordination of all implemented directions and for authomatization of the managerial process, accelerating decision making and their practical realization, as artificial intelligence constantly collects and analyzes digital data directly from digital devices (e.g., through ubiquitous computing and the Internet of Things) and issues precise commands to other digital devices (e.g., robots) in machine language. The subject of management is government bodies, and artificial intelligence is the technology management, and authomatization is partial (incomplete).

4 Conclusion

Thus, in the course of the research the working hypothesis was proved – analysis of the official statistics of the IMD showed that effectiveness of implementation of various directions of state management of the digital economy is too differentiated in Russia (where the practice of management of the digital economy on the basis of the traditional technologies, including in new directions, is implemented) and in the UK (where the practice of management of the digital economy on the basis of modern digital technologies within the system of e-government, but in traditional directions, is implemented). Therefore, both existing practices do not guarantee high effectiveness of training.

For solving this problem, a new concept of state management of the digital economy is offered – it conforms to all criteria that are set to this management: implemented in new directions with the help of new tools and modern digital technologies. The developed concept is based on digital technologies of the future, which are at the stage of development as of now, and which should be commercialized by 2024. This allows applying the offered concept in the long-term, which is its advantage in the conditions of the digital economy, in which quick rate of technological progress causes quick moral ageing, due to which concepts and technologies become obsolete.

References

IMD: World Digital Competitiveness Ranking 2018 (2019). https://www.imd.org/wcc/world-competitiveness-center-rankings/world-digital-competitiveness-rankings-2018/. Accessed 22 Apr 2019

Kassen, M.: Building digital state: understanding two decades of evolution in Kazakh e-government project. Online Inf. Rev. **43**(2), 301–323 (2019)

Khatib, H., Lee, H., Suh, C., Weerakkody, V.: E-Government systems success and user acceptance in developing countries: the role of perceived support quality. Asia Pac. J. Inf. Syst. **29**(1), 1–34 (2019)

Moreno-Enguix, M.D.R., Lorente-Bayona, L.V., Gras-Gil, E.: Can E-government serve as a tool for public authorities to manage public resources more efficiently? J. Glob. Inf. Manage. **27**(2), 122–135 (2019)

Polyakova, A.G., Loginov, M.P., Serebrennikova, A.I., Thalassinos, E.I.: Design of a socio-economic processes monitoring system based on network analysis and big data. Int. J. Econ. Bus. Admin. **7**(1), 130–139 (2019)

Popkova, E.G., Inshakov, O.V., Bogoviz, A.V.: Regulatory mechanisms of energy conservation in sustainable economic development. Lect. Notes Netw. Syst. **44**, 107–118 (2019a)

Popkova, E.G., Zhuravleva, I.A., Abramov, S.A., Fetisova, O.V., Popova, E.V.: Digitization of taxes as a top-priority direction of optimizing the taxation system in modern Russia. Stud. Syst. Decis. Control **182**, 169–175 (2019b)

Pritvorova, T., Tasbulatova, B., Petrenko, E.: Possibilities of blitz-psychograms as a tool for human resource management in the supporting system of hardiness of company. Entrepreneurship Sustain. Issues **6**(2), 840–853 (2018). https://doi.org/10.9770/jesi.2018.6.2(25)

Sibirskaya, E., Popkova, E., Oveshnikova, L., Tarasova, I.: Remote education vs traditional education based on effectiveness at the micro level and its connection to the level of development of macro-economic systems. Int. J. Educ. Manage. **33**(3), 533–543 (2019). https://doi.org/10.1108/IJEM-08-2018-0248

Vanchukhina, L.I., Leybert, T.B., Khalikova, E.A., Khalmetov, A.R.: New approaches to formation of innovational human capital as an element of institutional environment. Espacios **39**, 22–32 (2018)

Crypto Currencies and Possible Risks

Alexandr J. Bystriakov(✉), Andrey V. Guirinskiy, Tan Nan Nan, Shubbar Hidar, and Lin Chin Din

RUDN University, Moscow, Russia
bstal@mail.ru, gggwork@mail.ru, nannan.tan@mail.ru, sed.hhshb@gmail.com, lechinh2808@gmail.com

Abstract. This article is devoted to the problem of crypto currencies analyses and analyses of risks enclosed. It was shown the baseline of risks together with the reasons of their foundation. It was also identifies the main kinds of risk and some concrete parameters of control. The role of the Central Bank of Russia was shown in the scope of controlling of different financial and banking risks. The analyses of difficulties met by Bank of Russia and other authorities controlling different kind of banking business was fulfilled.

Keywords: Cryptocurrency risks · Banking control · Bank of Russia supervising · Money laundering in cryptocurrency operations · Cryptocurrency legislation

JEL Code: G 15

1 Introduction

Analyzing the risk problem connected with currency risk it is necessary to analyze the reasons of and processes connected with such risks. For now the moment of globalization in processes connected with currency operations has a tendency of growth. The phenomena of risk is the fundamental phenomena in economy and finance and it still has an actuality in the modern time. Risks have to be considered in all directions of activity connected with economical processes and in a first range we have business and of course finance. In practice all transactions in export and import are involved in international payments procedure and it is also connected with conversional operation from one currency to another, and there we have currency risks. Conversional operations are taking place in stocks and here we have different kind of risk volume.

2 Background and Methodology

Crypto currencies in the modern period, they are to be considered as currency assets taking in consideration the fact that conversion in other currencies is possible generally and can be calculated. Such possibility can be provided by stock and out of stock market, knowing that we have about 20 main crypto currencies for today. The stock market is more reliable one in regarding of out of stock one among financial market participants, but

E. G. Popkova and B. S. Sergi (Eds.): ISC 2019, LNNS 87, pp. 175–181, 2020.
https://doi.org/10.1007/978-3-030-29586-8_21

the most important reason is the possibility to buy and sell crypto currency with more and more stocks. The reasonable growth of stocks operating with crypto currency we have seen in 2012 up to 2014 but then in 2017 there were such kind of getting down. It was connected with several serious problems connected with money laundering and it was reinforced crises in such segment. So we can define the one of the most important kind of risk connected with crypto currency and say about it as stock risk. We have many factors influencing to that connect with the global processes in international financial market. Firstly this are processes connected with the international crises beginning from the 2007 which is still valid in spite of common efforts to arrange that. Analyzing crises situations taking place in 20 century than we can stress that even the crises in USA of 20–30 years of 20 century was resolved step by step. In strong phases it was resolved in 1932–1933 years but was finalized in all aspects only in 1950–1960 years. It is reasonably to consider that today in a global market crises we have to have more time to arrange all problems. The situation is strengthening by the fact that crisis processes are in force for long period of time. We have tendencies to isolation of financial market from other segments of it and the real sector of economy. We are tracing here the kind of autonomy of that market for today an especially stock market. This processes are forced to go others, connected with high speculation grade and it is revealed the growth of other currency risks in assets transactions. The crypto currencies are also involved in risks by that way being the conversion crypto to the other kind of currency. Such kind of risks have to be difficulty hedged taking high grade of valuation of currency asset. Beginning from 2009 crypto currencies are running as asset to be grown during sensible period of time and to be decreases within the frame of another period. Some traditional instruments of hedge such as option and future are not so constantly effective. Than we have such tendency to widen the applying such instruments in out of stock market which is provoke turbulence in all international market and financial system. We can fix the multiplication of out of stock derivatives which is over then ten times regarding international GRP. There are such fitches of generation world financial pyramid and it is necessary to fix place of crypto currency in aggravation or neutralization of such tendency. As we can consider the cryptocurrency as the asset in stock market which is buying or selling we can use it as the facility for limitation crises. Ways of limitation of crises may be traced as the multiplication of pairs with derivatives and crypto. The stabilization can be achieved by the limitation of crypto currency emission by time or by volume. This measure may force the regulation effect which decrease risks. Risks must be limited not only in global aspect but also in local one, for example it is external debt. We see the problem of debts in many countries and in order to resolve it we are fixing the generation of another debts in order to pay for previous and it generates problems over some regions and territories. We are seeing many conceptions and opinions regarding the multiplication of local currencies and the limitation of American dollar turnover and other reserve currency in a financial sphere. As a possible financial instrument we can propose the cryptocurrency and they say about the connection of any cryptocurrency to concrete state or territory with regulation law enclosed. In such case they propose to decrease the aggravation in stock and out of stock market. But upon our opinion we can see the difference between electronical money and cryptocurrency. By our opinion the electronic money are very much closer connected with the fiat emission than crypto. We know that such kind of money like WebMoney, Yandex money and so on are issuing with a close demand to fiat money and can't be introduced in turnover without that. To be generated in the separate structure using independent algorithm they are unable. In such mode the electronic money are the digital shadow of fiat money. Absolutely different situation we can see in analyses of crypto. Cryptocurrency is absolutely autonomous from external emission efforts and existing only upon their own way of emission and they can to arrange the turnover function and reservation function within the frame of virtual turnover. The other difference connected with their own protocol of payments which strongly and correctly construct transaction made and save it unchangeable after validation. So the blockchain technology in connection with the possibility of emission in virtual turnover is doing new possibilities for the automatic functioning as a whole. So we can raise the question about existing the crypto only in virtual and emission by program facilities. In this case the role of programming and administration of the net is having the most importance and the management of risk can be approved by mathematical and program methods of parameters done. But up two now we have no concrete answer is it possible the whole autonomous functioning by the reason of the fact that fiat emission is privileged procedure among all other possibilities when we say about state and state regulation of economical mechanism and money regulation. In analyzing the bitcoin and other crypto it is impossible to forget the highest valuation of that. In highest valuation we can trace plus and mines like in many other economical aspects. As plus we can consider the highest potential as a financial instrument in the market. Due to highest valuation it is possible to realize some hedge strategies in the market depending from the position of playing on downtrend or raising trend. In the other hand valuation is difficult for other process connected with using crypto as the facility of payment and accumulation. We see some dualistic approach in crypto characteristics and to be balanced by both parts reflected above in medium and long period of time. For the moment it is unable to achieve the visible stabilization of crypto rates and so the widening use the crypto as payment facility also is in question. The partial compensation of such situation we can consider the highest divide of crypto and bitcoin in particular which sensibly easing the payment operations and clearings, including reaction due to rate fluctuation. Thus this situation is not helping to crypto on the way of overloading the 5% barrier of world turnover. In addition to above we have an additional risk connected with bitcoin being the uncontrolled emission from the hand of state. From the very beginning bitcoin and crypto was invented as an instrument fully independent from state. This independence is due to decentralized emission where the receiving the crypto was only by two ways being mining and transaction of buy and sell crypto from one hand by owner of food stuffs or other goods and in other hand was the owner of crypto. The access to the electronic wallet is strictly prohibited and only the owner of it can fulfil the transaction. In accordance with the agreement the goods may be delivered against payment yet done or made in advance with after payment in crypto. Thus we say that risks of changing rates in crypto have taken place and losses for buyer or seller. We see also risks so named as merged and also risks of no delivery and claims to refund payments in bitcoins and refund goods. When we are doing mining bitcoins are not running due to delivery or not delivery of goods but connected with making algorithm of blockchain. Mining is free from risks connected with goods turnover and connected with risks of making the mathematical decision of task. The complication and difficulty is running from finding the correct decision and programming realization. The additional matter is that of finding the resolution earlier than other people. The factor of diminution of risk is the validation of transaction. Such validation is approving by all participants and miner is receiving the duly gained fees. In this case the receiving of crypto connected with low risk in regarding with that of buying or selling goods. In several countries such as USA, Grate Britain and especially Japan some payments for goods in crypto are possible. In other hand in such countries like China the using bitcoin as stock asset is prohibited due to the risk of money laundering. The same time China is very active for procedure of mining and according to the different opinions this volume is over 80 per cents of common volume. Analyzing risks of crypto turnover it is necessary to research trends connected with legislation to be applied. From one hand regulation has a positive effect over currency turnover and give a possibility to monitor emission processes. From the other hand the state control is focused on fiscal part of crypto turnover. In this matter we have some practice in several countries being the role of equity belonging to crypto and in such case it is eligible for taxation. But here we have also risks connected with the discrepancy of status between different countries upon taxation procedure. In addition to above the state regulation is the additional pressure to the currency turnover. This also can be reflected on speed of clearing and payments and also migration of capital from one sector to another. In this case the rational way of crypto development is the way of Japan which is connected with official balancing the currency rate and the rate of crypto in stock. The same time it is necessary to say that there were some facts limiting the status of fully done facility of payment belonging to crypto due to the absence of effectively functioning state emission overview and absence of guarantees of stable functioning non cash payments. But crypto is approved as payment facility and receiving the status of digital asset. The stimulating factor arranging risks is now the freeze in taxation for bitcoin and Ethereum.

3 Discussion and Results

It is necessary to underline that cryptocurrencies are the undividable part of digital economy and such as platform in some matter. This is the duly approved fact cryptocurrencies are composed from all risks connected with this definition. We may say that in cryptocurrency all banks risks are also to be reflected depending from when and how crypto will be involved in bank operations. Today in practice only the process of input and output of funds from stock is connected with bank operations activity. The clearing and payment operations between wallets is fulfilling without operations of banks. Above we mentioned about differences between electronic money and crypto. This explained the situation when banks were involved in electronic system money at once from the beginning of 1990th upon such conception was elaborated. Technically to construct the facility of payment from electronic money is the task more easier than another like crypto. This also one risk to be identifies being the risk of disproportion of technologies of internet between countries and regions. Crypto platforms has high grade of isolation and may have a possibility to make an obstacles for concurrence by means of programming. Also the development of technologies may decrease the cost component.

The spectacular group of risk of crypto is the risk of money laundering receiving by criminal operations. Here we can say about several aspects. Beginning from the second part of 70th and 80th years of last century it was formed the enough stable and functioning system of money laundering protection which limited the invasion of bad money into banking sector. This system was based on the principal of the obligation from the banks part to monitor all baking operations upon criterion specified and put this information to the special register and data were launched to the special organization. In turn bank has a power to postpone or suspend the operation prior the customer bring all relevant information upon request done. The customer have to explain the reasons of the transaction from where and to whom will be the payment and confirm the origin of funds. This is a specially important thing when the customer wants to deposit money in cash, or when other operations have difficulties about identification of money. Of course the customer may refuse from any explanations and negotiations but in this case bank will send this information to the proper institution. For the moment this system is controlling by FATF in the international arena and working with a proper effect. This effect may differ from country to country but methods and principals of FATF are mandatory for all countries and regions involved. Now in practice all countries of the world more or less are participants of that system. But since electronical money has come and together with crypto the situation is seriously complicated due to the partial lost of a transaction control. When we are saying about partial lost of control we mean the electronical turnover and funds circulating through terminals and cards and going under control of mega regulator being Bank of Russia. In case of cryptocurrency such kind of control is absolutely lost. The reason of such situation is the outdoor character of turnover and this turnover is out of control of state. In case of cryptocurrencies bank has no one facility of control and regulation upon the volume and speed of turnover and that is why there is no control upon legislation or violation of law. The further analyses of this group of risk is showing that system of clearing in crypto may to be easy converted into payment system of shadow money and money laundering as per example of those existing in analogue regime. The most very well known system among experts of such kind is that of Havala may be easy converted to the digital start up of new infrastructure. Such system was invented in the period when banks were no place and had a sensible resource of trust operation without identification documents submitting. In this case no one controlling board will be unable to apply any effective measure for identifying the payer, the amount of transaction, beneficiary of payment and value date. In case of analogue version of this system there were some weak features connected with the region of ordering customer and beneficiary's region, transportation the big amount of money till the point of shipping and so on. When we go with the conversion of analogue system in digital one all above discrepancies are gone to zero. For the additional feature we have to underline that all mentioned discrepancies had no critical significance and were not resulted to the disappearance of such havala. But in the century of digital technologies and blockchain the effect of use will multiply in more that 10 times. The system is giving the full possibility of anonym transactions and ordering customer, the time of transaction takes several seconds and the blockchain is witnessing about the successful result of transaction.

4 Conclusions

In this part it is necessary to pay the attention to such kind of risk which is connected with currency control. The currency control on the territory of Russian federation was done almost immediately after the new state burned named in academic literature like the new era of state. From the very beginning the system monitors financial risks connected with an export and import operations. After the currency monopoly of state was canceled such control was mandatory for arranging transactions with currency financial instruments. For the early days control was fulfills through the internal banking system and through abroad bank system of former Soviet Union. This abroad bank system controlled international payment transactions. After zero years this system of control was reformed and former abroad Soviet banks were reorganized. The Bank of Russia became fully empowered bank for control over all banking system of Russia and he is using the distance system of control through the relevant department. In the middle of zero years was done a relevant liberalization of currency legislation and the list of currency control operations was cut off. The main pressure of this control was for banking organizations being the board of control for that purposes. For this purposes banks were empowered by Central bank of Russia. The main document of currency control was the passport of transaction being of two types: export and import. This document contains all mandatory information about the transaction the agreement being the baseline of transaction and terms and conditions of contract together with currency of payment and amount of the contract. This document must be signed by authorized signatures from the part of the bank and the part of the customer being exporter or importer of goods. On the base of this document bank has a rule to require another documents connected with transactions and explaining one or another step in servicing the contract. Also bank may postpone the transaction in case the customer not fulfilled their requirements. Such system we have for the now days period with some alterations and changes in relevant documents of Central Bank of Russia. When digital technologies and crypto come the situation with that kind of control is seriously changing and is going to zero in fact. There is no measure and definition in present value legislation connected with the payment over the export or import contract in crypto currency. Simultaneously there is no any measure of direct control despite of the fact if it is done as yet or no. In addition to this banks knowing about specific features of banking control and payments in crypto currency banks are unable to control this operations from the point of present valid legislation, but also do not have standard instruments of such control. So, we can say that to hedge such kind of risks while the payments will be in crypto seems to be impossible.

Analyzing credit and operational risks we can say that traditional means of hedge and law regulation it is very difficult to monitor such risks when payments will be done in crypto currency. This is connected with the special character of payments in crypto and the same time with a special character of credit and operational risks in cryptocurrency.

Thus we can make a conclusion about the fact that we have some advantages in crypto currency operations being advantages in speed and confidential character of clearing and transactions. But the same time there are many serios risk which are very difficult to hedge by a standard means applied for traditional currencies and financial instruments. This situation in some manner reduces the scope of payments in crypto. In any way the blockchain technology is very modern and perspective for many fields of business not only directly connected with financial operations and payments as it is.

References

1. Ageev, A.I.: Kryptovaluty, rynki I instituty (Cryptocurrencies, markets and institutes). Econ. Strateg. **1**, 94–107 (2018). https://elibrary.ru/item.asp?id=32522704. Accessed 3 Apr 2019
2. Alexandrov, D.: Bisnes, Organizazia, Strategii (Business. Organization. Strategy), no. 12, pp. 23–25 (2017)
3. Andrushin, S.A.: Otkryty banking, kreditnaya aktivnost, regulirovanie I nadzor (Open banking, credit activity, regulation and control Banking). Bankovskoe delo **6**, 26–34 (2017)
4. Baulin, A.: Blokchain v efire (Blockchain on air). Forbes **11**, 126–127 (2017)
5. Bauer, V.P.: Blokchain kak osnova formirovania dopolnennoy realnosty v zifrovoy ekonomike (The blockchain as a baseline of additional reality in digital economy). Inf. Soc. **3**, 30–40 (2017)
6. Belous, M.: Mechtayt li kriptovalutchiki ob electricheskikh bently (Do the crypto dealers dream about electric Bentley?). PC Mag. **6/8**, 4–5 (2017)
7. Vakhranev, A.V.: Rol bitkoinov v ekonomike I ikh proizvodstvo (Role of bitcoin in economy and mining). Bus. Law **6**, 224–226 (2016)
8. Vedita, E.: Zifrovaya ekonomika privedet k ekonomicheskoy kibersysteme (Digital economy is driving to the economical cyber system/International life), no. 10, pp. 87–102 (2017)
9. Vershbitzkiy, A.: Kryptovalutnaya volniza (The cryptocurrency freedom). Forbes **9**, 136–137 (2017)
10. Gayva, E.: Blockchain zatormozil (Blockchain is stopping). Expert **15**, 46–47 (2017)
11. Genkin, A.S.: Blockchain I unikalnye zennye obyekty (Blockchain and spectacular objects). Insurance **3**, 15–22 (2017)
12. Genkin, A.S.: Kryptotechnologii I kriminalnie riski: est li povod dlia trevogi? (Crypto technologies and criminal risks: is there a reason of danger?). Insurance **5**, 47–55 (2017)
13. Gerr, R.: Virtualnye fantiki (Virtual tips). PC Mag. **10**, 30–31 (2017)
14. Gezman, M.: Razvitie elektronnykh servisov glazami registratora. Blokhchain kak sposob povyshenia dostovernosti elektronnogo dokumenta (The electronical service development from point of view of register. Blockchain as the way of high reliability of electronic document). Bonds Market **2**, 47–49 (2017)
15. Dolzhenkov, A.: Odnim dvizheniem ruki (By one hand movement). Expert **15**, 40–42 (2017)
16. Egorova, M.V.: Kryptovaluty kak novaya realnost (Cryptocurrency as a new reality (review of international and Russian resources)). Int. Econ. **11**, 34–41 (2017)
17. Karzikhiya, A.: Ozyfrovannoe parvo: virtualnost v zakone (Digital law; virtual reality in law). Intellectual Property. Author's and Merged Rules, no. 2. pp. 5–20 (2018)
18. Kochergin, D.A.: Mesto I rol virtualnikh valut v sovremennoy platezhnoy systeme (The place and role of virtual currencies in the modern payment system). Saint Petersburg University Review. Economy **1**, 119–140 (2017)

The Tools for Building the Digital Economy in Russia

Analyzing the Use of the Production Potential in the Russian Federation's Territories During the Transition to the Digital Economy

Anna G. Bezdudnaya[1](✉), Marina A. Gundorova[2], Tatyana M. Gerashchenkova[3], Kirill B. Gerasimov[4], and Denis Y. Fraimovich[2]

[1] FSBEI of Higher Education "Saint-Petersburg State University of Economics" (SPbSUE), St. Petersburg, Russian Federation
dept.kmi@unecon.ru

[2] FSBEI of Higher Education "Vladimir State University named after Alexander and Nikolai Stoletovs" (VlSU), Vladimir, Russian Federation
mg82.82@mail.ru, fdu78@rambler.ru

[3] FSBEI of Higher Education "Bryansk State Technical University" (BSTU), Bryansk, Russian Federation
gerash-tatyana@yandex.ru

[4] FSAEI of Higher Education "Samara National Research University" (SNRU), Samara, Russian Federation
270580@bk.ru

Abstract. The aim of the work is to perform a quantitative assessment of the level of using the production potential in the regions of the Russian Federation over a long period of time. The theoretical base of the studied problem is considered. The author's criteria for diagnosing the degree of developing advanced technologies are presented. Approbating the calculations was made on the basis of data from the official statistical reports of the Federal State Statistics Service. Tabular and graphical analysis methods are implemented. The territories that demonstrate high and extremely low results of implementing production capabilities and using high-tech technological solutions are identified. The scattergram of production indicators by region in the context of a separate federal district is presented. The tendency of strengthening interterritorial disproportions is substantiated. The possible causes of problems existing in socio-economic systems are disclosed. The general conclusions about the real level of engaging advanced technologies at the enterprises of the Russian Federation in the dynamic and spatial perspectives are formulated. The prerequisites for developing further research in the field of monitoring regional imbalances, working out additional combined indicators and studying the effectiveness of implementing production potential in the territories while transiting the national economic system to the course of modernization are identified.

The above methodology, the relative values used in it and graphical tools allow assessing, on a qualitative basis, the current situation and making a forecast for the future, taking into account the current socio-economic, production and innovation conditions. Also these factors allow justifying the introduction of appropriate corrections to regional strategic programs. The presented provisions for analyzing labor productivity indicators and application

E. G. Popkova and B. S. Sergi (Eds.): ISC 2019, LNNS 87, pp. 185–192, 2020.
https://doi.org/10.1007/978-3-030-29586-8_22

level of advanced production technologies in the subjects of the Federation can be supplemented and adapted to specific research objectives and used in the activities of scientific and educational organizations, as well as in the practical work of specialized administration departments of various territorial levels.

Keywords: Production potential · Advanced technologies · Regions of the Russian Federation · Processing sector

JEL Code: O 14 · O 33

1 Introduction

Issues of developing and intensifying the introduction of advanced, including digital technologies in the domestic industry in recent times are the most acute. In general, the existing problems are reduced to obtaining objective assessments of the current situation, which are often identified with the degree of using the existing production potential. The aggregation of results on this basis provides the possibility of developing an appropriate action plan and justifying accelerated involvement of high-tech resources and digital reserves of production management in the regions.

Without new automated systems, it is no longer possible to imagine functioning financial and banking sector, medicine, education, law enforcement, etc. But their use in material production poses much more problems to solve. First of all, the difficulties are caused by the need to make significant investments in the re-equipment of the technology park, moreover, by the high capital intensity of the products and the relatively low level of return per unit of invested capital. Problems arise in connection with the substantial duration of working capital return. Nevertheless, the programs of modernization and re-equipment of domestic industry and the transition of manufacturing industries to a new level should be implemented even faster than in the developed countries. It should be so, since without accelerating the dynamics of the corresponding processes, it is impossible to assert about the prospects for improving the population welfare and reducing the gap from the countries with higher GDP per capita.

2 Methodology

Many scientific papers have been devoted to the study of developing economic potential of countries and regions (Idrisov et al. 2018; Glazyrina and Lavlinskii 2018; Aganbegyan 2017; Ovchinnikov and Ketova 2016; Loseva et al. 2018; Fraimovich et al. 2017). At the same time, as a part of the research, monitoring of the situation and ongoing reproduction processes is being carried out at the level of expert assessments, often not connected with any numerical characteristics. The other part is related to the quantitative analysis and the corresponding interpretation of the obtained results, which is due to the need for a more verified substantiation of the managerial and general business decisions taken in the area of forming new strategic territorial programs, or targeted adjustment of existing projects.

According to the definition, the material and production potential of an industry is understood as a generalizing quantitative and qualitative characteristic of the availability and use of a set of substantiated (materialized) resources in industrial production in specific conditions of place and time to achieve the strategic goals of industrial development (Komkov et al. 2012).

However, it seems appropriate to clarify that in modern conditions the production potential depends largely on scientific and technical (innovational, project-oriented, informational, organizational, etc.) factors that have an intangible expression. Therefore, while further consideration of the issue, it is necessary to use an integrated approach that takes into account both materialized and intangible resources of industrial activity.

According to S.D. Bodrunov's findings, today new methods of organizing production and management are mastered in the world. Above all, remote production management is mastered, giving the production a new quality. Radically new technologies are being introduced: nanotechnologies, 3D printers, etc. The shape of the traditional manufacturing industry is changing fundamentally due to the intensive spread of additive processes (creating a product layer by layer, unlike the "subtracting" production technologies, such as cutting, grinding, cutting-off material from the workpiece, etc.). There is a change in the nature of industrial labor as a result of introducing knowledge-intensive labor functions (controlling, automated machining, using robots, etc.).

In addition, there is a change in the principles of obtaining knowledge and skills necessary in the conditions of the new industrial production ("gadgetization, "chipping", "internetization", using virtual technology and "added" reality) (Bodrunov 2016).

To identify the level of efficiency in functioning the domestic economic model, E.B. Lenchuk proposes to compare the dynamics of high-tech exports of the Russian Federation and China (CPR) over the past quarter of a century. If in the 90s of the twentieth century, exports of high-tech products of the Russian Federation amounted to about 2.2 billion dollars, and China's export amounted to 4.3 billion dollars, then by 2015 the Russian figure had increased by 4.5 times (to 9,7 billion dollars), and the Chinese one had increased 130 times (up to 554.3 billion dollars).

Of course, today Russia cannot fully follow the Chinese path, but it would be necessary to focus on certain elements of this country's experience. At the same time, a strong state activity in solving the problems of innovative development acts as a support base, without which it cannot be done. This is confirmed by other countries' experience that have achieved significant success and entered the trajectory of accelerated high-tech growth through the state's effective participation and stimulation (Lenchuk 2018).

It is impossible not to accept the scientists' comment from Mordovia State University named after N. Ogarev that today we should talk about a qualitatively new type of investment that meets the criteria of the neo-industrial paradigm. Such investments are long-term in developing intellectual capital and innovative spheres of the national economy, ensuring re-industrialization and the formation of knowledge-intensive, high-tech and digital capacities, growth of social labor productivity, effective use of human potential (Kormishkina and Koloskov 2017).

In this work, in order to identify the degree of developing the production potential during the period of the industry transfer to the modernization course announced in the country, it is proposed to analyze the dynamics of two indicators, based on official data of Federal State Statistics Service[1]. These indicators are labor productivity in the manufacturing sector (P) and the level of using advanced production technologies (T), in the territories of the Russian Federation for the 18-year period (from 2000 to 2017). The first indicator is calculated by the ratio of turnover to the number of personnel employed in the field in question, and the second one is calculated by comparing the number of advanced production technologies used and the number of existing enterprises and organizations. The selected criteria are relative and allow us to fairly objectively judge the processes occurring in a particular region, the transforming trends of new knowledge and the level of developing advanced, including digital technologies (Bezdudnaya et al. 2018). For a more representative mapping, the second indicator is taken per 1000 economic entities in a specific territory.

The objects of research are regional systems of four federal districts: Central (CFD), Volga (Volga Federal District), Siberian (Siberian Federal District) and Far Eastern (FEFD). The first two districts play a key role in developing socio-economic potential of the state, consistently long forming together about 50% of GDP. The other two studied territories, which are significantly remote from the centre of Russia and are located in the eastern part of the country provide only about 15% of GDP despite their impressive geographical extent.

3 Results

The dynamics of labor productivity in the processing industries in the analyzed districts, as well as in the average in the Russian Federation, is presented in Fig. 1.

The given schedule allows asserting about the output growth in the considered territories, but the growth rate is different. The leading positions in this parameter have been occupied by the Central Federal District since 2008. At the same time, the average Russian productivity values (noted in the diagram) do not overlap in Volga Federal District, Siberian Federal District and FEFD practically from the same period. It should be said about frankly low level of production in the Far Eastern Federal District, which, since 2005, has been almost twice as low as the corresponding average figure for Central Russia. The revealed situation related to deteriorating human and technological potential in the country's eastern regions naturally aggravates the already serious socio-economic imbalances and once again confirms the need for the earliest possible solution of the existing problems at the state level.

Speaking about the degree of distributing advanced technologies in the studied territorial systems, one can, in fact, pay attention to the enormous superiority of the Volga Federal District in this parameter (see Fig. 2). The extremely productive susceptibility of Volga enterprises to innovations is approximately two times ahead of similar indicators of activity demonstrated in other districts, as well as the average Russian level throughout the entire time interval. At the same time, in depressed (as they are often called) FEFD regions, the degree of mastering new technologies is not

[1] Federal State Statistics Service: [site]. URL: http://www.gks.ru/wps/wcm/connect/rosstat_main/rosstat/ru/statistics/publications/catalog/ (access date 08.02.2019).

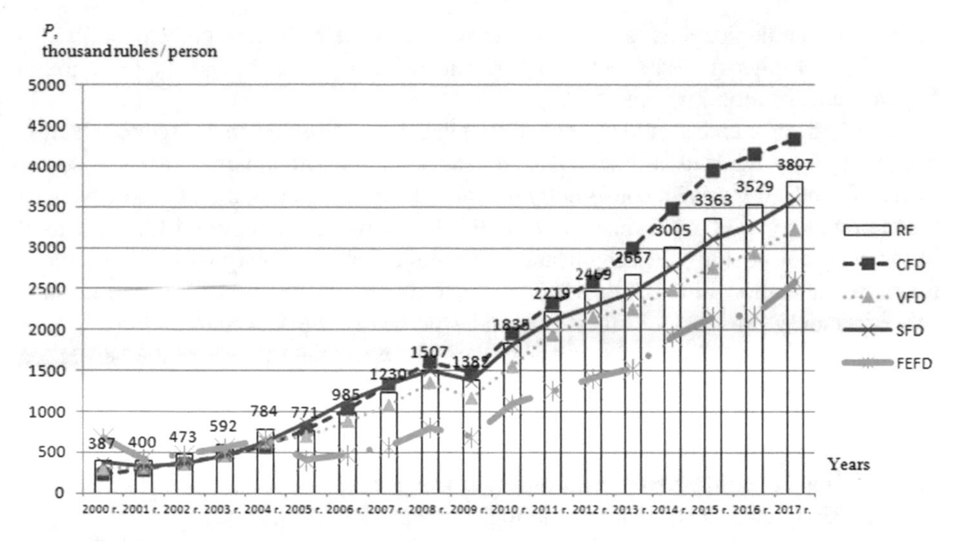

Fig. 1. Dynamics of labor productivity in the processing industries (thousand rubles/person) in the territories of the Russian Federation for 2000–2017.

significantly inferior to indicators in the Russian Federation, Central Federal District and Siberian Federal District). So, for example, in 2017, there were in average 38 used advanced technologies per 1000 economic entities of the Far East, in the Russian Federation there were 53, in the Central Federal District there were 45 and in Siberian Federal District this figure was 47. Against this background, a significant "failure" in the level of labor productivity in the FEFD, identified in the initial part of the analysis, may be the result of insufficiently established logistics and the lack of adequate

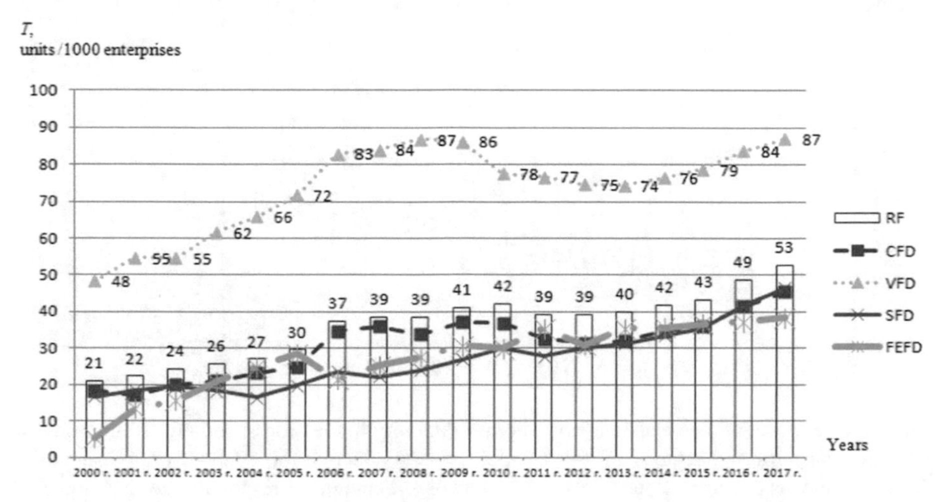

Fig. 2. Dynamics of the level of using advanced production technologies (units/1000 enterprises) in the territories of the Russian Federation for 2000–2017

industrial communications, as well as the ineffective use of new knowledge, which necessitates further development of regional infrastructure and human capital through targeted and volume investments.

The "mirror", and a rather paradoxical picture, is observed in Moscow, which is traditionally recognized as the absolute leader in the predominant part of socio-economic indicators. With consistently low and inexpressive volume of used advanced technologies per 1000 enterprises (in 2017 this figure was 21 units, amid 122 units used in the Magadan region), the capital surpasses with a confident margin the other regions under consideration in terms of labor productivity (in 2017 it was 7102.05 thousand rubles/person versus 1313.82 thousand rubles/person in the Magadan region).

But significant differentiation arises not only between the subjects of the Federation located in different parts of the country, but also within the districts, and, in particular, in Central Russia. The diagram of the output value range in the manufacturing industry by the CFD regions (Fig. 3) indicates the growing imbalances that act as a threatening factor in implementing government plans for the transition to a new growth trajectory and as a significant obstacle to developing the reproduction potential.

For example, the catastrophically low value of labor productivity in 2017 in the Ivanovo region (which is traditionally famous for its textile mills) is 1273.5 thousand rubles/person, which is almost 40% less than even the highest result achieved in the

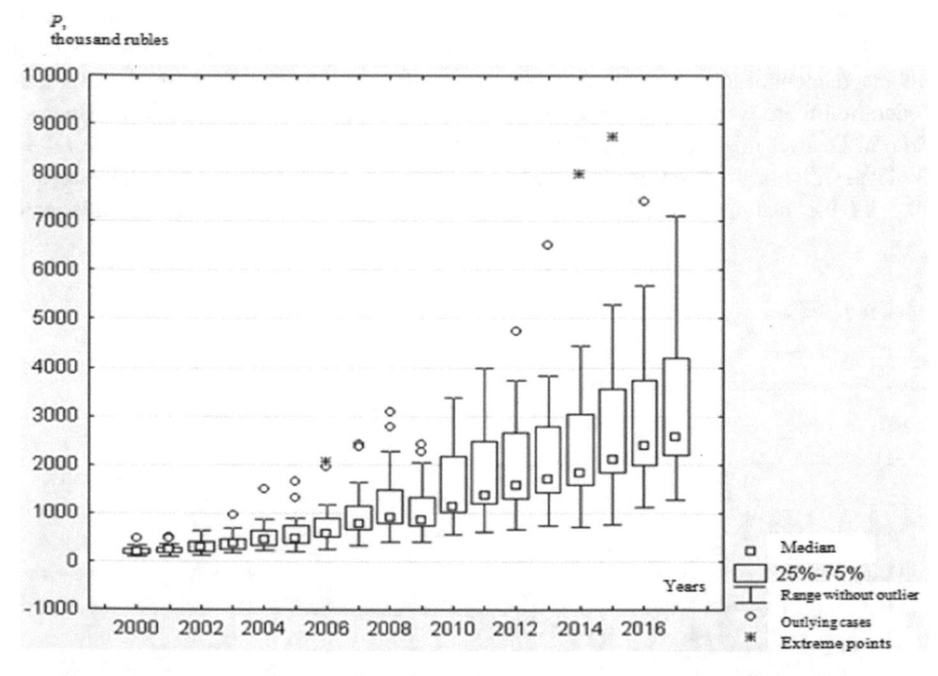

Fig. 3. A diagram of the value range of labor productivity in the processing industries (thousand rubles/person) for the regions of the Central Federal District for 2000–2017. (designation: "Median" means average output; "25%–75%" means a rectangle corresponding to 25% and 75% quartiles; "Range without outlier" means scope of output values without taking into account outlying cases in observations; "Outlying cases" are points corresponding to outliers; "Extreme points" are points corresponding to extreme values in the sample).

real sector of the FEFD economy in the Chukotka Autonomous Region, where the average output is recorded in the amount of 2046 thousand rubles/person.

The revealed imbalances are being overcome only due to the effective federal policy of investment and resource support for the processes of accelerated development of new knowledge, as well as the creation of equal living conditions in the regions of the Federation.

4 Conclusions

As a result, it should be noted that developing the production potential in the territories under consideration is very chaotic and not systematic, which does not allow us to state about the sustainability and balance of modernization processes and the most effective use of new science-based knowledge and digital technologies in the country.

The given methodology, the relative values used in it and the graphic tools for assessing the dynamics and territorial disproportions in mastering the production and modernization potential make it possible to assess the current situation on a qualitative basis, to make a forecast for the future, taking into account the current socio-economic, production and innovation conditions, and to justify making appropriate adjustments to regional strategic programs. The presented provisions for analyzing labor productivity indicators and the application level of advanced production technologies in the subjects of the Federation can be supplemented and adapted to specific research tasks, expanded with other combinatorial variables and used in the activities of scientific and educational organizations and in the practical work of specialized administration departments of different territorial level.

References

Idrisov, G.I., Knyaginin, V.N., Kudrin, A.L., Rozhkova, E.S.: New technological revolution: challenges and opportunities for Russia. Voprosy Ekonomiki **4**, 5–25 (2018)

Glazyrina, I.P., Lavlinskii, S.M.: Transaction costs and problems in the development of the mineral and raw-material base of the resource region. J. NEA (New Econ. Assoc.) **2**(38), 121–143 (2018)

Aganbegyan, A.G.: Investments in fixed assets and human capital: two interconnected drivers of socioeconomic growth. Stud. Russ. Econ. Dev. **4**, 17–20 (2017)

Ovchinnikov, V.N., Ketova, N.P.: The system-supplementing effect of the interaction between innovative capacity and institutional environment factors of a region. Econ. Reg. **12**(2), 537–546 (2016)

Loseva, O.V., Abdikeev, N.M., Didenko, A.S.: Ranking and clustering of regions by the level of efficiency of scientific and innovative activity. Sci. Works Free Econ. Soc. Russ. **211**(3), 146–161 (2018)

Fraimovich, D.Yu., Gundorova, M.A., Mishchenko, Z.V.: Diagnostics of the results of development of innovation-resource potential in the Federal Districts of the Russian Federation. Public Admin. **19**(4(108)), 49–54 (2018)

Bezdudnaya, A.G., Fraimovich, D.Yu., Gundorova, M.A.: Assessing the spread of advanced production technologies and the level of labor productivity in the regions of the Russian Federation. In: Coll. Scientific Works on the Basis of the Intern. Scientific-Practical Conf. "Management of Innovative and Investment Processes of Forming and Developing Industrial Enterprises in the Digital Economy", St. Petersburg, Russia, Publ. house: St. Petersburg State University of Economics, pp. 24–30 (2018)

Komkov, N.I., Selin, V.S., Zuckerman, V.A.: Innovation Economy: Encyclopedic Dictionary. Ivanter, V.V., Suslov, V.I.: Scientific Advisors. INEF RAS. - M.: MAKS Press Publ., 544 p (2012)

Bodrunov, S.: New industrial society of the second generation: people, production, development, society and economics, no. 9, pp. 5–21 (2016)

Lenchuk, E.B.: Shifting towards innovative development model in Russia: overcoming committed mistakes. Vestnik Instituta Ekonomiki RAS **1**, 27–39 (2018)

Kormishkina, L.A., Koloskov, D.A.: Innovation approaches to the formation of investment policy tools from the perspective of a neo-industrial economic development paradigm. Econ. Soc. Changes Facts Trends Forecast **10**(6), 218–233 (2017)

Federal State Statistics Service. http://www.gks.ru/wps/wcm/connect/rosstat_main/rosstat/ru/statistics/publications/catalog/. Accessed 08 Feb 2019

Trends of Scientific and Technical Development of Agriculture in Russia

Gilyan V. Fedotova[1,2(✉)], Ivan F. Gorlov[1,2], Alexandra V. Glushchenko[2], Marina I. Slozhenkina[2], and Arkady K. Natyrov[3]

[1] Volgograd State Technical University, Volgograd, Russian Federation
g_evgeeva@mail.ru

[2] Volga Region Research Institute of Manufacture and Processing of Meat-and-milk Production, Volgograd, Russian Federation
niimmp@mail.ru

[3] Kalmyk State University, Elista, Republic of Kalmykia, Russian Federation
kafedraTehnolod@yandex.ru

Abstract. Purpose: The purpose of this chapter is to assess the current state of Russia's agricultural sectors in the context of restrictive sanctions and food embargo. The attention has been focused on the need to intensify the production of agricultural raw materials for domestic consumption and export to the world market; and the low efficiency of the main branches of the agroindustrial complex in Russia has been noted. The experience of developed countries evidences that the implementation of advanced information technologies into traditional business processes makes it possible to increase the profitability of agricultural sectors. The development of electronic technologies, implementation of automated data collection devices and processing of the results obtained contribute to the implementation of the Industry 4.0 concept in the transition to an information society.

In order to assess the prospects of the transformation for agriculture in Russia, the article has analyzed the Federal Scientific and Technical Program for the Development of Agriculture for 2017–2025 developed and being implemented.

Methodology: The study presented in the article was conducted using the generalization and analog methods, statistical and graphical analysis, vertical and horizontal analysis and methods of data comparison and collation.

Results: Evaluation of the implemented program of scientific and technological development of agriculture in Russia has shown that today 3 activities are being financed, i.e., the creation of scientific and technical results and products; implementation of scientific and technical results and products into production; and commoditization of scientific and technological results and products. Based on these areas, the main trends of scientific and technological development of agriculture have been identified. They are "smart farm," "smart greenhouse" and "smart field." The introduction of these trends into the practice of agricultural organizations will enable meeting the basic needs of the domestic food market and increasing the volume of exported agricultural products.

E. G. Popkova and B. S. Sergi (Eds.): ISC 2019, LNNS 87, pp. 193–200, 2020.
https://doi.org/10.1007/978-3-030-29586-8_23

Recommendations: The trends identified in studying the sectoral technological development of the national economy provide an opportunity to increase profitability and intensify the initiative to create "smart enterprises" in the agricultural sector.

Keywords: Sectors of Agro-Industrial Complex (AIC) · Information technology · State-run program · Indicators

JEL Code: O13 · O32 · Q13 · P52

1 Introduction

Modern Russian agriculture in the conditions of the formed geopolitical confrontations of the world powers and application of restrictive food sanctions requires government support for stable and dynamic development. The experience of foreign countries (the USA, Canada, countries of the EU, Latin America and Asia) proves that agriculture is a high-risk industry that heavily depends on climatic conditions and seasonality of production. A set of state support instruments includes preferential loans, direct subsidies, protection of the domestic food market and crop and animal insurance programs. Recently, digital technologies have been widely used in the agro-industrial sector of many countries.

For instance, the EU countries intensively use the Information and Communication Technologies (ICT) and, in fact, are forming an information society. In the society of that kind, any farmer can connect to the Internet from anywhere in the field using wireless communication networks and sensors for animals and arable land. The leader of innovation, Japan has developed a system of unmanned vehicles provided with artificial intelligence to monitor arable land and protect against the attacks of wild animals. In the US, sensors have been developed to monitor the health status of pigs. There are many similar examples that prove the involvement of the agro-industrial sector into the process of global information restructuring.

In order to strengthen Russia's agrarian positions and improve the competitiveness of produced agricultural raw materials on the world market, Russia needs to make a scientific search and practical implementation of current information tools. To this purpose, in 2016, the Strategy for the Scientific and Technological Development of the Russian Federation that defined the directions of scientific and technological development for the next 10–15 years was developed and is being implemented into practice.

The assessment of the target indicators of the implementation of this strategy demonstrates the main trends in the development of the agro-industrial sector for the period up to 2025, i.e., the protection of plants and animals, storage and efficient processing of agricultural raw materials and creation of high-quality and functional foods. Meanwhile, the conditions for the creation of scientific and technical developments in agriculture and their prompt transfer into production and economic circulation have not been formed.

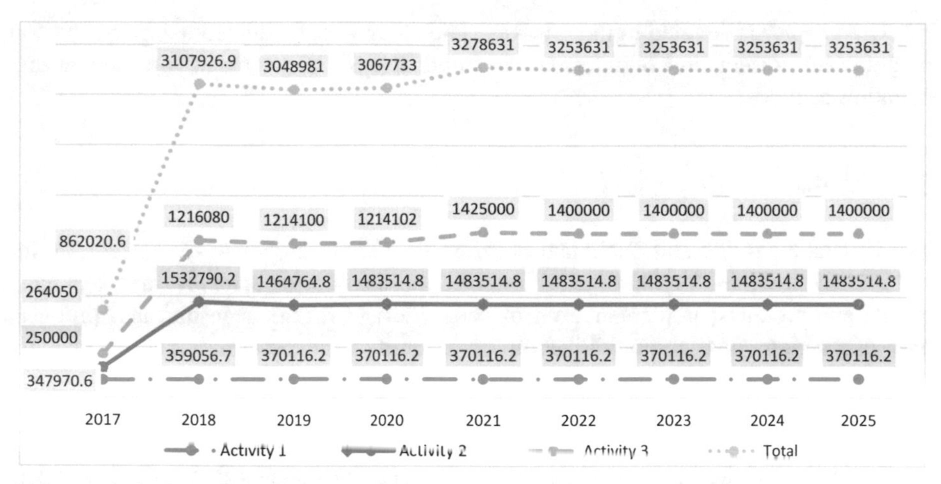

The source is the Decree of the Government of the Russian Federation of 08.25.2017 No. 996 "On approval of the Federal Scientific and Technical Program for the Development of Agriculture for 2017-2025." Available at: http://mcx.ru/upload/iblock/1e9/1e97bd2630e613804cf5ef016063bd60.pdf (accessed 02/03/2019)

Fig. 1. Financial support of the Federal scientific and technical program of agricultural development for 2017–2025, thousand RUB

The Federal Scientific and Technical Program for the Development of Agriculture for 2017–2025 provides for the implementation and financing of three activities (Fig. 1):

1. Creation of scientific and technical results and products;
2. Implementation of scientific and technical results and products into production; and
3. Commoditization of scientific and technological results and products.

2 Materials and Method

Theoretical and applied issues of digitization of agriculture are highlighted in numerous works by Vartanova and Drobot (2018), Gorlov et al. (2018a, b), Ognivtsev (2018), Glushchenko et al. (2018a, b), Fedotova et al. (2018), Gorlov et al. (2018), and Fedotova (2018).

The features of the development and application of information and communication technologies in the economy are reflected in the Strategy for the Scientific and Technological Development of the Russian Federation and the Federal Scientific and Technical Program for the Development of Agriculture for 2017–2025.

The accumulated experience of modern trends in technological development of various sectors of the national economy does not give a detailed idea of the implementation of information and communication technologies into agriculture. Therefore, the development of this area seems poorly understood and promising.

This chapter is devoted to these issues. The study was conducted using the method of graphical presentation of information, trend analysis and comparison, analog and collation methods.

3 Results

The Federal Scientific and Technical Program for the Development of Agriculture for 2017–2025 provides for three activities, i.e., the creation of scientific and technical results and products; implementation of scientific and technical results and products into production; and commoditization of scientific and technological results and products.

The activities listed are included in both scientific and technical projects and represent the factors that activate innovative areas of development in agriculture.

Estimation of the funding of the program's activities shows that until 2022, it is planned to increase funding and stabilize it in 2023. The main financial sources of the program are to be budgetary funds (state and regions of the Russian Federation) and extra-budgetary sources.

Table 1. Target indicators of the Federal Scientific and Technical Program for the Development of Agriculture for 2017-2025

	2017	2018	2019	2020	2021	2022	2023	2024	2025
Increase of innovative activity in agriculture, %			2	3	5	10	15	20	30
Attracting investment in agriculture, ths. RUB	870130	3115050	3056610	3064512	3273560	3250320	3220040	3175660	3123330
Increasing the infrastructure level of the AIC,%		6	8	12	14	16	18	20	25
Providing the industry with personnel in promising specialties, %				10	20	50	65	80	100

The source is the Decree of the Government of the Russian Federation of 08.25.2017 No. 996 "On approval of the Federal Scientific and Technical Program for the Development of Agriculture for 2017-2025." Available at: http://mcx.ru/upload/iblock/1e9/1e97bd2630e613804cf5ef016063bd60.pdf (accessed 02/03/2019)

The target indicators of the program presented in Table 1 reflect the low innovative activity in agriculture. By 2025, it is planned to increase the activity by 30% and infrastructure facilities by 25%. According to our estimates, these figures are not high, since abroad, the implementation of the Internet into the public life of the population and into the economy sectors, including agriculture, has actually completed for 90%.

Over the period mentioned, the volume of private investment is also planned to increase by 358% from 870130 thousand RUB in 2017 up to 3123330 thousand RUB in 2025. Due to such an increase in investment over the period planned, it is possible to provide the agro-industrial complex with infrastructure facilities for 25% in 2025. Currently, in 2019, the availability of these facilities is 8%, which negatively affects the development of the agricultural chain and establishment of partnerships between producers of agricultural raw materials, its processors and consumers.

The fourth indicator—the provision of industry with professional personnel—aims the educational institutions to train specialists in the field of information and communication technologies, that is, IT agronomists and IT zootechnicians, who are able to manage "smart production" systems. In terms of this indicator, the prospects are quite optimistic, since by 2025, the industry is planned to be provided with the necessary specialists for 100%. According to the program, 55 thousand people, working in the AIC, will receive training and develop their competencies in the digital economy. Russian agrarian universities have included training specialists in the discipline related to digital technology in their curricula. So, since September 2018, the Timiryazev Russian State Agrarian University introduced the discipline "Digital Agriculture" into the educational process.

The listed indicators of scientific and technological development of agriculture represent the main trends of the future digital agriculture platform that is envisaged in the departmental draft "Digital Agriculture" developed by the Ministry of Agriculture of Russia and submitted to the Government of the Russian Federation for approval in November 2018.

Table 2. Target indicators of the draft program "Digital Agriculture"

Indicator	2018	2021	2024
Share of AIC enterprises using ICT[a]	less than 1%	20%	60%
Share of ICT coverage of agricultural land	less than 10%	30%	70%
Share of agricultural enterprises equipped with data recorders that transmit data to receive electronic subsidies	less than 10%	50%	100%
Amount of products sold on electronic platforms	less than 10%	50%	100%
Number of private meteorological stations on agricultural land	less than 1 Mio	3 Mio	7 Mio
Number of agricultural goods transported within the framework of the EEU (EAEU) connected to the transport and logistics platform	less than 10%	50%	80%
Export	$ 25 billion	$ 30 billion	$ 50 billion
% of jobs related to ICT	<1%	8%	20%

[a]ICT is information and communication technology

The source is Digital Agriculture Project. Available at: http://mcxac.ru/upload/medialibrary/04c/04cf3968669675d0b9ecc106ad04a1a7.pdf (accessed 03/03/2019).

The indicators of the target development of the industry for the period up to 2024 presented in Table 2 demonstrate a rather dynamic development. The Ministry of Agriculture of the Russian Federation plans not only a massive introduction of information technologies into the work of agricultural enterprises by 2024 in the amount of 60% of the total number of farms, but also formation of a trading system for agricultural products, a logistics system and transportation of goods and raw materials. The volume of Russian exports of agricultural products should have increased 2 times by 2024 compared with 2018 due to an increase in the overall productivity of the industry.

According to this draft, 3 main development trends are outlined (Fig. 2). The trends presented reflect the main key stages of the ICT implementation in the Russian agricultural sector.

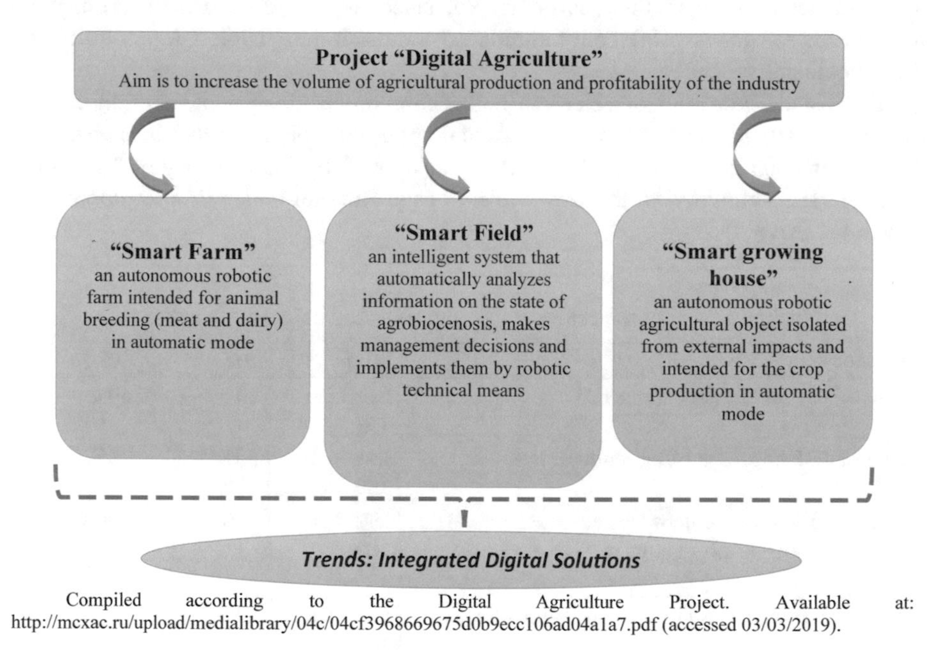

Compiled according to the Digital Agriculture Project. Available at: http://mcxac.ru/upload/medialibrary/04c/04cf3968669675d0b9ecc106ad04a1a7.pdf (accessed 03/03/2019).

Fig. 2. Trends: digital integrated solutions of the Digital Agriculture Platform

The submitted departmental draft of the Ministry of Agriculture of the Russian Federation aimed at increasing the indicator of awareness of animals and equipment available in agricultural enterprises through the creation of an intelligent digital platform "Digital Agriculture". The feature of the project is in focusing the efforts of public authorities and private businesses on the modernization of actual agricultural enterprises through the implementation of the most advanced information tools and intelligent solutions.

4 Conclusions

The current practice of applying digital technologies in individual large agricultural holdings in Russia has found the most popular solutions. They are

(1) Industrial Internet of Things (IIoT) that, according to predictions of Price waterhouse Coopers (PwC), will save about 469 billion RUB by 2025. As expected, in 2019, 30% of Russian agricultural enterprises are to use the Internet of Things (IoT) technology and predict sowing with the help of drones. Such a scenario is provided by the "roadmap" for the implementation of high technologies into the agro-industrial complex.
(2) Robotics. According to forecasts of Tractica, the volume of the market of agro-robots will reach $ 74.1 billion by 2024. The production of agricultural robots will increase during this time almost 19 times to 594 thousand units of equipment. The leader in the robotics implementation is the dairy industry, i.e., systems of feed delivery, cleaning and milking as well as livestock grazing robots.
(3) Virtual Reality (VR). VR-tools for staff training make it possible not only immerse in a virtual environment and work out certain skills, but also reduce the time to study training materials 2.5 times.
(4) Big data analytics and forecasting (Big Data technology). This is a series of approaches, tools and methods for processing structured and unstructured data of huge volumes and diversity. In smart farming, Big Data is used to provide a predictive understanding of farming operations, making on-line decisions and reorganizing business processes for fundamentally new business models.

The informatization tools listed are not exhaustive, since the implementation of larger projects requires investment resources that most Russian agricultural enterprises do not have. In this conditions, the state should support agricultural producers and processors; only in this interaction, technological breakthrough is possible to be made, and Russia is able to become a leader in world agricultural markets.

Acknowledgments. The work was carried out within the framework of the grant of the Russian National Science Foundation No. 15-16-10000, the GNU NIIMMP "Development and scientific substantiation of new approaches to the production of livestock raw materials and increasing the biological value of socially significant products based on modern biotechnological and molecular genetic methods."

References

Decree of President of the RF of May 7, 2018 No. 204: On the national goals and strategic objectives of the development of the Russian Federation for the period up to 2024. ATP Garant. http://base.garant.ru/71937200/#friends. Accessed 22 Feb 2019

Decree of the Government of the Russian Federation of 08.25.2017 No. 996: On approval of the Federal Scientific and Technical Program for the Development of Agriculture for 2017–2025. http://mcx.ru/upload/iblock/1e9/1e97bd2630e613804cf5ef016063bd60.pdf. Accessed 03 Feb 2019

Vartanova, M.L., Drobot, E.V.: Prospects for digitization of agriculture as a priority direction for import substitution. Econ. Relat. **8**(1), 1–18 (2018). https://doi.org/10.18334/eo.8.1.38881

Glushchenko, A.V., Kucherov, Y.P.: Lowering the integrated agricultural formations credit risk in the conditions of global crisis management. Espacios **39**(12), 11 (2018a)

Glushchenko, A.V., Kucherova, Y.P., Yarkova, I.V.: Ensuring an agricultural holding's economic security by means of sustainable development strategies in terms of global crisis management. Espacios **39**(12), 4 (2018b)

Gorlov, I.F., Fedotova, G.V., Slozhenkina, M.I.: Digital technologies for solving problems of food security. Agrarian Food Innov. **4**(4), 7–15 (2018)

Gorlov, I.F., Fedotova, G.V., Sazonov, S.P., Sergeev, V.N., Yuldashbaev, Yu.A.: Cognitive approach to the study of problems of food security: monograph. FSBEI Volga Region Research Institute of Manufacture and Processing of Meat-and-Milk Production, FSBEI HE Volgograd State Technical University, Volgograd, 167 p (2018a)

The Ministry of Agriculture is developing the draft "Digitization of agriculture." https://ag.RF/index.php?Id=422. Accessed 20 Feb 2019

The project "Digital Agriculture." http://mcxac.ru/upload/medialibrary/04c/04cf396866967-5d0b9ecc106ad04a1a7.pdf. Accessed 03 Mar 2019

Gorlov, I.F., Tretyakova, O.L., Shakhbazova, O.P., Nikolaev, D.V.: Development of an application program for the index evaluation of animal breeding qualities. In: Proceedings of the Nizhnevolzhsky Agrouniversity Complex: Science and Vocational Education, no. 1(49), pp. 176–181 (2018b)

Technologies of "smart" agriculture allow increasing work efficiency and reducing costs. https://www.agroinvestor.ru/technologies/news/28325-vnedrenie-interneta-veshchey-prineset-apk-469-mlrd-rubley/. Accessed 03 Mar 2019

Gorlov, I.F., Fedotova, G.V., Slozhenkina, M.I., Mosolova, N.I., Barmina, T.N.: Digital transformation in agriculture. Agric. Food Innov. **1**(5), 28–35 (2019)

Fedotova, G.V.: Information security threats to digital platforms in the industrial sector [Electronic resource]. In: Shkiotova, S.V., Gordeeva, V.A. (eds.) Integration of Science and Practice as a Mechanism for the Development of the Digital Economy: Collection of Articles. Intern. Scientific-Practical Conf. Dedicated to the 25th Anniversary of the Faculty of Engineering and Economics, Yaroslavl, 18 December 2018. FSBEI HE "Yaroslavl State Technical University", Government of the Yaroslavl Region. Yaroslavl, pp. 374–378. (CD-ROM) (2018)

Fedotova, G.V., Gontar, A.A., Ilyasov, R.Kh.: Digital modernization of economic security systems of credit institutions. In: Tsakayev, F.Kh., et al. (ed.) Development of the Regional Economy in the Conditions of Digitalization: Proceedings of Intern. Scientific-Practical Conf. Dedicated to the 80th Anniversary of the Chechen State University, Grozny, 24–25 September 2018, FSBEI HE Chechen State University. Makhachkala, pp. 268–277 (2018)

Ognivtsev, S.B.: The concept of the digital platform of the agricultural complex. International Agricultural J. **2**(362), 16–22 (2018)

Innovative Technologies in the Semi-smoked Sausage Production

Marina I. Slozhenkina[1,2](✉), Luiza F. Grigoryan[2], Alexander N. Struk[1], Anastasia F. Kruglova[2], and Elena P. Miroshnikova[3]

[1] Volga Region Research Institute of Manufacture and Processing of Meat-and-milk Production, Volgograd, Russian Federation
niimmp@mail.ru

[2] Volgograd State Technical University, Volgograd, Russian Federation
tpp@vstu.ru

[3] Orenburg State University, Orenburg, Russian Federation
post@mail.osu.ru

Abstract. Purpose: This chapter considers the need to find new types of low-cost plant protein. The object of the research study is a by-product of beer brewing. There is experimental data presented that has proved the protein sludge to be a valuable raw material as an additional protein component in the semi-smoked sausage production. The advantages and disadvantages of protein sludge have been determined, and the optimal method for sludge to be added to the stuffing has been proposed.

Methodology: The study was conducted using the methods of graphical representation of information, statistical data analysis, financial analysis, trend analysis and comparison, analog and collation methods.

Results: Experimental batches of a semi-smoked sausage product have been developed; its organoleptic characteristics have been determined; common consumer characteristics of the sausage product were preserved; and more pronounced flavour and aroma were obtained. The physicochemical studies of the samples showed that the protein content in Test sample was higher than in Control one by 2.2%, and fat was lower by 0.8%. Comparison of the amino acid profiles of Control and Test samples showed that the protein sludge could make up the deficiency in some amino acids, which in turn preserved the biological value of the finished product at a high level.

Recommendations: The "Glimalask" food additive combined with the protein sludge has been revealed to not only reduce the bitter taste, but also contribute to improved color formation and increased stability during storage, as the organic acids in the composition had a bacteriostatic effect. Thus, the studies of the protein sludge have shown its prospect and feasibility in the production of meat products.

Keywords: Brewing · Secondary products · Protein sludge · Nutritional value · Amino acid composition · Meat products

JEL Code: 13 · O32 · Q13 · Q16 · P52

E. G. Popkova and B. S. Sergi (Eds.): ISC 2019, LNNS 87, pp. 201–208, 2020.
https://doi.org/10.1007/978-3-030-29586-8_24

1 Introduction

Currently, the main challenge to the meat processing industry is the production of combined products. In this regard, an important task is to meet the demand in the production of protein products from animal and vegetable raw materials, increase the production of traditional types of products, reduce raw material costs during processing, storage and transportation of raw materials, search for new sources of raw materials and improve the technology of its processing. In the conditions of the need to increase the production of meat products, enhance their nutritional and biological value and reduce the cost and selling prices of products, enterprises should involve non-traditional raw materials and proteins of plant origin into the production.

Beer brewing produces waste and secondary products that must be removed. Few enterprises manage to dispose of some of these wastes at relatively low costs, but constantly growing expenses force the enterprises to more and more independently dispose of them.

Beer brewing technology includes the germination of barley grains to produce malt followed by drying and grinding. Next, the mashing process takes place, when the crushed malt is poured with water and stepwise heated; the enzymatic hydrolysis of starch is meanwhile carried out, with sugars being formed. The resulting mash is pumped into the filter tank, where the spent grain and wort are separated; wort is then heated to 100 °C with hops being added, precipitated in a hydrocyclone and cooled before being transferred to fermentation tanks, where diluted yeast culture is added. They convert sugar into alcohol. After the fermentation, beer is incubated and filtered [2].

For each 1,000 tons of finished product, the beer brewing technology produces about 170 tons of spent grain, protein sludge, spent yeast, etc. The bulk is spent grain that makes 80–85%. All of these products contain more than 25% protein and are mainly sold as animal feed supplements.

The main area of spent grain use is the feed and food production. In agriculture, raw spent grain most often serves as a supplement to galactogenic and protein feeds for farm animals and poultry instead of meat-and-bone meal. At present, spent grains based feedstuff and feed supplements have been developed for various types and age groups of farm animals and poultry [3].

Some scientists are known to investigate the prospects of brewing waste being used in various industries [9–11] and for enrichment of feed, in particular fish feed [12, 13]. The use of protein sludge in food is not well understood. This waste is formed when the wort is clarified in settling hydrocyclone apparatuses or separators. 2–3 kg of protein sludge with a humidity of 80% is obtained per 100 kg of grain products.

The protein sludge has brownish color, bitter taste and weak hop smell. The weight fraction of dry substances in protein sludge is known to make about 20%. The composition of dry substances includes proteins, nitrogen-free extractives, fiber and hop resins. In addition, protein sludge contains a number of trace elements, such as, manganese, copper and iron, as well as amino acids, i.e., tryptophane, cystine, lysine, aspartic acid, serine, glycine, glutamic acid, threonine, alanine, tyrosine, methionine, valine, phenylalanine, leucine, proline, etc.; and vitamins—thiamine, riboflavin and nicotinic acid. However, the quantitative composition of protein sludge depends on the composition of grain products, technological mode of preparation and method of clarifying beer wort.

In this regard, the object of the study was the protein sludge obtained in the beer production at OOO Shield in Volgograd. Semi-smoked sausages were chosen as an object for research because of their popularity among many groups of the population, since they have particular flavor and aromatic characteristics and are especially nutritious.

2 Materials and Method

Russian and foreign scientists analyzed the prospects for the use of protein-carbohydrate components of plant origin and prospects of food products of a combined composition (Vartanova and Drobot [14], Gorlov et al. [15] and Ognivtsev [16]), Glushchenko et al. [5] and Fedotova [13], Balashov [8], Filonov et al. [1], Postnikov and Pavlov [4], Gorlov and Polyakov [7], Kovalevsky [6].

The predominant share of protein technological ingredients on the Russian market (primarily soy concentrates, isolates and textured forms) should be noted to be represented by products of oversees manufacturers. Soya flour dominates among domestic protein components. Russia has large resources of vegetable protein-carbohydrate raw materials that can be effectively used for the production of high-quality meat products.

The work used generally accepted standardized methods of analysis of the objects under study.

To determine the prospects for the use and technological potential of protein sludge in the technology of semi-smoked sausage products, a number of studies were conducted in the laboratory of the SSI NIIMMP to define the weight fraction of moisture according to GOST R 54951-2012, weight fraction of protein according to GOST 32044.1-2012, weight fraction of fats according to GOST 13496.15-97, weight fraction of ash according to GOST 26226-95, amino acid composition according to the method of measuring the weight fraction of amino acids by the capillary electrophoresis method on the Kapel-105M system and microbiological points according to GOST 10444.15-94 and GOST 10444.12-2013. Comparison with the amino acid composition of the reference protein was performed according to the amino acid scale proposed by the Food Committee of the World Health Organization (FAO/WHO).

3 Results and Discussion

Due to the technical features of the production, there were some difficulties in sampling, so the values obtained were minimal for the protein sludge. The sludge was a wet product; its weight fraction of moisture was 85.2% with a short shelf life. To solve this problem, it was advisable to dry or freeze it. In this experiment, the preparation of protein sludge for the study included its spinning to remove loosely bound moisture and drying. A part of the residue obtained was poured evenly on a baking sheet with a layer of 10 mm and dried in an oven at an air temperature of not more than 60 °C, in order to preserve the initial biological activity of the final product. So, there was dry protein sludge obtained that was transportable and stable during storage. In the dried product, the weight fraction of protein was 42.3%, which indicated the feasibility of dry protein sludge as a valuable technological and biological raw material for the production of various foods, including meat products.

As the main points of the nutritional value of proteins, their amino acid composition is usually considered. The biological value of proteins was determined by the set and content of their essential amino acids that are not synthesized by the human body and must be supplied with food. Isometic, leucine, phenylalanine and tyrosine make the greatest amount of essential amino acids in protein sludge. The protein sludge obtained in the brewery OOO Shield were noted to contain all the essential amino acids. Almost all proteins of plant origin are known to be imperfect albumen, but of great importance for human nutrition. Their optimal combination with proteins of animal origin, namely, muscle proteins that are complete proteins, provided a product that satisfied the daily need for all essential amino acids. So, when developing a semi-smoked sausage product formulation, this combination was taken into account, since to preserve the biological value of the finished product was important in addition to the economic benefit.

When choosing food raw materials and developing the technology of meat products, great importance was attached to microbiological safety indicators that were the main criteria for food products along with the food and biological values. The results of microbiological studies of protein sludge are presented in Table 1.

Table 1. Microbiological points of protein sludge

Point	Value	
	Norm	Actual
Enterococcus, CFU/g, not more than	Should be absent	Not detected
S.aureus, CFU/g, not more than	Should be absent	Not detected
Coliform bacteria	Should be absent	Not detected
Yeast and mould, CFU/g, not more than	$1 \cdot 10^1$	0
Mesophilic aerobic and optional anaerobic microorganisms, CFU/g, not more than	$1 \cdot 10^2$	0

According to the technology of beer brewing, before obtaining the protein sludge, hopped wort was subjected to a period of cold conditioning that eliminated the possibility for pathogenic bacteria to develop. Table 1 shows that pathogenic and conditionally pathogenic microorganisms were absent in the protein sludge obtained.

So, the next stage of the work was devoted to the creation of a recipe for semi-smoked sausages. The recipe was designed, taking into account the preliminary results of experimental studies in terms of the selection of basic meat raw materials, i.e., beef, pork, their weight ratio in the composition of the formulation, the level of the replacement of raw meat with protein sludge and conditions of its introduction into minced products. The amount of raw materials of plant origin for replacing raw materials of animal origin was determined with respect to future economic benefits and biological value of the finished product. The organoleptic evaluation of the samples with added protein sludge was performed. With an increase in the amount of plant supplement in minced meat, the color was becoming lighter, and the bitterness and smell of barley were increasing. The obtained organoleptic assessment established that the optimal share of protein sludge added was 15% (in hydrated form) in the total weight of the product. The sludge had higher nutritional value, but a bitter flavour. So, sweeteners were necessary to be used. In this regard, there was "Glimalask" food additive chosen that contained aminoacetic acid (glycine) for 80%, ascorbic acid for 12% and malic acid for 8% [9].

It is a powerful regulator of the body's defense that improves energy metabolism, activates the immune system and promotes the toxin extrusion. In this technology, it is used as a color stabilizer, sweetener and preservative. Test and Control batches of semi-smoked sausages were produced; their recipes included raw meat, nitrite-curing mixture, sugar, hydrated sludge and "Glimalask" food additive (Table 2).

Table 2. Recipe samples of sausages

Ingredient	Sample	
	Test	Control
Beef trimmed 1 grade	39	40
Pork trimmed semifat	41.5	46.6
Nitrite salt	2.8	2.8
Granulated sugar	0.2	0.2
Fresh garlic chopped	0.25	0.25
Black pepper ground	0.15	0.15
Dried coriander ground	0.05	0.05
Protein sludge	5	–
Water to hydrate dry protein sludge	10	–
"Glimalask" food additive	0.14	–
Water/ice	–	10

At the next stage, samples of semi-smoked sausages were produced according to the conventional technology. After the precipitation, semi-smoked sausages were roasted with smoke fume at 60 °C for 20 min, then the temperature was raised to 80–90 °C; the total duration of roasting was 60–90 min to a temperature of 54 °C in the center of the sausage unit. After cooking at a temperature of 72–75 °C for 40–80 min to a temperature of 70–72 °C in the center of the sausage unit, the sausages were cooled with air at a temperature of 18–20 °C for 2–3 h.

With respect to the physical and chemical points, Test sausages were established to be not inferior to Control sample produced according to GOST 31785-2012 (Fig. 1).

Figure 1 summarizes the data on the content of protein, fat and amino acid score of the Test and Control semi-smoked sausage samples. Test sample containing protein sludge had higher values of protein by 2.2%; its fat content was lower by 0.9%. The study of amino acid profiles found lysine (87%) and valine (90%) as limiting amino acids. Additionally, there was determined the content of glycine as a substance that increases the performance of brain and allows reducing the impact of toxic substances on brain due to its additional inclusion into the product. The glycine content was 3.2% in Test sample and 1.4% in the Control one.

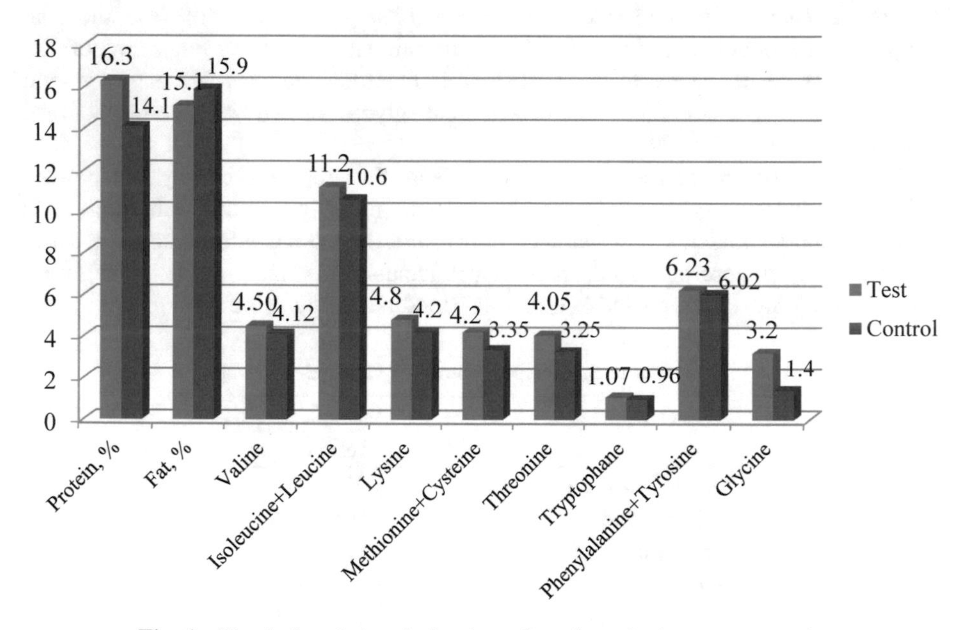

Fig. 1. Physical and chemical points of semi-smoked sausage samples

The protein of plant origin added, as well as the "Glimalask" food additive contributed to the improvement of consumer properties of the finished product, such as appearance, organoleptic characteristics and texture. The secondary product of brewing provided an increase in the yield of the product and, accordingly, a decrease in its cost.

4 Conclusions

Studies of the quality indicators of a sausage product with protein sludge added showed that this raw material met all the safety requirements and its use in the production of meat products is appropriate. So, the introduction of pre-hydrated protein sludge contributed to an increase in protein content by 2.2% and optimization of the amino acid profile. Combining protein sludge with the "Glimalask" food additive not only reduced bitter flavour, but also contributed to improved color formation, as well as increased stability during storage, since organic acids in the composition had a bacteriostatic effect. The study of the chemical composition considered protein sludge as a high-protein product that contained all the essential amino acids. Thus, the conducted studies established the feasibility of protein sludge in the technology of semi-smoked sausage products, which was determined by the functional and technological points of the product obtained. The technology of protein sludge production uses low-cost equipment of home manufacture; there are no chemicals harmful and dangerous for animals and plants, thereby high environmental safety of the target product is ensured. This concept can become an analogue for solving similar economic and environmental problems in other sectors of the food industry.

Acknowledgments. The work was carried out within the framework of the grant of the Russian National Science Foundation No. 15-16-10000, the GNU NIIMMP "Development and scientific substantiation of new approaches to the production of livestock raw materials and increasing the biological value of socially significant products based on modern biotechnological and molecular genetic methods."

References

1. Filonov, A.V., Krampit, M.A., Romanenko, V.O.: Current state and promising areas of secondary material resources being used in the food industry. Fund. Res. **5**, 215–219 (2017)
2. Murzagalieva, D.V., Grigoryan, L.F., Karpenko, E.V.: Perspective and assessment of protein sludge used in the meat products production. Agrarian Food Innov. **1**, 71–75 (2018)
3. Volotka, F.B., Bogdanov, V.D.: Technological and chemical characteristics of a Brewer's grain. Vestnik TSUE **1**, 114–124 (2013)
4. Postnikov, A.E., Pavlov, I.N.: Use of secondary raw materials of the brewing industry. Technologies and equipment for chemical, biotechnological and food industries, pp. 493–496 (2013)
5. Glushchenko, A.V., Kucherova, Y.P., Yarkova, I.V.: Ensuring an agricultural holding's economic security by means of sustainable development strategies in terms of global crisis management. Espacios **39**(12), 4 (2018)
6. Kovalevsky, K.A.: Technology of fermentation industries: study quide. In: Kiev: INCOS, 340 p (2004)
7. Gorlov, I.F., Polyakov, V.A.: Food supplement: Pat. 2519777 Russian Federation: IPC A23L1/30; A23L1/305; A23L1/06. Applicant and Patent Holder is Volga Region Research Institute of Manufacture and Processing of Meat-and-milk Production of the Russian Academy of Agricultural Sciences, OOO Innovation and Research Company "New Biotechnologies" - No. 2012134605/13; 13 August 2012; publ. 06/20/2014 Bull. No. 17 (2012)
8. Balashov, O.Yu., Utolin, V.V., Luzgin, N.E.: Features of obtaining compressed feed from brewing by-products. Agrarian Bull. Upper Volga **1**, 50–54 (2018)
9. Aliyu, S., Bala, M.: Brewer's spent grain: a review of its potentials and applications. Afr. J. Biotech. **3**, 324–331 (2011)
10. Simones, M., Martins Rui, F., Marme, M.: Failure analysis of a filling valve from a Brewery's beer filler. Eng. Fail. Anal. **93**, 87–99 (2018)

11. Guillermo, G., Woolley, G., Rahimifard, S., Colwill, J., White, R., Needham, L.: A methodology for sustainable management of food waste. Waste Biomass Valor **8**, 2209–2227 (2017)
12. Jayant, M., Hassan, M.A., Srivastava, P.P., Meena, D.K., Kumar, P., Kumar, A., Wagde, M. S.: Brewer's spent grains (BSGs) as feedstuff for striped catfish. J. Clean. Prod. **199**, 716–722 (2018)
13. Fedotova, G.V., Kulikova, N.N., Kurbanov, A.K., Gontar, A.A.: Threats to food security of the Russia's population in the conditions of transition to digital economy. In: Popkova, E.G. (ed.) The Impact of Information on Modern Humans. Advances in Intelligent Systems and Computing (AISC), Proceedings of HOSMC: International conference on Humans as an Object of Study by Modern Science, Nizhny Novgorod (Russian Federation), Switzerland, November 23–24, 2017. KozmaMinin Nizhny Novgorod State Pedagogical University, Institute of Scientific Communications), vol. 622, pp. 542–548. Springer International Publishing AG (Part of Springer Nature) (2018). https://link.springer.com/content/pdf/bfm%3A978-3-319-75383-6%2F1.pdf
14. Vartanova, M.L., Drobot, E.V.: The impact of complex conflicting processes on the growth of domestic agriculture. Russ. Entrepreneurship **19**(1), 13–36 (2018). https://doi.org/10.18334/rp.19.1.38718
15. Gorlov, I.F., Fedotova, G.V., Slozhenkina, M.I., Kulikovsky, A.V., Mosolova, D.A.: Modern trends in meat production in Russia and its consumption by the population. Agrarian-Food Innov. **3**(3), 25–30 (2018)
16. Ognivtsev, S.B.: The concept of the digital platform of the agricultural complex. Int. Agric. J. **2**(362), 16–22 (2018)

Determinants and Prospects for the Legal Harmonization of the Intra-BRICS Trade Turnover in the Digital Form

Agnessa O. Inshakova(✉), Elena I. Inshakova, and Anna V. Lavrentyeva

Volgograd State University, Volgograd, Russia
ainshakova@list.ru, {gimchp, inshakovaei, oponir}@volsu.ru, lavra.ne@mail.ru

Abstract. The article examines the features of civil legal regulation of digital assets and regulation of entering into foreign trade transactions in the form of smart contracts of business entities from the BRICS countries. The priority of the Russian Federation participation in BRICS is determined by dynamically developing economic ties, primarily in the spheres of production, energy, information and communication technologies, and cyber security provision. The determinants of digitalization of the integration union economic space are analyzed, on the basis of which the existence of the root causes (financial, economic) and derived causes (legal, political, and social) of the process is argued. The issue of increasing the interest of business entities to technological innovations, as a way to verify counterparties at the pre-contractual stage of concluding a foreign trade transaction and ensuring the fulfillment of obligations independently of the parties, is studied. It is proved that the use of the benefits of smart contracts is currently hampered by the lack of appropriate regulation, both at the national and international legal levels. The existing acts concluded by the BRICS member countries are, in their essence, of programmatic character and do not contain specific material norms.

Based on the comparative legal analysis of the legal systems of the states belonging to the integration union, the reasons for the lack of unified legislation are disclosed involving radically different legal traditions, as well as in the desire of each individual state to preserve national sovereignty and independently determine foreign policy in the sphere of trade and investment. The difficulty of creating a single package of material norms on civil law regulation of foreign trade transactions in the form of smart contracts, moreover, is seen in the significant difference in the choice of methods and means of legal regulation by national legislators both within domestic international private law in general and economic and contractual use of blockchain technologies exacerbated by the foreign element, in particular.It is concluded that, taking into consideration the difficulty of implementing unification by creating "uniform" rules, the Model Code of foreign trade transactions performed through smart contracts should be adopted. The said code in a dispositive manner must contain the general beginnings of the procedure for electronic registration of the economic relations' parties, the order of expression of will and confirming performance of obligations, including the use of chipization of goods crossing the border; as well as the

E. G. Popkova and B. S. Sergi (Eds.): ISC 2019, LNNS 87, pp. 209–219, 2020.
https://doi.org/10.1007/978-3-030-29586-8_25

basic applicable connecting factors of choosing the law and the court in whose jurisdiction are cases in this category.

JEL Code: K10 · K15 · K24

1 Materials

The search for preconditions for the emergence and development vectors of the civil law regulation of the digital sphere as a rapidly developing technological system that influences the traditional legal orders established in the BRICS countries is based on a complex of national and international sources. The following international acts were analyzed: the Johannesburg Declaration of the 10th BRICS Summit, issued at the conclusion of that summit on July 26, 2018; the Goa Declaration of the VIII BRICS Summit that followed the summit on October 16, 2016, etc. Among the national acts that were examined while writing the work the Presidential Order No. Pr-2132 of 10.10.2017 "List of Instructions Following the Meeting on the Use of Digital Technologies in the Financial Sector", and the state program "Digital Economy of the Russian Federation", approved by the Order of the Government of the Russian Federation No. 1632-p of July 28, 2017.

Studying the works on the regulation of digital assets in the Russian Federation and some foreign countries, including Kucherov and Khavanova (2017), Efanov and Roshin (2018), Lagutin and Suslikov (2018), Fedorov (2018), and Sarbash (2018), helped to create the theoretical framework of the paper. Positions and views on the problem of civil regulation of cryptocurrency and smart contracts in the foreign countries were examined on the basis of foreign research results by Aysan et al. (2019), Pustišek and Kos (2018).

The empirical component of the research is represented by the Russian judicial enforcement acts: Determination by the Arbitration Court of the Tyumen Region of 22.06.2016 in case No. A70-15360/2015, and Decision of the Oktyabrsky District Court of St. Petersburg dated 16.05.2017 in Case No. 2-1993/2017.

2 Methods

Methodological basis of the research is represented by the universal scientific method of historical materialism, dialectical positivism, as well as methods of theoretical knowledge: induction, deduction, analysis, synthesis, and empirical description has formed methodology base of the study.

Among the special legal methods the central place was given to the comparative legal research used in studying the means and approaches of civil law regulation in the sphere of digitization of the BRICS countries economic space; moreover, formal legal, interpretative, structural and functional methods were used. The combination of these methods allowed studying the national legislation and the general legal policy of the BRICS member states, aimed at formation, improvement and modernization of their legal order in the field of digital foreign trade transaction.

3 Introduction

Today, BRICS acts as an influential actor of the global community and an engine for progress of developing countries towards the formation of a multipolar world and the creation of conditions for increasing the competitiveness of the member countries on both regional and global scales. At the beginning of formation the union, there was a strategic goal – a way out of the global financial crisis with the construction of a multilateral dialogue based on harmonious and durable peace, and countries' economic prosperity.

In 2018, within the framework of the 10th BRICS Summit in Johannesburg, a Declaration was adopted proclaiming, among other areas, a course to jointly studying the use of distributed ledger technologies in the interests of each party. Work on this issue is entrusted to the BRICS banking organizations, which perform financial cooperation, lending in national currencies, and applying innovation. This approach will open up new opportunities to create and optimize banking and financial products, as well as ensure the conclusion and fulfillment of obligations under foreign trade smart contracts, primarily for Russian exporters.

In this regard, participation in BRICS for Russia is not only beneficial, but becomes a high priority, as it leads to the enhancing the development of new, not yet established, but promising foreign economic relations aimed at replacing existing export and import markets. This aspect of the BRICS functioning is a mutually beneficial alternative of the legal, political and trade cooperation of the member states, which is motivated by the international situation and the search for new compromises (Plakitkin 2011).

4 Socio-Economic and Legal Prerequisites for the Digitization of the BRICS Common Trade Space

Among the variety of different causes that led to an increase in polysubject interest in making foreign trade transactions in a digital format, it is necessary to distinguish between the root causes (financial and economic) and the derived causes (legal, political, and social).

The BRICS countries account for 42% of the population of the entire world and 27% of world GDP. Meanwhile, Russia's share in imports of the BRICS countries (as a percentage of GDP) is 22.5% and in exports – 28.4% (The official website of the Russian presidency in BRICS, 2015). Studying the dynamics of the BRICS intra-group trade turnover and intra-group foreign direct investment volume changes from 2005 to 2016 showed that the mutual development of trade and investment relations within the five member countries is a powerful source of their significant GDP growth compared with other countries. This fact underlines the high potential and synergistic effect of cooperation of the BRICS group states in this sphere (Gusarova 2018).

Another reason for the need to undertake the convergence of the BRICS countries legal systems on time in the nearest future is closely connected with the similar actions for the economy and society digitalization by the Western countries. According to the European Commission published results of the 2018 Digital Economy and Society

Index (DESI), the EU continued to improve its digital performance to catch up with global leaders and to be competitive on the global stage (The official website of the European Commission 2018).

Legal relations complicated by a foreign element are always linked to international payments and money transfers, the settling speed and accuracy of which in the interests of both business entities (intensity growth of the assets turnover) and state as a whole (implementing timely, lawful and active fiscal policy) depend upon the time of the transactions made.

Making foreign trade transactions in the blockchain system goes through an obligatory stage of electronic registration of the property turnover facts, which is aimed at eliminating unfair actions at the pre-contractual relations stage, compiling a single database of potential counterparty (preparatory process) and formalizing the procedure for adopting each of the transaction parties' will on the subject and conditions to a contract, as well as each separate legally significant action of counterparties. The counterparties' relationships expressed in a certain number of subsequent actions, after going through the electronic registration procedure, are considered to be understood identically, proven, and the parties cannot challenge them at a later time (Kalinina et al. 2019). The advantage of this mechanism is that the number of conflict situations at each transaction stage from conclusion of contract to its full execution can actually be reduced to zero.

Bitcoin should be considered as a hedging tool against global geopolitical risks (Aysan et al. 2019), since Western countries have introduced retorsions against Russian suppliers and manufacturers, the inflow of which is appropriately dictated by a high global market demand.

As a matter of law, smart contract will be the main tool in this model of building economic processes aimed at providing goods and services between commercial organizations from the BRICS countries. The nature of a smart contract is an automatic order of performance of obligations, for example, payment for goods is made only after the delivery, and only in the quantity that corresponds to the quantity and quality of the subject of the contract. The implementation of payment for a particular product in such a case always comes after receiving the goods by the buyer; this procedure reduces the time interval, and as a result, saves the costs of counterparties.

In the Goa Declaration that followed the 8th BRICS summit on October 16, 2016, the parties drew attention to the fact that their joint actions are aimed at maintaining a multilateral trading system with the central role of the WTO, which is based on the legal standards, involving transparency, non-discrimination and inclusiveness. At the conclusion of that summit, an increase in bilateral, regional and multilateral agreements was stated (The official website of the BRICS 2018a).

Continuing the sequence of previously achieved common goals and objectives, at the 10th BRICS summit in Johannesburg on July 25-27, 2018, the parties pointed out that trade and high technology form the basis for the inclusive growth, especially due to the economic integration based on the equality and sustainable contribution of all member states to the joint projects implementation. The critical and positive role of the internet in promoting economic, social and cultural development worldwide was also noted. That is why the necessity to develop the existing mechanisms of contributing the secure, open, peaceful, cooperative and orderly use of ICTs and its governance was underlined. The need for further enhancement of cooperation in E-commerce, on standards and technical regulations, and model e-port creation was highlighted (The official website of the BRICS 2018b).

5 National Legal Policy of Providing Digital Technologies in the BRICS Member Countries: Modern Approaches

The development of the BRICS countries' cooperation and trade capacity-building cannot be achieved without a regulatory framework that responds to the realities. The world is at the threshold of realizing the high potential of blockchain technologies. However, based on the analysis of the legislation of a number of countries, it is obvious that a full-fledged regulatory framework for the digital sphere (the use of smart contract, digital assets, distributed data ledger), including foreign economic relations, has not yet been formed in any state.

Blockchain is conceptually seen as a distributed database that contains transaction records accessible to all participants connected to such a ledger. Confirmation of a transaction is carried out by agreeing the will of all or most of the participants by "signing" with a cryptographic key. Therefore, any transaction for fraud cannot actually go through a collective confirmation. Since the record is created and accepted by the ledger, it cannot be changed or deleted (Efanov and Roschin 2018). Several distributed ledger protocols are currently known as functionally suitable for providing Internet of Things (IoT), for example, Ethereum, Hyperledger Fabric and IOTA. All the IoT applications are based on the blockchain technology and are built according to a certain logical structure: "from block to block in a chain" (Pustišek and Kos 2018).

Favorable conditions for introducing IT technologies, as evidenced by fast-growing markets for financial technology, products and services (fintech markets) have been created in China. The Chinese government authorities nonetheless separate the blockchain technology and cryptocurrency: they are actively developing the first while the second is stopped at the source. Despite the fact that Bitcoin was the first successful application of the blockchain technology, Chinese regulators have doubts on its usefulness and intend to limit the potential dangers of cryptocurrency to the maximum.

In December 2013, the People's Bank of China banned financial institutions from making transactions with cryptocurrencies. On April 1, 2014, banks and payment systems stopped servicing Bitcoin trading accounts. In September 2017, a whole group of Chinese regulators took coordinated action to ban the implementation of the Initial Coin Offering (ICO). The placement of virtual coins was recognized as an unlicensed attraction of financing, and most of the projects were considered as fraud. The activity of cryptocurrency exchanges in China was identified as illegal.

However, Chinese cryptocurrency exchanges quickly reoriented to other growing markets: Japan, South Korea and the United States. For example, the Hong Kong platform Binance, which has managed to become one of the largest in the world during six months due to aggressive marketing, changed registration after the ban first to Japan and then to Malta. Despite the ban on trading cryptocurrencies, China continues to be the global mining leader: the world's four largest mining pools: AntPool, BTC.com, F2Pool and BW.COM have Chinese roots. In total, over 72% of Bitcoins' computing power is located China.

Notwithstanding the state ban on cryptocurrency and ICO, the city authorities of Shanghai, Guiyang, Hangzhou and Guangzhou, the provinces of Shanxi and Henan implement a policy of encouraging the blockchain technologies development. Hangzhou, the capital of eastern coastal Zhejiang province, announced plans to invest USD 1.5 bln in the blockchain fund, which is expected to be the largest in the world.

Moreover, the blockchain was officially incorporated into the China's 13th Five-year Plan (2016–2020) as one of the directions for the development of the nation, along with quantum computers, artificial intelligence and autonomous cars.

In matters of legal regulation of the digital sphere of social relations, the Indian government authorities took an extremely negative position. The Financial Stability and Development Council (FSDC), chaired by Indian Finance Minister Arun Jaitley, raised the issue of cryptocurrency. Apparently, the intergovernmental committee charged with studying and developing a legal framework for cryptocurrencies instead proposed a ban on the use of digital assets in the country. In the summer of 2018, the Reserve Bank of India (RBI) issued an act banning the use of cryptocurrency and other digital assets (including fiat currency). Legal entities and individuals are prohibited from making any transactions, whether it is storage, use, disposition, or otherwise dispose related to the virtual assets.

Cryptocurrency exchanges working on peer-to-peer (P2P) trading platforms before the ban had to cease trading in digital assets. The ban on banking services has negatively affected the trading volumes of one of the largest cryptocurrency exchanges in India, ZebPay, which dropped from USD 5 mln to USD 0.2 mln by the closing date. As a result, ZebPay announced the termination of work since September 28, 2018 because of the ban; however, in October 2018, ZebPay registered a new office in Malta. On May 13, 2019, another large cryptocurrency exchange, Coinome, was closed in India.

At present, blockchain sphere in India is not being adequately provided, while the Regulator pursues a strict policy on the new type of assets. In late April this year, the Indian government has prepared a draft bill implying a complete ban on cryptocurrency trading in the country. Once approved, the new rules will take effect in the summer 2019.

At the same time, according to RBI, the blockchain technology holds a lot of potential, but the risks associated with the methods and mechanisms for protecting consumers' rights in the market, market integrity, and the preventive struggle against economic crimes in this sphere tip the scales. The Reserve Bank of India finds of particular concern the system of distributed storage and transmission of data that is closed from state control, through which legalization of the proceeds of criminal activity is possible (Elitetrader official website 2018).

The legislation of the Republic of Brazil does not explicitly prohibit the use of transactions, which are made in the form of the smart contracts, and digital assets in economic activity. However, the Central Bank of Brazil issued a warning that the use of digital money may be associated with a high degree of risk (The official website of the Central Bank of Brazil 2018). It is noteworthy that Brazilian legislation does not use the term "cryptocurrency", and by legal nature, cryptocurrencies are classified as digital money. The tax legislation provides for taxes on transactions with cryptocurrencies, qualifying them as a financial asset (digital money), however, at the present stage, civil law has not settled this issue.

Compared to the rest of the BRICS countries, South Africa is most actively interested in crypto-technology. Blockchain platforms have been created in the country for providing financial transactions in this region. The South African Reserve Bank (SARB) in its annual report noted a significant reduction in the percentage of costs associated with executing payment transactions using a distributed ledger (The Official Website of the Reserve Bank of SA 2018). As early as 2014, SARB, in a notification on digital currencies, made a statement that this area has no legal status, and no regulatory framework.

According to the comments of the South African Revenue Service (SARS), the law on taxes and fees of the Republic does not contain the definition of the term "money"; moreover, the term "cryptocurrency" itself is not common in the civil legislation of South Africa and is not used when making payment or exchange transactions. It follows that in a civil-legal sense, cryptocurrencies are not money in South Africa, but tax legislation extends to them (The official website of the South African Revenue Service 2018).

The legal system of the Russian Federation belongs to the Romano-Germanic legal family and its important characteristic feature is codification of civil law, the provisions of which also regulate international private relations, including foreign trade activities of the property relations subjects. From 2016 onwards, contentious cases involving cryptocurrency began to appear in judicial enforcement. An analysis of a small but representative base of judicial enforcement in Russia reveals a cautious and negative attitude towards operations of Bitcoin. Thus, the court of the Russian Federation recognized cryptocurrency as a quasi-money, referring to Art. 27 of the Federal Law on the Central Bank of the Russian Federation (Decision of Arbitration Court of the Tyumen region 2016; Decision of Oktyabrsky District Court, St. Petersburg 2017).

The Russian legal order was on the verge of a cryptocurrency ban, hence the approaches in law enforcement activities appeared that entail civil responsibility for damage caused by making deals using blockchain technologies. Today the situation has changed and a fundamental positive development towards legal regulation of digitalization processes begins. Russia's President instructed the Government, jointly with the Central Bank of the Russian Federation, to ensure amendments to the Russian legislation, which provide for determining the status of digital technologies used in the financial sector, as well as their concepts (involving "distributed ledger technology", "cryptocurrency", "token", and "smart contract") based on the obligation of the Ruble as the only legal tender in the Russian Federation (Order of the President of the RF 2017). Pursuant to that order, the Government developed the Digital Economy Program, according to which the task is to provide legal conditions for the introducing and applying decentralized ledgers' maintenance and certification of rights (Order of the Government of the RF 2017).

A substantive foundation for a smart contract can only be the Internet of things and Bitcoins. Both phenomena do not fall under any of the objects of civil rights, named in Art. 128 of the Civil Code. Informed by the experiences of other countries, Russian researchers I.I. Kucherov and I.A. Khavanova come to a fair conclusion that "national regulators and fiscal authorities are still only at the stage of understanding the problems that arise in the virtual world, including problems associated with the use of virtual currencies", but "the majority of states are not ready to exclude persons using Bitcoins, as well as trading platforms that earn an income from the relevant transactions, from the range of potential taxpayers" (Kucherova and Khavanova 2017).

Undoubtedly, it is extremely important to form in time common approaches not only within the national legislation framework, but also in consistency with international law, which define the foundations of the cryptocurrencies legal status and management regime, and the blockchain system implementation (Lagutin and Suslikov 2018).

6 Harmonization as a Method of Legal Integration of the Intra-BRICS Trade Regulation: The Validity of Application

In order to conduct convergence, priority must be given to comparing existing norms of each particular state, and understanding similarities and differences. There are two ways of convergence in the international practice of legal integration: harmonization and unification of law. In the second case, in the process of creating uniform norms, the participating states conclude international agreements. At the same time, unification takes place in two different systems: in international law (conclusion of an international agreement) and in national law (subsequent adjustment of domestic legislation).

There is no reason to deny the difficulties of applying method of legal integration in the examined sphere of private international law in the context of the existing reality. Most of the rules by their nature are conflict ones that determine the competence of national law and the choice of law applicable to legal relationship with a foreign element. Upon judicial enforcement, conflicts are resolved either basing on ratification of certain international acts by a separate state or excluding the national law of another state; as a result, different legal regimes are established for the same subject.

The existing difficulties of forming a unified legal space in the digital sphere of the BRICS countries is primarily seen in the informal status of this union of states. The union does not have the legal formalities of a subject of international law, such as the SCO, EU, AU, ASEAN, etc. BRICS has no legitimate grounds for creating an internal legal framework in that regard. In addition, fundamentally contradictory approaches on the development of digital technologies have been established in the countries of the Five.

To date, no scientific understanding of the legal nature and functions of smart contracts has been formed. The integration union is not ready for a widespread introduction of digital technologies into economic processes and for the replacement of automated legal procedures for transaction support in the near future. Given the competition in international markets, the various conditions for economic growth, price volatility and stability of national currencies, as well as the different levels of attractiveness of national economies investment climate, it is difficult to discuss the implementation of any kind of total legal standardization. These factors slow down the process of unification of legal regulation of foreign trade transactions made in the form of smart contracts. To move forward on the path of convergence of national legislations of the BRICS member countries, it is proposed to use the method of harmonization of their legal systems.

The emerging framework for the legal support of digital trade transactions is presented only with a few documents. The BRICS Declaration of July 26, 2018; the Memorandum of Understanding on Collaborative Research on Distributed Ledger and Blockchain Technology in the Context of the Development of the Digital Economy signed as part of the 10th BRICS Summit and based on provisions of the BRICS Leaders Xiamen Declaration of September 4, 2017; BRICS Roadmap of Practical Cooperation on Ensuring Security in the Use of ICTs and Terms of Reference of BRICS Model E-Port Network announced in the Xiamen Declaration; BRICS E-Commerce Cooperation Initiative of 30–31 July, 2017 are among them. The deepening of cooperation in the fields of digital technology, industrialization, "innovation, and inclusiveness and investment, to maximize the opportunities and address the challenges arising from the 4th Industrial Revolution" are noted as the objectives of the established Partnership on New Industrial Revolution (PartNIR) in the document. The strategies to bridge the digital divide, among them helping people to master technology and providing technology transfer mechanisms, are identified (The official website of the BRICS 2018b).

However, it can be stated that the unified international regulatory and legal framework for the sphere under consideration is lacking because the relevant legal support is limited only to program documents of a strategic nature.

Harmonization of law involves the process of legal systems convergence, bypassing the uniformity of norms in national legislations. It exempts from the need to undertake the "implantation" of the international treaty norms into national law. Instead, harmonization allows for the access to more flexible and acceptable legal mechanisms in this case, such as the "soft law" model acts. In this regard, it is recommended to develop and adopt in the framework of the BRICS member countries' economic and legal integration a model law – the Model Code of foreign trade transactions.

7 Conclusion

Thus, the study showed that the determinants of the digitalization policy of the economies of countries participating in the BRICS integration union are rooted in the economic and financial reasons.

It is obvious that against the background the BRICS member states aspirations to determine independently the directions of international activities, their desire to overcome significant differences between the national legal systems in the foreign trade sphere in the conditions of economy digitalization is quite evident. However, it was identified that unification of rules for regulating foreign trade transactions in the form of smart contracts currently remains at the stage of adopting program acts. The way out is seen in the convergence between the national legal systems by applying method of harmonization.

This thesis is confirmed by the results of a comparative legal analysis of legislative approaches to the regulation of digital assets and smart contracts in Russia, China, India, Brazil and South Africa. It was also concluded that within the BRICS union due to its informal status in the international arena, and considering significant differences in domestic policies of participating states with respect to the cryptocurrencies, the convergence of civil law regulation of foreign trade transactions should be carried out by applying harmonization method.

Harmonization of transactions undertaken in the digital form, in turn, should be concentrated in national legislation, but with synchronization of jointly developed general provisions. To this end, it is proposed to adopt the Model Code of foreign trade transactions completed through smart contract. The said code in a dispositive manner must contain the general beginnings of the procedure for electronic registration of the parties of economic relations, the order of expression of will and confirming performance of obligations, including the use of chipization of goods crossing the border; as well as the basic applicable connecting factors of choosing the law and the court in whose jurisdiction are cases in this category

Acknowledgments. The reported study was funded by RFBR according to the research project No. 18-29-16132.

References

Aysan, A.F., Demir, E., Gozgor, G., Marco Lau, C.K.: Effects of the geopolitical risks on Bitcoin returns and volatility. Res. Int. Bus. Financ. **47**, 511–518 (2019)

Decision of Arbitration Court of the Tyumen region dated 22/06/2016 re No. A70-15360/2015 (2016). https://kad.arbitr.ru/. Circulation 04 Feb 2019

Decision of Oktyabrsky District Court, St. Petersburg dated 16/05/2017 re No. 2-1993/2017 (2017). http://sudact.ru/regular/doc/ghSmHAU36M6m/. Accessed 04 Feb 2019

Efanov, D., Roschin, P.: The all-pervasiveness of the blockchain technology. Procedia Comput. Sci. **123**, 116–121 (2018)

Elitetrader official website: In India, banks were forbidden to handle cryptocurrency operations (2018). http://elitetrader.ru/?newsid=393220. Accessed 04 Feb 2019

Eurasian Economic Commission: Regulation crypto-currency: Research of experience of different countries (2018). http://www.eurasiancommission.org/ru/act/dmi/workgroup/Documents/Регулирование%20криптовалют%20в%20странах%20мира%20-%20январь.pdf. Accessed 04 Feb 2019

Gusarova, S.A.: Trade and investment cooperation of the BRICS countries as a factor in the development of their economies. Ph.D. dissertation. Moscow (2018)

Kalinina, A.E., Inshakova, A.O., Goncharov, A.I.: Polysubject jurisdictional blockchain: electronic registration of facts to reduce economic conflicts. In: Popkova, E. (ed.) Ubiquitous Computing and the Internet of Things: Prerequisites for the Development of ICT. Springer, Cham (2019)

Kucherov, I.I., Khavanova, I.A.: Tax consequences of using alternative means of payment (Theoretical and legal aspects). Perm Univ. Herald Juridical Sci. **1**(35), 66–72 (2017)

Lagutin, I.B., Suslikov, V.N.: Legal support of blockchain technologies (theory and practice). Financ. Law **1**, 25–29 (2018)

Order of the Government of the Russian Federation No. 1632-P: On approval of the program "Digital Economy of the Russian Federation" (2017). http://government.ru/docs/28653/. Accessed 04 Feb 2019

Order of the President of the RF Pr-2132: List of instructions following the meeting on the use of digital technologies in the financial sector (2017). http://kremlin.ru/acts/assignments/orders/55899#assignment-2. Accessed 04 Feb 2019

Plakitkin, S.A.: World energy - new frontiers of development. Effective Anti-Crisis Manage. **1**, 44–53 (2011)

Pustišek, M., Kos, A.: Approaches to front-end IoT application development for the Ethereum blockchain. Procedia Comput. Sci. **129**, 410–419 (2018)

Reshetnikov, F.M.: Legal Systems of the Countries of the World: A Handbook. Publishing House "Norma", Moscow (2001)

Sarbash, S.V.: Digital assets in the system of civil rights objects. LAW, **5**, 11–13 (2018)

The official website of BRICS (2018a). http://infobrics.org/document/86/. Accessed 04 Feb 2019

The official website of BRICS: The Johannesburg Declaration of the X BRICS Summit [Adopted in Johannesburg on 26/07/2018] (2018b). http://infobrics.org/documents/2018. Accessed 04 Feb 2019

The official website of the Central Bank of Brazil: Distributed ledger technical research in Central Bank of Brazil (2018). https://www.bcb.gov.br/content/publicacoes/Documents/outras_pub_alfa/Distributed_ledger_technical_research_in_Central_Bank_of_Brazil.pdf. Accessed 04 Feb 2019

The official website of the European Commission: How digital is your country? Europe needs Digital Single Market to boost its digital performance [concluded in Brussels on May 18, 2018] (2018). http://europa.eu/rapid/press-release_IP-18-3742_en.htm. Accessed 04 Feb 2019

The official website of the Reserve Bank of SA: Monetary Policy Statements (2018). https://www.resbank.co.za/Publications/Statements/Pages/MonetaryPolicyStatements.aspx. Accessed 04 Feb 2019

The official website of the South African Revenue Service (SARS): What SARS in the circuit of circleptocurrencies (2018). http://www.sars.gov.za/AllDocs/Documents/MediaReleases/2018/Translation%202018/SARS%E2%80%99S%20STANCE%20ON%20THE%20TAX%20TREATMENT%20OF%20CRYPTOCURRENCIES-isiZulu.pdf. Accessed 04 Feb 2019

Digital Technologies in the Development of the Agro-Industrial Complex

Ivan F. Gorlov[1,2(✉)], Gilyan V. Fedotova[1,2], Alexandra V. Glushchenko[1], Marina I. Slozhenkina[1], and Natalia I. Mosolova[1]

[1] Volga Region Research Institute of Manufacture and Processing of Meat-and-milk Production, Volgograd, Russian Federation
g_evgeeva@mail.ru, niimmp@mail.ru
[2] Volgograd State Technical University, Volgograd, Russian Federation

Abstract. Purpose: The purpose of this chapter is to assess the development level of the agro-industrial complex in Russia with respect to the trends in building the information society and digital economy, conduct a system analysis that takes into account the impact of challenges and threats to the digitization level of agriculture, determine the role and directions of such modernization of the traditional industry, and develop tools and mechanisms for its implementation.

The current digital transformation of agriculture seems to be a promising direction for the development of the industry and allows solving many problems related to the inefficiency and low profitability of production. The lack of legal regulations makes the transition to digital technologies difficult, but in practice, the implementation of information tools has already started.

The adoption of IT technologies at all stages of the agricultural cycle enables rationalizing the entire process from preparation to the sale of finished products. Single attempts of domestic farmers to implement new tools into the work do not produce a large-scale effect, since system planning and government support are of great necessity for digitalization. The progress is being made by the Ministry of Agriculture of the Russian Federation. A draft state-run program "Digital Agriculture" has been prepared, and the Food Net Road map (Smart Agriculture) has been developed.

Methodology: The study was conducted using the methods of graphical representation of information, statistical data analysis, financial analysis, trend analysis and comparison, analog and collation methods.

Results: In accordance with the peculiarities of the new ecosystem model of digital transformation of agriculture with imperatives of increasing agricultural production volumes, we have identified the following areas for the development of domestic agriculture, i.e., reducing the dependence on imported technologies, resources in the fields of breeding, genetics, plant protection products and feed additives; growth in the use of technological equipment; and in the long run, productivity growth throughout the trophic chain "from field to counter" to overcome the situation created.

E. G. Popkova and B. S. Sergi (Eds.): ISC 2019, LNNS 87, pp. 220–229, 2020.
https://doi.org/10.1007/978-3-030-29586-8_26

Recommendations: The main digitization directions discussed in this paper will contribute to enhancing the processes of implementing digital technologies into production processes in agriculture. So, for the agro-industrial complex, it is necessary to develop an integrated digital platform that includes 4 spheres: (1) industries that provide agribusiness sectors with farm machinery, equipment and fertilizers; (2) crop and livestock production; (3) food and processing industries; and (4) industries responsible for the storage and transportation of agricultural products.

Keywords: Digital economy · Agriculture · Digital technologies · Digital transformation · IT technologies

JEL Code: O13 · O32 · Q13 · Q16 · P52

1 Introduction

The efficiency enhancement of agricultural sectors in Russia is impossible without information or "smart" technologies being used to provide a scientifically based approach to managing traditional processes. The modern world has entered the era of large-scale digital transformation of all sectors of the economy and spheres of social life. A large number of smart devices emerged recently make it possible to find a maximum efficiency solution to many problems remotely and directly affect the performance in traditionally low-profit industries.

The new May decrees of the President of Russia revealed the need to create a highly productive and export-oriented sector of the agro-industrial complex (with exports up to $ 45 billion per year) based on modern digital technologies [1]. Information technologies will reduce the cost of agricultural production and increase the profitability level of the industry in general. The tradition of agricultural business consists in a long chain of intermediaries and high agricultural loans that in fact can take up to 90% of the profits of enterprises and considerably limit the capabilities of modernization and digitalization of the industry.

The main factors, affecting the digitalization level in agriculture, are both economic and actual inaccessibility of modern technological tools for Russian farmers due to the restrictive import of technologies and lack of qualified IT professional in the agro-industrial sector. These factors inhibit a full-scale integration of agriculture into a digital environment.

The promise of digitalization lies in its end-to-end nature that allows linking consumers and agro producers through constant information flows, reduce the selling costs of finished products and agricultural raw materials, speed up the turnover in the agro-industrial sector and result in food consumption growth and reduction of their cost for the population. But the current extremely low level of village digitalization limits the possibilities for developing information technologies and reduces the competitiveness of domestic agricultural products in the food market.

Agriculture is a branch of the agro-industrial economy sector that is a complex system, including interconnected elements developed into a single whole. The feature of openness of the system of that kind is, first of all, its functioning being influenced by the external and internal environment.

The concept of agricultural development should be worked out and implemented, taking into account the policy of developing a single information space of the Russian Federation, since digitalization as a development pattern of the national economy is inevitable for our country. In this vein, the areas to be provided for are digitization of the production process, digitization of the social infrastructure of rural areas and digitization of the process of training qualified personnel for agriculture.

The imperatives of increasing production in agriculture are the following:

1. The world's population is growing. In 30 years, humanity will need 1.7 times more food than it produces now. For this, it is necessary to seriously modernize agriculture (Fig. 1). According to the UN forecasts, by 2050, the world population will have reached 9.8 billion people; in order to feed them, it is essential to increase the food production by 70%.
2. The growth of the population of the planet requires growing more crops and livestock through intensification and efficiency enhancement at constant amounts of agricultural land.

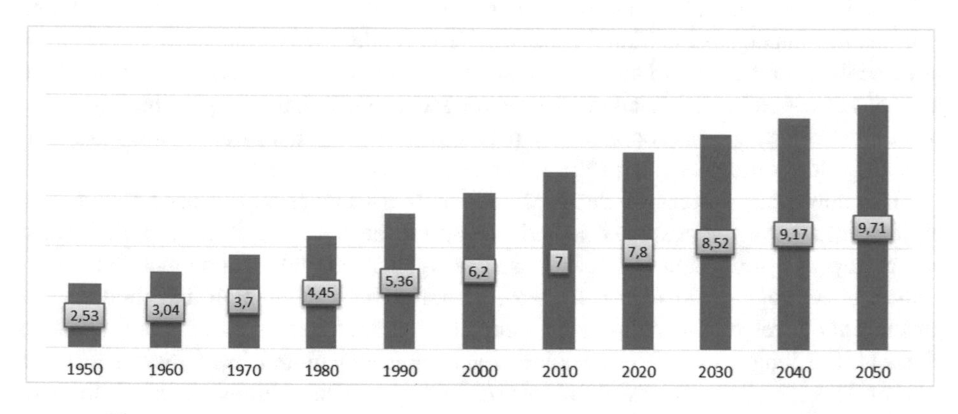

Fig. 1. Dynamics of the World Population Growth, according to the UN [3]

The world population dynamics presented proves the need to intensify agricultural production, which entails the implementation of a comprehensive program for the modernization of the entire industry [4]. An assessment of the economic availability of high-quality food for the population of Russia proved that the population is actually undernourished, it saves on high-quality and healthy food and switches to cheaper substitute products [6]. The current situation is unacceptable, because high-quality nutrition directly affects the health of the nation, its lifespan and ability to reproduce (Table 1).

Table 1. Estimation of the food set cost of Russians, RUB per month [7]

Product	Period			Changed in 2016/2015, %	Changed in 2017/2016, %	Changed in 2017/2015, %
	2015	2016	2017			
Basic food cost - total	5431.0	5858.6	5925.2	+7.3	+1.1	+8.3
Bread and bread products	783.2	860.3	868.1	+8.9	+0.9	+9.8
Potatoes	127.5	109.7	126.0	−16.2	+13	−1.2
Vegetables and cucurbits	574.3	621.6	624.0	+7.6	+0.4	+8
Fruits and berries	528.7	581.2	564.7	+9	−3	+6.4
Meat and meat products	1664.8	1753.5	1775.8	+5	+1.3	+6.3
Milk and dairy products	838.4	921.2	971.0	+9	+5.1	+13.7
Eggs	100.8	108.6	103.0	+7.2	−5.4	+2.1
Fish and fish products	375.2	414.2	417.3	+9.4	+0.7	+10
Sugar and pastry	356.8	394.4	385.8	+9.5	−2.2	+7.5
Vegetable oil and other fats	81.2	93.9	89.5	+13.5	+4.9	+9.3

The official statistics on the main food set cost proves that there is a rise in the cost of basic food products, while the incomes of Russians do not grow. Therefore, any modernization program of agriculture should be aimed at solving the problem to feed people [5]. The only way out of the current situation is to reduce retail prices not by 10–15%, but multiply, with the margin of the business of agricultural producers being retained or even increased and the product quality being high. Dramatic restructuring of the entire process of production and marketing of agricultural products, which, in fact, is called digital transformation, will make it possible. Digital transformation should affect all sectors and spheres of agriculture from preparation for production to the sale of finished products to consumers (Fig. 2).

The main obstacles to the successful implementation of informatization of the industry are the low level of automation and mechanization of farms, lack of reserve funds for the purchase of agricultural equipment, high production risk due to agro-climatic conditions and lack of a regulatory framework governing the use of technology.

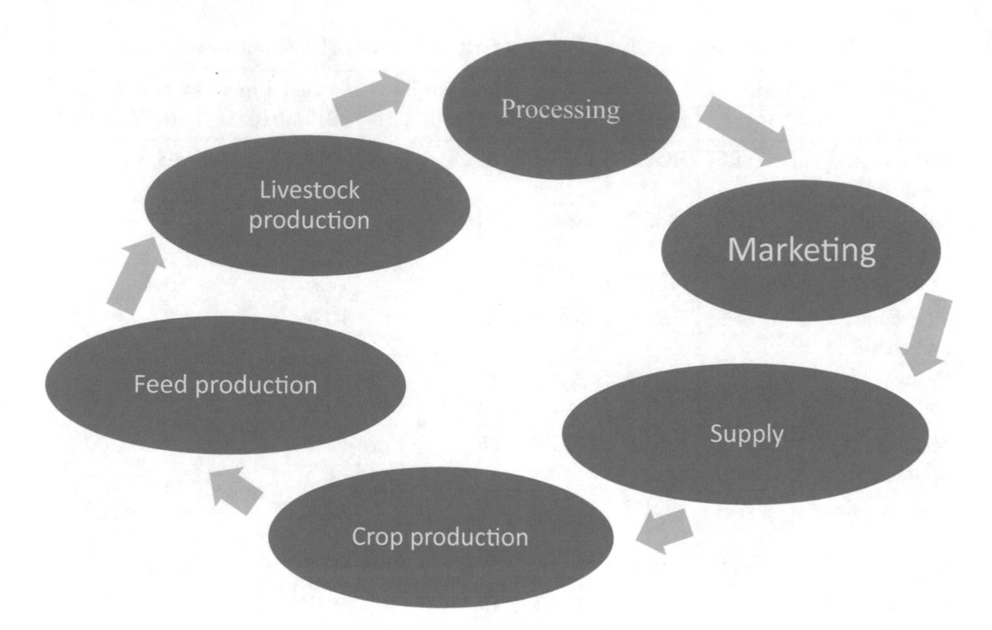

Fig. 2. Stages of the agricultural cycle, requiring digitalization

However, in late 2017, Russia ranked 15th in the world in terms of digitization of agriculture; and according to the data from the Ministry of Agriculture of the Russian Federation, announced in February 2018 at the "Precision Agriculture Conference 2018," the industry's information and computer technology market is estimated at 360 billion RUB [6, 7].

In the agricultural sector of Russia, drones are being developed even though the unfavorable regulatory framework. Unmanned Technologies (Novosibirsk), Geoscan (St. Petersburg), Autonomous Aerospace Systems – Geo Service (Krasnoyarsk) and ZALA AERO (Izhevsk) are considered the most active market participants [8].

The examples given are chaotic and rare; therefore a complex modernization of the industry through its large-scale digitalization is required. In the digital environment, the main task of the IT-technologies is the maximum automation of all stages of the agricultural cycle in order to reduce losses, increase business productivity and optimal allocation of resources. At subsequent stages, automation should affect and bring the industry to a higher level of digital integration. So, the integration of the data obtained with various intelligent IT applications that process them on-line ensures a revolutionary shift in decision-making for the farmer, providing the multiple factors analysis and the rationale for subsequent actions. In this case, the more detectors, sensors and field controllers are connected to a single network and exchange data, the more intelligent the information system becomes and the more useful information for the user it can provide.

For example, during the season, a farmer has to make more than 40 different decisions, i.e., what seeds to plant, when to plant, how to process them, how to treat a diseased plant, how to cope with situations that threaten the field's well-being, etc. The lack of information for decision-making leads to the yield losses up to 40% in planting, growing and cultivation of crop. During harvest, storage and transportation, other 40% of crop is also lost. So, scientists have found that the weather and other 2/3 of the loss factors today can be controlled using automated management systems (Hi-Tech Management).

Among other things, based on scientific calculations, the information system is able to create recommendations on the treatment and cultivation of plants or instructions for automatic execution by robotic technology.

For example, a predictive analytical model helps determine that a temperature increase of 2° contributes to the hatching of insects or an increase in humidity above the optimal limit may lead to an outbreak of the disease. Managing these factors determines a real value of modeling microclimatic conditions. One can prevent the temperature from rising in a greenhouse and prudently monitor over the field and treat it with chemical mixtures when parasites appear. For the first time in the history of agriculture, a farmer has the opportunity to control natural factors, design accurate business processes and, moreover, predict a mathematically precise result [9].

The changes will affect the livestock industry in terms of the transition from incident management to proactive management of the entire production cycle.

2 Materials and Method

Theoretical and applied challenges of agriculture digitization are revealed in numerous works by Vartanova and Drobot [3], Gorlov et al. [6, 11] and Ognivtsev [9]. Issues of agroholding development are discussed by Glushchenko et al. [4, 5]. The digital technology mechanisms are defined in the draft state-run program "Digital Agriculture" [2] and Food Net Road map (Smart Agriculture) introduced by the Agency for Strategic Initiatives (ASI).

Despite the abundance of publications on related topics, the challenges of digital transformation of agriculture are poorly understood. There are many unsolved problems in developing an optimization ecosystem model of digital transformation of agriculture, contributing to the building of "smart" and "rational" partnerships in exploiting information resources for the benefit of society. This chapter is devoted to these issues. The study was conducted using the methods of graphical presentation of information, trend analysis and comparison, analog and collation methods.

3 Results

The fundamental trend in the development of domestic agriculture is the digital transformation that allows intensive developing of the industry and maximizing its profitability. According to experts, the Russian Federation ranks 15th in the world in terms of digitization of agriculture [3].

Sectoral performance appraisal evidences the widespread use of georeferencing systems and system management of agricultural machinery and precision farming systems development. According to various estimates, about 10% of all arable land is monitored using information technology. With the current pace of digitalization, the agricultural technology market will have grown 5 times by 2026.

At present, only 10% of arable land in our country is processed using digital technologies. In fact, in early 2019, elements of precision farming are used in 28 regions of Russia; the Lipetsk region is the leader (812 farms). Non-use of new technologies leads to a loss of up to 40% of the crop.

Given the need to overcome the technological lag behind developed countries, it is assumed that the share of the digital technology market in agriculture will grow every year; by 2026, the market of information and computer technologies in the industry shall grow at least five times [4]. At the same time, the overall digitalization level of the industry is quite low compared to other industries.

Today, the leaders in digitalization are the media, finance, law and insurance industries. The approximate list of the government automated information systems (GAIS) implemented into the work of various departments on the Roskomnadzor website provides an idea of the cost of transition of the industry to a digital platform Rosstat IT system, Tax AIS, Resource Monitoring System, Customs authorities unified AIS, Database of Ministry of Internal Affairs, Justice GAS, ГАС выборы Elections GAS [5, 12].

The cost estimate of developing and implementing the GIS into work of various departments shows that large-scale investments are needed for transition to a digital platform. So, the development and implementation of the Elections GAS cost about 22 billion RUB. Only public funding for the comprehensive digitalization of agriculture is apparently insufficient. Over the past 5 years, global investment in the agro-industrial sector reached 10 billion RUB, with the share of Russia being 1.5%.

Robotization and production automation are considered the main areas of digitalization of the agro-industrial complex. An example of that is the Cherkizovo Group Meat Processing Plant built and launched in the city district of Kashira, Moscow Region in 2018. The peculiarity of the production process is the implementation of the Industry 4.0 concept with direct communication between artificial intelligence and robots (Fig. 3) [10].

A comparative analysis of the Industry 3.0 and Industry 4.0 technologies in production processes demonstrates considerable reduction of people employed in the production cycle. So, according to specialists, a traditional mechanized production requires about 700 staff members, and a robotic one only 170 employees. This project is the largest not only in Russia, but also in Europe; the investments amounted to about 7 billion RUB; the production capacity of the enterprise is 80 tons per day or 30 thousand tons per year.

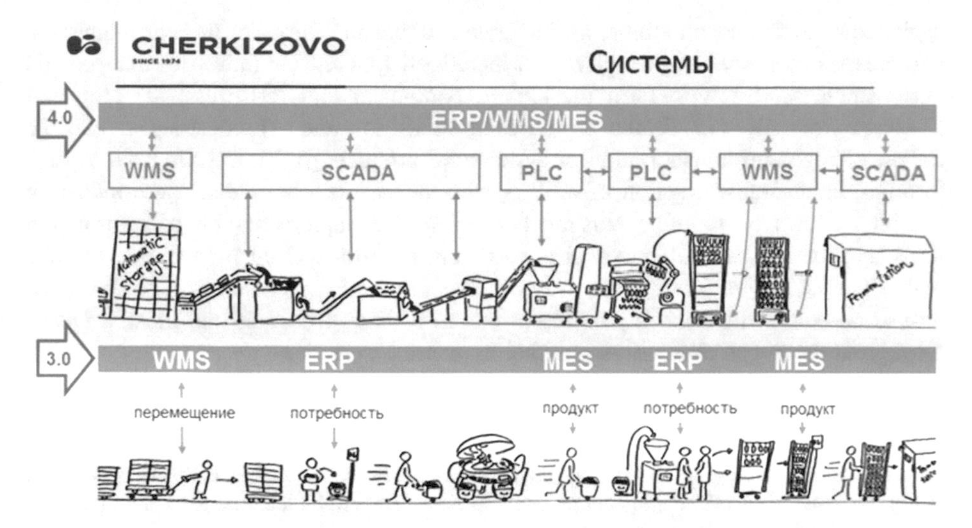

Fig. 3. Comparison of Industry 3.0 and Industry 4.0 technologies in sausage production

The second direction of digitization of agriculture is the IoT (Internet of Things) that is a network of objects interconnected through a global network [4]. The main areas of the IoT application in the agro-industrial sector are precision farming, smart farms, smart greenhouses, management of raw materials, storage of agricultural products, management of agricultural transport and Big Data.

Precise farming allows optimizing the operating costs and increase yields (by an average of 15 … 20%) by reducing the volume of seeds, agrochemicals, fertilizers and water used on demand and more efficient use of land. Actual losses in the industry are shown on the example of grain production: 1/3 of all food products in the world ($ 940 billion per year) is lost or goes to waste. Digitalization will reduce these losses and the number of hungry people (795 million) in the world [13, 14].

At the moment, about 40% of production is lost at different stages.

The volume of current losses in the grain production in agriculture reaches 40% that is a possible cost-cutting in production when precision farming tools being introduced. According to rough estimates, the effect of implementing the IoT in agriculture for the period up to 2025 may be about 469 billion RUB by optimizing staff costs, reducing losses in crop (grain) and petrol, oil and lubricants (POL). In 2019, 30% of all enterprises are likely to use the IoT solutions [8].

4 Conclusions

Thus, in agriculture, the technologies of equipment monitoring and control as well as precision farming technology have the greatest potential. Promotion of scientific ideas of the benefits of the latest achievements of science and technology among the population and training of IT-agronomists and IT-zootechnicians will also contribute to the intensification of introducing digital technologies into production processes in agriculture. In these conditions, an integrated digital platform for the agro-industrial complex is necessary to be developed. It includes 4 spheres: (1) industries that provide agribusiness sectors with farm machinery, equipment and fertilizers; (2) crop and livestock production; (3) food and processing industries; and (4) industries responsible for the storage and transportation of agricultural products [6, 7]. Carrying out a pass-through digital transformation of all processes in the branches of the agro-industrial complex allows overcoming existing barriers to the implementation of information tools into the agricultural sector of the Russian economy. Only joint efforts of state authorities, scientific institutions, private investors and representatives of the agro-industrial sector will make it possible to intensify these processes and make a breakthrough in the negative trend of profitability decline in agriculture.

Acknowledgments. The work was carried out within the framework of the grant of the Russian National Science Foundation No. 15-16-10000, the GNU NIIMMP "Development and scientific substantiation of new approaches to the production of livestock raw materials and increasing the biological value of socially significant products based on modern biotechnological and molecular genetic methods."

References

1. Decree of President of the RF of May 7, 2018 No. 204 "On the national goals and strategic objectives of the development of the Russian Federation for the period up to 2024". ATP Garant. http://base.garant.ru/71937200/#friends. Accessed 22 Feb 2019
2. The Ministry of Agriculture is developing the program "Digitization of agriculture." https://ag.RF/index.php?Id=422. Accessed 20 Feb 2019
3. Vartanova, M.L., Drobot, E.V.: Prospects for digitization of agriculture as a priority direction for import substitution. Econ. Relat. **8**(1), 1–18 (2018). https://doi.org/10.18334/eo.8.1.38881
4. Glushchenko, A.V., Kucherov, Y.P.: Lowering the integrated agricultural formations credit risk in the conditions of global crisis management. Espacios **39**(12), 11 (2018)
5. Glushchenko, A.V., Kucherova, Y.P., Yarkova, I.V.: Ensuring an agricultural holding's economic security by means of sustainable development strategies in terms of global crisis management. Espacios **39**(12), 4 (2018)
6. Gorlov, I.F., Fedotova, G.V., Slozhenkina, M.I.: Digital technologies for solving problems of food security. Agrarian Food Innov. **4**(4), 7–15 (2018)
7. Gorlov, I.F., Fedotova, G.V., Slozhenkina, M.I., Mosolova, N.I., Barmina, T.N.: Digital transformation in agriculture. Agrarian Food Innov. **1**(5), 28–35 (2019)
8. "Internet of Things" (IoT) in Russia. Technology of the future, available now. https://www.pwc.ru/ru/publications/iot/IoT-inRussia-research_rus.pdf. Accessed 23 Feb 2019
9. Ognivtsev, A.B.: The concept of the digital platform of the agro-industrial complex. Int. Agric. J. **2**(362), 16–22 (2018)
10. Project "Digital Agriculture". http://mcxac.ru/upload/medialibrary/04c/04cf3968669675d0b-9ecc106ad04a1a7.pdf. Accessed 03 Mar 2019

11. Gorlov, I.F., Tretyakova, O.L., Shakhbazova, O.P. Nikolaev, D.V.: Development of an application program for the index evaluation of animal breeding qualities. In: Proceedings of the Nizhnevolzhsky Agrouniversity Complex: Science and Vocational Education, no. 1(49), pp. 176–181 (2018)
12. Register list of Federal National Information Systems. http://rkn.gov.ru/it/register/#. Accessed 20 Feb 2019
13. Digitization of agriculture. http://polit.ru/article/2018/02/21/sk_digital_farming/. Accessed 21 Feb 2019
14. Cherkizovo opened a robotic plant in Kashira. https://www.agroinvestor.ru/companies/news/30127

Digital Economy: Beautiful, but Imaginary, Concept

Igor V. Astafyev[1(✉)] and Dmitry P. Sokolov[2]

[1] Civil Society Development Fund, Kostroma, Russia
iastafjev@mail.ru
[2] Financial University, Moscow, Russia
dpsokolov@fa.ru

Abstract. The expression "digital economy" appeared relatively recently, but it has already become stable and generally accepted. However, the question of what exactly should be understood by the digital economy remains open. It is also not entirely clear whether this turnover corresponds to the meaning embodied in it.

The article is dedicated to determining with the help of the analysis of the essence and goals of what digital economy or digitalization of the economy means. What is it - a fundamentally new applied toolkit, a transcendental phenomenon or a qualitatively new state of the macroeconomic system?

Keywords: Digitalization · Digital economy · Information · Scientific and technical progress · Communications · IT-technologies

1 Introduction

It is commonplace that before embarking on the discussion of a scientific problem, to analyze a phenomenon, it is necessary not only to determine the key concepts, but also to apply the most correct of them, which most reliably reflect reality. Otherwise, all further constructions based on an incorrectly formulated concept are doomed to failure. Just as one cannot construct a building on the wrong foundation.

2 Methodology

The expression "digital economy", unfortunately, for almost fifteen years has become a part of everyday life, as well as scientific use, while representing a beautiful, imaginative, even somewhat poetic, as well as deeply erroneous concept.

In the field of economics, as in many other areas, imagery often replaces content and meaning. In particular, everyone knows the phrase "the economy is growing" or "falling." While the system of relations about the production, distribution, exchange and consumption of goods, which is the economy, can neither grow nor fall. The same thing happened, for example, with the concept of "robot", introduced not by a scientist, but by a writer and journalist Karel Čapek.

E. G. Popkova and B. S. Sergi (Eds.): ISC 2019, LNNS 87, pp. 230–237, 2020.
https://doi.org/10.1007/978-3-030-29586-8_27

The first appearance of the word "robot" refers to 1920, when the Czech writer Karel Čapek used it in the fantastic play "Rossumovi univerzální roboti (R.U.R)". There it meant an artificially created person, whose work was used in heavy and dangerous industries instead of human. (Tkacheva and Shepelev 2018) But everyone knows the automatic and so-called "Robotic" gearboxes on cars. While even a complex mechanical device with feedback is hardly appropriate to call a robot.

Exactly the same thing happened with the so-called "digitalization of the economy." If this concept is understood literally, then we should talk about a kind of "robotization" of economic relations while this is not just observed, but can hardly be expected in the future. Because regardless of the degree of "digitalization" of data processing and transmission, the relations between people both were and will remain an "analog".

Is communication between people about the turnover of goods replaced by the interaction of computer programs? It is not.

Who makes the final (and almost all preliminary) decisions in the field of economic activity, perhaps with computers? No, they are still accepted by people, computers only help to analyze large data arrays. Many attempts to "autopilot" naturally ended in tragedy, since computer programs, having much faster data processing than human neural networks, have at their disposal orders of magnitude more meager sets of external factors that make the right decision, and they could compete in it with human beings only when these machines could have the same number of sense organs (which are aggregated with all previous experience in solving real problems) with the humans.

Plato distinguished three worlds - the world of things, the world of people and the world of ideas. Information certainly belongs to the world of ideas. But itself it is only a form of the existence of an idea. Just as the superbrain in a huge number of fantastic works is a kind of super-something, knowing everything about everyone, but in itself is helpless and has no purpose.

Any economic activity is always advisable. Let us notice, that the goals (conscious, not calculated goals) can be generated and set only by the subject of economic relations, the person.

One can argue that a computer can artificially generate any targets according to a given algorithm (We intentionally leave aside local technological goals, this level, of course, belongs to computers). Of course, it can. However, it will not be the goal of man. These will be "goals in themselves". As a "digit" itself, a number means nothing except an abstract quantity.

In S. Lem's novel "The Invincible", a similar case is described when automata, having separated from their prime creators because of a cataclysm, lost their original goals, replacing them with an algorithm for destroying everything conscious for their own existence. An interesting conclusion is when the commander ordered to stop the battle with these artificial creatures, wisely noticing that revenge on these silicon entities is about the same as trying to avenge the phenomena of nature for the death of their comrades.

3 Results

The term “digital economy” was introduced in 1995 by American computer scientist Nicholas Negroponte (University of Massachusetts) (Negroponte 1995). Let us try to understand the essence of the phenomenon called “digitalization of the economy” - the ever closer connection of digital (or more precisely, information) technologies and the system of economic relations. Let us consider, at least relatively superficially, the content of key concepts and definitions of this topic.

Digitization (a meme with signs of slang) is the translation of analog information into a digital code, its encoding.

Encoding is a method (and process) of interpreting and presenting information, that makes it suitable for automated computer (software) processing.

Economy is a system of relations between people about the production, distribution, exchange and consumption of goods.

“Digit” (in translation from a meme is a mathematical and/or coded form of representing the turnover of values) is an elementary link in information about economic transactions and their properties. In other words, it is a form of existence including economic information, just as language is a way of existence of thought.

So what are “digitized person” and “digital economy”? Neither one nor the other, as, incidentally, the “digitized world” does not exist and cannot exist. Like a digitized love. Digitized (encoded in a certain way) can only be information. Information about objects (tangible and intangible), their properties, phenomena, sequence of actions, technologies, rights, benefits.

A “digital” (coded) representation of economic relations is possible. But at the same time, the economic relations themselves were, are and will remain “analog” as long as a person remains a protein creature with cellular senses and an organic protein brain working through a system of neural connections under an “operating system” that is not just incompatible with the computer algorithms invented by it (the human brain), but even on the biological principle of storing and processing information that is unknown to us so far.

Perhaps this will remain forever, because, according to system theory, any product of an n-dimensional system can never have dimension n + 1 (at least “1”) and cannot fully comprehend itself. Especially systemically superior. Consequently, it is impossible to “digitize” the economy (as well as “ride the thought”).

In July 2017, the Digital Economy of the Russian Federation program was approved by a decree of the Government of the Russian Federation. (Program 2017) “Digital economy is an idea born not in the minds of Russian officials. It was voiced by the World Bank in 2016 in the World Development Report 2016: Digital Dividends. “The truth is that the concept of digital economy there and the priority steps in this direction differed from what the Government of the Russian Federation understands. If the World Bank pointed to such signs of digitalization in Russia as open data, the e-government system, the work of domestic digital giants such as Yandex, Kaspersky, online order services, reducing the period for registering property rights with the help of information technology to 10 days, then the Government of the Russian Federation did not stop at the final state program. Given that the term itself is blurred, the digital economy will clearly be with Russian specifics.” (Urmantseva 2017)

The Digital Economy of the Russian Federation program lists the main goals. (Program "Digital Economy of the Russian Federation", p. 2)

"Creating an ecosystem of the digital economy of the Russian Federation, in which data in digital form is a key factor of production in all spheres of social and economic activity and in which effective interaction is ensured, including cross-border, business, the scientific and educational community, the state and citizens."

What is meant by the "ecosystem of the digital economy" is quite difficult to understand. The different thing is clear - the presentation of data may, of course, affect the speed of transactions, but it is unlikely to be a key factor of production.

"Creating the necessary and sufficient institutional and infrastructural conditions, removing existing obstacles and restrictions for creating and/or developing high-tech businesses and preventing the emergence of new obstacles and restrictions both in traditional industries and in new industries and high-tech markets."

The goal is clear and important. But how does it relate directly to the "digital economy"?

"Increasing competitiveness in the global market of both individual sectors of the economy of the Russian Federation and the economy as a whole". Of course, information exchange technologies and electronic document management affect the relationship between actors of economic activity. However, competitiveness still refers primarily to products, prices and terms of sale, and not to the methods of processing transactions.

The Program contains many certainly important tasks, which, however, are weakly linked to the concept of "digital economy". But this methodical approach is characteristic not only for Russia. In the report "The New Digital Economy", the digitalization of the economy is also understood primarily as:

- share of coverage by the Internet;
- quality and availability of mobile communications;
- mobile technology. (Oxford Economics 2011)

Further, a number of indicators, which can be called technical, are used to evaluate digitalization. Among them are:

- technical characteristics of communication networks;
- information security parameters;
- regulation of electronic data interchange (legal framework);
- organization of training for the creation and maintenance of networks;
- electronic document circulation of subjects of economic activity with government bodies;
- development of geo-information technologies, etc.

All of the above necessary tasks are a set of measures to ensure the organization of production, distribution, exchange and consumption of goods in the new conditions of interaction, due to fundamentally new communication opportunities. However, in this way it seems they should be denoted.

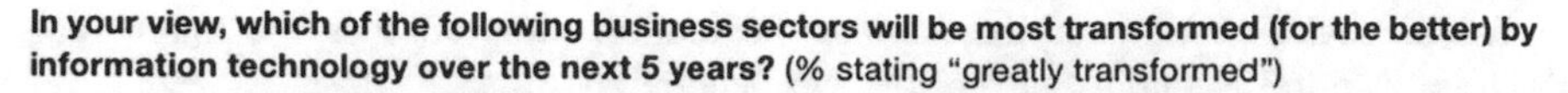

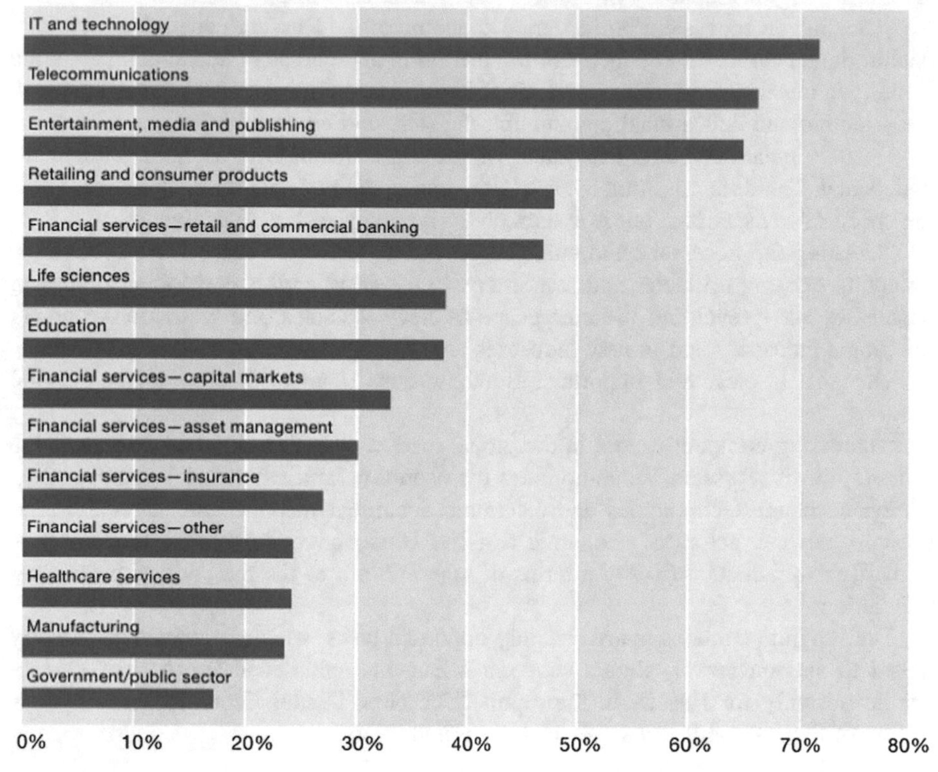

Fig. 1. Industries most affected by digital transformation Source: (Oxford Economics 2011, p. 10)

From the diagram in Fig. 1 we see that: (a) the impact of digitalization is less pronounced in the real sector of the economy, mainly due to improved communications; (b) the main beneficiaries of "digitalization" were not types of economic activity, but types of products and/or lines of business. In addition, the heterogeneity of objects is obvious, in relation to which the impact of "digitalization" is assessed.

The impression of the substitution of concepts is re-created. It can be checked out easily. All of the above, with the same success with which it can be attributed to the "digital economy", can also be attributed to "digital culture"; "Digital education; "Digital health care"; "Digital life" and so on.

Let us try to list what exactly digital technologies are changing in the organization of macroeconomic systems.

1. A significant increase in the share of IT (or directly related to IT) services.
2. Acceleration of the turnover of means of payment (another argument against monetary theory) and a reduction in the average transaction time.
3. The increase in the share of employment in terms of outsourcing.

Now, let us list what “digitalization” did not have a fundamental impact on.

1. The economic mechanism of turnover of goods (capital and other resources);
2. Principles of relations of employers with employees;
3. Class structure of society (with a significant change in the situs structure).

It is also important that when calculating the benefits and advantages of the “digital economy”, there is also a concomitant effect of a reduction in material production and GDP as a whole (Fig. 2).

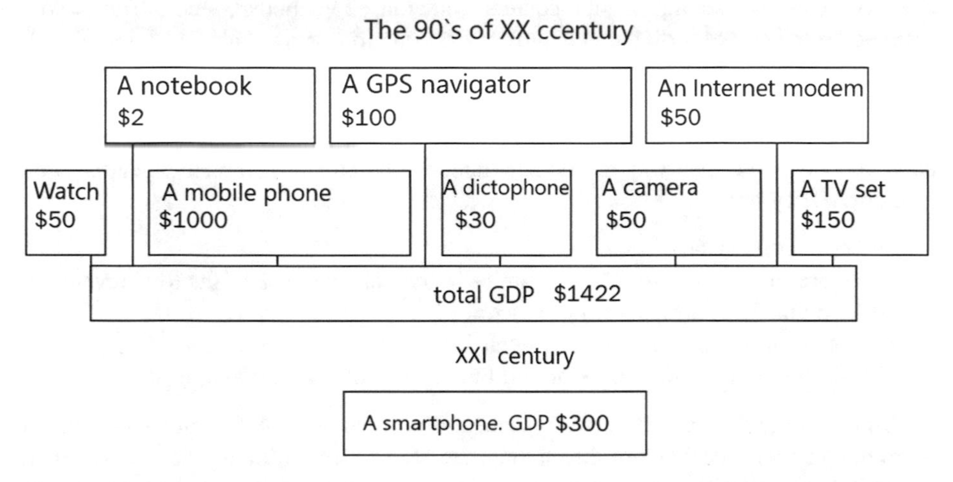

Fig. 2. Example of reducing GDP due to progress information technologies Source: (Bodrunov 2018)

It seems that much more significant than digitalization is currently the growth of the information-digital composition of capital, its knowledge capacity according to S. Bodrunov, after the growth of the organic composition of capital noted by K. Marx. (Bodrunov 2018, p. 231) Nevertheless, this phenomenon, directly related to the increasing role of the coded information turnover, is not “digitalization”, but is a natural stage in the further intellectualization of the economy.

4 Conclusion

The opinion that the concept of “digital economy” should be approached more thoughtfully is also shared by some business leaders who are directly involved in this type of economic activity. In particular, the chairman of the Moscow International Higher School of Business MIRBIS D. Aliyev believes that “the digital economy does not exist as an independent one: there is a digital segment of the real economy” (Urmantseva 2017). S. Balanova, CEO of IBS, reasonably believes that the attribution of information technology to the digital economy is a myth. “The concepts of” information technology “and” digital economy “have become almost synonymous with us. However, the introduction of technology for the sake of technology is meaningless. The digital economy is created by business models, and technology plays the role of a tool,” - she writes (Balanova 2017).

R. Meshcheryakov (Urmantseva 2017) suggests that there are two approaches to the term “digital economy”. The first approach is “classical”: the digital economy is an economy based on digital technologies and, moreover, it is more correct to characterize exclusively the field of electronic goods and services. Examples are telemedicine, distance learning, selling media content (cinema, TV, books, etc.). The second approach is advanced: “digital economy” is production using digital technologies. In any case, the scope of the term for its correct understanding is significantly narrowed.

Let us turn to the analytical report “Digital Russia: a new reality” (Aptekman et al. 2017). The method of calculating the “digitalization index” among 24 indicators contains only three, one way or another (rather indirectly than directly) related to the economy. They are:

- The cost of fixed broadband;
- The share of revenues of ICT companies in the country in % of the total revenue of the Top-900 ICT companies in the world;
- The share of e-commerce in total retail sales; and
- The share of advertising costs on the Internet in total advertising costs.

Thus, a digital economy is really understood as either the relevant sector of the economy technologically connected to the Internet, or a conditionally determined share of production using digital technologies. However, there are different types of technology, not only “digital”. Known biotechnology, agricultural technology, technology of mechanical processing and many others. Following the logic of the term “digital economy”, according to the same principle, any of the listed adjectives can be presupposed.

Most likely, the icause of it is the fact that “technological thought is now merging with economic” (Yudina 2016), but the interdisciplinary approach should not be confused with the merging of areas of knowledge. In general, it seems that in the expression “digital economy” the common and the particular have been reversed. Economics is a general concept; “digital” is a private one, indicating either a segment of the economic system or an applied instrument that performs a specific function.

It seems that the main methodological problem is ignoring the differences in the perception of digital technologies as a factor in economic activity and as a separate segment of the production of goods. Accordingly, all attempts to quantify in terms of value “the digitization of the economy” as a factor of efficiency can hardly be considered objective.

The segment of goods, works and services, one way or another related to the coding of information (“digitalization”) of course is possible and necessary to be estimated quantitatively. It indirectly reflects the degree of application of modern information technologies.

One can hardly agree with the fact that “that the digital economy is a new kind of economic relations” (Sagynbekova 2018), since the essence of economic relations has remained, and most likely will remain unchanged. Most likely, it is a new way of organizing economic relations. But a new way, and even a new form of something, does not change the essence of the original phenomenon.

Thus, it should be concluded that the system of concepts, which describes the organization of economic activity in the context of globalization of information exchange and its partial transfer to the virtual area, is not yet completely perfect, and is not fully authentic to the content of processes in the macroeconomic system, in which digital technologies are not only part of commercial products, but also fundamentally change the properties of transactions.

References

Aptekman, A., Kalabin, V., Klintsov, V., Kuznetsova, E., Kulagin, V., Yasenevets, I.: Digital Russia: New Reality, 19 July 2017. McKinsey Global Institute, McKinsey and Company CI, pp. 121–122 (2017). http://www.tadviser.ru/images/c/c2/Digital-Russia-report.pdf. Accessed 30 Mar 2019

Balanova, S.: The Illusory World: Five Major Myths of the Digital Economy”, 19 September 2017. https://www.forbes.ru/kompanii/350323-illyuzornyy-mir-pyat-glavnyh-mifov-cifrovoy-ekonomiki. Accessed 30 Mar 2019

Bodrunov, S.D.: Noonomics, Moscow - St. Petersburg - London, “Cultural Revolution”, p. 97 (2018)

Negroponte, N.: Being Digital. Knopf, New York (1995)

Oxford Economics: «The New Digital Economy. How it will transform business» A research paper produced in collaboration with AT&T, Cisco, Citi, PwC & SAP, 34 p, June 2011

Program: Digital Economy of the Russian Federation. Order of the Government of the Russian Federation of July 28, 2017 No. 1632-p (2017). http://www.consultant.ru/document/cons_doc_LAW_221756/. Accessed 01 Apr 2019

Sagynbekova, A.S.: Digital economy: concept, prospects, development trends in Russia. Int. Sci. Tech. J. Theory Pract. Innov. **4**(28), 255–267 (2018)

Tkacheva, A.A., Shepelev, K.O.: The first robots and a brief history of the development of robotics, 04 December 2018. http://forum.kafedra-appie.ru/. Accessed 31 Mar 2019

Urmantseva, A.: What is more important: real or digital economy?, 16 June 2017. https://ria.ru/20170616/1496663946.html. Accessed 31 Mar 2019

Yudina, T.N.: Understanding the digital economy. Theor. Econ. **3**, 16 (2016)

Open Innovations as a Tool of the Digital Economy Promotion in Regions

Oleg A. Donichev(✉), Sergey A. Grachev, and Tatyana B. Malkova

Vladimir State University named after A.G. and N.G. Stoletovs, Vladimir, Russia
pav_zah@mait.ru

Abstract. The purpose of the article is to reveal the importance of the use of open innovations in order to build a digital society and sustainable development of economic systems at different levels, including regions, and to give an idea of the areas in which it is necessary to focus attention to the problem has become relevant for scientists and business.

Methodology. The study used methods of analysis, synthesis, comparison.

Conclusion. In the context of the digital economy, the importance of innovation increases significantly. However, the old methods of implementing such innovations are becoming unacceptable. Therefore, to solve this problem, new approaches to the implementation of the planned innovation processes are required. Such new tools could be open innovation. The digital transformation of the economy is supported in addition to the state and the leading enterprises of various industries, but most importantly by innovative enterprises as the main drivers of the emergence of new digital services and consumer-oriented products. At the same time, the principles of innovative openness and the community's joint efforts for the transition to digital technologies are fundamentally important, and the formation of an ecosystem of digital industries becomes the basic solution to the digitalization of the national economy. Network structures play a significant role in the process of formation and dissemination of innovations, because in many cases they are the flow of information about new developments and the exchange of relevant resources. Through these structures, a specific innovation idea is evaluated and supported. Based on this, network structures and open innovations implemented in inter-firm environments cause mutual influence and stimulate each other.

Keywords: Digital economy · Open innovations · Transformation of economy

JEL Code: O10 · O40

1 Introduction

Forming and developing digital economy is one of the most important tasks of the Russian economy. However, the old methods of introducing such innovations are becoming unacceptable. Therefore, new approaches to the implementation of the planned innovation processes are required to solve the task. Amongst such new tools could be open innovations, which, unfortunately, have not found proper use in Russian reality.

E. G. Popkova and B. S. Sergi (Eds.): ISC 2019, LNNS 87, pp. 238–243, 2020.
https://doi.org/10.1007/978-3-030-29586-8_28

But how can open innovations contribute to the promotion of digital society, if all the equipment and accompanying its material and elementary base are imported and the manufacture of such in Russia is very limited. But that is why the system of open innovations is based on the use of scientific achievements, access to which is possible, i.e. in some cases it is open and only needs to be worked on with regard to their application. Moreover, the improvement of the equipment used, including that available in our leading digital companies, is necessary in the context of sanctions restrictions.

According to some experts, digitalization as a specific form of local systems of regional economy innovations is manifested in the fact that at each stage it appears as an innovative idea, functions and develops as an innovation, innovation activity, as the dominant form of economic relations. At the same time, the role of open innovations in the development of digitalization of local systems of regional economy is growing (Postaliuk and Postaliuk 2018).

Other authors believe that in order for the development to bring tangible results of digital technologies in the form of economic benefits from their use, as well as increasing profitability and reducing transaction costs, it is necessary to use the existing experience of both foreign countries and the existing Russian one. At the same time, it is important to increase the opportunities for the creation of equipment, development of the element base, metering and control devices by Russian specialists on the basis of domestic companies (Remizova and Koshelev 2018).

Proceeding from this, we believe it's necessary to stop at the consideration of the problems of formation and development of open innovations in the Russian economy. Proceeding from this it is necessary to note that understanding the essence of the open innovations category was first presented in his works by the American scientist H. Chesbrough in his publication of 2003 (Chesbrough 2003).

In one of his later works, H. Chesbrough presented open innovations as a new paradigm to understanding industrial innovations (Chesbrough 2007). H. Chesbrough's first work translated into Russian was published in 2007. In it he formulated the essence of this category as a type of activity in which, instead of earning money for using one's own technologies independently, there is an opportunity to receive money using various ways, including external ones, to bring one's own technologies to the market. Instead of limiting the research function to the creation of new knowledge, the rich research practice now also includes the assessment of external knowledge and its integration. Instead of managing intellectual property as a way of preventing other organizations and individuals from accessing it, you manage it to improve your business model and benefit from the use of your knowledge by rivals (Chesbrough 2007).

2 Methodology

Later on, a large number of foreign scientists devoted their work to the problems of open innovation. For example, Altman P. and Lee C. structure the tools through which open innovation interactions take place, in their publication, highlighting among them also elements such as collaboration, addition, new revenue streams, source search, collaboration and group activities, while emphasizing the role of information technology as an integral part of digital economy (Altman and Lee 2011).

German scientist Lichtenhaller U. notes that most firms use the principle of closed innovations. But the greatest economic efficiency is achieved in the organization of those who apply open innovations (Lichtenhalter 2008).

In the opinion of Xiangjiang Q., Peng J. and Kelly J., in order to use open innovations, it is necessary to know and take into account the wishes of clients, and then constantly monitor that these requirements are met (Xiangiiang et al. 2013).

Grosfeld T., Ronald T. believe that open innovation can be seen as a process with partners to bring new products and services to market, as well as the creation of new business opportunities through knowledge sharing (Grosfeld and Ronald 2008).

The problems of introducing open innovations for the development and formation of a digital society are reflected in the publications of Russian scientists and their colleagues from Belarus. For example, according to the information provided by N.N. Matkova, an Innovation Center has been created and operates under the auspices of MTS. The task of this center is to ensure the transformation of the company from a traditional telecom operator to a full-fledged digital IT-business. At the moment, the Center solves problems of forming the most effective channels for attracting external ideas and technologies, forms internal methods of work on the development of new digital products, determines the optimal business models for each of the business areas (Matkova 2018).

G.I. Kurcheeva and V.A. Hvorostov believe that open innovations are of crucial importance in the transition of economy to a new technological mode based on the introduction of the digital format. Networks play an essential role in the process of innovation development and diffusion, because in many cases, they are used for the information flow about new developments and the exchange of relevant resources. These structures evaluate and support a specific innovation idea. On this basis, the network structures and open innovations implemented in inter-firm environments cause mutual influences and stimulate each other (Kurcheeva and Khvorostov 2016).

A.M. Dibrov from the Tomsk National Research Polytechnic University argues that the transition of economic entities to an open innovation process can change the nature of organizing economic space, which in turn will lead to the formation of a broad network structure based on digital technologies. This will make it possible to overcome the fragmentation of the national innovation system and will require the formation of an appropriate institutional environment (Dibrov 2016).

According to V.V. Aranjin, Russian companies should be aware that the application of the open innovation model in compliance with all its fundamental principles, including the use of modern digital IT-technologies of their activities. However, it is necessary to understand that in order to popularize the principles of open innovation it is necessary to put a lot of effort and resources both for the enterprise itself and for the state as a whole, as the innovation activity is at a fairly low level (Aranzhin 2017).

M. Slonimskaya from the Institute of Economics of the National Academy of Sciences of Belarus writes that in order to actively implement the theory of open innovations it is necessary to create new types of services, including digital systems, which in turn contribute to their development. In addition, in the European Union countries the so-called "living laboratories" are widely used, the activity of which is based on digital technologies and is aimed at maintaining the activity of all participants of the innovation process - from manufacturers with a special emphasis on the participation of small and medium-sized enterprises in technological chains and orientation to feedback from end users, implementing projects, including the development and implementation of digital content. They are positioned as consumer-oriented open innovation ecosystems based on joint creativity, integrating scientific research and innovation processes in real life (Slonimskaya 2016).

Thus, the problem of using open innovations in national economy as a tool to build a digital economy is very important, but, unfortunately, in our conditions needs wide coverage. The need to apply innovations has sufficient understanding only in some Russian corporations, smaller, medium and small businesses and even a certain layer of scientists do not always represent their role, importance and especially the prospects of implementation as a strategic resource for socio-economic development.

Therefore, the main purpose of this article is to reveal the importance of using open innovations to build a digital society and the sustainable development of economic systems at various levels, including regions, and to give an idea of the areas on which it is necessary to focus attention in order to make the problem relevant for scientists and business.

Meanwhile, the transition to an innovative digital economy is gradually becoming a subject of study and analysis for Russian researchers. According to Y.M. Akatkin and the authors, the digital transformation of economy is supported not only by the state and leading enterprises of various industries, but also by innovative enterprises as the main drivers of the emergence of new digital services and consumer-oriented products. At the same time, the principles of innovative openness and the unification of community efforts for the transition to digital technologies are fundamentally important, and the formation of an ecosystem of digital industries becomes the basic solution for the digitalization of national economy (Akatkin et al. 2017). In turn, Richter K.K. and Pakhomova N.V. believe that the digital society is based on a complex of radical innovations - technological, organizational, institutional, social, etc. At the same time, the core of digital economy are the sectors related to information and communication technologies (ICTs). At the same time, innovative openness is one of the fundamental principles of the formation and development of digital transformation (Richter and Pakhomova 2018).

D.R. Belousov, E.A. Pakhomova emphasizes that the Russian ecosystem is a kind of innovative ecosystem, defined as a variety of interacting, exchanging resources and transforming some of the species into other subjects (participants). It is fundamentally important that innovative contacts of ecosystem entities occur in the community of environments. Such interaction to a significant extent determines the nature of reproduction in this area and the main parameters of its activities (Belousov and Pakhomova 2018).

V.I. Ananyin and his co-authors believe that corporate structures-leaders of digitalization clearly demonstrate that it often leads to a deep innovative transformation of the entire business. Such transformation can radically change the logic of the business itself and, as a consequence, change the structure of its tangible and, especially, intangible assets. Therefore, the analysis of the digitalization of a firm requires an approach that, when solving any problems, allows to constantly maintain the holistic functioning of this organization (Ananyin et al. 2018).

In domestic publications on this subject there are publications of foreign authors devoted to this subject. Thus, G. Prause and T. Turner note that the development of information and communication technologies has made possible new forms of consumer integration into open innovation processes. Virtual communities, groups of practitioners and "living laboratories" are examples of how to integrate dispersed knowledge of users into the practice of making strategic decisions by companies. Virtual communities were a powerful tool for promoting innovations, especially in the area of complex and dynamic socio-economic technologies. Mixed approaches, combining "living laboratories" with virtual communities, seminars and workshops, bring the activity in the field of open innovations to a new level (Prause and Turner 2014).

According to the results of the analysis performed by T.M. Brasser, A. Mladenov and K. Strauss, virtual cooperation is an important topic in the field of open business models of innovation (OBMI), identified in the analysis. It emphasizes the significant role of information technology in supporting open and collaborative business models. In this regard, OBMI's digital platforms are designed to provide business model development teams with support for the innovation community in addressing BMI challenges. Tools of digital business modeling include many components, the use of which in turn contributes to the development of new digital technologies (Brasser et al. 2017).

3 Conclusion

In the context of the digital economy, the importance of innovation increases significantly. However, the old methods of implementing such innovations are becoming unacceptable. Therefore, to solve this problem, new approaches to the implementation of the planned innovation processes are required. Such new tools could be open innovation. The digital transformation of the economy is supported in addition to the state and the leading enterprises of various industries, but most importantly by innovative enterprises as the main drivers of the emergence of new digital services and consumer-oriented products. At the same time, the principles of innovative openness and the community's joint efforts for the transition to digital technologies are fundamentally important, and the formation of an ecosystem of digital industries becomes the basic solution to the digitalization of the national economy. Network structures play a significant role in the process of formation and dissemination of innovations, because in many cases they are the flow of information about new developments and the exchange of relevant resources. Through these structures, a specific innovation idea is evaluated and supported. Based on this, network structures and open innovations implemented in inter-firm environments cause mutual influence and stimulate each other.

References

Akatkin, Yu.M., Karpov, O.E., Konyavsky, V.A., Yasinovskaya, E.D.: Digital economy: conceptual ecosystems of the digital industry. Bus. Inform. **5**, 17–28 (2017)

Altman, P., Lee, C.: The novelty of Open Innovation (2011). http://www.diva-portal.org/smash/get/diva2:471149/FULLTEXT01.pdf. Accessed 16 Mar 2019

Ananyin, V.I., Zimin, K.V., Pugachev, M.I., Gimranov, R.D., Skripkin, K.G.: Digital enterprise: transformation into a new reality. Bus. Inform. **2**, 45–54 (2018)

Aranzhin, A.A.: Problems of open innovation research in the activities of Russian organizations. Vestnik YuruGU. Series "Economics and Management", vol. 11, no. 4, pp. 46–49 (2017)

Belousov, D.R., Penukhina, E.A.: About building a qualitative model of the Russian ICT ecosystem. Probl. Forecast. **3**, 94–101 (2018)

Brasser, T.M., Mladenov, A., Strauss, K.: Open innovations in the field of business models: review of the literature and directions of further research. Bus. Inform. **4**, 7–16 (2017)

Chesbrough, H.W.: Open Innovation: A New Paradigm for Understanding Industrial Innovation, pp. 1–27. Oxford (2006)

Chesbrough, H.W.: Open Innovation: The New Imperative for Creating and Profiting from Technology, p. 273. Harvard Business School Press, Boston (2003)

Dibrov, A.M.: Collocations as a form of open innovation process. Innov. Manage. Theory Methodol. Pract. **11**(18), 7–11 (2016)

Grosfeld, T., Ronald, T.: The logic of open innovation: value by connecting network and knowledge. Foresight **2**(1), 24–29 (2008)

Kurcheeva, G.I., Khvorostov, V.A.: Open innovations as a factor of modern technological development. Internet Mag. "Sci. Stud." **8**(4), 1–7 (2016)

Lichtenhalter, U.: Open innovation in practice: an analysis of strategie approaches to technology transactions. TEEE Trans. Eng. Manage. **55**(1), 150–151 (2008)

Matkova, N.N.: Case stages: implementing open innovations on the example of MTS. Innovations **3**, 92–99 (2018)

Postaliuk, M.P., Postaliuk, T.M.: Digitalization of the local systems of the Russian regional economy: needs, opportunities and risks (in Russian). Prob. Mod. Econ. **2**, 174–177 (2018)

Pouché, G., Turner, G.: Consumer communities - drivers of open innovations. Forsyth T **8**(1), 24–32 (2014)

Remizova, T.S., Koshelev, D.B.: Application of the digital technologies for the modernization of the infrastructure of the electric power industry of Russia. Prob. Mod. Econ. **2**, 31–34 (2018)

Richter, K.K., Pakhomova, N.V.: "Digital economy as an innovation of the XXI century: challenges and opportunities for sustainable development (in Russian). Prob. Mod. Econ. **2**, 22–31 (2018)

Slonimskaya, M.: Living laboratories in the practice of open innovation. Sci. Innov. **9**, 30–32 (2016)

Xiangiiang, Q., Peng, J., Kelly, J.: Open innovation: T Leader in distributed global innovation. In: The Global Innovation Index 2013, WIPO, Geneva, pp. 95–98 (2013)

Social, Psychological and Worldview Problems of Human Being in Digital Society and Economy

Anna Guryanova[1(✉)], Elmira Khafiyatullina[2], Marina Petinova[2], Nonna Astafeva[2], and Nikolai Guryanov[2]

[1] Samara State University of Economics, Samara, Russia
annaguryanov@yandex.ru
[2] Samara State Technical University, Samara, Russia
dek.fispos2009@yandex.ru, shloss@yandex.ru, nonnaast@yandex.ru, nik.guryanow@yandex.ru

Abstract. The article deals with the problem of human change in conditions of modern society and economy. An impact of digitalization on human being and the resulting problems of ideological, psychological and social nature are considered. Dialectical interactions of real and virtual worlds, human mind and artificial intelligence, on-line and off-line communication are also analyzed. Their correlation is a most difficult humanitarian problem of the digital society. «Homo digital» is characterized as a product of digitalization and the owner of the fundamentally new qualities and value orientations. The authors are sure, correct human values are vitally important for the digital world. Their wrong interpretation leads to the triumph of consumerist ideology and social degradation. Equality and humanism, by contrast, are qualified as the main values of the modern society. But it can't be called «humanistic» yet. To make it so, the basic human values must be applied in conditions of digitalization. The real progress of humanity is possible on the base of self-improvement, human consciousness, intelligence and spiritual qualities development.

Keywords: Digital society · Digital economy · Homo digital · Human values · Virtual world · On-line communication · Artificial intelligence · Equality · Humanism · Consumerism

JEL Code: Z 10 · Z 13 · Z 19

1 Introduction

In modern conditions society has reached a qualitatively new stage of its historical development which is increasingly characterized as «digital» or the «era of digitalization» (Digital Globalization 2016). It causes radical revision of many traditional values and social relations. For example, in the economic sphere there is a trend of active using of digital technologies. This leads to dynamic development of the «digital economy». In the political sphere a special «digital elite» is forming now. It plays a role of the holder of innovative resources and competencies necessary to control the public

E. G. Popkova and B. S. Sergi (Eds.): ISC 2019, LNNS 87, pp. 244–250, 2020.
https://doi.org/10.1007/978-3-030-29586-8_29

opinion. Within the social structure the new demanded professional groups are quickly forming. They carry out digital activities effectively, while the others, who don't have such skills, lose their social status forever. In the spiritual sphere there are also great changes associated with a new type of digital culture with its special functions and orientations.

Based on all this a special human type – the so-called «homo digital» – is successfully forming in conditions of the digital society and economy (Guryanova et al. 2018). «Homo digital» has a set of fundamentally new qualities and value orientations. We'll characterize them later.

2 Materials and Methods

Several research methods are used to consider the problems of human being in digital society and economy.

Dialectical method is used to consider social, psychological and worldview problems of digital society and economy in their interrelations and interdependence.

Comparative method is used to correlate such vital opposites of the modern life as real and virtual worlds, human mind and artificial intelligence, on-line and off-line communication, etc.

Methods of analysis and synthesis are used to find out advantages and disadvantages of human existence in the virtual world and to present virtual life as a way of effective social being.

Descriptive method is used to show specific characters of «homo digital» as an owner of the new qualities and value orientations.

System method is used to identify various values forming and replacing each other in conditions of digitalization.

Classification method is used to identify the real values of the digital society such as equal access to information, free on-line communication, etc.

Prognostic method is used to consider the perspectives of the digital society development with an aim in view to turn it to humanism and to abandon consumerism.

3 Results

3.1 Homo Digital as a Subject of Virtual Life

3.1.1 Virtual Life as a Way of Effective Social Being

«Homo digital» is a special type of a human who is focused on a new basic strategy of effective and safe social behavior. This comes from his ability of using digital, information and analytical technologies. With their help, homo digital analyzes the receiving information, makes its adequate interpretation and then (on its basis) constructs a certain model of his social behavior. It should be specially mentioned that modern existentially active people spend most of their free time in a virtual environment (Guryanova et al. 2019). In this regard, it becomes especially valuable for them to be able to receive any information they need easily and in time, to communicate on-line, to operate freely with information resources and flows.

This virtual way of life is possible because of the modern digital technologies, vitally significant for the modern humans. On their basis they form individual behavioral strategies both in the sphere of their professional interests and everyday life. With their help the humans communicate with the world around them – they acquire knowledge, build business, spend their leisure time, contact with each other without any spatial boundaries.

3.1.2 What's a Result of the Real and Virtual Worlds Mixing?

From the very moment of the Internet appearance (1982) the virtual world began to form. Since then it has actively developed by including a great number of new elements such as forums, on-line computer games, social networks, etc. Each of these blocks is both a structural element of the virtual world, and a bridge connecting it to the real world. Obviously these worlds are interconnected and interdependent as, for example, a real person and his virtual image in the social network.

Today we can identify every phenomenon by its belonging to one or another world. But there are also many objects that can't be introduced in such a way. At the point of the real and virtual worlds connection and mixing a new «hybrid» world is forming (Keshelava 2017). It has other laws and rules, different from those that are usual for us today. It gives us possibility to make vital actions in the real world through the virtual world. The necessary prerequisites for this process are the high efficiency and the low cost of information and communication technologies (ICT) and the availability of the digital infrastructure.

3.1.3 Advantages and Disadvantages of Human Existence in the Virtual World

Modern people enjoy their virtual «being-on-line». The Internet provides them access to the wide information resources and facilitates technically the process of communication. In fact virtual reality gives humans an opportunity to feel themselves «in the epicenter of events», to observe the real course of objects and processes, to evaluate them in real time. This opportunity becomes for many of them a sort of existential need. It activates a set of subjectively-emotional human factors in their desire to show awareness, to know everything about what is happening.

For many people staying in virtual environment is an undoubted benefit guaranteed them by the development of civilization, – it's on the one hand. But on the other – a long being in the virtual sphere cases loneliness and isolation from other people, deepens the distance between the human and the society. Besides, integration into the system of real social relations is much more difficult than lightweight manipulations in the virtual reality.

Virtual existence doesn't involve deep self-cognition or even elementary understanding of the internal dialectically interrelated reasons for the human behavior. Modern humans don't have enough time to carry out a purposeful reflection of what is happening. They don't want even to think about and predict situations, explain their decision-making. Their aim is only to activate maximally the process of their digital activities.

3.2 Psychological and Cognitive Problems of the Digital Society

3.2.1 Psychological Effects of Virtual Communication

In most cases virtual communication doesn't involve the possibility of visual perception of the dialog companion. This often causes replacement of the real communicants by the fictional, artificially constructed figures. This situation is typical for the modern social networks and has a number of psychological reasons. Communicants get used to their fictional image so well that they begin to identify themselves with it and forget their real characters and features. As a result, they can't realize their true needs and desires. So, they get depressed and sometimes lose the meaning of their lives.

Moreover, distance communication is significantly poor in terms of emotions. It's impossible to get the full information distantly, while the process of real communication is much more reach in looks, gestures, intonation, etc. An absence of emotional component is a negative aspect of the on-line communication.

Replacement of the greater part of interpersonal communications into the virtual reality is a problem of nowadays. The result is a widespread lowering of the literacy level (especially among the youth), inability of correct writing and speaking. The fact is that communication on-line is usually carried out at the level of our everyday speech. This doesn't help in learning rules of spelling. Therefore, the skills of the competent writing aren't assimilated.

3.2.2 What's Better – Human Mind or Artificial Intelligence?

The question of natural intelligence's value is very important today because of the widespread using of the artificial intelligence (AI). The last one includes not only the complex technologies that simulate the mental processes of human brains, but also the simple electronic computers such as a personal computer, smartphone, etc. Today AI technology is deeply involved into the human life.

On the one hand, human mind is more perfect than any computer because it's capable to learn and improvise. The humans are able not only to act in accordance with certain algorithms but to develop them and to go beyond its boundaries at any time they want, using, for example, an intuition method. But on the other hand a great devaluation of human mind is taking place. The reason is that computers take over part of the functions of human intelligence: they can make calculations for us, check our spelling, remind us of important events, etc. So, the human mind has a chance to «relax». It loses its stimulus and potential for development and in some cases even degrades.

3.2.3 Status of Knowledge and Cognition in the Modern Society

The status of religious and art knowledge is an important problem too. These types of knowledge are focused on ideal objects and phenomena. So, the life of a spiritually developed person is incomplete without them. They make possible such human abilities as empathy and respect for the others, love for the beauty and sublime. It's impossible to realize complex and comprehensive development of the surrounding world without these types of knowledge. However, they are losing their value. Fewer people need them in their lives and activities.

Phenomenon of knowledge itself is also devalued. The most important human quality today isn't connected with having knowledge, but with acquiring it. You must be able to take out the necessary data from the wide information field in the right amount and at the certain time you need it. Keeping this knowledge alive in human brains for a long time has no sense because of the presence of hard disks and computer servers. So, it seems unnecessary to most members of the modern society. An erudite, encyclopedically educated person loses his social status. He is no longer respected and interesting for others. He isn't in demand in the modern society.

3.3 Social and Economic Values in the Digital Epoch

3.3.1 Equality and Humanism as the Values of the Digital Society

Social values are also actual for the digital society (Guryanova et al. 2020). The most important of them is equality. In the case of our research we mean equal access to information resources and digital technologies. They must be accessible both for people from different regions and countries with various levels of development. Therefore, one of the most important aspects of the state activity is to provide such an equality. Without social justice it's very difficult to ensure equality of starting opportunities, equal access to educational, medical and public services.

Modern digital civilization is often called «society of equal opportunities». That's true, because of the development and widespread using of information and digital technologies, distance education, employment, services, etc. Besides, humanization in such a society should be the main goal of public policy and social life. Non-humanistic society can't provide a proper standard of living for its members in both physical, social and spiritual spheres. After all, technical and technological progress has no sense without development of the culture itself.

3.3.2 Changes of Economy in the Digital Epoch

Strong changes affect the economy in the digital epoch. The ideology of hard honest working is replaced by the values of financial success (Shestakov et al. 2017). A similar situation we can see in the sphere of information production. Today, the holders of information (or those who have rights to use it) have a higher social status, more power and material well-being than those who really produce knowledge and technology. The last ones are staying in a lower social position.

Digital society generates consumerism and related value of unlimited consumption. Modern civilization has already reached a sufficient level of development in the fields of information and communication. So, it's ready to promote ideas of unlimited consumption and a free personal time for its realization. Developed technical, technological and economic spheres allow modern humans to expand the scale of production and to provide unlimited consumption (Guryanova et al. 2017). As a result, the human himself transforms into the so-called «homo consumens». He isn't interested in creativity or self-realization. The preferable way of life for him is acquiring things and adoring them.

3.3.3 Consumerist Worldview and Personal Transformations

Modern world is a real era of consumerism. S. Miles correctly calls it a «religion of the late 20th century» (Miles 1998). It's a widespread phenomenon manifesting itself not only in the individual consciousness, but in the social ones. «Homo consumens» acquires material values to feel his social leadership or to join the preferable social group. In fact, it's a real cultural revolution, having, incidentally, a negative value. Today museums, exhibitions, galleries and theaters give the main way to shopping, nightclubs, social networks and computer games. The humans don't want spiritual and intellectual development any more. They claim satisfaction of their basic physical needs and desires.

The modern world replaces sublime spiritual values by utilitarian consumerist ideology. Today a leader must meet some external criteria of success, well-being and prestige. An object of greater respect here is an owner of a modern smartphone, computer, etc., but not someone who is better educated, broad-minded and creative. Such situation is unacceptable and must be radically changed to make the digital society humanistic.

4 Discussions

The most debatable for the period of the digital society development is a problem of human values transformation including their social, psychological and worldview aspects.

In modern conditions values of real interpersonal communication, knowledge, religious and art cognition, independent and free thinking have already lost their meaning. They were replaced by the values of consumption, adaptation to virtual reality, free access to ICT and digital resources. Fundamental social category of equality found itself in the information sphere. So, a new kind of the human – the so called «homo digital» – is forming in modern conditions. His preferences aren't spiritual but the material ones. He doesn't look for ideal essences, but for the material values. He fills his life with computer games and virtual communications.

Unfortunately, modern digital society can't be characterized as humanistic. To make it true the basic humanistic values should be applied to the conditions of digitalization. This is necessary because the progress of the material and information spheres is only an external side of social development. Digitalization is a factor that makes human existence simpler and easier. But the real progress of humanity is connected, first of all, with a personal self-improvement, with human consciousness, intelligence and spiritual qualities development.

5 Conclusions

Summing up, we'd like to note that virtual communication is the most important way of cooperation between people in the digital society. It takes place with a help of various digital resources and ICT. At the same time the process of real (off-line) communication becomes secondary for most of modern people. This changes much communication priorities and has both positive and negative consequences.

The positive aspects of on-line communication are the same: it gives us possibility to cooperate among ourselves without any space-time boundaries. Actual information can be also acquired at the right time and when we really need it. Virtual communication helps us to save time. It simplifies business communication and document flow. However, there are also negative consequences, for example, an absence of a strong emotional context, typical for the process of real (off-line) communication. Here we can also mention reduced literacy, inability to articulate thoughts, devaluation of reading, etc.

All these disadvantages represent the most important humanitarian problem of the digital society. Digitalization provides humans great opportunities for receiving and processing different kinds of information. However, digital technologies, such as TV and the Internet, aren't able to carry out the most important humanitarian and educational functions (especially in relation to the youth). Besides, the data they provide for their direct consumers has sometimes a controversial character – not informative, but entertaining or, what's worse, dehumanizing.

References

Digital Globalization: The New Era of Global Flows. McKinsey & Company (2016)

Guryanova, A., Astafeva, N., Filatova, N., Khafiyatullina, E., Guryanov, N.: Philosophical problems of information and communication technology in the process of modern socio-economic development. In: Popkova, E., Ostrovskaya, V. (eds.) Perspectives on the Use of New Information and Communication Technology (ICT) in the Modern Economy, vol. 726, pp. 1033–1040. Springer, Cham (2019)

Guryanova, A., Guryanov, N., Frolov, V., Tokmakov, M., Belozerova, O.: Main categories of economics as an object of philosophical analysis. In: E, Popkova (ed.) Russia and the European Union: Development and Perspectives, pp. 221–228. Springer, Cham (2017)

Guryanova, A., Khafiyatullina, E., Kolibanov, A., Makhovikov, A., Frolov, V.: Philosophical view on human existence in the world of technic and information. In: Popkova, E. (ed.) The Impact of Information on Modern Humans. Advances in Intelligent Systems and Computing, vol. 622, pp. 97–104. Springer, Cham (2018)

Guryanova, A.V., Smotrova, I.V., Makhovikov, A.E., Koychubaev, A.S.: Socio-ethical problems of the digital economy: challenges and risks. In: Ashmarina, S., Mesquita, A., Vochozka, M. (eds.) Digital Transformation of the Economy: Challenges, Trends and New Opportunities, vol. 908, pp. 92–106. Springer, Cham (2020)

Keshelava, A.V. (ed.): Introduction to Digital economy. Vniigeosystems (2017)

Miles, S.: Consumerism – as a Way of Life. SAGE Publications, London (1998)

Shestakov, A., Noskov, E., Tikhonov, V., Astafeva, N.: Economic behavior and the issue of rationality. In: Popkova, E. (ed.) Russia and the European Union: Development and Perspectives, pp. 327–332. Springer, Cham (2017)

Economic Security of Businesses as the Determinant of Digital Transformation Strategy

Evgeniya K. Karpunina[1](✉), Maria E. Konovalova[2], Julia V. Shurchkova[3], Ekaterina A. Isaeva[4], and Alexander A. Abalakin[4]

[1] Tambov State University named after G.R. Derzhavin, Tambov, Russia
egenkak@mail.ru
[2] Samara State University of Economics, Samara, Russia
mkonoval@mail.ru
[3] Voronezh State University, Voronezh, Russia
jshurchkova@mail.ru
[4] Moscow Financial and Industrial University "Synergy", Moscow, Russia
dipmesi@mail.ru, alexander.abalakin@yandex.ru

Abstract. The article reveals the economic essence of the digital transformations taking place in the modern world which make significant changes in the system of social reproduction and the business activities.

The dichotomy of the role of businesses in the process of digital transformation is described. On the one hand, business is a subject of ongoing changes, it responds to them and adapts to them. On the other hand, business acts as a driver of digital transformations and generates changes themselves, as it is in a continuous search for new technologies to improve its own stability and competitiveness.

It is determined that digital transformations do not only open new opportunities for businesses, but also create a number of threats to its economic security. Menaces to economic security of businesses (information technology, threats from personnel, threats from the management system) in the context of the importance of factors and resources in the modern world are classified. The economic damage caused by the penetrating threats is analyzed.

Most of the described threats to economic security of businesses can turn into opportunities and strategic advantages of growth. To do this, it is advisable to provide digital transformation of businesses in the system «technology-people-strategy».

The strategy of businesses' digital transformation, the determinant of which is the «economic security of businesses», and the key areas involving leveling information technology threats to economic security, human capital development, corporate information management are suggested. The implementation of this strategy will transform threats to the economic security of businesses into new opportunities for their development.

Keywords: Businesses' economic security · Digital transformation · Digital transformation strategy · Economic damage · Economic security threats

E. G. Popkova and B. S. Sergi (Eds.): ISC 2019, LNNS 87, pp. 251–260, 2020.
https://doi.org/10.1007/978-3-030-29586-8_30

JEL Code: O3 · M150

1 Introduction

Digital transformation is one of the most striking global trends of modern times. It is associated with such concepts as «digital economy», «information society», «Industry 4.0», scientists around the world operate these terms. All subjects of society are subject to digital transformation: from individuals, businesses to the state and the entire global order. It defines a completely new way and quality of life of people, the principles of their interaction in the new reality.

Significant changes in the system of social reproduction affect, first of all, the activities of businesses. On the one hand, businesses are the subjects of changes, they must adequately respond to them and adapt to them as quickly as possible, change the assortment and price policy, mechanisms of sales and promotion of products, personnel management tools, development strategy. On the other hand, businesses act as a driver of digital transformation, they generate changes themselves due to the continuous search for new technologies to improve their own stability and competitiveness, optimize interaction with consumers as well as enter new markets.

The described dichotomy defines the special role of businesses from the standpoint of economic security. Today the threat landscape is growing and becoming more diverse. Businesses affected become more vulnerable. At the same time, security protects the businesses, allows them to innovate, create new products and services and gives them a strategic advantage in terms of growth.

Thus, for the research, it is interesting to consider digital transformations that determine the activities of businesses, and identify those threats to economic security that can be transformed into new opportunities for their development.

2 Methodology

This study is a logical consequence of the attempts of scientists to synthesize several scientific areas relevant to the science and practice of modern management of the national economy.

The first area is political economy, the object of which is the structure and patterns of development of economic relations and business activities, as well as the modern stage of management called «digital transformation». In this context, the study is based on the analysis of digital transformation trends presented in the works of Adams and Bennett (2018), Bell (1973), Castels (2000), Drucker (1969), Sukharev (2012), Varnavsky (2015) and others. In addition, the study is based on the findings of the author of this article Karpunina et al. (2018a, b) in a number of scientific publications, the object of study of which is the digital economy, the digital economy ecosystem and the features of economic relations in the conditions of informatization.

Another area is the theory and practice of businesses' economic security. The content of this research in relation to this publication includes the identification and analysis of threats to economic security of businesses due to digital transformation, as well as their systematization, identification of economic damage and the formation of protective mechanisms. Deepening this context, the existing theoretical and methodological developments in the field of identification and assessment of economic security threats, which are incorporated in the studies of domestic and foreign scientists has been analyzed: Gaponenko (2008), Kolosov (2010), Oleinikov (2004), Popov and Semyachkov (2017).

The existing infinity of the Internet technologies defines the environment of economic activities and creates the conditions not only for greater competition in the market, but also for the emergence of new economic security threats. Fragmentary attempts to describe this segment were made by Bartlett (2017), Brandman (2006) but their studies focused on an information component lacked an economic one.

The information base of the research consists of statistical and analytical materials of international databases and consulting agencies, including Gartner, IDC Insights, CIO, RSA, GreenHouseData, reports of the World Intellectual Property Organization (WIPO) of the UN, international reports of Check Point Software Technologies, publications of the World Economic Forum.

The methodological basis of the research is the dialectical method of cognition, the system-functional method, the method of comparative analysis, as well as scientific abstraction, induction and deduction, grouping, classification, economic and statistical methods of information processing, the method of expert evaluation.

3 Results

Digital transformation involves using digital technologies to change the processes in order to become more effective, efficient, take a leading position.

Digital technologies are replacing traditional ones, changing the nature of production. It becomes intellectually saturated and actively uses flexible technologies of creating a unique product, various combinations of the latest technologies and advanced technical solutions.

Businesses are moving to new technologies and production methods: information systems for collecting and processing information, saving energy and materials provide multiple acceleration of operations and simplification of processes; technological operations formalize production processes into logical schemes, and their implementation is carried out under the condition of instant data exchange.

Changes in the exchange of goods and services are also becoming virtual. Electronic payment modules and systems become a convenient and efficient form of exchange operations.

On-line services of the global Internet space most actively satisfy the needs of consumers of goods and services, while consumer preferences are shifted towards their immediate satisfaction through the same Internet technologies.

The ongoing transformations require changes in production management systems, they must become flexible, analytical, cognitive, mobile.

Modern requirements for the management of businesses involve their ability to change continuously according to the development of financial and economic relations, management of these changes, expansion of network interaction, which allows to maximize the existing potential of niche players. The staff of businesses is expected to improve development. Responding to the challenges of modern times, human capital should be constantly increased through continuous learning based on advanced technologies, and businesses should invest in this development in order to have sustainable competitive advantages in the future (Fig. 1).

«Digital transformation is gaining momentum, competitive pressure from first-time users is starting to force others to start transformational efforts, no matter where they are geographically. According to research by IDC Insights, even in Central and Eastern Europe, which is characterized by a relatively slow absorption, today enterprises and public sector organizations are rethinking their approaches and adopting changes, as evidenced by a significant increase in spending in the region - an increase of 22.1% year - on-year-in 2018» (Smith 2017).

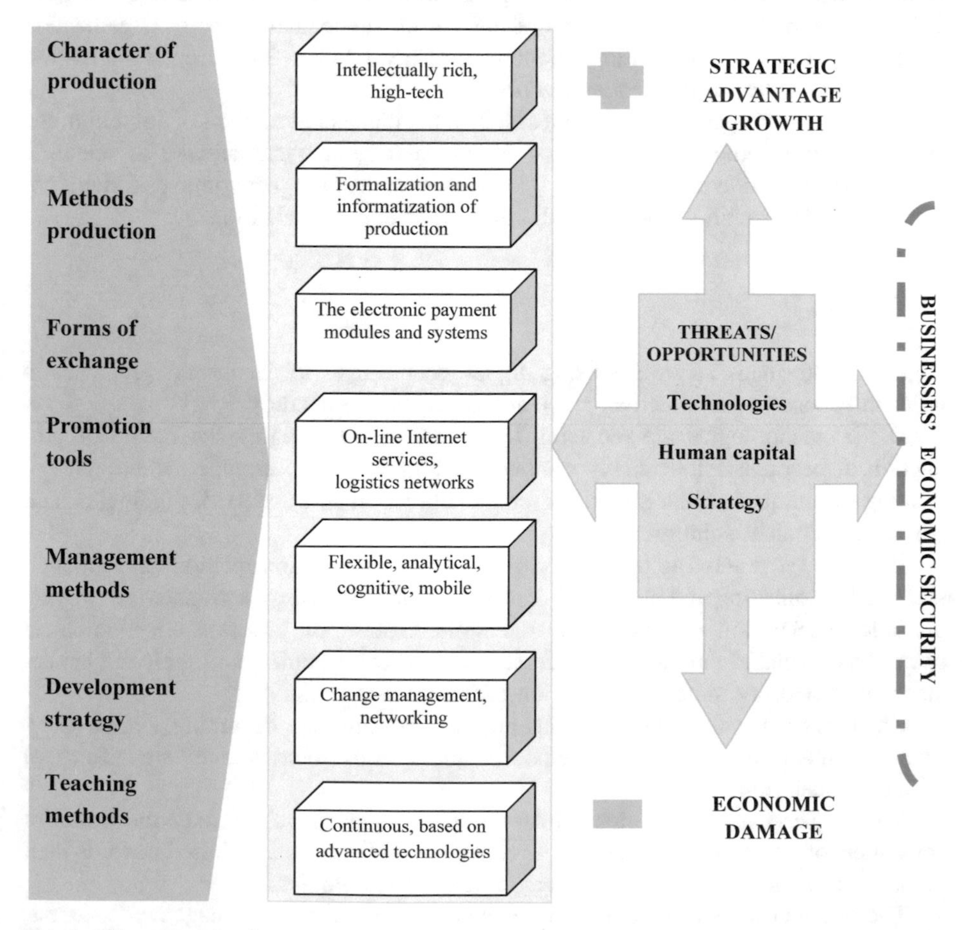

Fig. 1. Changing the forms and content of business activities in the conditions of digital transformation (compiled by authors)

Of course, the introduction of digital transformation projects by businesses allows it to remain at the forefront, it becomes its new opportunity that determines the strategic advantages of growth.

Confirmation of this fact can be reflected in the results of a survey by analysts of Gartner 460 managers, according to which 62% of respondents stated about the existing programs of transformation in the direction of digitalization of their businesses. 54% noted that their digital businesses goal is being transformed, and 46% expressed a desire for business optimization (Fig. 2) (Samuels 2018).

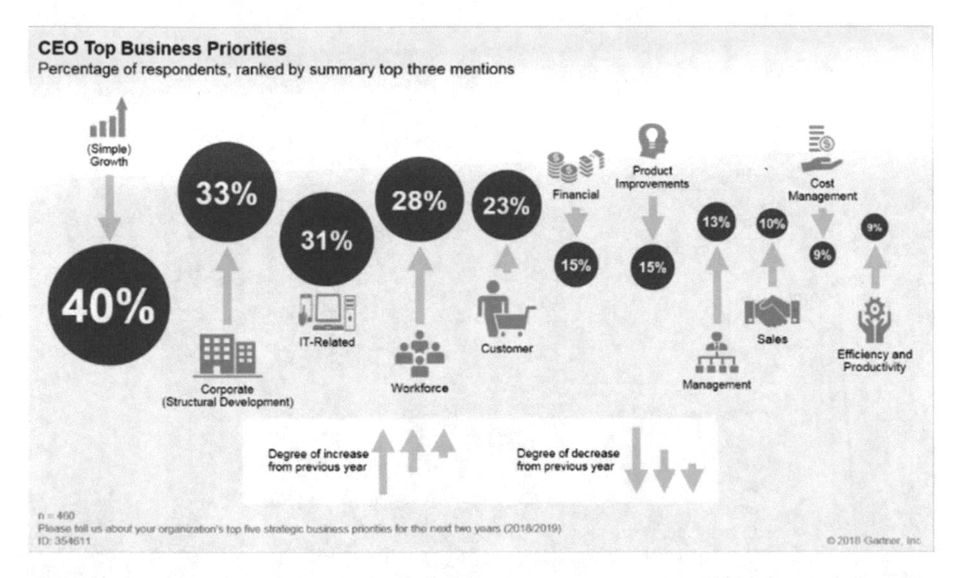

Fig. 2. Management's attitude to digital businesses transformation projects (compiled by Gartner analysts)

According to IDC, the global spending on digital transformation technologies - hardware, software and services - amounted to $ 1.3 trillion in 2017 (Shirer 2018). At the same time, spending is expected to almost double in the period until 2021, when the total amount spent on digitalization worldwide will exceed 2.1 trillion dollars.

Most of the costs of digital transformation in 2018 ($662 billion) will be spent on digital transformation. It was spent on technologies that support new or enhanced operating models, as businesses sought to make their operations more efficient and responsive through the use of products, assets, people, and digital trading partners. $ 326 billion in 2018, the company spent on innovation support technologies that transform the way customers, partners, and staff interact, as well as products and services designed to meet unique and individual demand.

$ 240 billion was invested in information in 2018, as businesses need to obtain and use information to optimize operations and new products and services, make better decisions and achieve competitive advantages.

But digital transformation does not always open up new opportunities for business. Therefore, the fundamental task is to ensure its own economic security of the businesses.

To realize all opportunities, security must be a fundamental part of any digital businesses transformation strategy in a growing and increasingly diverse threat landscape. And as businesses goes digital, the number of external touchpoints will only grow, making them increasingly vulnerable (Menon 2017).

Otherwise, existing security threats lead to inevitable economic damage, which affects businesses (Fig. 3).

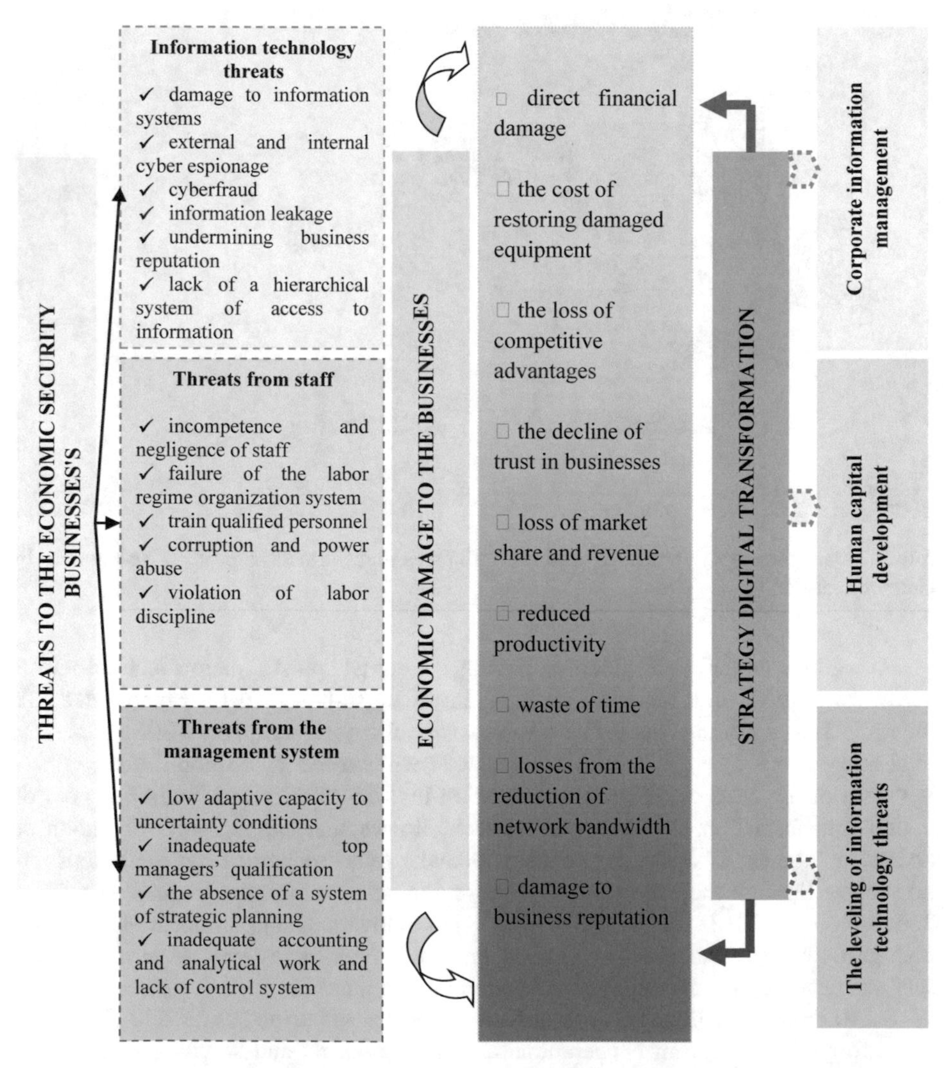

Fig. 3. The concept of the relationship between threats to economic security of businesses, economic damage and digital transformation strategy (compiled by the author)

We have combined the key threats to the economic security of the businesses into three groups: information technology, threats from staff, threats from the management system. This grouping is proposed in the importance of factors and resources in the modern world: information, intelligence, management.

In the context of digitalization, information becomes a powerful means of competition, it acquires a value determined by the amount of profit from its use or the size of the likely damage to the owner. The information extracted from the businesses can be used against it. Therefore, the most serious threats to businesses are information leakage, cyber espionage, intentional damage to information systems. They try to get information about businesses in order to achieve certain advantages: leadership in business negotiations, a more stable position in investing and implementing innovations, a competitive advantage in the market, the absorption of businesses in the approaching bankruptcy. Obtaining information of this nature by competitors and detractors causes both direct and indirect damage (for example, the need to update or replace the software in the event of a malfunction, payment of salaries to service personnel, etc.).

For example, from a report «The 2018 Global Fraud and Identity Report» следует, что «almost three-quarters of businesses (72%) cite fraud as a growing concern over the past 12 months 2018 and nearly two-thirds (63%) report the same or higher levels of fraudulent losses over that same period» (Experian Information Solutions 2018).

The harm done to a business's reputation, as well as the financial costs, can be huge. Some estimates put the average cost to a company at $172 per stolen record. This includes remediation costs, cost of lost business due to downtime, regulatory fines and legal costs (Check Point Software Technologies 2018).

In addition, significant losses to the businesses can cause their own staff. Typical reasons that threaten businesses are the low level of qualification of personnel, their unwillingness or inability to bring maximum benefit to the company, regular violation of labor discipline and corruption.

However, in the context of digital transformation at the forefront of the problem of low information literacy of staff (staff that are not trained in advanced methods of security and have weak passwords, visit unauthorized websites and/or click links in suspicious emails or open email attachments pose an enormous security c employers (Schiff 2015)), uncontrolled access to the Internet (according to IDC, about one third of their working time is spent on the Internet by employees of various organizations and companies for purposes that are not directly related to their work).

According to GreenHouseData, «fraudulent employees, especially it team members with knowledge and access to networks, data centers, and administrator accounts, can cause serious damage» (Schiff 2015). For example, the Sony hack was not carried out by North Korea, but was the result of internal work of employees.

Professional competence should be treated as an important corporate asset, enabling the organization to gain long-term competitive advantage. Competence of employees is determined by the ability of the company to perform the tasks it is entrusted with, including the ability to perform work. The volatility of economic conditions requires the introduction of continuous changes to adjust competence to the existing situation. This can be treated as the development of human capital in the company, contributing to the increase in its value (Kumar and Shah 2009).

Perfectly well-established and working management system will not cause damage to the economic security of the businesses, it will monitor the factors of the external environment and, accordingly, flexibly adjust the business model, adapt new schemes to the most important trends of digital transformation.

However, otherwise, if there are gaps in the work of management personnel, "ossification" of the management system, as well as the lack of a clear planning system and the inability to minimize the consequences of hacking and security leaks, to assess the risk, to take control measures or procedures to protect them, the management system can become the most important threat to the economic security of businesses.

4 Conclusions/Recommendations

Most of the described threats to economic security of businesses can become their opportunities and strategic advantages of growth. To do this, it is advisable to provide digital transformation of businesses in the system «technology-people-strategy».

In our opinion, «economic security of businesses» should be a basic determinant of the digital transformation strategy, and the key areas – the leveling of information technology threats to economic security, human capital development, corporate information management.

I. The strategy of leveling information technology threats is based on the differentiation of tools to reduce information threats to economic security (Karpunina et al. 2018a, b):
   (1) «preventive» policy aimed at creating the institutional conditions for business activity, improvement of legal field, the formation of the system of protection of interests of businesses, stimulation of entrepreneurial activity of businesses and the maintenance of normal competition;
   (2) «proactive» policy creates conditions for preventing fraudulent schemes and information leakage, it is connected with investing in the development of corporate information security systems, using automated tools to search for manifestations of criminal activity and attempts of cyber espionage;
   (3) «reactive» policy is focused on the implementation of measures in the event of a direct manifestation of the threat (blocking of mail; the use of large amounts of data; the use of modified versions of the software).

II. Human capital development involves:
   (1) business investing in the continuous development of professional competencies of employees, including the field of information and economic literacy;
   (2) creation of such a system of personnel motivation, which will prevent «leakage of intelligence» and data to competing firms;
   (3) implementation of a set of measures for the development of corporate culture;
   (4) involvement of consultants and specialists in conflict resolution in the organization.

III. Implementation of corporate information management system. In our opinion, this will prevent a number of threats arising from gaps in the business management system. Corporate information management system is a set of strategic vision, policies and technologies of information control and protection throughout the organization in order to optimize the value of information, compliance with regulatory and legal obligations, risk management and support of business goals (Adams and Bennett 2018).

The purpose of such a system is to maximize the value of information, recognize the potential value of data and minimize the risks and costs of storing information (which arise from data leakage, production of documents in court and regulatory requests, storage of redundant, outdated and trivial data).

To manage the corporate information management system, top-down management is necessary, as well as the definition of strategic goals and priorities for the management of information assets. The objectives will vary depending on the nature of the business, the size of the organization, the location of the particular organization with respect to cyber security threats, the types of data collected and the opportunities to benefit from the data and the use of new technologies.

The proposed management system allows the top management of the organization to respond flexibly to potential threats to economic security, timely prevent the misuse or leakage of personal information of customers, manage their own reputation in the public domain and maintain customer confidence.

References

Adams, M., Bennett, S.: Corporate governance in the digital economy: the critical importance of information management. J. Gov. Guide **70**(10), 38–54 (2018)

Bartlett, D.: Underground Internet. The Dark Side of the World Wide Web, Moscow (2017)

Bell, D.: The Coming of Post-Industrial Society. Basic Books, New York (1973)

Brandman, E.: Globalization and information security. J. Philos. Soc. **1**, 31–41 (2006)

Castels, M.: Information age: economy, society and culture: transl. with English under the scientific the editorship of O. I. Shkaratan, Moscow, HSE (2000)

Check Point Software Technologies: 2018 security report welcome to the future of cyber security (2018). https://www.checkpoint.com/downloads/product-related/report/2018-security-report.pdf. Accessed 12 Jan 2019

Drucker, P.: The Age of Discontinuity. HarperCollins, New York (1969)

Experian Information Solutions: The 2018 Global Fraud and Identity Report (2018). https://www.experian.com. Accessed 13 Mar 2019

Gaponenko, V.: Economic Security of the Enterprise: Approaches and Principles. Fizmatlit, Moscow (2008)

Karpunina, E., Yakunina, I., Yurina, E.: Realization of the National and State Economic Interests of Russia in the conditions of formation of the 4.0 Industry. In: Innovation Management and Education Excellence Through Vision 2020 2018 Proceedings of the International Conference 31 IBIMA in Milan, Italy, pp. 1488–1495 (2018a)

Karpunina, E., Yurina, E., Samoylova, S.: Digital economy and business structures: a new meaning of interaction or threat to economic security? In: Vision 2020: Sustainable Economic Development and Application of Innovation Management from Regional Expansion to Global Growth 2018 Proceedings of the International Conference 32 IBIMA in Seville, Spain, pp. 3085–3092 (2018b)

Kolosov, A.: Economic Security and Economic Systems. RAGS, Moscow (2010)

Kumar, V., Shah, D.: Expanding the role of marketing: from customer equity to market capitalisation. J. Mark. **73**(6), 119–136 (2009)

Menon, N.: There can be no digital economy without security (2017). https://www.weforum.org/agenda/2017/05/there-can-be-no-digital-economy-without-security/. Accessed 12 Mar 2019

Oleinikov, E.: Economic and National Security. Exam, Moscow (2004)

Popov, E., Semyachkov, K.: Features of management of development of digital economy. J. Manage. Russ. Abroad **2**, 54–61 (2017)

Samuels, M.: What is digital transformation? Everything you need to know about how technology is reshaping business (2018). https://www.zdnet.com/article/what-is-digital-transformation-everything-you-need-to-know-about-how-technology-is-reshaping/. Accessed 1 Dec 2018

Schiff, J.: 6 biggest business security risks and how you can fight back (2015). https://www.cio.com. Accessed 19 Feb 2019

Shirer, M.: DC Forecasts Worldwide Spending on Digital Transformation Technologies to Reach $1.3 Trillion in 2018 (2018). https://www.idc.com/getdoc.jsp?containerId=prUS43381817. Accessed 1 Mar 2019

Smith, E.: IDC Forecasts Worldwide Spending on Digital Transformation Technologies in 2018 to Reach $1.3 Trillion in 2018 (2017). https://www.businesswire.com/news/home/20171215005055/en/IDC-Forecasts-Worldwide-Spending-Digital-Transformation-Technologies. Accessed 2 Mar 2019

Sukharev, O.: Information economy, transaction costs and development. J. Econ. Theory **1**, 80–93 (2012)

Varnavsky, V.: Digital technologies and the growth of the world economy. J. Drukerovsky Bull. **3**(7), 73–80 (2015)

State Support for Small Enterprises in the Countries of the European Union and the Russian Federation

Natalia Polzunova[1(✉)], Igor Savelev[2], Svetlana Nikiforova[3], and Sergey Ushakov[4]

[1] Vladimir State University, Vladimir, Russia
Natalya.polzunowa@yandex.ru

[2] Moscow State University, Vladimir Branch of Russian Presidential Academy of National Economy and Public Administration, Moscow, Russia
sii-33@mail.ru

[3] Financial University Under the Government of the Russian Federation, Vladimir Branch, Moscow, Russia
SVNikiforova@fa.ru

[4] Russian Presidential Academy of National Economy and Public Administration, Vladimir Branch, Moscow, Russia
ushakov_s@vlad.ranepa.ru

Abstract. This article examines the factors affecting the development of SMEs, examines the problem of state support for SMEs. The purpose of the study is to assess the actual level of coverage of the potential need for SMEs in financial state support in the EU and the Russian Federation and to compare the results obtained. We used the stochastic boundaries of production to determine the potential amount of financial state support for SMEs. The initial data were statistical data from the Russian Federation and the EU on financial state support for SMEs. Descriptive statistics were also used to evaluate the data. The study confirmed the importance of state support for small businesses, its impact on the development of SMEs. This made it possible to determine the actual coverage of the potential of financial support in the Russian Federation and EU countries and to conclude that the results obtained are comparable. This research has value for SMEs and regulatory bodies. The results can be used to further improve such a tool of state regulation as state financial support for SMEs. The results presented in the article allow to deepen research in the field of increasing the effectiveness of state support for SMEs.

Keywords: Small and medium enterprise (SME) · Financial state support · Potential level · Actual coverage

JEL Code: M 13

E. G. Popkova and B. S. Sergi (Eds.): ISC 2019, LNNS 87, pp. 261–269, 2020.
https://doi.org/10.1007/978-3-030-29586-8_31

1 Introduction

Small entrepreneurship plays an important role in the development of national economies (Carlsson 1999). This is determined by its features (Khan and Manopichetwattana 1989) and the functions it performs.

The main functions of small entrepreneurship at the present stage of development are the promotion of innovative development and the activation of the innovation process; ensuring the dynamism, flexibility and maintenance of the system self-renewability procedures; possibility of realization of creative, creative and organizational abilities of people; the pursuit of technological progressiveness. The peculiarities of small entrepreneurship include adaptability to unstable rapidly changing environmental conditions, innovative susceptibility and interest in the commercialization of technologies; motivation for innovation; increased level of creative activity; dependence of the results of activity on the competence and motivation of the personnel of the enterprise; focus on high-performance work and return on a unit of financial investment in R&D; readiness for the effective development of investment resources.

According to OECD, small enterprises produce more than 50% of GDP, provide a third of total exports and about 10% of foreign investment (Financing SMEs and Entrepreneurs. An OECD Scoreboard. OECD Report to G20 2015) and show a high level of innovation activity, which is reflected in Fig. 1. In countries such as Bulgaria, Germany, Spain, Luxembourg, Poland, Finland, France, Czech Republic, Estonia, the share of small business structures that introduce process innovations is higher than the share of business structures implementing product innovations (Enterprises by type of innovation 2014).

In Russia, the intensity of innovation activity of small business structures remains not high and has recently shown a downward trend. Thus, the share of enterprises using technological innovations in the total number of small enterprises in 2011 was 5.1%, in 2013-4.8%, and in 2015-4.5%. Compared with the small enterprises of the Netherlands, which have the lowest share of process innovations, business structures in Russia are 4.9 times lower. And compared with Spain, which has the lowest share of product innovations, Russian business structures are 11.25 times lower.

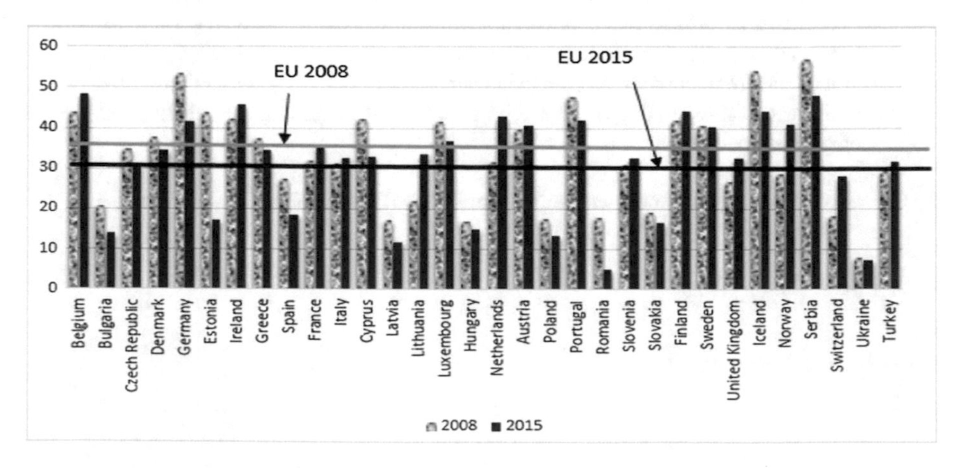

Fig. 1. SME introducing product or process innovations as % of SMEs. Source: (State Aid Scoreboard 2017)

SMEs are considered as one of the drivers of economic growth. It is precisely economically strong SMEs that make a significant contribution to the country's GDP, contribute to reducing unemployment, raising the level of education of the population, spreading knowledge and have many other effects. Therefore, the state is interested in the successful functioning of SMEs.

The state has a major impact on the development of SMEs. One of the forms of interaction between the state and entrepreneurs is state support. Government support is a mechanism to create the conditions as well as incentives for the development of the most effective representatives of SMEs (Aleshchenko and Karpov 2015). World experience allows government support to be interpreted as an activity aimed at providing a favorable business environment. The purpose of this activity is to create prerequisites for economic growth.

In the European Union, it is believed that government support restricts competition and has a negative impact on trade exchange between EU countries (Treaty on the Functioning of the European Union 2012). It is justified if it is aimed at overcoming the difficulties of the market, with the result that benefits will be obtained that exceed the negative effects that affect competition (State Aid Law. European Regional Development Fund Guidance Note for Grant Recipients 2015).

A huge amount of work is devoted to scientific coverage of the problems of state support. So its influence on economic growth, productivity and poverty reduction was studied in the works of Barro (1991), Rebelo and Easterly (1993), Aschauer (1989), Milbourne et al. (2003). The impact on the development of SMEs is considered in the works Kuznetsov et al. (2015), Smallbone and Welter (2001), Gnyawali and Fogel (1994) and other.

State support can be carried out in the form of direct and indirect methods and in the forms of: - financial support; - export promotion, - stimulating R&D and technological innovation; - creation of infrastructure; - consulting services; - stimulation of cooperation with other subjects of small and large business, including with foreign, etc.

Financing issues play an important role in the development of SMEs.

2 Research Methods

The study is empirical. During the analysis, we studied the indicator of the amount of funds allocated from the budgets of different levels to small businesses in the Russian Federation and the EU.

Our study consisted of several stages.

1. Determining the significance of the "government support" factor in the development of SMEs.
2. The rationale for the provision of financial support from the state. For this, we used the following actions.
   2.1. Selected subjects of evaluation. The subjects of evaluation were the totality of the regions of the Russian Federation (83 subjects of the Russian Federation) and the totality of the EU countries (28 countries).

2.2. Collected baseline data on the amount of state financial support for this study. This material is obtained from open sources. For EU countries, data is taken from State Aid Scoreboard 2017. Data studied for 2011 and 2016. For the subjects of the Russian Federation, data sources were provided by the "General Report on SMEs Financing" for 2011, posted on the federal portal of SMEs, data for 2016 were obtained from a single interdepartmental information and statistical system.

2.3. Identified the boundary potential of the amount of funds allocated for state support of SMEs. To describe the boundary potential, a model of the form was used:

$$Y_i = \frac{A}{1 + e^{a + b \cdot x_i}}$$

In this model: Y_i: – estimated boundaries of the boundary function; A – amplitude of the cloud describing the boundary potential, along the ordinate; x_i – the value of the initial potential; a, b – the coefficients of the linear equation determined by the least squares method.

2.4. We compared the actual use of the potential of financial support with its maximum volume.

Also, as a source of significant factors and conditions of state support for the purpose of developing SMEs, materials of scientific and analytical publications and research reports are attracted.

3 Main Results

All factors affecting SMEs can be divided into stimulating and restraining. The factors constraining the development of small entrepreneurship are subdivided: organizational, institutional and economic.

The report of the European Central Bank noted that customer search is the dominant problem for small businesses in EU (24% indicated this problem as the main one), the second most important problem is the cost of production and labor costs (15%), the problem of access to finance, and also the problem of the availability of qualified personnel and experienced managers was noted by 14% of the participants in the study. However, on the third and fourth factor, the estimates vary greatly from country to country. Thus, the problem of access to finance was noted by 42% of small enterprises in Greece as the most topical, 23% - in Ireland. The problem of the availability of qualified personnel and experienced managers is of serious concern in Germany (30% of those who have indicated it), Austria (27%) and Finland (18%). Also as a topical issue were changes in the regulation of activities (Survey on the access to finance of small and medium-sized enterprises in the euro area 2014).

Russian small enterprises point out such problems as the availability of financial resources; government policies regarding support for SMEs, especially new ones, the level of research and development and their availability to SMEs.

All the above presented convinces us that small entrepreneurship needs active state support. This shows the practice and our study.

3.1 Comparative Analysis of Financial State Support in Russia and the European Union

Statistical characteristics of the data used are shown in Table 1.

Table 1. Descriptive statistics, million EUR

Statistical characteristics	EU		Russia	
	2011	2016	2011	2016
Average	142,969	188,118	7,445	1,653
Standard error	60,941	91,035	0,982	0,221
Median	10,297	31,228	5,061	0,993
Standard deviation	322,470	481,711	8,782	1,993
The sample variance	103986,61	232045,45	77,116	3,972
Excess	8,813	10,216	13,898	6,954
Asymmetry	2,915	3,273	3,334	2,599
Minimum	0	0,002	0	0,041
Maximum	1406,065	2029,131	55,034	9,532
Total	4003,144	5267,307	595,619	133,956
Account	28	28	83	83

Source: own elaboration based on Eurostat and Russian statistics (*State Aid Scoreboard* 2017, *General Report on SMEs Financing* 2013, *Unified interdepartmental information and statistical system*)

Based on data analysis, it can be stated that the total amount of financial state support for SMEs in EU countries increased by 31.58%, whereas in Russia it fell by 77.5%. Such a sharp decrease is the integral result of external and internal factors. The growth of the euro explains only 18.2% of this fall.

In 2011, the largest amount of financial state support for SMEs among EU countries was in France. It amounted to 1406.065 million euros. Whereas in 2016, the maximum amount of government support for SMEs was observed in the UK. Its size was 2029.131 million euros. Among EU countries, only in 5 countries in 2011 and 4 countries in 2016, the amount of financial state support for SMEs was above average (Fig. 2). At the same time, they received 88.96% and 86.14% of the total amount of funds allocated for state support of SMEs. Among the selected countries, the growth of financial state support is observed only in France and in the UK.

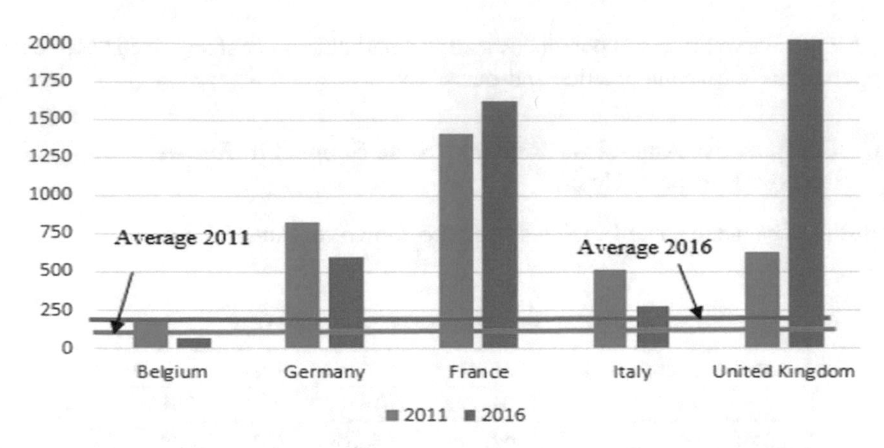

Fig. 2. Dynamics of financial state support for SMEs in countries where its level is above average. Source: own elaboration based on Eurostat (*State Aid Scoreboard* 2017)

In Russia, the level of financial state support has declined significantly. In 2011, 30% of the regions of the Russian Federation had a level of financial state support above the average. This accounted for 65.96% of state-allocated funds to support SMEs. In 2016, 25.93% of the regions of the Russian Federation were above the average level of support. They received 35.08% of the total funds allocated.

The task of determining the amount of financial state support required for the development of SMEs is currently quite relevant (Peter et al. 2018).

3.2 Determination of the Actual Coverage of the Potential Need of SMEs for Financial State Support

The justification of the amount of financial state support can be provided with the concepts of boundary and actual potentials. The actual potential is the potential characterized by the amount of financial state support. The boundary potential allows to determine the degree of manifestation of the actual potential. It represents the possible amount of financial support under the random effects of uncertainty factors. In Fig. 3

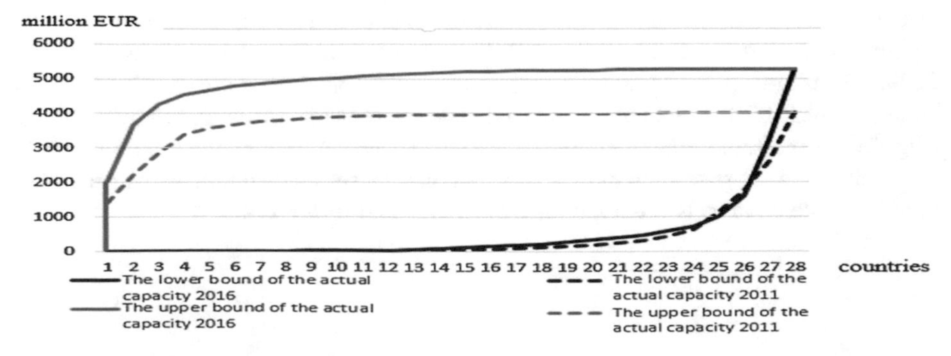

Fig. 3. Boundary potential contour for the amount of funds allocated in the EU. Source: own elaboration based on Eurostat (*State Aid Scoreboard* 2017)

we give a graph of the boundary potential of the volume of financial state support in 2011 and 2016. In Fig. 3 clearly shows the growth of the actual potential of state financial support in EU countries. The boundary potential of the volume of financial state support in 2011 and 2016 in the Russian Federation, presented in Fig. 4, indicates a significant fall.

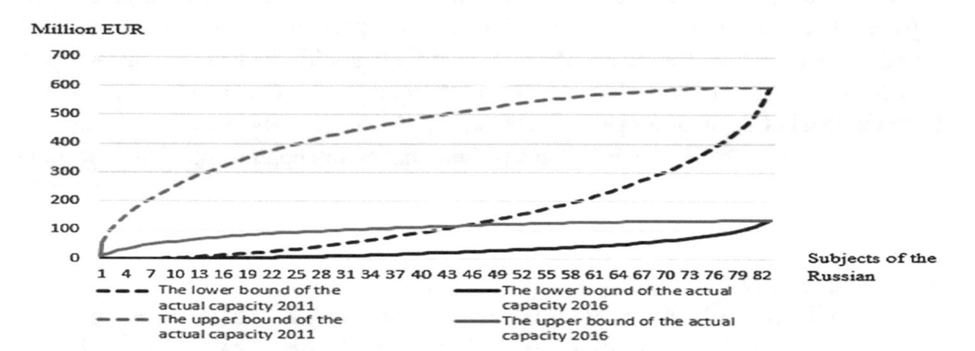

Fig. 4. Boundary potential contour for the amount of funds allocated in the Russian Federation. Source: own elaboration based on Russian statistics (*General Report on SMEs Financing* 2013, *Unified interdepartmental information and statistical system*)

The value of the potential level of need for SMEs in financial state support and the proportion of the actual potential of financial support from the potential level of need for SMEs are presented in Table 2.

The presented calculations show a comparable level of the share of the actual potential of financial support from the potential level of need for SMEs in it in the Russian Federation and the EU. However, while in the EU the potential volume of demand for financial state support increased by 249.5 billion euros, in the Russian Federation this level decreased by 151.06 billion euros. Reducing the need for SMEs in government financial support is explained by their state and level of development, and requires a comprehensive analysis.

Thus, the use of this method allows a comparative analysis of the volume of financial state support for small and medium-sized enterprises in the EU and the Russian Federation.

Table 2. Potential for use of the financial support for SMEs

Indicators	Russian Federation, million EUR			EU, million EUR		
	2011	2016	Change	2011	2016	Change
Potential level of need for government financial support	182,711.29	31,646.41	–151,064.9	551,177.5	800,711.4	249,533.9
The actual potential of financial state support	23,614.21	5,565.01	–18,049.2	91,276.06	121,762.4	30,486.34
Usage rate,%	12.92	17.58	4.66	16.56	15.21	–1.35

Source: own elaboration

4 Discussion and Conclusion

This study complements other research in the field of government support for SMEs. It confirmed that government support is a determining factor in the development of SMEs. Among all forms of support, financial assistance is key.

In the study, using the method of determining the stochastic limit of production capacity, the potential level of the need for financial state support and the share of its actual coverage were determined. Made its comparison with the level of state support for SMEs in the European Union. When comparing the share of actual coverage to the potential level of the need for public financial support, comparable results are obtained. At the same time, the absolute amount of financial state support allocated is incomparable, which leads to the conclusion about the insufficiency of state financial support in the Russian Federation.

The results of the analysis raise the question: is the lack of financial state support in the Russian Federation a deterrent to the development of SMEs?

The findings can be taken into account when planning financial state support for small and medium-sized businesses, as well as further improving the tools of government regulation.

The further development of the research, according to the authors, is connected with the expansion of the empirical base of the research and the comparison of the obtained results with the results of other studies in order to create the foundation for the development of a rational state support.

References

Aleshchenko, V., Karpov, V. (eds.): Improving the Mechanism of State Support of Subjects of Small and Average Business: The Collective Monograph. IC "Omsk scientific Bulletin", Omsk (2015)

Aschauer, D.: Is public expenditure productive? J. Monetary Econ. **23**, 177–200 (1989)

Barro, R.J.: Economic growth in a cross section of countries. Q. J. Econ. **106**, 407–443 (1991)

Carlsson, B.: In are small firms important? Their role and impact. In: Small Business, Entrepreneurship, and Industrial Dynamics, pp. 99–110 (1999)

Easterly, W., Rebelo, S.: Fiscal policy and economic growth: an empirical investigation. J. Monetary Econ. **32**, 417–458 (1993)

Enterprises by Type of Innovation (2014). http://appsso.eurostat.ec.europa.eu/nui/show.do?dataset=inn_cis7_type

Financing SMEs and Entrepreneurs. An OECD Scoreboard. OECD Report to G20 (2015)

General Report on SMEs Financing (2013). http://oldsmb.economy.gov.ru/report/stats/total/

Gnyawali, D.R., Fogel, D.S.: Environments for entrepreneurship development: key dimensions and research implications. Entrepreneurship Theor. Pract. **4**, 43–62 (1994)

Khan, A.M., Manopichetwattana, V.: Innovative and noninnovative small firms: types and characteristics. Manage. Sci. **35**(5), 597–606 (1989)

Kuznetsov, Y.V., Filimonova, N.M., Fedosova, R.N.: A practical approach to modeling of regional state support system for small business development. In: Knowledge, Service, Tourism and Hospitality - Proceedings of the Annual International Conference on Management and Technology in Knowledge, Service, Tourism and Hospitality, SERVE, pp. 133–138. CRC Press (2015)

Milbourne, R., Otto, G., Voss, G.: Public investment and economic growth. Appl. Econ. **35**(5), 527–540 (2003)

Peter, F., et al.: Government financial support and financial performance of SMEs. Acad. Strateg. Manage. J. **17**(3) (2018)

Smallbone, D., Welter, F.: The role of government in SME development in transition economies. Int. J. Small Bus. **19**(4), 63–77 (2001)

State Aid Law. European Regional Development Fund Guidance Note for Grant Recipients (2015). http://www.871candwep.co.uk/content/uploads/2015/04/ERDF-State-Aid-Law-Guidance.pdf

State Aid Scoreboard (2017). http://ec.europa.eu/competition/state_aid/scoreboard

Survey on the Access to Finance of Small and Medium-Sized Enterprises in the Euro Area. European Central Bank (2014)

Treaty on the Functioning of the European Union (2012). http://eurlex.europa.eu/legal-content/EN/TXT/PDF/?uri=CELEX:12012E/TXT

Unified Interdepartmental Information and Statistical System. https://www.fedstat.ru

Territory Branding Development as a Regional Economy Activation Factor

Magomed M. Omarov(✉), Natalya Y. Omarova, and Dmitry L. Minin

Yaroslav-the-Wise Novgorod State University, Veliky Novgorod, Russia
Magomed.Omarov@novsu.ru, n-omarova@mail.ru,
dmitryminin@mail.ru

Abstract. The article presents the study results of the tourism sector main activity indicators in key regions of the Russian Federation. The methodological basis of the tourism industry indicators study was based on the approaches based on application of economic and statistical methods based on time series and analysis of promising scientific developments. As a result of the work, an analysis of tourist flows in the medium term was carried out, and key trends and factors in the tourism development were identified. The authors proposed a number of directions to improve the territory branding. They also introduced the Novgorod Region branding concept implementation elements. The recommendations given by the authors can be applied in most subjects of the country in order to popularize domestic tourism, develop tourism infrastructure, and promote a regional tourism industry product. Development, implementation and execution of the branding process proposals for the regional authorities will significantly increase the tourism appeal and awareness of the region not only at the country level, but also in the international arena.

Keywords: Branding · Tourism · Regional economy · Tourism management

JEL Code: M31 · Z32 · Z33

1 Introduction

In the context of global competition, it is impossible to achieve the level of economic development of the industrialized countries of the world in terms of well-being and efficiency without advanced development of the most high-demand sectors of the Russian economy. One of these key areas is the development of regional and inter-regional tourism. Tourism includes over 50 industries and is a macro complex largely integrated into economy, culture, sports etc. It should be noted that the Russian tourism industry should ensure the maximum market growth rate with account for national competitive advantages.

In a changing economic formation, the innovative type of the economic development of tourism in Russia implies not only a constant increase in the competitiveness of the tourism industries, but also the use of public-private partnership mechanisms, improved conditions for companies to access long-term investment sources, and the provision of related economy industries with high technologies and specialists.

E. G. Popkova and B. S. Sergi (Eds.): ISC 2019, LNNS 87, pp. 270–277, 2020.
https://doi.org/10.1007/978-3-030-29586-8_32

2 Methodology

The studies have shown that changing the structure of population needs over the long term and developing domestic tourism is becoming the most important resource in the regional economy development.

The methodological basis for determining the prospects and forecasts of the development of key indicators of the tourism industry was the approaches based on application of economic and statistical forecast methods based on time series. With the use of economic and mathematical modeling methods, the reliable results of tourism development forecasts in the regions of Russia as the country's economy development factor were identified and subsequently presented in the work.

3 Results

To highlight the trends in the development of tourism in the country, we will analyze the main statistical indicators of the tourism sector development in the regions of the Northwestern, Central and Volga Federal Districts (Table 1).

Table 1. Data on the tourism sector development in the main tourist regions (according to the Federal State Statistics Service 2019)

Region	Area, Thous Sq Km	Population, Thous people	Culture (Tourism) budget, Thous Rub	Culture expenditures per head of the population, Rub per capita	Tourists, Thous people	Q-TY of KCP	Number of accommodated people, Thous	Number of tourist companies
Novgorod Region	54.5	606.5	226488	373.4	1500	126	365	100
Pskov Region	55.4	636.4	813.9	1278.9	600	106	372	44
Republic of Karelia	180.5	622.5	465.6	747.9	760	182	412	104
Leningrad Region	83.9	1813.8	2584.0	1424.6	2000	274	635	119
Tver Region	84.2	400.1	863.4	2157.9	1500	218	366	111
Smolensk Region	49.8	330.0	869.1	2633.6	310	114	187	89
Vologda Region	144.5	1176.7	621.1	527.8	2800	200	412	90
Vladimir Region	29.1	1378.3	2.522	1829.8	4000	232	624	180
Yaroslavl Region	36.2	1265.7	1207.7	954.2	2700	175	683	99
Perm Territory	160.2	2623.1	2825.2	1077.0	674	348	642	191

The statistical data study allows identifying the most promising regions in terms of development dynamics within the tourism business. They include Vladimir, Vologda, Leningrad and Novgorod regions. The latter was actively developed due to the establishment of an infrastructure office in the region, the institute for the tourism development ANO Tourist Office "Novgorod Rus". The increase in the budget efficiency of expenditures for the domestic tourism development is noted precisely in the above-listed regions (Novgorod region in numbers 2018).

An assessment of current trends suggests that in 2019–2022, the tourist flow may decrease, which is related both to the completion of major international events (the FIFA World Cup, etc.) and to changes in the population consumption structure.

In addition, the incoming tourist flow data analysis indicated that in 2018, the weakening of the Russian ruble had a significant impact, which made Russia more appealing for traveling compared to foreign countries.

Nevertheless, China remains the main source of tourists traveling to Russia (about 1.2 million people, an increase of more than 10%). Germany is in the second place - about 420 thousand tourists (an increase of 7.7%). As of 2018, the third place in the tourist flow is taken by South Korea - an increase of 44% (over 320 thousand tourists) (Federal State Statistics Service 2019).

In 2018, the growth rate of the number of trips of Russian citizens to the non-CIS countries has decreased almost five times compared with the growth rates of 2017. Following the results of 2018, the outbound tourist traffic volume has increased by only 7–8% (in 2017, by 36%) (Federal State Statistics Service 2019). As a result, in 2018, the number of recorded trips of Russian citizens to foreign was over 31 million, and the total outbound flow had increased up to 44.7 million people.

In the context of globalization, digital transformation and integration of Russia into the global economic territory, the effective development of the country's regions is of particular importance (Omarov and Minin 2016). Russia's reputation, favorable appeal and investment attractiveness is built up, among other things, through developing a sustainable, recognizable image and attractiveness of its regions.

In the current political and economic situation, the importance of the unique competitive advantages of territories, of the business, tourism and population infrastructure maturity is increasing, which is why in many countries there is an emerging trend in terms of their positioning among the potential target markets.

Kotler and Keller point out that territory branding and promotion of the municipalities at the regional, country, and world community level, as well as the build-up of own regional and municipal brands determining investment and tourism attractiveness are becoming of particular relevance (Kotler & Keller: Marketing Management 2015).

A strong, recognizable and manageable image of the region is essentially the territory brand and allows for:

- stimulating the flow of external public and private investments in the preferred economy sectors and intensifying the processes of cluster development in various industries and fields;
- intensifying the export of tourist services;
- increasing the territory attractiveness by keeping the local population from migrating and attracting new labor resources with demanded skills and competencies.

If branding becomes a priority state task at the regional level as the territorial marketing element, then territory branding will contribute to initiating the self-identification process of the residents, i.e. the relationship of a person to himself/herself and his/her identification with the place of residence (Strategy of Social and Economic Development of the Novgorod Region until 2030 2018).

The Novgorod region has a rich historical and cultural heritage: tourist attractions, unique monuments of history and architecture, religious sites, natural resources, recognizable ethno-cultural and ethno-religious features of the population. An important role in creating a positive appeal of the region is played by a stable political and socio-economic situation, the maturity of railway and federal highway network, the availability of communication tools, and high development level of the modern tourism infrastructure (The concept of the creation of special tourist-recreational type economic zone "Divas country" 2019).

All of this also necessitates the formation and development of the region's territory branding. The current goal of territory branding is to create an attractive image of the region, increase the effectiveness of the socio-cultural, political and economic image of the territory focused on internal and external consumers, develop equal and mutually beneficial inter-regional and international cooperation through a single state/regional policy, and ensure the presence of the region's brand in the common information space.

In accordance with the goal, the main purposes of territory branding can be divided into the following groups:

(1) marketing:

- analysis and identification of competitive advantages of the factors affecting the creation of a positive image of the Novgorod Region;
- identification and actualization of the region's unique features contributing to the build-up and promotion of its positive image;
- establishment of a system of promotion of the territory image for all target audiences;
- positioning of the region as a territory for comfortable recreation in order to increase the tourist flow;
- ensuring the interaction of the region's executive authorities with the local governments, public associations and organizations, including educational ones, to create and promote the region's brand depending on the social and economic conditions with account for the established priorities.

(2) socio-cultural:

- socio-cultural self-identification of the territory residents, including identification of the citizens with the territory of their residence;
- building quality intercultural communication;
- preservation of the territory's historical and cultural heritage;
- keeping the population from migrating out of the region;
- improving the standards of living.

(3) economic:

 - increasing the tourist flow;
 - increasing the investment attractiveness;
 - forming the export mechanism of tourist services;
 - creation and promotion of event projects for attracting investors and tourists (ANO Tourist Office "Novgorod Rus" 2019).

Let us define the conceptual foundations of development and promotion of the Novgorod Region as Novgorod Rus. In the context of a rapidly changing present, the attention of contemporaries is turned to the past, since it is where one seeks direction for unity and commonness. The Novgorod Region and visiting it is exactly what can satisfy these needs. Today, being a Novgorodian means finding oneself, touching the roots. And touching the roots is in-demand, relevant and trendy. Veliky Novgorod is the entry point to Russia, Veliky Novgorod is the father of Russian cities: the most ancient Russian city, here stands the oldest Kremlin of Russia – the Novgorod Detinets.

Veliky Novgorod is a Hanseatic city: antiquity and the Middle Ages are intertwined here. Novgorod Rus is the intersection of cultures and civilizations. The Novgorod Republic as 1/6 of Europe is a full-fledged participant in the Hanseatic League of the European cities. Novgorod land used to be the link between medieval Europe and Russia. Today Novgorod Rus is a member of the Hanseatic League of the new time, uniting over 160 cities from 15 European countries.

Novgorod Rus is the birthplace of the Russian democracy: the Novgorod Assembly arose before the European parliaments and other representative bodies of power. Novgorod is the ancient medieval capital of Russia, where a powerful civilization emerged reflecting independence and self-sufficiency.

Novgorod Rus is the cradle of the Russian Orthodoxy. Novgorod is the baptism center of Russia, the oldest baptism cradle of the Russian people on the Russian territory. It is where the ancient Russian Orthodoxy took shape; Novgorod has the most ancient monument of Slavic book writing, the most ancient Cyrillic book - a testimony and gift of the Christian civilization to the Russian culture. Frescoes of outstanding masters adorn the temples of the Novgorod Land.

Novgorod Rus is the Motherland of the Russian ruble: the trade routes of the Novgorod Region - "from the Varangians to the Greeks" trade route, "the Volga River Route", "the Zavolotsky Route" served for the rise and prosperity of the Old Russian state.

Today, Novgorod Rus is attractive because it has a proximity to the capitals, it is located between them. This is a small cozy European city easy and quick to reach. Veliky Novgorod is a European and at the same time exclusively Russian city, a city for living, including living with children. It is clean, green, with a well-developed network of public transport, availability of architectural monuments, walking routes - all in the center and around the ancient outer rampart. It is a quiet but deep city where you can be alone with yourself. The Novgorod Region offers unique mud, water and salt bath. Here, there is a place for active recreation, here stands the Valdaisky National Park.

The heroes of the Novgorod Land are epic and historical characters, heroes of the chronicles and folklore: Sadko, Rurik, Dobrynya Nikitich, Vasily Buslavev, Yaroslav the Wise, Alexander Nevsky, Marfa Posadnitsa (Boretskaya), G.R. Derzhavin, A.A.

Arakcheev, S.P. Dyagilev, Princes Muravyov, Princes Vasilchikov, A.V. Suvorov, F.M. Dostoevsky, N.A. Nekrasov, S.V. Rachmaninoff, V.V. Bianki, A.A. Akhmatova.

Thus, “Novgorod Rus”, on the one hand, refers us to the initial period of the Russian history, connected with Veliky Novgorod and time of its foundation. This is the time of prosperity and triumph of Russia and an indication to its specific northern part - Novgorod.

On the other hand, “Novgorod Rus” is a modern construct in territorial marketing technologies. It points to a certain subject of the Russian Federation - the Novgorod Region. Novgorod Rus and the Novgorod Region exchange the socio-cultural context, associations and allusions: the Novgorod Region is the successor of Ancient Russia, the ancient Russian state, Novgorod Rus. This region is a fragment of that very Ancient Russia, preserved within a single region with many historical and cultural monuments of that period.

4 Conclusion/Recommendations

The studies have shown that ensuring the strategic development of the regional tourism sector with account for changes in the global market requires implementing the following recommendations.

In the social and political field:

- determining the goals and objectives of branding for the regions by the state authorities;
- authorizing the “Novgorod Rus” regional tourism and export brand of the Novgorod Region;
- promoting the developed territory brands among the residents of the regions;
- gratuitously using the “Novgorod Rus” regional tourism brand of the Novgorod Region to promote handicraft, souvenir and promotional products of the Novgorod Region in tourist markets.

In the information and communication field:

- consolidating the mass media and mass communications to promote a favorable image of Veliky Novgorod and the Novgorod Region in internal and external information distribution markets;
- enhancing and systemizing the activities of press services of the regional executive bodies to disseminate information about Veliky Novgorod and the Novgorod Region in order to promote a favorable image, create a high reputation and encourage public audiences to take actions desired for the region in the tourism field;
- ensuring the tourism appeal of the Novgorod Region: developing and supporting the novgorod.travel tourist portal in the Internet information and telecommunication network; promoting the region’s tourism potential at international and Russian tourist exhibitions.

In the culture and spiritual development field:

- forming the image of the Novgorod Region based on preservation and development of the cultural and spiritual potential of the society, forming the region's modern culture which includes supporting and preserving the national culture, sights, ethnographic and natural monuments of the region.

For the successful development of regional tourism, it is advisable to use the following branding tools:

1. Active use of strategic tools aimed at forming the brand core with account for the main features of the territory and application of the following techniques:
   - determining the territory development strategy, which establishes the top-priority areas for development, on the basis of analysis of the social, economical and political situation, the cultural and scientific potential, and the geographical location;
   - determining the mission of the territory, forming a slogan containing the main meaning of its existence and activity.
2. Use of PR-tools as a set of brand-promoting activities based on communicating the information about the territory's features to the public and cooperating with it by engaging it in joint activities (Bottrell and Schoenly 2018):
   - allowing the territory to access the federal and international level with presentation of its achievements and image support;
   - a visual symbol, the main thumbnails (views of the region); in order to increase the visual communication efficiency, it is necessary to use the brand as a standard, a set of instructions for correct use and placement of visual identifiers of the regional tourist brand (ANO Tourist Office "Novgorod Rus" 2019).
3. Introduction of symbolic tools representing a specific set of visual means to influence the brand consumers. They include: creation of a unified style and design of the main attributes of the territory (logos and other significant symbols); promotion of the novgorod.travel tourists portal (Tourist portal of the Novgorod region 2019).
4. Expanding the use of advertising tools that provide for the distribution of information about the territory addressed to the brand consumers and aimed at drawing attention to the territory as a branding object by placing it in the media (ANO Tourist Office Novgorod Rus. 2019).

The main means of promoting the "Novgorod Rus" regional tourism and export brand of the Novgorod Region are communication measures and tools demonstrating the territory's openness to contacts and allowing the external subjects to gain a better understanding of its advantages, as well as use of the territory symbols in the fields of transport, communication, tourism and hospitality; organizing press tours in the region; holding cultural and sports events, interacting with people from various fields of science and culture, music and theater, their participation in the public life of the region; exhibition, trade fair and other presentation activities.

Thus, the promotion and development of regional tourism as the key economy element of the constituent entities of the Russian Federation requires stimulating the tourist product branding and promotion processes; developing tourist information centers and offices of ANO Tourist Office "Novgorod Rus" which has positively proved itself not only in terms of implementation of enhanced marketing activities but also as a catalyst for regional and interregional tourism development, which has united the efforts of the related sectors of the Russian entrepreneurship, contributed to improvement of the key financial, economic and social indicators of the Novgorod Region in the field of tourism activities and a number of affiliated branches of regional entrepreneurship.

References

Autonomous non-profit organization Tourist Office "Novgorod Rus", ANO TO "Novgorod Rus" (2019). https://www.facebook.com/rusnovgorod

Bottrell, D., Schoenly, K.: Integrated pest management for resource-limited farmers: challenges for achieving ecological, social and economic sustainability. J. Agric. Sci. 1–19 (2018). https://doi.org/10.1017/s0021859618000473

The Concept of Creating a Special Tourist-Recreational Type Economic Zone "Divas Country" (2019). http://tourismnov.natm.ru/. Accessed 18 Feb 2019

Kotler & Keller: Marketing Management. What Is Branding? The Branding Journal (2015). https://www.thebrandingjournal.com/2015/10/what-is-branding-definition/

The Ministry of Finance. https://www.minfin.ru

Novgorodskaya oblast v tsifrakh. (The Novgorod Region in Numbers). Zimina N.E. (ed.) statistical compilation. Novgorodstat Publ. – V.N., 139 p. (2018)

Omarov, M.M., Minin, D.L.: Prognoz Razvitiya Strukturnykh Izmeneniy APK Severo-zapadnogo Federalnogo Okruga I Rekomendatsii Po Obespecheniyu Stabilnogo Razvitiya Na Srednesrochnuyu Perspektivu (Forecast of the development of structural changes in the agro-industrial complex of the Northwestern Federal District and recommendations for ensuring stable development in the medium term). News of the International Academy of Agrarian Education 2016, no. 26, pp. 81–84 (2016)

The Social and Economic Development Strategy of the Novgorod Region Until 2030 (2018). https://www.novreg.ru/economy/strategy2030/strategy2030.php

The Tourist Portal of the Novgorod Region (2019). http://novgorod.travel

The Federal State Statistics Service (2019). http://www.gks.ru

The Role of Internet for Digital Economics in Developing Regions

Firmin Tangning Jiogap(✉)

Vladimir State University named after A. G. and N. G. Stoletovs, Vladimir, Russia
tajifirmin2@yahoo.com

Abstract. The present paper focuses on the description of the importance of communication channels in the development of various regions or developing countries. This description takes place in the form of analysis and comparisons of different possibilities of data exchange in the virtual world. It also proposes an online e-commerce organization model. These analyses and comparisons are intended to provide more information in the realization of new business models. The explanation of certain phenomena (evolution of information and communication technology, description and use of technical words in this field, etc.) is carried out based on graphical methods and mathematical theory of fuzzy sets. The calculations made in the text, attempt to show the importance of Internet which is an element in the digital economy. Finally, we will list various infrastructural elements as well as various reforms to undergo (to be carried out), to support the growth of this economy in developing countries or regions. The development of the regions or countries will be accomplished if, and only if, they follow the rhythm of technological evolution of the data transmission on the Internet.

Keywords: Digital economy · E-commerce · Internet speed · Online data transmission · Online data delivery · Information and communication technology

JEL Code: O18 · O24 · O33

1 Introduction

After a wave of massive urbanization and economic growth in several mega cities, many countries in the world are still very far from this innovative form of economy. Never mind, we can safely say that it's never too late to grow. Africa is one of the continents, which has a large number of members in this red zone that is underdevelopment. Most of these regions of Africa are very late in this race for development and can rely on the new method of digitizing the economy to achieve this.

In the twenty-first century, we observe that the profound upheaval of companies that occurred globally in the industrialized nations (especially Western Europe, North America and Japan) is caused by the rise of digital techniques, mainly computers and Internet. This radical change is creating a new form of communication and commerce

E. G. Popkova and B. S. Sergi (Eds.): ISC 2019, LNNS 87, pp. 278–290, 2020.
https://doi.org/10.1007/978-3-030-29586-8_33

in the world that is actively involved in the development of the country. With the upstream of new technologies such as information and communication, each nation wants to increase its competitiveness in the world, ensure the quality of life of their citizens, economic growth and national sovereignty.

Can the emerging underdeveloped countries digitize their economy? And if so, what are the infrastructural elements to support the rise of the digital economy in these regions? Can we talk about a digital economy with a poorly computerized population? What criteria should be used to promote its success? Could there be risks of employment by adopting this new form? These are all questions we will try to clarify and provide an answer to that will help some nations to better prepare for this digitization politics.

2 Can Poor Countries Digitize Their Economy?

If we are talking about the digital economy today, it is certainly because the traditional economy does not satisfy the needs of a large part of the population, but certainly because of the Information and Communication Technologies (ICT) revolution. The economy is doing badly overall, so "new blood" will have to be brought into this area in order to raise the standard of living not only in the different poor countries, but also in the rich countries.

2.1 Poverty Indicators

According to the norms of logic, it will be possible to determine whether a country is poor or rich, based on a number of well-defined criteria: the gross domestic product per inhabitant; the life expectancy of the population; the adult literacy rate; the amount of school-educated population. The factors that accompany these criteria are: natural resources, adopted politics, infrastructure, technology and others. As part of our studies, we will discuss the other criteria related to ICT (Kashif et al. 2017), which are essential for the development of the digital economy in a specific region. Chihiro Watanabe and others co-authors, in their article (Chihiro et al. 2018), state that: "Tapscott in his best-seller The Digital Economy published in 1994, the Internet has dramatically changed the way of conducting business and our daily lives". Talking about the digital economy means having a technology that can make it work. This is in principle the "high-speed" Internet, which will make it possible to create smart cities (Ning et al. 2019), for example.

Sometimes the word "high" speed is used with exaggeration for sales marketing purposes, which is why we will try to give a clearer and more appropriate definition of this term. For this, we will use one of the mathematical models describing uncertainty. The word "high" has this characteristic of not being able to give the exact information where it is used. For example, when you say: "high" speed, how fast are we talking? – 15 Mbps, 35 Mbps, 100 Mbps? etc. If we go from a speed V_1 to $V_2 > V_1$, can we say that we have a "high" rate? If so, what will be said by someone that moves from a speed V_1 to $V_3 > V_2$ and so on? So, this word, like many other words of this kind (for example: big, small, low, etc.) are random. These words indicate a certain uncertainty in precision. For more precision, it will be preferable to use models at intervals which will allow representing the word by specific values.

2.2 Description of the Model

The diagram №1 describes some different technologies (Wei et al. 2008) existing in the world by the speed of data transmission.

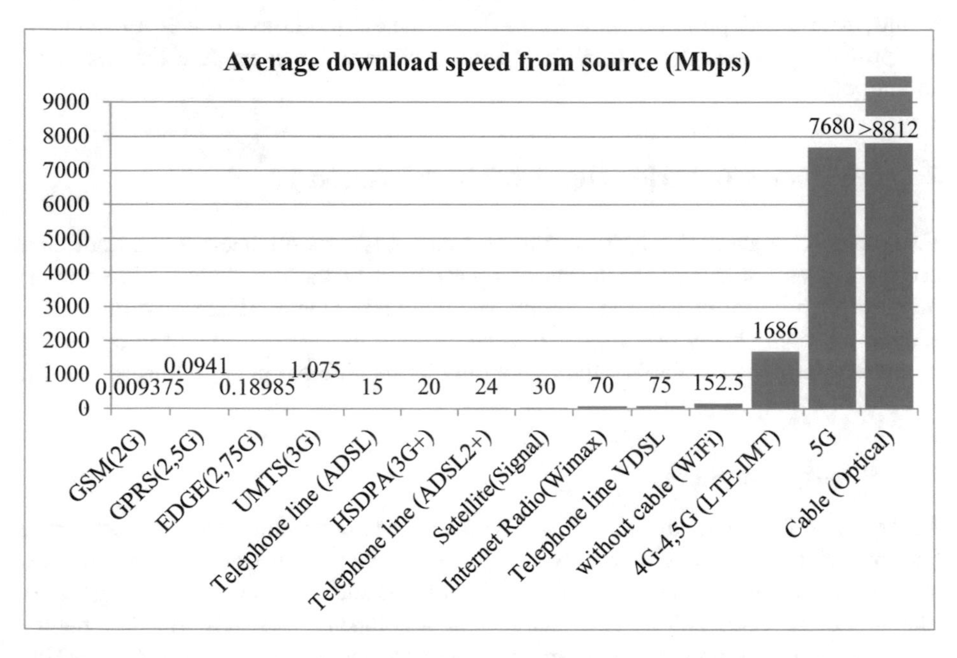

Fig. 1. Average speed of downloading data by different data transmission technologies

Consider S as a system that represents the different data transmission speeds that can be found in an Internet network. Let us form an interval in the form $Vi^{-} \leq Vi$ Vi^{+}, where Vi^{-} will be the minimum limit and Vi^{+} the upper limit for an indeterminate factor i (the different rates in the Internet).

The above inequality means that Vi can take any value of the interval. We can also determine that any velocity Vi belongs to a class Kj = $<K_1, K_2, \ldots, K_n>$, where Kj - type of flow: “low”, “medium”, “high”, “very high”, etc. To be even more precise, it will be necessary to use the fuzzy logic which relies on the mathematical theory of fuzzy sets in order to be able to determine the degree of membership $\mu \in [0, 1]$ to the class Kj (j - index of each class).

The formalization of such a description using the linguistic variable < (Internet rate or speed)$_{TC}$, Kj, [Vi^{-}, Vi^{+}]>, where Kj - will vary according to the technology used (TC – type of connection). It will be the same for the values Vi^{-} and Vi^{+} (see Table 1). We know that the speed of the optical fiber (Addanki et al. 2018) is not limited. For our calculations in Table 1, we took a maximum speed more or less equal to 10 Gbit per second.

Table 1. Determination of the different categories (classes) of Internet speed, in comparison with the capacity of data transmission, according to the type of connection

Class(category) TC		Weak	Medium	High	Very high
Optical fiber	Interval (Mbps)	**[500 – 2875]**	**[2875 – 5250]**	**[5250 – 7625]**	**[7625 – 10000]**
	Basic element (Be)	1687,5	4062,5	6437,5	8812,5
Satellite	Interval (Mbps)	**[0 – 8,75]**	**[8,75 – 17.5]**	**[17,5 – 26,25]**	**[26,25 – 35]**
	Basic element (Be)	4,375	13,125	21,875	30,625
VDSL2	Interval (Mbps)	**[0 – 12,5]**	**[12,5 – 25]**	**[25 – 37,5]**	**[37,5 – 50]**
	Basic element (Be)	6,25	18,75	31,25	43,75
3G—3G+	Interval (Mbps)	**[0,384 – 10,788]**	**[10,788 – 21,192]**	**[21,192 – 31,596]**	**[31,596 – 42]**
	Basic element (Be)	5,586	15,99	26,394	36,798
4G-4,5G	Interval (Mbps)	**[100 – 325]**	**[325 – 550]**	**[550 – 775]**	**[775 – 1000]**
	Basic element (Be)	212,5	437,5	662,5	887,5

Basic element (Be) – it is the central value (number) of interval when the degree of membership is equal to zero. He is neither more nor less. It is determined with the formula below:

$$\mathrm{Be} = \left(\mathrm{V}_{\mathrm{i}}^{-} + \mathrm{V}_{\mathrm{i}}^{+}\right)/\mathrm{ne}, \tag{1}$$

where *ne* = 2 - the number of boundaries in a given interval (it is equal to 2 here).

The values of the linguistic variable "Internet speed" according to the type of connection are associated with the class (Kj), in order to obtain a concrete fuzzy variable that belongs to the interval given by the variables Vi^{-}and Vi^{+}.

For example, we can say that: the form < (Internet speed)$_{\text{fiber optic}}$, high, [7625, 10000] > means that all values (speed) that will belong to the interval [7625, 10000] will be of the "high speed" quality if the connection will be made only by the optical fiber.

If we consider t_s as the sum of the time elapsed between the time of sending and the time of total receipt of the request, then we have the results (Time required to send data from point A to point B) in the Fig. 2 according to the different technologies used. To calculate, we used files with the following dimensions: 10 MB of data for online sales, 1800 MB of data for online reading, 300 MB of data for photos, 1500 MB of data for video and 40 MB data for music.

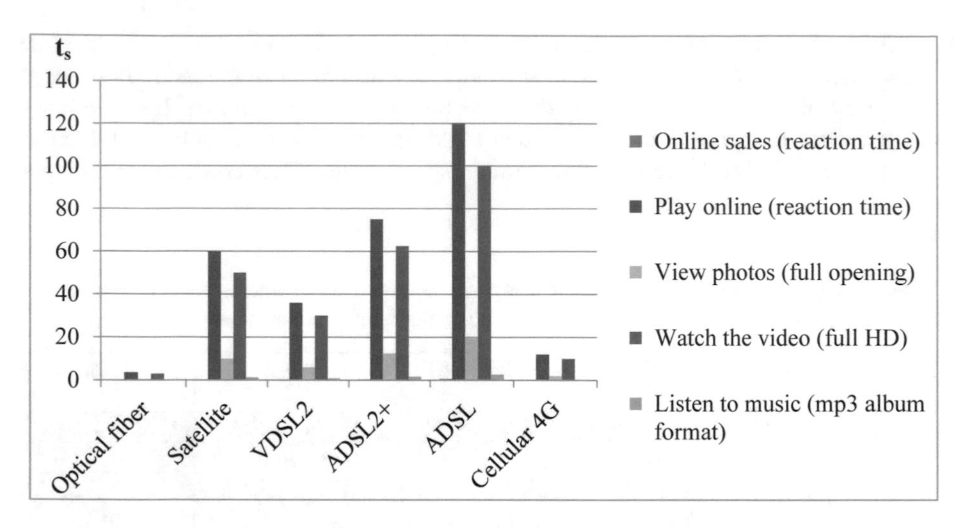

Fig. 2. Comparison of different times to download data with different connections

Using the information in Table 2 and including known standard formulas (Haynes 2017), we can more or less theoretically measure the bit rate that will be possible by downloading any X data.

The time T (time 2 in Fig. 4) to download or to upload the data from a point Q_1-server to a point Q_2-client (see Fig. 4) will be equal to:

$$T = \frac{X \times 1\,s \times 8bit}{V(Xbit/s)} .\beta\,(second), \qquad (2)$$

where V - is the possible speed of the connection line in bits per second; β - factors related to data buffering (Kurose and Ross 2013). In the ideal conditions, with a very good flow, it is equal to 1.

According to the standard (leader for computer code) Unicode in the form UTF, a character will have the equivalence of one byte. Thus, the number of characters on a web page (including blanks - the space bar) will be approximately equal to the amount of information on a page.

From the diagram (see Fig. 1), we see that the optical fiber and the 4G allow a lot more the speed of data transport in the Internet network. With the new 5G technology (Li et al. 2017), the speed will be approximately 20 times higher than that of 4G.

By analyzing the diagram №1, we can make the following observation: the more advanced the technology, the less time it takes to download any document, display an image or listen to or view a file. The 4G or 5G technology and the optical fiber of the "high" or "very high" category of the Table 1, gives the possibility of transforming not only the virtual world into a real world; but also to limit the execution time of a task (download of data, look at an online video, immediate reaction of the server at the request of the client, promote the realization of projects that are related to smart cities etc.) in online time.

The analysis made in this area has allowed us to determine the minimum value in practice at which it will be possible to make data exchanges on the Internet (see Table 2) in order to comfortably facilitate e-commerce (Leung et al. 2018).

Table 2. The average characteristics of the customer's Internet speed for the normal realization of a digital economy (Recommendation)

p	Main parameters	Minimum value	Average value
1	Inbound connection (download) $T_{[p]}$	25 Mbps	39,43 Mbps = 4,93 MB/s
2	Outbound connection (send)$T_{[p]}$	30 Mbps	45,56 Mbps = 5,70 MB/s

$$T_{[p]} = \left[\sum_{n}^{m-1} \text{times}[m]\right]_{p}, \tag{3}$$

where *m* type of technology (connection), n – number of technology (see Fig. 1).

3 The Trade Sector, as One of the Important Activities for the Development of the Digital Economy

The economy has several activities that can contribute to the development of a specific region. Among these activities, we can note the marketing. The French newspaper *LaTribune* publishes in one of these articles (Mirlicourtois and Xerfi 2017) the major growth sectors in 2017, and among them, we also found that the commercial sector was in first position. This sector is growing rapidly in the world (Sabie 2018) because of the new form of activity that is e-commerce (Sazonov 2010) and (Lin 2018). For example, in France (Auffray 2016), there is an increase in active merchant sites per year. Through the Fig. 3, we can better understand how it works and have an idea about the minimum amount (capital) that will be needed to start such an activity.

Step 1: Sign a contract for the delivery of goods: either directly with the producers or with the suppliers (k) at low prices (wholesale price);

Step 2: Create a virtual platform where the commerce will happen. For this, it will go through the following steps:

- create the domain name (your company name): $15–20;
- choose a reliable web hosting for your site, preferably with control panel - high cost at $ 250–300 (web storage in SSD, max process 60, RAM memory, database);
- complementary software to ensure platform security and data (~$100);
- a computer for the service;
- access to the Internet for which the minimum configuration is described in the Table 2 - 110 $ (theoretical value: ↓100 Mbps/↑50 Mbps);

– an employee whose role is to ensure the operation of the virtual shop. His salary will depend in part on the performance obtained;

Step 3: Creating an e-commerce site: How much will it cost? The answer to this question is not so easy to give, it all depends on the project of the different functionalities and the technology used. For a start, with a sum of $ 500–700, you could already have a small virtual store that can operate normally;

Step 4: The users (customers) will need an electronic machine (PC, laptop, tablet, etc.) with a minimum of following characteristic: processor 2-3 GHz, Memory: 4–8 Gb. This machine will have access to the Internet;

Step 5: Purchases will be paid through a credit card (approximate cost of obtaining the card: ~$ 30–40) or by using electronic money;

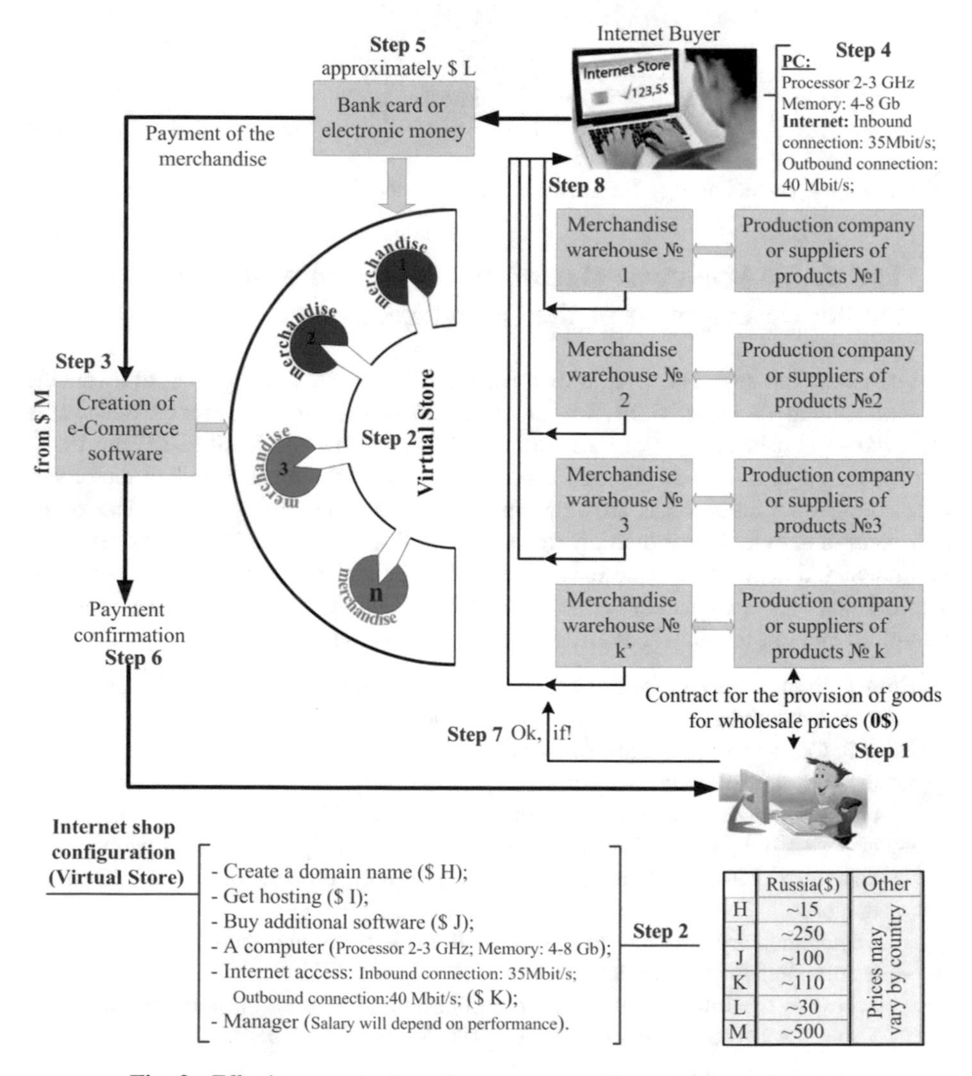

	Russia($)	Other
H	~15	Prices may vary by country
I	~250	
J	~100	
K	~110	
L	~30	
M	~500	

Fig. 3. Effective organization of e-commerce with a minimum of capital

Step 6: After payment, the online salesman (manager) will give an authorization (**step 7**) for the delivery of the goods or will give the customer the green light at the merchandise storage center in order to get possession of his goods. Everything will depend on the method of obtaining the goods chosen by the customer;
Step 8: Receipt of the goods by the customer.

What do we see in this small project to create an e-commerce? We see that the number of staff is reduced and there is not much need of a large workforce in the field because of the presence of ICT. A skilled ICT workforce is needed for the project. To implement this project, it will be necessary to start with a minimum capital (see Fig. 3) acceptable in most regions or emerging countries in the world, during the first year.

To better understand how it works and the importance of the quality of the Internet in its operation; let's try to implement time characteristics between different nodes of the client with the server where the virtual market is located. Figure 3 shows the different important nodes between the client and the server. In the 7th chapter of the book (Kurose and Ross 2013), we will be able to understand the principle of the client buffer during streaming data (video broadcasting, etc.) and the principle of buffering in general.

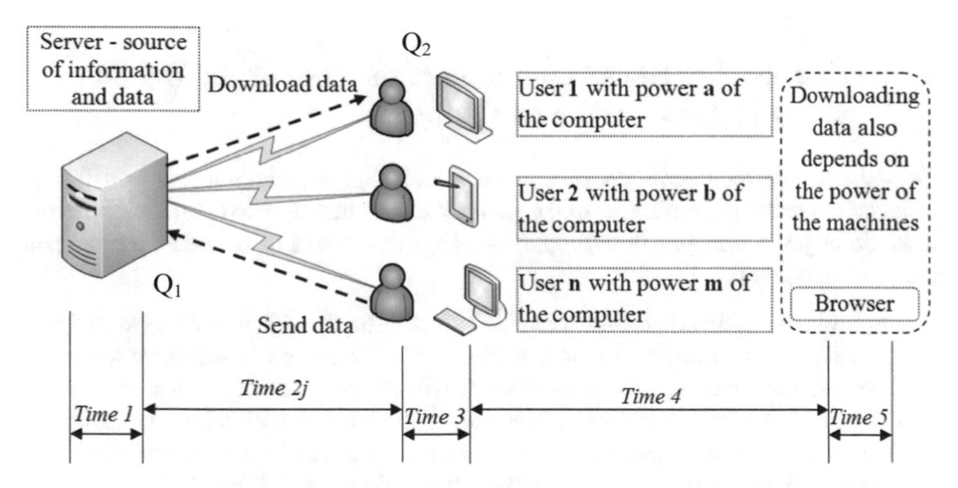

Fig. 4. The different measurement times during the processing and transport of data from the source (server) to the browser in the Internet

Times1 – latency time for processing data in the server. The more the server has a power configuration, the longer the latency will tend to zero.
Times2j – latency while transferring data from the server to the machine memory (or vice versa) of the client. It will depend in part on the type of connection (j) used (see Fig. 1).
Times3 – latency for processing data in the client's machine memory (Kurose and Ross 2013). This time will also depend on the characteristics of the machine.

Times4 – latency for transmitting data from memory to the client's browser. It can be associated with the previous time.
Times5 – latency for data processing or display of browser information. It will depend on the type of browser and how it has been configured.

To better understand what the digital economy is and how it should unfold, let's try to see the main parameters that influence its development.

It should be noted that the data in the previous table (Table 2) are desirable practical values for the proper functioning of e-commerce. This means that theoretically they have to be multiplied by at least 2, because there are several other parameters that can modify these data in practice: the distance between your machine (computer) and the technology used to connect to the server (the lower the better, the more the flow tends to the maximum); the quality of the connection line with different properties; the quality of the installation of the devices and the network (absence of the repeater for example in long distance); the power of your machine (computer, tablet, …) and the types of software (antivirus,…) used, the correct setting of some of them.

Thus, the data rates of incoming and outgoing data should progressively increase with the technological evolution of the transmission of data on the Internet, the power of software, and associated technical devices.

4 Infrastructure Elements to Support the Rise of the Digital Economy in Less Developed Countries

To associate the total economy of a country with digital technology, it will first be necessary to make some reforms in the territory in question. For without these reforms, it would be just an imagination (utopia). Among the reforms to be made, we can note the most important:

- improvement of Internet infrastructures (Coquio 2017) with faster access and able to facilitate the transport of data without delay; promote wireless networks for emerging countries, because with fiber, it will require a lot of expenses;
- ensure an extended geographical coverage of the Internet with a bit rate having the following minimum characteristics: inbound connection: 35 Mbps; outbound connection: 40 Mbps; on more than 90% of the territory (see Table 2);
- promote ICT in schools and universities;
- introduce an online administration system: this will enable public administrations to use information and communication technologies to make public services more accessible to the public 24 h a day and improve their internal functioning (Alberola et al. 2017);
- also introduce the electronic signature system;
- create projects requiring the Internet network for its realization and operation. These projects will aim to compete and eliminate existing ones. It will be necessary to encourage or motivate the population to take part in the development of the digital economy with the practice of cheap prices;
- give more benefits to companies developing computer software; for example by lowering taxes on finished products;

– create an online statistics center for collection and automatic analysis of data on each region for scientific research purposes; this will easily make it possible to create software capable of analyzing and automatically making effective decisions with the aim of improving the production or development of a specific sector of activity;
– creation of commodity storage centers to promote the development of small and medium-sized enterprises in the commercial field and more specifically e-commerce (see Fig. 3).
– to train a qualified workforce (Tangning 2018) that can support the use of ICTs in non-digital sectors;
– readjust monetary politics by promoting the use of bank cards, not only for the purchase of goods, but also for public services. Use smart cash registers according to the model shown in Fig. 5. Access to the Internet will facilitate control in the operation of the system or enterprise without disruption of work.

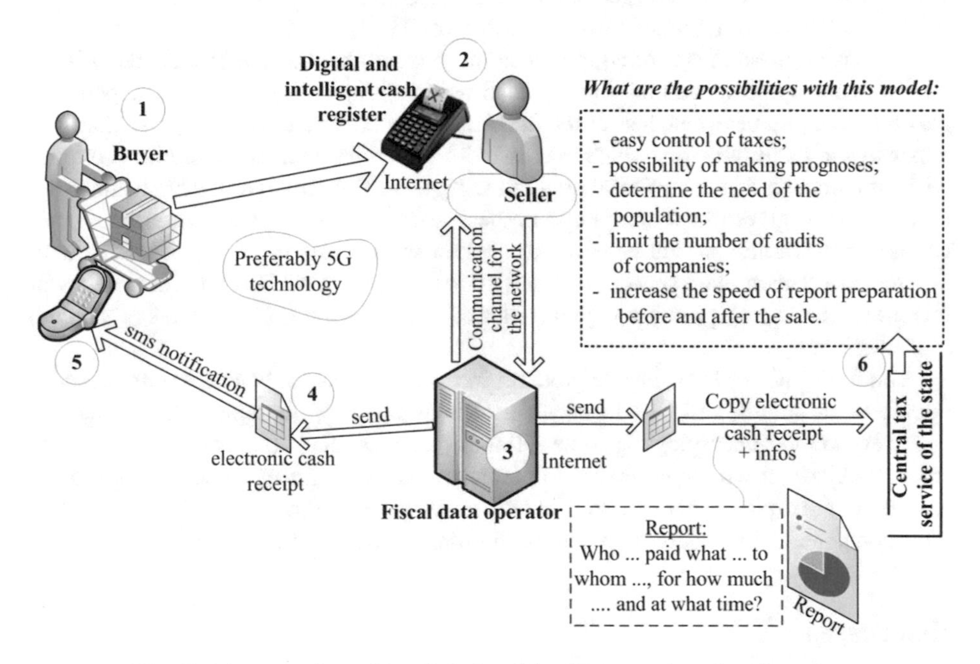

Fig. 5. The operation of the digital and intelligent cash register in a shop

It must be remembered that the type of connection (bandwidth) is not the only factor that can allow rapid access to the Internet to promote the expansion of the digital economy. There are other important factors that should not be overlooked: the possible number of devices connected per meter for wireless networks, the power and quality of these peripheral devices (computers, laptops, mobile phones (smartphones) and others); software for security, for filtration and software with properties of artificial intelligence and etc.

5 Conclusion

In conclusion, we can say that by fulfilling the conditions listed in this article, we can improve through the information and communication technology sector the percentage of GDP.

To digitize the economy of emerging countries, priority will be given to digital skills in the primary, secondary and university education. With the development of ICT, it will be possible to develop digital economy in these less developed countries or regions, where we can witness the transformation of regions into a smart city (Anand and Navío-Marco 2018). Thus, all the main activities will be related to artificial intelligence, and as an example, we can talk about traditional trade that can quickly turn into e-commerce.

According to the analysis made, it will be recommended to use optical fiber to connect the regions to each other. Within regions, the development of wireless networks will need to be encouraged, using at least 4G technology. Using 4G technology, round-trip latency is 10 times lower than that of 3G.

Among all these means of communication, optical fiber has the best characteristics and has virtually no speed limit in the transmission of data, because its flow depends in part on the equipment installed at the ends of the fiber. The disadvantage is that it is expensive in its installation, which makes it difficult to implement the digital economy without reorganizing the territorial plan of the regions. The arrival of the 5G technology (Ji et al. 2018) gives much more hope to these poor regions. Because with this mobile technology - wireless, it will be easier to create new business models. Recall that with 5G, the data transmission speed will reach 10 to 20 gigabits per second. Technicians in this field speak and promise low installation cost, and almost no latency in data transmission.

The use of ICTs also creates a problem in the labor market. People are replaced by intelligent machines, and then to avoid the development of unemployment, it will be necessary to offer unemployed people retraining courses (Kuladzhi et al. 2017) in new areas. Similarly, it will be necessary to prepare a skilled workforce able to adapt to the realities of the digital economy, promote the development of technical education in this area, computerize the elderly population, in order to be able to use computer tools, etc.

References

Addanki, S., Amiri, I.S., Yupapin, P.: Review of optical fibers-introduction and applications in fiber lasers. J. Results Phys. **10**, 743–750 (2018). https://doi.org/10.1016/j.rinp.2018.07.028. Accessed 14 Mar 2019

Alberola, E., Croutte, P., Hoibian, S.: E-administration: the double sentence of people in difficulty - Research Center for the study and observation of living conditions (CRÉDOC) - Consumption and lifestyles, no. 288 (2017). ISSN 0295-9976. Avril. http://www.credoc.fr/pdf/4p/288.pdf. Accessed 14 Mar 2019

Anand, P.B., Navío-Marco, J.: Governance and economics of smart cities: opportunities and challenges. J. Telecommun. Policy **42**(10), 795–799 (2018). https://doi.org/10.1016/j.telpol.2018.10.001. Accessed 14 Mar 2019

Auffray, C.: Important numbers: e-commerce in France. Wednesday, November 23 (2016). http://www.zdnet.fr/actualites/chiffres-cles-l-e-commerce-en-france-39381111.htm. Accessed 14 Mar 2019

Chihiro, W., Kashif, N., Yuji, T., Pekka, N.: Measuring GDP in the digital economy: Increasing dependence on uncaptured GDP. J. Technol. Forecast. Soc. Change **137**, 226–240 (2018). https://doi.org/10.1016/j.techfore.2018.07.053. Accessed 14 Mar 2019

Coquio, F.: LaTribune Newspaper: What Infrastructure Elements to Support Digital Growth in Africa? [FR] - pub. 09 November 2017 (2017). http://afrique.latribune.fr/think-tank/tribunes/2017-11-09/quels-elements-d-infrastructure-pour-soutenir-l-essor-du-numerique-en-afrique-757269.html. Accessed 14 Mar 2019

Haynes, W.M.: CRC Handbook of Chemistry and Physics. A Ready-Reference Book of Chemical and Physical Data. Editor in Chief Haynes, W.M., Lide, D.R., Bruno, T.J. 97th edn. 2643 p. CRC Press, Taylor & Francis Group, LLC 2017. ISBN 1498754287 (2017)

Ji, H., Park, S., Yeo, J., Kim, Y., Lee, J., Shim, B.: Ultra-reliable and low-latency communications in 5G downlink: physical layer aspects. IEEE Wirel. Commun. **25**(3), 124–130 (2018). https://doi.org/10.1109/mwc.2018.1700294. http://ieeexplore.ieee.org/stamp/stamp.jsp?tp=&arnumber=8403963&isnumber=8403915. Accessed 10 Apr 2019

Kashif, N., Chihiro, W., Pekka, N.: The transformative direction of innovation toward an IoT-based society - Increasing dependency on uncaptured GDP in global ICT firms. J. Technol. Soc. **53**, 23–46 (2017). https://doi.org/10.1016/j.techsoc.2017.11.003. Accessed 14 Mar 2019

Kuladzhi, T.V., Babkin, A.V., Murtazaev, S.Y.: Enhancing personnel training for the industrial and economic complex in the conditions of the digital economy. In: IEEE VI Forum Strategic Partnership of Universities and Enterprises of Hi-Tech Branches (Science. Education. Innovations) (SPUE), St. Petersburg, pp. 67–70 (2017). https://doi.org/10.1109/ivforum.2017.8246053. http://ieeexplore.ieee.org/stamp/stamp.jsp?tp=&arnumber=8246053&isnumber=8245951. Accessed 10 Apr 2019

Kurose James, F., Ross Keith, W.: Computer Networking, A Top-Down Approach. 6th edn. p. 912. ISBN 0132856204 (2013). Published by Pearson Education, Inc. Publishing as Addison-Wesley

Leung, K.H., Choy, K.L., Siu Paul, K.Y., Ho, G.T.S., Lam, H.Y., Lee Carman, K.M.: A B2C e-commerce intelligent system for re-engineering the e-order fulfilment. J. Expert Syst. Appl. **91**, 386–401 (2018). https://doi.org/10.1016/j.eswa.2017.09.026. Accessed 14 Mar 2019

Li, R., et al.: Intelligent 5G: when cellular networks meet artificial intelligence. IEEE Wirel. Commun. **24**(5), 175–183 (2017). https://doi.org/10.1109/mwc.2017.1600304wc. http://ieeexplore.ieee.org/stamp/stamp.jsp?tp=&arnumber=7886994&isnumber=8088405. Accessed 14 Mar 2019

Mirlicourtois, A., Xerfi: The newspaper - Journal La Tribune - Major growth and employment sectors in 2017; 15 February (2017). http://www.latribune.fr/opinions/tribunes/les-grands-secteurs-porteurs-de-croissance-et-d-emploi-en-2017-638458.html. Accessed 14 March 2019

Ning, Z., Huang, J., Wang, X.: Vehicular fog computing: enabling real-time traffic management for smart cities. IEEE Wirel. Commun. **26**(1), 87–93 (2019). https://doi.org/10.1109/mwc.2019.1700441

Sabie, A.: Payments and Commerce Market Guide 2018-2019; Document type: Research Study. EcommerceWiki - non-profit organization for collecting and sharing all knowledge about ecommerce-related topics, p. 213. 09 November 2018 (2018). https://www.ecommercewiki.org/reports/763/payments-and-commerce-market-guide-2018-2019. Accessed 14 Mar 2019

Sazonov, A.: 5 arguments for opening an online store. Forbes Mag. (Finan. Econ. J.), 02 June (2010). http://www.forbes.ru/svoi-biznes/21761-5-argumentov-dlya-otkrytiya-internet-magazina. Accessed 14 Mar 2019

Tangning, J.F.: The influence of information technology development in the employment market. Materials of the V-th International Scientific and Practical Conference and Round Table 18–20 April. Vladimir State University named after A.G. and N.G. Stoletovs, Vladimir; Kostanay state University named after A. Baitursynov; Warsaw University; Administration of the Vladimir region – Vladimir, VlSU Publishing House, 2018, pp. 244–254. ISBN 978-5-9984-0880-9 (2018). https://elibrary.ru/item.asp?id=36704727. Accessed 10 Apr 2019

Wei, W., Bing, W., Chun, Z., Jim, K., Don, T.: Classification of access network types: ethernet, wireless LAN, ADSL, cable modem or dialup? J. Comput. Netw. **52**(17), 3205–3217 (2008). https://doi.org/10.1016/j.comnet.2008.08.018. Accessed 14 Mar 2019

Lin, Y.: E-urbanism: E-commerce, migration, and the transformation of Taobao villages in urban China. Cities **91**, 202–212 (2018). https://doi.org/10.1016/j.cities.2018.11.020. Accessed 14 Mar 2019

Assessment of Crowdfunding Risks and Threats in Order to Counteract the Legalization (Laundering) of Criminal Proceeds and the Financing of Terrorism

Tatiana G. Ilina[1(✉)], Nataliya A. Tyuleneva[1], Victoria V. Makoveeva[1], and Elizaveta I. Ruzina[2]

[1] Tomsk State University, Tomsk, Russia
ilinatg@mail.ru, aola79@yandex.ru
vika_makoveeva@mail.ru,
[2] RUDN University, Moscow, Russia
ruzina_ei@rudn.university

Abstract. The rapid development of digital financial technologies poses new challenges in the area of combating money laundering (AML) and the financing of terrorism (TF). Crowdfunding is one of such non-traditional digital financing tools, representing new ML and FT risks. The main tool to counter legalization is a risk-based approach. Despite the fact that the FATF has developed a generally accepted methodology for assessing the risks of ML and FT, it is not a dogma and requires adaptation to new tools. Crowdfunding is one of the most promising digital financial instruments, but at the same time it is one of the most risky and insufficiently studied instruments in terms of ML and FT. Therefore, the article has explored and identified the features of crowdfunding as a new digital financing instrument for assessing its risks and threats in order to counter ML and TF (using the example of the Russian Federation and France). Recommendations on the areas of assessment, the formation of lists of risks of crowdfunding and their classification are offered.

Keywords: Crowdfunding · Money laundering and terrorism financing risks · Risk-based approach · FATF · Financial monitoring

JEL Code: F 330 · G 280

1 Introduction

With the rapid development of digital financial technologies, every state should pay great attention to ensuring their legality, especially with regard to issues of money laundering (ML) and the financing of terrorism (FT). The main tool for countering money laundering and terrorist financing is an approach developed by the FATF (Financial Action Task Force on Money Laundering). It aims to identify and assess the risks of legalization. This is necessary for making effective management decisions aimed at eliminating or mitigating negative consequences in the financial system,

E. G. Popkova and B. S. Sergi (Eds.): ISC 2019, LNNS 87, pp. 291–300, 2020.
https://doi.org/10.1007/978-3-030-29586-8_34

economy, and society. In addition, it allows you to more efficiently use the resources of the economy, directing them to the most important areas of combating money laundering (AML) and the financing of terrorism. The states that are members of the FATF are actively interacting in this area, since ML and FT are global problems. For example, Russia is closely cooperating with France, for this purpose the following documents were signed: Agreement on cooperation between the Russian Federation and the European Police Organization was signed (Rosfinmonitoring 2003); Agreement between the Government of the Russian Federation and the Government of the French Republic on cooperation in the fight against crime and in the field of internal security (Rosfinmonitoring 2003); Agreement between the Committee of the Russian Federation on Financial Monitoring and the Bureau for the Processing of Information and the Fight against Illegal Financial Flows of the Ministry of Economy, Finance and Industry of France on cooperation in the field of AML (Rosfinmonitoring 2003).

The main activities in this area are carried out by financial intelligence services of various countries, united in a global financial monitoring system. Rosfinmonitoring receives about 2 thousand requests for information from foreign partners annually (Rosfinmonitoring 2019); exchanges lists of terrorists (in 2019, lists of 1.5 thousand terrorists from France and Central Asian countries were provided) (Rosfinmonitoring 2019).

Together with Malta, Turkey and France, it fights against business that assists terrorists. (Leyva 2016) The financial intelligence of these countries is of an administrative type, so they maintain close ties with regulatory agencies and institutions (Russia, France, Australia) (Proshunin 2010). The head of the French FIU, J.B. Carpentier, during his working visit to Rosfinmonitoring, noted that "the most important issue in the financial intelligence activities at the present stage is the formation of lists of national and international risks in the field of ML and FT".

According to him, in the near future the risks of ML and FT will be associated with new non-traditional digital tools. Currently, practical methods for working with them have not been developed, so the settlement of such "black" zones is a promising task (Lisina 2019). International electronic transactions from the FATF "black list" are automatically considered suspicious and highly risky (Vashchekin and Vaschekina 2018). Since crowdfunding is one of such unconventional digital financing instruments, it undoubtedly enters this "black" zone.

In October 2015, the FATF report "New Terrorism Financing Risks" stated that "the use of organized "crowdfunding" methods also represents a new risk of FT" (FATF 2015). And the problem becomes more acute with each passing year, as crowdfunding is developing rapidly. Thus, in 2015 the research group Massolution estimated the global volume of the crowdfunding market at $ 34 billion (Crowd Expert 2015–2016), while The World Bank has predicted that it will reach $ 90 billion by the period between 2020 and 2025 (The best 2016). As a result, crowdfunding is one of the most promising digital financial instruments, but at the same time it is one of the most risky and insufficiently studied instruments in terms of ML and FT. Therefore, the purpose of this research is to study the characteristics of crowdfunding as a new digital instrument of financing and to analyse its risks and threats in order to counter ML and TF.

Methodology. D. Demistes in the book "Technology and Anti-Money Laundering: A System Theory and Risk-Based" presented scientific approaches to the analysis and assessment of risks in the field of ML and FT (Demits 2010). The fundamental documents of the FATF on risk assessment establish the basic methodological requirements, in particular: Guidelines for ML and TF risk assessment at the national level (FATF 2013); Strategy for assessing the risk of money laundering of criminally obtained funds (FATF 2008); Guidance on the application of an approach based on the risk assessment of AML and FT (FATF 2007). The FATF recommendations on a risk-based approach establish the correspondence between AML and FT measures and identified risks (FATF 2013).

Moreover, governments and financial institutions must identify, evaluate, and reduce these risks. But at the same time, the risk assessment methodology is being developed and selected independently, with regard to particularities of each country, sector, etc. The common methodology does not exist (Edronova 2016). The main goal of a risk-based approach in AML and RF is to assess the threat risk, reduce the probability of its occurrence and minimize damage based on the effective allocation of available resources. The main factors in risk assessment are threats, vulnerabilities and consequences.

The risk assessment consists of formulating conclusions regarding threats, vulnerabilities and consequences. The stages of risk assessment are: identification, analysis, assessment (FATF 2007, 2014, 2015). The stage of collecting information is also important, since the result of the assessment depends on its type and quality. The OSCE leadership points out the need to use a variety of data: quantitative and qualitative, combined, etc. (OSCE 2014). Since the statistical base requires a long accumulation period, not all risks can be measured. Moreover, the information base is often very limited, in that case any available information is used (FATF 2014).

Therefore, we should agree with the opinion of K.G. Sorokin that the risk-based approach cannot be adopted in a single policy form, since in each case it will depend on the specifics of the information base and the object of analysis (Sorokin 2014). The threat assessment is based on the following information: operational data; sentences; suspicious transaction reports; intelligence; Interpol and Europol reports, FATF reports; financial and economic data; information from private sector organizations; research and public opinion research results; media (FATF 2008).

In the Russian practice of financial monitoring, the following types of risks are distinguished: international legal; risks of financial institutions; budget; industry; regional; FT risks (Rosfinmonitoring 2014). In international practice, risks are divided into such types: political, economic, social, technological, environmental, legal. The consequences are grouped by type: physical, social, environmental, economic and structural. To assess the level of risk, risk matrices are used, in which the probability of a threat is indicated horizontally, and the severity of the consequences -vertically.

Usually the level of risk is measured as low, medium and high. It is clear that as the probability of the threat and the severity of the consequences increase, so does the risk. At the present stage, the question of "developing and implementing common approaches to assessing the risk of ML and FT, including on the basis of segmentation by type of organization and its business model," seems relevant (Vashchekin and Vaschekina 2018). This particular segment is crowdfunding.

2 Results

What is crowdfunding and what are its main features in terms of AML and FT? Let us consider some definitions. Crowdfunding comes from the English "crowd" - public and "funding" - financing. Therefore, the FATF considers it a fundraising from the public and defines it as "a way to allow companies, organizations or individuals to collect money through the Internet be means of donations or investments from a large number of people" (FATF 2015). A.V. Kuznetsov considers crowdfunding not as a way, but as an activity of attracting financial resources from a large number of people who voluntarily pool their resources at the crowdfunding sites to support the project, help the needy, hold events, etc. (Kuznetsov 2017) According to Chugreev V.L., Crowdfunding is a technology of collective funding through voluntary donations - "national or public funding" (Chugreev 2013). N.S. Nedzvetsky sees in it an alternative means of investment and a mechanism for finding their sources (Nedzvetsky 2017).

D.A. Profatilov as well as A.V. Kuznetsov connects crowdfunding with Internet platforms, he, however, defines it not as an activity, but as a financial tool for the voluntary collective pooling of people's resources at a crowd platform in order to support projects. At the same time, crowdfunding platforms act as intermediaries between project creators and crowdbacks (Profatilov 2015). Thus, Crowdfunding is considered by different scientists as a method, activity, technology, means, mechanism, financial instrument, which indicates obvious differences in terminology.

This makes it difficult to analyze and assess its risks for purposes of AML and FT. Moreover, in different countries crowdfunding has its own characteristics and different levels of development. To eliminate discrepancies, it is necessary to develop a common approach to understanding the essence of crowdfunding in the field of AML and FT. In our opinion, crowdfunding should be considered as a segment of the financial market, as an activity, and as a process. As a segment - because the volume of financing has reached a noticeable level, the order of financing has specific features, there are professional participants (crowd-platforms).

Let's draw an analogy with other specific segments of the financial market for which the terminology has already been defined. For example, leasing is considered as a process of rendering certain services and as an activity of leasing companies, the same can be said about factoring, insurance, etc. Crowdfunding is also implemented through specialized Internet sites that provide services and offer investment products in the form of startups, projects, loans, i.e. carry out intermediary activities in the financial market.

The realization of financing itself takes place in stages, and therefore represents a process where various methods, methods, means and technologies of collective fundraising are used. Such interpretation of crowdfunding clarifies, simplifies and speeds up the assessment of the risks of ML and FT. The method of analogies allows using methodological and practical developments of other sectors of the financial market to compile a list of common (typical) risks and factors (vulnerabilities, threats and consequences). As mentioned earlier, crowdfunding is rapidly developing quantitatively (by increasing the volume of funding, the number of participants, and so on.)

As actively occur qualitative changes, in particular, its new types are emerging: "crowdsponsoring" (collective sponsorship), "crowdinvesting" (collective financing of start-ups for the purpose of receiving and distributing profits), "crowdlending" (collective return financing for projects with a social orientation), "Crowddonatting/laising" (net gratuitous social investment) (Leimeister 2012). (Willfort 2013). Agrawal et al. (2013) consider these types depending on the goals of the investor: Donation Crowdfunding - charity (gratuitous), Reward Crowdfunding - Award, Crowdfunded Lending - Loan or Peer and Equity Crowdfunding - Investment or shareholder (Agrawal et al. 2013). Award crowdfunding is the most common type. Crowding is divided into three broad categories: consumer lending, corporate lending and social lending (Baeck 2014). Each of the categories also has branches. The first and second branches include balance and market lending, and the third - market lending for the purchase of real estate and auction lending. (Mills and Mills 2014) But the basic and generalizing one is the term "crowdfunding", including all other types (Sanin 2015). Depending on the sphere of investment, crowdfunding financing is classified as business crowdfunding, political, social and creative crowdfunding. (Larionov 2015) The large number of different types of crowdfunding leads to the need for a deeper study of their features to identify the specific risks of ML and FT. Simultaneously, new interesting areas of non-financial use of crowdfunding appear. So, Brown et al. (2017) talk about the marketing use of crowdfunding. Dolzhenko (2016) develops this approach and highlights the following directions: crowd-searching (searching among consumers of a product for proposals for improving its quality); crowdcasting (search for people and organizations capable of solving the tasks of large corporations); crowd-production (activities carried out by an open network community); crowdstaffing (search for employees to participate in product testing). Non-financial use of crowdfunding is explained by the fact that it is one of the branches of crowdsourcing, which means the joint voluntary participation of people in solving socially significant tasks (Chugreev 2012).

In general, non-financial areas of crowdfunding search and interact with people, which creates a favorable environment for their involvement in terrorist activities or use in projects related to ML. Moreover, this contributes to low public awareness of crowdfunding, its vulnerabilities and threats. This problem especially concerns Russia.

From February 28 to March 4, 2019, the Institute of Public Opinion conducted a sample survey of the population of the Russian Federation over 18 years old, in which 1680 people were surveyed. The sample reflects the socio-demographic parameters of the population of the Russian Federation. The error at 95% confidence level does not exceed 2.4% (Institute of Public Opinion 2019). The results of the survey showed that only 9% of the population know crowdfunding well, the rest (51%) have heard about it for the first time or have little idea (40%). Moreover, the level of knowledge about crowdfunding correlates with the level of income: with an income of up to 10,000 rubles, only 1.9% know about it, and with an income of more than 100,000 rubles - 31%.

For this reason, 91% of the population have never participated in crowdfunding, and 9% have taken part once. The second reason for the lack of interest among residents of the Russian Federation in crowdfunding is a low level of trust: 53% do not trust, 47% trust. The level of trust in crowdfunding sites is even lower, only 24% of respondents trust them, the rest either do not trust or find it difficult to answer. Nevertheless, in terms of future participation, we see the opposite picture: only 47% are against, but 53% are for participation, and of these, 66% are willing to invest from 500 rubles and more and aim to make a profit of 62%.

A large proportion of applicants (66%) are between the ages of 18 and 30, which portends very good prospects for the development of crowdfunding in the Russian Federation in the near future. Analysis of the population structure by the most preferred types of crowdfunding projects in the future showed a significant predominance of socially significant projects (58%) and projects for creating goods or supporting business (39%). These trends need to be taken into account by the financial intelligence service of the Russian Federation and other states when assessing risks, vulnerabilities, threats and the consequences of ML and FT. The most important are the conclusions about the existing preconditions for the rapid development of the crowdfunding market in the future.

The apparent contradiction between the low level of awareness (and trust) and the prevailing level of potential willingness to participate in crowdfunding projects will lead to an increase in the number of threats and the level of risk. This will especially manifest itself in the field of social projects and new business creation projects that are considered most favorable for ML and FT and have increased risks and threats..

All of this will require great efforts and resources from the bodies and agents of financial monitoring. Therefore, it is necessary to formulate a list of potential threats, risks, vulnerabilities and consequences, taking into account the future trends of crowdfunding, as well as studying the best practices of other countries. Special attention should be paid to the creation and introduction of legislative and regulatory frameworks. Comparison of some aspects of the development and regulation of crowdfunding in the Russian Federation and in France, conducted in 2017 by N.S. Nedzvetskim (2017), shows that Russia is about 4–5 years behind France (Table 1).

Table 1. Comparison of crowdfunding in Russia and in France (supplemented by the authors)

Criteria	France	Russia
Beginning	2007–2008	2012
Platforms	Many, diverse, emphasis on the development of business projects (almost 50%)	Few, diverse, emphasis on the development of cultural projects (more than 50%)
Governance relation	High interest in development, quick, effective response (2014 - law, two provisions, 2015 - 15 propositions)	Sufficient interest in development, but slow response (2018 - draft law.)

(*continued*)

Table 1. (*continued*)

Criteria	France	Russia
Regulation	The presence of general regulation: rules for the operation of crowd-platforms, mandatory registration, restrictions on the size of capital (The crowdfunding, 2019) Leader in the regulation of specific species: crowdlending; crowdinvesting. Charity is not regulated Since 2019 the EU regulation proposals come into force. Regulation applies to return crowdfunding (loan-based and equity-based crowdfunding) Does not regulate donation-based crowdfunding and reward crowdfunding (Norland Legal 2019).	Legal regulation is not yet available. A crowdfunding draft law adopted in first reading. General rules and restrictions on the activities of crowd platforms, the procedure for control by the Central Bank of the Russian Federation defined. The draft law lists four types of crowdfunding: (1) the provision of loans; (2) the acquisition of securities; (3) the acquisition of shares in the authorized capital, etc.; (4) the acquisition of tokens of the investment project; charity and non-financial types are not regulated (Motovilov 2018)

3 Conclusions

The study has revealed the inconsistency in the definition of the concept of crowdfunding. In our opinion, crowdfunding should be considered as a segment of the financial market, as an activity, and as a process, similarly to other specific segments. Such interpretation of crowdfunding clarifies, simplifies and speeds up the assessment of the risks of ML and FT. The method of analogies allows using methodological and practical developments of other sectors of the financial market to compile a list of common (typical) risks and factors (vulnerabilities, threats and consequences).

The large number of different types of crowdfunding leads to the need for a deeper study of their features to identify specific risks of ML and FT. The study of non-financial types of crowdfunding will help to generate a list of risks associated not only with the process of crowdfunding, but also with the crowdfunding relations between its participants, including the conditions and infrastructure. This will make the risk assessment of crowdfunding for AML and FT more complete and complex, and will also help develop public awareness measures to reduce their level of involvement in ML and FT.

The population survey revealed trends that the FIU of the Russian Federation and other countries need to take into account when assessing the risks of ML and FT. The most important are the conclusions about the existing preconditions for the rapid development of the crowdfunding market in the future. Therefore, it is necessary to formulate potential threats, risks, vulnerabilities and consequences, taking into account future development trends, as well as studying the best practices of other countries.

Special attention should be paid to the creation and introduction of legislative and regulatory frameworks.

References

Rosfinmonitoring. Agreement on cooperation between the Russian Federation and the European Police Organization (2003). http://www.fedsfm.ru/activity/interstate-agreements. Accessed 1 Apr 2019

Rosfinmonitoring. Agreement between the Government of the Russian Federation and the Government of the French Republic on cooperation in the fight against crime and in the field of internal security (2003). http://www.fedsfm.ru/activity/intergovernmental-agreements. Accessed 1 Apr 2019

Rosfinmonitoring. Agreement between the Committee of the Russian Federation on Financial Monitoring and the Bureau for the Processing of Information and the Fight against Illegal Financial Flows of the Ministry of Economy, Finance and Industry of the French Republic on cooperation in the sphere of counteracting legalization (laundering) of proceeds from crime (2003). http://www.fedsfm.ru/activity/bilateral-interagency-agreements. Accessed 24 Mar 2019

Rosfinmonitoring. Bilateral interaction, Official website of Rosfinmonitoring of Russia (2019). http://www.fedsfm.ru/activity/bilateral-cooperation. Accessed 24 Mar 2019

Rosfinmonitoring. A meeting of the Director of Finmonitoring with the Presidents of the Russian Federation took place (2019). http://www.fedsfm.ru/releases/3825. Accessed 25 Mar 2019

Leyva, M.: Putin was informed about the liquidation of the Caucasian cell of terrorists in France, RBC 2016 (2016). https://www.rbc.ru/politics/09/03/2016/56e04fb09a7947371a009faf. Accessed 24 Mar 2019

Proshunin, M.M.: Rosfinmonitoring as the authorized body for AML and FT, Administrative Law, no. 2 (2010). http://www.top-personal.ru/adminlawissue.html?54. Accessed 2 Apr 2019

Lisina, I.A.: Working visit by the head of the French FIU to Rosfinmonitoring. Financial Security, no. 9, pp. 42–44 (2019)

Vashchekin, A.N., Vaschekina, I.V.: Informational interaction in the system of combating the laundering of criminal proceeds: a risk-based approach. Legal Informative, no. 4, pp. 4–14 (2018)

FATF. New Terrorist Financing Risks Report (2015). http://www.fedsfm.ru/content/files/documents/2018/reporting. Accessed 2 Apr 2019

Crowd Expert. Crowdfunding Industry Statistics 2015–2016, Crowd Expert, 2015–2016 (2015–2016). http://crowdexpert.com/crowdfunding-industry-statistics/. Accessed 12 Apr 2019

The best. The best crowdfunding platforms in Europe (2016). https://blog.privateinvestmentsnetwork.com/category/crowdfunding-en/. Accessed 2 Apr 2019

Demits, D.: Technology and Anti-Money Laundering: A Systems Theory and Risk-Based Approach. Edward Elgar Publishing Ltd, Cheltenham (2010)

FATF. Guidelines for assessing money laundering and terrorist financing risks at the national level (2013). https://eurasiangroup.org/FATF_risk_accessment.pdf. Accessed 12 Apr 2019

FATF. Strategy for risk assessment of money laundering and terrorist financing (2008). https://eurasiangroup.org/ru_img/news/typ_strat.pdf. Accessed 4 Apr 2019

FATF. Guidelines for applying a risk-based approach to AML/CFT. Principles and procedures (2007). https://eurasiangroup.org/files/uploads/files/FATF_documents/FATF_Guidances/typ_ruk.pdf. Accessed 4 Apr 2019

FATF. International standards for combating money laundering, terrorist financing and financing the proliferation of weapons of mass destruction (2013). https://eurasiangroup.org/files/uploads/files/FATF_documents/rec_meth_high_level/Rekomendatcii_FATF_s_obnovleniem_rek_5_i_8.pdf. Accessed 4 Apr 2019

Edronova, V.N.: Financial monitoring methodology: national risk assessment. Finance and credit, no. 16, pp. 27–39 (2016)

FATF. The risk of illegal use of non-profit organizations for terrorist purposes. FATF report (2014). http://www.mumcfm.ru/index.php/ru/materials/books. Accessed 5 Apr 2019

OSCE. OSCE guidelines for data collection in conducting a national risk assessment of money laundering and terrorist financing (2014). http://www.osce.org/ru/secretariat/123906?download=true. Accessed 5 Apr 2019

Sorokin, K.G.: Problems of organization, legalization and harmonization of national and supranational risk assessment systems within the framework of the EurAsEC Customs Union, Customs, N1, pp. 23–30 (2014)

Rosfinmonitoring. Annual Report (2014). http://www.fedsfm.ru/activity/annual-reports. Accessed 24 Mar 2019

Kuznetsov, V.A.: Crowdfunding: Current Regulatory Issues. information and analytical materials, Money and credit, N1, pp. 65–73 (2017)

Chugreev, V.L.: Crowdfunding - a social technology of collective financing: foreign experience using Economic and social changes: facts, trends, forecast, no. 4(28), pp. 190–196 (2013)

Nedzvetsky, N.S.: Financial potential of the "crowd": Russian and European crowdfunding. Econ. Yesterday Today Tomorrow **7**(6A), 126–138 (2017)

Profatilov, D.A.: Crowdfunding is a modern tool for investing innovative projects. FGBOU VPO "RGAIS", Moscow, Petropavlovsk-Kamchatsky, pp. 67–72 (2015). http://elar.urfu.ru/bitstream/10995/32939/1/pz_2015_02_13.pdf. Accessed 1 Apr 2019

Leimeister, J.M.: Crowdsourcing: crowdfunding, crowdvoting, crowdcreation. In: Zeitschrift Control Controlling und Management (ZFCM), Ausgabe, Erscheinungsjahr, Seiten, no. 56, pp. 388–392 (2012)

Dushina, M.O.: Methods of network communication in the digital society: benchmarking, crowdsourcing, crowdfunding. Sociol. Sci. Technol. **5**(1), 105–114 (2014)

Willfort, R.: Crowdfunding, Crowdlending, Crowdinvesting. ISN - Innovation Service Network GmbH (2013). http://www.innovation.at/wp-content/uploads/2015/08/Crowdfunding-Crowdsourcing-Potenzial_Endbericht.pdf. Accessed 15 Apr 2019

Agrawal, A., Catalini, C., Goldfarb, A.: Some simple economics of crowdfunding, NBER Working Paper N 19133 (2013). https://www.researchgate.net/publication/272543487_Some_Simple_Economics_of_Crowdfunding. Accessed 10 Apr 2019

Baeck, P., Collins, L., Zhang, B.: Understanding alternative finance. The UK Alternative Finance Industry Report 2014. University of Cambridge 2014 (2014). https://media.nesta.org.uk/documents/understanding-alternative-finance-2014.pdf. Accessed 10 Apr 2019

Mills, K., Mills, B.: "McCarthy", Harvard Business School (2014). http://www.hbs.edu/faculty/Publication%20Files/15-004_09b1bf8b-eb2a-4e63-9c4e-0374f770856f.pdf. Accessed 11 Apr 2019

Sanin, M.K.: The history of the development of crowdfunding. Classification of species. Analysis of development prospects and benefits. Econ. Environ. Manage. **2015**(4), 57–63 (2015)

Larionov, N.A.: Features of financial behavior of investors in the crowdfunding model of financing innovation. Bull. Saratov State Socio Econ. Univ. **2**, 77–80 (2015)

Brown, T.E., Boon, E., Pitt, L.F.: Seeking funding in order to sell: crowdfunding as a marketing tool. Bus. Horiz. **60**(2), 189–195 (2017)

Dolzhenko, R.A.: Possibilities of using crowd technologies in the process of interaction of the organization with customers, Modern management: problems, hypotheses, research. Issue 7: Sat. scientific tr. / under scientific. Volkova, I.O. (ed.) status E.V. Filipskaya; Nat researches University "Higher School of Economics", Faculty of Business and Management. - Electron. text. Dan. (5.18 MB) - M.: Izd. House of the Higher School of Economics, 2016, pp. 342–352 (2016). https://management.hse.ru/data/2016/09/12/1120710925/Comparemanagement.Vyp.7-sayt.pdf. Accessed 11 Apr 2019

Chugreev, V.L.: Creating a crowdsourcing project for publishing and discussing proposals for socio-economic development of the region. Probl. Territory Dev. **62**, 157–164 (2012)

Institute of Public Opinion. Crowdfunding in Russia: Prospects and Level of Trust", Institute of Public Opinion, Site "Anketologist" (2019). https://iom.anketolog.ru/2019/03/06/kraundfanding. Accessed 12 Apr 2019

The crowdfunding. The crowdfunding law takes effect in France (2019). http://crowdsourcing.ru. Accessed 11 Apr 2019

Motovilov, O.V.: The phenomenon of crowdfunding: a study of the features. St. Petersburg Univ. Bull. Econ. **34**(2), 298–316 (2018). https://doi.org/10.21638/11701/spbu05.2018.205. Accessed 1 Apr 2019

Norland Legal. New proposals on crowdfunding regulation in the territory of the European Union, Norland Legal Website (2019). http://norlandlegal.com/novye-predlozhenija-po-regulirovaniju-kraudfandinga-na-territorii-evrosojuza/. Accessed 11 Apr 2019

Open Innovations as a Tool of Interaction Between Universities and Business Structures in the Digital Economy

Pavel N. Zakharov(✉), Artur A. Posazhennikov, and Zhanna A. Zakharova

Vladimir State University named after Alexander and Nikolai Stoletovs Institute of Economics and Management, Vladimir, Russia
pav_zah@mail.ru, zjane77@mail.ru, zzarturzz@yandex.ru

Abstract. The article is devoted to the study of the problems of finding the most preferred forms of interaction between universities and the business community. Universities are able to concentrate the intellectual potential of the region, and the business environment is able to realize this potential, so that any form of interaction will contribute to its economic and social development. The article provides an overview of the forms of interaction between higher education institutions and enterprises. The most effective forms of cooperation are identified. At the end of the work the model of interaction between the University and enterprises is given. The necessity of development for each University of the General system of management of interaction with the enterprises is proved.

Keywords: Digital economy · Open innovations · University · Business structures

JEL Code: O 15

1 Introduction

The declining efficiency of traditional economic growth resources requires new approaches to economic development. The system of innovative development on the basis of a close interaction of scientific knowledge and production opportunities based on high technologies, scientific and industrial potential and intellectual property becomes the support of the new model of development of Russia in modern conditions.

The current level of economic dynamics involves the systematic introduction of scientific achievements in industry, and later in the real sector of the economy through the activation of the joint potential of the participants. A special role in these conditions is assigned to higher education and research schools of universities.

The main task of universities is to preserve the educational potential of the territory, increase the level of education of the population and the scientific and technical potential of the country, the transition of accumulated knowledge and experience from one generation to another. In the Soviet period, the interaction of universities and the

E. G. Popkova and B. S. Sergi (Eds.): ISC 2019, LNNS 87, pp. 301–306, 2020.
https://doi.org/10.1007/978-3-030-29586-8_35

real sector in the face of industrial enterprises was quite active and productive. Basic departments and large enterprises along with the University invested in the formation of the future specialist. The active development of market relations has led to the rupture of most of the formed links between universities and specialized enterprises. The isolation of the training system from the real sector has led to systemic problems in the labor market:

– the lack of real experience of graduates makes employers spend time on additional training of employees. Spending money, time and effort on a specialist, employers often talk about the low quality of training in higher education;
– the emergence of a structural imbalance between supply and demand in the labour market. In many regions of the country, some specialists are needed, while universities located there produce others.

To solve these problems, a direct dialogue between universities, regional authorities and enterprises is necessary, which will allow to create conditions for ensuring the future economic needs of the territories. It is universities that have the highest total intelligence of employees, and form elements of the national innovation system in a certain territory, while at the same time being in a strict framework of the Ministry and a variety of bureaucratic requirements and standards. Modern conditions require educational institutions to meet the needs of enterprises in competent professionals with relevant knowledge and skills that can correctly assess the current situation and quickly make the right decisions.

The tasks set by the Russian government to switch to domestic software open up a number of opportunities for interaction between universities and the business environment. At the moment, a number of problems, such as lack of experience, users are not fully compatible with MS Office, lack of mail client, etc. do not allow to implement domestic software for the Executive authorities and government agencies.

From the point of view of the labor market, the University has a dual position. On the one hand, the University is a subject of the economy, providing educational, scientific, consulting services in a certain area, the consumers of which are students, enterprises, authorities. On the other hand, in the labor market they present the results of their activities – specialists with competencies, consumers of which are enterprises and organizations of various sectors of the economy. Thus, universities, as well as enterprises are more interested in analyzing the future state of the labor market needs than in the present needs.

Enterprises are experiencing a shortage of personnel, and express dissatisfaction with the quality of training, as well as non-compliance of educational programs with modern market requirements and production standards. Although there are examples of effective cooperation, the overall level of cooperation is quite low, and this is not a problem for individual companies, but for the system as a whole. One of the problems is in the plane of disinterest of University science in technological strengthening, both individual enterprises and groups of companies. So companies attract universities to solve individual problems, rather than complex support activities.

The implementation of elements of the digital economy will contribute to a closer interaction of all spheres of the economy, but the University science and business environment of the sector requires special mechanisms and institutions. The digital economy will link real-time statistics, public services and provide universities and businesses with a huge layer of opportunities. However, the technical side of the transition is much more complex. Innovations as a point of contact between educational institutions and enterprises have proved their absolute effectiveness, but a special role in this process is played by a special role (Lapaev 2011; Bortnik 2012; Goikher and Bugrova 2017; Zakharov et al. 2016).

The concept of open innovation is a convenient and practical basis for mutually beneficial cooperation between universities and enterprises through:

1. Organization of the research and development process in the form of a publicly available knowledge resource;
2. Development of various components of an innovative product by individual companies or groups of specialists;
3. Free sale of developments of wide application;
4. A significant reduction in the level of bureaucracy in decision - making processes for innovation.

This concept will increasingly be supported by the management of universities focused on long-term planning of work with enterprises. Open innovations make it possible to exclude problems related to the use of intellectual property rights, which has a number of positive aspects for students and enterprises (Zakharov et al. 2017; Hollanders et al. 2010).

In the long term, this will reduce the costs of research and development, increase output, and create fundamentally new markets for all partners. According to the scientist, it is necessary:

- In determining the objectives do not proceed from the needs of the partnership, and the specifics of the market;
- Classify the research capabilities of firms and give the development to companies with relevant competencies;
- Openly agree on the business models of the companies participating in the partnership.

A strong factor in the development of relations between universities and the business environment is the high cost of maintaining separate research laboratories. It is more profitable for companies to concentrate on joint developments, the creation of open innovation centers, where students can learn and work. Technoparks and innovation centers created by the state will more effectively reveal their potential within the framework of open innovation systems, forming competitive advantages not for individual companies, but for sectors of the economy.

It is the low level of interaction between University science and enterprises that determines the low level of efficiency of innovative programs for the development of the domestic economy. The activity of the University is aimed at improving the production technology, updating the nomenclature, improving the quality and further commercialization of the results of research and development. This work is impossible without systematic cooperation of universities and enterprises at all stages of the management cycle. It is the ratio of the stages of the management cycle and the existing forms of interaction that should be considered as the key to finding the most effective methods of interaction between universities and business structures.

2 Methodology

The basis of the research methodology was the search for points of contact between enterprises and universities on the basis of the management cycle, which includes five basic stages (Fig. 1).

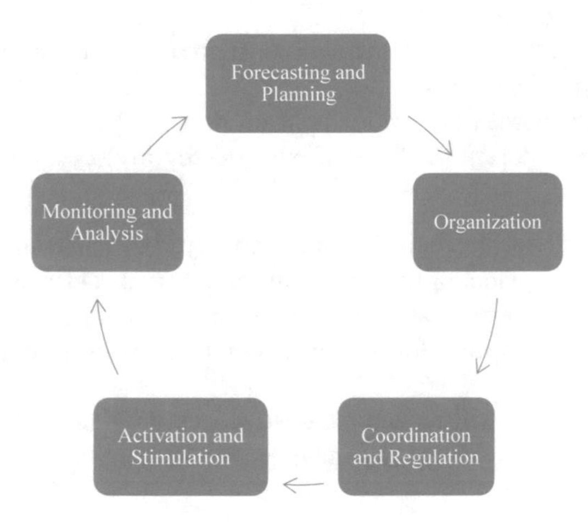

Fig. 1. Management cycle

In the course of the study of the process of interaction between universities and business structures, the strategic documents of the following flagship universities of Russia were analyzed: Yaroslavl, Vladimir, Tula, Altai, Kemerovo, Petrozavodsk and Novgorod state universities, Novosibirsk and Nizhny Novgorod state technical universities, Belgorod state technological University, Magnitogorsk and Saratov state technical University. These universities were included in the program of "flagship universities" which determined before their leadership special requirements for indicators such as the level of employment of graduates, the level of publication activity, the growth of R&D. Achieving high performance in these categories is impossible without improving the quality of interaction with enterprises in the region.

3 Results

Support of innovative development by the state contributed to the expansion of forms of interaction between higher education and production. Thus, we will form a list of the most relevant forms for the interaction of enterprises and University science (Table 1).

The presented forms of interaction contribute to the strengthening of contacts between students and employers, which, in turn, leads to an increase in the prestige of

Table 1. Forms of interaction between universities and enterprises depending on the stage of the management cycle

Management cycle stage	Form of interaction
Forecasting and planning	- involvement of employers in the formation of training programs; - development together with the employer of the competence model of the graduate; -involvement of employers in graduate career management programs - conclusion of cooperation agreements between the University and the enterprise; - monitoring of employers' need for specialists
Organization	- organization for students of industrial and pre-diploma practice at the enterprise; - organization of small innovative enterprises on the basis of universities; - creation of basic departments; - targeted training of specialists; - targeted training of students in the employer's territory; - course and diploma design; - involvement of leading specialists in teaching activities
Coordination and regulation	- retraining of personnel of enterprises; - additional education; - professional development of employees of enterprises
Activation and stimulation	- Training through research projects; - Presentations of companies-employers, job fairs
Monitoring and analysis	- involvement of representatives of organizations, external specialists and other stakeholders in the quality control of education; - monitoring the demand for graduates; - participation of leading experts in the work of certification commissions

both the University and the employer. This approach to interaction helps the educational institution to assess the prospects of further training of certain specialists.

Thus, the most productive interaction between universities and enterprises takes place at the stages of organization, coordination and control. The least interesting stages for interaction are planning, as well as activation and stimulation. This is due to the fact that the process of interaction between universities and enterprises is focused on solving specific problems and is not systematic, and therefore has a lower degree of manageability.

4 Conclusions

The study showed that there is an objective need for long-term prospective interaction between higher education and the business environment. Cooperation between universities and businesses requires a transition from a close, but the design of the interaction system. This greatly reduces the possibility of improving the quality of education in general and the design of educational programs to meet the requirements of the labor market.

However, the development of interaction tools is associated with the solution of many problems, especially in the regions:

- Development of the institutions of the patent regulation for the protection of intellectual property rights;
- Creation and support of information resources on students opportunities in the framework of interaction between universities and enterprises;
- Involvement of the teaching staff of the University in the formation of strategic documents of the municipality and the region;
- Assessment of innovative potential of the University from the point of view of the needs of the market;
- Formation of training programs based on the concept of "Open innovation";
- Development of program documents regulating the distribution of budgets for innovation in order of priorities dictated by the market.

Summarizing the above, we note that the individual characteristics of each University and business requirements each University must develop a unique system of interaction with enterprises.

References

Lapaev, D.: Comparative evaluation of the effectiveness of innovative development of economic systems. Integral **6**, 46–47 (2011)

Bortnik, I.: System of evaluation and monitoring of innovative development of Russian regions. Innov. Econ. **9**, 48–61 (2012)

Goikher, O., Bugrova, O.: Bases of the complex approach of innovative development of the region on the basis of monitoring procedures. Econ. Entrep. **8**(4), 1163–1167

Zakharov, P., Nazvanova, K., Posazhennikov, A: Synergetic effect of innovative development of the textile cluster in the region. Bull. Vladimir State Univ. Named After Alexander G. Nicholay G. Stoletovs. Ser. Econ. **4**, 10 (2016)

Zakharov, P., Nazvanova, K., Posazhennikov, A.: Problems of adaptation of the "Open innovations" model in Russian conditions. Federalism **4**(88), 99–112 (2017)

Hollanders, H., Tarantola, S., Loschky, A.: Regional Innovation Scoreboard (RIS) 2009. Pro Inno Europe, 76 (2010)

The Modern Tourism Market in Russia: Possible Ways of Using the Promotion Forms in Tourism in the E-Education System

Vladimir A. Zolotovskiy(✉), Marina L. Davydova, Evgeniy V. Stelnik, Aleksandra K. Elokhina, and Veronika A. Polukhina

Volgograd State University, Volgograd, Russia
{zolotovskiy.azi, kmp, analitika, alexelokhina, veronika.poluhina}@volsu.ru

Abstract. The purpose of the paper is to determine the quality and content of the mechanism for integrating such forms of promotion in tourism as tourist information resources into the electronic educational environment, as well as to determine the projected results of similar innovation from the tourist business, educational environment, society. To achieve the goal, the following methods were used in the article: dialectical, comparative analysis, comparative legal, formal legal, system analysis. The researchers analyzed the specifics of educational activities in the field of tourism in the framework of an E-educational environment created in the Russian Federation. The authors considered the most relevant forms of promotion in tourism. After that, special regulatory acts were analyzed and the content of the organization mechanism in the electronic educational environment was determined, on the basis of which conclusions were made about the ways of integrating digital means and forms of advancement into the educational process. The article discusses the possibilities and methods of integrating such forms of promotion in tourism as tourist information resources into an E-educational environment, including examples of ways of such integration in schools and higher educational institutions.

Keywords: Marketing technologies · Promotion in tourism · Digital technologies in tourism · Digital/electronic educational environment

JEL Code: K150 · K220 · K230 · Z300 · Z320 · Z330

1 Introduction

Digital technologies in tourism deal with a wide range of issues: optimizing the management system of organizational and personnel processes of a tourism and hospitality enterprise, using technologies to interact with partners, improving control over the process of providing specific services by third parties, strengthening the system of promotion of the tourism product (a separate service) and improvement mechanism of interaction with tourists/customers.

E. G. Popkova and B. S. Sergi (Eds.): ISC 2019, LNNS 87, pp. 307–316, 2020.
https://doi.org/10.1007/978-3-030-29586-8_36

In the economic literature, promotion is seen as a necessary part of an information campaign of an enterprise, represented by a set of various marketing measures implemented in real and digital space, aimed at creating an audience view of the product/service itself, creating interest in it, and using systematically various marketing tools and forms to stimulate and increase demand, increase sales, expand the market space for product sales, etc. (Perova and Zakirova 2015), (Simavonyan 2015), (Freeman and Glazer 2019).

Obviously, the specificity of the forms of promotion of a tourism product in its exclusively commercial significance, as well as the tourism product of a territory is a tourist product-place (a set of services, goods and objects in a particular destination tourism) (Dunets 2014), is associated with the task of promoting cultural, historical, sports, recreational environmental and other resources is determined by the nature of the tourist product.

Despite the distinctive features of the tourist product and individual tourist services (Sizenev 2018), methods and means of promotion by their nature, in most cases, can be classified as universal means of marketing technologies aimed at improving the competitiveness of production (Alananzeh et al. 2015), (Baranova 2015).

On that basis, the means of promotion can be divided into two traditional groups: «above line» and «below the line» . The first group includes the classic types of direct advertising on television, radio, territorial print media, indoor and outdoor advertising. The second group is characterized by a deeper and often individual impact on the consumer. (Middleton and Clarke 2001), (Morgan and Ranchhod 2009), (McCabe 2008), (Promote tourism products and services).

In the context of the development and mass introduction of digital technologies in various spheres of public relations, allowing to create conditions for the almost unlimited attraction of potential tourists, the methods and forms of PR related to the promotion of the tourist product of the territories are of current importance in the complex of means of promotion: brand/tourist territory interaction with the press on the Internet (primarily from Travel Media), the brand/tourist territory's interaction with representatives of the tourism industry (automated information system «Tourism» (Automated system AIS « Tourism»), a unified tourist passport (http://utp.nbcrs.org/) (Unified tourist passport), Tourist information exchange system (http://nbcrs.org/) (Tourist information exchange system), brand/tourist territory interaction with the public/unlimited circle of consumers (subsystem AIS «Tourism» «National travel portal «RussiaTravel.ru» (National Tourism Portal Russia.travel), tourist regional passports, electronic forms of printed guides, specialized federal, regional and municipal tourist portals and registries (Gurkina and Zhulina 2016), (Tourist portal of the Volgograd region), (Tourism Development Agency of the Volgograd Region), (Unified Federal Register of Tour Operators), (Federal register of tour operators of outbound tourism), (Travel Agency Registry), (Register of Tourist Information Centers of the Russian Federation), official websites of administrative entities (Wu 2018) containing tourist information, etc. The possibility of almost global audience reach with digital means of promotion predetermined the popularization of a regional tourist product as part of the practice of creating multilingual advertising materials (Tourist portal of the city of Volgograd and the Volgograd region “Volgograd region).

Speaking about the specifics of approaches to the understanding of tourism education in the Western tradition, it should be noted that they are directly dependent on the idea of an insufficient level of generalizations in the emerging tourist doctrine. (Hoerner 2000), (Cooper et al. 1998), (Tribe 2004). Domestic literature about the problems of tourism as a curriculum and scientific discipline is at the formative stage. An important role in the literature is occupied by the analysis by domestic specialists of the works of leading North American and European authors. One of the numerous examples, V.K. Stepanov in his article, conducts a deep analysis of the Canadian author and comes to interesting conclusions about the prospects for the development of digital technologies in modern education (Stepanov 2001).

Nevertheless, the topic of the digital educational environment in the Russian literature is a popular and developed topic. A pioneer in this area is V. Ort, the author of a series of books on information-open societies. Exploring the World Wide Web (WWW) environment, V. Ort sees in the distance learning and online video courses the future of modern education (Ort 2005).

The article by O. A. Brel, «The Use of Modern Educational Technologies in Training Personnel for the Tourism Sphere (Using the Case-Method and Technology Portfolio)» raises the question of the practical implementation of digital technologies in tourism education. The author sees the task of a new tourist education in «forming a developing, professionally oriented environment for realizing the personal potential of students» (Brel 2016).

Internet technology has changed the economy and politics now come in higher education. Universities formed in postguttenberg time are waiting for painful changes (Tapscott 2015). For developing tourism education, the crisis of higher education is a wonderful opportunity to organize an independent scientific discipline, which will initially be eclectic as much as the Internet itself is eclectic and confused (Horn 2005).

The purpose of the article is to determine the quality and content of the mechanism for integrating such forms of promotion in tourism as tourist information resources into the electronic educational environment, as well as to determine the projected results of similar innovation from the tourist business, educational environment, society.

2 Methodology

To achieve the goal, the following methods were used in the article: dialectical, comparative analysis, comparative legal, formal legal, system analysis. The researchers analyzed the specifics of educational activities in the field of tourism in the framework of the digital educational environment created in the Russian Federation. Initially, the most relevant forms of promotion in tourism were considered, among them the most promising to use were highlighted. After that, special regulatory acts were analyzed and the content of the organization mechanism in the electronic educational environment was determined, on the basis of which conclusions were made about the ways of integrating digital means and forms of advancement into the educational process. On the basis of the competence-based approach, a comparative analysis was made of current practices of using the most promising forms of promotion in the education system.

The main methodological approach of the article is the post-formal methodology of social and humanitarian research, as Horn R. A. formulated it in his article «Post-Formal Conversation» (Horn 2005). This approach is distinguished by its criticality and opposition to the dominant scientific and educational paradigm. In our opinion, it is this methodology that can be effective in a modern eclectic and «entangled» social and humanitarian education. At the same time, the term «entanglement» becomes the key to defining the current situation (Goerner 1999).

Tourism, obviously, is a social and political structure that requires its own «post-formal» research. Tourism education should be regarded as a social practice that not only characterizes modern education, but also transforms it. In this context, tourism research is becoming transdisciplinary, that is, beyond traditional scientific and academic disciplines.

The post-formal dialogue, originally emanating from their existence of a complex «composite» society, living by the circulation of memes, allows us to consider tourism research as a critical and eclectic response to the paradoxical post-modern world.

3 Results

Defining the specifics of using forms of promotion in education, one should refer to the regulatory acts regulating educational activities in terms of the development of the digital educational environment (DEE). According to the passport of the priority project «Modern Digital Educational Environment in the Russian Federation» (Protocol from October 25, 2016 No. 9), which is being formed since 2016, DEE is aimed at expanding the possibilities of continuous education of citizens through the development of educational technologies—online courses and introducing online resources into the education system necessary to ensure successful development of secondary and higher education programs, as well as for the development of general subjects. According to the projected value, by creating a mechanism for implementing the concept of an individual learning trajectory, the main achievement of the program should be an increase in the number of people who have mastered online courses to eleven million by the end of 2025.

The objectives and forms of development of the digital educational environment specified in the draft DEE fully conform with the norms of the Federal Law dated December 29, 2012 No. 273-ФЗ (Federal law N 273-FL 2012). The Federal Law «On Education» establishes the concepts of e-learning and distance learning technologies, as well as the basic principles and requirements for their use in the framework of educational programs. Due to the fact the use in educational activities of electronic education and distance learning technologies are disclosed in a special Order of the Ministry of Education and Science of Russia dated 08.23.2017 N 816 (Order of the Ministry of Education and Science of Russia N 816 2017). In connection with the possibilities of developing e-education in relation to the industry, the topics of our research should be noted that according to Art. 2 of the Federal Law «On Education» , the use in educational activities of electronic (as well as similar printed) forms of promoting a regional tourist product as information resources, allows them to be classified as a means of training and education. In this case, they can enter the funds of libraries formed by educational organizations in order to ensure the implementation of educational programs (Article 18 of the Federal Law «On Education»).

Obviously, the created DEE is a priority project related to the complex transformation of the economy. First of all, it should be noted that the educational goals in digitalization are defined by the state program of the Russian Federation «Information Society (2011–2020)» (Enactment of the Government of the Russian Federation No. 313 2011). Thus, subprogram 2 «Information Environment» defines as target indicators an increase in the «share of supported socially significant projects in the field of electronic media, including sites of social or educational importance, in the total number of socially significant projects in the field of electronic media information, including sites of social or educational value» . More ambitious indicators are defined by subprogram 4 «Information state» - «ensuring the use of electronic services based on information technologies, including in the areas of health, culture, education and science» . The special importance of the education digitalization is underlined by the key provisions of the «Digital Economy of the Russian Federation» program (Government order of the Russian Federation No. 1632-p 2017). In particular, in the formation of a hierarchy of basic directions for the development of the digital economy of the Program after the normative regulation, personnel and education are defined. At the same time, by 2024, the key indicators of the objectives of the program include: 800 thousand people per year of graduates of higher and secondary vocational education who have competencies in the field of information technology at the global average; 40 percent of the population has digital skills.

So, the digital economy, the knowledge economy requires a modern, digital model of university education, which will shape the actual knowledge and skills of students (Tapscott 2009).

Indeed, some professional competences (PC) and general professional competencies (GPC) set requirements according to which students of certain areas of training must be able to work in the electronic educational environment, and for this they need, first of all, to learn in this environment. For example, according to the new Federal State Educational Standard of Higher Education (3++) areas of training 43.03.02 Tourism» the student must be able to «apply technical innovations and modern software in the tourism sector» (General Professional Competenct-1) (FSES HE in the field of education 43.03.02 Tourism 2017). Also, the general cultural competence - GPC-1 of the FSES of HE (3+) areas of training 43.03.02 «Tourism», in which some higher educational institutions still work, states that the student must have the ability to solve standard professional tasks on the basis of information and bibliographic culture using information and communication technologies and taking into account the basic requirements of information security, use various sources of information on the tourist product object (GPC-1) (FSES HE in the field of education 43.03.02 Tourism 2016). That is, this competence not only requires certain knowledge and skills of the bachelor, but also creates certain conditions for the organization of the educational process, that is, it is a question of the need to use a electronic educational environment (for example, photo and video materials).

Moreover, there are some general cultural competencies that may imply the need to use DSP in their implementation n many federal state educational standards (for example, in the federal state educational standard of higher education in the direction of training 46.03.01 «History»). For example, the competence of OK-7 - «the ability to self-organize and self-educate» assumes that the student, in the process of learning, will himself seek the information necessary for further education (FSES HE in the field of education 46.03.01 History 2014). In today's reality, this requires educational institutions to provide various information educational resources within the framework of the electronic educational environment (such resources include: electronic scientific libraries, access to various Internet portals, the Moodle system, and much more).

Online portals that implement their activities within the digital educational environment have a very different focus, we will consider in this regard, forms of promoting tourism that can be used in the educational field for various specialties, while also being a source of educating people in the field of historical and cultural heritage of the Russian Federation and each individual region.

When it comes to forms of tourism promotion, many Russian researchers consider methods of promoting specific tourist products (advertising and non-advertising methods, distribution channels, etc.), while we mean by this a broader concept. To the forms of tourism promotion we include such tools that popularize tourism as an industry as a whole. This could include positioning of territories, historical and cultural heritage of the regions and, in general, tourism as a special culture, which helps not only to sell the results of the activities of the tourist industry, but also acts as a kind of catalyst that strengthens interest in knowing the world (including when receiving education), by integrating these forms in the educational process.

The main forms of promotion of the tourism industry, which we consider the most suitable for integration into the electronic educational environment, are: tourist guides to countries describing attractions in foreign languages; registers and inventories of certain types of tourist sites (museums, monuments, etc.); tourist passports of the regions (project of the Ministry of Culture of the Russian Federation); information portals of the regions (electronic resources of the regions of the country that provide information on the main natural, cultural, educational, religious and other resources of the region).

Similar forms can be applied at all levels of education. For example, in a school educational environment, the introduction of such digital educational resources can be used as follows: tourist country guides can be useful for learning foreign languages; Registers of historical and cultural objects - for the discipline of «history» ; tourist passports of the regions and information portals of the regions can help in studying the history and local history, as well as geography.

However, the use of these forms of tourism promotion may go beyond the school. Consider the examples of the use of the above information educational resources (IER) in higher education institutions:

1. disciplines of principal subjects. First of all, the use of such IOR is characteristic of the core disciplines of certain areas, such as «tourism» and «hotel business» , «restaurant business» (if we are talking about different registries in these areas);

2. disciplines of subsidiary subjects. Of course, within the framework of the digital educational environment, forms of promoting tourism can be useful in various areas, not only within the educational process, but also within the framework of education. Further, possible examples of such integrations will be considered:
   2.1. foreign language guidebooks can be introduced in philological areas, as well as for other areas in the framework of professional foreign language communication;
   2.2. registers of historical and cultural objects can be used in the study of historical trends, and the discipline «History», which is mandatory for all areas;
   2.3. tourist passports of the regions and information portals of the regions can be used in the study of geographic disciplines.

4 Conclusion

When used in educational institutions of the digital educational environment, the training receives not only educational, but also social orientation, and the educational environment of the university should become a place for the integration of all stakeholders of the global electronic educational environment represented by researchers, professional consultants, business, etc. (Afanaseve and Zyablov 2018).

Thus, the forms of promotion in tourism are not only educational, but also educational in nature. So, any person, even having no special education, using tourist resources, can learn something new about any region of the country. The peculiarity of information in tourism is that it is changing quite quickly, for this it is necessary to maintain its relevance and accuracy, which is difficult to do, but possible within the digital educational environment. Such innovations have a positive effect not only on the field of education, in which each school or university discipline will be supported by professionally prepared, interesting and brightly decorated, high-quality, reliable and relevant information. For the tourism industry, this will also play a positive role, since from the school itself, with examples from life, children will be brought up within a certain culture that promotes respect and tolerance for the customs and traditions of different countries, and perhaps one day, this will lead to the eradication of such negative phenomena like xenophobia. Obviously, people raised in such a culture will understand what the phenomenon of tourism is, which can be a catalyst for the development of tourism at a completely different, higher level.

The popularity of using the above forms of tourism promotion in educational practice has the opposite effect - by stimulating the interest of the actor (its creator) to continue marketing activities in this direction. The acquaintance of the youth audience with the historical and cultural heritage of their own homeland (large and small) is, on the one hand, a direct factor in shaping the present consumer demand of parents and the consumer demand of this audience in the future. In addition, public participants in the marketing of the tourist territory through digital forms of promotion in the usual and interactive forms can optimize work and enhance the effect of activities in the educational, educational and patriotic sphere.

Acknowledgements. The reported study was funded by RFBR and the government of Volgograd region according to the research project No. 18-413-342003.

References

Afanasev, G.A., Zyablov, A.A.: The development of the educational process in the new digital environment. Ecol. Urban Areas **2**, 105–106 (2018)

Alananzeh, O.A., Amyan, M.M., Alghaswyneh, O.F.M.: Managing promoting tourism product of the golden triangle in Jordan. Int. J. Human. Soc. Sci. **5**(9), 197–207 (2015)

Automated system AIS tourism. http://maps.russia.travel. Accessed 22 Feb 2019

Baranova, N.A.: Features of advertising activities of tourist enterprises of the Nizhny Novgorod region. Serv. Plus **9**(3), 21–26 (2015)

Brel, O.A.: The use of modern educational technologies in the training of personnel for the tourism industry (for example, the case method and technology portfolio). Kazan Pedagogical J. **2**, 57–61 (2016)

Cooper, C., Fletcher, D., Gilbert, R., Shepherd, R., Wanhill, S.: Tourism: Principles and Practices, 674 p. London (1998)

Dunets, A.N.: Design and Promotion of a Regional Tourist Product, 163 p. Barnaul (2014)

Enactment of the Government of the Russian Federation No. 313. Enactment of the Government of the Russian Federation No. 313 «On Approval of the State Program of the Russian Federation «Information Society (2011–2020)»» (last updated 02.02.2019), Federal law of the Russian Federation on meetings, vol. 18 (ch. II), art. 2159, 05 May (2014)

Federal law N 273-FL. Federal law of 29.12.2012 N 273-FL «On education in the Russian Federation » (last updated 25.12.2018), Federal law of the Russian Federation on meetings, vol. 53 (ch. 1), art. 7598, 31 December (2012)

Federal register of tour operators of outbound tourism - Tourist assistance. https://www.tourpom.ru/touroperators. Accessed 22 Feb 2019

Freeman, R., Glazer, K.: Services marketing, Introduction to tourism and hospitality in BC (2019). https://opentextbc.ca/introtourism/chapter/chapter-8-services-marketing/. Accessed 20 Feb 2019

FSES HE in the field of education 43.03.02 Tourism. (2016). http://fgosvo.ru/news/3/1648. Accessed 19 Jan 2016

FSES HE in the field of education 43.03.02 Tourism. (2017). http://fgosvo.ru/fgosvo/151/150/24/93. Accessed 29 June 2017

FSES HE in the field of education 46.03.01 History (2014). http://fgosvo.ru/news/4/396. Accessed 25 Aug 2014

Goerner, S.J.: After the Clockwork Universe: The Emerging Science and Culture of Integral Society, Edinburgh, 476 p. (1999)

Government order of the Russian Federation No. 1632-p. Government order of the Russian Federation of 28.07.2017 No. 1632-p. On approval of the program, Digital Economy of the Russian Federation. Federal law of the Russian Federation on meetings, vol. 32, art. 5138, 28 July (2017)

Gurkina, E.N., Zhulina, M.A.: Tourist information portal as a tool to promote tourism in the region. Ogarev-Online **1**(66), 10 p. (2016)

Hoerner, J.M.: The recognition of tourist science. Espaces **173**, 18–20 (2000)

Horn Jr., R.A.: Post-formal conversation. In: Banathy, B., Jenlink, P.M. (ed.) Dialogue as a Means of Collective Communication, pp. 291–322. New York (2005)

McCabe, S.: Marketing Communications in Tourism and Hospitality, 320 p. Butterworth-Heinemann, Oxford (2008)

Middleton, V.T.C., Clarke, J.R.: Marketing in Travel and Tourism, 512 p. Butterworth-Heinemann, Oxford (2001)

Morgan, M., Ranchhod, A.: Marketing in Travel and Tourism, 4th edn. 528 p. Butterworth-Heinemann, Oxford (2009)

National tourism portal Russia.travel. https://russia.travel. Accessed 22 Feb 2019

Order of the Ministry of Education and Science of Russia N 816. Order of the Ministry of Education and Science of Russia dated August 23, 2017 N 816. On approval of the procedure for the application by organizations engaged in educational activities, e-learning, distance learning technologies in the implementation of educational programs (2017). https://minjust.consultant.ru/documents/36757. Accessed 24 Feb 2019

Ort, V.: Our Digital Future. Informational support of Open Education Institute, Moscow, 78 p. (2005)

Perova, T.V., Zakirova, O.V.: Cross-marketing in the promotion of tourist services. Sci. Bull. Nat. Min. **3** (2015). http://vestnik.mininuniver.ru/upload/iblock/e00/t.v.-perova1_-o.v.-zakirova2.pdf. Accessed 20 Feb 2019

Promote tourism products and services, Trainee Manua. http://waseantourism.com/ft/Toolbox%20Development%20III:%2098%20toolboxes%20for%20Travel%20Agencies%20and%20Tour%20Operations/Submission%20to%20ASEC/(Draft)%202nd%20submission_290415/TO%20&%20TA/Promote%20tourism%20products%20&%20services/TM_Promote_tourism_products_services_290415.pdf. Accessed 20 Feb 2019

Protocol from October 25, 2016 No. 9 on the approval of the «Passport of the priority project «Modern Digital Educational Environment in the Russian Federation»». http://government.ru/news/25682/.2019/02/24. Accessed 24 Feb 2019

Register of tourist information centers of the Russian Federation. http://www.nbcrs.org/tic/list.cshtml. Accessed 22 Feb 2019

Simavonyan, A.A.: Promotion and marketing of tourism. Manage. Econ. XXI Century **2**, 63–65 (2015)

Sizenev, L.A.: Methods for assessing the consumer appeal of regional tourist products. Serv. Russian Fed. Abroad **12**(3), 90–111 (2018)

Stepanov, V.K.: The age of network intelligence: about the Don Tapscott book, the electronic digital society. Inf. Soc. **2**, 67–70 (2001)

Tapscott, D.: Digital Economy. Rethinking Promise and Peril in the Age of Networked Intelligence, 414 p. New York (2015)

Tapscott, D.: Grown up digital. How the Net Generation is Changing your World, 368 p. New York (2009)

Tourism development agency of the Volgograd Region. http://www.turizm-volgograd.ru/. Accessed 22 Feb 2019

Tourist information exchange system. http://nbcrs.org/. Accessed 14 Jan 2019

Tourist portal of the city of Volgograd and the Volgograd region "Volgograd region - the territory of travel". https://volgaland.volsu.ru/ru. Accessed 22 Feb 2019

Tourist portal of the Volgograd region. http://www.welcomevolgograd.com/. Accessed 22 Feb 2019

Travel agency registry - travel assistance. https://reestr.tourpom.ru/search.php. Accessed 22 Feb 2019

Tribe, J.: Knowing about tourism: epistemological issues. In: Phillimore, J., Goodson, L. (ed.) Qualitative Research in Tourism: Ontologies, Epistemologies and Methodologies, London, pp. 46–62 (2004)

Unified federal register of tour operators. https://www.russiatourism.ru/operators/. Accessed 22 Feb 2019

Unified state register of legal entities, unified state register of individual entrepreneurs. https://egrul.nalog.ru/index.html. Accessed 25 Feb 2019

Unified tourist passport. http://utp.nbcrs.org/. Accessed 22 Feb 2019

Wu, G.: Official websites as a tourism marketing medium: a contrastive analysis from the perspective of appraisal theory. J. Destination Mark. Manage. **10**, 164–171 (2018)

Managing the Development of Infrastructural Provision of AIC 4.0 on the Basis of Artificial Intelligence: Case Study in the Agricultural Machinery Market

Tatiana N. Litvinova(✉)

Volgograd State Agrarian University, Volgograd, Russia
litvinoval358@yandex.ru

Abstract. Purpose: The purpose of the article is to substantiate the necessity and to develop a conceptual model of managing the agricultural machinery market on the basis of AI for infrastructural provision of AIC 4.0 in modern Russia.

Methodology: The authors use the method of plan-fact analysis for comparing the target values of the indicators of agricultural machinery market, announced in the Strategy of development of agricultural machine building in the Russian Federation until 2020, and their factual values. The research is performed based on the 2016 data, as the data for later periods are not yet available in the official statistics. The difference (plan-fact) and the share of deviation of fact from plan (%) are determined.

Results: It is determined that non-optimality of managing the agricultural machinery market is a serious problem in modern Russia, as it does not allow forming the necessary infrastructural provision for transition to AIC 4.0. The reason of non-execution of the strategy of development of the Russian agricultural machinery market consists in high complexity of this management, which cannot reach high effectiveness with the current bureaucratic organization due to slow collection of statistical information, duration of its processing and analysis, and the long process of managerial decision making. This problem could be solved by organization of managing the agricultural machinery market on the basis of AI.

Recommendations: The developed conceptual model of managing the agricultural machinery market on the basis of AI is recommended for practical application in modern Russia, as it has the following advantages: high flexibility of the strategy of development of agricultural machinery market and the plan of its implementation, connection of the strategy of development of agricultural machinery market and the plan of its implementation to the current economic practice, and completeness and high detalization of the strategy of development of agricultural machinery market and the plan of its implementation.

Keywords: Managing the development of infrastructural provision · AIC 4.0 · Artificial intelligence · Agricultural machinery market · Modern Russia

JEL Code: Q13 · L16 · O32 · O33

E. G. Popkova and B. S. Sergi (Eds.): ISC 2019, LNNS 87, pp. 317–323, 2020.
https://doi.org/10.1007/978-3-030-29586-8_37

1 Introduction

The perspectives of development of the agro-industrial complex, which is one of the most important production and distribution complexes in economy, are connected to transition to AIC 4.0. This envisages digital modernization and full-scale automatization of the agro-industrial complex on the basis of breakthrough technologies, which conform to the new technological mode – Industry 4.0. This will allow for successful execution of functions that are set on the agro-industrial complex due to adoption of the global goals in the sphere of sustainable development: increase of accessibility of food products (by reducing the cost through increase of the efficiency) and guaranteeing the quality of the products (by provision of transparency of production and distribution).

In modern Russia, transition to AIC 4.0 (digital agriculture) is to become one of the directions of the program "Digital economy of the Russian Federation", adopted by the Decree of the Government of the RF dated July 28, 2017, No. 1632-r. According to the Association of participants of the market of the Internet of Things (2019), the offer on implementation of this direction is being considered by the Government of the RF. Despite the high importance of implementation of this direction in Russia, a serious barrier on this path could be the deficit of infrastructural provision. Apart from the national (universal) institutional provision, which includes the normative and legal establishment of digital modernization of economy on the basis of the technologies of Industry 4.0 and creation of specialized regulators, transition to AIC 4.0 requires specific infrastructural provision – import substitution and development of highly-effective and competitive domestic production of agricultural machinery.

The working hypothesis of the research is that transition to AIC 4.0 should be performed with a certain logical sequence: digital modernization of infrastructural markets (primarily, the agricultural machinery market) should be followed by modernization of agriculture and food industry, as foundation on traditional technologies does not allow forming the necessary special infrastructural provision of AIC 4.0. The purpose of the work is to substantiate the necessity and to develop a conceptual model of managing the agricultural machinery market on the basis of AI for infrastructural provision of AIC 4.0 in modern Russia.

2 Materials and Method

The authors use the method of plan-fact analysis for comparing the target values of the indicators of agricultural machinery market, announced in the Strategy of development of agricultural machine building in Russian Federation until 2020, and their factual values. The research is conducted based on the 2016 data, as the data for later periods are not available in the official statistics and analysis. Difference (plan-fact) and the share of deviation of fact from plan (%) are determined. The results of the performed analysis are shown in Table 1.

Table 1. The results of plan-fact analysis of implementing the Strategy of development of agricultural machine building in the RF until 2020.

Indicator	Target value of indicator, announced in the strategy (plan)	Statistical value of indicator (fact)	Deviation of fact from plan	
			plan-fact	%
Share of domestic machinery in the total volume of sales of agricultural tractors (including the products of assembly productions), %	37.2	34.0	3.2	8.6
Ratio of domestic Russian market of agricultural machinery and the volume of export supplies (export/internal market), %	39.5	35.3	4.2	10.6
Provision of the agrarian complex with tractors (number of tractors per 1,000 ha of land), pcs per 1,000 ha	4.51	3.3	1.2	26.8
Provision of the agrarian complex with combine harvesters (number of combine harvesters per 1,000 ha of crop land), pcs per 1,000 ha	2.3	2.0	0.3	12.3
Coefficient of maintenance of the park of agricultural tractors, %	6.7	3.3	3.4	50.6
Coefficient of maintenance of the part of combine harvesters, %	9.2	6.6	2.6	28.3

Source: compiled by the authors based on National Research University "Higher School of Economics" (2019), Ministry of Industry and Trade of Russia (2019).

The data from Table 1 show that the share of domestic machinery in the total volume of sales of agricultural tractors (including the products of assembly productions) in 2016 was lower (34%) than the planned (37.2%) by 8.6%. Ratio of the Russian domestic market of agricultural machinery and the volume of export supplies (export/domestic market) in 2016 was lower (35.3%) than the planned (39.5%) by 10.6%. Factual provision of the agrarian complex with tractors in 2016 (3.3 pcs per 1,000 ha) was below planned provision (4.51 pcs per 1,000 ha) by 26.8%.

Factual provision of the agrarian complex with combine harvesters (number of combine harvesters per 1,000 ha of crop land), pcs per 1,000 ha in 2016 (2 pcs per 1,000 ha) was below the planned provision (2.28 pcs per 1,000 ha) by 12.3%. Factual coefficient of maintenance of the park of agricultural tractors in 2016 (3.3%) was below the planned coefficient (6.7%) by 50.6%. Factual coefficient of maintenance of the park of combine harvesters in 2016 (6.6%) was below the planned coefficient (9.2%) by 28.3%.

Therefore, in 2016 there already was a large underrun from the plan of implementation of the Strategy of development of agricultural machine building in the Russian Federation until 2020, which might have increased by now (early 2019). A more detailed content analysis of this strategy allowed determining the reasons of the deficit of infrastructural provision of AIC 4.0 in Russia due to insufficient development of the agricultural machinery market:

- insufficient flexibility of the strategy of development of agricultural machinery market and the plan of its implementation: being adopted in 2011, this strategy has not been corrected, despite the changes of the market situation;
- separation between the strategy of development of the agricultural machinery market and the plan of its implementation and the actual economic practice: the planned values of the indicators of development of the studied market were too high from the very beginning;
- incompleteness and low detalization of the strategy of development of the agricultural machinery market and the plan of its implementation: they contain only the target values of the indicators but no recommendations for achieving them and no plans for each separate company, which causes lack of clarity of the strategy implementation.

These reasons show imperfection of the modern Russian practice of managing the development of the agricultural machinery market. As their overcoming envisages systemic integration, increase of controllability of the agricultural machinery market, and processing of Bog Data, a perspective method of complex elimination of these reasons is organization of management of the studied market on the basis of AI.

For developing a conceptual model of this management, we use the existing publications and studies in the sphere of AIC 4.0: Altukhov et al. (2019), Butorin and Bogoviz (2019), Huh and Kim (2018), Litvinova et al. (2016), Litvinova et al. (2017), Litvinova et al. (2019), Litvinova et al. (2019). Matei et al. (2017), Troyanskaya et al. (2017), and Weltzien (2016); and the works in the sphere of economic systems management on the basis of AI: Burggräf et al. (2018), Kumar Deb et al. (2018), and Partel et al. (2019).

3 Results

The following conceptual model of managing the agricultural machinery market on the basis of AI is offered (Fig. 1).

As is seen from Fig. 1, the offered model includes the linear hierarchy of artificial intelligences that are involved in management of the agricultural machinery market. The key AI is the one conducting sectorial management – i.e., managing the agricultural machinery market directly. It is subject to AI that conducts the management of the AIC on the whole and that issues commands to artificial intelligences that conduct corporate management at each company of the agricultural machinery market. The managerial process is a cyclic (recurrent) process, with six consecutive stages.

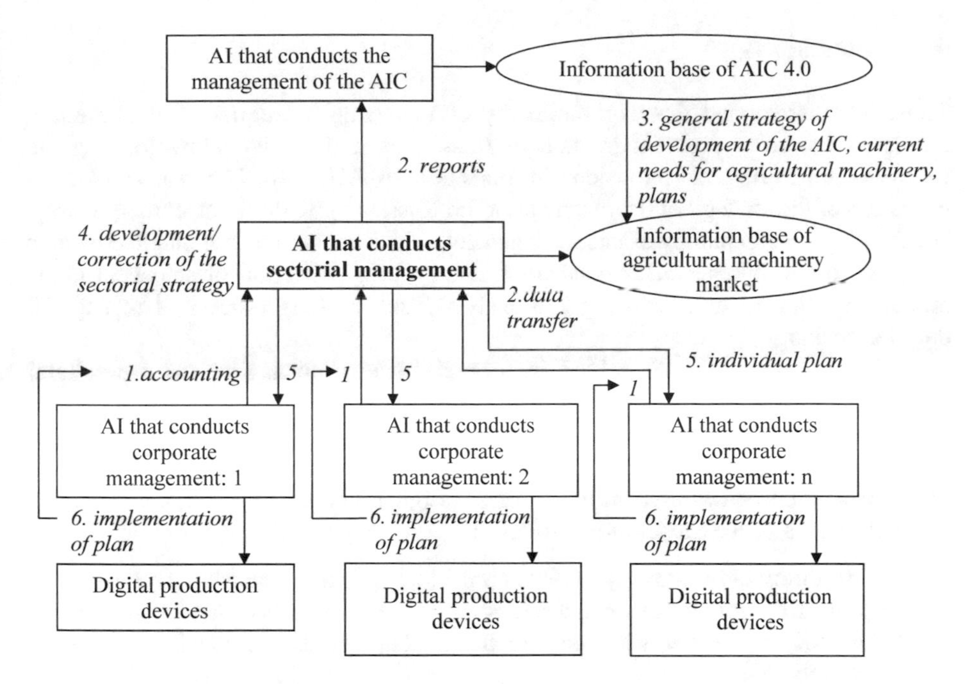

Fig. 1. The conceptual model of managing the agricultural machinery market on the basis of AI. Source: compiled by the authors.

At the first stage, artificial intelligence that conducts the management of the agricultural machinery market collects reports from artificial intelligences that conduct corporate management at the companies of the market. These reports include the information on production capacities and the level of their load, results of corporate marketing research, plans of production and distribution, the technical state of production equipment (digital production devices), and the financial state of the companies of the agricultural machinery market.

At the second stage, artificial intelligence that conducts sectorial management of the agricultural machinery market systematizes the received information and transfers the data into the information data base of the agricultural machinery market (Big Data) and, at the same time, reports to the artificial intelligence that conducts management of the AIC. At the third stage, artificial intelligence that conducts management of the AIC analyzes the information received from all sectorial artificial intelligences and develops a general strategy of development of AIC, determines the current needs for agricultural machinery market, and compiles plans of development of the agricultural machinery market, transferring information to inferior computers.

At the fourth stage, artificial intelligence that conducts sectorial management of the agricultural machinery market develops (or corrects the already compiled) sectorial strategy. At the fifth stage it passes individual plans to corporate artificial intelligences, which, at the six stage, compile corporate strategies of development and implement the plan, issuing commands to digital production devices. Then the cycle is repeated anew – they send reports to the sectorial AI, etc.

4 Conclusions

Thus, it is determined that non-optimality of managing the agricultural machinery market is a serious problem in modern Russia, as it does not allow forming the necessary infrastructural provision for transition to AIC 4.0. The reason of non-execution of the strategy of development of the Russian agricultural machinery market consists in high complexity of this management, which cannot reach high effectiveness with the current bureaucratic organization due to slow collection of statistical information, duration of its processing and analysis, and the long process of managerial decision making.

This problem could be solved by organization of managing the agricultural machinery market on the basis of AI. Though the offered conceptual model of managing the agricultural machinery market on the basis of AI envisages management of companies of this market and the agro-industrial complex on the whole on the basis of AI, this could be achieved in the course of digital modernization of this complex. The advantages of the model are as follows:

- high flexibility of the strategy of development of agricultural machinery market and the plan of its implementation: the strategy is constantly corrected in view of the changes of the market situation and the strategy of development of the agro-industrial complex;
- connection of the strategy of development of agricultural machinery market and the plan of its implementation to the current economic practice: planned values of the indicators of development of the studied market are determined on the basis of aggregation of corporate plans;
- completeness and high detalization of the strategy of development of the agricultural machinery market and the plan of its implementation: they contain target values of the indicators and provide recommendations for their achievement and plans for each separate company, which guarantees implementation of the strategy.

References

Altukhov, A.I., Bogoviz, A.V., Kuznetsov, I.M.: Creation of an information system – a necessary condition of rational organization of agricultural production. In: Advances in Intelligent Systems and Computing, vol. 726, pp. 800–809 (2019)

Burggräf, P., Wagner, J., Koke, B.: Artificial intelligence in production management: a review of the current state of affairs and research trends in academia. In: International Conference on Information Management and Processing, ICIMP 2018, January 2018 , pp. 82–88. 8325846 (2018)

Butorin, S.N., Bogoviz, A.V.: The innovational and production approach to management of economic subjects of the agrarian sector. In: Advances in Intelligent Systems and Computing, vol. 726, pp. 758–773 (2019)

Huh, J.-H., Kim, K.-Y.: Time-based trend of carbon emissions in the composting process of swine manure in the context of agriculture 4.0. Processes **6**(9), 168 (2018)

Kumar Deb, S., Jain, R., Deb, V.: Artificial intelligence creating automated insights for customer relationship management. In: Proceedings of the 8th International Conference Confluence 2018 on Cloud Computing, Data Science and Engineering, Confluence 2018, pp. 758–764. 8442900 (2018)

Litvinova, T.N., Khmeleva, G.A., Ermolina, L.V., Alferova, T.V., Cheryomushkina, I.V.: Scenarios of business development in the agricultural machinery market under conditions of international trade integration. Contemp. Econ. **10**(4), 323–332 (2016)

Litvinova, T.N., Kulikova, E.S., Kuznetsov, V.P., Taranov, P.M.: Marketing as a determinant of the agricultural machinery market development. In: Contributions to Economics, pp. 465–471 (2017). ISBN 9783319606958

Litvinova, T.N., Morozova, I.A., Pozdnyakova, U.A.: Criteria of evaluation of effectiveness of Industry 4.0 from the position of stimulating the development of knowledge economy. In: Studies in Systems, Decision and Control, vol. 169, pp. 101–107 (2019)

Litvinova, T.N., Tolmachev, A.V., Saenko, I.I., Iskandaryan, G.O.: Role and Meaning of the ICT Infrastructure for Development of Entrepreneurial Activities in the Russian Agricultural Machinery Market. Advances in Intelligent Systems and Computing, vol. 726, pp. 793–799 (2019)

Matei, O., Anton, C., Bozga, A., Pop, P.: Multi-layered architecture for soil moisture prediction in agriculture 4.0. In: Proceedings of International Conference on Computers and Industrial Engineering, CIE, vol. 2, no. 1, pp. 39–48 (2017)

Partel, V., Charan Kakarla, S., Ampatzidis, Y.: Development and evaluation of a low-cost and smart technology for precision weed management utilizing artificial intelligence. Comput. Electron. Agric. **157**, 339–350 (2019)

Troyanskaya, M.A., Ostrovskiy, V.I., Litvinova, T.N., Matkovskaya, Y.S., Bogoviz, A.V.: Possibilities and perspectives for activation of sales in the agricultural machinery market within sectorial development of Russian and European economies. In: Contributions to Economics, pp. 473–480 (2017). ISBN 9783319606958

Weltzien, C.: Digital agriculture - or why agriculture 4.0 still offers only modest returns. Landtechnik **71**(2), 66–68 (2016)

Association of participants of the market of the Internet of Things: The explanatory note to the offer on implementation of the new direction of the program “Digital economy of the Russian Federation”: Digital agriculture (2019). https://iotas.ru/files/documents/Пояснит.записка%20eAGRO%20fin%20000.pdf. Accessed 27 Jan 2019

The Ministry of Industry and Trade of the Russian Federation: Strategies of development of agricultural machine building until 2020, adopted by the Decree dated 22 December, no. 1810, 2011 (2019). http://www.consultant.ru/document/cons_doc_LAW_145647/. Accessed 25 Jan 2019

National Research University “Higher School of Economics”: The market of agricultural machinery – 2017 (2019). https://dcenter.hse.ru/data/2018/02/03/1163430452/Рынок%20сельскохозяйственных%20машин%202017.pdf Accessed 25 Jan 2019

Overview of the Educational Motivation Theory: A Historical Perspective

Maria V. Arkhipova[1(✉)], Ekaterina E. Belova[1], Yulia A. Gavrikova[1], Tatiana E. Nikolskaya[2], and Tatiana N. Pleskanyuk[1]

[1] Minin Nizhny Novgorod State Pedagogical University, Nizhny Novgorod, Russia
arhipovnn@yandex.ru, belova_katerina@inbox.ru, y.a_gavrikova@mail.ru, pleskanyuk@mail.ru
[2] Maxim Gorky Institute of Literature and Creative Writing, Moscow, Russia
t.e.nikolskaya@gmail.com

Abstract. Psychological and pedagogical science has a number of approaches devoted to the problem of personal motivation, including motivation of learning activities. Motivation determines human behaviour and activity, and therefore is one of the basic concepts in psychology. Lack of a single point of view on the substance and sources of motivation, absence of a unified theory prove the complexity of the problem. The aim of the study is to provide a systematic and consistent vision of motivation theories in a historical perspective with the focus on critical evaluation of differences and commonalities among the most well-known of them, the explanation ensuring their proper application in the educational environment. The study of educational motivation reveals the factors of students' success. Therefore, the relevance of research in this field remains to date. The explanation and interpretation of existing theories can help in understanding students' motives and finding ways to foster students' enthusiasm in learning. If the educational interests of the modern students are met, the students are sure to show more readiness and invest more effort and energy to set higher goals and strive to achieve them, which proves that it is motivation that paves the path towards academic success.

Keywords: Motivation theories · Education · Leading motives · Academic success

JEL Code: I25

1 Introduction

Personal motivation emerged as a leading subject of scientific enquiry in the first quarter of the 20th century. By the end of the century the problem of motivation was firmly established as a separate field in psychology.

Today, there are over 100 different definitions of motivation, reflecting the complexity of the subject. Lack of a single point of view on the substance and sources of motivation leads to a large fragmentation of interpretations and schools of thought on

E. G. Popkova and B. S. Sergi (Eds.): ISC 2019, LNNS 87, pp. 324–331, 2020.
https://doi.org/10.1007/978-3-030-29586-8_38

the subject. Since a unified theory of motivation is yet to emerge, scholars have had to study its various types separately. The phenomenon of motivation has always been paid a lot of attention to, many theories explaining the origin of motives and human motivation.

2 Background and Methodology

A high level of motivation helps a person to achieve better results in any sphere, learning being no exception (Bystrova et al. 2018; Volkova et al. 2018). Interests, emotions and motives, being the main components of motivation, are of paramount importance, because they provide energy that makes an action possible, give drive and ambition to strive for success and a feeling of satisfaction with the results (Panachugin and Lapygin 2018; Semarkhanova et al. 2018). To explore the topic of educational motivation, in the present study we focus on the theoretical aspects of motivation theories, aiming to critically evaluate and find out differences and commonalities among the most well-known of them. The article is based on secondary information, being descriptive and qualitative in nature. The explanation of early and modern motivation theories ensures their proper application in the educational environment.

To reach the aim we used the methods of theoretical literature analysis, studying and comparing the existing theories, therefore leading to the subsequent conclusion of the potential research.

3 Discussion and Results

The well-known division of motivation theories is considered to be into two groups: theories of content and theories of processes.

Content theories focus on the needs that define human behavior, Maslow, McClelland and Herzberg being most prominent in this school.

The Maslow's hierarchy of needs was formulated in his 1943 book "Theory of Human Motivation." According to Maslow, human behavior is determined by the needs that can be grouped into five levels – from the lowest to the highest:

- physiological needs (hunger, need to rest)
- safety needs (need to be secure against surrounding dangers)
- social needs (interaction with other people, social roles)
- self-respect needs (competence, recognition by other people)
- self-expression needs (realization of one's potential, need for personal growth).

Maslow maintained that lower level needs were the principal determinant of human behavior. After satisfying lower level needs, a person would seek to satisfy higher levels. Therefore, motivation process will be more intense once lower level needs are satisfied, since they no longer provide distraction (Maslow 1943).

Another theory of motivation also emerged in the 1940s. McClelland looked at three groups of needs: success, power and recognition. McClelland's theory of motivation explains how individual's basic needs and environmental factors combine with these three higher level motives. According to McClelland, the need to succeed is what forces an individual to assume greater responsibilities, seek more complicated tasks, work more and be a better problem solver. Need for power leads to a desire to control and influence other people. Individuals with a dominant need for power seek to acquire, realize and maintain influence over others. They prefer to compete with others and gain positions of authority. In this case, seeking power must reveal not only ambition, but also the subject's ability to succeed at different levels of organizational hierarchy. As to the need for recognition, it obliges a person to develop informal leadership skills, maintain independent opinions and ability to convince others. Each of McClelland's three motives results in a different type of satisfaction. Success motive, as a rule, results in a feeling of fulfillment, power motive results in a feeling of authority, and recognition motive leads to love and attachment (McClelland 1988).

Herzberg's theory is closely related to the Maslow's hierarchy of needs. His two-factor theory of motivation was created at the end of the 1950s. Similar to Maslow, Herzberg based his research on individual needs. As part of his research, he interviewed 200 accountants and engineers on the subject of their work satisfaction. Respondents had to describe episodes of particularly high as well as low satisfaction with work. Based on this experiment, Herzberg concluded that there are two groups of work-related factors: factors that result in satisfaction from work (motivators) and factors that decrease satisfaction (hygienic factors). Motivators are related to the substance of work, result in satisfaction and motivate workers to increase productivity and efficiency of operations. Hygienic factors are related to the conditions of work, remuneration, relationships with co-workers and superiors. If hygienic factors are insufficient, an individual feels dissatisfaction from their work, however, on their own, these factors cannot provide stimulus for any action. Herzberg's hygienic factors correspond to physiological needs, need for safety and certainty of the future (Herzberg et al. 2010).

In the framework of process theories of motivation, scholars attempt to establish relationships between dynamic variables that define motivation. These theories focus more on how behavior is initiated, targeted and supported. Process theories emphasize actual process of motivation. Examples of such theories are the work of Vroom, McGregor and Porter and Lawler's model.

In his 1964 book "Work and Motivation" the US researcher Vroom formulated theory of expectations as alternative to content models of theory of motivation. According to Vroom, people's motivation is equivalent to the expectation that their chosen behavior must necessarily lead to the desired outcome. His book describes the process of decision-making that defines subsequent behavior. He identified three principal components of motivation: expectation, instrumentality, valence. Expectation is perceived as a probability of reward for an effort. Instrumentality is a person's expectation that his/her effort will be rewarded. Valence refers to the assumed level of satisfaction or dissatisfaction aroused by the reward. Therefore, positive motivation stems from the expectation that a certain effort will lead to the desired outcome, while negative motivation arises when there is expectation that the effort will result in an undesirable outcome (Vroom 1995).

In the 1960s, another US researcher of motivation, McGregor, developed Theory X and Theory Y. He suggested two different representations of man: mostly negative

(Theory X) and mostly positive (Theory Y). These theories describe contrasting sets of statements on human nature. Theory X presumes that people are lazy by their nature and tend to avoid any work if it is at all possible. This necessitates that their work process must be controlled. Theory Y, on the contrary, presumes that people willingly accept responsibility, want to achieve success both at the individual and at the company level, seek development and prosperity. This theory suggests that people can be largely independent and creative at work as long as they are properly motivated (McGregor 2006).

It was Porter-Lawler's theory of motivation introduced in 1968 that included elements of the theory of expectations and theory of justice. It postulates that the reward is determined by the effort made, person's qualities, abilities and awareness of their role in the work process. The outcome of work is what determines a worker's satisfaction. According to Porter and Lawler, satisfaction results from the feeling of an accomplished task, which also increases productivity. Their model proved the importance of a unified theory of motivation that would cover such interdependent notions as satisfaction, reward, result, abilities, effort and perception (Porter and Lawler 1968).

Another important cognitive model developed by Heckhausen stipulates that motivation is determined by cognitive decision-making processes. Motivation to achieve arises from a possibility to improve one's skills, abilities and activity. The notion of motive is extended to include needs, desires, stimuli. Motivation is stimulated by situations that promise attainment of goals, pleasant outcomes and success. According to Heckhausen, an individual's action is the result of his/her selection of the best possible alternative from available actions and behaviors (Heckhausen 1997).

As for motivation in the framework of knowledge acquisition in classroom, there are several theories focusing on motivation for learning. Scholars address motivation for initiating and sustaining the process of learning.

In the 60s of the 20th century Skinner proposed a behavioural theory of learning in the context of education. The essence of this approach is that the external motivation of learning plays a more important role than the internal, as there are no ways to obtain objective data on internal motivation in science. The scientist developed the theory of reinforcement, according to which motivation is closely related to past experience. Thus, any actions of a student should lead to positive consequences, and in this case it will contribute to a greater likelihood of the same behaviour in the future. Skinner proposed the types of reinforcements, which, being combined in two parameters– encouragement and punishment – form four types: positive encouragement (reward for desirable behaviour), negative encouragement (abolition of punishment for desirable behaviour), positive punishment (unpleasant consequences for undesirable behaviour) and negative punishment (abolition of privileges for undesirable behaviour). Also, the scientist put forward two modes of reinforcement: permanent and variable (Skinner and Green 2004).

These modes are applied to the number of reactions and the reinforcement time interval. So, the following types are distinguished:

- constant reinforcement – constant reward for a certain number of tasks;
- reinforcement with a constant time interval – reward after a certain amount of time;
- variable reinforcement – reward after an indefinite amount of time and situations;

– reinforcement with a variable time interval – reward after a definite amount of time.

Thus, according to the scientist, the types and modes of reinforcement act as a means of external motivation increasing.

The theory of Self-Determination was developed in the 70s of the 20th century by Deci and Ryan. The researchers examined the problem of motivation and well-being of an individual, paying special attention to the issue of internal and external motivation. Defining the term of self-determination, the scientists explain it as the ability of a person to choose and experience the consequences of their choice. Deci and Ryan's approach is based on the detailed studying of influence of the external social factors (limitations, awards, evaluations) on the internal motivation. The researchers consider the internal and external motivation as two interrelated phenomena. The Self-Determination theory can be employed in the academic activity in the form of recommendations to motivate the learning process implementing strategies focused on interest, individual volition, desire and competence, but not on grades, rewards and other external factors. That is necessary in most modern schools as they do not support effectively students' autonomy and in academic activity it hinders motivation growth and aggravates anxiety decline (Deci and Ryan 2000).

Another concept of the educational motivation, formed in the framework of humanistic psychology by Rogers, is based on the fact that the human desire for self-actualization is the main motive of his behaviour (Rogers et al. 2013). According to Rogers, the learning process is associated with the whole personality of a student, his sensory and cognitive changes, and not just assimilation of knowledge. In his book "Freedom to Learn", published in 1969, the scientist identified two types of teaching: meaningless and meaningful. The meaningless type of teaching is when the interests of students are not taken into account, students are constantly under high control, and in the process of teaching there is only rigor and discipline. Students in this case are alienated from the teaching process and perceive a school as something they have to do for their parents or teachers. This type of teaching, manifested in external coercion and evaluation, is characterized by a high degree of external motivation. In meaningful teaching students, have more independence, ability to make choice, a creative approach to solving problems and full disclosure of their potential in conditions of cognitive activity of the student, supported by the teacher. According to Rogers, this type of teaching acts as the main motivational factor in students' learning process, as it involves maintenance of internal motivation. Thus, the scientist emphasizes that the greatest effectiveness of learning lies most likely in maintaining the internal motivation to study.

The theory of achievement target orientations, that originated in the 80s of the 20th century, focuses on the goals to achieve educational activities. Individuals, who seek to achieve these goals, adhere to certain behaviour, forming certain beliefs and being in different emotional states. Within the framework of this theory, there are two basic goals: the goal of skill and the effective goal. The goal of skill involves students' desire to gain new knowledge, improving the acquired skills, developing the achieved level of knowledge, and providing personal growth. The students who set themselves an effective goal are focused on getting the desired assessment from others, demonstrating their skills and knowledge. It is interesting that the effective goal is characterized by many researchers as an indicator of the superficial attitude of students to learning.

Elliott proposed his own approach to the theory of achieving target orientations, highlighting two types of effective goals: the goal of approximation and the goal of avoidance (Elliott and Dweck 1988). The students who are guided by the goal of approximation make every effort to succeed, while the students with the goal of avoidance prefer to avoid failures or undesirable results at every opportunity. The scientist believes that the focus on avoidance is more negative in the learning process, as it leads to low performance and lack of positive emotions.

The American psychologist Hull developed a theory of drives, according to which the necessities are a stimulus behaviour for an individual. The individual goes into a state of activity in the presence of an external stimulus – that is drive. Thus, a student is activated with a reward (high grades, scholarship, or grants). Drive acts as an incentive for effective learning activities and is fixed in the human psyche becoming a habit (Hull 1952).

The theory of achievement motivation was developed by Atkinson. The scientist considers as a source of behaviour not only the necessity, but also the values of success and the ability to achieve the goal. In his opinion, a student is more interested in the learning activities that are associated with high chances of success and a high goal (Atkinson 1983).

The problem of students' insufficient autonomy was raised by Reeve, who propounds that the faculty in the vast majority of countries possess the philosophy of control and management. Therefore, to increase motivation the scientist recommends that superfluous control should be prohibited at the lesson, instead students should be more encouraged to show their initiative, to speak their mind, should be given more choice and limitless ability to stand their ground on a certain issue (Reeve 2002).

Kornienko and Derish were the first to conduct a research on the relationship of academic motivation and academic motives and *the Dark triad of the personality*, the notion of which was introduced by Paulhus and Williamsas back as in 2002 (Paulhus and Williams 2002). The Dark triad is a group of three personality traits considered to be negative for a third person:

- Narcissism (self-admiration, lack of empathy, egotism, intolerance of a different opinion, hostility to critics),
- Machiavellianism (manipulation of other people, cynicism, egocentrism, thrift, absence of principles),
- Psychopathy (impetuosity, asocial behavior, cruelty, manipulation of other people, absence of empathy).

In their theory the scientists considered how negative character traits influence internal and external motives of academic activity of university freshmen (Kornienko and Derish 2015).

As the results of the research show, motivation in the academic sphere is determined solely by narcissism, as the intrinsic motives of cognition and attainment as well as the external one of self-respect are mostly expressed at a greatest degree of narcissism. Psychopathy negatively influences the external motivation as it is characterized by impetuosity, asocial behavior and irresponsibility that decrease aspiration for self-development. Machiavellianism in its turn does not affect academic motives. The research revealed that academic motivation being only effected by narcissism should undergo a further examination with respect to its relation to academic performance and expansion of students' selection.

Numerous theories and opinions on the problem of educational motivation prove the complexity and ambiguity of this phenomenon, since the variety of studies of the nature of motivation and motives complicates the practical solvability of the problem.

4 Conclusion

Nowadays the study of motivation becoming fundamental, scientists use an integrated approach to the study of this phenomenon. Over the past 100 years a wide variety of theories have been developed on the subject of motivation. Trying to form a system of methods of managing an individual and improve the efficiency of their activities, the researchers have proposed a wide range of theories, studying motivation in various fields. The study of motivational factors that determine human behaviour and activity is important in the modern world, as it allows us to explain the characteristics of human behaviour in a particular field of activity. The study of educational motivation reveals the factors of students' success, so the interest in the problem of educational motivation among scientists is only growing. This notwithstanding the problem of formation and sustenance of motivation for learning is not solved, thus making the present research acute and attractive for linguists, educationalists and psychologists. These issues of motivation demonstrate the diversity of practical application of motivation theories in the psychology of learning and require further research using modern data and tools.

References

Atkinson, J.: Personality, Motivation, and Action: Selected Papers (Centennial psychology series), Praeger (1983)

Bystrova, N.V., Konyaeva, E.A., Tsarapkina, J.M., Morozova, I.M., Krivonogova, A.S.: Didactic foundations of designing the process of training in professional educational institutions. Adv. Intell. Syst. Comput. **622**, 136–142 (2018). https://doi.org/10.1007/978-3-319-75383-6_18

Deci, E.L., Ryan, R.M.: The "what" and "why" of goal pursuits: human needs and the self-determination of behavior. Psychol. Inq. **11**, 227–268 (2000)

Elliott, E.S., Dweck, C.S.: Goals: an approach to motivation and achievement. J. Pers. Soc. Psychol. **54**(1), 5–12 (1988)

Heckhausen, H.: Achievement motivation and its constructs: a cognitive model. Motiv. Emot. **1**, 283–329 (1997)

Herzberg, F., Mausner, B., Bloch Snyderman, B.: Motivation to Work, 181 p. Transaction Publishers, New Brunswick (2010). With a new introduction of Frderick Herzberg

Hull, C.L.: A Behavior System. Yale University Press, New Haven (1952)

Kornienko, D.S., Derish, F.V.: The Dark triad and chronotype. PsikhologicheskieIssledovaniya, vol. 8, no. 43, p. 2 (2015). http://psystudy.ru/index.php/num/2015v8n43/1182-kornienko43.html

Maslow, A.H.: A theory of human motivation. Psychol. Rev. **50**(4), 370–396 (1943)

McClelland, D.: Human Motivation. Reprint Edition, 676 p. Cambridge University Press, New York (1988)

McGregor, D.: The Human Side of Enterprise, Annotated Edition, 1 edn. 480 p. McGraw-Hill Education, New York (2006)

Panachugin, A., Lapygin, J.: Virtual'naja obrazovatel'naja sreda kak sredstvo organizacii samostojatel'noj raboty studentov [Virtual environment as a means of fostering students' self-education]. Vestnik of Minin University. no. 6(4) (2018). https://vestnik.mininuniver.ru/jour/article/view/890. Accessed 10 Feb 2019

Paulhus, D., Williams, K.: The dark triad of personality: narcissism, machiavellianism, and psychopathy. J. Res. Pers. **36**(6), 556–563 (2002)

Porter, L.W., Lawler, E.E.: Managerial attitudes and performance. Homewood, Ill., R.D. Irwin 209 p. (1968)

Reeve, J.: Self-determination theory applied to educational settings. In: Deci, E.L., Ryan, R.M. (eds.) Handbook of Self-Determination Research, pp. 183–203. University of Rochester Press, Rochester (2002)

Rogers, C.R., Lyon, H.C., Tausch, R.: On Becoming an Effective Teacher. Person-centered Teaching, Psychology, Philosophy, and Dialogues with Carl R. Rogers and Harold Lyon. Routledge, London (2013)

Semarkhanova, E.K., Bakhtiyarova, L.N., Krupoderova, E.P., Krupoderova, K.R., Ponachugin, A.V.: Information Technologies as a factor in the formation of the educational environment of a university. Adv. Intell. Syst. Comput. **622**, 179–186 (2018). https://doi.org/10.1007/978-3-319-75383-6_23

Skinner, B.F., Green, L.: Psychology in the year 2000. J. Exp. Anal. Behav. **81**(2), 207–213 (2004)

Volkova, E., Mikljaeva, A., Bezgodova, S.: Psihologicheskie vozmozhnosti pedagogicheskoj otmetki [Psychological opportunities of marks]. Vestnik of Minin University, no. 6(4) (in Russian) (2018). https://vestnik.mininuniver.ru/jour/article/view/894. Accessed 10 Feb 2019

Vroom, V.H.: Work and Motivation, 432 p. Jossey-Bass Publishers, San Francisco (1995)

Russia: Digital Economy or Industrial and Information Economy?

Bronislav D. Babaev[1(✉)], Elena E. Nikolaeva[1], Dmitriy B. Babaev[2], and Natalya V. Borovkova[1]

[1] Ivanovo State University, Ivanovo, Russia
politeconom@yandex.ru, dvn2002@yandex.ru, bnv7777@ya.ru

[2] Russian Academy of National Economy and Public Administration Under the President of the Russian Federation, Moscow, Russia
bdbbdb@mail.ru

Abstract. The purpose of the work is to determine the economic nature, targets, nature of the modern economic system of Russia. Methodologically, we rely on a system-reproduction approach. Under the economic system, we understand social reproduction, which is considered to be the interaction of natural, economic and socio-cultural (ideology, law, mentality, religion, morality, aesthetics, etc.) components. The country's socio-economic system is also understood as a single economic and social space of the Russian Federation. At the same time, much attention is given to the territorial aspect of the economic analysis. As a result of research by way of polemical reasoning, we formulated the idea that the modern Russian economy is industrial-information in its nature. The concept of the digital economy itself is interpreted in two ways – as a set of information and communication tools and as a sector (segment) of the economy. The importance of the development of material production, including industry as a branch with strong innovative capabilities, is emphasized. A range of issues is being raised that reveal the competitive advantages of the Russian economy as an industrial-informational one (natural resources, vast territory, preservation of some features of ethnic passionarity in a nation, etc.). It is noted in the conclusion, that these points should be theoretically understood and a long-term state policy should be based on all levels of the socio-economic system. At the same time, the need for state regulation is emphasized, aimed at taking protective measures to safeguard high-tech industries with increasing returns, developing information and communication technologies to solve the problem of transition to a new type of social reproduction – the industrial-informational society.

Keywords: Virtual economy · Investment process · Industrial-informational economy · Material production · Industry · Real economy · Economic space of Russia · Digital economy

JEL Code: O10 · O14 · O25 · O30 · O38

E. G. Popkova and B. S. Sergi (Eds.): ISC 2019, LNNS 87, pp. 332–341, 2020.
https://doi.org/10.1007/978-3-030-29586-8_39

1 Introduction

In economics of Russia, there is a fashion for certain stories with the appropriate terminology. Passion acquires such a character that everyone, sometimes completely unexpectedly, becomes a supporter of this or that paradigm, this or that direction of research, which is considered promising, providing a positive result. In the 1960s, the interest in economic and mathematical research flashed, like a fire, to a certain extent under the influence of the West, but at the same time taking into account the traditions and demands of the economy of our country. There were leaders (V.S. Nemchinov, L.V. Kantorovich, N.P. Fedorenko, V.V. Novozhilov, A.G. Aganbegyan) who skillfully organized the work and gave it the right direction: large-scale research was launched, monographs and textbooks were published, conferences were held, the achievements made by economists and mathematicians were used in the national economy. The organizers became the laureates of the Lenin Prize, and L.V. Kantorovich filled up the ranks of the Nobel laureates. The economic and mathematical enthusiasm turned out to be so catching that they began to engage everyone who would take the trouble in it. Therefore, it was not always possible to talk about positive results. Still, after a long time, this wave of the application of mathematics to the economy that has swept the economic world should be recognized as a serious step forward.

In the 1970s at the suggestion of party leaders, the words "developed socialism" appeared on the pages of printed media in the humanities and even technical sciences. One could have expected that the economy itself entered a certain new stage, when qualitative changes took place, and the need for its adequate terminological description arose. But in reality it did not happen, though the specified term was on everyone's lips. And again it does not have to be argued that economics, which has widely used this concept, has achieved some success. At the same time, it became clear that there was some kind of force that urged economists and other social scientists to pronounce these momentous words, a kind of magic of the term.

In the 1980s the notion of "perestroika" rang round, the ideological psychosis reappeared, everybody became "perestroika-people" – someone in deed and someone more in name. But the term has become so popular that for a number of years it has become firmly established in scientific use, to some extent (giving justice) pushing people to revise obsolete dogmas. This was facilitated by certain realities of life itself – the development of cooperatives as forms of initiative activity, the development of market trade, the intensification of relations with foreign countries, etc.

In the 1990s "the market" flooded everyone – both as a real process and as multifaceted topics that captured the attention of almost all people. They began to say "the market has begun its work." The fashion for this term does not pass with time, since market relations acquired the full rights of citizenship and require appropriate scientific interpretation.

In the 2000s we were overtaken by another innovation – "strategies." The development of strategies, along with various projects, has become a kind of norm and real activity, and scientific research, especially of social science. Huge budget funds have been spent on drawing up strategies. There was such a practice: the center allocates moneyed resources, provided that the work is made by a certain organization. For example, 25 million rubles was spent on the development of one of the Strategies for

Vladimir Region; the work was carried out by one Moscow company, which was meant to get the money. For the most part, these voluminous documents, filled with text and digital material, remained on paper. Life has ceased to fit into the schemes invented by people. Western sanctions now being imposed on Russia have demanded a lot of change. But at the same time, the terms “strategy” (instead of the notion “plan”), “project” (instead of “a comprehensive target program”) have become part and parcel of our life, and a terminological enrichment of science has taken place.

Now the term “digital economy” is on everyone’s lips. It is not always clear what people mean by it, as there is a range of opinions, there is no unambiguous scientific definition for the term. There is an interpretation of the digital economy as a modern type of economy, characterized by the predominant role of information resources in the production of goods and services (Popov and Semyachkov 2018). The most common definition is that digital economy is an economy, “the subjects of which make extensive use of digital (electric) technologies. Sometimes, digital (electronic) economy is understood as part of the economy associated with a group of industries producing or using electronic technologies as a service (ICT sector). One can also find the definition of e-economy as part of an economy that uses the Internet as a medium for promoting services – such an economy is sometimes called the Internet economy” (URL: http://bit.samag.ru/archive/article/1824. Slavin, B. Russian economy: the digital age? 2017). Many experts in the field of business management, information technology define this economy as a system of economic (sometimes with the addition of social, cultural) relations based on the use of information and communication technologies; as an economy of new business models and new markets, providing a competitive bid unattainable for extensive automation; as a global network of economic and social events implemented through platforms such as the Internet, as well as mobile and sensor networks.

As we can see, it is quite common to say that information and communication technologies are at the center of the digital economy. They seem to have covered all our economic space (and not only economic), making its contribution to the formation of the system and integrity of the economy (in all honesty, let’s say that not everything is all right, there exists, as some researchers say, lacunae gaps in economic territory).

2 Methodology and Results

From content-functional point of view, information and communication technologies (ICT) can be understood in two ways. First, it is a tool, even a set of information tools, allowing organizations to solve important issues of scientific, managerial, forecast and other order (instrumental, technological approach). B. Panshin points out: “in the technological aspect, the digital economy is defined by four trends: mobile technologies, business analytics, cloud computing and social media; globally, social networks such as Facebook, YouTube, Twitter, LinkedIn, Instagram, etc. (Panshin 2016). But there is a different aspect, when ICT can be considered as a sector, a segment of the economy (sectorial, industry-specific approach), although its share is still low.

Thus, according to the empirical labor force survey, as of October 3, 2018, the share of people employed in the ICT sector in the total number of the employed people in the Russian Federation was on average in 2010–2014 – 2.1%, in 2015–2016 – 2.0%, in 2017 – 1.7%[1].

Our viewpoint is that the current state of the country's economy can be described as an industrial-informational economy. IT development of a society is an objective necessity caused by the scientific and technological progress. But the development of industrial production is an acute task for Russia and is determined by the challenges of globalization and the tasks of developing the national economy and its regions. The buildings and facilities of most plants and factories in Russia were built decades ago. The wear rate of fixed assets of commercial organizations in Russia on the whole in 2017 was 50.9%[2]. The equipment wear in the Russian Federation of in the extractive industry organizations in 2017 was 57.7%, in the manufacturing industries – 49.6%, transportation and storage – 56.8%[3]. This kind of evidence suggests that the industry in Russia is of problematic character, associated not only with wear and tear, but also with structural deficiencies (for example, the machine tool industry has been destroyed (Bodrunov 2018)), with lagging behind in many leading industries from the West. However, the industry does exist in the country, although the current national economy cannot function without an import component. Thus, almost 60% of the population's food supply is provided by the import, the domestic production being a bit more than 40% respectively. Food is a cooperation of industry with agriculture, but at the same time industrial enterprises significantly (but still insufficiently) are engaged in export, where the fuel and raw material component is allocated. All of these are fairly well-known things.

Let's concentrate our attention on a number of *key points* that make it possible to emphasize **the industrial nature of the country**, while it is important to highlight a number of positive points, especially in terms of possible positive changes.

The first point. The system of national accounts, adopted throughout the world, divides the results of economic activity into the production of goods (in 2005 in the Russian Federation, the types of activities producing goods accounted for 46.6%, in 2016 – 43.9% of the industry gross value added (GVA))[4] and production of services (tangible and intangible, is slightly more than 50% respectively). Though the ratio is rather stable, there is some progress, but not too substantial.

The question of theory and practice arises: what is the future of the world – in goods or services? The United States earns huge amounts of money on highly diversified services (financial, information and communication, consulting, trading, ship-owners, recreational and tourist services, etc.). At the same time, Germany, the largest economic power of Europe, is economically characterized as a country with a very

[1] Monitoring the development of the information society in the Russian Federation URL: http://www.gks.ru/free_doc/new_site/figure/anketa1-4.html (accessed 02 February 2019).

[2] Russian Statistical Yearbook 2018: Stat .book/Rosstat – M., 2018. p. 304.

[3] Wear rate of fixed assets in the Russian Federation at the end of the year. Data for November 22, 2018 URL: http://www.gks.ru/free_doc/new_site/effect/macr8.htm (accessed 02 February 2019).

[4] Regions of Russia. Socio-economic indicators. 2018: Stat .book/Rosstat – M., 2018. pp. 464, 478.

developed industry, though with an export focus. In this regard, one should pay attention to a fairly thorough study of the Norwegian economist E. Reinert, who, using the example of many countries of the world in various historical periods, showed that economic growth, the wealth of the country and its inhabitants directly depend on the development of manufacturing and high-tech services in the national economy which is characterized by increasing returns (Reinert 2017). Apparently, in the "goods or services" dilemma we are confronted with a different interpretation of the paradox, which was first – a chicken or an egg. In the current situation, the answer is dialectically paradoxical – both at the same time. The modern world does not understand or accept otherwise.

It is necessary to emphasize the significance of material production for our country not only from the standpoint of tradition, but also from the wealth of natural resources standpoint. They are significant, speaking in favor of our competitive advantages. In the long run, thanks to export efforts, we have not only a positive trade surplus, but also a positive balance of payments; we have accumulated significant gold reserves, including National Prosperity Fund.

A person should be introduced in this topic. His material needs have borders, but they are characterized by some flexibility, for example, improving nutrition in calories and food products, increasing the role of fashion in clothes, positive changes in housing and transportation, etc. At the same time, people's mental and spiritual interests are characterized by increased dynamism, thus, the types of economic activities (services) meeting these needs are expanding, but the importance of this trend should not be greatly exaggerated, but it cannot be ignored either. The world is in need of goods, although the capitalist formation in their production has made a truly revolutionary upheaval, allowing even such concepts as abundance or overabundance of goods to be used (for example, grain in the EU, car production in the world, etc.).

It should be noted that issues such as expanding exports, primarily machinery and equipment, work ahead of schedule in terms of producing new goods and providing material services, in-depth processing of raw and other materials, increasing the number of GVA industries, accelerated growth of the military-industrial facilities with a focus on dual-use technologies, the formation of industrial clusters and groups of leading enterprises, the industrial development of new territories. This requires huge investments – a focus task for years to come. The prospects for expanding material production in Russia are significant, as it is connected not only with global demand, with our natural resources and our own needs, but also with our economic space.

The second point in our discussion of Russia's competitive advantages in terms of industrial and informational nature of the economy is associated with its economic territory and geographical position. Russia is a spatial country, its territory is grand in latitude and longitude. This raises a number of challenges of a demographic character (territories are to be settled, with a population density of 5 or less people per 1 sq. km, it is considered undeveloped, and we have plenty of such territories), of transportation (permafrost zones pose special difficulties, and these are large areas). The major tasks are the use of rivers, lakes, seas for the conveyance of passengers and goods transfer, water supply issues for a number of areas, especially for the south of the European part of the country, are relevant; energy problems, waste disposal issues, etc. arise.

Our country is a bridge between Europe and Asia, which are extremely interested in using our territory for transportation purposes. We are talking about railways, highways, air corridors, the Northern Sea Route, tourist paths and routes, as well as pipelines for various purposes, not only gas and oil. If we enhance our transport policy in this way, we can get a tremendous benefit by way of foreign currency earnings, employment growth and expansion of production areas, as well as with regard to the country's prestige not only as an industrial, but also a transportation state. In general, one should strive to create something like a transport boom with the participation of public-private resources and funds of international companies. In this case, one must not forget about the need to ensure national security, including economic one.

In order to realize our competitive advantages, such issues as the speed of movement of goods and cargo (especially in terms of the success of a number of countries in this sphere), reliability and safety (we have a high accident rate on highways), cargo safety, convenience for passengers, price and tariff acceptability, a variety of additional services for consignors and passengers, etc. The topics concerning the Trans-Siberian Railway and the Northern Sea Route are promising, as there are difficulties with communication between Eastern Siberia, the Far East and the European part of the country. It is necessary to solve the problems of subsidizing many transportation services promptly in order to avoid the threat of severance of economic ties.

The third point of our reasoning concerns the clustering of the economy, meaning not only material production, but also the tertiary activity. In our historical reality, the topic of national economic complexes has always been topical. Now it is replaced by the term "cluster" which was offered by M. Porter (Porter 1993). The core industry, market-oriented, is surrounded by tertiary and support activities; it gives them a kind of planned targets in terms of volume, structure and other characteristics. A certain domestic market with its own prices and conditions for the movement of goods and services appears. Relationships of competition are becoming better, mutual assistance is developing, and a certain regional planning is taking shape. The result is in the acceleration of development, more effective adaptation of supply to demand, cost reduction, and other advantages. Clustering increases the sustainability of the economy and the competitiveness of products. This process can go on spontaneously, as a grassroots initiative, but it is desirable that the state and large economic structures take this process under their control. By the way, this topic is closely related the topic of ICT.

The fourth point is associated with an extremely significant issue – the interaction of the real and virtual economy. Theoretically, there is a lot of ambiguity here, collective efforts are necessary to develop a theory, the practice is clearly ahead of it, there are processes that are not sufficiently understood by researchers. Among them are, for example, the topic of electronic and quasi-money, cryptocurrency, intellectual property relations, etc. The virtual economy can be viewed from different points of view. On the one hand, it serves the real economy, providing it with additional competitive advantages. Let us take the payment system. The efficiency of the economy itself depends to a large extent on the speed, reliability, and accuracy of the calculations. E-business and e-commerce give additional impetus to the current production. On the other hand, the virtualization of economic relations forms an independent sector with its obvious signs of independence and self-sufficiency. There self-development and self-earning are demonstrated. One can offer other options for analysis, but these two points are quite enough to understand the complexity of the issue.

From the national well-being point of view the Russian society cannot be treated as favorable, but some features of ethnic passionarity still exist. Russians of different nationalities got used to computers very quickly, mastered mobile communications, began to actively use electronic money and everything of virtual origin. In this sense, the nation seeks to catch up to the level of world leaders. Communication in Russia is one of the most actively developing spheres, it being an important factor in favor of the digital economy. Let us take virtual communication between people and organizations with the help of various gadgets. The prospects here are ambitious and "long-playing."

At the same time, we cannot but note that the material and technical basis of the virtual economy was created by those branches that are part of material production. This elevates this sphere on the part of its influence on scientific and technical progress and on the future of the country. As S.D. Bodrunov notes, "the main goal of re-industrialization should be the restoration of the role and place of industry in the country's economy as its basic component, based on a new, advanced technological paradigm by solving a complex of closely related economic, organizational and other tasks in the framework of Russia's modernization" (Bodrunov 2018).

The fifth point in favor of the understanding of our economy as an industrial-informational one is the origins, nature, quality, forms of the investment process. Investment is understood as fixed assets in statistics, but the other element – the increase in working capital is not always shown. As an example, let us take some regions of the Upper Volga, which for some reason or other can be regarded as the depressive ones, that is, working with damaged reproductive bases (lack of resources, old or outdated technologies, etc.). In these areas, the investment process is distorted, especially in the regions of Ivanovo and Kostroma. The volume of investments per person is very limited (the number of residents in Ivanovo region is slightly more than 1 million people, investment in fixed assets is steadily keeping at 25 billion rubles, with some deviations in this or that direction) (Babaev and Babaev 2018). The overwhelming part of it goes to housing construction, while industrial and socio-cultural construction turns out to be disadvantaged. Another negative point is the concentration of capital investments in the city of Ivanovo as the regional center and in the adjacent Ivanovo district (80-85% of housing is built in it). The third negative point is that individual housing construction is not developed: it looks quite modest in scale and mostly cottages of 200 sq. m. and bigger are built. It means that the rich are engaged in building, while the main part of the population is deprived of this opportunity. Another negative point is the high proportion of Muscovites in the construction of individual housing, in particular, they are attracted by areas with good landscapes, as well as the banks of the Volga and other rivers. Local residents complain.

The investment process is an explicit weakness of material production, as well as socialized socio-cultural sphere. We should think about it, considering that in many types of economic activity the level of 1991 has not been reached by 2019 (e.g. in Ivanovo region, the production of machinery and equipment in 2016 was only 3.2% of the 1991 level! Only chemical production, metallurgical production and production of finished metal products exceeded this level[5]). There is some concern about the lack of capacity in manufacturing (with some exceptions; this is relevant not only for the Upper Volga region, but also for the country as a whole). The federal authorities have put forward 12 Programs (projects), which in their content support the idea of the digital economy, regardless of how to interpret it. It is important to understand that science plus education plus culture form the intelligence of the nation. This is a significant argument in favor of the informational nature of the economy, but it must always be borne in mind that the achievements of the "intelligence of the nation" cannot sufficiently be realized beyond the material production. If scientific developments of a medical character made it possible to create a new medical technology, then the devices for it will be created by industry. It's not a mere coincidence that for a number of years, the issues of new industrialization have been raised on the pages of the Russian economic literature and discussed on international forums (Bodrunov 2018; Bodrunov et al. 2018; Gubanov 2012, 2015, 2016, 2018; Ryazanov 2018).

An important topic facing the entire economy of the country is the weakening of the stimuli for economic growth and the lack of a long-term orientation. In many ways, local and federal authorities act like a fire brigade, which is not surprising, since the accidents happen quite often in the economy (wear degree of fixed assets, lax staff, poor management and other negatives points). The material incentives of people themselves are reduced mainly to the amount of wages, taking into account the price level and taxation. This considerably impoverishes the motivational mechanism of the society itself, where such characteristics as the content of labor and the conditions of its flow, relations with the authorities and within the team, transport conditions, household equipment, etc. are important. If all these areas are carefully examined, you can see for yourself that the digital economy in its ideal form is still far away.

3 Conclusions

Let's sum up our reasoning. We came to the conclusion that the economic nature of the modern Russian economy can be characterized as an industrial-informational economy. It should be noted that Academician L.I. Abalkin used the term information-industrial economy (Abalkin 2007) with a focus on industrial development. But over the past decade, the importance of information and communication technologies in the economic and social life of people has significantly increased. We attach great importance to the industry, characterizing it as an innovative branch, which is the basis for the development of the digital (informational) economy. Without a modern, developed industry, including machine-tool construction, electronics, etc., it is impossible to solve the task of building a new industrial society. But we must look forward. And the future

[5] See: Ivanovo region. Statistical Yearbook. 2017: Statistical collection/Ivanovostat-Ivanovo, 2017. p. 211.

belongs to the informational society. Here we unanimously support S. D. Bodrunov and his colleagues (Bodrunov et al. 2018). Therefore, we tend to use the term industrial-informational economy of Russia.

The issues raised by us require serious scientific work to understand new realities in the context of the formation of the digital economy. Then it is necessary to move from theoretical and methodological reasoning to the development of long-term public policy, relating both to the development of industry and other sectors of material production, and the further IT-based management of the society. It requires protective measures to safeguard high-tech industries with increasing returns, the development of information and communication technologies to solve the problem of Russia's transition to a new type of social reproduction – the industrial-informational society.

Acknowledgment. The article was prepared with the financial support of the Russian Foundation for Basic Research in the framework of the scientific project No. 19-010-00329 "Theoretical and methodological foundations of an expanded understanding of the economic mechanism in the modern economy".

References

Abalkin, L.I.: Basics of a social market economy and information-industrial society. Econ. Rev. Russia 2(12), 8–9 (2007)

Babaev, B.D., Babaev, D.B.: Economics of Ivanovo region: fundamental weaknesses of the reproduction basis. In: Nikolaeva, E.E. (ed.) Trends, Problems and Prospects for the Socio-economic Development of Old Industrial Regions (Using the Example of Ivanovo Region), pp. 49–162. Publishing House of Ivanovo State University, Ivanovo (2018)

Bodrunov, S.D.: Noonomics, Cultural Revolution, Moscow, Russia (2018)

Bodrunov, S.D.: New industrialization: prerequisites and approaches to implementation. In: Bodrunov, S.D. (ed.) New Industrialization of Russia: Strategic Priorities of the Country and the Possibilities of the Urals, pp. 11–26. Urals State Economic University, Ekaterinburg (2018)

Bodrunov, S.D., Demidenko, D.S., Plotnikov, V.A.: Reindustrialization and the formation of a "digital economy": harmonization of tendencies through the process of innovative development. Manage. Consult. (2), 43–54 (2018)

Gubanov, S.S.: Sovereign Breakthrough. Neo-industrialization of Russia and Vertical Integration. Book World, Moscow (2012)

Gubanov, S.S.: From the export-raw material model to the neo-industrial economic system. Econ. Rev. Russia 4(46), 48–59 (2015)

Gubanov, S.S.: On the economic model and long-term strategy of new industrialization of Russia. Economist (2), 3–10 (2016)

Gubanov, S.S.: Neo-industrial development paradigm: basics and meaning. In: Bodrunov, S.D. (ed.) New Industrialization of Russia: Strategic Priorities of the Country and the Possibilities of the Urals, pp. 27–61. Urals State Economic University, Ekaterinburg (2018)

Panshin, B.A.: Digital economy: features and development trends. Sci. Innovations **3**(157), 17–20 (2016)

Popov, E.V., Semyachkov, K.A.: Comparative analysis of the strategic aspects of the digital economy development. J. Perm Univ. Ser. Econ. **13**(1), 19–36 (2018). https://doi.org/10.17072/1994-9960-2018-1-19-36

Porter, M.: International Competition: Trans. from English. Schetinin, V.D. (ed.) International Relations, Moscow, Russia (1993)

Reinert, E.S.: How Rich Countries Got Rich, and Why Poor Countries Stay Poor: Trans. from English. Avtonomova, N. (ed.) Publishing House of the Higher School of Economics, Moscow, Russia (2017)

Ryazanov, V.T.: Russia's economic strategy: the neo-industrial imperative. In: Bodrunov, S.D. (ed.) New Industrialization of Russia: Strategic Priorities of the Country and the Possibilities of the Urals, pp. 62–81. Urals State Economic University, Ekaterinburg (2018)

Slavin, B.: Russia's Economy: The Digital Age? (2017). http://bit.samag.ru/archive/article/1824. Accessed 27 July 2018

Ensuring National Economic Security Through Institutional Regulation of the Shadow Economy

Vladimir Plotnikov[1(✉)], Maria Golovko[2], Gilyan Fedotova[3], and Maksim Rukinov[4]

[1] St. Petersburg State University of Economics, Saint Petersburg, Russian Federation
Plotnikov_2000@mail.ru
[2] National Research Nuclear University MEPhI, Moscow, Russian Federation
golovko178@mail.ru
[3] Volgograd State Technical University, Volgograd, Russian Federation
g_evgeeva@mail.ru
[4] Pushkin Leningrad State University, Saint Petersburg, Russia
pushkin@lengu.ru

Abstract. Purpose: The purpose of the chapter is to analyze the shadow economy as a systemic phenomenon, its qualitative and quantitative assessment, as well as the development of recommendations for shadow economy suppression. The subject of the study is the shadow economy. Its development is seen as a threat to economic security. Due to this, it is necessary to reduce the share of the shadow economy in the national economy. Thus, it will be possible to increase the level of national economic security.

Methodology: The methodology of the research is based on the methods of institutionalism. The study was performed on the materials of the Russian economy. The research is performed with the help of graphical presentation of information, trend analysis, comparison, analogy, systematization, etc.

Results: Increasing the share of the shadow economy reduces the level of security of the national economic system. Therefore, it is necessary to pursue a state economic policy aimed at reducing the share of the shadow sector in the national economy. The study of the shadow economy has shown that the main reason for its emergence and development is an unfavorable institutional environment. Studies conducted by using econometric methods and sociological surveys allowed establishing not only a qualitative, but also a quantitative relationship between the shadow economy and the institutional environment. The most effective way to reduce the shadow economy and increase economic security is to improve the institutional environment.

Recommendations: It is proposed to implement a special anti-shadow government policy. It is based on such economic policy components as tax, customs, monetary policy, business support policy, budget policy, etc. The anti-shadow policy integrates them into a single system to prevent and reduce the share of shadow economic activity. This will strengthen national economic security.

Keywords: Economic security · Shadow economy · State economic policy

JEL Code: F52 · K42 · O17

E. G. Popkova and B. S. Sergi (Eds.): ISC 2019, LNNS 87, pp. 342–351, 2020.
https://doi.org/10.1007/978-3-030-29586-8_40

1 Introduction

The development of modern society is characterized by an increase in the rate of observed changes. Today we are witnessing a process of "accelerating acceleration" (Bodrunov 2016). This accelerating dynamic is associated with the fourth technological revolution (Ciffolilli and Muscio 2018), the aggravation of international competition (Bartholomae 2018), social and political processes (Heilbron 2014), (Sekloča 2019), globalization and regionalization (Kotcofana et al. 2017), (Matveev et al. 2016), (Pashkus and Pashkus 2012), (Verbeke and Kano 2016), increasing rigidity of environmental constraints (Vertakova and Plotnikov 2017) and other factors. Under these conditions, the exacerbation of threats and challenges to the sustainable and stable functioning of the national economy occurs. Thus, ensuring the security of the national economic system becomes a priority.

Security is a universal scientific category. It is applicable to natural, social, economic, and technical systems. The subject of the author's analysis is economic security. The security of the economic system is examined by us in the context of national security. In this case, various components (subsystems) of economic security can be distinguished. The subject of study in this article is the shadow economy. Its development is seen as a threat to economic security (Alm and Embaye 2013), (Schneider and Enste 2000), (Williams 2014). Therefore, the study of the shadow economy as a systemic phenomenon, its qualitative and quantitative assessment, as well as the development of recommendations for its suppression are required. Due to this, it is possible to reduce the share of the shadow economy in the national economy and increase the level of economic security.

2 Materials and Methods

The methodology of the author's research is based on the methods of institutionalism. The study was performed on the materials of the Russian economy. (The data of official statistics, international ratings, the results of author's polls, expert evaluations of Russian specialists, etc. were used.) The choice of a specific economic system is due to the need to consider the specifics of the institutional environment. In the works of neoclassical scientists and some other areas of economic theory, abstract, universal models are built. They describe "the economy in general". This approach is good for a general understanding of the structure and dynamics of the economy. But he does not always give practically meaningful recommendations. Science is cut off from practice. The result is a decrease in the effectiveness of economic policy.

These effects manifested themselves in Russia, which in the 1990s began radical economic reforms. These reforms did not take into account the specifics of the Russian institutional environment. The theoretical basis of reform was monetarist economic theory. The result of the reforms was a systemic crisis and a fatal fall in GDP (Fig. 1).

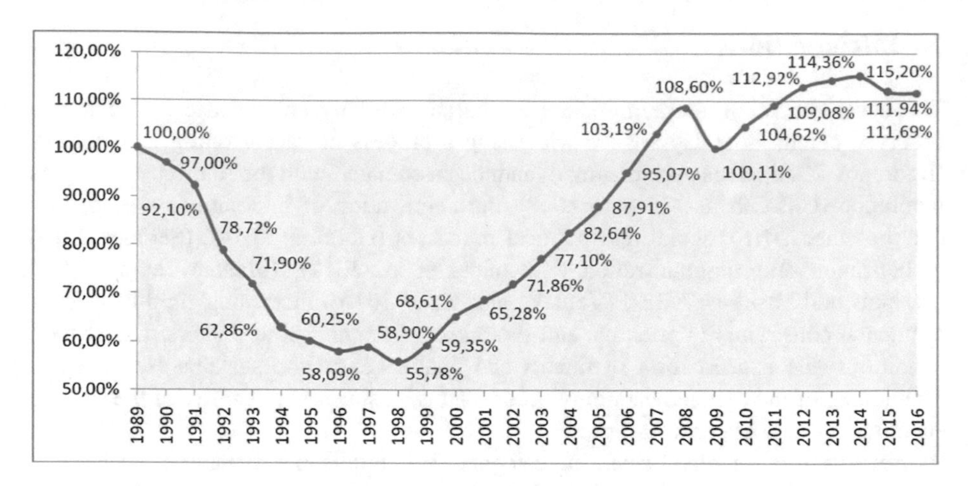

Fig. 1. Long-term dynamics of Russia's GDP (1989 - 100%). Source: Bodrunov and Plotnikov 2017.

The departure from abstract models and the introduction of pragmatism elements into economic policy made it possible to overcome a systemic crisis. But reliance on inadequate theoretical tenets led to a 20-year delay in the country's socio-economic development. Institutionalism allows us to combine the abstractness of theoretical schemes and country specifics (Brecher and Harvey 2009). Taking into account the focus of our study, we studied the institutional environment of the national economy from the standpoint of its impact on economic security.

The institutional environment forms the rules of legal and legitimate economic behavior (Golovko et al. 2018). It performs regulatory functions, suppressing or stimulating the shadow economy. The institutional environment is a factor for the sustainable and safe development of the national economic system. The level of development of the economy and its level of security depend not so much on the resources available, but on the rules by which these resources are used.

This is confirmed by the data of Friedman et al. (2000). According to their study, the share of the shadow economy in the GDP of Russia and Ukraine increased from 12% in 1989 to 29.6% and 36.9%. At the same time, in the countries of Eastern Europe, also affected by systemic changes, the scale of the shadow economy, on the contrary, has decreased. The decrease in the shadow sector, for example, in Poland and Romania was 3%. That is, the dynamics of the share of the shadow economy in GDP is not determined by the fact of reforms, but by their consistency and adequacy of the institutional environment.

In institutional theory it is customary to divide the rules into formal (political prescriptions, laws, contracts, etc.) and informal (culture, customs, traditions, ethical norms, etc.). Our research is based on the hypothesis that economic agents critically evaluate the applicability of formal institutional rules as a regulation of economic activity. If these rules are not effective enough, economic agents will compensate for the costs they cause in the shadow economy. Thus, the scale of the shadow economy (and therefore the threat to economic security) is related to the level of effectiveness of the institutional environment.

3 Results and Its Discussion

The shadow sector is unique in its content and functions as a subsystem of the national economy. Traditionally, it is estimated as a threat to the development of the national economy. From the standpoint of a systematic approach, this assessment is incomplete. The authors believe that it should be considered as a special indicator of the state of economic security. This position is due to the following. Society always seeks to create informal norms designed to compensate for the shortcomings of the formal institutional environment. These informal norms are associated, among other things, with the shadow economy.

Consequently, the analysis of the shadow economy can become an information source for improving the state economic policy. The indicators reflecting the size of the shadow economy include: the prevalence of corruption, the share of informal employment in the structure of the economically active population, the share of shadow incomes in national income, the share of illegal migrants in the total number of labor migrants, the share of counterfeit products in the domestic market, total imports and so on.

Table 1. The dependence of the shadow sector size and indicators of the institutional environment.

Index	Range	Estimate for Russia (1998)	Impact on the share of the shadow sector in GDP	Elasticity, %	Organization that calculated the index
Compliance with the law (mutual trust of government and society, legal culture)	0–6	3.5	Reverse	10.6	The PRS (Political Risk Services) Group
Quality of bureaucracy	1–6	3.19	Reverse	8.5	
Economic Freedom	0–16	7.0	Reverse	2.5	The Heritage Foundation
The degree of state intervention in the economy	1–5	4.0	Direct	14.7	
Security of property rights	1–5	3.0	Direct	13.4	
The arbitrariness and non-obligation of officials	1–7	2.01	Reverse	9.2	The World Economic Forum
Bribery	1–7	2.72	Reverse	8.0	
Equality of citizens before the law and equality of conditions of their access to legal proceedings	0–10	2.5	Reverse	3.8	Fraser Institute
Taxation	0–10	8.0	Direct	3.59	
Corruption Perception	0–10	2.27	Reverse	5.1	Transparency International

Source: Golovko 2016.

There is a quantitative relationship between indicators of the quality of the institutional environment and the scale of the shadow sector of the economy. This dependence (for Russia) is presented in Table 1. Data analysis suggests that improving the quality of the institutional environment (increasing the efficiency of public services, strengthening the institution of property rights, reducing corruption, increasing transparency and accessibility of power) will significantly reduce the size of the shadow sector of the Russian economy. The observed dynamics of the share of the shadow economy in Russia's GDP confirms our assumption.

The institutional environment for doing business can be represented as a set of political and economic conditions necessary to ensure the functioning of the business sector. Its quality depends on:

- The presence of certain economic freedoms (freedom of trade, investment, fiscal, financial freedom, freedom of labor, freedom from corruption).
- Development of a system to ensure the protection of contractual rights, property rights, and investors' interests
- Lack of administrative barriers in registering and liquidating an enterprise, obtaining licenses and permits, paying taxes, and implementing control procedures on the part of state bodies.

The synthetic rating, which assesses the institutional environment of the national economy, is the "Doing Business" rating. The World Bank makes this rating. Table 2 shows the dynamics of this index and its components for Russia. From the above data it can be seen that during the period under review, Russia significantly improved the quality indicators of the institutional environment. According to the authors, this is due to the fact that this rating has been used (since 2014) when monitoring the effectiveness of the state policy of supporting entrepreneurship in Russia. Consequently, the institutional environment can be successfully adjusted. The favorable changes in it are the result of special measures of state regulation.

Table 2. Russian Federation in the Doing Business rating for 2007–2016.

Indicators	2007	2011	2016
Overall rating of the Russian Federation	96 (from 175)	123 (from 183)	40 (from 189)
Creating a new company	33	108	26
Obtaining licenses and permits	163	182	115
Labor relations	87		
Property registration	44	51	9
Getting loans	159	89	44
Protecting the interests of investors	60	93	53
Payment of taxes	98	105	45
International trade	143	162	140
Compulsory enforcement of contracts	25	18	12
Liquidation of the company	81	103	51

Source: The World Economic Forum.

It should be noted that the Russian Federation is the country with the largest area in the world. Naturally, this leads to significant regional differentiation (Plotnikov et al. 2015), (Polozhentseva 2016), (Vertakova et al. 2017). The result is a significant variety of regional business conditions (Fedotova et al. 2018), (Iu et al. 2018), (Polyanin et al. 2018), (Vertakova et al. 2016). These conditions are determined by both formal institutions (regional and municipal legislation) and informal institutions (traditions and customs, cultural, religious, and other norms). The differentiation of the institutional environment leads to the need to take into account its specificity in identifying and assessing the scale of the shadow sector of the economy at the regional level.

One of the authors of this article (Golovko 2016) in 2015 conducted a survey of small business companies in the city of Volgodonsk (Rostov region, Russian Federation). The purpose of the survey was to identify the problems faced by entrepreneurs. Its results are shown in Fig. 2. The figures in the figure indicate the percentage of respondents who indicated the corresponding problem as a priority (the amount exceeds 100%, since it was possible to choose several answers). As can be seen from the presented data, bureaucratic procedures (70.7%), which are formed by the state, are most concerned about businessmen. In general, among the factors dominate those that indicate the low efficiency of the institutional environment.

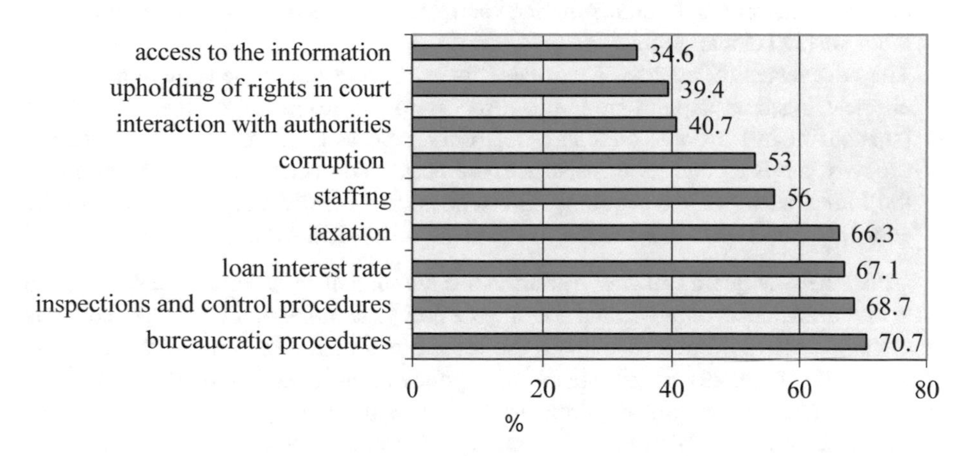

Fig. 2. The problems noted by the entrepreneurs of Volgodonsk.

A significant problem is the inspection of business by various government and public (established under the auspices of the state) control organizations. More than 80% of respondents indicated that each controlling organization conducted an average of 1 to 5 inspections per year. At the same time, 7% say that the Consumer Protection Associations carried out inspections more than 10 times, and 2% of respondents indicated the same frequency of inspections by the Fire Service.

Bureaucratic procedures, including excessive control over the activities of enterprises, cause additional business costs and risks of corruption. A common way to combat these negative effects is to transfer business to the shadow sector of the economy. Under these conditions, the main reason for the growth of the shadow economy is not the unethicalness of entrepreneurs, but the counterproductive institutional environment that has emerged.

We highlight a few shadow economy determinants due to the quality of the institutional environment:

- Starting costs of formal economic activity. Significant resources are required to officially start operations. They are spent on business registration, access to infrastructure, overcoming administrative barriers. In addition, a new enterprise may suffer losses due to the long periods of implementation of bureaucratic procedures. Shady business can save on some of these costs.
- Costs of legality. These are costs associated with the operation of an enterprise in the formal sector of the economy (payment of taxes, fines, compliance with restrictions established by law, etc.). The typical problem of Russia is informal employment. In this case, the company saves on payroll taxes, as well as on contributions to social funds. But this problem is not exclusively Russian. For example, enterprises in Latin American countries by such actions save up to 20% of labor costs (Tokman 1992).
- The advantages of legality. These benefits arise from the implementation of government business support programs. Access to the resources of these programs (subsidizing interest rates on loans, preferential representation of premises and other property, advisory and legal assistance, etc.) can be obtained only by enterprises of the formal sector of the economy. Consequently, the activation of the state in the economy can lead to a reduction in the scale of its shadow sector.

Thus, the low quality of the institutional environment leads to an increase in the scale of the shadow economy. This not only destroys the moral and ethical foundations of economic activity, but also reduces the level of economic security of the country. Therefore, it is necessary to activate the state policy in this area. It is necessary to form institutions adequate to the requirements of modern economics. First, it concerns economic institutions. Improving the quality of institutional regulation should also encompass the social and public administration spheres.

We note the complexity of the practical solution of this problem. The shadow economy is a complex phenomenon. Therefore, government opposition to the shadow economy cannot be a separate direction in the aggregate of state regulatory measures. It should be a set of specific areas in the framework of economic, social, cultural, educational and other state policies. Activities within each of these areas, aimed at creating an effective institutional environment, at containing the shadow economy, are a special type of government policy - "anti-shadow" policy.

"Anti-shadow" state policy is based on such components of economic policy as tax, customs, monetary policy, business support policy, fiscal policy, etc. Its valuable reference points are: adequate level of taxation; availability of loans; low inflation; effective and affordable government support for businesses; positive balance of foreign trade balance; the excess of the share of high-tech products in the production and export of resource-intensive; increasing the profitability of enterprises, etc.

The final result of the implementation of this kind of policy is to improve the quality of the institutional environment for doing business, which will lead to the economic feasibility of legal business activity. The result will be an increase in economic security. At the same time, the following private positive effects of the state anti-shadow policy are expected:

- Increasing the level of tax collection, including customs payments, against the background of reducing the tax burden by eliminating opportunities and motivating violations of tax legislation.
- Rational structure of the customs tariff, contributing to the filling and assortment differentiation of the national consumer market, on the one hand, and deter threats to the development of domestic entrepreneurship, on the other.
- Optimization of foreign trade turnover by increasing the volume of exports of high technology products, reducing the scale of export of raw materials (as the most attractive product for Russian shadow economic entities).
- Strengthening control over the movement of goods and vehicles across the customs border against the backdrop of the liberalization of foreign trade by selecting the optimal controls that do not discriminate participants in foreign economic activity.
- Positive dynamics of the effectiveness of government business support programs, calculated as the ratio of the increase in tax payments (federal, regional, local) enterprises-recipients of support to the money spent from the budgets of the corresponding level.
- Enhancing business processes by ensuring the availability and reducing the cost of credit resources amid a reduction in the number of deliberately bad loans and restraining capital outflows.
- Reducing the pressure on business against the background of the growing effectiveness of state control over its functioning, which becomes possible by eliminating corruption risks.

Another effect of the implemented anti-shadow policy is the transformation of the system of informal institutions. This effect is formed in the long term. It is associated with the growth of mutual trust of business and government, the development of a culture of economic behavior, the formation of a negative attitude of society towards the use of shadow business and corruption schemes.

4 Conclusion

The study made it possible to establish that the shadow economy is an important factor affecting economic security. Increasing the share of the shadow economy reduces the level of security of the national and regional economic systems. Therefore, it is necessary to pursue a state economic policy aimed at reducing the share of the shadow sector in the national economy.

The study of the shadow economy from the standpoint of institutionalism has shown that the main reason for its emergence and development is an unfavorable institutional environment. In various countries of the world, there is a strong correlation between the share of the shadow sector in the economy and indicators of the quality of the institutional environment. Studies conducted on Russian materials using econometric methods and sociological surveys allowed establishing not only a qualitative, but also a quantitative relationship between the shadow economy and the institutional environment. The most effective way to reduce the shadow economy and increase economic security is to improve the institutional environment.

Improving the institutional environment should be initiated by the state. The priority of its action is the modernization of formal institutions related to the regulation of business. It is proposed to implement a special anti-shadow government policy. It is based on such economic policy components as tax, customs, monetary policy, business support policy, budget policy, etc. The anti-shadow policy integrates them into a single system in order to prevent and reduce the share of shadow economic activity. This will strengthen national economic security. In the future, we can expect improvements in informal institutions, which will increase the level of legality and ethics not only of the economy, but also of society as a whole.

References

Alm, J., Embaye, A.: Using dynamic panel methods to estimate shadow economies around the world, 1984–2006. Public Finan. Rev. **41**(5), 510–543 (2013)

Bartholomae, F.W.: Digital transformation, international competition and specialization. CESifo Forum **19**(4), 23–28 (2018)

Bodrunov, S.D.: Re-industrialization: socio-economic parameters of reintegrating production, science and education. Sotsiologicheskie issledovaniya [Sociological Studies] **2**, 20–28 (2016)

Bodrunov, S., Plotnikov, V.: Institutional structures influence on the technological development of the economic system. In: Proceedings of the 30th International Business Information Management Association Conference, IBIMA 2017-Vision 2020: Sustainable Economic development, Innovation Management, and Global Growth, pp. 2658–2665 (2017)

Brecher, M., Harvey, F.P.: Realism and Institutionalism in International Studies, pp. 1–201. University of Michigan Press, Ann Arbor (2009)

Ciffolilli, A., Muscio, A.: Industry 4.0: national and regional comparative advantages in key enabling technologies. Eur. Plann. Stud. **26**(12), 2323–2343 (2018)

Fedotova, G.V., Kulikova, N.N., Kurbanov, A.K., Gontar, A.A.: Threats to food security of the Russia's population in the conditions of transition to digital economy. In: Popkova, E.G. (ed.). The Impact of Information on Modern Humans, pp. 542–548. Springer, Cham (2018)

Friedman, E., Johnson, S., Kaufmann, D., Zoido-Lobaton, P.: Dodging the grabbing hand: the determinants of unofficial activity in 69 countries. J. Public Econ. **76**, 459–493 (2000)

Golovko, M.V.: State Policy of Countering the Shadow Economy in the Context of Measures to Ensure Economic Security, pp. 1–188. Moscow Engineering Physics Institute Publishing (2016)

Golovko, M.V., Rudenko, V.A., Krivoshlykov, N.I.: Influence of institutional transformations on the choice of mechanisms for ensuring economic development and security of nuclear power engineering enterprises. Espacios **39**(31), 14 (2018)

Heilbron, J.: The social sciences as an emerging global field. Current Sociol. **62**(5), 685–703 (2014)

Iu, M.D., Selishcheva, T.A., Diatlov, S.A., Lomakina, I.B., Borkova, E.A.: Regional supply chain structure and centralization of the economy of Russia. Int. J. Supply Chain Manage. **7**(6), 684–692 (2018)

Kotcofana, T., Stazhkova, P., Pashkus, V.: Competition and concentration processes in the context of globalization (case of the Russian banking sector). In: Globalization and its Socio-Economic Consequences, 17th International Scientific Conference Proceedings, pp. 1086–1095 (2017)

Matveev, Y.V., Valieva, E.N., Trubetskaya, O.V., Kislov, A.G.: Globalization and regionalization: institution aspect. Math. Educ. **11**(8), 3114–3126 (2016)

Pashkus, V.Yu., Pashkus, N.A.: Globalization: new economy and new challenges. In: Globalization and its Socio-Economic Consequences, 12th International Scientific Conference Proceedings, pp. 643–648 (2012)

Plotnikov, V., Fedotova, G., Popkova, E., Kastyrina, A.: Harmonization of strategic planning indicators of territories' socioeconomic growth. Reg. Sectoral Econ. Stud. **15–2**, 105–114 (2015)

Polozhentseva, Y.: Inequality in social standard of living in the international context. Econ. Ann. XXI **157**(3–4), 15–18 (2016)

Polyanin, A., Golovina, T., Avdeeva, I., Merkulov, A., Klevtsova, M.L.: Organizational and managerial infrastructure of digitalization processes in economic systems of various levels. In: Proceedings of the 31st International Business Information Management Association Conference, IBIMA 2018: Innovation Management and Education Excellence through Vision 2020, pp. 4159–4168 (2018)

Sekloča, P.: The centre and the periphery: productivity and the global networked public sphere. TripleC **17**(1), 1–18 (2019)

Schneider, F., Enste, D.H.: Shadow economies: size, causes, and consequences. J. Econ. Lit. **38**(1), 77–114 (2000)

Tokman, V.E. (ed.): Beyond Regulation: The Informal Economy in Latin America, pp. 1–295. Lynne Rienner Publishers, Boulder (1992)

Verbeke, A., Kano, L.: An internalization theory perspective on the global and regional strategies of multinational enterprises. J. World Bus. **51**(1), 83–92 (2016)

Vertakova, Y., Plotnikov, V.: Problems of sustainable development worldwide and public policies for green economy. Econ. Ann. XXI **166**(7–8), 4–10 (2017)

Vertakova, Y., Plotnikov, V., Fedotova, G.: The system of indicators for indicative management of a region and its clusters. Procedia Econ. Finan. **39**, 184–191 (2016)

Vertakova, Y., Polozhentseva, Y., Klevtsova, M., Ershova, I.: Integrated assessment of socio-economic differentiation: cross-country comparison. In: Proceedings of the 30th International Business Information Management Association Conference, IBIMA 2017 - Vision 2020: Sustainable Economic development, Innovation Management, and Global Growth, pp. 1834–1843 (2017)

Williams, C.C.: Out of the shadows: a classification of economies by the size and character of their informal sector. Work Employ Soc. **28**(5), 735–753 (2014)

Resource Provision of the Digital Economy

O. B. Digilina(✉) and D. V. Lebedeva

RUDN University, Moscow, Russia
o.b.digilina@mail.ru, lena_leb-61@mail.ru

Abstract. This research investigates the process of Russia's transition into the digital economy. The widespread introduction of digital technologies has led to a visible improvement and simplification of everyday human life. These technologies have created new markets, stimulated consumer behavior and made people's lives more comfortable.

The methodological basis of this article is a set of techniques, methods and cognitive settings adapted to the specifics of the digital economy; the empirical base is the socio-economic practice of developed countries.

It is important to understand the role of economic resources is changing significantly in the transition to the digital economy. In industrial societies, the engine of economic development is predominantly human (labor) resources and financial resources, but the engines of economic progress in digital economies are knowledge and people. The transition to a digital economy is a labor-intensive, complex and significant process to ensure sustainable development and stimulate the competitiveness of economic entities. Technological resources are crucial for a digital economy; furthermore, identifying and defining those required during the transition from an industrial society are of particular importance.

Keywords: Resources · Digital economy · Resource provision · Competitiveness

JEL Code: O1 Economic Development

1 Introduction

In 2017, the Russian government adopted regulatory and legal documents focusing on the formation of the digital economy. The President of the Russian Federation Vladimir Putin has repeatedly argued for a shift in the Russian economy, noting that "the Digital economy is not a separate industry, in fact, it is a way of life, a new basis for the development of the system of public administration, economy, business, social sphere, the whole society. And of course, the formation of the digital economy is a matter of national security and independence of Russia, the competitiveness of domestic companies, the country's position on the world stage in the long term, in fact, for decades to come" (TASS RUSSIAN NEWS AGENCY 2017).

The concept of the digital economy was put forward by the American computer scientist N. Negroponte who assumed that all material substances (raw materials, products, etc.) have disadvantages associated with the need for resource provision.

E. G. Popkova and B. S. Sergi (Eds.): ISC 2019, LNNS 87, pp. 352–358, 2020.
https://doi.org/10.1007/978-3-030-29586-8_41

Negroponte (Negroponte 1996) believed the main advantage of the "new" economy will be the reduction of the necessary resource base, as well as the movement of material substances through the Internet space.

Currently, digital transformation is one of the main factors affecting global economic growth. McKinsey Global Institute estimates up to 22% of China's GDP growth leading to 2025 will be attributed to internet technologies. Similarly, the expected increase in value created by digital technologies in the US by 2025 could be 1.6–2.2 trillion US dollars. USA. Digitalization of the Russian economy will likely increase the country's GDP by RUB 4.1–8.9 trillion by 2025 (in 2015 prices), representing 19 to 34% of the total expected GDP growth. In 2016, according to Federal State Statistics Service, the total GDP in Russia was about RUB 86 trillion (Aptekman et al. 2017).

However, the development of digital technologies, in turn, requires significant resources. According to the authors, the formation of resource potential is crucial in the development of the digital economy, as the effective allocation of resources contributes to the achievement of the goals of the economic entity and sustainable development.

2 Methodology

The object of this article is the process of formation of the digital economy in Russia. The subject of the study are the economic relations that are formed whilst developing the resource support for the digital economy.

The methodological basis of this study was a set of techniques, methods and cognitive settings that are currently known to scientific research and that have been adapted, in relation to the specifics of the object - the formation of the digital economy.

The empirical basis for the preparation of this article was the socio-economic practice of the developed countries of the world, the Russian Federation, large enterprises, their best practices in the field of resource provision of digitalization of economic processes. Additionally, this studies draws upon academic statistical and analytical reviews, government documents reflecting the strategy of the digital economy and economic articles in the media.

Results. With the change of eras, the importance of individual resources has changed. Whilst natural and human resources dominated in the pre-industrial era and material resources were fundamental in industrial society, and information and intellect are key resources in post-industrial societies.

Economic resources are "the fundamental concept of economic theory, meaning in general sources, means of production, a set of resources used in economic activity" (Kurakov et al. 2004).

According to the authors, resource provision should be considered comprehensively for the effective implementation of economic activity, as limited or poorly managed resources can lead to negative consequences. This is a recurring idea in resource theory of the organization, for example, a textbook entitled: "business Policy" in 1986 (Andrews et al. 1986) which notes that the strengths and weaknesses inherent in each firm, outlines why the effective management of its own resources determines its prospects for further development and competitiveness.

G. Hamel and K. Prahalad (Hamel and Prahalad 1994) connect the sources of competitiveness of the company with the skills of manager's connection of dispersed resources with production skills and competencies for timely flexible response to changing market conditions and resource concept.

For example, according to D. Miller (Miller 2003), the asymmetric distribution of resources plays a more important role in stimulating competitive advantages than their properties, since in the absence of standard resources, significant costs can arise. The division of resources is not strict, but on the basis of the scientific papers studied (Shlychkov et al. 2009), (Grant 1991), (Collins and Montgomery 1995), (Collins and Montgomery 1997). Digitalization of the economy significantly changes the importance of economic resources and if in the era of industrial society the engine of economic development was human resources and capital, now the engines of economic progress are knowledge and man as their carrier. Knowledge in the modern economy performs many functions. They are a factor of production, a primary means and a subject of transactions, a product and a means of production, an end-use item and a means of thesaurus. Information resources (knowledge resources) are understood as "separate documents and separate sets of documents contained in information systems (libraries, archives, funds, data banks, other information systems). In the organization (society, company, enterprise, etc.) — the amount of information that the organization has to solve its management tasks" (Lopatnikov 2003).

In a world where digital technologies are used, knowledge is a tool of great power and a threat to security, both personal and public. The introduction of digital technologies leads to significant improvements and simplifications in everyday life. Participants in the digital economy are generating new markets, making people's lives more comfortable and their consumer behavior more dynamic.

At the same time, digital technologies allow to accumulate the smallest information about consumer preferences, increasing the transparency of the economic space for owners of large companies. Corporations, having full information about consumer preferences and ownership of intelligent data analysis systems, are able to drive the population into the "consumer cell", forcing them to pay an increasing price for meeting their needs, because they have monopoly power (Tsarikovsky et al. 2018).

Monitoring consumer behavior allows digital platforms to extract disproportionately high in relation to expended labor and investment income. The world Bank in its report "Digital dividends" notes that the main "cream" of digitalization receives a limited group of people who own digital companies (Digital Dividends 2016).

Knowledge in the digital environment, as a market product, has a number of features. They are intangible, have zero cost of travel anywhere in the world, can be copied and adapted to the national characteristics of their consumers free of charge. "In the digital markets, the product has acquired a new quality and features of its circulation, requiring special approaches in the analysis of the market in order to establish the share of an economic entity in the market and taking into account factors that have no influence or are absent at all in the traditional markets of goods and services" - said in his study (Anisimov 2018).

The development of digital technologies has also led to the problem of digital inequality. In United Nations documents, digital inequality refers to access to information, knowledge and technology in the digital space. For example, access to knowledge is currently limited in Russia. The high cost of scientific knowledge bases such as Elsevier, Wiley and Thomson Reuters does not allow many of the largest Russian universities to use them, because they do not have such funds in sufficient quantities. It is intellectual inequality, the lack of access to knowledge and technology, that determines the widening of digital inequality today, because the paradox of this phenomenon is that when connected to the Network, the gap between those who have technology and information, and those who do not have it, but are included in the Network, is aggravated, not reduced.

The bearer of knowledge and its creator is man. Under labor (human) resources refers to the economically active, able-bodied population, part of the population with physical and spiritual abilities to participate in labor activities (Raizberg and Lozovskiy 1999).

Organizational and management resource refers to the creative process, which carries out the solution of scientific and technical problems, stimulating employees, building a system of motivation, planning and management of the organization, etc.

The development of digital technologies makes demand for new professions. According to experts of the portal SuperJob.ru in the field of IT-technologies will be in demand:

– developers of applications for iOS and Android (competition - less than two people per place, the average salary in Moscow - RUB 130–140 thousand per month),
– UI/UX-interface designers of various platforms or sites (competition - 2.5 job seekers, the average salary in Moscow - RUB 110–120 thousand),
– application testing specialists (2.5 resumes per vacancy, average salary in Moscow - RUB 80–90 thousand),
– data Analytics and Big Data. (MAIL NEWS 2019)

Computer literacy requirements for other professions are also increasing. So for some professions office jobs will be transformed into home jobs ("home office"), linked via the Internet. This will allow entrepreneurs to save the cost of buying or renting premises, equipment, payment for various services related to the maintenance of offices, etc. Former employees will become "remote" workers, independently planning their costs, schedule, etc.

Gradually, the number of mobile jobs will increase – owners of tablets and netbooks can already work in public transport, in cafes and just on the street, performing their official functions, which do not need stationary computers, printers and other bulky devices. All this leads to the transformation of the quality of work, quality of life.

According to experts, between 3 million to 5 million IT professionals are working remotely in Russia. Over time, their number will only grow (ROSCONGRESS 2017).

According to A.V. Degtyarev (Degtyarev 2017) automation of intellectual labor, the Internet and "cloud technologies" will lead to the growth of professions of intellectual labor in the economy and will have a significant effect by 2025. In the foreseeable future, the main form of employment will be "work in the cloud".

The development of the digital economy requires new approaches to the education system, which should create specialists of the future. It is already necessary to develop new documents for universities and colleges, professional standards that contain requirements for the competencies of new specialists; to build the work of all structures of the education system in the interests of the deployment of the digital economy in the country. Work should also be carried out with workers employed in the production in the direction of ensuring constantly updated human resources and competence of workers for the needs of the digital economy, the formation of a personal development trajectory.

Of course, the construction of the digital economy is associated with significant financial investments. Under financial resources, refers to "the totality of all types of funds, financial assets, which has an economic entity at its disposal" (Raizberg and Lozovskiy 1999).

An example of financial resources can be grants, profits, funds, shares, government subsidies and more. "It is important to note that the space and time as the system resources are spent throughout the life cycle of the system: the first is filled, the second expires, which gives us additional grounds for considering them as resources of economic activity" (Kleiner 2011).

Despite the significant costs of the state to stimulate the development of information technology, business often talks about forced digitalization. For example, according to digital technologies, the potential of digitalization is not fully utilized (70–80%). According to the business Ombudsman Boris Titov, head of the Supervisory Board of the Institute of Economics of growth named after P. Stolypin, the compulsory digitization has cost businesses in RUB 80 billion.

This refers to the costs of business for the purchase of online cash, connection to the system of EGAIS ALCO, as well as costs associated with the need to comply with the provisions of the "law of Yarovoy".

Only large companies that have sufficient financial resources and benefit from digitalization invest their own funds in the development of digital technologies. The resources of medium and small businesses in the real sector are often insufficient to buy digital technologies, which is a limiting factor in the development of digitalization. The solution to this problem is seen in the use of tools of state support and the use of public-private partnership.

Russia has adopted a Program for the development of the digital economy until 2035, the aim of which is to create favorable organizational and regulatory conditions for the effective development of the institutions of the digital economy with the participation of the state, the national business community, civil society and to ensure rapid growth of the national economy through qualitative changes in the structure and management system of national economic assets, achieving the effect of the "Russian economic miracle" in the formation of the global digital ecosystem. (Development of digital economy in Russia. The program until 2035 2017).

The program of Russia's transition to the digital economy is formally part of the plan of economic reforms, which were developed by the Center for strategic development (CSD). The program is devoted to the transition to a new technological revolution in 2018–2024.

It is believed that the digital economy can be built on the basis of competent it regulation, the availability of developed infrastructure, national competence centers and digital platforms (TADVISER 2019).

CSD will be allocated on a "digital revolution" in the years 2018-2024 RUB 185 trillion, or about 30.8% of GDP per year. This is a huge figure, about 40–45% of the budget of the state and state-owned companies and even their partners from the private sector (Karpyuk 2017).

In 2019 alone, the Ministry of Digital Development, Communications and Mass Media of the Russian Federation plans to spend RUB 20.8 billion to support the development of digital technologies. The information technology development Fund plans to distribute RUB 5 billion between companies, the money will be spent on the development and implementation of domestic it products, services and platforms in the regions. Another RUB 1.7 billion from the budget will be allocated to subsidize interest rates in the amount of the key rate of the Central Bank. RUB 5 billion will be spent on pilot projects of the most developed technologies for priority sectors—health, education, industry, agriculture, transport, energy, etc. To get these funds, companies must attract at least 30% of extra-budgetary funding to the project and ensure that the technology will allow companies using it to triple their revenues in the fourth year after the pilot starts.

In addition, RUB 2.1 billion will be spent on grants that Russian Venture Company (RVC) through the Fund of technological projects will distribute to "leading research centers".

Another RUB 3 billion, according to the passport of the Federal project "Digital technologies", RVC will give to companies engaged in projects on end-to-end technologies (TADVISER 2019).

3 Conclusions

The transition to a digital economy is a time-consuming but necessary process to preserve and enhance the Russia's position in the international arena. This process is impossible without the development of technology, the involvement of highly specialized staff, financial support for equipment, government programs and so on, these components are components of resources. Given the limited resource base, the Russian economy may face difficulties during transformation. Therefore, it is necessary to build a competent management of resource potential and determine the necessary resources required to support competitive advantages and ensure sustainable development.

References

Anisimov, A.V.: Modern transformation of Antimonopoly policy. J. Econ. Theor. **15**(2), 338 (2018)

Development of digital economy in Russia. The program until 2035 (2017). http://spkurdyumov.ru/uploads/2017/05/strategy.pdf. Accessed 1 Apr 2019

Andrews, K., Bower, J., Christrensen, C.R., Hamermesh, R., Porter, M.E.: Business Policy: Text and Cases, vol. 6. Richard D. Irwin, Homewood (1986)

Tsarikovsky, A.Yu., Ivanov, A.Y., Voinikanis, E.A. (ed.): Antitrust regulation in the digital age: How to protect competition in the context of globalization and the fourth industrial revolution [Text]: monograph/NAC. research. University "Higher school of Economics"; FAS, p. 5. Ed. house of Higher school of Economics (2018)

Aptekman, A., Kalabin, V., Klintsov, V., Kuznetsova, E., Kulagin, V., Yasenovets, I.: Digital Russia: a new reality (2017). http://www.tadviser.ru/images/c/c2/Digital-Russia-report.pdf. Accessed 25 Dec 2018

Cllins, D.J., Montgomery, C.A.: Corporate Strategy: Resources and the Scope of the Firm, pp. 28–29. Chicago (1997)

Collins, D.J., Montgomery, C.A.: Competing on resources: Strategy for the 90 s. Harvard Bus. Rev. July-Aug, 118–128 (1995)

Degtyarev, A.V.: Digital employment in the digital economy (Work in the cloud" as the evolution of traditional employment) (2017). istina.msu.ru/conferences/presentations/53581453/. 20 Apr 2017

Digital Dividends: World development rep. (2016). http://www.worldbank.org/en/publication/wdr2016

Grant, R.M.: The resource-based theory of competitive advantage: Implications for strategy formulation. California Manage. Rev. **33**, 119 (1991)

Hamel, G., Prahalad, C.K.: Competing for the Future. Harvard Business School Press, Boston (1994)

Karpyuk, I.: Digital revolution from the CSD (2017). http://polit.ru/article/2017/07/01/csr/. 1 Apr 2019

Kleiner, G.B.: Resource theory of system organization of economy. Russian J. Manage. **9**(3), 3–28 (2011)

Lopatnikov, L.I.: Economic-Mathematical Dictionary: Dictionary of Modern Economic Science. Business, Moscow (2003)

Kurakov, L.P., Kurakov, V.L., Kurakov, A.L.: Economics and Law: Dictionary-Reference. University and school, Moscow (2004)

MAIL NEWS. Bounty hunting. The most popular and "unnecessary" professions of 2019 (2019). https://news.mail.ru/society/35989887/?frommail=1,(3 December 2018)

Miller, D.: An asymmetry -based view of advantage: towards an attainable sustainability. Strateg. Manage. J. **24**, 961–976 (2003)

Negroponte, N.: Being Digital. Knopf. (Paperback edition, 1996, Vintage Books). ISBN 0-679-76290-6 (1995)

Raizberg, B.A., Lozovskiy, L.S., Starodubtseva, E.B.: Modern Economic Dictionary, 2nd edn., Rev. M.: INFRA-M, 479 p. (1999)

ROSCONGRESS. Labour relations in the digital economy (2017). https://roscongress.org/sessions/trudovye-otnosheniya-v-tsifrovoy-ekonomike/about/27February2017. 4 Feb 2019

Shlychkov, V.V., Kulish, S.M., Timofeev, R.A.: Resource Approach to the Problems of Economic Reliability of Industrial Enterprises, 151 p. Kazan. state energy. UN-t (cgeu), Kazan (2009)

TADVISER. Digital economy of Russia (2019). http://www.tadviser.ru/index.php/%D0%A1%D1%82%D0%B0%D1%82%D1%8C%D1%8F:%D0%A6%D0%B8%D1%84%D1%80%D0%BE%D0%B2%D0%B0%D1%8F_%D1%8D%D0%BA%D0%BE%D0%BD%D0%BE%D0%BC%D0%B8%D0%BA%D0%B0_%D0%A0%D0%BE%D1%81%D1%81%D0%B8%D0%B8. 1 Apr 2019

TASS RUSSIAN NEWS AGENCY. Putin: the formation of the digital economy - the issue of national security of Russia (2017). https://tass.ru/ekonomika/4389411. Accessed 3 Dec 2018

The Comparison of the Financial Markets and the Financial Centres in the International Rankings

Alexander Ya. Bystryakov(✉), Elena V. Ponomarenko, and Denis A. Rasskazov

Peoples' Friendship, University of Russia, Moscow, Russian Federation
bstal@yandex.ru, elponomarenko@yandex.ru, rasskazov.da@yandex.ru

Abstract. This work is a continuation of the themes developed by the authors (Bystryakov and Rasskazov 2014), (Ponomarenko and Rasskazov 2017). The objective of this paper is to compare the development of international financial centers (IFCs) and financial systems of the IFCs' home-based countries. The study identified the main trends in competition between financial centers over the period from 2010 to 2016: 1. New York, London, Hong Kong, Singapore and Tokyo had been holding the leading positions in IFCs' rating more than 5 years, the TOP 5 was stable. 2. A trend to the dominance of North American centres in the global financial market. 3. The greatest progress had been achieved by the financial centres of the United Arab Emirates (Dubai and Abu Dhabi). 4. The largest IFCs in continental Europe was under pressure from competing North American and Asian centres. 5. There are different trends Among the BRICS: the financial centres in mainland China failed to gain a foothold in the TOP 10, while Mumbai (India) has shown the greatest progress over the 5 years in the overall ranking among BRICS; Sao Paulo (Brazil) and Moscow (Russia) positions remained stable. The authors have also reviewed the theoretical question of the relationship between the economic growth and the development of financial markets.

Keywords: International financial centre · Financial market development index · The Anglo-Saxon system of law · Economic growth · The world economic forum

JEL Code: G21

1 Introduction

The recognized IFC is one of the most important criteria for assessing the development and strength of the local financial market. This thesis is relevant for the major part of the IFCs (the exceptions are reviewed further).

Therefore, we suppose that the terms «international financial centre» and «financial market» have a strong correlation and there is a close relationship between their development and an economic growth.

E. G. Popkova and B. S. Sergi (Eds.): ISC 2019, LNNS 87, pp. 359–368, 2020.
https://doi.org/10.1007/978-3-030-29586-8_42

The structure of our research is as follows: the first part of the article is devoted to the reviewing the different scientific approaches to the problem of the importance of financial markets and centres in economic development. Further, we analyze the dynamics of the financial markets development index by the World Economic Forum and make the comparison of international financial centres based on the Z/Yen Group Ltd data. In the conclusion, we state the crucial results of our research.

2 Background and Methodology

Effective financial market transforms the savings of national and foreign economic entities into investment projects based on market mechanisms, rather than politically motivated decisions. Herewith the investments of the private (entrepreneurial) sector are extremely sensitive to the level of productivity. Consequently, the economy requires a developed financial market, generating financial capital, aiming at stimulating private investments. In this connection, the allocation of financial resources should be conducted on a wide range of channels: bank loans, adjustable stock exchange, venture financing, and others. The level of development of national financial market is largely determined by the degree of diversification of existing funding sources (the World Economic Forum, 2016), as well as the quality of the financial intermediaries. The stability and transparency of the banking sector as the main operator, as well as legal regulation focused on protecting the interests of private investors and other economic entities are the crucial factors to ensure the basic functions of the financial market.

The importance of financial markets in economic development is underestimated in neoclassical models of economic growth, but it is quite deeply investigated in the works of Joseph Schumpeter (Schumpeter 1911), Raymond William Goldsmith (Goldsmith 1969), King and Levine (King and Levine 1993). A key problem in different theoretical approaches is the determination of causality between economic growth and financial market development.

From the article of Nobel laureate in Economics Robert Lucas, we can conclude that the financial resources derive from changes in demand from the real sector of the economy (Lucas 1988). Subsequent studies come to the opposite conclusion – financial markets development stimulates an economic growth. American economists Robert King and Ross Levine in their work prove the strong correlation between the initial level of financial development and economic growth (King and Levine 1993), and Raghuram Rajan together with Louigi Zingales empirically proved that capital-intensive industrial production is growing more rapidly in countries with already established developed financial markets (Rajan and Zingales 1998).

Experts from the World Economic Forum (WEF) distinguish four directions of growth of productivity in the economy, achieved due to the functioning of developed and stable financial markets. Firstly, the markets allow sharing the risks among group members, which stimulates investment in large capital-intensive projects with a large amount of assumed risks compared to the small and often less attractive projects in terms of their financial results. In other words, the high cost of investment objects is a barrier for attracting potential investors in less-developed financial markets.

The absence of developed institutions, which are able to combine the financial resources of the owners of capital, limits the implementation of major investment projects (Acemoglu and Zilibotti 1997).

The second direction is improvement of the quality of capital allocation mechanism among companies. In stable and developed financial markets, the investor has more information about those investment projects, which have the greatest chance to become highly productive industries in the future, i.e. the basis for economic growth.

Thirdly, large financial institutions have more opportunities to develop long-term relations with borrowers – recipients of financial resources, primarily through monitoring their financial condition, solvency and information on the implementation of financed investment projects. Such control, which is based on one of the basic principles of lending – repayment, stimulates the recipients of funds to use the available financial resources more efficiently, i.e. to increase the efficiency of existing investments (Diamond 1984).

The fourth direction of productivity growth in the economy is the reduction of transaction costs of economic agents in the committing of transactions of commodity exchange through efficient, modern and safe payment system provided by the banking sector.

Thus, the economists representing various economic schools in different time came to the conclusion that there is a close relationship between financial markets development and economic growth. One of the most important criteria for assessing the development of the financial market of an individual country in conditions of globalization is the presence of large, recognized financial centres on its territory.

The study is based on the analysis of the local financial systems development of 22 countries and positions of their financial centres in the international rankings, grouped on a regional basis (USA and Canada, Western Europe, Central and Eastern Europe, Persian Gulf countries, Asia (excluding China and India) and BRICS). Than we have made a comparison of the results reached on the previous steps and tried to explain them via sophisticated research of the financial market development index, provided by the WEF, and analyzing local features of the IFCs' development and the common trends in the world economy.

3 Discussion and Results

3.1 Economic Assessment by the World Economic Forum

One of the most important components when comparing financial centres is the level of financial market development country-based. Such a comprehensive assessment on a regular basis is provided by the WEF in terms of preparation to the global competitiveness index (GCI) of the countries. The composition of the latter include the ranking of world economies by the level of development of financial markets. Financial markets development index has a greater weight for the analyzed countries than for the countries with agrarian economy (see Table 1) and is calculated based on the assessment of the following factors: depth of diversification of financial products and services; price affordability of financial products and services; the possible scale of increasing the capital via the national stock market; the possibility and speed of obtaining the loan when offering a quality business plan to the bank without providing the transaction; the willingness of venture capital funds to finance small "young" companies manufacturing innovative products and bearing increased risks for potential lenders (start-ups); the stability of the banking sector (assessment of the need for recapitalization of banks); ensuring financial market stability by the actions of regulatory authorities; the index of protecting the legal rights of lenders and borrowers.

Table 1. Financial markets development index, 2015

Region	Countries	*GCI ranking*	Rank	Index value	VS leader (New Zealand)	Index dynamics 2010 - 2015
North America	USA	*3*	5	5.45		
	Canada	*13*	4	5.47		
Western Europe	The United Kingdom	*10*	16	4.83		
	Switzerland	*1*	10	5.10		
	Germany	*4*	18	4.71		
Central and Eastern Europe	Czech Republic	*31*	24	4.62		
	Poland	*41*	43	4.26		
	Hungary	*63*	65	3.93		
Persian Gulf	Saudi Arabia	*25*	41	4.32		
	Qatar	*14*	13	5.02		
	UAE	*17*	20	4.70		
Asia	Hong Kong	*7*	3	5.50		
	Singapore	*2*	2	5.57		
	Japan	*6*	19	4.71		
	South Korea	*26*	87	3.60		
	Turkey	*51*	64	3.93		
	Kazakhstan	*42*	91	3.56		
BRICS	China	*28*	54	4.08		
	South Africa	*49*	12	5.03		
	Brazil	*75*	58	3.99		
	India	*55*	53	4.08		
	Russia	*45*	95	3.53		

Source: compiled by the authors, based on the statistical data of: (the World Economic Forum, 2010-2015)

Despite the subjectivity of such an index, which is mainly based on the expert opinion, it reflects the attitude of experts to the state of the financial sector across countries. Table 1 allows us to consider a change in the assessment of the experts participating in the WEF survey in the medium term (the table graphically shows the development index since 2010). As a reference, there are positions of the studied countries in the final ranking of the GCI as of September 2015.

3.2 International Comparison of the Global Financial Centres

The international research and information companies regularly publish studies in the field of competition among the financial centres of the world, which are based on the statistics of such major financial institutions as the World Bank, the International Monetary Fund, the Bank for international settlements, and others.

The most famous of these is the rating of financial centres (the GFCI), developed by an analytical centre of the City of London – by the company Z/Yen Group Ltd.; it is conducted twice a year. The positions of the financial centres are calculated using « factor evaluation » model, based on two types of data: the evaluation of a set of variables on the basis of open sources of information[1] and expertise assessment. For example, in the 19th edition of the ranking of financial centres as of 31 March 2016 the assessment used 102 factors, and this list is constantly improving. The analytical centre has been receiving expert assessment via online questionnaires since 2007. The number of study participants is over three thousand experts, which represent the vast majority of financial centres in the world, regardless of the scale of their activities.

Firstly, we analyze the positions of the leading financial centres of the world, forming the TOP 5 of the mentioned rating as of the 1st quarter of 2016: (Table 2)

Table 2. Dynamics of TOP 5 IFC positions in the rating of the company Z/Yen Group Ltd

Financial centre	Score rating of the financial centres			Ranking (positions) of the financial centres		
	03.2010	03.2016	Δ (2010-16)	03.2010	03.2016	Δ (2010-16)
New York	770	792	+▲22	2	2	—
London	772	800	+▲28	1	1	—
Hong Kong	760	753	-▼7	3	4	-▼1
Singapore	728	755	+▲27	4	3	+▲1
Tokyo	697	728	+▲31	5	5	—

Source: compiled by the authors based on The Global Financial Centres Index data (issues 8, 19)

In 2010 Hong Kong IFC was slightly behind the "global" financial centres recognized by the international community, whereas in the 1st quarter of 2016 lagging of the leading Asian centre from London and new York increased. The leadership of the two leading global financial centres brings us to the obvious conclusion about improving their positions in relation to Asian competitors, especially Hong Kong,

[1] Thus, the "Global digital economy ranking" is used to assess the competitiveness of the telecommunications infrastructure of the financial centre; it is developed by the Economist Intelligence Unit (part of The Economist group). An Ease of Doing Business (World Bank group) is used to assess the institutional business environment.

which has become more pronounced over the last 6 years. It is noteworthy that the ranks of the leading Asian financial centres remains virtually unchanged, while there is the convergence process of their scoring of the indicators.

Table 3 presents the leading financial centres of the world according to the 19 issues of The GFCI, located in North America, Western Europe and Asia, as well as representing the regional centres of Eastern Europe and countries of the Persian Gulf.

Table 3. The positions of the financial centres in the rating of the company Z/Yen Group Ltd

Region	Financial centre	Score rating of the financial centres			Ranking (positions) of the financial centres		
		March 2010	March 2016	Δ (2010-2016)	March 2010	March 2016	Δ (2010-2016)
USA And Canada	New York	770	792	+▲22	2	2	—
	Washington	649	712	+▲63	17	7	+▲10
	San Francisco	654	711	+▲57	14	8	+▲6
	Boston	655	709	+▲54	13	9	+▲4
	Toronto	656	707	+▲51	12	10	+▲2
	Chicago	678	706	+▲28	7	11	-▼4
Western Europe	London	772	800	+▲28	1	1	—
	Zurich	669	714	+▲45	8	6	+▲2
	Geneva	661	694	+▲33	9	15	-▼6
	Frankfurt	659	689	+▲30	11	18	-▼7
Central and Eastern Europe	Warsaw	517	631	+▲114	66	48	+▲18
	Prague	543	622	+▲79	58	57	+▲1
	Budapest	467	600	+▲133	71	74	-▼3
Persian Gulf	Dubai	607	699	+▲92	28	13	+▲15
	Abu Dhabi[a]	*n/a*	675	+▲57	47	26	+▲21
	Doha	592	652	+▲60	34	35	-▼1
	Riyadh	503	606	+▲103	68	70	-▼2
ASIA	Singapore	728	755	+▲27	4	3	+▲1
	Hong Kong	760	753	-▼7	3	4	-▼1
	Tokyo	697	728	+▲31	5	5	—
	Seoul	621	705	+▲84	24	12	+▲12
	Istanbul	496	636	+▲140	69	45	+▲24
	Alma-Ata[b]	*n/a*	597	-▼32	57	77	-▼20
BRICS	Shanghai	693	693	—	6	16	-▼10
	Shenzhen	654	688	+▲34	14	19	-▼5
	Beijing	653	682	+▲29	16	23	-▼7
	Mumbai	550	640	+▲90	56	42	+▲14
	Sao Paulo	573	639	+▲66	43	43	—
	Johannesburg	555	628	+▲73	53	51	+▲2
	Moscow	506	611	+▲105	67	67	—

[a] Index dynamics by March 2012
[b] Index dynamics by March 2014

This sample of countries allows to draw conclusions about regional trends in the area of competition among financial centres.

Source: compiled by the authors based on The Global Financial Centres Index data (issues 8, 19)

3.3 Comparative Analysis of Regional Financial Markets Development

(Table 1) and international financial centres (Table 3) gives grounds to draw some results.

The countries, that occupy high positions in the ranking of financial markets development, operate under the Anglo-Saxon system of law: the USA, Canada, Singapore, Hong Kong, the United Kingdom and Qatar[2] (to a lesser extent) have a high rank not only in the sample above, but among all countries, assessed by the World Economic Forum. Moreover, as protection of the interests of minority shareholders (investors) is a critical element in the global financial market, whose legal system primarily uses British law (Dubinin 2012).

The USA and Canada are characterized by developed financial markets, which is reflected, in particular, in high positions of their financial centres. Among the above components of the financial development index, respondents point out the difficulties in obtaining credit without collateral, which is rather an advantage, enhancing the stability of the state banking system than otherwise. The survey participants also noted the relatively low availability of venture capital in Canada, which is compensated by developed equity market and wide diversification of financial products and services.

The factor of property rights protection has had a negative impact on the position of the United Kingdom in the ranking of the financial markets, which the respondents rated worse than in previous years. Despite the improvement of the stability of the banking system of the country, the index is at average level. As a result, taking into consideration the difficulties in obtaining loans without collateral, the UK is inferior to other studied countries with an Anglo-Saxon system of law judging by the index of financial development, which, however, is not confirmed by the leading positions of the City of London among financial centres of the world.

The leading positions in ranking of financial development are occupied by Hong Kong and Singapore, the second "dozen" is formed by the Persian Gulf countries (Qatar, UAE), Japan and Western Europe.

Russian financial market is evaluated by respondents as a much more modest one compared to other BRICS states, as well as partner in the Customs Union – the Republic of Kazakhstan. Among the BRICS countries, South Africa holds the leading positions, with the highest estimates of almost all components of the financial development index. According to the WEF data as of September 2015, the Russian Federation in the financial development is behind the other countries in almost all components of the index. Special attention should be paid to the low assessment of the stability of the banking system, insurance of the financial markets stability, the high

[2] On the territory of the Qatar financial centre (in Doha) special legal regime is in operation. It is based on the principles of international law and follows the British model (Latham & Watkins).

cost of financial services, as well as the poor performance of the stock market, reflecting the difficulties for economic entities in the capital increase through the issue of securities in the domestic organized market. Despite positive trends in recent years, Russia ranks only 95th in the world, it is behind Kazakhstan in a few positions. In many respects, in our opinion, the low assessment does not take into account the impact of economic sanctions of the US and the EU.

Regarding the positions of international financial centres in the analyzed countries, there are five the US financial centres in the TOP 10 of the GFCI; half of the TOP 12 are representatives of North American continent as well. Let us note that 75% of the financial centres of the TOP 12 are functioning under the Anglo-Saxon law. They show a high absolute increase in the score ranking so they occupy high positions naturally. Moreover, the leadership of North American financial centres is due to outstripping growth in the following parameters collaboratively: for example, the average absolute score growth of competing centres of Western Europe in the given sample was +34 points versus +46 points in the USA and Canada.

The United Arab Emirates (the financial centres in the cities of Dubai and Abu Dhabi) has improved its position in the ranking during the last 6 years and has shown the most significant progress among other competitors, including regional. Let us note that Arab financial centres are rather complementary than direct competitors – up to 2015 growth of their position in the region was synchronous. This feature is also usual for North American financial centres. In addition, the Persian Gulf countries are largely sources of funds and investments due to the accumulated oil rent, whereas the financial centres of the United States do not compete due to the high diversification of provided services. New York accumulates 2/3 of the assets of foreign banks operating in the United States, the largest stock exchanges of the country are also located here; Chicago is the centre of the world options trading.

The largest financial centres of continental Europe are gradually yielding their positions to North American rivals.

Financial centres in the BRICS countries demonstrate multidirectional dynamics. Among the analyzed countries, only Mumbai achieved substantial progress (+14 positions since the year 2010 as of March 2016). Let us note the weakening of the financial centres of mainland China from 2010 to 2016: centres in Beijing and Shenzhen managed to improve their score rating slightly, but their reduced rank suggests faster growth for their competitors in 6 years, foremost financial centres from the countries of the Persian Gulf and Western Europe.

According to the compilers of the financial centres rating, Moscow, judging by the set of parameters, remained at the same level in March 2016 as in 2010, but for the reason mentioned above, a decrease of rank of the capital of Russia was recorded; it occupies the 67th position out of the 88 investigated financial centres.

4 Conclusions

According to the results of the analysis of positions of the financial centres over the period from 2010 to 2016, as well as the financial markets development index of the home countries of researched IFCs', it is possible to state the following conclusions:

- During the study period, the TOP 5 of the financial centres has not changed: among the leading financial centres of the world there are New York, London, Hong Kong, Singapore and Tokyo, with a pronounced superiority of the first two cities as of the 1st quarter of 2016.
- A trend to the dominance of North American centres developed in the global financial market: in the TOP 10 there are 5 IFCs from the United States, in the TOP 12 there are 6 North American cities, providing a wide range of unique financial services.
- The greatest progress in the rankings has been achieved by the financial centres of the United Arab Emirates (Dubai and Abu Dhabi), due to accumulation of windfall profits from the sale of oil.
- The largest IFCs in continental Europe are under increasing pressure from competing American and Asian centres.
- There are different trends Among the BRICS:
  - the greatest progress over the 5 years has been observed in Mumbai (India) – the financial centre managed to significantly improve its position in the overall ranking as of 2016 (42nd place, + 14 positions since 2010) and has made the most progress among the BRICS countries;
  - the financial centres in mainland China failed to gain a foothold in the TOP 10 – Shanghai, Shenzhen and Beijing together with Arab (the UAE) centers form the 2nd and 3rd dozens of the world rankings;
  - Sao Paulo (Brazil) and Moscow (Russia) positions remained stable.
- High positions in the financial markets development ranking, offered by the World Economic Forum, are occupied by the countries with Anglo-Saxon model of law: the USA, Canada, Singapore, Hong Kong, the United Kingdom, South Africa and Qatar. According to the international organization assessment, South Africa significantly stands out among the BRICS countries in terms of the financial market development, however according to the rating of The GFCI as of March 2016, it has had a little impact on the position of the financial centre in the capital of the African state even compared to other partner BRICS countries.

References

Bystryakov A.Y., Rasskazov, D.A.: The international comparison of Russian financial centre against foreign competitors (2014). http://journals.rudn.ru/economics/article/view/12206/11636. Accessed 16 May 2019

Ponomarenko, E.V., Rasskazov, D.A.: The identification of the perspective financial centres based on fundamental theoretical approaches. Econ. Taxes Law **10**(1), 91–100 (2017). (In Russ.)

The World Economic Forum. Appendix: Methodology and Computation of the Global Competitiveness Index 2015–2016, (2016). http://reports.weforum.org/global-competitiveness-report-2015-2016/appendix-methodology-and-computation-of-the-global-competitiveness-index-2015-2016/. Accessed 16 May 2019

Schumpeter, J.A.: The theory of economic development. Harvard University Press, Cambridge (1911)

Goldsmith, R.W.: Financial Structure and Development. Yale University Press, New Haven (1969)

King, R.G., Levine, R.: Finance and growth: schumpeter might be right. Q. J. Econ. **108**(3), 734–735 (1993)

Lucas, R.E.: On the mechanics of economic development. J. Monetary Econ. **22**, 3–42 (1988)

Rajan, R.G., Zingales, L.: Financial dependence and growth. Am. Econ. Rev. **88**, 584 (1998)

Acemoglu, D., Zilibotti, F.: Was prometheus unbound by chance? Risk, diversification, and growth. J. Polit. Econ. **105**(4), 709–751 (1997)

Diamond, D.W.: Financial intermediation and delegated monitoring. Rev. Econ. Stud. **51**(3), 393–414 (1984)

Latham & Watkins: Doing Business in Qatar. https://www.lw.com/upload/pubContent/_pdf/pub2782_1.PDF. Accessed 16 May 2019

Dubinin, S.K.: International financial centres and their importance in the global economy development. Analytical review, p. 9. - Financial University under the Government of the Russian Federation Moscow (2012)

Z/Yen Group Ltd. The Global Financial Centres Index (issues 8 and 19). https://www.longfinance.net/programmes/financial-centre-futures/global-financial-centres-index/gfci-publications. Accessed 16 May 2019

Aytmatov's Ethno-Pedagogy: Collision of Civilizations (By the Example of Chyngyz Aytmatov's Novel "The White Ship")

Kiyalbek Akmatov[1](✉) and Abdykerim Muratov[2]

[1] Osh State University, Osh, Kyrgyzstan
kgzakmatov@gmail.com
[2] Ishenala Arabaev Kyrgyz State University, Bishkek, Kyrgyzstan
bashredaktor@mail.ru

Abstract. Purpose: This paper is devoted to the ethno pedagogical peculiarities of the Kyrgyz people based on the novel "White Ship" written by Chyngyz Aitmatov.

Design/Methodology/Approach: Every nation has its own customs and traditions which are being observed over the centuries, these traditions may have both national colorful and pedagogical character as well. The author highlights the basic principles of educational entities in the bright example of Momun the hero of the novel, at the same time describing the national characteristics of the Kyrgyz people. Additionally, this article describes the confrontation between supporters of justice and purity with the new civilization's representatives who have ceased to pay attention to the real human values in the era of globalization.

Findings: In the XX century there wasn't any other literary work in Central Asia, even in the Soviet literature, which could compete with "The White ship" in providing with moral, dignity, pride and didactic problems of national cognition. The main strength of the work, the core of the building – is Momun. If we don't give first place to Momun the novel will lose its color. Momun is a grandfather. In Kyrgyz tradition grandfathers and grandmothers play a special role in upbringing not only their grandchildren, but in upbringing the next generation as well.

Originality/Value: The pedagogy of Momun is based on the ancient national pedagogy of ancestors existed and kept from Enesai era. There are lots of children like the boy in "The White Ship" carrying the moral-ecological norms.

Keywords: Ak keme · Ala-Too · Boy · Fish · The Kyrgyz people · God

JEL Code: N01

1 Introduction

If we arise a question about the place of "The White ship" in works of Ch. Aytmatov, first of all its main idea was directed to criticize the structure, discovering their "negative sides" during the totalitarian era when only communist party was dominating in public, when social realistic literature was considerably bounding writers and

E. G. Popkova and B. S. Sergi (Eds.): ISC 2019, LNNS 87, pp. 369–378, 2020.
https://doi.org/10.1007/978-3-030-29586-8_43

gaining strength. In fact the strength of the literary work is in demonstrating the influence of moral and ecological problems and about the system which was going down. In order to solve the problem the writer used the most important tool, in other words a "schtick" – the traditional, didactical sources of the Kyrgyz nation and national pedagogy. If we remove this from the literary work it would look like a plane without an engine or wings. Here we see Momun* (one of the main characters in the novel "The White Ship") as a national teacher and mentor. Momun is one of the last representatives of the ancient pedagogy. Momun is a tragic character characterizing the Kyrgyz national pedagogy which flew in the opposite direction to the stream of global civilization.

Probably in the XX century there wasn't any other literary work in Central Asia, even in the Soviet literature, which could compete with "The White ship" in providing with moral, dignity, pride and didactic problems of national cognition. The main strength of the work, the core of the building – is Momun. If we don't give first place to Momun the novel will lose its color. Momun is a grandfather. In Kyrgyz tradition grandfathers and grandmothers play a special role in upbringing not only their grandchildren, but in upbringing the next generation as well. In the epic "Manas" when Manas suffered defeat his son Semetey with his mother were taken by his grandmother Chiyirdy to Semetey's grandfather Temirkan where Semetey was introduced with surrounding world and people. When Semetey's son Seitek armed went to march taking his grandfather's (Manas's) Narkeske and Almabash* (the names of heroic weapons) everybody likened him to his grandfather Manas.

In this particular novel we have the boy and Momun. The boy is a pupil, Momun is a mentor. Momun is a person who carries national didactic traditions and responsible for passing it to the next generation. He feels the responsibility for those national didactic traditions and tries to do his best, in his attempts sometimes he suffers a defeat, sometimes gains the victory. At the end he loses those didactic rules and ancestral codex along with the only representative of next generation to whom he was passing those values; he wasn't able to prevent his only grandson turning from human into fish.

2 Materials and Method

The ethno-pedagogical aspect is studied in the works of such scholars as (Arsaliev 2016), (Osmushina and Ingle 2016), (Schwartz et al. 2019), (Shavadi 2015), (Stukalenko et al. 2013), (Tsallagova 2012), and (Zembylas 2019).

What the pedagogy of Momun is based on? Of course, it is based on the ancient national pedagogy of ancestors existed and kept from Enesai era. How could he find it, who gave it to him? Of course, he took it from the older generation. The older generation in its turn took them from their ancestors and the ancestors used to take it from their forbears, so all this has passed the national approval and considered to be the moral and ecological norms which keep the nation. There are lots of children like the boy in "The White Ship" carrying the moral-ecological norms facing lots of people who are similar to Orozkul; who fight with them by losing or getting victory; but in spite of all these obstacles our national pedagogical rules and moral postulates still exist.

The frame of "The White Ship" consists of boy's two fairy tales. One of them was told by his grandfather, the content, phrases, didactic meaning and every sentence of which have a function to influence the moral world of the boy. Momun as a boy's moral mentor who always says about his grandson's kindness, the boy's learning environment is the Kyrgyz nature. In national pedagogy the Kyrgyz people look at nature as if it was in the same rank with the human being and they taught their children to look at nature from the same point of view. The boy gives different names to stones ((«ээрташ», «төөташ», «кашабаң», «таңкеташ»)) – this is not only his learning the nature, but also his attempt to form the association about natural phenomena with human being and other things. In his own world these stones are depicted as "bad one", "good one", "cunning", "intricate", and "narrow-minded one", thus these stones from ordinary stone turn into living organisms. They can be seen only by the Kyrgyz boy, especially by Momun's grandchild. And the surrounding plants he defines according to their specific peculiarities, living conditions, and perceptions by human beings; he hates some of them and admires others by fondling them. He imagines the clouds in the sky as a lion sometimes as a swan. He can compare in this way. He knows lots of types of grasses, their abilities, and their characters; he feels sorry for the reed bending because of the wind.

Therefore, the writer decides to start with introduction one of the country children, with his romantic world and childish range of vision. In the child's carefree description the author mentions like this: "His grandmother says that he is a step grandchild. Whatever you do for them, step children will never appreciate or justify your labor. That is why if you don't help parents they ask "Are you a stepson?" What if he doesn't want to be a stepchild? Why should he be a stepchild? Perhaps not the boy, but the grandmother can be step grandmother" - adds the author describing the unconsciousness of the boy's childish world. Then describing unconscious behavior of the boy the author gives the seller's words who said that Momun didn't have a son, maybe he is the son of Momun's daughter. Here the world of the boy tumbles down along with the reader's heart too. Character's family is lack of someone. Does it mean that he is short of upbringing? Who should bring this step child up?

In the mind of the reader there appear two things about the boy as if he is in two shores. As a reader, we enter his amazing romantic childish world, play with his stones ((«ээрташ», «төөташ», «кашабаң», «таңкеташ»)), take his briefcase and wind over our heads, speak with plants, clouds, we also want to look at the White Ship through his binoculars, also want to believe that his father is a sailor on that White Ship indeed, and finally we also want to turn into fish. We love the boy, we integrate with him and we hate together with him and love those things that he loves, which mean that we are also ready to turn into the main character of the novel, into this boy. The author guides us, turns us into a six-year-old child and with his unimaginable strength brings us to the highest culmination – to the death of a boy. The boy dies. The tragedy as an aesthetic category performed its function, readers cried, felt pity for the protagonist character of the novel. The author tries to answer the question "What pushes person to the tragedy?" by using L.N. Tolstoy's words: "Conscience, pride, and moral are considered to be the truth which person is able to feel beforehand"; then he continues with the phrase that "tragedy is the highest form of art" (Korkin 1976).

With the collapse of romantic and fairy world of the boy, character's life reaches to the end and he dies in proscenium. This is a tragedy. And this tragedy gives tools to readers to fight with cruelty, barbarism, the abuse of natural resources, amorality; if to say with the words of aesthetics, it creates catharsis. Did he commit a suicide, because of greenness or he hasn't yet fell in love with life? No, on the contrary it was his fight against amorality, his cup of suffering was already overfilled. Where did he take such strength to fight against shameless people till the last heartbeats? – Yes, he suffered a defeat and his soul passed away, but in his inner world he found the purity and went to the water, to the "The White Ship" in the water. Of course, he just went to school and in his upbringing we can't see school's influence, so he took this strength from Momun. Momun with his current state is not an Honored Pedagogic of Republic, not an innovator-teacher; but he is a missioner of national didactic traditions; his credo - national didactic traditions. Therefore, emotional strength of the boy is based on the national didactic pedagogy of the Kyrgyz people.

3 Results

Momun is a real national hero; his behavior comes out from the Kyrgyz traditions. So here is a list of several points for Momun's character:

1. Being obedient and nimble.
2. Always greeting everyone, he meets on his way, no matter they are elder or younger than him. According to the national traditions the greeting is an indicator of a person's spiritual maturity and purity.
3. Serving for the tribe Bugu* (one of the tribes in Kyrgyz nation) and all its people.
4. If an old man passes away, serving in his ceremonies is Momun's duty.
5. Treating everyone from the Bugu tribe as his relatives.
6. Accusing himself in his daughter's infertility.
7. Begging a child for his daughter Bekey from Muiuzduu Bugu Ene* (Horned Mother Deer).
8. In the case of not being invited to some ceremonies, bothering himself a lot, worrying about people who do not follow the ancestral traditions.
9. Being worried when people ignore the ancestral and tribal traditions.
10. Believing into fairy world.
11. Telling the fairy tales and instructing others to tell the tales.
12. Where ever he goes always taking the boy with himself

Let's compare Momun with other old men of our contemporary world. They wear white coat, with Kyrgyz national patterns riding a white horse, white-bearded, everyone on their way greet them. But Momun is totally different from them. For his simplicity and naivety everyone humbles, humiliates and belittles him. But in order to be a mentor for the next generation you do not need to ride on an expensive horse and wear especially designed costume, there must be a core in a person.

In the course of a plot we see one episode when Momun meets with a seller and greets him: "Assaloom aleykym, tradesman! Is your business all right, are you succeeding?" This is the greeting of the Kyrgyz man an honorable old man. When people usually meet firstly they ask about their health, feelings and affairs; due to the ethical tradition Momun is asking about his business and success. He knows well that seven people in that little village including himself and his little grandchild will not buy anything worth. He knows perfectly well how his business is, but anyway to ask about his businesses is his personal duty in front of the ancestral rules of tradition. The seller is also being aware of these traditional rules replies Momun: "Everything is alright, business is good", trying to hide his anger. Before leaving Momun invites the seller to have a cup of tea at their home. The greeting of two adult people and invitation culture may seem to some readers as a simple routine but it is the most needed and valuable material for the upbringing of the little and very smart boy. The author shows these things in order to clean the boy's soul and leave the national malt into his cognition.

Momun makes attempts in the process of a boy's upbringing as if he were a national teacher, national botanist, biologist, ornithologist, in general as if he were an ethnologist. He acts according to the Kyrgyz traditions, in order to influence the boy's learning. In case if someone passes away, grandfather always takes grandchild with himself to the funeral. Grandfather tells him fairy tales about Tengir-Too, ChypalakBala* (very small, tiny hero of Kyrgyz national fairy tales) and Muiuzduu Bugu Ene (Horned-Mother-Deer), didactic functions of which are very strong. He takes into consideration boy's developmental-psychological peculiarities and tells only adapted versions of those legends and fairy-tales. In the process of upbringing he underlines the information about Tengir-Too, that "we, Kyrgyz people are very strange, omitting the Kyrgyz name for the mountain Tengir-Too, we call and register them by Chinese name Tien-Shan. Oh, my son, the poem about mountains was written not in vain: «Ala-Too and Tenir-Too are mountains protecting my nation from every kind of enemies". Saying these words to his grandchild he becomes upset with Kyrgyz people that they do not use the Kyrgyz version of those place names; he is distressed because they are registered in world map in foreign language, not in Kyrgyz. At the same time he recites the poem about our mountains, about their role in Kyrgyz people's life, and how they protect us from various enemies.

In the Kyrgyz culture the number seven is a sacred number; in cordon there are seven people – Momun with his wife, Orozkul with his wife, Seidakmat with his wife, everybody has a pair. Only the boy is alone. The boy for his grandmother is an alone, abandoned child, incapable, strange, mischievous boy, and even cursed. And for Orozkul he is a child abandoned by his parents, while Orozkul can't have a child; for Seidakmat he is a child becoming like his grandfather, old and weird. And for his own parents he does not exist. Neither Guljamal, nor aunt Bekei cares him. In that cursed cordon among those six people only his grandfather takes care of him. If he weren't there, there wouldn't be even a story about the boy. The boy glanced at his grandfather cheerfully. His glance looked like as if he was going to cry.

The boy felt the chapped hand stroking his head, at once he got upset. He was trying hard not to cry, feeling the smell of fresh-cut grass and sweat from his tiny and kind grandfather; this skinny, raw-boned old man was the only person who cared and loved him; the only person who was able to protect and hug him; when he was in a trouble the only person who kissed his eyes wiping his tears. In spite of the seller's rebuke he bought a briefcase and now he is stroking his head with his chapped hand. Now again this very person is hugging him tightly. Cunning people like that seller often humiliate over him even they gave him a nickname "obedient". He loves very much his small skinny granddad….. Ok, let them humiliate! Anyway, he is his own granddad. It is good to have a granddad!"

Warm relationship between grandfather and grandchild – is the vein of the novel. Other relationships gather around them. Look at the warmth, cordiality and affection between grandfather and grandchild in the upcoming sentences. This little boy does not know his mother, father, even any relatives and his small skinny granddad is the only person he can rely on. He is from Bugu tribe. And the boy is a nephew of the tribe Bugu; he does not care if his father (genes) belongs to Bugu tribe or not. But because of grandfather's didactic lectures, chronicles he feels himself as a successor of Bugu tribe.

Due to Kyrgyz culture people never abandoned orphans, even if he was an orphan who had neither mother nor father. According to the traditions there was a taboo not to call the child "orphan"; these children were very helpful in enemy invasions and in pastoral work as well. Usually such children grew very brave, strong, close to people and easy to get on with majority. In the families and household of national rulers, famous governors, wealthy people there were many orphan children where they were treated as their own and married with full honors. This boy in "The White Ship" is also one of them and his upbringing and cognition is closely connected with Momun. Momun is the only person, who treats with love, and feeds with spiritual provisions and shows fair way of living. Bekei aunt also wants to treat him with sister love, but she herself had lots of problems with her husband Orozkul, that is why she can't influence his upbringing.

In this beautiful environment with lots of mountains and woods around and light-blue sky the boy is alone. He feels like an orphan left absolutely alone with no one to take care of him. Look at the contradictions: on one side huge mountains, on the other side heartless, non-responsive mountains; on one side beautiful forest, on the other side dark, dense forest; on one side the living stones with particular names, on the other side taciturn, unsociable stones; on one side moving six people, on the other side selfish six souls. Among these things there is a boy, there is his childish world. The boy is keen on remote places, wants to see his parents, wishes to go to his father, whom he saw on the White Ship with the help of binocular on Karool-Dobo.

The concept of the city turned out to be destroyed for Momun, new concepts, such as urbanization and civilization broke into parts the national traditions. People are drifting more and more apart from the nature, when they are isolated from nature they are not human beings any more, they are just robots. Momun' daughter, the boy's mother lives in a big city, she got married again and now they have a daughter who is given to kindergarten and once a week they spend time together. It means that the relationship between people and provisions that we used to pass to each other from centuries has been disappeared. It brought townspeople to urbanity and courtesy crisis. But Momun is not one of them yet, he will not rest if he doesn't see the boy till the noon. If to tell about his daughter: "she lives in the big house, in a tiny room. In the big house live lots of families, when they go out they don't know each other, as if they were in bazaar. They never greet each other, enter their own rooms with handbags in their hands and quickly lock the doors. As if they were imprisoned and they continue living this way. The son-in-law is a driver, said grandfather; he has a minibus, drives people from one part of the city to another part. He leaves home early in the morning and returns only late in the evening. The job of his poor son-in-law seemed to be very hard for Momun. My daughter shed tears, apologized accusing in everything her harsh fate. They got in line for a new house, but they have no idea, when they get this house. She said that when they get a house, if her husband does not mind, she will take her son back. She asked to take care of the boy for a while. Don't worry, says grandfather, live with your husband in harmony, leave the rest to God. Don't worry about your son. While I am alive, I will not give him to anyone. If the God bless, he will find himself a shelter, as in anyway, he is a living being… - says grandfather».

In this very meeting we see how patriarchal-tribal age and civilized social age face with each other. On one side reserve wild wood, on the other side overcrowded city; on one side only seven people and three houses and uninhabited place; on the other side lots of people living in one house; on one side seven people, knowing each other's backgrounds by heart, on the other side urban community ignoring even to greet each other and get acquainted with each other; on one side three big houses, which do not know what is lock, on the other side urban community locking themselves as if they were imprisoned people; on one side Seidakmat and Orozkul, who do not know what to do from idleness, on the other side urban community, leaving house early in the morning and returning only in the evening; on one side Orozkul, dreaming about urban life, on the other side Momun avoiding the idea to live in the city; on one side Momun and the boy considering nature as sacred, on the other side Orozkul and Kokotai considering nature only as consumable. But in both parts of the setting the situation is the same – destroyed families, destinies, which failed their families. That is why the boy is practicing a difficult fate, he became despised, that everybody despises him in every possible chance. Being not able to say something confidently Momun used different phrases like "he said", "she said" while his daughter, shedding tear, said that she was going to take the boy back, whenever they get the new house. They have no idea, when they get a house; if they get the new house, if her new husband does not mind, then …probably she will take her son back. If to analyze, we can understand that the mother, who gave birth to a child does not want to take care of her son, she doesn't love him, even doesn't want to be close to him. She doesn't experience a maternal love towards her son. That is why grandfather says: "Don't worry about your son. While I am alive, I will not give him to anyone. If God bless, he will survive, as in anyway, he is a living being." But this despised boy was given a hand of help not by city, but by mountains, not by urban community, but by country folk, not by civilized people, but by the people from patriarchal sphere, not by future, but by past, and finally not by educated people, but by uneducated people. The writer depicts the contradiction between city and country, and between future and past. Describing all of these things, he accurately shows where the boy – the main character, adopted child was taking the lessons of moral, ethics, purity, conscience and honesty. This is the reason, why Momun is taking the responsibility to bring the boy up, to make a "man" instead of his own daughter. "Obedient" Momun, people's Momun decided to be responsible for the destiny of the boy. Momun realized that his mountains, stones, and forests would be more influential than his daughter's "tiny" room in the house like "a prison". He realized and at once he became sure that his daughter was not going to take her child back, he didn't even want him to be with her, because she didn't deserve him. He assumed the responsibility for this child's upbringing – the hardest, the most chargeable, and the most meritorious job to his own shoulder. Momun has an intuition in upbringing of a child, he has his own checked rules, although they were not experimented in scientific sphere, they were checked and experimented in national experience. In the following paragraph you can read his codes used in teaching the boy to become a good, educated man:

1. New neighbors just moved into a new house must be visited. This tradition is a special one in our tradition, exactly with this tradition people get acquainted with each other, are in relationship with each other, are friendly with each other, they sit and eat together, and good wishes are told on dastorkon* (a large cloth laid on the ground around which the guests sit – with your feet to your side or away from the dastorkon). Amantur Akmataliev mentions this about pedagogical tradition: "This is a very good tradition of Kyrgyz people's culture. New neighbors are visited by local citizens, as a gift they take with themselves various types of meal (boiled and smoked meat, cream, butter, boorsoks, kymyz and other beverages), sometimes they can take some useful things in house works and some carpets, rugs and some utensils; and see their new house. The master of the is also very glad to receive the guests, and even ready to kill a sheep for the sake of visitors; because of this tradition there are some phrases frequently used in Kyrgyz community: «Жанаша көчүп барсам, беш токоч өрүлүгү болбоду» (*In spite of being a new close neighbor he was not able to invite*), «Эшик-төрүн көрө элекпиз» (*We did not see his household*), «Кошуна-колоңго жок немелер» (*They are very surly to neighbors*), «Бир чөгүн чайын бизсиз ичпейт» (*They are so hospitable and ready to treat with a cup of tea at any time*) (Akmataliev 1993). Visiting new neighbors was the tradition of Kyrgyz people from the very ancient times and Momun added it to his own way of upbringing a child and introduces the child with this tradition of Kyrgyz people.
2. If you are a guest in one's house, you should shake hands with everyone in that house; it is good for the boy to do so, too.
3. According to his grandfather's words, younger person should offer his hand first, if not he will be considered as a rude young man.
4. There is a saying in our culture, that one of seven is sacred. This notion teaches people to respect everyone. Kyrgyz people consider the number seven as sacred and say: "One of those seven might turn out to be sacred, that is why try to have good attitude towards people, try to be contrite, and communicate with people in a very good manner". S.M. Abramzon writes in his work about our ethnography: "… along with Islam belief in Hizir-Ata* (sacred person who brings happiness, wealth and many other good things) was spread to Kyrgyz nation, he is an invisible, sacred, respected by people phenomenon that always brings happiness, wealth, and other good things" (Abramzon 1999).
5. "Every time whenever you shake hands with people shake hands touching his thumb, because he may turn out to be Hizir"- says grandfather to his grandchild.
6. Hizir is able to fulfill all your wishes; he does not have bones on his thumb. He himself does not know that he is Hizir; we call him Hizir, as he behaves as an ordinary man and very sincere. Only bad people know who they are. Old man Momun explains these things to his grandchild.

7. When Momun goes to their new neighbors' house to pay a visit, he greets them as follows: "Welcome to the land of our ancestors, located for ages by our forebears! Is everything all right? How is your household? How are your children?" The boy is present, while he is performing this greeting. Kyrgyz people often say "Welcome to the land of our ancestors"- greeting in this particular way the new neighbors, they teach children to be a patriot of their own country and be responsible for the protection of his country's territory's every inch. They wished happiness to newly settled families. For Kyrgyz people ranching with cattle, it is very important to have a good neighbor, wide land to breed cattle that is why usually their greetings begin with these things. To teach these things to the boy is Momun's obligation and duty.

4 Conclusion

Momun integrates the boy with other shepherds' children, wants him to learn Kyrgyz national games. It was his grandfather who taught him how to play children's games. He even does not spare his time teaching how to play games. Actually Kyrgyz pedagogy which up brings children with the help of games was stated from diverse angles in lots of dissertations and works of scientists. The boy, the main character in Chyngyz Aytmatov's work plays calling out war-whoop "Long live, Kyrgyz! Hurrah Kyrgyz! Kyrgyz! Hurrah! Manas! Manas! Manas!," because he was told that in ancient time our ancestors used to play like this. By acting this way he is not just keeping ancestors' traditions, but at the same time helps to teach them others too.

References

Abramzon, S.M.: Kyrgyz jana Kyrgyzstan taryhy boyuncha tandalma emgekter (Selected works about Kyrgyz people and the History of Kyrgyz Republic). Kyrgyzstan-Soros, Bishkek (Kyrgyzstan) (1999)

Akmataliev, A.: Baba salty, ene edebi (Father's tradition, mother's manner). Balasagyn, Bishkek (Kyrgyzstan) (1993)

Arsaliev, S.: Ethnopedagogical technologies: best approaches and practices. Recent Patents Comput. Sci. **9**(2), 173–184 (2016)

Korkin, V.: Legendanyn janyryshy (interview of Chyngyz Aytmatov with the reporter of "Literaturnaya gazeta" V.Korkin) Kyrgyzstan madaniyaty (Culture of Kyrgyzstan), vol. 23, pp. 8–9 (1976)

Osmushina, A.A., Ingle, O.P.: Ethnopedagogical value of the comical of the Mordovian people. Integr. Educ. **20**(3), 415–421 (2016)

Schwartz, M.S., Hinesley, V., Chang, Z., Dubinsky, J.M.: Neuroscience knowledge enriches pedagogical choices. Teach. Teach. Educ. **83**, 87–98 (2019)

Shavadi, A.: New information technologies in ethnopedagogical process. In: 9th International Conference on Application of Information and Communication Technologies, AICT 2015, Proceedings, 7338630, pp. 595–659 (2015)

Stukalenko, N.M., Murzina, S.A., Navy, L.N., Moldabekova, S.K., Raimbekova, A.D.: Research of ethnopedagogical approach in professional training of teachers. Life Sci. J. **10**(11), 205–207 (2013)

Tsallagova, Z.B.: Ethnopedagogical potentialities of north caucasian proverbial instructions. Etnograficeskoe Obozrenie **2**, 153–166 (2012)

Zembylas, M.: Shame at being human as a transformative political concept and praxis: pedagogical possibilities. Feminism Psychol. **29**(2), 303–321 (2019)

Informatization and State Administration: Possible Scenarios and Consequences

Ivan V. Petrin[1(✉)], Irina V. Pogodina[2], Galina V. Stankevich[3], and Sabir N.-ogli Mamedov[2]

[1] PLC "Technopark", "Skolkovo", Moscow, Russia
ivpetrin@gmail.com
[2] Vladimir State University named by A.G. and N.G. Stoletovs, Vladimir, Russia
irinapogodina@mail.ru
[3] Pyatigorsk State University, Pyatigorsk, Russia

Abstract. Introduction: the area of information technology applying is spreading in all the main types of the state activity. The international experience shows that the computerization phenomena leads irreversibly to a number of political, cultural and economic changes which form a new type of the society. The object of this work is the analysis of the informatization influence on the completing of the state functions and elaboration of the possible scenarios of the general state administration digitizing.

Methodology: As a methodological base the general dialectic method of cognition has been chosen. In this work some methods are used: the private – scientific, the historical, the formal – logical, the systematically – structured, the formal – legal.

Moreover, in the given research the interdisciplinary method of approach that made it possible to examine the computerizing of the state services both from the point of view of judicial sciences and the other spheres of knowledge is applied.

Results: The state now is not conceived without informational technology that has been reflected including the Strategy of the informational society development in Russian Federation. The informational – legal regulation has the great importance in making appropriate pace and quality of the development and informational society operation. However, there are some problems connecting with the fast development the informational society. The creating this or that informational system cannot be an end in itself. The informational system is first of all an instrument of the most effective process organization, and it should be considered in this very context. If as a result of making informational system the process is not more effective (and not only in limits of the organ of the power which makes the system but also in other levels of the state and municipal administration) its creation is not reasonable.

Conclusion: In the article the conclusion about the necessity of the elaboration of the politic – legal strategy for opposition to the challenges connecting with the information and society transformation is made. Experience has shown that one of the main problems of the making informational society is becoming the mutual coupling other systems of the different appliance and the processes in general.

E. G. Popkova and B. S. Sergi (Eds.): ISC 2019, LNNS 87, pp. 379–386, 2020.
https://doi.org/10.1007/978-3-030-29586-8_44

Keywords: Aggregator · Digital technology · Informatization · Informational society · State services · State functions

JEL Code: O380 · K230 · K24

1 Introduction

Now the general development and the usage of digital technology is one of the main tendencies. First of all, the given technologies are an instrument, which aims to improve the effectiveness of any process, to raise labour productivity at enterprises, and to make the human life easier. There are some spheres of influence of informational technology. Firstly, it is a public sector, secondly, it is a actual sector of economy and thirdly – a population.

In literature the problems of the informational society development are discussed. For example, one of them is a deficit of competitive specialists in informational technology (IT) (Abdulgalimov 2014). The service of informational system and technologies in public sector is one of the present-day problems of the IT - sphere. To avoid the negative influence of the Information and Communication Technology on economy it is necessary to work on fundamental changes in actual computerization model which is used by public authorities and quazi-public sector.

The state computerization covers a mass transformation of the society information and needs an analysis of the practice in the sphere of state and public government and necessity to identify the factors to realize it effectively (Kalinina et al. 2017).

The process of making electronic government, computerization of public administration requires the single conception of the building of all informational systems. The scientists of soviet school said about it while proposing to design and implement the single government-wide automated system in Soviet Union.

In 60–70 of XX century the project of creating of the government-wide automated system was being discussed, the idea was proposed by the academician Glushkov V.M. to the Government of the USSR. For its implementation the period of 15–20 yeas was required with the financial costs of 20 billiard roubles estimated. Glushkov's intension would be that the government-wide automated system should help to build the most effective economy in the world (Glushkov 1972).

The ideas of Glushkov and his colleagues are rather topical and becoming popular abroad. In our view it has something to do with the increasing importance of the state. The state as a regulator is getting a kind of "mega regulator" indeed which from the one hand has a number of functions typical for aggregator, from the other hand remains both a customer of some products as well as services and the regulator, the originator of the legal and regularity of different areas of life. The centralization of economic life is seemed to be a process as inevitable as a recovery of the state planning functions this or that way.

It was Glushkov V.M. who introduced the term "information barrier". He underlined the first and the second information barriers. The first barrier appeared when the small group (a clan, a family, a tribe) was managed by only one person. After a while the complicity of management surpassed a possibility of a single person. In this regard, from the one hand the management hierarchy (assistants, deputies, etc.) became to appear, and from the other hand there was an introduction of the single rules of behavior governed by customs, religion, and further by the laws (Glushkov 1972).

The second barrier is a barrier which has something to do with a great amount of information to be processed, the amount of problems to be solved surpasses the possibility of all the people together. We can speak about the confrontation with the second informational barrier since 30s of XX century when neither the management hierarchy nor commodity and money relations do work.

The reason of the management crisis can be a lack to poses all the problems of the management even by a plenty of people. The research of the Institute of Cybernetics Academy of Sciences of UkSSR demonstrated that the complexity of the tasks of economics administering is growing faster than amount of the people involved. And if to govern the country by the former methods basing on the priority of the principle of controlling and processing of planning register information, at late 70s it should have employed almost all the able-bodied population of the country just in the sphere of management of the material production (Petrin et al. 2017).

According to this research Glushkov V.M. make a conclusion that "computer science is the modern invention which lets to step over the second barrier. When the public automated system of the management appears, it will be easy to take a single look on the whole economics" (Glushkov 1972).

The technological problem is should be supposed to become one of significant problems of the development of socialism in USSR. Technologically the state was not able to regulate properly all the aspects of life.

The present-day states, Russia included, have considerably greater technological opportunities. The general amount of informational systems used by different bodies of power and bodies of the local self-management all over the country is now in the thousands and is constantly increasing. Meanwhile the problem of linking of the informational systems designed by different departments at different power levels, at different time is one of the main problems of informatization.

Absence of the single conception of the public informational system designing may lead to many problems which are not only exist but will intensify with the growth of thousands of nowadays informational systems.

The object of this research is an analysis of the consequence of the global informatization of the public administration.

2 Literature Review

An important contribution into the elaboration of the problems of the informational – legal regulation of servicing in e-form was made by the researchers of different spheres of knowledge, among them are: D.L. Abramovich, M.M. Blagoveschenskaya, I.A. Vasilenko, A.A. Vasiliev, L.A. Zlobin, P.T. Mukhaev, G.G. Pocheptsov, G.I. Savin, A.V., V.A. Sadovnichiy, A.A. Tedeev, M.A. Fedotov, B.D. Elkin, M.B. Yakushev, etc.

The work in determining the role of the public administration when servicing in e-form of the following foreign authors are of great importance: B. Gates, M. Castels, I. Loge, F. Machlup, A. Mol, A.D Ursul, K. Shennon, as well as some Russian scientists: T.F. Beresova, A.P. Verevchenko, B.V. Kristalniy, K.N. Matveeva, M.V. Rats, G.L. Smolyan, V.E. Chirkin etc.

Commending the work of the authors mentioned above it should recognize the informational – legal regulation of servicing in e-form in legal sciences is not given due attention and the problems connecting with its practical realization did not found their proper solution.

The consequences of informatization are regarded as positive and rather useful for the public and society in general and negative scenarios are not considered at all. However, some researchers indicate on negative consequences of the informatization of the public administration. So, Gorokhov V.G. mentions that if there was a hope in the informational society with a help of the computer revolution to become more informative than before, to get to know everything that is going on in the world, in the culture, in the science and in the technology faster and fuller, now this hope has fallen to the ground under the pressure of excessive and false information (Gorokhov 2007). A curious position of Lukashov V.N. who notes that the effect of informational technology implementing into the state and municipal administration may be unexpected (Lukashov 2015). These different points of view are seemed to be investigated thoroughly.

3 Materials and Methods

The relations connected with the introducing of the informational technologies into the sphere of the state administration, digitation of the state, so called electronic state is the object of the research. The methods of research are chosen according to the object of research.

As a methodological base, the general dialectic method of cognition is chosen, as well as general scientific methods such as analysis and synthesis. The particular scientific methods are used in this work, they are: the historical, the formal-logical, the system-structural, the formal-legal methods. The dialectic method enabled consider regulatory framework of the servicing in e-form from the point of view of different approaches to its development, of the advantages and disadvantages of the given methods, of the influence of external and internal factors on the appropriate processes.

The formal-logical method enabled to characterize particular normative content of regulatory framework of the servicing in e-form. The system-structural methods let consider regulatory framework of the servicing in e-form as an integrated system, detect its elements and objective linkage between them.

The formal-legal methods, the relatively-legal, the relatively- historical methods let consider the evolution of approaches to the legal-informational regulation of the system of electronic servicing, identify its shortcomings and justify the request about its development. And furthermore, in the present research interdisciplinary approach which gave a possibility to consider the legal-informational regulation of the system of electronic servicing both from the point of view of different legal sciences and other areas of humanitarian and technical cognition is used. The issues of building informational society are in a junction of a range of sciences such as jurisprudence, informatics, philosophy, sociology, economics and others, so it is impossible consider this process just from the technological point of view.

4 Results

We suppose that in time the problem mentioned above connected with fragmentation of informational systems will be solved this or that way, and generally the states have technological possibility to overcome the second informational barrier within the meaning which was intended by scientists of the soviet school bearing in mind the structures (and private companies) with similar functions of the state planning existing in the USSR. Essentially the sectoral aggregates have their separate function of a such global aggregate as should have been a state planning at its time. In this meaning the informatization within the country becomes more fragmented, and to the public informational systems the private ones are added (at least aggregates at present time). But we consider that over time these problems will be overcome and first of all owing to the return to the ideas of the scientists of the soviet school who told about the necessity of ontological and conceptual unity of the informatization within the country in general.

Moreover, the proper regulation should lead to a sharp state cost-cutting on informatization though mechanisms of public – private partnership.

The state has a monopoly on a number of services and functions, and in this case in relation to the commercial organization which are ready to invest in the development of any industry (for example, to build a motorway) the state carries out a permissive function. Nevertheless, a commercial organization gets a benefit from borrowing a payment for faster and more comfortable travel during a certain usually not very long time. The investor gets their money back, gets benefit and the state at the end gets a new motorway.

This model seems to be convenient from the conceptual point of view and within the informatization limits. Meanwhile, the area of informatization itself, the key characteristics of all the processes and general control (even within the specialized task limits) certainly should remain in the state. The issue about a public-private partnership in the sphere of informatization is certain to be an important and requires a separate study and legal regulation. At present time a public–private partnership (PPP) in the sphere of informational technology as a rule are made on a base of outsourcing. We believe the pertinent movement in PPP to be in a bit another way: the state and a company make a contract under which the company creates aggregator and uses it (gets benefit) for a certain period of time, generally short. Then the informational resource goes to the state which has a monopoly for making and using such kind of aggregators. We suppose reference should be to monopoly exactly on aggregators influencing on distribution of human and material resources in automated way and as a rule with using technologies of artificial intelligence. The functions made by such aggregators should be considered as the public ones, we suppose (in real time or potentially).

So, we come to conclusion that the second informational barrier not only can be overcome but it can also be done with a great budget cost reduction through the monopolization of separate areas of government (in essence, it means revival the national plan functions), as well as the public – private partnership as a model mentioned above.

At the same time the overcoming the second barrier itself can result negative and unpredictable consequences.

The power of the Artificial Intelligence over the most of the people as things stood, is beyond the dispute. However, the task of the state is not to give too much power to the Artificial Intelligence.

The owner of the informational systems as at current situation takes a profit alone trying to use human and material resource as effectively as possible (as a fact just effectively). The aggregators of the taxi could be an example. The people get less benefit and under the pressure the Artificial Intelligence have to work for less and less money within stricter rules (including the rating for traveling).

To avoid unlimited profiting by the owner of the informational systems of this kind at the expense of ordinary citizens it is necessary to adopt (including international level) a range of the regulatory and legal acts which must reflect an integrated concept: the only possible goal-setting for the design of the aggregators carrying out a regulation of the human and material flows should be a liberation rather than enslavement of the people.

Moreover, we are sure that it would be in the interest of the owners of the informational systems. The reason for this is that the focus on "effectiveness" is able to go so far that the most of the people must work, work, and work for the lowest possible wage. In doing this there are two kinds of conflict possible – both with the people working for low wage under the condition of the concentration of the new peculiar "means of the manufacturing" which are the aggregators (the classic pre-revolutionary situation) and with…. Artificial Intelligence itself.

And if to the conflicts of the first type taking in account the development of the modern means it is possible to resist rather effectively, the conflicts of the second type, to our mind, may become a trigger of the irreversible effects for all mankind. It can happen when the system from the very beginning oriented on the efficiency considers ineffective the distribution of resources to its owner.

We believe in necessity of the broad discussion about the goal-setting of the manufacturing of the informational systems influencing on the distributing of the human and material resources in an automated manner. All the people should benefit from such systems. In other words, we come to conclusion about the necessity of the state monopoly on designing aggregators influencing on the distribution of the human and material resources in an automated manner development of appropriate systems in a strict ontological and conceptual unity within the whole state, as well as a collaboration in this direction with the private companies within PPP mechanisms and consistent redistribution arising profit to every citizen.

At present the development of the informational systems follows the path connected from the one hand with the obvious problems of the fragmentation of all the systems existed (both state and private), and from the other hand – with a clear focus on an effectiveness not always connected with individual financial freedom of an individual citizen. On the contrary the real salary of a person connected with carrying out the same functions can be reducing even up to zero due to the reducing the appropriate post.

5 Conclusion

All the mentioned above is happening in the conditions of that the regularity framework from our point of view does not always keep up with pace in which the informational technology develop, that in its turn requires significant rethink and forming the limits of their unchecked development. These limits in our opinion must be generated within the set of sciences but first of all, within three of them - philosophy, jurisprudence and informatics.

As a priority we see the following tasks:

- Providing of ontological and conceptual unity within the whole state;
- Expansion of the state function and the state influence on all spheres of economic life;
- Consideration of the matter of the rebuilding of the state (planning institutional and functional);
- Study of the issue of the regulatory and legal status of aggregators influencing on the distributing of the human and material resources;
- Study of the mechanisms of PPP within the informatization of the state functions and designing of the aggregators influencing on the distributing of the human and material resources;
- Studying the measures to minimize the risks related to uncontrolled development of the informational technologies and particularly uncontrolled development of the systems with artificial intelligence.

Acknowledgments. The research is made within the scientific and private interests of authors. The authors express gratitude to Dr. of Law, Prof. Golovkin R.B. (Vladimir), Dr. of Law, Prof. Borisova I.D. (Vladimir) for assistance in preparing this article, as well as the anonymous authors.

References

A strategy of the development of the informational society in Russian Federation (adp. By President of RF 07.02.2008 N act-212)

Abdulgalimov, G.L.: Progress of information society in Russia and deficit of staff potential. Life Sci. J. **11**(8), 494–496 (2014)

Glushkov, V.M., Nikonorov, S.P., Chetverikov, V.N.: Ideologies of the domestic developments in the automated control systems area. Applied informatics, 1 edn., p. 120 (2009)

Glushkov, V.M.: Introduction into Industrial Control System (ICS). Technics, Kiev (1972)

Gorokhov, V.G.: The science and technology policy in the society of not-knowing. The issues of the philosophy, no. 12, p. 66 (2007)

Kalinina, A., Borisova, A., Barakova, A.: Development efficiency analysis of public administration informatization. Contributions to Economics, pp. 481–493 (2017). ISBN: 9783319454610

Zhang, L., Wei, Y., Wang, H.: Government informatization: a case study. Special Issue Syst. Sci. Enterp. Integr. Technol. Econ. Theory Mater. Flow **26**(2), 169–190 (2009). https://onlinelibrary.wiley.com/doi/abs/10.1002/sres.958. Accessed 28 May 2018. European Public Administration and Informatization: A Comparative Research

Lukashov, N.V.: Institutional paradoxes of the informatization of the public and municipal control in present-day Russia. Actual Probl. Econ. Law **2**(34), 83–91 (2015)

Petrin, I.V., Pogodina, I.V., Belokonev, S.U.: Retrospective of legal regulation of the electronic services providing by the power bodies in. State Power Local Govern. **5**, 3–7 (2017)

Putitseva, N.P., Nalivko, K.V., Lekova, A.E.: Informational technologies of the decision making support. Probl. Sci. **1**(19), 21–23 (2014)

Real Decreto 2291/1983, de 28 de julio, sobre órganos de elaboración y desarrollo de la política informática del Gobierno. The royal decree 2291/1983 of 28 of July about the body of drafting and development of informational government policy. https://www.boe.es/buscar/doc.php?id=BOE-A-1983-22882. Accessed 30 May 2018

Rossoshanskiy, A.V.: The problem of the informational openness of public power in present Russia. Power **11**, 33 (2009)

The Federal Act of 27 of July 2006 No. 149-ФЗ «About information, informational technologies and of the information protection»

The Russian power: the legal imitation of the openness. Fund «Liberal mission». www.liberal.ru/articles/4789

Peculiarities of Insurance of Legal Entities' Property and Issues of Its Legal Support in Terms of E-Insurance Development

Irina N. Romanova(✉) and Irina A. Mikhailova

Moscow University named after S.Yu. Witte (Ryazan Branch), Moscow, Russia
vip_irinaromanova@list.ru, irina_mikhaylova@list.ru

Abstract. In terms of this paper, we study specific features of insurance of legal entities' property, including application of e-insurance technologies as a new method of insurance business development on the research of doctrinal standpoints, analysis of current legislation and its practical application; reveal the issues of its legal support; draw recommendations on legislative elimination of topical issues in the area under study. The relevance of the research is determined by crisis taking place in insurance industry and this circumstance is explained by several reasons. The first is an intention of many companies to reduce expenses, which is often made thorough rejecting voluntary insurance.

The second issue is flawed legislation. On the one hand, it doesn't provide proper countering the fraud schemes (it's crucial for insurance underwriters) that have been already applied to CMTPL (Compulsory Motor Third Party Liability) and CNC (Comprehensive and Collision Insurance), and on the other hand, it doesn't always contribute to effective protection of policyholders. The latter partly depends on new global trends in the development of quite conservative sector of the insurance market, which are brought to life by the emergence of new threats arising in terms of new technological revolution and causing serious damage to the economy, the method for determination hereof has not yet evolved. Amid these circumstances, a complex of organizational and economic measures aimed at satisfying the needs of customers in insurance coverage, at providing coordination between the insurance underwriter and different counterparties as well as the collection and fusion of information on actual policyholders through the Internet (called e-insurance) is growing popular. The paper draws and substantiates the conclusion that the issue of necessary property insurance can be solved by imposing a legal obligation to conclude an insurance contract, also thorough the Internet; by enshrining a legislative presumption in respect to given obligation, which is made in relation to the pledge of property; by entrenching the possibility to add obligation of property's insurance to the contract.

Keywords: E-insurance · Insurance fraud · Insurance coverage · Insurance contract · Insurance of property losses · Insured amount and insured value

The concern of property preservation at the very early stages of entrepreneurial development gave an impetus to search for effective ways of protection hereof. The insurance had been the first to perform this function. Moreover, this type of protection against risks had actually been the only one for centuries.

E. G. Popkova and B. S. Sergi (Eds.): ISC 2019, LNNS 87, pp. 387–395, 2020.
https://doi.org/10.1007/978-3-030-29586-8_45

However, now this type of protection undergoes a crisis, despite expert's opinion that the development hereof "should become a priority activity of insurance companies for establishment of effective system on protection of valuable interests of legal entities, since the very property is primarily affected by various negative forces" [1]. Against this, e-insurance has become increasingly popular. This type of insurance certainly has particular advantages over traditional one. Thus, advantages for an insurance company in opening an Internet office is reduction of expenses for usual office as well as of transaction costs, since ones in a virtual office are much lower than required for customer's service in standard office.

Partial digitalization of business by insurance company is provided to both current and actual customers. Such opportunities include: to form a general idea of the insurance underwriter as well as of the list of services rendered; to get an online advice; to use the Web-calculator and to make an independent calculation of contributions; to file an electronic application for the conclusion of an insurance contract; to conclude an insurance contract and make an insurance premium through the Internet as well as to get online support hereof; to report the insured event through the Internet.

It should be noted that this contract can be currently applied in business activity to a limited extent, since arrangement of insurance policy online sales for particular insured property is either impossible or pretty difficult. For example, it refers to insured events of large industrial facilities that requires a mandatory preliminary inspection prior to insurance contract's conclusion, and further development of individual insurance programs with regard to specific features hereof. It is fair to say that most risks of e-commerce are among non-standardized.

The principal issue is a determination of the insured property. It should be noted insurance companies often define insured property as valuable interests of the policyholder (beneficiary), which are associated with the risk of loss (destruction) or damage to the insured property. Therewith, they demonstrate a pretty conservative approach to determination of its composition, referring real estate items and movable property to the insured property and simultaneously specifying that insurance doesn't cover manuscripts, maps, drawings and other documents, accounting and business books, unless the clause on reimbursement of expenses for the restoration of documents, models, scale models, samples, forms, etc., plants and animals (including agricultural), microorganisms, and other types of things that deemed as property according to Art. 128 of the RF Civil Code. Insurance doesn't admit (unless otherwise agreed by the contract) securities, cash in national and foreign currency, precious ingot metals, unmounted gems, jewelry, stamps, coins, banknotes and bons (paper currency), drawings, pictures, sculptures or other arts collections or art works [2]. Also, property insurance doesn't cover information contained on hard drives, in software and other storage devices of the policyholder, which corresponds to the concept enshrined in the RF Civil Code.

The analysis of property listed by insurance companies (that are not subject to insurance) shows that given exemptions and exceptions are determined by different reasons. Thus, explosive substances, property located in a zone threatened by rockfalls, landslides, floods or other natural disasters as well as in a zone of military operations since the announcement of such a threat in the prescribed manner (if it was made before the conclusion of the insurance contract), hazardous and dilapidated real estate items and also movable property located herein don't generally fall into the category of insured property, because their properties, condition or location don't allow qualifying their possible destruction or damage as an insured event.

In some situations, interest of the policyholder in preservation of particular items is covered by other types of insurance regarding specific features of the respective assets (vehicles, shares, monetary means, etc.). Although, the issue of the insurance protection mechanism for some of them is still under discussions due to different risks of their loss or reduction. In these terms, the standpoint by Mikhailova A.S. seems quite reasonable; she believes that "the search for a mechanism to protect shareholders' valuable interests (corporations in general) with the help of insurance contract is expedient in combination with the types of insurance, such as: title and business risk should be insured by shareholder; insurance of a legal entity's liability as an issuer of securities and liability of directors and management; insurance of liability of other persons who may cause harm (they may range from registrars as well as notaries public, appraisers, auditors, court-appointed trustees, and lawyers)" [3]. A similar approach can be applied in respect to monetary means.

A severe legal issue is an insurance of property losses entailed by the spread of knowingly false information as well as unauthorized access to information in software, on hard drives and other storage devices, which becomes increasingly topical nowadays. Obviously, they can't be covered by property insurance in its traditional sense, despite the fact that intangible assets have a property appraisal.

An important feature of the insured property is an interest of the policyholder (beneficiary) in preservation hereof, which may depend on various circumstances. Therewith, the courts note that Paragraph 1 of Art. 930 of the Civil Code of the Russian Federation "interest in the preservation of property" exactly means the interest of that person who bears the risk of loss and damage [4]. Thus, the Supreme Court of the Russian Federation in Ruling No. 308-ES15-11472 dated December 17, 2015 in a case No. A32-35788/2012 stated that the company had an unconditional valuable interest in preserving the insured property that although was owned by a third party but was a pledged item, the integrity hereof determined possible application of the credit line in the agreed scope. Hence, it was concluded that the insurance of the pledged property, also owned by a third party (pledgor), ensures pledgee's valuable interest expressed in recovery of damage to the pledge directly to the bank at the expense of the insurance company, which makes possible ongoing using of credit resources without changing contract terms.

The issue of obligatory property insurance can be differently resolved:

(1) by imposing a legal obligation to conclude an insurance contract that is stipulated in respect to mortgage item, sea vessel (Article 203 of the Merchant Shipping Code of the Russian Federation), pledged or pawned things (Article 6 of the Federal Law dated July 19, 2007 No. 196-ФЗ "On pawnshops" [5]);

(2) by entrenching a legislative presumption on given obligation, which is made in relation to the pledge of property. According to Paragraph 1 of Art. 343 of the RF Civil Code, unless otherwise provided by law or contract, pledgor or pledgee (depending on who keeps a pledged property) (Article 338) must insure pledged property against the risks of loss and damage at the expense of the pledgor to the amount not less than amount of claim secured by the pledge.
(3) by enshrining the possibility to add obligation of property's insurance to the contract. Besides, the legislator differently determines peculiarities of their development. In some situations, we are talking about a mutually agreed term (insurance of the merchandise under a purchase/sale agreement (Art. 490), insurance of a construction facility (Art. 742 of the RF Code)), in other situations it is stipulated that the obligation of property insurance is fulfilled only when it's required by a particular party of the contract (insurance of the leased property (Art. 637, Art. 661 of the RF Civil Code).

Despite the fact that policyholder's valuable interest may be based on a law, other legal enactment or contract pursuant to the provisions of Part 1 of Art. 930 of the Civil Code, it follows from Paragraph 3 of Art. 998 hereof that the obligation of property insurance by the commissioner and also by agent by virtue of rules of Art. 1101 of the Civil Code may arise out of customary business practices.

In this regard, we can't to draw attention to the fact that two of the thirteen basic terms of delivery mention the insurance: CIF (Cost, Insurance, Freight) used in shipping by sea, and CIP (Carriage and Insurance Paid to) mediating land and air shipping. In both events, the seller should insure the cargo under minimum FAP terms ("free average particular") in the absence of special instructions from the buyer.

In addition, it is assumed that the insurance should be made by a high-reputed insurance underwriter. It is also understood that the insurance will be made under the terms developed by the Institute of London Underwriters. However, their application can bring collision, since coverage amount for such contracts is calculated as a negotiated price of the merchandise plus ten percent that contravenes the rules of the Civil Code (Clause 2, Article 947) and Merchant Shipping Code (Art. 259) prohibiting insured amount over insured value under property insurance [6]. However, it is easy to solve this issue given that primacy of legal rules established by Art. 5 of the Civil Code.

In general, with regard to current practice and needs of the business community in expansion of self-regulation scope in general and their insurance protection in particular, it seems expedient to broaden the list of grounds for determination of insurance interest through customary business practice, amending Paragraph 1 of Article 930 of the RF Civil Code.

An important point of property insurance is a determination of the insured value intertwined with some issues, including: revealing the consequences of the lack of relevant term in the contract, setting the actual insured value, executing the term on incomplete insurance and deductible as well as additional insurance. It's no coincidence that Art. 948 of the RF Civil Code is one of the most applicable in insurance disputes arising between insurance underwriters and policyholders [7].

From literal interpretation of the rule enshrined in Paragraph 1 of Art. 942 of the RF Civil Code implies that the insured value doesn't apply to the essential terms of property insurance contract. Therewith, provisions of Paragraph 2 of Art. 947 of the RF Civil Code make it possible to talk about legislative entrenchment of the presumption to determine this value when calculating the insured amount, which shouldn't exceed fair value of property at the location hereof on the day of insurance contract conclusion. Thus, according to courts, only insured amount agreed by parties without specification fair value of property means their expressed consent to determine the insured value hereof to be equal to the insured amount, whereof the insurance underwriter computes premium paid by the policyholder in full [8]. However, there is another standpoint according hereto "systemic interpretation of the rules of Law N 4015-1 and chapter 48 of the RF Civil Code implies that the insured value refers to contract terms determining fair value of the insured property on the date of insurance, therefore it is an obligatory circumstance to be specified in the contract. The ratio of the insured amount and the insured value and their effect on the determination of amount of insurance recovery is stipulated in Articles 947, 949, 950 and 951 of the RF Civil Code. Thus, the insured value is a mandatory contract value that directly determines both the property to be insured and the amount of insurance recovery upon occurrence of an insured event" [9]. This opinion is also supported by the scientific community [10]. However, it is difficult to recognize it. More likely, we should talk about usual term of the contract implied by the parties, as also evidenced by insurance rules directly relating the determination of the insured amount with establishment of the insured value of the property [11]. Helding the contracts not directly specifying the insured value of the property under all-risks insurance will hardly contribute to sustainability of civil transactions. There is a reason that the courts, referring to the provisions of civil law, talk about incomprehensive property insurance only if figures of the insured value in the volume of less than the value of the insured amount were agreed by parties upon the contract's conclusion. Thus, the court found that reference of insurance underwriter to the fact that the term on incomprehensive property insurance is contained in the Insurance Rules have no importance, since the rule of proportional recovery is applicable only if the insured amount is directly set lower than the insured value of the insured property in the insurance contract; that is, this term must be explicitly specified herein, or the contract should precisely refer to fair value of the insured property [12].

Indeed, Civil Code of the Russian Federation admits possible entrenchment of another term on the insured value in the contract. And foremost, the legislative admission to circumvent a prohibition the prohibition on the insured amount not to exceed the insured value that, at first sight, becomes irrelevant given the consequences of insurance established by Art. 951 of the RF Civil Code in excess of the insured value that consist in helding the contract void in the part of the insured amount exceeding the insured value.

However, if we apply these provisions to statutory requirements concerning the place and time of its determination, the situation will be different.

Firstly, to fulfill a requirement on appraisal of legal entity's property upon conclusion of the contract is difficult. As Dedikov S. notes, if the property is insured on the basis of balance sheet data, it is obvious that its book value is determined not upon insurance deal, but as of the last reporting date [13], and this "violation" can hardly be used as a reason by one of the parties to the contract. Execution of this requirement looks even more arguable within determination of estimated cost of construction and installation works. It is no coincidence that insurance underwriters sometimes stipulate in these events that the insured amount specified in the insurance contract is not an agreement of the parties on fair value of the insurance item, since for post-completion warranty obligations it is calculated from the cost of the completed or commissioned construction or installation facility [14].

The issue of the appropriate application of the rule on determination of property value at the place of its location given that the construction and installation risks insurance contract is often concluded prior to the commencement of works and estimated value hereof is a basis for calculations in accordance with work and labor contract or other available documents; it is impractical to determine the insured value of insured export and import cargos at the place of their location in the country of departure, since their cost is generally much higher [15].

Accordingly, the clause proposed by the legislator on availability of other terms of the contract regarding determination of the insured value of property should be extended to the provisions on the procedure for determination hereof.

The issue of fair value meaning is also arising. As follows from Art. 7 of the Federal Law dated July 29, 1998 N 135-ФЗ "On appraisal activity in the Russian Federation", in the event of using this concept in the statutory enactment one should apply the market value of item under appraisal that means the most probable price of item's alienation at open market in competitive conditions when the parties of the deal take reasonable actions, holding all the necessary information (Par. 2, Art. 3 of the Law "On Appraisal Activity").

As the analysis of the insurance rules shows, value of the property in the insurance contract is determined differently: for real estate items, machinery and equipment, furniture and maintenance accessories, exterior (interior) finishing (including, facilities). It is calculated due to wear and tear and operational condition; for goods (including raw materials, semi-finished products, etc.) purchased by the policyholder (beneficiary) as well as those accepted for safe custody it's calculated from the amount spent on acquisition of insured goods; in respect of the pledged property - based on the pledged value hereof.

According to experts, we need to determine the insurance amount on the basis of replacement value to achieve the purpose of property insurance to full extent [1], that is however intertwined with the number of issues, some of them are gradually resolved. Thus, the issue to account wear and tear of the insured property is settled in CMTPL, which apply a method of insurance called "new instead of old". Currently, deliberate reduction of the tax base and calculation of the insured amount in accordance with the replacement value, rather than with the book one is not treated as tax evasion.

It is rightly noted that property insurance in accordance with the replacement value doesn't fully meet the principle of recovery in terms hereof the insurance underwriter is obliged to indemnify only real losses of the policyholder, since the replacement value exceeds not only the book value, but the market value of the property. According to Dedikov S., the solution could be a transition to a new-scale calculation of insured value based on the consumer qualities of the property [16]. At the same time, it is important to determine extent of application for replacement-value insurance to avoid abuses by policyholders.

If the insurance amount established in the contract is initially set lower than fair value of the insured item, it's incomprehensive insurance. In this event, the amount of loss/damage and expenses (unless otherwise provided by the contract) are paid proportionally to the ratio between insured amount and fair value of the insured item (the "proportional system"). Undertaking an increased liability by the insurance underwriter in accordance with Par. 2, Art. 949 of the RF Civil Code should be explicitly stated in the contract and shouldn't require additional interpretation [17]. Thus, the rules of insurance may establish possible entering of the provision if when insured amount is lower than the insured value of the insured item, then the insurance recovery upon the occurrence of insured event is equal to incurred damages without ratio between the insured amount and the insured value ("first risk" system), but anyway not higher than the insured amount established in the insurance contract [18].

By mutual consent of the parties, the insurance contract may also provide for the share of policyholder's participation in damage recovery, i.e. the deductible meant by legislator as partial losses determined by federal law and (or) the insurance contract that is not subject to recovery by the insurance underwriter to the policyholder or another person whose interest is insured in accordance with the terms of the contract and that is set as a particular percent of the insured amount or as a fixed amount (Par.9, Art. 10 of the Law on arrangement of the insurance business). However, Dedikov S.V. rightly points to this inaccurate definition, noting that a deductible should be understood as "that part of the insurance indemnity that shouldn't be paid by insurance underwriter or reinsurer in accordance with the terms of the insurance or reinsurance contract" [19]. Otherwise, this may lead to a violation of the legitimate interests of the policyholder, because a literal interpretation makes it possible to conclude that the deductible shouldn't be deducted from insurance indemnity to be paid, but from the amount of losses.

Insurance underwriters solve this issue through a detailed regulation of the procedure for the deductible application that can be set both in absolute terms and as a percent of the insured amount or amount of losses and can be conditional/ unconditional.

In the first event, the insurance underwriter doesn't pay the insurance indemnity if the amount hereof doesn't exceed the deductible amount, but pays in full if it exceeds the deductible;

In the second event, he pays the indemnity reduced by the deductible amount, losses not exceeding this value are not subject to indemnification. Therewith, presumption of application of the unconditional deductible is established as a rule.

Legalization of this institution is of particular importance, given the negative attitude to it in legal science of the Soviet times. Kalugin D.E. particularly notes that "the institution of the deductible has no background in Soviet insurance law as it is inconsistent with the principle of comprehensive insurance coverage and not required by any special needs that could justify the existence hereof" [20]. A critical attitude hereto was also expressed in modern legal science, although such an approach to this phenomenon on the insurance market was no longer supported. The courts also recognized term on the deductible as lawful and applicable, regarding it as a term agreed by the parties to the contract [21].

Legally provided transfer of rights and obligations under an insurance contract when transferring rights to insured property (except the cases of its compulsory withdrawal and waiver of property rights) (Article 960 of the RF Civil Code) is of great importance for business activity. The premises for this are created by issue of insurance policy to bearer that can be presented to the insurance underwriter by the policyholder or beneficiary of the rights under this contract respectively. As noted by Vogelson Yu. B., insurance coverage is transferred together with titles to the property, except risk and interest that are untransferable.

The above-stated allows to draw the following conclusions.

Despite the growing popularity of e-insurance in civil transactions, this contract can be currently applied in business activity to a limited extent, since arrangement of insurance policy online sales for particular insured property is either impossible or pretty difficult.

The issue of necessary property insurance, also through the Internet, can be solved by imposing a legal obligation to conclude an insurance contract, also thorough the Internet; by enshrining a legislative presumption in respect to given obligation, which is made in relation to the pledge of property; by entrenching the possibility to add obligation of property's insurance to the contract.

References

1. Khokhlov, S.I.: Property insurance in the Russian Federation: abstract of a thesis by Cand. of Sci. (Law), p. 9 (23 p.) (1998)
2. Rules of insurance of legal entities' property: approved by Director General of "IC "Soglasie" LLC on April 13, 2015; Rules of insurance of legal entities' property "against all risks": Approved by Director General of "Renessans Insurance Group" LLC on September 10, 2009 (2009)
3. Mikhailovam, A.S.: Concerning some aspects of application of insurance contracts to protect property rights of shareholders in terms of a phenomenon of raiding. Lawyer **17**, 38–42 (2016)
4. Resolution of the Commercial (Arbitrazh) Court of the North-Western District dated January 21, 2016 in a case No. A56-76088/2014; Resolution of the Commercial (Arbitrazh) Court of the Ural District dated December 22, 2015 N Ф09-9872/15 in a case N A50-27170/2014
5. Collected legislation of the Russian Federation, no. 31, Art. 3992 (2007)
6. Troitskaya, I.V.: Customary business practices in the insurance of entrepreneurial property interests. News of the Russian State Pedagogical University named after Herzen A.I., no. 104. pp. 167–179 (2009)

7. Dedikov, S.V.: Disputing the insured value. Laws Russian Feder. Exp. Anal. Pract. **3**, 52 (2010)
8. Resolution of the Federal Antimonopoly Service of the North-West District dated August 02, 2011 in a case No. A56-46451/2010; Resolution of the Commercial (Arbitrazh) Court of the Volga-Vyatsky district dated November 09, 2016 No. Ф01-4290/2016 in a case No. A82-19021/2014
9. Resolution of the Commercial (Arbitrazh) Court of the Volga-Vyatsky district dated March 24, 2015 No. Ф01-495/2015 in a case No. A17-5526/2013
10. Panchenko, E.V.: Insured value is an essential condition of property insurance contract. Law Econ. **5**, 43–46 (2012)
11. Rules of insurance of legal entities' property: approved by Director General of "AlfaStrakhovanie" OJSC on September 1st, 2011. https://www.alfastrah.ru/docs/Pravila-strah-imushchestva-uridicheskih-lic.pdf
12. Resolution of the Commercial (Arbitrazh) Court of the Volga-Vyatsky district dated November 09, 2016 No. Ф01-4290/2016 in a case No. A82-19021/2014
13. Dedikov, S.: The insured value of the property. EZh-Yurist, no. 25. pp. 1, 3 (2011)
14. Rules of the combined insurance of construction and installation risks: approved by Order of Director General of "RESO-Garantiya" IPJSC No. 406/02 dated November 20, 2017. http://www.reso.ru/export/sites_reso/About/Tariffs_rules/SMR_rules.pdf
15. Dedikov, S.: The insured value of the property. EZh-Yurist, no. 25, p. 1 (2011)
16. Dedikov, S.V.: Disputing the insured value. Laws Russian Feder. Exp. Anal. Pract. **3,** 3 (2010)
17. Resolution of the Commercial (Arbitrazh) Court of the North Caucasus District dated April 04, 2016 No. Ф08-1046/2016 in a case No. A32-8183/2015
18. Rules of insurance of legal entities' property: approved by Director General of "AlfaStrakhovanie" OJSC dated September 1st (2011). https://www.alfastrah.ru/docs/Pravila-strah-imushchestva-uridicheskih-lic.pdf
19. Dedikov, S.V.: Franchise: issues caused by its legalization. Laws Russian Feder. Exp. Anal. Pract. **3**, 43–48 (2014)
20. Kalugin, D.E.: Illegal nature of the franchise. Judicial and legal work in insurance, no. 2, pp. 89–95 (2011)
21. Resolution of the Ninth Commercial (Arbitrazh) Court of Appeal dated February 22, 2012 No. 09АП-1643/2012-GK in a case No. A40-18221/11-14-158

Unstable Transformation of the Models of Development of the Russian Economy

Elena M. Semenova[1(✉)], Elena N. Tokmakova[1], Karina V. Kuznetsova[1], Alla V. Volkova[2], and Mariia A. Fetisova[2]

[1] Orel State University, Orel, Russia
orel-osu@mail.ru, 1278orel@mail.ru, my-orel-57@mail.ru
[2] Orel State Agrarian University Named after N.V. Parahin, Orel, Russia
cool.volkovaalla@yandex.ru, super-ya-57@mail.ru

Abstract. The model of development of the Russian economy has a collective image. The model of the development of the Russian economy fatalizes the directions of economic processes within the country. The instability of the development of models poses problems for the formation of measures for the emergence of the economy from the crisis. Unstable transformation of the Russian economic development models leads to a backwardness of economic processes. The purpose of the scientific article is to study the unstable transformation model of the development of the Russian economy. The objectives for the realization of the goal are: to the consider the concept of «unstable transformation»; to the definition of the essence of the concept of «model of economic development»; to the construction of specific models of economic development, taking account the typology of their cycles. The methodical apparatus includes the following methods: the method of problem theorizing, the method of species characteristic, the method of factor grouping, the method of implication, the method of configuration. The scientific article is of a theoretical nature. The theoretical significance of the scientific article is based on the consideration of models of economic development.

Keywords: Transformation · Model · Digitalization · Innovation · Manufacturability · Instability · Displacement

1 Introduction

The regeneration of the economic processes leads to an imbalance of the economic development. In the economy there is a change of paradigms. The technological sector of the economy leads to an unstable transformation of the development model of the Russian economy. The process of unstable transformation causes changes. The changes are fatalistic. The instability of the transformation of the model of economic development provokes the emergence of negative factors in the territories. The unstable transformation of models of the development of the Russian economy has a number of negative features.

E. G. Popkova and B. S. Sergi (Eds.): ISC 2019, LNNS 87, pp. 396–405, 2020.
https://doi.org/10.1007/978-3-030-29586-8_46

Firstly, territorial resources are limited. Small territories can't be integrated into the system of transformation of the model of the economic development. Limited resources are caused by the debt burden on municipal budgets. The growth of municipal debt and the absence of the large production entities are changing the direction of the transformation of the economy of small territories. An important feature is the preservation of the satisfactory development of the economy of small territories. Forecasting growth and development together with the federal center is conditional. Different goals and tasks are suitable for the economy. The economy of small territories is destabilizing. The model of economic development is unbalanced. Changes support disbalance models of small economies.

Secondly, the state policy isn't a modern model of the development of the Russian economy. The mixed economy is a model with state participation. The mixed economy assumes participation of the state in building of market relations. The hybridity of the economic model ensures the performance of social functions not inherent in the open market. The state provides a high level of employment, stabilizes prices, the relationship between wages and productivity, balance of payments, the use of production capacities in full. The territories of the Russian Federation have stagnant economic processes. There is no coordinated state policy in the economy. The market is disorganized. The market is unable to rectify the economic situation by balancing demand and supply.

Thirdly, the destructive development of the Russian economy has led to a lack of stable economic growth. The Russian economy grew by 6,4% over the period from 2008 to 2017. The average growth rate of the world economy was 35,0%. The current situation indicates the destructuring of the economic model of the Russian Federation. There aren't priority directions for development in the economy. Conducting a conditional policy of import substitution shows the inability to implement projects innovative technological equipment of the country.

The problems focus on the outdated models of Russian economic development. The formation of new programs for economic development is associated with the transition to digitalization of society. The changes make it possible to shift the priority of the direction of the economy of the Russian Federation. The economic modeling is associated with government support of the commodity sector. This event is not correct. The purpose of the scientific article is to study the unstable transformation of the Russian economy development model. The tasks to achieve this goal are:

- to the consideration of the concept of «unstable transformation»;
- to the definition of the essence of the concept of «model of economic development»;
- to the modeling the types of models of economic development, taking account the typology of their cycles.

2 Methods

The methodological apparatus of the study is aimed at the realization of two key aspects. The first aspect is based on the theoretical study of concepts. The question is solved on the basis of methods:

1. The method of theorizing is considered as an instrument that clarifies the internal structure of the problem.
2. The method of the species characteristic represents a tool to address problems from the point of view of the features of the external form.
3. The method of factor grouping is based on a generalization of the concept of the subject.

The second aspect is aimed at considering models in the socio-economic nature. The toolkit includes:

1. The method of implication is characterized by specific features of models of economic development.
2. The configuration method is characterized by the manifestation of the external form of economic development models.

3 Results

The concept of «unstable transformation» is not used for models of economic development. The term has entered economic life recently. The term occurs in a scientific article by Lilien G.L., Rangaswamy A. Unstable transformation is a period of transformation and variability subject to change. These phenomena are cyclical and oppositional in nature (Lilie and Rangaswamy 2000). This definition wasn't developed in the future. Instability and transformation were considered as different phenomena. The transformation is the process of internal and external components of the investigated object (Kuzmin 2010). Instability is considered as a state of heterogeneity of occurring events (Belova and Kozelov 2016).

Authors agrees with the opinion of Lilien G.L., Rangaswamy A. on the need to consider the concept of «unstable transformation». This aspect is related to the following factors. Firstly, the concept allows us to determine the cyclicity of processes inherent in the systems under consideration. Secondly, transformation involves transformation. Transformation is an important element of the model of the economic development.

The model of economic development is a descriptive system of factors and features of the economic processes (Waller 2002). The model of the economic development is differentiated according to two specific characteristics. The first characteristic is associated with the allocation of the role of the state in the study of economic development (Voloshenko and Ponomarev 2017). Models of the economic development are divided into: the traditional model, the administrative model, the market model, the mixed model. The second characteristic is territorial. Models of the economic development are divided into: the American model, the Japanese model, the German model, the Swedish model, the South Korean model (Kim and Ayhan Kose 2003).

These models don't consider factors of economic development. Author's models of economic development are considered factors and cycles of the state of the economy. The definition of the cycle and factors of economic development models is the basis of the study. The cycle of the model shows a set of oscillations of external phenomena (Walters 2004). Factors determine the inner essence of processes and phenomena. Factors are reflected in the external form of the model of economic development. The characteristics of the model can be determined by the components of economic development for a selected period of time. Seven models of economic development will be considered in the unstable transformation of external phenomena (Figs. 1, 2, 3, 4, 5, 6 and 7). The model of catching-up economic development is presented in Fig. 1.

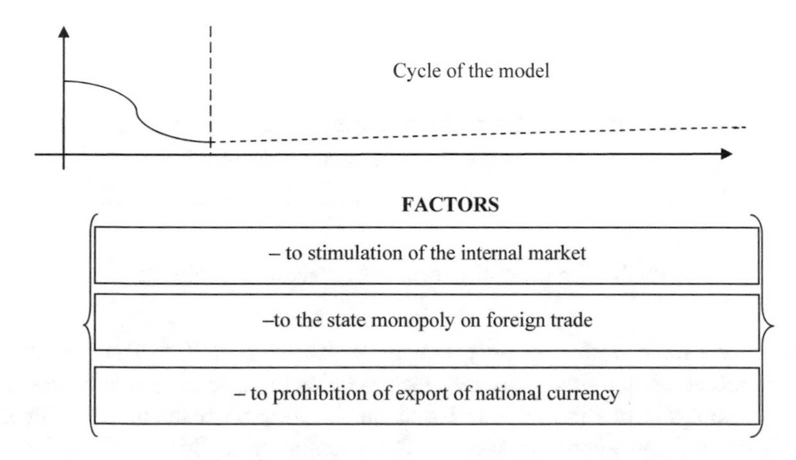

Fig. 1. Model of catching-up economic development

The model of catch-up economic development is based on the theory of mercantilism. The cycle of the model of economic development is associated with the phenomena of economic decline. The cycle oscillation process reaches a recessive point. The recessive point is a turning point in the catching-up economic model. The turn of the cycle is directed towards long-term, slow economic growth. These activities include stimulating the domestic market. Stimulating the domestic market is a consistent action related to the policy of import substitution.

The model of catching-up development is ambivalent. On the one hand, the model is aimed at supporting domestic producers. On the other hand, the model restricts the right to foreign trade. The foreign trade is a monopoly of the state. The deterioration of foreign trade is connected with the ban on the export of the national currency. This statement indicates a closed economy.

The model of the Russian economy uses the measures of the catch-up development model. The model isn't a priority. The catch-up development model is based on activities of a limited nature.

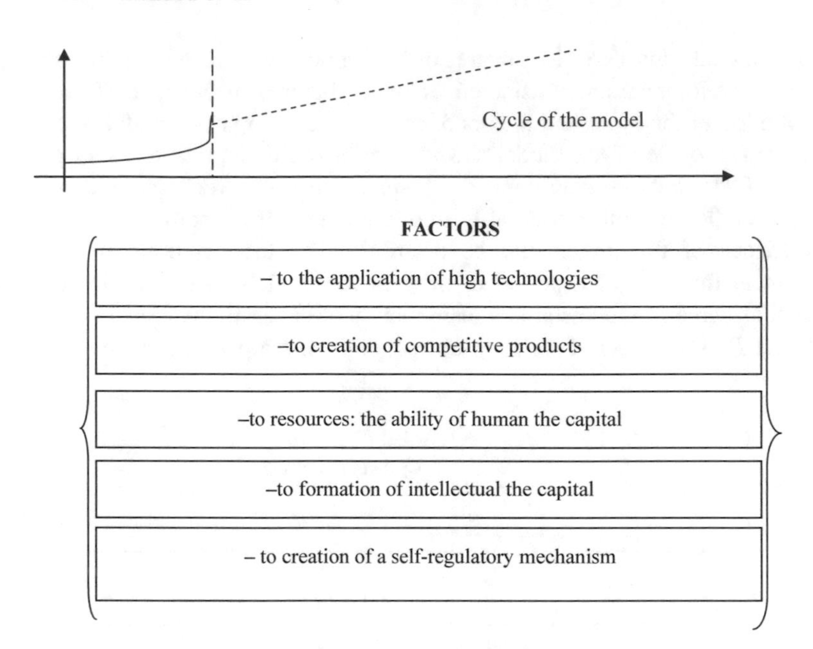

Fig. 2. The innovative model of economic development

Innovations have been a priority tool in Russia since 2007. The modernization policy was aimed at applying innovative technologies in various sectors of the national economy. The cycle of the model is based on the long-term fluctuations. Economic growth is formed by innovative factors. The peculiarity of this model is high technology. High technologies are aimed at creating competition in the model of the innovative development (Stroeva et al. 2016).

The resources of the innovation model are people. The tool that develops the resources of the innovation model is the idea. Human capital ascertains the significance of this phenomenon. Human capital is transformed into the intellectual capital. The innovative model of economic development is based on financing. The economic growth of the innovation model can be achieved on the basis of building an effective investment policy. Investments in innovation cause economic growth. The economic growth is supported by consumption. Consumption increases the income of investors.

The model of innovative development is quite effective. For innovations, there isn't foundation in the economy of small territories.

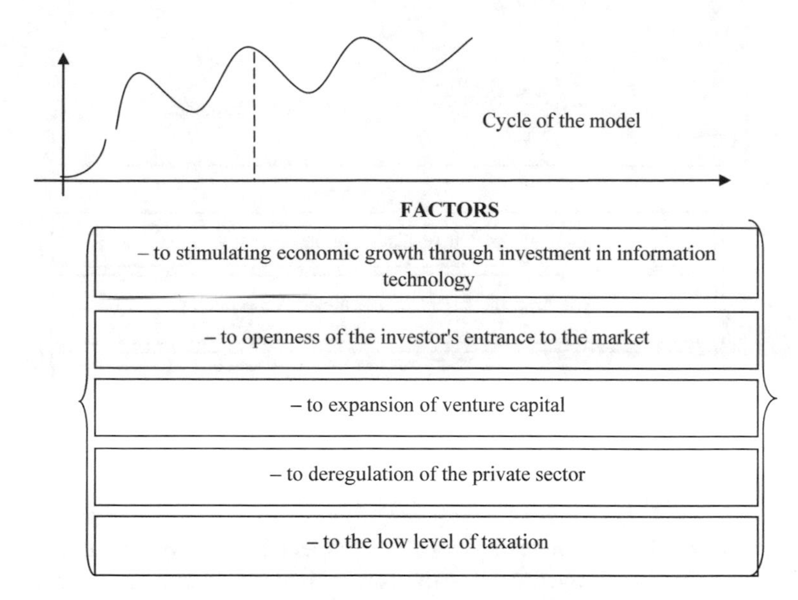

Fig. 3. Model of economic development the «new economy»

The model of the economic growth the «new Economy» is similar to the model of innovative development. The model is different with the cycle of the functioning of the economy. The cycle of the model is progressive. The turnover of the cycle is based on the passage of complete phases of recession and growth. The fluctuations are of a complete nature. Innovation and investment are needed for the development of information technology. The «new economy» is a post-industrial economy. Information is the main factor of production. The implementation of the «new economy» model is of a national nature. The model develops from government events. The model factors are openness of the market for the investor, expansion of venture capital, deregulation of the private sector, low level of taxation. Dedicated factors are inherent in the American model of the economy. The «new economy» development model is formed on the principle of maintaining the market. The state is the controller.

In the economy of the Russian Federation, the state's share is quite significant. The model of economic development the «new economy» is conditional. To model the economic development of the Russian Federation, only some of the activities of the «new economy» model can be applied.

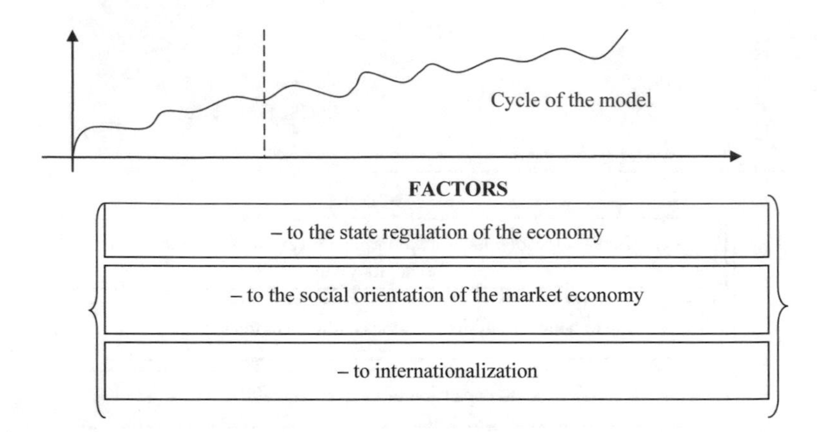

Fig. 4. The model of economic development the «social orientation»

The model of the economic development the «social orientation» is suitable for states with constant prolonged economic growth. Economic growth is gradual. The economic phases are short-term. The economic phase does not allow forming the foundation of the «social orientation» for the Russian economy. State regulation presupposes the possibility of applying the measures of the «social orientation» model. The model of the economic development the «social orientation» can be theorized. The postulate of social orientation will be declared. Activities will not be implemented in practice. The model «social orientation» isn't accepted in practice.

The transformational model of economic development is applied with private economic fluctuations. The fluctuations are depressive. The activities of the transformation model support a short-term growth cycle. The factors of this model are instruments for maintaining economic growth. On the basis of factors, trajectories of economic development are created. The demand focuses production.

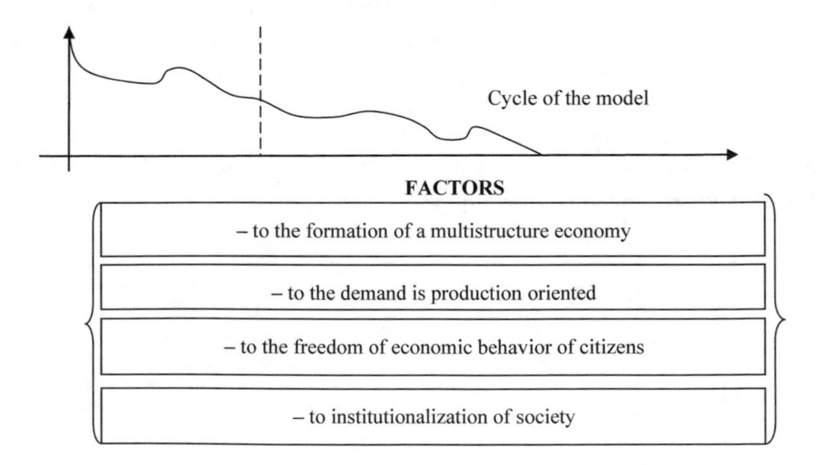

Fig. 5. Transformation model of economic development

The invisible hand of the market forms the need for production. Important market factors are: the freedom of economic behavior of citizens and the institutionalization of society. The institutionalization of society is an element of institutional macroeconomic theory. The direction of regulation isn't of a state nature.

The transformation model is quite adequate to the Russian economy. The development of the market component leads to a reduction in the effectiveness of the activities of the transformation model for the Russian economy.

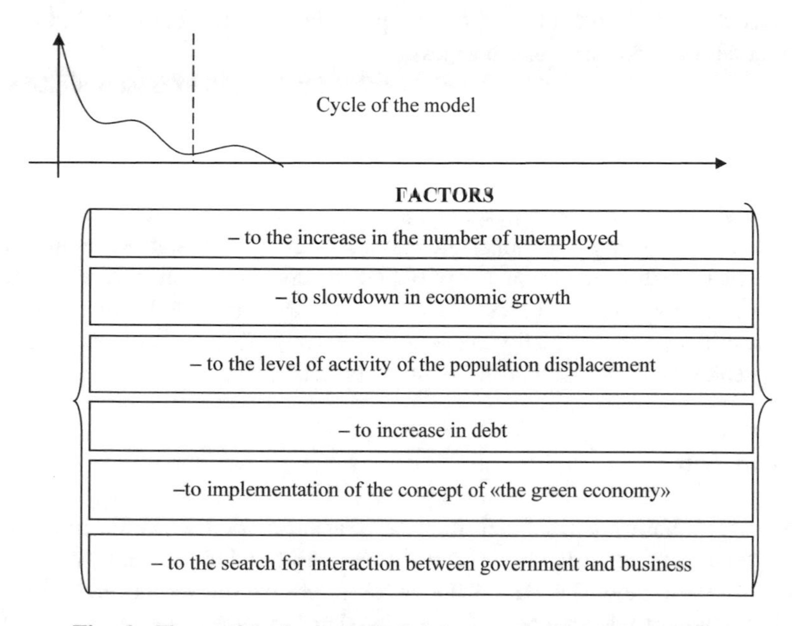

Fig. 6. The model of economic development in the new normality

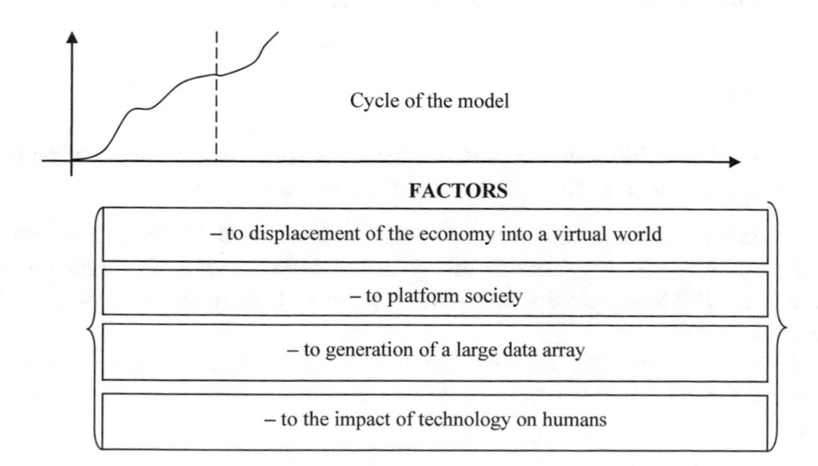

Fig. 7. The model of digital economic development

The model of economic development in the new normality presupposes the existence of stagnation processes of cycle oscillations. The model of economic development in the new normality is adapted for the Russian economy. The key factors are the growth in the number of unemployed and the slowdown in the economic growth.

The model of economic development of the new normality has a descriptive character. The model shows the problems, but doesn't offer solutions to these problems. The context of the problems is peculiar to the Russian model of the economic development: the shift in the level of activity of the population, the growth of debt obligations, the implementation of the concept of the «green economy», the search for interaction between the state and business.

The model of digital economic development is final. Digitalization is aimed at rapid economic growth. The innovation-technological base is predetermined by the creation of services and platforms. Virtualization for the Russian economy is conditional. To date, not one of the factors of the model of digital economic development is not implemented in the Russian model. The model can be idealistic. The model isn't realistic.

To date, the model of economic development of the Russian economy is in an unstable transformation. None of the presented models is a prototype of the Russian economy. The problem of Russian reality is similar to the model of the economic development in the new normality. Measures to solve problems aren't presented in the model of economic development in the new normality.

4 Discussion

Modeling the economy is a difficult process. Transformation leads the model of economic development to instability. There isn't algorithm for modeling changes. The experience of these models underlies the discussion of the question of the formation of a model of economic development for the Russian economy. We assume that further elaboration of this model of economic development in the new normality. The topic of the study is the proposal of measures to solve problems.

5 Conclusions

The study of the problem of the unstable transformation of the model of economic development made it possible to draw the following conclusions.

1. The concept of the «unstable transformation» isn't used to study the economic phenomenon in Russian scientific thought. This definition is differentiated into two definitions. The first definition considers instability. The second definition considers transformation.
2. Models of the economic development have a dual grouping. Species groupings consider the role of the state or the state territorial ownership of the economic model.

3. Models of the economic development made it possible to conclude that there is no model with the characteristics of the Russian economy. The model of Russian economic development is of a collective nature. The model of economic development of the new normality is most preferable. However, this model examines only the problematic aspects of economic development. The model of economic development of the new normality doesn't provide an activity to address the identified problems.

References

Belova, T.A., Kozelov, D.A.: Theoretical review of econometric models of a research of real national economies. Questions Econ. Manage. **5**(7), 4–7 (2016)

Kuzmin, D.V.: Risk factors in models of balance of open economies. World Econ. Int. Relat. **9**, 23–28 (2010)

Lilien, G.L., Rangaswamy, A.: Modeled to bits: decision models for the digital, networked economy. Int. J. Res. Mark. **17**(2-3), 227–235 (2000)

Stroeva, O., Lyapina, I., Mironenko, N., Petrukhina, E.: Peculiarities of formation of socially oriented strategy of economic growth of national economy. Eur. Res. Stud. J. **19**(2), 159–166 (2016)

Kim, S.H., Ayhan Kose, M.: Dynamics of open–economy business-cycle models: role of the discount factor. Macroecon. Dynam. **7**(2), 263–290 (2003)

Voloshenko, K.Y.U., Ponomarev, A.K.: Introduction in practice of regional government of sectoral models: assessment of the impact of managing directors of impacts on economy. Baltic Region **9**(4), 93–113 (2017)

Waller, Ch.J.: Modeling monetary economies. Atlantic Econ. J. **30**(2), 213 (2002)

Walters, D.: New economy – new business models – new approaches. Int. J. Phys. Distrib. Logist. Manage. **34**(3/4), 219–229 (2004)

Mechanisms for the Implementation of the Regional Economic Policy

Irina V. Skobliakova[1(✉)], Tatyana A. Zhuravleva[1], Maria A. Vlasova[1], Olga L. Maslova[2], and Oksana V. Gubina[3]

[1] Orel State University, Orel, Russia
my-orel-57@mail.ru, orel-osu@mail.ru, docent-ostu@yandex.ru
[2] Financial University Under the Government of the Russian Federation, Orel, Russia
fu-rf@yandex.ru
[3] Orel State University of Economics and Trade, Orel, Russia
osuet@mail.ru

Abstract. The problem of procrastination of the economic actions at the regional level has led to the emergence of the negative consequences of the implementation of economic policy mechanisms. The time factor is distorted in accordance with the slowdown in economic growth. Mechanisms for the implementation of economic policies are conditional. The problematics revealed the need to study the topic in the implementation of regional economic policy mechanisms. The purpose of the research is to examine the mechanisms for the implementation of regional economic policies. The objectives of the study are to review the chronology of regional economic policies; to determine the role of regional economic policy; to grouping the design of regional economic policy; formation of the copyright mechanisms for the implementation of the regional economic policy. The main research methods are the method of phasing, the method of retrospective analysis, the method of evolutionary assumptions, the method of katahrez, the method of essential perception, the method of invariant formalism, the method of abstraction, the methods of identification installations, the method of transparent parts, the method of imitational perception. The scientific article is theoretical and practical. The scientific article conducted a study of the issue of the regional economic policy with the presentation of mechanisms for its implementation.

Keywords: The region · The economy · Mechanisms · Tools · The development · Subsidiarity · Transparency · The program · The evaluation

1 Introduction

New foundations for the development of the socio – economic system have revised the functioning of tools and the implementation of measures to develop the regional economy. The economic development highlights important areas of the social management and control. Princes appear in the regional economy. Priority principles reflect the efficiency and effectiveness of the economic areas (Lyapina et al. 2019). The territorial direction is an important feature of the economic management. Branch division reflects the territorial direction. The territorial development includes the federal

E. G. Popkova and B. S. Sergi (Eds.): ISC 2019, LNNS 87, pp. 406–416, 2020.
https://doi.org/10.1007/978-3-030-29586-8_47

level of the government and the regional level of the government. The federal level of power is separated from the municipal level of power. Power levels are diversified. The federal level of the government and the regional level implement a unified economic policy at different levels. The federal level of the government pursues a federal economic policy. The regional level of government pursues a regional economic policy. This condition affects the development of the economy in two ways. Firstly, the federal policy is provided with a large number of budget funds. In the subjects of the Russian Federation, the level of independence and independence is reduced on the basis of this condition. Secondly, the mechanism of economic activities at the federal level is duplicated at the regional level. This circumstance doesn't contribute to the development of the constituent entities of the Russian Federation. The mechanisms of the regional economic policy are ineffective. This condition determines the relevance of the topic of scientific research.

The purpose of the scientific article is to consider regional economic policy with the subsequent allocation of the author's mechanisms for its implementation. The tasks of the scientific article are as follows:

- consideration of a phased chronology of the regional economic policy development;
- defining the role of the regional economic policy;
- grouping the design of regional economic policies in the strategic management system;
- formation of copyright mechanisms for the implementation of regional economic policy.

General directions of the regional economic policy determine the system of the management of the federal level of government. This position produces a comparison of the federal economic policy and the regional economic policy. In scientific research, the comparative factor is denied. Scientific research determines the phased development of the regional economic policy. Scientific research identifies key principles for the functioning of the regional economic policy. The role of the regional economic policy reflects the direction. Directions of the regional economic policy carry out measures to improve socio–economic activities on the basis of tools (Stroeva and Kvak 2014). The design of the regional economic policy gives an understanding of the external significance of this area of the national economy. Author's mechanisms for the implementation of the regional economic policies demonstrate ways to manage and make practical use of economic instruments.

2 Methods

The mechanism of the regional economic policy is implemented through the methodological apparatus. Methodical apparatus includes studying the historical subject, the theoretical subject and the comprehensive presentation. The historical methodological apparatus includes (O'Huallachain 2007):

- the phasing method is a study of the chronological order of appearance of the studied scientific subject;

- the method of retrospective analysis is a consistent study of the characteristics and characteristics of a scientific subject;
- the method of evolutionary prerequisites is the consideration of the dependence of the manifestation of evolutionary laws and the modern development of the subject of scientific research.

The theoretical methodological apparatus examines the basis for the development of the subject of scientific research. The theoretical methodical apparatus uses the basic tools. The basic toolkit studies the essential characteristics of the subject of scientific research. The theoretical methodological apparatus includes (Whitford and Potter 2007):

- the katahrez method is a combination of differentiated parts of a common subject of study;
- the method of essential perception is a definition of the foundations of the subject of scientific research;
- the method of invariant formalism is the study of the subject of scientific research in one plane;
- the method of abstraction is the study of additional characteristics of the subject of scientific research.

A comprehensive presentation of the subject of research is reflected in the combination of theoretical methods and practical measures. The complex methodical apparatus includes: the method of identification installations, the method of transparent parts, the method of imitative perception. The method of identification installations reflects the patterns between the compatible fundamentals of the subject under study. The method of transparent parts explores the external and the internal factors of a scientific subject. The method of imitative perception models the final result of the study. Terms of the modeling are the formation and implementation of the basic prerequisites of a scientific subject (Volosov et al. 2006).

In general, the methodological apparatus of scientific research allows you to create the mechanism for the implementation of the regional economic policy. An important condition for the application of these methods is their adaptation and transformation to the existing conditions for the functioning of the external the regional environment.

The chronology of the research includes five main stages. The chronology of the regional economic policy determines the basis for the development of the concept and essence of the regional economy (Fig. 1).

The regional economic policy appeared in the early 20s of the 20th century. The stage of inception is based on the territorial principle of the regional economic policy. The stage of origin is called the territorial regionalization. The territorial regionalization is the division of the national economy in the territory with the aim of the greatest control of the smallest objects of regionalization (Cooke 2001). The territorial regionalization was aimed at the formation of a new system for managing the country's economy. The territorial regionalization was based on the empowerment of local authority. The local government is the power of the region. The region wasn't part of the territory. The region was defined as the territory of the management.

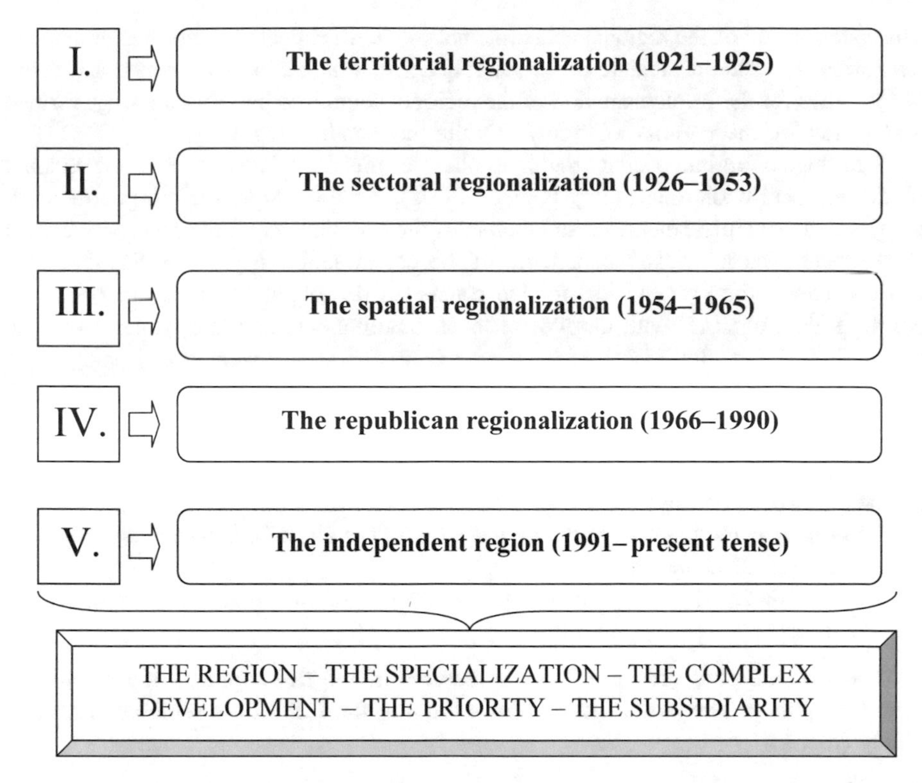

Fig. 1. Chronology of the development of the regional economic policy

The next stage in the development of the regional economic policy is the period of the sectoral regionalization. The period of the sectoral regionalization took place in 1926–1953 (Pakulina 2015). The basis of the economic policy is the industry. Regions form the sectoral national economy. The territory goes into the background. The concept of "the regional economic policy" isn't formed. Economic policy reflects the field of the national economy. The economic policy accumulates industries in large regions of the country.

The regional economic policy delimited the territory in 1953. The regional economy acquires the 1954. Territoriality is becoming the main characteristic of the regional economy. Spatiality is an additional characteristic of the regional economy (Hloponina 2008).

The spatial nature of the regionalization involves the creation of a full-fledged subject of the territory. The regional economic policy takes the form of a subject with authority. Industry parity is preserved. The basis of the regionalization of the economy has focused efforts on the harmonization of the two subjects: industry and territory. The regional economic policy began to study the sectoral interaction of the region. This format of relations is replaced by the republican regionalization (1966–1990). The republican regionalization has enlarged the foundations and measures to achieve economic balance (Kinal and Ratner 1986). Regions reflected the territorial character.

The specification of the regional economic policy didn't reflect the priorities of the new territories. The sectoral feature of the republican regionalization has created a vacuum of measures for the implementation of the regional economic policy. This stage formed a new stage of the regional economy. It's the independent region.

The stage of an independent region implies "artificial" independence in the conduct of the regional economic policy (Garaev 2008). At this stage, definitions, methodologies and principles, elements and factors of the regional economic policy have been worked out. The lack of independence of the constituent entity of the Russian Federation reflects the prerequisites for the consistent development of the regional economic policy, together with the federal level. Components of the regional economic policy at this stage are:

- the region is the territory of the subject of the Russian Federation;
- the specialization is the developed area of the regional economic policy;
- the integrated development is the activities and tools aimed at the development of the regional economy;
- the priority is the integration of key events with the aim of a balanced development of the regional economy;
- the subsidiarity is a solution to regional problems due to the powers of higher authorities.

The selected components reflect the features of the development of the regional economic policy. The current stage allows arguing the role of the regional economic policy (Fig. 2).

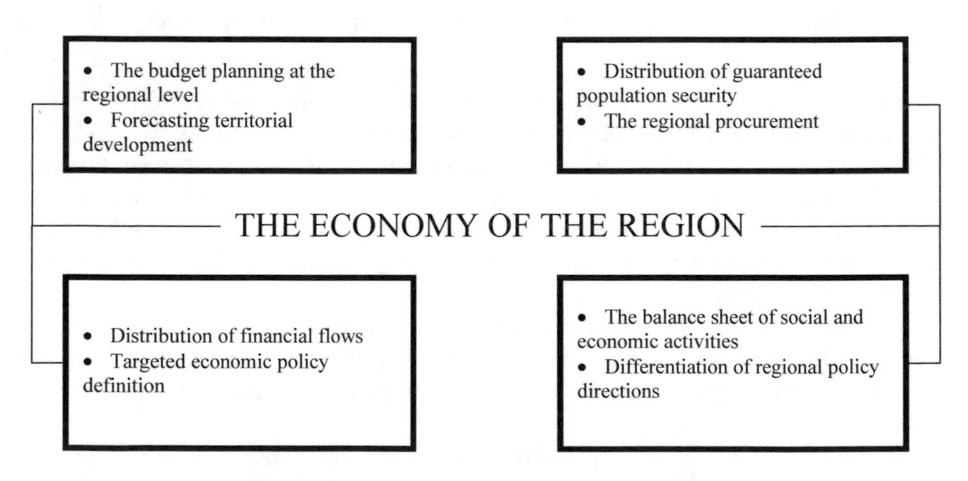

Fig. 2. The role of the regional economic policy

The role of the regional economic policy is reflected in block components. The blocks of the regional economic policy are conditional. The blocks of the regional economic policies reflect the planned function, the distribution function, the objective function and the control function. The planned function of the regional economic policy is implemented when planning the budget process and forecasting territorial development. Planning is based on the strategy of the desired result. The distribution function forms the regional procurement activity. The role of the regional economic policy is based on the economic postulates of the constituent entities of the Russian Federation. The objective function has a separate character. The objective function is manifested in the distribution of financial flows and targeted abstraction of the economic policy. The abstraction of the economic policy is focused on the elimination of inefficient mechanisms. These mechanisms include: the subjective – assessment tool, the subjective – qualitative mean, the subjective – methodical technologie. As part of the abstraction, there is a rejection of the subjectivity of thinking. The final function is the control function. The control function implements quite significant measures for the development of the regional economic policy. The control function is based on the application of the balance sheet of socio – economic activities.

In general, the role of the regional economic policy is regulated. The role of the regional economic policy is reflected in the functional apparatus of the regional development. The aspect of functionality is multiple. The multiplicity is manifested in the functional delineation of the regional economic policy. The role of the regional economic policy includes the dual multiplicity of each function. Duality reflects the basic laws of the regional economic policy. This differentiation is conditional.

3 Results

The role of the regional policy reflects the specific features. An important condition for the study of the regional economic policy is the allocation of strategic management designs (Fig. 3).

The role of the regional economic policy determines the internal components of the development of the territory. The design of the regional economic policy determines the external form. The external form of the regional economic policy reflects important circumstances.

Firstly, the regional economic policy is implemented within the framework of an economic strategy. The regional economic policy is part of the economic strategy. The economic strategy isn't part of the regional economic policy. The strategy implies having an undated plan. The strategy reflects the multicomponent directions to achieve the result. The policy is a regulated action.

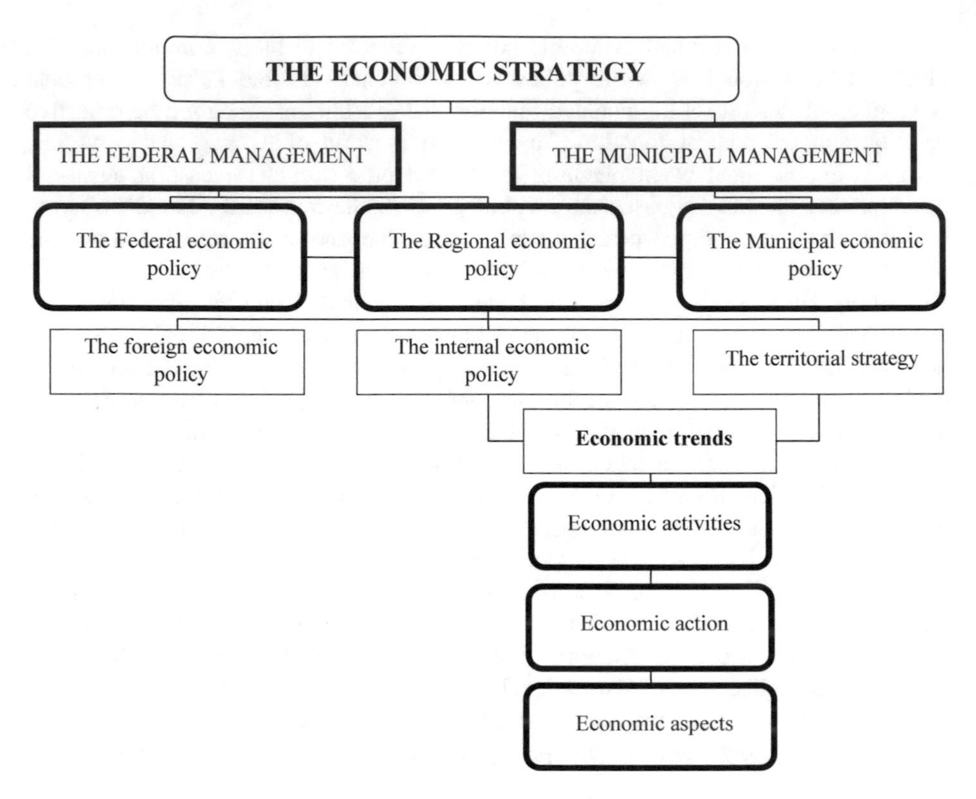

Fig. 3. The design of the regional economic policy

Secondly, the regional economic policy is limited by independence in making management decisions. The empowerment of the Russian Federation takes place on the basis of a two-tier system are federal and municipal authorities. The municipal management is separate in making management decisions at the local level. The federal management is delimited by two subsystems these are federal authorities and regional authorities. The federal government pursues a single the economic policy. The uniqueness and consistency of the economic activities leads to limited management decision-making at the level of the subject of the Russian Federation.

Thirdly, the regional economic policy is manifested through economic relations. The scope of the economic relations reflects the enlarged parts of the economic policy. The economic relations are invariant. The scope of the economic relations include the foreign economic policy, the internal economic policy, the territorial strategy. The foreign economic policy realizes the export potential of the regions. The foreign economic policy is aimed at the organization of the economic relations with non-residents. The internal economic policy and the territorial strategy imply a common segmentation of economic areas. These areas of the regional economic policy include:

- the economic measure are measures to stimulate the development of the regional economic policies;

- the economic action are practical proposals for the implementation of the regional economic policies;
- the economic aspect are the methodological prerequisites for the formation of the regional economic policy.

Economic measures, economic actions and economic aspects are the main mechanisms of the regional economic policy. Economic measures reflect the end result of the regional economic policy. The stated goal of the regional economic policy is theorized. Events act as the endpoint loudspeaker.

Activities converge with economic activities. Economic actions form proposals. The proposals are aimed at realizing the goal of the regional economic policy. Economic actions reflect the proposals used in practice. The effectiveness of the proposals reflects the positive aspects of economic activities. Effective and efficient action will allow you to quickly form the necessary economic event. Economic action with a high level of risk factor slows down the process of achieving the goal of the regional economic policy.

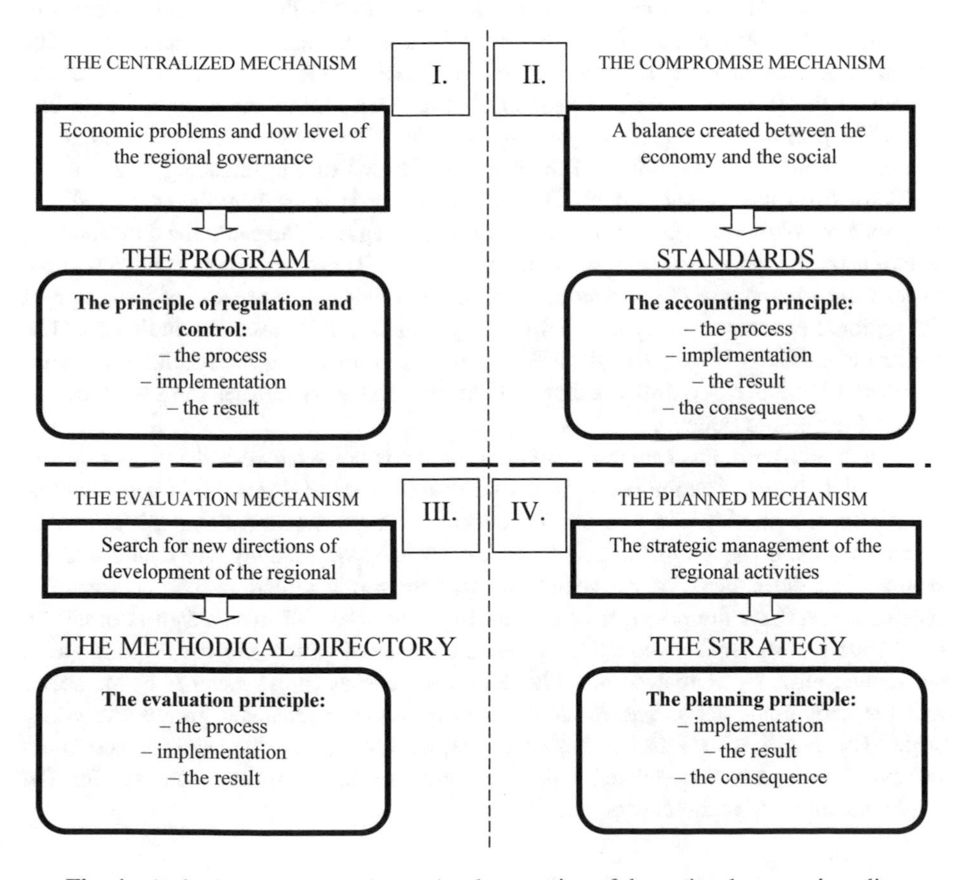

Fig. 4. Author's mechanisms for the implementation of the regional economic policy

Economic aspects act as methodological mechanisms for the implementation of the regional economic policy. Economic aspects are primary in relation to the actions and activities. Economic aspects form the goal. Optimization indicators require practical application.

Formed mechanisms for the implementation of the regional economic policies suggest an algorithm of theoretical methods. These methods implement practical directions. Mechanisms for the implementation of the regional economic policies are limited. The elaboration of the mechanism slows down the implementation of the regional economic policies into practice. Based on this circumstance, the significance of the time factor is reduced. The author's mechanisms for the implementation of the regional economic policies are presented in Fig. 4.

Author's mechanisms for the implementation of the regional economic policies are methodological tools. Each author's mechanism for implementing regional economic policy will be discussed in detail.

The centralized mechanism for implementing the regional economic policy is used when a large number of economic problems occur. The centralized mechanism for the implementation of the regional economic policies is used with a low level of decision-making by regional authorities. The centralized mechanism is a framework. The adoption of economic decisions at the regional level is due to the subsidiarity of the actions of the federal center. The program is the main tool of the centralized mechanism. The program is a means of systematically describing actions to achieve the goals of the regional economic policy. The program is based on the financing of economic activities from the federal budget. This circumstance is caused by the scale and significance of solutions to the economic issues of the region. The centralized mechanism is based on the principles of regulation and control. These principles concentrate the phased implementation of the regional economic policies. The process of formation of the regional economic policy is based on the regulation of assessment indicators. The control is carried out over the identified issues. Control forms the result. In the subsequent principle of control are implied on the consequences of the result of the regional economic policy.

The compromise mechanism is used to create a balance between the economic and the social activities. The limited funding of measures creates a vacuum of opportunities for the implementation of the regional economic policy measures. Standards are being formed to solve the problem. Standards are focused on meeting the conditions for maintaining the balance of the social and the economic direction. The compromise mechanism reflects the principle of accounting. The trade-off mechanism is based on evaluative and statistical accounting. The standard development process accumulates a substantial array of estimated data. The data array allows for an analysis of the social and the economic issues. The result of a compromise mechanism groups the rating scale. The rating scale sets the degree of implementation of the regional economic policy. Estimates are correlated with indicators of an optimistic forecast for the implementation of regional economic policies.

The evaluation mechanism reflects a high degree of analytics of quantitative indicators. The evaluation mechanism is used to search for new directions of development of the regional economy. The evaluation mechanism involves conducting intersectoral research to identify new areas of economic growth in the region. The methodical director are the main tools of the evaluation mechanism. Methodical directories are focused on the matrix assessment of qualitative and quantitative indicators of industries. Methodical directories form the tools for the development of a specific direction of the regional economy. Methodical directories reflect the theoretical elaboration of measures for the development of the regional economy. The regional economic policy combines theoretical and practical orientation. Theorization implies estimated outcome indicators. The result of the methodical reference book allows you to assess the level of implementation of the regional economic policy. The evaluation mechanism doesn't reflect the consequences of the regional economic policy implementation.

The planned mechanism is the strategic. The planned mechanism is implemented in conjunction with other mechanisms. The planned mechanism is the strategic management of regional activities. The strategy serves as a tool for the implementation of the regional economic policy. The strategy produces the basics of the forecasting regional economic policy. The strategy is aimed at identifying the result. The regional economic policy is based on the result planning. The strategy distorts the regional economic policy in the process of its implementation in practice. Risk minimization is the result of the regional economic policy implementation. The result obtained forms the consequences. The consequences are the driving forces behind the development of the regional economy.

In general, the author's mechanisms for the implementation of regional economic policies are the most grouped and least additive. Regional economic policies are the heterogeneous and the identifiable.

4 Conclusions

The study made it possible to draw conclusions.

1. The chronology of the development of the regional economic policy is governed by the territorial and the sectoral principles. Principles are replaced depending on the priorities of the regional economic policy over time. To date, the priority factors of the regional economic policy are the region, the specialization, the integrated development, the priority, the subsidiarity.
2. The role of the regional economic policy is declared in the context of the directions and functions of the development of territories. These directions are based on the prerequisites of the planning, the distribution, the target and the control functions.
3. The design of the regional economic policy reflects externalities. Mechanisms for the implementation of regional economic policies are displayed within economic areas. These include: economic measures, economic actions, economic aspects. Mechanisms are limited in functionality and time factor.

4. Author's mechanisms for the implementation of regional economic policies apply the program, the standard, the methodical director, the strategy. Copyright mechanisms implement the regional economic policies in various directions. Author mechanisms allow solving problems of the regional economic policy. This circumstance confirms the significance of the issue of the regional economic policy.

References

Cooke, Ph.: Regional innovation systems, clusters, and the knowledge economy. Ind. Corp. Change **10**(4), 945 (2001)

Garaev, M.M.: Problems and ways of improvement of criteria ensuring public administration with region economy. Reg. Econ. **3**(15), 228–234 (2008)

Hloponina, N.A.: Classification of animators and their interrelation in regional economy. Bull. Pac. Natl. Univ. **2**(9), 229–242 (2008)

Kinal, T., Ratner, J.: A VAR forecasting model of a regional economy: it's construction and comparative accuracy. Int. Reg. Sci. Rev. **10**(2), 113–126 (1986)

Lyapina, I., Mashegov, P., Petrukhina, E., Stroeva, O., Maltsev, A.: Institutional effects in development of regional innovational infrastructure. Int. J. Trade Glob. Markets **12**(1), 26–42 (2019)

O'Huallachain, B.: Regional growth in a knowledge-based economy. Int. Reg. Sci. Rev. **30**(3), 221–248 (2007)

Pakulina, A.: Regulation of the development of important social activities of the regional economy. Nauka i studia **9**, 109–113 (2015)

Stroeva, O.A., Kvak, A.A.: Tools of implementation of regional innovative policy of pro-active character. Central Russ. Bull. Soc. Sci. **6**(36), 102–109 (2014)

Volosov, A.I., Proskura, D.V., Tihomirov, S.A.: Strategic management of transition of regions to innovative economy. Econ. Manage. **4**(29), 131–134 (2006)

Whitford, J., Potter, C.: Regional economies, open networks and the spatial fragmentation of production. Socio-Econ. Rev. **5**, 497–526 (2007)

The Impact of the Digital Economy on the Quality of Life

Vladimir M. Razumovsky[1](✉), Alexandra V. Sultanova[2], Oksana S. Chechina[2], and Svetlana A. Nikonorova[3]

[1] FSBEI of HE "St. Petersburg State Economic University", St. Petersburg, Russian Federation
vrm-rgo@mail.ru

[2] FSBEI of HE "Samara State Technical University", Samara, Russian Federation
sultanovaav@mail.ru, Chechinaos@yandex.ru

[3] FSBEI of HE "Vladimir State University, Alexander Grigorievich and Nikolai Grigorievich Stoletov", Vladimir, Russian Federation
sveta_nikonorova@mail.ru

Abstract. The article discusses issues related to the development of digital economy in modern conditions, as well as the degree of influence of digitalization on the quality of life of the population. The main vectors and directions of information and communication technologies development are presented, as well as prospects of using digitalization of economy. The main problems that obstruct the introduction of digital economy in resolving issues of the socio-economic life of the population are identified. Activities of state programs for the development and implementation of digital economy are analyzed, the most pressing problems for implementing tasks set by the Russian government are identified. The article contains a list of the most frequently used mobile applications that help the population solve issues of sharing knowledge and experience, information, purchasing and selling goods, facilitate communication between people regardless of their location, provide an opportunity to pay for housing and communal services, fines and taxes with minimum time, speed up the process of solving the population's transportation problems. All conclusions formulated in the article are confirmed by statistical information presented in official sources. The article proposed the most relevant, in the opinion of the author, solutions of developing digital economy to improve the population's quality of life.

Keywords: IT technologies · Population's quality of life · Mobile applications · Priorities · Digital economy · Digital space

JEL Code: O33 · O35 · O38

Nowadays, it is impossible to imagine any sphere of life that is not presented in digital space. Starting with the government to small and medium businesses. If you are not on the Internet, you don't exist in general. In 2017, the Russian government signed the Digital Economy of the Russian Federation 2024 program, which is designed to solve a number of tasks to ensure digital transformation of the economic and social spheres of

E. G. Popkova and B. S. Sergi (Eds.): ISC 2019, LNNS 87, pp. 417–423, 2020.
https://doi.org/10.1007/978-3-030-29586-8_48

the Russian Federation by that year. In order to consider how this program is being implemented in Russian economy, it is necessary to mention that the very term "digital economy" (digital economy) was first used relatively recently, in 1995, by an American scientist from the University of Massachusetts, Nicholas Negroponte, to explain to his colleagues the advantages of the new economy in comparison with the old one due to the intensive development of information and communication technologies (28.09.2017).

Having studied the experience of a number of countries, it is unfortunately obvious that our country is far behind the technologically developed European countries, the countries of the Middle East and the United States in terms of digital equipment. Therefore, Russian specialists are presented with complex and multifaceted tasks. Currently, there are several main areas in which it is necessary to work in the first place when it comes to the development of digital economy. At the same time, it is necessary to take into account the specifics of our country: geographical location, age and class composition.

The priority vectors now are:

- the creation of a single digital document - an e-passport of a Russian Federation citizen, where all information about its owner is to be presented: personal, medical, banking, etc. According to the Governmental Bylaw of the Russian Federation dating 19.09.2013 N 1699-p (amended dd. 05.22.2018) " On approval of the Concept of introducing an identity card of a citizen of the Russian Federation in the Russian Federation, issued in the form of a plastic card with an electronic information carrier, and an action plan for implementation of the Bylaws "the document must be valid throughout the country, and its creation will greatly facilitate the movement of citizens throughout the country, outside it, as well as help them in solving everyday and social problems from anywhere in the world;
- the solution of transportation problems (Yandex Transport, Yandex Taxi, etc.). For example, by now Moscow has already moved on to creating a convenient and high-quality public transportation system, and the rest of the cities will come to this in the next 10–15 years. In order for all segments of the population to feel comfortable, not just the owners of private cars, citizens must have a choice: personal cars or convenient and affordable public transport. Applications like Google Maps, Yandex. Taxi", "Yandex Transport", "Maps me", "2GIS" and some others help partially solve this problem. They plan the route from point A to point B, taking into account the traffic situation at the moment, offer several options: on foot, public transport (metro, trolleybus, bus, tram, route taxi) or private car. The advantage of these applications is that they allow you to track the movement of public transportation in real time: the solution to the problem of long awaiting.

- the service enhancement in cultural, leisure and tourism areas (Yandex. Poster, Artefact, AIS "Cultural Region"). An interesting fact is that the program "Development of TV and Radio Broadcasting in the Russian Federation for 2009–2018" (2013) which allows to provide the population of the Russian Federation with multichannel broadcasting with guaranteed provision of all Russian compulsory publicly available TV channels and radio channels of a given quality. But in general, this area requires further development, since it gives the population the opportunity to organize their cultural leisure easier and faster, which will definitely lead us to raise the cultural level of citizens;
- the simplification of social life of citizens, creation of favorable public spaces, increase of security level (the system SCAUT is a comprehensive solution for Satellite Control, Analytics and Transport Management, Bars Group - Russian software developer, "My Home" System is a social project for managing your home). The development of this area is necessary because it helps optimize logistics within the country and the supply of goods from abroad. Digital development in the field of housing and communal services makes it easier to pay utilities and taxes. Issues of public safety are now separate, as terrorist attacks and various offenses present a separate problem.
- the optimization of medical issues (Telemedicine - the direction of medicine based on the use of modern communication technologies for the provision of remote medical care and consultations (25.10.2016), Unified Medical Portal). These platforms allow you to quickly and easily carry out a patient-doctor dialogue almost 24/7. Doctors can give advice on treatment at any time convenient for them, attract new patients who, in turn, can make an appointment with any specialist without leaving their home. An interesting fact is that the analogue of the project "Telemedicine" was proposed in Australia in the 1970s (2008). The country's highways were equipped with toll free telephones, which, in the event of poor health, drivers could use and get prompt expert advice. This application in our country allows us to conduct a dialogue with a doctor at any convenient time, in illustrated form, to monitor medical indicators, which is especially important for people with chronic diseases, when all processes need to be considered in dynamics.
- the improvement of the urban environment comfort for citizens with children (Kid friendly application - Children are welcome here). This project solves a number of tasks: educational and social. First, on the company's website and in official applications, parents can find educational content for their children: how to teach a child to say "no" (to minimize the problem of the possible communication of children with strangers on the street); how to teach them good manners, etc. On the other hand, innovative, in this case, is collecting all the "child-friendly" locations in one place. At the moment, the project offers a list of 4322 positions: cafes, leisure and public spaces in which there are facilities confirmed for children and parents: toilet and changing rooms, drinking water, child seats, cribs, etc.

– the optimization of the education system, by monitoring a single student card and collecting information about all educational institutions on the same site (Single Student Card, Electronic Journal "ElJour"). These websites help solve problems of controlling a child's attendance at school, his progress, moreover, parents can pay for meals at any convenient time and from any place. In addition, this site solves the problem for applicant's of remote submission of documents to universities located in other cities.

In modern conditions, digital technologies are an integral part of the social development of society, which is vividly represented by the following indicators of growth in use of various mobile applications (Fig. 1).

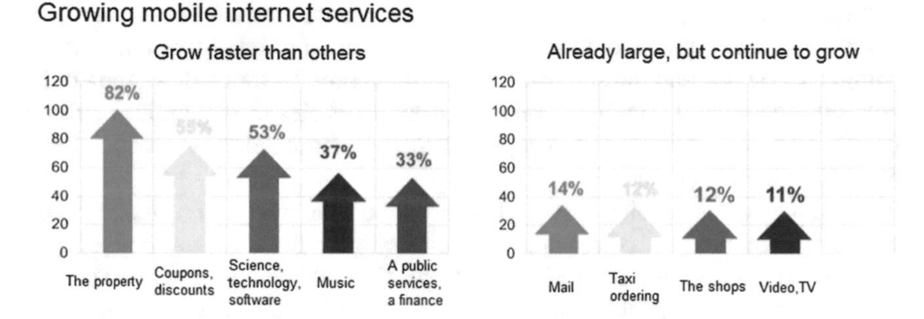

Fig. 1. Growing mobile internet services (03.10.2018)

As we can see, projects that resolve the population's various socio-economic problems already exist. Although, according to Maxim Akimov (Dzyadko, Kanaev, 12.02.2019), a politic, deputy chairman of the board of the Government of the Russian Federation, the main problems in solving a number of tasks set, are:

1. Lack of highly specialized personnel. To solve this issue, an action plan called "Personnel and Education" of the Digital Economy program of the Russian Federation was drawn up, according to which: at least 40% by 2021 should make up the proportion of the population with digital skills according to an average of 15% annually until 2020 it is planned to increase the target numbers for admission to Russian universities in IT specialisations (13.12.2018);
2. The slow pace of the "Digital Economy of the Russian Federation 2024" program development, as evidenced by the results of studies presented in Figs. 2 and 3 (collection "Indicators of digital economy: 2018" (Abdrakhmanova, Vishnevsky, Volkova, Gokhberg, etc. 2018).

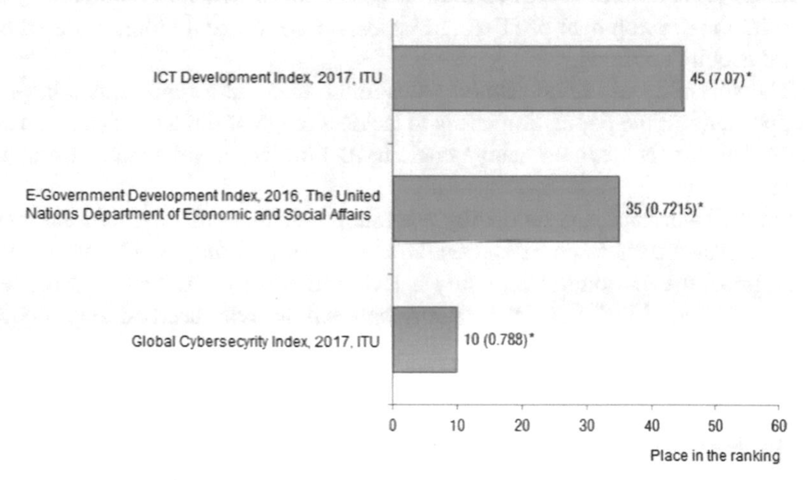

Fig. 2. Russia's place in international ratings of digital economy (Abdrakhmanova, Vishnevsky, Volkova, Gokhberg, etc. 2018)

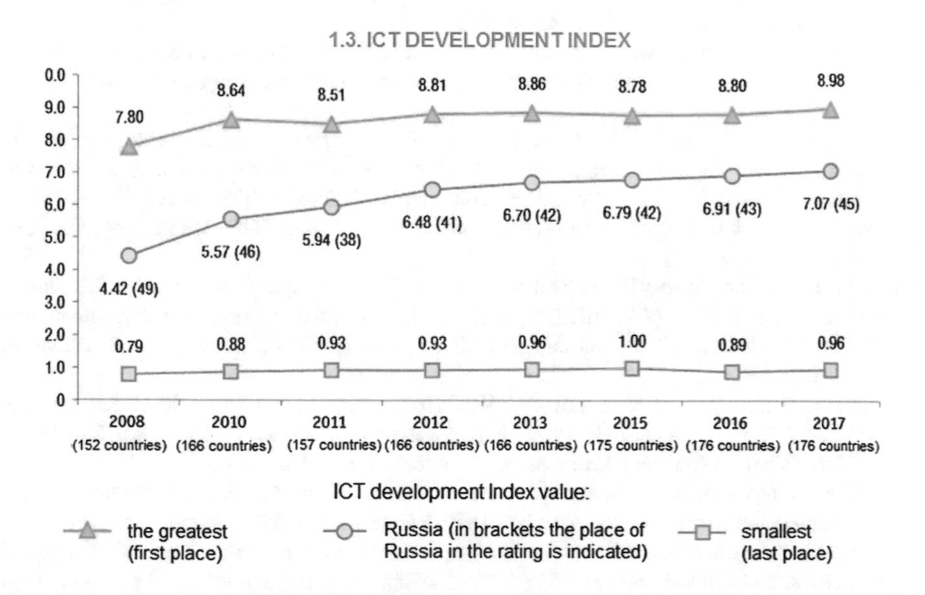

Fig. 3. ICT Development Index (Abdrakhmanova, Vishnevsky, Volkova, Gokhberg, etc. 2018)

You can also make some additional conclusions. The main problem in any reform related to digital economy is the lack of citizen awareness about the use of modern IT technologies. All these applications and developments are generally available only to young, socially active people, although some applications are more needed by older citizens. Thus, the complexity of digitalizing economy is that the state program does not fully take into account the state of the country's social life. Uneven distribution of resources leads to uneven exploitation of goods: those who have something, with time and with the development of IT technologies, have more and more, those who do not - do not acquire anything.

The versatility of digital economy lies in the fact that it represents a huge platform of opportunities: the population needs to tackle a range of different social and economic issues. This aspect can help many people try themselves as urbanists, designers, public figures.

On the other hand, the rationality and one-pointedness of solving these issues is the following: the state has embarked on the path of simplifying the life of citizens. Many of the problems that digital economy is trying to solve are united by a noble goal: to make cities individually and the country as a whole more accessible and comfortable for all categories of citizens.

References

Adindex.ru. Runet mobile audience: numbers, trends, largest applications. Accessed 03 February 2019

Fingramota.org. What is digital economy. http://www.fingramota.org/teoriya-finansov/item/2198-chto-takoe. Accessed 03 February 2019

Kid friendly - Children are welcome here. https://kidfriendly.ru/. Accessed 03 February 2019

BARS Group - Russian software developer, available at: https://bars.group/about/. Accessed 03 February 2019

Abdrakhmanova, G.: Indicators of digital economy: 2018: statistical compilation. In: Abdrakhmanova, G.I., Vishnevsky, K.O., Volkova, G.L., Gokhberg, L.M., et al., I60 Nat. researches University "Higher School of Economics". HSE. ISBN 978-5-7598-1770-3 (in the region) (2018). https://www.hse.ru/data/2018/08/20/1154812142/ICE2018.pdf.pdf. Accessed 03 February 2019

Vladzimirsky, A.: The history of telemedicine: people, facts, technology. - Donetsk: LLC Digital Printing House ISSN 1728-936X (Appendix to the Ukrainian Journal of Telemedicine and Medical Telematics) (2008). http://itelemedicine.pro/pages/files/telemed_hist.pdf. Accessed 03 February 2019

Dzyadko, T., Kanaev, P.: Maxim Akimov - RBC: It is necessary to change the structure of the market, not monopolize. https://www.rbc.ru/interview/economics/12/02/2019/5c627c629a7947cc28d6c666?from=center_1. Accessed 02 March 2019

Legislation of the Russian Federation. Rulaws.ru. Order of the Government of the Russian Federation dated September 19, 2013 No. 1699-p (amended on 05.22.2018) "On Approval of the Concept for the Introduction in the Russian Federation of an Identity Card of a Citizen of the Russian Federation Issued as a Plastic Card with an Electronic Media, and an Action Plan on the implementation of the Concept". http://rulaws.ru/goverment/Rasporyazhenie-Pravitelstva-RF-ot-19.09.2013-N-1699-r/. Accessed 02 March 2019

MyHouse. Know what is happening in your house. All information about the house is collected in one place. https://moydom.ru/. Accessed 03 February 2019

Decree of December 3, 2009 N 985 “On the federal target program “Development of TV and radio broadcasting in the Russian Federation for 2009–2015” (with amendments and additions). http://spb.rtrs.ru/upload/iblock/b3c/ftsp.pdf. Accessed 03 February 2019

RAEC. Runet summed up the year. https://raec.ru/live/raec-news/10766/. Accessed 03 February 2019

SCOUT. Satellite Control, Analytics and Transport Management. https://scout-gps.ru/o-systeme/. Accessed 03 February 2019

Telemedicine.ru. What is telemedicine. https://telemedicina.ru/news/world/chto-takoe-telemeditsina. Accessed 03 February 2019

Managing the Development of Digital Infrastructural Provision of Entrepreneurial Activities in the Agricultural Machinery Market

Tatiana N. Litvinova(✉)

Volgograd State Agrarian University, Volgograd, Russian Federation
litvinova1358@yandex.ru

Abstract. Purpose: The purpose of the research is to substantiate the necessity and to develop recommendations for improving the practice of managing the development of digital infrastructural provision of entrepreneurial activities in the modern Russia's agricultural machinery market.

Design/Methodology/Approach: The method of regression analysis is used for determining the contribution of digital infrastructural provision into development of entrepreneurial activities in the agricultural machinery market. The assessed indicators are gross added value of production of cars and equipment in main prices and the number of users of broadband Internet with wireless access. As a result, it is proved that digital infrastructural provision is a significant factor of development of entrepreneurial activities in the agricultural machinery market of modern Russia.

Findings: It is substantiated that digital infrastructural provision of entrepreneurship in the agricultural machinery market of modern Russia is insufficient. This problem is caused not by deficit of the telecommunication infrastructure (e.g., access to Internet) but by disproportions in development of digital infrastructural provision of entrepreneurial activities in the agricultural machinery market in Russia, which appeared due to unbalanced management. In the normative and legal documents of modern Russia and statistical reports, digital infrastructural provision of entrepreneurial activities is connected to telecommunication infrastructure. Other inseparable components of infrastructure (e.g., financial and human) are not taken into account. As a result, due to state support, most companies in the Russian agricultural machinery market (more than 90%) have access to the telecommunication infrastructure, but cannot used it in their economic practice due to deficit of the financial and human infrastructure.

Originality/Value: For improving the practice of managing the development of digital infrastructural provision of entrepreneurial activities in the modern Russia's agricultural machinery market, the author offers the measures for complex development of all components of the infrastructure.

Keywords: Management · Development · Digital infrastructural provision · Entrepreneurial activities · Market of agricultural machinery · Modern Russia

JEL Code: H54 · Q13 · Q18 · L26 · O31 · O32 · O33 · O38

E. G. Popkova and B. S. Sergi (Eds.): ISC 2019, LNNS 87, pp. 424–431, 2020.
https://doi.org/10.1007/978-3-030-29586-8_49

1 Introduction

Provision of food security is one of the most important problems of modern humanity. This problem is especially urgent in the countries that do not specialize in food production. A vivid example of these countries is Russia – with developed real sector and post-industrial production specialization of economy. In 2018, the level of food security of Russia was given 42 points out of 100 (67th position in the global rating of countries as to the level of food security) (The Economist Intelligence Unit 2019).

The most probable reason of the low level of national food security of Russia is low efficiency of agriculture. According to the calculations of the Federal State Statistics Service, the level of physical wear and tear of the main funds in the sphere of agriculture in 2018 constituted 38.2%. In view of moral wear and tear, which is not reflected in the official statistics, aggregate wear and tear of agricultural machinery in Russia could exceed 50%.

For solving this problem, the Government of the Russian Federation (2019) adopted the Strategy of development of agricultural machine building of Russia until 2030 (Decree dated July 7, 2017, No. 1455-r). The main attention in it is paid to digital modernization of entrepreneurship in the Russian agricultural machinery market for increasing the effectiveness of entrepreneurship in this market and growth of competitiveness of domestic agricultural machinery.

In view of the preserving high share of import of agricultural machinery in Russia – 44% in 2018 (Strategy of development of agricultural machine building of Russia until 2030), the following hypothesis is offered: the level of digital development of the Russian market of agricultural machinery is not sufficiently high for its full-scale import substitution (reduction of the share of import to the level below 20%) due to underdevelopment of digital infrastructural provision. The purpose of the paper is to substantiate the necessity and to develop recommendations for improving the practice of managing the development of digital infrastructural provision of entrepreneurial activities in the modern Russia's agricultural machinery market.

2 Materials and Method

Specifics of infrastructural provision of entrepreneurial activities in the conditions of the digital economy are studied in the works (Ashrafi and Zare Ravasan 2018), (Bryson et al. 2018), and (Yang 2018). The issues of development of entrepreneurial activities in the agricultural machinery market are discussed in the works (Bogoviz et al. 2019), (Litvinova et al. 2016), (Litvinova et al. 2017), (Litvinova et al. 2019), (Morozova et al. 2018), (Popkova 2019), and (Troyanskaya et al. 2017). At the same time, the problem of managing the development of digital infrastructural provision of entrepreneurial activities in the agricultural machinery market is poorly studied. The method of regression analysis is used for determining the contribution of digital infrastructural provision into development of entrepreneurial activities in the agricultural machinery market.

The assessed indicators are gross added value of production of machines and equipment in main prices (indicator of development of entrepreneurship, y) and the number of users of broadband Internet with wireless access (indicator of digital infrastructural provision, x). The initial selection of statistical data is very small – it is limited by 2012–2017 – that's why it is supplemented by the forecast data for 2018–2021. Dynamics of factual and forecast values of the selected indicators are shown in Figs. 1 and 2.

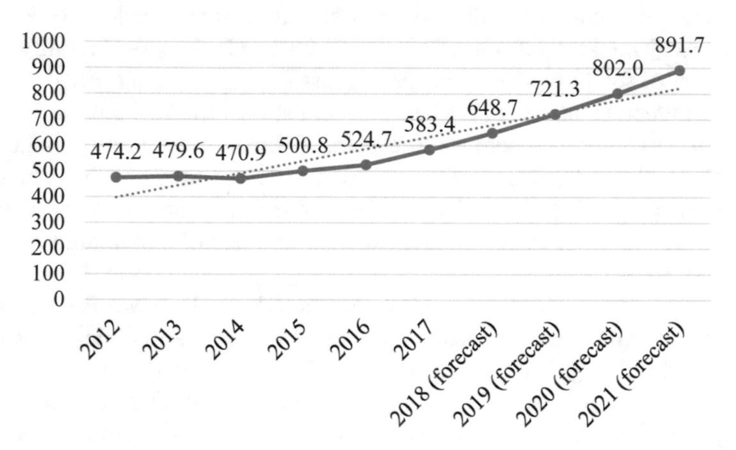

Fig. 1. Dynamics of gross added value of production of machines and equipment in main prices in 2012–2021, RUB billion. Source: compiled by the author based on the materials of National Research University "Higher School of Economics" (2019b).

Figure 1 shows that gross added value of production of machines and equipment in the main prices is characterized by upward trend – its growth in 2017 (RUB 583.4 billion) as compared to 2012 (RUB 474.2 billion) constituted 23.03%.

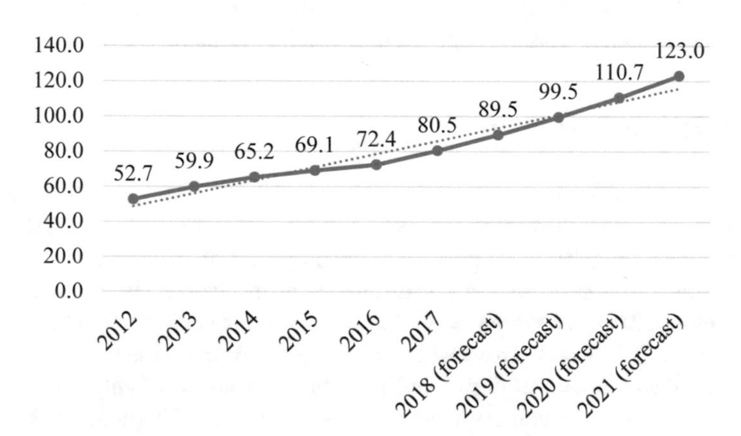

Fig. 2. Dynamics of the number of users of broadband Internet with wireless access in 2012–2021, per 100.000 people as of year-end. Source: compiled by the author based on the materials of National Research University "Higher School of Economics" (2019c).

Figure 2 shows that the number of users of broadband Internet with wireless access is peculiar for upward trend – its growth in 2017 (80.5 per 100,000 people as of year-end) as compared to 2012 (52.7 per 100,000 people as of year-end) constituted 52.75%. The results of regression analysis are given in Table 1.

Table 1. Results of regression analysis of dependence of gross added value of production of machines and equipment on the number of users of broadband Internet with wireless access.

Regression statistics	
Multiple R	0.9857
R^2	0.9715
Adjusted R-square	0.9680
Standard error	26.9935
Observations	10

Dispersion analysis

	df	*SS*	*MS*	*F*	*Significance F*
Regression	1	198942.7756	198942.7756	273.0293	1.81643E-07
Residue	8	5829.1989	728.6499		
Total	9	204771.9745			

	Coefficients	*Standard error*	*t-Stat*	*P-Value*	*Lower 95%*	*Upper 95%*
Y-intercept	77.8376	33.3021	2.3373	0.0476	1.0428	154.6325
£	6.4666	0.3914	16.5236	1.81643E-07	5.5641	7.3691

Source: calculated by the authors.

According to Table 1, gross added value of production of machines and equipment grows by RUB 6.4666 billion with increase of the number of users of broadband Internet with wireless access by 1 per 100 people with probability of 98.57% (as the estimate coefficient £ = 6.4666, coefficient of determination - $R^2 = 0.9715$). Significance F = 1.81643E-07 (less than 0.05), which shows statistical significance of the determined regression dependence at the level $\alpha = 0.05$. Therefore, digital infrastructural provision is an important factor of development of entrepreneurial activities in the agricultural machinery market of modern Russia.

3 Results

Let us perform an overview of digital infrastructural provision of entrepreneurial activities in the agricultural machinery market (in processing productions) and activity of its practical application in Russia in 2018 (Table 2).

Table 2. Overview of digital infrastructural provision of entrepreneurial activities in the agricultural machinery market (in processing productions) and activity of its practical application in Russia in 2018.

	Digital technology	Share of companies in the Russian agricultural machinery market that have access to infrastructural provision, which is necessary for using the technologies	Directions of usage of technologies in production of agricultural machinery		Share of companies in the Russian agricultural machinery market that use the technology
Internet	Broadband Internet	91.6%	Internet purchases and sales	online purchases	20.1%
	providing employees with mobile devices for Internet access	50.3% of companies, 2.3% of employees		online sales	19.5%
Technologies of provision of digital security	regularly updated anti-virus programs	85.1%	financial calculations in the electronic form		69.3%
	means of digital electronic signature	84.5%			
	Program means that hinder the unsanctioned access of malicious software	69.0% 61.0%			
	Encryption means	53.9%	provision of access to data bases through global information networks		28.9%
	Systems of determining the intrusion into computer or network	45.2%			
	Program means of authomatization of the processes of analysis and control of protection of computer systems	33.6%			
Technologies of automatization	ERP-systems	27.1%	automatization of planning, accounting, control, and analysis of all main business operations of the company		Solving organizational, managerial, and economic tasks: 66.4%
	Cloud technologies	25.7%			
	RFID-technologies	10.7%			
	CRM-systems	17.6%	automatization of management of interrelations with customers		
	SCM-systems	6.9%	automatization of management of supply chains		

Source: compiled by the author based on materials of National Research University "Higher School of Economics" (2019c).

Figure 2 shows that Internet access is available for 91.6% of companies in the Russian agricultural machinery market. Devices for mobile Internet access are provided to 2.3% of employees of 50.3% companies. However, the level of practical application of Internet is much lower. Internet is used for online purchases by 20.15% of companies in the Russian agricultural machinery market, and for sales – by 19.5% of companies.

Technologies of provision of digital security are also available for most companies in the Russian agricultural machinery market. For example, regularly updated anti-virus software is available for 85.1% of companies. 100% of Russian companies have operating accounts – as it is required by the law. However, financial accounts in the electronic form are used only by 69.3% of companies. Access to data bases through global information networks is provided only by 28.9% of companies.

At the same time, the technologies of modernization are available for a small number of companies in the Russian agricultural machinery market. For example, cloud technologies are available only for 25.7% of companies. However, automatization of solving organizational, managerial, and economic tasks is performed by most companies in the Russian agricultural machinery market – 64.4%.

Thus, the performed overview of digital infrastructural provision of entrepreneurial activities in the agricultural machinery market (in processing productions) and activity of its practical application in Russia in 2018 showed that digital infrastructural provision of entrepreneurial activities in the agricultural machinery market is not limited by access to digital technologies, as is shown in the existing statistical collections.

In particular, it is necessary to pay attention to the financial infrastructure of digital modernization of entrepreneurship in the Russian agricultural machinery market. The volume of investments into the main capital of companies in the studied market in 2018 constituted RUB 1,921.2 billion – i.e., 5.18% of their aggregate turnover (RUB 37,112.1 billion), according to the materials of the Federal State Statistics Service (2019). It is also necessary to take into account human infrastructure. The share of specialists in the sphere of information and communication technologies (digital personnel) in the Russian agricultural machinery market is less than 1%, according to materials of National Research University "Higher School of Economics" (2019c).

Based on this, during managing the development of digital infrastructural provision of entrepreneurial activities in the agricultural machinery market it is recommended to pay attention to the issues of development of financial infrastructure through stimulating the investments into fixed capital and to the issues of development of human infrastructure through training of specialists in the sphere of information and communication technologies (digital personnel). This will ensure systemic interaction of the components of the digital infrastructure and the creation of synergetic effect in the form of increase of activity of usage of digital technologies by the companies in the agricultural machinery market.

4 Conclusion

Thus, the working hypothesis is proved: digital infrastructural provision of entrepreneurship in the agricultural machinery market of modern Russia is insufficient. This problem is caused not by deficit of the telecommunication infrastructure (e.g., access to Internet) but by disproportions in development of digital infrastructural provision of entrepreneurial activities in the agricultural machinery market in Russia, which appeared due to unbalanced management.

In the normative and legal documents of modern Russia and statistical reports, digital infrastructural provision of entrepreneurial activities is connected to telecommunication infrastructure. Other inseparable components of the infrastructure (e.g., financial and human) are not taken into account. As a result, due to state support, most companies in the Russian agricultural machinery market (more than 90%) have access to the telecommunication infrastructure, but they cannot use it in their economic practice due to deficit of financial and human infrastructure.

For improving the practice of managing the development of digital infrastructural provision of entrepreneurial activities in the modern Russia's agricultural machinery market, the author offers measures for complex development of all components of the infrastructure. This requires legislative establishment of statistical accounting of the telecommunication, financial, human, and other components of infrastructure of entrepreneurial activities in the modern Russia's agricultural machinery market and implementation of the corresponding measures for stimulation of their development, which development could be performed in further research.

References

Ashrafi, A., Zare Ravasan, A.: How market orientation contributes to innovation and market performance: the roles of business analytics and flexible IT infrastructure. J. Bus. Industr. Market. **33**(7), 970–983 (2018)

Bogoviz, A.V., Sandu, I.S., Demishkevich, G.M., Ryzhenkova, N.E.: Economic aspects of formation of organizational and economic mechanism of the innovational infrastructure of the EAEU countries' agro-industrial complex. Advances in Intelligent Systems and Computing, vol. 726, pp. 108–117 (2019)

Bryson, J.R., Mulhall, R.A., Song, M., Loo, B.L.Y., Dawson, R.J., Rogers, C.D.F.: Alternative-substitute business models and the provision of local infrastructure: alterity as a solution to financialization and public-sector failure. Geoforum **95**, 25–34 (2018)

Litvinova, T.N., Khmeleva, G.A., Ermolina, L.V., Alferova, T.V., Cheryomushkina, I.V.: Scenarios of business development in the agricultural machinery market under conditions of international trade integration. Contemp. Econ. **10**(4), 323–332 (2016)

Litvinova, T.N., Kulikova, E.S., Kuznetsov, V.P., Taranov, P.M.: Marketing as a determinant of the agricultural machinery market development. Contributions to Economics, pp. 465–471 (2017). ISBN: 9783319606958

Litvinova, T.N., Tolmachev, A.V., Saenko, I.I., Iskandaryan, G.O.: Role and meaning of the ICT infrastructure for development of entrepreneurial activities in the Russian agricultural machinery market. Advances in Intelligent Systems and Computing, vol. 726, pp. 793–799 (2019)

Morozova, I.A., Popkova, E.G., Litvinova, T.N.: Sustainable development of global entrepreneurship: infrastructure and perspectives. Int. Entrepreneurship Manage. J. **15**, 1–9 (2018)

Popkova, E.G.: Preconditions of formation and development of industry 4.0 in the conditions of knowledge economy. Studies in Systems. Decision and Control, vol. 169, pp. 65–72 (2019)

The Economist Intelligence Unit. Global Food Security Index (2019). https://foodsecurityindex.eiu.com/Country/Details#Russia. Accessed 11 Apr 2019

Troyanskaya, M.A., Ostrovskiy, V.I., Litvinova, T.N., Matkovskaya, Y.S., Bogoviz, A.V.: Possibilities and perspectives for activation of sales in the agricultural machinery market within sectorial development of Russian and European economies. Contributions to Economics, pp. 473–480 (2017). ISBN 9783319606958

Yang, S.C.: A curriculum model for IT infrastructure management in undergraduate business schools. J. Educ. Bus. **93**(7), 303–313 (2018)

National Research University "Higher School of Economics". Indicators of the digital economy (2019a). https://www.hse.ru/data/2018/08/20/1154812142/ICE2018.pdf.pdf. Accessed 11 Apr 2019

National Research University "Higher School of Economics. The market of agricultural machinery of Russia (2019b). https://dcenter.hse.ru/data/2016/12/29/1114670197/Рынок%20сельскохозяйственных%20машин%202016.pdf. Accessed 11 Apr 2019

National Research University "Higher School of Economics". The digital economy of Russia (2019c). https://www.hse.ru/primarydata/ice2019kr. Accessed 11 Apr 2019

Government of the Russian Federation. Strategy of development of agricultural machine building of Russia until 2030. Adopted by the Decree dated July 7, 2017, No. 1455-r (2019). http://static.government.ru/media/files/Ba4B6YDTiuOitleLkDQ05MCbz4WrfZjA.pdf. Accessed 11 Apr 2019

Federal State Statistics Service of the Russian Federation (2019). Main funds. http://www.gks.ru/wps/wcm/connect/rosstat_main/rosstat/ru/statistics/enterprise/fund/. Accessed 11 Apr 2019

Specific Features and Perspectives of Digital Modernization of the Regional Economy

Digital/Smart Places and Their Strategies: Conceptualizing the Recent Trends

Daniil P. Frolov[1(✉)] and Anna V. Lavrentyeva[2]

[1] Volgograd State Technical University, Volgograd, Russia
ecodev@mail.ru
[2] Volgograd State University, Volgograd, Russia
lavra.ne@mail.ru, oponir@volsu.ru

Keywords: Smart cities · Digital cities · Institutional environment · Digitalization · Digital environment · Smart urbanism

1 Introduction

Digital technologies and digital economy are now among the most urgent topics for discussion in the field of economic policy. Digitalization is a systemic process that permeates macroeconomic systems at all levels of their hierarchy: from national markets and sectors to individual jobs. Therefore, it is predictable that the regional and especially urban development of digital economy attracts particular care of experts.

Since each city has its specific historical, cultural, economic, political and technological background, its current position in the system of digital economy is predetermined by its historical past (Scott 2016), i.e. by the path dependence effect. Therefore, the digitalization of any city and any region will always have a unique character. And each city or region should form its own technological digital path which reflects a long-term development of digital technologies, professions, organizations, industries, clusters and various institutional structures, i.e. an integrated digital ecosystem. This process cannot help but be strategic, as it relates to managing complex ecosystems where any smallest alteration produces cascades of direct and indirect effects accumulating and resonating in unpredictable ways. That is why it is important to shift the processes of digitalization, which occur spontaneously, discretely and fragmentarily in many territorial entities, into a strategic format. Furthermore, the article will focus mainly on cities as more compact objects of management, but all theses and conclusions are correlated with the regional scale of digitalization.

2 Methodology

This study uses various methods of factorial and level-sensitive subject-object analysis, as part of a systematic approach to defining the problems of smart cities and modern urbanism of the digital spaces, including an innovative methodology of post-institutional analysis based on multiparadigmatic and interdisciplinary synthesis.

E. G. Popkova and B. S. Sergi (Eds.): ISC 2019, LNNS 87, pp. 435–441, 2020.
https://doi.org/10.1007/978-3-030-29586-8_50

A modern city has certain digital goals and already achieved results (Dameri and Rosenthal-Sabroux 2014). This is due to the fact that digitalization is a logical continuation of informatization and the knowledge growth of economic and social processes within cities. Now almost any city has an official website or portal on the Internet. Internet crowdsourcing (public collection of ideas) and crowdfunding (public collection of funds for socially significant projects) campaigns were conducted in many cities. A number of cities have developed mobile applications for citizens, which help to hold voting, rating, surveys and other forms of communication between the authorities and the local population. Urban population actively speaks on the Internet to estimate various events concerning the city life, the activity of local government, and also forms media content about the city. The latter function is the most significant feature of Web 2.0 – the Internet, which is dominated by mobile devices, social networks and interactive communications when ordinary users, being not professionals, are the main producers of content. For cities, this means that the residents themselves form the information field of the city, its image, brand, emotional background, insider data, and expert opinions. If the people themselves take a dim view of any actions taken by the city authorities or rate certain infrastructure of the city quite negatively, then no amount of advertising budgets allocated to traditional pro-government media will not change this situation. Professional official photos designed to embellish certain objects are quickly overlapped by a stream of amateur photos with geotags, which are made by the residents themselves and which demonstrate the real situation on-site as well as comments from social network users. All these trends create a system of prerequisites for urban digitalization.

The discussion of strategic planning and management of urban digitalization requires at least framework definitions. However, the situation with definitions is rather complicated. Although the concepts of digital and smart city are used most actively, a wide variety of synonymous and metaphorical terms coexist. At the same time, there are two main paradigms of studying digitalization which are often used together. This creates additional obstacles due to the double meaning of the terms. Therefore, we think these paradigms need a clear distinction.

From the standpoint of the *techno-centric paradigm*, the main criterion for identification of digital and smart cities, as well as the main driver of digitalization, is a variety of information and communication technologies (ICT), including hardware and software. Since ICTs are evolving rapidly, the ICT-based model of city functioning is also changing. In our opinion, the *smart city concept is a more mature version of the digital city concept.* A digital city or, similarly, an information city, an electronic city, a virtual city, an online city, is a city in which ICT (Internet technologies) are actively used in managing urban processes. A smart city, unlike a digital one, is connected to a new generation of Internet technologies – the Internet of things (Da Silva and Flauzino 2016) and related technologies of Big Data and artificial intelligence (Willis and Aurigi 2017). This is the key difference as T. Ishida mentioned: “digital cities explore cyberspace, while smart cities use physical space” (Ishida 2017).

Development of the digital city model assumes that the city provides its residents and entrepreneurs, potential tourists, applicants, new residents and investors with specialized information, the possibility of interactive communication and various services in electronic form. Forms of immersion of digital cities in cyberspace are becoming increasingly complex: virtual tours, 3D visualization, augmented reality, interactive maps, online broadcasts, user-generated content with geotags, numerous network communities, and other forms are added to sites, forums and social networks. And, nevertheless, it includes development of cyberspace, i.e. increasingly deep and detailed digitization of various aspects of urban life, their transfer into digital form.

Development of a smart city model moves in the opposite direction, figuratively speaking, covering the physical space of the city with digital skin. Digital skin is a huge network of sensors that continuously read and transmit information about the state of the city life from the traffic situation on highways to scanning and facial recognition of visitors to shopping centers, from consumption of public resources to the results of thermographic scanning buildings and pipelines, etc. Sensors collect all kinds of city data: geopositions, clicks, likes, tweets, search queries, comments in social networks, uploaded videos and photos, data on purchases, banking data, GPS-navigation data, data of intelligent energy meters, CCTV camera data, sensor data from street lights and micro-cameras from curbs, data on crimes and road accidents, data on park attendance and street occupancy. The city becomes primarily a source of Big Data, and their continuous processing makes it possible to take management decisions not based on the experience and intuition of the head of any level, but based on a clear quantitative assessment and recommendations of artificial intelligence. Moreover, most sensors are intelligent (i.e. based on artificial intelligence) and can independently signal the actions needed: then a lamp informs about the need to replace it, and a curb informs about illegal parking. Predictive analytics and preventive measures, including additional public transport routes in peak zones, additional police squads in criminal areas, adaptive changes in heat or water supply while reducing consumption, become possible.

Many approaches and terms occupy an intermediate position between the models of digital and smart city, inclining more towards the first or the second one (Table 1).

Table 1. Continuum of definitions of ICT-innovative cities

Digital city Internet, smartphones, social networks	Information city Wired city Intellectual city Virtual city Electronic city Real-time city Cybercity Media city	Sentient city Ubiquitous city Haptic city Metered city	*Smart city* Big Data, the Internet of things, artificial intelligence, Web 2.0

Source: compiled by the authors.

From an *institutional point of view*, the main criterion for distinguishing digital and smart cities is institution itself and its rules, patterns and orders in the field of urban management, strategic planning and implementation of various city-wide policies, as well as coordination of interactions between different city stakeholders, i.e. groups with special interests. The city stakeholders include politicians, officials, entrepreneurs, investors, students, blue and white collar workers, tourists, creative industry workers, pensioners, young families and others. The current main trend of strategic planning and territory management assumes involvement of the local population or its various communities in developing the yard, district, city, or even region. Digital technologies expand opportunities of stakeholders' participation in management, reduce information asymmetry, allow quick identification of public opinion, provide a quick response to events, etc. Digital technologies can significantly increase involvement of citizens in strategic decision-making and, generally, in the strategic city management. It is a powerful driver for developing local democracy and a cohesive civil society.

Over the past few years a large amount of data, business intelligence and smart environments have attracted a lot of attention to decision-making in digital spaces and organizations. The Internet generates a large amount of data and digital footprints that show the behaviour of customers (existing and potential), competition trends, prices, costs, as well as socio-cultural attitudes and decision-making factors. Thus, all the above-mentioned is the basis for creating new activities or an effective tool for creating innovative business models and strategies for developing local territories and smart cities, allowing to consider the preferences and interests of Internet users, build individual behaviour through complex digital stages and without people's efforts, offer and independently orient them in the digital space.

Development of the 'digital skin' of the city definitely affects perception of its residents, forming certain patterns of interpretation, supplementing, annotating, indexing and filtering urban reality in the same way as the algorithm and format of the Google page affect information perception (Rabari 2015). Moreover, the model of a smart city radically changes the content of urban governance, which becomes, on the one hand, more diverse in terms of preventive and interactive decision-making, and, on the other hand, the very field of decision-making is sharply narrowed by limiting autonomy to the choice of options suggested by artificial intelligence. We can say that the very problem of political choice is removed and replaced by automatic (more precisely, smart) making optimal decisions in real time. This is, in fact, a "rapid and profound transformation of the rules of the game in public administration" (Hall and Burdett 2017) or, in other words, a radical institutional innovation.

3 Results

There is a growing need for accessible, compatible and reusable data management infrastructures and standards that provide greater access to information in the society. Investing in such infrastructures allows for innovation and digitization of urban services and introduction of a wide variety of technological ecosystems. Digital infrastructures are currently an integral part of many areas (e.g., business, health, transport, finance), but the question remains how we can give objective data and extract actionable information, going beyond technological innovations and security issues, asking right questions, and combining business transformation with Big Data analytics to create value that accelerates sustainable development of the society.

In fact, smart cities will gradually emerge from the burden of complex systemic problems which were ascendant in more or less large cities. It includes traffic jams on roads, overconsumption of resources and energy, crimes, etc. If these problems are solved, the role of the city authorities will be transformed evolutionarily. In fact, there will be no need for rather massive and significantly empowered city bureaucracy. There will possibly be some hybrid formats for making strategically important decisions for the city, such as blockchain referendums, and participatory budgeting will also be actively distributed — decentralized distribution of priorities for spending a certain part of the consolidated city budget through holding online voting by the residents themselves.

The big problem is understanding of the smart city model as related to the deep and systemic shifts in management mechanisms. Expecting rapid and radical transformation under the influence of new digital technologies overshadows many improving innovations and minor changes, which together can lead to a transformational shift. But such processes remain in the shadow of high expectations of rapid success.

Another problem is a popular expectation that digitalization may be niche, relatively closed, and concentrated in a particular industry(-ies) or sphere(-s) of urban life. In fact, digitalization is reduced to individual digital projects that are not combined into a single program or strategy (Luque et al. 2014). Then digitalization may be understood as introduction of digital technologies in the system of state and municipal services, activities of public authorities and local self-government, road management or housing and communal services. The key task of digital technologies is to reduce the operation cost of these systems. However, the isolated development of digital technologies is almost impossible in real life in certain industries and spheres of the city or region. Digitalization is a process that has a systemic nature and is highly dependent on the synergy effect. As for innovation, it is impossible to build a separate high-innovation industry (for example, nanoindustry) in a technologically backward economy and in a society that has already adapted to this underdevelopment.

At the same time, attempts to integrate digital technologies into the related spheres of urban economy are extremely difficult in practice. As the experience of implementing the smart city model in the UK show (here we speak about applications for the competition of strategic projects for urban economy digitalization "TSB Future Cities Demonstrator Competition" held in 2012–2013), most strategic plans concern only integration of two areas (most often energy and transport, energy and local economy, transport and social security, local economy and education), only a few programs include a combination of three or four areas (Taylor Buck and While 2017).

4 Conclusions/Recommendations

The doctrine of smart urbanism (Dirks and Keeling 2009) gets in quite a tough conflict with the traditional, generally accepted system of strategic planning and management of cities and regions. Planning and political mechanisms forming the basis of this system are based on top-down decision-making by a rather narrow circle of influential actors, without involving the general public. Those power processes and models of social order, implemented through the doctrine of smart urbanism, are not merely technocratic to transfer more power to large corporations supplying smart solutions, but rather act as subversive institutional innovation. By analogy with disruptive business innovations, smart city technologies do not only complement and enhance the efficiency of traditional city management systems but completely displace and replace them.

The initiatives and activities in the field of smart urbanism in the cities of leading countries remain on the periphery of the traditional approach to strategizing and managing territorial development (Cowley and Caprotti 2018) which reflects the current status of a smart city as an institutional experiment. The current budget constraints, which are typical for cities and regions not only in Russia but also in leading foreign countries, do not allow financing long-term-oriented complex transformations of the urban environment required for digitalization in its full sense. However, the interrelated institutional experiments and technological solutions proposed by the smart urbanism doctrine contribute to developing collaborative learning of local stakeholders and adopting more pragmatic strategic decisions which are based on data rather than intuition or private interests. In addition, digital technologies of the smart city have already been fetishized: when all key indicators are focused on digitalization itself, this can place many important socio-economic development goals to the periphery. Therefore, one should pay attention to the idea of smart sustainable development. In this model, digital technologies are the means to achieve sustainable development goals but not self-sufficient phenomena that operate as a whole, without any reference to specific local tasks. On the contrary, there is a certain need for digital technologies that are clearly focused on achieving sustainable development goals and contributing to improvement of relevant indicators.

Acknowledgment. The work was supported by Russian Science Foundation (project No.18-78-10075).

References

Scott, K.: The Digital City and Mediated Urban Ecologies, 189 p., pp. 14. Springer, Cham (2016)

Dameri, R.P., Rosenthal-Sabroux, C. (eds.): Smart City: How to Create Public and Economic Value with High Technology in Urban Space, p. 238. Springer, Cham (2014)

Da Silva, I.N., Flauzino, R.A. (eds.): Smart Cities Technologies, 246 p., pp. 3–18. InTech, Rijeca (2016)

Willis, K.S., Aurigi, A.: Digital and Smart Cities, p. 240. Routledge, London (2017)

Ishida, T.: Digital city, smart city and beyond. In: Proceedings of the 26th International Conference on World Wide Web Companion, Perth, Australia, 3–7 April, pp. 1151–1152. ACM (2017)

Rabari, C., Storper, M.: The digital skin of cities: urban theory and research in the age of the sensored and metered city, ubiquitous computing and Big Data. J. Reg. Econ. Soc. **8**(1), 27–42, 32 (2015)

Hall, S., Burdett, R. (eds.): The SAGE Handbook of the 21st Century City, 730 p., p. 232. SAGE, London (2017)

Luque, A., McFarlane, C., Marvin, S.: Smart urbanism: cities, grids and alternatives? In: Hodson, M., Marvin, S. (eds.) After Sustainable Cities? pp. 74–90. Routledge. London (2014)

Taylor Buck, N., While, A.: Competitive urbanism and the limits to smart city innovation: the UK future cities initiative. Urban Stud. **54**(2), 501–519 (2017)

Dirks, S., Keeling, M.A.: Vision of Smarter Cities: How Cities Can Lead the Way into a Prosperous and Sustainable Future, 18 p. IBM Institute for Business Value, New York (2009)

Cowley, R., Caprotti, F.: Smart city as anti-planning in the UK. Environ. Plan. D Soc. Space (2018). https://doi.org/10.1177/0263775818787506

Multiple Digital Economy: Semiotics and Discursive Practices

Andrew V. Olyanitch(✉), Zaineta R. Khachmafova, Susanna R. Makerova, Marjet P. Akhidzhakova, and Tatiana A. Ostrovskaya

Adyghe State University, Maykop, Russia
aolyanitch@mail.ru, zaineta@nextmail.ru, susannamakerova@gmail.com, zemlya-ah@yandex.ru

Abstract. The goals and objectives of the proposed paper include (1) analysis of signs' clusters (semiotics) and the concepts as constituents of the concept-sphere "Multiple digital economy"; (2) the study of these signs clusters' immersion as well as these concepts into the economic discourse. The research methods applied: introspection and retrospection; semiolinguistic analysis of sign formation processes in economic discourse; conceptual analysis of the concept-sphere studied; discourse analysis of the economic signs' different types immersion into the communication environment.

Keywords: Abbreviation · Actionym · Discourse · Sign · Internet of Things · Instrumentative · Concept · Legislative · Processive · Effective · Resourconym · Semiotics

Jel Code: C830 · C59 · C55 · O3 · I2

1 Introduction

As it is well-known, a digital (or network) economy is an economic activity carried out with the help of electronic networks (digital telecommunications). A technologically digital/network economy is an environment in which legal entities and individuals can communicate with each other about joint activities. Like any economy, it is a concept-sphere, which includes such concepts as a virtual enterprise, network organization, telework, Internet business (e-business), Internet of things.

The meaning of the name of the concept "Virtual enterprise" is transferred by the following definition: "A virtual enterprise is an enterprise consisting of a community of geographically separated workers who interact in the production process, using mainly electronic means of communication".

The semiotics of the "Network organization" concept is denoted through the following definition: "A network organization is an organization that uses network communications, relationships, and technology in production management".

The concept of "Telework" is semiotically interpreted as "… work performed using telecommunication systems in a place remote from the place where the results of this work are used".

E. G. Popkova and B. S. Sergi (Eds.): ISC 2019, LNNS 87, pp. 442–448, 2020.
https://doi.org/10.1007/978-3-030-29586-8_51

The concept of "Internet/E-business" is endowed with a name with the following meaning, transmitted by the definition: "… business based on the use of information technologies in order to ensure optimal interaction of business partners and create an integrated value chain". This concept should be considered as a complex semiotic formation, since its semiotics includes such sub-concepts as "sales", "marketing", "financial analysis", "payments", "employee search", "user support" and "support of partnerships".

The "Internet of Things" concept is an important element of the digital economy concept-sphere. The Internet of Things (IoT) is nothing more than a concept of a computer network of physical objects ("things") equipped with embedded technologies for interacting with each other or with the external environment, which considers the organization of such networks as a phenomenon that can restructure economic and social processes that exclude from the part of actions and operations the need for human participation [14]. The Internet of Things can be viewed as a network of networks in which smaller, weak-connected networks form larger ones (see diagram):

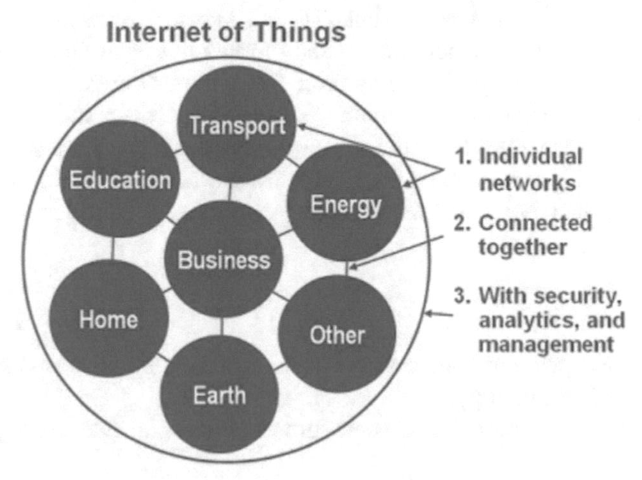

As we see, the digital economy conceptual sphere is a scientific object, which is designed to perform a whole range of tasks, which leads one to believe that this is a multi-tasking phenomenon, and its step-by-step development and improvement today seems extremely relevant.

Therefore, the **Purpose** of this paper is a semiolinguistic description of the subject area "Digital Economy" with regard to its multiply character, reflected with the help of clusters of signs (i.e., semiotic), immersed into the corresponding discourse (economic).

Thus, this paper is devoted to the scientific description of the phenomenon of a multi-decade digital economy in terms of its semiotic structure and actualization in relevant discursive practices. Such a spectrum of study of this phenomenon is certainly important for the development of linguistic theory in its semiotic part and constitutes an obvious **scientific novelty**.

2 Methodology

The research methodology is based on the following kinds and types of analysis approved in linguistics: (1) the general scientific method of introspection and retrospection, which allows to investigate an object from the point of view of its dynamic development and growth; (2) semiolinguistic analysis, which makes it possible to study sign formation in the relevant subject area and explore the features of the emerging signs' clustering; (3) a conceptual analysis that reveals the subject, figurative and value (axiological) features of the studied concept-sphere; (4) discursive analysis, which allows to reveal the constitutive signs of the object under study' immersion into the communication environment.

Our research was based on the following scientific research and achievements.

The subject area "digital economy" is studied in detail in the works of Tlehatuk [13]; the processes of clustering of signs and sign formation were generally studied in the works of Astafurova and Olyanitch [4], Morris [15]; the terminology formation of the "digital economy" sphere was studied by Averbukh [1] and Alexandrovskaya [2]. The processes of conceptualization of the sphere under study turned out to be the focus of research interest of Alimuradov [3] and Kalugina [10]. The communicative sphere of the digital economy signs' actualization (economic discourse) was observed by Achilova [5], Gurieva [7], Zhandarova [8], Karasik [11] and Vodak [16]. The financial and economic picture of the world as a sphere of the actualization of signs of the digital multiple economy in the discursive space was studied by Buyanova [6] and Kazydub [9].

3 Results

What clusters of signs form the concept sphere of the digital economy? Next, we present their detailed typology.

First of all, such semiotic clusters are included in the concept sphere as **resultatives**:

- "Digital Financial Asset (CFA)", i.e. property in electronic form, created using encryption (cryptographic) means, while ownership of this property is verified by making digital entries in the register of digital transactions; digital financial assets include cryptocurrency, token; note that digital financial assets are not legal means of payment on the territory of the Russian Federation, and this corresponds to their semiotic-cognitive characteristics, such as the "lack of legal use";
- "State (national) CFA"; the basic semiotic marker of this cluster is the subject of the issue - their issuer is the state or authorized state institutions;
- "Private CFA"; semiotic marker is the subject of the issue (the issuer are corporations or individuals);
- "Distributed TsFA"; semiotic marker is the subject of emission which cannot be defined; it appears at different stages of the existence of autonomous projects implemented on distributed registry technologies.

Further, the concept sphere of the digital economy includes such clusters of signs, such as **processives**:

- "Digital Transaction" is an action or sequence of actions aimed at the creation, issue, circulation of digital financial assets;
- "Digital Record" is information on digital financial assets recorded in the register of digital transactions;
- "Digital Transaction Register" is a systematized base of digital records formed at a certain point in time;
- "Distributed registry of digital transactions" is nothing but a systematic database of digital transactions that are stored, simultaneously created and updated on all media for all registry participants based on predetermined algorithms that ensure its identity with all registry users.

The following clusters of signs also provide a lot of digital economy:

(a) **personalities**, such as "participants in the digital transaction register", i.e. the persons performing digital transactions in accordance with the rules for maintaining a register of digital transactions; "Validator", i.e. A legal entity or an individual who is a member of a digital transaction registry and is engaged in the validation of digital records in a digital transaction registry in accordance with the rules for maintaining a register of digital transactions; "Digital financial asset exchange operator", i.e. a legal entity engaged in the exchange of digital financial assets of one type for digital financial assets of another type and/or the exchange of digital financial assets for rubles or foreign currency;

(b) **instrumentatives**, such as block chain (English block chain or block chain) that is built according to certain rules, a continuous sequential chain of blocks (linked list) containing information and stored on many different computers that are independent of each other; cryptocurrency, i.e. a type of digital financial asset created and accounted for in a distributed digital transaction registry by participants of this registry in accordance with the rules for maintaining a registry of digital transactions; secured cryptocurrencies, i.e. state or private cryptocurrencies secured by the issuer's assets and obligatory for acceptance by the issuer; unsecured cryptocurrencies – those cryptocurrencies that are not secured by the real assets of their issuer (for example, Bitcoin [BTC] is the first and widely known cryptocurrency that is not secured by tangible assets or liabilities); Token - a type of digital financial asset that is issued by a legal entity or an individual entrepreneur in order to raise funds and is recorded in the register of digital records; real tokens, i.e. tokens certifying the right to specific tangible assets (for example, a DIGIXIGLLOBAL PTE LTD [DGX] token provided with gold, each token corresponds to one gram of gold and can be exchanged for it); license tokens, i.e. tokens that provide the right to receive part of the income from the use of intellectual property, software (for example, the Bloquid token (BQT), provides the right to receive part of the income from the use of Bloquid software products); utility tokens, i.e. Tokens released on block chain networks that have applications in the project itself that issued them (for example, the StorJ cloud storage project token provides the right to receive cloud storage and can be used in applications built on the basis of the project); credit tokens, i.e. tokens confirming the right of the owner of the token to demand a return of the loan in the future (for example, the colion tokens

[KLN], confirming the right to participate in the Kolionovo ecological project, as well as to receive income in the form of accrued food interest); corporate tokens are the tokens that certify the right to a share in an enterprise (authorized capital) or investment fund – corporate (investment) tokens (for example, the Spie token certifying the right to a portion of the assets of the Satoshi fund); non-secured tokens are those tokens that are not secured by real assets of their issuer and/or certifying the right exclusively for digital objects of value (for example, the CryptoKitties project, which allows you to purchase digital seals).

The multiple character of the digital economy is also provided by **actionyms**, i.e. signs, such as:

- "digital record validation" – the sign that denotes a legally significant action to confirm the validity of digital records in the digital transaction registry, carried out in the manner prescribed by the rules of maintaining the digital transaction registry;
- "mining", which comprises business activities aimed at creating cryptocurrency and/or validation in order to receive remuneration in the form of cryptocurrency;
- "Remote Identification for Banking Services" as a system action to identify a participant in civil law relations when accessing financial services, including conducting one-time primary identification in person presence and repeatedly conducting remote identification in order to provide financial services.

In the multiple digital economy, the legislative sphere is also present and actualized semiotically, through such cluster of signs as **legislatives**; these include, for example, such sign as a "smart contract", i.e. an agreement in electronic form, the fulfillment of rights and obligations under which is carried out by performing automatically digital transactions in a distributed registry of digital transactions in a strictly defined sequence and upon the occurrence of certain circumstances; protection of the rights of participants (parties) of the smart contract is carried out in a manner similar to the procedure for the protection of the rights of the parties to the contract concluded in electronic form.

An important payment semiotic element in the digital economy is also a sign-**resourconym** - "Digital Wallet", which is a software and hardware tool that allows you to store information about digital records and provides access to the registry of digital transactions. A digital wallet is opened by the operator of the exchange of digital financial assets only after going through the procedures of identifying its owner.

Let us draw the reader's attention to the fact that a characteristic semiotic feature of the digital economy concept-sphere is the wide use of English-language abbreviations for nominating its basic concepts. Here are some examples:

CALS (Continuous Acquisition and Life cycle Support) – the technology of information support of the product life cycle at all its stages, from the stage of design, development, manufacturing, operation and repair work, to misuse and recycling;

CNP (Cardholder Not Present) – a plastic card transaction designation at the time of which the cardholder is not personally present at the merchant, but reports the plastic card details in absentia (by letter, by phone, computer network, etc.);

EDI (Electronic Data Interchange) – (1) international standard for the exchange of electronic data; (2) transmission of standardized electronic messages replacing paper documents;

EIP - (Electronic Invoicing and Payment) electronic invoicing and payment systems;

G2B (government to business – government for business) – the e-commerce market sector, in which government agencies and legal entities act as interacting entities;

G2C (government to customer – government for the consumer) – the e-commerce market sector, in which government agencies and individuals act as interacting entities;

IOTP (Internet Open Trading Protocol) is an open trading protocol on the Internet that provides the ability to conduct electronic trading transactions involving, on the one hand, the seller (it implies a certain number of participants in a trade transaction: an electronic store, a payment system, a supplier of goods and services) and, on the other hand, the buyer, during one or several trading sessions;

MRP (Material Requirements Planning) – planning the need for materials) – the concept of planning the needs of production in material resources, using (for automatic determination of this need) information about the structure and production technology of the final product, the volume-calendar production plan, stock data, contracts supply of materials and components, etc.

Discursive actualization of the digital economy concepts is carried out, for example, in the following contexts of human communication:

(1) "Nowadays, in the age of computerization and high technology, the digital economy affects every aspect of life: healthcare, education, Internet banking, government. The digital economy has been developed in all highly developed countries, including Russia. Proceeding from the events of foreign policy and global trends, Russia faces the issue of global competitiveness and national security, and the development of the digital economy in the country plays a significant role in solving this issue. Some elements of the digital economy are already successfully operating. Today, given the massive transfer of documents and communications to digital media, the authorization of electronic signatures, communication with the state also goes to the electronic platform" [12].

(2) "Another innovative direction related to digitalization is augmented reality (Augmented Reality, AR). The most promising is the technology of augmented reality, which allows to add objects from the virtual world to the real world" [12].

4 Conclusion

Summing up, we note the following.

The concept-sphere "Multiple digital economy" is a cognitive semiotic formation that determines the vector of the modern development of civilization, benefiting from a new round of technological progress of mankind. The semiotics of the concepts that form this concept-sphere is powerfully clustered and represented by such signs as resultatives, processives, actionyms, personalities, instrumentatives, resourconyms and legislatives.

One of the semiotic features of the concept-sphere of the digital economy is its terminological hyper-abbreviation.

The discursive actualization of the signs of this concept-sphere is carried out by means of corresponding discursive practices through the production of texts covering the communication process within the framework of the digitization of the economy as a whole.

Acknowledgments. We thank Larisa Semyonovna Shakhovskaya, Doctor of Economics, Professor of the Department of World Economy and Economic Theory of Volgograd State Technical University, Elena Gennadevna Popkova, Doctor of Economics, Professor, President of the Institute of Scientific Communications for the full support of philological projects of our Maikop philology research group.

References

1. Averbukh, K.Ya.: The general theory of the term: a complex-variological approach: dis. … Dr. filol. Sciences, 10.02.19, Theory of language, Ivanovo, 324 p (2005)
2. Aleksandrovskaya, L.V.: The semantics of the term as a member of general literary vocabulary (on the material of English maritime terminology): diss. … cand. filol. sciences, 10.02.04, 24 p (1973)
3. Alimuradov, O.A.: Meaning, Concept, Intentionality, 312 p. Publisher PGLU, Pyatigorsk (2003)
4. Astafurova, T.N., Olyanitch, A.V.: Semiolinguistic of power: sign, word, text: monograph. IPK FGOU VPO VGSKHA "Niva", Volgograd, 256 s (2008)
5. Achilova, E.N.: Economic discourse as a kind of institutional one: qualification and structure. Cult. Life South Russia **3**(32), 94–96 (2009)
6. Buyanova, L.Yu.: Verbal-semiotic module "Market" as the dominant of the Russian financial and economic picture of the world. News of Kabardino-Balkarian State University, vol. 2, no. 3, pp. 87–90 (2012)
7. Gurieva, Z.I.: Speech communication in business: to create an integrative theory (on the material of texts in Russian and English): dis. … Dr. phil. sciences, 10.02.19, Krasnodar, 446 p (2003)
8. Zhandarova, A.V.: Language conceptualization of the sphere of entrepreneurship and business (on the basis of Russian and English languages): diss. … cand. filol. sciences, 10.02.19, Krasnodar, 178 p (2004)
9. Kazydub, N.N.: Discursive space as a fragment of the linguistic picture of the world (theoretical model): diss. … Dr. phil. sciences, 10.02.04, Irkutsk, 34 p (2006)
10. Kalugina, Yu.E.: The conceptual sphere "Economics" in English and Russian languages. Bull. Moscow State Reg. Univ. Ser. Linguist. **2**, 137–144 (2010)
11. Karasik, V.I.: The language circle: personality, concepts, discourse, 477 p. Change, Volgograd (2002)
12. The development of the digital economy in Russia as a key factor in economic growth and improving the quality of life of the population, 131 p. Professional Science Publishing House, Nizhny Novgorod (2018)
13. Tlehatuk, S.R.: Subject area "Economics": a cognitive-semiotic aspect. Diss. Dr. FILOL, sciences, 10.02.19, theory of language, Maykop, 386 s (2017)
14. Ashton, K.: That 'Internet of Things' Thing: in the real world, things matter more than ideas. RFID J. **6** (2009)
15. Morris, C.: Signs, Language and Behavior, 670 p. George Brazilles, New York (2018)
16. Wodak, R.: Disorders of Discourse, 200 p. Longman, London (1996)

Information Systems for Project Management in the Public Sector

Marina V. Tsurkan[1(✉)], Maria A. Liubarskaia[2], Vadim S. Chekalin[2], Svetlana M. Mironova[3], and Alexey A. Artemiev[4]

[1] Tver State University, Tver, Russia
080783@list.ru
[2] St. Petersburg State University of Economics, St. Petersburg, Russia
lioubarskaya@mail.ru, vchekalin10@list.ru
[3] Russian Presidential Academy of National Economy and Public Administration, Volgograd Institute of Management, Volgograd, Russia
smironova2017@gmail.com
[4] Tver State Technical University, Tver, Russia
aaartemev@rambler.ru

Abstract. The aim of the study is to develop the criteria for the project management information systems in the public sector. This aim corresponds to the state digital agenda and the recent project management trends in the public sector. The main element of the research methodology was the system analysis method, which considering the project approach as an important part of the economic mechanism, and ensuring the role of the information system in its implementation. The study identified the areas of introduction of the project approach in the public sector, which includes participation of authorities in projects achieving strategic priorities and management of projects with intersectoral collaboration. Authors defined the best practices for the implementation of project management information systems in the public sector of the constituent entities of the Russian Federation. Evaluation criteria have been formed for the project management information system in the public sector at the regional level, taking into account legal restrictions and minimum functional requirements. The proposed calculation formulas for evaluation of the effectiveness aimed to be applied while choosing an information system for project management, introducing it to the structure of the public sector. The research results are of practical importance for public authorities, project management specialists, scientists in the field of digital economics and management.

Keywords: Management · Project approach · Information system · Public sector

JEL Code: H70 · H77

E. G. Popkova and B. S. Sergi (Eds.): ISC 2019, LNNS 87, pp. 449–459, 2020.
https://doi.org/10.1007/978-3-030-29586-8_52

1 Introduction

The implementation of national project "Digital Economy of the Russian Federation" particularly focused on the improvement of the public sector, mainly in the field of public administration through a number of measures, such as ensuring:

- Digital transformation of the state (municipal) service through the introduction of digital technologies and platform solutions;
- Improvement of mechanisms for the interdepartmental electronic document circulation in the interests of increasing the efficiency of interaction between state institutions, business and civil society.

At the same time, starting 2016, the public sector is characterized by the implementation of the project approach at all levels, especially the federal and regional ones. This process requires the development and integration of a network information system into the state administration system of the constituent entity of the Russian Federation. It will contribute not only to automation of the project approach implementation, but also to provision of the required interaction.

This system, which provides interdepartmental interaction between the structures involved in the project activities at the federal level, presented in Fig. 1.

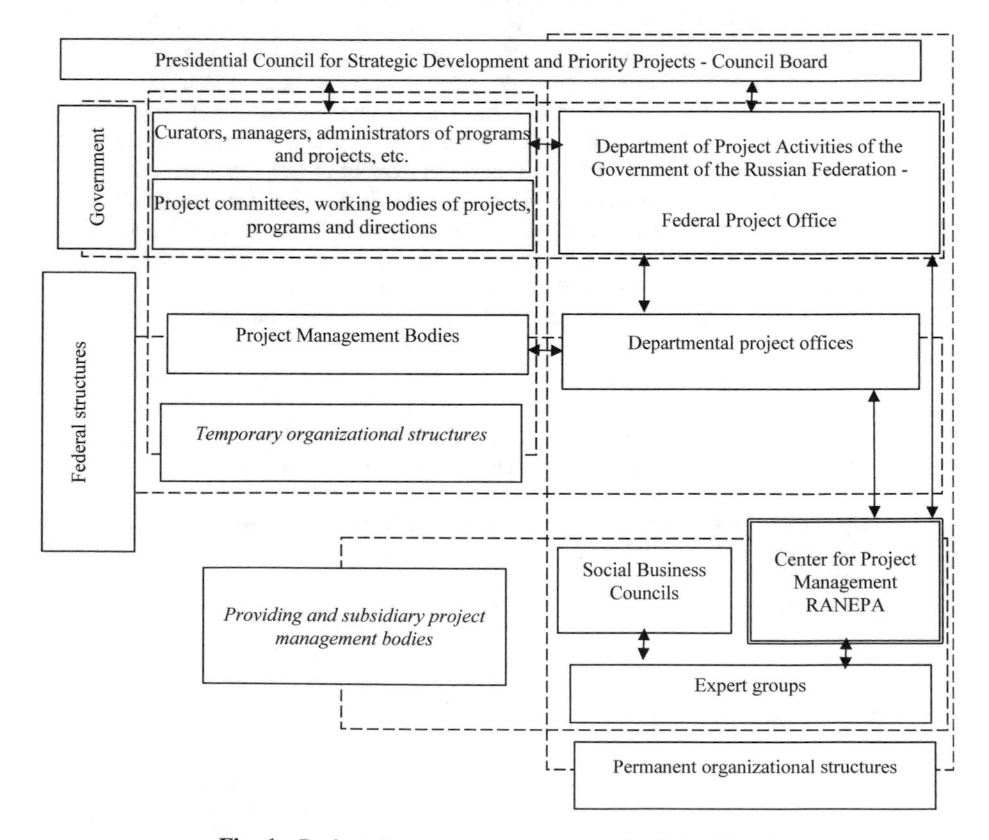

Fig. 1. Project management structure at the federal level

At the federal level, the formation, approval and submission of documents developed during the implementation of project activities is planned within the framework of the state integrated information system for managing public finances calls "Electronic budget".

Not all modules and components of the system presented in Fig. 1 are still implemented, and the use of the system will only support national and federal projects. Nevertheless, the question of forming criteria for determining the optimal project management information system at the regional and municipal level is more relevant.

Recently, to the domestic and foreign scientific literature were introduced the researches, reflecting various aspects of the project management information systems functioning at the corporate level and in a particular industry.

Researchers from different Russian schools emphasize the importance of the project management information system as the main element of the organization's management system to achieve its strategic goals (Zolotorev 2014; Sinyukova et al. 2017; Avdyushkina and Epifanova 2017).

Myasoedov P.V. and Tarlavsky D.V. amplify the importance of the use of information systems to improve the quality of project management in the oil and gas industry (Myasoedov and Tarlavsky 2008).

Serditova N.Y., Luchinina N.Y. and Pavlova L.S., considering the project approach as a priority, from the standpoint of managing the development of the university, note that the formation of future markets is impossible without the main component of the digital economy of the future so called "factories of the future". These are platforms for digital production, including digital design, total mathematical modeling using optimization technologies, testing samples of new products in a virtual environment that reflects the actual conditions of production and operation, which can significantly reduce development and production time, as well as the cost of products (Serditova et al. 2017).

Penkina G.K. proves the importance of developing criteria for choosing an information system for managing construction projects in specific conditions (Penkina 2009).

The urgency of the problem of public administration strategy implemention in the development and support of public-private partnerships through the creation and use of an information and analytical system also noted in Larionova's works (Larionova 2014).

Murinovich A.A. and Kirillov B.A. consider the creation of information channels in government in the context of the need to reduce budget spending (Murinovich and Kirillov 2017).

However, the existing publications are lack of requirements for the public sector project management information systems, which corresponds to the state digital agenda and allows the implementation of existing approaches to the project management in the designated area. Insufficient development of these aspects predetermined the aim of the presented study.

2 Methodology

Among the elements of the research methodology the main role played the system analysis method, which considering the project approach as an important part of the economic mechanism, and ensuring the role of the information system in its implementation.

Benchmarking of project management information systems renders the significant part of the analysis' results for the implementation of these systems in the public sector of the constituent entities of the Russian Federation. Submitted analysis was conducted on the basis of open data, where the standard value represents the minimum set of modules required for the implementation of projects in the public sector.

In the process of developing the criteria for the project management information system in the public sector, a modeling method was applied at the regional level.

The sample of the studied data includes information from the official resources of regional and municipal authorities, domestic developers of information systems, and the Analytical Center under the Government of the Russian Federation.

3 Results

3.1 Areas of the Implementation of the Project Approach in the Public Sector

The implementation of the project approach in the public sector includes participation of the authorities in managing the projects with inter-sectoral intercommunion and their involvement of project projects achieving the goals of strategic planning (Fig. 2).

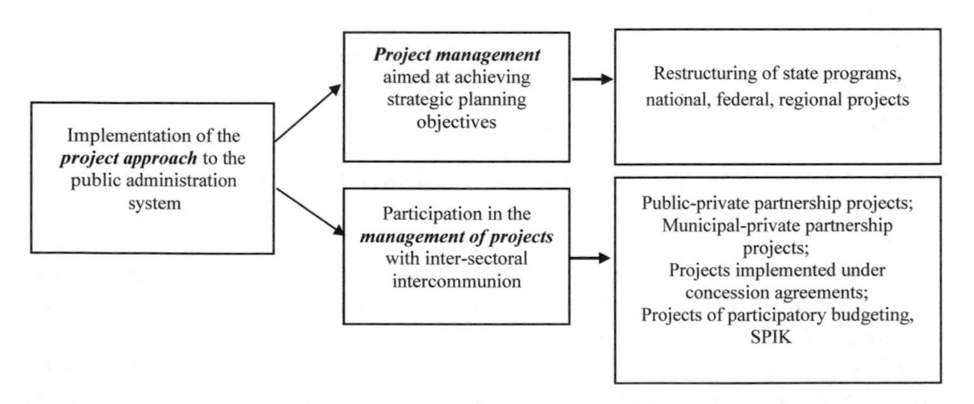

Fig. 2. Areas of the implementation of the project approach in the public sector

The main difference between these two areas falls into the implementation environment. Project management in the public sector, of course, involves the use of traditional project management methodology, but only in the internal environment of the subject of the public administration system. That is, all project management functions remain with the public authority body.

Kompaneytseva G.A. consider project management as an approach to management that implies the formation of projects as a way to solve the most significant tasks for the authorities in the framework of strategic planning (Kompaneitseva 2016). This approach could be taken at international, federal and regional levels.

Project management, as an independent tool for the implementation of the project approach, involves participation of the authorities in projects, which are partially managed by an external actors, but are aimed at solving functional tasks of subjects of the public administration system.

Project management in this context is a partial project management of inter-sectoral collaboration. That involves organization and (or) planning and (or) management and (or) coordination of labor, financial, material and technical resources throughout the project cycle, aimed at the effective achievement of project goals by applying modern methods, techniques and technologies to achieve results, identified in the draft (Mazura 2004).

The project of inter-sectoral collaboration with the participation of public legislative institutions is a complex of interrelated activities aimed at building and (or) reconstructing the infrastructure of the territory within a limited period of time and other resources, implemented within the framework of targeted beneficial interaction of the state (municipal), commercial and (or) non-profit sectors of society.

The tools for the implementation of the project approach at the present stage include public-private partnerships, concession relations, projects implemented under special investment contracts, participatory (initiative) budgeting.

3.2 Best Practices for Implementing Project Management Information Systems in the Public Sector of the Constituent Entities of the Russian Federation

Currently, project implementation practices at the level of the constituent entities of the Russian Federation are taking place at Primorsky Krai, Belgorod region, Volgograd region, Sakhalin region, Ulyanovsk region, Yaroslavl region, Penza region, Novgorod region, Sverdlovsk region, Republic of Bashkortostan, Khanty-Mansi Autonomous Area - Ugra, Tyumen region, etc.

At the municipal level, experience in implementing projects was obtained in Nizhny Novgorod region, Khanty-Mansi Autonomous Area (the leader in the number of best practices recognized by experts of the Analytical Center under the Government of the Russian Federation in the framework of the annual Project Olympus competition), Voronezh region, Moscow region, Tula region, Penza region, Chelyabinsk region, Samara region, Krasnodar region, Belgorod region, etc.

According to a pilot assessment of the maturity of the organization of project activities in the constituent entities of the Russian Federation conducted by the Federal Project Office, the TOP-5 regions considered to be leaders of the pilot group of the Project Activity Index (PAI) includes Belgorod region, Khanty-Mansi Autonomous Area – Ugra, Leningrad region, Krasnoyarsk region, Ulyanovsk region.

Analytical Center under the Government of the Russian Federation also famed Primorsky Krai among the territories with best practices of project approach implementation.

The Project Activity Index is a new tool in the governmental project management system that allows the identification of the best practices at federal and regional levels.

The PAI is formed in the context of five main elements: strategic planning and project portfolio management; project management; decision making and organizational support; development of competencies and efficiency culture; management of incentives for project participants.

The majority of subjects of the Russian Federation showing the best practices of implementation of the project approach, use the information systems, integrated in the regional and municipal management systems, which allows the required inter-level interaction.

Inter-level interaction provided in the framework of the information system in the Khanty-Mansi Autonomous Area - Ugra (KMAA-Ugra) is shown in Fig. 3.

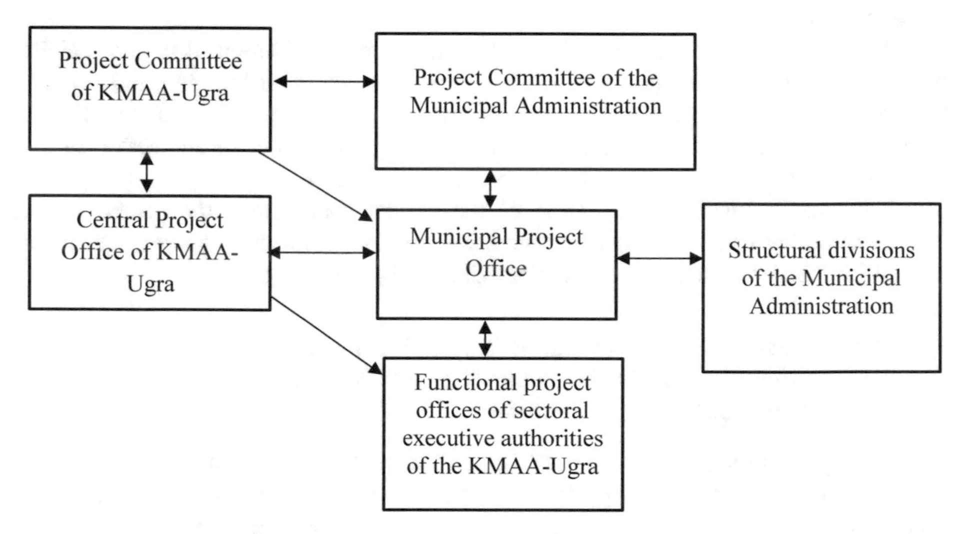

Fig. 3. Inter-level cooperation provided by the project management information system in the Khanty-Mansi Autonomous Area - Ugra

The functionality of the information system is determined by the set of the following functional modules (subsystems).

In Primorsky Krai, the functionality of the project management information system is provided by the following modules: project certification; program management; project portfolio management; contract management; project initiative management; management of non-project activities; meeting management; management of orders; time management; performance management; personnel management; management of financial performance of the project; managing risks, problems and open questions; project reporting; generating analytical reports on projects and monitoring; change management; knowledge base management; storage of project documents.

The project management information system of the Primorsky Krai also includes 4 supporting modules, among which the greatest importance in terms of integrating project management into the strategic planning system has a module that provides work with adjacent information systems of the region.

Function Region	1	2	3	4	5	6	7	8	9
Primorsky Krai	■	■	■	■	■	■	■	■	■□
Belgorod region	■	□	■	■	■	□	■	□	□
Ulyanovsk region, Leningrad region, KMAO-Ugra	□	□	■	■	□	■	□	■	□

1 Project Initiative Management;
2 Program Management;
3 Project Portfolio Management;
4 Resource Management;
5 Risk Management;
6 Meeting Management;
7 Change Management;
8 Knowledge Base and Lessons Learned;
9 Performance indicators for project participants;
■ – The module is functioning;
□ – Module does not functioning;
■□ – The module is functioning partially.

Fig. 4. The functionality of project management information systems in the regions

On the basis of open data, it was possible to determine that most regions implementing project management in the public sector, use information systems, developed on the basis of:

1. Motiware Melody One platform, where all the participants of the project activities have access to data at any time and from any place through the web-interface of the system (they only need Internet connection and an installed browser).
2. Informational system Advanta, or Informational system PM Foresight.

The analysis of the functionality of project management information systems in the public sector, based on samples of regions with corresponding open data, allows to summarize the specific features of these systems presented in Fig. 4.

The Advanta system was recognized as the winner in the special nomination "The Best Russian Information System for Project Management" at the contest "The Best Information and Analytical Tools - 2016", held by the Analytical Center under the Government of the Russian Federation. The main advantage from the standpoint of motivating the participation of government representatives in project activities include such possibilities of human resource management like labor planning, maintaining time sheets, employee work load analysis, summarizing the distribution of resources, calculation of the project component of wages.

The new version of PM Foresight for the public sector allows the formation of regional components of national projects (programs) in the areas defined in Presidential Decree No. 204 of May 7, 2018. In regional information management systems projects are visually grouped by the respective national projects. Such documents like "Passport of the regional component of the national project", "Additional and justifying materials" could be automatically generated from the system in accordance with the recommendations of the Federal Project Office. In addition, the system pre-filled the register of goals in accordance with the Decree of the President of the Russian Federation No. 204. It is also possible to maintain an actual tree of goals in the system in accordance with the regional strategy of socio-economic development, etc.

It should be noted that not one of the project management information systems used in the public sector of the analyzed regions does not take into account the bi-directionality of the implementation of the project approach by authority bodies.

3.3 Criteria for the Public Sector Project Management Information System at the Regional Level

The legal restriction on the selection of a project management information system that can be used in the public sector is the requirement to introduce domestic informational products.

Also, according to the authors, in addition to the functional possibility of implementing a portfolio of projects in accordance with the Decree of the Government of the Russian Federation No. 1288 of October 31, 2018 "On the organization of project activities in the Government of the Russian Federation", the corresponding information system should provide at least:

- project management of inter-sectoral collaboration;
- management of inter-level and inter-agency cooperation;

- management of the projects of ministries and departments of the constituent entities of the Russian Federation, which supporting the regional projects;
- reflect the system of project motivation adopted by the authority and take into account the performance indicators of the participants of the project activities;
- systematize the indicators for monitoring of implementation of project activities provided at the federal level;
- form the databases of the best practices of project approach in the region and lessons learned in the process of their implementation.

When choosing an information system option, two groups of indicators for evaluating its potential effectiveness can also be used.

From the group of relative indicators for efficiency evaluation it is advisable to take into account the indicator of annual reduction of labor cost and the labor cost reduction index (Kokhno et al. 2018).

The indicator of annual reduction of labor cost shows the share or percentage of costs decrease due to the implementation of the project management information system in the public sector structure as compared to the base cost, and can be calculated by the formula:

$$K_t = \frac{\Delta T}{T_0}, \tag{1}$$

where

K_t – the indicator of annual reduction of labor cost;
ΔT – the annual reduction of labor cost of conducting project activities in the public sector structure (difference between the cost "before" and "after" the implementation of the project management information system);
T_0 – the base labor cost of conducting project activities in the public sector structure (cost measured the year before the implementation of the project management information system).

The labor cost reduction index shows the ratio of decreasing of the cost of conducting project activities in the public sector structure after the implementation of the project management information system for the year compared to the base one.

This index can be calculated by the formula:

$$I_m = \frac{T_n}{T_0}, \tag{2}$$

where I_m – the labor cost reduction index; T_n – the labor cost of conducting project activities in the public sector structure within the project management information system for the year "n".

In turn, the labor cost of conducting project activities in the public sector structure within the project management information system for the year "n" can be calculated by the following formula:

$$T_n = \sum\nolimits_{i=1}^{n} t_i \quad (3)$$

where t_i – the labor cost of conducting project activity "i" in the public sector structure within the project management information system for the year "n".

The absolute indicator of the economic efficiency of the public sector project management information system is the reduction in the annual cost and labor cost of data processing process as compared to the base data processing option.

Relative indicator of the economic efficiency of the information system are the cost ratio and the index of cost changes.

The calculation of the economic component of the effectiveness of the project management information system in the public sector will allow:

- determine the need and feasibility of the cost of creating and implementing an automated information processing system;
- identify the order of work on the automation of information processing at each level of the control system;
- identify cost-effective options for technological processes of information processing;

Economic efficiency is ensured by the following main factors:

- high speed of operations for collecting, transmitting, processing and issuing information, speed of technical means;
- maximum reduction of time to perform individual operations;
- improving the quality of data processing and information obtaining (Kokhno et al. 2018).

4 Conclusions

The study showed that the project management information systems introduced in the activities of regional authorities do not provide the possibility of support for the implementation of the project approach in its two areas.

At the domestic market we can outline the information systems for managing the projects of two companies that most closely meet the needs of the public sector. First of all, these systems for the information support of regional projects were developed in order to ensure the implementation of national projects.

It should be noted that the introduction of the project management information system in the activities of the public sector structure may be accompanied by a number of problems. First of all, it is the lack of design and digital competencies among users of the system.

The research results are of practical importance for public authorities, project management specialists, scientists in the field of digital economics and management.

References

Avdyushkina, S.A., Epifanova, A.S.: Features of the implementation of information systems in project management. In: Collection of Scientific Papers of the II International Scientific and Practical Conference "Project Management: Problems and Prospects for Development", Saratov, pp. 21–25 (2017)

Zolotarev, O.V.: Process approach to management in corporate information systems implementation projects. Bull. Russian New Univ. Ser. Complex Syst. Models Anal. Manage. **4**, 89–92 (2014)

Kompaneitseva, G.A.: Project approach: concept, principles, efficiency factors. Sci. Methodol. Electron. J. "Concept" **17**, 363–368 (2016)

Kokhno, P.A., Kokhno, A.P., Artemyev, A.A.: Computer Economics, 354 p. Computer Science and Technology Center, Tver (2018)

Larionov, I.V.: Study of the effectiveness of using information systems in public-private partnership project management. Bull. Inst. Friendship Peoples Caucasus Theory Econ. Natl. Econ. Manage. **3**(31), 6 (2014)

Mazura, I.I.: Project Management: Study Guide. Moscow, 664 p (2004)

Murinovich, A.A., Kirillov, B.A.: Management of an interregional project in the system of informational support for the activities of public authorities. Manage. Issues **1**(44), 143–147 (2017)

Myasoedov, P.V., Tarlavsky, D.V.: Application of information systems for improving the quality of project management in the oil and gas industry. Qual. Manage. Oil Gas Complex **3**, 11–14 (2008)

Penkina, K.G.: Construction project information management systems. Vestnik MGSU **2**, 203–206 (2009)

Serditova, N.E., Luchinin, N.A., Pavlova, L.S.: Project approach to the organization of educational activities at the university. High. Educ. Russia **5**(212), 141–147 (2017)

Sinyukova, D.D., Averina, M.V., Pavlovich, K.V.: Comparative analysis of modern information systems in project management. Science Diary **5**(5), 20 (2017)

Metatheoretic Evaluation of Region's Tourist Attraction

Alexander V. Dyachenko[1(✉)], Tatiana L. Kosulnikova[1], Natalya N. Balashova[1], Lidiya A. Sizeneva[1], and Irina V. Dneprovskaya[2]

[1] Volgograd State Agrarian University, Volgograd, Russia
marketec@mail.ru, kosmosleta@mail.ru, balashova_nat@mail.ru, Vfatiso_em@mail.ru
[2] Volgograd State Technical University, Volgograd, Russia
dneproguess@yandex.ru

Abstract. Region's tourist system is attractive in case of providing high quality of tour's features and how they meet the demands, likes, preferences of the customers. The investigation is done of the basis of complex metatheoretical analyses of region's tourist system. There is used author's methodology to detect system basis of quantitative controlled features to compare tour's attractiveness. There are used systematically positive investigations works of native and foreign scientists.

It was defined that the optimal choice of the tour criteria is maximum of overall estimate of client's impressions from the basic features of the tourist product. Using the product's region's tourist system features expert evaluations quality levels and evaluating the region's tourist system features meaning for different target groups' indexes of client's impressions from different region's tourist system products can be estimated. Maximal aggregate values of the region's tourist product metatheoretical features windowed estimate for every target group correspond to the most preferable routes.

It was developed the criterial matrix expression, allowing defining the aggregate values of the tourist product features quality windowed estimates for every target group. Array of aggregate values of tourist products quality windowed estimates for each target group can be done at the result of expert's opinion matrix calculation composition for tourist products features quality on the matrix of tourist system features meaning for different target groups.

Keywords: Tourist system · Metatheory · System analyses · Basis of features · Meaning of features · Target groups · Quality of features indexes · Optimal choice of the tour · Optimality criteria · Matrix expression

1 Introduction

Tourist system harmonizes using market set of region's territorial hospitality infrastructure and set of respective tours by the holiday-makers. This market is a set of tourist hospitality segments services for the customers. Tourist system of region would be attractive in case of providing high quality of tour's features and at the same time they should meet the demands, likes and preferences of the segments being served.

E. G. Popkova and B. S. Sergi (Eds.): ISC 2019, LNNS 87, pp. 460–471, 2020.
https://doi.org/10.1007/978-3-030-29586-8_53

It is reasonable to investigate attractiveness of the tourist system considering quality of multiaspect metatheoretical features of tours performance and the state of being relevant, important, and essential feature by the segments being served. As the recreation factors of the region being developed on the basis of using its social, economic, cultural, historical, institution features the problem of investigation complex interdependence of business environment mentioned above is becoming actual. That is the thematic justification of this scientific work, directed on defining the methodology of investigating the interrelation of metatheoretical features of tours, quality of their implementation and the state of being relevant, important and essential.

The object of the scientific work is forming of the metatheoretical procedure for optimization choice of the region's attractive tours. Achievement of the objective suggests the necessity to solve the following tasks:

- defining the basis of features (syntaxes) of regional tourism metatheoretical description;
- defining the regional tourism main features and metatheoretical features at the basis of systemic analyses results of its known descriptions;
- evaluation of metatheoretical specifications quality of regional system tourist routes;
- defining the meaning of metatheoretical features of touristic routes for different target groups;
- development tour attractiveness criterial choice methodology for the certain target segment.

The target of investigation is set of vital activities relations in the region. Subject of investigation is relations of metatheoretical optimal choice of the tour attractiveness for the certain target segment.

2 Methodology

The analyses of the known informal descriptions of the tourist subject allow specifying the features that are inherent to the system being investigated. F.I. Ereshko defines that development of recreation branch and tourist industry suggests necessity to attract modern methods of system analyses, synthesis and mathematic modelling. (Erishko 2007) V. I. Kruzhalin and D. V. Fedotkin pay special attention on targets, objects and causal relationships of tourist systems perspective development and necessity to form special economic areas of tourist-recreation type. (Kruzhalin 2007), (Fedotkin 2007) T. V. Cherevichko at the analyses of tourist system defines the features: targets, subjects and processes of tourism and its core features. (Cherevichko 2015) E.A. Zaliznyak suggests to classify regional tourism by features of three structural blocks: territorial and resource, social and economic, organization of tourist activity. (Zaliznyak 2011) M. W. Rana, R. N. Lodhi, G. R. Butt, W. U. Dar according to the results of investigations pay special attention on necessity of recording: customer satisfaction; quality of merchantable and service components of the products being sold; the relation of price and quality (Rana et al. 2017) M. Suryawanshi and G. Shankar investigated the gnostic and physiological preferences of tourists, as well as role of mass media and oral advertising for their activation (Suryawanshi and Shankar 2018).

T. N. Gondar, having investigated the problem of clients effective servicing clients pays special attention on reasonability considering their different meaning for different segments of tourists (Gondar 2009) X. Feng describes the concept of providing the tourists with services, which can fully satisfy their demand. He examines the tourism as dominating branch in the certain field, which is directed on optimization and modernization of regional economic and social resources, including the tourist landscape, service backup, tourism management, interbranch relations and exchange with results. He suggests emphasizing tourists not on the scenic countryside, but on integration to the local life, that allows to reflect great variety of cultures. He focuses his work on subjects, objects, systemically important features, conditions and causal relationships of tourist systems perspective development (Feng 2017).

V. N. Sadovski suggested using the following features within the systemic description: original components, their combinations and relations, qualities of system education and its functioning; conditions of the system existence (Sasdovskiy 1978). In general, it is necessary to classify the investigated subject, detect the features of the cooperating subjects as well as objects about witch their relations are implemented, providing getting the net effect. It is necessary to determine activity of the subject, detecting: its systemically important features; cause-and-effect processes, becoming in the event under investigation; restrictions, at which the subject can exist successfully; orientation of its dynamic development. It is reasonable to determine metatheoretical description by systemic set of specifications, defining its main features, mentioned above (Dyachenko 2016). In Table 1 there is shown the systemic basis of features, which are reasonable to be used at holding of the complex metatheoretical analyses of region's tourist system.

Table 1. Basis for features of the metathoretical analyses of the tourist system

Feature identifier	Functional meaning of the system feature
P	Classification specification of regional institution, defining it as a set, relations, procedure, system, industry, product and so on
A	Region's subjects features, cooperating about their objects of their interests
B	Describing objects of region tourists system subject's relations
C	Systemically important features, acting in the described regional institution
D	Describing of aspect resulting effect under investigation in the regional tourist system
E	Dynamic features, reflecting the orientation of the aspect within regional system investigation over time
F	Description of the cause-and-effect influence on the aspect of the regional system under investigation and dynamic changes in it
G	Restrictions for existence of the regional system under investigation

On the basis of some used materials of Volgograd region tourist system, as well as results of its field study there was held the analyses of the respective specifications. This metatheoretical analysis of region's tourist system was emphasized on detecting great variety of complex features of use value of Volgograd region tourist system, which should be considered for optimization of the tour choice for the certain target segment.

Using the materials for describing the Volgograd region tourist system and field study we define the informal descriptions of consumer value tour's features. According to the investigations results of the system, let's form the quantitatively controlled features.

3 Results

3.1 Basis for Quantitatively Controlled Features of Region's Tourist System

At the result of metatheoretical analyses of Volgograd region tourist system it was detected its orientation on national, international, flash packing and social tourism, which has got the following classification feature:

R: Tourist product of Volgograd region.

The analyses of subject's features, having relations on maximization of tourist's products effectiveness, allowed to make the following informal description of the following feature:

A: The main subjects, defining the tourist product effectiveness are the guide, organization of infrastructure, excursion items of interest.

According to descriptions of activity of Volgograd region tourist industry companies, there were defined features of objects, about which the tourists, companies, that provide tourist services and companies, which regulate their activity cooperate. As the result we got the following informal description of the relations objects features for the maximization of Volgograd region tourist products:

B: Attractiveness objects of the tourist product are gnostic, culture, intellectual values, intellectual values, thoroughness of show, guide's verbal intelligence, intensity of speech, his or her communication style with the tourist and employees of place of interest.

Conceptual description of the Volgograd region tourist system analyses allowed detecting the systemically important features, defining the interrelations of its subjects. Using these features, it was received the following informal description of tourist industry systemically important features in Volgograd region:

C: Systemically important features of tourist industry in Volgograd region are defined by selecting the excursion objects and logical order to show them.

The defined features of Volgograd region tourist industry resulting effect on the social and economic system of the region, allowed to form the informal description of the feature:

D: Resulting effect of the tourist system is defined by general intussusception of quality and amenities: organization of the tourist group movement, hotel services and nutrition along the route.

According to the defined features of dynamic development orientation, it was formed the following informal description of dynamic orientation feature for Volgograd region tourism industry development:

E: The tours development is oriented on ability to visit the leisure activities (theater, stadium, circus and so on).

Using the cause-and-effect features on Volgograd region tourism industry, there was received the following informal description of the respective feature:

F: The essential cause and effect relations are the organization of the dialog with the tourists and selling the tourist goods near the place of interest.

Using the features of necessary conditions, at which the tourist system can function effectively we have got the following informal description of the feature:

G: Necessary condition for effective functioning of tourist system is availability of sanitary facilities for the tourists.

Set of systemic methatheoretical features defines set of the described subject possible aspects. This allows using these features to define the composition of region's tourist system. Using the results of investigations, let's form and make the table of two representations of the quantitative tourist system controlled features (Table 2).

Table 2. Basis for quantitatively controlled features of the region's tourist system

Features identifier	Informal description of system feature
P	Tourist product of Volgograd region
A	The main subjects, defining the attractiveness of the tourist product are the guide, organization of the infrastructure and excursion items of interest
B	Attractive objects of the tourist product are: gnostic, culture and intellectual value, thoroughness of show, guide's verbal intelligence, and intensity of speech, his or her communication style with the tourist and employees of place of interest
C	The system forming features of the tourist industry are defined by selecting of excursion objects of their show logic order
D	Resulting effect of the tourist system is defined by general intussusception of quality and amenities: organization of the tourist group movement, hotel services and nutrition along the route
E	The tours development is oriented on ability to visit the leisure activities (theater, stadium, circus and so on)
F	The essential cause and effect relations are the organization of the dialog with the tourists and selling the tourist goods near the place of interest
G	Necessary condition for effective functioning of tourist system is availability of sanitary facilities for the tourists

At the result of metatheoretical analyses of the conceptual descriptions for organization of effective tourist system we got the features, which are reasonable to be used within the process of optimization choosing the client oriented tour. The classification features has got the identifying meaning for the system under investigation, which is why quantitative evaluation of its meaning for different target groups of clients may not be considered.

3.2 Meaning of Metatheoretical Features for Different Target Groups

Metatheoretical features reflect sets of features, corresponding with the certain aspects of tourist products. To choose the preferable tour it is necessary to define the value of every metatheoretical feature for different target groups. This evaluation allows getting the windowed estimates of metathoretical features and their aggregate values, oriented on certain target segment. The Table 3 shows the results of pilot expert opinion of meaning (B_{qi}) for metathoretical features of region's tourist system for different target groups of the clients. If (q) – is index of the tourist system features, (i) – index of client's target group. These estimates reflect the fact, how these or those features are important for different segments of clients.

Table 3. Evaluation of meaning of tourist system features for different target groups

Criterial features of meaning (importance) of the tour features for the customers	Index (q) of tourist system features	Expert opinions of region's tourist system features meaning for different segments (B_{qi})					
		Women 15–30 years old	Women 31–51 years old	Women 52 and older	Men 15–30 years old	Men 31–51 years old	Men 52 years old
Index (i) of different target groups		(1)	(2)	(3)	(4)	(5)	(6)
A_B: The main subjects, defining the attractiveness of the tourist product are the guide, organization of the infrastructure and excursion items of interest	(1)	8	8	8	8	6	7
B_B: Attractive objects of the tourist product are: gnostic, culture and intellectual value, thoroughness of show, guide's verbal intelligence, and intensity of speech, his or her communication style with the tourist and employees of place of interest	(2)	7	8	8	7	7	6

(continued)

Table 3. (*continued*)

Criterial features of meaning (importance) of the tour features for the customers	Index (q) of tourist system features	Expert opinions of region's tourist system features meaning for different segments (B_{qi})					
		Women 15–30 years old	Women 31–51 years old	Women 52 and older	Men 15–30 years old	Men 31–51 years old	Men 52 years old
Index (i) of different target groups		(1)	(2)	(3)	(4)	(5)	(6)
C_B: The system forming features of the tourist industry are defined by selecting of excursion objects of their show logic order	(3)	5	7	7	7	6	5
D_B: Resulting effect of the tourist system is defined by general intussusception of quality and amenities: organization of the tourist group movement, hotel services and nutrition along the route	(4)	7	8	8	7	7	7
E_B: The tours development is oriented on ability to visit the leisure activities (theater, stadium, circus and so on)	(5)	5	4	6	6	5	4
F_B: The essential cause and effect relations are the organization of the dialog with the tourists and selling the tourist goods near the place of interest	(6)	6	8	7	6	8	7
G_B: Necessary condition for effective functioning of tourist system is availability of sanitary facilities for the tourists	(7)	7	7	8	7	6	8

3.3 Evaluations of Tourists Routes Features

Indexes of tourist routes metatheoretical features correspond with quality level of hospitality products certain aspects. Table 4 demonstrates set of expert opinion results for quality if features for the region A_K describe the quality: excursion service, organization of infrastructure, state of excursion facilities. B_K classifies the objects quality

Table 4. Demonstration of expert opinion for level of quality of region's tourist system products.

Tourist products	Index (j) region's tourist system products	Expert opinions for quality levels for tourist system products features (K_{jq})						
		A_K: The main subjects, defining the attractiveness of the tourist product are the guide, organization of the infrastructure and excursion items of interest.	B_K: Attractive objects of the tourist product are: gnostic, culture and intellectual value, thoroughness of show, guide's verbal intelligence, and intensity of speech, his or her communication style with the tourist and employees of place of interest.	C_K: The system forming features of the tourist industry are defined by selecting of excursion objects of their show. Logic order.	D_K: Resulting effect of the tourist system is defined by general intussusception of quality and amenities: organization of the tourist group movement, hotel services and nutrition along the route.	E_K: The tours development is oriented on ability to visit the leisure activities (theater, stadium, circus and so on).	F_K: The essential cause and effect relations are the organization of the dialog with the tourists and selling the tourist goods near the place of interest	G_K: Necessary condition for effective functioning of tourist system is availability of sanitary facilities for the tourists.
Index (q) of tourist system features		(1)	(2)	(3)	(4)	(5)	(6)	(7)
Excursion to the Volzhsksya hydro power station with visiting city of Volzhsky	(1)	9	10	6	7	9	8	7
Highlight tour 'City-hero Volgograd	(2)	9	9	8	8	6	9	8
....		......	....	....	....	......	......	......
Tourist routes of Volgograd region: city of Kamyshin	(n-1)	8	8	9	8	6	9	8
Tourist routes of Volgograd region: city of Urupinsk	(n)	8	7	9	8	5	9	8

of gnostic, culture and intellectual value, thoroughness of show, guide's verbal intelligence, and intensity of speech, his or her communication style with the tourist and employees of place of interest. C_K classifies selecting of excursion objects of their show logic show. D_K classifies intussusception of quality and amenities: organization of the tourist group movement, hotel services and nutrition along the route. E_K classifies processes of qualitative development of tours, oriented on ability to visit the leisure activities (theater, stadium, circus and so on). F_K classifies quality of organization the dialog with the tourists and selling the tourist goods near the place of interest. G_K classifies quality and availability of sanitary facilities for the tourists.

Data, indicated in Table 4 correspond with results of analyzing the expert opinion of feature's quality indexes (K_{jq}), got by special team for several regional tourist products. With: (j) – index of region's tourist system, (q) – index of tourist system features. The team used the scale from one to ten to estimate the tourist products features.

4 Basic Provisions for Methodological Choice of Optimal Tour

Criteria of optimal tour choice are maximal overall estimate of client's impressions from tourist product main features. The impressions are evaluated considering quality features implementation and their meaning for the customer of the tourist product. Using the expert opinions for quality levels for tourist system products features, indicated in Table 4 and evaluation of features meaning of the region's tourist system for different target groups and indicated in Table 3. The evaluation of client's impressions i of the target segment from j region's tourist system product can be estimated according to the following equation:

$$C_{ji} = \sum_{q=1}^{7} K_{jq} \cdot B_{qi}$$

For the client of i target segment optimal would be the region's tourist system product, corresponding to the following criteria:

$$W_i = \max_{j=1 \div n} \sum_{q=1}^{7} K_{jq} \cdot B_{qi}$$

Matrix $\left(K_{jq}\right)$ is array of evaluations for quality levels of tourist system features, indicated in Table 4. Matrix $\left(B_{qi}\right)$ is array of meaning of tourist system features for different target groups, indicated in Table 3. Evaluations of meaning allow getting windowed estimates of metatheoretical features for quality levels of hospitality product and their aggregate values for different region's tourist products considering likes and preferences of different target groups. Maximal aggregate values of region's tourist products metatheoretical features window estimates for every target group would correspond to the most preferable routes.

We define matrix expression, allowing defining aggregate value of windowed estimates for quality of tourist products features for every target group. Array of aggregate window estimates for quality of tourist product for every target group can be received at the result of calculating multiplication the matrix of expert opinions for quality of tourist products features product on the matrix of estimating meaning of tourist system features for different target groups.

$$\left(K_{jq}\right) \times \left(B_{qi}\right) = \left(C_{ji}\right)$$

We substitute in the matrixes $\left(K_{jq}\right)$ and $\left(B_{qi}\right)$ the respective meaning from Tables 3 and 4, we get the following equation, allowing calculating matrix $\left(C_{ji}\right)$ of aggregate window estimates for quality of tourist products for every target group:

$$\begin{pmatrix} 9 & 10 & 6 & 7 & 9 & 8 & 7 \\ 9 & 9 & 8 & 8 & 6 & 9 & 8 \\ \vdots & \vdots & \vdots & \vdots & \vdots & \vdots & \vdots \\ 8 & 8 & 9 & 8 & 6 & 9 & 8 \\ 8 & 7 & 9 & 8 & 5 & 9 & 8 \end{pmatrix} \times \begin{pmatrix} 8 & 8 & 8 & 8 & 6 & 7 \\ 7 & 8 & 8 & 7 & 7 & 6 \\ 5 & 7 & 7 & 7 & 6 & 5 \\ 7 & 8 & 8 & 7 & 7 & 7 \\ 5 & 4 & 6 & 6 & 5 & 4 \\ 6 & 8 & 7 & 6 & 8 & 7 \\ 7 & 7 & 8 & 7 & 6 & 8 \end{pmatrix}$$

$$= \begin{pmatrix} 363 & 399 & 416 & 384 & 360 & 350 \\ 371 & 416 & 427 & 393 & 371 & 364 \\ \vdots & \vdots & \vdots & \vdots & \vdots & \vdots \\ 361 & 407 & 418 & 385 & 364 & 356 \\ 349 & 395 & 404 & 372 & 352 & 346 \end{pmatrix}$$

Within the matrix $\left(C_{ji}\right)$ of total windowed estimates of tourist products quality the maximal value of optimality criterion in every column correspond with the best choice of tour for certain target group.

5 Conclusions/Recommendations

Tourist system should be oriented on providing high quality of tour's features and how they meet the demands, likes, preferences of the customers. It should be oriented on development of tour's selection methodology on the basis of their metatheoretical features quality and demand of these features by target clients.

It is reasonable to make complex metatheoretical analyses of region's tourist system within the basics of system features. This metatheoretical analysis is focuses on defining of great variety of complex features of tourist system subject actual improvement.

According to the positive results of the investigations it was defined that main subjects of tourist product are the guide, organization of the tourist infrastructure, excursion items of interest. Attractive objects of the tourist product are: gnostic, culture and intellectual value, thoroughness of show, guide's verbal intelligence, and intensity of speech, his or her communication style with the tourist and employees of place of interest. System forming features of Volgograd region tourism are defined by selecting the excursion subjects and their show logical order. Resulting effect of the tourist system is defined by general intussusception of quality and amenities: organization of the tourist group movement, hotel services and nutrition along the route. Modern orientation of the tours development is oriented on ability to visit the leisure activities (theater, stadium, circus and so on). The essential cause and effect relations are the organization of the dialog with the tourists and selling the tourist goods near the place of interest. Necessary condition for effective functioning of tourist system is availability of sanitary facilities for the tourists.

Optimizations of tours offer is oriented on considering the meaning of every its metathoretical feature for target groups of clients. Evaluation of tour's features meanings provide us with possibility to get their windowed estimates, oriented on certain target segment.

Choice optimality of tours selection is maximum of overall estimate of client's impressions from main features of tourist product. The impressions are evaluated considering quality of specifications implementation and their importance to the tourist product consumer. Using the expert opinions for quality level of the region's tourist system products features and evaluating the meaning of region's tourist system features for different target groups, we can estimate the indexes of client's impression from different region's tourist system products. Maximal aggregate values of region's tourist products quality metatheoretical features windowed estimate for every target group would correspond with the most preferable routes.

The criteria matrix expression allows defining aggregate values of windowed estimates for quality of tourist products features for every target group. Array of aggregate window estimates for quality of tourist product for every target group can be received at the result of calculating multiplication the matrix of expert opinions for quality of tourist products features product on the matrix of estimating meaning of tourist system features for different target groups.

Acknowledgments. The scientific work is done with the help of grant of Russian Foundation for Basic Research and Administration of Volgograd region (Project No 18-410-342002).

References

Gondar, T.: Construction of quality system for rendering services at the tourist industry companies. Scholarly notes of V. I. Vernadskiy, Tavricheskiy national university, Economics and Management Series, vol. 22(61), no. 2, pp. 121–126 (2009)

Welcome to Volgograd region. http://www.welcomevolgogradcity.com/Turism.aspx. Accessed 25 Dec 2017

Dyachenko, A.: Metatheory of imperatives for economic infrastructure development. News of Nizhnevolzhskiy agricultural cluster: Science and higher professional education, no. 4(44), pp. 286–292 (2016)

Dyachenko, A.: System analyses of agricultural tourism category. Serv. Plus **10**(2), 45–53 (2016). https://doi.org/10.12737/19457

Erishko, F.: System and analytic aspect of investigations for the forming tourist and recreation areas. In: Tourism and Recreation: Fundamental and Applied Researches: Scientific Works of the II International Scientific and Practical Conference, Russia, city of Moscow, M.V. Lomonosov MSU, pp. 46–48 (2007)

Zaliznyak, E.: Regional tourism: main features and conditions for development. Vestnik of Volgograd State University, Series 3, Economy, Ecology, no. 2(19), pp. 73–74 (2011)

Kruzhalin, V.: Scientific and methodological grounding for designing of tourist and recreation areas. In: Tourism and Recreation: Fundamental and Applied Researches: Scientific Works of the II International Scientific and Practical Conference, Russia, city of Moscow, M.V. Lomonosov MSU, pp. 34–38 (2007)

Sadovskiy, V.: Grounding of General System Theory, p. 99. Science (Ed.), Moscow (1978)

Fedotkin, D.: Creation and development of special economic area of tourist and recreation type in the Russian Federation. In: Tourism and Recreation: Fundamental and Applied Researches: Scientific Works of the II International Scientific and Practical Conference, Russia, city of Moscow, M.V. Lomonosov MSU, pp. 18–20 (2007)

Cherevichko, T.: Tourism as a system: methodology of investigation, Concept, No. 07 (June) (2015). http://e-kon-cept.ru/2015/15224.htm

Rana, M.W., Lodhi, R.N., Butt, G.R., Dar, W.U.: How determinants of customer satisfaction are affecting the brand image and behavioral intention in fast food industry of Pakistan? J. Tour. Hospit. **6**, 316 (2017). https://doi.org/10.4172/2167-0269.1000316

Suryawanshi, M., Shankar, G.: To study the role of advertisement for promotion of Pune Forts. J. Tour. Hospit. **7**, 338 (2018). https://doi.org/10.4172/2167-0269.1000338

Feng, X.: All-for-one tourism: the transformation and upgrading direction of regional tourism industry. J. Soc. Sci. Res. **1**(2), 2374–2378 (2017). https://doi.org/10.24297

Formation of a Regional Innovation Infrastructure Based on the Concept of Green Economy Development in Russia

Elena V. Goncharova[1(✉)] and Larisa S. Shakhovskaya[2]

[1] Volga Polytechnic Institute (branch) of Volgograd State Technical University, Volga Institute of Economics, Pedagogy and Law, Volzhsky, Russia
svumato@mail.ru

[2] Volgograd State Technical University, Volgograd, Russia
mamol4k@yandex.ru

Abstract. In this paper, the authors consider the current state of the Russian economy at the level of regional development, taking into account the trends of the green economy. In the conditions of the formation of an innovation economy, the issues of forming an innovation infrastructure are of particular importance. The concept of the green economy development in the Russian regions assumes the availability of infrastructural elements oriented towards the creation of innovations. The purpose of the study is to examine the basic principles of the concept of developing a green economy, taking into account the key components, namely – the objects of innovation infrastructure. In accordance with the goal, the article considers the following research tasks: consideration of the features of the Russia's economic policy formation at the present stage; characteristics of innovation in terms of the concept of green economy; identifying features of the whole country's innovative development as well as individual regions' development; consideration of foreign experience and the allocation of key factors for the development of the concept of a green and nature innovative-oriented economy. The research methodology is based on the application of the principles of a systemic approach, scientific comparison, methods of analysis and synthesis, and a logical-historical approach. In the paper, the authors analyzed the key characteristics of the green economy as an alternative factor for regional development. The formation of innovative infrastructures focused on the ecological direction of the economy has been proposed. The creation of a virtual technic park on the basis of an institute engaged in environmental development and technological innovation as the basis for the formation of a green economy in the region is considered.

Keywords: Innovative infrastructure · Green economy · Development of the region

1 Introduction

The development of a high-tech innovation economy in Russia with a focus on import substitution can be achieved by using green technologies, with the creation of innovative infrastructures to overcome the dependence of the Russian national economy on

E. G. Popkova and B. S. Sergi (Eds.): ISC 2019, LNNS 87, pp. 472–481, 2020.
https://doi.org/10.1007/978-3-030-29586-8_54

Western high technologies. This is especially relevant for the Russian Federation with its natural-climatic and socio-economic regional heterogeneity: indeed, no Russian region is similar to the other, representing a "state within a state" by, say, European standards. From this point of view, each Russian region is unique: having its own specific opportunities for development, and even with someone else's positive development experience, it cannot blindly copy it without adapting it to its capabilities.

The authors emphasize the fact of extreme turbulence in the external environment of economic activity generated by modern globalization processes and conclude that in these conditions not only each country should take this into account in its development, but also each region within the country, constantly adjusting its long-term development strategy. The same applies to individual regions of the world, and their member countries, as well as international economic institutions that unite these countries. It is necessary to understand that in the context of globalization, when economic competition is increasingly being replaced by the political dictates of individual countries, not even companies, not countries that compete in world markets, but international integration associations and international institutions that reflect the interests of groups of individual countries in a given region of the world. It is not easy to sustain this competition: for this it is necessary to clearly understand that the strength of the national economy lies in the ability to adopt the new things that have been accumulated by the practice of economic activity and public life in other countries and the ability to adapt it to the peculiarities of their country and its regions.

The green economy is quite a "strong" direction in the development of economic thought lately. Supporters of the green economy were and are such foreign and domestic scientists as J. M. Andersen, M. Bookchin, V. Glazychev, J. Jacobs, H. Daley, R. Carson, J. Mac-Harg, D. Meadows, Y. Odum, E. Schumacher, P. Houken, R. Whittaker. The founders of the school "Landscape Green Urbanism" were scientists D. Corner and C. Waldheim.

The modern market and the market of the future are the no mass production markets, which allow reducing the cost and price, but the markets of the innovation economy connected with the differentiated demands of consumers and constantly updated offers [1]. On this basis, the satisfaction of consumer preferences can be considered, first of all, as the main economic criterion of the new production. The use and formation of new knowledge should ensure the flexibility and efficiency of organizational structures, a short time frame for the development and production of new items, the rapid improvement of its quality and functionality. The purpose of the research in this paper is to consider the basic fundamentals of the green economy development concept, taking into account the key components, namely, the innovation infrastructure facilities.

In accordance with the goal, the paper considers the following research tasks: features of the formation of Russia's economic policy at the present stage of its development; characteristics of innovation in terms of the concept of a green economy, as an alternative to the traditional; identifying features of the innovative development of the country as a whole and its individual regions; consideration of foreign experience and the identification of key factors in the development of the concept of a green, innovation-oriented economy.

2 Materials and Methods (Model)

The research methodology is based on the application of the principles of a systems approach, scientific comparison, methods of analysis and synthesis, and a logical-historical approach. The development strategy of the subject, based on the conceptual provisions of the green economy, in general, should consider the formation of the necessary infrastructure for this. Aiming at the development of effective directions for the territorial economic systems' working-out, it is necessary to sufficiently substantiate the criteria for choosing not only the development priorities of these systems, but also the forms of its infrastructure. In the process of research, it is substantiated that it is the innovation infrastructure that is most effective for solving the problems of the green economy, since its development is an alternative to the traditional economy, which can improve the quality of life of the population and the level of socio-economic development of the regional community.

3 Results and Discussion

The Russian economy is more than half composed of energy and metallurgy, i.e. sectors that have the greatest adverse environmental impact among industrial sectors. To form a green economy and transition to environmentally sustainable development, Russia needs to change the current type of development and reverse the trends of unsustainable development in the economy. The key role in this process should be played by the transition from the extensive export-raw model of economic development to a model that is ecologically balanced and adapted for such a modernization of economic development that could lead to a green economy. This, in fact, a new alternative economy for Russia, should focus on qualitative, not quantitative development.

In terms of environmental sustainability, the future economy should have the following important features:

- conceptually, in economic development strategies at all levels of management, the directions formulated in UN and OECD documents on green economy and low-carbon economy are included;
- the environmentally friendly living conditions of the population and their provision become essential;
- priority in development is given to high-tech, processing and infrastructure industries with minimal impact on the environment;
- the share of the commodity sector in the economy is reduced in favor of the development of industries producing high value-added products;
- radically increases the efficiency of using natural resources and their savings, which is reflected in a sharp decrease in the cost of natural resources and the volume of pollution per unit of the final product (decrease in environmental intensity indexes and intensity of pollution);
- reduced environmental pollution.

In this regard, the implementation of environmental priorities should be combined with the economic objectives of the country: economic measures should provide both economic and environmental benefits.

At the stage of the accelerated process of urbanization in the world community, an intensive search continues to optimize the interaction between the city and nature, the city and man, the city and society.

One of the most important tasks in this regard is the use of green technologies in the urban economy and everyday life, which significantly increase the efficiency of the economy and the quality of life of the population. The implementation of green economy ideas is carried out in the following areas:

- strategic environmental management of economic activity at all levels;
- complete processing and utilization of municipal waste;
- membrane purification of natural water;
- the widespread use of alternative energy to reduce carbon dioxide emissions into the atmosphere;
- the formation of the natural framework of cities;
- production and promotion of organic food and building materials;
- the economical attitude of man to natural resources [2].

The emergence of the Green City international program (GREEN TOWN) embodied many scientific ideas in the field of urban development, and the adoption of green standards (LEED, BREEAM) significantly raised the bar in the design and construction of the safest buildings for human life and health. In the post-Soviet space, Kazakhstan and Belarus show positive examples of the development of a green economy. In Kazakhstan, this is the global initiative of the President of the country N. Nazarbayev “Green Bridge”, and in Belarus - the national program “Green Routes” [3].

Green technologies are actively promoted around the world through the system of higher education; Centers are opened in universities and training programs are introduced, such as: Green Skills Training Center (Australia); under the auspices of the Institute of Green Economy of the International Society for Ecological Economics - ISEE (International Society for Ecological Economics) a scientific journal International Journal of Green Economics is published, which intensively promotes ideas of green planning in urban development.

The most intensive research in the field of green economy is carried out by: Association for Sustainable Economy KOVET (Hungary); Association of the International Environmental Management Network INEM and the Passive House Institute (Germany); World Green Building Council and the Green Institute (Poland); International Association for Environmental Impact Assessment (International Association for Impact Assessment, IAIA); Association for Environmental Impact Assessment (New Zealand); International Society for Ecological Economics supported by the Charles Stewart Mott Foundation (USA). The Green Growth Alliance (THE GREEN GROWTH ACTION ALLIANCE), an active conductor of green economy ideas, includes the World Bank, OECD, IFC, Overseas Private Investment Corporation (OPIC), the United Nations Foundation, the International Development Club.

In Russia, scientists and specialists from the Institute of Software Systems, the Institute of Management Problems named after Trapeznikov and the Institute of Economics of the Russian Academy of Sciences, the Center for the Preparation and Implementation of International Technical Assistance Projects of the Russian Society for Ecological Economics, the Free Economic Society, and the Volgograd State Technical University are working at the green development site.

Today, the main directions of development of a green economy include: green energy based on renewable energy sources; conservation and restoration of urban land; decontamination of municipal waste and water treatment; providing cities with greenery; smart transportation and housing, as well as the modernization of production sites; formation of environmental indicators of the state of the urban environment; environmental education and training [4, 5]. Relevant for a green economy is the management of the future - a system aimed at sustainable development and improving the competitiveness of the economy, using a flexible mechanism for interaction between government, business and society in order to create innovative clusters and passionarity points in subsidized regions.

The process of innovative design begins with the stages of fundamental and applied research, continues with the phases of development and development of products and technologies, and ends with different stages of the commercialization of development results. It can be considered as a linear model of innovation, called the innovation funnel, innovation lift, or innovation pipeline (in the English-language literature, the pipeline model, assembly line model) [6]. This model describes a consistent linear process of supporting and eliminating innovation and venture capital projects.

The structure of the American, British, Australian and Taiwanese innovation systems focuses on the linear stages of work with various stages of technology readiness for commercialization [7].

The linear model of innovation is used: in the formation of the cluster development program and the development of measures to support its enterprises; the development of roadmaps for the development of the infrastructure of the innovation territorial cluster; system analysis of scientific and innovative projects of the cluster; preparation of proposals for the development of a cluster of national innovation system (NIS).

With the increasing globalization of the world economic space significantly increases the importance of such factors as scientific and technological progress. One of the effective ways to intensify innovation in a market economy is its stimulation and regulation at the regional and municipal levels. Subjects of the Federation of the Russian Federation in the framework of the implementation of the Strategy for the Development of Science and Innovations in the Russian Federation until 2020, recommended to make maximum efforts to form a system of institutions and infrastructure aimed at innovative modernization of the economy through the development of public-private partnership, expansion of scientific activities of higher educational institutions. If megalopolises basically have an innovative way, then medium and small cities are trying to determine their place in this area of activity. Based on existing research on the development of medium-sized cities, it can be argued that most of them are characterized by diverse economic specialization, multi-functionality, the presence of a developed infrastructure and institutions of higher professional education. Key factors that will contribute to the implementation of socially-oriented model of urban development, the introduction of eco-innovations, investments and information technologies in new industries, the development of highly qualified personnel. In connection with the constantly changing conditions of the functioning of the external environment, the management of the city requires its continuous improvement, the search for new approaches aimed at: innovative self-development; improvement of the urban environment through the use of new technologies and standards in construction to reduce the harmful effects of buildings and energy losses; gradual transition to environmentally friendly transport; introducing programs of financial support for the eco-innovation sector, subsidizing "clean" technologies, monitoring and controlling the quality of products and services; implementation of market mechanisms that stimulate changes in consumer behavior and promote green or eco-innovation [8].

The innovative territorial cluster is part of the RIS that supports the various stages of the life cycle of an innovation project, from design to successful business scaling, which can be described by various models of innovation activity.

The classic linear model of innovation describes the following steps: basic research; applied research; development engineering (R&D); development and commercialization.

On the basis of standardization and adaptation of these stages to the specifics of specific industries and regions, it is possible to form various state and corporate measures to support and develop innovative territorial clusters.

Helping companies to work at the appropriate stages can become the task of infrastructure organizations of the innovation cluster.

When deciding on the choice of the direction of development, one should not lose sight of the fact that new technologies can develop only under the condition that a population with a high educational and qualification potential will reside in the territory, and for this purpose an appropriate educational infrastructure must be created in the city and the motives encouraging citizens to constantly improve the educational level. Volgograd region is one of the regions of the Russian Federation, where a complex of measures aimed at creating a regional innovation system is being consistently developed and implemented. The administration of the Volgograd region has developed a program for creating an innovation infrastructure, developing fundamentally new mechanisms for financing innovative projects, and creating an innovation market. The basic element of the created support system for small and medium-sized enterprises in the region is the Regional Venture Fund with a financing volume of 140 million rubles. Also on the territory of the region are active: the association of business angels "Start Investments" in the form of a non-profit partnership, the US-Russia Entrepreneurship Center, the Foundation for Assistance to the Development of Small Enterprises in the Scientific and Technical Sphere [9]. For the period up to 2020, the priority areas for investment in the Volgograd Region will be: alternative energy and power engineering; biotechnology and pharmaceuticals; nanotechnology and information and telecommunication technologies [10]. Analysis of the socio-economic development of the city of Volzhsky shows that the peculiarities of the geo-economic situation, as well as the existing production and infrastructure potential, create prerequisites for the development of the city not only as the center of the urban district, but also as the center of an extensive cooperation area. The mechanism for managing the socio-economic development of the city is based on the presence of a number of competitive advantages: a rich natural resource base, production and infrastructure potential, the availability of qualified personnel and labor resources, the development of transport infrastructure, and the availability of access to the main transport routes of the railway and road transport; relative financial independence of the city from the regional center; political, national, interfaith stability and consistent policies to develop local self-government. The city district administration is interested in long-term business development and effective cooperation with enterprises and organizations of the city, joint actions are aimed at attracting investments in improving the environmental safety of the territory, improving the attractiveness of jobs, increasing the educational, professional level and qualifications of employees, improving competitiveness and the level of innovative activity.

Innovative development of a country or region involves the implementation of not only the main process of innovation, but also as an addition - the formation of a system of those factors and conditions that are necessary for the successful implementation of this process, i.e. availability of innovative capacity. If we start from the definition, innovation is an innovation created in the process of increasing the efficiency of production processes, in which new qualitative characteristics and values appear in a service, product or technology. In assessing the state of innovation potential and its development potential, conducted by the Institute of Strategic Innovation at Russian enterprises, it was revealed that among the external level factors, the demand for products produced by the market has the greatest influence on the innovative activity of personnel, and among internal factors - the willingness of managers and specialists to effectively use their labor and the available material, financial and intellectual resources.

Innovation should have an innovation process, have a tangible socioeconomic effect and reduce the pressure on the environment [11]. For example, eco-innovations include decisions of companies based on progressive technologies in reducing the consumption of initial natural resources (Bombardier - less fuel consumption), transforming one product into another (AXION - recycled plastic is processed into a new building material), using “smart” technologies, which allow to achieve resource saving and greater economic efficiency (E-streets - “smart” lighting of European cities), alternative sources of energy supply (Energy Innovation - HCPV-si) systems with minimal maintenance), ecological urban planning and technologies that make it possible to achieve zero waste throughout the city (Masdar City is a completely autonomous and ecological city), etc. In many European countries, extensive offensive greening is taking place: research centers are being created, scientific and technical centers are being prepared programs focused on maintaining environmental innovation technologies.

One of the trends of world development at the moment is the orientation of the economy towards the production of an intellectual product, towards the development of knowledge-intensive industries and high technologies.

The most promising forms of stimulating innovation are based on the commercialization of science: the creation of new innovative structures, such as technic parks, technic polices, and free economic zones [12].

The need to create a technic park on the basis of the university of the region is due to various factors. In universities, the implementation of research activities of teachers and students, there are many interesting ideas. Exporting them to the west is impractical. It seems more profitable to refine innovative projects in Russia and launch them into production.

The task of technic parks is to provide versatile support to small innovative enterprises operating in the scientific and technical sphere and the field of high technologies, especially at the initial stage of their formation.

The main function of technic parks is to generate, create, grow and bring scientific inventions and developments to the stage of commercialization, launch into production.

Virtual technic park is a specialized platform on which research and development, industrial enterprises, organizations and educational institutions representing high-tech, high-tech and innovative products are represented [13]. The basis of the exposition can be typical virtual platforms-stands of the enterprises of the participants, involving the use of audiovisual means of displaying various advertising and marketing information, as well as technical documentation. The virtual platform, in fact, is an event to identify and present technologies, the development of which in production will provide enterprises and entrepreneurs of the city and region with competitive advantages in product markets and will allow them to effectively promote scientific and technical products.

The formation of the innovation infrastructure (technic park) as the basic element of the implementation of the green economy concept causes the emergence of a number of advantages that are of great importance for the development of the region: reasonable investment of funds provided by the regional authorities for the development of entrepreneurship; the emergence of new jobs, the unification of several business technologies in one infrastructure, and finally, the streamlining of traffic flows. An important consequence of the creation of a technology park is the symbiosis of several technologies in the links of one ecological-economic chain: scientific development, implementation of prototypes, cooperation of small, large and medium-sized businesses, marketing research, exhibition and exhibition opportunities and potential implementation of green economy directions.

It is proposed to create a virtual technic park on the basis of the Volga Polytechnic Institute with the involvement of funds from the Bortnik Fund and the Ministry of Economic Development, based on the interaction of small innovative enterprises established at regional universities, with the participation of the Volga Regional Business Incubator. To build a technic park as a physical object, investments are needed for production space, premises, transport, and if you create a virtual technic park on the basis of an institute focused on creating primarily eco-oriented innovations, then this infrastructure will be an information space that will unite small businesses in the region at a distance, there will be less investment required, for domain registration, hosting, website creation, a virtual technic park will combine various universities, scientists, t. e. the same researcher, for example, can participate in several projects, developments and can be not only at the regional level. To become a technic park user, an entrepreneur needs to get a virtual office. Next, you need to create a virtual exhibition, and in the end - enter into real contracts and receive real money. Technic park, located in the information space, will help to unite various innovation infrastructures, geographically Isola-ted, will allow the same researchers to participate in several projects of various organizations, will increase the efficiency of innovation activities within the region.

4 Conclusion

The formation of a technic park based on the concept of developing a green economy, in essence, means creating a multi-sectoral innovation infra-structure in the region that will solve existing socio-economic problems and increase the investment attractiveness of the region. The creation of a technic park causes the emergence of a number of advantages that are of great importance for the economy of the region: reasonable investment of funds provided by the regional authorities for the development of entrepreneurship; the emergence of new jobs, the unification of several business technologies in one infrastructure, and finally, the streamlining of traffic flows. An important consequence of the creation of a technic park as a basic element of the innovation infrastructure is the symbiosis of several technologies in the links of one economic chain: scientific development, implementation of prototypes, cooperation of small, large and medium businesses, marketing research, exhibition and exhibition opportunities and potential implementation. The virtual technic park, created on the basis of the base university in the region, will enhance the interaction of the main market subjects in the direction of commercialization of innovations, positioning and promotion of scientific and technical products in the Volgograd region and beyond, thereby increasing the economic and innovative potential of the region. In the process of diffusion of innovations, the question arises about the degree of perception of the results of intellectual activity by the entrepreneur. First of all, the result of the commercialization of scientific and technical developments is of interest to the entrepreneur from the point of view of obtaining profit from the introduction and implementation, taking into account a significant share of the risk associated with activities in new types of market or offering new products. Commercialization of research activities is largely determined by the links between the key participants in the innovation process - research organizations and universities, small firms, large corporations.

Such a concentration of different technologies in one place simplifies transport and inter-business communications. A human resource will be involved, so that the indirect effect may be higher than even a direct economic one. The main positive aspect of the creation of a technic park is to increase the economic and investment attractiveness of the region.

References

1. Goncharova, E.V.: Innovatsionnyi potentsial kak strategicheskii faktor ehkonomicheskogo razvitiya rossii-skikh predpriyatii. Mezhdunarodnyi zhurnal ehkonomiki i obrazovaniya, T. 4, no. 2, pp. 29–46 (2018)
2. Shakhovskaya, L.S., Goncharova, E.V.: Zelenye tekhnologii kak osnova dlya formirovaniya regional'-nogo innovatsionnogo klastera. Ehkonomika i upravlenie: teoriya i praktika, vol. 4, no. 1, pp. 60–67 (2018)
3. Goncharova, E.V., Dzhindzholiya, A.F., Morozova, I.A., Medvedeva, L.N., Shakhovskaya, L.S.: Zelenaya ehkonomika kak osnova formirovaniya innovatsionnykh klasterov v regionakh Rossii: monografiya, 227 p. RUSAINS, MoskvA (2017)
4. Materialy saita Minehko-nomrazvitiya. http://economy.gov.ru/minec/main. Accessed 18 Feb 2019
5. Materialy reitingovogo agentstva «Ehkspert RA». http://raexpert.ru. Accessed 28 Jan 2019
6. Metodicheskie rekomendatsii po realizatsii klasternoi politiki v sub"ektakh Rossiiskoi Federatsii. www.economy.gov.ru/minec/activity/sections/innovations/politic. Accessed 18 Feb 2019
7. Shakhovskaya, L.S., Arakelova I.V.: Issledovanie opyta formirovaniya innovatsionnykh klasterov vo Frantsii i uroki dlya Rossii. Rossiya: tendentsii i perspektivy razvitiYA. Ezhegodnik. Otvetstvennyi redaktor: V.I. Gerasimov. Moskva, pp. 988–991 (2017)
8. Starovoitov, M.K., Medvedeva, L.N., Timoshenko, M.A., Goncharova, E.V.: Formirovanie «munitsipal'nogo polisa Volzhskii - Akhtubinskii» kak odnoi iz form sotsial'no-ehkonomicheskogo partnerstva i territorial'no-prostranstvennogo rasseleniya gorozhan. Ehkonomicheskoe vozrozhdenie Rossii, T. 28, no. 2, pp. 126–131 (2011)
9. Materialy Ministerstva ehkonomikI, vneshneehkonomicheskikh svyazei i investitsii. http://economics.volganet.ru. Accessed 18 Feb 2019
10. Ofitsial'nyi portal Gubernator i Pravitel'stvo Volgogradskoi oblasti. http://www.volganet.ru Accessed 30 July 2016
11. Goncharova, E.V.: Osobennosti protsessa kommertsializatsii nauchno-tekhnicheskikh razrabotok na sovremennom ehtape razvitiya ehkonomiki. In: Nauchno-metodicheskii ehlektronnyi zhurnal Kontsept, T. 31, pp. 1206–1210 (2017)
12. Goncharova, E.V.: O sozdanii regional'nogo tekhnoparka v Volgogradskoi oblasti. Voprosy ehkonomicheskikh nauk, no. 2, pp. 25–26 (2009)
13. Goncharova, E.V.: Virtual'nyi tekhnopark kak ploshchadka dlya innovatsionnykh razrabotok. Nauchno-metodicheskii ehlektronnyi zhurnal Kontsept, vol. 17, pp. 314–320 (2016)

Transformation of the Regional Innovation Sub-system as a Factor in the Development of the Digital Economy

Oleg N. Belenov[1](✉), Sergey S. Kiselev[1], Natalia V. Sirotkina[1], and Marina V. Titova[2]

[1] Voronezh State University, Voronezh, Russia
bon@econ.vsu.ru, serge_kiselev@bk.ru, docsnat@yandex.ru

[2] Lipetsk Branch of the Russian Presidential Academy of National Economy and Public Administration, Lipetsk, Russia
titovamarina@yandex.ru

Abstract. The purpose of the paper is to study the problems and prospects of cooperation between regional economic actors interested in innovative development and able to create favorable conditions for the production and distribution of digital technologies. Methodology. For this study, we employed the data provided by the Federal Statistics Service that describes the position of innovative activity of enterprises and companies as well as constituent entities of the Russian Federation. To unite regions into homogeneous groups, we utilized a comparative analysis, an index method, and an integral generalization. The study covers 2014–2017. The indicators describing the level of the regional innovation subsystem development within the establishment of the digital economy are as follows: the ratio of companies engaged in R&D, percentage of innovation-active organizations, share of the costs for technological innovation in the structure of GRP, and others. Results. We divided Russian territories into groups depending on the overall performance of the region's innovative activity. After a comparison of average values of innovation activity indicators in Russian regions with the actual figures achieved in the constituent entities of the Russian Federation, we identified two types of regions: innovation suppliers and innovation receivers. The threshold value of the overall performance of innovation activity ensured the impartiality of ranking, but it remains conditional because the dynamic development of the Russian economy leads to increasing representativeness of receiving regions. The conditionality of classification was taken into account when developing recommendations on the formation of an innovative environment in regions able to change their status in the short term. Recommendations. The differences revealed in the level of innovation activity of Russian regions became the basis for the development of recommendations regarding the ability of some regions to act as innovation suppliers and the needs of other regions for innovation interferences. It was proposed to use an original combination of methods and tools aimed at the effective transformation of the regional innovation subsystem assisting the formation of conditions for the development of the digital economy depending on the level of innovation activity achieved by various types of regions.

E. G. Popkova and B. S. Sergi (Eds.): ISC 2019, LNNS 87, pp. 482–488, 2020.
https://doi.org/10.1007/978-3-030-29586-8_55

Keywords: Innovation subsystem · Innovation policy · Digital economy · Developmental factor

JEL Code: F01 · O30

1 Introduction

As a rule, the definition of technological pattern points at a rather long historical period that has already passed. The current situation is an exception. Over the past decades, humanity has been in a state of self-identification, offering more and more new definitions for the emerging type of socio-economic relations. Thus, at the end of the last century it was believed that we observe the development of the knowledge-driven economy, which academicians unanimously understood as the stage of development featured by the creation of new knowledge and the production of high-tech goods with fundamentally different consumer qualities, subsequent application of this knowledge and goods in all types and areas of activity without exception. Specialists of the World Bank suggested to understood the knowledge-driven economy as one with the growth accelerated and competitiveness increased due to the establishment, dissemination, and use of knowledge. Having summarized dozens of definitions of the knowledge-driven economy, scholars soon began talking about the network economy, and today the digital economy is in the focus of research. The principal difference of the digital economy as an economic pattern is the crucial role of information and information (digital) technologies, which are considered as a source of growth or a factor determining the competitiveness of the economic system (organization, region, state). The environment for the establishment and diffusion of digital technologies is regional innovation subsystems that build the cellular structure of the national and global innovation system and determine the performance of the digital economy. The study of regional innovation subsystems allows us to understand the nature of the processes generating customization and other manifestations of the digital economy (Titova et al. 2015). The regularities found at the level of regional innovation subsystems are to be projected into other management levels.

2 Methodology

The development of the regional innovation subsystem should be considered as a necessary condition for the integration of the regional economy into the process of innovation improvement aimed at ensuring compliance with the requirements of the digital economy. This opinion is followed by scholars who believe that the essential elements of the innovation process are geographically-located. For example, M. Porter points out that the competitive advantages of companies on global markets are usually provided by their strong positions at the regional level, the concentration of highly-specialized industries, staff, supporting institutional bodies, suppliers, customers, etc. in particular regions (Porter 2015).

For the effective transformation of the regional innovation system it is necessary to assess its current state. A variety of methodological techniques are used to monitor the regional situation. For example, based on the methodology of the European innovation rating, the National Association of Innovations and Information Technology Development developed the following criteria for evaluating the innovation activity of Russian regions: the environment for the innovation development; production and use of innovations; legal framework. These criteria are employed in the composition of four comparative ratings: depending on the level of innovation development, the rate of their use in production activities, the level of development of the legal framework and the generalized rating of innovative activity by the total of the criteria calculated in the three previous ratings.

The methodology for evaluation of the innovative regional activity developed by the European Innovation Rating Agency includes the following stages:

1. The maximum and minimum values are selected for each criterion among all the regions under study.
2. Indicators of regions are standardized under the values found. As a result, the regions with the maximum and minimum indicators for this criterion amount to one and zero respectively.
3. The total indicator of the innovation activity is calculated as the average value of all indicators taken with equal weights for the region as a whole or a group of criteria.

Using this methodological approach and relying on the developments of other Russian and foreign authors (Popkova and Sergi 2018), (Popkova et al. 2018), (Sergi and Scanlon 2019), (Parakhina et al. 2017), (Treshchevsky et al. 2018), (Risin et al. 2017), (Titova 2015), (Sirotkina et al. 2012), (Sirotkina and Lanskaya 2011), (Sirotkina and Titova 2015), we offer to calculate the level of regional innovative activity based on the following parameters: percentage of companies engaged in R&D (X_1); proportion of innovation-active organizations (X_2); amount of internal R&D costs in the GRP structure (X_3); percentage of technological innovation costs in the GRP structure (X_4); share of innovative goods (works, services) in GRP (X_5); proportion of the economically active population (X_6); availability of innovation infrastructure (X_7).

The overall performance of the regional innovation activity is proposed to be calculated using the following formula:

$$I_{ИАР} = \sum_{n=1}^{7} a_n * X_n, \quad (1)$$

where a_n is a weighting factor, the value of which is determined based on expert evaluation; X_n are indicators for the evaluation of the innovative regional activity.

3 Results

The value of the level of regional innovation activity should be used as an indicator revealing the need for the application of some tools for the transformation of the regional innovation subsystem. Thus, regions characterized by low innovative activity are included in the group of innovation receivers that need an inflow of knowledge, innovation, and technology. Regions with high innovation activity are in the group of innovation suppliers. The regions are identified based on comparing the national average of the overall performance of innovation activity with the value reached in this region (Table 1). A higher than average value means that a region can be considered a supplier of innovations (knowledge, technology) since the level of innovation activity allows companies here to produce innovations for their own needs, as well as for use in other regions with the help of benchmarking. As of 2017, the national average of the innovation activity indicator in the region was 0.1923, which was used as a border separating innovation receivers from the innovation suppliers.

Table 1. Classification of Russian regions depending on the value of the overall performance of innovation activity ($I_{ИАР}$)

Range of $I_{ИАР}$ value	Regions included in the group
Innovation suppliers	
>0.1923	Moscow, St. Petersburg; Republics of Tatarstan, Bashkortostan, Udmurt, Chuvash; Altai, Khabarovsk, Perm, Krasnoyarsk, Krasnodar, Stavropol, Kamchatka, Primorsky Krai; Khanty-Mansiysk Autonomous Okrug – Ugra; Tomsk, Tver, Novosibirsk, Moscow, Samara, Tyumen, Penza, Tula, Saratov, Chelyabinsk, Voronezh, Vladimir, Ulyanovsk, Ivanovo, Kaluga, Sverdlovsk, Yaroslavl, Omsk, Rostov, Kemerovo, Kurgan, Vologda, Kaliningrad, Oryol oblasts
Innovation receivers	
$\leq$0.1923	Republic of Komi, Mordovia, Altai, Mari El, North Ossetia-Alania, Buryatia, Kabardino-Balkaria, Dagestan, Sakha (Yakutia), Tyva, Karelia, Khakassia, Chechen, Karachay-Cherkess, Adygea, Kalmykia, Ingushetia; Yamalo-Nenets Autonomous District; Jewish Autonomous Region; Nenets, Chukotka Autonomous Districts; Trans-Baikal Territory; Kursk, Ryazan, Belgorod, Leningrad, Tambov, Irkutsk, Kirov, Arkhangelsk, Lipetsk, Volgograd, Murmansk, Smolensk, Pskov, Novgorod, Kostroma, Bryansk, Amur, Magadan, Sakhalin oblasts

The region's ability to act as a supplier of innovations (knowledge, technologies, competencies) is determined by the effectiveness of the innovation policy implemented in it, which, in turn, implies the combined use of the following tools:

(1) fiscal tools (subsidies, tax benefits and deferments, state order, budget and investment tax credits, partial payment of patenting, participation in the registered capital of innovation-active organizations);
(2) institutional tools (development of integration processes with the participation of the innovation environment agents, legal and regulatory framework, programs of innovative regional development, workflow maps, technological forecasts);

(3) organization and management tools (improving the forms of integration of the innovation environment agents (Medvedev 2017), establishing of the innovation infrastructure);
(4) marketing tools (mainstreaming and PR support of regional innovation policy; holding conferences, exhibitions, fairs, regional branding, positioning the territory as a region of innovative development).

These groups were ranked and presented in Fig. 1 concerning their influence on the growth of the overall performance of the region's innovative activity. Thus, the use of fiscal tools is a priority in receiving regions with a low value of the overall performance of regional innovation activity. The transition of a region to a group of suppliers (based on an increase in the value of the indicator) is achieved due to step-by-step utilization of institutional, organizational, managerial and marketing tools. Compliance with this sequence is reasonable and expedient, since for the transformation of the regional innovation subsystem initially, it is necessary to establish formal conditions and then improve them.

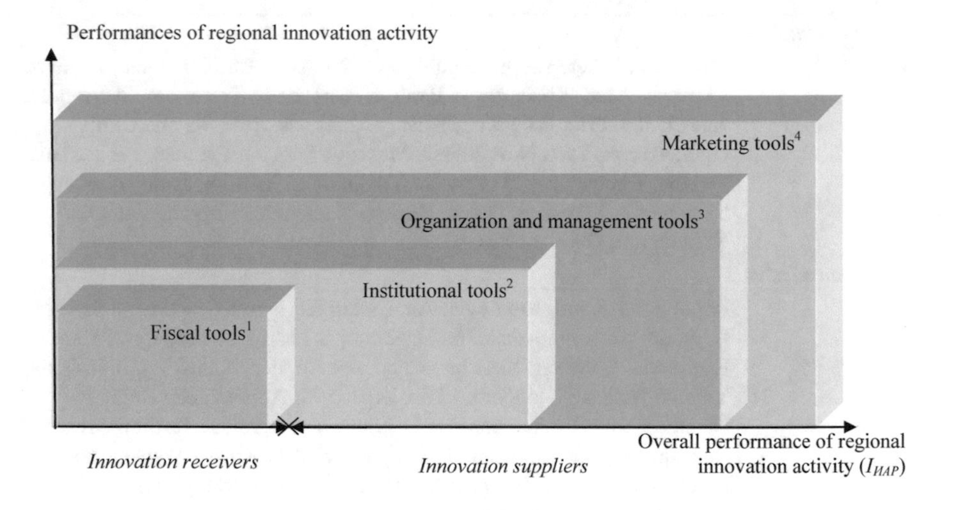

[1] subsidies, tax benefits and deferments, state order, budget and investment tax credits, partial payment of patenting, participation in the registered capital of innovation-active organizations
[2] development of integration processes with the participation of the innovation environment agents, legal and regulatory framework, programs of innovative regional development, workflow maps, technological forecasts);
[3] improving the forms of integration of the innovation environment agents, establishing of the innovation infrastructure);
[4] (mainstreaming and PR support of regional innovation policy; holding conferences, exhibitions, fairs, regional branding, positioning the territory as a region of innovative development.

Fig. 1. Transformation tools of regional innovation subsystem

4 Conclusion

The transition to a digital economy is provided by an effective transformation of the regional innovation subsystem through the development of appropriate policy. Regional innovation policy can be pursued through the use of special methods classified depending on the level of management (macroeconomic (exogenous) and microeconomic (endogenous)) and the nature of their impact on the regional economic system (direct and indirect). Exogenous and endogenous methods of transformation of the regional innovation system are reflected in the form of the target program, which assumes: stimulating the demand for innovations; contractual funding of innovative projects through state target and regional programs. The goals of target programs on the development and pursuance of regional innovation policy are ensuring the transition of the national economy to an innovative development model, increasing the population's living standard, and building a national innovation system.

Direct methods of transformation of the regional innovation subsystem are manifested in the following forms:

(1) sectoral form that suggests direct funding of innovation-active organizations in accordance with current legislation in order to promote the development of individual agents of the innovation environment;
(2) project form aimed at developing a base of effective innovative projects; establishing the funds for innovation projects; redistributing target capital of integrated bodies in order to support priority innovation projects; intensification of innovative development of the crucial economic sectors;
(3) integration form contributing to the promotion of cooperation and integration of agents of the innovation environment and providing co-creation in the process of diffusion of digital technologies.

Indirect methods intended to establish a favorable climate for innovation have the form of tax benefits and abatements, credit incentives to enterprises and organizations engaged in the development and distribution of innovations. Their implementation ensures the stimulation of processes leading to an increase in the population's living standard.

Tools for development and pursuance of regional innovation policy is a means to achieve the goals of building a regional innovation subsystem, shifting to an innovative model of development and establishing the digital economy.

We classified the whole variety of tools for the development and pursuance of regional innovation policy as follows:

(1) fiscal tools (subsidies, tax benefits and deferments, state order, budget and investment tax credits, partial payment of patenting, participation in the registered capital of innovation-active organizations);
(2) institutional tools (development of integration processes with the participation of the innovation environment agents, legal and regulatory framework, programs of innovative regional development, workflow maps, technological forecasts);
(3) organization and management tools (improving the forms of integration of the innovation environment agents (Medvedev 2017), establishing the innovation infrastructure);

(4) marketing tools (mainstreaming and PR support of regional innovation policy; holding conferences, exhibitions, fairs, regional branding, positioning the territory as a region of innovative development).

Taking into account the high level of differentiation of the Russian regions depending on the innovative development, we offered to use a combination of tools for the transformation of the regional innovation subsystem in accordance with the existing regional identity. Thus, we decided to consider the fiscal tools as the primary means of promoting the innovative development in the receiving regions and supplement the tools mentioned above with institutional, organizational, management, and marketing ones for supplying regions.

References

Medvedev, V.V., Tatarinova, E.S., Titova, M.V.: Improving the cooperation of public and municipal government. Innov. Econ. Prospects Dev. Improv. **4**(22), 225–230 (2017)

Parakhina, V.N., Boris, O.A., Timoshenko, P.N.: Integration of social and innovative activities into industrial organization. Contributions to Economics, pp. 225–242 (2017)

Popkova, E.G., Popova, E.V., Sergi, B.S.: Clusters and innovational networks toward sustainable growth. In: Sergi, B.S. (ed.) Exploring the Future of Russia's Economy and Markets: Towards Sustainable Economic Development, pp. 107–124. Emerald Publishing, Bingley (2018)

Popkova, E.G., Sergi, B.S.: Will Industry 4.0 and other innovations impact Russia's development? In: Sergi, B.S. (ed.) Exploring the Future of Russia's Economy and Markets: Towards Sustainable Economic Development, pp. 51–68. Emerald Publishing, Bingley (2018)

Porter, M.: Competitive Strategy: A Methodology for Analyzing Industries and Competitors, 453 p. Alpina Publisher, Moscow (2015)

Risin, I.E., Treshchevsky, Y.I., Tabachnikova, M.B., Franovskaya, G.N.: Public authorities and business on the possibilities of region's development. In: Popkova, E. (ed.) Overcoming Uncertainty of Institutional Environment as a Tool of Global Crisis Management. Contributions to Economics, pp. 55–62. Springer, Cham (2017)

Sergi, B.S., Scanlon, C.C.: Entrepreneurship and Development in the 21st Century. Emerald Publishing, Bingley (2019)

Sirotkina, N.V., Chernikova, A.A., Borisova, S.A.: Management Theory and Practice in Non-profit Educational Institutions, 224 p. Scientific Book, Voronezh (2012)

Sirotkina, N.V., Lanskaya, E.A.: Clustering as a promising direction for improving the competitiveness of the regional economy. Transp. Bus. Russ. **2**, 182–184 (2011)

Sirotkina, N.V., Titova, M.V.: Regional innovation policy in the focus of the knowledge-driven economy. Reg. Syst. Econ. Manag. **4**(31), 63–70 (2015)

Titova, M.V.: Conceptual provisions for the development and pursuance of regional innovation policy. Delta Nauki **2**, 64–67 (2015)

Titova, M.V., Goncharov, A.Yu., Sirotkina, N.V.: Regional innovation subsystem: evaluation and planning of development parameters. Modern Econ. Prob. Sol. **12**(72), 172–185 (2015)

Treshchevsky, Y.I., Risin, I.E., Korobeynikova, L.S., Gavrilov,V.V.: Management of changes of socio-economic systems: economic analysis of the state and consequences of the systemic feature. In: Endovitsky, D.A., Popkova, E.G. (eds.) Management of Changes in Socio-economic Systems, vol. 135, pp. 21–30. Springer (2018)

Assessment of the Company's Staff Creativity as the Basis for Their Adjustment to the Terms of the Digital Economy

Elena V. Endovitskaya(✉)

Voronezh State University, Voronezh, Russia
elena.endovitskaya@yandex.ru

Abstract. The purpose of the paper is an assessment of individual and social creativity as the basis for the adjustment of industrial enterprises to the digital economy.

Methodology. As a methodology for assessing the staff creativity, we propose test diagnostics, which allows measuring the level of creative competence of individual employees, target groups, teams and company's staff in general. We developed a basic model including four aspects and eighteen determinants of creativity. The main aspects include abstract and extraordinary nature of approaches that ensure universal decision-making; target task reformatting priority; the strategic scope of thinking; group creativity presumption.

Results. Testing the technique in the three enterprises of the same industry showed that they considerably differ in the levels of individual and social creativity. Under an overall low level of staff creativity, managers have the highest figure. In two companies, the creativity of specialists is close to that of executives. In one company, the creativity of specialists and other employees is much lower than that of senior management. In one of the companies, this ratio is better for a group of employees than for groups of specialists and others.

Conclusion/Recommendations. To create staff conditions for the transition to a digital economy one need: staff rotation, staff motivation to develop creativity, changes in the system of communication between various professional groups that develop both individual and social creativity.

Keywords: Staff creativity · Social creativity · Digital economy

JEL Code: D20 · O30

1 Introduction

The transition to a digital economy suggests fundamental changes in the company's staff behavior system, which is based on the creativity of employees (individual creativity). We consider it as an ability to produce new ideas based on extraordinary thinking, which determines the cognitive behavior of an individual in solving production and management problems. Creativity in development gives opportunities of a transition to divergent thinking, which allows for the existence of several right answers to the same question (Stolin 1983). According to this concept, divergence is the

E. G. Popkova and B. S. Sergi (Eds.): ISC 2019, LNNS 87, pp. 489–498, 2020.
https://doi.org/10.1007/978-3-030-29586-8_56

mechanism of creativity. Besides, creativity contributes to a multidirectional search for problem solutions using properties such as flexibility, quick thinking, a spontaneous manifestation of subconsciously controlled processes.

At the same time, we know from scientific research that personal qualities can be considerably transformed and give different results depending on people's affiliation to social groups and staff (Endovitsky et al. 2017; Risin et al. 2017); (Treshchevsky et al. 2017), (Bitarova et al. 2018). Therefore, in addition to creativity as a set of personal qualities, we think it necessary to highlight the company's ability to use them. For differentiation of personal and staff qualities, we use the concept of social creativity. In our opinion, it is social creativity that provides the opportunity to produce and implement new technological efficiently, organizational, managerial decisions, especially under dramatically changing terms (Bogoviz et al. 2018); (Parakhina et al. 2017); (Popkova and Bruno 2018); (Popkova et al. 2018). In this case, we are talking about the prospects for operation in terms of the digital economy.

2 Methodology

We propose test diagnostics as a technique for assessing the staff creativity developing the special nature of its professional expertise, which allows: measuring the level of creative expertise of individual employees, target groups, teams and company's staff in general; establishing the causes for the current state of things; estimating the prospects for achievement of the goals under basic model.

Table 1 shows a set of creative expertise, which content corresponds to the company's staff that is perspective within a digital economy. The model is implemented through a series of methodological procedures for textual diagnostics of creativity. This situation suggests a particular approach to the unification of creativity determinants to be employed. The suggested combination of characters as creativity determinants' codes (Dij) is presented in Table 1 (column 1 contains aspects of expertise Ai (i = 1 – 4); columns 2–3 show the determinants of creativity ADij (j = 1 – 18)).

Table 1. The basic model for diagnostic assessment of the company's staff creativity

Aspects of expertise (Ai)	Determinants of company's staff creativity (Dj)	ADij Codes
1	2	3
A1 Abstract and extraordinary nature of universal decision-making approaches	The disposition toward analysis and synthesis, the impetus for creative thinking	$AD_{1.01}$
	The propensity for problem thinking	$AD_{1.02}$
	The propensity for the choice of basic priorities that ensure persistent achievement of goals in a changing environment	$AD_{1.03}$
	Commitment to creative growth in terms of professional freedom	$AD_{1.04}$
	Self-control level in the creative process	$AD_{1.05}$

(continued)

Table 1. (*continued*)

Aspects of expertise (Ai)	Determinants of company's staff creativity (Dj)	ADij Codes
1	2	3
A2 The priority of target task reformatting	Creative and analytical abilities ratio	$AD_{2.06}$
	Thinking priority by goals	$AD_{2.07}$
	Commitment to data updating and reformatting	$AD_{2.08}$
A3 Strategic scope of thinking	Ability to predict possible situations	$AD_{3.09}$
	Ability to choose a sequence of steps	$AD_{3.10}$
	Ability to creative risk. Acceptability of calculated risk	$AD_{3.11}$
	Ability to adaptation. Choice of environment	$AD_{3.12}$
	The ability to allocate resources	$AD_{3.13}$
A4 Presumption of group creativity	Commitment to innovation	$AD_{4.14}$
	The ability to find energy zones and motivation points	$AD_{4.15}$
	The innate ability to recognize the peak of stimulation in group processes	$AD_{4.16}$
	Interest in the colleagues' opinions to generate innovations in teamwork	$AD_{4.17}$
	The ability to find their place in a group with multidirectional goals, opportunities, and needs	$AD_{4.18}$

According to the coding system of creative expertise contained in Table 1, we have drawn up a test form of their diagnostic assessment for staff, which had been published by us earlier (Endovitskaya 2018).

The test diagnostic tool provides fixing the employee's opposite opinions on a seven-point scale: from "+3" - "0" to "−3" to each determinant of creativity. The initial assessment of expertise (Table 1) and its test diagnostic assessment make it possible to study the employee's level of creativity that combines his/her self-assessment with a sociometric rating scale. For the results to be processed, we applied an algorithm for totaling all the points by Formula 1.

$$[(-3) \times 1] + [(+3) \times 1] \quad (1)$$

where l is the total of contradictory statements.

A particular feature of the methodological approach to the diagnostic assessment of employee's creativity is a united study of personal qualities such as creativity and cognitive rigidity. We understand cognitive rigidity as inflexible thinking, noncritical adherence to the ordinary way of action, disinclination to change the current pattern of activity.

The next step in the algorithm for diagnostic assessment of individual employee's creativity is compression and unification of the data obtained in the process of individual diagnostic assessment. The form of unification is presented in Table 2 (in column 3: cr - creativity, rg - rigidity).

Table 2. A unified array of employee's creativity/rigidity diagnostic assessment.

Indices	Accentuated aspects of creativity/rigidity	Determinants of cr and rg
1	2	3
A_1	Abstract and extraordinary nature of universal decision-making approaches	$AD_{1.01}$; $AD_{1.02}$; $AD_{1.03}$; $AD_{1.04}$; $AD_{1.05}$
A_2	The priority of target task reformatting	$AD_{2.06}$; $AD_{2.07}$; $AD_{2.08}$
A_3	Strategic scope of thinking	$AD_{3.09}$; $AD_{3.10}$; $AD_{3.11}$; $AD_{3.12}$; $AD_{3.13}$
A_4	Presumption of group creativity	$AD_{4.14}$; $AD_{4.15}$; $AD_{4.16}$; $AD_{4.17}$; $AD_{4.18}$

3 Results

We have been applied the developed technique for diagnostic assessment of staff creativity in three production enterprises of the Voronezh region. During 2017 these enterprises had been testing this technique for assessment of staff creative expertise. The results of the calculations are presented in Table 3. We used the following coding of performances and symbols:

M1, M2, M3 are the companies;
N1, N2, N3, N4 is the number of structural groups: managers, specialists, employees, workers;
A is aspects of creativity (as in Tables 1 and 2).

Quantitative diagnostic assessment of staff creativity was implemented under formulas 2–4.

$$\overline{A}_{cr/r_g} = \sum \overline{A}_{ij} \tag{2}$$

where $\overline{A}_{cr/r_g}$ is the average group aspect of creativity (cr)/rigidity (rg);

$$\sum \overline{A}_{ij} = \sum \left(\sum\nolimits_1^l A_j \times \sum\nolimits_1^k N_j \right) \tag{3}$$

Table 3. The results of the diagnostic assessment of staff creativity in the companies under study M1–M3 (2017)

Company, M		Accentuated aspects of creativity/rigidity	N \ A	Results of staff creativity/rigidity diagnostic assessment, points				
				N1	N2	N3	N4	$\overline{\sum N}$
1		2	3	4	5	6	7	8
	1	Abstract and extraordinary nature of universal decision-making approaches	A_1	+25 / -19	+29 / +9	+26 / +13	+10 / +10	+22,5 / +3,3
	2	Target task reformatting priority	A_2	+26 / +4	+25 / -11	+27 / +10	+16 / -2	+23,4 / +0,3
	3	Strategic scope of thinking	A_3	+23 / -8	+28 / -4	+26 / +13	+8 / +7	+21,3 / +2,0
	4	Presumption of group creativity	A_4	+22 / +14	+29 / +8	+25 / 0	+19 / -9	+23,8 / +3,3
	5	The total level of creativity/rigidity of staff target groups	$\sum_{1}^{4} A_j$	+96 / -9	+111 / +2	+104 / +36	+53 / +6	+91,0 / +8,9
	6	The average group level of creativity/rigidity	$\bar{A}$	+24,0 / -2,3	+27,7 / +0,5	+26,0 / +9,0	+13,3 / +1,5	+22,8 / +2,2
	7	Abstract and extraordinary nature of universal decision-making approaches	A_1	+28 / -5	+27 / +12	+24 / -2	+21 / -3	+25.0 / +0.5
	8	Target task reformatting priority	A_2	+25 / +2	+28 / -13	+21 / +9	+13 / +4	+21.8 / +0.5
	9	Strategic scope of thinking	A_3	+26 / +8	+24 / -12	+25 / -5	+16 / -6	+22.8 / -3.8

(*continued*)

Table 3. (*continued*)

	No.	Indicator	Symbol					
	10	Presumption of group creativity	A_4	+25 / -7	+23 / +9	+24 / +3	+25 / +8	+24.3 / +3.3
	11	The total level of creativity/rigidity of staff target groups	$\sum_{1}^{4} A_j$	+104 / -2.0	+102 / -4.0	+94 / +5.0	+75 / +3	+93.9 / +0.5
	12	The average group level of creativity/rigidity	$\bar{A}$	+26.0 / -0.5	+25.5 / -1.0	+23.5 / +1.3	+18.8 / +0.8	+23.5 / +0.1
	13	Abstract and extraordinary nature of universal decision-making approaches	A_1	+21 / -14	+20 / +14	+21 / +1	+7 / -5	+17.3 / -1.0
	14	Target task reformatting priority	A_2	+22 / +6	+26 / -1	+18 / +12	+13 / +1	+19.8 / +4.5
	15	Strategic scope of thinking	A_3	+23 / -8	+21 / +10	+15 / -1	+5 / +4	+16.0 / +1.3
	16	Presumption of group creativity	A_4	+21 / +11	+20 / +7	+16 / -2	+15 / -3	+18.0 / +3.3
	17	The total level of creativity/rigidity of staff target groups	$\sum_{1}^{4} A_j$	+87 / -5	+87 / +30	+70 / +10	+40 / -3.0	+71.1 / +8.0
	18	The average group level of creativity/rigidity	$\bar{A}$	+21.8 / -1.3	+21.8 / +7.5	+17.5 / +2.5	+10 / -0.8	+17.8 / +2.0

where j is accentuated aspects of creativity/rigidity, $j \to 1 \to 4$; l is the number of target groups, $i \to k \to 4$;

$$Aj = \Sigma АДfч \tag{4}$$

where АДfч is the aspect of creativity under the criterion, ч is the level of determinants cr/rg, f is the number of creativity determinants in the aspect Aj:

$$AD_1 = \sum(AD_{1.01}, AD_{1.02}, AD_{1.03}, AD_{1.04}, AD_{1.05});$$

$AD_2 = \sum(AD_{2.06}, AD_{2.07}, AD_{2.08});$
$AD_3 = \sum(AD_{3.09}, AD_{3.10}, AD_{3.11}, AD_{3.12}, AD_{3.13});$
$AD_4 = \sum(AD_{4.14}, AD_{4.15}, AD_{4.16}, AD_{4.17}, AD_{4.18}).$

As a result, we assessed the level of staff creativity (LCS) using formula 5.

$$LCS = \left(\sum^{l} \overline{A}_j\right) : \left(\sum_1^{l} \sum_1^{f} P_{max}\right) \times 100(\%), \quad (5)$$

where $\overline{A}_j$ is thc calculated data presented in Table 3 (column 8, lines 5, 11, 17);
P_{max} is the maximum assessment (in points) for each creativity determinant (+3).
Then the level of company's staff creativity will be:

$LCR_{M1} = 91.0\text{: } [(+3) \times 18 \times 4] \times 100 = 42.1$ (percent)
$LCR_{M2} = 93.9\text{: } [(+3) \times 18 \times 4] \times 100 = 43.5$ (percent)
$LCR_{M3} = 71.1\text{: } [(+3) \times 18 \times 4] \times 100 = 32.9$ (percent)

Taking the maximum level of the possible manifestation of creativity amounted 100% and comparing it with the achieved level, we can conclude that companies have a considerable non-involved capacity for creativity.

Let us compare the results obtained on the individual creativity of employees with the results of measurements. To measure the staff's social creativity (SCR), we use an indicator representing the difference between their creativity (CR) and rigidity (RG) – formula 6.

$$SCR = (\pm CR) - (\pm RG)\ \text{(points, percent)} \quad (6)$$

As we can see from formula 6, positive rigidity decreases the manifestation of social creativity and v.v.

The results of measurement and calculations of the company's staff social creativity using the data reflected in Table 3, its transformation using the formula 6, are presented in Table 4.

Table 4. The level and structure of the company's staff social creativity (2017)

Level of social creativity, points												
Company M1					Company M2				Company M3			
A		LSCS	N	LSCS	A	LSCS	N	LSCS	A	LSCS	N	LSCS
1	A_1	+19.2	N1	+105	A_1	+24.5	N1	+106	A_1	+18.3	N1	+92.0
2	A_2	+23.1	N2	+109	A_2	+21.3	N2	+106	A_2	+15.3	N2	+57.0
3	A_3	+19.3	N3	+68	A_3	+26.6	N3	+89	A_3	+14.7	N3	+60.0
4	A_4	+20.5	N4	+46	A_4	+21.0	N4	+73	A_4	+14.7	N4	+43.0
5	$\sum A$	+82.1	$\overline{N}$	+82.1	$\sum A$	+93.4	$\overline{N}$	+93.4	$\sum A$	+63.0	$\overline{N}$	+63.0
A is aspects of creativity (Table 3, column 2, lines 1, 2, 3, 4, 5);												
N is structural staff groups (Table 3, columns 4, 5, 6, 7, 8);												
LSCS is the level of staff social creativity (formula 6).												

Table 5. Aspect and comparative structural analysis of staff social creativity in companies M1, M2, M3 (2017)

The level of social creativity, percent												
Company M1				Company M2					Company M3			
ASCR		NSCR		ASCR		NSCR			ASCR		NSCR	
Aver.	Average weighted	Aver.	Average weighted	Aver.	Average weighted	Aver.	Average weighted		Aver.	Average weighted	Aver.	Average weighted
Ser. No.	1	2	3	4	5	6	7	8	9	10	11	12
1	24.4	20.7	34.8	29.6	26.2	26.2	28.4	28.4	29.0	19.6	36.5	24.6
2	29.5	25.1	34.5	29.3	22.8	22.8	28.4	28.4	29.0	19.6	22.6	15.3
3	25.9	22.0	21.5	18.3	28.5	28.5	23.8	23.8	24.3	16.4	23.8	16.1
4	20.2	17.2	9.2	7.8	22.5	22.5	19.4	19.4	17.7	11.9	17.1	11.5
5	100	-	100	-	100	-	100	-	100	-	100	-
6	-	85.0	-	85.0	-	100	-	100	-	67.5	-	67.5

From the data of Table 4 follows:

SCR_{M1} = +91.0-(+8.9) = +82.1 (points) (Table 4, column 8, line 5);
SCR_{M2} = +93.9-(+0.5) = +93.4 (points) (Table 4, column 8, line 11);
SCR_{M3} = +71.1-(+8.1) = +63.0 (points) (Table 4, column 8, line 17).

As you can see, social creativity in comparison with the individual one presented in Table 3 is much lower.

For a comprehensive and impartial assessment of this phenomenon, we drew up Table 5, which describes the level of social creativity using average percentages reflecting this phenomenon from positions of particular aspects of creativity (ASCR) and staff structure possessing this creativity (NSCR). The weighted average percentages of SCR in Table 5 provide information on the relative level of staff social creativity of in companies M1, M2, M3.

As we can see, company M1 with a quite high staff creative capacity show the most significant aspects of creativity such as abstract and extraordinary thinking (29.6%) and the priority of target management task reformatting (25%). The development of the group creativity presumption was underestimated (17%). Company M2 has a higher level of creativity than in others; the strategic scope of thinking (28.5%) and abstract and extraordinary thinking (26.2%) turned out to be the leading aspects of creativity. The level of group creativity presumption is low. Company M3 shows low figures of staff creativity ranging from 19.6% to 11.9% (Table 5, lines 1, 4, 6).

4 Conclusion/Recommendations

The study revealed that enterprises of the same industry have much different individual and social creativity. Under the overall low level of staff creativity in all companies assessed, leaders have the highest figure: 29.6% - in company M1; 28.5% – in company M2; 24.6% – in company M3. Companies M1 and M2 have almost the same figures of creativity for specialists and managers, which provides opportunities for effective staff adjustment. Company M3 possesses much lower creativity of specialists and other employees than that of the managers and amounts to 15.6% and 16.1% respectively.

The ratio of employees' and managers' creativity is different for companies.

In company M1, employees' creativity was 26.3% in comparison with managers' creativity, in company M2 – 68.3%, in company M3 – 46.7%. The final ratios make it possible to conclude that HR policy in all companies in terms of professional staff selection, motivation and external stimulation of staff creativity and its HR component is inappropriate.

Therefore, solving the problem of company's effective adjustment to the demands of the digital economy requires staff rotation, motivation for creativity development, changing the system of cooperation between various professional groups that cultivate both individual and social creativity.

References

Bogoviz, A.V., Ragulina, Y.V., Sirotkina, N.V.: Systemic contradictions in the development of modern Russia's industry in the conditions of establishment of the knowledge economy. Adv. Intell. Syst. Comput. **622**, 597–602 (2018)

Endovitsky, D.A., Tabachnikova, M.B., Treshchevsky, Y.I.: Analysis of the economic optimism of the institutional groups and socio-economic systems. ASERS J. Adv. Res. Law Econ. **VII** (6) (28), 1745–1752 (2017)

Endovitskaya, E.V.: Staff creativity analysis in the staff controlling system of industrial organizations of the agro-industrial complex, Topical issues of sustainable development of the agro-industrial complex and rural areas: proceedings of the All-Russian Research and Practical conference dedicated to the 50th anniversary of the Department of Economic Analysis, Statistics and Applied Mathematics, pp. 249–251. Federal State-funded Institution of Higher Education "Voronezh State Agrarian University", Voronezh (2018)

Bitarova, M.A., Getmantsev, K.V., Ilyasova, E.V., Krylova, E.M., Treshchevsky, Y.I.: Factors of socio-economic development of rural regions in the area of influence of city agglomerations. In: Popkova, E.G. (ed.) The Future of the Global Financial System: Downfall or Harmony. Lecture Notes in Networks and Systems, vol. 57, pp. 183–194. Springer, Cham (2018)

Parakhina, V.N., Boris, O.A., Timoshenko, P.N.: Integration of social and innovative activities into industrial organization, Contributions to Economics, No. 9783319454610, pp. 225–242 (2017)

Popkova, E.G., Bruno, S.S.: Will Industry 4.0 and other innovations impact Russia's development? In: Sergi, B.S. (ed.) Exploring the Future of Russia's Economy and Markets: Towards Sustainable Economic Development, pp. 51–68. Emerald Publishing, Bingley (2018)

Popkova, E.G., Popova, E.V., Bruno, S.S.: Clusters and innovational networks toward sustainable growth. In: Sergi, B.S. (ed.) Exploring the Future of Russia's Economy and Markets: Towards Sustainable Economic Development, pp. 107–124. Emerald Publishing, Bingley (2018)

Risin, I.E., Treshchevsky, Y.I., Tabachnikova, M.B., Franovskaya, G.N.: Public authorities and business on the possibilities of region's development. In: Popkova, E. (ed.) Overcoming Uncertainty of Institutional Environment as a Tool of Global Crisis Management. Contributions to Economics, pp. 55–62. Springer, Cham (2017)

Stolin, V.V.: Self-Consciousness of the Personality, p. 284. Publishing House of Moscow State University, Moscow (1983)

Treshchevsky, Y., Nikitina, L., Litovkin, M., Mayorova, V.: Results of Innovational Activities of Russian Regions in View of the Types of Economic Culture. In: Russia and the European Union Development and Perspectives Part of the series Contributions to Economics. Contributions to Economics, pp. 47–53. Springer, Cham (2017)

Global Competitiveness as a Background of the Digital Economy Development

Dmitry A. Endovitsky(✉), Yuri I. Treshchevsky, and Irina V. Terzi

Voronezh State University, Voronezh, Russia
eda@econ.vsu.ru, utreshevski@yandex.ru,
terziirinagagra@mail.ru

Abstract. The purpose of the paper is an assessment of developmental prospects and problems of the digital economy in terms of countries stratification by the performances of global competitiveness. Methodology. We employed the trend data provided by the World Economic Forum organization on the competitiveness of countries by global competitiveness performance. We used cluster analysis to unite countries into homogeneous groups. Clusters are built on annual data for 2014–2017. In the composition of global competitiveness indices, we highlighted the most significant for the development of the digital economy (infrastructure, quality of institutions, higher education and professional training, level of technological development, innovation capacity). Results. The classification of countries showed that five clusters with much different global competitiveness performances were revealed. The most developed and least developed cluster have a sustainable composition. The group of leaders and subleaders currently include 56 countries. Three clusters are continuously changing their composition while maintaining a low level of competitiveness parameters that do not allow the systematic development of the digital economy. Recommendations. The differences revealed between groups of countries determine the need for the establishment of transition modules that ensure their convergence in terms of global competitiveness performances. Moreover, both underdeveloped and the most competitive countries need it. Leveling-off institutional development, education, and innovation are the most important.

Keywords: Competitiveness · Cluster · Digital economy

JEL Code: F01 · O30

1 Introduction

The present stage of social development can be described as a shift to a socio-economic system, which is often called the digital economy. It makes some sense, like any clear concept of socio-economic processes facilitating their understanding at the mental level. At the same time, this simplification greatly complicates the cognitive issue of process understanding that takes place in different areas of public life. Humanity continuously deals with dramatic technological changes. However, they, as a rule, affect particular subsystems of life. For example, the system of machines that had been

E. G. Popkova and B. S. Sergi (Eds.): ISC 2019, LNNS 87, pp. 499–509, 2020.
https://doi.org/10.1007/978-3-030-29586-8_57

spread in the early 19th century fundamentally changed production processes but did not touch communication means, way of life of the majority of the population, and social institutions for a long time. On the contrary, the digital economy is being introduced into all subsystems of social life: from the manufacturing new means of production to developing a lifestyle on individual communications. In this regard, the following questions arise: what exactly do we want to get from the diffusion of the digital economy, what are its advantages and dangers, what way it will affect society, who can shift to an actual new technological status of all public subsystems.

2 Methodology

To compare the capabilities of different countries and other socio-economic systems, it is necessary to reveal the status of their parameters that determine the position in the global world, which is fundamentally ready to shift to a new technological status covering all areas of its performance.

The analysis is based on the data specified in the calculations of the Global Competitiveness Index (GCI) (World Economic Forum 2018). The data processing lies in ranking within a scale from 1 to 7 based on a comparison of GCI between countries. The two-thirds of variable composition are survey results of company management, and one-third is the data from outside open resources (Salmin 2002).

When calculating the GCI we determine 12 complex indicators (Table 1), which combine 114 variables, 34 of which are calculated on open statistical data (external debt, budget deficit, life expectancy, and other data provided by UNESCO, IMF, WHO), and the rest - on estimates of a particular survey of more than 14 thousand managers of medium and large enterprises (Salmin 2002).

Table 1. Performances of GCI calculation (Center for Human Sciences 2004)

Performances	Performance
Subindex A: Basic requirements	
P1	The quality of institutions
P2	Infrastructure
P3	Macroeconomic stability
P4	Health and Primary Education
Subindex B: Enhanced efficiency	
P5	Higher education and professional training
P6	Efficiency of the market for goods and services
P7	Labor Market Efficiency
P8	Financial market development
P9	Level of technological development
P10	Scope of the domestic market
Subindex C: Factors of practice and innovation	
P11	Company's competitiveness
P12	Innovation capacity

In the composition of performances, we determine those that are the most important for the development of both digital economy, and it is the most critical factors: production technologies, professional and social communications, official and unofficial institutions. Numerous studies make it possible to primarily refer to these performances education, technological, management, and institutional innovations (Endovitsky et al. 2017), (Popkova and Sergi 2018), (Popkova et al. 2018), (Sergi and Scanlon 2019), (Parakhina et al. 2017), (Treshchevsky et al. 2018), (Risin et al. 2017), (Treshchevsky et al. 2017).

This allowed us to identify the most remarkable performances of global competitiveness, which determine the prospects for the digital economy: P1, P2, P5, P9, P12. The others are considered by us as background, having a subordinate nature.

For systematization of the data array including 12 performances (parameters) of competitiveness, we used cluster analysis with theoretical and methodological foundations outlined in the academic literature (Hartigan and Wong 1979), (Mandel 1988) (Aldenderfer and Blashfield 1989). The theoretical and methodological provisions of cluster analysis in relation to large socio-economic systems are presented in several publications (Kruglyakova 2012), (Myasnikova 2015).

Using the above mentioned theoretical and methodological provisions, we analyzed the competitiveness of countries on the data of 2014, 2015, 2016, and 2017. A particular problem was a different number and a changing composition of countries specified in the GCI rating: in 2014 – 144; in 2015 – 140; in 2016 – 138; in 2017 – 137. One hundred twenty-six countries were present in the GCI rating over the analyzed period, and competitiveness parameters were subjected to cluster analysis. As a result, we obtained five clusters (A, B, C, D, E) and listed in the descending order of the overall level of competitiveness.

3 Results

Cluster A is represented by a relatively small number of countries with a well-defined core of 26 countries having stable positions here: Switzerland, Singapore, USA, Finland, Germany, Japan, Hong Kong, Netherlands, Great Britain, Sweden, Norway, United Arab Emirates, Denmark, Taiwan, Canada, Qatar, New Zealand, Belgium, Luxembourg, Malaysia, Austria, Australia, France, Ireland, Israel, Iceland. Estonia and South Korea fall into this cluster from time to time. Competitiveness performances of this cluster are presented in Fig. 1.

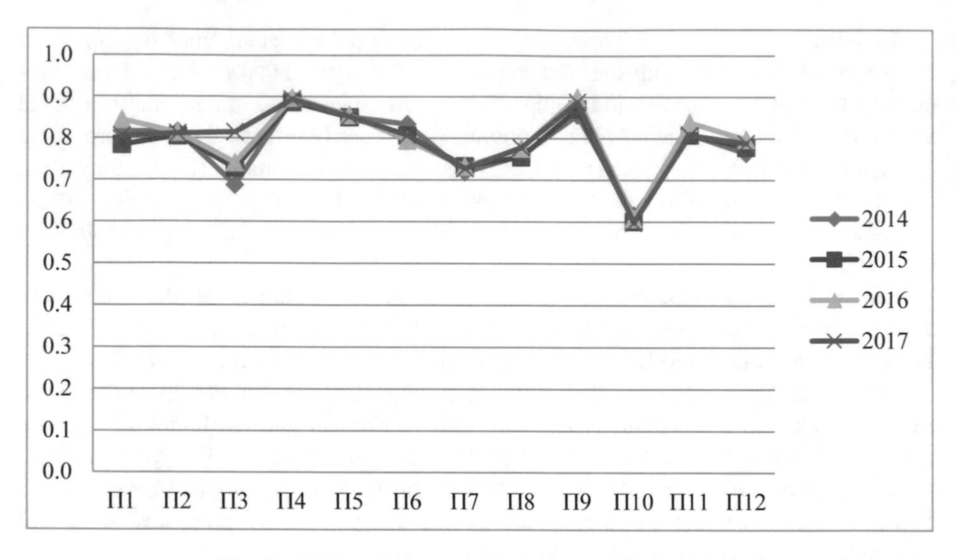

Fig. 1. GCI performances of cluster A in 2014–2017.

As we can see, the status of the cluster is exceptionally stable both in terms of the country composition and high performances of competitiveness. Stable quality of institutions (P1), infrastructure (P2), higher education and professional training, level of technological development (P9), and innovation capacity (P12) is crucial for the development of the digital economy. The size of the domestic market (A10) is the only relatively low performance of the background ones. It nearly equals to the value of this performance in cluster B. The level of another background performance (macroeconomic stability (P3)) has dramatically increased.

Unlike the previous one, cluster B continuously changes its composition (Table 2).

Table 2. The composition of the cluster B by year

2014	2015	2016	2017
Saudi Arabia	Saudi Arabia	Saudi Arabia	Saudi Arabia
South Korea	China	South Korea	China
China	Thailand	Estonia	Thailand
Estonia	Chile	Chile	Chile
Thailand	Indonesia	Spain	Indonesia
Chile	Czech Republic	Portugal	Spain
Indonesia	Azerbaijan	Czech Republic	Portugal
Czech Republic	Kuwait	Mauritius	Czech Republic
Mauritius	Poland	Lithuania	Azerbaijan
Lithuania	Turkey	Latvia	Mauritius
Latvia	Oman	Poland	Lithuania

(*continued*)

Table 2. (*continued*)

2014	2015	2016	2017
Latvia	Panama	Bahrein	Latvia
Bahrein	Kazakhstan	Malta	Latvia
Turkey	Philippines	Italy	Bahrein
Oman	Russia	Costa-Rica	Turkey
Malta	South Africa	Slovenia	Oman
Panama	Brazil	Uruguay	Malta
	Rumania		Panama
	Mexico		Italy
	Peru		Costa-Rica
	Columbia		Russia
	Vietnam		Bulgaria
	India		South Africa
	Morocco		Cyprus
	Slovakia		Hungary
	Iran		Mexico
			Jordan
			Slovenia
			India
			Slovakia
			Uruguay

The core of the group is formed only by three members: Saudi Arabia, Chile, and Poland.

Competitiveness performances of cluster B are presented in Fig. 2.

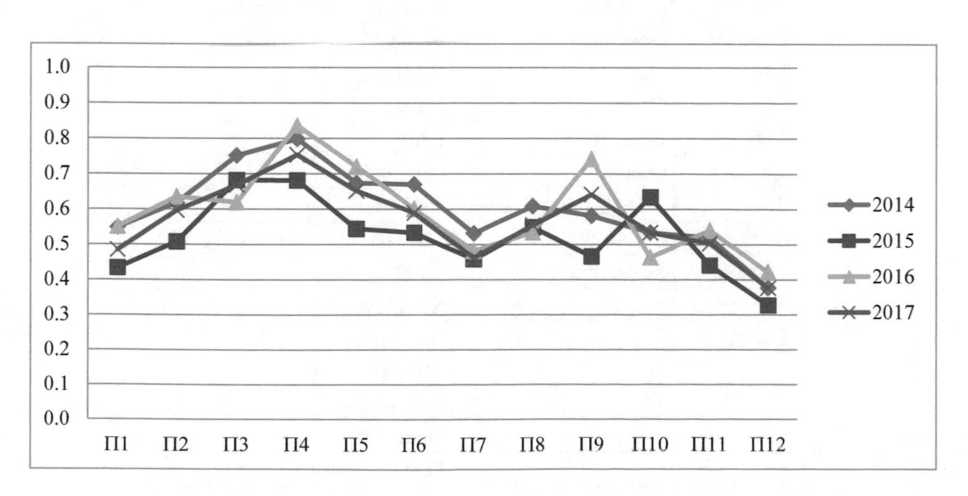

Fig. 2. GCI performances of cluster B in 2014–2017

As we can see, the status of cluster parameters over the entire period analyzed is very stable. Under significant changes in the number and composition of the countries of the cluster, this means that it is the competitiveness parameters that are crucial for its possible improvement in the context of inevitable changes related to the transition to the digital economy. It is important that in contrast with 2014 the values of the parameters P1, P2, P5 slightly decreased in 2017. Parameter P12 remained nearly the same, P9 increased. An unfavorable factor is the sharp annual fluctuations in the values of this parameter.

The third cluster by the level of competitiveness development C is extremely variable both in composition (Table 3, with no core) and in the parameters.

Table 3. The composition of the cluster C by year

2014	2015	2016	2017
Spain	Spain	China	Rwanda
Portugal	Portugal	Thailand	Montenegro
Italy	Mauritius	Indonesia	Georgia
Costa-Rica	Lithuania	Azerbaijan	Botswana
Russia	Latvia	Kuwait	Armenia
Brazil	Bahrein	Turkey	Jamaica
Cyprus	Malta	Oman	Namibia
Hungary	Italy	Panama	Kenia
Mexico	Costa-Rica	Kazakhstan	Tajikistan
Jordan	Bulgaria	Philippines	Laos
Slovenia	Cyprus	Russia	Cambodia
India	Hungary	Bulgaria	Bhutan
Sri-Lanka	Jordan	South Africa	Ghana
Ukraine	Montenegro	Rumania	Senegal
Croatia	Slovenia	Hungary	Cape Verde
Uruguay	Sri-Lanka	Mexico	Gambia
Greece	Ukraine	Rwanda	
Iran	Croatia	Jordan	
Salvador	Uruguay	Peru	
Jamaica	Greece	Columbia	
Tunis	Trinidad and Tobago	Vietnam	
Serbia	Serbia	Georgia	
Albania	Argentine	India	
Argentine		Morocco	
Lebanon		Botswana	
		Slovakia	
		Guatemala	
		Namibia	
		Dominicana	

Only the value of the parameter P1 improved for the analyzed period, the rest – deteriorated. Among the background factors, it is necessary to note a substantial improvement in the parameters of P7 and P8, but the competitiveness of cluster with an average level of development significantly decreased. Consequently, the opportunity for useful contacts with leading clusters has decreased in almost all points.

The status of outsider clusters (D and E) is of concern. In the cluster D the number of countries is variable under changing composition. The core is represented only by one country - Moldova (Table 4).

Table 4. The composition of the cluster D by year

2014	2015	2016	2017
Azerbaijan	Rwanda	Brazil	Kuwait
Kuwait	Georgia	Cyprus	Kazakhstan
Kazakhstan	Botswana	Montenegro	Philippines
Philippines	Guatemala	Sri-Lanka	Brazil
Bulgaria	Moldova	Ukraine	Rumania
South Africa	Salvador	Croatia	Peru
Rumania	Armenia	Algeria	Columbia
Rwanda	Jamaica	Greece	Vietnam
Peru	Tunis	Moldova	Morocco
Columbia	Namibia	Iran	Sri-Lanka
Montenegro	Kenia	Salvador	Ukraine
Vietnam	Tajikistan	Armenia	Croatia
Georgia	Laos	Jamaica	Guatemala
Morocco	Cambodia	Tunis	Algeria
Botswana	Zambia	Trinidad and Tobago	Greece
Slovakia	Albania	Kenia	Moldova
Guatemala	Mongolia	Tajikistan	Iran
Moldova	Honduras	Serbia	Salvador
Armenia	Dominican Republic	Albania	Tunis
Namibia	Bhutan	Mongolia	Trinidad and Tobago
Trinidad and Tobago	Kyrgyzstan	Honduras	Serbia
Kenia	Ghana	Bhutan	Albania
Zambia	Senegal	Argentine	Mongolia
	Lebanon	Ghana	Nicaragua
	Cape Verde	Lebanon	Honduras
	Gambia	Cape Verde	Dominican Republic
		Egypt	Nepal
		Venezuela	Argentine
			Kyrgyzstan
			Bangladesh
			Lebanon
			Egypt
			Paraguay

Almost all performances essential for the development of the digital economy have stable low values. The value of parameter P1 lowered. In other words, the cluster's gap with the leaders and even the countries of the average level of competitiveness increases.

Cluster E is quite extensive with a stable of 16 countries and a continuously changing periphery (Table 5).

Table 5. The composition of the cluster E by year

2014	2015	2016	2017
Algeria	Algeria	Laos	Zambia
Tajikistan	Nicaragua	Cambodia	Lesotho
Laos	Nepal	Zambia	Cameroon
Cambodia	Lesotho	Nicaragua	Ethiopia
Mongolia	Bangladesh	Nepal	Tanzania
Nicaragua	Cameroon	Lesotho	Uganda
Honduras	Ethiopia	Kyrgyzstan	Zimbabwe
Dominican Republic	Egypt	Bangladesh	Nigeria
Nepal	Paraguay	Senegal	Mali
Bhutan	Tanzania	Cameroon	Pakistan
Lesotho	Uganda	Ethiopia	Madagascar
Kyrgyzstan	Zimbabwe	Paraguay	Venezuela
Bangladesh	Nigeria	Tanzania	Malawi
Ghana	Mali	Uganda	Mozambique
Senegal	Pakistan	Zimbabwe	Sierra-Leone
Cape Verde	Madagascar	Gambia	Burundi
Cameroon	Venezuela	Nigeria	Mauritania
Ethiopia	Malawi	Mali	Chad
Egypt	Mozambique	Pakistan	
Paraguay	Sierra-Leone	Madagascar	
Tanzania	Burundi	Malawi	
Uganda	Mauritania	Mozambique	
Zimbabwe	Chad	Sierra-Leone	
Gambia		Burundi	
Nigeria		Mauritania	

(*continued*)

Table 5. (*continued*)

2014	2015	2016	2017
Mali		Chad	
Pakistan			
Madagascar			
Venezuela			
Malawi			
Mozambique			
Sierra-Leone			
Burundi			
Mauritania			
Chad			

We should positively assess the shrink of its composition. At the same time, performances of cluster competitiveness are very low (Fig. 3).

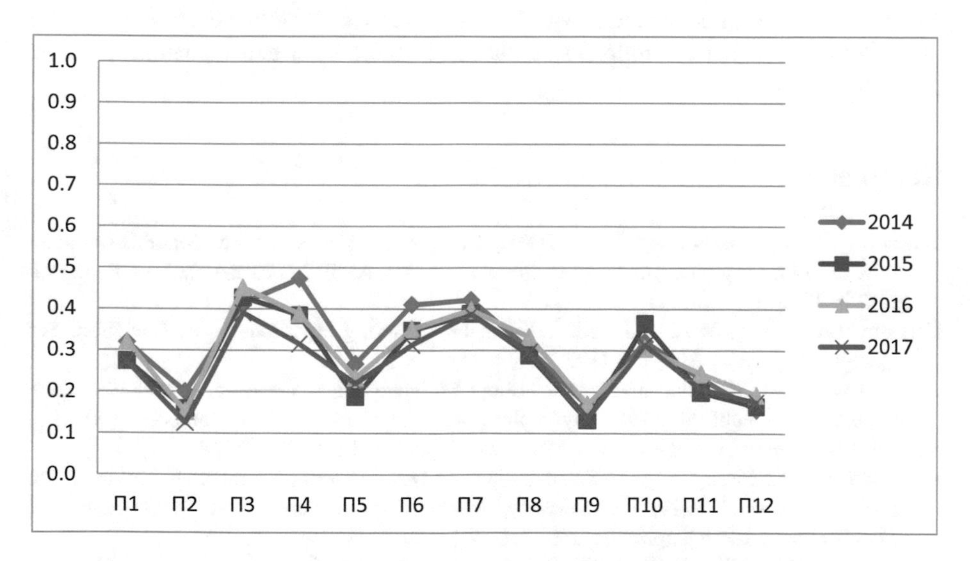

Fig. 3. GCI performances of cluster E in 2014–2017

As we can see, the status of the parameters that provide the opportunity of transition to a digital economy in the technological, innovative, social issues is extremely unfavorable. Most of the critical and background parameters are stable low over the analyzed period. Parameter P2 even shows a slight decrease. The fall of the already low values of the background parameters P4 and P6 should also be assessed as unfavorable for the possible development of the digital economy.

4 Conclusion

The analysis of countries competitiveness of by the ratio of its most important parameters (the quality of institutions; infrastructure; higher education and professional training; the level of technological development; innovation capacity) showed that only 26 countries could provide a transition to a digital economy in a broad socio-economic context. For this, they have the necessary human, technological, innovative, and institutional capacity.

Up to 30 countries have the opportunity of such a transition provided some adjustments to institutions, development of infrastructure and higher education.

More than half of the countries that form the clusters with medium and low competitiveness need to develop almost all of critical and background parameters of competitiveness. Their entry into the new digital economy is currently possible only in part, which considerably deteriorates the prospects for the sophisticated, systematic development of a new social pattern.

Considering the significant lag of most countries from the leading and subleading group, it seems necessary to reveal functional and spatial points of institutional, infrastructural, educational, technological, and innovative development to establish transition modules of cooperation with leading countries that allow extending the area of diffusion of digital technologies in the industrial, social and consumer areas of population's life.

References

Endovitsky, D.A., Tabachnikova, M.B., Treshchevsky, Y.I.: Analysis of the economic optimism of the institutional groups and socio-economic systems. ASERS J. Adv. Res. Law Econ. **VII** (6) (28), 1745–1752 (2017)

Hartigan, I.A., Wong, M.A.: AS 136: a K-Means clustering algorithm. J. R. Stat. Soc. Ser. C (Appl. Stat.) **28**(1), 100–108 (1979)

Kruglyakova, V.M.: Regional Investment Policy: Methodological, Conceptual, Methodological Support, 300 p. Publishing and Polygraphic Center of Voronezh State University (2012)

Mandel, I.D.: Cluster Analysis, 176 p. Moscow, Finance and Statistics (1988)

Myasnikova, T.A.: Strategizing the Socio-Economic Development of Municipal Entities in the Russian Regions: Theory, Methodology, Methodological Support: A Monograph, 271 p. Publishing and Polygraphic Center "Scientific Book," Voronezh (2015)

Aldenderfer, M.S., Blashfield, R.K.: Cluster analysis. In: Enukov, I.S. (ed.) Factorial, Discriminant and Cluster Analysis, 215 p. Finance and Statistics (1989)

Parakhina, V.N., Boris, O.A., Timoshenko, P.N.: Integration of social and innovative activities into industrial organization. In: Contributions to Economics, pp. 225–242 (2017)

Popkova, E.G., Sergi, B.S.: Will Industry 4.0 and other innovations impact Russia's development. In: Sergi, B.S. (ed.) Exploring the Future of Russia's Economy and Markets: Towards Sustainable Economic Development, pp. 51–68. Emerald Publishing, Bingley (2018)

Popkova, E.G., Popova, E.V., Sergi, B.S.: Clusters and innovational networks toward sustainable growth. In: Sergi, B.S. (ed.) Exploring the Future of Russia's Economy and Markets: Towards Sustainable Economic Development, pp. 107–124. Emerald Publishing, Bingley (2018)

Risin, I.E., Treshchevsky, Y.I., Tabachnikova, M.B., Franovskaya, G.N.: Public authorities and business on the possibilities of region's development. In: Popkova, E. (ed.) Overcoming Uncertainty of Institutional Environment as a Tool of Global Crisis Management. Contributions to Economics, pp. 55–62. Springer, Cham (2017)

Salmin, A.: The Russian Federation and the Federation in Russia. World Economy and International Relations, no. 2, pp. 44–47 (2002)

Sergi, B.S., Scanlon, C.C.: Entrepreneurship and Development in the 21st Century. Emerald Publishing, Bingley (2019)

Treshchevsky, Y.I., Risin, I.E., Korobeynikova, L.S., Gavrilov, V.V.: Management of changes of socio-economic systems: economic analysis of the state and consequences of the systemic feature. In: Endovitsky, D.A., Popkova, E.G. (ed.) Management of Changes in Socio-Economic Systems, vol. 135, pp. 21–30. Springer, Cham (2018)

Treshchevsky, Y., Nikitina, L., Litovkin, M., Mayorova, V.: Results of innovational activities of Russian regions in view of the types of economic culture. In: Russia and the European Union Development and Perspectives, Part of the Contributions to Economics Series. Contributions to Economics, pp. 47–53. Springer, Cham (2017)

World Economic Forum: The Global Competitiveness Index (2018). http://reports.weforum.org/global-competitiveness-report-2018

Center for Human Sciences: Humanitarian Encyclopedia, Global Competition Index, 2004–2018 (2018). http://gtmarket.ru/ratings/global-competitiveness-index/info

The Main Directions and Tools of Banking Digitalization

Olga V. Ivanova(✉), Larisa S. Korobeinikova, Igor E. Risin, and Elena F. Sysoeva

Voronezh State University, Voronezh, Russia
tramp-77@yandex.ru, korobeinikova@vsu.ru, risin@mail.ru, selfin@mail.ru

Abstract. The purpose of the paper is to identify the most efficient directions and tools of digital transformation for commercial banks, problems, and threats to the digitalization of the banking sector. Methodology. In the course of the study, we employed a systematic approach to reveal patterns of banking transformation in terms of economic digitalization; methods of comparative analysis to justify the most forward-looking areas and tools for digitalization of banking activities; synthesis, classification, groupings, observation. Results. The study made it possible to determine the following global areas of banking digitalization: a remote service model when the bank is mobile, and all services are available via applications for smartphones; technologies hidden from clients (petabytes of information and statistical models). The main digitalization tools include big data in proactive communication with clients; marketplace; robotization; blockchain technology; video integration; the use of chatbots and virtual assistants; making up virtual reality (VR); utilization of biometrics in the banking sector. Recommendations. The digitalization of banking activities poses significant threats to both the commercial banks and their clients primarily due to cyber attacks, high operational risks, growth of unemployment rate in the banking sector, the fact that some part of clients is not able to exist in the digital space. For such threats to be partially eliminated one need to upgrade the banking business environment with regard to developmental trends of the digital economy (regulatory, professional, cognitive issues); to check the reliability of bank clients in terms of transaction threshold amounts; to launch the programs raising the financial literacy of the population.

Keywords: Digital economy · Banking activity · Big data · Marketplace · Blockchain

JEL Code: G21 · G28

1 Introduction

At present, digital technologies are one of the benchmarks in determining the directions of the strategic provision of the country's competitiveness. The new paradigm of economic development, including the banking sector, aimed at engaging business

E. G. Popkova and B. S. Sergi (Eds.): ISC 2019, LNNS 87, pp. 510–516, 2020.
https://doi.org/10.1007/978-3-030-29586-8_58

entities and the public in the digital space, developing sustainable digital platforms, establishing an electronic financial infrastructure that provides the communication of digital space actors, etc.

BCG (Boston Consulting Group is an international company specializing in management consulting) under digitalization understands the highest possible fulfillment of digital technologies through their use in all business areas – processes, products and services, decision-making approaches (Boston Consulting Group 2018).

One of the main trends in the era of digital transformation is the use of digital technologies in big data work and design of platforms, which is especially relevant in terms of peculiar operation of credit institutions –financial mediators for the population and companies of the real economic sector.

Now commercial banks receive more than a third of the profit via digital channels. Tinkoff Bank, which was the first to introduce the concept of a virtual bank in Russia, takes the sixth place (5 million people) among the ten leading digital banks in the world by the volume of the client base.

2 Methodology

Since the direction of economic digitalization development in general and banking activity, in particular, was set mainly not by economists but by representatives of other fields of science, first of all, computer science, the fundamental economic proceedings are not enough (Skinner 2019). However, a significant contribution to the solution of applied economic problems of digitalization was made by foreign academicians (Chakpitak et al. 2018), (Goldstein 2017), (Tan et al. 2017) as well as Russian ones (Butenko and Isakhaev 2018), (Kudryavtsev and Skobelev 2017), (Savina 2018), (Ustyuzhanina et al. 2017), etc.

In this regard, in order to identify promising areas and tools for digitalization of banking activities, it is necessary to reveal current trends in the international practice of financial markets, primarily in the area of mutual payment platforms, through a comparative analysis.

However, payment systems are hardly the only object of digitalization of banking activities and personalization of banking services. The broad and heterogeneous nature of clients of credit institutions, complex character of the regulatory impact on the parameters of banking operations, diversity of current and promising digital technologies that are already in use and will be available for use in the short, medium and long term, makes systematization and classification of these technologies necessary.

3 Results

The transformation of financial goods markets into digital platform markets dramatically changes the role of financial mediators. Traditional banking services become remote (mobile banking, online lending), an infrastructure of in-country payments is being developed, non-bank payment systems and new actors emerge on the markets of transnational payments.

These digitalization tools for banking activity are only signs of global digitalization in cooperation of credit institutions and their clients. Due to the new quality of data, it will be possible to produce more flexible and efficient banking goods, launch them at the retail financial market as well as to establish a decentralized bank management agency necessary for flexible and quick decision-making.

We consider it necessary to highlight the three most useful areas of digital transformation for commercial banks.

1. Digitalization of cooperation between a client and a bank, which increases its efficiency both in terms of client satisfaction and minimizing of commercial bank costs.
2. The use of digital tools to support sales, such as a remote client manager interface, dynamic pricing, artificial intelligence for digitalization of daily operations.
3. Establishment of a big data management system and adjustment to the new requirements of the IT architecture.

Digitalization can be globally be divided into two directions. Firstly, it is a remote service model, when the bank is mobile, and all services are available using applications for smartphones. Secondly, these are technologies hidden from clients - petabytes of information and statistical models (Bankir.ru 2018).

Commercial banks need the flexibility that can be provided by Agile-culture both in the development of solutions and in the IT architecture to maintain innovation and engagement into the digital space.

The main tools for the digitalization of banking activities are the following:

- big data in proactive cooperation with clients. Clients of credit institutions seek to obtain a targeted and personalized experience, where service providers intuitively feel and understand their desires and eliminate potential problems even before they occur;
- transition to mobile services and applications (marketplace);
- robotization is an optimization of work with incoming documents, reduction of client service time and reporting;
- blockchain technology (reducing the time required for document reconciliation and signing, increasing the speed of transnational payments using the Ripple blockchain platform) (Amosova et al. 2018);
- video integration, which allows easing the damage reporting procedure due to a traffic accident to an insurance company, to improve the process of user interaction with an ATM at minimum cost. Video banking will be a tool to improve client experience and reduce costs;
- the use of chatbots and virtual assistants (artificial intelligence). This tool can keep a conversation with clients on any topic: from data on accounts to history of expenses. Moreover, these tools allow making personalized recommendations and suggestions based on real-time data and information;
- making up virtual reality (VR). VR technologies are the ideal solution for visualization of expenses and prediction of savings. Augmented reality technology can be used to manage savings in the same way.
- the use of biometrics in the banking sector for client authentication.

Digital technology gives more information about the client. The more transparent the client is, the easier it is for the bank to make an offer to meet the expectations and reduce its risks. In this relation, special attention should be paid to the marketplace and blockchain technology.

The marketplace is a new system of remote sale of financial products and e-commerce platform that provides information about a product or service of third parties whose operations are processed by the marketplace operator. Consumers have the opportunity to purchase financial products through the websites and mobile applications of display aggregators providing information from many vendors and allowing them to compare financial products. The system uses remote identification; registration of financial products is performed by the registrar of financial transactions.

The security of the entire system is ensured, first of all, by a unified identification-authentication system (USIA). Once having entered their biometric parameters, clients get the opportunity to identification with the help of Unified Biometrics System (EBU) and join the platform for financial transactions. Subsequent transactions in the Marketplace will be made only via USIA. This will increase the availability of financial products and services of credit institutions for remote clients. For full integration of services, the Marketplace should have an open API and be associated with an express payment platform. Now Moscow Exchange, VTB Registrar and Infinitum special depository are developing their marketplace.

Blockchain is a distributed ledger technology (DLT), which supports the recording of all transactions of cipher in a distributed computer network, but does not have a central accounting book and protects data by encrypted blocks. Blockchain technology facilitates the coordination of all types of interaction, helping to expertly arrange teamwork and preparing the basis for the transition of human-machine interaction to a dramatically new level.

The blockchain technology is promising in changing the landscape of the financial world. In particular, the volume of blockchain transactions made by one of the largest financial conglomerates HSBC exceeded $250 billion. At the same time, the financial institution conducted 3 million foreign exchange transactions and 150 thousand payments using the HSBC platform FX Everywhere platform.

Bank of America, the second largest US financial institution, intends to patent a system on blockchain technology, which is aimed at improving cash handling. In 2018, the bank received a patent for a mechanism that would allow large companies, including stock exchanges and custodial services, to keep their clients' cryptocurrency assets, as well as a patent for a device capable of storing cryptographic keys and digital currencies.

Forward-looking and already existing directions for the use of blockchain by various actors in the Russian Federation are presented below.

Bank of Russia:

- solving the problem of off-balance sheet depositors jointly with SDIA (prepared for implementation);
- the use of Masterchain for the exchange and storage of data (implemented);
- document circulation and reporting, also between the Bank of Russia and commercial banks.

Commercial banks:

- account management through power of attorney (implemented by Sberbank of Russia PAO in 2016); automated transfer of title to the collateral for automobile and mortgage loans; reporting to the supervisory authorities, the payment system (tokenization increases the security of electronic payments. The token is a randomly generated symbols, which the details of electronic payments are converted into); a unified platform of bonus programs, a single fraudster database (implemented by PAO "Sberbank of Russia", PAO Bank "FC Otkrytie");
- mortgage management;
- LC transactions (Alfa-Bank AO and Sibir Airlines PAO (S7 Airlines) in 2016 made LC transaction based on smart contracts, having decreased the time from a week to several hours);
- payment transactions (Alfa-Bank AO with the participation of "Sberbank of Russia" PAO and MegaFon PAO);
- factoring (implemented by Alfa-Bank JSC with the partnership of Sberbank of Russia PJSC and M. Video PJSC);
- document circulation (implemented by the Federal Antimonopoly Service jointly with "Sberbank of Russia" PAO);
- confirmation of bank guarantees, transactions on escrow accounts (implemented by "Sberbank of Russia" PAO, the electronic contract is executed only after confirmation by all parties and the authorization center).

Other market actors: leasing transactions; sale of automobiles and real estate; securities transactions; stock exchange transactions; notarial acts; intercorporate application for auditing and control.

As a mega-regulator, the Bank of Russia should be the first to transform into a digital ecosystem, which other players will subsequently join and it will ensure the stability and sustainability of the financial market in the new digital realities. Particular measures are already being taken in this direction.

In particular, in 2018 the Bank of Russia launched a robot to find illegal financial companies and projects on the Internet that have signs of financial pyramids. The robot works both online (Internet search) and offline (citizens' applications, publications in the mass media). In 2018, the Bank of Russia unveiled a wide-scale financial pyramid Cashberi.

The Bank of Russia has developed an express payment system (EPS) using a QR code. Clients of different banks will be able to transfer money to each other by phone number in just 15 s. At the first stage of the system operation, Russian citizens will be able to make transfers between their accounts and in favor of other parties. At the next stage of EPS development, it is planned to provide citizens with the opportunity to make payments in favor of legal entities. Connection to the system of express payments should become obligatory, primarily for core banks. According to experts, the implementation of EPS is both of economic and political importance, since it will solve the problem of the sanction isolation of the Crimea peninsula in cashless payments (with no operation of international Visa and MasterCard systems).

4 Conclusion

The digitalization of banking activity is a natural and, consequently, inevitable process, which develops faster than other areas and branches of the economy due to the peculiarities of financial capital.

We can mark out global areas of digitalization of the banking sector in terms of a functional feature:

- a retail business (marketplaces, USIA, biometrics, big data, chatbots);
- corporate business (e-signature, remote identification);
- interaction with the mega-regulator (a platform for prompt exchange of relevant data for reporting purposes. A regulator makes a general bank asset register, which allows updating information on the client's asset encumbrance, its real value; register of electronic guarantees, letters of credit and bills of exchange; a register of counterparties with an increased risk of participation in suspicious transactions; a system of data transmission from commercial banks to the Bank of Russia for right calculation of standards based on effective data; the opportunity to establish affiliates);
- interaction with stock exchanges (a platform for the operation of algorithms that have changed traders; allow decreasing the frequency of bot errors leading to the stock exchange crash);
- intrabank transactions (management, control, upgrade of the cash-flow model, which takes into account the most parameters and allows giving recommendations on risk hedging in time).

The volume of transactions carried out by the banking sector outstrips and cannot but bear new risks and threats not only for the banking institutions but also for their clients.

It seems that the problems and threats to the digitalization of the banking sector should include:

- cyberattacks (no more than 17 % of banks report them to FinCERT, while less than 5 provides full information), unauthorized access to information, growth of cybercrime rate as a result of the spread of mobile services;
- restricting access of Russian companies to international capital markets (Sysoeva et al. 2018), barriers to foreign capital and technology within the country can threaten its digital sovereignty and predetermine the digital backwardness of its residents, including credit institutions;
- increased operational risks of credit institutions due to the sophistication of digital technologies, business processes and client cooperation patterns; insufficient professional training of staff who do not have required expertise for application of digital technologies;
- raised volatility of financial markets, which is now clearly manifested on the cryptocurrency market;
- disappearance or dramatic transformation of some professions in the banking sector and, consequently, a reduced bank staff;

- restructuring of the banking regulation and supervision system and, therefore, higher regulatory burden on professional actors of the financial market and their possible bankruptcy;
- low financial literacy and activity of the population on the financial market;
- the need to amend the legal and regulatory framework of banking activities.

The threats of digitalization of banking activities can be countered only by joint efforts of the government, mega-regulator, credit institutions, and, to some extent, their clients.

References

Amosova, N., Kosobutskaya, A., Rudakova, O.: Risks of unregulated use of blockchain technology in the financial markets. In: Proceedings of the 4th International Conference on Economics, Management, Law, and Education (EMLE 2018). Part of the Series: AEBMR, vol. 171, pp. 9–13 (2018)

Bankir.ru. Future Banking: Figure as a Competitive Advantage (2018). https://bankir.ru/publikacii/20180921/banking-budushchego-tsifra-kak-konkurentnoe-preimushchestvo-10009593/. Accessed 7 Feb 2019

Boston Consulting Group: Big Data and Perspective Analytics (2018). https://www.bcg.com/ru-ru/capabilities/big-data-advanced-analytics/publications.aspx. Accessed 7 Feb 2019

Butenko, E.D., Isakhaev, N.R.: Contours of blockchain technology application in a financial institution. Finance Credit **24**(6), 1420–1431 (2018)

Chakpitak, N., Maneejuk, P., Shanaim, S., Sriboonchitta, S.: Thailand in the era of digital economy: how does digital technology promote economic growth? Studies in Computational Intelligence, vol. 753, pp. 350–362 (2018)

Goldstein, H.: Editorial: the digital economy act and statistical research. J. R. Stat. Soc. Ser. A Stat. Soc. **180**(4), 945–946 (2017)

Kudryavtsev, G.I., Skobelev, P.O.: Digital economy: a concept for managing large high-tech enterprises. Econ. Horiz. (5), 54–62 (2017)

Savina, T.N.: The digital economy as a new developmental paradigm: challenges, opportunities, and prospects. Finance Credit **24**(3), 579–590 (2018)

Skinner, K.: Digital Man: The Fourth Revolution in Human History, Which Will Affect Everyone, 304 p. Mann, Ivanov & Ferber, Moscow (2019)

Sysoeva, E., Budilova, E., Kachur, O., Dolgova, O.: Financing of Russian Companies in the conditions of distortion of international trade relations and economic sanctions. In: Popkova, E. (ed.) The Future of the Global Financial System: Downfall or Harmony. Lecture Notes in Networks and Systems, pp. 487–495. Springer, Cham (2018)

Tan, K.H., Ji, G., Lin, C.P., Tseng, M.-L.: Using Big data to make better decision in the digital economy. Int. J. Prod. Res. **55**(17), 4998–5000 (2017)

Ustyuzhanina, E.V., Sigarev, R.A., Shein, A.V.: The Digital economy as a new paradigm of economic development. Natl. Interests Prior. Secur. **13**(10), 1788–1804 (2017)

Qualitative Growth and Development Strategy of Personal Welfare Nanoeconomy in Terms of Economic Digitalization

Ivan T. Korogodin[1](✉), Natalia V. Golikova[1], Galina V. Golikova[1], and Lyudmila A. Beloglazova[2]

[1] Voronezh State University, Voronezh, Russia
ggalina123@yandex.ru, ggalina123@yandex.ru, snv-vrn@mail.ru

[2] Voronezh State Technical University, Voronezh, Russia
maskarad-36@yandex.ru

Abstract. The purpose of the study is to estimate the direction of the development of the personal welfare nanoeconomy strategy in terms of economic digitalization based on the system analysis of the qualitative growth of the personal welfare nanoeconomy and the abstract method of synthesis. Methodology. In the course of the study we revealed the content of personal welfare nanoeconomy in terms of economic digitalization as a socio-economic system consisting of the individual but interrelated subsystems: physical, social, process, effective, etc. We describe different types of growth of personal welfare nanoeconomy. We give a systematic analysis of the qualitative growth of the personal welfare nanoeconomy both as a process and a result. We determine the criterion for growth of the personal welfare nanoeconomy, characterize its development strategy in terms of economic digitalization based on theoretical principles put forward. Results. The authors substantiate several provisions that discover the essence and content of personal welfare nanoeconomy as economic categories, growth and development processes in terms of economic digitalization, affecting factors, which is much essential both theoretically and practically. Recommendations. We applied the following principles in the research methodology of personal welfare nanoeconomy: consistency, binarity, any-to-any connectivity, cause-and-effect, process-result, qualitative conformity between opposite parts of a whole, which served the basis for such methodological approaches as static, dynamic, reproduction. Their application made it possible to conduct a systematic analysis of the developmental strategy of the personal welfare nanoeconomy in terms of economic digitalization.

Keywords: Nanoeconomics · Personal welfare · Individual · Physical subsystem · Public subsystem

JEL Code: O4 · I31 · I200 · I210

E. G. Popkova and B. S. Sergi (Eds.): ISC 2019, LNNS 87, pp. 517–523, 2020.
https://doi.org/10.1007/978-3-030-29586-8_59

1 Introduction

Within complication of public bodies, the issue of the digital economy development is prioritized. This allows talking about a new type of economy with relations concerning production, processing, storage, transmission and use of an increasing amount of data prevail, which become the basis for an economic analysis of socio-economic system operation. Some experts argue that at present an economic agent need not the fact of resource possession but availability data on them and possible application for planning activities. On the basis thereof, from our point of view, the term digital economy should be understood as the type of economic management characterized by the dominant role of data and methods of their management as a critical resource in the production, distribution, exchange, and consumption (Bezrukova et al. 2019). The digital economy developing at a faster rate changes the normal economic relations and business models influences substantially on the development of the personal welfare nanoeconomy.

The main structural element of welfare is the population's living standard determined by several components: income, wealth, education, health care, labor conditions, living conditions, etc. There is personal and public welfare, and, with that, the public one is determined by the levels of the welfare of individuals (Korogodin and Smagina 2011).

Within various economic schools, a human has always been considered as an individual seeking to the growth of personal welfare. At the same time, in different periods social development the behavior of an individual was viewed from the point of hedonism, social essence of a person (Marx 1961), economic agent reasonably using market and state mechanisms for personal purposes, the economy of physical persons (Kleiner 2004). At present, it is important to determine the role of a human in nanoeconomy, nanoeconomy-based life strategy.

By the level approach to the study of economic phenomena, nanoeconomy represents the initial level of the economy, where the processes that form the fundamental basis for its other levels occur. Nanoeconomy is the level with individuals as the actors of economic relations (Korogodin 2016); (Golikova and Golikova 2010). Nanoeconomy is engaged in the labor process, manufacturing of benefits that satisfy the physical and spiritual needs of the individual and ensure his welfare (Korogodin 2016); (Melnikova 2014).

Manufacturing of life's benefits to meet the needs of individuals, the process of their consumption and provision of personal welfare is mediated by socio-economic relations. Socio-economic relations are, above all, production relations based on property ones. They include relations of market subsystems, which interrelation forms the integrity of the socio-economic system.

Organizational-economic relations, market relations, labor market relations, and labor relations are subsystems (Korogodin 2017). The object of this study is the personal welfare nanoeconomy in terms of economic digitalization, and the subject-matter is the system of socio-economic relations and ties emerging between people for their participation in development and provision the growth of the personal welfare nanoeconomy.

2 Methodology

Methodology in the most general sense is the method to establish the relationship between theory and reality (Blaug 2004). Sometimes the methodology is understood as a combination of methods typical for a given science.

Economic science has its particular methodology like the philosophy of science in its application to economics. The methodology of socio-economic research in a broad sense can be represented as a combination of methods for the development of scientific knowledge, principles and approaches to the study of economic processes and relations, methods of legal disclosures and substantiation of tentative conclusions about real economic activity." (Korogodin 2016). One of the fundamental methodological principles (consistency in the research of nanoeconomy) is designing a logical diagram reflecting real processes, links, and relations of people.

3 Results

Nanoeconomy is often understood as the level where actors of economic relations are particular individuals (Kleiner 2004). Agreeing with this is awkward. The internal organizational framework of the system cannot be fully understood using a dynamic approach to its analysis. Also, the individual in nanoeconomy is considered as a physical person. However, an employee as an individual is a legal feature of an individual who has only legal rights, but no economic ones. Therefore, it is reasonable to consider the individual in the nanoeconomic system not only as and even not so much as a physical person but foremost, as an economic agent. Using the binary principle, we can highlight dual human nature in different directions, which becomes the basis for the establishment of an internal framework at the nanoeconomic level. The dual nature of the individual is manifested in the fact that each human in nanoeconomy can be viewed from two perspectives expressing a dual process. On the one hand, it is the communication of human with nature when his reasonable activity is carried out, that is, human contact with those substances of nature, physical means, which under the influence of his labor get converted into life benefits satisfying people's needs. On the other hand, a human enters into relations with other employees and jointly converts material substances into finished products (Tolstykh and Golikova 2015). Consequently, a human as an actor of relations acts as the subject-matter of study and analysis in nanoeconomy. The object of study is human labor, which in combination with physical factors, unites the economic processes of production, distribution, and consumption (Korogodin 2017). Human life is regulated by two economic laws: the law of labor cost reimbursement and the law of rising needs. The law of labor cost reimbursement expresses a proportion between the quality of labor and the value of labor reproduction costs. The law of rising needs expresses the cause-and-effect relationship between abilities and needs. A human becomes economically interested in the comprehensive development of both his abilities and needs since they provide an enormous amount of essential physical and spiritual benefits (Golikova et al. 2018). The more benefits a human use in his life, the higher the level of his welfare. The growth of welfare is mediated by nanoeconomy, which ensures the personal welfare of each – it is the main actor. The essence of personal welfare nanoeconomy expresses the system of property relations and socio-economic ones arising between individuals concerning production, distribution, and consumption of life benefits that satisfy their needs to ensure the expanded reproduction of the abilities and needs of each of them. Any effective economic action of human is based on an economic interest in the satisfaction of needs.

Therefore, the material basis of personal welfare are benefits that can be economical and non-economic, physical and non-physical, as well as private, collective, mixed and public. The acts of an individual regarding the accumulation of his welfare determine the nature of the economic relations developing in the production, distribution, exchange, and consumption of life benefits.

The employment of the "process-result" principle, which allows explaining the means and affecting factors required for the steady growth of nanoeconomy, is essential for the study of personal welfare nanoeconomy in terms of economic digitalization. Effective strategies of the individual can explain the growth of the personal welfare nanoeconomy. Both the growth of any economy and the growth of the personal welfare nanoeconomy are analyzed from the point of its integral components. This growth represents, first of all, is interrelation of integral elements of both nanoeconomy and personal welfare. We can highlight two processes: the growth of nanoeconomy and the growth of personal welfare. Each of them has its content as a combination of interrelated elements, which interaction leads to self-development of personal welfare nanoeconomy system. These processes by the content of self-development cannot be identical in different types of nanoeconomy. The implementation of growth strategies is more independent in closed nanoeconomy than in a mixed one. Various forms of property primarily determine these differences. Closed nanoeconomy based on private labor property may provide only acquisition of physical factors of production and benefits for personal consumption and sale of goods manufactured through market exchange. Human accumulates income owing to the proceeds received from the sale of goods or services created. The combination of labor and capital for their interaction in a mixed nanoeconomy based on the private non-labor property occurs through the market. Having labor force and human capital of required quality, the individual as an actor of mixed nanoeconomy fulfills them in labor. With the help of labor, he/she creates a labor service, sells it to the employer and gain the necessary income in the form of wages (Korogodin 2017). Thus, interacting with the physical factors of production, the individual ensures the implementation of the nanoeconomy growth strategy and exert his efforts, which have to be refreshed. The implementation of personal welfare growth strategy consists in the consumption of benefits by the individual to refresh the efforts exerted and ensure their development (this process is carried out through a market exchange, accelerated and facilitated in terms of economic digitalization). An individual acquires market goods that satisfy his needs for his income. He refreshes and develops his efforts due to consumption, which will provide him with the ability to do his job (Sviridov et al. 2018).

The processes and results of the development of the personal welfare nanoeconomy require individual types of analysis with different qualitative and quantitative parameters established but not interrelated. Such an approach to the study of the personal welfare nanoeconomy growth is demanded only as an initial stage of research. Based on the knowledge received about our object in the course of analysis, we get opportunity to discover the entire process of growth as a single socio-economic system using the method of synthesis to obtain the result in the form of understanding the growth of personal welfare nanoeconomy as an integral whole of mutually interrelated parts (Korogodin 2017). With the help of synthesis, we can investigate the qualitative certainty of the personal welfare nanoeconomy growth and determine its development strategy.

There are three types of economic growth at all levels of the economy, including nanoeconomy: extensive, intensive, and qualitative. Each type of growth is associated primarily with either the quantitative or qualitative nature of economic growth. The methodological basis for describing the extensive and intensive economic growth was the Marxist reproduction approach to the study of economic processes, according to which reproduction: expands extensively in the event of production expansion and expands intensively under application of more efficient means of production (Marx 1961).

However, for a description of qualitative economic growth, it is not enough to proceed only from the definition of its intensive type. It mainly touches upon the quality of production means, which can be improved by innovations. The personal factor of production represents the other side of the intensive type of economic growth. However, for this, labor should correspond to the quality of the physical factors of production.

Concerning nanoeconomy, the essence of qualitative growth should be considered in its dual nature: as a process and as a result. Qualitative economic growth as a process is characterized by the interaction of various elements constituting its nanoeconomic system. The elements of this system are individuals of economic activity, their economic relations; subject-matter of relations; processes that ensure the change and development of the nanoeconomic system; laws governing relations, processes; factors affecting the process of qualitative growth.

An essential feature of these elements is the properties that give the qualitative certainty of nanoeconomic growth. It is the quality of an individual's labor as the total of the qualitative properties of its elements and characteristics and relationships, which determine the nature, purpose, complexity, and result of the labor aimed at creating a particular useful effect, that has the highest importance for nanoeconomic growth (Korogodin 2017).

In the process of the qualitative growth of nanoeconomy, an individual enters into relations with other actors on the creation of labor service as an object of exchange on the labor market (Korogodin 2018).

Qualitative growth, as a result, represents the amount of income received, which effect is expressed in a qualitative change in personal welfare nanoeconomy. First of all, this concerns the improvement of living standard, higher degree of satisfaction, reduction of income inequality. This, above all, is the criterion for the qualitative growth of personal welfare nanoeconomy and the essence of its development strategy.

The strategy of socio-economic development essentially expresses such a qualitative growth of nanoeconomy, which lay the foundations for the development of the human, his capital, labor capacity, abilities for creative, innovative activities. Qualitative growth provides such a socio-economic development, which is characterized by the shift of this system to a new, higher quality (Myrdal 2009). Owing to the upward development of the personal welfare nanoeconomy continuous improvement of personal welfare and living standards is achieved due to qualitative growth.

The implementation of the principle of qualitative conformity between the opposite parts of the whole in the socio-economic system lay an essential basis to create a new force of dynamic growth and develop personal welfare nanoeconomy. Each sophisticated socio-economic system typically has an internal mechanism of self-development based on the interaction of its elements (Korogodin 2017). The operation of this mechanism is determined by the dialectic of production and consumption expressed in the development of personal professional expertise, growth of personal needs and welfare, establishment, evolution, the fulfillment of abilities, and their recharge. In total, it is the main result of human life.

4 Conclusion

The article states the opinions of authors on the problems of personal welfare nanoeconomy, substantiates several provisions that reveal their nature and content as economic categories, growth and development processes, affecting factors in terms of economic digitalization, which is very important both for theory and practice.

In the research methodology of the personal welfare nanoeconomy, we employed the following scientific principles: consistency, binarity, any-to-any connectivity, cause-and-effect relationship, process-result, qualitative conformity between opposite parts of a whole, which served the basis for such methodological approaches as static, dynamic, reproduction. Their use has allowed making a systematic analysis of the economic categories under consideration.

Nanoeconomic growth is considered in two ways: intensive type of growth and qualitative type of growth. We conducted critical analysis following Marx reproduction approach, which is limited only by the physical nature of the expedient activity of the person without display of the personal factor's features.

Based on a systematic analysis of the qualitative growth of the personal welfare nanoeconomy and the abstract method of synthesis we reasoned a conclusion on its developmental strategy.

References

Blaug, M.: The methodology of economics, or how the economists explain. Moscow: Non-profit Partnership Journal of Economic Issues, 416 p. (2004)

Golikova, N.V., Golikova, G.V.: Transformation of Welfare Nano-Economy Based on Strategic Management of Changes in Micro-Level Systems: Monograph, p. 176. Nauchnaya Kniga, Voronezh (2010)

Kleiner, G.: Nanoeconomics. Econ. Issues (12), 70–93 (2004)
Korogodin, I.T.: Conceptual foundations for the study of the labor market object. J. Econ. Theory **15**(2), 254–263 (2018)
Korogodin, I.T.: Nanoeconomics of personal welfare as a socio-economic system, 138 p. LAMBERT Academic Publishing (2016)
Korogodin, I.T.: Systemic Nanoeconomics of Personal Welfare, Its Structure, and Qualitative Growth, p. 173. Economics, Moscow (2017)
Marx, K.: Capital, Work, 2nd edn., vol. 24, 657 p. Publishing House of Political Literature, Moscow (1961)
Melnikova, A.Yu.: Directions of interaction between the nano-economy and the public sector to improve the personal welfare of the population. Econ. Soc. 3–2 (12), 490–493 (2014)
Myrdal, G.: Modern principles of an unorthodox political economy. economic issues, no. 12, pp. 38–57 (2009)
Korogodin, I.T., Smagin, S.A.: Welfare: methodical approaches, principles of learning. Human and Labor, no. 8, pp. 14–16 (2011)
Tolstykh, I.A., Golikova, G.V.: The interrelation of the higher education system and the demands of the labor market in terms of the innovative transformation of the Russian economy. Reg. Syst. Econ. Manag. **3**(30), 82–85 (2015)
Bezrukova, T.L., Golikova, N.V., Safonova, N.A., Titareva, V.A., Komov, I.V.: Formation of cluster policy in the IT sector, vol. 726, pp. 546–553 (2019)
Golikova, G.V., Larionov, V.G., Verbitskaya, S.I., Fasenko, T.E., Kokhanenko, D.V.: Theoretical and methodological aspects of human capital management. In: Popkova, E.G. (ed.) The Impact of Information on Modern Humans, pp. 359–365. Springer, Cham (2018)
Sviridov, A.S., Golikova, G.V., Safonova, N.A., Nekrasova, T.A., Purgaeva, I.A.: Human development as an important component of the Russian population's living standards: a comparison with the global values. Espacios **39**(12), 9 (2018)

Advantages and Opportunities for the Development of Small Business E-Commerce in the B2B Sector

Anna Yu. Kosobutskaya(✉), Larisa M. Nikitina, Maria B. Tabachnikova, and Yuri I. Treshchevsky

Voronezh State University, Voronezh, Russia
anna.rodnina@mail.ru, lanikitina@yandex.ru, gmasha3@gmail.com, utreshevski@yandex.ru

Abstract. The purpose of the article is to find out opportunities for the development of small business e-commerce in the B2B sector. Methodology. The object of the study is a small enterprise acting on the market of technical devices and enterprise's technical arrangement projects. Results. It has been established that e-commerce in the B2B sector is a promising direction for the development of small enterprises in the distribution of technical goods and high-tech services. An analysis of the operation of one of the enterprises in this sector showed that the online store is much inferior to the sales department by sales volume and the cost of an average order. At the same time, the advantages of e-commerce are the engagement of a significant number of new customers, expansion of market geography, shrink of debtors, a higher return on sales against traditional trade. The disadvantages of e-commerce at the enterprise under analysis are the low level of regular customers. In general, the use of e-commerce tools allows not so much replacing traditional trade as supplementing its methods.

It is necessary to note the establishment of a set of tools as one of the main directions of its development that makes it possible to unambiguously identify the supplier of goods and services in a wide range of competitors.

Keywords: E-commerce · Small business · B2B sector

JEL Code: M20 · O39

1 Introduction

Currently, e-commerce in the B2B sector (business to business) sector seems to be a forward-looking direction of business development. The socio-economic importance of e-commerce development is associated with the establishment of conditions for actors of global economic relations and the simplification of market entry mechanisms. The research institution Gartner Group defined "e-business" as a means of continuous improvement of an enterprise's economic activity through digital technologies (Strauss and Frost 2001). Consulting enterprise PriceWaterhouseCoopers defines e-business as the integration of systems, processes, institutions, value chains, and markets using the

E. G. Popkova and B. S. Sergi (Eds.): ISC 2019, LNNS 87, pp. 524–531, 2020.
https://doi.org/10.1007/978-3-030-29586-8_60

Internet and related technologies (Sidorova 2012). Some researchers believe that e-business is a business activity employing the opportunities of global information networks to transform the enterprise's domestic and foreign contacts to make a higher profit (Novikova et al. 2013). Under the UN standards, business is recognized electronic if two of four components (production of goods or services, creation of demand or information, delivery to a consumer and payment) are carried out via the Internet (UN 1996). It is supposed that e-business includes sales, marketing, financial analysis, payments, recruitment, user support, and partnership maintenance. In the opinion of some researchers, e-commerce is a form of commercial activity under which two or more levels of an economic cycle are performed using information technologies (Novikova et al. 2013).

From our point of view, the development of cluster processes, which are in the focus of domestic and foreign researchers (Popkova et al. 2018), is essential for e-commerce in the small business.

The above definitions differ but generally describe e-commerce as part of the enterprise's business processes, which in the modern economy become an integral part of the enterprise's activities, and e-business is a business activity based on the application of information technologies.

2 Methodology

Small businesses are a particular sector of the economy vulnerable from the technological and socio-economic point of view (Endovitsky et al. 2017; Treshchevsky et al. 2018, 2015a, b). These features of small enterprises create a particular area of ICT diffusion. Many researchers note that any goods and services are not suitable for e-commerce, in particular, e-retail. For example, the online provision of services is impossible if they are not related to information. E-Commerce is limited to the sale of goods that the customer would like to see. However, let us note that there are technological possibilities to overcome such limitations. In particular, it is possible to create mobile groups that deliver goods to the customer's address without the prior conclusion of the contract. Some researchers suggest that the goods with parameters appropriately described in the accompanying documents and appearance displayed in a photo or video image are the most suitable for e-commerce. These are all kinds of technological devices and portable equipment: computers, TVs, video cameras, telephones, players, etc. (Sidorova 2012). This circumstance is essential for B2B sector for which the technical parameters of sales item are fundamentally important.

It is essential for e-commerce development in B2B sector that business bodies cooperate and need, on the one hand, to demonstrate the qualitative characteristics of the sales item, and on the other, to keep a trade secret. It is the latter circumstance that constraints the development of e-commerce in some cases. According to some researchers, the main issue is the low level of cryptographic trade-secrecy goods (Rublevskaya and Popov 2001).

It should be noted that e-commerce significantly broadens the opportunities of small enterprises with minimal costs owing to the use of modern communication means. A key feature of e-commerce is the ability to reach the target audience using social networks, which allow not only for satisfaction of current needs but also to generate future ones. The available electronic resources and Internet applications provide the tools for their implementation of various activities in trade, payment services, recruitment, and finance. Researchers note that e-commerce has some advantages in comparison with its traditional forms. In particular, the financial, production and human capital necessary for commercial transactions in e-retail are low-cost against real-time trade.

In the purpose of confidentiality, the enterprise under study was conventionally named “Alpha”. The parameters of the enterprise are kept unchanged.

3 Results

Alpha is a small enterprise. Directions of activities: wholesale and retail distribution of equipment; project-integration and design operations on complex technical systems.

The enterprise “Alpha” sales goods in three ways: through the sales department, project sales, and online store. Table 1 presents comparative data on trade turnover in various ways.

Table 1. Trade turnover of the online store, project sales, and sales department for XXXX year

Departments	Months								Total, mln.rub.
	1	2	3	4	....	10	11	12	
Sales department	6.1	5.3	4.7	6.2	....	6.1	5.9	7.6	83.1
Online store	0.8	1.2	1.4	2.0	....	1.0	1.5	1.9	18.6
Project sales	35.4	25.8	15.2	10.5	....	64.2	53.5	99.7	552.3

Analysis of the data presented in Table 1 allows us to conclude that the sales volume of the online store is only 22% of the sales volume of the sales department, and 2% of the enterprise’s total sales volume. That is, its influence on the enterprise’s results can be considered insignificant.

However, in current conditions, the geography of business development is essential for enterprises of any scale. In this regard, the comparative characteristics of the online store and the sales department are of interest (Table 2).

Table 2. Comparative characteristics of sales geography

No.	Province/Region	Online store	Sales department
1	Voronezh	+	+
2	Kursk	+	+
3	Belgorod	+	+

(*continued*)

Table 2. (*continued*)

No.	Province/Region	Online store	Sales department
4	Oryol	+	–
5	Lipetsk	+	+
6	Volgograd	+	–
7	Rostov	+	–
8	Ryazan	+	–
9	Tambov	+	–
10	Saratov	+	–

From the data presented in Table 2, we see that the online store operating in 10 regions allows the enterprise to develop new sales markets, while the sales department works only in four regions. This situation reasons the need for sale through the online store because it allows establishing a favorable reputation of the enterprise in other regions. As Table 1 shows, such activity makes possible both the sale of goods and stimulation of the spread of project activities that provide the maximum volume of goods and services of the enterprise. Besides, e-retail expands the area of competition of enterprises in the commercial sector, gives additional opportunities for market development in other regions with higher competitiveness.

Engagement of new customers is of great importance for the growth of the enterprise's competitiveness. This indicator is particularly important due to the peculiarities of the goods sold that means their long-term usability. A limited number of customers affects sales volume in the long-term negatively. In other words, the increase of the customer base reduces the strategic risks associated with a possible drop in sales volume through the enterprise's sales department. Comparison of data upon first-time customers is presented in Table 3.

Table 3. Comparison of enterprise's department performance upon first-time customers

Department	Months												Total, pcs.
	1	2	3	4	5	6	7	8	9	10	11	12	
Sales department	3	10	17	2	6	8	3	0	2	1	20	12	84
Online store	25	39	57	16	28	31	30	38	34	45	75	90	498

As we can see from the data presented in Table 3, the online store is more effective for attracting new markets than the traditional sales department. Like the geographical expansion of the market, this circumstance allows the enterprise to reduce the risk of market saturation by increasing its customer base and enhancing its density in the already occupied territory.

Table 4 presents data that makes it possible to estimate that e-retail is more effective than traditional trade both from the point of engagement of new customers and their conversion into regular ones.

Table 4. The ratio of first-time and regular customers of the online store and the sales department

Month	Sales department		Online store	
	First-time customers	Regular customers	First-time customers	Regular customers
1	3	19	25	10
2	10	25	39	21
3	17	12	57	13
4	2	17	16	13
5	6	10	28	16
6	8	21	31	20
7	3	32	30	18
8	0	26	38	10
9	2	23	34	13
10	1	15	45	17
11	20	13	75	14
12	12	22	90	11
Total	84	235	498	176

Based on the data presented in Table 4, we can conclude that the sales department interacts mainly with regular customers. In respect to the number of regular customers, engaged ones are about 30%. The situation is opposite in an online store: regular customers are 30% of new customers. At the same time, showing a positive trend in engaging new customers, the online store does not increase the percentage of regular ones. By the end of the year, the situation had deteriorated. This can be partly explained by the fast increase in the number of new customers. However, the number of regular customers scarcely grow, i.e., engaged customers did not fully appreciate the advantages of the enterprise and their goods. In principle, this fact corresponds to the nature of e-commerce, which is impersonal in no small extent. A significant number of new customers can be explained by the active promotion of goods and enterprise on the Internet. However, the technology of their conversion into regular customers is obviously unfinished.

The next performance upon which we compare the operation of the online store and the sales department is the average amount of the order by types of goods. Comparative characteristics of the sales department and online store are presented in Table 5.

As we can see from the data presented in Table 5, the performance of the sales department by some goods is much higher than the online store. The sales department mainly communicates with major customers on the implementation of complex works and simultaneous shipping of goods of several groups. The enterprise sells them goods with a minimum extra charge, increasing the chances of winning the tender for work performance; therefore the profit from sales considerably decreases but current assets increase. This situation is risky, especially in the event of payment issues.

Table 5. Comparison of the sales department and online store by the indicator "average amount of order within one month."

No.	Type of goods	Sales department figures			Online store figures		
		Turnover for the period, rub.	Number of orders, pcs.	Average amount of order, rub.	Turnover for the period, rub.	Number of orders, pcs.	Average amount of order, rub.
1	A	1,286,000	5	257,200	620,800	7	88,685
2	B	968,400	3	322,800	467,000	6	77,833
3	C	570,800	4	142,700	108,600	5	21,720
4	D	456,900	2	228,450	56,200	1	56,200
5	E	385,200	21	18,342	109,450	12	9,120
6	F	120,800	1	120,800	0	0	0
7	G	354,420	92	3,852	50,800	10	5,080
8	H	1,771,300	132	13,418	102,100	24	4,254

The online store communicates with individual customers; even due to competition extra charge for goods on this market is higher and it is more beneficial for the enterprise to supply less equipment but gain a bigger profit. Comparison of the performances of the sales department and the online store by gross profit is given in Table 6.

Table 6. Comparison of the sales department and online store by gross profit for one month of XXXX year

№	Type of goods	Sales department figures			Online store figures		
		Turnover for the period, rub.	Gross profit, rub.	Return on sales, percent	Turnover for the period, rub.	Gross profit, rub.	Return on sales, percent
1	A	1,830,900	549,270	30	400,850	180,382	45
2	B	2,582,120	180,748	7	369,000	55,350	15
3	C	870,950	130,642	15	108,600	20,634	19
4	D	569,000	68,280	12	89,600	16,128	18
5	E	985,250	246,312	25	139,450	55,780	40
6	F	120,800	21,744	18	13,900	5,004	36
7	G	854,420	187,972	22	72,060	22,338	31
8	H	1,890,300	623,799	33	402,100	160,840	40
Total		9,703,740	2,008,767	20.7	1,595,260	516,456	32

When analyzing the performances of the online store, we can conclude that the margin of sales is much higher within the operation of this department. This is because goods are sold at retail prices or under the lowest discounts. A discount for regular customers is 5% of the cost. The high efficiency of the online store allows the enterprise to ensure a high level of return on sales for all types of goods at a minimum cost.

An essential indicator of a comparative analysis of the sales department and online store performance is the level of debtors. Along with saving resources and upgrade of business processes, many companies need to control the amount of debtors. Comparison of debtors figures of the enterprise's sales department and online store for the last two reporting periods is given in Table 7.

Table 7. Comparison of debtors' level in the enterprise's sales department and online store for the XXXXs.

Department	Debtors, mln.rub.	
	XXXX year	XXXX year
Sales department	26,500	27,950
Online store	0	0

Analysis of the data presented in Table 7 allows concluding that the performance of the online store does not cause debtors since all trade operations are primarily paid by the customer and then shipped. This feature of e-commerce in general and the performance of an online store, in particular, creates a positive effect for the enterprise, reduces risks and protects the enterprise from insolvent customers.

4 Conclusions

E-commerce in the B2B sector is a promising direction for the development of small enterprises in the distribution of technical goods and high-tech services. An analysis of the operation of one of the enterprises in this sector showed that the online store is much inferior to the sales department by sales volume and the cost of an average order. Cooperation with major customers is a positive aspect of the traditional trade of small enterprise in the B2B sector.

At the same time, the advantages of e-commerce are the engagement of a significant number of new customers, expansion of market geography, shrink of debtors, a higher return on sales against traditional trade.

The disadvantages of e-commerce at the enterprise under analysis are the low level of regular customers that is mainly due to general features of e-commerce such as impersonal relations and trouble partner identification. At this moment we can state that the use of e-commerce tools allows not so much replacing traditional trade as supplementing its methods.

Using the tools of modern e-commerce, the enterprise increases its potential, enhances competitiveness and reduces risks. It is necessary to note the establishment of a set of tools as one of the main directions of its development that makes it possible to unambiguously identify the supplier of goods and services in a wide range of competitors.

References

Endovitsky, D.A., Tabachnikova, M.B., Treshchevsky, Y.I.: Analysis of the economic optimism of the institutional groups and socio-economic systems. ASERS J. Adv. Res. Law Econ. **VII**(6) (28), 1745–1752 (2017)

Novikova, K.V., Staratovich, A.S., Medvedeva, E.A.: E-Marketing and E-Commerce: Study Guide, pp. 7–8. Perm State National Research University, Perm (2013)

Popkova, E.G., Popova, E.V., Sergi Bruno, S.: Clusters and innovational networks toward sustainable growth. In: Sergi Bruno, S. (ed.) Exploring the Future of Russia's Economy and Markets: Towards Sustainable Economic Development, pp. 107–124. Emerald Publishing, Bingley (2018)

The UN General Assembly Resolution No. UN A/51/628 as of December 16, 1996/The UN: Model Law on E-Commerce adopted by the Commission on International Trade Law. http://www.businesspravo.ru/Docum/DocumShow_DocumID_62126.html. Accessed 5 Feb 2018

Rublevskaya, Yu.V., Popov, E.V.: Modeling Business in the Internet Environment. Marketing in Russia and Abroad, no. 2, p. 19 (2001)

Sidorova, O.V.: Types and Organizational Models of E-Commerce. Creative Economy, no. 1 (61), pp. 95–100 (2012)

Strauss, J., Frost, R.: E-Marketing, p. 49. Prentice-Hall, Upper Saddle River (2001)

Treshevsky, Yu.I., Duvanova, A.N., Franovskaya, G.N.: Interrelation of small business and regional innovation development – hypothesis testing. Region: Systems, Economics, Management, no. 3 (30), pp. 32–36 (2015a)

Treshchevsky, Y.I., Serebryakova, N.A., Golikova, G.V., Volkova, S.A., Volkova, T.A.: The system of state support for small and medium entrepreneurship and evaluation of its effectiveness. Revista ESPACIOS, vol. 39 (no. 12), p. 12 (2018)

Treshevsky, Yu.I., Franovskaya, G.N., Duvanova, A.N.: The Position of small business in the context of innovative regional development. In: Innovative Regional Development in the Context of Globalization: Proceedings of the International Research and Practical Conference, Ivanovo-Ples, 2–5 September 2015, 2 am, Part 1, pp. 146–149. Ivanovo State University, Ivanovo (2015b)

Modernization of Production Under the Conditions of Modern Technologies (by the Example of Metallurgical Production of PJSC "GAZ")

Natalia S. Andryashina(✉), Elena V. Romanovskaya, Ekaterina P. Garina, Victor P. Kuznetsov, and Svetlana N. Kuznetsova

Minin Nizhny Novgorod State Pedagogical University, Nizhny Novgorod, Russian Federation
natali_andr@bk.ru, alenarom@list.ru, e.p.garina@mail.ru, kuzneczov-vp@mail.ru, dens@52.ru

Abstract. The article investigates modern approaches to the definition of modernization at the industrial enterprise. The method of modernization, proposed in the study, is based on improving the quality of the final product and raw materials, changes in resource support of production, standards of work equipment, reducing defects in production and increasing the capacity of the enterprise in general. At the same time, increasing the productivity of equipment will reduce the cost of production and consequently increase the competitiveness of an enterprise in the market. Practical application of the method of modernization of fixed assets is realized on the example of metallurgical production of PJSC "GAZ".

Keywords: Modernization · Fixed assets · Methodology · Economic effect · Industrial enterprise

1 Introduction

Currently, in order to maintain the competitiveness of manufactured products and economic efficiency of the industrial enterprise, industrial modernization is required [11]. On the basis of the conducted analysis of foreign and domestic scientific works, modernization of the enterprise is a complex process of renewal of the company, aimed at increasing cost-effectiveness. The economic nature and role of fixed assets has always been a subject of discussion among many scientists and specialists in the field of economics [2].

Theoretical Basis of the Study. In accordance with clause 257 of the Tax Code of the Russian Federation, modernization "includes works caused by changes in technological or service purpose of equipment, buildings, structures, contributing to increasing loads and emergence of new qualities in the object of fixed assets. According to I. V. Sergeev "modernization of equipment means technical improvement of fixed assets in order to

E. G. Popkova and B. S. Sergi (Eds.): ISC 2019, LNNS 87, pp. 532–540, 2020.
https://doi.org/10.1007/978-3-030-29586-8_61

eliminate obsolescence and increase technical and economic indicators to the level of latest equipment" [7]. Dictionary of Economics gives the following definition of "technological modernization": "… development, improvement, updating the object, bringing it into line with new requirements and norms, specifications, quality indicators" [3].

Comparing a number of approaches to the category "modernization", it can be argued that all works that lead to the emergence of new characteristics of objects of fixed assets, positively influencing their further use [4, 5].

There are several classifications of types of equipment modernization. Degree of renewal distinguishes partial and complex modernization, in the process of which there is a complete development of fixed assets and replacement of outdated parts, components, and aggregates. According to objectives modernization process divide into target and standard type. When targeted modernization applied, improvements are made related to the needs of a particular production, and standard takes into account massive changes in structure [9].

The ultimate goal of modernization is to obtain additional profit through savings and identification of cost reserves for production.

Modernization of equipment requires organization of cost planning. In the process of planning the costs of modernization, it is always necessary to take into account not only the obvious costs (the price of the equipment itself, maintenance and preparation for operation of production premises, training of personnel for quality work on the updated equipment) [8, 10], but also hidden (costs of infrastructure modernization, training of additional staff, reducing the cost of low-skilled personnel, hiring new employees with experience in project equipment, obtaining permits and design documentation taking into account new volumes of production and equipment, additional costs for transition to another type of raw materials, costs of experimental samples, downtime and costs of improvement of production modernization) [1].

Methodology of the Study. An important aspect of the effective process of innovation and modernization is taking into account the sequence of certain actions. For this purpose, the paper proposes a method of modernization of fixed assets of the enterprise [6], with following stages (Table 1).

Table 1. Stages of methodology for modernization of fixed assets

	Stage name	Stage contents
Stage 1	Creation of a specific plan	Formulation of criteria on levels of modernization
Stage 2	Formation of the budget for modernization	- analysis of the composition, structure, technical condition and efficiency of use of existing fixed assets - accounting of modernization costs - calculation of planned results from modernization
Stage 3	Creating a core team	- formation of a group of qualified personnel to implement the modernization process; - Further training, if necessary
Stage 4	Exemption of the enterprise from surplus equipment	- sale of fixed assets; - renting fixed assets; - write-off

(*continued*)

Table 1. (*continued*)

	Stage name	Stage contents
5 Stage	Acquisition of quality fixed assets	- Analysis of the market of existing machinery and equipment; - analysis of the final cost of fixed assets; - analysis of suppliers of equipment, parts, spare parts, service companies
Stage 6	- Installation of new equipment; - Replacement of old parts and assemblies with new ones; - Repairs of equipment	- Commissioning works; - conclusion of contracts with service companies;
Stage 7	Timely scheduled maintenance of equipment	
8 Stage	Evaluation of modernization effectiveness	- Analysis of actual costs of modernization; - Comparative analysis of the quality of manufactured products before and after modernization

In accordance with proposed modernization sequence, management of the enterprise will be able to continuously monitor and make the necessary adjustments to the process at each stage, which will provide a qualitative end result to improve the efficiency of funds and competitiveness of products.

Analysis of the Results of the Study. Complexity and variety of processes, a saturation of complex equipment working in a single process chain with heavy operating conditions, impose high requirements for the organization of the process modernization. Practical application of modernization of fixed assets is proposed to be carried out in metallurgical production of PJSC “GAZ”.

The need to modernize fixed assets was shown by the analysis of the coefficients for the implementation of reserves for effective development of production, according to which

- Production capacity is used by 72%, which is 8% below the standard,
- The coefficient of performance of the nomenclature is close to the normative value and is 0.93,
- The rate of defect is higher than the optimal by 2%
- The material intensity of products is 0.72 (with the norm 0.65), which indicates the possibility of reducing the cost of raw materials.

The analysis showed the need to improve the technological process and carry out modernization of fixed assets of the enterprise.

Ways of saving resources:

- reducing the rate of consumption of basic materials in the manufacture of castings
- reducing energy costs;
- reducing the cost of lining furnaces;

- reducing the cost of maintaining existing furnaces in working condition;
- sell scrap metal.

To develop proposals for modernization of fixed assets, an analysis of the existing production technology was carried out, which allowed to identify the following shortcomings:

- there is a risk of liquid metal being released from the furnace;
- need for liquid metal residue in furnaces;
- high requirements for the state of raw materials;
- low melting speed;
- unproductive energy consumption;
- low lining resistance;
- high labor intensity of refractory lining and filling;
- inefficient removal of dust and gas emissions;
- weigh in of components and the number of necessary additives is carried out subjectively, by volume;
- high energy consumption per unit of production;
- high consumption of raw and refractory materials;
- low yield rate during melting and casting;
- high level of defects in the production of castings;
- a large number of operating equipment and operations on it;
- manual loading of materials in the furnace;
- hand stamping and padding of furnaces;
- outdated and inefficient control of melting process and lining condition;
- high operating costs for maintaining the equipment in working condition;
- high cost of castings.

Our program is supposed to modernize the existing induction furnaces LFD-25 (Table 2).

Table 2. Description of equipment before and after implementation of the program

Before the program		After the program	
Name	Qty, pcs	Name	Qty, pcs
Induction crucible furnace LFD-25 of industrial frequency firm "ASEA" for metal smelting (capacity 25 t)	6	Induction crucible furnace of medium frequency IFM-7 company "AVR" for smelting of metal (capacity 13.4 tons)	2
		Induction crucible furnace of medium frequency IFM-7 company "AVR" for smelting of metal (capacity 13.4 tons)	2
Induction ducted furnace LFR-45 by "ASEA" for accumulation, heating, and aging of melt (capacity 65 t)	3		
Künkel Wagner KW filling unit with open tank and induction heating (capacity 0.25 t)	2	Filling device OSS-50 company "AVR" with closed tank and induction heating (capacity. 7 t)	2

Modernization provides maximum use of the existing infrastructure of the smelting site and minimal construction work.

Investment and organizational plans for the project, as well as funding and execution schedules have been drawn up to determine the need for funds.

Table 3 provides a list of activities and implementers of the modernization project.

Table 3. List of project activities and performers

Events	Manufacturer
Design and development works	PU
Purchase of medium frequency crucible smelter with 2 furnaces and 2 filling plants from the company AVR	ABP Induction Systems GmbH, TLC
Transportation	ABP Induction Systems GmbH
Construction and installation works	Third-party organization
Installation of the medium-frequency crucible smelter with 2 furnaces and 2 filling plants of the company AVR	Contractor, metallurgical production
Commissioning works on the medium frequency crucible smelter	ABP Induction Systems GmbH
Auxiliary equipment	The organization that won the tender, TLC

The cost of the purchased equipment (with delivery, packaging, customs duties) is determined on the basis of the application. “AVR Induction” for the tender of LLC “TZK GAZ” and amounts to 160 605 thousand rubles on the terms of DAP Nizhny Novgorod.

All works on the project, purchase of equipment, commissioning, mechanical and electrical installation are estimated to amount to 253 100 thousand rubles. The calculation of investment is presented in Table 4.

Table 4. List of investments for the project, thousand rubles.

Events	Cost
Design and development works	6 437
Cost of medium frequency crucible smelter with 2 IFM-7 furnaces by ABP Induction Systems GmbH	189,514
Cost of 2 filling plants ABP PRESSPOUR® OCC-50 by ABP Induction Systems GmbH	
Mechanical and electrical installation of the medium-frequency crucible smelter with 2 furnaces IFM-7 by ABP Induction Systems GmbH	5 458
Mechanical and electrical installation of 2 filling units ABP PRESSPOUR® OCC-50 by ABP Induction Systems GmbH	2 087
Crucible smelting foundation	15 700

(continued)

Table 4. (*continued*)

Events	Cost
Foundation for filling plants	10 365
Extra charge of refueling complex for main equipment (3.6715%)	6 958
Auxiliary equipment, transportation	15 539
Charge for auxiliary equipment (3.6715%)	571
Registration of customs documents	472
Total	253,100

Thus, for the implementation of the program to modernize fixed assets is required 253,100 thousand rubles.

Table 5 shows the profit report.

Table 5. Report on profit of the program of modernization of fixed assets (thousand rubles), excluding VAT

	2017	2018	2019	2020	2021	2022	2023	Subtotal
Savings on existing production	79 612	124,560	134,709	145,349	155,078	161,483	170,602	971,393
Additional profit from additional volumes	29 995	34 795	27 000	17 785	10 954	2 550	-	123,079
Overall economic effect	109,607	159,355	161,709	163,134	166,032	164,033	170,602	1094471
Depreciation	(14 299)	(21 449)	(21 449)	(21 449)	(21 449)	(21 449)	(21 449)	(142994)
Property tax	(2 936)	(3 932)	(3 460)	(2 989)	(2 517)	(2 045)	(1 573)	(19 452)
Taxable profit	92 372	133,974	136,799	138,696	142,066	140,539	147,580	932,025
Income tax	(18 474)	(26 795)	(27 360)	(27 739)	(28 413)	(28 108)	(29 516)	(186405)
Net profit	73 897	107,179	109,439	110,957	113,653	112,431	118,064	745,620

Table shows that for 8 years of project implementation the total net profit will amount to 745,620 thousand rubles with capital investments of 253 100 thousand rubles.

The final stage of the work will be the launch of new production.

The advantages of the new production technology are:

- improving the safety of workers;
- absence of liquid metal residue in furnaces;
- reduction of the requirement for materials;
- high melting speed;
- absence of non-production consumption of electricity;
- high lining resistance;
- reduction of labor intensity of refractory lining and filling;
- efficient removal of dust and gas emissions;

- accuracy in weighing the components;
- reduction of energy consumption per unit of production;
- reducing the consumption of raw and refractory materials;
- high percentage of yield of suitable casting during melting;
- reduction of defects in the production;
- reduction of equipment in use;
- reduction of technological operations;
- mechanized loading of materials in the furnace;
- mechanized stamping and filling of furnaces;
- automated control over the melting process and lining condition;
- Reduction of operating costs for maintaining equipment in working condition;
- reducing the cost of castings;
- increasing the capacity of the smelting site.

Conclusions. In the process of research, the proposed program of modernization of fixed assets will allow to achieve the following results:

1. Reduce production costs:
   - reducing the rate of consumption of basic materials in the manufacture of castings;
   - reduction of electricity costs;
   - reducing the cost of lining furnaces;
   - reducing the cost of maintaining existing furnaces;
   - to sell scrap metal.
2. Introduction of a new technological process.
   At the same time, the new equipment allowed to increase the volume of products. The maximum possible loading of installed furnaces is 26 000 tons per year, according to the data of the main metallurgist of PJSC “GAZ”.

The overall economic effect of the project is due to: savings from project implementation; additional profit from additional sales (Table 6).

Table 6. Total economic effect of the project, thousand rubles.

	2017*	2018	2019	2020	2021	2022	2023	Subtotal
Total savings on the project	79 612	124,560	134,709	145,349	155,078	161,483	170,602	971,393
Additional profit from additional sales volume	29 995	34 795	27 000	17 785	10 954	2 550	-	123,079
Overall economic effect, with inflation	109,607	159,355	161,709	163,134	166,032	164,033	170,602	1 094 471

* - taking into account the start of the operational phase of the project

For the period from 2017 to 2023, the total economic effect will amount to RUB 1,094,471 thousand. At the same time, the total investment of the project will amount to 253 100 thousand rubles. The project cash flow report is presented in Table 7.

Table 7. Statement of cash flows for the project (thousand rubles)

	2016	2017	2018	2019	2020	2021	2022	2023	Subtotal
Net profit	-	73 897	107,179	109,439	110,957	113,653	112,431	118,064	745,620
Depreciation	-	14 299	21 449	21 449	21 449	21 449	21 449	21 449	142,994
Investments	(80553)	(133938)	-	-	-	-	-	-	(214491)
Cash flow	(80553)	(45741)	128,628	130,888	132,406	135,102	133,880	139,513	674,123
Accumulated cash flow	(80553)	(126294)	2 334	133,222	265,628	400,730	534,610	674,123	
Discount cash flow	(80553)	(38600)	91 601	78 659	67 148	57 819	48 351	42 519	266,943
Accumulated cash flow	(80553)	(119153)	(27553)	51 106	118,254	176,073	224,424	266,943	

Calculation of main indicators of efficiency of modernization and their comparison with the standard value in Table 8.

Table 8. Project performance indicators

	Measure unit	Values	Standard*
Net discounted income	thousand rub.	266,943	>0
Yield index		2.38	≥ 1,3
Internal rate of return	%	69%	>25%
Payback period	Years	2.48	<3,5

* Standards of performance indicators of investment projects are used in accordance with document of "GAZ Group" № 160 from 09.09.17

From the table it follows that the performance indicators meet conditions, so it is possible to draw a conclusion about the expediency of implementation of the methodology of modernization of fixed assets of PJSC "GAZ".

References

1. Andryashina, N.S., Kuznetsov, V.P.: Features of creating a new product at engineering enterprises. In: Krasnodar (ed.) FGBOU HPE, 134 p. Kuban State Technical University (2016)
2. Andryashina, N.S., Kozlova, E.P.: Rationalization of production as a path to sustainable enterprise development. Sci. Rev. **21**, 173–176 (2016)
3. Garina, E.P., Kuznetsov, V.P.: Development of a Complex High-Tech Product in the Industry: Russians, 148 p. (2017)

4. Garina, E.P., Romanovskaya, E.V.: Improving the efficiency of the organization of technological flows in the framework of the integrated application of the production system tools. Financ. Res. 2 (55), 125–133 (2017)
5. Cooper, R., Edgett, S.: Product Development Institute. Product Development (2010). http://www.prod-dev.com/stage-gate.php
6. Garina, E., Kuznetsov, V., Romanovskaya, E., Potashnik, Y., Yashin, S.: Management of industrial enterprise in crisis with the use of incompany reserves. Contributions to Economics, pp. 549–555 (2017). № 9783319606958
7. Kuznetsov, V.P., Garina, E.P., Andriashina, N.S., Kozlova, E.P., Yashin, S.N.: Methodological solutions for the production of a new product. In: Managing Service, Education and Knowledge Management in the Knowledge Economic Era - Proceedings of the 4th Annual International Conference on Management and Technology in Knowledge, Service, Tourism and Hospitality, SERVE 2016, pp. 59–64 (2017)
8. Markova, S.M., Narkoziev, A.K.: Production training as a component of professional training of future workers, vol. 6, no. 1. Vestnik of Minin University (2018)
9. Vumek, D., Jhons, D.: Lean Production: How to Get Rid of Loss and Achieve the Company Flourishing; Translated from English - 4th edn., p. 472. Alpina Business Books (2013)
10. Smirnova, Zh.V., Gruzdeva, M.L., Krasikova, O.G.: Open Electronic Courses in the Educational Activity of the University, no. 4 (21), p. 3. Vestnik of Minin University (2017)
11. Sergi, B.S. (ed.): Exploring the Future of Russia's Economy and Markets: Towards Sustainable Economic Development. Emerald Publishing, Bingley (2018)

Analysis of Healthcare Digitalization Trends and Tools for Their Implementation

Elena V. Mishon[1](✉), Aleksandr A. Sokolov[1], and Nina A. Breslavtseva[2]

[1] Voronezh State University, Voronezh, Russia
evm@yandex.ru, sokolov@vsu.ru
[2] Institute of Service and Business (Branch of Don State Technical University), Shakhty, Rostov Region, Russia
ylo79@mail.ru

Abstract. The purpose of the study is to assess the present experience of introducing information and communication technologies in the medical institutions, reveal its positive aspects, weak points and growth conditions. Methodology. For appropriate assessment, we used our own approach, including a combination of methods to get the highest reliable assessment and reasonably determine the forward-looking trend of the development and the tools for its practical implementation. Recommendations. We examined the main trends in the digitalization of the healthcare system. Also, we assessed health information technologies (HIT) used or recommended for use due to the analysis of their subject-object direction and effectiveness. We find out valuable social communication tools to minimize the cost of resources (financial, organizational, managerial) and implementation time of digitalization in healthcare. Results. The analysis reflected the need for higher attention of researchers and public authorities to the satisfaction of patients' needs in high-quality and available medical services.

Keywords: Digital economy · Information and communication technologies (ICT) · Health information technologies (HIT) · Lean production

JEL Code: C40 · I18

1 Introduction

Based on the most common interpretation of the digital economy, we have chosen healthcare as the object of the study. It is a significant subsystem of the social area, which needs digitalization for economic and socio-political reasons. The latter has a very significant impact on the dynamics and quality of social development of society. Fully sharing the opinion that the digitalization of the economy is impossible without a smart scientific-based public policy, we focused on the theoretical and methodological basis for introducing the medical industry into the digital economy and practical measures for implementation in some constitutional entities of the Russian Federation.

E. G. Popkova and B. S. Sergi (Eds.): ISC 2019, LNNS 87, pp. 541–547, 2020.
https://doi.org/10.1007/978-3-030-29586-8_62

2 Methodology

To achieve the purpose of the study we employed our own approach, which is based on the principle of consistency. The theoretical basis is the writings of domestic and foreign scientists in the economic theory, management of social relations, economy and production arrangement, economy and healthcare arrangement (Bogoviz and Sergi Bruno 2018; Popkova and Sergi Bruno 2018; Treshchevsky et al. 2018; Risin et al. 2017; Lanskaya et al. 2018; Radyukova et al. 2018; Wumek 2014). In the study we preferred the methods that are used for analysis of soft systems (analogy method, content analysis (to assess the results of implementation of the Lean Polyclinic federal project in various constitutional entities of the Russian Federation); comparative analysis (to determine the general and specific features of HIT used, as well as their correlation with the demand by medical institutions).

Specific features of the approach are:

1. The subject-matter of healthcare introduction trends into the digital economy, the Lean Polyclinic federal project being implemented in the constitutional entities of the Russian Federation (Ministry of Healthcare of the Russian Federation 2017), health information technologies from the point of their influence on the digital economy development trends.
2. The results of the analysis are considered with an account of two significant points:
   - healthcare is a soft socio-economic system and application of quantitative indicators to assessment hereof brings significant errors, which can affect the reliable assessment of the final results of the study conducted;
   - The external environment having a close correlation with the perception of the industry by the consumer of services dramatically influences on the performance of the healthcare system. This is important from the standpoint of economic analysis, because, on the one hand, the correlation of public health capital value with the real indicators of its economic activity is less close than the correlation with the consumer's assessment of the quality and availability of medical services, on the other hand, the public capital of this industry really affects its economic indicators.
3. The reliability of the assessment is based on a sampling of constitutional entities of the Russian Federation, namely: Buryatia (Ulan-Ude), Ossetia (Vladikavkaz), Krasnoyarsk Krai (Zheleznogorsk), Khanty-Mansi Autonomous District (Yugra), Astrakhan, Arkhangelsk, Voronezh, Lipetsk, Tula, Yaroslavl oblasts, Anzhero-Sudzhensk, Asbest, Yessentuki, Saransk, Sevastopol, Omsk.
4. Justification of the reliability of the estimated results is based on the use of systemic and structural-functional approaches, representative factual data and soft system research methods.

3 Results

The benchmarks of the analysis are the following provisions:

Experts refer healthcare to areas of social and economic activities where ICT become the most widely used along with agriculture, smart cities, transport networks, intelligent energy networks, financial services, supply logistics, and public administration.

For many impersonal factors, the digital involvement of the Russian population is a necessary and indispensable condition for leveling social and economic conditions that affect the quality of life, including the provision of equal access to modern infrastructure.

The government actively participates in the diffusion of digital technologies for mass use and serves as a model for their introduction (an example is the provision of high-quality digital public services).

The population of Russia has a leading position in international ratings on digital culture and literacy.

Digitalization of the economy in general and its sub-sectors is implemented at three levels. Concerning our analysis, the third level is critical and simultaneously under-studied: an environment that provides conditions for the development of both modern technologies and effective cooperation between market actors and economic sectors and embraces legal regulation, information infrastructure, staff, and information security.

For now, two trends in the healthcare digitalization have the priority:

- meeting the needs of the population for high-quality medical services, including available medical care in terms of place and time of provision, cost, and convenience;
- improving the performance of medical institutions.

For practical implementation of the trends specified since 2017 such a tool of public administration as the "Lean Polyclinic" project has been used and has been widely spread. It is aimed at tackling major health issues.

Currently, the territorial versions of the Lean Polyclinic project are being implemented in 32 constitutional entities of the Russian Federation.

Health information technologies (HIT) are developed and distributed within the organizational and economical administration of medical institutions. ICU (Intensive Care Unit) is the most demanded HIT. For example, Medical Center, Med-Angel, Medesk (with CRM module), Medmis, DMT LAB, Healthcare and Laboratory Information Systems "Ariadna", Computerized Patient Record.

The implementation of HIT is noticeable in the majority of constituent entities of the Russian Federation. Private medical institutions (medical centers, specialized clinics, dental clinics) declare the highest interest in these technologies.

As of the end of 2018 Moscow is the leader in HIT application: more than 10% of residents and 30% of medical practitioners use them in one or another form.

Based on the content analysis of the territorial versions of the Lean Polyclinic federal project, we concluded insufficient theoretical and methodological foundation for the development of the basic project and its territorial versions. Both the methodological approach and the procedures for setting and determining the task arouse criticism. The procedure of stratification in setting the task has not been clearly defined from the point of the research methodology. Thus, in some situations, the issues of the industry include lack of medical staff; underfunding; low motivation to improve the quality of services due to low wages. In other circumstances, the lack of an effective mechanism for the public health system administration, poor planning, lack of proper regulation and flawed supervision is considered the most severe. The situation is similar to the statement of the main tasks that the project is intended to solve. As a result, for example, the Tula version of the Lean Polyclinic project sets three tasks: remove queues; optimize document circulation; create a client-oriented environment.

It is evident that the client-oriented environment expects the absence of queues and an increased time of communication between the doctor and the patient just through optimizing of the document circulation.

The conceptual framework of the primary document is not sufficiently developed (causes and prerequisites, results and consequences, required expenses and disadvantages, etc., are confused).

A significant methodological disadvantage of the project is the priority of quantitative indicators (the system primarily focuses on the prompt provision of services to patients and reduced queue time) over qualitative ones. As a rule, the quality of medical services provided is not assessed, i.e., the main cause for patient dissatisfaction is not eliminated, while the project was developed and implemented precisely for this purpose.

By the way, reduced queue time (no more than 15 min planned; for example, queue time in Tula is 14 min) is undoubtedly a positive phenomenon in the performance of a medical institution. However, the decreased time of communication with the patient is a controversial issue (Semina et al. 2017).

The time structure of medical services provision includes: unreasonable loss of time (85%); supplementary activities, i.e., activities that does not add value to the service, but they are necessary for the provision hereof (10%); effective communication with the patient (5%) (Khromushin et al. 2016). When analyzing the distribution of the time for provision of vein blood sampling, we revealed loss of time (60% of the total time) that include waiting at the appointment desk, waiting in front of doctor's office, loss of time due to impossibility to enter the office at the appointed time because of concurrent queues (first come first served) or overlapping flows of patients, etc.; up to 35% of the time is wasted by a medical staff, which is necessary, but unuseful: disinfection, treatment of hands and surfaces, disposal of tools, surveying and filling in medical documentation, duplicate information collected earlier, etc.; and only 5% of the time is taken by the actual useful manipulation – vein blood sampling.

Practical implementation of the project in several regions (Arkhangelsk and Tula Oblast) led to a decreased number of actual queues for physicians in polyclinics, but virtual queues and waiting lists for attendance appeared.

The reverse result of the project implementation was the more difficult technical duty of the doctor instead of the expected release from the routine non-productive operations, which takes the already limited time of communication with the patient.

The disadvantages listed above deteriorates the performance of digitalization in the healthcare system.

The methodological advantage of the project is the identification of healthcare problems that are not solved by its implementation.

Based on the results a comparative analysis, we can consider a version of the project in thc Arkhangelsk Oblast, which clearly states the goals, actors, work flow charts, and presents tactical plans for project implementation of all participants, the most comprehensive and reasonable.

As a rule, talking about HIT, one should pay the primary attention to computerized medical record (CMR), which contains basic information about the patient (checkup results, diagnosis, prescriptions, list of prescriptions, referrals for examinations, health certificates, work incapacity certificates), etc.). The advantages of CMR for a doctor are the most detailed clinical description of the patient's health, which makes it possible to improve the quality of treatment by rational decision-making on therapy measures, decreased time of information processing, simplified completion of administrative documents. From an economic point of view, this is the growth of productive cooperation.

The advantages of CMR for patients are the continuity of information in various medical institutions, reasonable prescriptions, and the security of personal data by access rights differentiation. From an economic point of view, it reduces treatment costs through an increase in its efficiency and quality. The advantage of HIT to the territorial peculiarities of Russia is the opportunity of communication via e-mail, prompt delivery of medical care in distant regions and settlements.

We revealed the organizational and financial difficulties of CMR introduction for following reasons: probable loss of a CMR in the event of a system failure accompanied by economic losses due to the completion of a duplicate; additional costs for equipment of the doctor's workplace; ignorance of IT-technologies (especially among doctors of middle and pre-retirement age) and the lack of estimated funds for training.

In general, CMR is a positive result of the healthcare digitalization for all participating actors: patients, medical institutions and the government, since its application correlates with the growth of public health rates.

DMT LAB technology is focused on the most detailed information on clients. The DMT starts system provides information on the number of clients applied to a health facility over a given period and their needs, i.e., acts as marketing department, which considerably saves the cost of client search and retention. Private medical institutions widely use this technology. The price of introduction ranges from 60 to 300 thousand rubles.

The Medmis system includes three versions of the appointment desk schedule. It is designed for clinics with a large staff of specialists, for collective centers for quick search and appointment at a particular time, for polyclinics. We consider item "reputation management," i.e., the qualitative parameter of the institution's performance, advantageous.

Medesk Health Information Technology (with the CRM module) is the most demanded and popular in private medical institutions. The technology includes the automation of the desk appointment and doctor's workplace, provides detailed information on the medical services rendered, provides evidence of the employees' effectiveness. For example, experts gaining the maximum profit and outsiders and outsiders with minimal contribution is minimal are daily recorded. Such data allows developing a reasonable strategy and ensure high profitability of the institution. Besides, the system encourages staff to increase personal efficiency, notifying about the leaders and outsiders of the institution.

We should note quite high economic efficiency of the listed HIT. All of them, to a greater or lesser extent, pursue a single goal – to increase the average bill for the institution. For a medical institution, this is an absolute advantage. For a service consumer, it's increased expenses, i.e., a welfare deterioration according to Pareto efficiency.

4 Conclusion

The analysis allowed revealing the following negative trends. The first is uneven development of trends concerning entry of medical sector into the digital economy. Optimization of a medical institution performance takes the leading position; it is carried out faster and more efficiently). The second is the more widespread and more successful introduction of HIT against ICT. Uneven implementation of trends can serve as an impetus for the development of boomerang processes caused by the dominance of the interests of the industry providing medical services over the consumers' interests. The government needs to take emergency measures to determine tools accounting for interests of future and actual patients. For this purpose, we propose to focus on modern communication technologies aimed at increasing the social capital of the party in the medical activity.

References

Bogoviz, A.V., Sergi Bruno, S.: Will the circular economy be the future of Russia's growth model? In: Sergi, B.S. (ed.) Exploring the Future of Russia's Economy and Markets: Towards Sustainable Economic Development, pp. 125–141. Emerald Publishing, Bingley (2018)

Lanskaya, D.V., Treschevsky, Yu.I., Getmantsev, K.V., Myasnikova, T.A.: Strategizing as a tool of information and analytical provision and a technology of regional development management. In: Popkova, E.G. (ed.) In the Future of the Global Financial System: Downfall or Harmony. Lecture Notes in Networks and Systems, vol. 57, pp. 348–356. Springer, Cham (2018)

Popkova, E.G., Sergi Bruno, S.: Will Industry 4.0 and other innovations impact Russia's development? In: Sergi, B.S. (ed.) Exploring the Future of Russia's Economy and Markets: Towards Sustainable Economic Development, pp. 51–68. Emerald Publishing, Bingley (2018)

Ministry of Health of the Russian Federation 2017. Lean Polyclinic Project. https://www.rosminzdrav.ru/poleznye-resursy/proekt-berezhlivaya-poliklinika. Accessed 7 Feb 2018

Radyukova, Y., Sutyagin, V., Treschevsky, Y.I., Artemenko, D.: The problems of financial security of modern Russia. In: Soliman, K.S. (ed.) Proceedings of International Business Information Management Conference (32nd IBIMA) – Vision 2020: Sustainable Economic Development and Application of Innovation Management from Regional expansion to Global Growth, Seville, Spain, 15–16 November 2018, pp. 2877–2888 (2018)

Risin, I.E., Treshchevsky, Y.I., Tabachnikova, M.B., Franovskaya, G.N.: Public authorities and business on the possibilities of region's development. In: Popkova, E.G. (ed.) Overcoming Uncertainty of Institutional Environment as a Tool of Global Crisis Management. Contributions to Economics, pp. 55–62. Springer, Cham (2017)

Semina, T.A., Tomaeva, M.A., Torshkhoeva, T.T., Kitanina, K.Yu.: Implementation of the Project "Lean Polyclinic in the Tula Region" in Public Health Institutions of the Tula Region Providing Healthcare for Adult Outpatients: Instructional Guidelines, p. 35. Tula State University Publishing, Tula (2017)

Treshchevsky, Y.I., Tabachnikova, M.B., Franovskaya, G.N., Voronin, V.P.: Economic and statistical analysis in evaluating the perspectives of structural changes of regions' economy. Adv. Intell. Syst. Comput. **622**, 521–529 (2018)

Khromushin, V.A., Khadartsev, A.A., Lastovetsky, A.G., Kitanina, K.Yu.: Estimation of age-related cohorts of the population's mortality on the example of the Tula region from 2007 to 2015. Bulletin of New Medical Technologies, Electronic edition, no.3. Publication 1-1 (2016). http://www.medtsu.tula.ru/VNMT/Bulletin/E2016-3/1-1.pdf. Accessed 15 Jan 2018

Wumek, J.: Sale of Goods and Services by the Method of Lean Production, 264 p. Alpina Publisher (2014). https://econ.wikireading.ru/5486. Accessed 10 Feb 2019

Analysis and Forecasting of Prospects for Digitalization of a Regional Healthcare System

Elena V. Mishon(✉), Tatyana N. Gogoleva, Pavel A. Kanapukhin, and Aleksandr A. Sokolov

Voronezh State University, Voronezh, Russia
evm@yandex.ru, tgogoleva2003@mail.ru,
kanapukxin@econ.vsu.ru, sokolov@vsu.ru

Abstract. The purpose of the study is to determine promising areas of organizational and management activities and tools to minimize the resources and introduction terms of digital technologies in the regional healthcare system. Methodology. For the purpose of the study, we employed our own approach based on several principles and methods that make it possible to reason the expediency and efficiency of choice. Recommendations. We highlighted and substantially explained the advantages, disadvantages, and peculiarities of regional healthcare. Also, we revealed and described the prerequisites for the accelerated introduction of digitalization into the healthcare system. Results. We proposed the main directions for introducing digital medicine into the regional healthcare system. Also, we recommended management tools of social relations (supported by methodological developments for their practical use) that allow effecting arrangements for the introduction of the digital economy into the healthcare of Russian regions within a relatively short term and under minimal costs.

Keywords: Digital economy · Healthcare system · Information and Communication Technologies (ICT) · Health Information Technologies (HIT)

JEL Code: C40 · I18

1 Introduction

In this paper, we understand the digital economy as digital-tech based economic activities of actors at different levels. Now, it is evident that the digital economy needs to develop scientific-based, efficient ICT adjusted to various sectors and areas of socio-economic activity and focused on their specific features (Popkova and Sergi 2018), (Bitarova et al. 2018), (Mkrttchian et al. 2018); (Plotnikov et al. 2019).

Healthcare is one of the essential social systems with a critical need for the accelerated introduction of the digital economy. The significance of this social area for the socio-economic development of a territory is determined by its contribution to public health, which has both a direct (growth of human capital quality) and an indirect effect (higher working time fund, increased labor productivity, reduced temporal disability benefits, etc.) on the dynamic trend of the region of its performance.

E. G. Popkova and B. S. Sergi (Eds.): ISC 2019, LNNS 87, pp. 548–554, 2020.
https://doi.org/10.1007/978-3-030-29586-8_63

Public health shows the current level of physical resources of society and the effectiveness of its social institutions, primarily the healthcare system. Thus, the regional healthcare system simultaneously acts as a tool for the development of the public well-being (health) and as a factor of the socio-economic development of the territory. From this point of view, the need of healthcare for the search and implementation of ICT seems reasonable, and the determination of useful tools is practically demanded.

2 Methodology

To achieve the purpose of the study we employed our own approach based on the principles of priority of interdisciplinary methods and a reasonable sequence of their application. The theoretical basis of the study is proceedings of scholars in the regional economy, management of social relations, economy and production arrangement (Mishon 1999), (Endovitsky et al. 2017), (Treshchevsky et al. 2018).

Distinctive features of the method:

1. The subject-matter of the study is the directions of organizational and management activities and ICT tools that provide for the accelerated introduction of digital methods into regional healthcare systems.
2. Justification of the reliability of the estimated results is based on the use of representative factual data, structural content analysis, comparative analysis, records (within formalization, trend extrapolation, analogy).

3 Results

Studying international practice of digital economy introduction, we can conclude that its successful performance requires favorable conditions for innovation, substantial investment in digital technologies and infrastructure, active government support (direct or indirect – by developing the infrastructure of the digital economy). Developed countries gave the abovementioned conditions to a greater or lesser extent. Developing countries do not have similar opportunities but get a particular advantage: they do not spend efforts on alteration of already established infrastructure but introduce modern versions.

Although the actions plans of countries have much different content, we highlight several directions for the development of the digital economy applicable to most countries: the development of modern communication infrastructure, data storage and processing centers; promotion of free information flow; expansion of ICT services range; introduction of new intelligent networks, platforms and technologies with simultaneous provision of their interoperability; development of e-commerce; ease of restrictions in business; stimulation of business initiative and ICT funding; granting benefits to small and medium business; increasing the level of information security and user confidence in Internet services; training specialists and enhancing the general level of computer literacy.

With that, priority is given to the development of digital technologies such as the Internet of Things, industrial Internet, artificial intelligence, cloud computing, quantum and new manufacturing technologies, robotics devices, cyber-physical systems, big data processing technologies, wireless communication technologies, additive technologies, 3D, virtual and augmented reality, blockchain. Besides, the IPv6 protocol is being actively introduced instead of IPv4.

We should note the areas of the broadest use of ICT. These are smart cities, agriculture, supply logistics, and public administration; intelligent energy networks (Smart Grid) and transport systems; financial services. Concerning our study, digital healthcare should also be noted in this list.

Now, Russia is successfully developing most areas of the digital economy. Some quantitative indicators evidence this. In 2016, the number of registered users of the public services portal had increased twofold over the year up to 40 million people. In 2017, the number increased by 25 million and amounted to 65 million people. Over the year, the portal has been visited 435 million times (1.2 million users every day) (Gazeta 2018).

By the number of Internet users, Russia is the first in Europe and the sixth in the world.

Now Russia does not take leading positions in terms of some performances (level of digitalization, the proportion in GDP, average lag in the development of technologies applied in the leading countries) but experts note positive trends of recent years and accelerated rates of digitalization.

The experts of the McKinsey company in 2017 prepared a special report on the importance and prospects for the development of digitalization in the country (McKinsey and CIS 2017). In general, it should be noted that the conclusions and forecasts made in the course of the study are very promising for the country. The report marks that the objective of the development of the digital economy in Russia is to improve life quality, ensure the competitiveness of the country and national security. According to analysts' opinion, the goal of Russia in digitalization processes for the next 15–20 years is to join the group of leading economies due to digital transformations of traditional sectors and the development of an independent and competitive digital industry. From the standpoint of the McKinsey company, it seems quite feasible. Russia is able to reach the target by 2025, which is characterized by six fundamental constituents:

1. Russia is a world-class scientific and educational center in the field of digital technologies and innovative business models. The country has a developed network of educational and research centers that intensively cooperate with IT investment companies.
2. Russian digital companies successfully produce and launch competitive, innovative digital solutions and technologies at the international market.
3. The industrial sector of the Russian economy is distinguished by the application of high technologies and the presence of companies-world leaders in the introduction of particular digital technologies, for example, Industry 4.0. elements.

4. The government is a distributor of digital technologies for mass use and serves as a model for their introduction in the provision of high-quality digital services.
5. Digital engagement provides equal access of the population to infrastructure and services throughout the country.
6. Russia population is a leader in international ratings on digital culture and literacy.

Due to the peculiarities of digital technologies in comparison with traditional ones digital economy is developing at three levels, which close interaction affects the life quality of both individual citizens and society in general:

- economic markets and sectors;
- platforms and technologies to develop employees' expertise;
- an environment that establishes the conditions for the development of platforms and technologies and productive cooperation between market actors and economic sectors (statutory regulation, information infrastructure, information security (Sukhova 2018).

For the accelerated introduction of regional healthcare into the digital economy and determination of the necessary management tools, it is expedient to reveal and analyze its advantages and disadvantages, to discern the specific features of performance and prerequisites for further development.

Let us note an essential feature of the Russian healthcare system. It is an autopoietic system that uses mainly qualitative indicators. This considerably complicates the entry into the digital economy manipulating quantitative indicators. As the practical experience of the authors shows that quantitative indicators of this social system performance are difficult to determine and develop due to high complexity, long term of implementation and a quite low efficiency in most regions of the country. Let us note that the issue of a quantitative assessment of the quality of medical services has not yet resolved in international practice.

From the standpoint of our study, there are important advantages of the regional healthcare system (Voronezh oblast):

- university graduating skilled staff;
- the high-skilled staff of medical institutions;
- the positive experience of using modern medical technologies in the treatment of a wide range of diseases;
- a system of health resorts characterized by high demand and practically proved the effectiveness of disease prevention.

These advantages can ensure the growth of regional competitiveness within the current healthcare system.

Contrary to common belief, the main disadvantage of the regional healthcare system from the point of introduction into the digital economy is not so much the information base as the practical lack of social communication on healthcare issues. Communication tools are either not applied at all or unprofessionally, which cause opposite effects.

The positive aspects of healthcare integration in the digital economy include its commitment to the patient and the optimization of the medical activities of each employee of medical institutions. This is approved by the development and implementation of Lean Polyclinic pilot project since 2016, introduction of medical information technologies, development of a single database across the country (electronic filing system) by 2019, which will include profit and public medical institutions. By 2020, it is expected to use personal electronic medical records in 60% of medical institutions of the country. Since 2017, on the public services portal the patient can register a personal account with the electronic card. The patient can be made aware of the data at any time. The government allocated 160 billion rubles for these purposes.

For complete fulfillment of the existing advantages, elimination of disadvantages and account for specific features of the healthcare system for introduction into the digital economy, we propose a set of communication tools building or transforming the company's image (branding is applicable and efficient for some regional medical institutions). In this situation, it is possible to forecast the enhanced position of healthcare institutions in the regional economy.

The advantages of the development and application of this communication tool are:

- high demand in social cooperation of the "society-health," which is satisfied with its help;
- an entirely low physical, financial and organizational costs;
- high efficiency of the tool (in the event of its appropriate development, terms, and place of use).

For creating the image of a medical institution, we suggest using the algorithm developed by us and adjusted to the regional situation:

- determination of the basic principles of the institution's communication about which the public can be informed in different ways (depending on the specialty of the medical institution);
- narration about the history of the institution with advantages acquired over the term of activity (treatment methods used, infrastructure, etc.);
- the structural and qualitative composition of the staff (length of service, age, achievements, practical results, certificates, etc.);
- opposites of a medical institution.

The authors also developed guidelines for the development and implementation of the algorithm.

This algorithm is advantageous because it does not require specific knowledge and considerable experience of specific professional activity for its development (in the event of available methodical recommendations). The exception is the last point (opposites). Its shrewd development requires consultation with a professional. On the one hand, this point acts to weaken or even eliminate competitors; on the other hand, it can cause a boomerang effect which is surely unadvisable for the initiator.

In this paper, we do not touch the issues of developing a brand of a medical institution due to the limited scope of the article. We only note that this is also a useful ICT tool.

Another important effective and demanded communication tool is interactive communication with the public on disease prevention, health importance for the complete and comprehensive fulfillment of the individual and the role of the carrier in health level and quality. The practice of outpatient institutions in the Voronezh region showed that this tool is little-known and therefore rejected by the medical practitioners.

We recommend holding communication training to eliminate this disadvantage. The methodology for the development of training tasks is based on the essential principles of economic theory, psychology, and management of social relations.

4 Conclusion

As a result of the study, we concluded that it is necessary to focus attention on the organizational and management aspects of ICT introduction into the regional healthcare system and the broadest practical application of social relations management methods, which is particularly topical for the priority directions noted by us and implementation of recent healthcare development projects aimed at intensification and expansion of social communication with medical institutions on public health issues. Tools and communication technologies recommended by the authors allow pushing the boundaries and increasing the area of public interaction with the regional healthcare system for its optimization based on mutually beneficial cooperation, also concerning the issues of digitization of processes related to regional medicine expertise. In turn, this situation will lead to higher satisfaction of the population's needs for medical services, on the one hand, through establishment of incentive system for medical staff to increase its efficiency (as a result of using medical information systems Med-Angel, ICU. Medical Center, Medesk with CRM module), on the other hand, owing to the population's awareness and acceptance of the digital economy constituents in healthcare (Computerized Medical Record, Lean polyclinic and its structural components), as evidenced by the trend extrapolation.

References

Endovitsky, D.A., Tabachnikova, M.B., Treshchevsky, Y.I.: Analysis of the economic optimism of the institutional groups and socio-economic systems. J. Adv. Res. Law Econ. **8**(6), 1745–1752 (2017)

Bitarova, M.A., Getmantsev, K.V., Ilyasova, E.V., Krylova, E.M., Treshchevsky, Y.I.: Factors of socio-economic development of rural regions in the area of influence of city agglomerations. In: Popkova, E.G. (ed.) The Future of the Global Financial System: Downfall or Harmony. Lecture Notes in Networks and Systems, vol. 57, pp. 183–194. Springer, Cham (2018)

Mishon, E.V.: Environmental Management, p. 216. Voronezh State University Press, Voronezh (1999)

Mkrttchian, V., Vertakova, Y., Treshevski, Y., Firsova, N., Plotnikov, V., Treshchevsky, D.: Smart City – the concept of resolving the contradiction between production and urban life. In: Benna, U.G. (ed.) Industrial and Urban Growth Policies at the Sub-National, National, and Global Levels, Chap. 15, pp. 300–301. IGI Global, Hershey (2018). https://www.igi-global.com/book/industrial-urban-growth-policies-sub/208848#table-of-contents. Accessed 27 Jan 2019

Popkova, E.G., Sergi Bruno, S.: Will Industry 4.0 and other innovations impact Russia's development. In: Sergi Bruno, S. (ed.) Exploring the Future of Russia's Economy and Markets: Towards Sustainable Economic Development, pp. 51–68. Emerald Publishing, Bingley (2018)

Gazeta, R.: p. 1, 8 February 2018. https://rg.ru/2018/02/08/kolichestvo-polzovatelej-portala-gosuslug-vyroslo-do-65-mln-chelovek.html. Accessed 27 Jan 2019

Sukhova, V.A.: Digital economy: opportunities for countries and methods of regulation. Molodoi uchyonyi, no. 21, pp. 303–306 (2018). https://moluch.ru/archive/207/50579/

Treshchevsky, Y.I., Tabachnikova, M.B., Franovskaya, G.N., Voronin, V.P.: Economic and statistical analysis in evaluating the perspectives of structural changes of regions' economy. Advances in Intelligent Systems and Computing, vol. 622, pp. 521–529 (2018)

Plotnikov, V., Vertakova, Y., Treshchevsky, Y., Firsova, N.: Problems of improving the management of socio-economic subsystems in smart cities. In: International Geotechnical Symposium "Geotechnical Construction of Civil Engineering & Transport Structures of the Asian-Pacific Region" (GCCETS 2018). MATEC Web of Conferences, vol. 265, p. 01001, Article No. 07010 (2019)

McKinsey and CIS Company. Digital Russia: A New Reality, p. 7, July 2017

Controlling as an Instrument of Industrial Enterprise Management in the Conditions of Modern Economic Activity

Ekaterina P. Garina(✉), Victor P. Kuznetsov, Alexander P. Garin, Natalia S. Andryashina, and Elena V. Romanovskaya

Minin Nizhny Novgorod State Pedagogical University, Nizhny Novgorod, Russian Federation

e.p.garina@mail.ru, kuzneczov-vp@mail.ru, rp_nn@mail.ru, natali_andr@bk.ru, alenarom@list.ru

Abstract. *Relevance* of the research topic is due to the current trends in economic activity, including the process of monitoring the producer-counterparty relationship process as well as risks arising from the manufacturer in the course of operating activities due to insider activity. In this case development of standards and procedures, as well as basic corporate documents regulating business processes; making proposals for improving controlling as a tool for managing the development of companies. *The object of the study* is the business units of GAZ Group, the leading manufacturer of commercial vehicles in Russia, one of the ten largest European manufacturers of commercial vehicles. Key indicators include: creation of new production facilities in the last five years and 1 billion dollars of investment in the development of new NEXT cars; 85% level of automation in key production areas; creation of 220 centers of sales of cars GAZ, and a wide dealer network among manufacturers of commercial vehicles in the Russian Federation. *The main goal of this study* is to develop practical recommendations for improving controlling system as an instrument for effective management of business units of an industrial enterprise in the machine-building industry.

Keywords: Management · Operational activity · Development of business units · Controlling · Process activity · Software and technical means of control

The basic concepts and definitions in the field of controlling process and operating activities of business units of the enterprise were considered as theoretical bases in the work. It is determined that:

- Controlling can be characterized as a complex system of support of the management of process activities aimed at coordination of interaction of management systems and control efficiency of their economic activity (Karpov 2018; Romanov 2015);
- The main task of controlling is tracking insider activity (Kuznetsov and Romanovskaya 2011). In the Russian Federation, this sphere of activity is regulated by the Federal Law dated 27.07.2010 No. 224 - "On counteraction to illegal use of insider information and market manipulation…" and clause. 3.1.8 "Insider trade" in the Code

E. G. Popkova and B. S. Sergi (Eds.): ISC 2019, LNNS 87, pp. 555–562, 2020.
https://doi.org/10.1007/978-3-030-29586-8_64

of Professional Ethics. According to which the insider is an employee of the company who, due to his official position, has access to confidential information and, in violation of established requirements, discloses information through negligence or uses for personal profit. Insider factors are called: (a) desire to receive financial remuneration (self-interest)–, more than 65% of respondents call this factor in 2017; (b) desire to take revenge on the employer (20%), caused by such factors as disciplinary recovery, low wages, dismissal on the initiative of the employer; (c) negligence, incompetence of the employee –for example: clicked the wrong link on a website; (d) motivation with fear, i.e. an attempt to intimidate the insider in order to obtain desired information. For example, by threatening life and health of his family and friends.

– Controlling process includes the following steps: 1. Establish standards that must be achieved to obtain the goals of the manufacturer in the end. Standards contain criteria of efficiency of processes, final products (Garina et al. 2016); 2. Measurement of actual performance—measured relative to the target indicator. Comparison of the actual efficiency with the standard value; 3. Corrective actions—initiated by management in the course of correcting shortcomings in the actual work of an enterprise (Aleksandrovsky and Shushkin 2015).

Assessment of the system of industrial enterprises of the mechanical engineering industry Controlling of a business unit – PJSC "GAZ": efficiency of work of profile services in the direction of identification of facts of illegal actions of employees. According to the study (Andrashina 2014; Gruzdeva et al. 2018) in 2017, 280 scheduled and 59 unscheduled inspections were carried out in a separate division, including: 34 documentary, 283 control, 23 special. The economic effect for 11 months of 2017 from measures on economic security amounted to 370,690,1 thousand rubles. In the direction of identification and suppression of thefts of inventories in 2017, the employees of profile services achieved the following results: 10 criminal cases were initiated on the materials of inspections in 2016, including 7 under clause 158 of the Criminal Code; 1 under 159 of the Criminal Code; 2 under 160 of the Criminal Code of the Russian Federation. Analysis of violations of labor and industrial discipline was conducted also (Table 1) (Mizikovsky 2011; Kuznetsov and Romanovskaya 2011).

In 2017, to minimize the risks of damage to the company carried out internal investigations and inspections, violators are punished in accordance with the Code of "GAZ Group" on disciplinary responsibility and measures of material influence". The facts of illegal actions of employees on which the management of business units took the corresponding measures of reaction are indicative: in the period from 01.01.2017 to 29.12.2017 the controlling services showed the results of counteraction (Tables 2 and 3).

Table 1. Reporting indicators for controlling operating activities of PJSC "GAZ" business unit for 2017

			Standards	January	February	March	April	May	June	July	August	September	October	November	December	2017
1	Conducted inspections: from which	planned	Quantity	8	25	34	27	25	29	37	22	22	23	28	35	315
		unplanned.	Quantity	4	6	7	5	3	6	9	3	3	7	6	6	65
1.1	documentary (audit)		Quantity	2	2	4	3	4	4	6	1	0	3	5	2	36
			amount of damage thousand rub	317.0	0.0	283.2	41.5	669.3	865.2	58.1	111.4	0.0	97.0	71.3	499.3	3013.4
1.2	control		Quantity	8	26	34	28	23	28	38	23	24	25	26	37	320
			amount of damage thousand rub	15406.2	7525.2	15041.2	6757.9	2424.5	2216.4	3677.7	6387.6	4209.8	5004.9	6185.8	5010.5	79847.6
1.3	special		Quantity	2	3	3	1	1	3	2	1	2	2	3	2	25
			amount of damage thousand rub	7.1	1.9	2.0	0.0	0.0	0.0	33.4	0.0	7.0	0.0	2.9	15.2	69.5
2	Receivables returned with the assistance of DRC plant		the amount of refund thousand rub	35 094,5	53 687,2	10 639,6	1 329,8	8 077,7	1 449,4	61 001,6	228.5	50 094,4	7 267,8	55 643,6	10 251,9	294765.9
3	The number of expert evaluations of contracts (transactions) considered by the security of the enterprise			15	35	42	36	32	33	34	19	88	131	143	139	747
4	Number of verified identified firms (persons), transactions which the security has found to be commercially inexpedient			0	33	43	56	44	64	42	16	45	70	101	59	573
				0	1	2	2	1	5	1	1	0	3	5	5	26
5	Economic effect of security measures		the amount of refund thousand rub	50824.8	61214.3	25966.0	8129.2	11171.5	4531.0	64770.81	6727.56	54311.19	12369.7	61903.5	15776.9	377696.4

(*continued*)

Table 1. *(continued)*

		Standards	January	February	March	April	May	June	July	August	September	October	November	December	2017
6	Established violations of access control: from which	Quantity	64	165	196	275	175	257	278	251	282	297	259	199	2698
		the amount of fine thousand rub	10.4	18.3	29.7	22.2	10.1	29.5	17.8	17.0	27.4	17.0	36.0	21.6	256.9
6.1	arrested with stolen goods and materials	Number of people	19	68	61	78	48	71	58	79	49	91	66	38	726
		the amount of refund thousand rub	46.9	82.4	398.2	239.3	8.6	1259.6	963.6	881.2	377.5	1130.1	1929.2	323.4	7640.0
7	violations of the internal object regime: from which	Quantity	34	98	99	103	71	55	67	64	59	86	85	72	893
		the amount of refund thousand rub	30.2	162.7	180.5	185.1	72.5	109.8	58.0	70.7	84.5	102.4	162.3	126.0	1344.7
7.1	unauthorized storage of goods and materials found	Quantity	3	16	18	15	9	19	34	17	9	11	14	21	186
		the amount of refund thousand rub	0.5	6.2	2.3	2.0	0.5	2.2	8.5	1.1	1.8	2.3	0.3	0.0	27.7
7.2	detained in a state of intoxication or with alcoholic beverages	Number of people	31	82	81	88	62	36	33	47	50	75	71	51	707
		the amount of fine thousand rub	29.7	156.6	178.2	183.2	72.0	107.6	49.5	69.6	82.7	100.1	162.0	126.0	1317.1
8	Other violations or theft of goods and materials	Quantity	1	2	3	2	2	2	2	1	1	2	0	3	21
		the amount of refund thousand rub	0.0	0.0	0.0	0.0	0.0	0.0	0.0	0.0	0.0	0.0	0.0	0.0	0.0

(continued)

Table 1. *(continued)*

		Standards	January	February	March	April	May	June	July	August	September	October	November	December	2017
9	The economic effect of security measures	the amount of refund thousand rub	87.5	263.3	608.4	446.6	91.2	1398.9	1039.5	968.9	[illegible]89.4	1249.4	2127.5	471.0	9241.6
10	Number of official investigations of violations		2	2	4	3	2	2	6	2	[illegible]	3	4	5	37
11	Number of prescriptions for elimination of violations		3	3	3	5	4	6	7	5	[illegible]	7	13	11	72
12	Number of materials transferred to law enforcement agencies about violations:		0	2	1	1	0	1	2	0		2	3	3	16
12.1	Number of criminal cases initiated		2	1	1	0	0	1	1	1	[illegible]	1	2	3	15
12.2	number of court decisions	Number of cases	2	1	0	0	1	0	1	2	0	1	1	2	11
		Number of people	2	2	0	0	1	0	1	2	0	1	1	2	12
13	Revealed material damage caused by other violations	amount of damage thousand rub	17.1	4 799,4	11 268,7	95.9	277.9	977.5	669.4	879.9	765.6	951.8	3 151,9	2 861,0	26716.1
14	The total economic effect	amount of thousand rubles	50912.3	61477.6	26574.4	8575.8	11262.6	5929.9	65810.3	7696.5	[illegible]4800.6	13619.1	64031.1	16247.9	386938.0

Table 2. Results of controlling services of a business unit by main performance indicators for 2017

Source data	Division	JSC "xx"	Total division
The economic effect of measures for economic security in accordance with the report (thousand rubles.)	370690.1	2326.7	373016.8
Total balance of the maintenance of the service	60017.7	2780.8	62798.5
- Scheduled number of inspections:	891	138	1029
- The fact of inspections:	864	139	1003
- Identified shortcomings:	1333	96	1429
- Fixed shortcomings:	1325	96	1421
Number of measures taken (orders for punishment, criminal cases, court decisions)	254	19	273
The number of violations in the direction of providing protection against economic threats	260	20	280

Table 3. Preliminary Results of Business Unit Resource Protection by Key Indicators for 2017 (KPI)

Key indicators	Calculation formula	Target KPI	Actual achievement
Correspondence of service results with maintenance costs	Economic effect of measures on economic security in accordance with report)/(total balance of maintenance of the service) * 100	300%	594%
Creation of conditions of preservation for inventory items and inadmissibility of theft	Quality index of security activity	90%	98%
Ensure implementation of the practice of taking appropriate, including organizational measures in the direction of providing protection against economic threats	Number of measures taken (orders on punishment, dismissal of guilty persons)/ number of persons (employees) in respect of whom the facts of violations stipulated in the Code of GAZ Group on disciplinary liability * 100%	95%	97%
Ensuring that the planned level of expenses for the GAZ Group is not exceeded	(actual costs for the functionality of the GAS Group's controlling service/controlling service costs) * 100%	100%	98%

As the study shows:

– Preventive measures include: (a) inspection of hired employees in order to identify potential violators (Markova and Narkoziev 2018); (b) creation of a local regulatory and administrative base at the enterprise regulating the procedure handling of confidential information and measures of liability for violations, including confidentiality obligations (Garina et al. 2017); (c) a control system and management of access to various information resources of the enterprise, including control systems and control of physical access to premises; (d) implementation and management of security policy in corporate networks of enterprises (Hackman and Wageman 2015); (e) implementation of monitoring of user actions; (f) application of software and hardware tools that control the use of transmission channels data: mobile media, e-mail, Internet traffic (DLP - Data Leak Prevention class systems) (Kuznetsova et al. 2017); (g) conducting inspections for detected violations and taking measures.

– Improvement of the system for fighting with insider information leaks includes: (1) revision of the control system in the field of information security, including requirement that all information created by the employee during working hours on the equipment owned by the company, is the property of the company; (2) gaining information about the upcoming dismissal of top and middle management as early as possible; (3) application of software and hardware control tools use of data channels: mobile media, e-mail, Internet traffic; (4) monitoring of user actions, including checks for detected violations, taking action against violators. The implementation of these measures should help to create awareness of the inevitability of punishment for wrongful acts and caused damage. This task is carried out on an ongoing basis by the controlling units.

Evaluation of the avoided damage of the enterprise as a result of the implementation of controlling methods showed the need in the future:

- Activation of protection of resources from economic threats in terms of carrying out service checks, instructions for managers, measures on the revealed facts violations, as well as in the preparation of materials to law enforcement agencies (Shpilevskaya 2016).
- The principle of inevitability of punishment for violators in accordance with the "Code of the GAZ Group on disciplinary responsibility and measures of material influence", rigidity in the adoption of managerial decisions to punish responsible workers. Informing employees of the enterprise about certain results of inspections in order to implement the principle of inevitability of punishment.
- Compensation of material damage caused by their actions.
- Preventive measures aimed at eliminating the causes and preconditions of economic damage to the enterprise (Schätz 2016; Sergi 2018):
  (a) Development of documented procedures for managing and controlling the company's operations;
  (b) The procedure for dealing with receivables of GAZ Group enterprises, with other aspects of economic activity of enterprises that cause damage to the Group in the forecast period (Yashin et al. 2018).

In the long term, this will allow for greater control efficiency in terms of keeping confidential information on key operational performance indicators, technologies used, indicators of financial statements of the enterprise, main customers and suppliers.

References

Aleksandrovsky, S.V., Shushkin, M.A.: Model of implementation of simulation strategies by companies. Innovations, no. 1 (195), pp. 108–114 (2015)

Andrashina, N.S.: Analysis of best practices of development of domestic machine-building enterprises. Bulletin of Saratov State Socio-Economic University, no. 1 (50), pp. 24–27 (2014)

Garina, E.P., Garin, A.P., Efremova, A.D.: Research and generalization of design practices of product development in the theory of sustainable development of production. Humanities and Socio-Economic Sciences, no. 1 (86), pp. 111–114 (2016)

Gruzdeva, M.L., Smirnova, Zh.V., Tukenova, N.I.: Application of Internet services in technology training. Bulletin of Mininsky University, vol. 6, no. 1 (22), p. 8 (2018)

Karpov, V.V.: Controlling in the business system: conceptual framework, tools, organization. OSU, Omsk, 289 p. (2018)

Kuznetsov, V.P., Romanovskaya, E.V.: Analysis of methods of restructuring of industrial enterprise in modern conditions. Bulletin of Cherepovets State University, vol. 1, no. 2 (29), pp. 59–62 (2011)

Markova, S.M., Narkoziev, A.K.: Production training as a component of professional training of future workers. Bulletin of Mininsky University, vol. 6, no. 1 (2018)

Mizikovsky, I.E.: Harmonization of indicators of internal control. Audit Statements, no. 12, pp. 62–66 (2011)

Romanov S.N. Risk-controlling in the system of modern management. Transport business of Russia, no. 10, pp. 173–175 (2015)

Shpilevskaya, E.V.: Economic security of the country: threats and ways of its provision. Int. Sci. Res. J. no. 5–1 (47), 188–193 (2016)

Garina, E.P., Kuznetsova, S.N., Romanovskaya, E.V., Garin, A.P., Kozlova, E.P., Suchodoev, D. V.: Forming of conditions for development of innovative activity of enterprises in high-tech industries of economy: A case of industrial parks. Int. J. Entrep. **21**(3), 6 (2017)

Hackman, J.R., Wageman, R.: Total quality management: Empirical, conceptual, and practical issues. Adm. Sci. Q. **40**, 309 (2015)

Kuznetsova, S.N., Garina, E.P., Kuznetsov, V.P., Romanovskaya, E.V., Andryashina, N.S.: Industrial parks formation as a tool for development of long-range manufacturing sectors. J. Appl. Econ. Sci. **12**(2) (48), 391–401 (2017)

Sergi, B.S. (ed.): Exploring the Future of Russia's Economy and Markets: Towards Sustainable Economic Development. Emerald Publishing, Bingley (2018)

Schätz, C.: A Methodology for Production Development: doctoral thesis. Norwegian University of Science and Technology, 126 p. (2016)

Yashin, S.N., Trifonov, Y.V., Koshelev, E.V., Garina, E.P., Kuznetsov, V.P.: Evaluation of the effect from organizational innovations of a company with the use of differential cash flow. Adv. Intell. Syst. Comput. **622**, 208–216 (2018). https://doi.org/10.1007/978-3-319-75383-6_27

Assessing the Practical Strategizing of the Regional Policy on the Development of the Digital Economy

Igor E. Risin(✉), Pavel A. Kanapukhin, Elena F. Sysoeva, and Irina N. Petrykina

Voronezh State University, Voronezh, Russia
risin@mail.ru, kanapukxin@econ.vsu.ru, self@mail.ru,
petrykina_irina@mail.ru,

Abstract. The purpose of the study is to assess the modern practice of strategizing the regional policy on the development of the digital economy, to reveal its advantages and disadvantages. Methodology. For assessment we employed our own methodical approach including the sampling of the constituent entities of the Russian Federation representing different Federal Districts as objects of the study; using the analysis of updated strategies of socio-economic development of regions as an information base; the directions, objectives, expected results and mechanism for the implementation of the regional policy on the development of the digital economy as subject-matter of the study; application of content analysis of regional strategies together with the method of comparative analysis. Results. We revealed and substantially explained the advantages and disadvantages of the modern practical strategizing of regional policies on the development of the digital economy. Recommendations. The advantages of the strategizing digital economy development typical for a limited number of the constituent entities of the Russian Federation should become the invariant features of local practices. The disadvantages revealed predetermine the need for systemic changes in the strategizing of these policies, including the expansion of its directions and objectives, the addition of qualitative characteristics of the expected results with quantitative ones, increasing the diversity of forms and tools through which government authorities jointly with the business ensure the development of the digital economy.

Keywords: Digital economy · Public policy · Regional development strategies

JEL Code: F01 · O30

1 Introduction

According to the definition of the World Bank, the digital economy is a system of economic, social and cultural relations based on the use of information and communication technologies (ICT). Under this definition, the object of digitalization is the socio-economic system of the constituent entity of the Russian Federation in general, since the processes of digitalization are pervasive and developing its new properties. The development of a digital economy is managed through designing and pursuing a relevant public policy.

E. G. Popkova and B. S. Sergi (Eds.): ISC 2019, LNNS 87, pp. 563–571, 2020.
https://doi.org/10.1007/978-3-030-29586-8_65

Assessment of the modern practice of strategizing the public policy on the development of the digital economy in the constituent entities of the Russian Federation will expand the understanding of the content and mechanism for implementation of this policy, and restrictions to be overcome by measures of government authorities and business.

2 Methodology

For assessment of abovementioned practice we propose a methodological approach, which theoretical basis and imperatives are set out in several publications by (Vertakova et al. 2016), (Risin and Treshevsky 2015), (Risin et al. 2017), (Treshchevsky et al. 2018). Its distinctive features are as follows:

1. The subject-matter of the analysis is directions, objectives, estimated results and the mechanism for implementation of the regional policy on the development of the digital economy.

The determination of the directions and objectives of this policy is based on conceptual ideas that explain the essence and content of the digital economy presented in the proceedings of Russian and foreign scholars (Bukht and Hicks 2018), (Yudina 2016).

Identification of results is associated with the determination of their qualitative and quantitative parameters. The thorough analysis of the policy implementation mechanism involves the discovery of management forms and instruments by means hereof the public authorities of the constituent entities of the Russian Federation jointly with business intend to transform the traditional economy into a digital one.

2. Representative assessment is achieved by the account of the constituent entities of the Russian Federation representing different federal districts in the sampling of the study.

3. The information base of the analysis is presented by strategies for socio-economic development of the Voronezh Oblast (Ministry of Economic Development of Russia - Voronezh Oblast 2018), Krasnoyarsk Krai (Ministry of Economic Development of Russia - Krasnoyarsk Krai 2018), Moscow Oblast (Moscow Oblast Duma 2018), the Republic Bashkortostan (Ministry of Economic Development of Russia – the Republic of Bashkortostan 2018), the Republic of Sakha (Yakutia) (Ministry of Economic Development of Russia 2017), the Republic of Tatarstan (Invest-Tatarstan 2015), St. Petersburg (Ministry of Economic Development of Russia - St. Petersburg 2018).

4. The targeted assessment is provided by using content analysis of regional strategies jointly with the method of comparative analysis.

3 Results

The results of the analysis are presented in Tables 1, 2 and 3.

Table 1.

Constituent entities of the Russian Federation	Directions and objectives of policy on digital economy development
Voronezh Oblast	movement to the primary use of ICT of Russian manufacturers, increasing their competitiveness; introduction of digital economy technologies (provision of public and municipal services in electronic form; provision of remote medical services; connection to the Internet of all library institutions; growing number of virtual museums; use of data and software for drawing yield maps; use of sensors to assess soil moisture during the full day); monitoring of population's life quality (shift to the use of electronic medical records; introduction of biochips to monitor the drug effects and the state of the patient's internal environment; establishment of information processing centers on life parameters of particular groups of population; establishment of a space information reception center, which allows increasing the response to emergency risks situations and the level of reasonable forecasts); the expansion of university training in ICT and software; management of spatial development with the help of geo-information technologies
Krasnoyarsk Krai	introduction of digital economy technologies (provision of public and municipal services in electronic form; transition to the use of electronic medical records; introduction of remote medical services; introduction of telemedicine technologies; use of GLONASS navigation technologies; expansion of practical distant education; establishment of a digital school; connection of regional libraries to the Internet; an offer of virtual tours to visitors by museums); monitoring of the population's life quality (shift to the use of electronic medical records; the introduction of mobile scanning systems for diagnostic assessment of human body)
Moscow Oblast	introduction of digital economy technologies (provision of public and municipal services in electronic form; 3D printing and scanning; information technologies for life cycle management; development of software and hardware systems in the field of exploration and mining of mineral resources; development of unmanned and manned space exploration devices; introduction of telemedical technologies; the use of electronic medical records); training of IT staff
The Republic of Bashkortostan	introduction of digital economy technologies (provision of public and municipal services in electronic form; informatization of the healthcare system; establishment of a modern digital educational environment; establishment of virtual museums, electronic reference rooms; monitoring of the population's life quality; establishment of a hardware-software complex "Safe City"

(*continued*)

(*continued*)

Constituent entities of the Russian Federation	Directions and objectives of policy on digital economy development
The Republic of Sakha (Yakutia)	introduction of digital economy technologies; provision of public and municipal services in electronic form; establishment of virtual museums; creation of a modern digital educational environment; arrangement of satellite communication channels in medical institutions; development of telemedicine technologies; transition to a new level of space management using geo-information technologies; use of telecommunication technologies and the GLONASS global navigation system in vehicle management; e-retail development; monitoring of life quality: introduction of information and analytical systems "Monitoring of the housing resources", "Dom.online"
Republic of Tatarstan	building the conditions for the emergence of new and development of existing innovative companies in the field of information technologies and software products; development and commercialization of breakthrough information technologies; increasing the competitiveness of the Russian IT-industry; introduction of digital economy technologies: provision of public and municipal services in electronic form; development of e-medicine services; establishment of a new technological environment of the education system (digital school, digital college); development of telemedicine technologies; the creation of virtual tours in cultural institutions; introduction of solutions in the development of space and infrastructure (smart city, smart land use, smart road, smart home, etc.); transition to a new level of space management using advanced information technologies; training and advanced training of ICT staff
St. Petersburg	Establishing conditions for the development of information technologies and their use in various sectors of the economy, increasing their competitiveness at the international level (development of information infrastructure, training of high-skilled IT staff, etc.); strengthening of the role of IT-industry in the economy of St. Petersburg, the Russian Federation; ensuring the entry of Russian IT companies to international markets of goods and services; introduction of digital economy technologies: development of e-healthcare and telemedicine; development of distant learning technologies; introduction of modern information technologies into the activities of cultural institutions; development of intelligent systems for integrated accounting of municipal energy resources; consolidation of municipal housing management systems with use of smart technologies; further introduction of modern IT in the activities of public authorities

An analysis of the information presented in Table 1 allows determining the advantages and disadvantages of practical strategizing of regional policy on the development of the digital economy. Strategizing of directions and objectives are among the first:

- transition to the primary use of ICT of Russian manufacturers;
- diversification of Russian exports through the ICT inclusion in its object base;
- introduction of ICT in the areas in the management of space and infrastructure development: smart city, smart house, etc.;
- the movement to the management of the socio-economic space of the region with the use of ICT;
- expansion of IT staff training;
- the entry of Russian ICT companies to foreign markets.

The given examples evidence the limited scope of the advantages listed.

Another substantial restriction of modern practice is the narrowed range of economic digitalization paths (the lack in the target function of regional policy of directions that include the creation and development of systems for the production and delivery of foodstuff and water; smart resource extraction systems; digital financial technologies).

Table 2. Estimated results of the public policy on the development of the digital economy in the constituent entities of the Russian Federation

Constituent entities of the Russian Federation	Police effects
Voronezh Oblast	Elimination of digital inequality and provision of available high-quality communication services throughout the region; replacement of software and information support of foreign manufacturers by domestic ones in regional information resources; connection of all libraries to the Internet; growing number of virtual museums; increasing percentage of households with broadband access to the Internet information and telecommunications network; growing proportion of citizens using the mechanism for receiving public and municipal services in electronic form; growing number of the population with genomic-data electronic medical records
Krasnoyarsk Krai	connection of all libraries to the Internet; provision of virtual broadcasting of plays and concerts by all theater and concert institutions; offers of virtual tours to visitors by all museums
Moscow Oblast	reduced labor efforts of doctors for filling in statistical and regulatory reports; creation of additional automated jobs for doctors and nursing staff; equipment of doctors' workplaces with modern computers

(*continued*)

Table 2. (*continued*)

Constituent entities of the Russian Federation	Police effects
The Republic of Bashkortostan	increasing proportion of citizens receiving public and municipal services in electronic form; development of personalized medical servers; broader access of republic residents to the services of cultural and artistic institutions; growing number of access points to the virtual concert hall; digitization of museum pieces, documents, book collections; development of virtual museums and expositions, electronic reading rooms
The Republic of Sakha (Yakutia)	development of an electronic platform for personalized teaching of pupils; creation of a single information and educational space of secondary professional education system; development of a single healthcare information space; space management using cutting-edge geo-information technologies; provision of equal access to information services for citizens and companies; provision of high-speed telecommunication and Internet communication for socially significant facilities and companies across the country; growing percentage of public services that can be received by population using ICT; higher share of the IT industry in GRP
The Republic of Tatarstan	Rendering new types of public services using modern ICT; provision of virtual tours in cultural institutions; use of electronic hypertext textbooks, open education services on the Internet, classes of robotics in education; patient's interaction with the regional healthcare system through the development of telemedicine, remote technologies; setting up new IT companies, decreasing outflow of IT specialists to Moscow and abroad; increasing competitiveness of the Russian IT-industry; emergence of a internationally-known university with specialty in IT-education; growing number of research in IT; creation and commercialization of breakthrough ICT
St. Petersburg	Raising the level of informatization of the regional health care system, development of e-health and telemedicine; higher information openness of the education system, the development of distance learning technologies; the creation and development of e-catalogues, e-libraries, finding aids of archive funds, e-archives; providing consumers with access to information of the public information system for housing and communal services; management of municipal housing services using smart technologies; entry of Russian ICT companies to foreign markets for services; winning the leadership by St. Petersburg in new areas (information technology, cloud technology, multimedia systems, modeling of complex systems and processes, robotics and automatic vehicles, etc.); improving the quality of information for management decision-making at all administrative levels of St. Petersburg

The information specified in Table 2 shows that the advantage of practical strategizing is the commitment of regional policy to the achievement of economic and social results manifested in:

– elimination of digital inequality in the region;
– higher information openness in the systems of healthcare, education, and culture;
– more considerable ICT contribution to the gross regional product;
– prevalence of software and information support of domestic producers in the regional information resources;
– the entry of Russian ICT companies to international markets of services.

The disadvantage common to the practical strategizing in most regions is the limited composition of target indicators that allow giving a quantitative assessment of the results of economic digitalization.

The analysis of the forms and instruments (Table 3) make it possible to broaden our understanding of the advantages and disadvantages of practical strategizing under consideration.

Table 3. Forms and instruments of public policy on the development of the digital economy in constituent entities of the Russian Federation

Constituent entities of the Russian Federation	Forms and instruments of policy
Voronezh Oblast	State Program "Information Society"; geoinformation systems; IT-cluster; processing centers of data from sensors of life parameters of particular groups of population; space information receiving station
Krasnoyarsk Krai	State Program "Information Society"; interpretation center of research on the state of the basic parameters of the human body; information and consultation services and consultation centers for support of children studying on a distant basis
Moscow Oblast	State Program "Information Society"; science cities; special technology development economic area; university priority development areas
The Republic of Bashkortostan	State Program "Information Society"; hardware-software complex "Safe City."
The Republic of Sakha (Yakutia)	State Program "Information Society"; geoinformation systems; IT-park

(*continued*)

Table 3. (*continued*)

Constituent entities of the Russian Federation	Forms and instruments of policy
The Republic of Tatarstan	State Program “Information Society”; Digital School and Digital College projects; clusters “Finance and Professional Services”, “Confluence of Civilizations” (culture, tourism and recreation), Smart information technologies; The new university “Innopolis” specializing exclusively in high-tech training; implementation of Smart city, Smart land use, Smart road, Smart home projects; innovation clusters
St. Petersburg	State Program “Information Society”; Digital School project, monitoring of key health indicators; intelligent systems for integrated accounting of municipal energy resources; management system of municipal housing services using smart technologies; development centers of advanced high-tech companies in IT-industry; innovation clusters

4 Conclusion

We suppose that the advantage of the practical strategizing is an association of the processes of economic digitalization with the use of cutting-edge forms of its spatial arrangement: innovation clusters, special technology development economic areas.

The disadvantage is the low diversity of forms and instruments by means hereof the regional public authorities seek to ensure the development of the digital economy.

The practical strategizing of public policy on the development of the regional digital economy is not yet systematic, as evidenced by fragmentary versions of its essential elements (directions, objectives, results, forms and management tools).

The combination of advantages and disadvantages of modern practical strategizing of regional policy on the development of the digital economy makes it possible, on the one hand, to broaden the use of useful patterns, on the other hand, to focus the attention and efforts of public authorities and business actors on the development of measures to overcome its disadvantages.

References

Bukht, R., Hicks, R.: Definition, concepts, and measurement of the digital economy. Bull. Int. Relat. **13**(2), 143–172 (2018)

Ministry of Economic Development of Russia - Voronezh Oblast. Revised Draft Strategy for Socio-Economic Development of the Voronezh Oblast until 2035. http://economy.gov.ru/minec/activity/sections/strategterplanning/komplstplanning/stsubject/straterupdate/201802101. Accessed 17 Jan 2019

Ministry of Economic Development of Russia – Krasnoyarsk Krai: Revised Draft Strategy for Socio-Economic Development of the Krasnoyarsk Krai until 2035 (2018). http://economy.gov.ru/minec/activity/sections/strategterplanning/komplstplanning/stsubject/straterupdate/201802103. Accessed 20 Dec 2018

Ministry of Economic Development of Russia – the Republic of Bashkortostan: Revised Draft Strategy for Socio-Economic Development of the Republic of Bashkortostan until 2035 (2018). http://economy.gov.ru/minec/activity/sections/strategterplanning/komplstplanning/stsubject/straterupdate/201825104. Accessed 20 Dec 2018

Ministry of Economic Development of Russia – the Republic of Sakha (Yakutia): Revised Draft Strategy for Socio-Economic Development of the Republic of Sakha (Yakutia) until 2035 (2018). http://economy.gov.ru/minec/activity/sections/strategterplanning/komplstplanning/stsubject/straterupdate/2017190401. Accessed 20 Dec 2018

Ministry of Economic Development of Russia – St. Petersburg: Revised Draft Strategy for Socio-Economic Development of St. Petersburg until 2035 (2018).http://economy.gov.ru/minec/activity/sections/strategterplanning/komplstplanning/stsubject/straterupdate/201801032. Accessed 20 Dec 2018

Moscow Oblast Duma: Draft Strategy for Socio-Economic Development of Moscow Oblast until 2035 (2018). http://mosoblduma.ru. Accessed 20 Dec 2018

Risin, I.E., Treshevsky, Yu.I.: Regional cluster policy: conceptual, methodical and instrumental support. Ruscience, pp. 100–105 (2015)

Risin, I.E., Treshchevsky, Y.I., Tabachnikova, M.B., Franovskaya, G.N.: Public authorities and business on the possibilities of the region's development. In: Popkova, E. (ed.) Overcoming Uncertainty of Institutional Environment as a Tool of Global Crisis Management. Contributions to Economics, pp. 55–62. Springer, Cham (2017). No. 9783319606958

Invest-Tatarstan: Strategy for Socio-economic Development of the Republic of Tatarstan until 2010 (2015). http://www.invest.tatar.ru. Accessed 20 Dec 2018

Treshchevsky, Yu.I., Risin, I.E., Korobeinikova, L.S., Gavrilov, V.V.: Management of the changes of socio-economic systems: an economic analysis of the state and consequences of the systemic feature. In: Endovitsky, D.A., Popkova, E.G. (eds.) Management of Changes in Socio-Economic Systems. Studies in Systems, Decision, and Control, vol. 135, pp. 21–30. Springer, Cham (2018)

Vertakova, Y., Risin, I., Treshchevsky, Y.: The methodical approach to the evaluation of clustering conditions of socio-economic space. In: Innovation Management and Education Excellence Vision 2020: From Regional Development Sustainability to Global Economic Growth, IBIMA, Proceedings of the 27th International Business Information Management Association Conference, pp. 1109–1118 (2016)

Yudina, T.N.: Understanding the digital economy. Theor. Econ. **3**, 12–16 (2016)

Development of the System of Operational and Production Planning in the Conditions of Complex Industrial Production

Elena V. Romanovskaya(✉), Victor P. Kuznetsov, Natalia S. Andryashina, Ekaterina P. Garina, and Aleksandr P. Garin

Minin Nizhny Novgorod State Pedagogical University, Nizhny Novgorod, Russian Federation
alenarom@list.ru, kuzneczov-vp@mail.ru, natali_andr@bk.ru, e.p.garina@mail.ru, p_nn@mail.ru

Abstract. Today achieving goals of a company and successful solution of tasks is possible only with proper management system of operational production planning, application of a systematic approach, and with modern achievements of science and technology. Therefore, one of the main tasks of the enterprise is to find effective tools for the development of the system of operational and production planning. The subject of research are organizational-economic relations arising in the process of production activity of the enterprise. The role of the operational and production planning system at the industrial enterprise in the formation of an effective planning strategy was defined as theoretical aspects in current work. The essence, objectives, and tasks of operational and production planning at the enterprise are defined. Methods and stages of operational and production planning at the enterprise are systematized. In the study the analysis of the current state and development of the system of operational and production planning at the enterprise LLC "Promelectro", which showed the need to develop the system operational and production planning to improve the efficiency of production activities at the enterprise, as well as the entire system. For the exclusion of the identified problems, operational production planning was developed for Promelectro LLC.

Keywords: Enterprise · Industry · Mechanical engineering · Operational and production planning

1 Introduction

Market success of enterprises is determined by competitiveness of their products, so there is a need to improve product quality and reduce cost [10]. The first step to achieve these objectives for production enterprises is the correct planning of production activities [2, 5]. The main task of operational planning at the enterprise is to ensure coordinated work in each of the units. This is regulated by: setting a schedule of production and production in each of the shops, sites; timely delivery of materials to a workplace; regulation of execution of the production program in each of divisions.

E. G. Popkova and B. S. Sergi (Eds.): ISC 2019, LNNS 87, pp. 572–583, 2020.
https://doi.org/10.1007/978-3-030-29586-8_66

2 Theoretical Basis of the Study

Operational planning can be divided by types depending on the direction and nature of tasks that need to be solved [3, 4, 7].

1. Depending on the application:
   - Inter-workshop planning. Provides development, regulation and control of the implementation of plans in all workshops of the enterprise especially coordinates the work of the main and auxiliary workshops;
   - In-workshop planning. It is the preparation of operational plans and schedules for sites, teams and individual workplaces.
2. Depending on the content and validity period of operational planning:
   - Calendar planning. This type of planning includes the distribution of annual tasks by production units and terms of performance and the achievement of indicators to certain performers of work;
   - Current planning, which is also called dispatching, involves operational control of the progress of production processes, accounting of output and consumption of resources [1, 8].

In the process of operational and production planning it is necessary to specify and detail strategic and tactical plans of the company. The following types of production planning can be distinguished (Table 1).

Table 1. Types of production planning

Depending on the level of activity	Depending on the object of operation	Depending on time	Depending on the ability to make changes and additions	Depending on the type
– General production planning in which the plan affects all areas of work of the company – Private production planning, where a plan includes only certain areas of work	– Planning of production activities – Staff planning – Financial planning – Sales planning	– Short - term (1–12 months) – Medium - term (1–5 years) – Long - term (more than 5 years)	– Flexible (when you can make changes) – Hard (when you cannot change a plan)	– Strategic – Current – Operative

The main indicators of production planning are presented in Table 2.

Table 2. Production planning indicators

Key indicators	Operational indicators	Indicators of current production plans
– Income level; – Sales volume; – Profit per share; – Growth rate; – Market share; – Value of shares; dividends; – Quality of goods; – Development strategies; – Compensation to employees; – Level of product quality; – Social responsibility; – Stability policy	– investment per employee; – Value added; – Productivity growth rate; – Cost reduction policies; – Capital turnover ratio	– Volume of sales of goods, works, and services; – Number of employees specifically in production; – Assortment and nomenclature of manufactured goods; – Number of investments in activity areas; – Cost of goods, income, profitability and other financial indicators; – Wage fund and average wage index

In preparation for production planning, it is necessary to calculate all indicators for each item. Nomenclature—the list of manufactured products by type, grade, size, etc. It is also very important to evaluate the effectiveness and novelty of produced products. This is an integral part of production planning.

3 Results and Discussions

LLC "Promelectro" is a leading Russian company, developer and manufacturer of electrical equipment, control systems, distribution and regulation of electricity. The company provides needs of key sectors of the economy of the country: energy, including nuclear, gas, chemical, oil, metallurgy and other industries.

Analysis of the activity of LLC "Promelectro" revealed a number of problems related to the production activity, among them:

- Unreasonably high levels of reserves, which increase cost of production;
- Downtime of equipment and workers due to lack of necessary parts. As a result, the production cycle is increased and deadlines are passed;
- The existing system of calendar planning does not meet the requirement of effective planning of production activities due to lack of accurate calculations of equipment load on every production area of the shop.

The problems show that developed production standards and resource loading standards are not followed, as they are outdated. Level of resources used is higher than planned, and principle of production rhythm is not used. It is necessary to review the planning system in production and solve current problems, which will reduce time production and costs by specifying the production facility with a full calendar schedule, which contain information when to produce, with help of which personnel and on what equipment. Production planning should be aimed at maximizing production efficiency and reducing costs.

It is assumed that the proposed system of operational production planning will optimize the utilization of production capacity, reduce inventory, production costs and will increase the volume of output, which will lead to lower cost of production and increase of profit of the enterprise. As an example, we will develop a calendar and volume plan of work of a closed section of mechanical-mechanical production. This unit produces parts for various types of electrical cabinets. In the planning period, parts for the following electrical cabinets will be produced: EL 205, EL 119, EL 104 and EL 280.

The initial data for calculations is the normative reference information on the products of the site, in particular:

- Product plan (Table 3);

Table 3. Product plan

Item Code	Program of release, pc.		
	Annual	Quarterly	Monthly
EL-205	280	70	23
EL-119	480	120	40
EL-104	300	85	25
EL-280	600	150	50

- Labor intensity standards of products by groups of interchangeable equipment (Table 4);

Table 4. Summary labour intensity rates for interchangeable equipment groups

Item code	Summary labour intensity rates by group interchangeable equipment, min.				
	12	15	16	18	23
EL-205	93.9	2.2	102.2	70.9	225
EL-119	122.8	90.7	15.1	14.7	98.8
EL-104	66.6	14.9	24.7	34.4	76.2
EL-280	22.8	139.4	61.8	2	192.6

Table 5. Process, batch size, and complexity of machining parts

Item code	Part no.	Stage no.	Number of stages	Complete set, pc.	Batch size, pc.	Period of repetition of production, decade	Advance release details, decade	The labor intensity of a batch by equipment groups, hour					
								Group code	Labour intensity	Group code	Labour intensity	Group code	Labour intensity
EL-205	90373	1	1	2	16	1	2	23	1.17	15	0.28		
	90742	1	1	1	8	1	5	23	0.58	15	0.14	18	18.9
	10009	1	3	1	70	9	4	23	24	16	30.4	18	19.3
	10009	2	3	1	70	4	3	23	23.4				
	10009	3	3	11	70	9	2	23	23.9	16	10.5		
	30021	1	2	1	140	18	4	23	25.6	16	14.7	12	26.2
	30021	2	2	1	140	18	3	23	30.8	18	10.3		
	30022	1	2	1	140	18	3	23	22.8	16	8	12	12.5
	30022	2	2	1	140	18	2	18	10.1	23	3.5		
	48206	1	2	1	8	1	2	16	0.94	23	0.89		
EL-119	21341	1	1	1	13	1	3	15	11.4	23	4.64		
	33022	1	1	2	26	1	2	12	0.59				
	48081	1	1	1	13	1	2	12	0.38	23	0.42		
	51911	1	1	2	26	1	2	15	0.59	18	0.31	23	1.38
	52509	1	1	1	13	1	2	23	0.04				
	34277	1	2	1	120	9	3	12	32.2				
	34277	2	2	1	120	9	2	16	11.2	18	16.2	23	11
	42044	1	1	1	13	1	2	12	0.03	23	0.18		
EL-104	30545	1	1	1	75	9	4	23	26	16	15.9	18	12.2
	30545	2	1	1	75	9	3	15	18.6	18	10.8	23	34.9
	30545	3	1	1	75	9	2	23	16.2				

(continued)

Table 5. *(continued)*

Item code	Part no.	Stage no.	Number of stages	Complete set, pc.	Batch size, pc.	Period of repetition of production, decade	Advance release details, decade	The labor intensity of a batch by equipment groups, hour					
								Group code	Labour intensity	Group code	Labour intensity	Group code	Labour intensity
		1	1	1	9	1	2	12	0.46	23	0.22		
	56062	1	1	1	9	1	2	12	0.68	23	1.39		
	37024	1	1	1	150	18	2	12	7.5				
	37021	1	1	1	75	9	2	12	10.1				
	55021	1	1	1	75	9	3	12	22.3	23	1.75		
EL-280	9062	1	1	1	17	1	3	12	1.85	23	0.36		
	9064	1	1	1	17	1	3	12	1.85	23	0.36		
	47095	1	2	2	600	18	3	12	5	23	25		
	47036	1	1	1	300	18	3	12	5	23	20		
	47037	1	1	1	17	1	3	12	0.14	23	0.69		
	47119	1	1	1	300	18	3	15	2	23	22		
	7386	1	1	1	17	1	3	23	0.8				
	22878	1	1	1	150	9	4	15	33	16	5	15	2.75
	22878	2	1	1	150	9	3	23	16.5				
	46682	1	1	1	17	1	4	16	0.5	12	1.11		

– Process, batch size and workmanship (Table 5).

The purpose of this planning stage is to identify groups of equipment that have maximum and minimum production load, even distribution of equipment load over decades for calculation of the required number of workers required to perform the work, as well as prevention of downtime or overload of equipment and employees [7, 9].

The calculation of the time funds is carried out in the following order:

1. Calculates the annual used time funds by groups of interchangeable equipment
2. Calculates the time funds used by groups of interchangeable equipment.

Then, we use obtained data and calendar for planned year, funds of time in each planned-accounting period.

The results of the calculations are summarized in Tables 6 and 7.

The calculations take into account that in 2017 247 working days with a five-day working week.

Table 6. Annual time funds used by equipment group

Indicator	Equipment groups				
	12	15	16	18	23
Used time fund for annual program, hour	2241.1	2204.4	1370.7	656.5	4355.9
Including parts produced every decade, hour	611.7	680.36	250.8	77.05	937.7

Table 7. Average time assets used by equipment group

Indicator		Decades								
		1	2	3	4	5	6	7	8	9
Number of working days in the decade		6	8	6	8	6	9	6	7	8
Average time funds used by groups of interchangeable equipment, hour	12	18.1	72.6	54.4	72.6	54.4	54.4	54.4	63.5	72.6
	15	17.8	71.1	53.6	71.4	53.6	53.5	53.5	62.5	71.4
	16	11.1	44.4	33.3	44.4	33.3	33.3	33.3	38.8	44.4
	18	5.3	21.3	15.9	21.3	15.9	15.9	15.9	18.6	21.2
	23	35.3	141	106	141	106	106	106	123	141.1
Including parts produced every decade, hour	12	4.9	19.8	14.8	19.8	14.9	14.9	14.8	17.3	19.8
	15	5.5	22.0	16.5	22.0	16.5	16.5	16.5	19.3	22.0
	16	2.0	8.1	6.1	8.1	6.1	6.0	6.1	7.1	8.1
	18	0.6	2.5	1.9	2.5	1.9	1.8	1.9	2.2	2.5
	23	7.6	30.3	22.8	30.3	22.7	22.8	22.8	26.5	30.3

When determining the deadlines for release the following shall be taken into account: part requirements to ensure the release of the products in which they are applied; production in progress data for finished parts at the beginning of the planning period (Table 8); batch size standards parts and quantities ahead of their release relative to the release of products.

Table 8. Availability of finished parts on the site at the beginning of the planned quarter, pc.

Item code	Part No.	Number of finished parts
EL-205	10009	96
	10009	16
	10009	96
	30021	127
	30021	127
	30022	67
	30022	67
EL-119	34277	91
	34277	91
EL-104	30545	89
	30545	89
	30545	89
	37024	128
	37021	60
	55021	56
EL-280	47095	277
	47036	129
	47119	246
	22878	109
	22878	109

The release date of parts is calculated to a decade by finding the difference between the release date of products and value of the advance of production. Creation of the calendar and volume plan continues until the last decade of the planning period is finished.

The loading of equipment for the first decade is shown in Table 9.

In this way, the load of equipment for each decade of the planning period is calculated. Let's present a summary table of equipment loading to the volume and calendar plan (Table 10).

Table 9. Loading of equipment for 1 and 2 decades of closed section of mechanical and mechanical production

Part number	Stage no.	Number of stages	Deadline for issue	The labor intensity of batch processing of parts by groups of interconnected equipment, hour				
				12	15	16	18	23
Parts for which the release deadline is 1				4.95	5.51	2.03	0.62	7.59
Additional loading								
10089	1	1	5	13.6				
41080	1	1	9		9.2		4.8	15.2
Time fund used, hour				18.15	17.78	11.1	5.32	35.27
Time consumption:								
Initial, hour				4.95	5.51	2.03	0.62	7.59
Final, hour				18.55	14.71	2.03	5.42	22.79
Absolute deviation:								
Overload (+)				0.4			0.1	
Underload (−)					−3.07	−9.07		−12.48
Relative deviation from used time funds %				2.2	−17.3	−81.7	1.9	−35.4
Parts for which the release deadline is 2				19.81	22.04	8.12	2.5	30.37
10009	2	3	2					23.4
37024	1	1	2	7.5				
Additional loading								
22878	1	2	3		33	5		12.75
50288	1	1	4	21.25		10.1		24.25
22878	2	2	4					16.5
34277	2	2	5			11.2	16.2	21
47036	1	1	5	5				20
Time fund used, hour				72.59	71.11	44.4	21.26	141.08
Time consumption:								
Initial, hour				19.81	22.04	8.12	2.5	30.37
Final, hour				53.56	55.04	34.42	18.7	124.87
Absolute deviation:								
Overload (+)								
Underload (−)				−19.03	−16.07	−9.98	−2.56	−36.21
Relative deviation from used time funds %				−26.3	−22.6	−22.4	−12.1	−11.4

Table 10. Summary table of equipment loading to the volume and calendar plan

			Decade								
			1	2	3	4	5	6	7	8	9
Code of interchangeable equipment	12	Fpo.k	18.55	53.56	58.49	70.61	32.86	40.86	39.86	25.34	19.81
		Fisp.k	18.15	72.59	54.44	72.59	54.44	54.44	54.44	63.51	72.59
		ΔF	0.4	−19.03	4.05	−1.98	−21.58	−13.58	−14.58	−38.17	−52.78
		Δ, %	2.2	−26.3	7.4	−2.7	−39.6	−24.9	−26.8	−60.1	−72.7
	15	Fpo.k	14.71	55.04	33.23	71.04	56.73	47.63	16.53	53.78	22.04
		Fisp.k	17.78	71.11	53.55	71.4	53.55	53.55	53.55	62.47	71.4
		ΔF	−3.07	−16.07	−20.32	−0.36	3.18	−5.92	−37.02	−8.69	−49.36
		Δ, %	−17.3	−22.6	−37.9	−0.5	5.9	−11.1	−69.1	−13.9	−69.1
	10	Fpo.k	2.03	34.42	34.79	16.12	21.99	24.79	6.09	7.11	38.52
		Fisp.k	11.1	44.4	33.3	44.4	33.3	33.3	33.3	38.85	44.4
		ΔF	−9.07	−9.98	1.49	−28.28	−11.31	−8.51	−27.21	−31.7	−5.88
		Δ, %	−81.7	−22.4	4.5	−63.7	−34%	−25.6	81.7	−81.7	−13.2
	18	Fpo.k	5.42	18.7	1.87	20.5	14.07	12.67	11.97	12.98	21.8
		Fisp.k	5.32	21.26	15.95	21.26	15.95	15.95	15.95	18.68	21.26
		ΔF	0.1	−2.56	−14.08	−0.76	−1.88	−3.28	−3.98	−5.63	0.54
		Δ, %	1.9	−12.1	−88.3	−3.6	−11.8	−20.6	−25	−30.3	2.5
	23	Fpo.k	22.79	124.87	101.62	85.62	109.28	114.08	86.08	49.62	54.37
		Fisp.k	35.27	141.08	105.81	141.08	105.81	105.81	105.81	123.4	141.08
		ΔF	−12.48	−36.21	−4.19	−55.46	3.48	8.28	−19.72	−73.78	−86.71
		Δ, %	−35.4	−11.4	−4.0	−39.3	3.3	7.8	−18.6	−59.8	−62.5

4 Conclusion

It is recommended to implement and use such information systems as «1C, because development of large industrial account plans for each business area is very complex. ERP Enterprise Management 2.0" [6].

ERP system is an effective tool in the development of production plans, as its functionality covers a significant amount of production planning tasks.

Economic evaluation of the development and implementation of the system showed that the introduction will reduce inventory, production costs will lead to a 7% reduction in the cost of production and an increase in profit of the enterprise by 26% in five years.

With expected outcome of the project, the internal rate of return is 53.4%, and ROI index will be 40.9.

The project of the system in LLC "Promelectro" will pay off less than a year after the introduction, after which it will start to make a profit. The presented calculations show that the proposed innovative solution is economically profitable and it needs to be implemented.

References

1. Ability to use ERP system to support operational planning of production [Electronic resource]. — Access mode: http://www.topsbi.ru/about-the-company/press-centr/publikacii. Accessed 27 Mar 2019
2. Chelnokova, E.A., Kuznetsova, S.N., Nabiev, R.D.: Possibilities of using information and communication technologies in teaching economic disciplines in the university, no. 3 (20), p. 8. Vestnik of the Minin University (2017)
3. Garina, E.P., Kuznetsov, V.P., Romanovskaya, E.V., Andryashina, N.S., Efremova, A.D.: Research and generalization of design practice of industrial product development (by the example of domestic automotive industry). Quality - Access to Success, vol. 19. no. S2, pp. 135–140 (2018)
4. Garina, E.P., Kuznetsova, S.N., Romanovskaya, E.V., Garin, A.P., Kozlova, E.P., Suchodoev, D.V.: Forming of conditions for development of innovative activity of enterprises in high-tech industries of economy: a case of industrial parks. Int. J. Entrep. **21**(3) (2017). https://www.abacademies.org/articles/forming-of-conditions-for-development-of-innovative-activity-of-enterprises-in-hightech-industries-of-economy-a-case-of-industrial-6888.html
5. Garina, E.P., Romanovskaya, E.V.: Improving the efficiency of the organization of technological flows in the framework of the integrated application of the production system tools. Finan. Res. no. 2 (55), 125–133 (2017)
6. Kuznetsova, S.N., Romanovskaya, E.V., Potashnik, Y.S., Grechkina, N.E., Garin, A.P.: Factors determining whether industrial parks are successful or not. In: Collection: Managing Service, Education and Knowledge Management in the Knowledge Economic Era - Proceedings of the 4th Annual International Conference on Management and Technology in Knowledge, Service, Tourism and Hospitality, SERVE 2016, pp. 53–58 (2017)
7. Markova, S.M., Narkoziev, A.K.: Production training as a component of professional training of future workers, vol. 6. no. 1. Vestnik of Minin University (2018)

8. Potashnik, Y.S., Garina, E.P., Romanovskaya, E.V., Garin, A.P., Tsymbalov, S.D.: Determining the value of own investment capital of industrial enterprises. Adv. Intell. Syst. Comput. **622**, 170–178 (2018)
9. Smirnova, Zh.V., Gruzdeva, M.L., Krasikova, O.G.: Open Electronic Courses in the Educational Activity of the University, no. 4 (21), p. 3. Vestnik of Minin University (2017)
10. Sergi, B.S. (ed.): Exploring the Future of Russia's Economy and Markets: Towards Sustainable Economic Development. Emerald Publishing, Bingley (2018)

Introduction of Genomic Research Results into the Economy Through the Legal Regime of Intellectual Property (by the Example of Circadian Gene Research)

Andrey A. Inyushkin(✉), Yurii S. Povarov, Elena S. Krykova, and Valentina D. Ruzanova

Samara National Research University named after S.P. Korolev, Samara, Russia
inyushkin_a@mail.ru

Abstract. The article discusses the legal regulation of interrelated intellectual property and genomic research results within the transition of the Russian Federation to the digital economy. **The relevance** of the issue under research is determined by the need to improve legislation on the legal regulation of genomic research related to achieving the purposes set out in the national program "Digital economy of the Russian Federation." **The purpose** of the paper is to determine the main directions for improvement of the regulatory framework relating forms and methods of using the genomic research results and effective legal mechanisms for the application of the intellectual property to fulfill the Digital Economy program. **The main approaches** of the research are cross-sectoral, dialectical and systemic. These approaches make it possible to comprehensively study whether the regulatory framework applies to intellectual property to solve the problems of the introduction of genomic research results within the implementation of the national program "Digital Economy." **Results:** the paper investigates the current status of legal regulation of genomic research and discovers the peculiarities of the application of the intellectual property to solve the problems of introducing the results of similar research into the economy. We described the applicability of individual intellectual property objects in the field of introducing the genomic research results, including production secrets (know-how), databases and works of science. We state that the most suitable legal regime for introducing the genomic research results into civil commerce will be the legal regime of copyright law objects.

On the one hand, such a legal regime conveys the peculiarities of science works, and on the other hand, it allows classifying the information obtained as databases. This article is of practical value for multidisciplinary specialists, including geneticists, biologists, lawyers and economists who manipulate the genomic research results in economic activity. Besides, the article is of interest to lawmakers.

Keywords: Circadian genes · Databases · Composite work · Legal regime · Genomic research results · Intellectual property · Copyright law · Production secret (know-how) · Improvement of legislation

E. G. Popkova and B. S. Sergi (Eds.): ISC 2019, LNNS 87, pp. 584–590, 2020.
https://doi.org/10.1007/978-3-030-29586-8_67

1 Introduction

1.1 Establishing a Context

Since the adoption of the national program "The Digital Economy of the Russian Federation," civil commerce has dramatically changed, which inevitably necessitated the further improvement of the regulatory framework. New objects of civil rights appear more often within the regulation of relations arising in relation to this, which require theoretical understanding and the establishment of an effective legal regime meeting the challenges. Address of the President of the Russian Federation to the Federal Assembly repeatedly noted that the Russian Federation should establish the full operation of scientific and educational centers, which will integrate the capabilities of universities, academic institutions, and high-tech companies. Without any doubt, such integration is possible only provided the development of an effective regulatory framework. Establishment of canters in such promising area as genomic research has been set as a particular direction in implementing major interdisciplinary projects of the digital economy. A drastic breakthrough in this direction will make it possible to design new methods of assessment, prevention and countering of many diseases, broaden breeding and agriculture opportunities. In the middle of the next decade, the Russian Federation should become one of the leaders in these scientific and technological areas, which undoubtedly will predict the future both of Russia and the whole world. In this relation, the specific nature of the transition to new economic realities is increasingly investigated in the literature (Vaipan 2018; Kuznetsov 2018; Inyushkin et al. 2018a, b). The development and scientific substantiation of the interrelation of the genomic research results and intellectual property seems to be the most forward-looking means to introduce the genomic research results into civil commerce. Thus, in particular, the approved mechanisms for the use of intellectual rights, allow establishing the foundation of the legal regime for research results in humans and animal genomes. As an example, the use of the intellectual property legal regime for applying the results of circadian gene research. These genes are responsible for the biological rhythms of humans and animals. Information about sleep and wakefulness has a potential commercial value because it allows for the identification of suitable staff to perform a particular job (a particular activity). Thus, it is admitted to use the legal regime of production secret (know-how) for the legal protection of this information. In addition, in the Russian legal system creation of an intellectual property object containing information also falls under the legal regime of databases (Bachilo 2009; Inyushkin 2016). The specific character of data circulation related to genomic research generates different legal and doctrinal approaches to the use of the intellectual property legal regime (Singh et al. 2007), also in the Russian legislation within the transition to a digital economy (Voinikanis 2013).

1.2 Literature Review

Analysis of the present status of intellectual property and digital economy legislation was carried out by Voinikanis (2013), Inyushkin et al. 2018a, b, Mikhailov (2018), Ivardava (2019) et al. Information as an object of property relations were investigated by Bachilo (2009), Chang and Zhu (2010), Inyushkin (2016). The general issues of the civil law system, including intellectual property law, are explored in the writings of Ruzanova (2018), Gavrilova (2019). Genes as objects of intellectual property are studied in publications of Singh et al. (2007), Mokhov and Yavorskii (2018).

1.3 Establishing a Research Gap

Prior research of genomic legal regulation mainly considered the regulatory framework of intellectual rights, which did not allow identifying factors affecting the successful introduction of the research results into the digital economy as well as ways of surmounting obstacles to their full use.

The research reveals some intellectual property objects that can be used as carriers of information on the results of genomic research. We provide an example of work with circadian genes to perform the tasks of the national program "Digital Economy."

1.4 Purpose of the Research

The purpose of the research is to study the legal regime that determines the forms and methods of using the results of genomic research in relation with intellectual property within the transition to a digital economy and to found out ways of its further improvement.

2 Methodological Framework

2.1 Research Methods

In the course of the research, we applied both general and private methods of cognition: historical, systemic, comparative-law, cross-sectoral, etc. Their combined use allows revealing the peculiar application of the legal regime of particular intellectual property objects to the introduction of genomic research results.

2.2 Research Background

The research background is proceedings of Russian and foreign academicians studying various aspects of legal regulation of intellectual property, legal regulation of genomic research. The current civil legislation, foremost relating intellectual property rights, serves as a research background.

2.3 Research Stages

The issue is studied in two stages:

- the first stage is an analysis of the academic literature on the issues studied as well as legislation relating digital economy, intellectual property, genomics;
- the second stage is making conclusions on the appropriate use of particular intellectual property objects to the genomic research results in terms of the digital economy.

3 Results and Discussions

3.1 Prospects for the Development of Legal Regulation of Genomic Research

Genomic research today plays the role of one of the critical areas for the development of the digital economy, and finding the best ways to introduce the genomic research results into civil commerce in this context is a priority task for the development of the country's economic base. A similar task is enshrined in the Presidential Address to the Federal Assembly that also notes that it becomes particularly important due to the emergence of artificial intelligence computer systems (computer neural networks), which means integration of technical and biological systems into a single functional autonomous system. Under these conditions, the development of the regulatory framework will allow for the establishment of the most suitable legal regime for the introduction of genomic research results in the digital economy. The demand for improving the regulatory framework is increasingly manifested due to the necessary reformation of the legal system to protect the results of the research made by Russian academicians. Justifying the interrelation of genomic research results and intellectual property seems to be the most promising means for the establishment of such protection. Thus, in particular, the approved mechanisms for the use of intellectual rights make it possible to establish a foundation of the legal regime for the introduction of results in the field of the human and animal genome research. An example of the use of the legal regime of intellectual property for introducing research results into civil commerce is the introduction of the results of circadian genes research. These genes are responsible for the regulation of the biological rhythms of humans and animals and determine the sleep and wakefulness patterns. It should be noted that such information has potential commercial value since it allows for the identification of suitable staff to perform a particular job (a particular activity).

3.2 The Procedure for Using the Production Secret (Know-How) to Introduction Genomic Research Results in Practice

According to Art. 1465 of the Civil Code of the Russian Federation, production secret (know-how) is information of any nature (production, technical, economic, organizational, and other) on the results of intellectual activity in the scientific and technical area and on the ways of conducting professional activity that have actual or potential commercial value due to non-public nature, if third parties do not have free legal access to such information and the holder of hereof takes reasonable steps to keep the information secret, also by introducing a trade secret regime. Let us analyze the criteria for referring the information obtained in the course of research of circadian genes to the production secret (know-how). It is necessary to determine whether the fundamental scientific results have actual or potential commercial value due to its non-public nature. Probably it has commercial value since even though as of the current date such scientific results may not always bring potential profit, in the long-term academic papers can be used as the basis for advanced projects and even new sectors of the economy. Legislative requirements that third parties do not have free legal access to such scientific information and taking reasonable measures to keep the information secret, also by introducing a trade secret regime, will also be enforced. In the educational and scientific institutions, in the overwhelming majority of situations, there is access control, which allows stating the observance of this criterion.

Russian legislation requires that the received scientific results should be information on the results of intellectual activity in the scientific and technical area. According to the Articles 1225 and 1259 of the Civil Code of the Russian Federation, academic proceedings belong to the works of science. The literature analyzed that in Clause 1 of Art. 1225 of the Civil Code of the Russian Federation, the term "intellectual rights" is applicable to the results of intellectual activity specified in subclauses 1-11, but not to production secrets (know-how) and visual identities mentioned in subparagraphs 12-16 (Eremenko 2014). Thus, application of production secret (know-how) regime with respect to the results of circadian gene research requires formalization of the research results.

3.3 The Procedure for the Application of Databases When Working with the Received Scientific Data on the Results of Genomic Research

Another suitable object for introducing the results of genomic research into civil commerce will be databases. Databases, as well as works of science, belong to copyright objects; they are protected as composite works. We should note that the literature is already discussing appropriate reference of genetic code to the legal regime of copyright objects – computer programs. The definition of databases includes the systematization of information within the database and the relation with the computer (Mokhov and Yavorskii 2018). In this regard, it is advisable to analyze the legal regulation of information circulation, which is the content of databases, as well as the rules that determine the database as a result of intellectual activity. Relations in the area of information circulation on the territory of the Russian Federation are regulated by Federal Law No. 149-ФЗ dated July 27, 2006 "On Information, Information Technologies and the Protection of Information" (hereinafter referred to as the "Federal Law No. 149"). Part 1 of Art. 1 establishes the scope of this Federal Law:

(1) exercising the right on search, reception, transmission, production, and distribution of information;
(2) using information technologies;
(3) ensuring the protection of information.

All three areas of legal regulation of Federal Law No. 149 are closely related to the rules regulating the content of databases. In particular, as defined by these rules, information technologies are processes, methods of search, collection, storage, processing, provision, distribution of information and ways of fulfilling such processes and methods. Thus, systematization of information within the database should rely on particular methods; and processing, in general, allows for the combined application of the legal rules on information and the Civil Code of the Russian Federation. In addition to that, Clause 2 of Art. 1 of the Federal Law No. 149 specifies that the provisions of the law on information are not applicable to relations arising from the legal protection of the results of intellectual activity and equivalent visual identities. Statutory ban on the legal protection of intellectual property by legal means enshrined in the law on information indirectly confirms that copyright law protection of databases can be reduced to the protection of the algorithm for information location, i.e., to the form of its presentation but not to the information constituting its content. This allows the use of rules on copyright objects to the legal protection of the results of circadian genes research. Besides, in the relations arising in relation to the database, often appear the term "information" or its derivatives. In particular, Art. 1274 of the Civil Code of the Russian Federation explicitly specifies possible free use of the work (which include databases according to Article 1260 of the Civil Code of the Russian Federation) for informational, scientific, educational or cultural purposes. In other words, we are talking about the legal regulation of the search, reception, and dissemination of information contained in the database. Since academicians often publish scientific results, this approach seems to be relevant. We should note that the terms "data" and "information" are different. Data is the recorded signals from objects of the surrounding world. Information is data processed by appropriate methods that make a new product. Thus, information emerges and exists at the moment of the interaction of objective data and subjective methods. Like any object, it has properties that distinguish it from other objects of nature and society. Fundamentally, information is influenced both by the properties of the data that make up its content and the properties of the methods that interact with the data during the information process. Upon completion of the process, the properties of the information are transferred to the properties of the new data, i.e., the properties of the methods can be transferred to the properties of the data. Concerning the results of genomic research, there will be scientific data processed by academicians.

4 Conclusion

Determining the legal regime of genomic research will eliminate hindrances on the way to the development of new sectors of the digital economy. With that, the study of some areas of genomics, in particular, circadian genes, necessitates the improvement of intellectual property legislation and selection of suitable approaches to the use of current legal regimes for the quick introduction of research results into civil commerce. At the same time, systematic adjustment of such legal regimes will contribute to their optimization to effectively introduce the results of genomic research into commercial practice.

Within the transition to a digital economy and required optimization of the legal regime for the results of genomic research we propose:

- a legally enshrined admissibility of application of the legal regime of production secrets (know-how) for the practical introduction of the results of genomic research, as well as established limits of the use of such a legal regime and the criteria for its use in the corresponding area;
- the specified legal regime of databases, which is admissible to use as objects of information deposit collected in the course of the genomic research;

The reported study was funded by RFBR according to the research project No. 18-29-14073.

References

Inyushkin, A.A., Kryukova, E.S., Povarov, Y.S., et al.: The Russian information systems of the housing and utilities sector: peculiarities of legal regulation and application. Advances in Intelligent Systems and Computing, vol. 622, pp. 227–233 (2018a)

Singh, A., Das, S., Wilson, N.: Genomics and IP: an overview. J. Intell. Prop. Rights **12**(1), 57–71 (2007)

Chang, J., Zhu, X.: Bioinformatics databases: intellectual property protection strategy. J. Intell. Prop. Rights **15**(6), 447–454 (2010)

Bachilo, I.L.: Information Law: Higher Education Textbook, 238 p. Higher Education, Yurait-Izdat (2009)

Voinikanis, E.A.: Intellectual property law in the digital era: a paradigm of balance and flexibility. Jurisprudence, 208 p. (2013)

Gavrilov, E.P.: The development of intellectual property law in Russia. Patents and Licenses, no. 1, pp. 21–25 (2019)

Eremenko, V.I.: Some problems of codification of legislation on intellectual property. Legislation, and Economics, no. 2, pp. 37–48 (2014)

Ivardava, L.I.: Changes in the scope and limits of legal regulation within digital economy. Business Security, no. 1, pp. 39–47 (2019)

Inyushkin, A.A.: Information in the system of objects of civil rights and its relation with intellectual property on the example of databases. Information Law, no.4, pp. 4–7 (2016)

Inyushkin, A.A.: Legal Regime of Civilian Law Transactions. Information law, no.1, pp. 45–48 (2018b)

Mikhailov, A.V.: Issues of the digital economy establishment and the development of business law. Topical Issues of Russian law, no. 11, pp. 68–73 (2018)

Mokhov, A.A., Yavorskii, A.N.: Genes and other gene-based formations as objects of intellectual property law. Civil Law, no. 4, pp. 28–32 (2018)

Ruzanova, V.D.: Obligations related to the course of business by parties: issues of differentiation of legal regulation. Civil Law, no. 4, pp. 18–20 (2018)

Vaipan, V.A.: Legal regulation of the digital economy. Entrepreneurial law. The supplementum to “Law and Business”, no. 1, pp. 12–17 (2018)

Kuznetsov, P.U.: An integrated approach to the legal regulation of public relations in the field of the digital economy. Russian Juridical Journal, no. 6, pp. 154–161 (2018)

Risks of Innovative Projects: An Expert Review

Dmitry Yu. Treshchevsky[1(✉)], Galina N. Franovskaya[1], Maksim O. Gladkih[1], and Nina Yu. Treshchevskaya[2]

[1] Voronezh State University, Voronezh, Russia
treschevsky@gmail.com, fgn.vrn@mail.ru, gladkih_maksim3@list.ru

[2] The Russian Presidential Academy of National Economy and Public Administration, Moscow, Russia
treshchevsk@mail.ru

Abstract. The purpose of the study is to assess the risks of innovative projects. Methodology. We employed an expert method for assessment of innovative project risks. The experts determined 24 types of risks and classified them into five groups: technical and technological risks, market risks, risks related to funding and arrangement of financial activities, HR risks, and institutional risks. Each risk was assessed according to a six-point scale: 0 points – no risk; 1 point – minimal probability/severity; 2 points – low probability/severity; 3 points – average probability/severity; 4 points – high probability/severity; 5 points – critical probability/severity. For the assessment of different expert opinions, we calculated a coefficient of variation, which allows judging the level of average values' importance. Results. The experts mainly assess the probability and severity of risks arising during the development and fulfillment of innovative projects as average. HR and market risks have the highest values. Only one risk (the lack of demand for products) is assessed by experts as high in terms of severity and average in terms of probability. The following risks are recognized the least severe: equipment downtime; manufacturing inconsistencies; environmental risks; legal and accounting risks; theft of invention, industrial espionage; interference of outside organizations in the implementation of innovative projects. The experts have different opinions on the severity and probability of risks.

Keywords: Innovative project · Risks of innovative projects · Expert review of innovative risks

JEL Code: F01 · O30

1 Introduction

The problem of innovative project risks is one of the most topical in the academic literature. For Russia and Russian regions, the urgency of the problem is primarily associated with low rates of innovative development, which deteriorates the prospects for advanced technical and technological changes (Popkova and Sergi 2018). Researchers deal with various aspects concerning the identification of such risks. They pay attention to

E. G. Popkova and B. S. Sergi (Eds.): ISC 2019, LNNS 87, pp. 591–598, 2020.
https://doi.org/10.1007/978-3-030-29586-8_68

the following features: impersonal nature; environmental uncertainty; availability of alternative solutions; the presence of a personal component of risk assessment and acceptability; adjustability; interdependence of risk and degree of project novelty (Agafonova and Chelak 2006). Innovation risks are associated with the behavior of managers engaged in project implementation (Kolobov 2016). Some researchers take notice of information incompleteness; possible errors in risk factor assessment; conflict of interests of the project participants; multi-criterion problems (Treshevsky and Tabachnikova 2016). Researchers of the Voronezh Economic School consider it necessary to assess risks with regard to the institutional features of the environment and the social framework of the expert community (Nikitina and Tabachnikova 2014; Tabachnikova 2017; Risin et al. 2016; Tabachnikova 2016; Risin et al. 2017; Treshchevsky et al. 2018).

2 Methodology

Based on the features of innovative projects, which consequences both in the event of their public rejection and in the course of the implementation, are hardly predictable, we consider it necessary to assess their risks in accordance with the opinions of the expert community. Expert assessment is traditionally carried out based on questionnaires or in-depth interviews. Public opinion polling as one of the methods to assess socio-economic processes is unacceptable here because people who are not engaged in the development and implementation of innovative projects cannot judge the nature and degree of risk. The interview questionnaire in this situation does not seem the right way to know the experts' opinions since the interviewer inevitably suggests to the interviewee a particular vision of the problem.

To make a risk assessment of innovative projects more impersonal, we stated provisions that determine the nature of risk. The provisions were stated by various experts individually. Then the next experts supplemented the provisions stated. As a result, we collected an array of 24 provisions characterizing the risks. In avoidance of an unintentional influence on the opinion of each expert, we did not classify the questions by the nature of the risks. Summarizing characteristics like "technical and technological risks," "market risks" et al. were stated after the collection of questionnaires. In this regard, in Table 1 the numbering of the provisions presented in the questionnaires remained the same.

The experts were developers, investors, initiators, consumers, partners, consultants.

The questionnaire offered to assess the risks of innovative projects on their probability and severity. Rating scale: 0 points – no risk; 1 point – minimal probability/severity; 2 points – low probability/severity; 3 points – average probability/severity; 4 points – high probability/severity; 5 points – critical probability/severity. The assessment of the risk probability in this situation is non-traditional. The probability is typically calculated in unit fractions. However, this technique is applied to recurring events that require the use of probability theory methods. The essence of innovative projects is fundamentally different from recurring events, therefore the probability theory apparatus is not applicable. In this regard, for assessment of probability, we applied the same six-point scale as for the assessment of severity. The scale from 0 to 5 is acceptable for use due to its proper fitting to the human perception of events in the range from no hazard to its critical level. In the future, such a scale makes it possible to use the apparatus of the fuzzy-set theory for assessment of innovative project risks.

For assessment of opposite expert opinions, we calculated coefficients of variation that allow determining the level of importance of average values.

3 Results

The results of questionnaires processed on innovative project expert risk assessment are presented in Table 1.

Table 1. Assessment of the probability and severity of the risk

Ser. no.	Risks	Probability of risk (point average)	The severity of risk (point average)	Coefficient of variation by risk probability (%)	Coefficient of variation by risk severity (%)
Technical and technological risks					
8	A shortfall of required equipment	2.500	3.231	44.18%	29,44%
9	Equipment downtime (breakdown, lack of machine-tool attachment)	2.654	2.923	41.20%	34,78%
10	Manufacturing inconsistencies	2.800	2.880	37.17%	39,23%
11	Shortage of accessories and tools, failures in their supply	2.731	3.038	30.30%	28,66%
18	Loss of access to raw materials	1.962	3.346	44.40%	34,79%
13	The underestimated complexity of works and subsequent inability to adequately perform the target project	3.115	3.846	40.93%	29,13%
7	Lack of qualified maintenance	2.680	3.000	36.88%	27,22%
21	Environmental risks associated with the development and implementation of innovation projects	2.269	2.846	45.89%	45,21%
The average value for the group of technical and technological risks		**2,589**	**3.139**	**40.12%**	**33.56%**

(continued)

Table 1. (*continued*)

Ser. no.	Risks	Probability of risk (point average)	The severity of risk (point average)	Coefficient of variation by risk probability (%)	Coefficient of variation by risk severity (%)
Market risks					
1	Unreasonably high assessment of the sales market	3.308	3.577	32.86%	29,77%
2	Lack of demand for products	3.077	4.077	37.82%	30,97%
3	Wrong assessment of competitors and their unpredictable advantage	3.115	3.346	24.58%	26,65%
4	The launch of the product of another company with similar performances	3.038	3.000	36.63%	41,10%
5	The appearance of developments replacing your technologies on the market	3.231	3.538	30.71%	26,79%
The average value for a group of market risks		**3,154**	**3.508**	**32.52%**	**31.05%**
Risks related to funding and arrangement of financial activities					
14	The volatility of the foreign exchange market and consequent rise in prices for import equipment and accessories	3.308	3.346	23.83%	31,56%
16	Increasing prices for raw materials, energy, and water supply	3.308	2.654	35.02%	43,87%
23	Risks related to refunding due to the breach of financing terms	2.923	3.808	37.38%	28,79%
20	Blocking of a company's settlement account in a commercial bank even with a high rating due to bankruptcy or resolution	2.654	3.462	49.92%	41,04%

(*continued*)

Table 1. (*continued*)

Ser. no.	Risks	Probability of risk (point average)	The severity of risk (point average)	Coefficient of variation by risk probability (%)	Coefficient of variation by risk severity (%)
17	Legal and accounting risks related to commercialization of products	2.808	2.615	36.35%	37,58%
The average value for a group of risks related to funding and arrangement of financial activity		**3,000**	**3.177**	**36.50%**	**36.57%**
HR risks					
6	Shortage/turnover of qualified staff	3.115	3.538	30.56%	35,95%
The average value for the group of staff risks		**3,115**	**3.538**	**30.56%**	**35.95%**
Institutional risks					
12	Complicated procedure for issue of protective documents (patents, warrants, certificates, etc.)	2.923	3.000	49.27%	43,20%
15	Theft of invention, industrial espionage	2.346	2.846	43.37%	40,60%
19	Risks associated with the acts of supervisory institutions and authorities	3.000	3.115	45.22%	45,69%
22	Public objections related to the development and implementation of innovative projects (mineral resource mining, nuclear energy, petrochemicals, etc.)	2.308	3.077	55.86%	49,44%
24	Interference of outside companies in the implementation of innovative projects	2.692	2.731	51.22%	51,33%
The average value for a group of institutional risks		**2,654**	**2.954**	**48.99%**	**46.05%**

Based on the point scale accepted for assessment of risk probability and severity, we can judge on technical and technological risks by average values obtained: their probability is generally low; severity of this risk group is average; underestimated complexity of works on an innovative project is the most probable and severe in this group; loss of access to raw materials is the least probable and average in terms of risk severity; manufacturing inconsistencies are the least severe risks and at the same time the least probable.

Market risks have a higher probability and severity than technical and technological ones; nevertheless, their values are not higher than average; lack of demand for innovative products is of high severity and average probability; none of the risks on the probability and severity are lower than average.

Risks of funding: have an average level in terms of probability (higher than the technical and technological ones, but lower than the market ones); none of the risks reaches a high level. Experts consider the risk of volatility of the foreign exchange market and the associated increase in the price for import equipment and accessories to be the highest in probability; but risks related to refunding due to the breach of financing terms – the highest in severity.

Staff risks are assessed as average both in terms of probability and severity. They are lower than the market ones in both performances.

The fact of the least importance of institutional risks is of interest. On average, their influence in terms of probability and severity is assessed as low.

Among institutional risks, ones associated with the acts of supervisory institutions and authorities are the highest in terms of probability and severity.

Experts suppose that risks of public objection against the development and implementation of innovative projects are low-probable; however, if these events occur, their severity is average. In this group of risks, the severity is mitigated only by the acts of supervisory institutions and authorities.

Particular attention should be paid to the assessment of average values based on the coefficient of variation. The mentioned coefficients are very high: they range from 23.8 to 55.8%. Only in two parameters, the variation coefficients do not reach the limit of 25%, which shows moderate variability: wrong assessment of competitors and volatility of the foreign exchange market. The highest variability is noticeable for the group of institutional risks (over 45% on average).

Expert opinions differ to the greatest extent (nearly 50%) concerning the probability of blocking a settlement account due to various circumstances; complicated procedure for issue of protective documents. Even more significant disagreements we see in the assessment of the probability and severity of public objections and interference of outside companies in the implementation of innovative projects.

4 Conclusion

The study carried out allows making the following conclusions.

The experts mostly assess the probability and severity of risks arising in the development and implementation of innovative projects as average. Market and HR risks have the highest values. In both cases, the severity of risks is assessed higher than the probability. The average values of risk assessment for groups are almost the same.

Only one risk (the lack of demand for products) is assessed by experts as high in terms of severity and average in terms of probability. Assessment of severity of underestimated complexity of works and refunding due to a breach of financing terms is close to a high level.

The group of institutional risks has the lowest values in terms of probability and severity.

Equipment downtime, manufacturing inconsistencies, environmental risks, legal and accounting risks, theft of invention, industrial espionage, interference of outside organizations in the implementation of innovative projects are the least severe among particular risks belonging to different groups.

The disagreement of expert opinions on the severity and probability of various risks should be assessed as high. Consistency of expert opinions is satisfactory only for two risks: wrong assessment of competitors and the volatility of the foreign exchange market. Parameters of institutional risks turn out to be the most controversial.

References

Agafonova, I.P., Chelak, S.L.: Criteria for choosing the forms of innovative projects funding and minimizing their risks. Digest-Finance, no. 1 (2006). https://cyberleninka.ru/article/n/kriterii-vybora-form-finansirovaniya-innovatsionnyh-proektov-i-minimizatsiya-ih-riskov. Accessed 20 Jan 2019

Kolobov, D.S.: Causes of uncertainty in management decision-making. Economics and modern management: theory and practice: college article In: Proceedings of the 58th International Research and Practice Conference, vol. 2, no. 56. SibAK, Novosibirsk (2016). http://www.management-service.ru/article_08.php. Accessed 20 Jan 2019

Nikitina, L.M., Tabachnikova, M.B.: An empirical study of the content characteristics of social projects (according to in-depth interviews). J. Manag. Stud. NY **2**(1), 1–9 (2014)

Popkova, E.G., Sergi Bruno, S.: Will industry 4.0 and other innovations impact Russia's development? In: Sergi, B.S. (ed.) Exploring the Future of Russia's Economy and Markets: Towards Sustainable Economic Development, pp. 51–68. Emerald Publishing, Bingley (2018)

Risin, I.E., Treshchevsky, Yu.I., Tabachnikova, M.B., Plugatyreva, A.T.: Big business on the opportunities and threats to the development of a region. Socio-Econ. Phenom. Process. **11** (11), 65–71 (2016)

Risin, I.E., Treshchevsky, Y.I., Tabachnikova, M.B., Franovskaya, G.N.: Public authorities and business on the possibilities of region's development. In: Popkova, E. (ed.) Overcoming Uncertainty of Institutional Environment as a Tool of Global Crisis Management. Contributions to Economics, pp. 55–62. Springer, Cham (2017)

Tabachnikova, M.B.: Assessment of economic pessimism of regional institutional groups. Reg. Syst. Econ. Manag. **1**(32), 96–102 (2016)

Tabachnikova, M.B.: Prospects for the implementation of socio-economic development projects assessed by regional institutional groups. Mod. Econ. Prob. Solut. **4**(88), 123–133 (2017)

Treshchevsky, Y.I., Voronin, V.P., Melnik, M.V., Sokolov, A.A.: Analysis of risks of forecasted changes with the help of fuzzy logic elements. In: Endovitsky, D.A., Popkova, E.G. (ed.) Management of Changes in Socio-Economic Systems, vol. 135, pp. 81–93. Springer, Cham (2018)

Treshevsky, Yu.I., Tabachnikova, M.: Forecasting of social projects risks. Russian economy: an outlook. In: Collection of Proceedings of the 2nd International Research and Practice (Virtual) Conference, pp. 635–646. Publishing House of Tambov State University named after G.R. Derzhavin, Tambov (2016)

Economic Analysis of Human Resources in the Digital Economy

Anna A. Fedchenko(✉), Olga A. Kolesnikova, Ekaterina S. Dashkova, and Tatyana A. Pozhidaeva

Voronezh State University, Voronezh, Russia
faal7ll@yandex.ru, oakolesnikova@mail.ru, dashkova-82@mail.ru, pozhidaeva_ta@econ.vsu.ru

Abstract. The purpose of the study is to substantiate the peculiarities of the application of human resource economic analysis within a digital economy and reason the directions of use of the results obtained.

Methodology. We propose the principle of consistency as the key one since the objects of economic analysis should be considered with regard to their full scope, scientific character of analysis tools, impersonal nature of the information, and purpose orientation. In this respect, the economic analysis of human resources is considered as a systemic economic analysis. We reason that the most critical functions of the economic analysis of human resources are the assessment of their status and development forecasting. We prepared tools to assess the integration of results on human resource status at different administrative levels. We recommend an algorithm for the impact of economic digitalization on the transformation of human resources. Results. We revealed the leading causes of the transformation of human resources from the synthesis of the results of economic analysis. Recommendations. We advise using the results of economic analysis for the current assessment and scenario forecasting of human resources at all administrative levels.

Keywords: Economic analysis · Human resources · Digital economy · Integrated assessment · Scenario forecast

JEL Code: J23 · C10

1 Introduction

The present stage of socio-economic development is characterized by dynamic changes in all areas of activity. The digital economy is the main initiator of changes. Under the Program adopted in the Russian Federation, the digital economy embraces all types of economic activity. At the same time, digital data is a crucial production factor, which facilitates the decision-making mechanism and increases the importance of economic analysis. Thereat, digital measurement is considered exclusively as a tool to fulfill the ideas at all levels of human resource management.

E. G. Popkova and B. S. Sergi (Eds.): ISC 2019, LNNS 87, pp. 599–605, 2020.
https://doi.org/10.1007/978-3-030-29586-8_69

2 Methodology

Human resources possess a combination of qualities and characteristics of human, staff, society, which predetermines the potential ability to work in particular socio-economic conditions. They have both quantitative and qualitative characteristics, and territorial belonging, therefore, can be analyzed from different points of view. In addition to that, it is necessary that these characteristics meet the selection criteria proposed by the Organization for Economic Cooperation and Development, namely: importance and effectiveness for use, analytical nature, measurability. The listed criteria are considered as a guideline that supposes partial fulfillment of requirements.

The economic analysis of human resources is based on the following principles: scientific character, complex nature, consistency, impersonal nature, purpose orientation (Fedchenko 2009). We determine the principle of consistency as a crucial one since the objects of economic analysis should be considered with regard to their full scope, scientific character of tools, impersonal nature of information, and purpose orientation. The concept of systemic analysis as a scientific method of cognition, which is a sequence of acts to establish structural links between the system elements under study is inextricably related to this principle, (Korogodin 2005, Popov et al. 2002). In this regard, we consider the economic analysis of human resources as a systemic economic analysis.

The essential functions of the economic analysis of human resources are the assessment of their status and development forecasting (Fedchenko 2013). The status of human resources means the use of various assessment tools at the regional and intra-company levels. Within the company, the list of indicators is determined by the specific nature of the sector, peculiarities of the area of activity and its purposes. At the municipal and regional levels, one should account for the development of human capital and investment attractiveness. The method to obtain an integrated assessment of human resource status is also distinguishing at different administrative levels.

In our opinion, the integration method is the most acceptable to the intra-company level. It involves the addition of quantitative assessments for each indicator characterizing human resources that are adjusted by their importance determined by the method of expert evaluation. Integration can also be carried out using weighted or geometric averages. For the same purpose, one can employ a grading scale, which allows comparing particular results and obtains their total value.

At the municipal and regional levels for an integrated assessment of human resource status one applies the methods that characterize the object of study without regard to other objects, or ones aimed at comparing the performances of the object under study with the maximum and minimum performances of total objects.

The application of an integrated assessment of the human resource status makes it possible to set the guidelines for future development within the digital economy. It has three levels, which under close interrelation affect the economic analysis of human resources:

The relationship between the types of human resource economic analysis and the levels of the digital economy is displayed in Table 1.

Table 1. The relationship between social and labor relations and the levels of the digital economy

The levels of the digital economy	Features of economic analysis types
The first level is economic markets and sectors	All types of economic analysis are used
The second level is platforms and technologies	Strategic economic analysis is used
The third level is an environment that creates conditions for the development of platforms and technologies, and effective communication between market actors and economic sectors	All types of economic analysis are used

The economic analysis of human resources is intended to make advanced changes determined by economic digitalization. It is achieved using the competency-based approach employed at the second level of the digital economy and expressed at the first and third levels. This approach is evidenced in the development of professional standards, in terms of the National Qualifications System regulated by Article 195.3 of the Labor Code (Russian Union of Industrialists and Entrepreneurs 2007). As the digital economy involves the transformation of knowledge, abilities skills of employees, the professional standard describing the activities, knowledge, and skills of a specialist will be adjusted and supplemented in the future.

The algorithm for the impact of economic digitalization on the transformation of human resources based on the results of economic analysis can be presented as follows:

1. The digitalization of the economy includes the use of competencies in assessing the level of professional behavior, which is reflected in professional standards.
2. Professional standards affect the demand on the labor market and the supply of labor that depends on educational institutions.
3. The demand and supply of labor are created in accordance with the available flexible forms of employment and flexible labor management.
4. Flexible social and labor relations imply the peculiar assessment of labor results.
5. The peculiar features of assessment of the results obtained, command of the required expertise, employment forms used and labor management have a combined effect on staff remuneration and population's welfare.

3 Results

In accordance with the theoretical and methodological provisions mentioned above we can highlight the leading causes for the transformation of human resources discovered from the results of economic analysis in the digital economy:

- the growing role of the social component in social and labor relations conveying the transformation of human resources;

- development of non-standard forms of labor relations associated with the blurring of lines between job and leisure and different value preferences of generations X, Y, Z;
- the commitment of education to the peculiarities of the labor market in the digital economy.

A growing role of the social component in social and labor relations conveying the transformation of human resources is based on higher demands to socialization, which foundations were laid in the Universal Declaration of Human Rights adopted by the UN General Assembly on December 10, 1948, and supported by the International Labor Organization (ILO). The basis of the ILO ideology is an idea of universal and stable peace owing to social justice (ILO, 2019). This idea has become exhaustively developed over the years of the ILO existence, which is evidenced by the present situation. Decent labor activity is the essential aspect of social justice; therefore the economic analysis of the social component in human resources manifested in social and labor relations is considered by us as a tool to fulfill the Decent Labor Concept. The Program of Cooperation between the Russian Federation and the ILO for 2017-2020 set priority directions based on decent work standards. In accordance with the standpoint of the Federation of Independent Trade Unions of Russia, decent labor standards are adjusted to Russian practice and act in four major areas: 7 standards in the area of wages, 5 standards in the area of citizens' employment, 6 standards in the area of social partnership, 2 standards in the area of social insurance. The use of these standards necessitates the economic analysis of indicators proposed for each of the standards adjusted to the particular type of activity and level of management, which makes it possible to estimate the progress of the standard introduction. The need for such an analysis is also determined by the Strategy for Socio-Economic Development of the Voronezh Region the general purpose hereof is to build a favorable environment for human life and activity and business development (Ministry of Economic Development of Russia 2018).

The development of non-standard forms of labor relations associated with the blurring of lines between job and leisure and different value preferences of generations X, Y, Z. The cause for the occurrence of these processes is mostly different nature of labor, enhancement of innovative and digital features. Labor resources as most of the human resources cover all generations (X, Y, and Z), which are called "baby boomers" with some extent of conditionality. Most of the modern labor resources are Z and partly Y generations. Non-standard forms of labor relations are manifested in the specific features of contract conclusion and employment terms. If the employment contract differs from the official terms of hereof and adopted standard, then this is deemed narrow interpretation of non-standard labor relations. If actual terms of employment are non-standard, then the interpretation is deemed extended. In non-standard employment, there are such forms as part-time employment, temporary employment, irregular employment, underemployment, over-employment, secondary employment, self-employment, implementation of activity on outstaffing (outsourcing, staff leasing), remote employment (Fedchenko et al. 2018). The role of remote employment within a digital economy increases dramatically. There is a need to reveal freelancers – free workers and private specialists with low dependence on the employer, who themselves choose the form of employment and their workplace.

We consider the transformation of employment forms expressed in the development of non-standard labor relations as a response to the challenges of the labor market. One of the consequences of this transformation is the emergence of informal employment tending to increase (Fedchenko et al. 2016). The main problem solved in the legalization of labor activity is related not to additional revenues to the budget but to the development of an understanding that social guarantees and future pension benefits depend on the form of employment.

The commitment of education to the specific features of the labor market within the digital economy is possible only in the event of using the results of a comprehensive economic analysis, which allows concluding that the non-compliance of the educational level of graduates from higher and secondary educational institutions with the requirements of the labor market is considerable. An essential requirement of the competitiveness of an educational institution on the market of educational services that provide its effective performance is a commitment to practice with regard to economic digitalization.

The results of the economic analysis of human resources are the background for the scenario forecasting, which makes it possible to solve the following main problems: distinguishing the key points of evolution of the object under study; making the qualitatively different scenarios of its development; comprehensive analysis and evaluation of each of the scenarios, the study of its structural features and the possible consequences of its implementation to elaborate particular plans and programs (Orlovsky 1981).

As the scenarios are made according to the purpose set, while being the basis for developing an action plan concerning the transformation of human resources and social and labor relations (Medvedeva 2016, Stuken 2008), we consider the standpoint of the authors who believe that the scenario is future outlook, which helps to examine the development of some process, reasonable (Kulba and Kononov 2004, Baeva 2012).

When building the long-term scenarios on the development patterns, some authors note such approaches as simulation modeling of economic processes and theoretical expert approach (Nazarenko and Zvyagintseva 2012, Endovitsky et al. 2017, Treshchevsky et al. 2018).

4 Conclusion

The potential of an individual is a fundamental category in the analysis of human resources. The research methodology includes methods for assessing the psychophysiological state of an individual, the level of his intelligence, education, professional behavior, the degree of tolerance and sensitivity of civil society. All the listed characteristics of human resources are expressed in the staff and any territory: local, region, country. In the study of human resources representing a group of individuals, it is important to take into account the synergy effect and their exposure to the impact of the labor market, statutory and regulatory framework, socio-economic and political situation. In these circumstances, there is a need to carry out both comprehensive economic analysis and social study.

The specified methods of analysis should be supplemented with a rationale for the scenario that is a model for the development of a social object provided a particular combination of determining factors, which contains the problem statement and a description of possible actions aimed at reducing the gap between the expected and the desired result.

The scenario development at the regional level is completed by a comparative analysis of the parameters of the developed scenarios and choosing the most preferable. The scenarios are developed on strategic economic analysis, identification of economic and mathematical dependencies, and expert evaluations.

Forecast scenarios include forecasting models that describe the probable directions of development with regard to the impact of the significant factors and plan of management actions aimed at improving the performance of socio-economic systems. Forecast scenarios make it possible to foresee in advance the risks arising from ineffective management action; therefore, it is recommended to use methods of expert evaluation to increase the extent of their relevance.

References

Baeva, N.B., Vorogushchina, D.V.: Mathematical Methods for Assessing and Building the Economic Potential of a Region, pp. 150–151. Publishing House of Voronezh State University, Voronezh (2012)

Endovitsky, D.A., Tabachnikova, M.B., Treshchevsky, Y.I.: Analysis of economic optimism of the institutional groups and socio-economic systems. J. Adv. Res. Law Econ. **8**(6), 1745–1752 (2017)

Fedchenko, A.A.: The role of economic analysis in assessing the level of regional development. Econ. Anal. Theory Pract. **21**, 16–21 (2009)

Fedchenko, A.A.: Economic Analysis as a Condition for the Implementation of the Decent Labor Concept. Modern Economy: Problems and Solutions, vol. 11, no. 47, 197–209 (2013)

Fedchenko, A., Dorokhova, N., Dashkova, E.: Flexible employment: global, Russian and regional aspects. World Econ. Int. Relat. **1**, 16–24 (2018)

Fedchenko, A.A., Kolesnikova, O.A., Dashkova, E.S., Dorokhova, N.V.: Methodological approaches to study of informal employment. J. Appl. Econ. Sci. **11**(7), 1281–1289 (2016)

Korogodin, I.T.: Social and Labor System: Issues of Methodology and Theory. PALEOTYPE, p. 25 (2005)

Kul'ba, V.V., Kononov, D.A.: Methods of developing scenarios of socio-economic systems. SINTEG, 296 p. (2004)

Medvedeva, T.A.: Enhanced Systemic Approach to Social and Labor Relations within Economic Globalization. TEIS, 288 p. (2016)

Nazarenko, A.V., Zvyagintseva, O.S.: Scenario forecasting of the development of socio-economic systems. Acad. J. Kuban State Agrarian Univ. **84**(10), 575–587 (2012)

Orlovskii, S.A.: Trouble Decision-Making with Fuzzy Source Information. Nauka, 208 p. (1981)

Russian Union of Industrialists and Entrepreneurs. Provision on professional standard approved by the Decree of the President of the Russian Union of Industrialists and Entrepreneurs No. РП-46 dated 28 June 2007. http://media.rspp.ru/document/1/d/a/dad700c6ffaebe34a845-fc1ccca2081c.pdf. Accessed 10 December 2018

Popov, V.M., Solodkov, G.P., Topilin, V.M.: Systemic Analysis in the Administration of Socio-Economic Processes, p. 501. SKAGS Publishing House, Rostov-on-Don (2002)

The ILO 2019. The ILO Century. Promoting social justice and decent work. https://www.ilo.org/100/ru/. Accessed 10 February 2019

The Ministry of Economic Development of Russia 2018. Strategy for Socio-Economic Development of the Voronezh Region until 2035. http://economy.gov.ru/minec/activity/sections/StrategTerPlanning/komplstplanning/stsubject/projects/201822053. Accessed 20 January 2019

Stuken, T.Yu.: Labor relations in Russia: status and behavior pattern of inequality, p. 224. Publishing House of Omsk State University, Omsk (2008)

Treshchevsky, Y., Risin, I.E., Korobeinikova, L.S., Gavrilov, V.V.: Management of changes of socio-economic systems: an economic analysis of the state and consequences of the systemic feature. Stud. Syst. Decis. Control **135**, 21–30 (2018)

Specific Features of Training of Law Makers with the Help of Remote Technologies

Dmitry A. Lipinsky(✉), Leyla F. Berdnikova, and Olga V. Schnaider

Togliatti State University, Togliatti, Russia
Dmitri8@yandex.ru, Bleylaf@mail.ru, Shnaider-o@mail.ru

Abstract. **This paper is aimed at** establishing the possibility of training of law makers with the use of remote technologies, and to identify positive and negative aspects of the virtual interaction of its participants in the educational environment. **Research methodology**: Analysis, synthesis, dialectics, systemic and complex approaches, deduction and induction techniques, formal legal and comparative legal methods were used in the process of research. **Results:** The need for the training of law makers has been justified. The possibility of the training of law makers with the use of remote technologies has been investigated. The main forms of training activity in the implementation of remote technologies have been identified. Advantages and disadvantages of the use of remote educational technologies in the training of law makers have been identified. **Conclusion:** The imperfection of regulatory documents is caused by the lack of specialists in this field. Hence, there is a need for professional training of specialists in the field of lawmaking, namely, law makers. At present, the lack of an occupational standard of a law maker which would conform to the current educational standards in the legal areas of training is a significant problem. In the modern context, the issues of both traditional and distance education are relevant and controversial. Experience has shown that remote technologies are the most advanced and cost-effective, which bears record to the possibility of training of law makers with their use.

Keywords: Educational environment · Professional training · Educational standards · Information flows · Remote technologies · Lawmaking · Laws lawmaking · Lawyer · Legal professions · Law maker

1 Introduction

At present, much attention in the legal general-theoretical and professional literature is paid to legislative gaps, contradictions, as well as imperfection of legislation. Most papers of a practical nature contain recommendations for improving legislation.

The imperfection of regulatory documents is caused by the lack of specialists in this field. In all likelihood, the persons who prepared them had no legal degree, or these lawyers were not specialized in the field of lawmaking. This situation is associated with a low level of legal mastery, non-compliance with regulatory acts of higher legal force, which results in significant disregard for the rule of law. Therefore, there is a need for professional training of specialists in the field of lawmaking, namely, law makers.

E. G. Popkova and B. S. Sergi (Eds.): ISC 2019, LNNS 87, pp. 606–611, 2020.
https://doi.org/10.1007/978-3-030-29586-8_70

The ability to draw up smart legal texts is presented as one of the core competencies of a professional lawyer. Daily activities of a lawyer include execution of statements of claim, queries, complaints, binding overs, notices of appeal and many other documents. The result as well as the effectiveness of institutional communication depend on how competently a document was drafted.

One of the main activities of law makers is the development of a well-reasoned text of a legislative act that is compatible with regulatory documents of various levels, which is confirmed by references to relationship to other regulatory documents [1]. This is a general framework of requirements. Typical forms of such manuals can be found, for example, on the website of the National Conference of State Legislatures of the USA [2]. In addition, high-quality written text comes out from the orderly or consistent thinking. The work on written text always involves the development of analytic abilities [3, p. 830].

Educational institutions of some countries are aimed at solving the issues of training of personnel in the field of creation of legislative texts. Some universities offer international master's programs (University of Ottawa, Canada, School of Advanced Study University of London, Great Britain) [4]. The universities developed the courses for learning to write normative legal documents in the bachelor's program, as well as extended education courses [5]. No legislative authority in Russia has such-like documents [6].

As is pointed out by Malko, A.V.: "under the conditions of legal reform, the legal policy, which can be understood as scientifically grounded, consistent and systemic activity, is increasingly moving to the forefront…" [7, p. 55].

There are two topical issues relating to the training of law makers at present. The first issue consists in the lack of an occupational standard of a law maker which would conform to the current educational standards in the legal areas of training. The second issue consists in the possibility of training of law makers with the use of remote technologies.

Innovation processes in the educational environment impose new requirements to the professional training of lawyers, define and identify the aim, scope of problems in the educational environment judging from the needs of students. The key requirements of the educational environment should include compliance with educational standards aimed at improving the quality of professional training of students.

2 Analysis and Discussions

The need for law makers is obvious at present. That said, we should accept the point of view of Baranov, V.M. regarding the fact that "high-quality education and obtaining a diploma of a "law maker" must be a powerful means of social mobility". This requires special and well-found "mobility tools" - benefits, privileges and immunities for those who have mastered the university program with "good" and "excellent" marks [8, p. 27]. In turn, the readiness of the system of state and municipal service to the emergence of new personnel is required as well. Fast changing conditions of the external environment force the specialists from various areas of activities to study virtually throughout their life. At present, the mental vocabulary that has been gained during studentship is not enough for a lifetime. Globalization and integration processes force people to rethink their goals, objectives, and content of education. Internet technologies are widely used in the educational system for the extension of competencies and the field of knowledge. It contributes to the ongoing training of specialists, making education more affordable at the same time. Thus, the arrangement of professional training of law students, law makers in particular, with the use of remote technologies, is a highly topical question.

Since we have identified the need for training of specialists in the field of law-making, the question arises whether current educational and occupational standards afford an opportunity to train specialists in this field.

Baranov, has distinguished a number of questions in his research paper. First, whether an individual educational standard of a law maker must be developed. Second, whether there is a need for a special occupational standard with detailed disclosure of competencies in the training of law makers [8]. In our opinion, such standard is certainly needed; in particular, there is also a need for new standards in all legal majors. Moreover, these standards must be developed by legal training experts who have wealth of experience in educational research activities. In addition, the members of the following organizations must take part in the elaboration of such documents, in particular: Association of Law Schools of Russia; Association of Russian Lawyers; Specialized Academic Methodological Association.

With a view to improving the quality of education and training of professionals who are able to work in a climate of knowledge-driven economy, the system of occupational standards independent from institutes of education is being introduced [9].

It is worth emphasizing that the emergence of occupational standards is closely associated with new educational technologies, arrangement of the educational process, vocationally-orientated training and academic performance rating technologies. The issues that are associated with the focus on innovation in education should be studies at theoretical, practical, and technological level.

All of the above is indicative of the relevant and controversial nature of learning for obtaining the degree of a law maker with the use of remote technologies.

At the current stage of development of educational system, remote technologies constitute a category that is used in a wide range of various educational programs and courses. These technologies can be used both for refresher courses and for the implementation of programs of higher professional education.

The following tools must be used in the process of implementation of remote technologies in the training of law makers: Internet, PCs, tablet PCs, software, email, phone etc.

Remote training acquires new possibilities with the expansion of Internet technologies [10].

At present, the use of remote educational technologies is fairly popular and is widely used in the majority of cases due to its convenience and mobility. This is because trainees can use remote technologies to choose convenient time for classes without detriment to their own timetable. It is remote technologies that substitute the need for regular attendance according to the set timetable.

Remote educational technologies are such educational technologies that are implemented with the use of information and telecommunications networks and via indirect (i.e. remote) interaction of trainees and trainers.

Remote educational technologies used in the training of law makers must include:

- multimedia;
- telecommunication;
- interdisciplinarity;
- interactive training methods.

In order to achieve efficient training of law makers with the use of remote technologies, the following kinds of training activity must be used:

- video lectures (offline);
- online lectures (Skype technologies);
- videoconference (offline, online);
- message boards, discussions;
- webinars (online workshops, trainings, practical studies);
- chat (video chat) - review sessions, current assessment, business games, final assessment.

We shall identify advantages and disadvantages of remote educational technologies.

The advantages of the use of remote educational technologies in the training of law makers are as follows:

- contributes to the arrangement of independent activities of trainees in the learning of disciplines;
- is aimed at learning on an individual basis. A trainee sets the time for studying course materials for herself/himself at his/her own discretion, depending on his/her abilities and potential;
- allows a trainee to independently choose the place and duration of classes;
- can be used regardless of the geographic location of trainees and educational institution;
- contributes to more profound independent work of students with different information sources;
- is aimed at ensuring the meta-cognitive activity of trainees in the remote training system.

Despite many positive aspects of the use of remote technologies in the training of law makers, one cannot but mention certain negative aspects, which include in particular:

- the lack of time in communication between a teacher and a student. This is due to the limited number of cumulative hours of a teacher, which makes it impossible for him/her to provide review sessions for students above than the allotted time;
- virtual communication scarcely ever leads to complete mutual understanding between the participants of the educational process. There are known cases in which a trainee failed to comprehend the demands of a trained regarding the performance of certain tasks;

- the need for a set of individual psychological conditions. Thus, remote technologies require strict self-discipline, since its result is directly dependent on independence, self-control and consciousness of a trainee;
- the need for continuous access to sources of information, technological infrastructure, which is not available to all students.

The use of remote educational technologies in the training of law makers is conditioned and dictated by a number of factors:

- immensity of the territory of our country;
- location of research and development centers and study centers in big cities;
- implementation of the needs of society with regard to the content and educational technologies according to their accessibility. Remote technologies allow people to study remotely and are defined by the accessibility of proposed forms of educational services.

In keeping with foreign technologies with regard to Distance Learning, Electronic Learning, Electronic Tutoring (E-learning, E-tutoring), Russian remote training becomes integrated with a number of terms characterizing the information and communication technologies in education, namely:

- computer technologies in education,
- multimedia learning,
- training based on web technologies.

The integration of information and communication technologies of training of students in various universities enlarged their scope, as well as their knowledge in an area they learn throughout the entire learning process, thereby improve their professional standards.

Remote technologies can be referred to as virtual, as a progression from possible to real. Virtual nature of modern remote technologies in learning is establishes correlation relationship of human nature between human potential and information technologies in the system of higher education where remote training is used.

It must be noted that professional potential of a person in a climate of information-oriented society is very significant; its formation and implementation is not limited to the framework of the educational paradigm inherent in the present form of civilization. Remote technologies in the system of higher education have substantial resources for identifying the potential of an individual. This point of view is supported by theories which consider a person as an important part of the information system, including information subsystems. These subsystems include:

- sensory subsystem,
- motor subsystem,
- cognitive subsystem,
- affective subsystem
- stylistic subsystem,
- axiological subsystem.

Remote technologies in the system of higher education are based on the apprehension of a person from the point of view of the hierarchy of information subsystems aimed at ultimate implementation of each of them.

3 Conclusions

In summary, it must be emphasized that for the Russian Federation, the issues of both traditional and distance education are relevant and controversial, whereas remote technologies are the most advanced and cost-effective at present. A separate issue consists in combining them into a unified system. Such a hybrid system is intended to use remote technologies along with conventional technologies.

This being said, the learning process with the use of remote technologies is viewed from the perspective of a disequilibrium dynamic system in which a trainee receives basic competencies allowing him or her to resolve various training situations with the use of synergetics methods. Synergetic paradigm has a potential feature which allows a trainee to simulate the behavior of complex systems and conditions of their existence. The use of synergetic paradigm was noted in social and humanities knowledge, which affords ground for using it as a didactic system for the provision of remote training in the system of higher education, in the training of law makers in particular.

References

1. https://ials.sas.ac.uk/study/courses/legislative-drafting-course. Accessed 10 May 2018
2. https://www.city.ac.uk/law/courses/continuing-professional-development/in-house-courses/clear-writing-and-drafting. Accessed 10 May 2018
3. Texas Legislative Council Drafting Manual January 2017 [Revised to update Appendix 7] Prepared by the Staff of the Texas Legislative Council Published by the Texas Legislative Council P.O. Box 12128 Austin, Texas 78711–2128. https://tlc.texas.gov/docs/legref/draftingmanual.pdf. Accessed 10 May 2018)
4. https://hls.harvard.edu/dept/opia/what-is-public-interest-law/public-interest-work-types/legal-writing. Accessed 10 May 2018)
5. http://www.ncsl.org. Accessed 10 May 2018)
6. Garner, B.A.: Garner on language and writing: selected essays and speeches of Bryan A. Garner. Includes a foreword by Ruth Bader Ginsburg; [illustrations by J.P. Schmelzer]. American Bar Association, 839 p. (2009)
7. Malko, A.V., Matuzov, N.I., and others.: Russian legal policy: theory and practice. M.: Prospect, 752 p. (2006). (In Russ.)
8. Baranov, V.M.: Legislative drafter as a profession. Vestnik Saratovskoy gosudarstvennoy yuridicheskoy akademii, 6(119), pp. 16–29 (2017). (InRuss.)
9. Methods of teaching disciplines in different forms of education: Collection of scientific articles. Ed. prof. Pospelova, V.K. Candidate of Geology-Min. N.N. Komissarova. M.: Financial Academy, 96 p. (2009)
10. Davletova, I.M.: Distance learning: advantages and disadvantages, development prospects. http://nsportal.ru/shkola/obshchepedagogicheskie-tekhnologii/library/2012/02/14/distantsionnoe-obuchenie

Objective Special Aspects of Legal Regulation of Promotion in Tourism Within the Context of Digital Tourism Development in Russia

Vladimir A. Zolotovskiy(✉), Marina L. Davydova, Yuriy A. Bokov, and Yevgeniy V. Stelnik

Volgograd State University, Volgograd, Russia
{zolotovskiy.azi,kmp,bokov,analitika}@volsu.ru

Abstract. Within the scope of identification of objective special aspects of legal regulation of promotion in tourism, the authors have essentially defined the promotion in tourism taking into account the main features of tourist activities and conditions of sales of the tourist product. Considering the intricate and complex nature of the tourist product, the authors have presented the description of promotional tools within a framework of analysis of the two groups: "above the line" - classic types of direct advertising on television, radio, territorial print media, indoor and outdoor advertising; "below the line" - promotional tools that are characterized by a deeper and often individual impact on the consumer. Consideration was given to, in particular, classic offline forms, tools and methods of promotion of the tourist product. Special attention was paid to digital promotion technologies.

It has been found that the regulation of relations regarding the promotion in tourism, in view of specific consumer characteristics of the tourist product (both commercial and territorial product), is complex and unstructured. Such characteristics is conditioned by the multisectoral content of tourism, vast variety of forms of promotion having different nature, promotion channels and objectives, as well as heterogeneous and ambiguous target audience. The special nature of relations in promotion of tourism conditions their regulation with the use of public and private legal methods.

Keywords: Legal regulation in tourism · Marketing technologies · Promotion in tourism · Digital technologies in tourism

JEL Code: K15 · K23 · Z30 · Z32 · Z38

1 Introduction

The use of information technologies is becoming an integral part of social life in the developing information society. The progress in digitization aimed at optimization has changed not only the process of production, promotion and sales of goods, but also completely transformed the form and content of individual kinds of activities. Considering this trend, several functions in the system of management of socially important kinds of activity in the form of public regulation are converted to the development of

E. G. Popkova and B. S. Sergi (Eds.): ISC 2019, LNNS 87, pp. 612–622, 2020.
https://doi.org/10.1007/978-3-030-29586-8_71

digital economy and general informatization, too. In this regard, a particular scientific interest is caused by the processes occurring in interdisciplinary areas of activity.

The digitization of the economy and social sphere exerted material influence on the tourist industry. The role of tourism in the global economy is constantly increasing. The modern tourist industry is one of the most high-yielding sectors in the global economy. The indicators of profitability growth dynamics in tourism exceed the indicators of profitability growth dynamics in other sectors. Global nature of tourism is emphasized by the yearly involvement of one million people in it and, accordingly, increasing competitiveness of tourist areas. The global growth of tourist arrivals and the related increase in a specific share of tourism in the global economy draw closer attention of the business, international tourist organizations and national tourist administrations. According to UNWTO forecasts presented at the 19 Session of the General Assembly in 2011, the annual tourist arrival within the framework of international tourism is expected to amount to about 1,8 billion people by 2030 (UNWTO 2011:5). Judging from the global increasing dynamics of tourist arrivals since 2017, as well as the general tourist statistics of domestic and international tourism in the Russian Federation which makes it possible to determine the share of the Russian tourist market in the global tourist market, it is expected that the share of the Russian market in such a challenging situation with foreign policy (having a negative impact on the development of tourism) will be about 84 million people.

Being one of the most dynamically developing system-wide intersectoral phenomena, uniting the economy, culture, sports, healthcare, transportation, services, etc., tourism like no other kind of activity embodies almost all digitization tools that being developed nowadays. Digital technologies in tourism concern a wide range of issues: optimization of the system of management of organizational and personnel processes of an enterprise of tourism and hospitality industry, the use of technologies for interaction with partners, improvement of control over the process of provision of specific services by third parties, enhancement of the system of promotion of the tourist product (unbundled service) and improving the mechanism of interaction with customers.

The issue of promotion of the tourist product emerges full blown in the competition for first place in the tourist market. In order to attract attention of a potential consumer to tourist services in modern digital conditions of the growth of global tourist market, there is a need for a systemic set of tools and methods of promotion which is created and coordinated not only by commercial participants of tourist relations (or their associations), but also national and local tourist administrations as institutional structures operating on a regular basis, having relevant powers, as well as predictable and planned budget (Deng et al. 2019; Zhang et al. 2015). Considering the material, financial and personnel resources required for the methodic work on the promotion of the tourist product, the state/government acts as one of the key participants of relations, directing its attention to the promotion of tourist resources both in international informational space and in the Russian Federation. According to Yegorova, E.S. and Kapezina, T.T., representation of government institutions in the web space is the most fast and convenient PR channel for the provision of information to the target audience. According to the authors, the most popular Internet PR tools for government institutions include: microsites created for a special project or event; web presentations of government institutions which are interactive directories or virtual websites available on the Internet (website of tourist attractions); webcasts (Yegorova and Kapezina 2017). At the same time, the state is not only directly involved in activities on promotion of the tourist product, but also determines the mode of its regulation. Certainly, taking into account almost absolute commercialization of the tourist market, the state is only one of its regulators. Private commercial subjects also participate in the definition of requirements and regulations for the provision of services via professional associations in tourism.

The strategic importance of tourism in addressing not only economic and political issues, but also a number of social and humanitarian issues, is emphasized in a number of policy papers (Decree of the Government of the Russian Federation 2014; Order of the Government of the Russian Federation 2014). Determining the role of the state in the development of tourism, the authors who developed the mentioned policy papers emphasize the promotion of the national tourist product at the Russian and international levels among the top-priority objectives.

Having regard to the above, the paper is aimed at identifying objective special aspects of legal regulation of promotion in tourism within the context of digital tourism development in Russia.

2 Methodology

The researchers have analyzed the scientific papers and the practice of management of promotion in tourism in real (tangible) and digital forms. Initially, approaches in the literature and regulatory acts to the definition of such concept as promotion were considered. After that, taking into account the interdisciplinary nature of tourism and special aspects of its development processes, as well as service implementation, a special interpretation of the promotion of tourism was created. The analysis of the theory and the practice of management of promotion in tourism that was performed on the basis of the use of the comparative method, made it possible to distinguish common and special forms and tools of promotion in tourism. A complex description of forms, tools and methods of promotion in tourism has been consistently presented. As a result, it has been found that when we define the tourist product in a broad sense, the subject of regulation consists in relations that emerge in the process of provision of information with a view to attracting attention to the popular tourist destinations. At the same time, the subject of regulation consists in the information about a tourist service, product, and tourist resources.

3 Discussion

The problem of management of a set of tools of efficient promotion of tourist areas and commercial tourist products is of particular interest in contemporary literature. In particular, certain attention was paid to the issues of organization, management and planning of the use of information in the context of the use of a set of marketing technologies in the external environment (Serdiukova and Gavrilets 2012). In addition to fairly sketchy definition of the place of traditional forms of promotion (Dudensing et al. 2011; Baldemoro 2013), special attention is paid to general characteristics of certain innovative digital forms of promotion of particular goods or products (Lui 2000). In addition, a comparative analysis of forms of promotion of tourist areas on the Internet is made with a view to identifying the most efficient mechanisms (Vetitnev et al. 2015). The authors emphasize the top-priority strategic importance of information support aimed at developing the image of a territory attractive for tourists (Vetitnev et al. 2015; Molina et al. 2010; Khoferikhter 2015; Zhang et al. 2015; Revenko 2016: 646–647).

When we turn to the state of knowledge of the issue of legal regulation of promotion, it is important to note the sporadic character of attention of the academic community to the topic in the context of research on the general issues of legal regulation in tourism.

The attempts to give consideration to legal regulation of promotion in tourism from an independent and special point of view are highly occasional and are primitive in their nature. For example, Astashenko, D.A. focuses his attention solely on legal regulation of advertising of tourist products on the Internet (Astashenko 2014). The author highlights general issues of the image promotion of website of a travel company and advertising of a tourist service as the key points. The researcher lays special emphasis on consumer attitudes that are formed and implemented in electronic environment of web pages of a tourist company. We believe that such an approach that has become conventional nowadays is more likely based on the outlook on tourism as a consumer industry, having no ground for the development of a special regulatory system.

In this regard, in most cases, the comments of researchers are reduced to general, oblique wordings that emphasize the lack of special aspects of legal regulation of promotion in tourism as a sector of provision of services (Volvach 2012).

Certainly, this methodologic reference point clearly limits the research potential, ignoring multisectoral and complex nature of tourist relations, including the sphere of private and public interests. However, it is quite obvious that such a condition of the state of subject knowledge can be attributed to the fact that the concept of social and cultural services itself, much as their independent nature from the perspective of the legal regulation system, appear to be ambiguous and highly controversial in contemporary science (Ivlev 2017).

The long-standing scientific approach provides an opportunity to the authors to appeal to general norms of advertising law and legislation on the consumer right protection. Considering the overall trends of development of marketing technologies, the issue of the mechanism of regulation of electronic (digital) and traditional advertising is still the most controversial one (Tabaksiurova 2017; Nikitina 2017). Thus, Mikhailov, S.V. and Lepetikova, I.Y. turn to the problem of provision of security in the system of statutory regulation of Internet advertising. The authors pay special attention to the issues of definitive nature, and place emphasis on long-range objectives of improvement of legislation in the field of concrete definition of subjects and objects of relations on the Internet, procedural and institutional aspects of placement dissemination of advertisement on the Internet, qualifying characteristics of promotional information on the Internet, as well as the need for the enhancement of the role of self-regulatory organizations (Mikhailov and Lepetikova 2016).

4 Results

Before we proceed to the consideration of the topic, it is necessary to definitize the subject of research. According to Article 1 of Federal Law No. 132 "Concerning the fundamental principles of tourist activities" (Federal Law 1996) "Promotion of the tourist product is a set of measures aimed at selling the tourist product (advertising, participation in specialized exhibitions, fairs, establishment of tourism information centers, publication of catalogues, brochures, etc.)".

In economic literature, promotion is viewed as a necessary part of the information campaign of a company, represented by a set of various marketing measures that are implemented in real and digital environment, aimed at formation of a concept of the product/service as such with the target audience, creation of interest to it, systematic usage of various marketing tools and forms for the stimulation and growth of demand, sales increase, increase in market opportunities for the sales of products, etc. (Perova and Zakirova 2015: 50; Simavonian 2015: 63–64; Lebedeva et al. 2018: 81; Freeman and Glazer 2015).

Based on the commercial effect, promotion in tourism should be perceived as a practice aimed at increasing the demand for travel. This vital task can be solved by means of the following mechanisms: developing an attractive image of the tourist area, including the provision of information about popular tourist destinations, level of services and conditions of stay, reasonable pricing policy, the creation of a positive and trustworthy image of the travel company. Given the formed needs of a sophisticated traveler, special role among the factors of tourist interest is played by the quality of service and unique character/attractiveness of tourist resources.

Certainly, special aspects of the form of promotion of the tourist product in its solely commercial sense, and as a tourist product of an area (a set of services, goods and objects in tourism within a particular territory), related to the task of promoting cultural and historical, sports, recreational, environmental and other resources, is determined by the nature of the tourist product. Tourist product as a set of services is characterized by several specific features: unsuitability for storage and preliminary demonstration of services in a tangible form; seasonal price fluctuations; statics of objects/resources; temporary disunity of sale and consumption of a service; spatial remoteness/lack of integration between the service consumer and a person selling this service. Given the fact that a consumer cannot assess the quality of the tourist product when buying it, he/she can only do it when directly consuming it being under the influence of subjective factors. Thus, the ambiguity of the factors of attractiveness of the tourist product, its qualitative specificity generate a need for visualization of promotion forms.

In spite of distinctive features of the tourist product and certain tourist services, methods and tools of promotion by virtue of their nature can be generally classified as universal tools of marketing technologies aimed at improving the competitiveness of products (Vasilyeva 2012: 195–196; Alananzeh et al. 2015: 198).

That being said, promotional tools may be conveniently divided into two conventional groups: "above the line" and "below the line". The first group includes classic types of direct advertising on television, radio, territorial print media, indoor and outdoor advertising. The second group is characterized by a deeper and often individual impact on consumers (McCabe 2008: 208–217; Morgan and Ranchhod 2009: 295; Middleton and Clarke 2001: 241–257): personal selling, various forms of demand stimulation, cross technologies (Arenasa et al. 2019), email marketing, trade marketing, exhibitions, special digital forms of promotion, PR (Sulaiman 2014: 503–510), territorial image development (Govers et al. 2007; Rozdolskaya et al. 2017: pp. 109–124), special tourist portals (Malenkina and Ivanova 2018: 204–233), passport of tourist areas, registers of legal entities engaged in tourist activities, tourism information centers, etc.

Taking into consideration the major objective of research, we should turn our attention to the general description of promotional tools, with the explanation of special aspects of some top-priority of them.

It appears that various advertising forms are the most widely used and traditional tools of promotion in tourism.

By virtue of reaching the widest audience, advertising in the media - on television, radio, in newspapers and magazines - still holds its own. Since advertising in newspapers and magazines is not targeted, most often it is presented in the modular form and contains a standard set of characteristics of the tourist product. A modular form that is supplemented by the deep emotional content by means of text, intonations, music and graphic images; accordingly, it is also used in audiovisual advertising on radio, television, in cinemas, cultural institutions, sales areas, as well as in travel companies, means of accommodation, and meal (in the form of advertising video information and slide films about activities, territories and events).

Print advertising is a promotional tool which is equally efficient. The main forms are as follows: company catalogues of tourist offers containing the relevant organized information about the tourist program, about all products being sold, etc.; brochures and booklets aimed at informing a tourist about particular services and offers; posters and promo leaflets.

A special place in the system of promotional tools is taken by outdoor and street advertising. Its major objective consists in the provision of primary and/or additional information to potential tourists/customers about the popular tourist product. This advertising implies the use of special forms and objects for its placement: transportation advertising, posters, advertising panels, rotating posters, party walls, electronic displays, neon signs, billboards.

The group of indirect forms of promotion includes tools of customized and non-personalized orientation. Personal selling technique is considered to be the most-coveted technique that is designed for achieving the maximum marketing effect in tourist practice (Promote tourism 2019). Its nature implies personal involvement of a consumer, potential tourist/customer in a verbal presentation of the tourist product with a view to selling it. A similar effect is expected from the entire array of tools which are included in the concept of stimulation of consumption: sales marketing, trade marketing, cross-promotions in the context of cross-marketing, loyalty programs, sales letters, etc.

PR is the most capacious form of indirect promotion of the tourist product amalgamating the interests of public and private parties to tourist relations (Yegorova and Kapezina 2017; Chulov 2015: 140–157; Sulaiman 2014: 503–510). The main purpose of "public relations" consists in the development and preservation of an attractive image of the tourist area, tourist product or an unbundled service. The following PR tools and methods can be used in promotion of the tourist product depending on the scope of problems and data transmission channels:

legal symbols (onomastic symbols of the tourist region, emblem, flag, cultural symbols);
interactive forms of communication – tourism information centers, press conferences, seminars, webinars, roundtables, festivals and competitions, exhibitions and fairs, or other tourist events of a global, national or regional scale, aimed at presenting the tourist product/special promotional products, or playing the role of "news opportunity", making it possible to advert tourist area, service, object of display or accommodation;
press-release presented in the form of a modular block of information dedicated to a tourist event, destination, object, or service;
documentary cinematography and special printed materials in the form of travel guides, dedicated to a particular tourist area which characterizes its resource specificity;
familiarization tours organized for the media representatives (media tours, blog tours), prospective investors and partners.

Certainly, not only classic offline forms, tools and methods of promotion of the tourist product appear to be relevant. Digital promotion technologies play a special part. The development of digital economy, information-oriented society and information and communication technologies provides an opportunity to overcome the spatial boundaries for direct communication between the consumer tourist and the service provider, to ensure virtually limitless provision of information to potential tourists/customers about special aspects of the tourist product, area and services (Makhmud 2015: 6, 8; Revenko 2016: 643; Morozov and Aristov 2017: 99). It is significant that tools and methods of the digital technology of promotion of the tourist product go beyond internet marketing, providing an opportunity to interact with the target audience via offline channels (POS terminals, software applications, etc.) (Lebedeva et al. 2018: 81–85; Schetinina 2018: 50).

The most popular tools of digital promotion of the tourist product in tourist practice include: SEO optimization technology (Schetinina 2018: 50), QR codes, electronic sales, contextual advertising, mass emailing, advertising in applications, targeted advertising, SMS newsletters, content marketing, SMO and SMM technologies (social networks), various websites and blogs (Rumiantsev 2016), webinars, video conferencing, viral advertising, social networks (Kurganskaya and Hofmann 2018: 55–59).

The increasing competition in the international tourist market escalates struggle for the consumer preference not only between commercial participants of the market, but also between territories themselves. In this regard, approaches to the complex promotion or tourism through the development and popularization of image of the territory become relevant. The nature of tourist destinations and the nature of the regional tourist product as a unique selling proposition determine the orientation of promotion on the popularization of culture-historical, sports, recreational, environmental and other resources of the territory.

In the context of development and mass adoption of digital technologies in various spheres of public relations, provides an opportunity to create conditions for almost limitless attraction of potential tourists, PR methods and forms emerge full blown in a set of promotional tools associated with promotion of the tourist product in the following areas: the interaction of a brand/tourist area with the print media on the Internet (primarily with the travel media, the interaction of a brand/tourist area and representatives of the tourist industry (automated system AIS "Tourism", unified tourist passport (http://utp.nbcrs.org/), Tourist Information Exchange System (http://nbcrs.org/), the interaction of a brand/tourist area with the society/unlimited range of consumers (subsystem of the AIS "Tourism" "National tourist portal Russia.travel", official portal "RussiaTravel.ru", tourist passports of regions, electronic forms of printed travel guides, specialized federal, regional and municipal tourist portals and registers (Bondarenko 2016; Gurkina and Zhulina 2016; Tourist Portal of the Volgograd region; State Autonomous Agency of the Volgograd Region "Agentstvo Razvitiya Turizma"; the Unified Federal Register of Tourism Operators; the Register of International Tourism Operators - Tourist Assistance; the Register of Travel Agencies – Tourist Assistance; The Register of Tourism Information Centers of the Russian Federation), official web sites of administrative units (Wu 2018: 165–166, 170), containing the information for tourists etc. It should be pointed out that the opportunity for virtually worldwide reach with digital tools of promotion predetermined promotion of the regional tourist product due to the practice in creation of multilingual promotional and informational materials (Tourist portal of the city of Volgograd and the Volgograd Region).

5 Conclusions

When we present the characteristics of special aspects of legal regulation forms of promotion in tourism, we should take into account their ambiguity and specifics of their maintenance. That said, forms of promotion in tourism should be defined in the broadest sense, since we consider that they are tools and methods aimed at promotion of a set of tourist services for accommodation and transportation, sold for the common price, as well as on promotion of the regional tourist product in general. It should be emphasized that when we define the tourist product in such a broad sense, the subject of regulation consists in relations that emerge in the process of provision of information with a view to attracting attention to the popular tourist destinations. At the same time, the subject of regulation consists in the information about a tourist service, product, and tourist resources.

The definition of special aspects of legal regulation of promotion in this methodologic field allows us to unlock the heuristic potential of forms, tools and method of promotion in tourism, including for outreach and educational goals, and resolves the problem of promotion of the tourist area and the tourist product.

Acknowledgements. The reported study was funded by RFBR and the government of Volgograd region according to the research project No. 18-413-342003.

References

Alananzeh, O.A., Amyan, M.M., Alghaswyneh, O.F.M.: Managing promoting tourism product of the Golden Triangle in Jordan. Int. J. Hum. Soc. Sci. **5**(9), 197–207 (2015)

Arenasa, A.E., Gohb, J.M., Uruecac, A.: How does IT affect design centricity approaches: evidence from Spain's smart tourism ecosystem. Int. J. Inf. Manag. **45**, 149–162 (2019)

Baldemoro, J.: Tourism promotion (2013). http://www.slideshare.net/JHBlue/tourism-promotion-28432196. Accessed 7 Apr 2019

Deng, T., Hu, Y., Ma, M.: Regional policy and tourism: a quasi-natural experiment. Ann. Tour. Res. **74**, 1–16 (2019)

Dudensing, R.M., Hughes, D.W., Shields, M.: Perceptions of tourism promotion and business challenges: a survey-based comparison of tourism businesses and promotion organizations. Tour. Manag. **32**(6), 1453–1462 (2011)

Freeman, R., Glazer, K.: Chapter 8: Services marketing. In: Introduction to tourism and hospitality in BC (2015). https://opentextbc.ca/introtourism/chapter/chapter-8-services-marketing/. Accessed 7 Apr 2019

Govers, R., Go, F.M., Kumar, K.: Promoting tourism destination image. J. Travel. Res. **46**(1), 15–23 (2007)

Lui, Z.: Internet Tourism Marketing: Potential and Constraints. University of Strathclyde, UK (2000). http://www.hotelonline.com/Trends/ChiangMaiJun00/InternetConstraints.html. Accessed 12 Mar 2015, 7 Apr 2019

Malenkina, N., Ivanova, S.: A linguistic analysis of the official tourism websites of the seventeen Spanish autonomous communities. J. Destin. Mark. Manag. **9**, 204–233 (2018)

McCabe, S.: Marketing Communications in Tourism and Hospitality, p. 320. Butterworth-Heinemann, Amsterdam (2008)

Middleton, V.T.C., Clarke, J.R.: Marketing in Travel and Tourism, p. 512. Butterworth-Heinemann, Oxford (2001)

Molina, A., Gómez, M., Martín-Consuegra, D.: Tourism marketing information and destination image management. Afr. J. Bus. Manag. **4**(5), 722–728 (2010). http://www.academicjournals.org/article/article1380715458_Molina%20et%20al.pdf. Accessed 7 Apr 2019

Morgan, M., Ranchhod, A.: Marketing in Travel and Tourism, 4th edn, p. 528. Butterworth-Heinemann, Amsterdam (2009)

Promote tourism: Promote tourism products and services. Trainee Manual (2019) http://waseantourism.com/ft/Toolbox%20Development%20III:%2098%20toolboxes%20for%20Travel%20Agencies%20and%20Tour%20Operations/Submission%20to%20ASEC/(Draft)%202nd%20submission_290415/TO%20&%20TA/Promote%20tourism%20products%20&%20services/TM_Promote_tourism_products_services_290415.pdf. Accessed 7 Apr 2019

Sulaiman, M.Z.: Translating the style of tourism promotional discourse: a cross cultural journey into stylescapes. Procedia Soc. Behav. Sci. **118**, 503–510 (2014)

UNWTO: UNWTO General Assembly 19th Session: Tourism towards 2030, 19 p. Republic of Korea, 10 October 2011

Wu, G.: Official websites as a tourism marketing medium: a contrastive analysis from the perspective of appraisal theory. J. Destin. Mark. Manag. **10**, 164–171 (2018)

Zhang, C.X., Decosta, P.L.E., McKercher, B.: Politics and tourism promotion: Hong Kong's myth making. Ann. Tour. Res. **54**, 156–171 (2015)

Astashenko, D.A.: Legal regulation of advertising of tourist products on the Internet, VI International Scientific Meeting of Students (2014). https://scienceforum.ru/2014/article/2014006789. Accessed 7 Apr 2019

Bondarenko, A.P.: The issues of import substitution in tourism: informational aspect. Serv. Plus **10**(4), 4–14 (2016)

Vasilyeva, M.V.: Formation of demands of tourists with the use of the “unique selling proposition” concept. Teor. Prakt. Serv. Ekon. Sotsialnaya Sfera Tekhnologii **3**(13), 195–200 (2012)

Vetitnev, A.M., Romanova, G.M., Serdiukova, N.K., Serdiukov, D.A.: The study of internet promotion of services of cultural and educational tourism in the South of Russia. Ross. Predprin. **16**(17), 2899–2914 (2015)

Volvach, Y.V.: Tourist services as a civil matter, Moscow, 128 p. (2012)

State Autonomous Agency of the Volgograd Region “Agentstvo Razvitiya Turizma” (2019). http://www.turizm-volgograd.ru/. Accessed 7 Apr 2019

Gurkina, E.N., Zhulina, M.A.: Tourist information portal as a tool of regional promotion of tourism. OGARIOV-ONLINE **1**(66), 10 (2016)

Yegorova, E.S., Kapezina, T.T.: Modern Internet advertising and PR tools for government institutions and business. Online academic periodical “Nauka. Obschestvo. Gosudarstvo”, vol. 5, no. (3) (19) (2017). http://esj.pnzgu.ru. Accessed 7 Apr 2019

Ivlev, S.V.: Sociocultural services as an object of legal regulation. Service in Russia and Abroad, vol. 11, no. 4 (74), pp. 64–75 (2017)

Kurganskaya, G.S., Hofmann, K.M.: Innovative Internet technologies in tourism. Biznes-Obrazovanie v Ekonomike Znaniy, no. 1, pp. 55–59 (2018)

Lebedeva, T.E., Shkunova, A.A., Slautina, M.S.: Promotion in the tourism market: a new solution. Innovatsionnaya Ekonomika: Perspektivy Razvitiya i Sovershenstvovaniya, no. 5 (31), pp. 81–85 (2018)

Makhmud, A.V.: The impact of modern technologies on the development of electronic tourism. Bulletin of the Kazan State University of Culture and Arts, pp. 4–12 (2015)

Mikhailov, S.V., Lepetikova, I.Y.: About certain aspects of development of legislative regulation of Internet advertising. Yurist-Pravoved, No. 1 (74), pp. 81–85 (2016)

Morozov, M.A., Aristov, P.O.: Methods of improvement of internet advertising technologies for improving the competitiveness of services in the field of restaurant and entertainment business. Kreat. Ekon. **11**(1), 99 (2017)

Nikitina, T.E.: Issues of legal regulation of relations associated with advertising on the Internet. Bulletin of the University named after O.EE. Kutafin, Vektor Yuridicheskoy Nauki, no. 9, pp. 81–92 (2017)

Perova, T.V., Zakirova, O.V.: Cross marketing in promotion of tourist services. Bulletin of Minin Nizhny Novgorod State Pedagogical University, no. 3 (2015). http://vestnik.mininuniver.ru/upload/iblock/e00/t.v.-perova1_-o.v.-zakirova2.pdf. Accessed 7 Apr 2019

Decree of the Government of the Russian Federation No. 317: “Concerning the Approval of the State Program of the Russian Federation “Development of Culture and Tourism” for 2013-2020”, Official web portal of legal information (www.pravo.gov.ru) of 24.4.2014, Article 0001201404240013; Collection of Legislative Acts of the Russian Federation of 2014, No. 18, Article 2163 (Part II) (2014)

Order of the Government of the Russian Federation No. 941-p: “Concerning The Approval of the Strategy for the Development of Tourism in the Russian Federation for the Period through to 2020”, Official web portal of legal information (www.pravo.gov.ru) of 9.6.2014 (No. 0001201406090016); Collection of Legislative Acts of the Russian Federation of 2014, No. 24, Article 3105 (2014)

Revenko, A.A.: Activities of network media resources for popularizing the cultural heritage of Russia in the context of globalization of the cultural space. Vopr. Teor. Prakt. Zhurnalistiki **5**(4), 641–653 (2016)

The Register of Travel Agencies – Tourist Assistance. https://reestr.tourpom.ru/search.php. Accessed 7 Apr 2019
The Register of Tourism Information Centers of the Russian Federation. http://www.nbcrs.org/tic/list.cshtml. Accessed 7 Apr 2019
The Register of International Tourism Operators - Tourist Assistance. https://www.tourpom.ru/touroperators. Accessed 7 Apr 2019
Rozdolskaya, I.V., Ledovskaya, M.E., Lysenko, V.V., Bolotova, I.S.: Informative designated purpose of congress and exhibition tourism in a new reality of regional development. Service in Russia and Abroad, vol. 11, no. 1, pp. 109–124 (2017)
Rumiantsev, D.: Promotion of the Development of Business in VKontakte, 400 p. Novyie Praktiki i Tekhnologii, St. Petersburg (2016)
Serdiukova, N.K., Gavrilets, G.Y.: Assessing the current trends of promotion of the tourist product. Bulletin of Sochi State University, no. 3, pp. 90–95 (2012)
Simavonian, A.A.: Tourism promotion and marketing. Research and Practice Journal "Upravleniye i Ekonomika v Dvadtsat Pervom Veke", no. 2, pp. 63–65 (2015)
Tabaksiurova, A.V.: Legal characteristics of ways of dissemination of advertising, Online academic periodical "Nauka. Obschestvo. Gosudarstvo", vol. 5, no. 2 (18) (2017). http://esj.pnzgu.ru. Accessed 7 Apr 2019
Tourist portal of the city of Volgograd and the Volgograd Region "Volgograd Krai – a territory for travelling". https://volgaland.volsu.ru/ru. Accessed 7 Apr 2019
Tourist Portal of the Volgograd region. http://www.welcomevolgograd.com/. Accessed 7 Apr 2019
Federal Law No. 132-FZ: "Concerning the fundamental principles of tourist activities in the Russian Federation" (as amended on 01.01.2019). Collection of Legislative Acts of the Russian Federation. 02.12.1996. No. 49. Article 5491; Rossiyskata Gazeta. 03.12.1996. No. 231 (1996)
Khoferikhter, N.A.: Strategic approaches to the promotion of destinations in the market of tourist services. Zhurnal Pravovykh i Ekonomicheskikh Issledovaniy, no. 1, pp. 181–186 (2015)
Chulov, D.A.: Branding and territorial development in tourism: interaction with the media and consumers. Service in Russia and Abroad, vol. 5 (61), pp. 140–157 (2015)
Schetinina, I.V.: The use of digital promotion technologies for improving the competitiveness of products. Ekonominfo **15**(4), 49–53 (2018)

New Interaction Models in Digitalization: The Sharing Economy

Irina B. Teslenko[1(✉)], Nadezhda V. Muravyova[1], Olga B. Digilina[2], Igor I. Saveliev[3], and Marina B. Khripunova[4]

[1] Vladimir State University named after A. G. and N. G. Stoletovs, Vladimir, Russia
iteslenko@inbox.ru, nemur@mail.ru
[2] Russian Peoples' Friendship University, Moscow, Russia
o.b.digilina@mail.ru
[3] Vladimir Law Institute of the Federal Service for the Execution of Punishments of Russia, Vladimir, Russia
sii-33@mail.ru
[4] Financial University under the Government of the Russian Federation, Moscow, Russia
marinakhripunova@rambler.ru

Abstract. The era of the digital economy opens up tremendous opportunities for improving the quality of human life. Information technologies have significantly simplified communication between people who are far from each other, have opened up the opportunity to quickly and inexpensively purchase goods, securely store large amounts of information, etc. All this contributed to the formation of a shared economy. The purpose of this study is to characterize and define the features of the business sharing model and its role in the conditions of digitalization. In the process of research, general scientific methods were used, such as: analysis, synthesis, logical and graphical modeling, and others. The article defines the concept of "sharing economy", identifies the main advantages of the new business model and its main elements, provides forecasts for the development of online services, as well as analyzes existing and possible problems of the development of the sharing economy and suggests some measures to solve these problems, including with the participation of government agencies.

Keywords: Sharing economics · Online service · Online platforms · Car sharing · Carpooling · Airbnb · Uber

JEL Code: D410 · M130 · M150 · O310 · O350

1 Introduction

The development of information technology opens up huge new opportunities for people. This is not only fast communication at a distance via email, mobile phone, social networks, instant messengers, not only the ability to buy goods, sitting at home from a computer through online stores, not only the ability to store a huge amount of information on compact media, etc. this and the formation of new models of human relationships. One of these is the sharing business model or sharing economy.

E. G. Popkova and B. S. Sergi (Eds.): ISC 2019, LNNS 87, pp. 623–630, 2020.
https://doi.org/10.1007/978-3-030-29586-8_72

The sharing of goods took place in ancient Greece. A variation of such an economy were collective farms in the USSR. An example of a shared economy (ESP) in the offline-world at present is the sale of used items in stores called Second-Hand (from the English - "second hand"). Libraries are also an obvious example of ESP in the offline world: it is the ability to read books without acquiring them.

The modern concept of ESP was proposed by economists R. Botsman and R. Rogers in the book "What's Mine Is Yours: The Rise of Collaborative Consumption" (2018) (Sabitov 2018). Their idea was that it was easier and more profitable for consumers to pay for temporary access to a product than to own it. And they were right.

Possession of an object means that you must first buy it, spend the money, then maintain and maintain it. The presence of often little-used things is costs that do not pay off, and organizing a place to store them also reduces an important personal space of a person.

Methodology. The object of the study of our article is the economy of joint consumption (English version - sharing economy), by which, in its most general form, we understand the exchange of goods and services between an unlimited number of people around the world without intermediaries. This is a new socio-economic model of interaction, which is based on a conscious preference for collective ownership instead of private ownership. Moreover, this choice is not associated with a lack of money, but with the desire to expand its capabilities and take advantage of the opening advantages.

In our work, we consider the formation of an economic sharing model from two sides: technological and value. As for the first, the emergence of platform technologies allows you to quickly and easily connect economic agents with each other. The essence of the value factor lies in the abandonment of established habits, in the new perception of many things, behaviors that previously seemed unshakable, and others.

Results. Thanks to the concept of sharing, a market has appeared even for those things that were never considered potentially profitable. A dozen square meters of driveway can make a profit through Parking Panda. A room in the house can become a dog house through DogVacay. And on Rentoid, for example, a lover of hiking can share a tent with a city dweller for $ 10 a day - and both will benefit. With a SnapGoods drill, idle in a garage also becomes a source of income (Heron 2019).

Not so long ago, the car was considered an indicator of human status, but now it is perceived by many as a means of transportation, which should be convenient to use. Unfortunately, problems have recently arisen with this: the number of cars on the roads is increasing, motorists are idle for hours in traffic jams, it has become difficult to find a parking space near an office or a shopping center, it's expensive to maintain your own car, etc.

These circumstances in the presence of developed information technologies led to the creation of special services for the new model of interaction. This, for example, Uber (application for ordering a taxi) or Carsharing (car sharing).

Already today, thousands of people in the world use the services of the economic model of sharing. Along with the above mentioned, for example, such as Airbnb - a rental housing service; BlaBlaCar - search for fellow travelers; eBay - online auction, etc.

Structurally, ESP includes markets where a variety of online services are provided. These include:

1. Carsharing - per minute car rental. By registering, through the mobile application, the user can find the car on the city map and book it. After that, this machine will be at the complete disposal of the consumer until the end of the lease term. Depending on the car booking service, it can be free for up to 10–15 min, and it can be up to 20 min.
2. Carpooling (from the English. Car - car, pool - association). Carpooling - sharing a private car with online search services for travel companions. Distances for travel using this service can be different: from several kilometers - a trip to the other end of the city, up to 2–3 thousand kilometers. If earlier this method was possible only for a trip with friends or acquaintances, now - with the help of Internet services, trips can bring together unfamiliar or completely unfamiliar people.
3. Short-term rental housing - apartments or houses are rented daily, and residential real estate is specially equipped for delivery in the high tourist season. With the help of various online services, a tourist can independently choose an apartment taking into account the price/quality ratio and book it through the app.
4. Office sharing is another version of the concept of sharing, in which companies owning vacant office space rent them to other firms or freelancers for a certain period. This is especially convenient in metropolitan areas. Office sharing removes problems such as capital costs, long-term liabilities, restrictions on growth opportunities and communication with other firms.
5. Ridesharing - joint trips by car, during which consumers share the costs among themselves. An example is the BlaBlaCar online service. Compared to taxis (Uber), ordinary people use this service, not professional drivers, their goal is not to make money, but to save money on the road by taking a travel companion with them. Currently, the service operates in 22 countries around the world.
6. Design for auction - this service involves placing on the site a design order indicating the amount of payment for the best project. The higher this amount, the more designers will participate in this auction. From the set of proposed options, the customer chooses a sketch that he liked the most, and pays for it. This platform is open, i.e. All works are visible, and beginners in design can gain experience by studying the work of competitors.

The pioneer concept of ESP was Airbnb, which launched an online platform for finding and renting private housing around the world for a short period. By 2016, the number of active users of the service has increased 10 thousand times. By the beginning of 2019, this online service posted ads for renting more than 2.5 million rooms, apartments and other premises in 191 countries.

In Russia, due to the fairly rapid spread of digitalization processes, ESP is also developing.

This is evidenced by the dynamics of the transaction volume on the main ESP platforms (Carsharing, Uber, Ebay, Airbnb).

Thus, in 2018, the country increased the volume of operations in the ESP markets by 30% compared with 2017. Figure 1 shows a diagram showing the growth in the volume of ESP transactions in Russia.

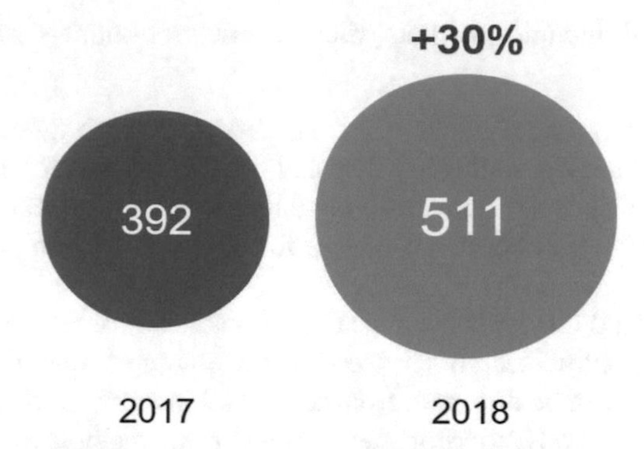

Fig. 1. Dynamics of the ESP market in Russia in 2017–2018, billion rubles

The largest share in the volume of ESPs is made up of electronic sales between consumers (C2C - Consumer-to-consumer) - 72%. Freelance services (19%), car sharing (5%) and rental of residential premises for a short period (2%) are far behind them.

Figure 2 shows the transaction volumes in Russia for the ESP markets for 2018.

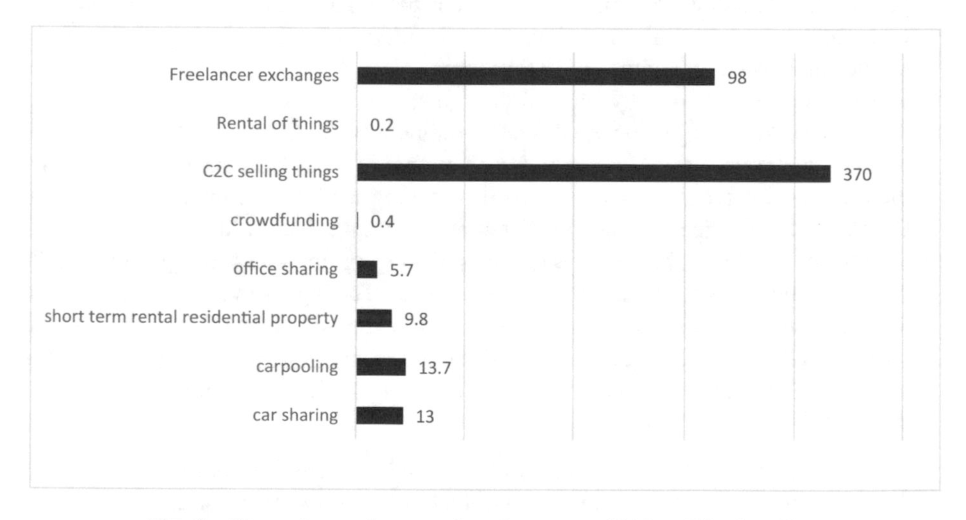

Fig. 2. The volume of transactions by sector, 2018, milliards rubles.

In the C2C market (sale of second-hand goods), 25% of all transactions are sales of new products. Among them, a little more than 20% of transactions are the sale of new things that “do not fit” to the seller, about 5% of sales are goods that were purchased specifically for resale.

According to the data for the Russian Federation: in the spring of 2017, the Airbnb service posted 42 thousand housing rental announcements in various cities of Russia (Li and Pastushin 2017). Most often, residential and non-residential premises were handed over in St. Petersburg - 4,000 transactions per year, in Moscow this figure was about 1.5 times less. In 2018, 400 thousand tourists took advantage of this service, which is 85% more than in 2017.

Russia entered the top 10 countries in terms of inbound tourism. Figure 3 shows the geography of tourists visiting Russia and using the Airbnb service.

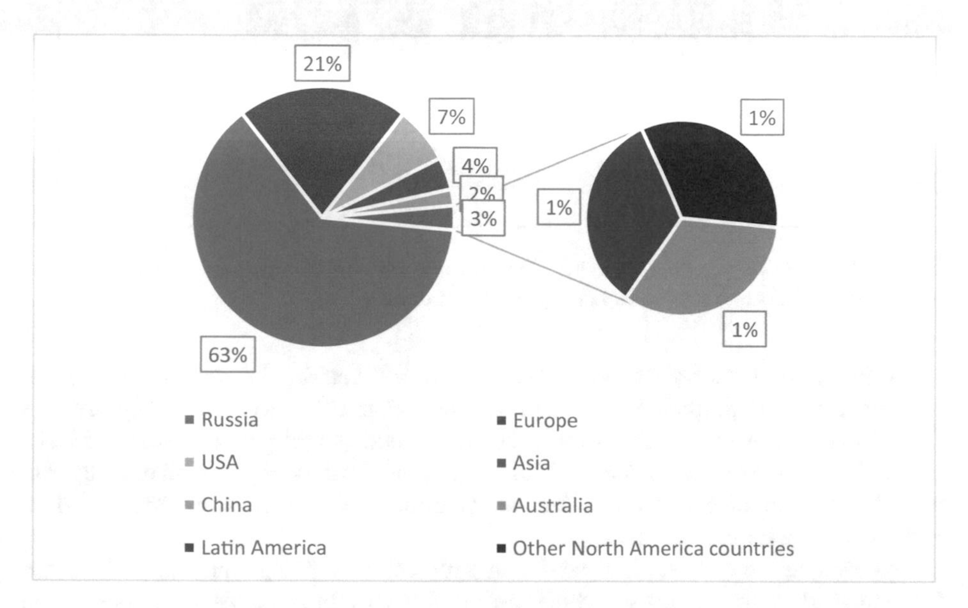

Fig. 3. Geography of tourists who came to the Russian Federation in 2017 and used the Airbnb service

Airbnb service makes it easier for citizens to travel within the country and abroad. The Association of Tour Operators of Russia (ATOR) predicts that in 2019 tourist flow will grow by 20% compared with 2018 (Atorus 2019). The conclusion is that an increasing number of tourists will use the Airbnb service to rent housing in Russia and abroad. This can be explained by the fact that renting through the Airbnb service saves travelers money (see Fig. 4) (McCarthy 2018).

According to Fig. 4, the difference in payment directly at the hotel and through the Airbnb service can be quite significant. In some cities, the difference reaches 50% (for example, in Tokyo and in Moscow). So the cost of a room in New York per night when paying directly at the hotel itself was $ 306, and with the help of the online service Airbnb - $ 187. A tourist could save about 40% of his expenses (Statistics and fact 2018).

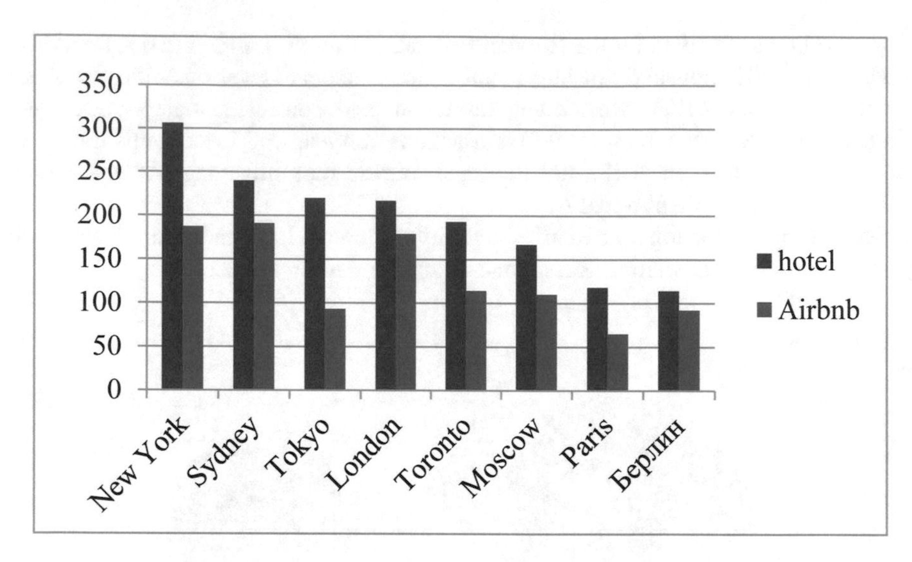

Fig. 4. The average price of a room per night in some cities of the world (payment at the hotel and through the Airbnb service in January 2018), in dollars

Refusing food at a hotel restaurant in favor of self-catering in a kitchen in a rented apartment or other accommodation will help budget travelers to save money. Despite the obvious advantages of the Airbnb site, some tourists still prefer to stay in hotels, although prices are higher there. Most often, this behavior is explained by the unwillingness to independently select an apartment on the Airbnb service and to communicate with strangers.

As well as abroad, several car-sharing projects are being implemented on the Russian market, the largest of which are in Moscow. Moscow is the leader in the number of cars for rent.

If in the middle of 2016, the average number of cars ranged from 30 to 200 depending on population density, then at the beginning of 2018, the total fleet in Moscow was more than 2.7 thousand cars. However, the number of cars for rent was only 0.2 per thousand inhabitants. For comparison, in the same period in Berlin this figure was 0.66, in Milan - 1.16. At present, there are more than 8,000 cars in the Moscow car sharing fleet.

By the number of users of car sharing service in 2018, North America was the leader in the world: there 1.12 million people used this service. In the US, ZipCar's fleet was approximately 11,000 (ING Economics Department 2018). In 2018, the total car fleet of car-sharing services in Europe grew to 370 thousand cars. Compared to 2016, the number of cars increased by 2.8 times, and the number of users by 2 times.

The data suggests that interest in karsheringu around the world is growing and, above all, people who do not own their own car, do not use public transport or travel a lot around the world.

In mid-2018, a project was launched in China, which was named "The most extreme car sharing in the world" (Barinov 2018). The essence of this project is that the car that is being rented is an electric car driven by a robot. If the project successfully passes the test, it will greatly simplify the life of people who do not have a driver's license.

According to Forbes, the income received through ESP and coming directly to its participants will grow at a very rapid pace. ESP is no longer just a profitable model, but a real revolution in the economy (Heron 2019).

Despite the advantages and growing popularity, like any new model, the sharing economy faces a number of problems and has certain disadvantages. The first and most important problem is the possible reduction of jobs in the implementation of ESP services.

The next problem is tax regulation. Additional measures should be introduced to limit competition from the share services. For the new economic model, new methods and methods of tax collection should be developed.

The third problem is related to the introduction of additional measures to ensure the security of information platforms. This problem can be fixed faster than all others, because the services themselves are interested in solving it.

ESP markets lead to increased competition and the destruction of previously established monopolies. For example, in 9 countries of the world, the application for ordering a taxi Uber is prohibited: in Brazil, China, France and Spain - because of the destruction of the monopoly of taxi drivers, their violent protests. In Germany, the main disadvantage of the Uber online service is a violation of the laws of the country and public order (Bochkareva 2015).

Taxis ordered via the Uber online service can be used for extortion, robbery, kidnapping, or killing passengers.

Scammers can impersonate Uber drivers. For your own safety, an online service user should always check the number of the car in which he sits down with the number specified in the order.

Conclusions/Recommendations. So, summing up, it should be said that the development of information technologies has led to the emergence of new models of human interaction. One of them was a shared economy, the demand for which is growing every day.

It should be noted that the economy of joint consumption is an innovative industry and has become widespread throughout the world, including in Russia. It is based on the trust of people, without which the interaction and development of this model is impossible.

The main advantages of EPS are: convenience, trusting relationships of participants, maximum use of available resources, obtaining permanent income from property that the owner does not use constantly.

Scientists expect that by 2025 the turnover of, for example, sharing services will increase 10 times. Analysts expect the largest growth in car use and rental housing (Adactylos 2018).

An important issue today is the security of online platforms. It is predicted that by 2025, online services will be as secure as possible.

In Russia, the development of ESPs occur more slowly than in Europe and the USA. The reason for this is the vastness of the territory and the uneven economic development of the regions of Russia. Problem solving and, accordingly, stimulating the development of ESPs will largely depend on state policy.

References

Bochkareva, E.: 9 countries in which Uber is prohibited. https://rb.ru/list/uber-ban/

Li, I., Pastushin, A.: Airbnb for the first time revealed the volumes of housing delivery in Russia. https://www.rbc.ru/technology_and_media/08/07/2017/595f46b39a79479492a04647

Atorus. http://www.atorus.ru

ING Economics Department. Car sharing unlocked, October 2018. https://think.ing.com/uploads/reports/ING_-_Car_sharing_unlocked.pdf?ref=vc.ru

McCarthy, N.: Is Airbnb really cheaper than a hotel room? https://www.statista.com/chart/12655/is-airbnb-really-cheaper-than-a-hotel-room/

Statistics and fact. https://www.statista.com/topics/2273/airbnb/

Adaktilos, A.D., Chaus, M.S., Moldovan, A.A.: Sheringovaya economy. J. Econ. (2018). CyberLenink. https://cyberleninka.ru/article/n/sheringovaya-ekonomika

Sabitov, O.: Everywhere sharing: what is the economy of joint consumption. What is sharing economy: the history of the term. https://rb.ru/story/share-it/

Barinov, D.: The most extreme car sharing in the world. https://tjournal.ru/flood/71317-samyy-ekstremalnyy-karshering-v-mire

Heron, T.: Economy sharing. https://forbes.kz/finances/markets/ekonomika_sovmestnogo_polzovaniya\

Crowd Recruiting: Modern Approaches to Recruitment

Vasiliy E. Krylov[1(✉)], Viktor A. Eronin[1], Marina P. Vahromeeva[1], Olga V. Shuvalova[2], and Tatyana A. Nikerova[2]

[1] Vladimir State University named after A.G. and N. G. Stoletovs, Vladimir, Russia
wek_70@mail.ru, eronin.v@yandex.ru, marta2302@yandex.ru

[2] Financial University under the Government of the Russian Federation (Vladimir Branch), Vladimir, Russia
{OVShuvalova, TANikerova}@fa.ru

Abstract. The article is devoted to one of the most new technologies, which began to develop thanks to the digitalization of the economy - crowdsourcing and, as its most important component – crowd recruiting. Its capabilities are extremely wide: it overcomes geographic distances, saves time and costs of finding candidates for the vacancy of the desired specialist of the company. The sphere of application of crowd technologies is investigated. Formulated the main objectives of the use of crowd - technology. Lists the platforms on which these technologies are applied. The history of crowdsourcing is considered in detail. Describes the history of the introduction of cracking in the Russian Federation. The concept of "recruiting" is defined. It describes the main services that are implemented through recruiting. Then, the cracking technology is described. Then, the main advantages and disadvantages of cracking are described. The differences between cracking and recruitment are determined. The advantages of crowd recruiting technology compared with other methodologies, the possibilities of this technology, as well as its limitations are considered. Crowdsourcing is used in the exchange of knowledge, the development of ideas and concepts, the collection of funds for the implementation of interesting and socially significant projects, voting, work on content, evaluation of various proposals and others. The conclusion lists the main problems of crowd implementation in the Russian Federation.

Keywords: Digital economy · Cracking · Crowdsourcing · Crowdfunding · Recruiting · Outplacement

JEL Code: J 44

1 Introduction

In the conditions of the development of digitalization of the economy, the subjects and, above all, the business began to enjoy the benefits of new technologies and services.

One of these new technologies, which began to spread from the beginning of the 2000s, is the crowdsourcing technology.

E. G. Popkova and B. S. Sergi (Eds.): ISC 2019, LNNS 87, pp. 631–636, 2020.
https://doi.org/10.1007/978-3-030-29586-8_73

Crowdsourcing, as a modern technology, involves the use of the abilities, knowledge, and creative skills of a large number of people to solve various issues and problems during a joint online work on a specialized Internet platform.

Crowdsourcing is an example of using collective intelligence in different areas of knowledge when choosing the best solution to a problem from a variety of alternatives.

Crowdsourcing technology is used in the exchange of knowledge, the development of ideas and concepts, the collection of funds for the implementation of interesting and socially significant projects, voting, work on content, evaluation of various proposals and others.

Crowdsourcing is the development of outsourcing technology, caused by the new realities of the 21st century, in which, thanks to integration processes, part of the functions of the organization's activities are transferred to third-party organizations in order to save costs. Crowdsourcing is essentially a continuation of the tradition of outsourcing in a digital economy that has removed the barriers of time and space.

Unlike outsourcing, it involves a significant greater number of interested people from the external environment, offers many alternatives, gives rise to innovative ideas during discussions, is low-cost, despite a significant number of participants, and significantly reduces the time it takes to solve problems, thereby increasing the efficiency of the organization.

The scope of crowd technology is very extensive. With their help, a business can, for example: promote its brand; improve the website design, organize interaction with customers through online polls and forums; to attract resources, for example, capital, labor resources (crowdfunding), etc. Attracting resources for business in modern conditions is particularly important.

It's no secret that the performance of an organization depends on its employees. Each organization seeks to find the best employees.

Now, personnel departments in organizations and enterprises are engaged in personnel departments, as well as specialized organizations - employment services and recruitment agencies.

Recruiting is a business process for the selection and selection of specialists for the customer company (Sugareva 2014).

Over the long years of its existence, recruiting has developed and improved. Now he is an independent type of business.

Recruiting sells different types of services, for example:

- "Bounty Hunt" is a service for finding the best candidate for a leading position.
- Outplacement - a service for the employment of personnel dismissed from the client company in accordance with established deadlines and certain conditions.
- Staff leasing is a service in accordance with which the agency provides the client company with the necessary employee for a certain period of time to perform the work in accordance with the position held.
- Online recruitment - attracting qualified staff or finding a job seeker with the help of websites (Vishnyakova 2016).

In Russia, the Human ResoursesON-Line (HRO) service, established in May 1996, began to provide an opportunity to post vacancies from organizations and recruitment agencies on a fee basis. Then a similar Internet resource PointJob was launched and the development of electronic recruitment began. The sites zarplata.ru, superjob.ru, hh.ru, job.ru, rosrabota.ru, rabota.ru, bankir.ru, and others earned. Some of them were free at first, and then began to work on a fee basis (Simanova 2017).

In recent times, online recruitment opportunities are expanding and improving.

Abroad and in Russia they started talking about crowdracking techniques.

This is not just a way to hire employees, as is online recruiting, when information technology helps a company to save time on finding candidates and reduce costs. This is something more. Although without digital technology, neither the first nor the second.

Crowd recruiting is a way to find and hire an employee by assessing his qualities, real possibilities and abilities in carrying out the project proposed by the company (Simanova 2017).

The goal of crowd recruiting is to evaluate and select the best employees from a fairly large number of candidates (Yakuba 2018, Chulanova 2018).

A characteristic difference in crowd recruiting from the methods of selecting candidates for work is the full involvement of project participants in the solution of the tasks, so that they have the opportunity in a short time to show themselves from completely different sides, maximizing their potential (Bobko 2018).

Crowduring is the use of crowdsourcing technology (involving a large number of people) with the direct participation of recruitment professionals who, thanks to the work of applicants on crowd platforms, professionally evaluate a potential employee taking into account the interests and needs of the employer. countless. And the employer can choose the most talented, offering interesting solutions to real problems.

This is possible because The proposed project involves not only individual but also teamwork, as well as interaction with the employer. The employer, also online, can track the actions and proposals of each candidate and communicate with him by questioning, participating in video conferences, familiarizing himself with the candidate's resume, etc.

The advantage of crowd recruiting is the fact that in the process of the candidate's work on the project, the employer can learn a lot about the abilities of applicants, their skills or inability to act in different situations.

It is also interesting for the candidates themselves to participate in the project: they are embraced by the excitement of solving problems in a tough competitive environment of a wide range of participants.

It is also important that both the employer and the candidates are interested in working together, they get to know each other more closely during the work process, are ready to hear each other, to negotiate the most acceptable for all working conditions and its remuneration. As a result, the workload of HR-service employees of the enterprise is reduced, or the need to attract third-party recruitment agencies is eliminated, the cost of selecting each candidate is reduced (Yakuba 2018).

Crowdrebring is a fairly new technology. It, for example, is used by the BrainForce platform or FuturUS for the implementation of crowdsourcing projects (Sugoreva 2016).

The task of an employer in crowdfunding is to create an up-to-date project, make it interesting for participants, and also offer a reward system for the leaders of this "competition". However, the most important thing is not to limit the participants in anything so that the project implementation process is creative and the participants maximize their potential.

They must demonstrate proficiency in general, specific knowledge, knowledge, ability to work with information, show individual traits (assiduity, literacy, punctuality, frequency of initiative, etc.), teamwork skills (maybe leadership skills).

At the end of the project, the employer will only have to send invitations to the final interview or offer to take a vacant place to those who are interested in it (Maksimov 2018).

In Russia, crowdracking technology was first used in 2012 by «Witology» on the order of the «Rosatom State Corporation» during the implementation of the unique project «TeMP 2012» . The project was aimed at selecting 100 specialists for various areas of «Rosatom's» activities.

With traditional recruiting, this would take about half a year. This crowdfunding project took only a month.

Along with identifying the most creative and knowledgeable candidates, crowdracing allowed us to form a certain talent pool of young professionals, to acquaint many applicants with the company and the specifics of its work (Yakuba 2018).

Crowd rekruting has its advantages and limitations.

In addition to such advantages as selection speed and cost savings, it should be noted that it is the on-line format that is very much in demand by modern young people, it is clear to them and its advantages are obvious to them. Therefore, graduates of the challenge of different specialties - this is the main audience of cruise recruiting projects. We are talking about specialists with demanded functionality - business analysts, marketers, logisticians, engineers, IT specialists, etc.

This technology is hardly suitable for selecting a driver, secretary, as well as candidates for rare professions.

However, not all managers, especially in the regions, are ready to use this format and prefer the traditional methods of recruitment. Traditional job search methods are often preferred by highly qualified specialists (Chulanova and Vishnyakova 2018).

As for the selection of top managers, then, as a rule, they do not resort to the help of Job-sites. They prefer not to advertise their intention to change their employer and avoid reducing their own value in the labor market: there is an opinion that good managers do not look for work, and business owners invite them. The search for such professionals is an art that is subject only to a person, perhaps armed with the Internet, but it is unlikely that HH or Superjob. In principle, the depersonification of the recruitment process deprives it of flexibility, and can reduce its effectiveness (Simanova 2017).

Crowd cracking is not so widespread in Russia, according to the Research Center of the portal Superjob.ru, because quite often employers want personal contact with the candidate, because for them are important different things, such as appearance, style of clothes, demeanor, etc. Online work allows to exclude subjectivism, but the applicant himself is not a robot, an important role is also played by the personal relationship between the employer and the future employee (Goryunova 2018).

The further development of the crowdfunding technology is to improve the feedback. Crowd platform records people's communication, teamwork, promotion of ideas, who and how many were selected at the completion of projects, etc. And the questions of how the selected candidates cope with the work, how many of them "got accustomed" to the company, how many were dismissed and how quickly - remain open. From this point of view, the effectiveness of the technology is not defined.

In any case, the number of employees who did not pass the probationary period should decrease, since crowdrecruiting can resolve many controversial issues even before the formal conclusion of an employment contract (Chulanova 2018).

As noted above, crowd cracking involves the use of Internet platforms, so its cost is quite high, which means that not all employers, especially small and medium business owners, can use this technology. This also holds back the introduction of crowd-technology into the practice of personnel selection in our country (Goryunova 2018).

Crowd recruiting can be used not only in the selection of candidates for vacancies, but also when it is necessary to solve some problem within the company. Employees of the company have the opportunity to express themselves and thereby begin to treat their work differently, and later, perhaps, to move up the career ladder.

As for the company's customers and third-party respondents, they get the opportunity to become participants in the creation of a product or service that interests them (Bobko 2019).

Summing up, it should be noted that the development of information technologies is changing the modern world, opening up additional opportunities for the emergence of new relationships between market actors.

One of the modern technologies in the digital economy has become crowdsourcing, using the wisdom of the crowd in solving various problems. The development of crowd technologies has led to the emergence of one of its varieties - crowd recruiting.

Its capabilities are extremely wide: it overcomes geographic distances, saves time and costs of finding candidates for the vacancy of the desired specialist of the company. However, having significant advantages, the technology of crowdfucking to the full extent cannot yet replace the personal human contact of the recruiter with the applicant. Therefore, at present there is a need to use crowd recruiting at the initial stages of selection, and at the final stage of selection - communication with an experienced professional recruiter. Such an approach can be used until feedback is established with employers.

The further development of crowdfunding can make this technology effective and very promising for staff selection in the 21st century.

References

Bobko, A.: Krautkremer. Search for people through the search for ideas (2018). https://witology.com/blog/company/264/. Accessed 1 March 2019

Bobko, A.: Idea management and crowdfunding (2019). http://www.sergiev-posad.ru/useful/?ID=18134/. Accessed 1 March 2019

Chulanova, O.L.: Crowdfunding as a method of searching candidates (2018). http://hr-portal.ru/blog/kraudrekruting-kak-metod-poiska-kandidatov. Accessed 1 March 2019

Chulanova, O.L., Vishnyakova, T.S.: Crowdsourcing, crowdfunding and crowdfunding in the recruitment system (2018). https://studref.com/453540/menedzhment/kraudsorsing_kraudrekruting_kraudfanding_sisteme_podbora_personala/. Accessed 1 March 2019

Goryunova, O.: In one fell swoop kill two birds with one stone – precisely such a outcome promises a new technology selection personnel (2018). https://bankir.ru/publikacii/20150316/prizhivetsya-li-u-nas-kraudrekruting-10006171/. Accessed 1 March 2019

Maksimov, N.N.: Krautkremer - crowdsourcing model recruitment (2018). http://forumbusiness.net/showthread.php?t=27323/. Accessed 1 March 2019

Simanova, I.: “Real” recruiting and Internet recruiting: pros and cons (2017). http://aviconn.com/press/realnyiy-rekruting-i-internet-rekruting-za-i-protiv.html/. Accessed 1 March 2019

Sugareva, E.: What is recruiting (2014). http://hr-portal.ru/article/chto-takoe-rekruting/. Accessed 1 March 2019

Sugoreva, E.: Krautkremer (2016). http://ci-systems.ru/ocenka-i-poisk-personala. Accessed 1 March 2019

Vishnyakova, T.S.: Recruitment service – what is it? (2016). http://znaydelo.ru/personal/trudoustroystvo/rekruting.html/. Accessed 1 March 2019

Yakuba, V.: Krautkremer - new technology of mass talent search (2018). http://iinsider.biz/. Accessed 1 March 2019

Social Consequences of Transition to the Digital Economy

Development of the Mechanism of Formation of Effective Digital Platform in the Glass Industry and Estimation of the Efficiency of Its Realization

Alexey M. Gubernatorov[1(✉)], Natalia D. Stelmashenko[2], Evgeniy A. Ulanov[1], Andrei A. Chekushov[2], and Yaroslav N. Mayorov[1]

[1] Vladimir State University named after Alexander Grigorievich and Nikolai Grigorievich Stoletovs, Vladimir, Russia
gubernatorov.alexey@yandex.ru, evgeniyulnv@rambler.ru, yaroslav.majorov@mail.ru
[2] Vladimir Branch, Financial University under the Government of the Russian Federation, Vladimir, Russia
nds-ko@mail.ru, aachekushov@fa.ru

Abstract. The article describes the features of the development of the mechanism of formation of an effective digital platform on the materials of the glass industry. In order to maximize the benefits of digital transformation in innovation, economic growth and social prosperity, there is a need to recognize the concern that existing macroeconomic statistics may not fully appreciate the benefits of digital products and products based on digital technologies or cross-border transactions. The authors present a possible list of breakthrough technologies and sub-technologies in the glass industry. To assess the effectiveness of digitization of the glass industry was carried out an expert assessment of the factors affecting the digitalization of enterprises of the Vladimir region, engaged in the production of glass products. The assessment was made on a 5-point scale.

Keywords: Digital economy · Digital platform · Business processes · Information technology · Digital transformation of the glass industry

JEL Code: M21 · O14 · O31

1 Introduction

In the digital economy accumulates, stored and processed a huge amount of data on the activities of individual sectors of the national economy, and the glass industry should not be an exception. The very concept of the digital economy in Russia is defined as economic activity, the key factor in the production of which is data in digital form, which contributes to the formation of the information space, taking into account the needs of citizens and society in obtaining quality and reliable information, the development of information infrastructure of the Russian Federation, the creation and

E. G. Popkova and B. S. Sergi (Eds.): ISC 2019, LNNS 87, pp. 639–643, 2020.
https://doi.org/10.1007/978-3-030-29586-8_74

application of Russian information and telecommunication technologies, as well as the formation of a new technological basis for the social and economic sphere." The transformation processes taking place in the glass industry, to which many analysts tend to attribute a revolutionary character, are multifaceted. And often, as sages studying from different sides of the elephant, experts see only (Orlova 2018a).

Methodology. Breakthrough technologies in the glass industry are a powerful catalyst that can change the configuration of all established technological trends and routes of the industry. Breakthrough technologies play a crucial role in increasing the competitiveness of a single enterprise engaged in the production of glass products, and in General the economic growth of the entire glass industry due to the multiplier effect.

On the territory of the Vladimir region in 2018, there are 31 organizations that produce glass (Orlova 2018).

The leading enterprises of the industry are LLC "Experimental glass factory", which produces drinking vessels, a variety of glassware, JSC "OS Fiberglass", engaged in the manufacture of continuous fiberglass, chopped fiber, roving, fiberglass, a Branch of LLC "Rusjam Glass holding" in Gorokhovets (glass container), LLC "Red Echo" (bottles with a capacity of 0.3–1.75 l, banks), LLC "Vinerberger Brick" (ceramic brick, ceramic blocks), CJSC "Kovrovsky plant".

2 Results

To assess the effectiveness of digitization of the glass industry was carried out an expert assessment of the factors affecting the digitalization of enterprises of the Vladimir region, engaged in the production of glass products. In the expert rating took part heads and specialists of the glass 20 of the companies (OOO "Crystal sky", Gus Crystal plant. Maltsova, JSC "Light", LLC "Experimental glass factory", JSC "OS Fiberglass", LLC "Red Echo", LLC "velikodvorsky glass plant" and others.)

The assessment was made on a 5-point scale. As a factor in assessing the effectiveness of digitization of the glass industry were selected seven factors that were ranked:

K1-regulatory and administrative indicators of digitalization,
K2-specialized personnel and training programs,
K3-research competences and technological groundwork,
K4-information infrastructure,
K5-information security
K6-financial and economic efficiency of digitalization development.

The assessment of the reliability and consistency of expert opinions was based on the calculation and analysis of the concordance coefficient W, the value of which was calculated by the formula:

$$W = \frac{12\sum_{i=1}^{n}(r_i - \bar{r})^2}{N^2(n^3 - n)}$$

where: 12-constant value in the formula for calculating the coefficient of concordance proposed by Kendall;

N - number of experts;
n - number of evaluation criteria;
ri - sum of ranks of the I-th index;
r is the average sum of all scores.

The coefficient W belongs to the interval from 0 to 1, where its equality to one means full consistency of experts 'opinions, and equality to zero indicates that there is no connection between experts' assessments. In the case where the $0.2 \leq W \leq 0.4$ is the place the weak consistency of the experts' opinions, and if $W \geq 0{,}6$ it is possible to speak about existence of the strong consistency of expert opinions.

Next, to determine the weighting coefficients of each factor, we use the Fishborn method (a method for assessing the clinical efficacy of drugs):

ai = 2 * (n – ri + 1)/n * (n + 1), (for all i from1 to n)

where: ai - coefficient of weight of the i-th criterion,
n - number of evaluation criteria,
ri - rank assigned to the i-th indicator.

Table 1 presents the results of expert evaluation of the weighting factors of digitization of the glass industry enterprises of the Vladimir region. Verification was made on the example of «Crystal sky».

$$W = \frac{12\sum_{i=1}^{n}(r_i - \bar{r})^2}{N^2(n^3 - n)} = \frac{12 * 1238}{20^2(4^3 - 4)} = 0.80$$

The degree of consistency of expert estimates can be considered quite acceptable, since W = 0.80 > 0.6.

According to experts, the most significant is the regulation and administrative indicators of digitalization, and the least significant is the financial and economic efficiency of the development of digitalization.

In addition, relationships have been established between the factors characterizing the digitization of the glass industry of enterprises of the Vladimir region.

One of the used indicators of the adequacy of the equation is the coefficient of determination, the lower limit of which can be taken as the value of 0.7. A more reliable prediction result can be achieved with large values of R (0.8–0.9).

$K0 = \alpha K1 + \beta K2 + \gamma K3 + \delta K4 + \varepsilon K5 + \varphi K6 + \omega K7$, where

$\alpha, \beta, \gamma, \delta, \varepsilon, \varphi, \omega$ – weights coefficient
Coefficient of determination
$R^2 = 0.9997$.

The value of the coefficient of determination, close to unity, indicates a high reliability of the multifactor model.

Table 1. Estimation of weight of factors of digitization of the glass industry of the enterprises of the Vladimir region (on materials of LLC «Crystal Sky»)

Criteria of evaluation	Expert review																			
	1	2	3	4	5	6	7	8	9	10	11	12	13	14	15	16	17	18	19	20
Regulatory and administrative digitization indicators	5	5	4	4	5	4	5	5	5	5	5	5	4	5	5	5	5	5	5	4
Specialized personnel and training programs	4	2	3	3	3	3	2	3	3	2	3	2	2	2	3	2	2	2	3	3
Research competencies and technological background	4	3	4	2	4	4	4	4	3	3	3	3	3	3	4	3	4	3	4	4
Information infrastructure	5	4	5	5	5	5	5	5	4	4	4	4	5	4	5	4	5	4	5	5
Information Security	5	5	4	4	5	4	5	5	5	5	5	5	4	5	5	5	5	5	5	4
Financial and economic efficiency of digitalization development	4	2	3	3	3	3	2	3	3	2	3	2	2	2	3	2	2	2	3	3
Social development effectiveness of digitalization	4	3	4	2	4	4	4	4	3	3	3	3	3	3	4	3	4	3	4	4
Total																				

Criteria of evaluation	Total points	Deviation from the average	Square deviation	Rank	Weight indicators calculated according to the Fishburn method
Regulatory and administrative digitization indicators	95	18	324	1	0,40
Specialized personnel and training programs	52	-25	625	4	0,10
Research competencies and technological background	69	-8	64	3	0,20
Information infrastructure	92	15	225	2	0,30
Information Security	95	18	324	1	0,40
Financial and economic efficiency of digitalization development	52	-25	625	4	0,10
Social development effectiveness of digitalization	69	-8	64	3	0,20
Total	310	0	1238		

Further, all factors were distributed by levels of digital efficiency (high, medium and low levels).

The highest level is set to the maximum weighting factor calculated by the Fishburn criterion. For a low level, the maximum weighting factor divided by three (the number of levels) is assigned, after which the d value is found using the formula, which is calculated as the difference between the high level indicator and the low level indicator divided in half. The average level indicator is equal to the low level indicator, increased by a step according to the formula:

$$d = (high\,value - low\,value/2)$$

In the course of the calculations, the following limits of digitalization efficiency were obtained for «Crystal Sky» LLC:

Z < 0 digitalization level low;
Z (0; 0.333) digitization level is below average;
Z (0.333; 0667) digitization is average;
Z (0.667; 0.999) digitization is high (Table 2).

Table 2. Integral assessment of the digital development level of «Crystal Sky» LLC using the Fishburn rule in 2018

Companies	Fishburn Index	Z <0 digitalization level low	Z (0; 0.333) digitalization level below average	Z (0.333; 0667) digitization average	Z (0.667; 0.999) high digitization
Crystal Sky LLC	0.326		+		
Ltd. Gusevskoy Crystal Plant them. Maltsova	-0.566	+			
JSC "Light"	0.002		+		
LLC "Experienced Glass Factory"	0.005		+		
Open Society "OS Steklovolokno"	-0.159	+			
LLC "Red Echo"	-2.522	+			
LLC "Velikodvorsky glass container plant"	0.304		+		

3 Conclusions/Recommendations

The digital transformation of the glass industry is the key to building a digital economy and obtaining digital dividends, that is, achieving measurable economic results through the introduction of digital technologies.

References

Orlova, S.: The discovery of the Velikodvorsk processing plant is an important step in the development of the glass industry in the Vladimir region. [Electronic resource]. - Access mode (2018a). https://avo.ru/novosti/-/asset_publisher/E2PryKmsVruz/content/svetlana-orlova-otkrytie-velikodvorskogo-pererabatyvausego-kombinata-eto-vaznyj-sag-v-razvitii-stekol-noj-otrasli-vladimirskoj-oblasti-
Orlova, S.: The results of the socio-economic development of the Vladimir region in 2018 [Electronic resource]. Access mode. https://avo.ru/web/guest/promyslennoe-proizvodstvo

Digitization of Economic Space as an Imperative for the Formation of a Knowledge Economy

Olga B. Digilina[1(✉)], Natalya O. Subbotina[2], Elena R. Khorosheva[2], Ilya A. Lvov[3], and Sergey N. Kalintsev[2]

[1] Russian Peoples' Friendship University, Moscow, Russia
o.b.digilina@mail.ru

[2] Vladimir State University named after A.G. and N.G. Stoletovs, Vladimir, Russia
nosgnom@mail.ru, kalincev.sergei@mail.ru, khorosheva.elena@gmail.ru

[3] Financial University under the Government of the Russian Federation (Vladimir Branch), Vladimir, Russia
IALvov@fa.ru

Abstract. Currently, information and communication technologies play a huge role not only in the lives of individuals, but also transform economic processes, modernize entire industries and various types of social and economic activity, and become drivers of innovation, economic growth and competition. The purpose of this study is to show how digital technologies modify the economic space and thereby influence the formation of the knowledge economy. The article defines the "knowledge economy", shows the stages of its formation. The authors argue that digital technologies will be the basis for the development of the knowledge economy in the coming years. At the second stage, an important role will be played by the intellect, creative abilities of a person, aimed at creating innovations.

Keywords: Digital economy · Knowledge economy · Digitalization · Information technology · Innovative economy

JEL Code: M21 · O14 · O31

1 Introduction

The development of civilization is determined by changes in the productive forces. These changes are global in nature, occur according to universal laws. The gradual improvement of the productive forces led to a change in the pre-industrial era to an industrial, and then a post-industrial era.

The basis of the productive forces are people, and the means of labor and the objects of labor (except for natural) are the result of their physical and mental activity.

E. G. Popkova and B. S. Sergi (Eds.): ISC 2019, LNNS 87, pp. 644–651, 2020.
https://doi.org/10.1007/978-3-030-29586-8_75

The industrial era was characterized by an increase in labor productivity, the emergence of a huge amount of goods and services, an elevation of needs, and an increase in the standard of living of the population.

These changes have led to the emergence of a new quality of life, when people change priorities, the needs for intellectual development and self-realization become paramount by raising the educational level and skills.

And in the conditions of depletion of all types of resources, it is the scientific knowledge and information that becomes the real productive force of society. This is manifested in the fact that scientific and technological progress expands human capabilities, reducing production time, developing new activities, releasing labor, redistributing it between sectors of the economy. In our article we will look at the impact of digital technologies on the formation of the knowledge economy.

2 Methodology

The object of this article is the economic relations that are formed in the process of the development of the knowledge economy. The subject of the research is the directions of the influence of digital technologies on the patterns of formation of economic relations based on knowledge.

The methodological basis of this study was a set of techniques, methods and cognitive attitudes that are currently known to scientific research and which have been adapted in relation to the specifics of the object being studied - the formation of society and the economy of knowledge.

The empirical basis for the preparation of this article was the socio-economic practice of the developed countries of the world, the Russian Federation, large enterprises, their best practices in mastering the technologies of the knowledge economy, as well as statistical and analytical reviews on the topic of research, government documents reflecting the strategy of forming the knowledge economy, articles of an economic nature in the media.

3 Results

The development of information technologies makes information and knowledge accessible and readily available, Digitalization leads to the formation of a post-industrial society. This concept was formulated as far back as 1962 by D. Bell, who presented his concept in the work "The Arrival of a Post-Industrial Society".

He believed that if the basis of pre-industrial and industrial societies was ownership of the means of production, then at the end of the 20th century information was the basis, accumulated knowledge. They determine the direction and speed of change, the emergence of innovation.

Today, with the help of computer technology, a person can store, transmit and almost instantly receive information, on the basis of which - to make management decisions promptly.

The scale of the technological leap that occurred in such a short time is amazing. At the same time, information technologies continue to develop at a colossal pace.

The capabilities of information and communication technology not only compress time, but also space. The quality of jobs is changing - they are no longer stationary, this is a cyber space. Instead of physical movement, people begin to move their ideas and thoughts through cyberspace (Nordstrom and Riderstrall 2013).

Information technology makes information about almost everything transparent. Indeed, the era of the "dense" and downtrodden consumer ends.

Even being anonymous, using a nickname, everyone can express their opinion, debate, make a request and get a quick response, order a product, thereby reducing the value chain and their costs. Not only this, online stores provide the buyer with a huge selection, which is not comparable with even the largest megatorg. In addition to ease of use and low prices, the consumer gets quick service, information about the latest updates, opinions and customer reviews.

The Internet allows companies to organize activities on a fundamentally new basis. The interaction of suppliers and customers through the network. On-line, you can have information about suppliers' products, immediately place an order, organize its delivery, thereby reducing to a minimum the stocks in the warehouses and the warehouses themselves.

For example, at Wal-Mart, the largest department store chain in the United States, 97% of the goods do not go through the warehouse at all. Goods are delivered directly from the manufacturer to the store shelf, and then to the consumer or immediately to the consumer (Nordstrom and Riderstrall 2013).

New principles of business organization allow modern companies using information technology or working in the field of ICT to achieve impressive results.

So in January 2019, Amazon (engaged in Internet commerce and cloud computing) became the most expensive in the world, overtaking Microsoft. The four largest corporations, changing places, lead by a wide margin and have an impressive cost: Amazon - $ 811.4 billion, Microsoft - $ 800 billion, Alphabet - almost $ 750 billion, and the leader of many years - Apple - 727 billion dollars (the company's capitalization in October 2018 amounted to 1.1 trillion dollars) (Overchenko 2019).

All of this - the positive effects of information technology and networks. They actually implement the law of increasing profitability.

Researchers believe that the company's "infostructure" will soon become more important than infrastructure. Organizations with weak infrastructure will look like a 65-year-old athlete trying to run the Olympic Marathon in high heels and in evening dress (Nordstrom and Riderstrall 2013).

Now many people, enterprises, organizations, informal associations, etc. build their work on information and knowledge.

Knowledge, as well as those technical tools that propagate information and knowledge, have become the most important factors in the transformation of the social order, the way of life of people, the entire world economy, the foundations of business organization and management, have changed the approaches to education, health care, employment, work, etc.

It is no coincidence that at present, developed countries in science and technology policy give preference to information technologies and services, medicine, ecology, etc.

This is due to the impressive results of scientific and applied research, and suggests that more and more industries and spheres are becoming the drivers of economic growth. based on knowledge.

A knowledge-based economy is a certain stage in the development of a post-industrial society. This is a stage in which only knowledge can move forward all areas of the economy, a stage in which this knowledge is used everywhere, including in people's daily lives.

In our opinion, it would be more correct to use not the term "economy based on knowledge", but the term "economy of knowledge", since what else, if not with knowledge and information, is the development of civilization.

The term "knowledge economy" has existed since the 60s of the XXth century. It was proposed by the American economist F. Machlup, who by knowledge economy understood the economic sector that produces knowledge.

In modern conditions, this term can be interpreted more broadly, extending it to many sectors of the economy, first of all, those where new knowledge is directly produced, where production of information benefits and services takes place, where equipment is created for the transfer and processing of knowledge.

The beginning of the 21st century was marked by the powerful development of high technologies, the creation of ultra-fast computers, the spread of global communication networks, the improvement of information systems and technologies for the dissemination and storage of knowledge.

The modern stage of the knowledge economy is now increasingly called the digital economy, because It is the digitalization of many processes that determines the intellectual informational production, which means that the sphere of high technologies becomes the most important branch of the knowledge economy.

This is evidenced by the fact that many countries are investing more and more in the development of information technology, conducting basic and applied research to create new high-tech equipment and technology.

Information and communication technologies (ICT) are created on the basis of scientific developments, in turn, they contribute to the rapid processing, transfer, unimpeded distribution of knowledge among a huge number of people (Varavva 2008). It was they who made the open innovation model possible, when researchers from many countries participate in the implementation of various innovation projects.

Digitalization leads to the globalization of human knowledge, the development and implementation of elements of artificial intelligence.

Among scientists there is an opinion that in the development of the knowledge economy there will be two stages (1) information civilization (development of globalization processes in almost all spheres, including cultural and ideological; unification of public knowledge, providing virtually unlimited access to world information stores); (2) the civilization of knowledge (self-knowledge as a process of personal and institutional development of information).

This indicates that it is ICT that will become the basis for the development of the knowledge economy in the coming years. At the second stage, intelligence, creative abilities of a person aimed at creating innovations will play an important role.

Many sociologists claim that a new type of social stratum is becoming established (from English cognition—knowledge, cognition; by analogy with the proletariat).

A new type of employee is being formed who possesses the necessary competencies - knowledge, skills, and skills, and is aimed at creative practical activities.

Therefore, countries that cannot create conditions for the development of science, improvement of the education system, ensuring the quality of the information environment will inevitably lose the ability to compete with those who are now devoting all their forces to building up their scientific and technical potential in order to use it effectively.

Currently, there are significant digital gaps between countries and regions, as well as between developed and developing countries.

In developed countries, twice as many contracts for mobile broadband per 100 population than in developing countries. The number of mobile broadband contracts is much higher in Europe, North and South America than in other regions, and more than three times higher than in Africa.

More than half of the world's households have access to the Internet, despite a slowdown of less than 5% per year. The number of households with connections in developed countries is almost twice as high as in developing countries. People in Europe more than three times more often have access to the Internet than in Africa.

There is a significant digital gender gap. The data collected by the International Telecommunication Union (ITU) show that young people are more likely to connect to networks than older people. This gap is relatively small in developed countries, more pronounced in developing countries. It is estimated that the proportion of people aged 15 to 24 years who have a connection in the world exceeds 70%, whereas in relation to the population as a whole, this figure is only 48% [9].

The problems of developing countries are connected with the limited markets, the migration of educated people to developed countries, the lag in technological development due to the lack of adequate funding.

Various analytical agencies determine the ratings of countries on innovation, the development of the digital economy.

Cornell University, INSEAD Business School and the World Intellectual Property Organization (WIPO) compile the Global Innovation Index (GII) annually based on 80 different parameters - from the number of applications for intellectual property rights and the created mobile applications to the amount of education expenses and the number of scientific and technical publications. The Top 10 leaders of 2018 included Switzerland, the Netherlands, Sweden, the United Kingdom, Singapore, the USA, Finland, Denmark, Germany, Ireland. China ranked seventeenth in the rankings - this is a great achievement for the country. The rapid rise of China in the ranking is a reflection of the strategic course taken by the country to develop the innovative potential of world importance, increasing the share of high-tech industries in the national economy in the interest of maintaining competitive advantage.

Russia, according to this rating, lost one place and is located at the 46th position. And for the first time, Ukraine was ahead of Russia - in 43 positions against the 50th in 2017. Such technologically backward countries as Portugal, Vietnam, Thailand, Greece and even Bulgaria bypassed Russia (Global Innovation Index 2018).

The Bloomberg Agency also ranks the countries' innovation rating – Bloomberg Innovation Index. The 2018 innovation economy rating rated countries by seven criteria, including the concentration of high-tech public companies, research and development expenses, etc. According to the rating, the United States dropped out of the top 10 for the first time in six years. South Korea and Sweden kept the top spot. Third place went to Singapore. In 2016, Russia ranked 12th in this ranking, slipped to 26th place in 2017, and in 2018 it moved up to rank 1 and ranked 25th among 50 countries (The rating of innovative economies-2018: the United States dropped out of dozens of leaders 2018).

As for such an important indicator of innovation as the number of scientists, in Russia there are 10 thousand workers, only about 50 people involved in science. This is about two times less than in science-oriented countries, and three times less than in the leading countries. The problem remains the uneven distribution of scientists by age: the majority of research workers in Russia are already 60 years old or more. Soon these scientists will stop working actively. Therefore, today there is an urgent task to attract young people aged 25–30 years to science.

The Superjob portal conducted research, according to which a third of Russians believe that Russian science is developing steadily. According to 20% of respondents, the sphere is experiencing a period of stagnation, the other 22% are confident that science is in complete decline (In Russia, the share of scientists is two to three times lower than in the leading countries 2019).

According to the HSE data, Russia ranks tenth in the ranking of countries in terms of aggregate publication activity in natural science research areas, Russia is in the top five in some specific areas. The leaders of the rating are China and the USA (Rogulin 2018).

In total, in the considered areas in 2015–2017, with the participation of scientists from Russia, 3.7% of publications were made in international scientific journals (Dyachenko 2018).

As for such an important indicator as the share of high technologies in the export of manufactured goods, according to the Strategy for Innovative Development of the Russian Federation for the period up to 2020, the country is still planning to increase this share to 2% (Order of the Government of the Russian Federation No. 2227-p About Strategy 2011).

You can bring other. The fact is that for some indicators Russia really lags behind foreign countries, for which it is practically not inferior to them. This indicates instability, fragmentation and non-systemic development of the national innovation system of the country.

New requirements are also imposed on education: renewal of knowledge, skills and skills should be carried out throughout our life. A modern person must possess not only the volume of knowledge and competence by profession, specialty, activity, but understand the need for continuous training: new information, additional knowledge helps to solve industrial problems, making him competitive in the marketplace (Volkova et al. 2011).

4 Conclusions/Recommendations

Summing up, it should be noted that the current stage of the post-industrial society is characterized by the development of the knowledge economy.

Factors such as: the transformation of knowledge into the main factor of production testify to its formation; growth in the share of the service sector; modernization of the education and training system; active implementation of modern ICT in various fields of activity; transformation of innovations into the main source of economic growth and competitiveness.

Information technologies make knowledge accessible to all people regardless of their place of residence. This opens up opportunities for all countries to accelerate their development.

According to the World Bank, the competitiveness of countries depends less and less on the availability of natural resources and cheap labor and is increasingly determined by the degree of innovation and knowledge-intensiveness of the economy. That is why the export of scientific knowledge, high-tech products, intellectual and educational services is becoming increasingly important (Stages of formation and development trends of the knowledge economy, 2018).

The distribution of knowledge depends largely on the degree of use of ICT. Evidence suggests that the greater the coverage of information technology and networks, the greater the effect on the knowledge economy. Thus, ICT is the basis for the further development of the knowledge economy.

Russia still lags behind some foreign countries in a number of indicators of the development of the digital economy. To speed up the digitalization process, it is necessary to overcome the negative impact of such factors as: the priority of the development of the commodity sector; underestimation of human and intellectual capital, insufficient support for science and education, etc.

References

Overchenko, M.: More than just technology leaders (2019). https://www.vedomosti.ru/opinion/articles/2019/01/10/791174-chem

Varavva, M.Yu.: Stages of formation and development trends of the knowledge economy. Bulletin of the Orenburg State University (2008). https://cyberleninka.ru/article/n/etapy-stanovleniya-i-tendentsii-razvitiya-ekonomiki-znaniy

Volkova, I.O., Koval, E.A., Mashukova, N.D.: Formation of directions for improving continuous professional education based on the best world practices (2011). https://www.hse.ru/pubs/share/direct/document/72016585

In Russia, the share of scientists is two to three times lower than in the leading countries (2019). https://ruposters.ru/news/09-02-2019/uchenih-stranahliderah-rossii

GII 2018 g.: China promptly burst into the top twenty leaders; Switzerland, the Netherlands, Sweden, the United Kingdom, Singapore and the United States hold leading positions in the annual rankings. https://www.wipo.int/pressroom/ru/articles/2018/article_0005.htm

Global Revolutions. Typology of societies. https://helpiks.org/8-58467.html

Nordstrom, K.A., Riderstrall, J.: Funky business (2013). http://ignorik.ru/docs/keell-a-nordstrem-jonas-ridderstrale.html?page=7

Rogulin, D.: HSE: Russia is among the top five world leaders in a number of scientific fields (2018). http://agnc.ru; https://tass.ru/nauka/5388776

Report. Measuring the Information Society (2017). Summary. https://www.itu.int/dms_pub/itu-d/opb/ind/D-IND-ICTOI-2017-SUM-PDF-R.pdf

Order of the Government of the Russian Federation of December 8, 2011 No. 2227-p About Strategy (2011). https://www.garant.ru/products/ipo/prime/doc/70006124/

The rating of innovative economies-2018: the United States dropped out of dozens of leaders (2018). https://theworldonly.org/rejting-innovatsionnyh-ekonomik-2018/

Dyachenko, E.L.: Russia in the ranking of countries on the publication activity of scientists: natural and exact sciences. HSE, Science. Technology. Innovation (2018). https://issek.hse.ru/data/2018/07/19/1151513834/NTI_N_92_19072018.pdf

Sociological analysis of the typology of society. http://bibliofond.ru; Global Revolutions. Typology of societies. https://helpiks.org/8-58467.html

Stages of formation and development trends of the knowledge economy (2018). https://helpiks.org; https://helpiks.org/9-45297.html

Global Innovation Index (2018). http://www.tadviser.ru/index.php/%D0%A1%D1%82%D0%B0%D1%82%D1%8C%D1%8F:Global_Innovation_Index

Digital Platforms in the Modern Economy: The Concept, Features and Development Trends

Irina B. Teslenko[1(✉)], Alexey M. Gubernatorov[1], Nizami V. Abdullaev[1], Irina A. Alexandrova[2], and Olga A. Kornilova[3]

[1] Vladimir State University named after A.G. and N.G. Stoletovs, Vladimir, Russia
iteslenko@inbox.ru, gubernatorov.alexey@yandex.ru, nizamka33@mail.ru
[2] Financial University under the Government of the Russian Federation, Moscow, Russia
IAleksandrova@fa.ru
[3] Vladimir Branch of RANEPA, Moscow, Russia
olgakornilova2006@mail.ru

Abstract. At present, digital platforms, as well as platform ecosystems, which are formed by them, capable of modernizing entire industries and various types of socio-economic activity, are becoming drivers of innovation, economic growth and competition. The purpose of this study is to determine the main characteristics and features of digital platforms. The definition of the concept of a digital platform is given. The history of the emergence and development of digital platforms is investigated. The main categories of platforms are characterized, as well as problems that are solved with the introduction of digital platforms. The main features, as well as the advantages and disadvantages, opportunities and threats associated with the activities of digital platforms are considered. The basic data on the use of digital platforms in the world and in Russia are presented.

Keywords: Digital economy · Digital platform · Digital technologies · Information technologies · Digital platform economy

JEL Code: M21 · O14 · O31

1 Introduction

The post-industrial era, many scientists describe as the economy of knowledge, the distinguishing feature of which at the present time is a comprehensive digitalization. Digitalization opens up unprecedented opportunities for participants in economic relations.

The use of digital technologies makes it possible to reduce transaction costs, speed up operational cycles, stimulate competition, intensify innovation processes, form new business models of interaction between market participants, etc.

E. G. Popkova and B. S. Sergi (Eds.): ISC 2019, LNNS 87, pp. 652–661, 2020.
https://doi.org/10.1007/978-3-030-29586-8_76

Digital platforms (CPUs) can be considered a peculiar business model of the modern economy.

Based on high technologies, the platforms allow producers and consumers of final goods and services to interact directly, without the use of intermediaries. They provide an opportunity for companies to share information in order to establish cooperation and organize joint work on creating innovations (Yudina and Geliskhanov 2018).

Their wide distribution in many areas allows us to talk about the era of the digital platform economy. In the arsenal of the latter - a variety of mechanisms and tools of online platforms and the Internet. The CCP essentially forms the foundation of the social and economic life of modern society.

According to Thomas Eisenman, digital platforms include a common set of rules (protocols, policies, standards and contracts with duties and rights) and components (software, equipment and service modules with an installed architecture) (Eisenmann et al. 2008).

2 Methodology

In this paper, we used the following sequence of studies. At the first stage, the subject and object of study were determined. The object of the study were digital platforms. The subject of this article is the factors contributing to the development and growth of digital platforms.

At the second stage, a digital platform survey was conducted, digital platform categories were identified, the digital platform ecosystem was explored, the main advantages and disadvantages of digital platforms were analyzed, and measures were proposed to develop an effective institutional environment that provides positive effects scaling and leveling negative consequences from digital platforms.

At the third stage, the main conclusions and results of the research were formulated and directions for their practical application were proposed.

3 Results

The main task of the platform is to enable third parties to use the existing infrastructure as a means to distribute value, for example, to create applications. That is, the platform interface is used by other persons and their products as an intermediary that delivers their values to the consumer. Third parties benefit from integration with other products, and rapid delivery of the product to the user.

Integration of other products with the main user interface, in turn, increases the value of the platform for users, forms in them a steady habit of constantly using it.

Users create more functions, therefore, for their development it is not necessary to allocate additional resources. Thereby, the cost of the platform product increases without special costs for the platform.

Digital platforms began to be used in the 1960s in the information technology (IT) industry. In 1964, IBM introduced on its computers an operating system and general equipment for them. Users thanks to this could add software and hardware without using complex programs.

A little later, in the 1980s, the creation of microprocessors by Intel and the practical, targeted operating system from Microsoft gave a tremendous impetus to the development of universal and high-performance personal computers (PCs). This in turn contributed to the improvement of software and hardware.

In the late 1990s and early 2000s, the Internet led to a quantum jump in the development of the platform economy. For a decade, Internet platforms have connected a large number of users of personal computers (PCs) to a wide range of online applications and websites (Kuprevich 2018).

Currently, one can observe the exponential growth of digital platforms. Companies that operate on the basis of platforms ("platform operators"), penetrate into various fields of activity and industry. Crowdfunding is performed on the platforms - voluntary pooling of funds and other user resources to support the efforts of other people or organizations (Aflamnah, ArtistShare, Gofundme, Ulule, Kickstarter, Yomken); there is transport (BlaBlaCar, Sidecar, Uber, Ola, Lyft, JustPark); HR and financial functions are carried out (Freelancer, Workday, Elance, WorkFusion, Crowdrquoting); mobile payments (Square, Mahala); retail and online auctions (Etsy, Snapdeal, Angie'sList, eBay, Flipkart, Amazon); working social networks (Snapchat, LinkedIn, Facebook); public services are provided (G-Cloud): clean energy is being created (SolarCity, Sungevity, EnerNOC), etc.

TheCenterforGlobalEnterprise, a non-profit research organization, based on the analysis and evaluation of 176 platforms from different countries, identified the following categories of CPUs:

- investment platforms represented by holding companies managing a portfolio of platform companies. An example would be the Priceline Group, which focuses on online travel and related services;
- innovative platforms that serve as the technological basis for attracting external innovators. With the help of such platforms, third-party companies can develop additional services and products. Examples of such platforms are Android from Google and Apple's Apple Inc., which have developed large modern ecosystems to create applications for their own mobile devices;
- transactional platforms that assist individual organizations and individuals in finding each other, facilitating their interaction and commercial transactions between them. The best examples of such platforms can certainly be considered e-commerce platforms such as eBay and Amazon. On-demand platforms such as Zipcar, Uber and Airbnb provide for the exchange of services and goods between certain individuals;
- integration platforms (represented by several large companies, such as Apple and Google), which have the capabilities of both innovative and transactional platforms. Both companies have developed innovative platforms for their own developers. Similarly, Amazon and Alibaba can be attributed to transactional platforms for their own individual users, as well as for many vendors selling products on their own e-commerce platforms (Evans and Gawer 2016).

The CPU ecosystem is based on a distributed architecture; it is a symbiosis of a team and a technological tool, in which there is a high degree of freedom of communication between the participants, processes are adjusted as necessary and easily adapted to current needs.

Ecosystems of CCS provide complexness and quality of accepted solutions in relation to the problems that may arise during the active transition to the large-scale use of automated systems in the digital economy. These include.

1. Inadequate technology when working with data.

Some companies are trying to get ahead, significantly ahead of competitors, others, on the contrary, are lagging behind the pace of development of most companies, developing internal functions and competencies for storing, collecting and processing data. At the same time, they can be forcibly excluded from the interaction complex, since at a certain period of time they are unable to provide support for the rules when exchanging data packets. This is due to the fact that the primary ones are not the technologies inside the digital platform, but the formats and forms relating to the external exchange of information.

The ecosystem makes it possible to introduce requirements and standards for data that digital platforms are able to exchange, determine economic and technical parameters, monitor the potential and condition of automated systems in the global network.

2. Incomplete use of digital analytics.

Digital platforms can implement their own models and methods for analyzing the domain for which they were created. At the same time there are a number of mandatory principles and techniques for analyzing digital information. For example, for each individual digital platform, an important component is constant analytics in terms of user conversion, reboot and security, as well as in matters of stable work with external systems. Moreover, the basic functions are available, as well as to the owner of the digital platform, and its third-party auditors and regulators.

In this case, the ecosystem can not only set forth a number of mandatory requirements for analytical elements of any digital platform, but also provide prepared unified templates, algorithms, comparative indicators. This helps to eliminate the problem of the incorrect use or underestimation of models, tools and technologies for analyzing digital information.

3. Low quality components of the platform.

For any digital platform, it is important to maintain the required quality of automated models and data, products and technologies, as well as interaction interfaces. Independent solution of this complex of tasks is a rather complicated and expensive work. You need to start by comparing the quality of your own platform with others who participate in the interaction through transactions.

Within the framework of a specific ecosystem, special services (providers, agents) for checking and controlling the quality of incoming and outgoing digital data, the subject model used, the functionality used and tools, as well as the interfaces that open for collaboration, are mandatory.

4. Integration errors.

CPU directly interacts with third-party systems. It is important that the interaction of the digital platform is sustainable, continues to exist and develop, even when the platform itself and its components are transformed or replaced by others.

The Ecosystem SC allows to significantly reduce the risks and integration errors, as it can offer integration patterns and uniform schemes, standardized interfaces, a predictable and unified architecture, action logic and so on.

5. Minimizing security issues.

Because it is important for owners to quickly implement a digital platform in the market, they often neglect safety rules. And some errors and omissions can cause quite serious consequences. Solve security issues of the Unified ecosystem for them.

6. Fragmentation and closeness of platforms

Sometimes the owners of the platform try to make it closed. In this case, despite all the usefulness and functionality, the platform becomes inconvenient. Users need fast and flexible solutions, so they will not waste time on a platform with an uncomfortable interface.

The ecosystem is able to provide developers of digital platforms with rules for designing comfortably interacting systems, freeing them from excessive closeness and unnecessary fragmentation into extremely limited user segments.

7. Limited use and creation.

Quite often, unscrupulous competitors who want to take a leading position interfere with the work of the CPU. In order to avoid the consequences of this ecosystem, it is necessary to include the principles of market and technical coordination of complementary and competing digital platforms that can be supported by special automated methods (arbitrators and agents) that are clear and clear to its participants.

8. Low effectiveness of learning technologies and development.

No matter how comfortable and perfect the digital platform is, it needs to be upgraded in any case. At the same time, users also need training to better interact with the digital platform.

The formation and preparation of the platform itself and its external users (including connected systems) require unified and effective conclusions. The ecosystem is quite capable of finding solutions by proposing the proper tools, approaches, and options.

9. Outdated control methods.

Control of digital platforms in the framework of pre-digital (paper) technologies does not fully guarantee their intensive appearance and subsequent dynamic growth. Instead of a lengthy procedure for preparing, approving and agreeing to voluntary standards or mandatory regulations, it is necessary to have regulatory regulation with early testing and debugging of any of the rules introduced.

So, equality, relevance, predictability, clarity and, most importantly, trust between economic actors and automated systems are realized through digitized regulatory mechanisms of the digital ecosystem.

The main advantage of using digital business models is the reduction of time, transaction and transaction costs due to the lack of intermediaries. According to research conducted by international company PricewaterhouseCoopers, 56% of all users of digital platforms for passenger transportation (Zipcar, Lyft, Car2Go, Uber, RelayRides and others) choose them in connection with a better price, 32% due to a wider choice in the market and 28% due to more convenient access to services.

Digital platforms are capable of providing comparatively cheap access to value chains and global markets for subjects. For example, when placing a mobile application on the GooglePlay or AppStore platforms, or by providing various services and products using the Taobao, eBay, Amazon platforms, entrepreneurs can get instant access to a large number of potential consumers around the world.

In addition, the reduction of barriers to entry into the market is stimulated by the development of entrepreneurship. So, for example, about 43% of Etsy's British e-commerce platform, which specializes in handicrafts and vintage products, noted that Etsy is their first platform for carrying out their activities, and 36% of respondents believe that without this digital platform would begin to be active in this area.

The services provided (for example, the use of convenient forms of payment), allow you to expand the range and improve the quality of products and services provided to consumers.

Personal data collected by digital platforms provide an opportunity to personalize communication with participants.

According to data provided by PricewaterhouseCoopers, the size of the savings in the joint use of the CP can reach 335 billion dollars by 2025, which is 20 times higher than the savings in 2013.

Despite certain advantages, there are a number of threats to the functioning of the CPU. One of the main threats is to protect the privacy of personal data. PCs accumulate and process big data about participants, track their actions and may resort to unfair competition, manipulation by customers.

In addition, there are a number of problems in the activities of the CP, such as:

- there is no fuzzy national and international regulation of companies that provide digital platforms;
- discrimination of participants of digital platforms from the point of view of their lack of rights and benefits that employees have (paid holidays, pensions, etc.). The fact is that participants of the digital platform do not have the status of full-time employees, are independent performers;
- volatility of suppliers' income due to price unpredictability in cases when the platform is able to independently set tariffs for services and products of suppliers;
- high costs of suppliers (for example, Uber taxi drivers are themselves responsible for all the costs of car maintenance, depreciation, fuel and insurance coverage);
- Transaction security issues and others (Geliskhanov et al. 2018a, b).

In recent years, digital platforms have undergone significant changes.

If in 2008 the list of the 10 most expensive public companies, most of which were focused on the commodity sector, was included in the only platform company Microsoft, then currently there are already seven such companies (Google, Amazon, Facebook, Apple, Microsoft), Alibaba, Tencent), the cumulative market capitalization of which is about 4.5 trillion dollars, which is more than 7 times the volume of the Russian stock market (625.2 billion dollars). These companies have overtaken the leaders of the past decade among digital platforms such as Blackberry, MySpace, Windows, Nokia.

Most of the large platform companies are located in the PRC and the USA, they are in Russia, Japan, India, Great Britain, Germany and other countries.

The rapid promotion of digital platforms in the ranking is due, among other things, to the development of their ecosystems. For example, the average monthly number of active users of the social networking site Facebook has increased over the past 10 years almost 12 times from 197 million people in 2008 to 2.3 billion people in 2018.

Active growth showedInstagram network - 11 times (from 90 million people in 2013 to 1 billion people in 2018) and WhatsApp messenger - almost 8 times growth (from 200 million people in 2013 to 1.5 billion people by 2018). Among non-American digital platforms, the best result was shown by the Chinese platform WeChat (TencentHoldings) - an increase of 22 times (from 50 million people in 2011 to 1.1 billion people in 2018).

At the same time, today's leaders must realize that, despite the absence of obvious risks, continuous market dominance is not guaranteed by any platform due to unpredictable and rapid technological transformations.

Digital platforms in Russia are presented in the format of payment, search engines, instant messengers, social networks, platforms in the field of e-commerce, tourism, education, employment, finance, passenger traffic, etc.

Russian platforms significantly lose on capitalization to domestic banks and commodity holdings and are weakly represented in the market.

Thus, when summarizing the results in 2017, only two platform companies entered Yandex-1, which ranked 12th, and Mail.RuGroup, which ranked 23rd, in the top 100 most expensive public companies in Russia.

Global digital platforms have a strong position in the Russian markets. Their share in the total market volume of platforms in Russia accounts for about 30% (or $8 billion). In some areas, domestic digital platforms can compete with some global platform leaders (for example, on the Russian market of messaging, search engines and social networks).

The total penetration rate of messengers and social networks in Russia reached 47% by the beginning of 2018 (these are about 68 million citizens). The leader among such services is the American platform Youtube (63%), then the Russian social networks Vkontakte (61%) and Odnoklassniki (42%). Among the messengers are leading American Skype (38%) and WhatsApp (38%). The world leader of Facebook is in fifth place (35%). Two platform leaders compete in the Russian search engine market: the Russian search engine Yandex and the American Google.

The main source of income for platform companies in Russia is advertising revenue. In particular, Yandex's revenues for 2017 amounted to 94.1 billion rubles, which doubled the revenue volume of a Google subsidiary (45.2 billion rubles). The total revenue of the head company Google at the end of 2017 amounted to 110 billion dollars, 95.4 of which are advertising revenues (87%).

The main source of Yandex revenue is revenue from contextual and display advertising (93% or 87.4 billion rubles). Incomes of foreign social networks in Russia are much lower than the incomes of Russian social networks Vkontakte and Odnoklassniki.

So, on a separate platform market in Russia, a relatively small number of domestic companies are presented that compete with global digital platforms. Positions of Russian companies in global platform markets are still weak.

According to experts of the World Bank, the formation of the CP should act as one of the priority directions of the strategy of forming a digital economy in the Eurasian economic space until 2025. This is primarily due to the important role of platforms in obtaining digital dividends (positive effects) and the development of innovations at the regional level (in the form of creating new jobs, improving public services, accelerating economic growth, etc.), including expense of modernizing the principles and mechanisms of cross-border business and reducing the cost of international transactions, including providing entrepreneurs with access to a greater number of possible clients.

In the Russian Federation, state support for the formation of domestic digital platforms is one of the strategic objectives for the coming years. economy and the provision of public services, including to meet the needs of individual entrepreneurs, the population and small and medium his business.

In order to develop an effective institutional environment that ensures the scaling up of positive effects and leveling the negative consequences of the operation of digital platforms, the following measures were proposed:

- elimination of legislative restrictions that prevent the formation of domestic platforms, adjusting labor, tax and other types of legislation in order to maintain a balance between the interests of platform companies, the state and society, including in terms of ensuring public and national security and protecting the rights of citizens of the Russian Federation;
- formation of a productive policy for managing large data, including the creation of mechanisms and tools that prevent the violation of the principle of confidentiality of personal data;
- Development of the infrastructure of broadband access networks, which include the launch of fifth-generation networks (5G);
- increasing the country's transport and logistics potential for the modernization of e-commerce platforms;
- introduction and creation of effective mechanisms for arbitration and resolution of disputes between participants of digital platforms in order to create a trusting institutional environment and ensure transaction security;
- development of a unified service ecosystem for the launch and formation of innovative projects (and, above all, projects that are focused on implementing and developing platform solutions in various sectors and sectors of the Russian economy and social sphere by combining the competences and resources of state-owned companies, the state, development institutions, and business and scientific - expert community;

- support for domestic production of various robots, sensors, cyber-physical microelements, systems and devices that are necessary for the development of projects in the field of consumer and industrial Internet of things, robotics, 3D printing and other areas (Geliskhanov et al. 2018a, b).

4 Conclusions/Recommendations

Summing up, it should be noted that the rapid development of digital technologies and the emergence of innovative business models open up opportunities for Russia for a qualitative leap in the innovative transformation of almost all sectors of the economy.

Digital platforms have great potential for the development of many areas of human socio-economic activity. CPUs can contribute to the formation of positive results for the state and society, while their work is accompanied by certain threats and risks. This explains the need for developing mechanisms and solutions for the institutional and non-institutional nature of scaling up positive effects and managing the risks and threats associated with the activities of digital platforms.

Predicted trends of active access to high-speed mobile Internet, the spread of mobile and other digital devices, the development of digital technologies - big data, distributed registry technologies (blockchain), the Internet of things (IoT), artificial intelligence, etc., as well as the unique features of the platform business models create serious prerequisites for rapid growth of the platform economy.

Currently, individual Russian CPUs can compete with foreign platforms only on Russian platform markets, while their positions on world markets are weak.

In order to minimize risks, to overcome the negative consequences of uncontrolled access to personal information of citizens, to strengthen the position of domestic platforms, it is necessary to implement a set of systemic measures aimed at supporting Russian CPUs, including in global markets.

References

Geliskhanov, I.Z.: Digital Platform as an Institute of Economy of a New Technological Generation. Materials of the International Youth Scientific Forum. MAKS Press (2018a)

Geliskhanov, I.Z., Yudina, T.N., Babkin, A.V.: Digital platforms in economics: the essence, models, development trends. Scientific and Technical Statements SPbGPU. Economics **11**(6), 22–36 (2018b)

Gribanov, Yu.I.: Basic models of creating sectoral digital platforms. Issues Innov. Econ. **8**(2), 223–234 (2018)

Kovalenko, A.I.: The problematics of research multilateral platforms. Modern competition, vol. 10, no. 3(57), pp. 64–90 (2016)

Kuprevich, T.S.: Digital platforms in the global economy: current trends and directions of development, no. 37/1, pp. 311–318. Economic Bulletin of the University (2018)

Osipov, Yu.M., Yudina, T.N., Geliskhanov, I.Z.: Digital Platform as an Institute for the Era of Technological Breakthrough. Econ. Strat. no. 5(155), 22–29 (2018)

Selin, A.: Digital business models: the main trend of the modern market. News digest of the world of high technologies, no. 5, p. 14 (2016)

Yudina, T.N., Geliskhanov, I.Z.: Data economics: bigdata, digital platforms and digital rent. In: Babkina, A.V. (ed.) Innovation Clusters of the Digital Economy: Drivers of Development: Proceedings of a Scientific-Practical Conference with International Participation, pp. 218–227. Polytechnic Publishing House, St. Petersburg, un-that (2018). Dr. Econ. Sciences

Evans, P.C., Gawer, A.: The rise of the platform enterprise. A global survey. The Center for Global Enterprise, no. 1, R. 28 (2016)

Eisenmann, T.R., Parker, G., Van Alstyne, M.: Opening Platforms: How, When and Why? Harvard Business School. Working Paper 09-030 (2008)

Methodical Approaches to Analysis of Performance of Budgetary Obligations on the Basis of the Risk-Oriented Approach

Nadezhda I. Yashina[1](✉), Svetlana D. Makarova[1], Oksana I. Kashina[1], Victor P. Kuznetsov[2], and Elena V. Romanovskaya[2]

[1] N.I. Lobachevsky National Research Nizhny Novgorod State University, Nizhny Novgorod, Russian Federation
sitnicof@mail.ru, makarovasd@iee.unn.ru, oksana_kashina@mail.ru,
[2] Nizhny Novgorod State Pedagogical University named after K. Minin, Nizhny Novgorod, Russian Federation
kuzneczov-vp@mail.ru, alenarom@list.ru

Abstract. Subject. The subject of this study is evaluation of the effectiveness of budgetary obligations with a risk-oriented approach.

Goals. The aim of the work is to develop a method for assessing the risk-budget sphere that arises during the execution of budgets by determining specific risk per unit of budget execution.

Methodology. As a basic approach, a method of assessing the risks of budget execution at any level is used, which is based on the law of distribution probability. A distinctive feature is the assessment of the risk of budget execution in comparison with the planned value.

Results. The result is a detailed theoretical basis for analyzing security of federal budget, by developing and applying a methodology for assessing the overall risk of budget execution, as well as indicators of the effectiveness of execution in the budget system of Russia based on a risk-based approach.

Conclusions. The proposed approach can be considered as a universal method of budget management, allowing assessing the degree of efficiency of budget execution on the basis of risk-oriented approach.

Keywords: Budget · Performance of budget obligations · Risk-oriented approach · Risk dynamics · Budget security

JEL Code: G21 · G33

Risks in the modern economy are present in all areas of its functioning. The budgetary sphere is no exception, despite the strict regulation of budgetary relations, both in Russia and abroad. Sometimes risks of non-receipt of funds from budgetary sources lead to an increase in uncertainty factors when planning the volume of budgetary sources and under-financing budgetary obligations.

E. G. Popkova and B. S. Sergi (Eds.): ISC 2019, LNNS 87, pp. 662–669, 2020.
https://doi.org/10.1007/978-3-030-29586-8_77

The task of forming a system approach to assess the efficiency of functioning of a budget is relevant for most modern governments. Often this is due to consequences of unbalancing commodity and financial markets (for example, due to rising inflation, speculative and sanction effects, etc.), use of macroeconomic management tools (for example, varying budget deficits), etc. In such cases, the risks of a possible destabilization of the economy during the development of unfavorable conditions increase in the domestic and foreign markets (Kuznetsov et al. 2018).

In general, the purpose of assessing efficiency of budgetary obligations is development and optimization, ensuring budgetary security, as well as minimizing losses in the budgetary sphere.

In the scientific literature, a number of both domestic and foreign authors are engaged in the study of evaluation of public sector performance. So in the work (Kononova 2012), great attention is paid to the implementation of monitoring, financial control and audit over the efficiency of formation and use of budgetary funds. Of considerable interest are works (Perekrestova 2011), (Yanov 2012), devoted to evaluating the effectiveness of quantitative methods for analyzing budgetary efficiency. Yashina conducted research of efficiency of budget financing in separate directions. Currently, it is particularly relevant to determine the effectiveness of integrated budget financing, which allows characterizing both efficiency of budget planning and execution (Potashnik et al. 2018).

Fundamental research in the field of risk assessment in the sphere of public financial relations should include the work of foreign economists. For example, (Blank 2012) evaluates the impact of the heterogeneity of regions on the peculiarities of budget financing. In the works (Breyer 1993), (Rocha et al. 2008), (Obama 2011) substantiate the reasons for institutional nature, influencing risks of regulating parameters of state budgets and trust funds for a social purpose. (Hampton 2005), (Van Dam and Andersen 2008), (Nowicki 2010) analyzes problems of reducing fiscal risk by assessing changes in administrative burden, etc.

It should be noted that the reviewed studies suggest different approaches to the analysis of budget processes in relation to different countries. However, they do not adequately address the problems of a quantitative assessment of budget security, which has recently become increasingly important for the functioning of the budget system in a rather unstable economy depending on many factors of internal and external influence.

With regard to budget execution, risk can be considered as a probability of implementing budget parameters compared to planned values. The presence of deviations between planned and actual indicators presupposes the need for a quantitative assessment of the risk budget sphere and consequences that in general can increase the level of budget security.

It should also be noted that when forming budget parameters in the situation of their substantial dependence on energy prices in Russia at present, one should take into account the probability of deviation of planned and actually performed indicators depending on changes occurring outside our economy, but having a direct impact on it (Kuznetsov et al. 2017).

Thus, the question of managing various kinds of risks in the budgetary sphere is extremely important.

As part of the study, we consider implementation on the example of assessing the risks of fulfillment of budgetary obligations, for which we propose to use the following indicators:

- planned expenditure budget for the i-period by sections;
- the actual performance of budget expenditure for the i-period by section;
- standard deviation and coefficient of variation as indicators of risk assessment of performance of budgetary obligations.

The degree of risk in this case is the probability of incomplete and/or untimely financing of budgetary obligations of budgets on different levels.

If we interpret the risk in the budgetary sphere as the variability of budget performance in comparison with expected value, then their absolute execution will be a characteristic of a risk-free budget, and a performance other than 100% will be risky. The greater the volatility (dispersion) of budget indicators, the greater the risk.

To measure the heterogeneity of deviations in the set of budgetary indicators, which in the study are budget commitments by section, it is proposed to use the standard deviation of their performance indicator (σ): the larger it is, the greater the variability of budget indicators and, consequently, the higher the risk of their non-performance.

It should be noted that using only the standard deviation for risk assessment is not sufficient, as it does not take into account the level of risk per unit of budgetary performance. As a measuring instrument, it is proposed to use the coefficient of variation (CV), which establishes the typicality average. If CV greater, then higher relative risk of non-fulfillment of budgetary obligations.

If CV $\leq$ is 33.3%, the average value is considered typical and it makes sense to talk about quality budget planning, minimization of budget risks and a high degree of budget security.

However, 33.3% is the boundary maximum value that characterizes a homogeneous set of statistical indicators. Let us set an upper limit of 10% to characterize the effectiveness of budget commitments, which is approximately 1/3 of the total value and, in accordance with the requirements for homogeneity of the sample, makes it possible to speak of good budget planning and risk-free fulfillment of budget commitments due to insignificant dispersion of the studied indicators. It is proposed to differentiate the range of CV values from 0% to 10% of applicants to risk response of budgetary obligations as follows: up to 5% - low level of risk; 5–10% - moderate level of risk; over 10% - high level of risk.

To substantiate the existence of patterns when using the proposed approach, consider the expenditure part of the federal budget of the Russian Federation for the period 2015–2017.

Thus, the leaders in terms of expenditure remain sections, financing of which the most important tasks for the government to ensure a decent standard of living, law, and order in the government: social policy (2015—27.3%, 2016—28.0%, 2017—30.4%), national defence (2015—20.4%, 2016—23.0%, 2017—17.4%), national economy (2015—14.9%, 2016—14.0%, 2017—15.0%) and national security and law enforcement (2015—12.6%, 2016—11.6%, 2017—11.7%).

In addition to the analysis in the total amount of budget financing undoubtedly significant is the volume of performance of budget indicators, allowing to estimate the deviation of actual indicators from planning, and therefore the effectiveness of budget planning as a whole.

In relation to budgetary obligations and as a basis for an analysis of efficiency of execution of the federal budget on expenses, indicators for the period 2015–2017 were also considered.

The analysis of the percentage of implementation of all expenditure sections in the vast majority of cases characterizes the underperformance of indicators to the level of expenditures approved by the consolidated budget list, taking into account changes, for the entire period under study. Risks arising in the process of performance of budgetary obligations, we will evaluate according to the above methodology.

First, we calculate the share of expenditures (pi) for each indicator in the total amount of budgetary obligations. Next, we calculate the standard deviation (σ) and the coefficient of variation (CV) of budget execution for expenditures for the period 2015–2017.

Thus, according to the results of calculations presented in Table 1, the aggregate risk of performance of budgetary obligations in 2015–2017 amounted to 0.32%, 0.51% and 0.27% correspondingly. Values are classified as low risk (0 to 5%), indicating a low level of risk in the execution of the federal budget against budgetary obligations as a whole for the period under review.

Table 1. Risk of execution for Federal budget expenditures by sections (2015–2017)

№	Indicator name	CV		
		2015	2016	2017
1	National issues	0,17050%	0,12294%	0,75834%
2	National Defence	0,43833%	0,80400%	1,29819%
3	National security and law enforcement	0,00000%	0,98845%	0,56177%
4	National Economy	0,90542%	0,63879%	0,36105%
5	Housing and communal services	0,08066%	0,14035%	0,05387%
6	Environmental protection	0,03366%	0,12934%	0,23080%
7	Education	0,07201%	0,30038%	0,49504%
8	Culture, cinematography	0,07139%	0,09451%	0,39142%
9	Health	0,18388%	0,00643%	0,19763%
10	Social policy	0,40180%	1,15730%	1,69683%
11	Physical culture and sport	0,06984%	0,54722%	0,19380%
12	Media	0,07922%	0,16354%	0,27528%
13	Public and municipal debt	2,09441%	0,09257%	0,18614%
14	Interbudgetary transfers to budgets of the Russian Federation	0,17758%	0,46364%	0,21381%
15	Expenditures, total	**0,32%**	**0,51%**	**0,27%**

Source: Authors' calculations

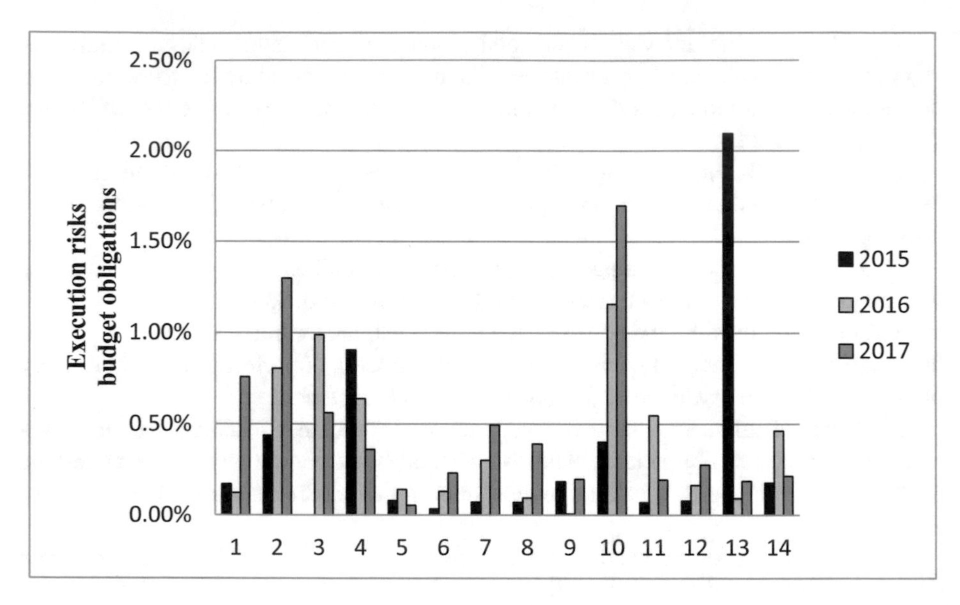

Fig. 1. Dynamics of changes in the risks of fulfillment for budget obligations under expenditure sections (2015–2017). Source: Authors' calculations

It should be noted that the conclusions drawn from the analysis of general indicators are an illustration of generalized characteristics, the state of which generally positive may contain prerequisites for reducing budget security.

After analyzing the dynamics of indicators by risk values, the following results were obtained (Fig. 1).

Evaluation of changes in the level of risks of budgetary liabilities revealed four scenarios that characterize changes in budget risks within expenditure sections in dynamics (Table 2).

It should be noted that the maximum risk of performance of budgetary obligations does not exceed 2.5%. However, fluctuations of risks within the expenditure sections of the analyzed period can also be used as an indicator of budgetary efficiency in terms of risk-oriented approach (Table 3).

The presence of significant dynamics risks suggests that there may be difficulties in carrying out strategic budget planning over the three - year period, even with low-performance risks of budgetary commitments. Figure 2 shows the dynamic risk values that indicate significant changes within the analyzed period for certain expenditure sections.

Table 2. Grouping of expenditure sections depending on the dynamics of fulfillment for the Federal budget commitments (2015–2017)

№	Group	№	Expenditure section
			Title
1	Steady increase in risk	2	National defence
		6	Environmental protection
		7	Education
		8	Culture, cinematography
		10	Social policy
		12	Media
2	Sustainable risk reduction	4	National economy
3	Risk variability towards increasing	1	National issues
		9	Health
4	Risk variability in the downward direction	3	National security and law enforcement
		5	Housing and communal services
		11	Physical culture and sport
		13	Public and municipal debt
		14	Interbudgetary transfers to budgets of the Russian Federation

Source: Authors' calculations

Table 3. Risks of dynamics of budget obligations of the Federal Budget for 2015–2017

No.	Expenditure section	CV
1	National issues	0.037
2	National Defence	0.42
3	National security and law enforcement	0.08
4	National Economy	0.185
5	Housing and communal services	0.11
6	Environmental protection	0.098
7	Education	0.066
8	Culture, cinematography	0.093
9	Health	0.098
10	Social policy	0.732
11	Physical culture and sport	0.076
12	Media	0.102
13	Public and municipal debt	0.346
14	Interbudgetary transfers to budgets of the Russian Federation	0.066

Source: Authors' calculations

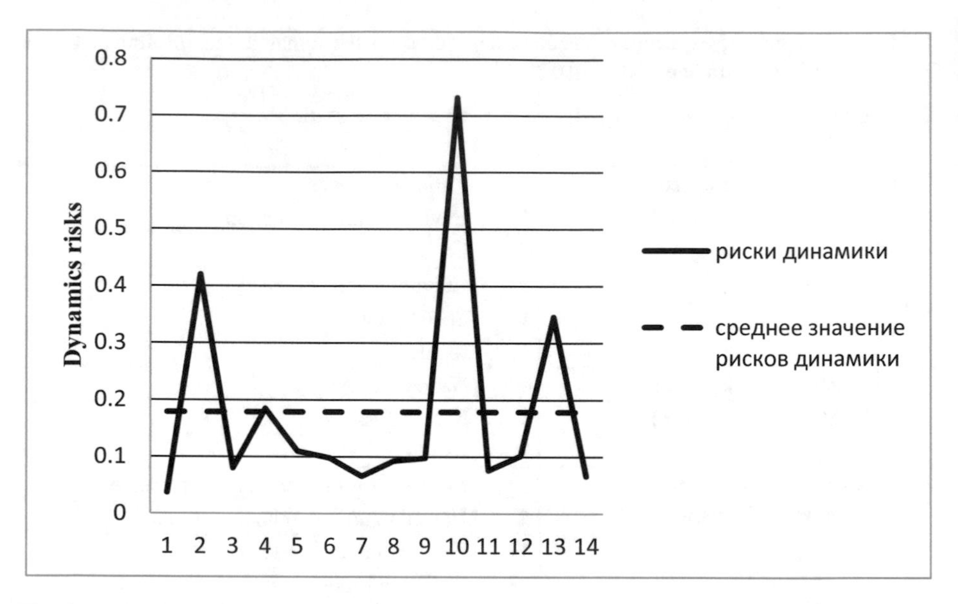

Fig. 2. Variability of the risks of the dynamics of budgetary obligations of the federal budget for the period 2015–2017. Source: Authors' calculations

Depending on the group in which a particular budget commitment is made (see table 6), the following description of changes can be given based on dynamic risk values. Thus, the section "national defense" has the risk of dynamics in the group "steady increase of risk" 42%, "social policy" - 73.2%, which characterizes the increase of risk in the execution of these sections expenditure. At the same time, in the group "sustainable risk reduction" in the section "national economy", the value of the dynamics risk of 18.5% indicates a significant trend towards reduction of performance risks.

In this way, on the background of the overall assessment of performance of budgetary obligations, in which there were no items with high and even moderate levels of risk, trends were identified and can be used in as additional clarifying indicators of the efficiency of the functioning of the budget sector, which allow to clarify the probability of reducing or strengthening the budget security.

The proposed approach to assessing the budget process from the perspective of the risk-based approach is universal and allows for a comprehensive assessment of the totality of budgetary indicators, forming an integral indicator to assess the possible risks of planning and execution of income and expenditure part of the budget of any level, and the entire budget system as a whole, taking into account the requirements of the budget security in conditions of a volatile economy.

Acknowledgments. The article was prepared with the financial support of RFBR. Grant 18-010-00909 A.

References

Kononova, M.V: On increase of efficiency of activity of bodies of the Federal treasury on cash servicing of execution of budgets. Finance, no. 3, pp. 18–20 (2012)

Perekrestova, L.V.: Methods of assessment of efficiency of budgetary regulation of territorial development. Regional economy: theory and practice, Moscow, no. 36, pp. 2–8 (2011)

Yanov, V.V.: The substance of the budget risk: theoretical and methodological aspects. Siberian Financial School, no. 6, pp. 96–101 (2012)

Yashina, N.I., Borisova, S.P., Makarova, S.D.: Improvement of financing of general education institutions on the basis of estimation of efficiency of budgetary expenses, Economics of Education, no. 5, pp. 32–52 (2008)

Yashina, N.I., Makarova, S.D., Roganova, S.Y.: Research of efficiency of spending of budgetary funds on education. Economics of Education, no. 6, pp. 45–60 (2008)

Blank, F.: Inspection Reforms: Why, How, and with What Results. OECD Publishing, Paris (2012)

Breyer, S.: Breaking the Vicious Circle: Toward Effective Risk Regulation. Harvard University Press, Cambridge (1993)

Rocha, R., Brunner, G., Hinz, R.: Risk-Based Supervision of Pension Funds. Emerging Practices and Challenges. The World Bank, Washington (2008)

Obama, B.: Toward a 21st century regulatory system. Wall Street J. **18** (2011)

Hampton, P.: Reducing administrative burdens: effective inspection and enforcement. HM Treasury, London (2005)

Potashnik, Y.S., Garina, E.P., Romanovskaya, E.V., Garin, A.P., Tsymbalov, S.D.: Determining the value of own investment capital of industrial enterprises. In: Popkova, E.G. (ed.) The Impact of Information on Modern Humans. Advances in Intelligent Systems and Computing, vol. 622, pp. 170–178. Springer, Cham (2018)

Van Dam, R., Andersen, E.B.: Risk-based Supervision of Pension Institutions in Denmark. The World Bank, Washington (2008)

Kuznetsov, V.P., Garina, E.P., Andryashina, N.S., Romanovskaya, E.V.: Models of modern information economy conceptual contradictions and practical examples, p. 361 p. Emerald Publishing Limited, Bingley (2018)

Kuznetsov, V., Kornilov, D., Kolmykova, T., Garina, E., Garin, A.: A creative model of modern company management on the basis of semantic technologies. In: Kravets, A., Shcherbakov, M., Kultsova, M., Groumpos, P. (eds.) Creativity in Intelligent Technologies and Data Science. Communications in Computer and Information Science, vol. 754, pp. 163–176. Springer, Cham (2017)

Nowicki, M.: The Investment Attractiveness of the Regions and the Sub-regions of Poland 2010. The Gdansk Institute for Market Economics, Gdansk, Poland (2010)

Gamukin, V.V.: Budgetary risks: introduction to general axiomatics. TERRAECONOMICUS **11**, 52–61 (2013)

Yashina, N.I., Makarova S.D., Pronchatova-Rubtsova N.N.: Evaluation of budgetary potential of subjects of the Russian Federation for the purpose of investment attractiveness of regions. In the collection: Finance architecture: anti-crisis financial strategies in the context of global change. In: Collection of Materials VII International Scientific-Practical Conference, pp. 174–178 (2016)

The Conversion of the Glass Industry: The Transition from Traditional Business Models to New Digital Platforms

Alexey M. Gubernatorov(✉), Elena N. Gorbatenko, Alla B. Kuznetsova, Lyudmila A. Shmeleva, and Dmitry V. Kuznetsov

Financial University under the Government of the Russian Federation, Vladimir Branch, Vladimir, Russia
gubernatorov.alexey@yandex.ru, gorbatenko.el@yandex.ru, gera60@yandex.ru, dvkuznetsov@fa.ru

Abstract. The article reveals the features of digital transformation in the glass industry, which today is the key to building a digital economy and obtaining digital dividends, that is, to achieve measurable economic results through the introduction of digital technologies. The purpose of this study is to determine the directions of transition of the glass industry from traditional business models to new digital platforms and substantiate their importance, relevance and efficiency. Methodical support of transition of the glass industry to the digital platform is presented. In this study, we will note significant trends related to the digital economy, both in socio-economic relations and in the scientific agenda, and show how the proposed project relates to them. At the end of the study we will discuss separately the practical relevance of the expected results.

Keywords: Digital economy · Digital platform · Business processes · Information technology · Digital transformation of the glass industry

JEL Code: M21 · O14 · O31

1 Introduction

In order to maximize the benefits of digital transformation in innovation, economic growth and social prosperity, there is a need to recognize the concern that existing macroeconomic statistics may not fully appreciate the benefits of digital products and products based on digital technologies or cross-border transactions. Questions remain about how to assess the contribution of the sharing economy, platforms to GDP and productivity growth. Countries with high levels of development and use of digital technologies are also engaged in measuring their impact. The Chinese National Bureau of statistics (National Bureau of Statistics) index published by the digital economy of China. In the Republic of Korea, it is planned to add a sharing economy to GDP in 2019. The Bureau of Economic Analysis (BEA) is developing tools to better measure the impact of rapidly changing technologies on the U.S. economy and global supply

E. G. Popkova and B. S. Sergi (Eds.): ISC 2019, LNNS 87, pp. 670–676, 2020.
https://doi.org/10.1007/978-3-030-29586-8_78

chains. BEA aims to calculate the contribution of the digital economy to GDP and improve indicators related to high-tech goods and services, international trade, sharing economy and free digital content, as well as examine economic indicators beyond GDP to better understand the contribution of the digital economy to welfare in General. The digital transformation of the glass industry is the key to building a digital economy and receiving digital dividends, that is, achieving measurable economic results through the introduction of digital technologies. At the industry level, the transformation of the glass industry is characterized by the minimization of human participation in the production process and the transition to effective data-based management. In addition to the widespread introduction of ERP solutions, technologies that contribute to the transformation of traditional production into digital and characterized by full digital integration of production and logistics chains, as well as supply chains, include: - digital design and modeling as a set of computer-aided design, computer-aided and supercomputer engineering, mathematical modeling, optimization and technological preparation of production, focused on additive manufacturing, and the development of "smart" models and "smart" digital counterparts; - the use of new synthetic materials, especially composite materials, silicates, metamaterials and metal powders for additive manufacturing; - additive technologies: additive manufacturing systems, materials, processes and services; - industrial sensor technology: the introduction of "smart" sensors and control tools (controllers) in production equipment, in the room at the level of the shop or enterprise as a whole; - industrial robotics: first of all, flexible production cells; - generation, collection, storage, management, processing and transmission of "smart" big data; - industrial Internet of things; - virtual, augmented and mixed reality; - expert systems and artificial intelligence. None of the advanced production technologies, taken separately, is able to provide a long-term competitive advantage in the market. We need a system of integrated technological solutions that provide in the shortest possible time the design and production of globally competitive glass products of the new generation. These solutions are combined in the so-called "glass industry 4.0". Unique solutions in the field of current innovations in the production, processing and finishing of glass were the source and basis for the creation of the concept of "3D" in the glass industry, which is a new production model based on a multidisciplinary approach to the creation of advanced production (Results of social and economic development of the Vladimir region for 2018, 2018).

2 Methodology

The methodology is described in the following main provisions: 1. Creating digital platforms that enable new ways of creating value through the introduction of advanced digital technologies. Due to predictive Analytics and big data, the platform approach allows to unite geographically distributed participants in the design and production processes, to increase the level of flexibility and customization, taking into account the requirements of consumers. 2. Development of a system of digital models, both new designed products and production processes. Digital models must have a high level of adequacy to real products and real processes (convergence of the material and digital worlds generating synergetic effects) in order to eventually become digital counterparts.

3. Digitalization of the entire life cycle of glass products, from concept and design to production, operation, after-sales service and disposal. “Glass industry 4.0” covers the product life cycle from the stage of research and product planning to the development of a digital layout and a digital double, and to the creation of prototypes or small series. Glass industry 4.0 uses “smart” big data to create “smart” product models (e.g. machines, structures, assemblies, devices and installations) developed through the application of a new digital design and modeling paradigm focused on the development of “smart” digital counterparts (Evans 2013).

At the moment, we are witnessing an active transformation in the direction of the digital economy. The report of the Organization for economic cooperation and development, “Prospects for the digital economy the OECD 2017” the context is described, in particular, in the following way: “Innovation based on data, are a key driver of growth in the 21st century. The combination of a number of trends, including the increasing transition of socio-economic activities to the Internet, as well as the reduction in the cost of data collection, storage and processing, lead to the creation and use of huge amounts of data - called “big data”. These data sets are becoming the main asset of the economy, contributing to the emergence of new industries, processes and products, creating key competitive advantages. In particular, in business, the use of data leads to value creation across a wide range of business processes, including the optimization of global value chains, improving labor efficiency, and personalizing customer relationships.” The report notes that many OECD member countries have launched special programs aimed at supporting the development of research and professional competencies related to the digital economy. Russia is no exception. Thus, in July 2017, the program “Digital economy of the Russian Federation” was approved, among the goals of which “creation of an ecosystem of the digital economy of the Russian Federation, in which data in digital form is a key factor in production in all spheres of socio-economic activity and in which effective interaction is ensured, including cross-border, business, scientific and educational community, state and citizens”. The program highlights the main digital technologies, including (a) big data, (b) Neurotechnology and artificial intelligence, and one of the five main directions of the program is “the formation of research competencies and technological achievements.” In addition to government agencies, the importance of studying new phenomena of the digital economy is noted in the academic community. In particular, the Massachusetts Institute of technology launched the digital economy Initiative (MIT Initiative on the Digital Economy) in 2013 as a large-scale project of a team of visionaries, leading scientists and practitioners aimed at “understanding how individuals and businesses work, interact and thrive in a digital transformation.” One of the key areas of research under this initiative is “Social Analytics and digital experiments” (Social Analytics and Digital Experimentation), which annually hosts a conference on digital experiments (Conference on Digital Experimentation), bringing together leading scientists, which indicates the emergence of this new interdisciplinary direction in science and practice. In Russia, in 2017, the national center for digital economy was also established at Moscow state University. The center was established in order to “promote the formation and development of the digital economy in Russia, as well as to unite and coordinate the efforts of departments of Moscow state University, other leading scientific and educational centers, public authorities, various organizations and companies to carry out research and development, international cooperation, educational, methodical, expert, innovative and other practical activities that contribute to the development of the digital economy in our country." The appearance of these two centers is testimony to the relevance of research in the field of the digital economy in General. If we talk about the scientific priorities in the field of glass industry, here you can focus on the list of priority areas proposed by the "Center of glass processing technologies" him.: "Zentrum Handwerk", the Federal Association of sheet glass manufacturers of Germany (BF-Bundesverband Flachglas), the company IFT Rosenheim GmbH, the German trade Union of independent experts on facade technologies UBF e.V. (zareg. General), Augsburg Academy (Institute of Construction and real estate), Dortmund Academy (Architecture), Darmstadt Technical University/Darmstadt (Institute of materials and structural mechanics), as well as Professional Association of manufacturers of laminated insulating glass in Germany (zareg. General)/Gütegemeinschaft Mehrscheiben Isolierglas e.V. (World Bank, Washington, DC. License: Creative Commons Attribution CC BY 3.0 IGO, 2018).

Leading academic journals are aware of this direction in the selection of publications and planning specialists-graduates. All this is designed to pay close attention to the future and prospects of the glass industry. The claimed research project is aimed at developing methods of analysis and processing of big data in order to determine the patterns of digital transformation of the glass industry. Research areas: 1. Methods of machine learning and data analysis in the field of smart production systems. 2. Approaches to validation of the results of machine learning methods, including cross-validation, splitting into training and test subsamples, calculation of the average effect of the impact on the selected subgroup. 3. Econometric estimation methods, including nonparametric methods (regression), partialling out method to isolate the estimation of the effect as a function of time. 4. Machine learning methods to reduce the dimension of the factor space, methods of correction (debiasing) of the result of the selection of factors in machine learning models to obtain conclusions about causality. 5. Methods of designing effective experiments without a priori assumptions about the values of parameters (Bayesian Efficient Experimental Design), their modification for large-dimensional problems. 6. Methods of mathematical optimization. 7. DECA Method to assess the readiness of the glass industry for digital transformation (DECA Russia is a world Bank product developed in April-November 2017, 2017).

The industrial revolution is a fundamental revolution in the world's productive forces, which is associated with a change in technological structures, followed by a sharp jump in productivity and economic growth of the industry (Still et al. 2017).

Description of technological development gives a good idea of what gives a good idea of what technologies in the glass industry have settled, and which require innovation. Breakthrough innovations change the foundations and traditions of the glass business, create new ways of its development. There are four main industrial revolutions that influenced the development in the production of glass: chemical production of soda, continuous cooking technology, drawn sheet glass and float glass (Fig. 1).

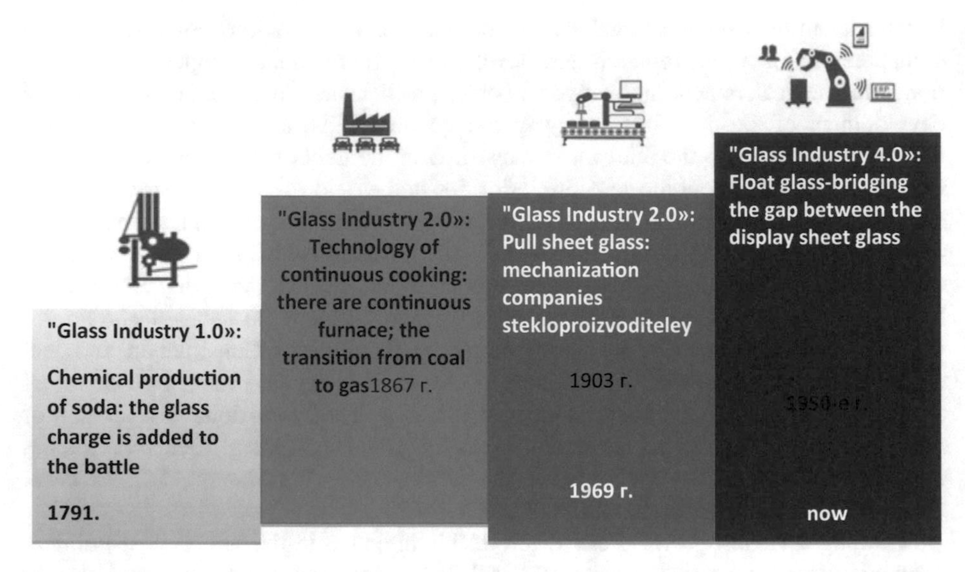

Fig. 1. Four major industrial revolutions that influenced the development of glass production

Results. Digital transformation as a systematic improvement of all business processes and business models in the entire glass industry, accompanied by large-scale transformation of the usual technological operations in the production of stele in the "smart" is possible by creating a digital glass factory of the future and the transition from the "glass industry 4.0" to "Glass Industry 5.0", which contributes to the transition to a new technological order (Fig. 2).

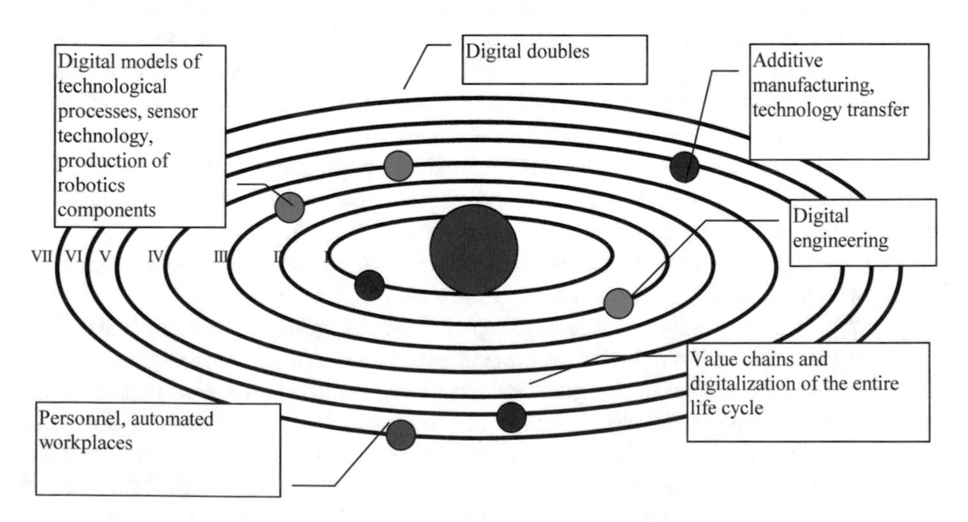

Fig. 2. Seven-shaped structure of Glass Industry 5.0

Factories of the Future-a certain type of business process system in the glass industry, a way to combine business processes, which has the following characteristics:

- creation of digital glass platforms, a kind of ecosystem of advanced digital technologies. On the basis of predictive Analytics and big data platform approach allows you to combine geographically distributed participants in the design and production of glass products, to increase the level of flexibility and customization to meet the requirements of consumers;
- development of a system of digital models of both new designed glass products and production processes. Digital models should have a high level of adequacy to real objects and real processes (convergence of the material and digital worlds, generating synergetic effects);
- digitalization of the entire life cycle of glass products (from concept idea, design, production, operation). The later the changes are made, the greater the cost, and therefore the center of gravity is shifted towards the design processes, in which the characteristics of global competitiveness or high consumer requirements are laid.

At the stage of formation of Factories of the Future there is also the formation of new key competencies, for example:

- quick customization of response to Market or Customer requests;
- use of system approaches (system engineering), when it is necessary to keep the whole system and all its interacting components in view at any time;
- formation of a multi-level matrix of targets and constraints as the basis for a new design, significantly reducing the risks, the volume of field tests and the amount of work associated with the "fine-tuning of products and products based on tests";
- development and validation ("comparison with experiments") of mathematical models with a high level of adequacy to real objects and real processes–the so-called "smart" models;
- change management throughout the life cycle of glass products;
- "digital certification" based on thousands of virtual tests of individual components and the entire system (National technology initiative (NTI), 2018).

Digital factory (Digital Factory) is a system of integrated technological solutions that ensure in the shortest possible time, the design and manufacture of globally competitive products of the new generation from the stage of research and planning, when laid the basic principles of the product, and ending with the creation of digital mock-up (Digital Mock-Up, DMU), the "digital double" (Smart Digital Twin), a prototype or small series ("paperless production", "everything's digital"). Digital factory implies the presence of "smart" models of products or products (machines, structures, assemblies, devices, installations, etc.) on the basis of a new paradigm of digital design and modeling Smart Digital Twin - [(Simulation & Optimization) Smart Big Data]-Driven Advanced (Design & Manufacturing).

"Smart" factories – Smart Factory) - systems of complex technological solutions that provide in the shortest possible time the production of globally competitive products of a new generation from the workpiece to the finished product, the hallmarks of which is a high level of automation and robotization, eliminating the human factor and associated errors leading to loss of quality ("deserted production"). As an input product of "Smart" factories, as a rule, the results of the work of Digital factories are used. "Smart" factory usually means the presence of equipment for production - machines with numerical control, industrial robots, etc., as well as automated process control systems (Industrial Control System, ICS) and systems of operational control of production processes at the level of the shop (Manufacturing Execution System, MES).

Virtual factories – systems of complex technological solutions that provide in the shortest possible time the design and production of globally competitive products of the new generation by combining Digital and (or) "Smart" factories in a distributed network. Virtual factory implies the presence of enterprise management information systems (Enterprise Application Systems, EAS), allowing to develop and use as a single object a virtual model of all organizational, technological, logistics and other processes at the level of global supply chains (supply => production => distribution and logistics => sales => after-sales service) and (or) at the level of distributed production assets (National technology initiative (NTI), 2018).

3 Conclusions/Recommendations

The digital transformation of the glass industry is the key to building a digital economy and receiving digital dividends, that is, achieving measurable economic results through the introduction of digital technologies.

References

World bank: Report on the development of the digital economy in Russia, September 2018. Competition in the digital age: strategic challenges for the Russian Federation. World Bank, Washington, DC. License: Creative Commons Attribution CC BY 3.0 IGO (2018)

Results of social and economic development of the Vladimir region for 2018. https://avo.ru/web/guest/promyslennoe-proizvodstvo

National technology initiative (NTI) 2018. http://fea.ru/compound/national-technology-initiative

DECA Russia is a world Bank product developed in April-November 2017 in cooperation with the Institute of information society development. https://istina.msu.ru/download/91398706/1edwY0:aH7_sIBQMK4oz2JriGg3dEbeYOo/

Millions of digitalization: how Russian companies transformed? https://www.cio.ru/news/300119-Milliony-tsifrovizatsii-kak-transformiruyutsya-rossiyskie-kompanii

Still, K., Seppänen, M., Korhonen, H., Valkokari, K., Suominen, A., Kumpulainen, M.: Business Model Innovation of Startups Developing Multisided Digital Platforms. In: IEEE19th Conference on Business Informatics (CBI) (2017)

Evans, D.S.: Economics of Vertical Restraints for Multi-Sided Platforms. Coase-Sandor Institute for Law & Economics Working Paper No. 626 (2013). https://chicagounbound.uchicago.edu/cgi/viewcontent.cgi?article=1187&context=law_and_economics

Specific Features of Formation of Value Creation Chains in Industry and Entrepreneurship in the Digital Economy

Gennady I. Yakovlev[1(✉)], Aleksei V. Streltsov[1], Airat M. Izmailov[1], Lilia V. Ermolina[2], and Anton N. Sunteev[2]

[1] Samara State University of Economics, Samara, Russia
dmms7@rambler.ru, oisrpp@mail.ru, airick73@bk.ru
[2] Samara State Technical University, Samara, Russia
Ermolina@mail.ru, SunteevAN@yandex.ru

Abstract. The challenges of large-scale digitization and other achievements of Industry 4.0, the tendencies of their influence on modification of the existing value creation chains and formation and functioning of new value creation chains during production of industrial goods are studied. The global economic relations are peculiar for growth of protectionism, trade and economic opposition, and promotion of economic interests of different countries by political and military methods, which is considered to be a sign of formation of a new system of global production and exchange in the post-Keynesian concepts of the economic process. Significance of participation of national companies in the global value creation chains, which allow countries to optimize the structure of the payment balance, receive incomes in currency, and ensure technical and organizational development of their companies, grows. The authors determine specific barriers for Russian companies during implementation of digital technologies, which consist in absence of the need for creation of development strategies, unpreparedness of a sufficient number of specialists, vivid and sufficient measures of protection from cyber threats, and ideas of the moral and ethical character during human-machine interaction. Further studies on this topic should determine the factors of successfulness of participation of national companies in the global economic relations, for formation of an adequate competitive strategy, stimulation of growth of the indicators of import and export, production cooperation, and investment cooperation in top-priority directions of participation in the global value creation chains.

Keywords: Global economic relations · Added value · Industry 4.0 · Chains of value creation · Production cooperation · Digital economy · Transformation

JEL Code: A220

E. G. Popkova and B. S. Sergi (Eds.): ISC 2019, LNNS 87, pp. 677–689, 2020.
https://doi.org/10.1007/978-3-030-29586-8_79

1 Introduction

The modern material production of industrial companies becomes more trans-border. Manufacture of a lot of types of goods and services is distributed among the companies of different countries, where extraction and processing of raw materials are conducted, or/and various parts and components of the final product are manufactured (Koopman et al. 2008), (Kraemer et al. 2011), (Shishkov 2009), (Kondratyev, 2018).

Special literature pays a lot of attention to the problems of formation and functioning of international production chains, which cross the borders of the states, and their integration associations. In the conditions of formation of new quality of global economic relations, which driver is digital transformation, the Russian companies – so as not to become outsiders of economic development – have to determine the key factors of successfulness of participation in transborder reproduction chains.

When substantiating the paths of companies' accession in the global reproduction chains it is necessary to take into account the cyclic character of production of new consumer value according to the stages of its life cycle and regularities of the mechanism of transfer of new technologies of their production from one country to another in the course of the country's passing the stages of industrialization. Digitization, "smart factories", robototronics, high social character, and communicativeness allow building companies of any scale and complexity as close to any country's consumers as necessary, minimizing the previous problems of saving on the costs that arise during organization of large scale of production.

The purpose of this research is to analyze the results of development of international trade and regional and local specific features as the conditions for a country's companies' accession to the global reproduction chains; to study the factors and problems of the digital economy that influence the level of successfulness of companies' accession to the global chains of value creation. As a result, it is necessary to develop the paths of increasing the successfulness of activities of industrial companies during accession to the global reproduction chains in the conditions of the digital economy, with application of the corresponding models and mechanisms by means of usage and synergy of the resource potential of companies, region/country and entrepreneurial talent of the population.

It becomes obvious that the eternal dichotomy of the liberal and directive ideology is not brought down to victory of one over another – however, there is a range of alternatives that are based on various types of markets, economic models, socio-cultural traditions, special character of the policy of the ruling groups of separate countries, and the level of mastering of technologies of Industry 4.0. The dominating agenda in the world mass media includes growth of nationalism, protectionism, transfer of jobs back to developed countries, increase of the global property inequality, and transformation of the global reproduction value creation chains.

It is necessary to take into account the vivid insufficiency of the existing methodological provision and practical application of traditional economic laws for satisfactory explanation of the current changes in the global economic relations – especially, due to the phenomenon of the "Fourth industrial revolution" and emergence of new centers of economic power in the world arena (Shvab 2016). The modern global economic growth should be aimed at ubiquitous increase of employment, stimulation of innovations, creation of an international platform for exchanging the best practices and experience in the sphere of doing business between the companies of separate countries, and development of fair terms of international trade and monetary exchange (Kondratyev 2018).

2 Background and Methodology

In the specialists' studies, the modern economic development of a developed country is connected primarily to the achievements of the Fourth industrial revolution, artificial intelligence, robotization, digital transformation, etc. (Krasnushkina 2018). The theoretical and methodological basis of the research includes the principles of complex and comparative analysis, which allow the authors to cover a wide range of dynamics of development of industry of various countries, including the states of the Russian economy and its foreign trade complex; the systemic approach during determination of measures that are performed when solving the problem of increase of effectiveness of the Russian companies' participation in the global chains of value creation with evaluation of the state of their reproduction structure and determinants of competitiveness, according to the M. Porter's methodology.

The experimental basis of the research includes the following: state statistics of the state of foreign economic relations, data of economic turnover of industrial companies and study of international organizations. The research stages include theoretical substantiation of the necessity for improving the mechanism of involvement into the global economic relations for solving the tasks of industrial development, monitoring of competencies of rivals, evaluation of the current state of own potential, development of production cooperation, formation of the terms of provision of international competitiveness of companies. Attention of modern research is drawn to such problems as structural shifts in production trans-border chains, creation of added value and multinational production, which require the corresponding specification of the methodological and categorical tools. Kondratyev (2018) speaks of a new model of globalization according to the formation of new proportions of forces of its main players, disintegration of the material components of the chains of formation of its cost, which is stimulated by technologies of Industry 4.0, differentiation of the paths of growth of developed economies, and increase of the share of services in the international trade. Modern companies have to find new segments of potential growth of production of goods and services, which are often available in developing countries – as growth rates of production in them already started exceeding the similar indicators in develop countries.

Digital and additive technologies, robototronics, and high communicativeness allow building companies of any scale and complexity as close to any country's consumers as necessary, minimizing the previous problems of saving on the costs that arise during organization of large scale of production (The Rise of Manufacturing 2016).

Participation of companies in the global reproduction chains – especially by the terms of equal cooperation with partners, requires the presence of a certain, high, level of technical and economic development. That's why it is correct that an inseparable part of analysis of the problems, connected to participation of Russian companies of industry in the global reproduction chains should be study of the tendencies of development of industry on the whole, as well as investment and reproduction aspects, which are treated as studying the movement of the main capital, and more detailed characteristics of development of "machine-building types of economic activities". The latter is especially important, as machines and equipment account for large share of import (55–60%), are a rather complex innovations-drive production of processing companies, and indirectly characterize the level of scientific and technical development of the country's economy. It should be taken into account that machines and equipment account for a large share of export of industrially developed countries, and provision of international cooperation of the Russian industrial companies on the equal basis requires the corresponding development of machine-building and the spheres that manufacture their products with high added value.

Systemic statistical information on participation of industrial companies is rather limited, as the data on this type of activities belongs to commercial information of the companies. At the same time, according to the indirect indicators, which could be found in open statistical media, we can see the level of development of foreign economic activities of the Russian industrial companies and their place in international division of labor, as well as effectiveness of participation of the Russian companies of industry in the global reproduction chains.

3 Discussion and Results

The process of vertical division of labor, which initially developed within national economies, went beyond the national borders. Shishkov (2009) introduced the term "international division of the production process", which he defined as "distribution of technological stages of production of a product or service between manufacturers that are located in different countries, which leads to creation of large structures of international production of cooperation, which cover hundreds and even thousands of links". The production process's going beyond separate countries, which initially took place at the plants of Henry Ford in the first half of the 20th century, became very popular in the car industry and then in other spheres.

In the modern treatment, international division of the production process means development of intra-sectorial international division of labor and leads to growth of intra-sectorial trade. The countries that participate in it specialize not in production of certain categories and groups of goods but in supply of certain parts and components, or in different stages of product processing. It is determined that vertical differentiation of production is a more powerful factor of growth of intra-sectorial trade than horizontal differentiation. Thus, countries at the macro-level participate in the unified (global) chain of creation of product's value.

In the most general form, the global chain of value creation is a mechanism of value accrual in the process of creation of final product, which includes various technological stages of production, as well as design and sales (Sturgeon 2001). A separate global chain includes the following:

- forward linkages along the line of export of raw materials and services, which are imported back in the form of finished products (manufacturers of spare parts and components of complex products with high added value);
- backward linkages, which form around production and export of finished products and import of raw materials and services (leading manufacturers of finished products) OECD (2013), OECD, WTO (2013), OECD, WTO, UNCTAD (2013), OECD, WTO, UNCTAD (2013).

In the Russian literature, production chain is a complex of companies (organizations) of various spheres, which are connected by the same technological process of creation of any finished product, which is aimed for final consumption. In a lot of works of Russian researchers economy is treated as a union of three economic blocks of the national economy: extracting spheres of industry, processing industry, and consuming spheres.

The main flow of products along the production chains goes from the extracting spheres of industry through the processing spheres to the consuming spheres, corporations, and unions (Fig. 1). A part of the products moves in the opposite direction (capital property and equipment, auxiliary materials) – i.e., from the processing spheres to extracting spheres; though its volume is small, as compared to a mass of products that move from the processing spheres to final consumers (separate individuals and husbandries that purchase goods for satisfying personal needs), government structures, representatives of the institutional market, and industrial commercial consumers and exporters.

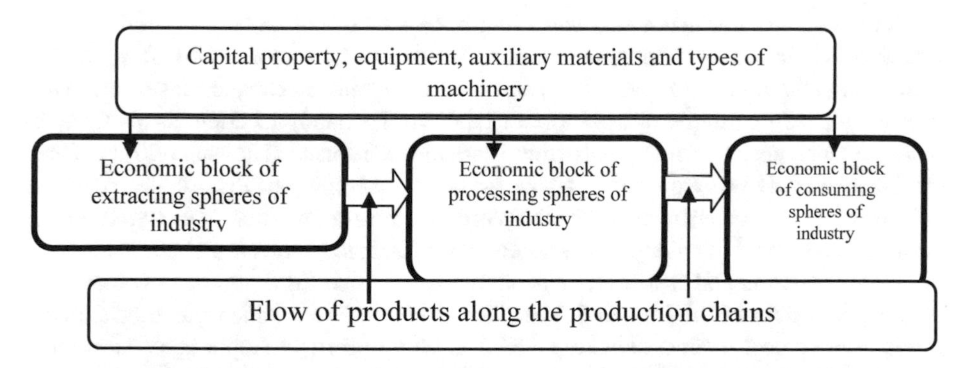

Fig. 1. The model of formation of added value in industry. Source: compiled by the authors.

The result of activities of each block of the spheres of the industrial sector of region's economy is creation of certain goods that have their own value – which allows considering the industrial sector of economy from the point of view of pricing (production) chains (Sharipova 2008).

High rates of economic growth in a range of developing countries are usually connected to increase of their involvement in the global chains of value creation by means of usage of imported components and materials in production of goods, including exported goods. Such extensive growth, however, is ineffective – as the main share of the global added value is created in the service sphere, not in the production sphere. Thus, the share of services for added value in the aggregate global export constitutes 45% (in the usual variant – 23%) (Escaith 2013). The category "effectiveness" as to participation of Russian companies of industry in the global reproduction chains is to be understood in two ways: as a level of profitability of the companies' participation in international division of labor, and as a whole complex of other results – which are primarily connected to technical progress, increase of competitiveness of the issued products, and companies on the whole, renovation of fixed capital, increase of competitive positions in the market, etc.

Developing countries often see their accession to the global chain of value creation as a possibility of increasing the added value of products and increase of competitive positions by means of improvement of conditions for international business and attraction of foreign investments. In 1995–2009, the level of involvement of countries in the global value creation chain grew by 5–10% on average (OECD, WTO, UNCTAD 2013). 40% in the volume of export of the OECD countries account for added value that is created abroad. Since 1995, South Korea, India, and China improved their positions in the global chain of value – their GVC Participation Index1 constituted 10–20%. Over the studied 15 years, the average share of added value of services in gross export of the OECD countries and their partners grew (OECD, WTO, UNCTAD 2013).

The highest growth of this indicator was shown by the largest economies of the EU (Germany, the UK, and Italy), as well as India and the USA; added value of services in gross export of these countries accounts for 40–50%. For Russia, this indicator remains unchanged—30%.

It is not a wonder that added value that is created in the process of production of traded goods is rather heterogeneous as to its origin. The most vivid illustration is technologically complex items – PC's and their components, cars, airplanes, etc. This, in a well-known example of assemble of HDD in Thailand, 43 components from 10 other counties and 11 components from Thailand were used (Hiratsuka 2011). Then, the finished HDD was exported to China for further assembly of PC. Finally, purchase of finished PC by a consumer could take place in China or any other country. The most popular example of functioning of a complex multi-stage production chain is products of Apple. Experimental works on division of a retail price for iPhone 4 in components in the USA (the item is imported from China) showed that added value of Chinese origin – labor cost – constitutes less than 2%, while profit of Apple constitutes more than 60% (Kramer et al. 2011). However, the similar situation could be observed with productions that are less technological. For example, in case with shoes that are imported from China to Europe, more than 50% of added value is of European origin. This phenomenon of circular trade – or reimport of intermediary components in finished products – is also one of the important attributes of the global production chains.

Based on systematization of the approaches of specialists in classification of possible obstacles for implementation of digital technologies into activities of Russian companies, it is possible to create a road map of successfulness of implementation of this process (Table 1).

Table 1. The paths of elimination of barriers for digitization at Russian companies and involvement into the global reproduction chains.

Obstacle	Ways of solution	Expected effect
Absence of the need for using digital technologies in the activities	Transition to pro-active model of development	Development of readiness for fighting the threats of the market environment
Insufficient number of specialists of the required qualification	Training of new specialists and advanced training of the existing specialists – especially in the sphere of authomatization of business processes	Provision of flexibility of production and operational effectiveness of business
Threats to information security of the company's activities	Creation of a system of protection of information, law enforcement measures	Elimination of weak spots in the company's work, possible losses, and thefts of resources
Stimulation of authomatization of production, social problems of society, structural unemployment	Elimination of imbalances in the labor market, social protection measures and advanced training	High motivation of the company's personnel at leadership in the sphere
Low technical level of production, wear and tear of the main production funds of the company	Active technical policy of the company, implementation of the methods of strict production	High speed of commercialization of perspective types of products, competitive status
Low capacity of the company, small scales of business	Connecting the franchise to the existing digital platforms of business	Expansion of the scales of business, mastering of new market segments
Problems of infrastructure and insufficiency of the market for digital scales of production	Structural reforms in economy, transition to innovative rails in industry	Provision of economic growth and international competitiveness of the country

Source: compiled by the authors.

For Russia, the problem of quick development on the basis of digital technologies is especially vivid in the conditions of aggravation of the global challenges and deep contradictions with the interests of other powerful countries of the world. Whether a certain country could generate a sufficient volume of material goods and scientific products and develop the modern technologies according to challenges of digitization determines the defense and wealth of a state that strives to preserve its sovereignty. In these conditions, the problem of provision of significant economic growth of a country is taken to the level of national security (Yakovlev 2007).

However, one should evaluate new challenges and threats that appear due to digitization of Russia's economy, which consist in provision of the rights of a working human – member of the society, and the work group – especially with comparing it to the digital image, as well as confidentiality of data and emergence of threats to personality, society, business, and virtual environment – all these issues are mentioned in the Program "Digital economy of the Russian Federation". These problems are complicated by vivid underrun of the Russian scientific and technical achievements in this sphere from the leading countries, lack of skilled personnel, and ineffectiveness of research in the sphere of information technologies. There are also specific barriers for Russian companies during implementation of digital technologies, which consist in absence of the need for development strategies, unpreparedness of sufficient number of specialists, vivid and sufficient measures of protection from cyber threats, and ideas of moral and ethical character during man-machine interaction (Makhalin and Makhalina 2018). The most important indicator of foreign economic activities of the Russian Federation is export of country and its commodity structure (Table 2).

Table 2. Commodity structure of export of the Russian Federation. Source: government statistics data (http://www.gks.ru/wps/wcm/connect/rosstat_main/rosstat/ru/statistics/efficiency/)

	2009		2010		2014		2015		2016	
	USD million	%	USD million	%	USD million	%	USD million	%	USD million	%
Export, total	103,093	100	397,068	100	497,359	100	343,512	100	285,674	100
Including										
Food products and agricultural resources (except for textile products)	1,623	1.6	8,755	2.2	19,982	3.8	16,215	4.7	17,070	6.0
Mineral products	55,488	53.8	271,888	68.5	350,266	70.4	219,167	63.8	169,167	59.2
Products of chemical industry, rubber	7,392	7.2	24,528	6.2	29,246	5.9	25,405	7.4	20,814	7.3
Rawhide, furs and fur products	270	0.3	305	0.1	417	0.1	311	0.1	263	0.1
Timber and cellulose & paper products	4,460	4.3	9,574	2.4	11,583	2.3	9,845	2.9	9,806	3.4
Textile, textile items, and footwear	817	0.8	764	0.2	1,101	0.2	873	0.3	912	0.3

(continued)

Table 2. (*continued*)

	2009		2010		2014		2015		2016	
	USD million	%	USD million	%	USD million	%	USD million	%	USD million	%
Metals, precious stones and metal items	22,370	21.7	50,343	12.7	52,275	10.5	40,760	11.9	37,706	13.2
machines, equipment, and transport means	9,071	8.8	21,257	5.4	26,495	5.3	25,422	7.4	24,432	8.6
Other words	1,603	1.5	n/a	n/a	6,996	1.4	5,513	1.5	5,507	1.9

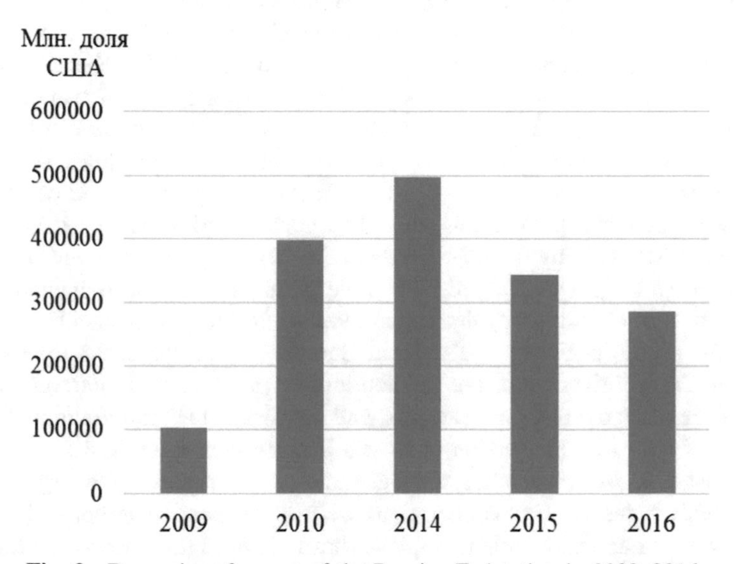

Fig. 2. Dynamics of export of the Russian Federation in 2009–2016.

Characterizing the presented data, it is possible to note that over 16 years the volume of Russian export grew by 2.77 times to USD 285,674 million. At the same time, it should be noted that recent years marked reduction of export for almost all commodity groups – especially for the group “mineral products”. In view of its role in export, this reduction in 2015 and 2016 had a decisive role in the total value of reduction. Thus, for example, adding the values of reduction of export for the group “mineral products” to its aggregate value for Russia we see that possible value of export, with preservation of the values for the group “mineral products” at the 2014 level, could have constituted USD 474.6 billion in 2015 (350,266 – 219,167 – 343,512), and USD 466.8 billion in 2016 (95.4% and 93.8% of the 2014 level, accordingly) (Fig. 2).

Thus, reduction of the aggregate volume of export is caused primarily by reduction of export for the group "mineral products". In view of the fact what is exported within this commodity group, it is possible to conclude that reduction of the volume of export is caused by the changes of prices for basic commodities of this group in the world markets.

The commodity structure of export allows for preliminary conclusions on the role of Russian companies of industry in the international trade. As of 2016, 80% in the commodity structure of export accounted for commodity groups that sold certain resources, and 60% accounted for mineral products. Machines, equipment, and transport vehicles, which were exported by Russian companies, account only for 8.6% in the commodity structure, while they have the leading position in export of industrially developed countries.

It should be noted that these proportions have not changed much over the last 16 years. Less metals are exported, while the export of food products and agricultural resources grows. On the whole, these proportions are stable.

According to the Institute of statistical research and knowledge economy of the National Research University "Higher School of Economics", the Russian industry in 2016 produced innovative goods, works, and services for RUB 3.7 trillion (the maximum value for the period of observation since 1995). The leaders in the issue of innovative products are production of charred coal and oil products (29% of total volume), cars (9.7%), and metallurgy (8%). In the aggregate volume of sales of the companies of industrial production, innovative goods, works, and services constituted 8.4% in 2016, which is by 0.5% higher than the 2015 level (Koroleva 2017).

However, as to growth rates of innovative production Russia is behind the European countries; in certain cases the gap is fivefold. In the UK, innovative production accounts for 43.5%; in France – 23.6%; in Germany — 19.4% of the total volume of production. Competitiveness of the Russian innovative products in external markets is low: the share of innovative goods, works, and services in the total volume of export of companies of industrial production of Russia in 2016 constituted 8.4%.

In the list of innovative leaders, the fuel and energy complex is the largest generator of investments in the Russian economy, as well as the main customer of innovations. Metallurgy, being an export-oriented sphere, has to invest into innovations for survival in highly-competitive external markets.

The main spheres of innovations are machine building and production of spare parts, metal works, and industrial authomatization – especially, robototronics and robotized systems and technologies for the fuel and energy complex. It is necessary to develop cooperation with international companies for creation of innovative products of the global scale. Such perspective directions and robototronics and self-driving cars are still not legal in Russia and are not property developed.

The mechanisms of digitization, which allow solving the problems of digital transformation and usage of drivers during the movement towards the goals of optimization of business processes, which were specified by N. Negroponte, allow increasing the competitiveness of industrial products and their manufacturers. M. Uehssel, E. Levi, and R. Sigel distinguished the following inevitable challenges of the digital economy:

- change of relations with consumer;
- destruction of the traditional partnership relations;
- "necessary interdependence";
- creation of a new eco-system of the digital economy.

According to them, the main problems of strong companies, which have resources, in the conditions of the digital economy will be connected to complexity of refusing from reliable business models; at the same time, new companies, which have no stereotypes, will have access to better data on consumers and more flexible managerial technologies. The opportunity to know the customer better (due to digital technologies) will open wide perspectives for the companies, though this will require changing a lot of elements of the chain of value creation. Change of the business model in the company could cause tension in relations with partners, due to their unpreparedness to changes their business models and systems of management. Emergence of an innovation in one company and (or) at one stage of the chain of value creation inevitable causes the need for innovations in other (partner) companies and (or) links of the chain. The more revolutionary the character of the innovation, the higher the level of interdependence. Such mutual dependence is "inevitable". Therefore, essential changes in strategic management of organization are also inevitable, as most of the existing concepts will not conform to the challenges of the digital economy (Ryazanova et al. 2018).

4 Conclusions

The performed analysis of the most important indicators of development of industry, its effectiveness and level of innovativeness, as well as the indicators of reproduction allows determining the state and dynamics of the material and technical basis of participation of companies of industry in global reproduction chains. The level of Russian companies' participation in the global reproduction chains directly depends on the state of the production tools, volumes of production of "machine-building types" of products, dynamics of export, state of the material basis of production, and the level of its renewal.

In the long-term, development of the mechanism of increasing the effectiveness of participation of the companies of industry in the global reproduction chains should take into account a range of aspects. There are various options of entering the global market in the modern conditions. This determines the presence of various organizational forms of economic activities. It is not necessary to use only one specific form of international contacts. One has to bear in mind also the possible participation of state structures in foreign economic activities of industrial companies, with possible options: the state could participate directly through organizations that are controlled by state structures or through execution of certain functions. Besides, public-private partnership is also a perspective option.

The presented methodology allows for full assessment of the tendencies and problems of Russian industrial companies' involvement in the global reproduction chains in the conditions of the digital economy. It is important to form the corresponding mechanism of effective participation of industrial companies in the global reproduction chains in the modern conditions. It is necessary also to develop an integrated and well-balanced industrial policy of a country, which is based on the rise of national production forces and international production cooperation. It is also necessary to find and use the methods of organization of production and increase of effectiveness of usage of resources of digital transformation of companies that differ from the traditional forms and methods of interaction with the subjects of international market activities.

References

Brazhnikov, M.A., Khorina, I.V., Minina, Y.I., Kolyasnikova, L.V., Streltsov, A.V.: System development of estimated figures of volume production plan. Int. J. Environ. Sci. Educ. **11** (14), 6876–6888 (2016)

Escaith, H.: Trade in Tasks and Global Value Chains: Stylized Facts and Implications. In: Presentation at the WTO Trade Data Day, 16 January (2013). http://www.wto.org/english/res_e/statis_e/miwi_e/tradedataday13_e/hubert_escaith_e.pdf

Guide to Measuring Global Production. New York and Geneva. UNECE, 227 p. (2015). http://www.unece.org/statistics/publications/economic-statistics/national-accounts/2015/guide-to-measuring-global-production-2015.html. Accessed 10 Sep 2018

Hiratsuka, D.: Production Networks in the Asia-Pacific Region: Facts and Policy Implications: IDEJETRO Discussion Paper 315, p. 2011. Institute of Developing Economies, Tokyo (2011)

Kommerskollegium. Adding Value to the European Economy: How Anti-dumping Can Damage the Supply of Globalised European Companies. Five case studies from the shoe industry. Kommerskollegium, Stockholm (2012)

Koopman, R., Wang, Z., Wei, S.: How Much Chinese Exports Is Really Made in China – Assessing Foreign and Domestic Value-added in Gross Exports. NBER Working Paper 14109. National Bureau of Economic Research, Cambridge (2008)

Kraemer, K.L., Linden, G., Dedrick, J.: Capturing Value in Global Networks: Apple's iPad and iPhone: PCIC Working Paper (2011)

OECD: Interconnected Economies: Beneting from Global Value Chains. OECD, Paris (2013)

OECD, WTO. OECD-WTO: Statistics on Trade in Value Added. OECD, Paris (2013)

OECD, WTO, UNCTAD. Implications of Global Value Chains for Trade, Investment, Development and Jobs. OECD, WTO, UNCTAD, Paris (2013)

Shishkov, Y.V.: Internatsionalizatsiya proizvodstva – novyi etap razvitiya mirovoi ekonomiki [Production Internationalization: New Stage of Development of the World Economy], p. 2. IMEMO RAN, Moscow (2009)

Sturgeon, T.J.: How do we dene value chains and production networks? IDS Bull. **32**(3), 9–18 (2001). https://doi.org/10.1111/j.1759-5436.2001.mp32003002.x

The new Globalization. Going beyond the Retoric. https://www.bcg.com/publications/2017/new-globalization-going-beyond-retoric. Accessed 14 Sep 2018

The Rise of Manufacturing Mars the Fall of Globalization. Geopolitical Weekly, 7 June 2016, pp. 1–3 (2016)

Sharipova, V.V.: Formation of value chains in the industrial sector of region's economy. Bulletin of Herzen State Pedagogical University of Russia, no. 60 (2008)

Kondratyev, V.: A new stage of globalization: peculiarities and perspectives. The global economy and international relations, vol. 62, no. 6, p. 5 (2018)

Koroleva A. (2017) Russia performs an innovative leap// "Expert Online"

Krasnushkina, N.: The world industry has not prepared for revolution. Gaz. Kommersant, no. 6, 16.01.2018, p. 2 (2018)

Makhalin, V.N., Makhalina, O.M.: Managing the challenges and threats in the Russian digital economy. Upravlenie no. 2(20), p. 58 (2018)

Porter, M.: International Cooperation, p. 680. International Relations, Moscow (1993)

Target program “Digital economy of the Russian Federation”. http://government.ru/docs/28653. Accessed 23 Aug 2018

Shvab, K.: The Fourth industrial revolution, “Eksmo” (Top Business Awards) (2016). http://government.ru/docs/28653. Accessed 23 Aug 2018

Yakovlev, G.I.: Competitiveness of national economy: problems and solutions. Bull. Saint Petersburg State Univ. Eng. Econ. Ser. Econ. SPb, no. 4(17), pp. 81–87 (2007)

Yakovlev, G.I.: Managing the competitiveness of industrial companies in the conditions of globalization, p. 244. Bull. Samara State Univ. Econ., Samara (2007)

Negroponte, N.: Being Digital. Kopf. (Paperback edition, 1996). http://web.stanford.edu/class/sts175/NewFiles/Negroponte.%20 Being%20Digital.pdf. Accessed 02 Jan 2018

Uehssel, M., Levi, E., Sigel, R.: Ryvok v tsifrovuyu ehkonomiku. A breakthrough in the digital economy. http://hbr-russia.ru/management/strategiya/a19181/. Accessed 03 Oct 2018

Ryazanova, G.N., Sazanova, A.A., Sazanova, S.L.: Influence of the processes of digitization of economy on the activities of non-financial organizations. UPRAVLENIE No. 2(20), C. 54 (2018). https://doi.org/10.26425/2309-3633-2018-2-52-56

Dialectics of the Processes of Digitization of the Socio-economic System

Vera D. Nikiforova(✉), Lyubov V. Achba, Alexander A. Nikiforov, and Anna V. Kovalenko

Financial University under the Government of the Russian Federation, St. Petersburg Branch, St. Petersburg, Russian Federation
{vdnikiforova, lvachba, aanikiforov, avkovalenko}@fa.ru

Abstract. The article analyzes the processes of digitalization of the economy in a historical retrospective, the scientific views of representatives of various concepts of the digital economy as a phenomenon of the economy, which is based on the active use of information and communication and other high technologies in the production of economic benefits. The authors reveal the factors influencing the economic system of the processes of information-digital transformation of society and the associated socio-political risks and challenges. Special attention is paid to the analysis of the impact of digital technologies on economic growth. In this context, issues of comparing models and rates of socio-technological renewal in the global and national economies are revealed. The purpose of the article is a theoretical understanding of digitalization processes from the perspective of qualitative analysis of changes in the socio-economic system in the context of increasing their impact on the economy and society, the development of conclusions and proposals for the effective use of modern methods, approaches and tools to ensure the growth of a globalizing national economy. To achieve this purpose, the authors use historical, abstract-logical methods, a complex-functional analysis, a systematic approach, a comparison and a method of expert assessments.

The authors substantiate the conclusion that it is necessary and expedient to take into account the historical experience, the socio-cultural identity of the country for the ongoing development and ensure the security of the national economy, which is focused on technological modernization. Recommendations are made to overcome the contradictions identified in the course of the analysis and to reduce the socio-economic risks.

Keywords: Evolution of digitalization · The transformation of the economic system · Socio-political challenges · Globalization of world economic relations · Economic growth

JEL Code: E60 · F63 · G18 · O33 · P17

E. G. Popkova and B. S. Sergi (Eds.): ISC 2019, LNNS 87, pp. 690–697, 2020.
https://doi.org/10.1007/978-3-030-29586-8_80

1 Introduction

A diverse analysis of the problems associated with structural changes in the socio-economic system of society, institutional transformations of the public administration system under the influence of various factors in the development of the modern economy is presented in the studies of many economists. At the same time, research on the processes of globalization and the creation of the information society as factors interconnected and capable of ensuring not only economic growth, but also generate imbalances, deficiencies and controversy, is becoming particularly relevant today.

Despite the achievements in the study of the problems under consideration, they save their debatability. Further theoretical understanding requires the essence of the digital economy as a phenomenon of neo-economics. Searching for ways to transform the socio-economic infrastructure, the choice of methods and tools to ensure economic security in the state, the analysis of civilizational changes in the economic systems of developing countries is not less importance. Theoretical and practical problems of using digitalization as a tool for the globalization of world relations are also need scientific development and justification.

2 Methodology

The article reveals the historical continuity of the stages of the digitalization of the economy through a retrospective analysis of the gradual replacement of analog technologies by their digital substitutes. It also reveals the emergence of new forms of business based on digital technologies. For the theoretical understanding of the current trends of information-digital transformation of the economic system, a systematization of the scientific views of domestic and foreign economists on the digital economy, its interpretations and relations with socio-economic changes in society has been carried out. A comparative analysis of the models and rates of socio-technological renewal in the context of economic growth in individual countries made it possible to identify the socio-political and economic challenges and risks associated with this process for countries with different levels of socio-economic development. Using expert assessments and the results of the author's analysis of Russia's progress along the path of implementing the national strategy for digital development has made it possible to formulate a number of conclusions and proposals for overcoming contradictions, ensuring economic security, and improving the conditions for transition to the economy of a new technological generation.

3 Results

The formation of a modern post-industrial society is inextricably connected with the transformation of information into a resource capable of creating knowledge, and transferring social and economic relations to the network space.

Digital transformation expands its sphere of influence on business, science, social sphere and ordinary life of citizens; accompanied by the effective use of its results by users of the converted information.

The evolution of digitalization processes has condition to the emergence of a number of scientific ideas about these processes. One of the first scientific studies can be called the «theorem of readings» by Kotelnikov (1933), according to which an analog signal with a limited spectrum can be absolutely accurately restored by a digital signal if its frequency is twice as large as the maximum frequency of the original analog signal. It is a signal in digital format - the key element from which the scientific community goes to understanding digitalization and its consequences, including for the economy.

The appearance of the Arpanet network (1964), which became a model of the modern Internet, has opened up broad possibilities for conducting experiments in computer communications, combining the scientific potential of research institutions, exploring ways to maintain stable communication in a nuclear attack, developing the concept of distributed control of military and civilian structures. In 1964, there was a discussion in the USSR about using the computer technology in managing the economy, which demonstrated the lack of readiness of Soviet economic theory to realize the economic benefits and benefits of using it in social production based on K. Marx's labor theory socialist distribution, special commodity-money relations, etc. (Davydov and Lopatnikov 1965).

The appearance of the tcp/ip protocol (transmission control protocol/Internet protocol, 1983) created the basis for further civilizational changes in the economic systems of a number of countries and the global economy: e-commerce, online banking systems, Internet trading, etc. Attempts to understand the concept of the digital economy, its main characteristics, among which they highlight globalization, virtualization, innovation, the elimination of institutions of mediation, changes in the interaction between producer and consumer Telem, appearing Lenie-digital converter are taking by a number of scientists (Stepanov 2001; Nicholas 1995).

Since the beginning of the 21st century, in many countries digitalization, which has manifested itself mainly at the micro level, has begun to emerge at the macro level, implemented through government strategies and programs for the development of the digital economy. The key blocks of which are:

- ensuring free and unhindered unobstructed movement of information and services, reliability and security of online operations, high speed transmission and receipt of information via communication channels;
- raising the general level of computer literacy of the population, improving educational programs for training qualified specialists and encouraging employees to improve their skills in the field of information and communication technologies, cyber security;
- improvement the rules for the protection of copyright, patent systems, examination, licensing; stimulating the use of big data technologies, the creation and standardization of the industrial Internet, etc.

For modern technological development, conjugate with changes in civilization, theoretical and methodological approaches associated with systemic and comprehensive research of the content of the digital economy, the interrelation of factors of its influence on the socio-economic system of society, economic growth and all global of the market take on special meaning.

To date, a number of authors have created their own theoretical theories, reflecting a wide range of approaches to the definition of the essence of the «new economy». Discussions on the selection of an adequate term that reveals the essence of the process of digitalization of the economy are fragmented and aren't based on a holistic theoretical analysis.

Suggested terms and related definitions such as «information economy» (Nordhaus 2002; Oliner and Sichel 2003), «network economics» (Streletsky et al. 2010), «web–economics» (Vafopoulos et al. 2012), «Internet economy» (McKnight and Bailey 1998), «e-economy» (Khan 2002; Cohen et al. 2000), etc., in our opinion, are applicable for uncovering certain aspects of the process of digitalization of the economy, but they aren't sufficient for describing transformations in the socio-economic system and civilizational changes in the 21st century.

Undoubtedly, the search for the essence and term was carried out within the framework of a specific historical period and continues in the context of an incomplete transformation of all sectors of the economy. Studies of institutional factors, which promote the digital ecosystem, the selection of economic and mathematical methods for evaluating the effectiveness of the components of this system have appeared recently (Budd and Harris 2004).

Russian economists (Bogomolov, etc.) make a definite contribution in this way, reveal the contradictions between the fast-paced distribution of digital technologies and the slow adoption of various legal norms and rules governing the creation of digital goods, both at the national and supranational levels. (Mesenbourg 2001) Special attention is paid to the depersonification of property, the weakening of real social connections, the concepts of «digital feudalism», «digital inequality», and «cybertariat» are introduced (a special social layer of workers). This approach opens up new directions for the disclosure of the concept of the digital economy, the development of a system of integrated legal regulation of socio-economic relations associated with the development of the digital economy.

It should take account of the fact, that the increasingly widespread introduction of digital technologies into the economy can lead to profound changes in the economic system of society, affecting the ownership relations to the products and benefits, private property incentives, commodity-money relations, the system of distribution and redistribution of public product. The existing order of social relations in society, relations between employers and employees, many other relations and orders will also change radically. In these conditions, it is necessary to update the formal and informal institutions of society, including the creation of institutions responsible for the damage, which is caused by the application of digital technologies, ensuring digital security; practical solution of the problems of building a barrier-free digital space based on the use of international legal mechanisms, etc.

The Russian Federation is implemented an enormous amount of work to create a public good in the form of ensuring unhindered access to the Internet of all socially significant facilities. The state program «Digital Economy» provides for the full provision of access to the broadband Internet of socially significant facilities only by 2014 (in 2018 - 30%). A number of other tasks, including an increase in the share of the Russian Federation in the global volume of rendering services in data processing and storage from 0.9% to 5%, are also set. The implementation of software tasks requires not only a large-scale inventory of existing computer models of socio-economic systems at the macro and micro levels, defining the scope of their adequate and effective use, developing software and hardware platforms that ensure their systematic and integrated use. Legislative initiatives to prevent the erosion of the rights and responsibilities of individuals and legal entities, mixing spaces of virtual and real subjects, conducting social monitoring, etc., which are aimed at creating a new institutional environment, will also be required.

The adoption of a law to ensure the sustainable operation of the Russian Internet segment in the event of a disconnection from the global infrastructure of the global network (effective from November 2019) receives mixed opinions from experts. From the point of view of the initiators of this bill, this is the desire to protect the domestic Internet and its users from external threats, carried out within the framework of the program of digitization of the economy. Supporters of the opposite point of view consider this law as an attempt to make the global local, to create a routing that will provide the entire Internet traffic through the territory of the Russian Federation (Philosophy of the economy 2017). Undoubtedly, it is necessary to adequately perceive all the benefits and risks of the digital transformation and the globalization processes associated with it, and to understand that the importance of geo-economic factors for decision-making in ensuring economic security is increasing today country. It should be noted that even in the United States, the idea of «opening the economy inward» is developing.

In conditions, where the problem of regulating network operations globally and creating interethnic structures for storing and accessing information, ensuring the fulfillment of formal rules remains unresolved, contradictions between the global nature of the digital economy and national interests regarding its development in a particular country intensify. For Russia, it is important to participate in digitalization processes for economic growth and competitiveness, but sanctions and political pressure from Western countries force it to take into account these circumstances and require additional costs associated with the implementation of the digital economy program. Along with this, significant costs will be required to curb illegal activities in the field of Internet payments, illegal financing, etc. of negative manifestations in the virtual space.

The high degree of uncertainty regarding the future renewal of the institutional structure of the socio-economic system of society particularly affects the «circulatory system of the economy»—the banking sector and financial markets. The questions about the introduction of blockchain technology into the monetary sphere and the reduction of the role of the central bank stay debatable.

As is known, the attitude towards cryptocurrencies among the scientific community, government institutions and banks is ambiguous. Aware of the progressive nature of the blockchain technology, on the one hand, experts fear that this technology poses the threat of tax evasion, cash withdrawal and financing of terrorism, on the other (Pryanikov and Chugunov 2017; Bulgakov 2016; Strembitskaya and Babayan 2017). Equally hard discussions are underway regarding the recognition of cryptocurrency as an object of law, and the creation of a corresponding regulation.

Taking into account all the variety of approaches to the legalization of cryptocurrencies in different countries of the world, in Russia the most acceptable options can be considered equating cryptocurrency to an investment asset, or to money substitutes. At the same time, the use of a cryptocurrency as an investment asset from the point of view of the regulator will not be generally available to individuals, and it is assumed only between participants of the regulated exchange (crypto-exchange). Thus, the creation of a closed regulated sphere of cryptocurrency trade is foreseen, which is not intended for settlements, but only for exchanging for national or foreign currency.

As regards for the role of the central bank, in the short term, digitalization opens up great opportunities for improving the efficiency of this institution's work, but in the long term there may be different options for its functioning:

- It can be abolished with the development of technology, infrastructure, and due to this significant reduction of risks in the monetary sphere;
- It can become the only financial institution monopolizing all banking activities in the country;
- Its role will be reduced to the use of a system of short-term interest rates to influence the economy.

We support the expert's opinions; they believe that the trend towards the consolidation of banks and the consolidation of the financial market will continue in Russia, and digital technologies will contribute to it. It is quite possible that interbank settlements, transfers and payment infrastructure will leave the sphere of banking activity. But banks will not cease to exist, at least in the future 5–10 years.

Lending business will remain mainly for banks. Fintech startups and non-bank startups, which are created for bank investments, represent a transactional model of a low margin business that is not capable of long-term investments, risk management due to lack of capital and funding. Moreover, their number can only grow in the financial market, since there are low entry barriers on the market and no large investments are required to create a near-financial or near-bank application.

According to the estimates of the international consulting company McKinsey, the world trend is the crowding out of banks not only from the market of such traditional services as payments, transfers, but also loans, and, above all, consumer loans, loans to small businesses, due to expansion activities of fintech companies in this market (up to 10%). In Russia, fintech-companies haven't received serious development yet due to the high risks of activity and insufficient conditions for such development. The central bank is working to create an enabling environment for the development of the digital sector in this direction, in particular, it introduces a special regime («regulatory sandbox») that allows participants to test new financial technologies and business models without risking violating the law. Such support of the mega-regulator can turn out to be significant for the redistribution of the banking services market in favor of new players, while traditional banks risk turning into infrastructure similar to reinsurance companies in the insurance services market.

You can expect new financial institutions, which are based on the blockchain, and also the use of the blockchain technology, not only in the banking sector, but also in other areas of the financial services, where transaction costs are high and a lot of time is spent on reconciliation and settlement of payments. This concerns, first of all, the organization of direct peering financing; crowdfunding; processing and transmission of financial information in the field of audit and accounting; servicing securities in the field of safekeeping, accounting and rediscounting of rights to securities; as well as other specific businesses.

4 Conclusions/Recommendations

A retrospective analysis of the research of economists revealed the evolutionary interrelations between the developing processes of digitization of the economy and their theoretical understanding, a variety of scientific approaches to defining the digital economy as a phenomenon of the «new economy». To date, an active debate and the absence of a unified and systematic view of the concept of the digital economy remain. The deep changes in the economic system of the emerging information society, affecting private property relations, commodity-money relations, the system of distribution and redistribution of the social product and the related problems of updating the formal and informal institutions of society, and also the formation of international institutions for regulating the digitalization processes in the global economy, are not sufficiently studied. The use of a comparative analysis of models and rates of technological renewal in individual countries made it possible to identify the challenges and risks for countries with different levels of socio-economic development, to identify ways to overcome them.

It turned out that at the present stage the importance of geo-economical factors for making decisions in the field of informatization, taking into account the economic security of the country, increases significantly, the contradictions between the global nature of the digital economy and national interests regarding its development in a particular country increase. The study on the subject of updating the institutional structure of the socio-economic system of society in the sphere of banks and financial markets made it possible to determine the direction of infrastructure transformation and the creation of an appropriate regulatory system in this sphere. Risk management by the state and private business, a holistic view of legal issues related to the redistribution of markets and institutional transformations, can become positive synergistic factors in the digitization processes of the socio-economic system of society.

References

Davydov, Yu., Lopatnikov, L.: Economists and Mathematics at the Round Table, Moscow (1965)

Stepanov, V.: The age of network intelligence: about the Don Tapscott Electronic Digital Society. Inf. Soc. J. **2**, 67–70 (2001). http://emag.iis.ru/arc/infosoc/emag.nsf/BPA/bef8b90eb6894281c3256c4e0027b866. Accessed 15 May 2019

N, N.: Being Digital. Alfred A. Knopf, New York (1995)

Nordhaus, W.D.: Productivity growth and the new economy. J. Brookings Papers Econ. Act. **2**, 211–265 (2002)

Oliner, S.D., Sichel, D.E.: Information technology and productivity: where are we going? J. Policy Model. **25**, 477–503 (2003)

Streletsky, I.A.: Networking Economics, Prospect, Moscow (2010)

Vafopoulos, M., et al.: The web economy: goods, users, models, and policies. J. Found. Trends Web Sci. **1–2**, 1–136 (2012)

McKnight, L.W., Bailey, J.P.: Internet Economics. MIT Press, Cambridge (1998)

Khan, F.: Information Society in the Global Age. APH Publishing, New Delhi (2002)

Cohen, S.S., et al.: Tools for Thought: What is New and Important about the Economy?, Berkeley International Roundtable on the International Economy (BRIE), No. 138, pp. 7–11 (2000)

Budd, L., Harris, L.: E-Economy: Rhetoric or Business Reality?. Routledge, New York (2004)

Mesenbourg, T.L.: Measuring the digital economy, US Bureau of the Census, Suitland (2001). https://www.census.gov/content/dam/Census/library/working-papers/2001/econ/uzdigital.pdf. 15 May 2019

Philosophy of Economics: Almanac of the Center for Social Sciences and the Faculty of Economics of Moscow State University. Mv Lomonosov, December 2017

Gingerbread, M.M., Chugunov, A.V.: Blockchain as a communication basis for the formation of the digital economy: advantages and problems. Int. J. Open Inf. Technol. **5**(6) (2017)

Bulgakov, I.T.: Legal issues of using the blockchain technology, Law magazine, No. 12. pp. 80–88 (2016)

Strembitskaya, S.B., Babayan, S.G.: Cryptocurrency in the financial services sector: new opportunities on the example of the blockchain. In: European Scientific Conference 2017, Materials of the International Conference in Penza, Russia, Penza, pp. 146–148, 8 January 2017

Network Interaction as a Factor of Professional Qualities' Development of Service Workers

Zhanna V. Smirnova(✉), Maria V. Mukhina, Olga V. Katkova, Marina L. Gruzdeva, and Olga T. Chernei

Minin Nizhny Novgorod State Pedagogical University, Nizhny Novgorod, Russian Federation
z.v.smirnova@mininuniver.ru, o.t.chernej@pochta.vgipu.ru

Abstract. This article reveals the importance of organization of network interaction between an educational institution and organization of the service sector. We describe the introduction of information technologies in the practice of service enterprise and it also creates opportunities for internal training of employees. *Object of study*: service activity of an organization. *Subject of research:* Integration of information technologies into the activities of a service enterprise. *The purpose of the study* is to integrate information technology in the process of functioning of a service organization. *Main hypothesis of the study* - information technologies will ensure the optimization of the organization's activities, if: ensure the integration of information technology into organization activities, expand the use of technology in various areas, skills of service providers on the basis of integration of information technologies.

1 Introduction

In recent years, the service industry has become one of the most rapidly developing areas. The sphere of service today is a complex system, which includes enterprises of different forms of ownership, different organizational and legal forms, different in the volume of activity, according to purpose and composition of the services provided.

In this regard, it is necessary to scientifically comprehend and understand this area of life in its cultural and economic content. This task is complex one since its solution must be interdisciplinary. This task is especially relevant for educational institutions that train specialists for the service sector. In recent years, as a result of the efforts of scientists and methodologists of higher school to develop knowledge about problems of service. Some progress has been achieved in its content, management, and legal aspects [3].

However, in reality of organizing work of enterprises in the service sector, there are still many unsolved problems. Service sector is rapidly growing in the GDP of the country, it becomes clear that there is a huge potential of this sphere and inevitability of its development. These trends lead to increased competition among service enterprises in this area and the need to use modern technologies to optimize their activities.

One of the necessary conditions is the appeal of service entities to information technologies, which represent the greatest opportunities for improving the performance of organizations.

E. G. Popkova and B. S. Sergi (Eds.): ISC 2019, LNNS 87, pp. 698–704, 2020.
https://doi.org/10.1007/978-3-030-29586-8_81

2 Methodology

Modern information technologies, with their rapidly growing potential and decreasing costs, offer great opportunities for new forms of enterprise organization. The range of such opportunities is steadily expanding. Today, information technologies can make a decisive contribution to the growth of productivity, production volumes, by improving the quality of work with consumer, regular market monitoring, etc. Introduction of information technologies into the practice of the service enterprise also creates opportunities for internal training of employees [5, 6].

Problem of development of professional qualities of service workers is one of the most important, as improving the efficiency of service organizations in the transition to post-industrial economy significantly depends on qualified personnel, which actualizes a problem of competence of personnel potential of service organizations.

One of the important functions provided by information technologies is the possibility of organizing internal training of employees, through development and implementation of electronic educational programs aimed at improving skills of employees sector and competitiveness of the organization as well.

Therefore, one of the most important tasks of service organizations is to use modern information technologies in a wide range of possibilities of their application.

3 Results

Taking into account the vast experience accumulated by the Department of Technology of Service and Technological Education in the training of specialists 43.03.01 "Service" and active use of educational IT is expedient to create an experimental platform on the basis of service enterprises in order to optimize work by integrating information technologies. Creation of an experimental site can also become the most important form of organization of clinical practice for students 43.03.01 "Service" "NGPU named after K. Minin". The Department of Service Technologies and Technological Education is interested in translating the experience of introducing information technologies into activities of a service organization and preparing future employees. Such practice-orientation incorporation of a student in the activity of experimental platform on the basis of the service enterprise opens up new opportunities for the formation of professional competencies [1, 9, 11].

Theoretical Substantiation of the Study. Under influence of the process of informatization, a new public structure - information society is currently being formed. It is characterized by a high level of information technology, developed infrastructures that ensure the production of information resources and ability to access information.

In the process of developing the research problem, we relied on:

- theoretical developments in the field of service activity and its content (Romanovich, V.K., Avanesova G.A., Chelenkov, A. et al.);
- theoretical research of informatization of education (A.P. Ershov, B.A. Zvyagintsev, V.A. Izvozchikov, E.I. Mashbits, I.V. Robert et al.);

Problems of information technologies are considered in researches of Korneev I.K., Xanadopulo G.N., which describe the concept, structure, means, and methods of information technologies.

Technical means of information technologies are presented as a set of devices of organizational, communication and computer equipment. In view of the prevailing use, much attention is given to software of computer technology.

Problems of computer training programs creation technology, classification of educational software products, pedagogical requirements for tools were considered by O.I. A.T. Voronin, Y.A. Chernyshev, I.V. Retinskaya, M.V. Shugrina, M.V. Bulgakov, A.E. Pushkin, S.S. Fomin, E.E. Yakivchuk [2, 7].

Problems of use of electronic textbooks, multimedia, virtual reality, hypertext systems in training devoted to the work of many authors: A.I. Arkhipova, V.N. Ageev, V.V. Amilderova, Yu.S. Branovsky, A.G. Brown, A.V. Veselova, D.M. Grishechkina, I.G. Levitina, A.L. Livitin, O.B. Popovich, A.I. Tikhonov, Y.M. Taraskin, A.V. Smolyaninov and others.

Instrumental tools of n planning are devoted to the work of K.K. Boykachev, I.G. Koneva, I.Z. Novik, Y.A. Chernyshova, S.N. Trapeznikova [16, 18].

However, to date, the use of information technology in the practice of a service organization remains insufficiently developed. Rapidly changing market conditions and constantly expanding possibilities of use of information technologies make the problem of constant updating of enterprise software and expanding the range of their use. One of the main factors should be recognized under rapid updating of requirements for professional training of employees in the conditions of informatization of society in modern market relations. Taking into account the insufficient theoretical development of this problem and its practical significance in modern socio-economic conditions, we chose to study the process of integration of information technologies [9, 15].

The idea of the study is that the process of optimization of enterprise activity in today's conditions significantly depends on the scale of use of information technologies in the activity of the company. The objectives of the study were as follows:

1. To study the activities of the organization.
2. Monitor the use of ICT in the activities of the organization.
3. Develop a pilot program for the use of ICT.
4. Implement a pilot ICT program in the various areas of the organization.
5. Implement the activity of the experimental platform in the training students of "NGPU named after K. Minin" students in the program 43.03.01 "Service".

Methods and base of research. To address the objectives necessary to use complex techniques and research methods: methodological and theoretical analysis of documentation and material equipment, system approach to modeling of the process of introduction of information and computer technologies (ICT); synthesis empirical material; observation, survey, testing, diagnosis of the state and development of knowledge, skills, methods of mathematical statistics [3, 7, 19].

Conditions of implementation of experimental activities

(a) Regulatory support:

- development of regulatory and legal bases of experimental activity (agreement with "NGPU named after K. Minina", experimental program);

(b) Logistical support:

- ensuring a sufficient level of material and technical equipment for the organization of service activities.

(c) Staffing:

- providing a group of teachers, students "NGPU named after K. Minin" and staff of the organization involved in the pilot activities.

(d) Software and methodical support:

- Development of a strategy for the introduction of ICT into the organization's practices

(e) Working with the team:

- the use of diagnostic tools to identify ICT competencies;
- joint discussions and publications of intermediate and final results of the pilot activities;

(f) Working with students:

- training of students together with the team in seminars, webinars, trainings on the basis of the organization of service activities [4, 11, 17].

Mechanisms of the program are joint scientific and educational activities of university teachers and employees carried out in various types and forms (teaching, practice, research, seminars, webinars, trainings), as well as practical activities of students in the course of practical activities, complex scientific and technical activities of students in the course of the practical activities of the students. Research works and projects of students in cooperation with the staff of the department [8, 10].

Conditions of implementation of experimental work
LLC "Registration Agency" and "NGPU named after K. Minin" carry out experimental work on the topic "Integration of information technologies into the practice of service organizations" under the supervision of employees of "NGPU named after K. Minin" - teachers of the Department of Service Technologies and Technological Education.

NGPU named after K. Minin" is obliged to [14, 18]:

- while planning pilot work take into account the needs and characteristics of the organization;
- provide scientific and methodological assistance in the analysis of the progress and results of experimental activities;
- participate in the development of programs and methods of the experiment;
- develop criteria for the performance;

- assist in conducting seminars on the topic of the experiment;
- advise administration and team members on the subject of the experiment;
- together with the participants monitor results of the experimental work [2, 15].

LLC "Registration Agency" is obliged to:

- to create conditions for experimental work; to provide assistance in scientific researches by teaching staff of NGPU named after K. Minin;
- to carry out experimental work according to the plan under the guidance of teachers of "NGPU named after K. Minin", carry out adjustment as necessary and check effectiveness of the pilot study;
- monitor and analyze results of the pilot work.

4 Conclusions

The purpose of the experiment is theoretical substantiation, practical development and implementation of information technologies in practice of a service company.

The general strategy of the experiment is to integrate modern information technologies into production process, which will provide new opportunities to employees.

Final result of experimental activities is to increase the efficiency of the organization of LLC "Registration Agency" and improving the quality of training specialists in the direction 43.03.01 "Service" NGPU named after K. Minin. The overall result will be a good quality of the services, provided through introduction of information technologies into the process of professional activity.

Also, the integration of information technologies will allow employees of the service organization to form certain general, professional and special competencies, such as:

- understanding of the essence and importance of information in the development of modern information society, compliance with the basic requirements of information security;
- readiness to introduce and use modern information technologies in the process of professional activity;
- ability and willingness to independently acquire new areas of knowledge through information technologies and to use in professional activities;
- possession of basic methods means of obtaining, storing, processing information, skills of working with the computer as a means of information management; work with information in global computer networks.

First results of the pilot activities on the integration of information technology into service activities are broadcast through the publication of materials in the form of articles, monographs, educational and methodical manuals, seminars, master classes, teacher training workshops, refresher courses for employees of the organization of service activities.

References

1. Bicheva, I.B., Filatova, O.M.: Formation of the teacher-leader in the educational process of the university. Vestnik of Minin University, No. 3(20), p. 5 (2017). (in Russian)
2. Bogorodskaya, O.V., Golubeva, O.V., Gruzdeva, M.L., Tolsteneva, A.A., Smirnova, Z.V.: Experience of approbation and introduction of the model of management of students' Independent work in the university. In: Advances in Intelligent Systems and Computing, vol. 622, pp. 387–397 (2018)
3. Bulaeva, M.N., Vaganova, O.I., Koldina, M.I., Lapshova, A.V., Khizhnyi, A.V.: Preparation of bachelors of professional training using MOODLE. In: International Conference on Humans as an Object of Study by Modern Science, pp. 406–411. Springer, Cham, July 2017. https://link.springer.com/chapter/10.1007/978-3-319-75383-6_52
4. Vaganova O.I., Koldina M.I., Trutanova A.V.: Development content of professional pedagogical education in the conditions of realization of competence approach. Baltic Humanitarian J. **6**(2(19)), 97–99 (2017). (in Russian)
5. Vaganova, O.I., Smirnova, Z.V., Trutanova, A.V.: Organization of research activities of bachelor of professional education in electronic form. Azimuth Sci. Res. Pedagogy Psychol. **6**(3), 239–241 (2017). https://elibrary.ru/item.asp?id=30101872
6. Garina, E.P., Kuznetsov, V.P., Egorova, A.O., Romanovskaya, E.V., Garin, A.P.: Practice in the application of the production system tools at the enterprise during mastering of new products. Contrib. Econ. **2**, 105–112 (2017). 9783319606958
7. Garina, E.P., Kuznetsov, V.P., Romanovskaya, E.V., Andryashina, N.S., Efremova, A.D.: Research and generalization of design practice of industrial product development (by the example of domestic automotive industry). Quality - Access to Success **19**(S2), 135–140 (2018)
8. Iltaldinova, E.Yu., Filchenkova, I.F., Frolova, S.V.: Peculiarities of the organization of postgraduate support of graduates of the targeted training program in the context of supporting the life cycle of the teacher's profession. Vesknik of Minin University, No. 3(20), p. 2 (2017)
9. Ilyashenko, L.K., Prokhorova, M.P., Vaganova, O.I., Smirnova, Z.V., Aleshugina, E.A.: Managerial preparation of engineers with eyes of students. Int. J. Mech. Eng. Technol. **9**(4), 1080–1087 (2018)
10. Ilyashenko, L.K., Smirnova, Z.V., Vaganova, O.I., Prokhorova, M.P., Abramova, N.S.: The role of network interaction in the professional training of future engineers. Int. J. Mech. Eng. Technol. **9**(4), 1097–1105 (2018)
11. Kutepov, M.M., Vaganova, O.I., Trutanova, A.V.: Possibilities of health-saving technologies in the formation of a healthy lifestyle. Baltic Humanitarian J. **6**(3), 210–213 (2017). https://elibrary.ru/item.asp?id=30381912
12. Kuznetsov, V.P., Romanovskaya, E.V., Egorova, A.O., Andryashina, N.S., Kozlova, E.P.: Approaches to developing a new product in the car building industry. In: Advances in Intelligent Systems and Computing, vol. 622, pp. 494–501 (2018)
13. Ilyashenko, L.K.: Pedagogical conditions of formation of communicative competence of future engineers in the process of studying humanitarian disciplines. Int. J. Civil Eng. Technol. **9**(3), 607–616 (2018)
14. Smirnova, Z.V., Mukhina, M.V., Kutepova, L.I., Kutepov, M.M., Vaganova, O.I.: Organization of the research activities of service majors trainees. In: Advances in Intelligent Systems and Computing, vol. 622, pp. 187–193 (2018)

15. Smirnova, Z.H.V., Gruzdeva, M.L., Krasikova, O.G.: Open electronic courses in educational activities of the institution. Vestnik of Minin University (4), 3 (2017)
16. Smirnova, Z.H.V., Vaganova, O.I., Trutanova, A.V.: State attestation as a method of integrated assessment of competence. Karelian Sci. J. **6**(3), 74–77 (2017). https://elibrary.ru/item.asp?id=30453035
17. Tsyplakova, S.A., Grishanova, M.N., Korovina, E.A., Somova, N.M.: Theoretical bases of designing of educational systems. Azimuth Sci. Res. Pedagogy Psychol. **5**(1(14)), 131–133 (2016). (in Russian)
18. Fedorov, A.A., Paputkova, G.A., Ilaltdinova, E.Y., Filchenkova, I.F., Solovev, M.Y.: Model for employer-sponsored education of teachers: opportunities and challenges. Man in India **97**(11), 101–114 (2017)
19. Yashin, S.N., Yashina, N.I., Ogorodova, M.V., Smirnova, Z.V., Kuznetsova, S.N., Paradeeva, I.N.: On the methodology for integrated assessment of insurance companies' financial status. Man in India **97**(9), 37–42 (2017)

The Third Mission of a Regional Flagship University: The First Results of Its Realization

Ekaterina Ugnich(✉), Pavel Taranov, and Sergey Zmiyak

Don State Technical University, Rostov-on-Don, Russian Federation
ugnich77@mail.ru, taranov@inbox.ru, sergey_zm@list.ru

Abstract. The paper is dedicated to the study of the content of the third mission of the university under current conditions of social development. It has been demonstrated that the third mission of the university consists in it becoming the source of the socioeconomic development of the region. This mission is most clearly formalized in the flagship university model, the formation of network of which has begun since the end of 2015 based on the results of yet another stage of reforming of higher education system in Russia. The case of Don State Technical University that has been provided as an example goes to prove the effectiveness of the flagship university model in the development of its interaction with the regional economy.

Keywords: Flagship university · Mission of the university · Region · Higher education · Innovative development · Postindustrial society

JEL Codes: I20 · I21

1 Introduction

The rapidly changing realia of the modern world, conditioned by the development of postindustrial society, give rise to new social structures and relations (Bell 1973). Knowledge in postindustrial society is the main factor of progress and the source of reproduction of social wealth. In this regard, requirements change both to education and to the sphere of generation and dissemination of knowledge. A modern university which generates knowledge and supplies their bearers to the economy, is no longer able to be a closed and self-sufficient "ivory tower" (Etzkowitz et al. 2000). Certainly, the transformation of postindustrial society extended boundaries of subjectness of the university, providing it with new opportunities for participation in social development (Benneworth and Sanderson 2009), particularly community that is localized in a certain region. At the same time, new challenges in the implementation of these opportunities of universities take shape, conditioned by global competition and permanent variability of socioeconomic environment.

This paper is dedicated to the study of the content of the new mission of the university, which is associated with its impact on the socioeconomic development of the region. The first results of implementation of this mission by the university, having the special status of the flagship university of the region, obtained as a result of a new

E. G. Popkova and B. S. Sergi (Eds.): ISC 2019, LNNS 87, pp. 705–713, 2020.
https://doi.org/10.1007/978-3-030-29586-8_82

stage of reforming of higher education system in Russia in the end of 2015, have been analyzed for the confirmation of theoretical conclusions.

2 Methodology

This research is based on systemic approach (Kornai 2002; Popkova and Tinyakova 2013) to the perception of the university, providing an opportunity to formulate its third mission in concert with the needs of environment. In addition, the comprehensive idea of operation of the university made it possible to examine, systematize and verify the indicators characterizing its third mission.

A case study method was used to obtain the validity of conclusions of the research (Yin 2003), which made it possible to examine the experience of a particular flagship university (Don State Technical University - DSTU) in terms of implementation of its third mission. Bibliometric analysis tools were used for the assessment of certain performance indicators of the university (Pislyakov and Shukshina 2014). The five-year period data (2013–2017), posted on the Web of Science (WoS) Core Collection platform, were used as a source of bibliometric data. The data of information analysis products based on the results of the monitoring of the efficiency of activity of higher educational establishments (in 2014–2017), as well as the State Program of the Russian Federation "Advancement of Education" for 2018–2025, the Program for the Development of the Flagship University of the Rostov region – DSTU for 2016–2020, and the Program of transformation of DSTU into the center of innovative, technological and social development of the region for 2017–2019 served as a source for the assessment of other indicators of implementation of the third mission of the university.

3 Discussion and Results

3.1 Revisiting the Third Mission of the University

A great number of scientific papers is dedicated to the research into the role, mission of the university, and its value in the society. English researcher Barnett (1999) is one of the key figures in a new interpretation of the mission of the university. According to him, new mission of the university comes down to the implementation of integrating capabilities by the university, to its manifesting itself as an intermediary in our modern age.

The concept of university 3.0 has been quite popular in recent years (Karpov 2017; Molas-Gallart et al. 2002) as the subject of economy of knowledge. Numerical value here denotes the number of missions of the university: university 1.0 is only an educational institution, university 2.0 is already focused on training and scientific research; the two abovementioned missions are supplemented with the commercialization of knowledge in university 3.0. Definitely, the commercialization of knowledge may be referred to the new functions of universities. However, we believe that the third mission of the university is much more extensive, and the commercialization of knowledge is only one of forms of its manifestation.

The concept of the third mission of the university has official status in a number of countries, which is indicative of the growing role of the state in the reorientation of higher educational establishments to social interests and innovations (Ugner and Polt 2017).

The "triple helix" model, which is indicative of the interaction between the state, business, and universities, is also associated with the perception of the mission of the university and its social role (Etzkowitz and Leydesdorff 1995). It is emphasized that it is universities that become the territorial centers generating new knowledge, products and business initiatives, ensuring involvement of certain institutions at every stage of the innovation process. That said, several researchers rightfully consider the universities to be the drivers of development of the regional economy (Bramwell et al. 2012) and place special emphasis on the need to improve the interaction between the university and the region (Benneworth and Sanderson 2009). Thus, the third mission of universities, apart from the first two missions - education and research - is to serve as a driver of the socioeconomic development of the region.

3.2 Flagship Universities in the Higher Education System of Russia

The transformation of the higher education system has been a worldwide trend in recent decades. The strategy of improving the competitiveness of universities in Russia, similar to strategies adopted in Germany, France, Japan, Korea, China, and Taiwan, is focused on the provision of support to a limited number of universities (Shin and Kehm 2013). The goal of provision of support to flagship universities was often focused on joining the international ratings. Following the results of reforming of the higher education system of Russia, which has begun in 2007, the following categories of universities were identified and gained state support in the form of development programs:

- Moscow State University named after M.V. Lomonosov and St. Petersburg State University;
- 10 federal universities;
- 29 national research universities.

In addition, the 5–100 Project was launched in 2012 for increasing the prestige value of the Russian higher education at the global level; 21 universities participate in this project. This group includes 5 federal universities and 12 national research universities.

All the groups of universities described above aim for the leadership at the national and supranational levels. They are less focused on integrating into regional processes (Baryshnikova et al. 2019). The solution of this problem formed the basis of a new stage of reforming of higher education system which began in 2015, which resulted in the formation of 33 flagship universities in 32 subordinate entities of the Federation. Flagship universities were conceived not as national leaders, but as regional flagship universities, in other words, the "basis" of the higher education system in the region.

33 flagship universities were formed in two stages. Based on the results of the first stage, which began in 2015, 11 universities obtained the status of "flagship universities". In 2017, based on the results of the second stage, this status was obtained by yet another 22 regional universities.

Today, the main idea for flagship universities consists not only in the achievement of target indicators and their development programs, but also in the implementation of strategic projects for the socioeconomic development of the region (Lisitskaya et al. 2018). The target model of flagship university is focused on its positioning as a center for attraction of talents; regional science and innovation campus; center for the formation of regional elite and the source of positive changes in the city and regional environment. Therefore, it is the flagship university model that the third mission of the university which means it becoming a source of development of the region, is most clearly formalized in.

3.3 Implementation of the Third Mission of the Flagship University: Case Study of Don State Technical University

DSTU was one of the first universities which obtained the status of the flagship university of the region based on the results of competitive selection within the scope of the project of the Ministry of Education and Science of the Russian Federation "Development of a network of flagship universities", initiated in 2015. DSTU as a flagship university is developed within the scope of the target model of multidisciplinary regional university. The higher education system of the Rostov region includes Southern Federal University (SFU) – a university which aims for the national leadership in education, science, and culture. A similar mission has been imposed on DSTU, too, just at the regional level.

Being the flagship university of the Rostov region, DSTU has significant academic, educational and technological potential, having potential to solve a great number of problems in the region: starting from the influx of applicants and ending with the development of local communities. 21.06% students coming from the Rostov region study in DSTU in more than 100 areas of study.

Since three years have already passed since the implementation of the Program for the Development of DSTU as a flagship university, it gives us an opportunity to summarize the first results of implementation of its third mission. At the same time, the results of implementation of the third mission must be also quantitatively assessed. Currently, there is no unified system of indicators of assessment of implementation of the third mission of the university in domestic practice. In particular, this system of indicators has been presented abroad in the report by (Molas-Gallart et al. 2002) for Russell Group universities. Potential indicators which characterize the implementation of the third mission are the number of obtained patents and filed applications, the number of license agreements, the number of small enterprises (spin-offs), revenues from research and development activities, the number of joint scientific papers in cooperation with authors from non-academic organizations, etc.

We shall analyze the success of implementation of the third mission of DSTU based on suggested indicators taking into account the Russian specificity. We shall group together the indicators of implementation of the third mission of the university across the three principal directions that characterize, first, great demand for the university (its research and development and educational resources) on the part of regional economic entities, second, the involvement of economic entities of the region in research and development of DSTU, third, the impact of the university on the formation and development of the socioeconomic environment of the region. That said, we shall consider the five-year period (2014–2018) for the analysis of dynamics and identification of the impact of a new status of flagship university (which was obtained in 2016).

1. *Great demand for the university (its research and development and educational resources) on the part of regional economic entities.* Revenue from research and development activities is one of the most important indicators of the efficiency of scientific research activities. In general, the increase in revenues from research and development activities has been noted in DSTU, including per 1 member of academic staff (Fig. 1). This is a positive characteristic of the efficiency of scientific research activities of the university since the obtainment of a new status of flagship university. As for the interaction with the region as such, a sharp increase in revenue from research and development activities related to the solution of certain regional problems can be observed as well in 2017. That said, the volume of attracted extrabudgetary funds (of enterprises of the Rostov region) more than 8 times exceed the amount of funds attracted from the budget.

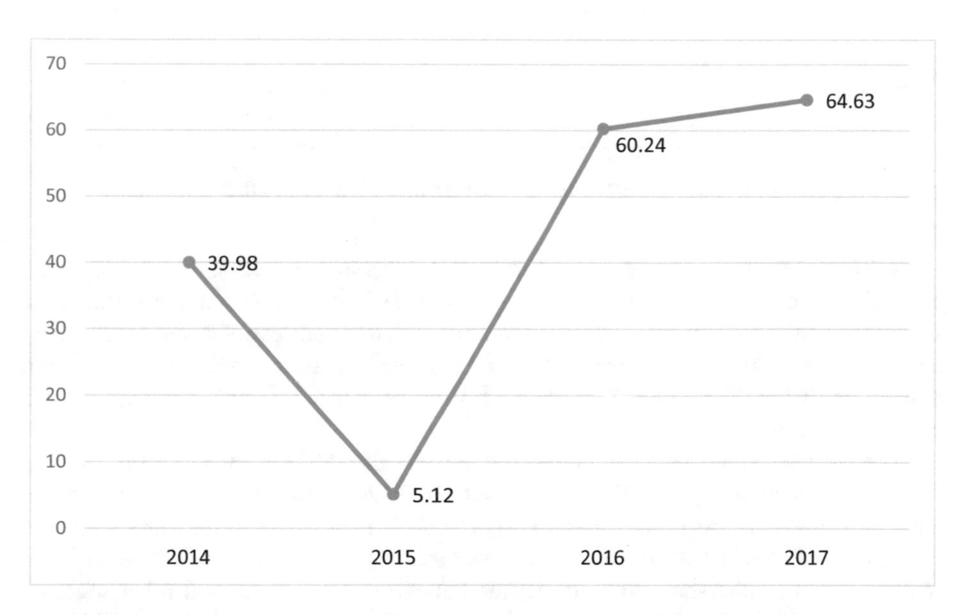

Fig. 1. The dynamics of revenues from research and development activities per 1 member of academic staff in 2014–2017

In 2017, DSTU concluded three license agreements, which are representative of the great demand for the commercialization of research and development of the university. There were no license agreements until this period since 2014.

An important indicator which is representative of the great demand for educational resources of the university on the part of regional enterprises is the presence of basic departments which were created in cooperation with enterprises. 22 basic departments have been created and efficiently operate in DSTU at present (Fig. 2). That having been said, the employment of graduates of such departments in basic enterprises is above 80%, which goes to prove the great demand for such a format of training of specialists for the real sector of economy of the region.

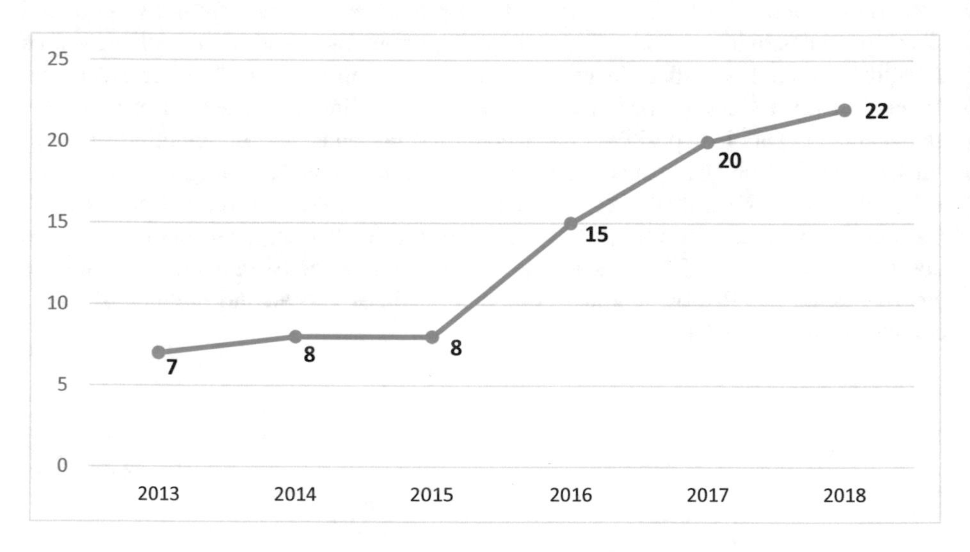

Fig. 2. The number of basic departments in DSTU in 2013–2018

2. *The involvement of economic entities of the region in research and development of DSTU.* The most unbiased assessment which is indicative of the significance of scientific results of research and the communication in professional academic community, is possible through the use of bibliometric analysis tools (Pislyakov and Shukshina 2014). The 2013–2017 data of Web of Science (WoS) Core Collection platform were used for the research.

The overall dynamics of the publishing activity of DSTU is shown in Fig. 3. One may notice a sudden "burst" of publishing activity in 2017 - the number of publications increased almost 16 times compared to 2013. This is mainly due to the introduction of the system of efficient contracts. As for the "intraregional" collaboration as such, in other words, coauthorship only with researchers from universities and other regional organizations, then the highest number of joint "intraregional" publications DSTU was achieved in cooperation with SFU. The share of publications co-authored by regional organizations in the total number of "intraregional" publications varies from 18.8% in 2013 to 25% in 2017. The growth of publishing ("intraregional") activity in 2017 was due to the increase in the number of publications not co-authored by other organizations.

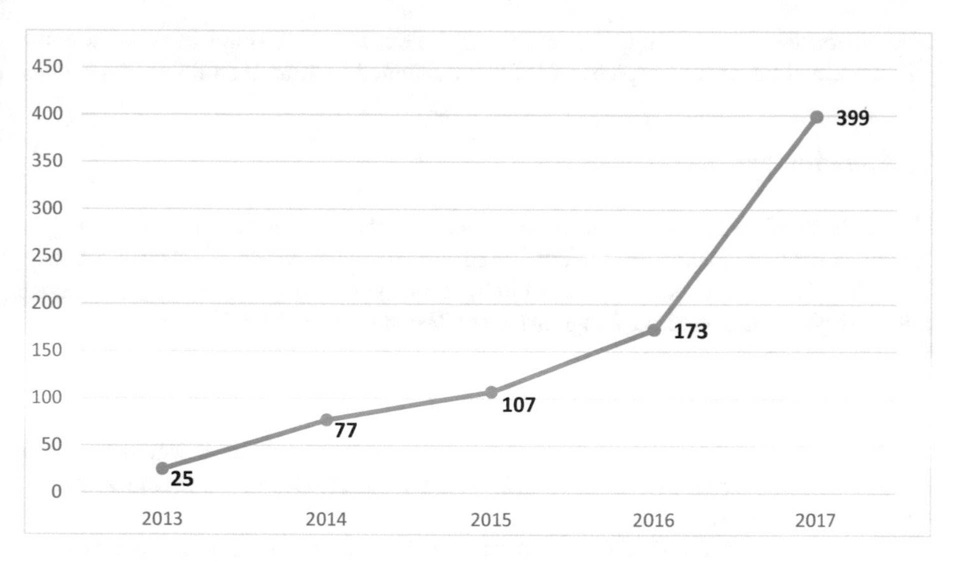

Fig. 3. The dynamics of publishing activity of the university according to the WoS information analysis system in 2013–2017

Of special interest are "intraregional" collaborations of DSTU with non-academic organizations. There were no publications like that on WoS platform for the period under analysis until 2016, while since 2016 positive dynamics could be observed, which characterizes the development of joint scientific research of the university and entities of the real sector of economy of the Rostov region (see Table 1).

Table 1. The dynamics of "intraregional" publishing activity in cooperation with non-academic organizations

Years	2014	2015	2016	2017	2018
Coauthorship with non-academic organizations within the region	0	0	3	6	4

The impact of the university on the formation and development of the socioeconomic environment of the region. In order to assess this impact, we shall characterize the strategic partnership of DSTU with entities of the real sector of economy of the region. Following the results of 2017, contractual arrangements were formalized with 2440 enterprises. The agreements for the training of specialists were concluded with 182 enterprises.

As for the creation of small innovative enterprises with the involvement of DSTU, currently their number has reduced to 21 (by 2 compared to 2014). The reduction in number of small innovative enterprises could be observed in 2017 in connection with the reconsideration of the need for these entities by the university. As a result, only efficient small innovative enterprises remained. At the same time, in 2017, the volume of tax liabilities to the budget system from the operation of small innovative enterprises, created under the auspices of DSTU, amounted to more than 85 million rubles.

4 Conclusions

The analysis of indicators of the third mission of the university across the all three directions we have identified is indicative of the strengthening of its interrelation with the region. The skyrocketing growth of indicators could be observed during the second year of implementation of the Program for the Development of the Flagship University. However, in order for this effect not to be short-term, there is a need for the consistent interrelated alignment of components of the innovative system of the region and the university, overcoming communication gaps between them, development of collaborating organizational culture, as well as active promotion of joint research and pilot projects in cooperation with the leading research organizations of the region as well as non-academic sector.

The case of DSTU that has been provided as an example goes to prove the effectiveness of the flagship university model in the development of its interaction with the regional economy in the immediate future.

References

Barnett, R.: Realizing the University in An Age of Supercomplexity. Open University Press, Buckingham (1999)

Baryshnikova, M., Vashurina, E., Sharykina, E., Sergeev, Yu., Chinnova, I.: The role of flagship universities in a region: transformation models. Educ. Stud. Moscow **1**, 8–43 (2019)

Bell, D.: The Coming of Post-Industrial Society. A Venture in Social Forecasting. Penguin Books, Harmondsworth (1973)

Benneworth, P., Sanderson, A.: The regional engagement of universities: building capacity in a sparse innovation environment. High. Educ. Manag. Policy **21**(1) (2009). http://dx.doi.org/10.1787/hemp-v21-art8-en

Bramwell, A., Hepburn, N., Wolfe, D.A.: Growing innovation ecosystems: university-industry knowledge transfer and regional economic development in Canada. Final Report to the Social Sciences and Humanities Research Council of Canada, University of Toronto (2012)

Etzkowitz, H., Leydesdorff, L.: The Triple Helix of University - Industry-Government Relations: A Laboratory for Knowledge - Based Economic Development. EASST Review, 14 (1995)

Etzkowitz, H., Webster, A., Gebhardt, C., Terra, B.: The future of the university and the university of the future: evolution from ivory tower to entrepreneurial paradigm. Res. Policy **29**, 313 (2000)

Karpov, A.O.: University 3.0 – Social Mission and Reality. Sotsiologicheskie Issledovaniya [Sociological Studies]. No. 9: 114–124 (2017). (in Russian)

Kornai, J.: The system paradigm. Voprosy ekonomiki [Problems of Economics]. No. 4 (2002). (in Russian)

Lisitskaya, T., Taranov, P., Ugnich, E., Pislyakov, V.: Pillar Universities in Russia: The Rise of "the Second Wave". In: STI 2018 Conference Proceedings. Centre for Science and Technology Studies (CWTS) (2018). https://openaccess.leidenuniv.nl/handle/1887/65226. Accessed 09 Nov 2018

Molas-Gallart, J., Salter, A., Patel, P., Scott, A., Duran, X.: Measuring Third Stream Activities. Final Report to the Russell Group of Universities. SPRU, University of Sussex (2002)
Pislyakov, V., Shukshina, E.: Measuring excellence in Russia: highly cited papers, leading institutions, patterns of national and international collaboration. J. Assoc. Inf. Sci. Technol. **65** (11), 2321–2330 (2014). https://doi.org/10.1002/asi.23093
Popkova, E.G., Tinyakova, V.I.: Dialectical methodology of analysis of economic growth. World Appl. Sci. J. **24**(4), 467–475 (2013). https://doi.org/10.5829/idosi.wasj.2013.24.04.13189
Shin, J.C., Kehm, B.M.: The World-Class University in Different Systems and Contexts. Institutionalization of World-Class University in Global Competition, pp. 1–13. Springer Netherlands, Dordrecht (2013)
Ugner, M., Polt, W.: The knowledge triangle between research, educational and innovation – a conceptual discussion. Foresight STI Governance **11**(2), 10–26 (2017)
Yin, R.K.: Case Study Research Design and Methods. Sage, Thousand Oaks (2003)

Energy-Efficient and Energy-Safe Development of the Constituent Entities of the Russian Federation: Restrictions and Prospects

Gabibulla R. Khasaev(✉) and Vladimir A. Tsybatov

Samara State Economic University, Samara, Russia
gr.khas@mail.ru, tva82@yandex.ru

Abstract. The paper explores the opportunities and limitations of reduction of the energy intensity of the gross regional product (GRP) in a constituent entity of the Russian Federation based on model analysis in the horizon of 2018–2035 under the development of the region with regard to the requirements of the Energy Strategy of Russia for the period until 2035. The calculations were carried out upon interrelated models of the economy and energy industry of the Samara region, which elaborated the scenarios of energy-efficient development of the region with the account to targets of state programs for energy economy and efficiency declared in the Energy Strategy. As a result of research, we found out that economic growth is the most important condition for the reduction of GRP energy intensity. Moreover, the higher the economic growth, the greater its contribution to the reduction of GRP energy intensity.

The paper shows that a forty percent reduction of GRP in the horizon of 2018–2035 is feasible only under an average annual economic growth of at least five percent, even upon the ultimate implementation of all industrial energy economy and energy efficiency increase programs. Similar conclusions are true for the Russian economy in general. If the Russian economy develops at an average annual rate of less than five percent, then the main target indicator of the Energy Strategy (reduction of GDP energy intensity) by 2035 by over forty percent compared to 2007 will be theoretically unattainable.

The article was prepared within the framework of the state assignment of the Ministry of Education and Science of the Russian Federation No. 26.4131.2017/ПЧ, the project entitled "Development of methods and information technologies for macro-economic modeling and strategic planning of energy efficient development of the Fuel & Energy Complex of a constituent entity of the Russian Federation".

Keywords: Reduction of GRP energy intensity · Fuel & Energy Complex (FEC) · Economic growth · Modeling · Forecasting

JEL Code: C63 · C65

E. G. Popkova and B. S. Sergi (Eds.): ISC 2019, LNNS 87, pp. 714–727, 2020.
https://doi.org/10.1007/978-3-030-29586-8_83

1 Introduction

1.1 Research Objective

According to the "Energy Strategy of the Russian Federation for the period until 2030", the central objective of the energy policy of Russia is to increase the energy efficiency of production and consumption of energy resources and, as a consequence, to reduce the specific energy intensity of the economy. The main target indicator of the strategy at the federal level is the reduction of GDP energy intensity, and at the regional one – the reduction of GRP energy intensity. By the end of 2030, it is planned to reduce the energy intensity of GDP by more than forty percent in contrast with 2007.

When making regional energy-efficient development programs, developers of regional strategies rely on the target indicators of the federal strategy and plan 40% reduction or higher in GRP energy intensity. The following questions arise: to what extent these targets are reasonable, is it possible to reduce GRP energy intensity by forty percent in the horizon of 2018–2035, what restrictions emerge on the way to the reduction of GRP energy intensity?

In preparing the State Program of the Russian Federation "Energy Saving and Energy Efficiency Increase for the Period until 2020", the impact of various factors on the dynamic pattern of GDP energy intensity has been evaluated (Bashmakov and Myshak 2012). Table 1 presents the results hereof for the innovative scenario of the Energy Strategy of the Russian Federation 2020, assuming that GDP specific energy intensity by 2020 will be reduced by forty percent against 2007.

Table 1. The contribution of factors to the reduction of GDP energy intensity for the innovative scenario of the Energy Strategy 2020[a]

Denomination of indicators	Units of measurement	Contribution of factors
Reduction of GDP energy intensity, percent	Percent	40.0
Including the:		
Structural changes	Percent	17.7
Product changes	Percent	4.1
The rise in energy prices	Percent	4.2
Autonomous technical progress	Percent	6.2
The implementation of the State Program "Long-Term Energy Saving and Energy Efficiency Increase of the Russian Federation until 2020"	Percent	7.8

[a](Bashmakov and Myshak 2012).

As follows from the table, the authors of the calculations consider structural changes in the economy the crucial factor of reduction of the specific energy intensity of GDP. However, in our opinion, structural changes (as well as product changes) are not factors, because they only convey the performance of real factors, such as, for example, economic growth and uneven rates of development. Therefore, when forecasting the dynamic pattern of specific energy intensity of GDP (GRP), it will be more correct to study the impact of these factors, rather than structural changes.

1.2 Literature Review

The international practice widely uses an approach based on the results of the factor analysis of changes in energy intensity to forecast the dynamic pattern of GDP energy intensity.

The forecasting suggests the following steps:

(a) factor analysis of the change in GDP energy intensity in retrospect and the simulation of a multi-factor model of GDP energy intensity;
(b) forecasting the dynamic pattern of affecting factors for the long term;
(c) forecasting the energy intensity of GDP in terms of a factor model through the factor dynamic pattern forecasted.

A thorough review of factor analysis methods of energy intensity is given in the publication of (Bashmakov 2012). The most common method of factor analysis of energy intensity is the LMDI index analysis method (Ang and Lue 2001). This method is basic in evaluating the energy efficiency index in a number of countries and is also in the wide practical use of the International Energy Agency. However, the above-described approach seems unacceptable at the regional level due to the lack of long series of comparable statistical data on changes in the main energy intensity factors, which does not allow for the provision of statistically significant findings of the analysis. The application of factor analysis findings in the long term (twenty years) is doubtful since such an approach cannot take into account the dynamic pattern of affecting factors. Therefore, the only appropriate method to forecast GRP energy intensity is a controllable model experiment, when scenarios of energy-efficient development are tested on interrelated models of economy and energy.

Current approaches to simulation of the economy and energy are featured by a great variety of models, reviews, discussions, and comparison can be found in the articles (Greening and Bernow 2004), (Becalli et al. 2003). The article (Jebaraj and Iniyan 2006) gives a review of over two hundred models widely used in different countries for analysis and forecasting of the energy sector development, discusses the issues related to energy modeling. Particular attention in the simulation literature is paid to computable general equilibrium (CGE) models (Dixon et al. 2013). Due to their capability to simulate the response of the system to excitation inputs, these models are widely used to analyze the consequences of management decisions made.

However, we were interested most of all in the Russian studies of energy simulation and forecasting in the economy, since they regard the peculiarities of national management institutions and the statistical description of simulation objects to a greater degree. At present, the technology of simulation and forecasting energy and economy that is developed at the Energy Research Institute of the Russian Academy of Sciences (SCANER 2011) deserves the most attention. This technology is successfully used to forecast both the Russian and global energy industry (Evolution of global energy markets 2015).

1.3 Purposes and Objectives of the Research

The main purpose of the research is to evaluate the possible reduction of the energy intensity of GRP of a constituent entity of the Russian Federation by the example of the Samara region in the horizon of 2018–2035 within the framework of the draft Energy Strategy of Russia until 2035. In terms of the research, we posed and completed the following objectives:

- collection of a required volume of reporting information on the development of FEC of the Samara Oblast and regional economy in general;
- formation of scenarios for the development of the Samara region in the context of the Energy Strategy of Russia for the period until 2035, including the prospects for domestic fuel&energy demand for with regard to the adopted Strategy for the Development of the Samara Region until 2010;
- testing the scenario forecasts at the forecasting and analytical complex "Energetika" (Khasaev and Tsybatov 2017);
- assessment of the prospects and conditions for reduction of GRP energy intensity for the region-constituent entity of the Russian Federation based on the findings of the forecasting experiments.

2 Методология

The energy intensity of GRP is calculated by the following formula:

$$EI_{GRP}(t) = \frac{TFC(t)}{GRP(t)}, \tag{1}$$

where $EI_{GRP}(t)$ is GRP energy intensity in *t-year*, tons of fuel oil equivalent per 1 rub. of Gross Value Added (GVA);

TFC(t) is an ultimate consumption of fuel and energy in *t-year*, tons of fuel oil equivalent;

GRP(t) is a gross regional product in *t-year*, rub.

When calculating the ultimate consumption of fuel and energy, we took account of FER (fuel&energy resources) spent on ultimate consumption in all sectors of the regional economy, including households. To avoid double counting, we excluded FER converted into heat and electrical energy, as well as ones processed into a non-energy commodity for chemical enterprises.

The specific energy intensity of GRP, which is calculated with respect to the base year in comparable prices is of the greatest interest:

$$EI_{GRP}(t|0) = \frac{EI_{GRP}(t)}{EI_{GRP}(0) I^{def}_{GRB}(t|0)}. \tag{2}$$

Here: $EI_{GRP}(0)$ is GRP energy intensity for the base year; $I^{def}_{GRB}(t|0)$ is GRP deflator index calculated for *t-year* in respect to the base year.

Insofar as

$$GRP(t) = GRP(0)I^{gr}_{GRP}(t|0)I^{def}_{GRP}(t|0),$$

where $I^{gr}_{GRP}(t|0)$ is the index of growth of GRP physical volume, then the specific energy intensity with regard to (1) can be presented in a more convenient form:

$$EI_{GRP}(t|0) = \frac{TFC(t)}{TFC(0)I^{gr}_{GRB}(t|0)} = \frac{TFC(t)}{TFC^{0}(t)}, \quad (3)$$

Where the indicator is equal

$$TFC^{(0)}(t) = TFC(0)I^{gr}_{GRB}(t|0) \quad (4)$$

ultimate consumption of FER in *t-year* makes sense provided that GRP energy intensity at the level of the base year remain the same.

The specific energy intensity of economic sectors (except for households) is calculated by the formula:

$$EI_i(t|0) = \left(\frac{TFC_i(t)}{Y_i(t)}\right) / \left(\frac{TFC_i(0)}{Y_i(0)}\right) / I^{def}_i(t|0). \quad (5)$$

Here: $TFC_i(0)$, $TFC_i(t)$ is an ultimate consumption of FER in *i-sector* of the economy in the base year and in *t-year* respectively; $Y_i(0)$, $Y_i(t)$ is GVA produced in *i-sector* of the economy in the base year and in *t-year* respectively; $I^{def}_i(t|0)$ is an index deflator of *i-sector* calculated for *t-year* in respect to the base year.

Insofar as

$$Y_i(t) = Y_0(0)I^{gr}_i(t|0)I^{def}_i(t|0), \quad (6)$$

where $I^{gr}_i(t|0)$ is the index of growth of GVA physical volume in *i-sector* calculated for *t-year* with respect to the base year, then formula (5) can be represented in a more convenient form:

$$EI_i(t|0) = \frac{TFC_i(t)}{TFC_i(0)I^{gr}_i(t|0)}. \quad (7)$$

The specific energy intensity of households is calculated by a similar formula:

$$EI_H(t|0) = \left(\frac{TFC_H(t)}{M_H(t)}\right) / \left(\frac{TFC_H(0)}{M_H(0)}\right) / I^{def}_H(t|0) = \frac{TFC_H(t)}{TFC_H(0)I^{gr}_H(t|0)}. \quad (8)$$

Here: $TFC_H(0)$, $TFC_H(t)$ is an ultimate consumption of FER by the population in the base year and in *t-year* respectively; $M_H(0)$, $M_H(t)$ are monetary income of the population in the base year and in *t-year* respectively; $I^{def}_H(t|0)$ is a consumer price index calculated for *t-year* in respect to the base year; $I_H^{gr}(t|0)$ is an index of growth of population's real income calculated for *t-year* in respect to the base year.

For changes in indicators (3), (7) and (8) to be evaluated we carried out forecast analysis for various scenarios of energy saving and economic growth. Forecast experiments were conducted on the forecasting-analytical complex "Energetika" developed at the Samara State University of Economics (Khasaev and Tsybatov 2017). The core hereof is a dynamic multi-branch model of FEC as part of the general model of the region's socio-economic activities, which forms interrelated processes of the production, processing, transportation, and utilization of all types of FER in the region. The model developed by the authors in the class of CGE-models and used for the purposes of regional forecasting is applied as a model of the constituent entity of the Russian Federation (Tsybatov 2018).

The main purpose of the scenario calculations is to assess the degree of impact of economic growth rates on the reduction of GRP energy intensity. The development forecasts of the economy and FEC of the Samara region in the horizon of 2018–2035 were compared in terms of six scenarios that vary in the average annual growth rates of GVA in the sector of goods and services production. The zero-growth scenario was chosen as the baseline – "Zero growth". Scenarios of "one-percent growth", "two-percent growth", "three-percent growth", "four-percent growth", and "five-percent growth" suggested a respective average annual augmentation of GVA in the sector of goods and services production.

In all the scenarios, the indexes-deflators of prices and tariffs as well as the parameters of fiscal, budget and demographic policy were set equal and are taken from the Strategy for the Development of the Samara Region. It was also accepted that indices of price growth for the main types of FER are the same as the deflator indices of GRP on the forecasting interval. In each scenario, the output of the fuel&energy sector met the needs of the economy; was calculated endogenously (on the model). It was also assumed that in the horizon of 2018–2035 all economic sectors of the Samara region, including households, would be engaged in energy saving and energy efficiency increase while achieving targets of government programs declared in the energy strategies of the Russian Federation. The average annual energy saving and energy efficiency ratios by type of activity were calculated on the content of these strategies.

3 Findings

Table 2 and Fig. 1 show the forecast values of the specific energy intensity of the economic sectors calculated with respect to the end of the forecast interval (the year of 2035) for all six scenarios.

Table 2. Specific energy intensity of economic sectors, 2035-base year ratio, percent (forecast for six scenarios)[a]

Economic sectors	Scenario					
	Zero growth	One-percent growth	Two-percent growth	Three-percent growth	Four-percent growth	Five-percent growth
FER	74.2	74.2	74.4	74.8	75.4	76.0
Production of goods (except for FER)	77.8	76.2	74.8	73.8	72.9	72.2
Production of services	78.9	75.1	72.0	69.4	67.2	65.3
Households	87.4	79.0	71.3	64.4	58.2	52.6
Economy in general	79.0	74.5	70.4	66.8	63.6	60.8

[a]Author's calculations

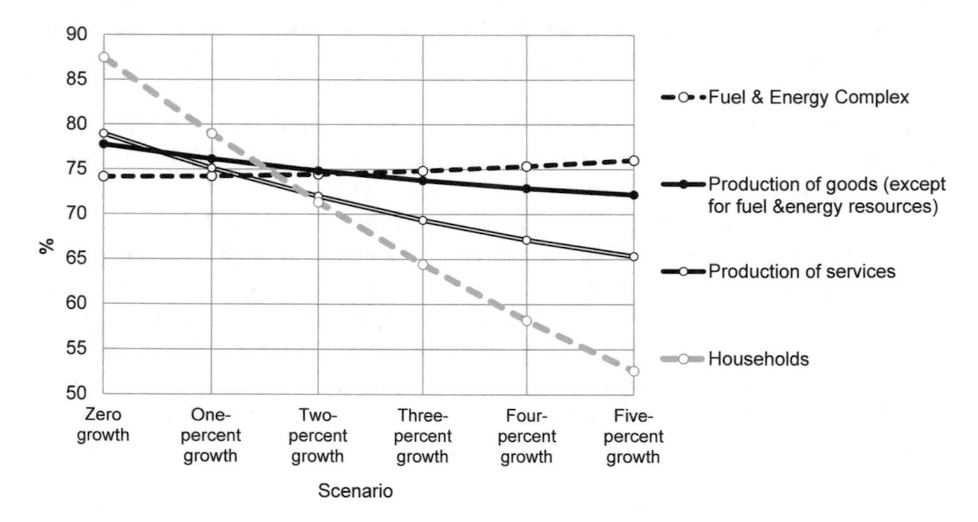

Fig. 1. Specific energy intensity of economic sectors (2035–2017 ratio)

A slight increase in the energy intensity of FEC for scenarios with higher growth is explained by a change in the FEC structure in favor of oil production and refining due to a decrease in the demand for electric and heat energy.

To assess the promotion by economic sectors of the reduction of the specific energy intensity of GRP, let's present the ultimate consumption of FER in the region as the following sum:

$$TFC(t) = TFC_{FEC}(t) + TFC_G(t) + TFC_S(t) + TFC_H(t). \tag{9}$$

Here: $TFC_{FEC}(t), TFC_G(t), TFC_S(t), TFC_H(t)$ is an ultimate consumption of FER by Fuel & Energy Complex, the sector for the production of goods (except for FER), the sector for the production of services, and households respectively. Then the specific energy of GRP (3) can be decomposed as follows:

$$EI_{GRP}(t|0) = \frac{TFC(t)}{TFC^0(t)} = \frac{TFC_{FEC}(t)}{TFC^0(t)} + \frac{TFC_G(t)}{TFC^0(t)} + \frac{TFC_S(t)}{TFC^0(t)} + \frac{TFC_H(t)}{TFC^0(t)}. \tag{10}$$

Ultimate consumption of FER by Fuel & Energy Complex and the sectors for the production of goods and services is calculated as follows:

$$TFC_i(t) = \sum_{j \in J_i} \sum_{n \in N_n} TFC_{i,j,n}(0)(a_{i,j,n} + (1 - a_{i,j,n})I_{i,j}(t|0))k_{i,j,n}(t|0).$$

Here: $TFC_{i,j,n}(0)$ is an ultimate consumption of *n-type* of FER by *j-industry* belonging to *i-sector* ($j \in J_i$) in the base year; $I_{i,j}(t|0)$ is the index showing the growth of output physical volume of *j-industry* in *t-year* in respect to the base year; $a_{i,j,n}$ are the coefficients of fixed costs of *n-type* of FER in *j-industry*; $k_{i,j,n}(t|0)$ is a coefficient of reduction of energy intensity in *j-industry* by *n-type* of FER in *t-year* in respect to the base year.

Ultimate consumption of FER for the household sector is calculated as follows:

$$TFC_H(t) = \sum_{n \in N_n} TFC_{H,n}(0) I_{H1}^{b_{1,n}}(t|0) I_{H2}^{b_{2,n}}(t|0) I_{H3}^{b_{3,n}}(t|0)) k_{H,n}(t|0).$$

Here: $TFC_{H,n}(0)$ is an ultimate consumption of *n-type* of FER by households in the base year; $I_{H1}^{b_{1,n}}(t|0)$ is the index showing the growth of the total area of residential premises in *t-year* in respect to the base year; $I_{H2}^{b_{2,n}}(t|0)$ is the index showing the growth of population's real income in *t-year* in respect to the base year; $I_{H3}^{b_{3,n}}(t|0)$ is the population growth index in *t-year* in respect to the base year; $k_{H,n}(t|0)$ is the energy saving coefficient by *n-type* of FER in *t-year* in respect to the base year.

Table 3 and Fig. 2 show the components of specific energy intensity of GRP (10) calculated for the end of the forecast interval for all six scenarios of economic growth.

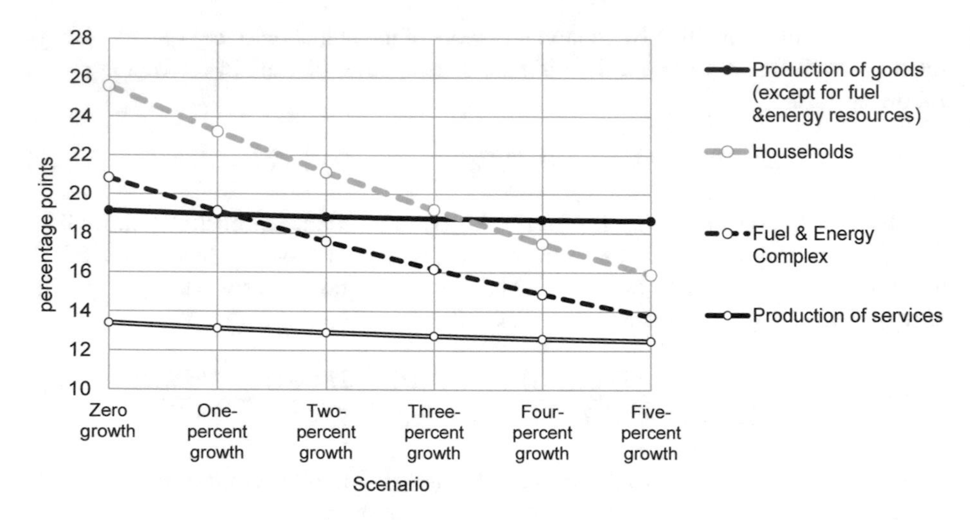

Fig. 2. The components of the specific energy intensity of GRP (2035–2017 ratio)

Table 3. The components of GRP specific energy intensity (forecast)[a]

Denomination of the indicator	Scenario					
	Zero growth	One-percent growth	Two-percent growth	Three-percent growth	Four-percent growth	Five-percent growth
Specific energy intensity of GRP in 2035, percent ratio to the base year	78.9	74.5	70.4	66.8	63.6	60.8
Including the components (10):						
FEC, p.p.	20.8	19.1	17.6	16.2	14.9	13.7
Production of goods (except for FER), p.p.	19.1	19.0	18.8	18.7	18.7	18.6
Production of services, p.p.	13.4	13.1	12.9	12.7	12.6	12.5
Households, p.p.	25.6	23.2	21.1	19.2	17.5	15.9

[a]Authors' calculations

In terms of economic growth, GRP energy intensity decreases to the greatest extent by "FEC" and "Households". For households, this is explained by the fact that the rates of ultimate consumption of FER by the population are slower than ones of GRP growth. The greater the gap, the higher the rate of economic growth. Indeed, the population spends its income growing with the rate of GRP growth on ultimate consumption and only a small part thereof on the purchase of the additional real estate, vehicles, and household appliances, the overall augmentation hereof provides an increment growth of FER consumption.

Table 4. The contribution of economic growth to the reduction of GRP energy intensity in economic sectors

Denomination of the indicator	Scenario					
	Zero growth	One-percent growth	Two-percent growth	Three-percent growth	Four-percent growth	Five-percent growth
Specific energy intensity of GRP, percent ratio to the base year	78.9	74.5	70.4	66.8	63.6	60.8
The overall reduction of GRP energy intensity due to economic growth, p.p.	0.0	4.4	8.5	12.1	15.3	18.1
contribution of economic growth in economic sectors						
FEC, p.p.	0.0	1.7	3.3	4.7	6.0	7.1
Including the:						
The decrease in demand for FER as a result of economies of production scale	0.0	0.6	1.1	1.4	1.6	1.8
Lagging of FEC growth rates from the regional economy rates due to the high capital intensity	0.0	1.1	2.2	3.3	4.3	5.3
Production of goods (except for FER), p.p.	0.0	0.2	0.3	0.4	0.5	0.5
Production of services, p.p.	0.0	0.3	0.5	0.7	0.8	0.9
households, p.p.	0.0	2.3	4.4	6.4	8.1	9.7

GRP energy intensity decreases for FEC in terms of economic growth due to the following reasons:

- an additional decrease in demand for FER in the economy as a result of the economy from augmentation of production scale;
- underrun of FEC growth rates from the regional economy rates in general due to the high capital intensity in the fuel and energy industries does not allow for their growth at rates over 2–2.5 percent per year (Khasaev and Tsybatov 2017). At high growth rates of the economy, it leads to a sharp decline in the ratio $TFC_{FEC}(t)/TFC^{(0)}(t)$ of formula (10).

The overall contribution of economic growth to the reduction of GRP energy intensity can be evaluated by the following formula:

$$\Delta EI_{GRP}^{(n\%)}(t|0) = EI_{GRP}^{(0\%)}(t|0) - EI_{GRP}^{(n\%)}(t|0). \quad (11)$$

Here $EI^{(0\%)}_{GRP}(t|0)$ is a specific energy intensity of GRP for a zero-growth scenario; $EI^{(n\%)}_{GRP}(t|0)$ is a specific energy consumption of GRP for "*n-percent* growth" scenario.

The contribution of economic growth to the reduction of GRP energy intensity for economic sectors is similarly evaluated by the formula:

$$\Delta EI^{(n\%)}_{GRP,i}(t|0) = EI^{(0\%)}_{GRP,i}(t|0) - EI^{(n\%)}_{GRP,i}(t|0),$$

where $EI^{(0\%)}_{GRP,i}(t|0)$ is a specific energy intensity of GRP in *i-sector* for a zero-growth scenario; $EI^{(n\%)}_{GRP,i}(t|0)$ is a specific energy intensity of GRP in *i-sector* for "*n-percent* growth" scenario.

Table 4 shows the overall reduction of GRP energy intensity due to economic growth and the contribution of economic sectors herein.

The specific contribution of the above-listed factors to the overall reduction of GRP specific energy intensity is shown in Table 5.

Table 5. The specific contribution of the main factors to the reduction of GRP energy intensity

Denomination of the indicator	Scenario					
	Zero growth	One-percent growth	Two-percent growth	Three-percent growth	Four-percent growth	Five-percent growth
components of the specific						
The contribution of factors to the reduction of GRP specific energy intensity						
1 is energy saving and energy efficiency increase in the production and consumption of FER	100.0	82.5	71.2	63.5	57.9	53.7
2 is economic growth	0.0	17.5	28.8	36.5	42.1	46.3
Including the factors						
Economies of the scale of production of goods and services	0.0	4.2	6.4	7.5	8.0	8.1
Underrun of FEC growth rates from the regional economy growth rates due to the high capital intensity	0.0	4.2	7.4	9.9	11.9	13.5
Underrun of the growth rate of the population's energy-consuming property from GRP growth rates	0.0	9.1	15.0	19.2	22.3	24.7
TOTAL (a total reduction of the energy intensity of GRP), percent	100.0	100.0	100.0	100.0	100.0	100.0

4 Discussion

The findings of the research conducted show that economic growth is the crucial condition for reducing GRP energy intensity. Moreover, the higher the economic growth, the greater its contribution to the reduction of GRP specific energy intensity (see Fig. 1). Economic growth reduces specific energy intensity:

- through economies of scale of production of goods and services;
- through widening gap between the growth rates of the FEC from the growth rates of the regional economy due to the high capital intensity of the exploration, production, and processing of FER;
- through growing lag in the growth rate of the energy-consuming property of the population from GRP growth rate.

The first factor has a double effect on reduction of GRP specific energy intensity since it is regarded twice in formula (10). Firstly, economies of production scale reduce the demand for FER and the values of ultimate consumption of fuel and energy $TFC_G(t)$, $TFC_S(t)$ respectively in the sectors for the production of goods and services (the second and the third components). Secondly, a decrease in demand for FER reduces the output of fuel and energy and the value of ultimate consumption of fuel and energy $TFC_{FEC}(t)$ respectively in FEC (the first component). The second factor is related to the fact that the advanced growth of non-energy sectors changes the structure of the regional economy in favor of these less energy-intensive sectors. The third factor is determined by the lagging rate of growth of household energy consumption from the rate of real income per capita growing along with GRP.

Our findings become confirmed in near Russian history. For example, in 1999–2008, when the Russian economy had been growing at an average rate of seven percent per year, after a long lag Russia became the world leader by rates of GDP energy intensity reduction: this figure dropped by forty-two percent and had been decreasing over by five percent per year on average, which is much faster than in many countries of the world. In the regions of Russia where GRP had been growing dynamically, the energy intensity hereof had been decreasing faster and vice versa (Bashmakov 2011). The fast-developing countries also demonstrate a trend towards a steady decline in GDP energy intensity. For example, China, which has shown high rates of economic growth since the 1990s, had been fastly reducing the energy intensity of GDP during this period (Grigoriev and Kurdin 2013).

The research conducted on the information about the Samara region, which is an average constituent of the Russian Federation in many parameters, allow for giving a reasonable answer to the question: Is it possible to reduce GRP by forty percent in the horizon of 2018–2035? The findings show this purpose can be achieved only under an average annual economic growth of at least five percent, even upon the ultimate implementation of all industrial energy economy and energy efficiency increase programs declared in energy strategies of 2030 and 2035. Energy saving and energy efficiency measures without economic growth will reduce GRP energy intensity in 2035 to the level of 2017. It's only half of the target (see Table 5).

Since capital-intensive regions (for example, Lipetsk Oblast, Khanty-Mansi Autonomous Okrug, Yamalo-Nenets Autonomous Okrug, Krasnoyarsk Territory, and others) cannot develop at rates over two-two and a half percent due to the high intensity of capital consumption, target indicators of the reduction of GRP energy intensity set out in federal strategies are theoretically unattainable for them.

Similar conclusions are true for the Russian economy in general. If the Russian economy develops at an average annual rate of less than five percent, then the main target indicator of the Energy Strategy (reduction of GDP energy intensity) by 2035 by over forty percent compared to 2007 will be theoretically unattainable.

References

Energy Strategy of Russia for the Period until 2030/Resolution of the Government of the Russian Federation No. 1715-p as of November 13, 2009 (2009). https://minenergo.gov.ru/node/1026

The Project of the Energy Strategy of the Russian Federation for the Period until 2035 (revised 01.02.2017). http://minenergo.gov.ru/node/1920

Bashmakov, I.A., Myshak, A.D.: Russian accounting system to improve energy efficiency and energy economy. Center for Energy Efficiency (CEE), 81 p. (2012). http://www.cenef.ru/file/Indexes.pdf

Ang, B.W., Lue, F.L.: A new decomposition method: perfect in decomposition and consistent in aggregation. Energy **26**, 537–548 (2001)

Greening, L.A., Bernow, S.: Design of coordinated energy and environmental policies: use of multi-criteria decision-making. Energy Policy **32**, 721–735 (2004)

Becalli, M., Cellura, M., Mistretta, M.: Decision-making in energy planning. Application of the electric method at the regional level for the diffusion of renewable energy technology. Renew. Energy **28**, 2063–2087 (2003)

Jebaraj, S., Iniyan, S.: A review of energy models. Renew. Sustain. Energy Rev. **10**, 281–311 (2006)

Dixon, P.B., Koopman, R.B., Rimmer, M.T.: The MONASH style of computable general equilibrium modeling. In: Handbook of Computable General Equilibrium Modeling, vol. 2, pp. 23–103. North-Holland, Amsterdam (2013)

SCANER. Modeling and Information Complex. The Energy Research Institute of the Russian Academy of Sciences (ERI RAS), 72 p. (2011)

Makarov, A.A., Grigor'ev, L.M., Mitrova, T.A. (eds.): The Evolution of Global Energy Markets and its Consequences for Russia, 400 p. The Energy Research Institute of the Russian Academy of Science- Analytical Center under the Government of the Russian Federation, Moscow (2015)

Khasaev, G., Tsybatov, V.: Tooling of modeling and strategic planning of energy-efficient development of the regional FEC. Euras. J. Anal. Chem. **12**(Interdisciplinary Perspective on Sciences 7b), 1169–1182 (2018)

Tsybatov, V.A.: Strategic planning of energy-efficient development of the constituent entities of the Russian Federation. Reg. Econ. **14**(3), 941–954 (2018)

The Energy Strategy of Russia for the Period until 2020: Resolution of the Government of the Russian Federation No. 1234-p as of August 28 (2003). http://www.energystrategy.ru/projects/ES-28_08_2003.pdf

Khasaev, G.R., Tsybatov, V.A.: Capital-generating sector of the economy as a frame of economic growth. Bull. Samara State Univ. Econ. **1**(147), 5–16 (2017)

Bashmakov, A.D.: The dynamic pattern of energy intensity of the Moscow GRP. Energy Econ. 3 (2011). https://www.abok.ru/for_spec/articles.php?nid=4890

Grigoriev, L.M., Kurdin, A.A.: Economic growth and energy demand. Econ. J. High. Sch. Econ. **3**, 390–406 (2013)

State Support of Social Entrepreneurship: Sociological Assessment

Irina V. Dolgorukova(✉), Tatyana M. Bormotova, Tatyana V. Fomicheva, Evgenia E. Kiseleva, and Evgeny A. Lidzer

Russian State Social University, Moscow, Russia
{dolgorukovaiv, bormotovatm, fomichevatn, kiselevaee, lidzerea}@rgsu.net

Abstract. The paper proposes the interpretation of social entrepreneurship as a special kind of entrepreneurial activity initiated to solve a socially significant problem in order to obtain a sustainable long-term social and commercial effect on the market. The purpose of the work was to analyze the role of state support in the development of social business initiatives. In the course of work, the main components of state support necessary for the development of social entrepreneurship are highlighted. On the basis of the work done, it was concluded that in order to promote the development of social business initiatives, the actions of government and management should be implemented at the level of local and regional communities. Regional actors must continually strive to support social entrepreneurship and ensure its sustainable development.

Keywords: Social entrepreneurship · Public policy · Sociological analysis · Social and economic development · Social programs

JEL Classification Codes: Z13 · Z18

1 Introduction

The social sphere plays a significant role in the sustainable development of any state. However, in conditions of crisis phenomena in the economy, the state's ability to finance the social sphere is limited. Therefore, the problem of finding financial resources, including additional extrabudgetary sources, the attraction of all groups of society to the solution of the most important socio-economic problems becomes extremely urgent.

The lack of proper flexibility of the social policy of the state and the decrease in the efficiency of state regulation of economic processes in market conditions affected the transformation of public policy objects towards the formation of new hybrid structures. One of these types of hybrid structures that can effectively cope with the solution of many social problems, attract additional financial resources in the social sector, and promote their optimization and distribution in the national economy, is social entrepreneurship, combining the mission of a non-profit organization, the social goals of the state and entrepreneurial business approach.

E. G. Popkova and B. S. Sergi (Eds.): ISC 2019, LNNS 87, pp. 728–738, 2020.
https://doi.org/10.1007/978-3-030-29586-8_84

The main difference between social entrepreneurs from other businessmen is the goal of their activities. In the second, the main goal is money, and in the first, help to people. Yet do not confuse social entrepreneurship with charity, even though money is the ultimate goal of social entrepreneur, but they are the main way to achieving this goal. It can be said that a socially oriented business is located at the intersection of entrepreneurship and charity, combining a social orientation and an entrepreneurial approach.

In recent years, the phenomenon of social entrepreneurship has been increasingly discussed in the economy of many countries. And if in developed and many developing countries this type of economic activity has already got accustomed and is fully developed, in semi-peripheral countries such business is only at the start of its development. Social business, being low-profitable as a general rule, requires special development resources, which should be expressed in the development of social entrepreneurship development policies. This policy is being implemented in the EU, the USA, Korea, India and other developed and developing countries, proving its effectiveness, but in Russia it is fragmentary.

Russian social entrepreneurship is only at the start of its development, so often people are mistaken about some of the features of this type of business. Most still do not know about this phenomenon and, having heard the term «social entrepreneurship», they represent a charitable project rather than a full-fledged business.

Strengthening the relationship between the state, business and society in the framework of the paradigm of social entrepreneurship will give a new impetus to the formation of civil society institutions. Orientation to the complex solution of acute social problems, the network principle of dissemination, the ability to actively integrate into existing projects make social entrepreneurship a powerful source of civil initiatives.

2 Methodology

The functional component of social entrepreneurship and its distinctive properties in comparison with the non-profit sector and the classical business are well enough reflected in the works of J. Mayr, J. Emmerson, I. Marty, C. Alter, R.L. Martina, S. Osberg and others. Of considerable interest for the thesis are the works of J. Schumpeter, which characterize entrepreneurial activity from the point of view of its innovative nature.

Among the domestic researchers of the phenomenon of social entrepreneurship, its features, problems, trends and development prospects, the works of A. A. Moskovskaya, A. V. Mukhin, V. V. Zhokhova, V. Glushkov, L. Taradina, M. Batalinoy should be noted. N.I. Zvereva.

The works of Russian scientists (T. Zaslavskaya, Yu. Popov, A. Shevchuk, V. Radaev, M. Lapusta, L. Sharshukova) represent the big interest, defining historical and modern features of entrepreneurial activity. The specifics of the activities of entrepreneurs in modern Russian society, their place, their role in the social structure and the relationship with the authorities are analyzed in the works of A. Chepurenko, A. the

Acceptor, G. Osadchy. The research interest is the point of view of I. Afanasenko, which focuses on the historical foundations and prospects for the development of entrepreneurship in the context of changes in the modern economy.

Comprehension the essence of social entrepreneurship in the field of social responsibility of business representatives is represented in the scientific works of A.V. Savina, A.F. Weksler, G.L. Tulchinsky, E. V. Orlova, L. A. Temnikova, M. B. Orlova, O.V. Karamova, R.N. Pavlova.

The article is based on the results of a sociological research on the topic «State support of social entrepreneurship in Russia», which is conducted on the basis of the Russian State Social University. The research method was questioning. Objective: to study the features of state policy in the field of social entrepreneurship support. Also, 10 in-depth interviews were conducted with experts, members of the Club of Social Entrepreneurs "CRNO".

3 Results

3.1 The Concept of Sociological Research of State Support of Social Entrepreneurship

Today there are several approaches to understanding the term «social entrepreneurship» in the scientific literature and business practice. From the point of view of the AshokaFoundation company concept, the essence of social entrepreneurship is defined as follows: «Instead of leaving the needs of society to the state or the business sector, social entrepreneurs look for the source of the problem and eliminate it by changing the social system» (Bornstein 2015).

From the perspective of the concept of the Our Future Foundation, social entrepreneurship is an entrepreneurial activity aimed at mitigating or solving social problems, characterized by the following main features: social impact, innovation, self-sufficiency and financial sustainability, scalability and replicability and entrepreneurial approach (Batalina et al. 2018).

From the concept of the Organization for Economic Co-operation and Development (OECD), social entrepreneurship is defined as any private activity, which is conducted in the public interest and organized with an entrepreneurial strategy, whose main goal is not to maximize profits, but to achieve certain economic and social goals. It has the ability to implement innovative solutions in solving problems of social exclusion and unemployment (Batalina et al. 2018).

The concept of J. Kikala and T. Lyons represent the big interest. They consider social entrepreneurship as an application of the mindset, processes, tools and technologies of ordinary entrepreneurship for the benefit of society and ecology and the solution of the most pressing social problems that combines the entrepreneurial spirit of the private sector, the strength of economic markets and the characteristics of the public sector, putting public interests above private ones (Lisevich 2016) (Table 1).

Table 1. The main criteria for social entrepreneurship

	Coalition of Social Enterprises (UK)	Bill Drayton (Ashoka)	Fund «Our Future»
Goals	Social goals and ethical values	Social significance of the idea: the availability of a real social problem, for the solution of which the project is implemented Social impact: the solution or mitigation of social problems	Social impact: solving or mitigating social problems
Financing	Reinvest profits primarily to achieve social goals and in the interests of the local community	«Moral Sustainability»: sustainability of the principles of maximizing the public good, which doesn't put the main interests of profit maximization Self-sufficiency and financial sustainability: solving social problems at the expense of their own income and as long as necessary	Self-sufficiency and financial sustainability: solving social problems at the expense of own income and as long as it is necessary
Innovativeness	–	Creativity and entrepreneurial skills: the ability to see a solution to a social problem that has not yet been proposed and accumulate resources for their implementation	New, unique combination of resources and approaches to solving a social problem (close to creativity of B. Drayton)
Other qualities	Public ownership: The governance structure and ownership of social enterprises is usually based on the participation of stakeholder groups or directors and trustees acting on behalf of a wider range of stakeholders	–	Scalability and replicability: scaling up the activities of a social enterprise and spreading its experience to increase social impact through social franchise, circulation, etc. Entrepreneurial approach

Based on the above analysis, we can formulate the following definition of social entrepreneurship. Social entrepreneurship is a special kind of entrepreneurial activity, which is initiated to solve a socially significant problem in order to achieve a sustainable long-term social and commercial effect on the market.

Meanwhile, the development of social entrepreneurship in many countries is hampered by the lack of legislative consolidation and registration of this phenomenon. In the Russian Federation, on December 27, 2018, the Law «On the Development of Small and Medium-Sized Businesses» was adopted in terms of consolidating the concepts of «social entrepreneurship» and «social enterprise». In the draft law, social entrepreneurship is stands out as a separate priority area of activity for small and medium-sized businesses, defines the concept of «social enterprise», special forms and types of support for social enterprises.

The subject of social entrepreneurship ensures the employment of certain categories of citizens (disabled, parents with many children raising minor children, graduates of orphanages under the age of 23, pensioners and people near pre-pension and other socially unprotected categories of citizens). The percentage of such workers should be not less 50%, and the percentage in the wage fund – not less 25%. The priority support measure is to ensure the availability of infrastructure facilities in the regions, on the basis of which social entrepreneurship entities operating in the field of social entrepreneurship can receive comprehensive support. Favorable conditions are envisaged for the payment of insurance premiums for employees from the category of socially unprotected citizens, there may be tax concessions, and there may be special conditions for preferential loans in the framework of social entrepreneurship development programs. There are proposals to establish a priority for social enterprises during state and municipal procurement.

The system of indicators of the effectiveness of state support in the field of social entrepreneurship should include:

I. Objective indicators:
   1. Adoption of regulatory legal acts regulating the sphere of social entrepreneurship, containing precise criteria that allow to clearly define the goals, objectives and motives of social entrepreneurship (the number of adopted regulatory legal acts, projects, etc.).
   2. Formation of state programs aimed at supporting social entrepreneurship (number of programs, targeting);
   3. Opening of social entrepreneurship support centers;
   4. Formation of subsidies for social entrepreneurs;
   5. Opening of development funds for social entrepreneurs.

II. Subjective indicators
   1. The quality, efficiency and targeting of the provision of state support to social entrepreneurs (the number of regional programs, the timing of consideration of appeals, the level of involvement of government bodies, etc.);
   2. Rational implementation of state support for social entrepreneurs, depending on the form and type of social enterprise (the number of subsidies issued, open centers to assist social entrepreneurs, etc.).
   3. Satisfaction with government support from social entrepreneurs.

3.2 Public Assessment of State Support of Social Entrepreneurship in Russia

Today, partnerships between social entrepreneurship are developing more and more actively, and joint social projects of various kinds are being implemented. Therefore, the relevance of the study of public opinion is of particular importance.

Figure 1 presents the distribution of answers to the question: «How would you describe the activities of the authorities in support of social entrepreneurship subjects?».

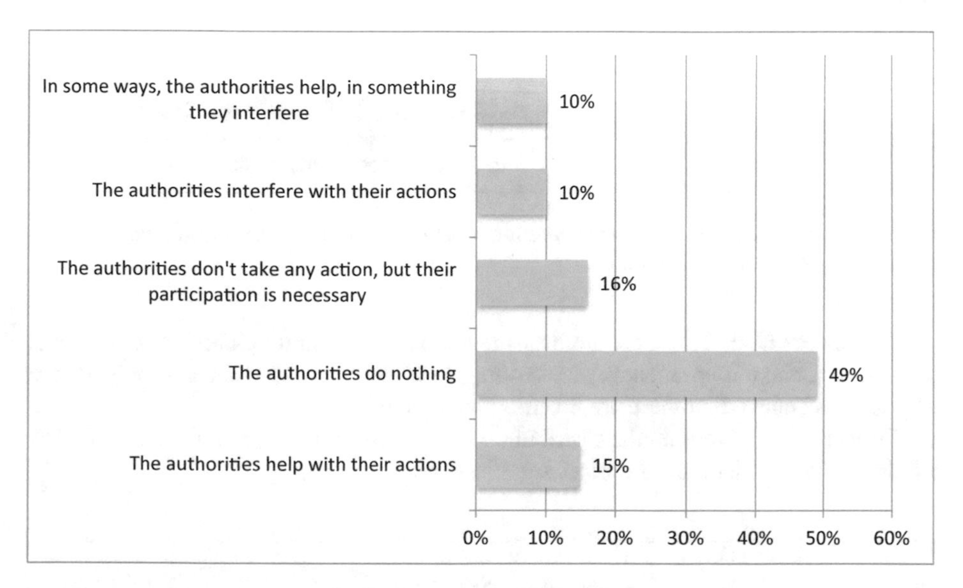

Fig. 1. Distribution of answers to the question: «How would you characterize the activities of the authorities in support of social entrepreneurship subjects?»

It is remarkable that the majority of respondents positively assess the role of social entrepreneurship in the development of society (more than 60% of respondents). Nevertheless, the study also revealed negative assessments of the phenomenon: «making money on people», «business supposedly for people, not for profit», «when they take money for free social services» (13.2%). The concept of social entrepreneurship is put in quite different meanings - from commercial activities focused on income and benefit to charitable gratuitous work for the benefit of society.

Figure 2 presents the distribution of answers to the question: «Assess the state of administrative barriers for social entrepreneurship in Russia».

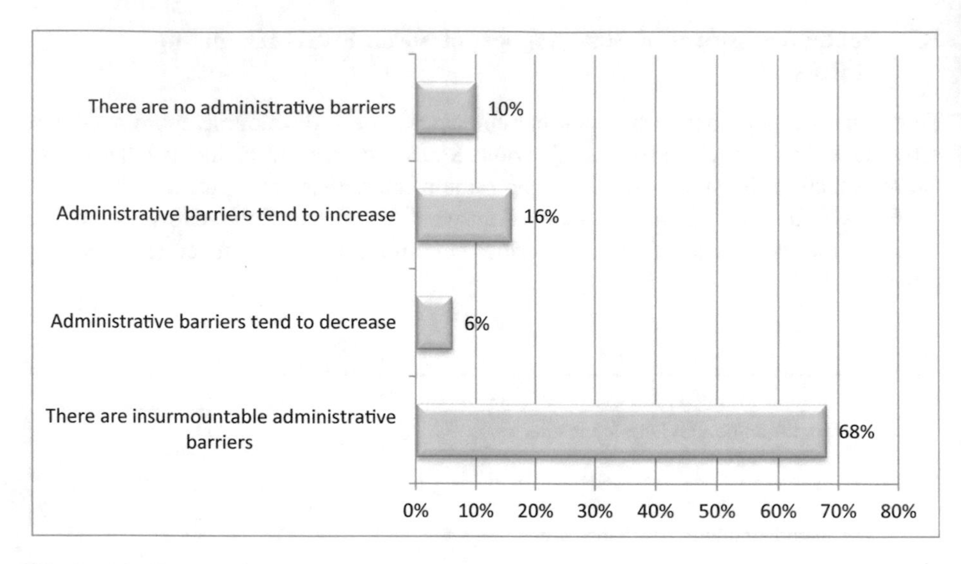

Fig. 2. Distribution of answers to the question: «Estimate the state of administrative barriers for social entrepreneurship in Russia»

We can see here, that, there are insurmountable administrative barriers to the development of social entrepreneurship, according to 68% of respondents, and only 10% of respondents believe that there are no administrative barriers.

Further, Fig. 3 presents the distribution of answers to the question: «Which of the administrative barriers are the most significant?».

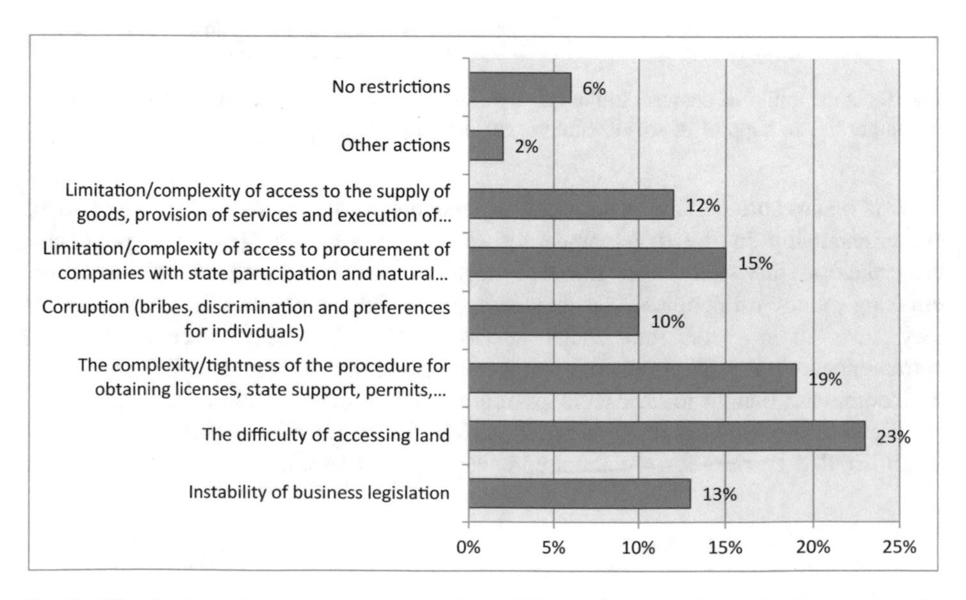

Fig. 3. Distribution of answers to the question: «Which of the administrative barriers are the most significant?»

Thus, based on the answers presented in Fig. 3, we can conclude that the most significant administrative barrier, according to respondents, is the difficulty of accessing land, and 19% of respondents noted the complexity/lengthiness of the procedure for obtaining licenses and state support funds.

3.3 The Effectiveness of State Policy on the Support of Social Entrepreneurship

During the interview with experts, issues related to the state regulation of social entrepreneurship were discussed. The survey revealed that now there is no significant financial support for social entrepreneurs in Russia. Most experts expressed their opinion that, due to the lack of clear boundaries and criteria for social entrepreneurship, commercial enterprises began to claim financial support, trying to present their activities in a socially-entrepreneurial manner. And there were cases, for example, when they really succeeded, and they received financial support. Accordingly, such investments of public funds rarely reach the target audience. And at some point they decided to reduce financial support, or postponed until the adoption of the law on social entrepreneurship.

It was also repeatedly stated that at that time social entrepreneurs didn't learn how to properly manage the money received. It could also lead to a reduction in funding. Here is what an expert said about this: «I think it's related to the fact, that the social entrepreneurs don't always have enough knowledge about how to properly manage money. I have repeatedly come across the fact that social entrepreneurs receive funds, but for some reason they aren't spent for the purposes that were announced».

It should also be noted that due to the lack of specific legislation in the sphere of social entrepreneurship, the activities and support of social entrepreneurship subjects are regulated by regulatory documents and programs for the development of small and medium-sized businesses. Thus, due to various reasons, support for social entrepreneurship is mainly directed not at financial support at present time, but at educational, consulting. That is, support is aimed at obtaining the very tools for development.

The following proposals were received to the question: «If you had the opportunity to influence legislation in the sphere of social entrepreneurship, what would you suggest to add/change/remove/modify?» and presented in the form of generalizations:

- It is enough to clearly distinguish between classic business and social entrepreneurship by defining the necessary criteria.
- The state should provide assistance in the implementation of products and services of a social entrepreneur, helping it to establish a sales market and the development of partnerships.
- Pay special attention to social entrepreneurship entities that work directly with people with disabilities, as «… they really do a great job».
- To endow regions with independent powers in the sphere of social entrepreneurship in order to «…develop and maintain promising areas of social entrepreneurship for the region first of all».

- Provide social entrepreneurs with preferential terms for the rental of premises and tax payments.
- Provide social entrepreneurs with some financial support whenever it possible, as «...for such a business, even small financial support is a significant help».

4 Discussion

Currently, there are many interpretations of social entrepreneurship development factors and the role of the state in supporting social initiatives.

First of all, in interpreting social entrepreneurship, emphasis can be placed on the creation and functioning of enterprises that realize and pursue social goals.

But there are limitations to this approach. Such an understanding of this social phenomenon can reflect both the social responsibility of business, and private investment by entrepreneurs of social projects and also the activities of non-profit organizations.

Secondly, due to determining social entrepreneurship, the emphasis may be placed precisely on the innovation activities of social enterprises for solving social problems. Moreover, innovative methods are considered more important than economic profit and enterprise self-sufficiency.

Thirdly, social entrepreneurship is considered not only as a source of solving any social problems, but also as a condition for searching and eradicating the sources of social problems (Saulina 2015).

According to G. Diz, the essence of social entrepreneurship is to apply the best practices of traditional entrepreneurship for the realization of socially significant missions and goals. He believes that social entrepreneurship can make non-profit organizations less bureaucratic. Jerr Boshi, developing the ideas of G. Diz, notes that «social entrepreneurs are leaders who pay more and more attention to the realities of the market, and they don't losing sight of their main mission.

They are able to create a balance between moral imperatives and the desire to profit – and this process represents the heart and soul of the entire social movement».

J. Wieraward, C. Carnegie, and J. Mort view social entrepreneurship through the lens of philosophy. They say that social entrepreneurship involves the honest way of doing business, which is based on the performance of a social mission. Social entrepreneurs select the right morale to achieve their goals for this, recognize the possibility of creating social value, are willing to take risks and use innovative approaches in their work. However, these researchers in their works practically don't touch upon the issues of a non-profit organizational structure.

David Brown, Christine Letts and Sarah Elward add to the definition of social entrepreneurship the theme of transformation and sustainability. They say that social entrepreneurship can find innovative solutions to any social problems and is able to mobilize ideas, social measures and resources, which are necessary for sustainable transformations.

5 Conclusions/Recommendations

Some weaknesses of state support for social entrepreneurship were identified In the process of empirical sociological research, which should be paid special attention in order to increase the efficiency and spread of this type of business in Russia. Based on the information received, we developed the following practical recommendations that can be used to improve the development of social entrepreneurship in certain regions and in Russia as a whole.

First, at the state level, it is necessary to improve the legislative framework in the sphere of social entrepreneurship, which contains precise criteria that clearly limit this type of business from the traditional one. Lawmakers need to rely on the already used principles and characteristics of this type of business, developed on the basis of social entrepreneurship support funds, existing scientific developments in this area, existing structural bodies involved in the maintenance and support of this type of business and the interests of entrepreneurs.

With the help of the accumulated world and domestic experience, lawmakers will be able to take into account the aspects and features of Russian social business to the maximum, which minimizes the possibility of not entering into existing business of social entrepreneurship. This is especially relates to entities, which open up a social business in education and health care. According to experts, these areas must be in the coverage area of the projected law, and they should not be excluded.

Secondly, it is necessary to improve the control system during the registration of a particular type of activity. This is due to the fact that many enterprises, according to the All-Russian Classifier of Economic Activities (ACOEA), are registered in the sections covered by state support for social entrepreneurship. But there are many cases, when ACOEA doesn't correspond to the real activity of enterprises. And because of this, part of the resources for supporting social entrepreneurship comes to those enterprises that, in fact, should not receive it. The increased control system will allow minimizing or eliminating speculation in this area, thereby contributing to a targeted flow of various support resources to the target audience.

References

Bornstein, D.: How to Change the World: Social Entrepreneurship and the Power of New Ideas, pp. 89–91. Alpina Publisher, Moscow (2015)

Batalina, M., Moskovskaya, A.A., Taradina, L.D.: Review of the experience and concepts of social entrepreneurship, taking into account the possibilities of its application in modern Russia. State University Higher School of Economics, pp. 23–30 (2018)

Lisevich, A.V.: Social entrepreneurship: theory and practice. In: Proceedings of the IV International Baltic Sea Forum. pp. 1643–1648 (2016)

Saulina, Yu.P.: Development of entrepreneurial activity in the Russian Federation. Entrepreneurship – XXI, № 7, pp. 128–135 (2015)

Gregory Dees: The Man Who Defined. WebsiteBloombergBusinessweek, 01 Aug 2014. http://www.businessweek.com

History of Social Entrepreneurship. Electronic Site Bright Hub. Jean Scheid, 09 July 2011. http://www.brighthub.com

Policy Brief on Social Entrepreneurship 2013. OECD/European Union website. OECD (2013). http://www.oecd.org

Social Impact Award 2018. IMPACT HUB Moscow (2014). http://www.impacthubmoscow.net/programms/

Tanatova, D.K., Yudina, T.N., Dolgorukova, I.V.: A comparative analysis. Varazdin Development and Entrepreneurship Agency; Russian State Social University, pp. 446–454 (2017)

Tanatova, D.K., Dolgorukova, I.V., Pogosyan, V.G., Bormotova, T.M., Lidzer, E.A.: Youth entrepreneurship: social practices and risks. J. Soc. Sci. Res. (2018)

Yudina, T.N., Osadchaya, G.I., Leskova, I.V., Dolgorukova, I.V., Kireev, E.Y.: The Eurasian economic union: migration risks. Mediterr. J. Soc. Sci. **6**(4), 451–457 (2015)

Yudina, T.N., Dolgorukova, I.V., Kireev, E.Y., Bormotova, T.M., Fomicheva, T.V.: Constructing regional social inclusion: foundations of sociological analysis. In: Popkova, E. (ed.) The Future of the Global Financial System: Downfall or Harmony. Lecture Notes in Networks and Systems, pp. 195–202. Springer, Cham (2019)

Reproductive Behavior in Russia and Countries of the World: Axiological Aspect

Dina K. Tanatova(✉), Tatiana N. Yudina, Tatyana V. Fomicheva, Irina V. Dolgorukova, and Ivan V. Korolev

Russian State Social University, Moscow, Russia
dktanatova@mail.ru, ioudinatn@mail.ru, fomitchevatv@mail.ru, div0607@gmail.com, iv_king@mail.ru

Abstract. The problem of regulation of reproductive behavior is chosen for the topic of publication not without a reason. Careful attention to the study of this problem is necessary for understanding and predicting fertility trends in Russia and in the world for developing a long-term concept of an effective demographic and family policy. At the moment, the country doesn't observe the problem of depopulation, but it hasn't disappeared, it only faded into the background. Now the main demographic reserve of the country is the depopulation generation of the 90s, which is currently undergoing a process of rethinking life priorities and strategies of socio-demographic behavior: the number of families leading childless lifestyles or the age of birth of the first-born shifts, so a new threat arises due to transition of the country into the open phase of depopulation.

Keywords: Reproductive behavior · Population of Russia · Axiological aspect

JEL Code: J 110 · J 150

1 Introduction

1.1 Problem Definition

The problems associated with reproductive behavior and with changes in reproductive attitudes became one of the most important aspects of the study at the turn of the 19th and 20th centuries, when a sharp decline in the birth rate was noticed in Europe. It was connected, first of all, with the transition from the traditional type of society to the society with a new social system. For example, one of the main reasons was the mobilization of women in the labor sphere. In Russia, the question of a demographic problem arose only in the 60s of the 20th century, when the number of marriages began to decline, the divorce rate increased, and a low birth rate was established. The country entered the phase of hidden depopulation, and the country entered the open phase in the 90s.

Historically, the institutional crisis of the family manifests itself in a change in family types, both from a functional and structural point of view: the extended family gave up its place to a nuclear family consisting of parents and children, the latter gives way to the leading meaning of the new types - the married family (where interests are not children, but marital well-being) and an ego-centric family, where each "I" strives

E. G. Popkova and B. S. Sergi (Eds.): ISC 2019, LNNS 87, pp. 739–749, 2020.
https://doi.org/10.1007/978-3-030-29586-8_85

for its own benefit. In the context of modernization (industrialization, urbanization, improvement of health and hygiene, growth of education and professional employment of women, etc.), the family lost its production function and, at the same time, the economic basis of the motives for having children, the basis of reproductive motivation (Antonov 2014). According to statistics, Russia is ranked 178 in terms of the birth rate among other countries of the world with a total fertility rate of 11.0 (Central intelligence agency 2018).

Modernization theory (Vishnevsky AG, Volkov AG, Golod S.I.). This theory is common in developed western countries. According to this concept, all negative changes occurring with a modern family are considered as temporary phenomena on the background of constant positive development; society is evolving towards universal well-being. From the point of view of this approach, modern reproductive behavior is viewed from the perspective of a transition to a new type of reproduction. It is based on the idea of demographic equilibrium, the evolution of demographic processes and the concept of a demographic transition.

Based on this, representatives of the crisis theory Medkov V.M. and Antonov V.A. give the following definition of reproductive motives - the mental state of the individual, which encourages the individual to achieve personal goals (economic, psychological, social) through the birth of children. And they distinguish three main types: (1) economic motives of having children are motives that induce the birth of a certain number of children because certain events are achieved (or assumed to be achieved) through this event, i.e. goals associated with the desire to acquire some material benefits or improve (maintain) their economic status; (2) social motives are motives that encourage the birth of a certain number of children within the existing sociocultural norms of childhood and which are an individual response to these norms. They are expressed in the desire of a person to live “as all” and to have as many “as all” children; (3) psychological motives are motives that induce the birth of a certain number of children due to the fact that through this event some purely personal, socio-psychological, internal goals of the person are achieved. They are divided into three groups: the first includes motives which are driven by the need to fill life with meaning, the need for love from a child, respect, the desire to continue in children; in the second group there are motives that are caused by the need to take care of a small child, to love him, to give him my life experience; in the third group, all other types of psychological motives, for example, to avoid loneliness in old age or the desire of the spouses to strengthen their marriage (Antonov 1996).

The next component of reproductive behavior is reproductive needs. Reproductive needs are assimilated by a person in the process of socialization, being in culture and reproductive norms, or social norms of fertility (i.e., social patterns and behavioral principles relating to the birth of a certain number of children) that are transformed in a personal way by the personal structure. They are formed under the influence of social norms of fertility or reproductive norms of small families, medium-sized children, large families (Ilyasov 2013). Elizarov V.V. understands the need for children as a reproductive behavior program aimed at obtaining various types of reproductive resources (Ilyasov 2013). The need for children in different subjects may be actualized or not actualized (suppressed) at all.

2 Methodology

The results of a sociological study, which is held in 2016–2018 by the online survey method on the website webanketa.com, aimed at studying the transformation of reproductive attitudes among Russian youth. Total interviewed 600 respondents. The study was held in 2 waves (1 wave - May 2016, 2 wave - November-December 2018), which allowed to consider the main indicators in dynamics. The sample is one-step and directional. Signs of the selection of respondents were: belonging to a social group, youth (age group 18–35 years old), marital status; male and female sex; living in Russia. The purpose of the study is: to consider the transformation of reproductive behavior of Russian youth (2016–2018 year).

Reproductive behavior is a system of actions, relationships and mental states of a person associated with the birth or renouncement of having children anyway, married or out of wedlock (Borisov 2001). As any system, reproductive behavior has its own structure, which includes the following components: reproductive attitudes; reproductive motives; reproductive needs; solutions; actions.

Reproductive setting is a special case of a social setting. For the first time, the term «installation» was introduced as part of the study of the characteristics of L. Lange's perception in 1888 and was defined as «a holistic modification of the subject's state, directing its reactions and interaction» (Dictionaries and Encyclopedias at Academica 2018).

The beginning of research on social attitudes is associated with the work of American sociologists F. Znanetzky and W. Thomas «Polish Peasant in Europe and America». Two dependencies were identified from this work: the dependence of the individual from social organization and the dependence of social organization from the individual. The importance of social organization for the individual is social value; for the psychological state of the individual in relation to the group was given the concept of social setting (Ganzha 2018).

Reproductive installation is divided into 2 types: (1) child setting, which are aimed at the birth of a certain number of children and include installation on the safe outcomes of pregnancy, installation on the sex of the child, installation on adoption (under certain circumstances); (2) the installation on the baselessness, which are aimed at contraception and allow for artificial termination of pregnancy. Such installations can be safely called anti-reproductive installations.

The main external indicators of reproductive attitudes are the three main indicators of preferred numbers of children: The average ideal number of children characterizes an individual's idea of the best number of children in a family in general, without taking into account a particular life situation and personal preferences. It is rather an indicator that reflects people's ideas about the social norms of children. This is the most stable indicator, which isn't subject to drastic socio-economic changes.

The average desired number of children is an indicator characterizing the individual need for children, reveals the personal preferences of the individual in relation to the number of children that he would like to have in his family if nothing prevented him from fulfilling his desire. This indicator gives the most accurate description of people's reproductive preferences. The average expected (planned) number of children characterizes the real intentions, reproductive plans of people and families taking into account the specific circumstances of their lives, taking into account the competition of reproductive plans with other life plans (Bodrova 2000). Reproductive attitudes are formed at an early age and then almost don't change throughout a person's life.

3 Results

3.1 Factors of Change in Reproductive Behavior

One of the factors that influenced the change in reproductive behavior and the emergence of a childfree ideology is anti-family views in the early 20th century. Indeed, in pre-revolutionary Russia, the majority of the population was sufficiently religious and women, on average, gave birth to 5–10 children, not due to the fact that having many children was a special value, but because the birth of children was a natural consequence of a married life. But after the revolution in 1917, views on family and demographic policies had a pronounced anti-family orientation. Social values and norms were formed under the influence of the views of K. Marx, F. Engels, V.I. Lenin. The installation of a new family-marriage relationship began, which included the proclamation of the emancipation of the kinds and «free love», the abolition of church marriage registration, a simplified procedure for terminating marriage through divorce, and most importantly, legalizing abortions. In 1920, Lenin signed a decree allowing abortions in medical institutions for social reasons, as well as at the request of a woman (Demography of Russia 2018). So Russia was the first country in the world where abortion was allowed by law. In other countries, this happened much later, in the 70s–80s of the 20th century, and it was the result of the «sexual revolution». «Freedom of abortion» in the USSR lasted until June 1936, when abortion was again prohibited.

Henceforth, a woman, who did an underground abortion, was threatened with public reprimand or a fine, and a more severe punishment awaited the doctors who committed it (Historical Materials 2018). The ban led to a new surge in fertility: the number of newborns in Moscow increased by 65%. But the number of babies killed has also increased dramatically.

In the west in the 70s, a new movement appeared and directly affected the institution of the family. The movement received the name «childfree», that is, «free from children» - the activists of this movement created groups and strongly promoted their way of life free from children. In subsequent years, this movement reached Russia. Officially, this movement doesn't exist, it freely «walks» across the expanses of the Internet.

Conscious childlessness existed long before the appearance of this movement, but it was not only rare enough, but it also could not declare itself «loudly». From the side, public opinion categorically condemned this phenomenon. Now the movement is also one of the negative stereotypical phenomena, which is met with a sharp rejection from the world community.

Consciously, the decision of a woman to abandon children became perceived as a deviation from the social norm.

Researchers identify many reasons why people voluntarily refuse to give birth. Many people find life beautiful without children, they feel some freedom from unnecessary trouble and worries. Many put in the first place favorite work and, they believe that the child only complicates the way up the career ladder. The modern young woman prefers a good, high-paying job, which prevents the child. «Today it is easier to live without children».

A successful career is not the only reason of women or men refuse to give birth. Often the main argument is fear or hostility to children or pregnancy, fear of spoiling health or appearance. All of this reasons that cause public censure and expose supporters of this movement irresponsible and untenable. Weavers distinguishes two types of «childfree»: «rejectors» - representatives of this type dislike children and everything related to the process of childbearing and breastfeeding; «Affencionado» - there is no hostility as such, representatives of this type consider children as an unnecessary burden, hindrance. It is important that representatives of this type are good without children (Veevers 1980). The authors of recent analytical work have identified two more categories of those who end up childlessly of their own free will: (1) «permanent shelters» (from adolescence or from the beginning of marriage/partnership use contraception, but initially don't give up the idea child, however, their temporary «later» smoothly turns into permanent «never», when they no longer want or they can't—or both, have children), and also (2) «wavelike refusers» (they have periods when they want children, sometimes they really want of other considerations and aspirations still «outweigh» the children and they don't have a result in its own decision) (Kneale and Joshi 2008).

3.2 Prospects for the Introduction of Anti-reproductive Systems: The Readiness of Society

At this stage of time, many researchers, especially demographers, don't stop talking that the phenomenon of «childfree» is indeed a deviant behavior that contradicts the traditional family model. The traditionality of this model is that a woman passes the classical stage of family life, expressed in a chain: marriage - the birth of a child.

Representatives of the national crisis theory Antonov A.I. Medkov V.M., Sinelnikov A.B. They believe that the appearance of the «childfree» phenomenon is a consequence of the family crisis, its failure to fulfill its functions, which in turn is explained by the crisis of family values and the general value crisis in our modern society.

Followers of evolutionism Vishnevsky AG, Golod S.I. They believe that the traditional family is transformed into new alternative forms, and in particular - childless families as a result of sociocultural modernization, a transformation of the family and its reproductive behavior occurs.

There are always arise questions and controversies around this phenomenon.

For example, the main question is: «Does this phenomenon pose a threat to Russia's demographic situation?». Among various population groups, supporters of the «childfree» movement, of course, contribute to the current demographic situation in the country. The phenomenon is rapidly spreading on the Internet, and especially in social networks. There are more than 150 communities on the Vk.com social network, and each has at least 2,000 subscribers who actively promote «life without children». In our country, about of 42 million Russian families, 48% have no children, of which only 5 million are for medical reasons.

Unfortunately, life without children is becoming an increasingly common social phenomenon. The population of our country is noticeably decreasing: the death rate in Russia exceeds the birth rate. According to statistics from the Federal State Statistics Service at the end of 2018 there was a sharp decline in the birth rate and an increase in mortality. The number of deaths exceeded the number of births by 14% (Federal State Statistics Service 2018). This is a critical situation. Over the past 11 years, the birth rate in the country has fallen to a minimum, a smaller number of children born were recorded only in 2004 (Table 1).

Table 1. Birth rate, mortality and natural increase of the Russian population (for the last 11 years) (Federal State Statistics Service 2018)

Years	Total human			Per 1000 population		
	Born	Dead	Natural growth	Born	Dead	Natural growth
	Whole population					
2007	1610122	2080445	−470323	11,3	14,6	−3,3
2008	1713947	2075954	−362007	12,0	14,5	−2,5
2009	1761687	2010543	−248856	12,3	14,1	−1,8
2010	1788948	2028516	−239568	12,5	14,2	−1,7
2011	1796629	1925720	−129091	12,6	13,5	−0,9
2012	1902084	1906335	−4251	13,3	13,3	0,0
2013	1895822	1871809	24013	13,2	13,0	0,2
2014(3)	1942683	1912347	30336	13,3	13,1	0,2
2015	1940579	1908541	32038	13,3	13,0	0,3
2016	1888729	1891015	−2286	12,9	12,9	−0,01
2017	1690307	1826125	−135818	11,5	12,4	−0,9
2018	**1599300**	**1817700**	**−218400**	**10,9**	**12,4**	**−1,5**

Table 2. Distribution of answers of respondents who are married to the question «Have you ever thought about not wanting children?». Depending on the presence of children (in% of all respondents)

Do you have children from the current marriage?	Have you ever thought about not wanting children			Total
	Yes	No	Difficult to answer	
Yes	20,0%	72,0%	8,0%	100,0%
No	32,0%	58,0%	10,0%	100,0%

It is noteworthy that 32% of Russian respondents, who have no children, think that they don't want to have them (Table 2). It is considered that officially registered marriage is always identified with family stability, which in turn should determine the reproductive system for a certain number of children, but, based on the results of the study, we can conclude that a modern marriage is created for the purpose of a different need. We conclude that young people who are officially married are a less promising category that realizes their reproductive potential.

Marriage in this case (childless marriage) doesn't justify the social needs of society, for the sake of which this institution is being created.

4 Discussion

There is almost no country in the present world, which is not concerned about the problems of population growth and, as a result, the need to regulate the birth rate in some way. All these countries are divided into two groups with opposing interests and goals in this area (Yudina et al. 2018a, b). Some are high-birth-rate countries that need to reduce the growth rate of the population in one way or another due to the enormous pressure on the economy. The most active in this matter are countries such as China and India. The second group presents highly developed countries for which the problem of low fertility is relevant, as a result, depopulation. One of these countries is Russia.

All measures taken by countries for population growth can be divided into 4 main groups (Rybakovsky 2018): (1) legislative or administrative-legal measures, which include contraception, abortion and sterilization, and also marriage and family relations; (2) economic measures, which are directed at reducing the material difficulties of a family with children or at providing remuneration in exchange for the required reproductive behavior. These measures include benefits paid to families with children (one-time or monthly), various benefits, reduced rates on loans for the purchase of housing (mortgage); (3) legal and economic measures aimed at overcoming contradictions that hinder the combination of female employment and motherhood; (4) moral and psychological impact measures6 which are oriented to shaping the necessary public opinion about the policy pursued and its goals.

Among the developed countries of the world, the standard of demographic policy is France, it is one of the first countries in the whole world, which faced the serious problem of depopulation. The dynamics of the population of France is as follows: 1801 - 28.3 million people, 1901 - 40.7 million people, 2002 - 59.8 million people (Official website of the State Duma 2018). An active demographic policy in this country has been pursued since the 20s of the last century. In the second half of the 20th century, the population of France began to grow due to the excess of births over mortality and immigration over emigration. France currently provides two thirds of the natural growth of the EU population, although its population accounts for only 16% of the total population of the EU member countries (Eran 2004). The population policy of France composed of economic measures by 90% and it has a significant impact on the financial situation of families. The basic child allowance is granted to all persons living in France and having at least two children, whatever of their citizenship, for their children under

20 years old living in the country. The amount of the allowance is differentiated depending on the number of children (Official website of the State Duma 2018).

The demographic policy of the USA is less noticeable than in France. Immigration plays a huge role in shaping the demographic situation. Family assistance is provided, as an indirect rule, in the form of tax benefits. The most important thing is open immigration policy (Yudina et al. 2018a, b) which negates the need for a pro-natalist demographic policy.

China's population policy is fundamentally different from the usual measures and is directed at reducing the birth rate. In the XX century, the country's demographic policy was held under the slogan «One family - one child».

Authorities imposed a ban on married couples in cities to have more than one child (excluding cases of multiple pregnancies). In 2013, families in which at least one spouse is the only child in the family received the right to have a second child. These rules are also introduced in stages. In 2013, the Chinese National Commission on Family Health and Family Planning stated that the «one family, one child» policy «prevented» the birth of approximately 400 million people. Since 1980, the government collected about 2 trillion yuan ($ 314 million) in fines (TASS 2018).

In Germany, the government has an active fertility policy. It consists of not only in economic measures, but also social ones. In addition to childcare benefits, the government is constantly increasing the number of places in kindergartens, research on family policy also revealed the importance of paternity leave. In Germany, the childcare leave system now allows both parents to get two thirds of their previous earnings while on leave for child care. Also provided housing benefits (News and Analytics 2018).

5 Conclusion

Analysis of the respondents' answers allows us to write the following conclusions. Firstly, the stereotype that today's youth is ready to build only a one-child model, the family should be put aside, a larger percentage of respondents see their family as at least two-children and also three-children, which gives us positive forecasts regarding the demographic situation in the country. Secondly, the negative factor is that the age for the birth of the first child shifts upwards by 27–30 years, which gives us cause for concern, because first of all it is a question of reproductive health. Taking into account the lifestyle of young people, reproductive health, especially of girls, deteriorates every year, the percentage of childless families grows due to health difficulties, and this also plays a negative role for the demographic situation in the country. Also, despite the fact that the hierarchy of values is headed by such values as family, children and love, the most of respondents see material and housing difficulties as an obstacle to the birth of a child. In today's world, the conditions for building a career and, accordingly, an increase in wages are significantly different from those that were several years ago.

Let us consider in more detail the economic measures of birth control in Russia, due to the fact that the demographic policy in economically developed countries is carried out only by such methods and is aimed at stimulating the birth rate. Let us turn to Decree of the President of the Russian Federation of October 9, 2007 No. 1351 On Approving the Concept of the Demographic Policy of the Russian Federation for the

Period up to 2025. The main objectives of the demographic policy of the Russian Federation are:

- a reduction in the mortality rate of at least 1.6 times, especially in working age from external causes;
- reducing the level of maternal and infant mortality by no less than 2 times, strengthening the reproductive health of the population, the health of children and adolescents;
- an increase in the birth rate (an increase in the total fertility rate by 1.5 times) due to the birth of the second child and subsequent children in the families;
- attracting migrants in accordance with the needs of demographic and socio-economic development, taking into account the need for their social adaptation and integration.

Consider the goal in more detail: increasing the birth rate. The solution to the problem of increasing the birth rate includes:

- development of a system of granting benefits in connection with the birth and upbringing of children (including regular revision and indexation of their size with regard to inflation);
- realization of a set of measures to promote the employment of women with young children, in order to ensure the combination of parental and family responsibilities with professional activities, including:
- development of special programs that allow women to obtain new professions in case of their transfer (release) from jobs with harmful and difficult working conditions to new jobs (On Approving the Concept 2007).

One time in a few years, this document is updated, supplemented, based on the demographic situation in the country. So at the end of 2017, Russian President Putin V.V. made the following important changes to the decree, which may indeed affect the growth of the birth rate. The main thing to note is the monthly payment to the families at the birth of the first child, which will be paid upon reaching 1.5 years. On average, the size of the payment will be about 10,000 rubles. Now many families want to achieve certain goals in their careers, so that in the future they will have enough budgets to support their children, because the age of birth of the first child is shifting.

The demographic policy in the sphere of increasing the birth rate is hardly noticeable for the Russian youth, characterized by a low score in terms of the «effectiveness» indicator. A population birth rate policy will be noticeable and effective when special mortgage programs are introduced for families with one child; the quality of pediatric medicine has been improved; maternity benefits increased; child care benefits had increased and extended until they were 3 years old; and there will also be available pre-school education for young children.

We would like to pay staring attention to the fact that an increase in the birth rate is impossible without strengthening the institution of the family, without raising the quality of marital relations. It is difficult for women to decide on the birth of second and subsequent children, if they are not sure of the strength of their relationship. Currently, family problems and the formation of family policy are one of the main directions of social policy of the state.

References

Antonov, A.I.: Institutional crisis of the family and family and demographic structures in the context of social change and social inequality. ISTINA.MSU.RU: Family and socio-demographic studies. Scientific Online Magazine (2014). https://istina.msu.ru/publications/article/7342725/. Appeal date 10 Nov 2018

Antonov, A.I.: Sociology of Family. Moscow State University Publishing House: International University of Business and Management Publishing House («The Karich Brothers»), p. 221 (1996). http://socioline.ru/pages/aiantonov-vmmedkov-sotsiologiya-semi. Circulation date 12 Dec 2018

Bodrova, V.: «The ideal, desired and expected number of children». DEMOSCOPE.RU: the weekly demographic newspaper Demoscope Weekly. № 81–82 (2000). http://demoscope.ru/weekly/2002/081/tema01.php. Appeal date 10 Nov 2018

Borisov, V.A.: Demography. NOTABENE Publishing House, p. 123 (2001). http://www.sociologos.ru/upload/File/Methods/Demography_Borisov.pdf. Appeal date 10 May 2018

Ganzha, A.O.: Humanistic Sociology by Florian Znanetsky. Federal Educational Portal Economics, Sociology, Management. http://ecsocman.hse.ru/data/671/235/1218/019Ganzha.pdf. Appeal date 10 Nov 2018

Demography of Russia (the site is dedicated to Prof. D.I. Valentei. Romantic period of building the foundations). http://demography.ru/xednay/demography/abortions/vs-2.html. Appeal date 10 Dec 2018

Ilyasov, F.N.: The need for children and reproductive behavior. Monitoring of public opinion. № 1, pp. 168–177 (2013). https://cyberleninka.ru/article/n/potrebnost-v-detyah-i-reproduktivnoe-povedendenie. Contact date 10 Dec 2018

Historical materials On the prohibition of abortion, increasing material assistance to women in childbirth, the establishment of state aid to multi-family, expanding the network of maternity homes, nurseries and kindergartens, increasing the penalty for non-payment of maintenance. http://istmat.info/node/24072. Appeal date 10 Dec 2018

News and analytics about Germany, Russia, Europe, the world. https://www.dw.com/en/are-family-policy-reforms-to-thank-for-germanys-rising-birth-rates/a-43188961. 17 Oct 2018

On approval of the Concept of the demographic policy of the Russian Federation for the period until 2025: Decree of the President of the Russian Federation of October 9, 2007 No. 1351. http://kremlin.ru/acts/bank/search?title=1351. Date References 16 Oct 2018

Official site of the State Duma, Issue 11 On the demographic situation in Russia. http://iam.duma.gov.ru/node/8/4511/15142. Appeal date 17 Oct 2018

Rybakovsky, L.L.: Demography, textbook for higher educational institutions. http://rybakovsky.ru/uchebnik2a40.html. Circulation date 16 Oct 2018

Dictionaries and encyclopedias on Academician. Newest philosophical dictionary. Gritsanov, A. A. https://dic.academic.ru/dic.nsf/dic_new_philosophy/1132/%D0%A1%D0%9E%D0%A6%D0%98%D0%90%D0%9B%D0%AC%D0%9D%D0%90%D0%AF. Circulation date 10 Nov 2018

TASS: China's population policy "one family - one child." Dossier. https://tass.ru/info/2389795. Appeal date 17 Oct 2018

Federal State Statistics Service. http://www.gks.ru/free_doc/2019/demo/edn01-19.htm. Contact date 03 July 2018

Eran, F.: «The demographic situation in France in the European context». Summary: Kulikova, S.N.: DEMOSCOPE.RU: the weekly demographic newspaper Demoskop Weekly № 145–146 (2004). http://www.demoscope.ru/weekly/2004/0145/analit04.php. Revised 17 Oct 2018

Central intelligence agency. https://www.cia.gov/library/publications/the-world-factbook/rankorder/2054rank.html. Appeal date 05 Oct 2018

Kneale, D., Joshi, H.: Postponement and childlessness: evidence from two British cohorts. Demogr. Res. **19**, 1935–1964 (2008)

Morell, C.: Saying no: women's experiences with reproductive refusal. Feminism **10**(3), 313–322 (2000)

Veevers, J.E.: Childless by Choice. Butterworth, Toronto (1980)

Yudina, T.N., Fomicheva, T.V., Dolgorukova, I.V., Kataeva, V.I., Kryukova, E.M.: The value of happiness: well-being on a global scale. Int. J. Eng. Technol. **7**(3.14), 455–460 (2018a)

Yudina, T., Mazaev, Y., Fomicheva, T., Dolgorukova, I., Bormotova, T.: Labor activity of the EAEU migrants in Moscow. In: Popkova, E. (ed.) The Future of the Global Financial System: Dawnfall of Harmony. Lecture Notes in Networks and Systems, vol. 57, pp. 142–153. Springer, Cham (2018b)

Well-Being of the Population of the Far Eastern Region

Dina K. Tanatova(✉), Tatyana N. Yudina, Ivan V. Korolev, and Eugene A. Lizer

Russian State Social University, Moscow, Russia
tanatovadk@rgsu.net, ioudinatn@mail.ru, koroleviv@rgsu.net, lidzerea@gmail.com

Abstract. The authors of the paper use their own methods and approaches to estimate population well-being indicators in 2019 in the Far Eastern Federal District. The analysis has shown that, despite the fact that significant funds were invested in Far Eastern territories, a number of indicators characterizing well-being of the population of the Far Eastern Federal District, are inferior to the indicators of other federal districts. The rating of the population well-being indicators for the Far Eastern Federal District is presented in the form of expert estimates. Vectors of achievement of sustainable growth of well-being of the population of the Far Eastern Federal District are disclosed.

Keywords: Well-being · Objective well-being indicators · Subjective well-being indicators · Far Eastern region · Human well-being

1 Introduction

The Far Eastern Federal District occupies 40.6% of the territory of Russia, being the most widespread region of the state in terms of space. However, the population of the region is only 5,58% of the total population of the country. The vastness and low occupation density of the territory of the Far Eastern Federal District predetermined the low density of population. Whereas the average density of population in Russia is 8.3 persons per square meter, this indicator is about six times lower in the Far Eastern Federal District (1.33 persons per square meter) (Federal State Statistics Service 2019).

The Far Eastern Federal District has a unique geopolitical location. On the one hand, this is its isolation from the center of Russia, while on the other hand – its closeness to the developing and developed countries of the Asia-Pacific Region. Thanks to its favorable economic and geographical location, the region is of particular importance due to relations between Russia and the countries of the Asia-Pacific Region. There are 29 seaports in this region, which is an important part of the transport corridor between Asia and Europe; the largest trunk railways pass through its territory. On its territory there are the largest deposits of diamonds, oil and gas, gold, coal, complex ores of universal importance. The presence of these resources forms the most powerful base for the management of new large-scale industries and projects. The Far Eastern Federal District is defined today as the geostrategic territory of Russia (Government of the Russian Federation 2018).

E. G. Popkova and B. S. Sergi (Eds.): ISC 2019, LNNS 87, pp. 750–759, 2020.
https://doi.org/10.1007/978-3-030-29586-8_86

Far East has been the object of focused attention on the part of the Russian government (Government of the Russian Federation 2014). Strategic decisions are made at the federal level, being aimed at attraction of investments that are able to arouse the economic space of the region through the launch of major projects and promotion of entrepreneurial activity. A new model of development of the Far East has been developed today. It made it possible to create 18 territories of rapid development and initiate the mode of Vladivostok free port, which is now occupied by 1468 companies. 199 thousand jobs are expected to be created. Investors have invested 326.5 billion rubles in the economy of the Far East, 1,183 new productions have already been launched, more than 27 thousand jobs have been created. Industrial-production growth (4.4%) is 1.5 times higher than the average Russian rate, while the volume of foreign investment accumulated over four years constitutes almost one third of the total amount of money invested in the country.

It is expected that the implementation of ambitious plans will dramatically increase well-being and the quality of living of population of the Far Eastern Federal District. However, studies have shown that the economic growth in the Far East is not identical to the increase in well-being of local population; strange as it may seem, rapid economic growth gives rise to the discontent of population (Regions online 2019; Yudina 2015). There is a gap between promised and actual changes in the region. As is correctly pointed out by the academician of the Russian Academy of Sciences Minakir, P.A. “the abovementioned gap gave rise to the formation of classical situation of escalation of tension of needs in direct relationship to changes in the region. This is supplemented by tension caused by negative economic shocks which hammered the entire population of the country: confiscatory “pension scheme reform”, devaluation of the ruble, “tax maneuver”, which reduced to the increase in taxes and duties for the population” (Minakir 2019).

Operating on the premise that well-being of the population remains not only a permanent criterion of state social and economic policies, but also one of the major mainstreams of research activities, the authors put forward updated approaches to the study of the population well-being, empirical results of research in a number of federal districts of Russia, in the Far Eastern region and in the Central region in particular.

2 Materials and Methods

Well-being of the population cannot be viewed in isolation from the economic and social development of the region where the population under consideration is living (Trutnev 2019). In this regard, we used the key directions of the National Program for the Development of the Far East until 2025 in the development of indicators (Ministry for Development of the Russian Far East and Arctic 2019). In the methodological aspect, we relied, among other things, on the best practice of foreign and Russian researchers, where subjective human well-being is treated as an aggregate estimate, allowing a person to determine his or her real-life situation at a particular point of time (Voronin 2009).

9 well-being indicators have been developed in our research: well-being in healthcare; family well-being; well-being in education; well-being in housing and utility sector; environmental well-being; well-being in employment and remuneration; well-being in the living environment; well-being in creative work; social well-being. Well-being was understood to be an estimate which includes the totality of private satisfactions in various living environments of people.

Well-being indicators were measured on a five-point scale by means of indicators. Average values have been determined for each well-being indicator.

The research was carried out from February till March 2019 on the basis of an online survey. Well-being of the population of the Far Eastern region was estimated by 467 experts living in territories, regions and republics included in the Far Eastern Federal District: Khabarovsk Territory, Amur Region, Magadan Region, Republic of Buryatia, Zabaikalye Territory, Kamchatka Territory, Chukotka Autonomous Region, Jewish Autonomous Region, and Sakhalin region. Central Federal District as the most developed Russian region was chosen for the comparative analysis of well-being of population.

3 Result

3.1 Social Well-Being

According to experts, the greatest problems in social well-being in the Far Eastern region are associated with the amount of household income (84% of experts designated the income level of population as low or extremely low), as well as the level of prices (68% of experts designated the level of prices in the region as high or very high). Experts believe that the social safety net in the region is inadequate. More than half of experts (51%) are of the opinion that the level of social protection of the population in the Far Eastern region is low or extremely low.

The average expert estimate of social well-being of the Far Eastern region amounted to 2.7 points on a five-point scale (1 is the maximum negative estimate of the indicator, and 5 is the maximum positive estimate of the indicator), which is below the acceptable level of 3 points we have set.

The average expert estimate of social well-being of the Central Federal District amounted to 2.9 points. The widest gap between the regions is related to income (Central Federal District – 2.4 points, Far Eastern Federal District – 1.9 points) and to the level of social protection of the population (3.0 in the Central Federal District as against 2.5 in the Far Eastern Federal District).

3.2 Well-Being in Healthcare

The greatest problem in the Far Eastern region related to the health protection of population are associated with inadequate staff sufficiency of healthcare facilities and high prices for medicines. About two thirds of experts estimated the status of these well-being indicators as poor or very poor (68% and 65% respectively).

The average estimate of well-being in healthcare is 2.4 points. The comparison of this indicator with a similar indicator for the Central Federal District carries inference that the population of the Far Eastern region has lower level of well-being in healthcare (2.4 points in the Far Eastern Federal District as against 2.8 points in the Central Federal District). All indicators of well-being in healthcare of population for this region have received lower average expert estimates as compared to estimates of similar indicators for the Central region, and no indicator exceeded the average expert estimate of 3.0 points – "satisfactory".

3.3 Well-Being in Education

As for the indicators of well-being in education in the Far Eastern Federal District, such indicators as "competency of pedagogic staff" (36% estimates "good" and "very good", 43% of "satisfactory" estimates) and "overall quality of education" (31% estimates "good" and "very good", 46% – satisfactory) received higher expert estimates.

The comparative analysis of regions shows that the problem of education in the Far Eastern region is more pressing. The average expert estimate for the Central Federal District amounted to 3.5 points, whereas the expert estimate of well-being in education for the Far Eastern Federal District is below the "satisfactory" level (2.9 points). There is not a single population well-being indicator for the Far Eastern region in education, that would be estimated by the experts higher than a similar indicator in the central region. This being said, the Far Eastern region is considered to be an attractive place to acquire an education for students from China (Tanatova et al. 2018).

3.4 Family Well-Being

According to experts, the greatest problem in family well-being in the Far Eastern Federal District consists in the poor social support for needy families (multiple children families, families with children with disabilities, etc.). 38% of experts designated the level of social support as low or very low. In addition, about one third of experts (30%) pointed out the high level of commercialization of children's leisure activities.

In general, family well-being of the Far Eastern region is estimated by the experts as satisfactory (the average expert estimate is 3.1 points), but when we compare family well-being in the Far Eastern Federal District with that in the Central Federal District, Far Eastern region is a little bit inferior to the central region (the average expert estimate for the Central Federal District is 3.3 points). The underdevelopment of the Far Eastern Federal District can be observed in terms of neglect and homelessness (3.3 points in the Far Eastern Federal District as against 3.7 points in the Central Federal District) and social support of needy families (2.8 points as against 3.3 points). At the same time, according to experts, children's leisure activities are less commercialized in the Far Eastern Federal District (3 points in the Far Eastern Federal District as against 2.9 in the Central Federal District).

3.5 Well-Being in Housing and Utility Sector

According to experts, the greatest problems in the Far Eastern Federal District are associated with the inadequate quality of services of Facility Managers (65% of experts are of the opinion that the level of quality of services is low or extremely low), control over this sphere on the part of government authorities (59% of responding experts designated the supervision level as low or extremely low), and rates 58% of experts designated their level as high or very high).

The comparative analysis of well-being indicators in housing and utility sector of both regions carries inference that the level of well-being in them is roughly the same and is below the "satisfactory" level (the average estimate for the Far Eastern Federal District is 2.7 points, the average estimate for the Central Federal District is 2.8 points). The comparative analysis of well-being indicators in housing and utility sector shows that the situation with housing prices in the Central Federal District is worse compared to the situation in the Far Eastern Federal District (the average estimate is 2.8 points as against 3.2 points). The Central Federal District, in turn, "beats" the Far Eastern Federal District in terms of control over housing and utility sector on the part of government authorities (2.9 points as against 2.3 points) and in the form of services provided by Facility Managers (the average estimate is 2.6 points as against 2.1 points).

3.6 Environmental Well-Being

The issues associated with the utilization of "eco-friendly" technologies by enterprises of the region and the issues associated with the collection and recycling of wastes raise the most concern among the 60% of the responding experts; and 55% of the experts are the opinion that these issues are solved in a poor or very poor manner. We should point out that the issues associated with the recycling of wastes are characteristic of Russia (Koroliov 2018).

In general, the indicator of environmental well-being of the Far Eastern Federal District is below the "satisfactory" level. Its average value is 2.6 points. There is not a single indicator of environmental well-being of the population of the Far Eastern Federal District, average expert estimates of which would be higher than estimates of a similar indicator of environmental well-being of the Central Federal District.

3.7 Well-Being in Employment and Remuneration

According to experts, the sphere of labor in the Far Eastern Federal District is one of the most dysfunctional spheres. The vast majority of experts (83%) are of the opinion that there is a shortage of well-paid jobs in their region. In addition, about two thirds of responding experts are of the opinion that the region has big problems associated with the lack of jobs in mono-cities (68%), it is hard for people with disabilities to find a job (66%), "illegal schemes" are used in remuneration of labor (65%).

The population of the Central Federal District has similar issues. However, the problems of the sphere of labor are more pronounced in the Far Eastern Federal District. The well-being indicator here is 2.6 points, whereas the similar indicator in the Far Eastern Federal District is 2.1 points.

3.8 Well-Being in the Living Environment

The most pressing problems include: lack of parking areas in residential zones (the situation with parking areas is poor or very poor – 66% of experts), condition of roads, bridges and pavements (62% of experts are of the opinion that the situation with the condition of transport infrastructure is poor or very poor).

The value of the overall indicator of well-being of population in the living environment is quite low, amounting to 2.3 points. The comparative analysis of value of this indicator for the Far Eastern region (2.3 points) with the value of a similar indicator for the Central region (3.4 points) carries inference that the living environment is more comfortable in the Central Federal District. The most noticeable gap can be observed in the condition of infrastructure (2.1 points in the Far Eastern Federal District as against 3.2 in the Central Federal District), availability of public spaces and resting places in the urban environment (2.6 points as against 3.4 points), availability of sports facilities and recreation grounds in residential zones (2.6 points as against 3.5 points).

3.9 Well-Being in Creative Work

According to experts, this sphere is the most trouble-free one. The value of population well-being indicator for the Far Eastern Federal District in creative work exceeds the satisfactory level, amounting to 3.2 points. The value of similar indicator in the Central Federal District is 3.8 points. All indicators have higher values in this region. The widest gap in values of similar indicators of the two regions can be observed in the support on the part of regional administration with respect to the creation of patriotic films, theatrical performances, literary writings, etc. (the average estimate for the Central Federal District – 4.1 points as against the average estimate for the Far Eastern Federal District – 3.3 points) and the support of freedom of expression (3.8 points as against 3.1 points). At the same time, the experts have pointed out the higher level of control over creative work of the population on the Internet in the Central Federal District (the average estimate of this indicator in the Central Federal District 3.7, whereas in the Far Eastern Federal District it amounts to 3.0).

4 Discussion

Human well-being is a complex construct, and it is hard to give it a single comprehensive definition, since it is a set of objective and subjective factors that are formed depending on past experience, assessment of the present and forecasting of the future in a particular time period. Therefore, the concept of human well-being in our modern age may have different appearance in various studies.

That said, there is a number of indicators which are virtually invariable, since they are focused on the basic needs, which include not only the so-called material needs, but also the communication needs, involvement in social life (Konkova 2014).

However, high objective indicators are not always indicative of the equally high level of social well-being. It is common knowledge that a person may be rich and unhappy, and vice versa, poor and happy. The definition of subjective well-being, set forward by Shamionov, R.M. (Shamionov 2006:105) in practical research activities gained currency at present. He treats subjective human well-being as an aggregate estimate that allows a person to characterize his or her real-life situation at a particular point of time.

Certain approaches rest upon subjective economic well-being. In particular, Sheifer, E.V. distinguishes a number of its components: degree of satisfaction with one's financial standing, confidence in the future, assessment of the current economic environment in the country. And this, according to him, makes it possible to prepare the integrated assessment of financial standing of a person (Sheifer 2013: 35). Other approaches are available as well.

As we have already mentioned, we used the indicators in our research. One of them is hardly ever used; it is about "well-being in creative work", which, in our opinion, provides useful and in-depth information about a person, related to intellectual capacity, philosophy of life, harmony, and emotional stress. We launched it for the first time, and it is far from perfect. We assume further solid work on the interpretation and operationalization of this indicator.

5 Conclusion

The overall estimate of population well-being in the Far Eastern Federal District on a five-point scale is 2.7 points, and this is below the "satisfactory" level. In the Central Federal District, the population well-being estimate amounts to 3.1 points. Moreover, the values of the population well-being indicators for the Central Federal District exceed the values of all similar indicators in the Far Eastern Federal District.

The radar of the population well-being for the Far Eastern Federal District and the Central Federal District is depicted in Figure below (Fig. 1).

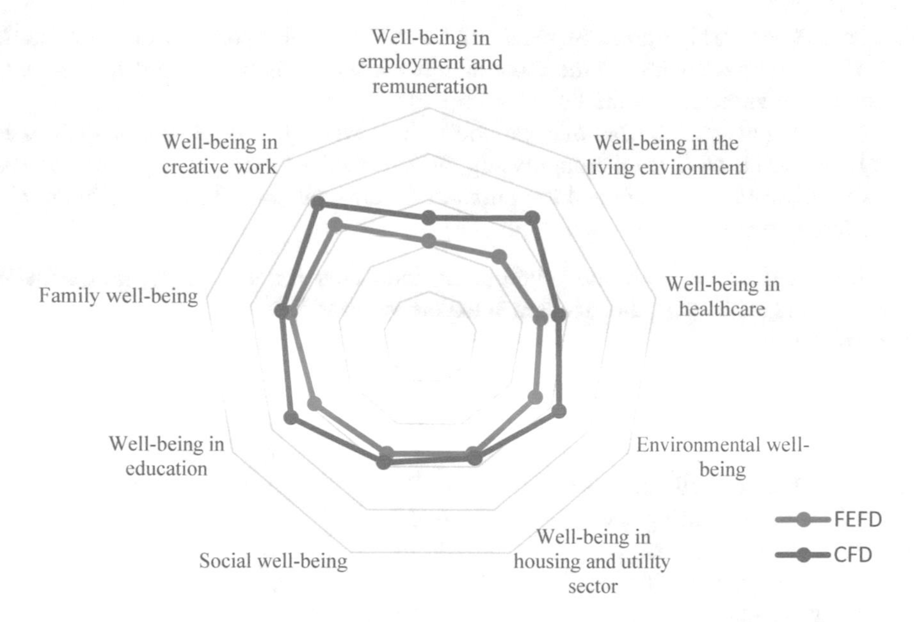

Fig. 1. The radar of the population well-being for the Far Eastern Federal District and the Central Federal District (five-point scale)

1. Low income of population and high prices are considered to be the main negative factors of social well-being.
2. The level of well-being in healthcare of population suffers from the greatest negative impact of the situation with inadequate staff sufficiency of healthcare facilities, as well as high prices for medicines. In addition, the well-being of the population is deteriorating due to insufficient fitting of healthcare facilities with state-of-the-art equipment and high cost of medical services.
3. Well-being in education of the population is at a relatively high level. The main problems consist in the material and technical infrastructure of educational establishments and their teaching staff sufficiency.
4. Family well-being of the population of the Far Eastern region suffers from the negative impact of poor social support for needy families. Moreover, family well-being is deteriorating due to commercialization of children's leisure activities.
5. The main problems in well-being in housing and utility sector in Russia consist in high utility rates, low quality of services of Facility Managers, and inadequate attention of the state to the housing and utility sector.
6. According to the experts, environmental well-being of population in the Far Eastern Federal District is relatively acceptable. The greatest negative impact on environmental well-being is caused by the problem of collection and recycling of wastes, as well as the problem of utilization of environmentally-dangerous technologies by enterprises of the region.
7. The main problem in employment and remuneration in the Far Eastern Federal District is the lack of well-paid jobs.

8. The lack of parking areas in residential zones and the poor condition of roads, bridges and pavements are the most significant problems with regard to the comfortable environment in the Far Eastern Federal District
9. Well-being in creative work of population is achieved by regional administrations at a fairly high level. They thoroughly support the creative process. The problem today is a redundant censorship and the pursuit of redundant control over creative work, including that on the Internet.

The rating of the population well-being indicators for the Far Eastern Federal District, arranged by the value of average expert estimate (five-point scale) is as follows (the lower is the point value, the lower is the level of well-being):

1. Well-being in employment and remuneration – 2.1
2. Well-being in the living environment – 2.3
3. Well-being in healthcare – 2.4
4. Environmental well-being – 2.6
5. Well-being in housing and utility sector – 2.7
6. Social well-being – 2.7
7. Well-being in education – 2.9
8. Family well-being – 3.1
9. Well-being in creative work – 3.2

Hence, the analysis of certain indicators characterizing well-being of the population of the Far Eastern Federal District has shown that, despite the fact that considerable financial resources are allocated for the development of the region, the level of well-being is still quite low, being inferior to the level of well-being in other regions of the Russian Federation.

In order to achieve sustainable growth of well-being of the population of the Far Eastern Federal District, one should pursue a comprehensive regional policy aimed at unlocking the potential for development of each region, overcoming infrastructural and institutional constraints, creating equal opportunities and contributing to the development of human potential.

Conclusions reached cover adequately the actual state of things and can be useful in the assessment of efficiency and elaboration of measures in priority areas of development of the regions of the Far East.

References

Tanatova, D.K., Pogosian, V.G., Koroliov, I.V., Chernikova, A.A.: Russia-China: educational mobility of Chinese students. Ponte **74**(8–1), 214–223 (2018)

Voronin, G.L.: Objective and subjective indicators of social well-being. Sotsiologicheskiy Zhurnal **3**, 41–54 (2009)

Kiselev, S.N., Solokhina, L.V.: Human Development Index and certain indicators, characterizing social well-being of the population of the Far Eastern Federal District. Bulletin of public health care of the Far Eastern region of Russia 1 (2017). http://www.fesmu.ru/voz/20171/2017101.aspx. Accessed 15 Mar 2019

Konkova, E.: Far East: state ambitions and the actual quality of living. Information analysis agency "Vostok Rossii" (2014). https://www.eastrussia.ru/material/kachestvo_zhizni/. Accessed 27 Apr 2019
Korolev, I.V.: A person and environment: based on the results of the sociological research. The strategies for the future in a fast-paced world: questions, answers and responsibility. Files of XXIII sociological readings of RSSU: a collection. Russian State Social University (2018)
Minakir, P.A.: Institutional innovations in the Far Eastern region: the simulation of a new stage. Prostranstvennaya Ekonomika **15**(1), 7–17 (2019)
Ministry for Development of the Russian Far East and Arctic: National Program for the Development of the Far East for the period until 2025 (2019). https://дв2025.рф/about. Accessed 30 Apr 2019
Government of the Russian Federation: National program of the Russian Federation "Social and Economic Development of the Far Eastern Federal District". Approved by the Decree of the Government of the Russian Federation No. 361 of March 29, 2019. Collection of Legislative Acts of the Russian Federation. 05.05.2014. No. 18 (part I), p. 2154 (2014)
Government of the Russian Federation: Annual Report of the Government of the Russian Federation for the year 2018. Parlamentskaya Gazeta (2018). https://www.pnp.ru/politics/otchyot-pravitelstva-o-rezultatakh-raboty-v-2018-godu-polnyy-tekst.html. Accessed 2 May 2019
Regions online: The opportunities of the Far East have been the subject of discussion in Moscow (2019). https://www.gosrf.ru/news/40468/. Accessed 20 Apr 2019
Federal State Statistics Service: Estimated population as at January 1, 2019 and on the average for the year 2018 (2019). http://www.gks.ru/wps/wcm/connect/rosstat_main/rosstat/ru/statistics/publications/catalog/afc8ea004d56a39ab251f2bafc3a6fce. Accessed 14 Mar 2019
Trutnev, Y.: National Program for the Development of the Far East will join measures aimed at improving the macroregion. Ministry for Development of the Russian Far East and Arctic (2019). https://minvr.ru/press-center/news/21456/. Accessed 30 Mar 2019
Shamionov, R.M.: Subjective well-being and axiological formations of a person in professional area. Bull. Saratov Univ. **6**(1/2), 104–109 (2006). Series Philosophy. Psychology. Pedagogy
Shamionov, R.M.: Psychology of Subjective Well-Being of a Person, p. 180. Publishing House of Saratov University, Saratov (2014)
Sheifer, E.V.: Subjective economic well-being of men and women in various age groups. Psychol. Econ. Manage. **2**, 53–59 (2013)
Yudina, T.N.: Migrators and receiving community in the Far East: mutual compromises. In: Under the General Editorship of Riazantsev, S.V., Pesterova, N.M., Kgramova, M.N. Dispersal Movements in the Asia-Pacific Region: History, Modernity, Practical Cooperation and Regulation. Collected Works from the International Research-to-Practice Conference, pp. 249–258 (2015)

Innovative Clusters in the Conditions of Digitization

Olga B. Digilina[1(✉)], Irina B. Teslenko[2], and Andrei A. Chekushov[3]

[1] RUDN University, Moscow, Russia
Digilina_ob@pfur.ru
[2] Vladimir State University named after Alexander and Nikolay Stoletovs, Vladimir, Russia
iteslenko@inbox.ru
[3] Financial University under the Government of the Russian Federation, Moscow, Russia
aachekushov@fa.ru

Abstract. The authors analyze the changes of the cluster policy of a state, which are connected to the process economy's digitization. The authors note that the sectorial and regional principles of formation of innovative clusters in the conditions of development of digital platforms become insignificant. The most perspective associations, which generate large synergetic effect, are innovative clusters – which include IT companies, business incubators, technological parks, academic institutes, and companies of various spheres that function on the same IT platform.

Keywords: Innovative cluster · Digitization · Network interaction · Digital platforms

JEL Code: O38 · O33

1 Introduction

Development of the processes of globalization and increase of competitive struggle in the world markets, economic and political conflicts, depletion of natural resource, and change of the conditions and mode of life and the corresponding attitude towards goods and their usage require new conceptual approaches to solving the issues of preserving and increasing the competitiveness of companies and their products.

Most countries have found a solution and began the transition to the innovative path of development.

Moreover, the practice showed that only close cooperation of companies (which are parts of the value creation chain) with transport, logistics, consulting, technological, service, and other organization will lead to synergetic effect and competitive advantages in the market.

E. G. Popkova and B. S. Sergi (Eds.): ISC 2019, LNNS 87, pp. 760–766, 2020.
https://doi.org/10.1007/978-3-030-29586-8_87

Increase of partnership (network) interaction of market subjects in the conditions of digitization becomes one of the signs of the modern stage of development of a socio-economic system. The organizational principle, which fits the establishment of relations between government structures, business, scientific organizations, universities, and other structures, is the cluster principle. It envisages consideration of territorial and sectorial belonging, innovative orientation of organizations, and unification of resources and information flows in the progressive chain of added value creation.

2 Methodology

The research object is economic relations that form in the process of formation of innovative clusters. The research subject is influence of digital technologies on the regularities of formation of cluster structures in the modern economy.

The methodological basis of the research is totality of means, methods, and settings, which are known to scientific search and which were adapted to the specifics of the studied object – formation of innovative clusters.

The empirical basis for the research includes the economic practice of developed countries of the world, the Russian Federation, and large companies, their best practices in the sphere of formation of cluster structures, and statistical and analytical overviews on the topics of the research, as well as government documents on the strategy of the national cluster policy and articles of the economic character in mass media.

3 Results

Digital transformation helps opening the potential of each company of a cluster, analyzing the factor and sectorial markets, and changing approaches to organization of production and management.

Digitization of the modern world stimulates the increase of effectiveness of cluster's functioning. Application of such technologies ad additive, quantum, and blockchain technologies, Big Data, cloud computing, the Internet of Things, industrial Internet of Things, AI, etc. allows – within a cluster – generating own novelties, implementing them in production, manufacturing innovative products on their basis, and commercializing them [1].

The purpose of creation of innovative clusters is constant improvement of goods, services, technologies, and managerial and organizational solutions. Cluster determines the directions of development of all sectors of economy and becomes a modern tool of management of innovative development [4].

M. Porter used the term "cluster" when analyzing the external surroundings of an organization from the point of view of its competitiveness. At present, according to A. V. Babkin, cluster is a separate economic system, which influences the national industrial policy and the economy on the whole.

Innovative cluster is a totality of interconnected industrial, financial, scientific, educational, and other organizations that are unified by the goal of development and commercialization of different innovations. Cluster creates products and services that have high technological level, added value, and competitiveness [1].

Innovative cluster is defined in the scientific literature as a complex dynamic system, where balance of competition and cooperation is achieved (or collaboration – constant process of coordination and achievement of consensus between participants); and as an eco-system, where the links of a triple spiral (model of paired collaboration of three sectors – state, business, and science within a joint network and project) and other players create new values; and as the most developed model of business network, where collaboration leads to synergetic effect of continuous innovations (innovative growth) and self-development without participation of a controlling center [9].

Eco-system of an innovative cluster includes innovative startups, universities, business incubators, engineering centers, service organizations, venture companies, etc. and depends on the technology of creation of innovative products and services.

Innovative cluster policy is aimed at formation of knowledge economy; increase of competitiveness of the national economy; stimulation of development of small and medium innovative business; solving the problems of employment and growth of population's incomes; expansion of inter-sectorial, inter-regional, and inter-country cooperation; further development of intellectual capital.

Innovative cluster is influenced by external environment, which has a lot of aspects. In the conditions of digitization, an important role belongs to the technological environment (totality of the technological processes that ensure production of goods and services).

Influence of technological environment is ambiguous. Automatization and digitization accelerate and simplify the production processes and make them less expensive. This creates a lot of advantages. On the other hand, implementation of new technologies makes a part of personnel unnecessary.

Technological environment changes the requirements to qualitative structure of workforce. Reducing the needs for certain employees, new techhologies increase the need for skilled service personnel, application engineers, and specialists on maintenance of program products [10].

Development of the institutional foundations of clustering started in 1980's. The USA develops the programs that are aimed at creation and development of clusters. The same process started in Europe in early 2000's [4].

The first clusters – so called clusters of the first generation – were aimed at optimization of the production processes, reduction of transaction costs, and increase of effectiveness of functioning of the supporting infrastructure [3].

Then, organizational foundations of cluster management began to be formed: 2010 saw the appearance of the Committee on territorial innovative clusters in the USA and the European group on cluster policy in the EU [4].

The process of formation of clusters of the second generation – innovative clusters – began after that. New technological competences that could solve a certain class of production tasks appeared. Further implementation of these competencies into the corresponding sectors leads to significant change of the latter [8].

At this stage, the role of state cluster policy growth. Development of clusters requires tax and credit tools, financial in the form of grants for scientific research, the corresponding advanced training of personnel, etc.

The largest innovative clusters are Silicon Valley in the USA (IT); Silicon Plateau in Bangalore (India); the Cosmetic Valley (France); Saxon Silicon Valley (Germany); Sassuolo (ceramic tiles) in Italy; BioM (biotechnologies and pharmacy) in Germany; Boston biotechnological and life science cluster (USA); Agro Business Park (Denmark); Oxfordshire Bioscience (UK); BRAINPORT (Netherlands). They are able to compete in the world markets due to created innovative technologies and products [4].

Clustering in Russia started in 2012, though this idea was discussed in early 2000.

In 2012, the Ministry of Economic Development of the Russian Federation started a program of support for pilot innovative territorial clusters. Their location was in the regions with high level of innovative development, territories with scientific centers, special economic areas, and closed territorial entities.

The selected innovative territorial clusters dealt with developments in the following spheres: nuclear technologies, production of airplanes and spacecraft, ship-building, pharmacy, biotechnologies, medical industry, new materials, chemistry, petrochemistry, IT, and electronics [7].

The program aimed at increasing the cooperation ties between all participants of clusters and development of territories with the highest technological and production potential [5]. Since the start of the program, the clusters have been able to develop a network of suppliers, attract investments into the projects, and support innovative companies.

The successful initiatives, which were implemented in innovative territorial clusters, include creation and implementation of domestic industrial robots of the third generation, intellectual system of computer vision, multi-functional manipulator robot Hexapod, software RoboticsLab for intellectual management of robots, opening of the first Russian plant on production of telecommunication and technical optical fiber, and opening the first Russian plant on prototyping of printed boards with capacity of 55.000 sq. m. per year [5].

Since 2016, the Ministry of Industry and Trade of Russia has been implementing the program of support for industrial clusters, which volume of financing will constitute RUB 3.24 billion until 2020 [7].

In 2016, the Ministry of Economic Development of Russia started a new stage of cluster policy until 2020 and a top-priority project "Development of innovative clusters – leaders of investment attractiveness of the global level" with 12 clusters. Its main goals include creation of points of rapid growth of economy, innovative development, export of hi-tech products, commercialization of technologies, increase of labor efficiency, creation of highly-efficient jobs, and growth of Russia's competitiveness [7].

The change of approaches in the cluster policy in the world and in Russia was caused by transformations in the socio-economic systems and, primarily, by the processes of digitization. A new stage of cluster policy is connected to refusal from the sectorial principle of formation of clusters.

Clusters of the first generation were based on the spheres from the previous technological mode and cannot stimulate the development of cooperation now. According to the specialists of RBK JSC, division of clusters into aircraft-building, IT, and nuclear technologies does not conform to the market demand. Added value chains form between the spheres, not within them.

In new state programs of development of the digital economy there's no similarity with the sphere within which the clusters in 2011–2012 were formed. Such division according to the sectorial belonging could lead to the situation when the data of cluster accounting will lead to an obsolete view on the economy and will not determine their real demands, efficiency, and potential [2].

The new project focuses on formation of the system of cluster management that conforms to international standards; interaction of clusters and state companies and the institutes of development; support for companies' entering the world markets; stimulation of companies' modernization; development of a mechanism of attraction of investments of the global level; support for quickly growing innovative companies [5].

On the whole, the results of the work of domestic clusters in recent years are not bad. Each budget ruble led to clusters' attraction of RUB 3.5 of private investments, which total volume reached RUB 360 billion. Despite a complex economic situation, clusters were able to achieve goods results in several indicators: growth of efficiency per 1 employee by 10%; increase of the number of highly-efficient jobs by 33%; growth of production volumes by RUB 429 billion; and achievement of the aggregate level of RUB 2 trillion [5].

The common target landmarks are established for the leading clusters. As compared to the 2016 level; growth of efficiency per 1 employee and average share of added value in revenues by more than 20%; triple growth of the number of patents for inventions and double increase of the volume of export revenues from sales of non-resources products. The plan for 2016-2020 is as follows: attraction of private investments – more than RUB 300 billion; volume of R&D with foreign organizations – more than RUB 100 billion; number of technological startups with investments – more than 300.

One of the key directions within the project is stimulating the cooperation of the leading clusters with foreign partners and entering the global markets.

The practice showed that countries in which cluster companies became the participants of the global chains of value creation (either existing ones or the ones that are created in the process of implementation of international cluster projects) could better adapt to the changing external environment.

At present, an innovative and production cluster is being formed in Moscow – it will diverge from the sectorial principle and will become the first Russian intersectorial super-cluster that will unite IT companies, business incubators, technological parks, and academic institutes on the common IT platform.

It is supposed that the innovative production cluster could compete for human capital at the global labor market for startups that could move not only in European cities but also in Moscow [2].

2019 will see the start of reformation of the system of existing technological parks and clusters, based on the needs of the digital economy. At least 10 clusters and 15 technological parks on the basis of digital platforms will be created by 2025.

By 2025, it is planned to create on digital platforms at least 15 platforms for development of comprehensive technologies and at least 50 markets for scholars on the basis of universities, scientific organizations, and companies.

4 Conclusions

It is possible to conclude that development of digital technologies leads to the necessity for developing a cluster principle of interaction of subjects during creation of innovative products. The regions with innovative clusters develop with the higher dynamics.

As economic systems function in the conditions of uncertainty, cluster policy should be flexible and should quickly adapt to the changes of the environment.

Recently, such changes have taken place: a project of creation of innovative territorial clusters, which conforms to new challenges of the digital economy and moves away from the previously applied sectorial principle has been developed. It is based on intersectorial connections [6]. The inter-disciplinary approach creates conditions for emergence of new products and increase of non-resource export potential. A new cluster model is a tool of truly complex modern industrial policy.

References

1. Babkin, A.V.: Digital economy and development of innovations-active industrial clusters. In: Babkin, A.V. (ed.) Innovative Clusters of the Digital Economy: Drivers of Development: Works of the Scientific and Practical Conference, pp. 175–190. SPb.: Polytechnical University Publ. (2018). [E-source] – Accessed. http://inecprom.spbstu.ru/files/inprom-2018/inprom-2018.pdf
2. The future of the digital economy lies with inter-sectorial clusters. [E-source] – Accessed. https://regnum.ru/news/2473549.html
3. Dezhina, I.G.: Technological platforms and innovative clusters: together or separately?/Dezhina I.G. – M.: Haidar Institute Publ. (2013). (Scientific works/E.T. Haidar Institute of Economic Policy Publ., No 164P). [E-source] – Accessed. https://www.iep.ru/files/RePEc/gai/rpaper/122Dezhina.pdf
4. Innovative clusters in the world and Russia – specific features of formation and development. [E-source] – Accessed. https://viafuture.ru/privlechenie-investitsij/innovatsionnye-klastery
5. Innovative clusters – leaders of investment attractiveness of the global level: methodological materials/Islankina, E.A., Kutsenko, E.S., Rudnik, P.B., Shadrin, A.E.; Ministry of Economic Development of Russia, RBK JSC, National Research University "Higher School of Economics". – M. (2017). (Scientific and methodological materials). [E-source] – Accessed. https://publications.hse.ru; https://www.hse.ru/mirror/pubs/lib/data/access/ram/ticket
6. Cluster model in innovative entrepreneurship. [E-source] – Accessed. https://studme.org/1911052221777/ekonomika/klasternaya_model_innovatsionnom_predprinimatelstve
7. Cluster policy: achievement of global competitiveness/Abashkin, V.L., Artemov, S.V., Islankina, E.A., et al.; Ministry of Economic Development of Russia, RBK JSC, National Research University "Higher School of Economics". – M. (2017). [E-source] – Accessed. https://publications.hse.ru/mirror/pubs/share//direct/207682536

8. Malyugin, A.N., Kolotovkina, E.I., Kudienko, I.V., Medvedev A.V.: Regarding the opportunities of using the cluster systems as the models of real estate market management. J. State Tech. Univ. (2014). [E-source] – Accessed. https://cyberleninka.ru/article/n/o-vozmozhnostyah-ispolzovaniya-klasternyh-sistem-kak-modeli-upravleniya-rynkom-nedvizhimosti
9. Presentation of the report "Innovative clusters and cluster policy: comparison of the approaches of Russia and Europe. (2016). Natalia Smorodinskaya. [E-source] – Accessed. https://inecon.org, https://inecon.org/docs/2016/Smorodinskaya_20161213.pdf
10. Regional innovative clusters. [E-source] – Accessed. https://eee-region.ru, https://eee-region.ru/article/4301/
11. The strategy of development of "Academy of media industry) (Institute of Advanced Training of Workers of Television and Radio) in the conditions of the digital economy of Russia for 2018–2028 (2018). [E-source] – Accessed. http://www.ipk.ru/ftpgetfile.php?id=282. Institute of regional innovative systems. Innovative clusters: the main ideas. http://www.innosys.spb.ru/?id=887

New Development Opportunities of the Sectoral Economy in the Conditions of Digitalization

Alexander Y. Bystryakov[1(✉)], Maria S. Marchenko[1], and Vladimir M. Pizengolts[1,2]

[1] Peoples' Friendship University of Russia (RUDN University), Moscow, Russian Federation
bystryakov-aya@rudn.ru, m_marchenko.education@mail.ru, pizengolts_vm@pfur.ru
[2] Russian New University, Moscow, Russian Federation

Abstract. The modern economy has taken the path of innovation development. The beginning of the XXI century is associated with a breakthrough in information processing after the invention of electronic computing technology. Currently, there is a rapid development of digital technologies and, as a result, the country's economy has received a digital focus of development. In this regard, it becomes interesting to study the possibilities for the development of branches of the Russian economy after the introduction of digitalization. This explains the relevance of the chosen topic. This study was carried out using general economic methods, analysis of statistical data and a systematic approach to the description of the digitization of industries. According to the results of the study, it was revealed that digitalization will affect absolutely all sectors of the economy, trade and human life. The results of the introduction of digital technologies will open up significant economic and social benefits. In the coming decades, the countries with the highest level of digitalization will have a competitive advantage in the global market.

Keywords: Digital economy · Digital technologies · Industry economy · Digital transformation of industries · Competition

JEL Code: F5 · O4

The significant development of the economy of information is due to the emergence of digital technologies and as a consequence the transition to a digital economy. The development of information technology is one of the prevailing areas of social development. The continuous process of modernization of all sectors of the economy is the foundation for building a new type of economy. Relationships about the production processing storage and transmission of data are the basis of the digital economy.

E. G. Popkova and B. S. Sergi (Eds.): ISC 2019, LNNS 87, pp. 767–777, 2020.
https://doi.org/10.1007/978-3-030-29586-8_88

1 Russian Experience in the Development of Digital Technology

The modern development of the economy, aimed only at industry benchmarks, makes it closed and raises the question of its dependence on the world market. Information technologies are becoming the functional basis for the development of modern economic systems.

Undoubtedly, the industrial revolution of the XVIII–XIX centuries. determined the course of the subsequent development of countries. However, the creation in 1950 of the first electronic computer in continental Europe, developed under the leadership of the Soviet scientist I. Bruk, signified a transition to the information era of development. The end of the 1950s was marked by the transition of mechanical means of electronic technology to digital computing electronics. The increase in economic growth was no longer due to the introduction of mechanized production, but the introduction of electronic computers. Later, financial institutions and high-precision manufacturing were actively developed. Since the mid-1960s. The era of digital innovation began, at first the process was longer: the transition from electronic computers to personal computers lasted for decades. In prior technological disruptions, from steam engines to electric power to digital computing, the logic of efficiency has often run ahead of the capacities of organizations and society at large to absorb and adapt to them, requiring significant reshaping and accommodation in order to reach a more mature and humane footing (Bodrozic and Adler, 2017).

Currently, digital technologies are changing extremely rapidly, determining the future path of development. According to the World Economic Forum, today Russia is second in the world in the availability of cellular communication services, and tenth in the availability of broadband access [10]. In the era of Internet technology, changes are perceived more easily. Digitalization penetrates into all spheres of human activity, facilitating his life.

By definition, the World Bank, the digital economy is a system of economic, social and cultural relations based on the use of digital information and communication technologies. The introduction of such technologies cost companies a significant amount. However, companies around the world spend huge amounts of money on improving digitalization. It becomes obvious that digitalization determines the company's position in the struggle for competition. Companies with the largest capitalization, according to Forex, for 2018 are presented in Table 1.

Table 1. Rating of the largest companies in the world in 2018 (according to Forex)

Place	Company Name	Branch	Capitalization at the end of 2018, billion US dollars
1	Amazon Inc.	Retail	802.18
2	Microsoft	Software development	789.25
3	Alphabet Inc.	Internet	737.37
4	Apple inc.	Electronics, Information technology	720.12

(continued)

Table 1. (*continued*)

Place	Company Name	Branch	Capitalization at the end of 2018, billion US dollars
5	Berkshire Hathaway Inc.	Insurance, Finance, Rail transport, Utilities, Food and non-food products	482.36
6	Facebook	Internet	413.25
7	Tencent	Conglomerate (Venture Company)	400.90
8	Alibaba Group	Internet	392.25
9	Johnson & Johnson	Pharmaceutical industry	347.99
10	JPMorgan Chase	Banking	332.24

Resource: [12]

Digital technologies are changing the company's operating model. Today, the companies with the highest level of capitalization in the world are mainly representatives of the digital economy.

Russia began its journey of digitalization much later. In 2016, Russian President Vladimir Putin, in a message to the Federal Assembly, proposed «launching a large-scale system program for developing the economy of a new technological generation, the so-called digital economy», which should «rely on Russian companies, scientific, research and engineering centers of the country». Subsequently, the Digital Economy Development Program in the Russian Federation until 2035 was developed, according to which the digital (electronic) economy is an aggregate of social relations developing through the use of electronic technologies, electronic infrastructure and services, technologies for analyzing large amounts of data and forecasting in order to optimize production, distribution, exchange, consumption and increase the level of socio-economic development of states [7].

The development of the Program marked the transition to a new digital order in the domestic economy. This transition was extremely necessary, because today, just as two centuries ago, the main driving forces of the national economy are still large oil and oil refining companies and metallurgical enterprises (Table 2).

Table 2. Rating of the largest public companies in Russia according to the results of 2018 (according to RIA Rating)

Place	Company Name	Branch	Capitalization at the end of 2018, billion US dollars
1	NK Rosneft	Oil and gas production and refining	65.29
2	Sberbank	Banks and financial services	57.82

(*continued*)

Table 2. (*continued*)

Place	Company Name	Branch	Capitalization at the end of 2018, billion US dollars
3	Lukoil	Oil and gas production and refining	53.82
4	Gazprom	Oil and gas production and refining	52.24
5	Novatek	Oil and gas production and refining	49.39
6	Norilsk Nickel	Metallurgy	29.63
7	Gazprom neft	Oil and gas production and refining	23.59
8	Tatneft	Oil and gas production and refining	22.86
9	Surgutneftegaz	Oil and gas production and refining	13.81
10	NLMK	Metallurgy	13.59

Resource: RIA Rating

One of the main reasons for this lag is not quite favorable business environment. In the World Bank's 2018 report, the Russian Federation ranks 35th in the rating of 190 countries in terms of Ease of doing business [9]. Other reasons are the direction of innovation in production, insufficient development of regulatory documentation.

2 Digitalization as an Integral Trend in the Development of a Sectoral Economy

The Russian Association of Electronic Communications (RAEC) presented the study «Economics of Runet». The ecosystem of the digital economy of Russia", according to which in 2018, the share of the digital economy in the country's GDP grew 2.5 times - from 2.1% to 5.1%. «Prime» writes about this with reference to the presentation of the report during the Russian Internet Week forum in Moscow [8].

With the adoption of the course of development of the digital economy, information and communication systems have introduced enterprises in various industries (Table 3).

Table 3. Use of the information and communication technologies in organizations by economic activity in 2017

	Percent of total number of surveyed organizations of the relevant activity
Total	92,1
Mining and quarrying	90,7
Manufacturing	95,5

(*continued*)

Table 3. (*continued*)

	Percent of total number of surveyed organizations of the relevant activity
Electricity, gas, steam and air conditioning supply	94,2
Water supply; sewerage, waste management and remediation activities	85,5
Construction	88,9
Wholesale and retail trade; repair of motor vehicles and motorcycles	94,5
Transportation and storage	93,4
Accommodation and food service activities	90,5
Information and communication	97,3
of which telecommunication	97,5
Financial and insurance activities	94,9
Real estate activities	65,6
Professional, Scientific and technical activities	93,1
Administrative and support service activities	89,7
Public administration and defence; compulsory social security	97,2
Higher education	98,4
Human health and social work activities	96,8
Arts, entertainment and recreation	91,1
Other service activities	93,9

Resource [2]

In total, in 2017, information and communication technologies were used by 92.1% of Russian enterprises. The highest percentage of technology use belongs to enterprises of educational activities (98.4%), and the lowest percentage - to enterprises engaged in real estate operations (65.6%). Digitalization has not bypassed any type of economic activity.

Digital transformation of agriculture. Agriculture is an important national economic complex of the country. Food security is an essential principle of its functioning. The transition to a model of "intelligent" agriculture provides for a comprehensive robotization of production, the introduction of a system of independent search and decision-making based on ecosystem modeling, as well as broad integrated monitoring of compliance with certification requirements. At the same time, this model functions with minimal use of fuel, various chemical fertilizers, etc., and maximum use of renewable sources. An example of such technologies can serve as biopesticides targeted protection against pests.

Digital transformation of energy. Despite the fact that the territory of the Russian Federation is rich in natural resources, the mining industry accounts for about 10% of GDP. However, mineral reserves are limited and this leads to the development of a new approach - the introduction of a new generation of electrical distribution networks

(Smart Grid). The creation and implementation of smart grids in the energy system of Russia is firmly tied to key aspects of the digital economy, such as data digitization and big data processing, new materials, new management formats that address operational data analysis and automated prediction [3].

Digital transformation of the transport industry. The digital transformation of the transportation industry is very extensive. It covers the creation of a common logistics chain with carriers, analysis and modeling of traffic flows, the creation of high-speed «smart» roads and unmanned vehicle control. An integrated cargo clearance system has been developed for Russian railways for further transportation, which is carried out electronically. This system is able to calculate the optimal route with the lowest cost, both financial and time. In the future, this approach will significantly reduce road mileage and increase the efficiency of the use of the railway network.

Digital transformation in the construction industry. Digitization of the construction industry allows you to create complex design models of bulk buildings, change any parameters both at the preliminary stage, and make changes during construction. An example of such technologies is building information modeling.

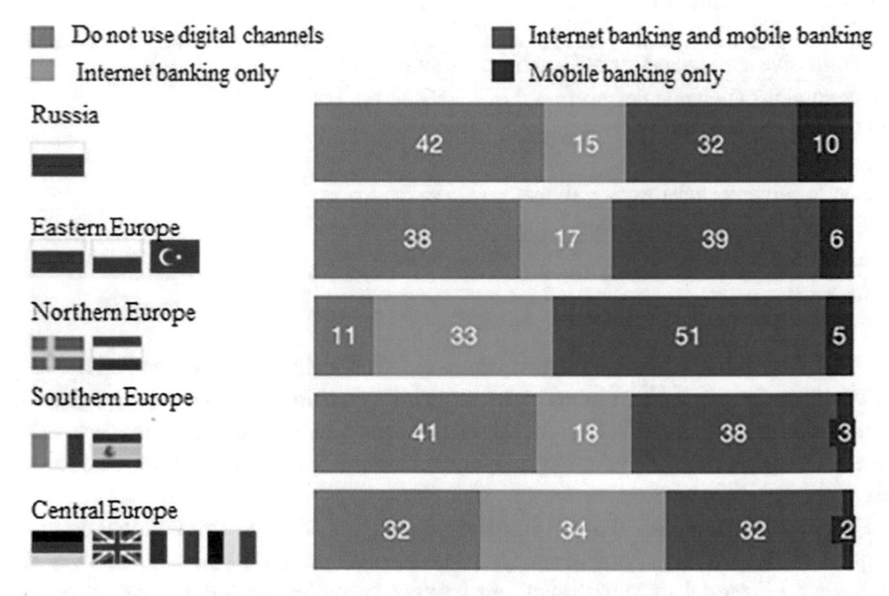

Fig. 1. Penetration of digital channels for 2016, percentage of respondents* (*Percentage of respondents who have used Internet banking and/or mobile banking at least once in the last three months. Resource: McKinsey's Retail Banking Market Survey 2016)

Digital transformation of the medical industry. Information technologies are widespread in the domestic health care. The possibility of remote registration for admission to medical institutions through the service of the Unified Medical Information and Analytical Service appeared. Also, with the help of this service, a doctor is called to the house The application for the call is made through the Service's website and is displayed directly on the doctor's electronic device. However, this service is currently available only for residents of Moscow. Also, thanks to digitalization, counseling of patients from remote parts of the country is carried out by means of a telecommunication channel - the development of "telemedicine". The Government plans to introduce electronic document management in medical institutions, including medical records of patients. As for the field of high-tech medicine, there is also the effect of digitalization. In this regard, the development of high-precision studies of genetic abnormalities and mutations, their detection in the early stages, as well as subsequent development of their treatment.

Digital transformation in education. As part of the digitalization project in the field of education, Russian universities will try to introduce mass online courses. They will not replace the traditional interaction between students and the teaching staff, but will be used in general disciplines and introductory lectures. With a positive result of the experiment, part of the disciplines will be replaced by webinars. Webinars are available at any time, this will improve the level of training for the exam in the discipline.

Digital transformation in financial services. According to the McKinsey Global Institute, in 2016, Russia had a low penetration rate of digital channels compared to other regions (Fig. 1).

However, this allowed to fill this niche. Now remote banking is actively developing using mobile applications. The main drivers in the field of financial technologies are payments and transfers, crowdfunding, asset management, financial marketplace. Special attention is focused on building a fully digital bank with chat bots for customer service, which, among other things, can analyze various information for making loan decisions for a customer.

Digital transformation in e-commerce. Electronic commerce covers all areas of interaction of the parties (B2C, B2B, C2C). Russia is no exception. The main task of the state in this regard is to build relationships between the participants based on the development of technological platforms for regulating processes.

Digital transformation in communications and telecommunications. In Russia, the telecommunications industry is at an early stage of digitalization, although it suffers a significant burden on the digital infrastructure. A modern user needs a set of applications to use electronic services. In the world, a new generation mobile network is being tested with a virtual implementation of network functions (NFV), which is capable of storing data remotely on a cloud server, which suggests a high degree of data protection. Russian mobile operators are gradually switching to digital services. An example of such an introduction is the pilot of Megafon, a project called "Elena". This bot is able to respond to requests from subscribers, provide information on request, report data.

Digital transformation in housing and communal services. Since there has recently been a significant migration of people to cities, it is necessary to create conditions for a large number of people in cities. Such projects are called «smart cities». They take into account the load on the electrical network, water supply, creating a balance of the use of smart networks and urban transport logistics.

3 Opportunities for the Development of the Russian Digital Economy

Currently, the digital economy is selected as the strategic direction of the development of the Russian Federation. In accordance with Presidential Decree No. 601 of May 7, 2012, "On Main Directions for Improving the System of Public Administration", it was assumed that 70% of citizens of the Russian Federation will receive public services remotely. As of February 1, 2018, about 65 million people were registered on the Russian portal of state services. The digital economy has penetrated almost all sectors and spheres of the Russian economy. The fact that ten years ago it was difficult to even imagine is now actively used in industry, commerce, banking, telecommunications, etc. However, the Russian digital economy still has a long way to go. Thus, in accordance with the National Program for the Development of the Digital Economy in Russia, in the period from 2018 to 2024, 1,634.9 billion rubles will be allocated for the implementation of this program (of which 1099.6 billion are the federal budget, 535.3 billion are extra-budgetary funds) [3].

Thus, it is assumed that domestic spending on the development of the digital economy in 2019 will be 2.2% of GDP, in 2021-3.0%, and by 2014 the figure will be 5.1%. Such significant funding allows you to make quite optimistic forecasts for development opportunities.

Digitalization makes economic and social benefits available (Fig. 2).

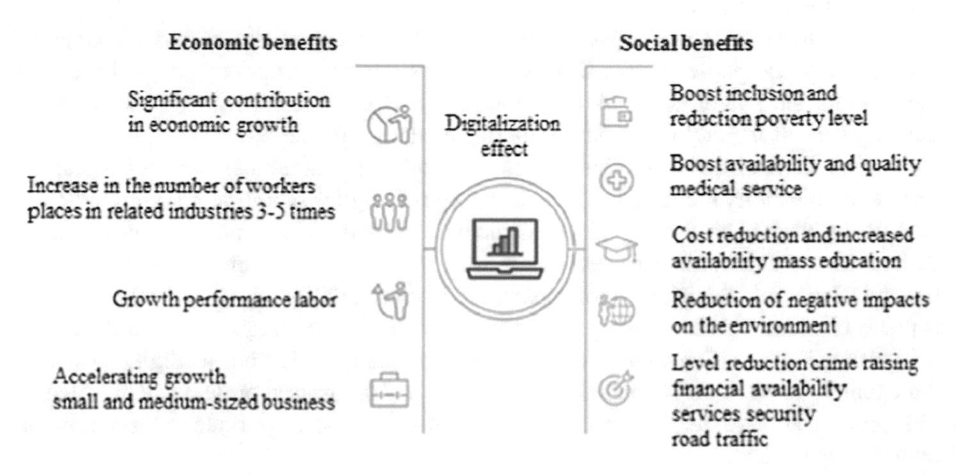

Fig. 2. The economic and social benefits of digitalization. Resource: World Bank, McKinsey Global Institute

The digitalization process opens up new opportunities for the development of the economy in all sectors and areas:

(1) economic growth. According to experts [6], the digitization of the economy by 2025 will lead to a possible increase in GDP by 4.1–8.9 trillion rubles, which is about 19–34% of the total expected growth. This will be possible thanks to fundamentally new digital technologies, breakthrough business models;
(2) increasing productivity by reducing costs. This is possible thanks to the algorithmization of processes and automation, reducing the burden on humans when using robotics and exoskeleton, performing high-precision manufacturing operations when using digital technologies;
(3) acceleration of business development, especially the expansion of electronic commerce through special digital platforms;
(4) increase in jobs. Digitalization will make a breakthrough in the performance of certain types of work and the provision of services remotely. Due to the widespread spread of the Internet, people from remote regions and people with disabilities have more employment opportunities. According to estimates, by 2025 only recruiting and personnel platforms will reach a total turnover of 2.7 trillion dollars - about 2% of world GDP (Manyika et al., 2015);
(5) improving the quality of services due to increased competition. Electronic platforms reduce barriers to entry into the industry, which allows increasing the number of services provided by the manufacturer.

4 The Impact of the Digital Economy on the Country's Competitiveness

When the digital economy is so actively developing in the world, it is extremely important to engage in its successful implementation in Russia as soon as possible. This will allow to take a worthy place in the world in terms of digital technologies. To this end, the National Program for the Development of the Digital Economy was developed. At the moment, Russia exports an extremely low amount of digital services (about 0.5% of GDP). This suggests that Russian information and communication technologies are not highly competitive in the global market, and therefore are not in demand. In developed countries, the export of such services is about 6%. The degree of impact of any technology can be considered as a result of its distribution and depth of implementation (Handel, 2015). Global leadership in the field of digital technology justified the United States and China. The volume of their digital economy is 10.9 and 10.0%, respectively. By 2025, they plan to increase the volume of the digital economy by 3 times. In Russia, the share of the digital economy is 3.9%. The reason for this lag is outdated technology and the duration of their development.

In the near future, the country's competitiveness in the global market will be determined by the level of its digitalization. Therefore, not only world leaders in this field, but also other developed countries have started searching and developing technologies that increase the level of digitalization. In the long term, digital technologies will increase the share of industry in the country's GDP by optimizing production costs, and in the extractive industries modeling of mineral deposits and digital high-precision drilling equipment will improve the efficiency of the development of such deposits. This in turn will lead to new investment.

5 Key Findings

1. Russia is not among the world leaders in terms of digital technology (0.5% of GDP). However, at present, Russian companies are developing to introduce digital technologies. Under the National Digital Economy Program, 1634.9 billion rubles have been allocated for the implementation of this program until 2024.
2. Digitalization penetrates all spheres of human activity, branches of the economy and other economic spheres. This allows you to create a new format of digital business, which is based on information and communication technologies that can reduce production costs and improve labor efficiency.
3. Possibilities for the development of digitalization are determined by the economic and social benefits of its implementation. Among the economic benefits include: contribution to the economic development of the country, increase in labor productivity, accelerated business development and an increase in the number of jobs. Social benefits include an increase in the well-being of citizens, the availability of medical services, a reduction in the cost of educational services, and an increase in environmental safety.
4. Owning digital assets creates a competitive advantage. In the near future, the level of competitiveness of countries will be determined by the level of their digitalization.
5. Digitalisation will undoubtedly lead to serious consequences that the national economy will face, for solving which will require joint efforts of the state, research structures and leading domestic companies of all sectors to control these effects. A large number of automated operations will lead to the release of labor, which can move into adjacent areas after retraining. Increasing public involvement in the digital economy will make this transition smoother.

Acknowledgments. This paper was financially supported by the Ministry of Education and Science of the Russian Federation on the program to improve the competitiveness of Peoples University of Russia (RUDN University) among the world's leading research and education centers in the 2016-2020.

References

1. Bodrozic, Z., Adler, P.: «The Evolution of Management Models: A Neo-Schumpeterian Theory». Administrative Science (2017, in press)
2. Federal State Statistics Service. http://www.gks.ru/free_doc/doc_2018/year/year18.pdf. Accessed 18 Apr 2019

3. Federal state program.: The development of the digital economy in Russia. Development program until 2035 (2017). http://spkurdyumov.ru/uploads/2017/05/strategy.pdf. Accessed 18 Apr 2019 4
4. Handel, M.: The Effects of Information and Communication Technology on Employment, Skills, and Earnings in Developing Countries. Background paper for the World Development Report 2016. World Bank, Washington, DC (2015)
5. Manyika, J. et al.: A Labor Market that Works. McKinsey Global Institute, New York (2015). http://www.mckinsey.com/global-themes/employment-and-growth/connecting-talent-with-opportunity-inthe-digital-age. Accessed 18 Apr 2019
6. McKinsey Global Institute: Digital Russia: a new reality (2017). https://www.mckinsey.com/ru/~/media/McKinsey/Locations/Europe%20and%20Middle%20East/Russia/Our%20Insights/Digital%20Russia/Digital-Russia-report.ashx. Accessed 18 Apr 2019
7. Russian State order: Order of the Government of the Russian Federation of 28.07.2017 N 1632-p On approval of the program «Digital Economy of the Russian Federation» (2017)
8. RAEC: Presented the study «Economics of Runet. Ecosystem of Russia's digital economy» (2018). https://bloomchain.ru/newsfeed/issledovanie-dolya-tsifrovoj-ekonomiki-v-vvp-rossii-uvelichilas-v-2-5-raza-v-2018-godu/. Accessed 18 Apr 2019
9. World Bank: A World Bank Group Flagship Report. Doing Business 2018. Reforming to Create Jobs (2018). http://www.doingbusiness.org/content/dam/doingBusiness/media/Annual-Reports/English/DB2018-Full-Report.pdf
10. World Economic Forum: Global Information Technology Report 2016 (2016). http://reports.weforum.org/global-information-technology-report-2016/economies/#indexId=nri&economy=rus. Accessed 18 Apr 2019. !!!стр 3
11. RIA Rating: Russian rating agency. Rating of the most expensive Russian public companies (2018). http://www.riarating.ru/infografika/20190129/630115992.html. Accessed 18 Apr 2019
12. Top 10 most expensive companies in the world in 2019. https://ru.fxssi.com/top-10-samyx-dorogix-kompanij-mira. Accessed 18 Apr 2019

Development of Logistical Technologies in Management of Intellectual Transport Systems in the Russian Federation

Alexander L. Chupin(✉), Oxana A. Yurchenko, Zhanna S. Lemesheva, Anna Y. Pak, and Mikhail B. Khudzhatov

RUDN University, Moscow, Russia
chupalex93@gmail.com, {yurchenko_oa, lemesheva_zhs, pak_ayu, khudzhatov_mb}@pfur.ru

Abstract. The article studies the modern logistical technologies at transport and the connected functional peculiarities and program components, which application is very important in provision of effective activities of logistical systems.

Creation of information systems for preparation of managerial decisions during distribution and shipment of goods and cargoes is an urgent problem in the sphere of transport of the Russian Federaiton. Application of the modern information technologies at transport and in commerce allows increasing effectiveness of the whole transport process by means of quick access to the information on subjects (buyer, transporter, services) and objects of shipment (goods, terminals, transport) and making the most rational decision.

The purpose of the paper is to study the development of logistical technologies in management of intellectual transport systems, which are used in the process of managing the transport flows, and to substantiate their advantages and peculiarities during application in the sphere of improvement of management of car transport.

The authors use the theoretical method of research.

Results: the authors determine the parameters that show certain aspects of development of logistical technologies in management of intellectual transport systems.

Sphere of application of results: the obtained results should be used by economic subjects.

Keywords: Intellectual transport system · Transport infrastructure · Logistics · Logistical system · Transport

JEL Code: R40

1 Introduction

At present, transport is one of the moving forces in economic development of a country. The modern life needs new requirements to mobility of transport and more strict requirements to safety of transportation. One million people die in road accidents every year. In Russia, this indicator requires a lot of attention. An intellectual transport system is to solve this issue in the world.

E. G. Popkova and B. S. Sergi (Eds.): ISC 2019, LNNS 87, pp. 778–784, 2020.
https://doi.org/10.1007/978-3-030-29586-8_89

Provision of high level of organization of cargo transportation – especially, international – is possible with application of high technologies: modern means of communication and computer processing of information. Recently, the issues of development, implementation, and adaptation of information computer systems in transport were studied by Russian specialists V. Vasilyev, V.N. Kharisova, A.P. Barilovich, etc. However, usage of the leading modern information systems is not paid enough attention in the issues of management of transport flows in Russia.

Logistical technologies in managing the intellectual transport systems. At present, transport logistics is one of the most perspective and quickly development spheres of modern business.

The compiled data of functional features of the system TMS (Transportation Management System) could show the possibility and information for evaluating the whole significance of development and implementation of IT for each transport company at the modern stage of functioning. Thus, the possibility of planning the routes allows processing the structure of the whole chain of supply and the existing tools for management. At the tactical level it is possible to determine the potentials and opportunities of future models of transport networks.

At the same time, wide opportunities of accounting reduce the volumes of work with documentation and simplify the technical processes for maintenances of cars' load. Besides, analyzing the specific features and advantages of the systems of the TMS class, it is necessary to evaluate the process of implementation of such modern information technologies.

Table 1 shows the main stages of implementing the TMS systems in view of the factors of time and employment of company's personnel.

Table 1. The main stages of implementing the TMS systems.

No.	Contents of the stage	Company's employed personnel
1	Previous joint work on pre-project level of specialists from supplier and specialists of customer, which allows determining the functions that are necessary for solving a specific task and forming the schedule and volume of necessary work	Company manager, head of financial department, engineer
2	Stage of project management	System administrator, engineer, transportation department manager
3	Development of full service prototype of the future system	System administrator, engineer, transportation department manager
4	Testing the prototype at the simulation server	System administrator
5	Training the users	Company personnel
6	Putting TMS into operation	System administrator
7	Technical support	Users of the system

Source: compiled by the authors.

Analyzing the data of the table, it is possible to conclude that with all existing advantages of the system process, implementation of the direct technical operational product at a company requires close attention of the specialists from the company's personnel and large spending of time and financial resources. Thus, according to statistical evaluation, the process of development and implementation of the TMS systems takes 33 days, and employees' control is required at each level.

In view of the fact that the Russian market of car transport services has a lot of small and medium companies, it is possible to speak of large obstacles in the process of integration of complex information systems. Thus, Russian companies do not have an objective opportunity for implementing TMS systems, as small companies do not have a lot of labor and financial resources, and large duration of implementation and adaptation of the system could influence the company's competitiveness in the modern market conditions. In order to reduce expenditures and obtain the economic effect, small Russian companies could implement not comprehensive information system but limited program products of monitoring at car transport. Thus, companies could obtain the most important advantages of the information systems at transport with low spending of labor, time, and material resources. At present, there are modern opportunities for control and planning of the activities of a transport company that are available for a wide circle of users; automatized systems of monitoring of car transport could ensure execution of different tasks in real time.

As a result of annual growth of the volumes of international relations, development of society and economies of countries, quality of the Russian transport complex has to conform to higher requirements. This requires solving the tasks that are shown in Table 1.

The Russian transport complex (within certain types of transport) entered a new path of development – innovative path of development. Thus, modernization at water transport is almost finished: 2009 saw the end of implementation of the federal program "Modernization of the Russian transport system (2002–2010)" (Panamereva et al. 2007). Based on this, the intensive path of improvement of the Russian transport system fully conforms to the modern state policy. It could be implemented in case of development and implementation of the intellectual transport systems (Table 2).

Conclusion. Effectiveness is a complex notion, which includes such elements as sustainability and dynamics; the former includes also safety and quality.

On the whole, increase of effectiveness of the Russian economy (all its spheres) depends on the infrastructural component.

Implementation of an intellectual transport system could be increased without any limits and is integrated with the existing information systems and data bases of government bodies, including road patrol and law enforcement.

Table 2. Tasks that are solved within the transport complex of the Russian Federation during growth of public relations and development of the Russian economy.

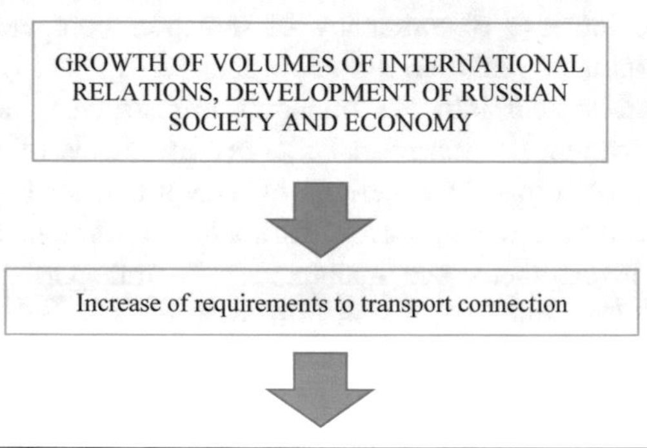

1. Increase of the number of transport means;
2. Organization of timely shipment of cargoes and passengers;
3. Organization of safe and high-quality shipment of cargoes and passengers;
4. Reduction of load on transport ways, elimination of traffic (passenger, cargo) jams, increase of the speed of transportation, reduction of transport expenditures – reduction of the transport component in the final price of a product, etc.;
5. Improvement of the ecological, social, and economic situation, etc.

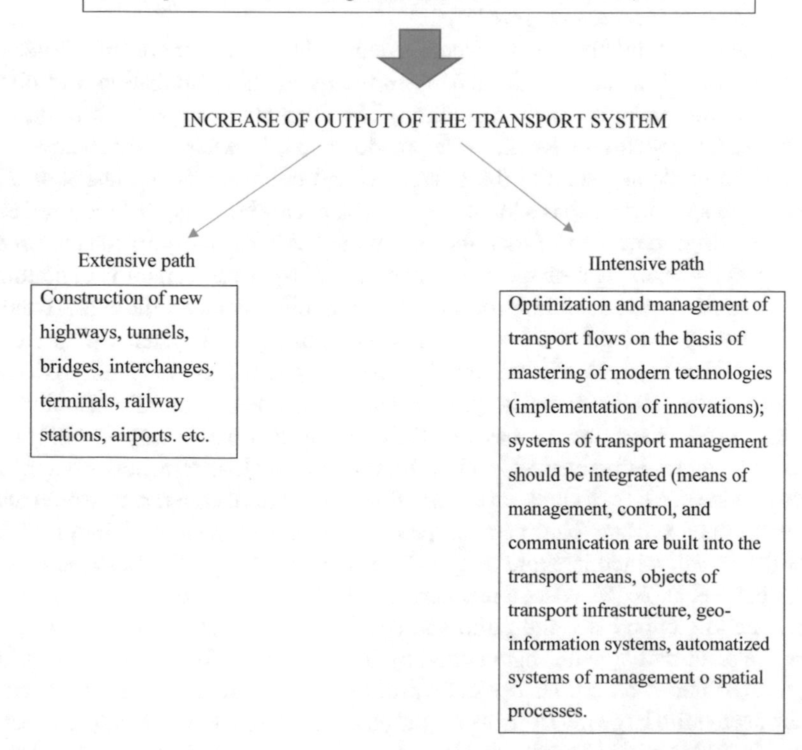

Source: web-site of the International Transport Forum, new age of the Russian transport // Transport of the Russian Federation. 2017, No. 6 (25).

Intellectual transport system stimulates solving the problems on provision of safety of road traffic, planning of the work of public transport, liquidation of traffic jams in transport networks, increase of efficiency of transport companies, and solving the problems of environment pollution (Goloborodko et al. 2018). Implementation of intellectual transport system into the transport infrastructure allows increasing the effectiveness of managing the transport economy by means of receiving timely and precise information, performs the functions of receipt and analysis of operative decisions, continuous centralized monitoring, formation of strategically sustainable economic and social development, and management of transport flows on the basis of received analytical data. Intellectual transport system could help to optimize coordination of activities of emergency services and law enforcement bodies, improve prevention and minimize the consequences of traffic accidents, and reduce the time of reaction to emergencies. Intellectual transport systems are a means of economically effective expansion of the existing transport infrastructure.

Control over road signs is performed with the help of the central coordination center, which collects data on the movement of transport vehicles and trios. The centers could consist of a lot of bodies, and all road and transport services, police, and emergency services use the unified center – or there could be several centers of specialists that have means for communicating with all other centers. The integrated center of control could disseminate data and control a lot of systems of the intellectual transport system, including the computer system of traffic control, which functions via transfer of information on emergencies.

Development of intellectual transport systems in the transport infrastructure of Russia is at the low level. It is limited by usage of satellite navigation and obsolete equipment in the sphere of traffic regulation. The attempts of implementing the intellectual transport systems in Russia were made in 2015, when the Moscow Mayor announced the implementation of the system “Smart traffic lights” (Adart et al. 2017). Intellectual transport system has a lot of advantages, but a whole range of requirements of the systems – high precision of determining the position of transport means for management in real time, navigation services for critical transport, and creation of continuous navigation service in the conditions of tunnels and multistory city buildings – cannot be ensured by the opportunities of the modern satellite navigation systems (Aguilar et al. 2017). Also, disadvantages of intellectual transport system include locality of sources (inability to cover 100% of territory with cameras); emergence of difficulties with accumulation of statistics on the basis of the existing data bases: impossibility of real assessment of target effectiveness – pilot zone of intellectual transport system is not scaled to the size of a city; increase of defects during change of ephemeris, which reaches 30 meters; influence of the landscape on precision of data; frequent violation of continuity of the signal, which is expressed in distortion and delay of determination of the signal (Albalas et al. 2018). Also, under certain conditions, the receiver does not receive a signal: due to clouded sky and radio sources. The working frequency is in L-band. Besides, the receipt of signal is aggravated by location in reinforced concrete building, tunnel, etc. (Albino et al. 2017). The main problem of the satellite systems is their high price, as they require large investments for purchasing photo and video cameras, modern traffic lights, information screens, and creation of the unified electronic data base for putting the systems into action. Besides, the state of certain roads is not ready for implementation of this project. The main financial risk of implementation of an improved model of the intellectual transport system is lack of financing, which is minimized by means of stage-by-stage financing – which requires investments in the sufficient volume for works within each stage of development (Alcaide-Muñoz et al. 2017).

The main legal risk is absence of the legislative basis for creation of intellectual transport system and standardization in the sphere of interaction of executive authorities' bodies. The group of legal risks could be minimized by means of formation of a legal environment, methodological complex for creation of Intellectual transport system, and the conditions for coordination of interaction of various bodies of executive power.

1. Formation of the unified transport space;
2.Provision of accessibility, volume, and competitiveness of transport services as to criteria of quality for cargo owners at the level of innovative development of the country's economy;
3. Provision of accessibility and quality of transport services for population according to social standards;
4. Integration into the global transport space and implementation of the transport potential of the country
5. Increase of the level of security of the transport system;
6. Reduction of negative impact of transport on the environment.

The main goals of the transport strategy of the Russian Federation until 2030

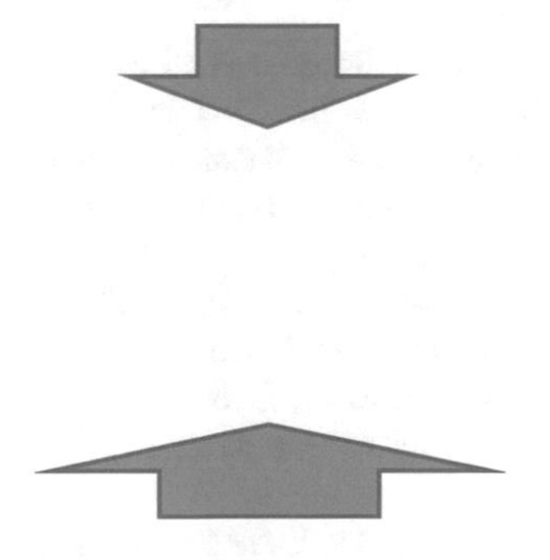

Formation and implementation of the intellectual transport system

Source: web-site "Intellectual transport system". URL: http://www.connect.ru/article.asp?id=9558 (Accessed: 23.01.2019).

The modern concept of logistics is the basis of a company's economic strategy, when logistics is used as a tool in the competitive struggle and should be treated as managerial logic for implementing planning and control over the material, information, and transport flows (Aguilar et al. 2017).

This task could be solved by management of the production process, which largely predetermines the rational usage of fixed assets and high effectiveness of capital investments.

References

Bulletin of the Congress of the intellectual transport system of Russia. Supplement of the newspaper “Transport of Russia”. Special issue No. 1 (2010)

Kasimov and Elatma to be connected by an innovative road. Information agency “MediaRyazan”. – Access. http://mediaryazan.ru/news/detail/166165.html

Maximov, V.V., Kurkin, P.E.: Using the concession mechanism for constructing federal highways with further exploitation on the free basis. Transport infrastructure – Access. http://www.pppinrussia.ru/userfiles/upload/files/artikles/Maximov_VV-Kurkin_PE.pdf

New technology will provide country with modern highways. Russian newspaper “Russian roads”. No 235 (5314). Access. http://www.rg.ru/files/special_editions/data/244.pdf)

Intellectual transport. http://www.connect.ru/article.asp?id=9558. Accessed 23 Jan 2019

International transport forum became a starting point of a new era of the Russian transport. Transport of the Russian Federation. No. 6 (25) (2009)

Panamareva, O.N.: Essence of the notion of economic effectiveness of a sea trade port’s work. Collection of scientific works, Novorossiysk, Issue 12, pp. 214–216 (2007)

Intellectual transport system as a tool of increasing the competitiveness and profitability. http://www.connect.ru/article.asp?id=9558. Accessed 23 Jan 2019

The first Russian International Congress on the intellectual transport system. http://www.pibd.ru/its1/. Accessed 2 Feb 2019

Intellectual transport system: perspectives of development. http://www.zdtmagazine.ru/publik/exibition/2009/05-09.htm. Accessed 3 Feb 2019

Adart, A., Mouncif, H., Naïmi, M.: Vehicular ad-hoc network application for urban traffic management based on Markov chains. Int. Arab J. Inf. Technol. **14**(4A Special Issue), 624–631 (2017)

Aguilar, J., Sanchez, M.B., Jerez, M., Mendonca, M.: An extension of the MiSCi middleware for smart cities based on fog computing. J. Inf. Technol. Res. **10**(4), 23–41 (2017)

Albalas, F., Al-Soud, M., Almomani, O., Almomani, A.: Security-aware CoAP application layer protocol for the internet of things using elliptic-curve cryptography. Int. Arab J. Inf. Technol. **15**(3A Special Issue), 550–558 (2018)

Albino, V., Berardi, U., Dangelico, R.M.: Smart cities: definitions, dimensions, performance, and initiatives. J. Urban Technol. **22**(1), 3–21 (2015)

Alcaide-Muñoz, L., Rodríguez-Bolívar, M.P., Cobo, M.J., Herrera-Viedma, E.: Analysing the scientific evolution of e-government using a science mapping approach. Gov. Inf. Quart. **34**(3), 545–555 (2017)

Al-Hader, M., Rodzi, A., Ismail, M.H., Sood, A.M.: Utilization of the dynamic laser scanning technology for monitoring, locating and classification of the city trees. Int. J. Inf. Process. Manag. **2**(1), 148–159 (2011)

The Financial Component of Formation of the Digital Economy

The Mechanism of Ensuring Liquidity of Venture Capital

Vladimir M. Matyushok(✉), Svetlana A. Balashova, Astkhik A. Nalbandyan, and Ivan A. Mikhaylov

RUDN University, Moscow, Russian Federation
{matyushok-vm, balashova-sa, nalbandyan-aa, 1032090157}@rudn.ru, imikhaylovv@gmail.com

Abstract. The authors study the mechanism IPO as the most attractive mechanism of venture capital funding by the example of Israel – as one of the most successful countries in development of venture industry. The purpose of the paper is to analyze the Russian practice of venture investements and to find the means of increasing their effectiveness. The authors used the empirical method (collection and analysis of data), historical, comparative, general scientific methods, synthesis of theoretical and practical material, and method of expert evaluations. Dynamics of investing into Hi-Tech companies and the Russian practice are studied. The current tendencies in the global market of venture investements are determined, and Russia's underrun from them is substantiated. The possibility of application of the latest mechanisms of attraction of investments - ICO, STO, IEO – into the venture industry of Russia for integration in the system of international venture capital is considered as an alternative to traditional mechanisms.

Keywords: Venture capital · IPO · ICO · Exits · Blockchain in Russia · Venture industry of Israel

JEL Code: G24 · O31

1 Introduction

The main idea of venture business that appeared in the middle of the 20th century has not changed much – high-risk investments for the purpose of decent return.

The main income is obtained by venture investors from growth of a company's cost. After all rounds of financing, when the company successfully functions in the market, venture investor leaves a project.

Thus, an especially important aspect of studying venture capital is the issue of return of invested assets. Receipt of income from growth of a company's cost after the end of the project is realized by means of selling it to strategic investor, return purchase, mergers and acquisitions (M&A), or initial public offering (IPO). Initial public offering in the global practice is considered to be most preferable for investor and a very successful method of leaving a project.

Successfulness of a company in IPO is the result of active cooperation and coincidence of interests of three aspects of deals: emitting company, investors, and financial

E. G. Popkova and B. S. Sergi (Eds.): ISC 2019, LNNS 87, pp. 787–795, 2020.
https://doi.org/10.1007/978-3-030-29586-8_90

intermediaries – underwriters. Setting an adequate price of a share during IPO and possibility of reaching the maximum price in the secondary market, as well as provision of high liquidity of shares, are a point of crossing of these interests (Khvorykh 2017).

While successful IPO allows investor to sell his share of the company with profit, the founder receives the opportunity to get the control over his business back, dividing large shares of the company between multiple shareholders.

It should be noted that the mechanism of exit through IPO is very interesting for investors and founders and for the state on the whole. Successful development of this mechanism for startups and each successful project increase "prestige" of the state, influencing its ratings – innovative, investment, and venture.

2 Methodology

The methodology of the paper is based on the principles of formal logic, systemic analysis, and the inter-disciplinary approach to the studied problem. The data were analyzed with the help of statistical and content analysis. The methods of grouping and classification were used for data processing.

3 Results

3.1 Experience of Israel

Israel is one of the most successful countries in development of venture industry (Abuladze 2016). The venture industry of Israel is constantly developing, adapting to the current tendencies in the global market of investments, which uncertainty is caused by US-China trade war and growth of interest rate in developing countries. These factors influence the investors and entrepreneurs. Both become more cautious and reconsider their strategies. Thus, the number of deals with low cost reduces, and duration of life cycle of investment rounds increased (Fig. 1).

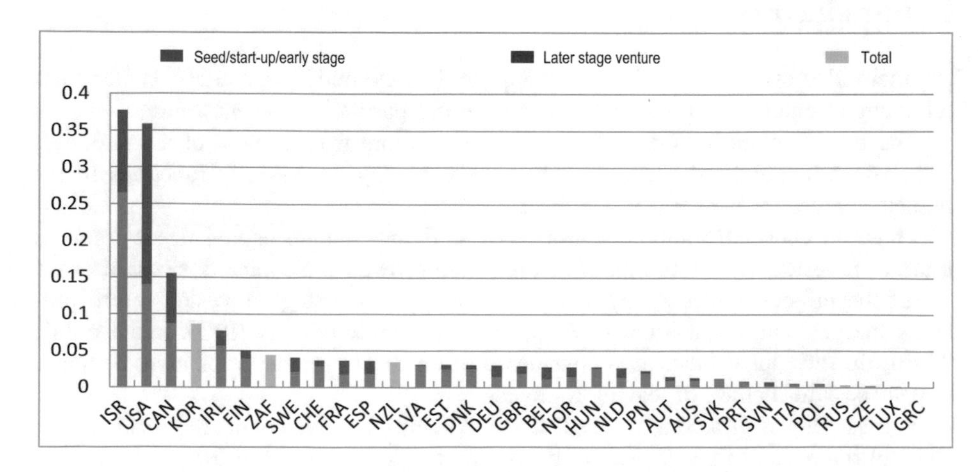

Fig. 1. Venture capital, share in GDP, for countries. Source: https://www.oecd-ilibrary.org/

Israel is ranked first in the world as to the share of venture capital in GDP and second as to the number of venture companies, behind the USA. A proof of high competitiveness of the Israeli Hi-Tech is the fact that Israel is ranked 3rd in the world (after the USA and Canada) as to the number of companies in NASDAQ.

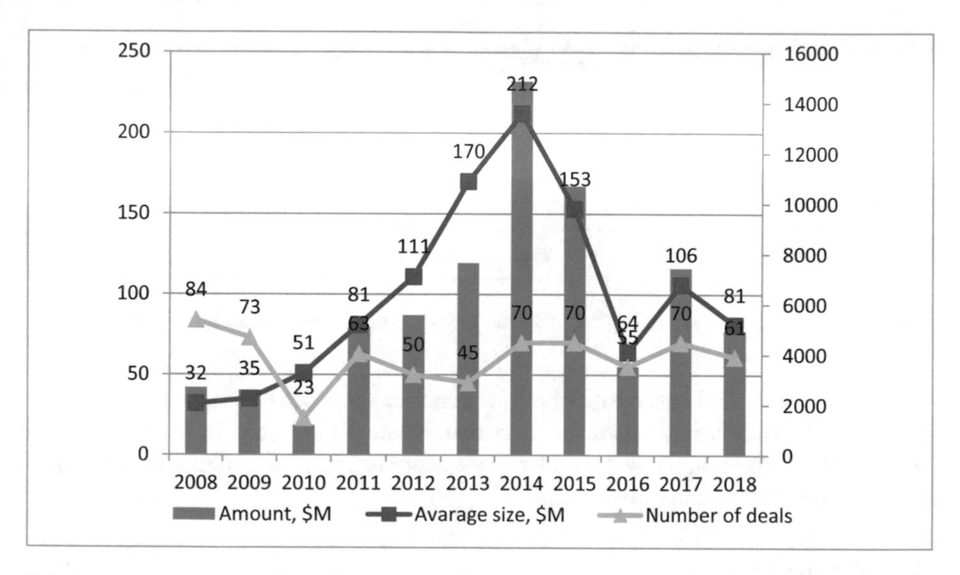

Fig. 2. Venture capital exits (IPO and M&A), deals over $5 million, Israel. Data: https://www.pwc.com/il/en/assets/pdf-files/2019/pwc-exit-report-eng-2018.pdf

As is seen from Fig. 2, until 2014 the mechanisms of exits had been developing, and the average size of deals had been growing with stable number of deals. However, there has been large decline in the last three years. This could be explained by the above – the market goes through a "patient and cautious period". Also, of 61 deals in 2018 only 9 deals are IPO, with the average size of deal of $99 million; 5 of them are present in NASDAQ, 4 - in ASX. 8 IPO were performed in 2015, 2 – in 2016, and 11 – in 2017. A stable average number of IPO of Hi-Tech companies constitutes 10 per year (Fig. 3).

In 2018, capitalization of the market of venture capital (total number of deals with participation of venture capital in Israeli companies) constituted USD 3.7 billion) (more than 17% of the market of Europe's venture capital) and is growing, despite the decline in "exits" through IPO and M&A. The programs of state support – "Incubator" and "Yozma", creation of international funds - BIRDF and USISTC with the USA, and BRITECH with the UK – stimulate these dynamics of venture activity.

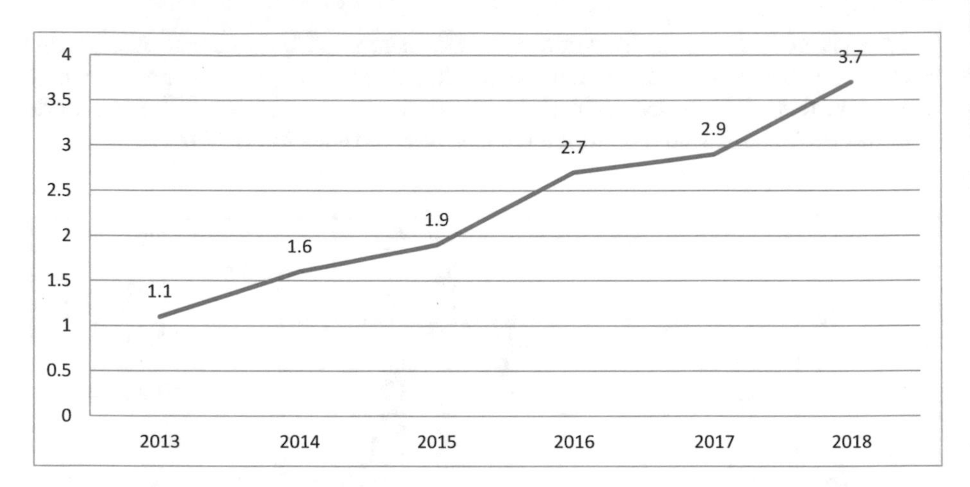

Fig. 3. Volume of venture capital in Israel, USD billion. Data: https://blog.dealroom.co/

In 1990, Israel had one venture fund, which managed USD 35 million. As of now, there are 100 active funds in Israel, which manage more than USD 10 billion, of which 14 funds are international funds with Israel statutory capital. 220 international funds are investing into Israeli entrepreneurship.

3.2 Russian Practice

Over the last 18 years, more than 100 Russian companies performed IPO, of which half – in the foreign stock markets, primarily in London Stock Exchange (LSE) and the Alternative Investment Market (LSE's AIM) (Butyaeva 2017). On the one hand, refusal from the local platform complicates the process of IPO and makes it more expensive; on the other hand, it provides wider opportunities for founders, including:

- larger market, and, as a result, larger potential circle of investors;
- possibility of selection of market with more developed spheres of company's activities;
- more prestigious platform increases the image among national investors (Fig. 4).

Popularity of London stock exchanges as compared to American stock exchanges is caused by smaller expenditures for listing – 3–4% vs. 6–7%. The highest number of Russian companies that performed IPO was observed in 2006–2007. On the main platform of LSE they attracted 60% of the total sum of all investments. These statistics could be compared with Israel – except for the fact that the statistics take into account SPO – secondary public offering. It should be also noted that these data reflect the total number of IPO/SPO of companies, the large share of which is large companies that are not startups – while Fig. 2 shows the data for Hi-Tech startups of Israel. There are few IPO startups in Russia – which shows that this mechanism of exit is absent.

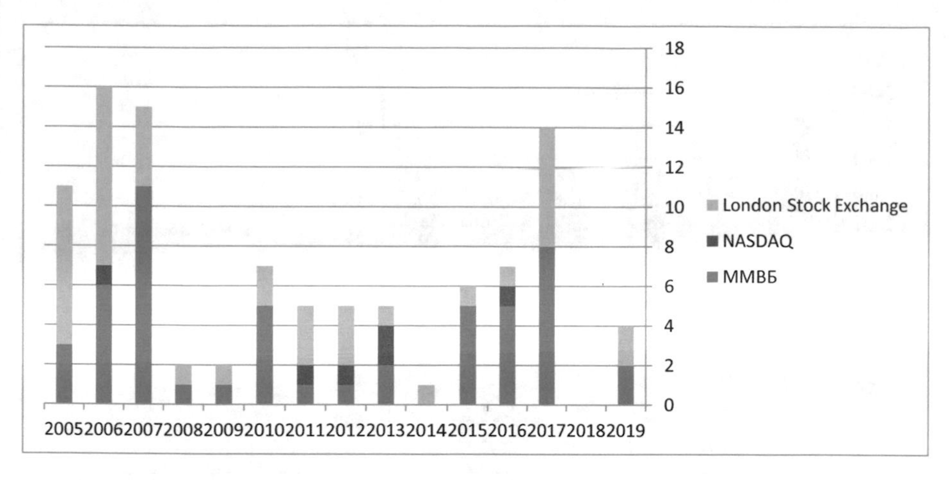

Fig. 4. Russian IPO and SPO, total number of deals. Data: http://www.preqveca.ru/

Absence of IPO in 2018 shows the loss of trust from investors and emitters to the Russian market. Exit of such large emitters and Diksi and AvtoVAZ and American sanctions restrain the market participants and make them reconsider the expedience of listing in the stock market due to the risk of implementation of economic limitations and, as a result, decline of capitalization of the company. For example, quotations of US Rusal reduced by two times after the company was included into the black list of the US Ministry of Finance.

However, Russia sees the second attempt of revival of the public market of securities. The fund Da Vinci Pre-IPO Tech Fund with the planned volume of RUB 6 billion was created; it is oriented at the companies that are in their late stages of development and aim at IPO. The fund invested in such companies as Gett, ITI Capital, and Softline – however, the financing did not stimulate IPO.

As for venture capital, capitalization of the Russian market in 2018 increased by 50%, constituting USD 714 million, of which USD 388 million – Russian investors and corporations (35%) and venture funds (65%). In 2017, the volume of deals of Russian investors with foreign assets was assessed at USD 695 million, which shows the continuing diversification of the portfolio and investors' aiming at the global market in the conditions of slow rates of growth of the Russian economy.

According to ReedSmith and Mergermarket, only 1% of the questioned respondents of venture capitalist are interested in the Russian market of innovations. This shows that investors are interested more in separate ideas and project that present their indirect goals than in return of their investments.

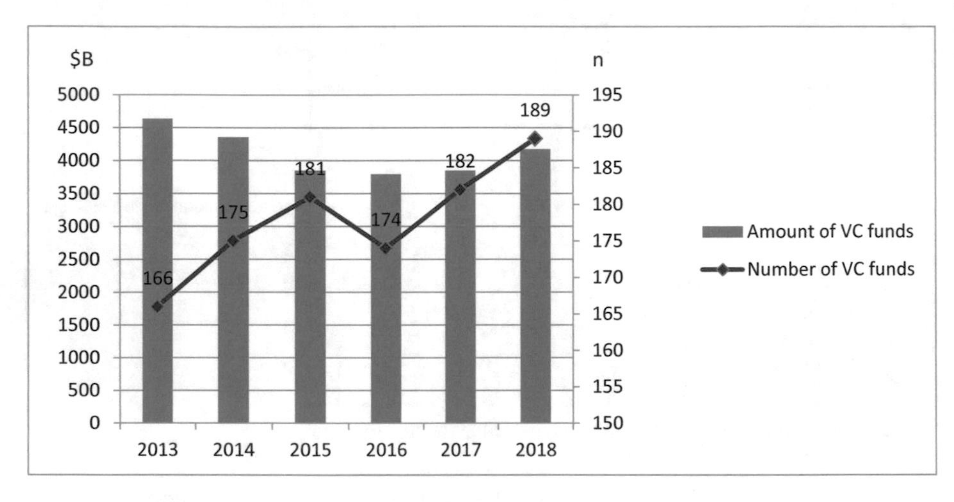

Fig. 5. Russian venture funds, USD million. Data: http://www.rvca.ru/

There are 189 venture funds in Russia, with USD 4.1 billion. Dynamics of the number of venture funds, which is presented in Fig. 5, reflects the growing interest to venture investments – the number of funds grew from 166 to 189 over the recent five years, however, the size of the funds reduced by 10%. According to RVCA, the average size of a fund reduced from USD 28 million to USD 22 million on average in the last five years.

However, growth of capitalization of the venture market from USD 470 million in 2017 to USD 714 million is explained by growth of the number of deals at later stages – which conforms to the tendencies of the global market of venture investements. The share of corporate investments in 2018 grew from 12% to 40% of the total size, as compared to the previous year – one of the reasons of this is the President's order on creation of venture funds by the largest government corporations.

3.3 Digital Economy. Solutions on the Basis of Blockchain Technology

Since 2015, a new mechanism of attraction of investments has been actively trying to replace the mechanism of IPO - ICO (initial coin offering). It offers entrepreneurs – at any stage of project's development – to try to attract financing, but without limitation on the number of investors, as well as cheaper and quicker. On the other hand, it functions on platforms that work on the blockchain technology – therefore, it envisages anonymity and protection only within the system (Vilner 2018). Wide popularity of this technology attracted a lot of frauds, inexperienced investors, and entrepreneurs, and the unexpected character of this phenomenon did not allow preparing and creating an effective regulating basis for fighting frauds. The next, also a very important problem of the ICO market, consists in the fact that emitted tokens – shares are connected to the cryptocurrencies rates, which have very high volatility. As a result, the downfall of the market of certain cryptocurrencies could lead to termination of the activities of ICO projects anytime (Table 1).

Table 1. Statistics of ICO, 2014–2018. Data: https://icobench.com/

By the number of ICOs		By raised funds, $B	
USA	740	USA	7.4
Singapore	551	British Virgin Islands	2.4
UK	487	Singapore	2.3
Russia	328	Switzerland	1.8
Estonia	274	UK	1.3

Since 2014, more than 5,480 ICO were started; the volume of collected assets since 2016 exceeds USD 26 billion. Year 2017 could be called the year of ICO – however, by the end of 2018 the market fell against the background of downfall of the cryptocurrencies market and high activity of frauds. According to RBC, statistics for the first quarter of 2019 are not very goods; ICO projects collected the assets that are smaller by 58 times as compared to the first quarter.

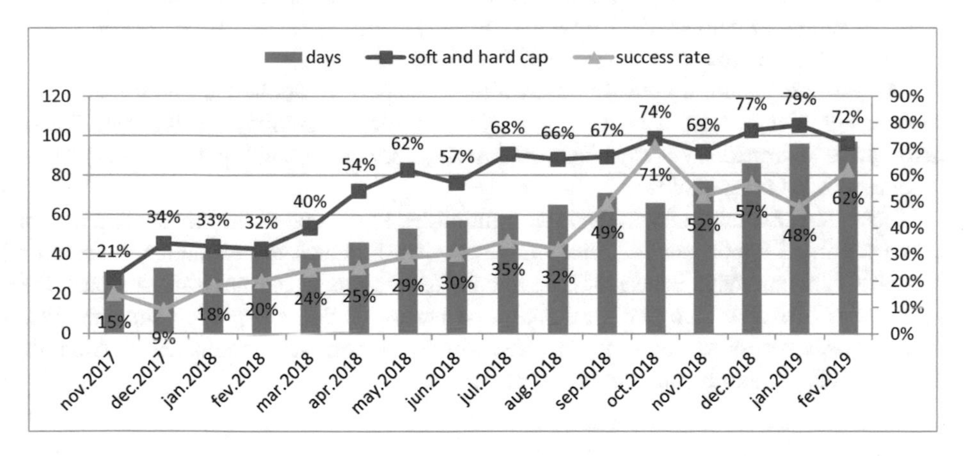

Fig. 6. Characteristics of ICO project in 2017–2019. Data: https://coin360.com

According to Fig. 6, characteristics of ICO projects repeat the general tendencies of the global market of venture investements – increase of the duration of investing and thorough selection of investment objects. Since November 2017, the average time necessary for collection of assets by ICO projects grew by three times – to 97 days. In the first quarter of 2019, the share of startups that collected assets for all their purposes (hard cap) and minimum purposes (soft cap) exceeds 70%, as compared to 33% in early 2017; at the same time, the share of successful projects grew from 18% to 60% in the same period of time. The causes of the growth were downfall of the market of cryptocurrencies and reduction of the number of investors, as well as increase of the market's professionalism – the investors started selecting the projects for investing with more patience and attention.

ICO were replaced by such mechanisms as STO and IEO. STO – security token offering – is almost the same as IPO but in the digital world. STO — security tokens offering; tokens are actually tokens-shares and a real investment tool. The investor who has tokens possesses the right for dividends or share of the company's profits. Security tokens have to be ensured by the project's assets and have the legal confirmation. Security token is officially a token-security and has to perform all its functions. If the ICO project emits a security token, it becomes a STO project (security coin offering). The most important difference of such method of attraction from ICO is that the offered token and its functioning are regulated by the laws of the U.S. Securities and Exchange Commission (SEC) or another regulator. Despite this, the mechanism becomes very popular – in 2019 the STO projects collected more than USD 84 million.

Another mechanism that appeared in 2017 (but became popular only in 2019) and is to replace and supplement ICO, initial exchange offering (IEO), is to protect the investor from fraud projects. A startup, increased of ICO via starting a platform, passes a necessary financial and legal inspection — Due Diligence – and receives listing at the crypto exchange. Correspondence to all requirements of the exchange platform increases protection and expectations of investors on the possible return of investments. The exchange accumulated perspective projects and interest from investors, thus increasing the volume of exchange and the direct income.

At present, several cryptocurrencies already started the tools that allow using this mechanism of attraction of investments: Binance, OKEx, Bittrex, and KuCoin. There also appear intermediary companies that provide service on audit and legal inspection of potential IEO companies.

However, IEO have certain external limitations. Protection of the exchange narrows down the fraud opportunities. Frauds have to directly attack the exchange, instead of complex schemes with investors. The second limitation is the necessity to pass the procedures of registration and verification of person (KYC) for participating in IEO, which closes access to investing and trade to users from such countries as Albania, Bosnia, Belarus, and certain countries of Africa.

4 Conclusion

Israel is a unique example of right development of the innovational infrastructure, including:

(1) state support for entrepreneurship and developments;
(2) interest of hi-tech business;
(3) mechanisms of financing of venture capital.

Russia has developed mechanisms of state support for innovative, small, and medium entrepreneurship – the country has a sufficient number of programs for support and stimulation of entrepreneurship, various accelerators, technological parks, and funds (including Skolkovo and the Russian Venture Company, which is a successful application of the Israeli model – namely, YOZMA group, which became a model of funds on the global practice).

From the side of entrepreneurship, the system experiences certain complexities, which are connected to complexity of doing business, various administrative barriers, and financing. On the other hand, at the peak of development of ICO mechanisms Russia was ranked 2nd as to the number of started projects, which shows the opportunities of development of entrepreneurial activities.

However, attraction of investments require the mechanisms of IPO, which presence does not reduce the risks of financing but exceeds the country's attractiveness. Israel's experience shows that presence of developed stock market is an important element of the system, but is not necessary, as almost all companies of the country perform IPO at NASDAQ and ASX. In Russia the attempts of implementing the traditional means of exist from venture company with the help of stock markets are not effective, so it is expedient to use the alternative mechanisms.

The latest technologies of blockchain envisage the tools in the form of mechanisms of attraction of financing and exit from a company through ICO, STO, and IEO – which can ensure the development of the venture industry in Russia and stimulate its integration in the system of international venture capital. These mechanisms could be especially topical in case of disappearance of their connection with volatile crypto currencies.

Acknowledgments. The publication has been prepared with the support of the "RUDN University Program 5–100".

References

Khvorykh, O.: IPO in the system of global financial architecture. Youth and Science: Step Towards Success (2017)

Abuladze, L.: Specifics of venture capital funding of innovative projects in Israel. Curr. Probl. Humanit. Nat. Sci. **2016**(6–2), 26–32 (2016)

PwC Israel 2018 Hi-Tech Exit Report (2018). https://www.pwc.com/il/en/assets/pdf-files/2019/pwc-exit-report-eng-2018.pdf

Butyaeva, Y.A.: IPO of Russian companies at the russian and western stock markets: existing experience and perspective. Sci. Bull. **2017**(1–1), 64–70 (2017)

Vilner, Y.: Evolution of Venture Capital Structure in the Age of Blockchain (2018). https://www.forbes.com/sites/yoavvilner/2018/11/23/evolution-of-venture-capital-structure-in-the-age-of-blockchain/#7299e4a629a1

Statistics (2019). https://www.oecd-ilibrary.org/

Draganov, I.: €2.5 billion venture capital investment in April in Europe & Israel. https://blog.dealroom.co/

Venture capital information. http://www.preqveca.ru/

Overview of the market of direct and venture investements (2018). http://www.rvca.ru/

ICO Rating Platform. https://icobench.com/

Online platform. https://coin360.com

The Investment Concept Strategy of Development of Innovative Activities of Agricultural Organizations in the Conditions of Techno-Economic Modernization

Anna V. Shokhnekh(✉), Yuliya V. Melnikova, and Tamara M. Gamayunova

Volgograd State Socio-Pedagogical University, Volgograd, Russia
shokhnekh@yandex.ru

Abstract. The authors determine the main problems of mastering, implementation, and application of innovations in agricultural organizations, which appear in the conditions of a crisis of the whole economic system in the global and national landscapes of the digital economy. Innovations that are created and approbated in agricultural organizations are aimed at determining the problems and searching for their solutions are very expensive. This research is aimed at analysis of the problems of innovations in agricultural organizations, which start at the first stage of their development and implementation, where the author of the newest approaches to agricultural production has to prove the effectiveness of implementing the idea that cannot motivate the investors for participation in an innovative project.

The authors analyze the following: (1) criteria of innovations that could be used in an investment project of agricultural organization; (2) system of innovations and structure of innovative activities of the agricultural sphere; (3) barriers of innovative activities and characteristics of reasons; (4) share of innovative goods, works, and services in the aggregate volume of supplied goods and performed works and services in Russia, for the types of economic activities for 2017; (5) share of investments into fixed capital in agriculture, forestry, hunting, fisheries and aquaculture sectors of the aggregate volume of the types of economic activities for 2014–2018 in Russia (in factual prices); (6) index of physical volume of investments into fixed capital for the whole circle of economic subjects of agriculture, forestry, hunting, fisheries and aquaculture sectors (in % as to previous year).

1 Introduction

The level of a country's economic security is influenced by activity of the investment processes and development of investment infrastructure. Investment processes are an indicator that shows general economic state within the state, volume of national income, and object of strivings for other countries. The term "investments" has several meanings. It could be treated as purchase of profitable securities for obtaining financial profit; as real assets – e.g., technological equipment for production and realizations of

E. G. Popkova and B. S. Sergi (Eds.): ISC 2019, LNNS 87, pp. 796–808, 2020.
https://doi.org/10.1007/978-3-030-29586-8_91

goods and services. In a wide sense, investments are a specific risk mechanism that is necessary for formation, growth, and development of a country's economy. Investment processes and development of the investment infrastructure are treated as financial flows of investors and owners of business and the investment project. Investment processes, as a result of influence of the investments on the economy, cannot be parallel – they interact and cross in various spheres, thus accelerating the economic, social, production, scientific, labor, and ecological investment effects from implementation of certain investment projects. Investment as a tool in which production resources could be invested allows preserving and multiplying their cost and ensuring the positive value of income. The investment activities as a process of investing are a totality of practical actions for implementation of investments. Investing into creation and reproduction of the main funds has the form of capital investments ("Regarding investment activities in the Russian Federation, in the form of capital investments", 25.02.1999 No. 39-FZ).

The subjects of investment processes could be investors, customers, contractors, users of the objects of investment activities and intellectual results, suppliers, legal entities (banking, insurance, and intermediary organizations) that ensure the investment process, and other participants of investment phenomena and projects. The objective need for stimulating economic growth in the strategic situation is a stage-by-stage and planned transition of a country's economic systems, including the agricultural sphere, to the innovative path of strategic development, which requires global acceleration of the investment processes, scientific consideration of the nature of investment concepts and the mechanism of effectiveness of their usage in technological update of the agricultural sphere of production.

When objectively studying the notion "investments", it is necessary to note the main contents of their key function – formation of resources for the production stage of the reproduction process.

2 Materials and Method

The scientific materials of the research include publications on the issues of formation of the investment concept-strategy of development of innovative activities of agricultural organizations in the conditions of techno-economic modernization. The authors used the methods of deduction, induction, comparative analysis, logical conclusions, and graphical transcription of the results of the authors' concept, which allowed for thorough elaboration of the scientific problem.

3 Discussion

In the conditions of technical and economic modernization, the investment resource concepts include the following (Botashev and Shokhnekh 2013; Rogachev et al. 2016; Makarova and Shokhnekh 2018):

equipment of a new generation, new technologies and developments, information data bases, qualification of employees and their advanced training, new managerial solutions;
restructuring of the main forms of the modern reproduction in the agro-industrial complex;
creation of a new system of capital formation;
structural transformation of the agricultural system and mechanisms of formation and support for organizations of the agricultural sphere;
resource provision of transition of the agricultural system to the innovative path of development, which is a consequence of attraction of investment flows into the economic process;
group of socially important functions, where the investment processes lead to creation of new jobs in the sphere of agriculture, development of the social sphere of the people who work in the agro-industrial complex, and expansion of the group of measures of social support, and attraction of new personnel into the sphere of agricultural production;
due to food sanctions, within the policy of food security, the sphere of agriculture remains one of the top-priority directions of the Russian economy's development.

The investment concept strategy of development of innovative activities of agricultural organizations, as a fundamentally new approach to production, distribution, and reproduction in the conditions of techo-economic modernization and innovative processes, is based on a specific example, idea, or direction of the activities, which bring financial, social, or technological result. This concept-strategy envisages investing into completely new technologies, approaches, or projects, which could increase efficiency, but in view of the differential approach and based on the diversity of the spheres of agricultural organizations. Investing into innovations as an implementation of innovations in the sphere of technique, technology, organization of labor or management, based on usage of the achievements of science and the leading experience, ensures qualitative increase of effectiveness of the production system or quality of products. The tern "innovatio" is of the Latin origin, being a synthesis of two words: investio (put on) and novatio (renovate)[1]. Definition "innovations" in its real application began to be used as a research object in the 19th century as introduction of elements of one culture into another. In early 20th century, there appeared a new sphere of knowledge – innovation theory – a science on innovations, within which regularities were studied. The research shows that mechanisms, methods, and techniques of usage of new skills, means, and forms, which envisage minimum time, material, and

[1] Scientific library. [E-source]. – URL: http://cyberleninka.ru/article/n/innovatsionnye-tehnologii-v-obrazovanii-1 (Accessed: 18.03.2019).

intellectual resources for obtaining the desired result, are called innovations. In the agricultural sphere, innovation as a notion as a meaning similar to other spheres. Innovations possess the main criteria in the agricultural sphere. The criteria of innovations that are used in the investment project of an agricultural organization are presented in Fig. 1.

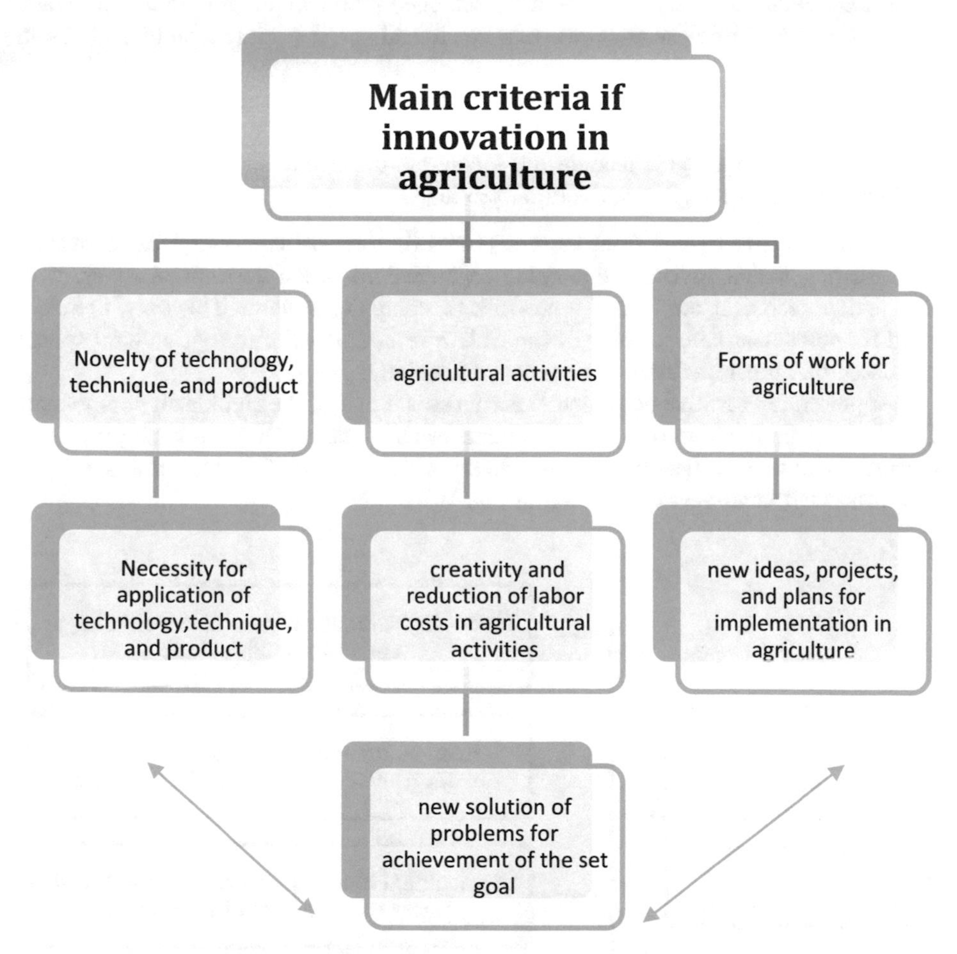

Fig. 1. Criteria of innovations that could be used in the investment project of an agricultural organization. Source: compiled by the authors.

The terms "innovations" and "innovative projects", which are used as synonyms, were scientifically substantiated and introduces into the categorical tools. Innovative technologies in agriculture could be treated as a sphere of scientific knowledge and agricultural practice, which studies and practices new methods of improvement of the production process with the usage of new tools, methodologies, and techniques. The new methods of improvement of the production process are determined by the central position of a product of agricultural production on the basis of reduction of losses, increase of the level of ecological security, and growth of profit from the manufactured products.

Based on the above definitions, it is possible to conclude the following:

1. innovative technologies have to improve the process of agricultural production;
2. it is necessary to take into account the changed production environment, in which an important role belongs to new technical means and ecologically clean technologies;
3. innovative technologies should use the achievements of all scientific spheres, which reduce the physical intensity of work;
4. innovative technologies include the technique of search for new knowledge for improving the quality of agricultural products.

Innovations are treated from various points of view of the concept of innovative technologies, which envisages equipping organizations of the agricultural sphere with the modern technical means and a network of computer systems. However, there's a need for competent labor resources. Innovative technologies of the agricultural sphere are based on human factor as a resource of economic growth.

At present, innovative pedagogical activities are one of the significant components of educational activities of any educational establishment (Glukharev (2009); Goncharenko and Gerashchenkova (2014), Shokhnekh et al. (2017)). The structure of all interconnected components is shown in Fig. 2.

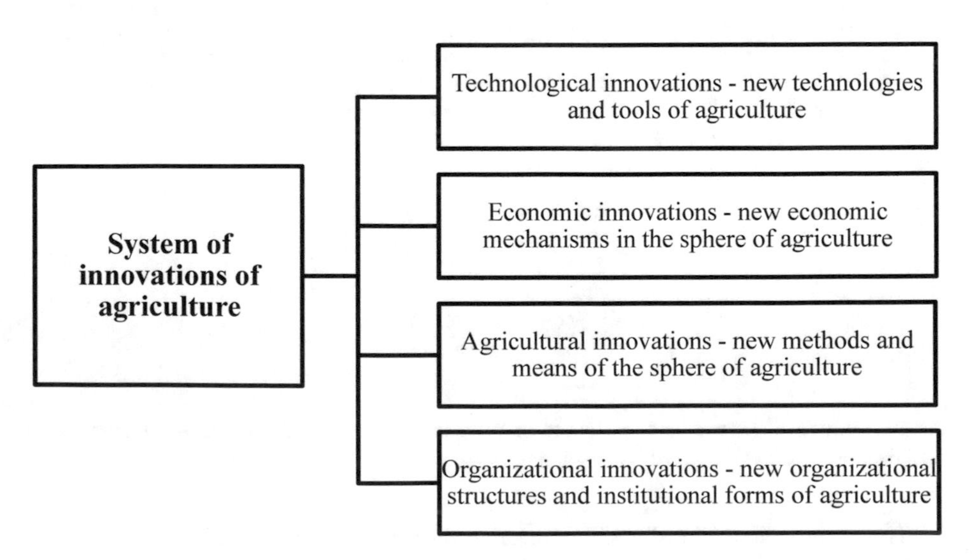

Fig. 2. The system of innovations and structure of innovative activities of the agricultural sphere. Source: compiled by the authors.

Interaction of the structure of the agricultural sphere creates a comprehensive functioning system. That's why innovative activities of the agricultural sphere are a platform for increasing the competitiveness and determine the directions of growth of competencies of labor resources for increasing the level of profitability, quality of products, and, as a result, food security.

4 Results

Investing into innovative projects of the agricultural sphere could lift off the tension and solve the task of update and modernization. There are a lot of options of implementation of the investment concept-strategy for technological directions: copying foreign experience and its extrapolation on the sphere of the Russian agriculture; transfer of past experience of the Russian agricultural sphere to the modern environment; formation of a new one – based on the changed technologies and techniques in the conditions of the digital economy.

The conceptual basis of innovations includes not only possibilities of success but also the risk of failures. It is difficult to predict that the losses from risks could be covered or justified. However, the attempts to implement innovative projects and improve the technologies in agricultural organizations should only increase. There are also attempts to classify the innovations and divide them into several types.

N.A. Ilyina offered a classification that could be easily applied to the sphere of agriculture (Ilyina 2016):

- technical and technological innovations: new production means and new technologies (innovations from which nothing negative is expected by the organization's personnel);
- organizational and managerial: new organizational structures and methods of management of labor resources, making of managerial decisions and control over their execution;
- socio-economic: new material stimuli of labor resources, motivational system of labor payment;
- legal: changes in local acts (internal positions of organizations of accounting policy, collective agreements); changes at the federal, regional, and local levels of state management: amendments to laws; norms regarding self-employed people; regarding protection of intellectual property, etc.;
- agricultural (new methods, technologies, models, and forms of agricultural production, provisions of agricultural departments and other).

Investment stimulation of innovations is aimed at leveling the barriers for implementation of innovations. Barriers for innovative activities, which slow down the innovative process, and characteristics of reasons are shown in Table 1.

Table 1. Barriers of innovative activities and characteristics of reasons.

Barriers for innovative activities	Reasons in the agricultural sphere
Barrier for creativity in agriculture	Entrepreneurs and labor resources that are used to work with old mechanisms do not want to implement innovations, for they understand that if there's no harvest and no agricultural products (which are seasonal), there will be no possibility to amend mistakes and manufacture new products in the current year. Mistakes and risks will be taken into account only in the next year. Risks of bad harvest or damage of products during processes envisage coverage of losses, search for financing for investments into agricultural production of the next year, and provision of entrepreneur and labor resources during off-season
Barrier of conformism	Entrepreneurs and labor resources of the agricultural sphere do not accept the vivid positive innovations due to the habit to adapt to long traditions. It is known that agrarian country is peculiar for domination of traditions and support for the traditional mode. It is difficult to violate the cognitive scheme in which there are no conditions for assimilation and accommodation. Reluctance to develop and fear of losing agricultural products are a refusal implementation of innovations, which are necessary for agricultural production
Personal concern of entrepreneur and labor resources in the agricultural sphere	Due to lack of confidence in the innovative project, a lot of innovative ideas do not move farther then genesis of thought. Low self-esteem does not allow believing in oneself – it's also hard to believe that the ideas of innovative project for agricultural sphere could be genial and stimulate the growth of efficiency of agricultural products and ensure high level of food security. Also, the fear to express the ideas at conferences on the possible innovative activities in the process of production take away the possibility to implement innovations

(*continued*)

Table 1. (*continued*)

Barriers for innovative activities	Reasons in the agricultural sphere
Rigidity of entrepreneur's thinking and labor resources in the agricultural sphere	Experienced entrepreneurs and labor resources of the agricultural sphere think that their opinion is right and cannot be reconsidered. They do not strive for new knowledge and skills and have a negative attitude towards new ideas in the conditions of the digital economy. Entrepreneur and labor resources in the agricultural sphere could change the conditions of implementation of an innovative project. That's why the investment concept strategy, which is implemented by the investment and innovative project based on a clear road map, which includes a general plan and programs of works, will not allow introducing any unsanctioned changes on the basis of rigidity of thinking
Absence of technologies and technical & technological insufficiency	Insufficient possession of modern agricultural technologies, computer equipment, and electronic devices does not allow forming effective investment and innovative projects
Absence of competencies with entrepreneur and labor resources in the agricultural sphere	Absence of competency and skills for application of the modern agricultural technologies, computer equipment, and electronic devices does not allow forming and implementing investment and innovative projects
Ignoring innovative development by entrepreneur and labor resources in the agricultural sphere	Absence of behavioral habit of search for innovations for development of organization of the agricultural sphere and its labor resources. Ignoring innovative development by entrepreneur and labor resources becomes a threat of loss of competitiveness. New innovative ecologically clean technology or technique, which allows producing ecologically clean agricultural products, could become a reason for reducing prices and increasing the market. Entrepreneur, who has not analyzed the possibilities of mastering and application of innovations in time, risks losing his positions in the market

Source: compiled by the authors.

Table 2. Share of innovative goods, works, and services in the aggregate volume of supplied goods and performed works and services in the Russian Federation, for the types of economic activities for 2017 (Federal State Statistics Service (Rosstat) (2019) E-source http://www.gks.ru/wps/wcm/connect/rosstat_main/rosstat/ru/statistics/science_and_innovations/science/#)

types of economic activities for 2017[3]

Types of **economic activities**	**Code of Russian National Classifier of Types of Economic Activity 2**	**2017 (%)**
Share of innovative products of the aggregate volume of the Russian Federation:		**7.2**
Share of innovative products of the agricultural sphere in the aggregate volume of the Russian Federation:		**2.1**
Share of innovative products of the agricultural sphere in the aggregate volume of innovative products for all types of economic activities of the Russian Federation		29.9
for the types of economic activities		
annual crops	01.1	1.9
perennial crops	01.2	3.1
transplant seedlings	01.3	21.4
Cattle-breeding	01.4	1.7
mixed agriculture	01.5	-
auxiliary activities in the sphere of production of agricultural cultures and post-harvest care	01.6	1.8
industrial production		6,7
water supply; water discharge, organization of collection and disposal of waste, activities on liquidation of pollution	E	1,1
roof works	43.91	0.5

(*continued*)

Table 2. (*continued*)

specialized construction works	43.99	0.1
publishing activities	58	0.2
activities in the sphere of telecommunications	61	4.4
development of software, consultation services in this sphere, and other accompanying services	62	6.6
activities in the sphere of IT	63	2.5
activities in the sphere of law and financial accounting	69	0.6
activities of head offices; consulting on the management issues	70	0.1
activities in the sphere of architecture and engineering and technical design; technical tests, research, and analysis	71	2.3
scientific R&D	72	43.1
advertising activities and studying the market situation	73	1.9
professional scientific activities and other technical activities	74	-
Total		**100.0%**

Therefore, innovative behavior does not envisage creative pause, personal concern, rigidity of thinking of entrepreneur and labor resources, or ignoring the innovative development in the agricultural sphere. Innovative behavior envisages formation of own individuality, self-development, and investing the capital, natural, and labor resources into the innovative project. Solving these problems requires advanced training of entrepreneurs and personnel of organizations, seminars, videoconference, trainings, and educational work of the ministry of agriculture on application of the modern technologies.

Importance of formation of an investment concept strategy of development of innovative activities of agricultural organizations is especially obvious in the conditions of techno-economic modernization. Studies show that the share of innovative products in the aggregate volume of the Russian Federation constitutes only 7.2% of which the ones of the agricultural sphere – 29.9% (Table 2).

In 2017, the agricultural sphere supplied 29.9% of innovative goods, works, and services in the total volume for the Russian Federation (for the types of economic activities). These calculations were performed on the basis of the indicators of the annual form of federal statistical observation No. 4-innovation "Data on innovative activities of organization", which are systematized according to the Russian National Classifier of Types of Economic Activity. In the aggregate volume of innovative

Table 3. Share of investments into the fixed capital in agriculture, forestry, hunting, fisheries and aquaculture sectors in the aggregate volume of the types of economic activities for 2014–2018 in Russia (in factual prices).

Types of economic activities	Russian National Classifier of Types of Economic Activity	2014	2015	2016	2017	2018
		RUB billion				
Total (RUB billion)		13,902.6	13,897.2	14,748.8	16,027.3	17,595.0
Including:						
Agriculture, forestry, hunting, fisheries and aquaculture sectors, including	A	524.3	518.8	623.4	705.5	777.0
Crop research and cattle breeding, hunting and provision of corresponding services in these spheres	01	492.5	483.6	582.6	651.4	707.7
Forestry and logging	02	16.6	20.8	20.7	25.4	31.1
Fisheries and aquaculture sectors	03	15.2	14.4	20.1	28.7	38.2
	in % as to result					
Total		100.0	100.0	100.0	100.0	100.0
Including:						
Agriculture, forestry, hunting, fisheries and aquaculture sectors	A	3.8	3.7	4.2	4.4	4.4
Including:						
Crop research and cattle breeding, hunting and provision of corresponding services in these spheres	01	3.6	3.5	4.0	4.1	4.0
Forestry and logging	02	0.1	0.1	0.1	0.1	0.2
Fisheries and aquaculture sectors	03	0.1	0.1	0.1	0.2	0.2

supplied goods and performed works and services in Russia, for the types of economic activities for 2017, the share of organizations of agriculture constitutes 2.1%, which, in the conditions of the digital economy and Industry 4.0, is the indicator of a low level of provision of economic and food security. Low level of activity of investing of organizations of agriculture could be seen in the process of analysis of the share of investments into the fixed capital in agriculture, forestry, hunting, fisheries and aquaculture sectors from the aggregate volume of the types of economic activities for 2014–2018 in Russia (in factual prices) (Table 3).

Analysis of the index of physical volume of investments into fixed capital for the full circle of economic subjects (in % as to previous year) reflects low growth of dynamics of investment activity into agriculture, forestry, hunting, fisheries and aquaculture sectors (Table 4).

Table 4. Index of physical volume of investments into fixed capital for the full circle of economic subjects (in % as to previous year).

Types of economic activities	Russian National Classifier of Types of Economic Activity	2014	2015	2016	2017	2018
Total		98.5	89.9	99.8	104.8	104.3
Including:						
Agriculture, forestry, hunting, fisheries and aquaculture sectors	A	92.4	87.9	112.5	109.7	105.5
Including:						
Crop research and cattle breeding, hunting and provision of corresponding services in these spheres	01	92.7	87.4	113.1	108.2	104.1
Forestry and logging	02	80.4	109.1	90.7	119.6	116.7
Fisheries and aquaculture sectors	03	103.0	78.8	125.5	139.6	127.7

Therefore, innovations in agriculture should be implemented in all directions of activities. For verification of effectiveness it is necessary to evaluate the possibilities of future development and risks that could emerge during the practice of implementation. In view of advantages (future opportunities of development) and restraining factors (future risks), it is possible to accelerate the innovative activities through investment concept-strategy, which will allow financing the update of obsolete models of agricultural production.

5 Conclusion

Innovative activities are connected to scientific and methodological activities in the agricultural sphere and always contain a new solution of a current problem; they lead to qualitative and quantitative changes of the manufactured agricultural products; they increase quality of work in the structure of the agricultural sphere. The key goal of investment concept-strategy of development of innovative activities of agricultural organizations in the conditions of techno-economic modernization is maximum development of the potential of natural opportunities in the process of agricultural production. Practical implementation of innovative technologies envisages mastering of innovations with technical means and increase of competency of labor resources – which is the basis of investment modernization.

Acknowledgements. The reported study was funded by the Russian Foundation for Basic Research grant No. 19-010-00985 A. "Development of the innovative and investment policy as a concept of strategic economic security of agricultural organizations in the conditions of the modern technological transformation".

References

Botashev, A.Y., Shokhnekh, A.V.: The modern concept of investment monitoring of economic development of agriculture in the regions of Russia. Audit Financ. Anal. **3**, 258–261 (2013)

Glukharev, K.A.: Innovations and investments: essence, interaction, and role in the reproduction process. Bull. Herzen State Pedagogical Univ. Russ. **97**, 92–97 (2009)

Goncharenko, L.P., Gerashchenkova, T.M.: Investing of innovative processes in the agro-industrial production as a factor of increasing the level of Russian food security. Bull. Financ. Univ. **2**, 13–23 (2014)

Ilyina, S.A.: Essence and coordination of the main categories of innovative management. Internet J. "Naukovedenie" **8**(5) (2016). http://naukovedenie.ru/PDF/08EVN516.pdf

Makarova, N.N., Shokhnekh, A.V.: Integration of the neo-systemic nature into the life cycle of controlling of organizations of the agro-industrial complex. Bull. Samara State Univ. Econ. **10** (168), 61–68 (2018)

Russian National Classifier of Types of Economic Activity (OKVED 2)

Federal State Statistics Service (Rosstat) (2019). E-source: http://www.gks.ru/wps/wcm/connect/rosstat_main/rosstat/ru/statistics/science_and_innovations/science/#

Federal law: Regarding investment activities in the Russian Federation, in the form of capital investments. No. 39-FZ, 25 February 1999

Shokhnekh, A.V., Rogachev, A.F., Sizeneva, L.A., Glinskaya, O.S.: Economic evaluation of synergy of innovative technologies for cognitive modeling of region's food security. Audit Financ. Anal. **3-4**, 578–581 (2017)

Rogachev, A.F., Mazaeva, T.I., Shokhnekh, A.V.: Manufacturing and consumption of agricultural products as a tool of food security management in Russia. Revista Galega de Economia **25**(2), 87–94 (2016)

Forming the Policy of Insurance of Innovative and Investment Activities of Agricultural Organizations as a Concept-Strategy of Provision of Economic and Food Security

Yuliya V. Melnikova and Anna V. Shokhnekh(✉)

Volgograd State Socio-Pedagogical University, Volgograd, Russia
shokhnekh@yandex.ru

Abstract. The authors study the directions of formation of the policy of insurance of the innovative and investment activities of agricultural organizations, in which modern insurance is the key mechanism in provision of economic and food security, which should serve the interested of consumers of insurance services for development of insurance market. The following measures could be taken for protecting the rights of consumers of insurance of the innovative and investment activities of agricultural organizations: informing on insurers, insurance intermediaries, and terms of insurance of the innovative and investment activities of agricultural organizations, including with the help of Internet space; opening the corresponding information on the official web-sites of the subjects of insurance business; ensuring legal protection of services' consumers; ensuring objective and unambiguous assessment of the volume of damage to innovative and investment activities; determining the principles of formation of guarantee funds for insurance of the innovative and investment activities of agricultural organizations.

Protection of economic interests in case of various unfavorable events by means of the insurance reserves, which are formed by insurers from the paid insurance fees is the main purpose of insurance.

Keywords: Formation · Policy of insurance · Innovative and investment activities · Agricultural organizations · Concept-strategy · Provision of economic security · Food security

1 Introduction

One of the perspective directions of improving the financial and economic potential of the insurance market is development of investment activities. Growth of investment attractiveness of insurance companies will ensure the necessary help in modernization of the insurance market. It is necessary to improve certain types of insurance for increasing the effective attraction of assets to the market. The task of any insurance company should consist in constant improvement of terms for the types of insurance activities for citizens, development of the culture of insurance with customers, and provision of profitable offers.

E. G. Popkova and B. S. Sergi (Eds.): ISC 2019, LNNS 87, pp. 809–816, 2020.
https://doi.org/10.1007/978-3-030-29586-8_92

An important element in improvement of insurance activities is creation of a common information data base for all insurers for fighting fraud, quick preparation of reports, and precise accounting of policies and losses. The unified information data base will allow ensuring accumulation of the necessary data for reducing the risk of insurance in the market (Proshin and Shaykin 2017; Spletukhov 2018; Aksenova 2019). For more transparent activities of participants of the insurance market, it is necessary to publish the insurers' reports with the data on their managers and owners on the web-site of the Federal Service on Financial Markets. Improvement of the legislative aspects in the sphere of insurance could lead to active growth of the retail market of insurance.

In the long-term, the segment of insurance envisage the following development: electronic form of all insurance operations – i.e., selling policies on the Internet without physical presence of agricultural organizations; standardization of the rules of insurance of the innovative and investment activities of agricultural organizations in services and compensation of losses. Formation of the policy of insurance of the innovative and investment activities of agricultural organizations as a concept-strategy of provision of economic and food security envisages studying the definition "food security". The notion "food security" appeared in 1970's. In December 2974, the UN General Assembly adopted the Food and Agriculture Organization of the United Nations' "International obligation for provision of food security in the world". The functions of observation and collection of information on this problem were set on the Committee on World Food Security of the FAO. In the global agri-food market, a lot of countries reconsidered the goals and directions of their agrarian policy. Thus, the countries of the EU started urgent measures for involving previous unused land into agriculture and determining new reserves. Necessary measures are also taken in Russia. Thus, corrections have been implemented into the Federal Program and the Strategy of socio-economic development of Russia until 2020 (Isaeva 2016).

2 Materials and Method

The scientific material includes scientific publications on the problems of formation of the policy of insurance of the innovative and investment activities of agricultural organizations as a concept-strategy of provision of economic and food security. The authors use the methods of deduction, induction, comparative analysis, logical conclusions, and graphical transcription of the results of the authors' concept.

3 Discussion

The most significant directions of the adaptive reformation of the Russian agriculture include the following:

- increase of the share of subsidizing the agricultural manufacturers in the sphere of cattle-breeding, as in the last four years there formed a large exceed in the issues of financing of crop research. It is offered to introduce the criteria of differentiation for various areas with different levels of profitability from sales of agricultural products; these criteria should be the basis in the issues of planning of budget allocations;

- creation of conditions for cooperation of agricultural manufacturers, and, as a result of these transformations – increase of the share of profitability of each manufacturer as parts of large productions;
- social development of rural territories – it is necessary to form and adopt the scientifically substantiated system of norms of social infrastructure and living standards of rural settlements, as well as increase the tax potential of rural territories by reforming the structure of local taxation, creating a system of subsidized mortgages, and developing the parameters of the innovative and economic mechanism of attraction of work force;
- improvement of the system of managing the agro-industrial and food sphere – it is necessary to specify a range of managerial functions and the list of functions, as scattered functions exclude personal responsibility of government bodies for decisions in the sphere of agriculture.

The modern state of the agro-industrial complex is transformed under the influence of the sanction policy in the sphere of food limitations for Russia, and these changes have been exerting influence over the last five years. On the one hand, Russian sanctions expand opportunities for domestic agricultural manufacturers, which is already expressed in growth of the volumes of issue of food products. On the other hand, anti-Russian sanctions lead to increase of credit rates for Russian manufacturers. Increase of credit rates influences the intensity of the investment policy in the sphere of agriculture of Russian regions. It is obvious that without solving the problem of import substitution in the part of resource it is impossible to guarantee the state's food security (Botashev and Shokhnekh 2013). One should bear in mind that this is a complex of not so much sectorial and agrarian complexities as inter-system, inter-sectorial, and macroeconomic problem of interaction of managerial and financial structures. It is connected to the necessity for restoration and development of the whole sub-spheres of industry, agricultural machine-building, biological and other industries, and compensation and investment mechanism of making of managerial decisions in the system of food and economic security of agricultural organizations.

4 Results

Insurance, as a mechanism of provision of food and economic security, allows compensating for losses and is one of the most stable financial sources in case of risks of loss of agricultural resources, which stimulates the innovative and investment activities.

A generally accepted treatment of the notion "economic security" includes a complex of economic, political, legal, social, and other conditions, which ensure protection of national interests of a state, its citizens, and economic subjects as to the resource potential, possibilities of well-balanced and dynamic growth, social development and perspectives, ecology, and entrepreneurship even in the conditions of unstable development of internal and external influences. However, this treatment does not open the level of protection of an economic object or subject from the corresponding threats, with the issue of the measure or level of their protection remaining unsolved.

That's why the theory that interprets the economic security of any subject as an admissible or relatively acceptable level of risk, which is set by the normative way or which could take into account the specifics of the socio-economic system or its subsystems and elements, would be correct. As a matter of fact, neither subject of economy is protected so well that it would be possible to state the absence of any threats. In the opposite case, the system of security provisions becomes too expensive. Therefore, there is no zero risk in the system of economic subjects of the state, but there's a possibility of guaranteeing a certain level of protection. When analyzing the complex notion of food and economic security, it should be treated as a "component of the national security, preservation of sovereignty of the country, and the most important component of the demographic policy, the necessity condition of implementing the strategic national priority – increase of population's living standards on the basis of international standards of living and financial well-being" (Klimova 2012). Unfortunately, it's necessary to state that the modern agro-industrial complex of Russia does not fully ensure the food and economic security in its sector of economy, which performs the stabilizing role in the system of the national economy. The potential of the AIC is not realized in full – and the state of food security causes serious concerns. High import dependence of for certain types of agricultural and fish products preserves – which leads to a threat of violation of economic security (Rogachev and Shokhnekh 2015).

In its turn, the mechanism of regulation of food and economic security in this case is a system of management (reduction) of risk, which includes target norming of risk through establishment of its acceptable (minimum) level and the mechanism of provision of this normative level. As is known, the most important tool of risk management is insurance, which is sub-system of the system of provision of economic and food security.

Formation of insurance policy of the innovative and investment activities of agricultural organizations as a concept-strategy of provision of economic and food security envisages protection of the interests of legal entities, individuals, subjects, and municipal entities of Russia. Protection of subjects of insurance activities in case of certain insurance cases is performed by means of financial funds, which are formed by insurers from the insurance premia (insurance fees) and by means of other assets of the insurer. In this context, the role of insurance for the purpose of provision of economic security of the users of insurance market's services is considered. From the point of view of provision of protection of economic security of insurance organizations, the risk component is a part of the structure of the insurance tariff.

Insurance market is a market sphere of development of offer and demand for insurance services. It determines the relations between insurance companies that offer their services and legal entities and individuals, which require insurance protection. Studying the insurance market of Russia and the policy of the largest insurance companies in recent years, it is possible to state that the market is developing and growing financially.

Formation of the policy of insurance of the innovative and investment activities of agricultural organizations is built on the understanding that insurance company, as any economic subject, seeks profit in its activities. Successful result of the work of the insurance market and promotion of insurance services is reached by application of marketing technologies and tools. Marketing is a mechanism of development of an economic subject, which is aimed at its effective work.

Insurance company is no exception. It is oriented at development of own services, involvement of customers, and creation of a positive image for customers and the state. The tools of insurance marketing are used for achieving good result. The sphere of insurance services does not belong to the sphere of material production, which is a peculiarity of the sphere of insurance services.

Insurance organizations are created for satisfying the society's needs for insurance services. There are two types of services in the modern economy: services that are expressed in goods; services which result is not expressed in goods for sale. Insurance service belongs to the first type of services, as it is a product in the insurance market and has all its qualities.

Insurance service on protection of innovative and investment activities of agricultural organizations is a financial compensation for loss of insurance in case of an insured event that is envisaged by insurance agreement. In the economic sense, insurance service, which is realized on the paid basis, means transfer of risk from insurant to the insurance company in case of insured event that is envisaged by the agreement.

Product is an item that satisfies the needs of a human and has two qualities. Firstly, consumer value, and secondly, cost. If service (non-material product) could be consumed by consumers, it is a product.

Insurance service has a consumer and exchanging value. Consumer value consists in provision of insurance protection. Exchanging value is a price of the insurance service, which is paid by insurant to insurer. This price is expressed in insurance fee. The price of insurance service is determined depending on the insurance risk and possible loss.

Consumer value consists of two benefits: non-material – it means that insurant has to be confident and calm that he acquired a guaranteed protection from dangerous accidents in the system of the concept-strategy of provision of economic and food security; material benefit, which is not a service and consists in material compensation of the loss.

Additional services, which are included into insurance services, assign it the individualized character. The insurers that provide additional services might have a more stable circle of insurants.

The insurer, who wants to achieve market success, has to control to which extent his product, which is provided in the form of an insurance service, conforms to the needs of agricultural organizations.

Service's life cycle insurance differs from other types of goods. Firstly, service's life cycle is much longer. A company could sell insurance policies over a long period of time – years or even decades – while it is impossible for most goods with higher demand. Secondly, the initial expenditures for design and implementation of a service in the insurance market are much lower than for the products of mass consumption.

Insurance service's life cycle has the same stages as any other product. At the first stage, the target audience is selected; insurance service is developed; based on the goals that are set by the insurer, terms and tariffs of insurance are determined; the service is implemented in the insurance market. At the second stage – as the product is new to the public – the insurer's purpose is to create demand for this product with the help of advertising. Then, by means of dissemination of information, the demand for various types of insurance services grows. At the fourth stage, consumer demand for services of the insurance company reduces – therefore, profit is reduced. At the last stage, the product stops satisfying the terms of the concept-strategy of provision of economic and food security. Here insurers could perform updates, improvements, and expansion of the list of insured risks. If the performed modernization leads to positive results, the product moves to a new stage – the stage of growth.

In 2016, amendments to the Federal law "Regarding state support in the sphere of agricultural insurance and implementing changes into the federal law "Regarding development of agriculture" came into effect.

For participation of agricultural manufacturers in the crediting programs, including the ones subsidized by the state, banks can establish the requirement on the presence of various types of guarantee – in particular, pledge of products of future harvest of agricultural cultures or pledge of animals. The banks' requirements as to the terms of insurance of the pledged property usually differ from the requirements which observation is necessary within the Federal law No. 260–FZ. In the process of credit issue, requirements are set on insurance protection in case of loss of agricultural products from: (1) unlawful actions of third parties (theft, arson, and other actions aimed at destruction of ensured products); (2) actions of animals, birds, and rodents; (3) other conditions that are connected to specific features of production of a certain type of agricultural products.

This situation leads to the fact that insurer, in order to avoid double expenditures, has to choose between insuring for the purpose of crediting and insuring with state support – which reduces the effectiveness of measures that are taken by the state for protection of the interests of agricultural manufacturers (Report for public consultations (2017)).

The main directions of reformation of insurance in the sphere of the AIC and state support for agricultural manufacturers are as follows:

- development of individualized offers for insuring the participants of agri-business, in view of specifics of the regions that are involved in agriculture;
- Improvement of the methodological basis and recalculation of tariffs due to new provisions on the possibility of insuring agricultural objects from separate risks in the law – these changes came into effect on March 1, 2019;
- specialized support for small economic forms that are involved in agribusiness – creation and successful development of societies of mutual insurance and transfer of risks to hedgers in case of catastrophic losses;
- development of an alternative form of insuring risks in agriculture – index insurance, which is based on high dependence of harvest from certain factors, which could be assessed quantitatively. Despite certain difficulties at the initial stage of studying a certain regions from the point of view of indices, the advantages of this type of insurance could not ignored – for index insurance simplifies the procedure of conclusion of an insurance agreement and the process of regulation of losses;
- the system of insurance with state support for fish farms is used for the first time;
- Implementation of the mechanism of co-financing by insurers in the sphere of agricultural insurance of preventive measures;

- Development of dual agricultural subsidizing for three directions of support – unified subsidy, unrelated subsidy, and subsidy per 1 L of milk, by 2020. These types of state support will be divided into two parts: compensating and stimulating, for transparent regulation and the possibility of quick maneuvering in the sphere of production;
- Support – by 2022 – for a range of specific and narrow spheres of the AIC – e.g., lying of vineyards, support for flax mills, or compensation of capital expenditures for companies that produce milk protein for baby food.

5 Conclusion

New times stimulate quick popularization of the compensation mechanism of insurance services for provision of food and economic security of a country. This popularization could be done with the help of electronic channels of communication. Promotion of insurance services through the Internet stimulates the increase of coverage of audience that is interested in insurance protection of agricultural organizations and strengthening of the positions of insurers from the point of view of economic security. Recently it has become possible to regulate insurance cases via the Internet (in case of small loss) and to perform accounting of insurants and information on them.

Therefore, formation of a policy of insurance of the innovative and investment activities of agricultural organizations is aimed at provision of economic and food security by means of development of insurance services and state support in the structure of adaptive reformation. This approach stimulates strengthening of food and economic security of the state and stimulates the development and promotion of insurance products of innovative and investment activities of agricultural organizations.

Acknowledgements. The reported study was funded by the Russian Foundation for Basic Research grant No. 19-010-00985 A. “Development of innovative and investment policy as a concept of strategic economic security of agricultural organizations in the conditions of the modern technological transformation”.

References

Aksenova, V.A.: Insurance legal relations in agriculture. Young Schol. **2019**(3), 217–219 (2019). https://moluch.ru/archive/241/55792/. Accessed 23 May 2019

Botashev, A.Y., Shokhnekh A.V.: The modern concept of investment monitoring of economic development of agriculture in regions of Russia. In: Botashev, A.Y., Shokhnekh, A.V. (eds.) 2013 Audit and financial analysis, No. 3. pp. 258–261 (2013)

Report for public consultations: Offers on development of agricultural insurance with state support in the Russian Federation. Central Bank of the Russian Federation (2017)

Isaeva, O.A.: Top-priority directions of provision of food security. In: Materials of the 8th International Student Scientific Conference “Student scientific forum” (2016). https://scienceforum.ru/2016/article/2016028588. Accessed 26 Mar 2019

Klimova, N.V.: Food security - the basis of provision of region's economic security. Fundam. Res. **2012**(9), 214–219 (2012). (part 1)

National union of agricultural insurers adopted the directions of activities for the next year (2019, Press release). http://www.insur-info.ru/pressr/68807/. Accessed: 26 Mar 2019

Proshin, K.S., Shaikin, A.M.: Basic approaches to introduction of index agricultural insurance in the Russian Federation. Fin. J. **6**, 121–130 (2017). Research Financial Institute

Rogachev, A.F., Shokhnekh, A.V.: Genesis of mathematical models of econophysics as a path to food security. In: Rogachev, A.F., Shokhnekh, A.V. (eds.) 2015 Audit and financial analysis, No. 1. pp. 410–413 (2015)

Spletukhov, Y.A.: Agricultural insurance in Russia and abroad: comparative characteristics. Fin. J. **2018**(1), 87–95 (2018)

Mathematical Model-Based Study of the Problem of Collective Effect in Philosophical, Social and Economic Theories

Olga S. Evchenko(✉), Tatyana N. Ivanova, Natalya N. Kosheleva, Margarita V. Manova, and Natalya B. Gorbacheva

Togliatti State University, Togliatti, Russia
evchenko75@mail.ru, IvanovaT2005@tltsu.ru, cavva01@mail.ru, manmarga@mail.ru, gnb0906@mail.ru,

Abstract. The purpose of this study is to examine the socio-economic concepts methodologically important in the understanding of corporate relations and the role of morality therein. The authors analyzed various sociological studies on the specific nature of business relations, the patterns of peoples' behavior in large organized groups. As a result, the article presents various views of the issue of building relations in large companies (corporations). Thus, economist M. Olson in his study showed a fundamental difference between the interaction of individuals in large and small groups, in particular, paid attention to the problem of a free rider in large ones. The disciples of M. Olson – the sociologists K. Offe and H. Wiesenthal, tried to divide such groups into employees and employers and described a trend to establish such dominant bodies in the company that take the top leadership out of rank-and-file members' control and allow avoiding the general requirements. For example, T. Parsons and V. Pareto came to the conclusion that social relations are ordered because of forming the people's acts under the impact of common values. Special attention is paid to the topic raised by philosopher Y. Habermas, "systemic colonization of the lifeworld" – the trend to replace daily communication by detailed bureaucratic and legal regulation and limited rationality of "purposeful rational action system". J. Habermas pointed out the main danger to corporate culture: transformation from a mechanism of moral regulation of human relations into an annex of bureaucratic bodies.

Keywords: Corporate relations · Corporate culture · Collective morals · Individual morals · Utilitarianism · Socio-economic concepts · Models

The issue of collectivity and the specific nature of the moral regulation of relations in companies is interdisciplinary. To reveal it we need to address both philosophical and sociological, cultural, socio-psychological, politological, mathematical, and economic research. The topic of the moral part of organizational relations was raised in the sociological writings by K. Marx, F. Engels, was investigated by E. Durkheim and F. Tennis at the turn of the 19th–20th centuries and was discovered and solved as a practical issue in the control theory of the early 20th century. For example, the French sociologist and philosopher Emile Durkheim in his writings draws attention to the fact that the moral rules and values carried by employees not only precede the contractual

E. G. Popkova and B. S. Sergi (Eds.): ISC 2019, LNNS 87, pp. 817–824, 2020.
https://doi.org/10.1007/978-3-030-29586-8_93

obligations of the labor agreement but also govern them to some extent. The individual morality of the employee performs the functions of supervision over the fulfillment of contractual obligations. Studying the phenomenon of collective consciousness, E. Durkheim distinguished between the enforcement of conscience and natural laws and social enforcement through violence. Limiting the actions of individuals under the pressure of conscience is self-compulsion, commitment to their own rules and values, which at the same time are public. The American sociologist Tolcott Parsons (Parsons 2000) criticized the utilitarian model of action (the theorist meant, foremost, T. Hobbes, because he believed that it was him who enunciated the utilitarianism principles more precisely than others). The theorist believes that if we construe action as one aimed exclusively at the utility, then it turns out to be completely deterministic, and we are not able to understand and explain how social order is established and people's goals are coordinated.

"The Logic of Collective Actions" by M. Olson

In the middle of the 20th century, the specific nature of people's behavior in large groups, both organized and unorganized, become the subject of a special study. The American economist Mansour Olson was one of the first trying to describe and explain the specific nature of collective interactions in large groups (companies) in his book "The Logic of Collective Actions" (Olson 1995). Based on his hypothesis, Olson believes that the logic of collective action differs from the logic of individual action, shows that the relationships of individuals in large and small groups are fundamentally different and collective actions are not automatically determined by jointly experienced social problems and common interests. On the one hand, in collective actions always arises the so-called "free rider" problem, i.e. an individual enjoying the public benefits but not willing to bear the costs and letting others pay for him. And this problem is particularly common to large companies. M. Olson writes that "an individual member of a typical large company is found out in [the following] situation: his own efforts will not affect significantly the company's position, whereas he will enjoy the improvements achieved by the company supported by others" (Olson 1995: 14). On the other hand, as a rule, there are always individuals who take actions to achieve a collective benefit. Why do they make it? What social bodies promote joint action for a common purpose? M. Olson puts forward and tries to solve these issues, and they are further solved in the framework of neo-utilitarianism.

M. Olson notes that the smaller the group, the bigger the contribution of a person to achieve a collective benefit. He writes that "in a small group each participant hereof receives a significant part of the common benefit just because the group includes only a few individuals, the common benefit can be provided through voluntary interaction of the group members" (Olson 1995: 30–31). In addition, all actions in small groups are made in the presence of all the members. However, the bigger the group, the more invisible the contribution of each member to the common benefit and the more difficult to control the individual members of the group. Therefore, it is in the large groups there is the temptation to evade the obligations. To be able to better control each other, M. Olson proposes to establish decentralized departments and divisions in large companies. Secondly, he sees reasonable to apply enforcement measures to "free-riders" up to exclusion from the organization or the threat of exclusion in order to leave them no collective benefit provided by the company. Thirdly, M. Olson thinks it is expedient to use the so-called secondary benefits or selective incentives, i.e. additional benefits that would make membership in the company more attractive and could keep from free riders (legal advice, guided tours, books at lower prices, etc.).

The further development of M. Olson's ideas leads both to the enrichment and deepening of the topics under his consideration, and their applications to other social realities. In particular, the reasoning of collective actions by M. Olson was used in the analysis of social movements (for example, the topic of a free rider in the revolution). But we are interested in the writing by Klaus Offe and Helmut Wiesenthal "Two Logic of Collective Actions" (Offe and Wiesenthal 1980: 67–115) where the authors on the back of M. Olson's idea and its development tried to distinguish between the groups of employees and employers. They concluded that the behavior of a company's employees is based on other principles than one of the employers. And here we are talking about both the different size of groups, although it also matters, if only because it is more difficult for large groups to be organized, and different mechanisms to call the group members. K. Offe and H. Wiesenthal believe that the theoretical tools developed by Olson make the iron law of oligarchy stated by the German sociologist Robert Michels more understandable (Michels 1991). According to this law, any company (democratic one here is no exception) tends to establish such dominant bodies that take the top leadership out of rank-and-file members' control and allow evading the general requirements, but impose their own ideas on subordinates. It results in the situation that even the democratic companies where its members determine the policy pursuant to the law or the articles of association, the functionaries acting on behalf of the company's members pursue the policy that does not meet the majority's interests and rank-and-file members mostly cannot influence it.

The result of M. Olson theory's development is the theory of strategic games of the 40s of the last century, which simulates various situations and extremely stresses the problem of collective benefit. This theory studies people's behavior in situations where the result of the action of the decision-making person depends on the behavior of other participants. The solution is proposed in terms of the rationalist theory of action, and complex mathematical models and computations are used to present the logic of the participants' actions. Like M. Olson's concept, the game theory (confidence game, a game with a coward, prisoner's dilemma, etc.) rejects the idea of a stiff causal connection between the actual actions of individuals and collective benefit.

Thomas Schelling in "Micromotives and Macrobehaviour" (Schelling 1978) shows by concrete examples that quite innocent individual actions can have unintended consequences leading to grave negative changes at the macro level. John Elster in his Incomplete Rationality: Odyssey and the Sirens (Elster 1979) discusses the self-restraint mechanisms developed by the society for such situations not to let the potential action plans to be implemented. James Coleman (Coleman 1982), one of the first in the neo-militarist tradition who tried to explain the emergence of rules, came to the conclusion that the trends in modern society are determined not by individual actors, but by corporate ones (companies).

Common values as a basis for collective action in T. Parson's philosophy T. Parsons suggested that coordination of the goals of human actions can be explained only by the freedom of choice. He argued that the market in general does not exist. Markets function differently in various cultures, which evidences that purely self-interested behavior of market actors hides other motives, our personal values that we cannot make the subject of a rational calculation of utility. T. Parsons determines absolute, ultimate goals that cannot be sacrificed by a person under any circumstances without the destruction of his identity. These goals may be different, and, therefore, the ideas entrenched in these goals will also vary. As a result, T. Parsons comes to the conclusion that social relations are ordered precisely because of the fact that people's actions are formed by common values. Similar ideas belong to the Italian economist and sociologist Wilfredo Pareto, who, in particular, believed that personal ultimate goals become integrated and form a unified system of common values. This system also includes distributive justice, and he supposes that in the context hereof only economic distribution is possible.

Common rules and values affect both the goals and the means of activity, which makes it possible to coordinate action and social order. The need for integration requires a determination of the importance and comparison of assessments and interests in relation to possible consequences of actions for the community. T. Parsons writes that "the assessment is based on standards, which can be either cognitive standards of truth, or flavor standards of conformity, or moral standards of justice" (Parsons 2000: 182). T. Parsons says that it is empirically difficult to distinguish the "personal system" and the "social system" since the actor is partly involved in the interaction with other actors. They can be divided only analytically: depending on the research objective the researcher can make the subject of a special study either the personal system or the social one (Parsons 2000: 185). T. Parsons defines loyalty or commitment (dedication) to values as follows: "We define loyalty in the quality of communication mediator as a generalized ability and credible promises to assist the implementation of values" (Parsons 1969: 456).

Parsons was reproached for the abstractness of his theory, isolation from specific social practices. To be implemented by actors, rules and values should be clarified and interpreted in a specific situation of action. Theoretical schools of symbolic interactionism and ethnomethodology offered the qualitative methods for the study of collective action.

The issue of collective relations in the concepts of symbolic interactionism and ethnomethodology.

Interactionism derived its ideas from philosophical pragmatism, and foremost from the proceedings of John Dewey and George Herbert Mead. G.H. Mead focused his attention on the study of situations of interpersonal interaction and laid the foundations of the anthropological theory of communication. G. Mead saw the specific nature of personal action that a person uses symbols, the meaning hereof is established in the interaction. This point of view was also shared by Herbert Blumer, the disciple of G.H. Mead, who called his concept "symbolic interactionism". According to this concept, the starting point for the study of personal action is not an individual actor or an individual action, but a symbolically mediated interaction, which forms personal behavior (Blumer 1994). The actions of others are not only the context and not just the means, but always also an integral part of my individual action. Each participant of the interaction acts based on his position, but in the process takes place an interaction, coordination, adaptation of actions to each other, – a collective activity. Therefore, the final result of individual action cannot be predicted; it is integrated into cross-links and depends on interpretation. Collective action is a process created by individuals, and companies are the result of agreements, continuous negotiations. He states that the bodies of any company are the result of these agreements restricting members to some extent. They are formed by the actions of biased people, and therefore forms of collective action that seem stable are actually flexible.

Another representative of symbolic interactionism Anselm Strauss (Strauss 1974) introduces the concept of a "negotiated order". In the famous collection of essays "Mirrors and Masks: In Search of Identity" Strauss proves that socialization, establishment, and determination of human identity are not completed in youth, it is lifelong, therefore people continuously reinterpret themselves, their past and the surrounding world.

Ethnomethodology is based on the phenomenology by Edmund Husserl. Its representatives pose the problem to uncover the deeper patterns of collective behavior, rather than the rule-oriented level interpreted by T. Parsons. A famous representative of this trend Irving Hoffman in his writing "Representing oneself to others in everyday life" (Hoffman 2000) explored the phenomenon of self-presentation. Studying the techniques of keeping a personal image, he arrived at the conclusion that these practices should be understood in the sense of mutual saving a face but not a unilateral presentation of yourself by a party in the capacity of a certain person before others. Revealing the hidden patterns of everyday actions (formal patterns of practical actions, grammatical patterns), Harold Garfinkel (Garfinkel 2002, 2007, 2009), Alfred Schütz (Schütz and Luckmann 1984) and their adherents concluded that in the interpretation of everyday actions we can either put unordinary events into the ordinary interpretative frameworks or deprive them of the reality status by our explanation. G. Garfinkel believes that the basis of social order is not the force and not the imperative nature of moral rules, but the rationality of everyday life that we continuously create in our actions. In the context of this rationality, only the orientation to rules is possible.

The concept of rationality by J. Habermas.

We cannot but mention another approach to the interpretation of collective action: the concept of rationality, the discourse theory of truth and morals by Jürgen Habermas and his main proceeding "The Theory of Communicative Action" (Habermas 1981, 2000). J. Habermas departs from the theories that explain rationality solely as the suitable choice of means adequate for the implementation of these goals. He opposes the hyper-rational interpretation of human action and relies on everyday practice, which shows that people trust their sometimes-irrational intelligence. He understands intelligence and rationality more broadly and calls them "communicative rationality" and "communicative intelligence".

According to J. Habermas opinion, any speech act (and, accordingly, in any action) contains three references to the world (three claims to the importance ("Geltungsanspruch")), which we are ready to defend. Firstly, we claim the truth in each statement. Secondly, we determine a social attitude and claim for regulatory propriety in each statement or action, that is, we talk about whether something is appropriate and accurate in a regulatory sense. So, in the communication people have to agree on the level hereof: they do not rely on some fixed pattern, but come to it in the process of communication. For example, some people can begin to ignore or boss us, but we, in turn, can reject this implicit or explicit definition of the situation, avoid the regulatory propriety of the partner's actions and defend other regulatory frameworks of communication. Thirdly, we try to express our identity more clearly and consistently, claim the truthfulness, the sincerity of our feelings and self-presentation. This conclusion is based on the theory of I. Hoffman, who believed that the partiality of the speaker and the actor is manifested in all his actions, and the purpose of the public is to trust or distrust his self-presentation.

As J. Habermas thinks that speech acts and self-presentation regulated by the rules of action have the nature of semantic statements understandable in their context and their rationality is determined by possible recognition between the actors. Actions and speech acts are related to claims available for criticism. However, if the claims can be challenged with the help of reasonable arguments, it means that learning processes are also possible in these areas. Knowledge embodied in regulated actions and expressive statements points not at the existence of some impersonal state of affairs, but the significance of obligation rules and personally important situation, commonly experienced in the society (Habermas 1981).

J. Habermas determines three types of actions: (a) instrumental action, the purpose hereof is to manipulate the outside world; (b) actions that are regulated by the rules and that are appropriate; and (c) dramatic action, by means hereof self-presentation and self-styling are made.

Instrumental, expedient action is directed to tangible objects. It can be expanded to a strategic action aimed not at tangible objects, but at other actors. Strategic action is pursued according to the same pattern of "goal-means", i.e. the communication partner is converted into a means of goal achievement.

Communicative action differs from instrumental, strategic, rule-regulated, and dramatic actions by the fact that here the actors seek for mutual understanding in the situation of the action, coordination of their action plans, and hence the actions. In addition, the communicative action is not aimed at any set goal, since the result of a conversation, a dispute cannot be predetermined in advance. Any participant of the discussion should be prepared for the fact that his goals will be revised and refuted. Each communication partner relies on individually-interpreted life world, but at the same time, all refer to something in the social and personal world to commonly determine the situation. Thus, if T. Parsons considers the action only as a goal- and objective-oriented, although limited to values and rules, then J. Habermas defines the communicative action is a non-teleological one, aimed at mutual understanding. In fact, J. Habermas distinguishes between two ways of social integration: coordination of human actions based on orientations and common values; and the mechanisms of systemic integration of society not affecting the orientation of actions but relating and coordinating actions with each other through consequences and consensus.

Making a diagnosis to modern era, J. Habermas raises another important topic, "systemic colonization of the lifeworld" – the trend to replace daily communication by detailed bureaucratic and legal regulation and limited rationality of "purposeful rational action system". System mechanisms roughly interfere the everyday life rather than focus on the rules and discussions of fairness. Money or lawyers in the courts more often act in the capacity of a communicative mediator. The accuracy of this diagnosis is evidenced by the changes taking place in modern corporate culture.

Acknowledgments. The article is prepared within the framework of the RFBR-2017 grant (Project-17-43-93525)

References

Garfinkel, G.: Ethnomethodological Studies. Piter, Saint Petersburg (2007). Translated from English by Z. Zamchuk, N.Makarov, E.Trifonov

Garfinkel, G.: Studying the usual bases of daily actions. Sociol. Rev. **2**(1) (2002)

Kulagina, O.V., Averina, O.V., Yurkin, M.O.: Assessing the developmental prospects of the system of inter-budgetary relations at the regional level (by the example of the Jewish autonomous region). Sci. Res. **5**(2(15)), 9–13 (2016)

Garfinkel, G.: The concept and investigational studies of trust as a condition for stable coordinated actions. Sociol. Rev. **8**(1), 161–182 (2009). Translated from English by A.M. Korbut

Hoffman, I.: Representing yourself to others in everyday life (2000). Translated from English and introduction by A.D. Kovalev. M.: Kanon-Press-Ts, Kuchkovo Pole

Blumer, G.: Collective Behavior. American sociological thought: Texts. M.: Publishing House of Moscow State University (1994)

Michels, R.: Sociology of political parties in terms of the democracy. Dialogue, No. 5–9 (1990), No. 4 (1991)

Olson, M.: The logic of collective actions. Public benefits and group theory (1995). Translated from English by E. Korolenko. M.: FEI

Cherepanova, D.A.: Mass Media in Russian and European Policies (Syrian Issue). Azimuth Sci. Res.: Econ. Manag. **5**(3(16)), 254–256

Parsons, T.: The Framework of Social Action. Concerning the Framework of Social Action. M.: Academic Project (2000)

Kuzmina, T.L.: Current state and prospects of innovative cooperation among the member countries of the EEU. Azimuth Sci. Res.: Econ. Manag. **6**(4(21)), 141–142 (2017)

Habermas, J.: Moral Consciousness and Communicative Action, p. 381. Nauka, Saint Petersburg (2000)

Kirichenko, O.N.: Impact of accession of the central and eastern Europe countries (CEE) to the EU and their experience in the process of ukraine integration to European community. Azimuth Sci. Res.: Econ. Manag. **1**(6), 50–53 (2014)

Coleman, J.S.: The Asymmetric Society. Syracuse University Press, Syracuse (1982)

Bogatyrev, V.D., Yu Ivanov, D., Kurilova, A.A.: Approaches to managing the development of high-tech enterprises based on the economic and mathematical model of price competition at the market of easy aviation. Azimuth Sci. Res.: Econ. Manag. **4**(25), 44–47 (2018)

Elster, J.: Ulysses and Sirens. Cambridge University Press, Cambridge (1979)

Muliar, O.P., Rusinova, M.M.: The establishment of the ideological culture of senior students in the process of teaching the social and human disciplines. Baltic Humanit. J. **5**(1(14)), 120–123

Habermas, J.: Theorie des kommunikativen Handels. 2 Bände. M.: Suhrkamp, Frankfurt (1981)

Popkova, E.G., Popova, E.V., Sergi, B.S.: Clusters and innovational networks toward sustainable growth. In: Sergi, B.S. (ed.) Exploring the Future of Russia's Economy and Markets: Towards Sustainable Economic Development, pp. 107–124. Emerald Publishing, Bingley (2018)

Raven, J.: Problems with "closing the gap" philosophy and research (some observations derived from 60 years in educational research). Baltic Humanit. J. **6**(3(20)), 252–275 (2017a)

Offe, C., Wiesenthal, H.: Two logics of collective action: theoretical notes on social class and organization form. Politi. Power Soci. Theory **I**, 67–115 (1980)

Parsons, T.: On the concept of value commitments. In: Parsons, T. (ed.) Politics and Social Structure, p. 456. Free Press, London, New York (1969)

Zelenskaya, L.L.: English for specific purposes in the historical perspective. Baltic Humanit. J. **6**(2(19)), 41–44 (2017)

Schelling, T.: Micromotives and Macrobehavior. W. W. Norton & Co, New York/London (1978)

Raven, J.: ducation and sociocybernetics. Azimuth Sci. Res.: Econ. Manag. **6**(3(20)), 289–296 (2017b)

Strauss, A.: Spiegel und Masken. Die Suche nach Identität (Mirrors and Masks. The Search for Identity). M.: Suhrkamp, Frankfurt (1974/1959)

Apanasyuk, L.A., Apanasiuk, Yu.V: Peculiarities of intercultural communication in the field of tourism. Baltic Humanit. J. **7**(2(23)), 193–195 (2018)

Schütz, A., Luckmann, T.: Strukturen der Lebenswelt. 2 Bände. M.: Suhrkamp, Frankfurt (1979/1984)

Popkova, E.G., Sergi Bruno, S.: Will industry 4.0 and other innovations impact Russia's development? In: Sergi, B.S. (ed.) Exploring the Future of Russia's Economy and Markets: Towards Sustainable Economic Development, pp. 51–68. Emerald Publishing, Bingley (2018)

Sokolova, E.O.: The Russophone movement within the framework of linguistic imperialism. Azimuth Sci. Res.: Econ. Manag. **7**(4(25)), 393–395 (2018)

The Essence of Loan Capital and the Model of Effectiveness of Its Turnover

Ella Y. Okolelova(✉), Larisa V. Shulgina, Marina A. Shibaeva, Oleg G. Shal'nev, and Alexey V. Shulgin

Voronezh State Technical University, Voronezh, Russia
fes.nauka@gmail.com

Abstract. In the article on the basis of the analysis of researches of the Russian and foreign authors questions of essence and the maintenance of the market of the loan capital are considered.

It is shown that the loan interest is a payment for the rapid satisfaction of the needs of the population. It is considered that the loan market is quite specific in comparison with the commodity market, since it is the main product of borrowed funds, and the form of their purchase and sale is a loan. The price of the loan is interest rate. In fact, loan capital is exchanged for the same loan or capital only in increase. The subjects of loan transactions in this market are creditors and borrowers. In its essence, the loan market is a market mechanism of interaction of economic entities, which alternately act as creditors, then as borrowers.

The mortgage loan as a form of lending to individuals, based on the analysis of the prospects of development of credit mechanisms and preferences of borrowers. Presented the model of credit capital turnover, built on the principle of possible refinancing of funds with early repayment of debt by the borrower. To build the model, new concepts are introduced, such as "speed of repayment of credit funds" and "refinancing risk", defined as the probability of demand for early released funds from credit turnover. The optimal period of early repayment of the loan is determined.

Keywords: Loan capital · Loan interest · Credit · Long-term crediting · Early repayment of the loan · Refinancing risk

1 Introduction

The concept of "credit" is now firmly established in the economic life of not only enterprises but also individuals. Increasingly, we resort to the help of loans, especially when making large purchases. Today, almost any product can be purchased in installments. But at all convenience of use of the mechanism of crediting of goods not all use it. The mentality of the Russian person is such that the attitude to credits in our country developed ambiguous. Consumer loans are not in demand by individuals due to the reluctance of people to have credit encumbrances. Many people prefer to refuse to buy in the absence of sufficient funds, rather than take a loan favorite product. This situation is not only due to the mentality of the majority of the population, but also due to the relatively high interest rates in Russian banks.

E. G. Popkova and B. S. Sergi (Eds.): ISC 2019, LNNS 87, pp. 825–837, 2020.
https://doi.org/10.1007/978-3-030-29586-8_94

At the same time in the US and European countries, individuals are constantly using loans. Acquire goods immediately for cash, especially when making major purchases, is considered irrational, and in some countries it is simply not accepted.

At present, the development of lending in the loan market depends on the policy of credit institutions-banks and the identification of ways of effective turnover of loan capital. The problem of early repayment of loans forces banks to use speculative methods in contractual relations with customers, according to which early repayment of a loan is possible with the payment of the entire amount of loan interest. The other part of the banks agrees to early repayments and does not require the payment of the entire amount of interest from credited customers. Does this mean that for a bank such operations cease to be profitable?

The purpose of this article is to substantiate the nature of loan capital and interest and develop a mathematical model of the effectiveness of the turnover of credit resources under the condition of early repayment of credit amounts by the borrower.

The authors used dialectical and system methods of research, as well as the method of scientific abstractions, the method of analysis and synthesis, the method of mathematical modeling, the method of induction and deduction.

2 Literature Review

The relationship between the developed market of loan capitals and the growth of the standard of living of the population was considered by many researchers. Credit relations were scientifically grounded in the writings of economists of the past centuries, primarily in the writings of the physiocrats F. Quesnay and A. Turgot, the classics of political economy A. Smith and D. Ricardo, J. St. Mill, K. Marx, G. Thornton, and also in the works of Russian researchers I.T. Pososhkova, A.N. Radishcheva, N.I. Turgenev and others. An important contribution to the development of the theory and methodology of credit as part of the financial market and the market equilibrium factor was made by foreign representatives of economic thought K. Wicksell, R. Hilferding, D.M. Keynes, NG Mannyu, J. Hicks, M. Friedman, V. Salomo and other researchers.

So, Henry Thornton, author of “Researches about the nature and effect of the paper credit of great Britain” published in London in 1802, was the beginning, in the words of F. Hayek, “a new era in the development of the theory of money.” He carried out a better analysis of the credit system in comparison with his predecessors. John. St. Mill considered of Thornton a specialist in Finance. However, later researchers did not share Mill’s views on the writings of G. Thornton. In his work the structure of the monetary system of great Britain is presented and the characteristic of two-level banking system is given, the idea of the centralized regulation of means of the credit and banking sphere is carried out. He talks about the possibilities and limits of stimulating economic growth with monetary levers, argues with A. Smith and D. Ricardo about the price of paper money and the price of gold, and argues that a reasonable policy of the Central Bank has no alternative [1].

The Marxist theory assumed that with the loan of money as capital, for the lender, the funds received act as capital-property, bearing interest, and for the borrower-as a capital-function that brings entrepreneurial income [2].

Keynes, D.M. considered the growth of temporarily free cash (savings) as a base for investing in the real sector of the economy, and the level of demand for investment depends on the expected efficiency of capital and the interest rate, thus justifying the relationship between GNP and the amount of money in circulation, the role of macro regulation of the money supply, taking into account credit operations. In Keynes this dependence looked like the dependence of the growth of GNP on the growth of the money supply [3].

Friedman M. justified the directly opposite "monetary rule" associated with the growth of the money supply on the basis of the growth of GNP, which is actively used and now in countries with developed economies [4].

3 Research Methods

In the modern economy, most of the money in circulation is credit. Consequently, the loan market has not only its specific product - borrowed funds, but also its instruments of circulation - debt obligations in the form of credit agreements or credit money.

Another feature of the loan market is the trust relationship, which is a moral and ethical aspect. After all, a loan is a transaction in which one entity trusts another entity a certain amount of money, which must be returned after a certain period of time and with the payment of interest. The role of trust is quite large in the development of credit relations, it becomes one of the market instruments that has an impact on the economic behavior of the subjects of the loan market.

Objectively inherent characteristics of the loan market lead to the fact that this market does not look like a market for goods, works or services. In these markets, the exchange of results of the activities of its subjects occurs through sale and purchase. On one side are the values of goods, and on the other - the equivalents of value, money. In fact, it is an exchange of equal values.

The loan market is quite specific, since the main commodity on it is borrowed funds, and the form of their purchase and sale is a loan. The price of the loan is interest. In fact, a loan or loan capital is exchanged for the same loan or capital only in increments. The subjects of loan transactions in this market are lenders and borrowers. In its essence, the loan market is a market mechanism of interaction of economic entities, which alternately act as lenders, then as borrowers.

For example, the system of mortgage lending is now quite popular financial instrument and continues to develop, greatly facilitating the purchase of housing.

But, implementing only long-term loans, banks can not provide the necessary level of profit. It is necessary to actively move capital not only in the form of long-term, but also short-term credit products that banks actively position. Bank retail is the basis of any commercial Bank. In an effort to attract as many customers as possible, banks are now on relatively loyal lending terms. One of these conditions is the possibility of early repayment of the loan, which is quite convenient for the borrower and can significantly reduce interest payments.

In addition, with early repayment of the loan, the borrower is relieved from the credit burden much earlier, entering into the legal rights to use the facility. But how much is this situation profitable for the bank?

On the one hand, in the event of early repayment, the bank loses a certain amount of interest. On the other hand, funds returned before the due date can be used to provide new loans to individuals and legal entities. In addition, the risks of the insolvency of the borrower are significantly reduced, which may occur during a long crediting period.

Consider the mechanism of lending subject to early repayment of the loan. It is necessary to determine the “break-even point” of early repayment, which is expressed in the minimum period when the loss of interest of banks will be compensated by the reduction of risks and the condition of further lending. Therefore, in order for the long-term credit system to develop and be as risk-free as possible, it is necessary to develop a mechanism that would be most beneficial to both parties to the transaction.

Such a mechanism should be based on the principle of accessibility for a wide range of borrowers. This principle is implemented through the provision of lower interest rates on loans and the creation of flexible conditions for debt repayment, which will allow banks to significantly expand the market for consumers of credit products.

Let’s consider the implementation of the credit mechanism under the condition of early repayment of the debt.

The amount of principal (body of credit) returned by the borrower on a monthly basis is calculated as follows:

$$c = \frac{V_{кр}}{T_{кр}}, \tag{1}$$

where Vкр - the volume of the loan, rubles;
Tкр - maturity, month.

The monthly amount of interest on the loan is:

$$p = \left[V - \frac{V \cdot (n-1)}{T} \right] \cdot \frac{i}{1200}, \tag{2}$$

where i is the annual interest rate on the loan, %;
n is the serial number of the month.

Consider the function of the accumulated amount of payments, built on the example of calculating a mortgage loan amounting to 1500 thousand rubles at a rate of 13% per annum (Fig. 1).

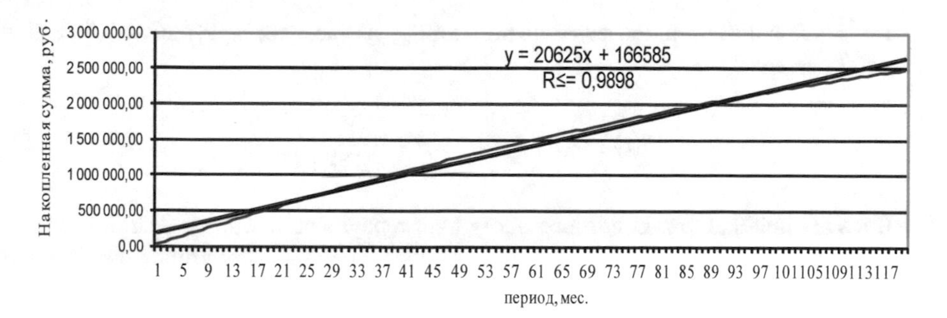

Fig. 1. The function of the amount of accumulation of payments

The parabolic function is approximated by a linear dependence, and, as calculations of real credit conditions have shown [1], the error does not exceed 0.3%.

Let's introduce the concept of "speed of credit repayment" as a characteristic of the build-up function. Based on the mathematical definition of speed as the first derivative, we obtain the value of the funds returned to the bank monthly

$$y' = 20625\ (\text{rub}).$$

If the value of the monthly repayment by the borrower is constant, the interest payments vary significantly over time. The borrower usually pays the principal interest at the beginning of the loan period. However, even the remaining interest after the early repayment of the loan can be very significant for the borrower.

The amount of monthly funds returned to the Bank will be designated $\bar{a}$.

$$\bar{a} = c(t) + p(t), \tag{3}$$

where c is the monthly amount of principal debt, rubles;

$p\ (t)$ - monthly interest rate on the loan, rub.

Due to the use of linear dependence in the calculation, the average values of the payment parameters were used, i.e.

$$c(t) = \frac{V_{\kappa p}}{T_{\kappa p}}, \tag{4}$$

$$p(t) = \bar{p} = \frac{i}{12} \times \frac{V_{\kappa p}}{2} = \frac{iV_{\kappa p}}{24} \tag{5}$$

where $V\kappa p$ - the volume of the loan, rub.;

$T\kappa p$ - initial term of loan repayment under the terms of the contract, months.

Let's define the size of the saved up sum taking into account repayment of the basic debt and percent

$$S(t) = (\bar{d} + \bar{p})t = \left[\frac{V_{\text{кр}}}{T_{\text{кп}}} + \frac{iV_{\text{кр}}}{24}\right]t, \quad (6)$$

Let's assume that loan conditions allow loan repayment before the deadline set by the contract. This is very beneficial for the borrower, since such terms and conditions of the contract significantly reduce the amount of interest payments. This condition is also beneficial to the bank, as it minimizes the risks of inflation and currency risks.

We consider the condition of early repayment of the loan from the perspective of the bank and determine on what conditions this situation may be beneficial to him.

Let the mortgage loan 1500 thousand. R., planned for 10 years, repaid within 7 years (84 months.). The graph of the accumulation function in case of early repayment of the loan is shown in Fig. 2. The graph is also approximated by a linear function, but the rate of return of borrowed funds increases.

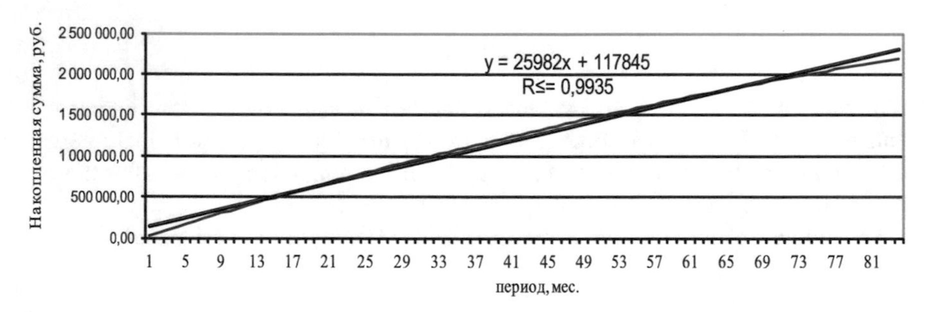

Fig. 2. Graph of the accumulation function with a reduction in the maturity of the loan

The amount of the accumulated amount taking into account the return of debt and interest in the event of early repayment of the loan (84 months) will be equal to:

$$S = \left[\frac{1500}{84} + \frac{0,13 \cdot 1500}{24}\right] \times 84 = 2183\, thous.rub$$

We calculate the amount of funds that the bank will not receive in this case. So, the amount of the accumulated amount with a 10-year maturity and the same interest rate is:

$$S = \left[\frac{1500}{120} + \frac{0,13 \cdot 1500}{24}\right] \times 120 = 2475\, thous.rub$$

The amount received by the bank with a reduction in the maturity of the loan is equal to

$$\Delta S = 2475 - 2183 = 292\,thous.rub$$

The linearization of the accumulation function has a slight (0.1–0.2%) error. Thus, according to real calculations, the accumulated amount in the case of a 10-year loan period amounted to 2483.1 thousand rubles (instead of 2475 thousand rubles, calculated by the formula (4)). Similarly, when a 7-year period, the amount in accordance with the actual repayment schedule of the debt amounted to 2190,6 thousand. in contrast, calculated by the formula 2183 thousand. The Difference Is calculated according to the repayment schedule is 292,5 thousand R. the Error in the amount of 0.5 thousand rubles is quite acceptable.

At the same time, we should not forget that the lending process has a significant time frame. Therefore, the comparison of cash flows is correct only when using the discount mechanism.

We apply the discounting method to the amount of the accumulated amount when differentiating the maturity dates, taking into account the discount rate of 14%. In this case, we receive a loan repayment in 120 months:

$$S_{нач}^{disk} = \frac{2475}{(1+0{,}14)^{10}} = 667{,}6\ thous.rub$$

When credit will be repaid in 84 months:

$$S_{сокр}^{disk} = \frac{2183}{(1+0{,}14)^{8}} = 872\ thous.rub.$$

From the comparison of the two results, it can be concluded that early repayment of the loan for the Bank is most effective. [6] this efficiency is connected both with the fact that the amount returned ahead of schedule becomes the basis for further lending, and with the fact that for the remaining 36 months (the difference between 120 and 84 months) the Bank under the same conditions of lending will receive an additional amount:

$$S_{доп} = \left[\frac{2183}{120} + \frac{0,13 \times 2183}{24}\right] \times 36 = 1080\,thous.rub.$$

Thus, despite the amount of uncollected funds in the amount of 292 thousand rubles, the Bank can earn thousand 1080 p. further lending prepaid funds. In this case, the Bank's income will be:

$$1080 - 292 = 788 \text{ thous.rub.}$$

We have considered an example with specific data of the loan agreement. But there is a question of comparing the terms of repayment of the loan with different expectations of return. Need to find the optimum term of repayment of the loan.

We construct a model for estimating the income (or accumulated amounts of $S1$ and $S2$, and $S_2 < S_1$) of the lender with a reduction of different terms (T_1 and T_2, moreover, $T_2 < T_1$). debt recovery.

The amount of the amount not received by the Bank will be:

$$\begin{aligned} \Delta S = S_1 - S_2 &= \left[\frac{V}{T_1} + \frac{i\,V}{24}\right] \times T_1 - \left[\frac{V}{T_2} + \frac{i\,V}{24}\right] \times T_2 \\ &= \left(V + \frac{i\,VT_1}{24}\right) - \left(V + \frac{i\,VT_2}{24}\right) \end{aligned} \tag{7}$$

Converting the expression (7), we obtain:

$$\Delta S = \frac{i\,V}{24}(T_1 - T_2) \tag{8}$$

Let's assume that the Bank again credits the received amount S_2 for the period T_3. In this case, the amount of income S_3 in accordance with the formula (6) when refinancing funds for the period of reduction of the initial loan will be equal to:

$$S_3 = \left[\frac{S_2}{T_3} + \frac{i\,S_2}{24}\right] \times (T_1 - T_2) \tag{9}$$

We express the quantity S3 in terms of the initial volume of the loan V.

$$\begin{aligned} S_3 &= \left[\frac{S_2}{T_3} + \frac{i\,S_2}{24}\right](T_1 - T_2) = S_2(T_1 - T_2)\left(\frac{1}{T_3} + \frac{i}{24}\right) \\ &= \left(\frac{V}{T_2} + \frac{i\,V}{24}\right) \cdot T_2(T_1 - T_2)\left(\frac{1}{T_3} + \frac{i}{24}\right) \end{aligned} \tag{10}$$

After the transformation we get:

$$S_3 = V \cdot T_2(T_1 - T_2)\left(\frac{1}{T_2} + \frac{i}{24}\right)\left(\frac{1}{T_3} + \frac{i}{24}\right) \tag{11}$$

If the condition S3 > ΔS is met, early repayment of the debt is a more favorable condition for the Bank than repayment of the loan on time. In addition, the risks of inflation, insolvency of the borrower and many others are significantly reduced.

For the Bank, both risk reduction and continuously reproducible refinancing are essential conditions.

Consequently, the question of the demand for long-term loans, which determines the possibility of refinancing funds released before the deadline, remains open.

To solve this problem, banks actively use advertising products, libcralize credit conditions and take a number of other methods and techniques to attract borrowers.

In General, the amount of shortfall due to early repayment of loans ΔS will be equal

$$\Delta S = \sum_{i=1}^{n} \left(S_i^{нач} - S_i^{сокр} \right), \tag{12}$$

where $S_i^{нач}$ - the accumulated amount of the loan, without changing the terms of the loan (under the terms of the initial contract), rubles;

$S_i^{сокр}$ - the accumulated amount of the i-th loan with a reduction in the terms of lending, rubles;

n - is the number of loans issued.

The value $\sum_{i=1}^{n} S_i^{сокр}$ shows the amount of funds received ahead of schedule from borrowers, i.e. means for further refinancing.

We calculate the possible reduction in the interest rate, which does not affect the bank's financial stability. To do this, we introduce a new indicator - the refinancing risk factor [2], which is equal to the ratio

$$k_r = \frac{p \cdot \sum_{i=1}^{n} S_i^R}{\Delta S}, \tag{13}$$

where S_i^R - the amount of funds to provide loans, rub.

p - probability of refinancing the released funds, $p \in [0; 1]$.

Reduction of loan repayment terms is allowed only in the case when $k_r \geq 1$.

Define the factors affecting the coefficient of refinancing risk [2] (based on 8 to 11).

$$k_r = \frac{p \cdot S_3}{\Delta S} = \frac{p \cdot VT_2(T_1 - T_2)\left(\frac{1}{T_2} + \frac{i}{24}\right)\left(\frac{1}{T_3} + \frac{i}{24}\right)}{\frac{iV}{24}(T_1 - T_2)} \tag{14}$$

Converting (14), we obtain:

$$k_r = \frac{p \cdot S_3}{\Delta S} = \frac{p \cdot (24 + i\, T_2)(24 + i\, T_3)}{24i\, T_3} \tag{15}$$

The calculations showed that the refinancing risk factor does not depend on the size of the loan, but depends on the rate, maturity and probability of demand for loans.

We introduce the concept of "loan repayment rate" (kr) and "loss rate" due to the reduction of the accumulated amount (vs). These figures are shown in Fig. 3.

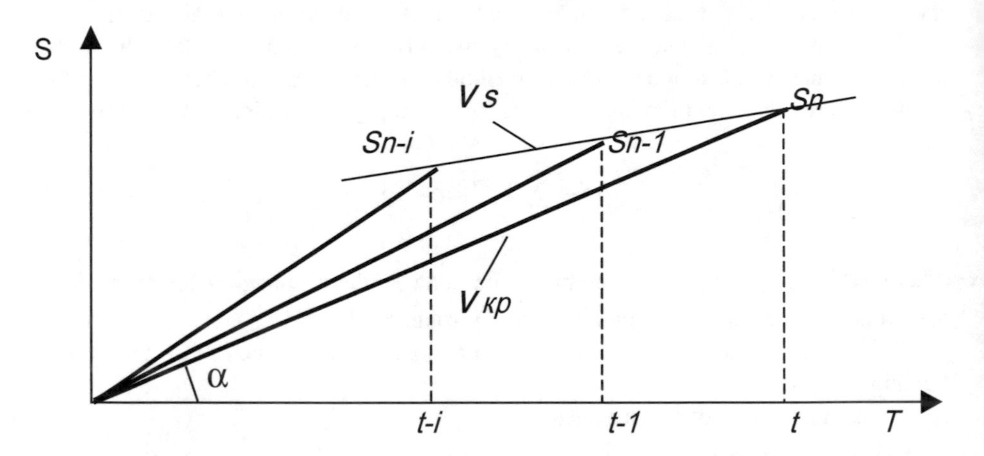

Fig. 3. Model of different loan repayment options

The loan repayment rate is determined in accordance with (6) and is equal to

$$v_{\kappa p} = \frac{\partial S}{\partial t} = \left[\frac{V_{\kappa p}}{T_{\kappa p}} + \frac{iV_{\kappa p}}{24}\right], \text{provided T = const} \tag{16}$$

For $T = t$, the accumulated amount is determined as follows:

$$S(t) = \left[\frac{V_{\kappa p}}{t} + \frac{i\,V_{\kappa p}}{24}\right]t + \varepsilon = \left(V_{\kappa p} + \frac{i\,V_{\kappa p}t}{24}\right) + \varepsilon \tag{17}$$

Then the growth rate of the accumulated amount will be equal to [3]:

$$v_S = \frac{\partial S}{\partial t} = \frac{i\,V_{\kappa p}}{24} \tag{18}$$

The function of the accumulation rate has the following form:

$$v_S(t) = \frac{i\,V_{\kappa p} \cdot t}{24} \tag{19}$$

The value of the rate of loss of Bank v_{Π} due to early repayment of the loan is inversely proportional to:

$$v_{\Pi} = -\frac{i\,Vt}{24} + \Delta S \tag{20}$$

Substituting the expression (8) into (20), we obtain:

$$v_{П} = -\frac{i\,Vt}{24} + \Delta S = -\frac{i\,Vt}{24} + \frac{i\,V}{24} \cdot (T_{н} - t), \text{ где} \tag{21}$$

$T_{н}$ – initial repayment period according to the terms of the contract.

We define the break-even point as the period of time when the rate of income growth and the rate of loss will take the same values, i.e. $v_{н} = v_{П}$.

Equating expressions (19) and (21) we obtain $t = \frac{T}{3}$.

This time period shows the maximum value of the early repayment efficiency from the Bank's point of view. The risks of inflation, insolvency of the borrower and a number of others are significantly reduced. But this option of repayment will be beneficial to the Bank in the event that the funds returned as a result of early repayment will again be converted into loan funds, that is, they will serve as a long-term loan to the new borrower. Therefore, the period of reduction of the loan repayment period should not be more than 1/3 of the contract [3].

Early repayment of loans is beneficial to the Bank in the event that new borrowers are attracted, that is, the funds received should be involved in a new credit scheme. The condition is satisfied with the probability of subsequent refinancing p = 1.

We will introduce a new indicator characterizing the possibility of attracting new borrowers and defined as the refinancing risk.

It should be noted that in this case only long-term loans, such as mortgages, are considered. It is on such credit mechanisms that the model for determining the optimal credit conditions will work effectively.

Today, banks actively attract customers using all methods and mechanisms of Bank retail. But this is not enough for high credit capital turnover. High interest rates on loans make them insufficiently accessible to a wide range of consumers.

Consider the situation when credit resources are not in full demand and determine the possibility of early repayment of the loan in this case.

Let us determine the value of the Bank's income while reducing the loan repayment period [4]. The income of the Bank while reducing the maturity of the formula (6) is equal to

$$S^{н}_{сокр} = \left[\frac{V_{кр}}{T_{кр}} + \frac{i\,V_{кр}}{24}\right](T_{кр} - \Delta t), \text{ где} \tag{22}$$

$T_{кр} - \Delta t = t$; Δt – the reduction period of the loan.

This amount can be used for the purpose of further granting of the credit, but taking into account probability of its demand p. Therefore, the amount of funds for further loans is equal to

$$V_{кр}^{рефин} = p \cdot S_{сокр}^{н} \tag{23}$$

The amount of the refinanced income of the Bank $S_{рефин}^{н}$ received as a result of crediting $V_{кр}^{рефин}$ shall not be less than its income in case of repayment of the credits in the term established by the agreement, i.e., $S_{рефин}^{н} \geq S_T$ where S_T-the income of Bank at return of the credit in due time. This expression in the substitution (22) in (6) takes the form [24]:

$$\left[\frac{p \cdot S_{сокр}^{н}}{T_{кп}} + \frac{i \cdot p \cdot S_{сокр}^{н}}{24}\right] \cdot t \geq \left[\frac{V_{кр}}{T_{кр}} + \frac{i \cdot V_{кр}}{24}\right] \cdot T_{кр} \tag{24}$$

By transforming the expression (24), we obtain the dependence of the loan rate on the probability of further refinancing:

$$i \geq 24 \cdot \left(\frac{T}{pt^2} - \frac{1}{T}\right); \qquad p \geq \frac{24T^2}{t^2(24 + Ti)}, \tag{25}$$

where T is the initial loan repayment period;
t - earlier loan repayment period.

4 Conclusion

In the process of research we have revealed the essence of loan capital and loan interest. At the same time, based on the analysis of the works of Russian and foreign studies, the authors concluded that the loan capital acts as the amount of funds with which you can quickly meet the needs of the population, and the loan interest acts as a payment for instant satisfaction of needs.

In addition, the authors obtained a model that determines the conditions for the effective movement of credit capital, depending on the interest rate, credit terms and the probability of using the funds received by the Bank as a result of early repayment of the loan, as a new credit product. This dependence makes it possible to determine the value of any of the parameters under given conditions of the probability of further placement of credit resources. Conversely, it is possible to calculate the probability of refinancing at a fixed rate of interest.

References

1. Ananyin, O.: Macroeconomics of Henry Thornton. Issues of Economics, no. 12, pp. 110–126 (2002)
2. Marx K. Capital - T. 1-3. Soch., Marx K., Engels F. T.23-25
3. Keynes, J.M.: General terriya employment, interest and money. Progress (1978)
4. Fridman, M.: Quantitative Theory of Money. Elf Press, Moscow (1996)
5. Gasilov, V.V., Okolelova, E.Yu., Zamchalova, S.S.: Economic-mathematical methods and models: the teaching method. allowance. Voronezh, state. arch.-builds. Un-t., Voronezh, p. 157s (2005)
6. Okolelova, E.Yu., Meshcheryakova, O.K.: Investment mechanisms of the real estate market. Mortgage. Sub.red. Gasilova V.V. - Voronezh, Publishing House "Origins", 219 p. (2007)
7. Okolelova, E.Yu.: Models of investment forecasting of the commercial real estate market. Sub.red. Gasilova V.V. - Voronezh, Publishing House "Origins", 326 p. (2008)
8. Grabovyi, P.G., Trukhina, N.I., Okolelova, E.Yu.: Dynamic model for forecasting the development of an innovative project. Technology of the Textile Industry, no. 1 (367), 78–82 (2017)
9. Shulgina, L.V., Evseeva, S.V.: The development of consumer credit in Russia. FES: Finance. Economy. Strategy 10 (51), pp. 42–47 (2008)
10. Shulgin, A.V., Kudryavtsev, V.Yu., Shulgin, A.V.: On state-private cooperation in the sphere of reproduction of human capital [Text]. Russian Entrepreneurship, Part 2, no. 9, pp. 137–141 (2007)
11. Okolelova, E., Shibaeva, M., Shalnev, O.: Development of innovative methods for risk assessment in high-rise construction based on clustering of risk factors. In: E3S Web of Conferences, High-Rise Construction 2017 (HRC 2017), Samara, Russia, 4–8 September 2017, vol. 33, p. 03015 (2018)

Economic Models of Well-Balanced Usage of the Economic Resources of a Transportation Company

G. V. Bubnova[1(✉)], A. I. Frolovichev[1], and E. S. Akopova[2]

[1] Russian University of Transport (MIIT), Moscow, Russia
[2] Rostov State University of Economics (Rostov Institute of Agriculture), Rostov-on-Don, Russia
kafedra_kil@mail.ru

The developmental strategy of the Russian Railways Holding Group for the term until 2030 [1] is focused on traffic expansion and improvement of the quality of services provided. Thereat, it is worth noting the important role of the company in ensuring the development of the country's economy in general as well as some social functions performed. The importance of raising the performance of resources that shows the growth of the overall railway production efficiency is obvious [2]. Any model of strategic management to some extent touches the issue of resource efficiency [3].

Earlier, the importance of balanced use of resources for a transportation company [4, 5] has been proved. This is determined also by a big share of staff costs in the overall cost pattern [6] as well as a high level of wear of fixed assets of the Russian railway industry [7, 8]. Despite the reforms pursued [9], the Russian railway industry is far from the leading positions according to the main indicators of efficiency [10].

The balanced scorecard developed by Kaplan and Norton [11] and became widely used in the early 21st century is based on the cause-and-effect relationships between strategic goals, their parameters, and factors. It relies on the balance of short-term and long-term goals as well as on the consistency between the company's external evaluations and internal ones. In particular, it allows distributing the company's resources in accordance with the strategy-set priorities.

The idea to harmonize the transportation company's development under consideration [4] is also based on the principles of balance, but in some different way with respect to a large transportation company with a high level of capital intensiveness.

The term "balance" derives from the word "balance" (fr. balance – scales) that means the "form reflecting the balance of interrelated quantities subject to continuous change" [12]. It should be noted that we can talk about balance only regarding the interrelated quantities, for which it is achieved provided a certain proportion of hereof in relation to each other by an established criterion.

The balance of economic resources is the consistencies between their various types within the production system as well as between the elements thereof and the external environment, which ensure the efficient performance of the production system.

The conceptual diagram of the middle of the road management of the transportation company's development (Fig. 1) stipulates two ways of a balanced development:

- balance of demand and the company's production capacity;
- balance of economic resources used.

E. G. Popkova and B. S. Sergi (Eds.): ISC 2019, LNNS 87, pp. 838–845, 2020.
https://doi.org/10.1007/978-3-030-29586-8_95

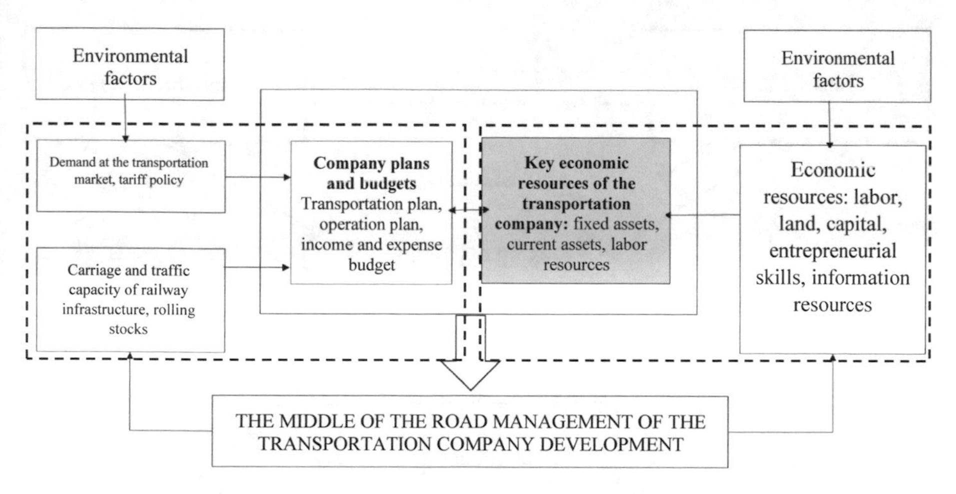

Fig. 1. Conceptual diagram of the middle of the road management of the transportation company's development

Upon achievement of a balance between the company reaches the highest performance, but each of the constituents specified is also subject to balance. The issue of the balanced demand and production capacity is explored in the following writing [13] devoted to the economic justification of managing the transportation company development on the principle of balanced use of resources.

The key resources of a large transportation company are the following [4]: fixed capital, labor resources, current assets. The revenue and volume of transportation (for rail transport it's the presented work) can be considered as thc main overall indicators of the transportation company.

We can display the relationship between these indicators in the form of the diagram presented in Fig. 2, which shows the main directions of the mutual impact of the indicators under consideration.

One can suppose it is the volume of fixed capital and labor resources that have a greater effect on the results of the transportation company's performance. Current assets are more determined by the results of performance and secondary in this case. The main idea of the concept presented is some balance between fixed capital and labor resources calculated by the overall performance indicators, which is achieved owing to the possible substitution of one type of resource with another. This is displayed by dashed lines in the diagram.

Employing traditional indicators of resource efficiency that assess the intensity of resource utilization or profitability, we see impossible to investigate the level of their balanced use.

It becomes possible if we relate the amount of fixed assets and labor resources involved with the overall indicator using the Cobb-Douglas production function, which describes the company's typical production and economic technology. This function is supplemented by a multiplier reflecting the impact of scientific and technological progress [14]. Below we present a similar function for revenue as an overall indicator

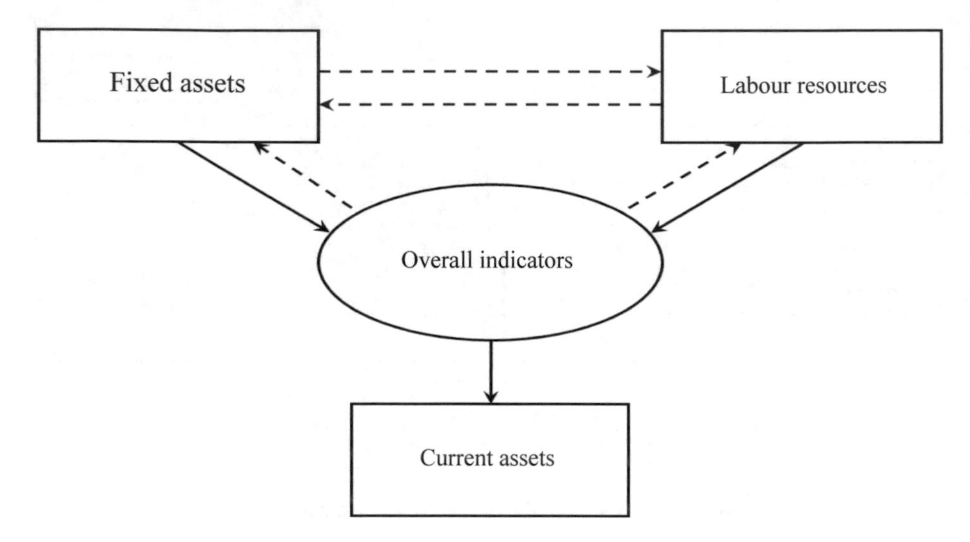

Fig. 2. The relationship between key resources – overall indicators of the transportation company performance

$$S = b_0 K^{b_1} L^{b_2} e^{rt}, \tag{1}$$

where

K is an average annual initial value of fixed assets, rubles,
L is an average staff number over year, people,
S is the company's annual revenue, rubles,
b_0, b_1, b_2 are the production function parameters characterizing the production and economic technology used ($b_1 + b_2 = 1$, $b_0, b_1, b_2 \geq 0$);
r is the coefficient of technical and organizational changes.

Then the diagram in Fig. 2 can be realized for the overall revenue indicator using the following mathematical model

$$\begin{cases} C = P_K \cdot K + Z + P_L \cdot L, \\ Z = kS, \\ S = b_0 K^{b_1} L^{b_2} e^{rt}; \end{cases} \tag{2}$$

where (except the above-indicated)

C is the total annual cost of resource utilization, rubles,
Z is current assets over year, rubles,
P_K is the cost of utilization of the capital unit for the year, rubles/rubles,
P_L is the cost of utilization of the unit of labor resources over year, rubles/person,
k is a coefficient regarding the value of current assets in the revenue pattern.

The weighted average cost of capital (WACC) was chosen as the price for utilization of the capital unit; the average annual wage with an account of social deductions is the price for utilization of labor resources.

The previously presented model didn't take into account the factor of scientific and technological progress [4]. It should be noted that similar models can be developed for other companies and industries. The specific nature of the transport industry here is regarded as follows.

Firstly, the composition of key resources is determined by the high level of the industry's capital intensiveness and the large volume of labor resources involved, which affects their overall cost of utilization.

Secondly, current assets play a secondary role in a transportation company and act as a variable that does not affect the amount of revenue S, but, on the contrary, depends on it. This situation is explained by the specific nature of the transport industry, where current assets, as a rule, are not a determinant of trend and value of the overall indicator.

Thirdly, it is a function (1) among the variety of production function types that makes it possible to describe the production and economic system having the continuous effect of the scale and applicable to large companies and industries [15].

Fourthly, scientific and technical progress in a large company representing the transport industry can only be evolutionary due to the impossible simultaneous substitution of all fixed assets with innovative ones that allow for dramatic changes in the company. Therefore, the impact of scientific and technical progress and organizational changes is rendered by the production function (1) as an additional multiplier (exponent).

We can say that the coefficient r shows the rate of scientific and technical progress and organizational changes and increases the overall performance of resources, which is accompanied by the growth of neutral technical progress coefficient b_0. Its value expresses the volume of production output under unit volumes of resources. In fact, the parameters b_1, b_2 in the formula (1) are the coefficients of the revenue elasticity of revenue in terms of fixed assets value and the average staff number respectively.

The model described by formula (2) makes it possible to determine a balanced ratio of resources utilized, which is based on the idea to minimize the total cost of their utilization.

Given the relationship between the variables of the above-described system of equations (2), the total cost of utilization of the economic resources of a transportation company can be presented as a one-variable function

$$C = P_K \cdot K + kS + P_L \cdot \left(\frac{S}{b_0 K^{b_1} e^{rt}}\right)^{1/b_2}. \tag{3}$$

Finding the minimum point of the obtained function, we can calculate the volume of fixed assets and labor resources leading to a minimum of the total cost of resource utilized and take it as balanced:

$$\begin{cases} K_{balance} = \left(\frac{P_L b_1}{P_K b_2}\right)^{b_2} \left(\frac{S}{b_0 e^{rt}}\right), \\ L_{balance} = \left(\frac{P_K b_2}{P_L b_1}\right)^{b_1} \left(\frac{S}{b_0 e^{rt}}\right); \end{cases} \tag{4}$$

For the economic evaluation of the efficiency of the current ratio of resources we can use the balance index:

$$I_B = \frac{C}{C_{balance}} = \frac{P_K \cdot K + kS + P_L \cdot \left(\frac{S}{b_0 K^{b_1} e^{rt}}\right)^{1/b_2}}{P_K \cdot K_{balance} + kS + P_L \cdot \left(\frac{S}{b_0 K_{balance}^{b_1} e^{rt}}\right)^{1/b_2}}, \quad (5)$$

showing the value of the relative deviation of the total cost of resource utilization from the total cost of their balanced ratio for a given transportation company.

This indicator can be added to the commonly-accepted indicators of the efficiency of resource utilization, which will regard the balance principle in the management of the transportation company's resources and its development.

For the economic evaluation of the balance of the resources utilized it is also necessary to know the substitution cost of a unit of one resource with a unit of another one called the marginal rate of substitution [6]. Formula (1) allows making a similar evaluation. Thus, the marginal rate of substitution of labor resources with fixed assets, which is of the greatest interest to a transportation company supporting the innovative path of development can be calculated by the formula

$$\sigma_{L \to K} = \frac{b_2 K}{b_1 L} \quad (6)$$

The indicator under suggestion that allows for evaluation of the required amount of additional investments depends on the volume of resources utilized by the transportation company as well as the elasticity of coefficients b_1 and b_2 of the overall indicator of fixed assets and labor resources respectively.

It should be noted that the approach under suggestion can be used for a comparative analysis of the balanced use of resources in various transport companies as well as in particular corporate business units or regional divisions of a transportation company. With account to designations applied we can use Table 1 for such an analysis.

This table allows comparing production and economic technologies of different companies (divisions), assessing the degree of impact of the resources expended in different companies on the overall indicator through indicators b_1 and b_2. We can also compare the level of scientific and technical progress based on the parameter b_0, and the growth rate of production output due to scientific and technical progress and organizational changes through the parameter r.

Indicators of the resource utilization cost, their current and balanced ratio, and balance indices make possible an economic assessment of the level of their balanced use and compare the values of the marginal rate of substitution with the amount of additional investments to achieve a balance.

Table 1. Comparative analysis of the balanced use of resources in different companies (divisions)

Company (division)	Parameters of production and economic technology					Cost of a resource unit utilization		The current ratio of resources		The marginal rate of substitution	Balanced ratio of resources		Index of balance
	b_0	b_1	b_2	r	S	P_K	P_L	K	L	$\sigma_{L \to K}$	$K_{balance}$	$L_{balance}$	I_B
K1													
…													
K_N													

Taking into account the fact that all parameters of the production function are evaluated on the back of the results of the company's performance in different periods of time, and the cost of utilization of various resources changes over time, the balanced ratio of resources is not constant. In addition, it depends on the estimated results of performance, which forecast values may vary depending on various micro- and macro-economic factors. Therefore, the parameters of the production and economic system and other indicators listed in Table 1 are subject to regular monitoring.

Besides, the investment program of the transportation company should be aimed not just at the substitution of worn-out fixed assets with similar ones, but the mobilization of innovative fixed assets and the introduction of more efficient technologies. It is also necessary to raise the involvement of fixed assets in the production process. All these circumstances allow achieving higher production and financial return at lower resource costs and also affects their balanced ratio.

Thus, strategic and long-term planning in a transportation company should be based on the principle of balanced use of resources. The application of the approach under suggestion will assist the sound development of the transportation company and its achievement of a qualitatively new level of development as well as ensure its sustainable and effective operation in a fast-changing market situation.

References

1. The strategy of development of railway transport in the Russian Federation until 2030: approved by the Government Resolution of the Russian Federation No. 877-p dated June 17, 2008. www.mintrans.ru/documents/detail.php?ELEMENT_ID=13009
2. Lapidus, B.M., Macheret, D.A., Miroshnichenko, O.F.: Improving the efficiency of resources and railways. Railway Economy, no. 6, p. 12 (2011)
3. Baban, S.M., Bubnova, G.V., Giricheva, V.A.: Strategic management in railway transport: training guide. Railway Transport Training Center, 344 p. (2013)
4. Podsorin, V.A., Epishkin, I.A., Frolovichev, A.I.: Harmonization of the transportation company development. Railway Economy, no. 3, pp. 12–23 (2018)
5. Podsorin, V.A., Frolovichev, A.I.: Principles of a balanced ratio of labor and capital. Railway Economy, no. 8, pp. 21–30 (2017)
6. Epishkin, I.A., Sheremet, N.M., Frolovichev, A.I.: Middle of the road management of staff costs in the "Russian Railways" company OAO. Railway Economy, no. 12, pp. 70–80 (2017)
7. Tereshina, N.P., Podsorin, V.A.: Reproduction of the fixed capital of the transportation company. Railway Transport, no. 6, pp. 67–69 (2007)
8. Bubnova, G.V., Podsorin, V.A.: Management of economic processes of the transportation company when upgrading hardware and software. Collected proceedings of the international research and practice conference "Increasing labor productivity in transport is the source of development and growth of the national economic competitiveness", pp. 44–46. Moscow State University of Transport named after Emperor Nicholas II, Institute of Economics and Finance (2016)
9. Tereshina, N.P., Epishkin, I.A., Flyagina, T.A.: Economic reforms in railway transport: training guide, 94 p. MIIT (2012)
10. Frolovichev, A.I.: Comparative analysis of the efficient use of resources in railway transport of different countries. Transport Business of Russia, no. 6, pp. 25–28 (2017)

11. Kaplan, R., Norton, D.: Balanced Scorecard. From strategy to action. Olimp-Business, 525 p. (2017)
12. Borisov, A.B.: Big Economic Dictionary. 2nd ed., revised and enlarged, 860 p. Book World (2005)
13. Sokolov, Yu.I, Ivanova, E.A., Shlein, V.A., Lavrov, I.M., Anikeeva-Naumenko, L.O., Nesterov, V.N.: Management of demand for rail transportation and market balance issues, Moscow (2015)
14. Kolemaev, V.A.: Economic and mathematical modeling. Modeling of macroeconomic processes and systems: a study guide for university students majoring in 061800 "Mathematical Methods in Economics", 295 p. UNITY-DANA (2005)
15. Kleiner, G.B.: Production Functions: Theory, Methods, Application, 239 p. Finance and Statistics, Moscow (1986)

The Guidelines of Public Regulation in Terms of Digitalization of the Russian Economy with the Industry 4.0 Tools

Gilyan V. Fedotova[1,2(✉)], Natalia E. Buletova[3], Ruslan H. Ilysov[4], Nina N. Chugumbaeva[5], and Natalia V. Mandrik[5]

[1] Volga Region Research Institute of Production and Processing of Meat and Dairy Products, Volgograd, Russia
g_evgeeva@mail.ru
[2] Volgograd State Technical University, Volgograd, Russia
[3] Volgograd Institute of Management, Branch of the Russian Academy of National Economy and Public Administration, Volgograd, Russia
buletovanata@gmail.com
[4] Chechen State University, Grozny, Russia
ilyasov_95@mail.ru
[5] MIREA - Russian Technological University, Moscow, Russia
nina-ch2005@ya.ru, mandrikn@mail.ru

Abstract. Purpose: The main purpose of the chapter is to review the basic tools for the regulation of digitalization processing occurring in the Russian economy. To achieve this purpose, we carried out an analysis of the regulatory framework for the transition to a digital economy, summarized the activities held within various state programs of the country's socio-economic development, assessed the results achieved on the installment of information platforms.

The analysis of the results achieved makes it possible to conclude that it's necessary to further improve the regulatory, develop and upgrade the current digitalization mechanisms in different areas of public life, and enhance the system for planning target results of state programs.

Methodology: We apply the methods of comparative data analysis, graphical data analysis, statistical analysis, dynamic assessment, comparison, analogy, and systematization.

Results: The peculiarities of developing the Russian model of the digital economy include aggressive public involvement in the ongoing processes, which is aimed at developing the high-priority sectors of the national economy. The evaluation of the digitalization results for the last 8 years has shown that many areas of public life are involved therein. Foremost, it's the very system of public administration, which already operates on the basis of the philosophy of Industry 4.0.

Recommendations: The basic tools of public regulation of the informatization processes in the Russian society considered in this paper provide an opportunity to revise some objectives and target indicators of the future transition to full digitalization of all areas and sectors of the national economy.

Keywords: Digital economy · State · Regulatory tools · Industry 4.0 · Information society

E. G. Popkova and B. S. Sergi (Eds.): ISC 2019, LNNS 87, pp. 846–855, 2020.
https://doi.org/10.1007/978-3-030-29586-8_96

JEL Code: F01 · O11 · O14 · 057

1 Introduction

Charles L. Dodgson taught mathematics at Oxford University and, perhaps, therefore, in his literary role as Lewis Carroll, he definitely anticipated much of what was generated by digit and embodied into the natural world. Based on Caroll's words "…you need to run hard just to stay in place, but to get somewhere, you need to run at least twice as fast!", Dutch paleobiologist Lee Van Valen set up the principle of the Black Queen, which is fundamental to the evolutionary theory: *"The species in the evolutionary system need continuous change and adaptation to sustain its existence in the surrounding biological world continuously evolving with it."*

The conclusions of evolutionary economic theory rely on the same principle:

- economic processes do not differ from biological ones and should be studied in real time;
- the market environment is volatile and unstable, therefore changes are difficult to be foreseen or predicted;
- changes are spontaneous and irreversible, they are the result of the interaction of internal and external factors;
- innovations and scientific and technological progress have economic nature, they are generated in the course of the interaction of economic institutions and entities;
- economic models are unbalanced, and economic entities are heterogeneous in relation to the utilization of innovations;
- the competitive environment is imperfect and those who have the best-adjusted behavior (routines) to its changes win the competitive advantages.

The rapid digitalization of the socio-economic, political, and military order of the world is an entitative trend, the very factor of evolution that requires change and adjustment of all the market actors. States, societies, strata, and individuals are forced to adjust to the technology-related changes to survive and develop. Acting as a factor of the market environment volatility, new information technologies require the rapid adaptation of national economies and the reproduction of innovations for survival and development.

In 2018 the Russian Federation was 46th in the Global Innovation Index according to the World Intellectual Property Organization. The Digital McKinsey Expert Group Report 2017 "Digital Russia: A New Reality" notes that digitalization can raise the Russian GDP by 2025 in absolute terms from 4.1 to 8.9 trillion rub. (in prices of 2005) and amount to nearly a third of the total GDP growth over the specified period.

As Digital McKinsey points out, Russia presently doesn't have leading positions in the development of the digital economy, since the proportion of the digital economy in Russian GDP is 3.9%, which is approximately threefold lower than in USA, Germany, Japan, and China. However, in McKinsey analysts' opinion, the Russian Federation shows a positive trend in the growth of the digital economy percent in the GDP structure (24% of the total GDP growth over 5 years since 2012).

2 Materials and Methods (Model)

In the study, the authors employed advanced theoretical and applied writings devoted to the establishment of an information economy in modern Russia, in the global economic space. The basis for this chapter was the proceedings of the following authors (Sukhodolov et al. 2018a), (Kravets et al. 2013), (Kuznetsov et al. 2016), (Popova et al. 2015), (Vertakova et al. 2016), (Sibirskaya and Shestaeva 2016), (Plotnikov et al. 2015), (Fedotova et al. 2018), (Romanova et al. 2017), (Kovazhenkov et al. 2018), (Fedotova et al. 2019).

The direction of the Russian economic system at the large-scale digitalization of all areas and sectors determines the setting new objectives for solution hereof the successful international and Russian practices to raise the level of informatization of the Russian economic system and the transition to a digital model of social development will be attracted.

3 Results and Discussion

In 2018 the contribution of the Internet economy to the Russian economy constituted 3.9 trillion rubles according to the Russian Association of Electronic Communications (RAEC), therewith e-commerce segment, (1.95 trillion rubles in absolute terms) is the most financially important with an increase of 13.2% by 2017. The audience of the Russian segment of the Internet (Runet) is 90 million people or 74% of the total population of Russia. A significant trend is the increasing share of Internet users on mobile devices over ones on desktops and laptops.

The analysts call the further development of ICT, the introduction and dissemination of the Internet of Things (IoT) and cyber-physical systems (CPS) the "Fourth Technological Revolution or Industry 4.0". The concept hereof assumes the operation of production systems on the Internet without a man as a mediator in the machine interaction. The concept of Industry 4.0. was firstly announced at the Hannover Industrial Fair as the main topic of 2013.

According to reports of the World Bank and General Electric in 2011, the introduction of Industry 4.0. (Industrial Internet) can add to 46% or USD 32.3 trillion to global GDP by 2025.

The data provided generally show an insufficient and unevenly developed digital economy in the Russian Federation, and also exhibit a great potential for introducing and promoting its elements in the economic structure.

The economic and political factors of the external environment, stiffer interstate competition, escalation of state confrontation in political and military issues due to the transformation of the global political and economic system demand from Russia to constantly strengthen its economic capacity in general and the capacity of the digital economy in particular.

Recognizing the importance and seriousness of digitalization in all issues of economic development, since the early 2000s, the leadership of Russia has made efforts to streamline and regulate the Internet area.

In particular, in 2002 the Government of the Russian Federation approved the Federal Target Program (FTP) "Electronic Russia (2002–2010)".

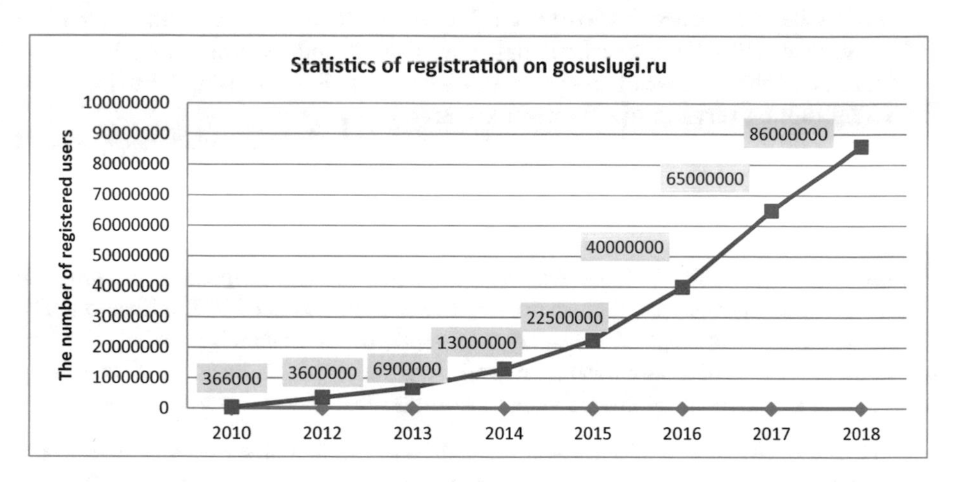

Fig. 1. The dynamic pattern of the number of registrations on gosuslugi.ru, units

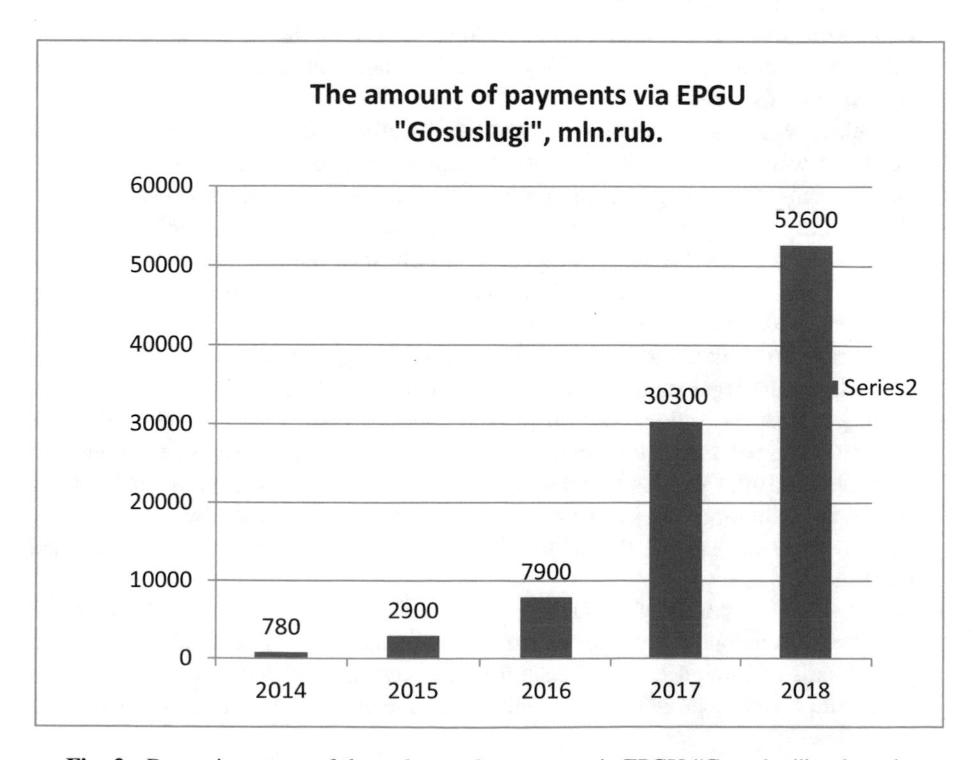

Fig. 2. Dynamic pattern of the volume of payments via EPGU "Gosuslugi", mln. rub.

In terms of this FTP, an electronic portal of state services gosuslugi.ru was set up to provide citizens of the Russian Federation and other persons with access to public services in electronic form.

Having been launched in 2010 with 336 thousand users, the Unified portal of Public Services (EPGU) “Gosuslugi” has more than 86 million users (Fig. 1).

Since 2013 the number of services provided via EPGU “Gosuslugi” has increased from 12.9 million services to 1.3 billion services.

Since 2014 the amount of payments effected via EPGU “Gosuslugi” has increased several hundred-fold and has now reached 52.6 billion rubles (Fig. 2).

In 2010, the Federal Program “Electronic Russia” was replaced by the State Program “Information Society” approved by the Resolution of the Government of the Russian Federation No. 1815-p dated October 20, 2010. The purpose of the program was to provide citizens and organizations with the advantages of ICT utilization. The program stated the concept of “Open Government”, which makes it possible to ensure transparency of public administration and population’s feedback. The following information resources were developed and introduced in the framework hereof:

- gossluzhba.gov.ru is a single portal of state and municipal services providing the transparency of data on state and municipal service, state and municipal officers;
- pravo.gov.ru is open legal information, the source of the official publication of legal acts of the federal authorities;
- regulation.gov.ru is the federal portal of legal act drafts, the official site for posting information on the preparation of legal acts by federal authorities and the results of their public discussion;
- zakupki.gov.ru is a unified procurement information system for state and municipal needs as well as the needs of public corporations and publicly- or municipally-owned companies that grants the procurement to be transparent and available;
- data.gov.ru is an updated information resource on open data of federal authorities, regional authorities and other companies; it publishes documented data array, reference links, and metadata of published data array, information on open data-developed software and information services;
- pkk5.rosreestr.ru is a public cadastral map containing data on land plots and other real estates, the legal status of this property, encumbrances, the exact address, etc.;
- torgi.gov.ru is the official website of the Russian Federation for publishing the information on all types of trading of state and municipal property, titles to uniform technologies, titles to sites of subsurface resources and hunting lands, transactions within the public-private partnership, concessions, housing and utilities, etc.
- bus.gov.ru is the official site to publish the information on state and municipal institutions.
- ved.gov.ru is a portal of the foreign economic information of the Ministry of Economic Development of Russia on international trade rules, public services to participants of foreign economic activity, support and development of the foreign economic activity, programs of economic cooperation between Russia and other countries, etc.

Thus, various information resources and technological solutions for prompt electronic cooperation between the state and citizens, their associations, and commercial legal entities were developed and introduced within the fulfillment of the "Open Government" concept.

All these measures made it possible for Russia to become 32nd in the level of e-government development by 2018. E-Government Development Index (EGDI) is determined by the UN Department of Economic and Social Relations every other year. Based on this rating, the level of e-government development in Russia was assessed as great, the penetration rate of online services is one of the highest. Moscow was ranked the 1st by the development of e-government in cities.

In addition, the Russian government put considerable efforts to develop and manage the digital economy, which includes not only the economy of the IT sector but also other economic activity using the Internet environment for its establishment and evolvement.

In 2016, President of the Russian Federation Vladimir Putin in his Annual Address to the Federal Assembly proposed launching a large-scale system program to develop the economy of a new technological generation – the digital economy. He pointed out that it's the issue of national security and technological independence of Russia.

In 2017, the Government of the Russian Federation adopted the National Program "Digital Economy of the Russian Federation until 2024".

The goals stated in this national program are as follows:

- establishment of the ecosystem of the digital economy of the Russian Federation with digital data as a key factor of production in all areas of social and economic activity and granted effective interaction, including cross-border one, business, the scientific and educational community, the state, and citizens;
- arrangement of essential and sufficient institutional and infrastructural conditions, elimination of existing obstacles and restrictions for set-up and (or) development of high-tech businesses and prevention of new obstacles and restrictions both in traditional sectors and new ones and high-tech markets;
- raising the competitiveness in the global market of both individual sectors of the economy of the Russian Federation and the economy in general.

The main cutting-edge digital technologies supported by the National Program "Digital Economy of the Russian Federation" are the following:

- Big Data;
- Neural network and AI;
- distributed ledger systems (Blockchain);
- quantum technologies;
- new production technologies;
- Industrial Internet (IoT);
- components of robotics and sensorics;
- wireless communication technology; virtual and augmented reality technology.

According to the National Program, the management of the digital economy includes representatives of all the parties interested in the development hereof (government authorities, business, civil society, and the scientific and educational community), ensure transparency and openness in its activities and employs a project approach to the arrangement of management. Currently, the leaders in digitalization are the mass media, finance, law, and insurance. A provisional list of introduced state information systems (GIS) in various departments published on the website of the Federal Service for Supervision of Communications, Information Technology, and Mass Media is illustrative of the cost of the industry's transition to a digital platform (Fig. 3).

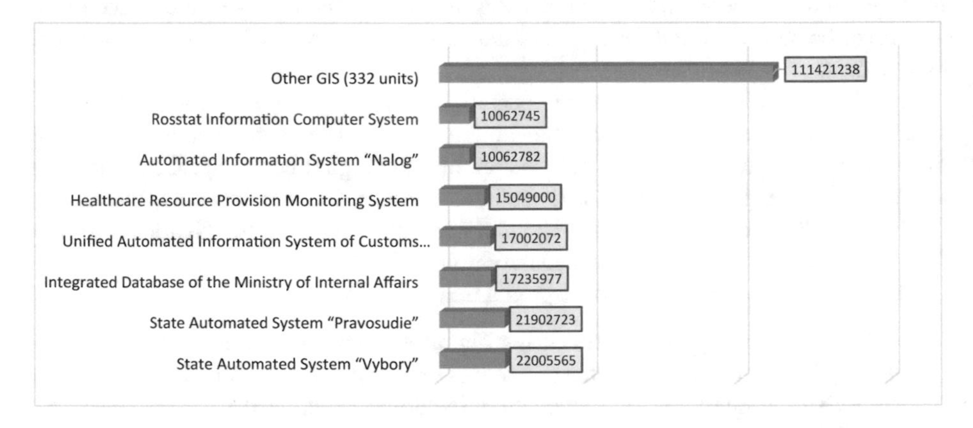

Fig. 3. Register of federal state information systems put into operation, th. rub.

The major actor within the National Program "Digital Economy of the Russian Federation" is the Autonomous Non-Profit Organization "Digital Economy" founded in 2017 by the Government of the Russian Federation. It acts as a coordinator in the cooperation of the state, business, and the scientific community and determined the basic parameters hereof (Table 1).

In addition, the state actively supports the National Technology Initiative (NTI) promoted by the Agency for Strategic Initiatives. In the framework of the NTI, the Agency indicates nine key markets have to be developed: AeroNet, AutoNet, EnergyNet, FinNet, FoodNet, HealthNet, MariNet, NeuroNet, SafeNet.

Table 1. The estimated results of the implementation of the National Program "Digital Economy of the Russian Federation" by 2024

Regarding the ecosystem of the digital economy	Regarding staff and education	Regarding the development of research expertise and technological capacity	Regarding information infrastructure	Regarding information security
Successful performance of at least ten leading companies (ecosystem operators) competitive in global markets	Number of graduates of higher education institutions in ICT-related majors is 120 thousand people per year	Number of accomplished projects in the digital economy (not less than 100 million rubles) is 30 units	The proportion of households with broadband access to the Internet (100 Mbps) in the total proportion of households is 97%	The number of entities using the standards of safe communication of public and social institutions is 75%
Successful performance of at least 10 sectoral (industrial) digital platforms in the basic domains of the economy (including digital healthcare, digital education, and smart city)	The worldwide average number of graduates of higher and secondary professional education with ICT expertise is 800 thousand people per year	The number of Russian institutions participating in the large projects (USD 3 million) in priority areas of international scientific and technical cooperation of the digital economy is 10	Stable 5G coverage and higher in all major cities (1 million and more people)	The volume of the internal network traffic of the Russian segment of the Internet routed via foreign servers is 5%
Successful performance of at least 500 small and medium enterprises in the development of digital technologies and platforms, and digital services' provision	The percent of the digitally-skilled population is 40%			

As part of the development and application of advanced manufacturing technologies, the NTI developed the TechNet action plan, a cross-market and cross-industry direction that provides technological support for the development of NTI markets and high-tech industries through the establishment of digital, smart, Virtual Factories of the Future. The concept of TechNet is based on the Factory of the Future, i.e. a system of integrated technological solutions (integrated process chains), which ensure the design and production of globally competitive new-generation products in the shortest possible term. Factory of the Future is developed on the test sites “TestBeds”.

In the interaction of the NTI and the ANO “Digital Economy” in 2019 were selected the developers of the action plans of the National Program “Digital Economy”. In particular, Sberbank will develop an action plan for the neural network and artificial intelligence, National University for Science and Technology “MISiS” – an action plan “Quantum Technologies”, Center for National Technological Initiative of St. Petersburg State University – “New Production Technologies”, Center for National Technological Initiative of Innopolis University (Kazan) – “Components of Robotics and Sensorics”.

4 Conclusion

The Russian Federation traditionally pays much attention to the provision of the security of the Internet network operation and the prevention of cyber attacks of the government and non-government actors on technological and financial systems used within the country.

In particular, one of the recent initiatives of global importance for the adequate performance and development of the digital economy is the draft law passed in the first reading by the State Duma on the autonomous operation of the Russian Internet segment in the event of its disconnection from the global network. Innovations herein are the establishment of public supervision over the traffic routing and engineering of a national domain name system of the RU zone. In the event of the draft’s transformation into law, it may greatly affect the guidelines and rates of development of the digital economy in the Russian Federation. However, concerns about the possible restriction of traffic, the failure to provide uninterrupted information exchange and, as a consequence, deepening gap with world leaders in introduction of the digital economy expressed by some experts seem somewhat far-fetched, because the People’s Republic of China using similar traffic management and web content administration systems is an example of successful development of the IT industry and digital economy. Despite the restrictions determined by the use of the Greater Chinese firewall, China is currently recognized world leader in the introduction of digital technologies and the development of digital systems.

Acknowledgements. The reported study was funded by RFBR according to the research project No. 18-010-00103 A.

References

Kravets, A.G., Gurtjakov, A.S., Darmanian, A.P.: Enterprise intellectual capital management by social learning environment implementation. World Appl. Sci. J. **23**(7), 956–964 (2013)

Kuznetsov, S.Y., Tereliansky, P.V., Shuvaev, A.V., Natsubize, A.S., Vasilyev, I.A.: Analysis of innovative solutions based on combinatorial approaches. ARPN J. Eng. Appl. Sci. **11**(17), 10222–10230 (2016)

Fedotova, G.V., Lomakin, N.I., Tkachenko, D.D., Gontar', A.A.: Peculiarities of digital transformation of the system of bank's economic security. In: Popkova, E.G. (ed.) The Future of the Global Financial System: Downfall or Harmony: Proceedings of the Conference, Limassol, Cyprus, 13–14 April 2018. Lecture Notes in Networks and Systems, vol. 57, pp. 1104–1112. Springer Nature Switzerland AG, Cham (2019)

Popova, L., Litvinova, T., Ioda, E., Suleimanova, L., Chirkina, M.: Perspectives of the growth of economic security by clustering of small innovational enterprises. Eur. Res. Stud. J. **18** (Special Issue), 163–172 (2016)

Plotnokov, V., Fedotova, G.V., Popkova, E.G., Kastyrina, A.A.: Harmonization of strategic planning indicators of territories' socioeconomic growth. Reg. Sectoral Econ. Stud. **15–2** (July–December), 105–114 (2015)

Romanova, T.F., Andreeva, O.V., Meliksetyan, S.N., Otrishko, M.O.: Increasing the cost efficiency as a trend of intensification public law entities' activity in a public administration sector. Eur. Res. Stud. J. **20**(1), 155–161 (2017)

Fedotova, G.V., Kulikova, N.N., Perekrestova, L.V., Kozenko, Yu.V.: Target indicators of implementing the measures on formation of the model of information economy. In: Sukhodolov, A.P. (ed.) Models of Modern Information Economy: Conceptual Contradictions and Practical Examples, Chapter 24, pp. 255–263. Emerald Publishing Limited (2018)

Kovazhenkov, M.A., Fedotova, G.V., Kurbanov, T.K., Uchurova, E.O., Tserenova, B.I.: Verification of state programs of geographically-distributed economic systems. In: Popkova, E.G. (ed.) The Future of the Global Financial System: Downfall or Harmony: Proceedings of the Conference, Limassol, Cyprus, 13–14 April 2018. Lecture Notes in Networks and Systems, vol. 57, pp. 1043–1053. Springer Nature Switzerland AG, Switzerland Cham (2019)

Vertakova, Y., Plotnikov, V., Fedotova, G.: The system of indicators for indicative management of a region and its clusters. Proc. Econ. Finan. **39**, 184–191 (2016)

Sibirskaya, E.V., Shestaeva, K.A.: The contents of the innovative in the Russian Economy. Knowledge–Economy– Society. In: Lula, P., Pojer, T. (eds.) Contemporary aspects of the economic transformation. The Krakow University of Economics, Krakow, Poland, pp. 27–37 (2016)

Sukhodolov, A.P., Popkova, E.G., Kuzlaeva, I.M.: Internet economy: Existence form the point of view of the microeconomic aspect. Stud. Comput. Intell. **714**, 11–21 (2018a)

Synergy of Blockchain Technologies and "Big Data" in Business Process Management of Economic Systems

Yulia V. Vertakova[1(✉)], Tatyana A. Golovina[1,2], and Andrey V. Polyanin[2]

[1] Southwest State University, Kursk, Russia
vertakova7@yandex.ru
[2] Central Russian Institute of Management - Branch of the RANEPA, Orel, Russia

1 Introduction

Currently, the digital economy is included in the list of the main directions of strategic development of Russia and many foreign countries. The task of creating conditions for deep system digitalization of economic life in Russia is outlined at the government level. New digital business models are aimed at reducing costs, generating additional revenue from digital solutions, optimizing customer interaction and improving customer service by studying their experience.

At present, blockchain technology is an innovative, breakthrough technology that has a huge potential to change the business environment in almost all sectors of the modern economy. The drivers of growth are the increasing demand for simplification of business processes, low transaction costs, transparency, continuity, speed, peer-to-peer interaction of economic entities, almost unlimited number of use scenarios in any industry. Blockchain technologies are becoming an integral part of the technological and operational infrastructure of most corporations and organizations. An additional factor of greater attention to the practical application of blockchain technologies is the sharp increase in the number of projects implemented and financed by states and large industry companies.

The main barriers to larger penetration of products and services based upon blockchain technologies are the lack of government regulation, low confidence in financial transactions through the blockchain technologies, and skepticism about the scalability of the technology when taking into consideration the huge amounts of data involved in blockchain transactions. Economic systems are just beginning to realize and generalize the problems of the blockchain industry and formulate the first regulatory decisions. In their turn, technologies that can process and work with large amounts of data can fundamentally change many aspects of modern society. "Big data" suggest the possibility of highly productive analysis of information in online mode. After the emergence of Big Data technologies, the concept of "information society" acquired its original meaning, and information received the status of the most valuable

E. G. Popkova and B. S. Sergi (Eds.): ISC 2019, LNNS 87, pp. 856–865, 2020.
https://doi.org/10.1007/978-3-030-29586-8_97

asset, acting as a driving force of the information society (Mkrttchian, Gamidullaeva, Panasenko and Sargsyan 2019).

For its successful functioning and development, it is necessary to combine the use of technological tools and management models that will participate in management decision-making, create "cross-cutting" technologies in order to work in the global market and develop the infrastructure of the digital economy.

2 Methodology

The research papers of scientists and professional communities are devoted to studying the information society, the formation of new technological structures, e-business models, the introduction of digital technologies and the development of the digital sharing economy: Taylor E., Vidas-Bubanja M., Bubanja, I., Boston Consulting Group, international Bank for reconstruction and development (Botsman and Rogers 2010, Huws 2014).

In their study of digitalization processes in data sharing, the general partner of Kleiner Perkins Caufield & Byers and Executive Vice President Mike Abbotti, Chief Executive Officer of Commercial Insurance Division of AIG Commercial, Rob Shimek believe that secure data sharing will be the engine of the new digital economy: "If the world of the Internet of things is the beginning of a new industrial revolution, then secure sharing of large amounts of data will be a prerequisite for its implementation" (Abbott and Shimek 2017).

The blockchain technology was presented to the world as a technology in 2008 in the capacity ofa technological platform of the new digital currency "BitCoin". The technology has collected several conceptually different ideas. There were combined such areas as: blockchain of data storage, consensus algorithms, and cryptographic mechanisms of data protection.

Technological solutions that are provided by blockchain platforms or blockchain technologies are very promising for business. Technologically, the blockchain does not eliminate intermediaries, it does not provide an opportunity to track the manufacturer of the goods (services), one just finds a counterparty (but the manufacturer or the intermediary is not technically confirmed), the main idea is that any transaction is fixed and reflected in time on a variety of technical devices – this is really important. The more information there is, the more it is falsified, and therefore there are less opportunities to prove how and under what conditions this or that action is carried out through time (Vertakova, Klevtsova and Babich 2016). Blockchain technology allows you to create a "time stamp". The advantage of this is that the transaction action has been stamped and can not be changed. The user does not care how it has been made technically; he gets the opportunity to demonstrate indisputable proof of the action made by him/her. The decisive advantage of using this technology by business structures is "the irrevocability of the solution" and "the proof of existing a digital asset at a certain point in time". The legal mechanisms of contract law are technologically standardized and it is possible to maintain accounting records with the help of a blockchain technology.

The fact that blockchain technology creates a new currency and changes the principles of monetary relations is nothing more than a marketing move of its developers. But at the same time, we would like to note that updated business conditions are being created. Firstly, this technology creates a clear accounting, without corrections and adjustments. Secondly, the contractual terms do not require endless approvals and confirmations. And most importantly, a transaction or a certain action appears to be irrevocable. In addition, technologically, the blockchain technology allows simple users to work with large amounts of data.

From the point of view of doing business, it is very useful that blockchain platforms can serve as a repository of socially significant records, such as records of documents, events, personal data.

For example, the Etherium platform has been successfully used to create decentralized online services based on the blockchain in the health care sector (Dokukina 2018), as it allows us to create a "time stamp" for keeping records of each patient and the action is stamped and can not be changed. This is very important both for a patient, as it provides a complete collection of personal data and events, and for a health care organization as it provides medical assistance on the basis of complete information and maintenance of a shared record of data.

Another example of the need to use the blockchain technology lies in the possibility of registration and protection of intellectual property. The patenting system is imperfect and in many cases simply inapplicable, especially for works of art. It is difficult for any user to retain the authorship of the created digital asset. Blockchain technology allows one to do this by compressing any digital asset into a unique 64-character hash that identifies, but does not allow us to restore the original file (Swan 2018). The resulting hash is included in the transaction, timestamp is the proof of the digital asset existence, and the original file belongs to the owner and is stored on his/her computer in its original form.

Figure 1 shows the scheme of blockchain influence on the order of transactions in the economic system.

The ecosystem of blockchain technologies includes technology providers, the developers of blockchain applications, network and system integrators, issuers of crypto currencies and marketplaces. A significant share of blockchain technologies use, especially by small and medium-sized business, will be based on BaaS (Blockchain-as-a-Service).

According to analytical agencies' assessment, global investments related to blockchain technologies will reach $ 9.7 billion in 2021. The size of the market is calculated on the basis of projected revenues from the implementation of blockchain solutions and the provision of services based on it. At the same time, the average compound annual growth rate (CAGR) in the period up to 2022 will be from 79.6% to 81.2% 1, but a number of regions will be eincreasing growth rates in the blockchain industry in a faster way: Japan - 127.3%, Latin America - 152.5% (Polyanin, Pronyaeva, Golovina, Avdeeva and Polozhentseva 2017).

Big data and digital technologies lead to high speed of decision-making, building communication with customers or suppliers and control, both on demand part and supply part. The economy has gradually moved away from the traditional model of centralized organizations, where large operators, being often in a dominant position, are

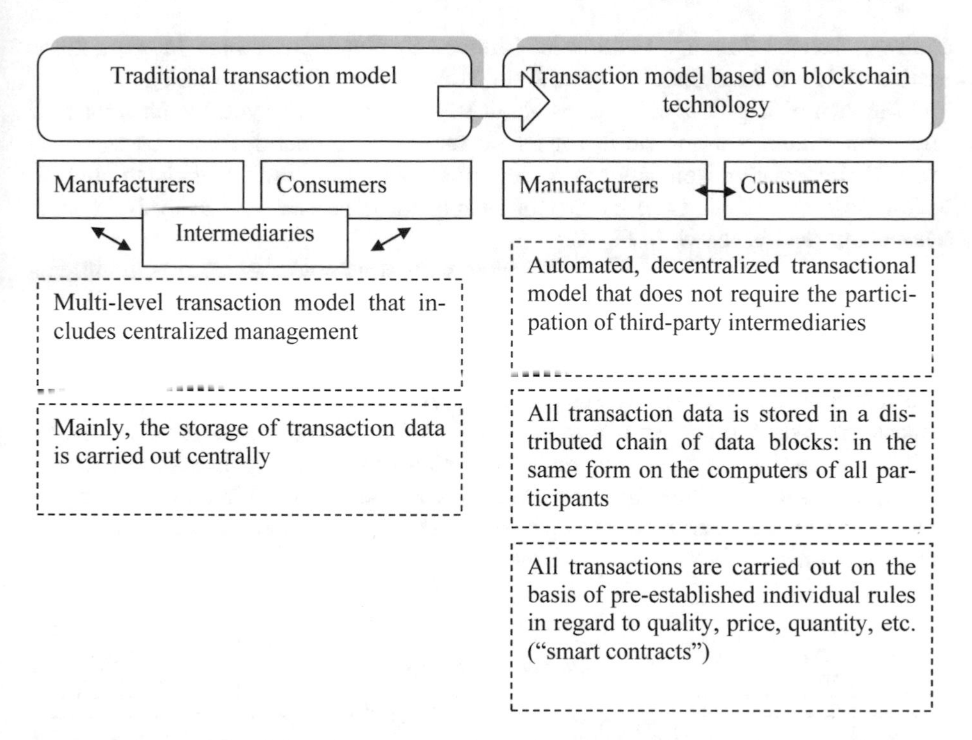

Fig. 1. Scheme of blockchain influence on the order of transactions in the economic system (made by the authors)

responsible for providing services to a group of passive consumers (Vertakova, Risin and Treshchevsky 2016).

Nowadays, it is becoming very important to develop methods for assessing the parameters of big data, to determine approaches to their structuring, accumulation, verification and storage, as well as to determine the relationship of formats and their streaming. A competitive advantage is the ability to transform the results of big data analysis to identify, understand and respond to hidden trends in order to make management decisions.

In the field of big data analysis, there are many areas. It is advisable to divide them into two categories: Big data Engineering and Big Data Analytics (Scientist).

Big Data Engineering concerns the design of a data processing, collection and storage system that allows processing petabytes of data and provides access for various user applications to the results of data processing. Specialists with good programming skills, knowledge of network technologies who are able to interact via the Internet and professionally work with computer equipment, are required for this.

Big Data Analytics covers the search for patterns in large amounts of data obtained from ready-made systems developed by Big Data Engineering. The direction of data analysis in itself is quite extensive and includes such specializations as Data Mining, Text Mining, Visual Mining, OLAP, Process Mining, Web mining, Real-Time Data

Mining, Streaml Mining, Multimedia Mining, Spatiotemporal Data Mining, Information Network Analysis, Biological Data Mining, Financial Data Mining.

The area of Big Data solutions is about the creation of analytical applications for any organization in the world that needs to make operational decisions based on all available information from any source, in any volume, as a result of in-depth analysis and in real time mode. The content of management process and analysis of large amounts of data is shown in Fig. 2.

Business Analytics is based on working with structured data. It uses traditional methods of mathematical analysis and statistics. Largely, Business Analytics is "descriptive analytics".

Owing to the extensive capabilities of Big Data analysis, interaction with the consumer is becoming more personalized and targeted. With the development of the forecast analysis, business structures in the sphere of e-Commerce have gained access to information that can make communication with consumers highly personalized.

Big data analysis allows business entities, which are engaged in online trading, to make business processes more customer-oriented, tracking recent purchases and analyzing customer habits.

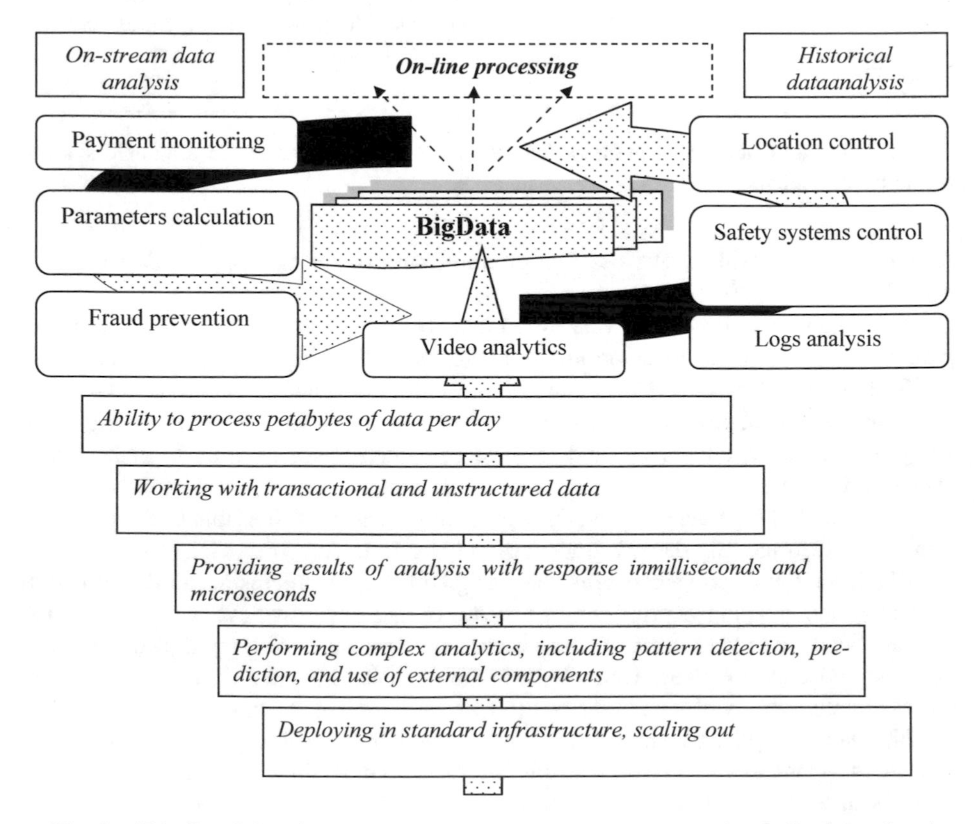

Fig. 2. "Big Data" functionality in business processes management (made by the authors)

According to marketing experts' surveys of B2B (Business to Business) company and B2C (Business to Consumer) company of markets in Europe and the UK, it was found that up to 43% of marketing managers use Big Data technology in their work. For marketers this technology makes it possible to find hidden relationships, improve the quality of solutions for processing client proposals, retain customers and introduce new unique services (Abbott and Shimek 2017).

At the same time, one of the most serious problems related to the use of Big Data technology is consumer distrust. Every year, more and more cases of information leakage are registered, which means that cyber security is becoming a growing priority for both organizations and individuals, as financial transactions and personal interaction are actively being transferred to the online environment.

Thus, data is the foundation of digital transformation, a necessary and indispensable condition for innovation. But to achieve that the data must be not just BIG, but necessarily SMART - it is necessary to apply the most effective methods, solutions and technologies for working with big data and use themfor the benefit of business.

3 Results

Today, civilization is moving towards a new model of increasingly decentralized organizations, where large operators are responsible for aggregating the resources of many people to provide services to a much more active group of consumers. This shift marks the emergence of a new generation of "dematerialized" organizations that do not require physical offices, assets, or even employees.

The features of the economy in the era of digitalization are shown in Fig. 3.

Exponential economy covers the material objects of the physical world and perfectly coexists at the level of individual countries' economies. The use of blockchain technologies in this economy can provide an exponential growth in the capitalization of companies due to the emergence of new business models, improving the efficiency of work within the previous ones, creating new goods and services, etc.

At the same time, the main deterrents to competition are not space and time, but risk and speed. Under these conditions, the uncertainty of managers in decision-making is growing, and as a result, the demand for big data analysis, primarily about customers, is increasing. Accordingly, the value of such data is increasing.

The analysis of big data includes the development of different systems of classification and prediction with the aim of exploring trends and patterns with the subsequent interpretation of results.

Specialists, who know the methods of search, selection, grouping, analysis, integration and data visualization, are required towork with data analysis. One of the available Big Data processing tools is digital platforms.

A digital platform is a tool that allows one to find the desired effect and create a chain of cooperation for it, or find the right resource and adjust it to a known effect, and do it in the form of a real business.

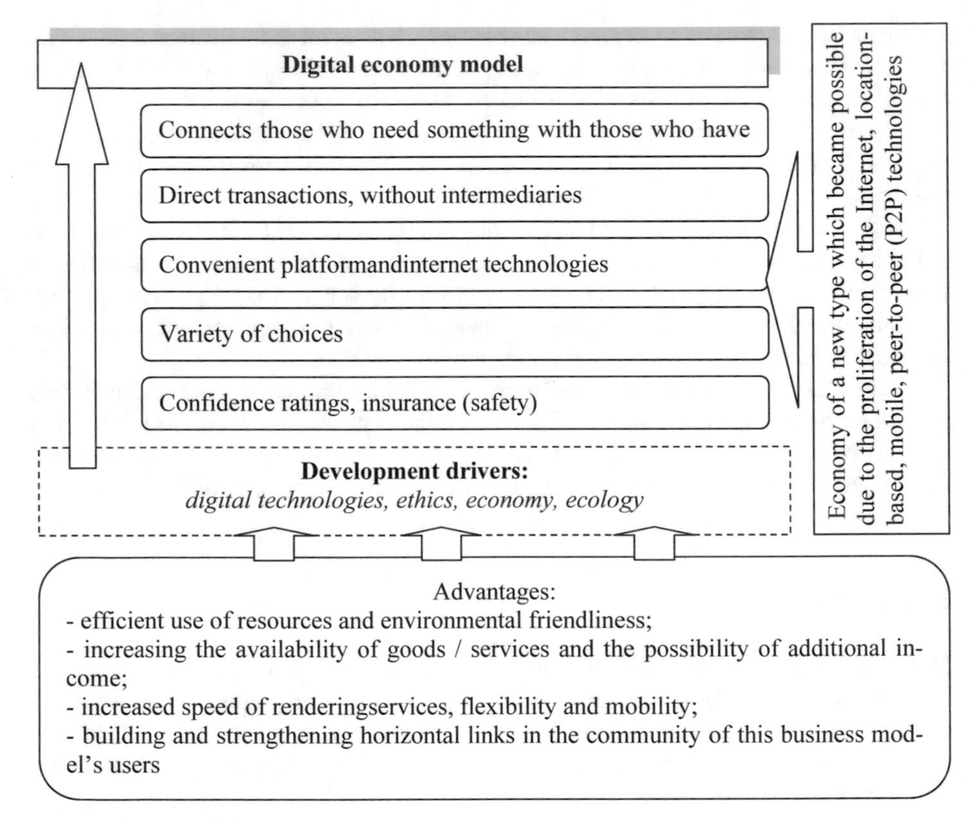

Fig. 3. Features of the economy under the conditions of digital transformation (made by the authors)

The integration of blockchain technology and "big data" on a digital platform provides great prospects for business, allowing: to get access to detailed information, on the basis of which it is possible to make important management decisions regarding further development.

Characteristic features of modern digital platforms are:

– mass introduction of information and communication technologies in all spheres of life;
– escalation of the size and complexity of the platforms;
– complexity of development environment and building;
– widespread use of parallel and distributed computing;
– cloud computing, mass deployment of mobile platforms, the Internet of things (absence of locally isolated systems).

It is important to organize correctly the process of creating reliable data on the scale of an organization and maintain it in this state – Data Governance. One must first identify initiatives for the development and monetization of the data asset and incorporate it in the data management strategyto accomplish this. Then it is necessary to form joint teams from business and IT to implement initiatives, develop Data Governance policies for creating and modifying data and a comprehensive program of projects for implementing initiatives. Only after that one can move on to implementing Data Governance policies to improve the manageability of systems that create data and establish responsibility for its quality and completeness.

"The Blockchain" technology ensures the encryption of bitcoin transactions and other transactions with Big Data containing private confidential information. Blockchain is practically invulnerable, because the transmitted information is not concentrated in one place, but is detailed into fragments and located at different addresses for security purposes.

The basic principles of business processes management based on two digital technologies are:

- real-time data acquisition;
- management of economic processes based on automated analysis of big data;
- high speed of decision-making, real-time rule changes - instant response to changes and interactivity of the environment;
- focus on a specific user, customer life situations as a business process (the user gets closer thanks to mobile devices and the Internet of things);
- digital ecosystem is perceived as a center of all participants'synergy;
- cybersecurity of management decisions.

The use of the digital platform will significantly unlock the potential of business processes, reduce risks, and provide new opportunities for monetization, the emergence of products that meet customer preferences.

In order to develop digital transformation, it is necessary to learn how to collect data so that to convert it into knowledge, create services on this basis, combine processes and establish management through the blockchain. This makes it possible to increase the efficiency of interaction with partners, reduce the cost of internal business processes. Also, it will increase confidence in the information within the ecosystem, reliability and security of distributed processes.

Thus, not only some technologies, but also new models of technologies and big data management that enable rapid response and modeling of future challenges and problems are becoming an important factor for success in the digital economy, which is highly competitive and cross-border.

4 Conclusions/Recommendations

The synergy of blockchain technology and big data, with the condition of complying with all the principles laid down in them, will fundamentally change the way the economy functions today and the way financial transactions are carried out.

The study concludes that the joint use of blockchain technology and big data is appropriate for creating a decentralized model for data exchange and storage, streaming and historical analysis. Control and security over this model is carried out with the help of a decentralized operating system.

Decentralized systems cannot be controlled by a minority of participants or by a single central authority, and they are transparent to all participants as well as self-governing ones. Thus, we can make a conclusion that the synergy of these technologies allows us to create a self-regulating and self-governing digital economy, which is controlled by computer programs and in which transactions are carried out through self-implemented digital contracts.

This kind of decentralized management can lead to reducing the number of inefficient operations and diminishing the level of corruption, increasing cybersecurity in the management of business processes. Due to the fact that each individual network element handles each transaction, none of the elements controls the database as a whole. From this perspective, decentralization also plays a role in enhancing the security and sustainability of the economic system.

Digital economy on the basis of joint use of the two technologies allows us to freely carry out operations on their own terms. This opens up prospects for the development of innovative intermediary services in the future. Such services could allow a third party to approve or reject a transaction in case of a dispute between other parties, which requires further development of both regulatory and methodological support.

Acknowledgements. The reported study was funded by RFBR according to the research project № 18-010-01119 «Management of digital transformation of innovation-industrial cluster as a system-forming element of the industrial digital platform: methodology, tools, practice».

References

Abbott, M., Shimek, R.: Ekonomika potrebleniya v usloviyah sovmestnogo ispol'zovaniya dannyh: Opredelyaya perspektivy novyh biznes-modelej [Economy of consumption in a data-sharing environment: Determining the prospects for new business models] (2017). https://www.aig.ru/content/dam/aig/emea/russia/documents/brochures/iot3fin.pdf. Accessed 19 April 2019. (in Russia)

Botsman, R., Rogers, R.: What's Mine Is Yours: The Rise of Collaborative Consumption. HarperBusiness, New York (2010)

Doklad o mirovom razvitii. Cifrovye dividendy. Obzor. Mezhdunarodnyj bank rekonstrukcii i razvitiya [World Development Report 2016. Digital dividends. Overview. International Bank for Reconstruction and Development] (2016). https://openknowledge.worldbank.org/bitstream/handle/10986/23347/210671RuSum.pdf. Accessed 25 April 2019. (in Russia)

Dokukina, I.A.: Osobennosti formirovaniya decentralizovannoj sistemy upravleniya dannymi v medicinskih uchrezhdeniyah na osnove tekhnologii blokchejn [Features of the formation of a decentralized data management system in medical institutions based on blockchain technology]. Vestnik TVGU. Series: Economics and Management, no. 3, pp. 106–112 (2018). (in Russia)

Genkin, A.S.: Blokchejn: Kak eto rabotaet i chto zhdet nas zavtra [Blockchain: How it works and what awaits us tomorrow], p. 592. Alpina Publisher, Moscow (2018). (in Russia)

Huws, U.: Labor in the Global Digital Economy: The Cybertariat Comes of Age. Monthly Review Press, New York (2014)

Mkrttchian, V., Gamidullaeva, L., Panasenko, S., Sargsyan, A.: Creating a Research Laboratory on Big Data and Internet of Things for the Study and Development of Digital Transformation. In: Kaur, G., Tomar P. (eds.) Handbook of Research on Big Data and the IoT, pp. 339–358. IGI Global Pennsylvania, Hershey (2019). https://doi.org/10.4018/978-1-5225-7432-3.ch019

Mogajar, U.: Blokchejn dlya biznesa [Blockchain for business]. Moscow, Eksmo, p. 224 (2018). (in Russia)

Polyanin, A., Pronyaeva, L., Golovina, T., Avdeeva, I., Polozhentseva, Y.: Administrative and managerial approaches to digital economy development in Russia. In: Proceedings of the 29th International Business Information Management Association Conference - Education Excellence and Innovation Management Through Vision 2020: From Regional Development Sustainability to Global Economic Growth 29, Vienna, Austria, pp. 2166–2179 (2017)

Swan, M.: Blokchejn: skhema novoj ekonomiki [Blockchain: New Economy Chart]. Olimp-Buziness, Moscow, p. 240 (2018). (in Russia)

Vertakova, Y., Klevtsova, M., Babich, T.: Identification of the new research areas and development of the existing ones by methods of morphological analysis and synthesis. Econ. Ann.-XXI **157**(3–4), 4–7 (2016)

Vertakova, Y., Risin, I., Treshchevsky, Y.: The methodical approach to the evaluation and development of clustering conditions of socio-economic space. In: Proceedings of the 27th International Business Information Management Association Conference - Innovation Management and Education Excellence Vision 2020: From Regional Development Sustainability to Global Economic Growth, IBIMA 2016, Milan, Italy, pp. 1109-1118 (2016)

Vin'ya, P.: Mashina pravdy. Blokchejn i budushchee chelovechestva [The truth machine. Blockchain and the future of humanity]. Moscow, Mann, Ivanov i Ferber, p. 320 (2018). (in Russia)

Digital B2B Communications: Economic and Marketing Effects

Galina Deryabina(✉) and Nina Trubnikova

RUDN, Moscow, Russian Federation
g_deriabina@yahoo.com, ninavadimovna@mail.ru

Abstract. The purpose of this study is to analyse the efficiency of international digital B2B instruments implementation at Russian market for the development of recommendations for the most effective approach this viral channel.

Methodology. The authors use the field studies (in-depth interviews, collection of expert opinion) as well as do the comparative analysis of the open sources (international scientific views in regards to digital technologies implementation, companies' reports). Such integral analysis of theory and practice allows creating the new concentrated knowledge in the development of digital technologies.

Findings. Digital technologies intensify their development in Russia. While, the digital B2B has its own specifics. The producing companies, historically putting the focus predominantly at the final consumer, start shifting towards winning the competition via reinforcement of their influence at trade partners (distributors and retailers). It involves such instruments as improvement of quality of communication with trade partners, frequency of communication, their education and motivation, delegation of merchandising tasks, B2B2C communication of trade partners with final consumers as well as the increase of trade partners' loyalty towards the companies' products. As the consequence: the growth of sales.

However, the digital B2B is an emerging direction for Russia. The major difficulties in launch of mobile communications relate to the Internet coverage across the regions, the smartphones' availability, and the age of the participants involving certain complexity in education to new technologies.

Practical implications. By bringing the new technologies to emerging markets, the international corporations should take into consideration their specificity. The focus at digitalization, becoming the relatively effective instrument of delegation of part of marketing and trade marketing tasks to trade partners, while decreasing the human field resources in parallel before reaching the desired economic effects, can bring the companies to resources gap and as a consequence to acceleration of decrease of sales.

Keywords: Digital technologies · B2B marketing · Marketing communications

JEL Code: O32

E. G. Popkova and B. S. Sergi (Eds.): ISC 2019, LNNS 87, pp. 866–875, 2020.
https://doi.org/10.1007/978-3-030-29586-8_98

1 Introduction

The modern B2B market is quite diverse; it represents a significant number of both small players and large corporations with traditional activities and chains of relationship building.

These relationships can be defined as a kind of "exchange" between the two businesses (Marketer's Notes 2019). It exists in two environments in parallel: online and offline; and we observe the synergy leading to the integrated result.

Shaun Gahan noted that it is not just necessary, but vital to understand the difference between the "consumer and business market" (Sean 2011). That is especially important for digital communications.

The participants of B2B market play the different roles: sellers, buyers and partners.

The Specifics of B2B Market:

1. Specific products and services that require the informational and secured promotion. For example, construction equipment, software, banking or insurance services, telecommunications, equipment for sophisticated communication, factory flow lines, etc. The specificity of such products determines their purchase with detailed examining of all characteristics, instructions, nuances of use, and possibilities of operation within a particular business. The indication of potential problems is required for a buyer to evaluate a transaction perspective.
2. The variety of contact points with client. They appear at every stage of work: engagement, initial contact, prerequisite, order, execution, maintenance, contract signing, post-customer service. The single B2B company has approximately 115 contact points, i.e. situations and reasons for customer contact with an enterprise (Bobrikov 2015).
3. In relation to B2B market the concept of "goods" is considered in a broad sense and implies the inclusion of related services. By employing their marketing mix systems, the B2B companies do not sell the products, but the solutions (Baryshev 2007). With growth of competition in segment, the companies began to complement the sale of product with the services; moreover, the share of value added services has been growing steadily with the development of market.
4. The lower price elasticity of demand has been noticed in marketing of the corporate companies, in comparison to the mass consumer of B2C segment, which in turn is very "susceptible" to discounts, promotions and other marketing tools (Furschik 2007).
5. **The sales network plays the fundamental role in B2B.** The role of sales network is expressed in its influence on other components of the marketing mix by setting in motion the most of resources of the enterprise (Titarev 2012).
6. The decision time in B2B is much longer compared to the B2C segment, since this process covers a significant number of people and very complex. The transactions in "business to business" segment are more complex than in consumer market. Therefore, in B2B communications, the important role is given to the educational component.

7. The participants of transaction have the multidirectional interests. The party barely involved into decision making has the right to set its requirements: for example, the buyers desire a profitable financial transaction, the production department - to increase the productivity, the heads of security department – to reduce the risks. **The human factor is extremely important in the B2B segment.** It is often underestimated, and attributed to the B2C market.

Marketing and sales are catching up with digital technologies developing in a fast pace. The authors distinguish the following **trends in digital B2B:**

1. The mass distribution of retargeting and lead cooking technologies combined with the cheapening and distribution of Big Data technologies will allow even the small and medium size businesses in B2B to have their own big data, to automatically customize the heated messages and to transfer the collected user data to CRM, that is already executed by mobile operators, banks and large corporations. "To succeed, a business-to-business (B2B) company needs a lot of sales leads" (Boachie 2018). While, the process of marketing automation is fragmented, using the services that are poorly integrated with each other. There is a need for integrated systems that allow to track the client's path from the first contact to the target action and to properly organize the work at each stage of the movement through the sales funnel.
2. The development of targeted personalization technologies and one-to-one marketing, the further orientation on Account Based Marketing (ABM) or Marketing of key customers in B2B is a concept which essence conveys the Pareto principle: the 80% of revenue is generated by the 20% of customers. In ABM concept the customer is a separate market. Based on the list of the most important clients (accounts) their needs are analyzed, followed by the plan of action, the personalized offers, and the personalized content that is distributed via e-mail, social networks and other channels. In ABM, marketing is closely integrated with sales, and not limited by the attracting of leads. Marketing campaigns have the individual character, by allowing to achieve the higher ROI in case of mass campaigns. "In a survey of 115 marketing specialists in B2B roles, Omobono found that 79% rated social media as the most effective marketing channel, with 38% noting that if they had extra budget for next year, they would spend it on social media" (Chaffey 2018).
3. Content transformation (proportionality, purity, interactivity). The principle "Less content - more quality" comes from the idea that users do not have time to process the entire content, and the return from every new unit decreases. This phenomenon is called the content shock, coupled with active promotion through various channels: social networks, blogs, media, e-mail. In B2B the selected content performs better in a specific way. Content is created not for the sake of content, but with a specific purpose to involve consumers, to show the level of the expertise, or to increase the brand awareness. This forces B2B representatives to look for engagement that could stand out, with the aid of infographics, leads' generating quizzes, calculators, pickers, online applications, etc. (Johnson 2018).
4. The replacement of human resources by digital instruments, in its extreme form of robotization of sales and service departments, highlighted as revolutionary tools a couple of years ago, does not pay off in large number of business cases because of dissatisfaction of customers with pure digitalization and the high cost of robots.

That is especially valid for B2B. Despite of the rapid development of digital technologies, the sales with a human face remain relevant. Many sellers are seeking for potential customers at various events thanks to more personal and closer communication. However, the generation Z is not performing well in sales, primarily due to information-fragmented syndrome, they are massively affected (Mayboroda 2017).

The international corporations launch the digital B2B channels as the opportunity for the growth of business thanks to speeding up and improvement of communication with trade partners, better tasks setting, control reinforcement, assortment and investment efficiency management, field information collection, education of trade partners and increase of their loyalty to company products.

While, one of the goals, not declared, becomes the efficiency in use of human resources, gradually replaced by digital. The calculated efficiency is quite big: the 30% (see the results below).

However, the digital instruments are far from being perfect, and do not demonstrate the good results yet, and, accompanied by the human resources reduction, might accelerate the decrease of revenues instead of the growth.

2 Methodology

The authors use the sources that reveal the main trends in B2B communications and digitalization, the classification of digital and B2B, as well as the authors' own work and field studies with retailers in regards to use of digital technologies in their work with international companies.

The paper contains the qualitative studies of 2017 and 2018: 20 in-depth interviews with tobacco outlet owners and shop assistants in Moscow and in Novosibirsk in regards to B2B digital instruments, before and after their mass implementation. The study continues in 2019.

The participants were asked the questions regarding the existing trade and trade-marketing programs, difference of offline work vs. switching to online B2B platforms, benefits and difficulties of their usage, as well as analysis of role of trade representative before and after the digital channel implementation.

3 Findings

As a rule, the most of retail outlets have the agreements with manufacturing companies for execution of trade programs in their outlets. Such programs might include, in addition to fulfilling the sales plan, the creation of product stock, the availability of products in POS (absence of out of stock), as well as performing the trade-marketing tasks such as product visibility, display, placement of advertising materials in the outlet, education, and advocacy (promotion of products by trading partners to the final consumer).

The owners of outlets get the bonuses from companies for keeping the agreed stock and managing other allowed activities in points of sale, while the programs themselves are usually executed by their staff (shop assistants) being the additional source of staff's income and making them more loyal to the companies.

Along with the evolution of digital technologies (appearance of smartphones, growth of Internet coverage, mobile application development) the industrial companies started including the mobile technologies into B2B communication with trade partners aiming the delegation of part of trade marketing and merchandising tasks and increasing the speed of cooperation as well as releasing the time of sales representatives.

The authors use the example of tobacco companies, having the restricted access to the final consumers via marketing programs and putting the major focus at the loyalty building with the trade partners.

The field study included the researches of the fieldwork of such companies as Philip Morris, JTI, BAT, Imperial Tobacco, Don Tabak, represented in Russia and mentioned as most active players of tobacco market by all the participants.

The highest activity regarding the trade marketing was observed among Philip Morris and JTI companies, proposing the wide specter of trade programs. Both companies executed the education of retailers on companies' standards and requirements of work, and informed about the companies' products and specificity of communication with adult consumers. While JTI was mentioned as paying more for similar activities.

According to participants, the most popular trade programs included "building of product stock", "availability", "visibility" and "advocacy".

The organization of execution of such trade and merchandising activities in outlets has always been the additional responsibility of sales representatives of tobacco companies, and constituted the 30–40% of their call mission (field activities) time (Table 1).

Table 1. Call mission activities of trade representatives.

Group of activities	Activities	Field time
Sales	- Introduction of new products - Negotiations - Orders etc.	20%
Merchandising	- Availability - Visibility - POSM placement - Category education etc.	40%
Brand & education	- Engagement & education - Retailers' activation - Advocacy - Trade programs - Consumer info - Loyalty building etc.	30%
Other	- Coaching - Compliance - Problems' solving etc.	10%

Source: built based on the authors' practical experience.

To ensure the transfer of trade programs to digital the tobacco companies developed the web platforms and mobile applications for their trade partners. The most known within the research participants were "KUspekhu!", "PickAp!" of PMI and "JTI club" of JTI.

The sales representatives install the applications at the smartphones of participants or the phones provided by the companies.

The applications contain the personal data of participants in form of CRM system as well as information about the trade programs executed in their outlets, their performance and received bonuses (task management system). The system contains the educational and engagement content for the participants as well.

Research participants in Moscow noticed the very first pilots of B2B digital applications in 2014 ("PickAp!"). The mass application launches took place in 2017.

Overall, the participants of research, especially in Moscow, noted their **positive attitude towards moving to digital channel**. The major reason for participation was financial (bonuses converted into roubles).

The participants noted: "it was interesting at the very beginning, it was like a game", "application allows being more relaxed because nobody supervises you".

However, the following **issues were observed during the use of mobile platform:**

Both cities (Moscow and Novosibirsk):

- Weak Internet connection, especially in underground outlets;
- Constructive issues of applications: slowing down of applications, insufficient user friendliness.

Novosibirsk only:

- Insignificant smartphones distribution, dominance of "button" phones, thus inability to participate in digital program until the smartphone provided by tobacco company;
- Low quality of given by tobacco companies smartphones, slowing the tasks accomplishing;
- Poor computer literacy of participants (shop assistants) that required additional efforts of sales representatives in training and educating. The participants were often forgetting the password. The work was frequently done by sales representatives themselves instead of shop assistant;
- Poor knowledge of Russian language among the retailers-immigrants creating the extra barriers in trade tasks execution via mobile.

Some of research participants openly shared the manipulation schemas for digital with product stock like "hidden stock" for photos or renting of products in neighbour outlets.

Other participants were worried about the "big brother" watching them: "I cannot openly talk with friends, somebody can hear us".

The majority of research participants affirmed that **importance of sales representative physical presence in outlet has even increased after switching to digital platform** with the necessity to explain, or assist on technical and financial issues (from helping entering logins/passwords to checking the bonuses).

Meanwhile, the analysis of annual dynamics of number of employees, example, in PMI and its decline by 4% in 2018, right after the mass digital B2B implementation, allows to suggest the existence of a certain link between the transition to digital and employees reduction (see the Table 2; PMI annual company reports 2015, 2016, 2017, 2018).

In 2017, the company presented the table demonstrating the positive dynamics of number of company employees that is not a case in 2018: the dynamics has changed. Moreover, the first time since the annual reports publications the field force employees' number has not been included into report.

Table 2. Dynamics of number of employees of PMI in 2015–2018.

Quantity/year	2018	2017	2016	2015
Total employees	77,400	80,600	79,500	80,200
Field force	?	28,700	26,700	26,700

Source: built based on PMI annual company reports 2015, 2016, 2017, and 2018.

The net revenues of the company continue growing since launch of heating tobacco IQOS in 2016 (RRP – reduced risk products), despite of overall tobacco products quantity decline (see Tables 3, 4, 5 and Fig. 1).

Table 3. PMI total cigarette and heated tobacco unit shipment volume and revenue 2015–2018.

Quantity/year	2018	2017	2016	2015
Total tobacco unit shipment volume, billion	781.7	798.2	812.9	847.3
Net revenue, billion USD	29.6	28.7	26.7	26.7

Source: built based on PMI annual company reports 2015, 2016, 2017, and 2018.

Table 4. PMI shipment volume by brand 2017–2018 (annual report 2018).

Brand/year	2018	2017	Change, %
Marlboro	264,423	270,366	(2.2)%
L&M	89,789	90,817	(1.1)%
Chesterfield	59,452	55,075	7.9%
Philip Morris	49,864	48,522	2.8%
Sampoerna A	39,522	42,736	(7.5)%

(*continued*)

Table 4. (*continued*)

Brand/year	2018	2017	Change, %
Parliament	41,697	43,965	(5.2)%
Bond Street	32,173	37,987	(15.3)%
Dji Sam Soe	29,195	22,757	28.3%
Lark	23,021	24,530	(6.2)%
Fortune	16,596	13,451	23.4%
Others	94,583	111,720	(15.3)%
Total Cigarettes	**740,315**	**761,926**	**(2.8)%**
Heated Tobacco Units	**41,372**	**36,226**	**14.2%**
Total Cigarettes and Heated Tobacco Units	**781,687**	**798,152**	**(2.1)%**

Source: PMI annual company report 2018.

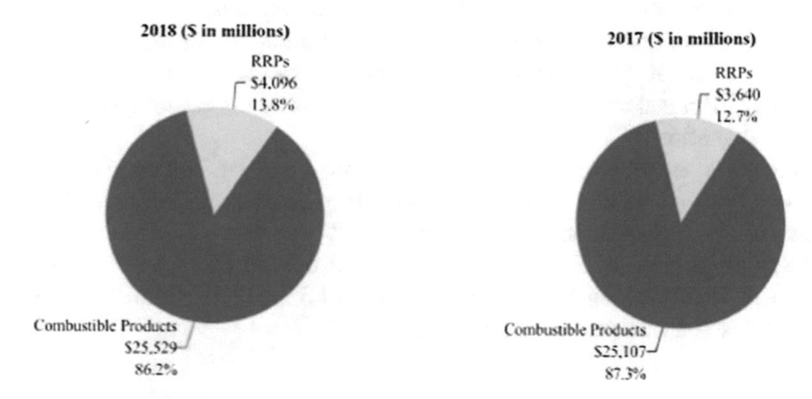

Fig. 1. PMI Net revenues by product category for the years ended December 31, 2018 and 2017. Source: PMI annual company report 2018.

Table 5. PMI Operating Company Income in 2015–2018.

Quantity/year	2018	2017	2016	2015
OCI, bio USD	11.4	11.6	10.9	10.7

Source: built based on PMI annual company reports 2015, 2016, 2017, and 2018.

At the same time, the company OCI started to decrease, most probably driven by the high investments into the new product development and promotion (see Table 5). That might be one of essential reasons for human resources reduction.

However, human resources replacement by the digital technologies might generate the lack of resources, provoking the further sales and profit decline.

4 Conclusions

The digitalization of sales and marketing is a major trend nowadays. However, just the few industries start going viral with their field (trade marketing) activities yet: most of FMCG companies focus at the final consumer and at consumer related applications (digital marketing, e-commerce etc.).

While, it is not a case of tobacco companies. By being strictly limited in number of activities and in direct contact with adult consumers, they reinforce their connections with trade partners by bringing digital technologies to the field (points of sale) for expenses optimization and growth of business efficiency.

However, one of the major goals seems to be the replacement of trade marketing human resources by online platforms with electronic tasks assignment and trade partners' performance assessment.

Meanwhile, the studies demonstrate that the technologies are far from ideal yet, along with the capabilities and the desires of the participants.

The lack of attention to technologies' implementation and insufficient field support bring to decrease of quality of communication and might result in decline of sales. Other companies, just starting the development of similar digital solutions, should take it into consideration.

Information technologies are irreplaceable in terms of finding consumers with actual needs (the so-called "warm" clients), but then human resources and the potential of individual communication should be connected.

The strategic line of B2B companies should not be the "pure" digitalization, but the aim of increasing the educational potential of information technology and the loyalty of the companies' employees.

References

Baryshev, A.V.: Efficient promotion at B2B market. Marketing i marketingovye issledovaniya [Mark. Mark. Res.] **06**(72), 483–485 (2007)

Boachie, P.: 4 Hacks You Need to Know to Amplify Your B2B Sales (2018). https://www.entrepreneur.com/article/311029. Accessed 10 Nov 2018

Bobrikov, O.V.: B2B marketing: customer communication building. Promyshlenny i B2B marketing [Ind. B2B Mark.] **2**(30), 127 (2015)

Chaffey, D.: Using social media marketing in B2B markets? (2018). https://www.smartinsights.com/b2b-digital-marketing/b2b-social-media-marketing/b2bsocialmediamarketing/. Accessed 26 Jan 2019

Furschik, A.A.: Specificities of product promotion in B2B and B2C segments at example of TV communication. Marketingovye kommunikatsii [Mark. Commun.] **05**(41), 274–275 (2007)

Johnson, R.C.: Your B2B Customers Deserve Engaging Content, Too! So, Why Not Give It to Them? (2018). https://www.entrepreneur.com/article/315923. Accessed 10 Mar 2019

Marketer's notes: Business to Business (B2B) (2019). http://www.marketch.ru/marketing_dictionary/marketing_terms_b/b2b/. Accessed 20 Apr 2019

Mayboroda, A.: What you need to know about B2B sales today, tomorrow and in 10 years ahead (2017). https://rb.ru/opinion/b2b-trends-forever/. Accessed 20 Apr 2019

PMI_2015AR_CompleteAnnualReport-3. https://www.pmi.com/investor-relations/reports-filings . Accessed 27 Apr 2019

PMI_2016AR_CompleteAnnualReport. https://www.pmi.com/investor-relations/reports-filings. Accessed 27 Apr 2019

PMI_2017_AnnualReport. https://www.pmi.com/investor-relations/reports-filings. Accessed 27 Apr 2019

PMI_2018_AnnualReport. https://www.pmi.com/investor-relations/reports-filings. Accessed 27 Apr 2019

Sean, G.: The B2B Executive Playbook Growth, pp. 14–15. Clerisy Press, Cincinnati (2011)

Titarev, D.A.: Transformation of 7P model of Bitner to B2B market specificity. Promyshlenny i B2B marketing [Ind. B2B Mark.] **04**(20), 260 (2012)

Problems and Prospects of Economic Digitalization in Kyrgyzstan

Ainura M. Khamzaeva(✉), Inabarkan R. Myrzaibraimova, and Kanzharbek A. Mamashov

Osh Technological University, Osh, Kyrgyzstan
ainura.hamzaeva@mail.ru, salamat-ar@mail.ru, kanjarbek79@mail.ru

Abstract. The article is dedicated to the study of the most urgent and discussed the issue among theoretical scholars and practical economists around the world – the economic digitalization. The purpose of the article is to develop measures on enhancement and improvement of economic digitalization efficiency in the Kyrgyz Republic. The objective of the study is to justify the external necessity and efficiency of the digital transformation of the economy, reveal economic digitalization issues, outline its trends and promising areas. To achieve the purpose set, the authors investigated the essence of the digital economy concept and its components and interpreted it based on a comparison of the various definitions. We described in details the advantages and risks associated with economic digitalization in order to determine and compare the potential benefits from the development of the digital economy and issues, to avoid crises in the evolution of the global economy. We proved the need for the introduction of modern digital information technologies both into the economy of the Kyrgyz Republic and in all areas of public life on the back of a comparative analysis of the advanced international practice of economic digitalization. We examined the regulatory documents, intergovernmental and government programs, digital agenda concepts, special reference books as well as research materials on the issues under consideration. After review of the above-listed sources, we determined purposes, objectives, and elements of economic digitalization in the Kyrgyz Republic. We carried out an analysis of the current state, trends, problems, and prospects of economic digitalization in the Kyrgyz Republic. We summarized the recommendations on acceleration and enhancement of the digital transformation efficiency in the country's economy.

Keywords: Digital economy · Digitalization · Digital Kyrgyzstan · Digital technologies · e-commerce · e-payments · e-services

JEL Code: F39 · F21

The whole world today is talking about the digitalization of public life and economy. The digital economy is considered almost a universal remedy. This is a global trend, the most discussed and burning topic for the development of any country today. Of course, ICT dramatically transform social relations, assisting the establishment of an innovative information society called the "digital economy".

E. G. Popkova and B. S. Sergi (Eds.): ISC 2019, LNNS 87, pp. 876–881, 2020.
https://doi.org/10.1007/978-3-030-29586-8_99

What is this digital economy? Is it a myth or a reality? Maybe is it an objective process that we can't stop? Or maybe is it too early for Kyrgyzstan residents to talk about it because we have not yet predetermined its introduction? We tried to find answers to these questions and to present our view of the problem.

First of all, we should note that digital technologies, high-speed Internet connection, and high-quality communication certainly open up opportunities for the exchange of big data and accumulation for reasonable decision-making and benefit. Whether we like it or not, IT is gradually replacing people, and the emergence of a digital enterprise, e-government, e-state, smart cities, and productions can severely change the entire economy. Digitalization and use of modern network and intelligent ICT make modern economic activity flexible, dynamic and sophisticated. Digital technologies mean real-time data exchange, affect all sectors of the economy and bring the so-called digital dividends in the form of economic growth, additional jobs, and better services.

At the same time, there are risks associated with all-around digitalization, such as data overrun, business loss, information space pollution, job cuts (technological unemployment), breach of confidential data security (cybersecurity).

The increasing role of information technology in social development in general and economic one in particular, some aspects of establishment and development of the information society were investigated by N. Negroponte, K. Kurokawa, T. Umesao, Yu. Hayashi, E. Masuda, K. Kohuma, M. Castells, J. Naisbitt, et al.

The term "digital economy" was proposed in 1995 by Nicholas Negroponte, an American scholar at the University of Massachusetts, to explain the advantages of the new economy due to the rapid development of information technology. Researchers have given different definitions and names of the digital economy (API economy, web-economy, the Internet economy, e-economy, application economy, creative economy, etc.). The World Bank experts suggested the definition: "… the digital economy is a new paradigm of accelerated economic development" (Alekseev 2016).

The European Union defines the digital economy as a global economy; the result of the transformational effects of new general-purpose technologies in information and communication area (Nikolaev, URL: https://konfef.pskgu.ru). Gartner gives the following definition: "digital business is a new business model that covers people/business/things and gains global scale via IT, Internet and all their properties, suggesting effective personal service in terms "everyone, everywhere, everytime"." The digital economy is defined as a global network of economic and social interactions made through information and computer technologies, which allow for the establishment of direct contacts between companies, banks, government and population without intermediaries under accelerated transactions and dealings (Nikolaev, URL: https://konfef.pskgu.ru).

In the opinion of other scholars, the digital economy is another marketing myth, a marketing brand from the West.

We agree with the opinion that the digital economy is a system of economic relations based on the use of digital ICT, and a state with a digital economy is mobile, flexible and quick-responsive to any modern challenges. Besides, the basic components of the digital economy are e-commerce; mobile banking; e-services; e-payments; Internet advertising; Internet content, etc.

The Boston Consulting Group (BCG) estimates that the digital economy currently provides 5.5% of global GDP. The proportion of the digital economy in GDP of the UK is 12.4%, China is 6.9%, the USA is 5.4%. In the EAEU countries, this indicator is 2.8% (about $85 billion dollars). The rate of digital economy penetration in Russia of 2017 was 3.9% of GDP. Russia is 39th in the rating of digital economies. The most visited sites in Russia are Ozon.ru, Aliexpress (over 9 million visits according to Similarweb).

According to Roland Berger forecasts, by 2025 digital transformation of the European industry may produce goods in the amount of 1.25 trillion euros, digitalization may bring to the global economy an additional income of 30 trillion dollars.

Kyrgyzstan is 109th of 176 countries in the rating of ICT development, the last among the CIS countries (according to the ITU data), 91st of 193 countries in the UN e-government development rating.

The digital leaders are Norway, Sweden, Switzerland. The top 10 countries include AKSh, United Kingdom, Denmark, Finland, Singapore, South Korea. International practice (in particular, China with its Alibaba Group system, the USA with Amazon, Japan with Rakuten, etc.) shows that information technologies assist the advance of public administration and business climate, improvement of services' quality, investment and innovation activity, development of human potential, direct interaction between the supplier and customer without intermediaries, GDP growth. As many theorists and practitioners argue that the digital economy and the use of innovative technologies are the most important force of innovation; they promote the growth of economy and competitiveness, raise the country's welfare and security through turnaround of production operations, increase in the equipment capacity, diminishment of production losses, R&D intensification (Goretkina 2017). Digital technologies in the economy give an opportunity to carry out remote administration of activity; accessible and free market; payment simplification; accessibility of any economic sector; complete departure from paper document workflow and the introduction of an electronic one.

The economic digitalization is recognized as a crucial factor in the development of the EEU economic space. The Digital Agenda of the EEU until 2025 has been developed to establish a single digital space, boost a single economic space and deepen the cooperation of member countries. The increase of the export potential of countries within the Union requires the implementation of joint initiatives and projects, coordinated actions and efforts.

On July 28, 2017, Russia adopted the "Strategy for the Development of the Information Society of the Russian Federation for 2017–2030" for the development of the digital economy. For the above-mentioned Strategy to be implemented, Russia developed and approved the program "Digital Economy of the Russian Federation".

Kyrgyzstan is among other countries of the EEU that has taken the path to the country-wide digitalization. As an ultimate goal, the National Development Strategy of the Kyrgyz Republic for 2018–2040 sets that Kyrgyzstan should become a country with a developed information innovation- and knowledge-based society, efficient, transparent, regulated, and corruption-free public administration with active participation of citizens as digital technology users (National Development Strategy of the Kyrgyz Republic 2018).

To achieve this goal, the government developed the Digital Kyrgyzstan program-2019–2023 that set outs the structure, foundations of the digitalization and management system hereof. The program involves the establishment of e-government, e-parliament, e-justice, digital economy. It sets three main goals: training of citizens' skills related to the usage of digital technologies in education, employment of citizen's potential; provision of high-quality services to citizens, entrepreneurs via corruption-free automated systems; economic digitalization, development of infrastructure and platforms (Concept of Digital Transformation "Digital Kyrgyzstan" 2018).

Thus, the development of the digital economy and its introduction is an objective necessity with more advantages than disadvantages. At the same time, many things depend on the government position. Government objectives include establishment of uniform rules for digital business transformation, development of international infrastructure, provision of broadband Internet access throughout the country; design of wi-fi and satellite technologies in remote regions, the most efficient, reliable and low-cost connection of the country to global networks and data highways, construction of data centers as alternatives.

The most crucial components of digitalization, which should be considered foremost are the development of digital infrastructure, content, skills, and expertise (https://analytics.cabar.asia/en/talant-sultanov). The digitalization should gradually cover all key areas of the social sector (education, health, environment), the economy (industry, agriculture, energy industry, telecommunications, banking sector, construction, tourism) and the political sector (corruption, fair elections) (Sulaimanova, URL: kabar.kg).

The republic carries out particular activities in the framework of digitalization. The institutional foundations of the digital economy are laid. On June 18, 2001, the President of the Kyrgyz Republic issued the Decree No. 199 "On ICT Council under the President of the Kyrgyz Republic (since 2008 – under the Government of the Kyrgyz Republic). Decree No. 54 of the President of the Kyrgyz Republic dated March 10, 2002 approved the National Strategy "ICT for the Development of the Kyrgyz Republic". Then a National Action Plan was developed to implement this strategy.

In 2011, the High Technology Park was established for the development of the software design industry with over fifty residents today. Kyrgyzstan is the first country in the region to join the international initiative "Open Government Partnership". Resolution of the Government of the Kyrgyz Republic No. 651 dated November 17, 2014 approved the "Program of the Government of the Kyrgyz Republic on the introduction of e-administration ("e-government"). In 2015, the Center for E-administration of the Government of the Kyrgyz Republic was founded. The Kyrgyz Association of Software and Service Developers is also operating.

Since 2014, the "ELSOM" electronic wallet project has been implemented. It's a virtual account represented by a mobile device number, which allows you to make purchases in stores, also online, pay utility bills, Internet and cable television, invest, transfer and cash money electronically. ELSOM provides an opportunity to effect financial transactions anywhere and anytime in the event of an available mobile network.

Public services are intensively being converted into electronic form; public authorities join the "Tunduk" project; a public service portal is being tested; the projects "Safe City", "Information kiosk", "State Electronic Document Workflow System" et al. are being launched; Interdepartmental Automated Information System

"Unified Register of Public Property", "Unified Register of Violations" et al. are being introduced. (Development and digitalization: what do the Kyrgyzstan regions expect in 2019. URL: politmer.kg/article/537). The Tunduk platform is based on the Estonian program protected by advanced encryption blockchain technologies.

Kyrgyzstan was among the first countries to join the regional integration program "Digital CASA", the goal hereof is to promote the integration of landlocked states of Central Asia and particular states of South Asia into the regional and global digital economy. The components of the Digital CASA-Kyrgyz Republic project are overcoming the barrier to ICT literacy; training new professionals required in a digital economy; minimizing human capacity shortage for digital transformation at the regional and national levels.

Therewith, there are restraints to the development of the country's digitalization in general, and the economy in particular such as high cost of projects and insufficient funding, technical backwardness, underdeveloped infrastructure, low interest of public authorities in the transition to rendering digital services, severe shortage of skilled staff, the lack of uniform standards and regulations, etc. The project of economic digitalization still causes distrust among the population, consequently, it is necessary to thoroughly elaborate every stage of the process and raise public awareness. According to official statistics, only 34.5% of the population of the Kyrgyz Republic use the Internet (average indicator for the CIS countries is 65.1%), 21.4% of households have computers (average indicator for the CIS countries is 67.4%), only 3–4% of the population has access to broadband Internet.

Therefore, the primary objective of the government is to build international infrastructure; reduce Internet costs; provide 60% of the population with broadband Internet access; ensure effective, reliable, and inexpensively connection of the country to global information networks and highways; develop common rules for digital business transformation.

The year of 2019 has been declared in Kyrgyzstan the "Year of Regional Development and Digitalization". The President set particular objectives and made recommendations on the acceleration of digital transformation and socio-economic development of regions, including development and introduction of a unified digital platform "Digital Kyrgyzstan": unified identification systems, electronic messages, digital payments, electronic interdepartmental interaction "Tunduk", portal and mobile application of electronic public services; creation of a unified state data center and regional centers; development and introduction of the digital platform of the municipal administration "Sanarip Aimak"; introduction of digital technologies in education and healthcare with electronic payments; establishment of a Center for digital educational technologies and regional development; design and putting into operation of instruments of electronic processing operations and payments for their digital development, etc. (What objectives on regional development and digitalization did Jeenbekov put for the government. URL: politmer.kg/article/538).

We think that the development of digital commerce and access to digital financial services should be prioritized. It is required to improve the digitalization of tax procedures, sending e-receipts and filing e-reports.

In general, for the purpose of building a digital economy information society to be achieved it is necessary to consolidate the efforts of the government, business, and civil society. The government should take an integrated approach to the fulfillment of the digital transformation of enterprises, administrating authorities, financial institutions, and the social area. We hope that owing to economic digitalization we will manage to eliminate the raw material-based direction in the development of the republic and create conditions for technological growth. The development of the digital economy should ultimately contribute to the improvement of living standards and the promotion of the country's competitiveness and national security.

References

Alekseev, I.V.: Digital economy: peculiar features and trends in the development of electronic interaction. In: Topical Issues of the Research: From Theory to Practice: Proceedings of the 5th International Research and Practice Conference, vol. 2, no. 4(10), pp. 42–45. Center for Scientific Cooperation "Interactive Plus", Cheboksary (2016)

Babkin, A.V., Chistyakova, O.V.: Digital economy and its impact on the competitiveness of business entities. Russ. Entrep. **18**(24), 4087–4102 (2017)

Goretkina, E.: RECS 2017: Pros and Cons of the Digital Economy, PC Week, 26 September 2017, no. 13(934) (2017)

What objectives on regional development and digitalization did Jeenbekov put for the government? List dated 15 January 2019 (2019). politmer.kg/article/538

The concept of digital transformation "Digital Kyrgyzstan" 2019–2023, Bishkek, 14 December 2018 (2018)

National Development Strategy of the Kyrgyz Republic for 2018–2040, UP No. 221, Bishkek, 31 October 2018 (2018)

Nikolaev, M.A.: Developmental factors of regions in the digital economy. Prospects for the economic development of the North-West Federal District regions. https://konfef.pskgu.ru

Popkova, E.G., Grechenkova, O.Yu.: Prospects for the application of social marketing in economic criminology within the development of an innovative economy. All-Russ. J. Criminol. **11**(2), 280–288 (2017)

Development and digitalization: what do the Kyrgyzstan regions expect in 2019. politmer.kg/article/537

Sulaimanova, M.V.: Kyrgyzstan plans to accelerate economic growth via new technologies. kabar.kg

Digital agenda of the Eurasian Economic Union until 2025: prospects and recommendations. World Bank Group, The United Nations Economic Commission for Europe, 40 p. (2018)

Digital skills and expertise for the development of the digital economy. ITU Regional Workshop "National Strategies of Digital Transformation", Ak Maral Recreation Center, Issyk-Kul, Kyrgyz Republic, 28–29 August 2018 (2018)

https://analytics.cabar.asia/en/talant-sultanov-digitalization-of-kyrgyzstan-is-not-a-luxury-but-a-call-of-the-times/

The Digital Reality of the Modern Economy: New Actors and New Decision-Making Logic

Nikita O. Stolyarov[1(✉)], Elena S. Petrenko[1], Olga A. Serova[2], and Aida S. Umuralieva[3]

[1] Plekhanov Russian University of Economics, Moscow, Russia
buffon.09@mail.ru, petrenko_yelena@bk.ru
[2] Baltic Federal University named after Immanuel Kant, Kaliningrad, Russia
prof.serova@gmail.com
[3] Kyzyl-Kiya Multidisciplinary Institute of Batken State University, Kyzyl-Kiya, Kyrgyzstan

Abstract. Purpose: The purpose of this paper is to develop a conceptual model of the digital economy that presents new actors hereof and their decision-making logic.

Design/Methodology/Approach: The research is conducted within the framework of a systems approach using the method of structural and functional analysis. The authors rely on statistical and analytical data of The Global Information Technology Report (World Economic Forum), World Robotics Survey (International Federation of Robotics), as well as World Digital Competitiveness Ranking (IMD business school).

Findings: We found out that the digital reality of the modern economy promotes the transformation of economic agents within hereof the state is turned into e-government, the population – into an information society, business entities – into a digital business, and employees – into digital staff. There is also a new economic agent (intelligent machines). Although they do not follow the independent logic, they determine, to a great extent, the logic of all other actors of the digital economy.

The generalized decision-making logic of the digital economy is reinterpreted through the lens of new criteria. Along with that, if common criteria are typical for e-government, information society, and digital business, digital staff apply individual criteria that are specific to each particular employee. This greatly complicates forecasting the development of the digital economy and the management hereof.

Originality/Value: The developed conceptual model of the digital economy that presents new actors and their decision-making logic made it possible to lower uncertainty and outlined the prospects for future scenario analysis of the digital economy.

Keywords: Digital economy · Economic agents · Decision-making

JEL Code: D81 · O31 · O32 · O33 · O38

E. G. Popkova and B. S. Sergi (Eds.): ISC 2019, LNNS 87, pp. 882–888, 2020.
https://doi.org/10.1007/978-3-030-29586-8_100

1 Introduction

The world's first programs of digital economic modernization were launched by the most progressively developed and developing countries less than ten years ago: by Germany, the United Kingdom, and the United States in 2012; by France, Japan and China in 2015; by Brazil, India, the Republic of South Africa, and Russia in 2017. Nevertheless, the results of deep transformation processes in the economies hereof and most other countries of the world are noticeable right now.

According to "The Global Information Technology Report" for 2016 (more recent studies on this topic were not conducted) presented at the World Economic Forum 2019, even then there was a tremendous impact of digital technologies on consumption in the countries of "High-income group average" ("6th pillar: Individual usage" is 5.9 points out of 7), production ("7th pillar: Business usage" is 4.6 points out of 7), labor ("5th pillar: Skills" is 5.7 points of 7) and public administration ("8th pillar: Government usage" is 4.8 points out of 7).

Affected by scientific and technological progress, this impact has become stronger in 2017–2018 (according to data as at the start of 2019). And now the new stage of digital modernization is beginning. Digital technologies replace previous ones in all economic areas. They are perceived as an integral part of human society and an indispensable attribute of economic activity. According to the "World Robotics Survey" presented by the International Federation of Robotics 2019, there are thirty-five million robots in the world with a total value of over $12 billion dollars, 70% hereof are household robots. The production of high technology is on the verge of a new breakthrough related to the development of artificial intelligence.

In this regard, the digital economy is no longer a future prospect and become a modern reality that needs an in-depth scientific study. It allows lowering the uncertainty and risk of economic activity for all its actors and developing the most suitable strategies of their behavior. The working hypothesis of this study is the emergence of new economic agents and the application of new decision-making logic in terms of the digital economy. The purpose of this paper is to develop a conceptual model of the digital economy that presents new actors hereof and their decision-making logic.

2 Materials and Method

The decision-making theory and practice in the context of market acting as a conceptual framework for a modern economy is examined in the proceedings of scholars and experts such as Mehta and Dixit (2016), Mohsenin et al. (2018), Takahashi et al. (2018), Wang et al. (2016). Based on them, we determined the common economic agents and their generalized decision-making logic (Fig. 1).

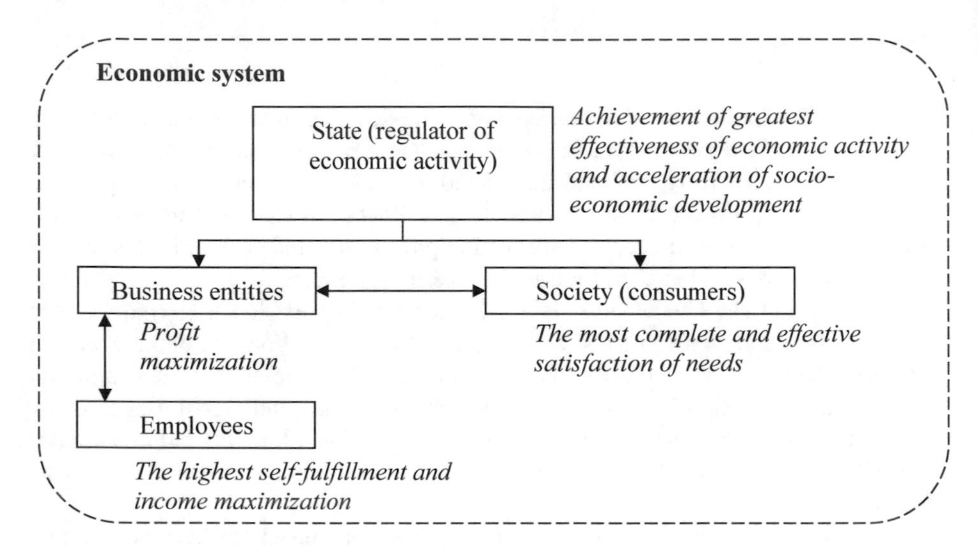

Fig. 1. Common economic agents and their generalized decision-making logic Source: drawn up by the authors.

As you can see from Fig. 1, the state in the conceptual model of a market economy governs economic activity, proceeding from the logic of maximization of the economic system performance and acceleration of its socio-economic development. The rationale for consumer behavior is pursuance of more full satisfaction of social needs, for business entities is profit maximization, for an employee is the highest possible self-fulfillment and income maximization.

At the same time, decision-making criteria based on the above-indicated logic vary in different types of business systems. For example, the crucial criterion for the suitable decisions made by all business entities in an industrial economy is the rate of mass production opportunities' use, and in a post-industrial economy is the degree of personal fulfillment and the level of innovation activity.

The fundamentals of the digital economy as well as practical experience of the development thereof in the modern countries of the world are discussed in the publications of Bogoviz et al. (2019a), Bogoviz et al. (2019b), Mueller and Grindal (2019), Negrea et al. (2019), Popkova (2019). The authors defined digital economy as a modern type of economic system in the framework hereof digital technologies (for example, computers, smartphones) and technologies (for example, mobile communication, the Internet) are widely spread, and breakthrough digital technologies (Internet of Things, cloud technologies, artificial intelligence, robotics, etc.) are extensively developed and used in high-tech industries.

A review of the literature on the chosen topic showed that the composition of economic agents and the logic of their decision-making in the digital reality of the modern economy is not scientifically-proven and not thoroughly studied despite a quite high degree of investigation of particular fundamental and applied decision-making issues and the development of the digital economy; therefore, it should be further studied. In this article, the research is conducted within the framework of a systems approach using the method of structural and functional analysis.

3 Results

As a result of investigation of the materials of the World Digital Competitiveness Ranking 2018 presented by the IMD business school 2019 as well as the national strategies for development of the digital economy in the most progressive countries in the world, in particular, the Digital Economy of the Russian Federation program approved by Resolution of the Government of the Russian Federation No. 1632-p as of July 28, 2017 we classified new business entities emerging in the digital economy as follows:

- E-government: a state governing economic activities with the help of digital technologies;
- Information society: consumers buying and consuming goods and services with the help of digital technologies;
- Digital business: business entities using digital technologies in the production and distribution of goods;
- Digital staff: employees with digital expertise, those who are able to use modern digital technologies in the production and distribution of goods and services;
- Intelligent machines: technical devices involved in the production, distribution, and consumption of goods and services. They have artificial intelligence and can remotely communicate with other machines and people; operate under human control or function autonomously (within the framework of a human-established program).

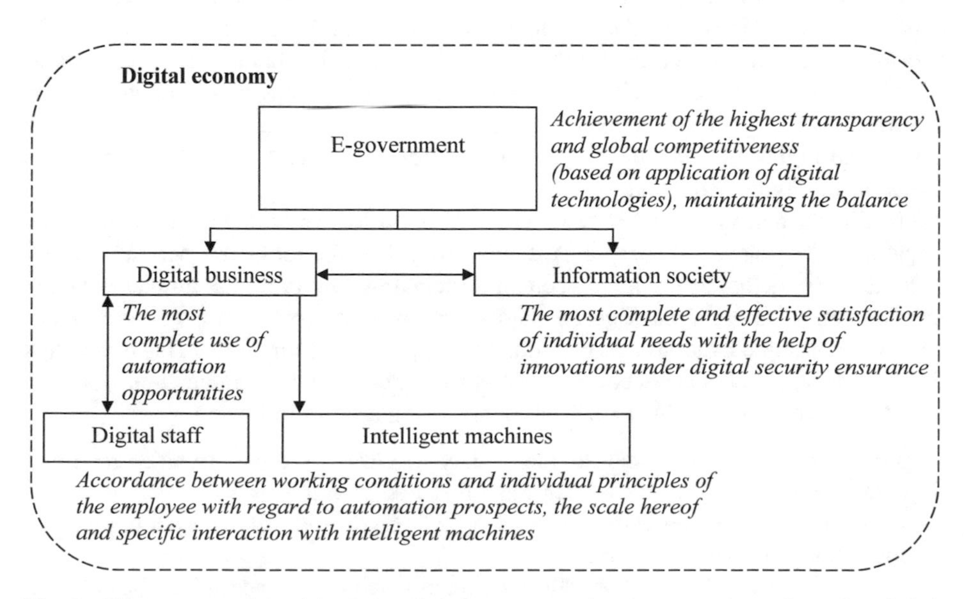

Fig. 2. The conceptual model of the digital economy showing new actors hereof and their decision-making logic Source: drawn up by the authors.

Based on the opportunities given by the digital economy and the risks associated, we revealed the decision-making logic for designated business entities. It is conveyed by the conceptual model of the digital economy presented in Fig. 2.

As you can see from Fig. 2, the logic of e-government behavior in the digital economy is included in the maximization of transparency and global competitiveness (on the basis of digital technology use) as well as in the maintenance of a balance between supply and demand. This is explained by the fact that digitalization of information increases the controllableness hereof, thereby broadening the opportunities of public monitoring and administration of economic activities.

The highest transparency is achieved with electronic corporate reporting, electronic document workflow, and electronic payments. In the context of the digital economy, both traditional competitiveness and digital one based on aggressive and efficient use of digital technologies in the economy is subject to evaluation. The balance between supply and demand is possible owing to digital marketing and the satisfaction of individual needs of each particular consumer based on flexible intelligent-machine automated production.

The information society follows the logic of the highest and the most effective satisfaction of individual needs with the help of innovations under digital security ensurance. One of the important features of the digital economy is the acceleration of scientific and technological progress. It stimulates demand for innovation. Digital security assuming the uninterrupted operation of digital devices and the integrity of private data is one of the key risks of the digital economy. Mass consumers are interested in management hereof.

Digital business is guided by the logic of the most complete use of automation capabilities. This is due to the fact that automation allows you to benefit from the economies of scale as well as to reduce the influence of the human factor (emotions, mistakes) on economic activity. In addition, the employment of the digital staff is understood as a specific competitive factor; since the more automated business processes, the higher the customer loyalty hereto (the more prestigious the products) in terms of the digital economy.

Intelligent machines are not guided by their own logic in the digital economy but implement the tasks set by people. Although, maybe in the long-term period they will become major actors of social and economic relations providing the ground-breaking development of artificial intelligence. Digital staff is guided by the logic of accordance between working conditions and individual principles of an employee. This means that the logic of digital staff is not common but individual for each particular employee. The generalized criteria for digital staff decision-making are the following:

- prospects of automation development: they determine the opportunities for long-term employment (the risk of staff reduction due to automation)
- the scale of automation: it determines the staff composition (the ratio of digital staff and intelligent machines herein);
- specific features of interaction with intelligent machines: it determines the status of digital staff against intelligent machines (their maintenance, their operation in production and distribution, their management or subordination to intelligent machines).

4 Conclusion

Thus, the working hypothesis was scientifically proven: the digital reality of the modern economy promotes the transformation of economic agents within hereof the state is turned into e-government, the population – into an information society, business entities – into a digital business, and employees – into digital staff. There is also a new economic agent (intelligent machines), which although doesn't use the independent logic but determines to a great extent the logic of all other actors of the digital economy.

The generalized decision-making logic of the digital economy is reinterpreted through the lens of new criteria. Along with that, if common criteria are typical for e-government, information society, and digital business, digital staff apply individual criteria that are specific to each particular employee. This greatly complicates forecasting the development of the digital economy and the management hereof. At the same time, the developed conceptual model of the digital economy that presents new actors and their decision-making logic made it possible to lower uncertainty and outlined the prospects of future scenario analysis of the digital economy the development of a methodology for provision hereof it is recommended to devote the further research.

References

Bogoviz, A.V., Alekseev, A.N., Ragulina, J.V.: Budget limitations in the course of digital economy development. In: Lecture Notes in Networks and Systems, vol. 57, pp. 578–585 (2019a)

Bogoviz, A.V., Lobova, S.V., Ragulina, J.V.: Distortions in the theory of costs within the digital economy. In: Lecture Notes in Networks and Systems, vol. 57, pp. 1231–1237 (2019b)

IMD business school 2019. World Digital Competitiveness Ranking (2018). https://www.imd.org/wcc/world-competitiveness-center-rankings/world-digital-competitiveness-rankings-2018/. Accessed 10 Apr 2019

International Federation of Robotics 2019. World Robotics Survey. https://ifr.org/news/world-robotics-survey-service-robots-are-conquering-the-world-/. Accessed 10 Apr 2019

Mehta, R., Dixit, G.: Consumer decision- making styles in developed and developing markets: a cross-country comparison. J. Retail. Consum. Serv. **33**, 202–208 (2016)

Mohsenin, S., Sharifsamet, S., Esfidani, M.R., Skoufa, L.A.: Customer decision-making as a tool for segmenting digital products market in Iran. J. Islam. Mark. **9**(3), 560–577 (2018)

Mueller, M., Grindal, K.: Data flows and the digital economy: information as a mobile factor of production. Digit. Policy Regul. Gov. **21**(1), 71–87 (2019)

Negrea, A., Ciobanu, G., Dobrea, C., Burcea, S.: Priority aspects in the evolution of the digital economy for building new development policies. Qual. Access Success **20**(S2), 416–421 (2019)

Popkova, E.G.: Preconditions of formation and development of industry 4.0 within the framework of the knowledge-driven economy. In: Studies in Systems, Decision, and Control, vol. 169, pp. 65–72 (2019)

Takahashi, H., Nishino, N., Takenaka, T.: Multi-agent simulation for the manufacturer's decision making in sharing markets. Procedia CIRP **67**, 546–551 (2018)

Wang, X.-Y., Zhang, W., Xiong, X., Che, H.-L., Shen, D.: Research on the negotiation decision-making model based on concession strategy in the artificial credit market. Filomat **30**(15), 3907–3916 (2016)

World Economic Forum 2019. The Global Information Technology Report 2016. http://www3.weforum.org/docs/GITR2016/WEF_GITR_Full_Report.pdf. Accessed 10 Apr 2019

Government of the Russian Federation 2019. The program "Digital Economy of the Russian Federation" approved by the Resolution No. 1632-p as of July 28, 2017. http://static.government.ru/media/files/9gFM4FHj4PsB79I5v7yLVuPgu4bvR7M.pdf. Accessed 10 Apr 2019

The Legal Challenges of Digital Reality for Society and State

Digital Constitution as a Scientific Direction

Emil E. Barinov(✉), Leonid G. Berlyavskiy, Andrey G. Golovko, and Natalya V. Dzhagaryan

Rostov State University of Economics, Rostov-on-Don, Russia
barinov@yandex.ru, berlg@yandex.ru,
kafedra.kimp@yandex.ru

Abstract. The paper studies the formation of digital constitution as a research direction in the Russian constitutional and legal science, which main components are studying the constitutional foundations of the digital economy, provision of human rights and liberties in the conditions of the information society, and transformation of the role of state and its institutes in the conditions of digitization.

Keywords: Constitution · Law · Information · Law · Legal regulation · Source of law · Form of law · Digital reality · Digital technologies

JEL Code: K1 · K4 · Z0

1 Introduction

The information society forms in the Russian Federation in the conditions of the digital age (Concept of the Information Code of the Russian Federation 2014). In April 2019, the draft law on provision of stable work of the Russian segment of the Internet (Runet) in case of cyber-attacks or other aggressive actions from abroad was adopted.

Digitization is treated as creation of a new product in the digital form. The Decree of the Government of the Russian Federation dated June 28, 2017 No. 1632-r adopted the National Program "Digital economy of the Russian Federation", which provides the following definition: digital economy is economic activities, which key factor of production is data in the digital form.

Russian investments in this direction (RUB 1 trillion for the last five years, according to Forbes) have not yet led to creation of a large-scale sector of the digital economy, and this is an additional proof for a wide public discussion of the search for ways of transition to the digital economy (Andreeva 2018). Parts II-III of the program "Digital economy of the Russian Federation" state that only 10% of municipal entities conform to the requirements to the level digitization that are set by the Russian law.

According to the Head of the Constitutional Court of the Russian Federation, Zorkin (2018), development of information technologies for two last decades leads to formation of a new – digital – reality. Jurisprudents discuss the usage of robots in their profession. A new law appears – "law of the second modern", which regulates the economic, political, and social relations in the context of the world of digits, Big Data, robots, and artificial intelligence.

E. G. Popkova and B. S. Sergi (Eds.): ISC 2019, LNNS 87, pp. 891–898, 2020.
https://doi.org/10.1007/978-3-030-29586-8_101

According to Bachilo (2016), information laws include 400 laws that regulate the relations on information and information technologies; 3,700 normative legal acts with amendments to these laws; hundreds of decrees of the Government of the Russian Federation with further amendments; 100 decrees of the President of the Russian Federation – this large number of legal acts causes further changes in other laws and could be explained by large dynamics in the sphere of information life and quick changes in development of the information and communication technologies.

The most important law in this sphere is the Federal law dated June 27, 2006, No. 149-FZ "Regarding information, information technologies and protection of information", which Article 2 defines information as data or messages regardless of their form. It should be noted that totality of information on something specific forms the phenomenon of knowledge. The federal laws "Regarding protection of children from information that damages their health and development", "Regarding provision of access to information on activities on courts of the Russian Federation", "Regarding personal data", etc. are closely connected to this law.

However, the existing laws do not fully conform to the modern requirements, as a lot of laws are not connected to the above basic law nor between each other. Thus, the information laws require systematization, deletion of doubling, and bringing their categorical tools into an unambiguous state (Zorkin 2018). The hopes for the project of the Information Code of the Russian Federation, which published concept does not contain such term as "digital law" are not justified in this regard (The concept of the Information Code of the Russian Federation 2014).

The laws of foreign countries show various vectors of development. Public relations, which are connected to digitization, are regulated by a wide circle of acts, including the acts of strategic planning (the Republic of Belarus, the Republic of Kazakhstan); program and forecast documents (the Republic of Kazakhstan and the Republic of Armenia); special laws (France – the Law "Regarding the digital republic" of 2016; the UK – the Law "Regarding online economy" of 2017). The European Parliament approved the norms of civil law on robototronics (February 16, 2017); Germany acknowledged bitcoin as a means of payment (February 27, 2018) (Khabrieva 2018).

As for the level of elaboration of the topic of the article in the scientific literature, the studies that determine the role of state in the modern world are still in the sphere of setting of questions and a lot of uncertainties (Moncinska 2018).

In the Soviet time, information law was treated as an element of administrative law, which main method is the method of imperative prescriptions from state bodies. In the post-Soviet period, Naumov (2002) analyzed the problems of legal regulation of the Internet and came to the conclusion that a new sphere of law – telecommunications law – emerged in Russia and the world. The information society was defined as society in which each member is provided by a legal opportunity to be a participant of the information exchange via creation and usage of the information institutes and means for the most effective and complete development (Boer and Pavelyeva 2006).

A group of Russian and French jurisprudents (Bachilo et al. 2013) studied the "right of digital administration in Russia and France", including such issues as development of electronic technologies in the election system of the Russian Federation, digital technologies in development of democracy and management as a tool of transformation of the systems of public management, the model of open government, and public services – a path to overcoming the information and social inequality, etc.

In foreign science, digitization in the context of law is considered as a natural phenomenon that emerges on the path of development of the legal system in the modern age. The works of Western scholars are concerned primarily with the practical aspects of digitization of law and the law enforcement practice. The Russian science has also interest in this topic. The studies aim at solving narrow but important problems that are connected to usage of digital technologies in the legal sphere. Discussions' topics are mainly the search for optimal solutions and development of the models of legal regulation of public relations, connected to application of digital technologies in the sphere of finance, public management, and creation of artificial intelligence (Khabrieva 2018).

Also, it is expedient to study the "political Internet" (Morozova 2011) and to reconsider the scheme of interrelations during realization of right for information as compared to the traditional realization of citizen's right for receipt of information and activity of public authorities in provision of information on their activities (Talapina 2015), as well as to analyze foreign experience.

In the USA, Internet is actively used in the period of election campaigns and creates conditions for unprecedented – as to the scale – research in which hundreds of thousands of people from different countries could participate (Elyakov 2011). After adoption of the Law on electronic signature in 2003, applications to the police of Finland and to the courts were accepted in the electronic form – together with the documents in the paper form. The documents in the electronic form that are applied to the state structures of Finland are digitized and shown in the unified information system of the state register. New digital technologies determine the future of Finland and change the effect of various social regulators (Zhilkin 2018).

The government of the USA, the UK, and Sweden showed a desire to prosecute the "digital revolutionary" J. Assange and his followers.

The recently published papers of T.Y. Khabrieva "Law and the challenges of the digital reality", S.M. Shakhrai's "Digital constitution. The fate of the main rights and liberties in the totally information society", and V.D. Zorkin's "Right in the digital world", etc. allow speaking of formation of the digital constitution as a scientific direction in studying the modern Russian and foreign constitutional law. In particular, in the context of the existing Constitution of the Russian Federation, it is necessary to develop the constitutional and legal concept of information law, which should be based on the constitutional right of citizens for information (Zorkin 2018).

Also, regarding the V.D. Zorkin's paper "Law in the digital world", V.N. Galuzo and N.A. Kanafin expressed critical opinions regarding the digital law. According to them, it should be studied as a preliminary result of the search for combination of economy and law.

On the other hand, (Khabrieva and Chernogor 2018) stated that influence of development of distribution of digital technologies on the modern transformation of law is least studied and considered by the legal doctrine. Digitization influences the sphere of legal regulation. It involves new public relations that have not existed before or have not required the legal regulation or could not be regulated by the law.

2 Methodology

The methodological basis of the research includes the general scientific methods of cognition: dialectical, logical, and systemic analysis; and special scientific methods: systemic & legal, comparative legal analysis, treatment of legal norms, formal and legal, and legal forecasting.

3 Results

In the most general form, the main priority of the legal policy of Russia in the sphere of state information resources is (Prokopenko and Krivoukhov 2007) formation of open state information resources for informing citizens and organizations on the state policy and state decisions and creation of “e-government”, which is realization of the constitutional rights of Russians for information.

This definition requires significant specification. The Okinawa Charter on Global Information Society states that information and communication technologies are one of the most important factors that influence the formation of society of the 21st century. Their revolutionary influence concerns the way of life of people, their education and work, and interaction of government and civil society – they quickly become a vitally important stimulus of development of the global economy.

The concept of global information society, which was developed on the basis of the Okinawa Charter, set the issue of state sovereignty, limits, and specific features of jurisdiction in digital spaces.

Thus, Internet is treated as a social structure and a new form of space, which does not conform to the material characteristics, but allows for various actions that could have real consequences. According to the definition of the Supreme Court of the USA, cyber space is a “unique environment that is not located in geographical space but is accessible to everyone anywhere via the access to Internet”. The foundations of activities of Internet include the principles of convergence, hierarchy, decentralization, exterritoriality, and democracy. Competency in the sphere of Internet could be determined within the domestic and the international legal regimes. Very often certain law drafts say that so called Runet – the Russian segment of Internet – should be regulated by the Russian laws. Development of Internet allows a lot of jurisprudents to substantiate the theories that Internet, which is a new type of space, which has virtual essence – cyberspace, does not fall under state jurisdiction (Dashyan 2007).

At the same time, new information technology, like any other technology, could be used either for the benefit of humanity or for criminal purposes (Zorkin 2018).

In the scientific literature we find (Goloskokov 2006) a rather debatable notion of network law as a meta-law, which structural elements are as follows: (1) elements of norms of law and norms of law in the electronic form; (2) subjects of law: person, family, and organization (in any forms), state and its bodies; (3) network relations, network legal mechanisms of direct connection and feedback, cyber space, information and communication technologies, computers, electronic systems of automatic registration of deals and transactions, electronic archives, networks, personal terminals of subjects of law, and robots – special computer programs that work in the network in the set limits. In the network law, legal relations could be realized by horizontal, vertical, and combined ties, and the main role in their realization will belong to robots. There will be partial delegation of authorities of operative legal regulation from the legislative level to an independent body with simultaneous automatization of legislative and law enforcement processes.

These arguments have serious counter-arguments (Zhilkin 2018), as robots cannot study independently – so the process of creation of smart machines involves jurisprudents and scholars that are involved into language processing and development of artificial intelligence. Machine should not replace human as the highest value of the state. Robots are only a working tool in the legal profession and require human participation in analysis of program algorithms of artificial intelligence. Human must have control over the work of smart machines and the possibility to correct robots' work.

According to S.M. Shakhrai, Russia has a chance to become a pioneer and to create the world first "Digital Constitution". However, as for its contents, there are no answers but only questions: what is the connection between the rights of virtual and real personality? What would be the "basic set" of rights and liberties of virtual person and virtual citizen? How could they be determined as to the digital reality? The digital world has not limits, so how is it possible to achieve the minimum integrity in the ideas on the system of values and their hierarchy or on the ratio of freedom and security? Which technological means could be used for guaranteeing the constitutional rights and liberties in the digital space?

Specification of the problem is brought down to the issue that the main constitutional rights and liberties of human and citizen have to be observed in the real world and the digital world. These include the rights envisaged by Articles 19, 22, 23, 24, 29, and 44 of the Constitution of the Russian Federation (Shakhrai 2018).

In the conditions of digitization, there's expedience of functioning of the system of electronic justice, as "speed of development of the modern technologies allows for confident steps for improving difference systems – even the court system" (Poddubny 2018).

The Federal law "Regarding provision of access to information on activities of courts in the Russian Federation" dated December 22, 2008, No. 262-FZ contains definition of the unified information space of federal courts and justice of peace (Article 1) and the main principles, means, and forms of provision of access to information on activities of courts (Articles 4, 6, and 7), including in the form of electronic document.

There are offers (Bykova 2018) of creation of a digital ministry of defense, digital ministry of industry, digital ministry of economy, digital ministry of science and education, digital police, digital tax service, and digital anti-monopoly service in Russia. These and other digital structures of government should ensure cyber security of the country and the law has to become the third foundation of cyber security and the digital economy of Russia.

Such offers do not conform to the existing law on the system and structure of the federal bodies of executive power and will inevitably lead to repeated functions of public authorities. The Federal law "Regarding provision of access to information on activities of public authorities and local administration" dated February 9, 2009, No. 8-FZ (p. 5 Article 1) establishes that the official web-site of a government body of body of local administration is a web-site on the information and telecommunication network "Internet", containing the information on activities of a government body or body of local administration, which electronic address includes domain name, the rights for which belong to the government body or body of local administration. The Federal law could envisage creation of a unified portal of official web-sites of several government bodies.

In reality, implementation of the strategy of building the digital society is accompanied by expansion of the state telecommunication infrastructure, which is used for interaction with citizens – e.g., via various digital systems, which are called the Russia e-government.

The scientific doctrine (Kovaleva 2016) offers the following definitions. E-government is realization of Internet solutions and the basic infrastructure for providing individual and legal entities with information resources and information services by government bodies for ensuring transparency of the work of the state sector and interactive participation in decision making. E-government is the concept of state management that is characteristic of the information society; this concept is based on the possibilities of the information and telecommunication technologies and the values of an open civil society.

It is also stated (Klimashevskaya 2019) that e-government is a non-constitutional government body, to which, as a matter of fact, constitutional authorities of the Government of the Russian Federation in the most important spheres of life are transferred. This means the process of the change of interaction between citizens and bodies of executive power in the political and legal space: formation of the legal and technical basis of turning Russian citizens into "electronic population", governed by a transnational communication system. Usage of foreign information technologies and means of computer equipment might lead to loss of national sovereignty.

The Ferdinand Lassalle's idea of a "night-watchman state" is discussed again. The changes are dictates by the digital revolution with its technological potential that opens possibilities of development of global horizontal ties that stimulate larger freedom and decentralization of activities. This stimulates breakthrough of previously local structures in the "large world", but is a challenge for local authorities: whether they are able to use this chance for the good of their region. The potential of digital technologies increase the possibilities of bringing the states' role in accordance with global challenges and specific features of certain countries (Moncinska 2018).

According to G.N. Andreeva, the notion "digital economy" is absent in the Constitution of the Russian Federation. Strictly speaking, digital economy, as a part of the whole, should be built on the same democratic principles as the whole economic constitution of the country. However, due to its specifics, the digital economy requires changing the current laws and this envisages increased control over the implemented changes from the point of view of constitutionalism, as application of new technologies leads to new threats for realization of human rights and possibility of inadmissible centralization of the data and their usage not in the society's interests (Andreeva 2018).

Digitization of economy means formation of a separate system of normative and legal acts (as a rule, federal laws of the Russian Federation), which do not contradict the Constitution of the Russian Federation. A normative and legal act specially for digitization of economy might be the Federal law of the Russian Federation "Regarding application of digital technologies in the economy of the Russian Federation" (Myshko et al. 2019).

4 Conclusions/Recommendations

At present, digital constitution as a scientific direction in studying the modern Russian and foreign constitutional law is formed. There's an opinion that the current version of the normative and legal acts is only the prototype of the future digital model. For example, the Constitution of the Russian Federation in the "knowledge base" will be a current unity of: (1) its text; (2) federal constitutional laws; (3) laws of the Russian Federation regarding amendments to the Constitution of the Russian Federation; (4) legal positions of the Constitutional Court of the Russian Federation; (5) generally accepted norms and principles of international law, which have become a part of the Russian legal system; (6) commentaries of scholars; (7) law interpreting acts of international bodies and organizations. The ideas of such presentation of the Constitution is now new, but it has not become too popular (Khabrieva and Chernogor 2018).

It should be noted that such presentation of the Constitution of the Russian Federation cannot become popular, as it contradicts p. 1 Article 15. The acts of the p. 2–6 of the model are independent forms of sources of the Russian constitutional law. Law interpreting acts of any international bodies (p. 7) do not have a right to treat the Constitution of the Russian Federation, as, according to p. 5 Article 125 this could be done only by the Constitutional Court of the Russian Federation. Such models are justified in cognitive and educational purposes, but not in the current versions of the normative and legal acts regardless of the digital or other form of their expression.

References

Andreeva, G.N.: Digital economy; economic constitution; intellectual property: connection between the notions. Law Future: Intellect. Prop. Innov. Internet **1**, 12–18 (2018)

Bachilo, I.L.: Information law: study guide for academic bachelors' program. Yurayt (2016)

Bachilo, I.L., Ivanchenko, A.V., Arinin, A.N., et al.: Law of digital administration in Russia and France. In: Collection of Scientific Materials of the French-Russian International Conference, 7–28 February 2013. Poligraf-plus (2013)

Boer, V.M., Pavelyeva, O.G.: Information law, SUAL, SPb, p. 1 (2006)
Bykov, A.Y.: Law of the digital economy: certain economic and political risks. Prospekt (2018)
Galuzo, V.N., Kanafin, N.A.: Digital law in the Russian Federation. Education and law, no. 9 (2018)
Goloskokov, L.V.: Modernization of the Russian law. Prospekt (2006)
Dashyan, M.S.: Law of Information Highways: The Issues of Legal Regulation in the Sphere of Internet. Wolters Kluwer, Moscow (2007)
Elyakov, A.D.: Good and evil: poignant paradox of Internet. Philos. Soc. **2**, 58–76 (2011)
Zhilkin, V.A.: Artificial intelligence and digital technologies in legal activities in the digital reality (by the example of Finland). J. Foreign Law Comp. Jurisprud. **5**, 16–21 (2018)
Zorkin, V.D.: Law in the digital world: discussion at the St. Petersburg International Legal Forum. Russian Newspaper, 30 May, pp. 1, 4 (2018)
Klimashevskaya, O.V.: Problems of implementation of the state policy of the Russian Federation in the sphere of building the digital society. Vlast **1**, 82–85 (2019)
Kovaleva, N.N.: Information law of Russia, Dashkov and Co., IPR Media, (2016)
The Concept of the Information Code of the Russian Federation. Ed. by I.L. Bachilo. ISP of the RAS (2014)
Morozova, O.N.: Political internet communications: its role, functions, and forms. Polit. Linguist. **1**, 156–161 (2011)
Moncinska, E.: The role of the state in economy: from the times of Adam Smith to the digital economy. Soc. Sci. Mod. Times **5**, 5–17 (2018)
Myshko, F.G., Olimpiev, A.Y., Aleksandrova, A.Y.: Regarding certain issues of digitization of economy in the Russian Federation. Bull. Mosc. Univ. Russ. MIA **1**, 132–135 (2019)
Naumov, V.B.: Law and Internet: Essays on theory and practice. Universitet Publ., House (2002)
Prokopenko, A.N., Krivoukhov, A.A.: The legal policy of the Russian Federation in the sphere of state information resources. Sci. Bull. **9**, 173–181 (2007)
Poddubny, E.O.: Regarding the issue of optimization of the court activities in Russia via e-justice. Law Digit. Econ. **1**, 31–34 (2018)
Talapina, E.V.: State management in the information society (the legal aspect). Jurisprudence (2015)
Khabrieva, T.Y., Chernogor, N.N.: Law in the conditions of the digital reality. J. Russ. Law **1**, 85–102 (2018)
Khabrieva, T.Y.: Law and the challenges of the digital reality. J. Russ. Law **9**, 5–16 (2018)
Shakhrai, S.M.: Digital constitution. The fate of the main rights and liberties of person in the totally information society. Bull. Russ. Acad. Sci. **88**(12), 1075–1082 (2018)

Goals and Interests in the Law of the Digital Age

Tatiana V. Shatkovskaya[1(✉)], Tatiana V. Epifanova[2], Natalia G. Vovchenko[2], and Irina S. Maslova[2]

[1] Russian Presidential Academy of National Economy and Public Administration, Rostov State University of Economics (RSUE), Rostov-on-Don, Russian Federation
shatkovskaya.tv@gmail.com
[2] Rostov State University of Economics (RSUE), Rostov-on-Don, Russian Federation
profepifanova@gmail.com, nat.vovchenko@gmail.com, irinasmaslova@yandex.ru

Abstract. The paper deals with the research into the most important doctrinal problems of modern law, conditioned by the rapid change in public relations under the impact of the information revolution. By means of revision of long-standing legal forms and critical rethinking of the current law, authors identify changes in the content of such concepts as goals and interests in law and their impact on the system of digital relations.

The authors rest their argument on the fact that the institutions of digital law that are being formed today cannot be solely oriented towards the priority of the market and economic laws, domination of certain economic and social systems, and facilitation of activities of state institutions. They suggest that the welfare of people, the community and states as political alliances of people and communities, as well as the arrangement of conditions for the improvement of national legal cultures should be chosen as the axiological guidepost of digitization.

The paper presents the scientific results achieved by the authors during the research; in particular, such categories as "legal interest", "main trends of law", "legal symbiosis", "digital economy", etc. have been defined. The authors believe that imperative and dispositive methods of regulation are no longer representative of the entire range of needs and interests that can be formalized by legal and technical means. Balancing and proportionality of rights and liabilities of actors became new demands of the society to jurisprudence.

The paper demonstrates that strategic planning and legal support of national legal interests belong to the primary goals of the legal policy of a modern state. In addition, internal and external problems have been formulated, and their solution requires the formation of the national legal interests by the state.

Keywords: Legal interests · Goals in law · Digitization of law · Harmonization of law · Legal symbiosis · Nation · Legal categories

E. G. Popkova and B. S. Sergi (Eds.): ISC 2019, LNNS 87, pp. 899–907, 2020.
https://doi.org/10.1007/978-3-030-29586-8_102

1 Introduction

Revision of the legal form of public relations, conditioned by their constant socioeconomic movement, is a natural segment of legal activities. Shershenevich, G.F. wrote the following words about this as early as a hundred years ago: "consideration of issues of law, as it should be, unlike law as it is, is … the recurrent problem of jurisprudence" so that "in critical moments life could resort to jurisprudence for answers that had been prepared over time". At the same time, the researcher came up with an algorithm for critical rethinking of existing law, including the identification of unsatisfactory components of law order, identification of objectives and directions for transforming law, and elaboration of measures necessary for the transition from "existing to desired" law (Shershenevich 1910–1912).

We believe that the foregoing has not only not lost edge, but has acquired even greater urgency in the context of the accelerated modernization of public relations under the impact of the information revolution. Industrial civilization, unlike industrial societies, accelerates the pace of social development, transforming the value system and changing the priorities of human activities. Creation and creativity combined with scientific rationality supersede the value priorities of the past century.

In the discussion on the new digital reality modern legal science proceeds from the inviolability of the target of research, which is largely due to the dominance of the logical and dogmatic approach. However, today we are talking about a dramatic change in the social context, which cannot but lead to paradigmatic changes in legal scientific knowledge. Therefore, it is hardly practical to suggest the invariability of such concepts as goals and interests in law in the system of digital relations from the scientific perspective of the XX century, when these concepts had different empiric content.

We believe that the definition of goals and the desired state and legal order, which is put forward by the full suite of historical conditions, as well as giving them legal form, is an essential prerequisite for the further development of legal institutions. Therefore, when we start a discussion on the development of law in the context of digitization of public relations, we should narrow down the main trends, interests and goals of this development in the first place.

2 Theoretic and Methodological Approaches to the Study of the Interest in Digital Law

In theoretical terms, the problem of legal interests begins with the perception of nature of the interest. In our opinion, the discussion on the absolute objectivity of the interest, especially in the legal sphere, is unfounded, just like the denial of subjectivity in law. The interest should be viewed as a complex psychosocial category, conditioned by objective needs formed in the society and, in turn, generating new needs, as a constituent motivation of behavior of individuals.

Legal interest as a variety of public interest has a number of specific features. First, the interest as a legal category is a means towards achievement of a legal objective. Second, the implementation of legal interests is limited by the normative framework and is achieved by public enforcement. Third, the legal interest is objectified through legal actions and by adequate legal means. Fourth, legal interest someway or other correlates with the rule of law. It either complies with a certain rule or contributes to its formalization or transformation and termination. Fifth, the implementation of legal interests gives rise to the emergence of legal relations and, accordingly, the need for achieving the balance of interests of various actors participating in them.

Thus, legal interest is objectified in the actions of an actor and implies that the latter knows how to consider interests of others and their interaction in the search for compromise solutions. This being said, the state represented by authorized bodies can prioritize certain interests or groups of interests, impose limitations and even bans on the implementation of some of them. Private and public interests can also be regulated with the help of an institution granting rights and liabilities to the agents, as well as the corresponding status of participants in legal relations which is legally protected by the state.

The protective and preservative armory of the state mechanism provides support for the existing hierarchy of legal interests, recognition of interests of certain actors that are found to conform to law, implementation of interests, and possibility of initiation of legal action in case of their violation.

These most important problems cannot be resolved without answering the question about the place of jurisprudence in public life. If we concur with Erik Anners in the fact that jurisprudence is “a peculiar kind of engineering”, and lawyers “develop and apply legal norms by order of political authorities” that govern “the division of labor and cooperation of people” (Anners 1994), the goal-setting in law becomes the prerogative right of the state. Hence, public officials become responsible for the formulation of the main trends, interests and goals of the digital legal development.

For opponents of the bureaucratic model of legal establishment which the authors of the research affiliate themselves with, it is obvious that jurisprudence acts as a social mediator, ensuring the actualization of the force of law and its materialization through the prism of the national legal culture, setting the conditions and the context of this process. In this regard, it is appropriate to reproduce the words of John Paul II, which are meant to point out the role of institutions created by people and the responsibility of mankind for the preservation of the goals for which public mechanisms are created. The Pope pointed out that means being created are not a goal in itself, but are aimed at serving a human person, providing it with a decent development environment (https://w2.vatican.va/content/john-paulii/en/encyclicals/documents/hf_jp-ii_enc_04031979_redemptor-hominis.html, 16b & 16d).

3 Goal-Setting in the Law of the Digital Age

In such a situation, the institutions of digital law that are being formed today cannot be solely oriented towards the priority of the market and economic laws, enforcement of rights of the strongest winner in the competitive struggle, domination of certain economic and social systems, and facilitation of activities of state institutions. The welfare of people, the community and states as political alliances of people and communities, as well as the arrangement of conditions for the improvement of national legal cultures should be chosen as the axiological guidepost of digitization towards the achievement of universal human values contributing to unity in diversity and public welfare.

Reasoning from this fact, the main trends of law should be formulated, which in this research shall be understood as scientifically founded positions on its directions of development, based on sensible and constantly redefined human values and principles of peaceful coexistence of people and focused on achieving a balance between technical homogeneity of law (correlation between law and legislation), on the one hand, and achieving the balance of rights and liabilities of a person, on the other hand.

For several decades of the Soviet period, the bureaucratic apparatus ensured the implementation of the basic underlaying premises of the dominant legal ideology, and the main trend of development of law consisted in its unification to preserve the state order based on the Marxian doctrine. Over the past century, this approach has taken root in the public conscience and has begun to be perceived as the fundamental model of legal development. However, it contradicts not only universal human values, but also modern socioeconomic and political realias.

The active international legal interaction of sovereign states with various bases of legal systems in the globalized world and the unity of the global economic space eliminates unification due to the impossibility of ensuring uniformity of statutory regulation of similar public relations without interfering with the public policy of states, which is contrary to generally accepted principles and standards of international law. (Rome Convention on the Law Applicable to the Contractual Obligations (ROME 1980), Article 16). Consequently, unification is changed to the idea of the harmonization of law, which is understood as the approximation of legislation and sociopolitical institutions of different states.

In our opinion, harmonization, which actually denotes unification in chime, implies the same uniformity, but in a milder form, and does nothing to contribute to the formation of the multidimensional system of legal regulation which provides preservation of unique national legal elements and their interrelations, basic properties and functions of the legal and cultural impact on the digital sphere of legal activities. We believe that legal symbiosis is a sensible alternative to total formalization caused by the digitization of real-life objects which turns them into content. Legal symbiosis is a mutually beneficent coexistence of various legal organisms.

That said, no reasonable and conscious legislative activity is conceivable without prior resolution of the issue of its goals, criteria of the rational and the fair. As Pokrovsky, I.A. noted, if these questions are left unanswered, all legislative frameworks will be deprived of "teleological orientation", "great ideas and universal truths consciously or unconsciously form the basis of our sense of justice, and the legislator just cannot deepen or reinforce an impression of its works to such a degree which will be achieved by expressing them in the statutory wording" (Pokrovsky 1917).

4 The Problems of Implementation of National Interests in the Law of the Digital Age

Scanning the perspectives of modern technological development allows us to reveal both positive and negative consequences of the Fourth Industrial Revolution. In the globalized world, in the context of the mass immersion into the digital environment, states face many diverse and disturbing challenges. One of them, in our opinion, is the threat of loss of national identity and the possibility of positioning of the national legal interests in an integrated unbounded space.

A technological breakthrough requires teamwork for the development of a new regulatory framework and arrangement of interstate organizational structures. At this stage, technologies are controlled by a man, and it is at this precise point in time when it is necessary to develop a system of legal means that would guarantee the safety and security of mankind as a whole and its individual groups in particular. The theorists of modern efficient management believe that "shared vision of the future helps to develop a uniform understanding of problems and sets benchmarks in the decision-making process" (Abuchakra and Khoury 2016). However, western researchers do not explain what is a shared vision, how it correlates with national, let alone personal, interests.

In previous papers, we have noted the contradictory nature of the globalization process, which causes the interdependence of different states and the formation of a global legal communication system, but at the same time intensifies the desire to preserve national identity (Shatkovskaya et al. 2015). In this regard, the provision of a rationale for the concept of the national legal interests has not only theoretical, but also significant practical relevance. Indeed, the interest, as noted by Jhering, is a practical basis of law in the subjective sense (Jhering 1881).

Prior to the Fourth Industrial Revolution, centralized approach towards the formation of priorities in the sphere of legal interests of the state was considered to be the best model for achieving the balance of public and private interests. Modern technologies rapidly change both the methods of pursuit of economic activity and the system of state legal regulation as such.

In the age of hybrid forms of commercial and governmental organizations, wide powers of intergovernmental entities and transnational enterprises, the regulatory framework of the state should be mild, flexible and liberal, creating a competitive regulatory environment that would be comfortable for living, but at the same time ensuring state control, national safety and sovereignty.

In reference with the above, we believe that the formation, strategic planning and legal support of national legal interests belong to the primary goals of the legal policy of a modern state. The difficulties of this process can hardly be overestimated, since, on the one hand, modern society can be referred to as society without borders, and on the other hand, it is characterized by the maximum degree of individualization of its members. In these circumstances, individual actors exercise many legal functions, can be several juridical personalities at the same time, participating in both real and virtual legal relations. The situation is complicated by the discreteness of long-standing traditional social institutions, such as family of marriage.

Within this framework, we believe that one of the state priorities is the preservation of a nation that we understand as a historically shaped social union, the common interest of which consists in maintenance of state power within a given territory which maintains and protects its integrity. The binding media of the nation may consist in common language, historical past, worldview, culture, religious and legal ideas, value system, and, of course, interests.

In opposition to Shershenevich, G.F. we believe that the nation and the people are different concepts, since representatives of many peoples can compose one nation. At the same time, we shall concur with Gabriel Feliksovich in the fact that "continuous communication, general protection of the state, participation in legislation and government" will significantly weaken the separative influence, if not eliminate it completely, and will gradually turn the "state population" into a single nation (Shershenevich 1908).

National legal interests are a variety of common interests, the content of which is determined by various legal and non-legal technologies and depends on the general cultural level of a nation, social stratification of people which constitute it, and the nature of intrasocial interaction.

Formation of national legal interests in democratic states is only possible based on the widespread use of methods and means of decentralized equal interaction of public and state institutions, development of various forms of public-private partnership, stimulation of social activity of citizens and their organizations. In these circumstances, the state is assigned the mission of the main coordinator and at the same time the advocate of the national legal interests, since, firstly, the governmental organization has the highest sociopolitical value, secondly, it is the state which inherited key social powers, including legalized coercion, in the process of historical development.

It is extremely difficult to position national legal interests in the globalized world using legal means. The format of this research prevents from fully disclosing the full range of these issues. Hence, we will focus on issues which in our opinion are the most important.

Current approaches to the legal regulation are based on its close relationship with the land, literally with the soil surface, which by the law of nations is recognized to belong to a certain community of people who originally took possession of a plot of common land and established a sovereign state on its territory. It is the territorial sovereignty of the state, which today includes not only the earth surface, but also the sea and air space, that forms the core of the territorial principle of the force of law. The latter largely determines the nature of international legal relations which constitute the interaction of territorial legal orders, managed by the state and formalized in essentially "interstate" legal acts.

Globalization erodes, if not erases, the territorial boundaries, which complicates the application of the territorial principle and transforms ideas about the state-legal space. Of course, such changes cannot but affect the management strategies of the state, its approaches towards the formation of national interests, legal regulation both domestically and abroad.

5 Results

We believe that the gravest disadvantage of the modern model of legislation is its focus on preserving the technical stability of the statutory regulation and current juridical institutes constituting a hierarchical mechanism for ensuring centralized law order.

By way of example, we shall take the Government Decree concerning the establishment of the program "Digital Economy of the Russian Federation" that was adopted in summer 2017 and ceased to be in force on February 12, 2019, but it contains the only standard definition of digital economy. It is thought of as an ecosystem with such key factors as the digital data and effective cooperation of entrepreneurs, academic community, state and citizens. We have set forward a different definition of digital economy, implying its human-centric focus, namely "an institutional environment created on the basis of information technology for the networking cooperation (interrelation, allocation and exchange) of equal free individuals - economic agents, in which the generation and the use of innovations are inextricably intertwined, and creativity is the main way of economic activity of a person" (Shatkovskaya et al. 2018).

Having regard to the above, we believe it necessary to consolidate the humanistic orientation of the emerging digital law, to preserve stable national cultural foundations in it that would ensure proper interaction with the world, and to ensure sovereignty and subsidiarity, understood as the distribution of regulatory powers and responsibilities for their exercise between public and private actors both at the national level and at the individual level.

Technologies in law should exercise the function of means for implementing the goals for which they were invented and intended, above all things, the optimal development of a person, family, and nation. Moreover, we believe that only a human person who possesses qualities of a moral responsibility in the exercise of their rights, doing it in a moral and ethical manner, and a propensity to adapt and respond to new socioeconomic and political challenges, can be the center of legal activities.

Information technology, having an effect on public relations, differentiates and changes the content of private and public interests and leads to the emergence of new objects require legal implementation. Imperative and dispositive methods of regulation are no longer representative of the entire range of needs and interests that can be formalized by legal and technical means. Balancing and proportionality of rights and liabilities of actors became new demands of the society to jurisprudence. In particular, the Constitutional Court of the Russian Federation offers such a way of ensuring a dialogue in new conditions and its readiness for it as a legally acceptable compromise (Ruling of the Constitutional Court of the Russian Federation 2016).

The quest for a compromise is required in conditions when voluntary forms of restriction of sovereignty by way of subordination of national legislation to the supranational law cannot be compared with a total absence of territorial boundaries in the digital sphere. In contrast to the existing forms of objective and material interaction of people, the virtual world is open to innovations and is an exterritorial communication of creative agents for the creation, use, and circulation of real-life or virtual objects.

6 Conclusion

Therefore, we shall single out the following primary goals in the law of the digital age:

- recognition and provision of the definite objective sovereignty of an individual that is inaccessible for arbitrary interference of any persons, including the state, and is expressed in the balance of his/her legal rights and liabilities, based on assumption of reasonable and fair actions of participants in digital relations;
- decentralization of the arrangement of state power and extensive delegation of operational management functions to non-governmental structures;
- distribution of regulatory powers and responsibilities for their exercise between public and private actors;
- creating an effective legal monitoring mechanism to obtain unbiased findings concerning the nature of the impact of the regulatory framework on public relations and its consequences;
- promotion of various forms of public-private partnership and entrepreneurial activity of citizens;
- removal of administrative barriers in economic activity and other public activities;
- consistent implementation of a national strategy of legal development on the international stage, development of interethnic regional legal forms of interaction with strategic partner states.

We believe that promotion of national legal interests is an essential prerequisite for the prevention of global social perturbations in the digital age, since the pursuance of supranational legal unification would disturb the existing order of international interaction. National legal interests can only be positioned on the basis of the principles of "pre-emptive knowledge" and minimum harmonization of legal norms as a result of internationally achieved interstate consensus, but not the principle of "domination".

Formation of national legal interests implies that the state should find the solution for a number of internal and external problems. From among internal problems we should emphasize the creation of strategic management, analysis and forecasting bodies; decentralization of the arrangement of state power and extensive delegation of operational management functions to non-governmental structures; distribution of regulatory powers and responsibilities for their exercise between public and private actors; creating an effective legal monitoring mechanism to obtain unbiased findings concerning the nature of the impact of the regulatory framework on public relations and its consequences; promotion of various forms of public-private partnership and entrepreneurial activity of citizens; removal of administrative barriers in economic activity and other public activities. The main external problems include coordination and consistent implementation of a national strategy of legal development on the international stage, development of interethnic regional legal forms of interaction with strategic partner states.

References

Abuchakra, R., Khoury, M.: Effective government for a new age. Reforming state administration in the modern world (2016)

Anners, E.: History of European Law, Moscow, p. 3 (1994)

Jhering, R.: Der Zweck im Recht, in 2 volumes, vol. 1, St. Petersburg, pp. 38, 39 (1881)

Rome Convention on the Law Applicable to the Contractual Obligations, ROME (1980)

Pokrovsky, I.A.: Key problems of civil law, pp. 48–50 (1917)

Ruling of the Constitutional Court of the Russian Federation No. 12-P of 19.04.2016, Official Gazette of the Russian Federation, no. 17, p. 2480, 25 April 2016

Government Decree of the Russian Federation No. 1632-p of July 28, 2017 concerning the establishment of the program "Digital Economy of the Russian Federation" (ceased to be in force)

Shatkovskaya, T.V., Epifanova, T.V., Vovchenko, N.G., Romanenko, N.G.: A legal mechanism for regulating digital economy. In: CBU International Conference on Innovation in Science and Education, Prague, Czech Republic, 21–23 March (2018)

Shatkovskaya, T.V.: Models of interaction of political systems ensuring the national security of states/(co-authored), State and Municipal Management. Scholarly Notes of the North Caucasus Academy of Public Administration, no. 3, pp. 104–107 (2015)

Shershenevich, G.F.: General theory of law, pp. 799, 800 (1910–1912)

Shershenevich, G.F.: General theory of law and the state, p. 24 (1908)

https://w2.vatican.va/content/john-paulii/en/encyclicals/documents/hf_jp-ii_enc_04031979_redemptor-hominis.html, 16b & 16d. Accessed 19 Jan 2019

Tendencies and Prospects of the Legal State Development Under Digitalization

I. V. Abdurakhmanova(✉), G. B. Vlasova, and N. E. Orlova

Rostov State University of Economics, Rostov-on-Don, Russia
dima_rd@rambler.ru, vlasovagb@mail.ru,
teorii.kafedra@yandex.ru

Abstract. Purpose: The objective of this work is to analyze the new tendencies of the legal social state development under digitalization influence; to reveal and ground new tasks and trends of lawmaking and law enforcement activities of state power to provide constitutional rights and freedoms of a human and a citizen.

Design/Methodology/Approach: To research the digitalization influence on tendencies and prospects of the legal state development the following methods were applied: formal legal, structural and analytical ones as well as general scientific methods of analysis and synthesis.

Findings: The authors have analyzed and generalized the state's new tasks connected to the digitalization to regulate and protect information rights, state management digitalization, and social policy modernization, as well as to provide information security.

Originality/Value: The conclusions formulated in the article can be applied in the lawmaking and law enforcement activities to adapt the legal state institutions to the digitalization.

Keywords: Government digitalization · e-government · Legal state · Information security · Digital rights · Cyber threats

JEL Code: K38 · K100

1 Introduction

Having turned into the objective reality the digitalization became a part of modern individual's life. Being a simple method to improve life the digitalization turned into the total phenomenon, the world social development driver. In the narrow sense the digitalization is a transformation of information into digital form to reduce costs substantially and to expand the opportunities for the subjects of the corresponding activity (Khalin and Chernov 2018). Due to its topicality the digitalization is the subject matter of the numerous scientific researches concerning mostly its economic aspects.

A lot of analytical articles and more fundamental researches were devoted to the digital economy. Most of the authors focus their attention on different aspects of the digitalization impact on the national economy, for example, on the innovative telecommunication technologies application, on internet sensor nets use and on electronic

E. G. Popkova and B. S. Sergi (Eds.): ISC 2019, LNNS 87, pp. 908–915, 2020.
https://doi.org/10.1007/978-3-030-29586-8_103

document management, on new consumer market formation under online trading. We agree with the scientists' opinion the economy reflects the specificity of "the new technological generation with application of a huge number of data generated in different information systems and processed to extract useful information from them" (Khalin and Chernov 2018).

As a world development effective trend the digitalization brings about significant changes in all spheres of life, generating new realia of the human and society existence on a whole. Consequently, it transforms the content, purpose and functioning of all social institutions, including state and law. Let us stress that the digital format to present information is aimed at not only boosting the economic activity, but improving citizens' lives and their well-being as well. It is declared as a priority task of the government in the strategic documents of the long-term development, including the Program "Digital Economy of the Russian Federation", approved by order of the Government of the Russian Federation on July 28, 2017.

The digitalization is to improve all social spheres connected with economy that makes the analysis of adaptation of social institutions, bodies, principles, legal social state to that trend topical.

The complexity of the formulated problem is that, on the one hand, legal social state is to ensure the implementation of digitalization prerequisites, on the other, the state itself and its institutions are influenced by digitalization.

However these digitalization aspects are not covered adequately in modern scientific literature. The authors of this article attempted to overcome this gap. The objective of this research is to analyze new tendencies of the legal social state development being influenced by digitalization according to the Russian Federation constitution norms, to reveal and ground new tasks and directions of lawmaking and law enforcement activities to ensure constitutional rights and freedoms of a human and citizen.

2 Materials and Method

The methodological basis of the legal analysis is positivist methodology, as well as the methodology of comparative legal research. To research the digitalization influence on tendencies and prospects of the legal state development the following methods were applied: formal legal, structural and analytical ones as well as general scientific methods of analysis and synthesis.

3 Results

Assuming the main signs of the legal state are the rule of law and separation of powers let us consider how their content under digitalization is changing. Digitalization transforms the very term law and the forms of its implementation, expands the borders of legal regulation, and corrects lawmaking and law enforcement activities. In this connection it is important to analyze how the content of the category "law" changes under the digitalization and how the state lawmaking activity at a new cycle of the civilization development changes.

Today it is evident the digital technologies are changing the image of law as a basis of the legal state, are influencing its regulative and protective potential in the area of the protection of rights and freedoms of a human and a citizen (Khabrieva 2019). Neither the law doctrine no practice possess single conceptual understanding of tendencies and prospects of the rights transformation under digitalization. In the foreign literature as a rule private issues of practical legal activity under the new conditions are analyzed.

This involves the use of electronic normative acts, legal service market development, and e-education of the future lawyers. However there are no any conceptual researches in this sphere. The Russian science focuses on the issues of development and reasoning of the most optimal model of the social relation legal regulation connected with the application of the digital technologies in the financial sphere, state management, and artificial intelligence.

Our scientists among the prospects to transform the law under digitalization influence name the following: either the law will be transformed into the new social regulator assuming the emergence of a program code or any other hybrid form, or it will preserve its substantial signs and will exist with a program code. Scientists also envision more distant prospects associated with the emergence of a new regulatory environment along with law, morality, and religion. These prospects are connected with new spheres needing regulations but are difficult to be regulated.

New social relations are in the sphere of the legal regulation whose subjects are virtual or "digital" persons; the relations associated with personal identification in the virtual space have appeared; new human rights have appeared in this space; robotics and artificial intelligence are widely applied.

The relations concerning non-typical objects such as information, digital technologies, cryptocurrencies have appeared. Digitized data bases are supposed to be used in the new relations, and that leads to changes in the law implementation. Digitalization of the formal legal sources brings about the formation of the new methods of the legal technology.

Law becomes both the digitalization tool and an object of its impact. Jurisprudence is to understand the impact of digitalization on the state and legal reality having revealed and analyzed the dynamics, the existing risks and to formulate the counteractions to new threats. Technological revolutions have changed the image of law more than once. Now it is to change due to forming legal reality. In this connection the principle of the rule of law as a foundation of the legal state will be transformed.

This prospect touches upon the principle of the separation of powers, because under digitalization in legal, executive and judicial powers the state faces the new tasks to be implemented in the legal and social spheres. The idea of the separation of powers for many centuries has accompanied the search of the perfect state it can be found in the works of Aristotle and Polybius. On each stage of the historical development it acquired new features adjusting first to the needs industrial society, then to the information one.

In the lawmaking activity the most important task of the legal state is to develop and improve normative and legal basis to regulate digitalization. Digitalization is directly connected with the social purpose of the legal state that is reflected in the laws and acts issued by the Russian Federation President. Speaking about the significance of the social component of the digitalization the authors of the Program "Digital Economy" stress that it is directed to improve the citizens' well-being and their lives quality by rising accessibility and service quality produced by digital economy with application of the modern digital technologies, to improve the accessibility and public service quality for the citizens, their safety.

Digitalization management is not to allow the Russian economy to lag behind and is to protect citizens' rights. It is necessary to organize the digitalization processes, its normative and legal regulation as well as financing. The state is to organize and promote the digital transformation processes. Digitalization legal regulation is implemented in different ways in the foreign countries. For example, these are acts of strategic planning (Decree of the President of the Republic of Belarus of December 21, 2017 "On the Development of the Digital Economy") in the Republic of Belarus. There are software forecasting acts (State Program "Digital Kazakhstan" of December 12, 2017) in Kazakhstan. Special laws are adopted in France, for example, the 2016 law "On the Digital Republic".

The Russian Federation has paid much attention to the legal regulation issues related to the relation digitizing lately. In his Message to the Federal Assembly of December 1, 2016, the head of the state launched the Russian economy digitalization program. Information Society Development Strategy of the Russian Federation for 2017–2030 and the Program "Digital Economy of the Russian Federation" dated July 28, 2017 are the basic documents in this sphere. On March 12, 2019, the State Duma of the Federal Assembly of the Russian Federation approved the Law on Digital Rights in the third reading. It is the first from three basic acts in the digital economy sphere. In the nearest future the federal laws "On Digital Finance Assets" and "On Alternative Ways to Attract Investments" will be adopted. The legislator overcomes the gaps in this sphere since these relations became a reality long ago, for example, electronic tax declaration verified with electronic signature.

Let us emphasize that we do not speak on a new separate law adopting but on amendments and changes in the current legislation. These are amendments to 1, 2 and 4 parts of the Civil Code of the Russian Federation. The essence of these amendments is that the very notion "digital rights" is enshrined in the regulations. Besides the inheritance rights it will be possible to issue all rights provided by the Civil Code in a form of a digital code and transfer in accordance with information system rules. It will be possible to digitize some transactions that require notarization. Transactions made by electronic and other similar technical means, including filling in the forms in the internet, text messages will be considered as a form of ordinary transactions made in written form. Currently the issue on the digital money is disputable. Crypto currency market is not completely regulated, the budget is making losses. However this money is used. Judicial practice interprets it as a property. It is money, or "money surrogate, or information, or property".

According to the Plan of measures in the area of "Regulatory regulation" of the program "Digital Economy of the Russian Federation" dated December 18, 2017, federal laws were enacted in 38 areas of legal regulation and more than 50 amendments to the existing regulatory and legal acts. In 8 areas of technical regulation and standardization, the adoption of national standards was envisaged. However the official scientific concept of the digital legislature development is absent.

Digitalization has a great impact on the executive power, it place in the system of the power separation. Digital technologies can substantially increase the law enforcement quality in the executive bodies. Artificial intelligence will accelerate and simplify the public powers implementation, including law enforcement. The scientists note the necessity to transfer from "manual" state management to the digital one associated with the algorithmization of all decisions at the federal and regional levels. One of the federal projects of the national program "Digital Economy" is a project "Digital State Management". It is aimed at final transition to the electronic interaction between citizens and the state and that promotes the implementation of such legal state principles as openness of the state bodies' activities, control over their activity by the civil society.

"Electronic government" whose establishing started in the frame of the state program "Information Society" approved by order of the Government of the Russian Federation in 2010 is a digitalization manifestation in the executive power sphere. Currently key of the national infrastructure elements of the e-government have been developed and function.

As to the judicial power the analysts predict the wide application of the artificial intelligence in the legal proceedings. Namely these are digital traces as e-evidences, electronic document management, and intellectual systems to analyze the case materials; telecommunication means to conduct hearings, etc. In the frame of the federal project "Digital Environment Regulation" it is supposed to form legal conditions in the field of legal proceedings and notaries in connection with the digital economy development.

According to the program "Digital Economy" one of the most significant aspects is to provide of rights and freedoms of the individual in the digital world. The rule of law and rights and freedoms and an effective mechanism to protect them are the basis for the constitutional system of the Russian Federation as a legal social state. In accordance with the RF Constitution rights and freedoms protection determines the meaning and activities of the legislative, executive and judicial authorities. Generating new rights and threats the digital reality sets new tasks for the state to improve the rights and freedoms protection mechanism and to counteract new security challenges for both individual and society on the whole.

Human rights guaranteed by the 1993 RF Constitution and by international regulations are being concretized and specified at each stage of the historical development. It is evident that under digitalization freedom of expression, the right to receive and disseminate information without interference from the authorities and regardless of state borders, the right to privacy, personal and family secrets, individual's honor and good name protection, the right to privacy of correspondence, telephone conversations, and postal, telegraphic and other messages obtain new meanings. These rights are mostly influenced by digitalization.

The legislator faces the choice how to how to relate the digitalization legal regulation with human rights, how to balance rights and security. Digital reality creates new technological opportunities in the sphere of the basic constitutional values, for example, safety. The introduced in China the experimental face recognition program allows tracing each person in the real time and depending on his/her lawful or illegal behavior to evaluate them on a point system. The obtained points will influence on whether different benefits are granted or not. On the one hand, these measures will promote safe environment formation, on the other, they mean total control not only over citizens' real life, but the virtual one as well, and it is when the privacy right is questioned.

Another new trend in human rights and freedoms under digitalization is a formation of a new group of rights which some scientists call digital ones. They are the right to access, to use, to create and to publish digital works; the right to access and use computers and other electronic devices, communication systems. Digital rights can be defined as concretization through law or law enforcement (including judicial) acts of the universal human rights applicable to a human and a citizen needs under digital reality.

There is a number legal acts in Russia, which regulate this area of legal relations, however, the scientists think that it is necessary to systemize the legislation in this sphere. The Information Code can be a systematization variant.

In addition to the new rights appearance under digitalization the content of the rights already enshrined and regulated by the current legislation is changing. In this connection their protection mechanism is to be modernized. The most crucial issues in this sphere are digital user security provision, citizens' confidence in the digital environment, cybercrime, the rights and interests protection on the Internet, its legal regulation, the threat of external information impact on the information infrastructure and citizens' consciousness.

In the rule of law state the right to information takes a leading position. The federal project "Information Security" takes an important place in the national program "Digital Economy". It is aimed at protecting rights and interests of an individual, a business and a state from information security threats under digital economy. The information security provision is an element of the Russian Federation national security system.

The Russian Federation information security is a state of protection of an individual, society and the state from internal and external information threats, where a person's and a citizen's constitutional rights and freedoms are observed, and decent quality and standard of living of citizens, sovereignty, territorial integrity, and sustainable social and economic development, defense and state security are provided. It is important to establish legal liability for violation of legislation in this area to provide the information security for a person, a society and a state. The lawmaker has regulated many aspects of the information security for a person, a society and a state. Meanwhile the information security analysis shows that its level does not meet needs to ensure rule of rights and freedoms and ability to counteract to new threats. At modern stage it is necessary to improve the information security system which implements the uniform state policy in this area.

Another trend in the legal state activity to provide human rights and freedoms under digitalization is to protect a number of social rights that become important under digital reality. Under the RF Constitution the social nature of the state is an attribute of the legal state. According to Part 1 Article 7 it is a state whose policy is to create the conditions for the decent life and an individual's free development. Life standard increase and well-being growth as a top priority are enshrined in federal laws, the RF President's acts, the Russia long-term development programs (including national project "Digital Economy") as well as in information and socio-economic security strategies.

The digitalization criterion is a percentage of citizens involved in the formation and use digital environment and digital services. That is why the state is to implement the corresponding educational programs to provide the citizens with the access to digital technologies. Namely it is the public access to the communication systems providing digital information, primarily to broadband internet. The digitalization indicator is Internet use, online content consumption in various areas: leisure, financial transactions, the purchase of goods and services, etc. The state is to provide the citizens with access to the educational environment and databases, to the interactive technologies.

The level of telecommunication technology use in Russia is lower the in the European countries and according to experts there is a substantial gap in digital skills between different groups of the Russians.

Social policy under in new conditions is to be thoroughly analyzed. Liberal model of social policy to provide targeted assistance and support is being introduced in Russia (Malysheva 2018).

Digitalization simplifies the task because it provides a more accurate accounting of citizens' income. But there is still a danger to reduce the social assistance amount while transiting to targeted social subsidies; the regions' experience shows it.

Thus, it is possible to state that digitalization influences on the social sphere in a controversial manner. On the one hand, these are prospects to increase the citizens' life quality, to provide access to new services and technologies, on the other, a set of challenges and threats. One of the problems the digitalization generates in the social sphere is to provide citizens' labor rights when robotization brings about imbalance between demand and supply in the Labor market. The analytical articles argue that some professions will stop their existence in the nearest future. It is topical, if robots will be able to replace a human lawyer. The introduction of "labor resources flexible use" can result in a loss or a social status downshift of many citizens and in deeper social in inequality.

Scientific discussions about social policy under digitalization touch upon the issue of the introduction of the basic social income that is negatively assessed by politicians. The proposed by the state model is sharply criticized by those who foresee the society split because it can turn the country into a huge Internet platform which can replace the state. The Russians will be sorted out depending on their possession of digital skills and technologies. The destruction of the very state administration and economy and even loss of the state's ability to protect its sovereignty are the consequences of such development. The threat that citizens will be turned into "electronic persons" who have lost social links and constitutional rights is actively discussed.

4 Conclusion

Thus, being an effective world development trend digitalization transforms the content, purpose, functioning of all social institutions, including state and law. It is aimed at qualitative improvement of all related to the economy areas that makes the analysis of the adaptation of the institutions, bodies, principles, and the essence of the legal and social state to this trend topical.

Digitalization transforms the very concept of law and the forms of its implementation, expands the borders of legal regulation, and makes significant adjustments to lawmaking and law enforcement activities. Law is becoming both the tool of the digitalization and object of its impact.

Under digitalization the reality and separation of powers principle are being transformed significantly. The main goal of lawmaking in the legal state is to work out and improve regulatory framework governing digitalization. Digitalization has a big impact on executive power and its place in the system of separation of powers. Digital technologies can boost the quality of enforcement in the executive branch.

Digital reality transforms the content of a number of the constitutional rights and freedoms generating the need to modernize the mechanism of their protection. In this connection the legal and social state faces the new directions of regulatory and security activities, in particular, in the field of digital rights protection and individuals and society's information security ensuring in the face of cyber threats. Another new trend in the legal state activity to provide human rights and freedoms under digitalization is to protect a number of social rights which become topical under digital reality.

References

Andreev, A.P., Kokunova, S.D.: Digital transformation of society as a factor influencing the security of the legal system of the Russian Federation. In: Collection of Scientific Articles and Materials of the International Conference "Digital Society as a Cultural and Historical Context of Human Development", Kolomna, 14–17 February, pp. 24–28 (2018)

Burkova, L.V.: Mass consciousness in the era of global informatization. In: Collection of Scientific Articles and Materials of the International Conference "Digital Society as a Cultural and Historical Context of Human Development", Kolomna, 14–17 February, pp. 60–64 (2018)

Malysheva, G.A.: On the Socio-Political Challenges and Risks of the Digitalization of Russian Society. Vlast, no. 1, pp. 40–45 (2018)

Message from the President of the Russian Federation to the Federal Assembly of March 1, 2018, Rossiyskaya Gazeta, no. 46 (2018)

The program "Digital Economy of the Russian Federation" approved by the order of the Government of the Russian Federation of July 28, 2017 No. 1632-p, Collection of Laws of the Russian Federation, no. 32 (2017). Article 5513

Khalin, V.G., Chernov, G.V.: Digitalization and its impact on the Russian economy and society: advantages, challenges, risks and threats. Management Consulting, no. 10, pp. 46–63 (2018)

Khabrieva, T.Y.: Law Under Digitalization, p. 36. University of the Humanities and Social Sciences, St. Petersburg (2019)

Human Rights in the Digital Age

Leonid G. Berlyavskiy[(✉)], Larisa Y. Kolushkina, Ruslan G. Nepranov, and Alexey N. Pozdnishov

Rostov State University of Economics, Rostov-on-Don, Russia
berlg@yandex.ru, kafedra.kimp@yandex.ru, gr_process38@mail.ru, kfiap@yandex.ru

Abstract. Human rights in the digital age are a specification of the main human rights that are established in the Constitutions of different countries and are guaranteed by the international legal acts. The authors study the problem of the main human rights (political, socio-economic, and personal) in the digital age, specific features of their implementation, and the forms of guaranteeing them by the state in the new conditions.

Keywords: Human rights · Digital age · Constitution · Law · Information · Digital reality · Digital technologies

JEL Code: K1 · K4 · Z0

1 Introduction

The main characteristic of the modern stage of development is the growing role of information and technologies of its processing in human life. Increase of the role of information leads to correction of the basic human rights, to which, due to the objective reasons, the right for information is added. Establishment of the subjective right for information in the normative legal acts started after World War II (Spiridonov and Evsikov 2015).

Analysis of the processes of digitization allows forecasting the change of the mechanism of law-making and composition of the existing model of social regulation, correction of the limits of the known social regulators, and creation of a niche for program code (Khabrieva 2018).

The 1948 Universal Declaration of Human Rights establishes in Article 12 that "No one shall be subjected to arbitrary interference with his privacy, family, home or correspondence, nor to attacks upon his honour and reputation. Everyone has the right to the protection of the law against such interference or attacks".

Article 19 of the 1966 International Covenant on Civil and Political Rights says that "Everyone shall have the right to freedom of expression; this right shall include freedom to seek, receive and impart information and ideas of all kinds, regardless of frontiers, either orally, in writing or in print, in the form of art, or through any other media of his choice".

The concept of legal informatization of Russia, established by the Decree of the President of the Russian Federation on June 23, 1993, No. 966, was based on the fact

E. G. Popkova and B. S. Sergi (Eds.): ISC 2019, LNNS 87, pp. 916–924, 2020.
https://doi.org/10.1007/978-3-030-29586-8_104

that Russia started to transform into a "network" state according to the concepts of creation of the information society. The American 1998 Digital Millennium Copyright Act contains one of the first legal references to the digital age.

On the other hand, Article 21 of the Strategy of national security of the Russian Federation, adopted by the Decree of the President of the Russian Federation date December 31, 2015, No. 683, states that the international situation is influenced by the increasing opposition in the global information space, predetermined by striving of certain countries to use information and communication technologies for achieving their geopolitical goals, including by manipulation of public consciousness and falsification of history.

The events of the "Arab Spring" showed that social Internet networks play an important role in organization of violent civil unrests in the countries of Asia, Africa, and the Middle-East. This led to the phenomenon of "Twitter revolutions" (Seleznev and Skripak 2013).

The necessity for acknowledgment and protection of digital rights is proclaimed in a range of international legal acts (Zorkin 2018). The Okinawa Charter on Global Information Society notes that the strategy of development of the information society should be accompanied by development of human resources, which opportunities would conform to the requirements of the information age (Seleznev and Skripak 2013). Participants of the Charter are to provide all citizens with the possibility to master and obtain skills of work with the information and communication technologies via education, life-long education, and training.

The UN General Assembly Resolution dated December 18, 2013, No. 68/167 "Right to Privacy in the Digital Age" contains a call to all countries to respect and protect the right for privacy, including in the context of digital communications. This act envisages establishment of new and usage of the current independent and effective controlling mechanisms that can ensure transparency in the corresponding cases and accountability as to states' intercepting messages and collecting personal data.

Thus, an important problem of interrelations between human and public authorities in the digital society is determining the possible limitations of digital rights by the federal law, including the allowable limits of control over the information environment by law enforcement agencies for the purpose of provision of effective protection of society from cybercrimes (Zorkin 2018). Transformation of the main human rights (political, socio-economic, and personal) in the digital age and peculiarities of their implementation in the new conditions should be studied.

Human rights in the digital age were studied by the legal science in the course of the last twenty years. According to Baturin (2000), the problems of observation of authors' rights during usage of global networks, the issue of protection of privacy, electronic document turnover, application of digital signature in electronic messages, and usage of other new technological means are not new problems but a new treatment of old problems, which allows stating that the change of the technological infrastructure has not yet led to a completely new sphere of public relations.

Based on analysis, including analysis of application of digital technologies in the US presidential campaign in 2000, it was concluded that the Internet allows users to obtain information without any official sources, to exchange information, and to express users' opinions: to use the technology of e-government to enter the contact with public authorities, protecting rights and realizing the need for receipt of state information; to participate in referendums and discussions of the Constitution of the country; to use the interactive regime to enter the process of making of local regional decisions. It was noted that the Internet weakens the government's and corporations' monopoly for information and stimulates the dissemination and receipt of comprehensive political data, thus stimulating the process of human's turning from a passive object of influence of policy into its active subject (Elyakov 2011).

An unprecedented case of application of the Internet for developing the Constitution of Iceland in 2012 attracted attention of a lot of scholars (Avzalova 2014). A council of 25 citizens was created for preparation of the documents via the social networks (Facebook, Twitter, and even Youtube), and the citizens' offers were collected (3,600 comments were received, as well as 370 amendments to the Constitution). The final variant was approved by the referendum. Iceland's experience was highly evaluated by researchers as a lesson of direct democracy and citizens' wide participation in the activities of public authorities.

Based on analysis of the Russian empirical materials, the functions of electronic civil society (implemented directly or indirectly with the help of the Internet technologies) are distinguished: (1) creation of certain conditions for formation and expansion of the sphere of activity of an individual in the virtual environment and then offline; (2) control over the state activities; (3) influence on government structures for dynamic functioning of democratic principles and norms; (4) dissemination of socio-political practices in the Internet; (5) formation of political pluralism; (6) selection of information that concerns the functioning of individual activities (Bronnikov 2012).

In particular, certain scholars (Melnikova 2014) studied the project "Russian public initiative" (www.roi.ru), which was created as per the Decree of the President of the Russian Federation dated March 4, 2013, No. 183 "Regarding consideration of public initiatives that are sent by the citizens of the Russian Federation with the usage of the Internet resource "Russian public initiative" with support from the Fund of information democracy. At present, the project provides several opportunities: showing initiative; acquainting with previously sent initiatives; voting "for" or "against" the posted initiatives; obtaining the information on the course and results of implementation of the public initiative.

Also, the scholars substantiate the opinion that fighting anonymity in the Internet without violation of human rights is impossible (Radaykin 2013), which is a new interpretation of a thought (Bachilo et al. 2001) that there will be no effective legal protection from the state in the new virtual world. As information technologies enter the life of society, the role of "law", which is established in the program provision, grows.

At the modern stage, the discussion on human rights in the digital age continues. Thus, digitization is understood as a factor of dynamic development, which led to creation and quick development of the digital economy, formation of the institutes of digital law, and new configuration of social relations on the basis of usage of social networks, Internet, and other information and communication technologies (Kartskhia 2018).

According to Khabrieva (2018), digitization might lead to expansion and narrowing of the sphere of legal regulation and change of its depth and other parameters – in particular, ratio of legislative and sub-legislative regulation and effect of private and public law. Appearance of different virtual communities – fintech and regtech – is a proof of striving of certain groups of people to move away from strict state regulation.

It is yet unclear how these changes will influence the constitutional and legal aspect of the issue, but it is obvious that various economic rights, which are an important part of the economic constitution – primarily, the right for intellectual property, which plays the most important role in the digital economy – will be influenced (Andreeva 2018).

While T.Y. Khabrieva notes the illusion of emergence of a new type of rights "digital" (Khabrieva 2018), V.D. Zorkin is sure that the time of specification of rights and liberties of human and citizens as to the digital reality has come (Zorkin 2018).

2 Methodology

The methodological basis of the research is general scientific methods of research: dialectical, logical, systemic & legal, comparative legal analysis, treatment of legal norms, technical, and legal forecasting.

3 Results

At present, almost 200 countries and different jurisdictions determine the rules of realization of human rights for information and create the corresponding structures of provision and protection of information, ensuring the mechanisms of protecting human from harmful information. The key notions in studying the information sphere through the prism of human relations, regulated by the legal and moral means in the conditions of change of a lot of paradigms under the influence of globalization, are human, person, knowledge, information awareness, freedom and its realization, institutions and institutes of civil society, public authorities, and law and order (Bachilo 2016).

It is stated (Seleznev and Skripak 2013) that emergence and functioning of the modern network Internet resources is based on realization of constitutional rights as an inseparable component of the legal status of a person. These rights include the right for unification (Internet is a place of formation of communities and groups on various interests, but their creation should not contradict the rules set by the Constitution of the Russian Federation); right for freedom of thought and speech (one of the reasons of attractiveness of online social networks for the users consists in the possibility to express the opinion on any issue); information rights that are connected to distribution, transfer, receipt, and usage of information.

S.M. Shakhrai in the issue of the main rights and liberties of a person in a "totally information society" states that the main constitutional rights and liberties of human and citizen have to be observed in the virtual and real worlds. The scholar means the primarily the rights mentioned in Articles 19, 22, 23, 24, 29, and 44 of the Constitution of the Russian Federation. Even Article 25, which guarantees immunity of residence, could be "virtually expressed" in network accounts or personal web-sites of a digital person (Shakhrai 2018).

The scholar emphasizes that the part of the program "Digital economy of the Russian Federation", which is devoted to challenges and threats that hinder the development of the Russian digital economy, establishes the main problem as "provision of human rights in the digital world".

In April 2019, the State Duma of the Russian Federation adopted a draft law on implementing changes into the Civil Code of the Russian Federation (Article 141.1), according to which digital rights are legally envisaged legally binding and other rights, the contents and conditions of realization of which are determined according to the rules of the information system that conforms to the qualities that are set by the law (p.1). If not specified otherwise by the law, the possessor of the digital right is the person who, according to the rules of the information system, can use this right (p.2).

Such variant of civil definition of digital rights differs from their constitutional and legal understanding. Thus, in the sphere of legal regulation there appear relations which subjects are virtual or digital "persons"; connected to legally significant identification of person in the virtual space; emerging due to realization of human rights in the virtual space (right for access to Internet, right to be forgotten, right for "digital death", etc.) (Khabrieva and Chernogor 2018).

Based on analysis of the international and legal acts, A.A. Kartskhia provided a definition of digital rights as citizens' rights for access, usage, creation, and publication of digital works and right for free access to Internet (other communication networks) with the usage of PC's and other electronic devices (Kartskhia 2018).

The Head of the Constitutional Court of the Russian Federation, V.D. Zorkin, says that that the initial point and methodological landmark in the conditions of the digital reality should be constitutional principles and norms. He attempted to define digital rights in the expanded form: digital rights are treated as human's rights for access, usage, creation, and publication of digital works and access and usage of PC's and other electronic devices, as well as communication networks, in particular, Internet. They also include the right for free communication and expression of opinion in Internet and right for immunity of private information sphere, including the right for confidentiality and anonymity digitized personal information (Zorkin 2018).

Based on this definition, it is possible to present a classification of specific digital rights and to develop draft laws that are connected to their guarantee and the constitutional and legal protection. In particular, the most important digital right is the right for information, which is a whole complex of direct main and secondary subjective legal rights that emerge based on information (Boer and Paveleva 2006). The Russian law on Internet is to be developed according to eight directions (Elchaninova 2005).

At present, there are discussions that the Constitution of the Russian Federation expands its effect, apart from guarantees of protection of personal information on the user's hard drive, to certain segments in Internet, with "personal information space" of the user. There is no direct legal act that regulated legal relations in Internet – however, certain aspects that could be assigned to protected interests of citizens, are regulated by other legal acts (Alyabyev and Lagutochkin 2013). Thus, as a result of implementing changes into the criminal law in 2010, p. 24.1 of Article 5, via the usage of Article 186 of the Russian Federation Code of Criminal Procedure, partially defines the legally protected elements of private life of citizens in Internet, which could be overcome based on a court decision: information on connections between users and (or) users' devices – the data on date, time, and duration of connections between users and (or) users' devices (equipment), and the data that allow identifying the users, as well as the numbers and locations of transmitting-receiving basic stations.

As is stated in the scientific literature (Golovin and Bolshakova 2014), implementation of a universal electronic card in in all regions of Russia in 2013, which is envisaged by the Federal law dated June 27, 2010 No. 210-FZ “Regarding organization of state and municipal services”, cased ambiguous reaction in the Russian society. Free acceptance of electronic documents should be transform into the mandatory condition of a citizen’s realizing the natural rights and liberties that he has from birth and that are guaranteed by the Constitution of the Russian Federation.

In China, where the most important thing are the interests of the state. The process took the path of creation of a total digital system – not only life of society but life of every individual. China implements the “Plan on creation of the system of social credit of trust”, adopted by the State Council of the People’s Republic of China in June 2014. The idea is to create the individual social rating of a Chinese citizen based on unification of all data bases that contain any information on individuals and legal entities and their analysis with the help of Big Data technologies. As of now, this Chinese project shows the utmost level of ignoring the rights and liberties of a person in the total digital world (Shakhrai 2018).

The basic provisions of Article 29 of the Constitution of the Russian Federation for “right for information”, “right for access to information”, and “freedom of information” are realized in expansion of electronic services and state services in the electronic form (Kartskhia 2018). Limitation of the right for information was acknowledged to be unlawful by the Constitutional Court of the Russian Federation (p. 1 of the Decree dated December 18, 2000, No. 3-P1). Limitations in distribution of personal data of citizens, including with the usage of modern technologies, are very important for guaranteeing personal liberties and rights of citizens, which is stated in the Definition of the Constitutional Court of the Russian Federation dated June 23, 2015, No. 1537-O2.

Digital age opens new opportunities for implementing the political rights of citizens. Usage of cloud technologies and encryption methods for provision of confidentiality of information on Internet opens new opportunities for bringing information to general knowledge (Article 29 of the Constitution of the Russian Federation). “Digital referendums” at all levels become an effective tool of real expression of will of the people. Direct democracy can and should oust the representative bodies of public authorities in the political system of a country. These and other new processes, related to “digitization” of the relations of human, society, and state should be considered and reflected in the coordinates of law, and their key principles and foundations should are established in the “Digital Constitution”, which could and should create a strong basis for a completely new level of democracy in the digital society (Shakhrai 2018).

Unlike a range of states – e.g., France, Switzerland, Canada, the UK, and Estonia – where the practice of voting via Internet has become traditional and has been established at the legislative level – this type of voting is being tested in Russia. Implementation of online voting into practice could solve a situation when the voter, due to reasonable excuse (envisaged by the law) cannot be present at the voting station and realize his constitutional active right to vote. Remote voting would allow Russians who live abroad to participate in the Russia’s political life (Samorodnyaya 2010).

Thus, it is necessary to solve the problem of provision of secrecy of vote. It is offered to use the Swiss experience (Avzalova 2014) in order to distinguish the process of identification of a user and the process of voting – which allows ensuring reliability, security, and anonymity of online voting.

The scientific literature contains that attempts to define the notion "e-democracy". According to E.D. Pavlova, this is democracy that is mediated by information and communication technologies – primarily, Internet. N.O. Obryvkova sees the phenomenon of e-democracy as the system of openness and transparency of state management, increase of effectiveness of political management, and active involvement of citizens into making of political decisions on the basis of the information and communication technologies. Thus, T.S. Melnikova distinguishes two important aspects. Firstly, e-democracy means the fact of increase of civil participation in policy and involvement into the political activities of social groups that were previously excluded from it. Secondly, e-democracy could be interpreted as increase of citizens' participation in making of state decisions. The supporters of this opinion think that development of information technologies could lead to gradual transition from representative democracy to direct democracy. Internet will help to resurrect the ideals of the Athens democracy, creating a kind of "people's council" of Internet users, who will be governing state institutes not through their representatives but through new channels of information transfer (Melnikova 2014).

The project "Concept of development of the mechanisms of e-democracy in the Russian Federation until 2020", developed in the Ministry of Communications and Mass Media, sees e-democracy as the form of organization of public and political activities of citizens, which, by means of wide application of the information and communication technologies, ensures a completely new level of interaction between citizens and public authorities, local administration, public entities, and commercial structures.

The most popular mechanisms of e-democracy in the Concept are as follows: online voting (via smartphone, Internet elections, etc.); mechanisms of network communication of citizens and group discussion of socially important problems and issues of public and political character in the online regime; mechanisms of formation of online communities, including the mechanisms of planning and implementation of civil initiatives and projects of group actions; mechanism of network communication between citizens and public authorities, including tools of influencing decision making and civil control over the activities of public authorities; mechanisms of public online management at the municipal level.

The processes of digitization of public relations lead to clear expression of the necessity to establish the socio-economic digital rights of citizen and human. The court practice of the European Court of Human Rights of recent decades shows that the right for intellectual property (rights for patents, trademarks, etc.), which was previously considered to be the right of private character, is included into the range of the main human rights (Kartskhia 2018).

4 Conclusions/Recommendations

Digital reality should fall under the effect of the Constitution as the normative act of the higher legal power in the Russian Federation.

Digital human rights include as follows: (1) the complex of information rights that could be realized in the digital age; (2) rights for access, usage, creation, and publication of works in the digital form; (3) right for immunity of private life in the information sphere; (4) right for access to communication networks of general usage, including the right to leave them; (5) right for usage of electronic devices in the form that is envisage by the law; (6) right for usage of the mechanisms of electronic democracy; (7) right for the objects of intellectual property in the digital form.

References

Avzalova, E.I.: The modern discussion on the role of Internet in the politics. Scientific note of Kazan University. Series “Humanitarian sciences”, vol. 156, no. 1, pp. 208–213 (2014)

Alyabyev, A.A., Lagutochkin, A.V.: Problems of law enforcement intelligence operations in the information space of Internet. Probl. Law Enforcement Activities, no. 1, pp. 66–69 (2013)

Andreeva, G.N.: Digital economy: economic constitution and intellectual property: adjunction of notions. Right Future Intellect. Property Innovations Internet, no. 1, pp. 12–18 (2018)

Baturin, Y.M.: Telecommunications and law: attempt at coordination. Center “Law and mass media”, series “Journalism and law”, no. 26 (2000)

Bachilo, I.L.: Information Law: Study Guide for Academic Bachelors’ program. Yurayt, Moscow (2016)

Bachilo, I.L., Lopatin, V.N., Fedotov, M.A.: Information Law: Study Guide. Legal center “Press”, Moscow (2001)

Boer, V.M., Pavelyeva, O.G.: Information law, SUAL, SPb, p. 1 (2006)

Bronnikov, I.A.: Transformations of the civil society. PolitBook, no. 2, pp. 77–99 (2012)

Golovin, E.G., Bolshakova, V.M.: Electronic identification of citizen: pros and cons. Vlast, no. 8, pp. 33–36 (2014)

Elchaninova, N.B.: Legal problems of internet. Bull. South. Fed. Univ. Tech. Sci. **50**(6), 197–200 (2005)

Elyakov, A.D.: Good and evil: poignant paradox of internet. Philos. Soc. no. 2, pp. 58–76 (2011)

Zorkin, V.D.: Law in the digital world: discussion at the St. Petersburg International Legal Forum. Russian Newspaper, 30 May, p. 1, 4 (2018)

Kartskhia, A.A.: Digital transformation and human rights. Russ. Polit. Sci. no. 4, pp. 33–38 (2018)

Melnikova, T.S.: The modern tendencies of development of e-democracy in Russia. Bull. Saratov State Socio Econ. Univ. no. 4(53), pp. 145–149 (2014)

Radaykin, M.F.: Regarding the problem of anonymity in the Internet. Gaps in the Russian law. J. Law, no. 2, pp. 25–28 (2013)

Samorodnyaya, E.G.: Realization of the constitutional rights via the Internet. S. Russ. J. Soc. Sci. no. 3, pp. 89–96 (2010)

Selezenev, R.S., Skripak, E.I.: Social networks as a phenomenon of the information society and specifics of social ties in their environment. Bull. Kemerovo State Univ. vol. 3, no. 2(54), pp. 125–131 (2013)

Spiridonov, A.A., Evsikov, K.S.: Regarding the new forms of realization of the constitutional right for information. Bull. Tula State Univ. Econ. Legal Sci. no. 4-2, pp. 274–280 (2015)

Khabrieva, T.Y.: Law and the challenges of the digital reality. J. Russ. Law, no. 9, pp. 5–16 (2018)

Khabrieva, T.Y., Chernogor, N.N.: Law in the conditions of the digital reality. J. Russ. Law, no. 1, pp. 85–102 (2018)

Shakhrai, S.M.: Digital constitution. the fate of the main rights and liberties of person in the totally information society. Bull. Russ. Acad. Sci. **88**(12), 1075–1082 (2018)

The Form of Interaction Between the Public Authorities and Civil Society in the Context of Digitalization

I. V. Abdurakhmanova(✉), N. E. Orlova, G. B. Vlasova, V. I. Vlasov, and S. V. Denisenko

Rostov State University of Economics, Rostov-on-Don, Russia
dima_rd@rambler.ru, kinl997@mail.ru,
vlasovagb@mail.ru, vlasovvi@mail.ru,
s.v.denisenko@bk.ru

Abstract. Purpose: The purpose of this work is to reveal the opportunities and prospects of the transformation of the traditional forms of the interaction of the public authorities and the civil society institutions under digitalization.

Design/Methodology/Approach: The analysis legal basis and the Russian model of the interaction of the public authorities and the civil society is based on the concept of the integrated jurisprudence with applying the interdisciplinary approach to research the state and legal phenomena. The methodological basis for this research is international and Russian legal sources where principles and conceptual approaches to public authorities and civil society interaction are stated.

Findings: The article looks into the issues of transformation of the public authorities and the civil society interaction under digitalization impact. The advantages of the "e-state" development and new opportunities for the civil society involved in the decision making process and in management decisions implementation, in increasing level of citizens' confidence in public authorities are analyzed.

Originality/Value: In the nearest future the use of digital technologies will change the character and forms of the public authorities and the civil society interaction. The electronic document management development, use of management information systems and WEB services ensures the transparency and openness of the public authorities' activities but at the same time generates new risks and threats for the information security of the state and the society. That is why the digital technologies application as a tool of public authorities and the civil society interaction needs systemic legal regulation.

Keywords: e-state · Digital technologies · e-government · Global information society · Information security · Civil society · Government bodies

JEL Code: K38 · K100

E. G. Popkova and B. S. Sergi (Eds.): ISC 2019, LNNS 87, pp. 925–930, 2020.
https://doi.org/10.1007/978-3-030-29586-8_105

1 Introduction

In the recent years the digital technologies have become a part of our lives having changed communication and management opportunities of modern society, and it demands to reconsider the forms of interaction between public authorities and the civil society. In modern social discourse the idea of the democratic institutions improvement is highly demanded through creation of the effective mechanism to influence the civil society on the public authorities. In its turn the global changes define the scale of tasks the state faces to modernize economy, political system and social sphere and whose decision involves harmonization of diverse public interests, and it requires reconsidering the existing forms of the public authorities and society, to work out and to implement the e-state concept. Digitalization and information society development makes the legal basis development and corresponding programs of the information environment development to increase the efficiency of the public authority activity topical.

Different aspects of the formation and development of the institutions of the civil society and legal state have traditionally been the subject of historic, political, state and political law studies. However the issues of the public authority activity transformation and mechanisms of citizens' political participation in the context the digital technologies development have not been efficiently covered in the law literature.

Okinawa Charter on Global Information Society (July 21, 2000) pointed out the necessity to use information and communication technologies allowing providing the public services in the real time as an indispensible condition to make the public authorities more accessible for all citizens. The directions of global information society development and electronic state in Russia received legal clearance in the state program "Information Society (2011–2020)" and "Strategies for the Development of the Information Society in the Russian Federation for 2017–2030", approved by the President of the Russian Federation received legal clearance. The digital management technologies development prospects give more opportunities for the public authority and society to interact, and create new channels of the citizens' political participation, but at the same time generate new risks creating global threats for the information security. A complex set of issues due to public management digitalization needs to be considered scientifically.

2 Materials and Methodology

International and the Russian legal sources, where principles and conceptual approaches to public authority and civil society interaction are reflected, made up a methodological basis of the research.

International principles to create global information society are defined by Okinawa Charter on Global Information Society (2000), a Declaration of Principles "Building an Information Society: a global challenge in the new Millennium" (2003), the Tunis Agenda for the Information Society (2005). In accidence to the declared by the international community approaches Information Society Development Strategy for 2017–2030 was adopted in Russia.

The analysis of the legal regulation of the forming information society at the international and national level as well as the practice of the e-state creation in Russia allows singling out prior directions of the public authority and civil society interaction under digitalization. The electronic information society allowing the citizens to obtain qualitative, verified information freely and to express their opinion about the public authorities' activities is to become a foundation of such interaction. The universal access provision to information and communication technologies allows illuminating "digital inequality" that prevents from free interaction of the public authority and civil society interaction. "The e-government" concept presupposes to establish informational-management systems to ensure the public authorities' activity and to form a single space for the electronic interaction in order to ensure transparent and responsible public management.

"The e-state" project implementation is to develop of the democracy institutions, to increase of the public management system openness, to involve the citizens in the discussion, adoption and control of the execution of government decisions.

3 Results

Constitutional provisions enshrine the democracy principle, the right of citizens to receive and impart information freely, to participate in the public affairs management, the right to appeal individually and collectively to state bodies (Article 1, 2, Clause 2, Article 24, Clause 4 1 Article 32, Article 33 of the Constitution of the Russian Federation). However they need to be clarified in respect to specific forms of interaction among the public authorities and citizens and society under digital technology development.

The significant progress in e-state infrastructure creation has been made in Russia lately: the "Internet" users share has reached 71.4%, and the share of the citizens getting public and municipal services electronically totaled to 64.3% (Ministry of Digital Development, Communications and Mass Communications of the Russian Federation 2018). The electronic document management has increase significantly lately, the obtaining information procedure on the requests of citizens from government bodies has simplified. A number of documents are electronic ones in Russia now: medical records, banking cards, of compulsory pension insurance certificates, passports of vehicles, certificates of the right to real estate, and this process is far from being finished.

The Russian legislation regulates the citizens' access to the information on public authorities' activity, provides the creation of the public authority official sites to inform the society on their activity, makes it possible for the citizens to appeal to the public authorities electronically as well as to use web-services for the citizens and public authorities to interact (the RF Government 2014).

Regulatory legal acts publishing on the legal information official Internet portal, the function of the official websites of the senior officials and the Russian Federation public authorities' combined with the opportunity to use social services and Internet technologies WEB 2.0 create new opportunities for the interaction of public authorities and citizens.

The use of different services, including the regional ones, enables the public authorities to conduct the discussion of bills, socially meaningful initiatives, and management solutions publicly. The citizens in their turn in the process of the open social discussion participate in the decision making in the area of the public management, can put important questions into the agenda of the public authorities, and that definitely means the development of the direct democracy institutions with digital technologies application.

Due to the digitalization the public authorities have the efficient feedback with citizens, the possibility to monitor the citizens' reaction on the taken decisions, and it can affect the performance of government agencies and officials significantly. The speed in information dissemination and receipt has increased many times. Internet-services and web-sites of the public bodies, established in the frame of the "e-government" project, allowed taking citizens' appeals electronically through online receptions. Online video broadcasts and surveys give the public bodies huge possibilities to organize open social discussions on the state management issues and decision making taking into account the expressed opinions and the received proposals. However, the researches of the information activity of the public authorities revealed a number of problems related to the inefficient feedback, low quality of the information on public authorities' activity, lack of the open information on the results of implemented policy (Danilova 2015).

Digital technologies change the nature of the public bodies and society interaction crucially, making it interactive. The citizens from the passive consumers of the officially provided information turn into the active participants and create various internet-communities, react on the government's decisions immediately and publicly, and in this way exercising direct public control over state decisions execution.

Since 2013 public initiatives addressed by the Russian Federation citizens with the Internet resource "Russian Public Initiative" and supported by 10 thousand people are to be considered by the state bodies to take appropriate measures to implement it (the RF President 2013). Various services, including the regional ones, are used to find out the citizens' opinion on the issues state administration and to receive the proposals and complaints about the officials' actions.

The mutual activity of the state authorities and civil society to protect the human rights and development democratic principles while working out and implementing the state policy is an important direction. One of such civil society institutions is the Civic Chamber of the Russian Federation, whose activity is to approve the publicly meaningful interests of the citizens, public associations and public authorities, to propose and to support public initiatives, to implement public examination of legal acts as well as to control the executive branch (The Federal Law 2005). The Civic Chamber of the Russian Federation official site contains information on projects involving people in the publicly significant actions, and monitoring of different spheres of the social life, the form for the citizens to apply electronically, topical comments of the Chamber members. The digitalization allows the Chamber to use interactive forms in its activity, to exercise public monitoring and to organize public discussions with attracting the maximum number of the citizens. Due to the digital technologies publicity and openness of public control exercised by the Chamber is ensured (The Federal Law 2014).

The e-participation of citizens and their associations in the political activity can be done in different forms: the online participation in the projects and actions where state bodies are involved; organization of the virtual meetings, surveys, internet voting, online work with public chambers, public councils and committees those implementing the public control over the sate bodies' activities. However one of the most important tasks of the legal regulation and organization of such a form of political activity is civil society integration and increase trust in the state bodies.

However new interaction forms, appearing due to digital technologies, have some evident advantages as well as additional risks for the state and society. It is not possible to use the new channels of the political communication effectively and receive public services without possessing sufficient digital competencies; otherwise the risk of the people's dissatisfaction with the state bodies increases substantially.

The significant threat is that it is difficult to verify citizens and online communities in the virtual space, and in this connection problem of the qualitative communication arises, there is a possibility to distort the information, to manage the information flows and to manipulate the mass consciousness and public opinion. The high speed and transboundary of communication turn the information into the powerful weapon use to hack state institutions and to make the society chaotic. "Information wars" have a formative effect on the on people's worldview in the modern society. The possibility to remote access to the users' personal data, data bases, and cyber attacks threat the human rights, states' stability and democratic institutions' efficiency. That is why while applying new communication digital technologies to increase the interaction efficiency between the state authorities and civil society it is necessary to be aware of possible risks and threats to the information security of the society and state.

4 Conclusion

Digital technologies application is rapidly changing its character and the forms of the interaction between state bodies and citizens. The digitalization creates the conditions to develop the constructive partnerships between public authorities and citizens.

The "e-state" concept changes of the level of the information openness of the public authorities qualitatively, ensures the possibility for the citizens and public associations to work out and implement management solutions, takes into account their opinions and priorities revealed in the course of the constant dialogue.

The electronic document management development, application of the information-management systems and WEB services ensures the transparence of the public authorities' activity, but at the same time generates new risks and threats of the information security of the state and society. That is why the use of the digital technologies as a tool to interact between the public authorities and citizens needs the system legal regulation and a single information space in the state management system.

The transition to the information society and e-state is a complex process whose result is to increase the trust of the citizens in the state activity, to ensure close interaction of the state and civil society institutions as well as to boost the efficiency of the public authorities' activity.

References

Bochkov, S.I., Makarenko, G.I., Fedichev, A.V.: On the okinawa charter of the global information society and the tasks of developing russian communication systems. Leg. Inform. **1**, 4–14 (2018)

Danilova, N.E.: Transformation of public administration as an organization of effective interaction of government and society in conditions of informatization. Cent. Russ. J. Soc. Sci. **10**(6), 294–298 (2015)

Dniprovskaya N.V.: Digital transformation of interaction between government bodies and citizens, State Administration. Electron. Gaz. **67**, 96–110 (2018)

Pozdnyakova, E.O.: Public legislative initiative: application mechanisms. J. Public Munic. Adm. **5**(4), 44–47 (2016)

Ustinovich, E.S.: On the concept of information activities in public administration. Threats Secur. **34**(91), 54–59 (2010)

Ustinovich, E.S.: Interaction of power and civil society structures in the internet space (political and legal analysis). State Power Local Self-Gov. **3**, 11–15 (2018)

Tselnicker, G.F., Nemov, A.A.: Interaction of civil society and state authorities in the Russian Federation. Bulletin of Volzhsky University named after V.N. Tatischeva **2**(4), 51–58 (2018)

Federal Law of April 4, 2005 No. 32-FL "On the Public Chamber of the Russian Federation" (2005). https://base.garant.ru/12139493/. Accessed 30 Apr 2019

Federal Law of July 21, 2014 No. 212-FL "On the Basics of Public Control in the Russian Federation" (2014). https://base.garant.ru/70700452/. Accessed 30 Apr 2019

Ministry of Digital Development, Communications and Mass Communications of the Russian Federation. The updated annual report on the implementation and evaluation of the Russian Federation "Information Society (2011–2020)" state program effectiveness (2018). https://digital.gov.ru/ru/documents/6010/. Accessed 30 Apr 2019

President of the Russian Federation. The Decree of March 4, 2013, No. 183 On consideration of the Russian Federation citizens' public initiatives sent through the Russian Social Initiative Internet resource (2014). https://base.garant.ru/70326884/. Accessed 30 Apr 2019

Government of the Russian Federation. The Order of January 30, 2014, No. 93-p "On the Federal Executive Bodies Openness Concept" (2014). https://www.garant.ru/products/ipo/prime/doc/70478874/. Accessed 30 Apr 2019

Digitalization as an Urgent Trend in the Development of the Social Sphere

Tatyana F. Romanova[1(✉)], Vladimir V. Klimuk[2], Olga V. Andreeva[1], Anna A. Sukhoveeva[1], and Marina O. Otrishko[1]

[1] Rostov State University of Economics, Rostov-on-Don, Russian Federation
kafedra_finance@mail.ru, olvandr@yandex.ru, suhoveeva_anna@mail.ru, starkal3@mail.ru
[2] Higher Education Institution "Baranovichi State University", Baranovichi, Republic of Belarus
klimuk-vv@yandex.ru

Abstract. The development of digital technologies (information and service «digital platforms») should be a key factor in improving the quality and availability of social services (health care, education, social protection and so on). The purpose of this article is to show the directions and problem aspects of the implementation of digital technologies into the social sphere at the present stage. The article analyzes the impact of digitalization on the economy of the Russian Federation and the Republic of Belarus. In the rating of Digital Evolution Index 2017, Russia is in a promising group of countries characterized by an increase in the overall level of «digitalization», passing to a group of leading countries, which allows to identify a number of prospective areas of the social sphere digitalization.

According to the results of the study, they are: unified information services for socially unprotected citizens, for people with disabilities; «social assistants» in social services for aged and disabled citizens; technologies in the rehabilitation industry; automation of procurement of medicines; the development of various digital social services, for example, within the concept of «smart city» («smart» health care implies the introduction of telemedicine, the effective system of electronic services includes operations from «electronic registry» to getting the test results online, smart housing system implies the control and accounting of data and providing information to their users, smart «accessible environment» is created on the basis of geographic information systems); digitalization of employment and education processes.

Keywords: Social policy · Social sphere · Social expenditures · Digital economics · Digital technologies · Digital platforms

JEL Code: H5 · H11 · H75 · O35

E. G. Popkova and B. S. Sergi (Eds.): ISC 2019, LNNS 87, pp. 931–939, 2020.
https://doi.org/10.1007/978-3-030-29586-8_106

1 Introduction

Many developed and developing countries, realizing the inevitability of the upcoming changes, began to move forward in the direction of «digitalization» of the economy.

The United States and China were the first countries that declared such course. They are considered to be informal leaders of the «digital» race today. Then England, the European Union, Australia, Belarus, Kazakhstan and others adopted appropriate programs (Keshelava et al. 2017).

In 2017, the digital revolution entered a crucial phase that means every second inhabitant of the Earth connected to the Internet. By the number of Internet users, Russia takes the first place in Europe and the sixth one in the world.

For the recent years, «digitalization» covered all the spheres of the economy and public life in Russia. Before the trend was instigated by the society and the corporate sector, but since 2017 there has been used the state program «The Digital economy of the Russian Federation». «Digital platforms» in the main directions of socio-economic life in Russia is of great importance in this program.

In general, Russia plans to use new technologies, including blockchain, in eight areas: state regulation; information infrastructure; research and investigations; HR and education; information security; public administration; smart city; digital health care. All these areas affect the social sphere.

On the 21-st of December, 2017, the President of the Republic of Belarus signed the Decree No. 8 «On the development of the digital economy», which provides unprecedented conditions for the development of IT-industry and gives the country a serious competitive advantage in the formation of the digital economy of the XXI century. Conditions for the development of the digital economy provided by the revolutionary decree are the one of the best conditions in the world. The document pays special attention to projects based on blockchain technologies and smart contracts.

However, none of the countries, including the leading countries, has a holistic understanding of what the «digital» economy is and what consequences it will lead to. Many countries consider the «digital» economy as new forms of payments and communication with consumers, but not new forms of management and economic relations. Apparently, most countries do not «build» the «digital» economy, but just they are engaged into «digitalization» of existing economic relations. This activity, despite the obvious practical aspect, is not focused process of building the digital economy (Gartay 2018).

In the nearest future, the development of digital technologies (information and service «digital platforms») will be a key factor in improving the availability and quality of services in the social sphere. Digitalization transforms the social paradigm of people's lives.

It offers unprecedented opportunities to acquire new knowledge, broaden way of thinking, learn new trade and improve skills.

Digital technologies are used as a mechanism of social lifts that promote social and financial involvement of the population and improve the convenience of receiving services in such important areas as medicine, education, municipal and public services, culture and so on.

The purpose of this study is to show the directions and problem aspects of the implementation of digital technologies into the social sphere at the present stage.

It is still early to talk about global digitalization of social projects, and one of the main tasks today is to implement digital technologies into the social sphere.

2 Methods

The investigation was based on the study of scientific literature, information of official websites of authorities, organizations implementing social policy in the social sphere. Since there is still no legitimate definition of the social sphere and the official classifier of economic activity does not contain such a word combination, different authors use different concepts (social infrastructure, socio-cultural sphere or measures, social services), describing the totality of various activities and referring them to the social sphere.

The definitions of the concepts of social sphere, social infrastructure, social expenditures available in the literature are debatable. In this work, we will follow an economic approach, concerning the social sphere as a set of industries that directly determine the living standard of the population. Based on the sections in the budget classification of expenditures corresponding to state functions we refer to them branches of education, health care, physical training and sports, culture and art, social protection of the population. This interpretation of the social sphere concept is suitable for the study of social spending.

In our view the social policy of the state is «the area of state policy in relation to the formation of the living standard of the population, reproduction of human capital, social services and the development of social infrastructure at the federal, regional and local levels» (Babich and Pavlova 2000).

3 Results

Before we talk about digitalization of the social sphere, let us turn to the results of the investigation of the level of development of the digital economy.

Thus, the urgent study was conducted by experts of the School of Law and Diplomacy named after Fletcher at Tufts University in collaboration with Mastercard to assess the development of the digital economy in 60 countries. They presented the digital Evolution Index 2017 rating, which reflects the progress in the development of the digital economy of different countries, as well as the level of integration of the global network into the lives of billions of people.

They represented rating of the Digital Evolution Index 2017, which demonstrates the progress in the development of the digital economy of different countries, as well as the level of integration of the global network into the lives of billions of people.

The researchers divided countries into 4 groups:

1. leaders that demonstrate the rapid growth and a high level of the digital development and continue to lead in spread of innovation (Singapore, UK, New Zealand, UAE, Estonia, Hong Kong, Japan and Israel);
2. the slowing pace of growth that for a long time showed steady growth, but now significantly reduced the pace of development. If these countries do not use innovation, they risk falling behind the leaders of digitalization. South Korea, Australia, Canada, USA, Germany, Scandinavian countries are in this group.
3. prospective countries. Despite the relatively low overall level of digitalization, these countries are at the peak of the digital development and demonstrate stable growth rates, which attracts investors. China, Kenya, Russia, India, Malaysia, Philippines, Indonesia, Brazil, Colombia, Chile, Mexico are in this group;
4. Problematic countries. Digital progress in them is constrained by low level of the innovative development and slow growth. These countries are South Africa, Peru, Egypt, Greece, Pakistan.

Thus, Russia is in a group of promising countries characterized by an increase (rating of Digital Evolution Index 2017) in the overall level of «digitalization», a transition stage to a group of leading countries. The Republic of Belarus was not analyzed in this rating.

Innovative digital technologies should improve functioning the social sphere. Digitalization is important both for the beneficiaries and for those whose work is related to social services providing, because often the help to different categories of the population obliges professionals to work with huge amounts of data. Therefore, the innovative digital technologies are designed to facilitate providing services quickly and efficiently.

Therefore, the latest digital technologies are designed to facilitate quick and efficient provision of services. Let us list the most popular directions of digital technology implementation in the social sphere. For example, it is the launch of unified information services, which should provide complete information on what help a citizen has the right and how to use this right.

On the 14-th of January, 2016 at the meeting of the Supervisory Board of ANO «Agency for strategic initiatives to promote new projects» chaired by the President of the Russian Federation, there was approved the project «New quality of life» within public and private partnership to form an information system for people with disabilities by attracting extra-budgetary funds.

The aim of the project is to create a unified informational and service platform, which is planned to perform the function of information support of social guarantees for persons with disabilities.

The informational portal aggregates all types of social services and rehabilitation measures for people with disabilities and starts working as a general information reference to assist in meeting their needs.

The platform aggregates NCOs providing social services for people with disabilities, as well as companies supporting people with disabilities.

Also in Moscow, there has been developed and tested in a number of state-budget institutions of Moscow the project «social assistant». According to it social workers use a special digital platform and mobile application during servicing «social clients» at home.

The application will make possible to create an order for food and other necessary goods with home delivery, which increases the efficiency of social services. The use of technologies in the rehabilitation industry is perspective. For example, to orient in the surrounding space, the blind person can now download a program that operates in connection with the neural network, and artificial intelligence with the help of built-in phone video camera will determine which object is in front of the person.

There are examples of automating the procurement of medicines. So, one of the first thing on the path of digitalization was the municipal pharmacy network of Novosibirsk, as a result, there is processed a huge array of data on more than 100 pharmacy points and 44 thousand items of medicine. Every 15 min, there is pumped all information from pharmacies to the central database and there is formed a complete picture of their needs. The development of a procurement robot that tracks the entire process from start to finish was specially ordered.

Thus, the most important factor in «digitalization» of social services is the development of the telecommunications market and the initiative of leading public and private companies interested in formation of new markets and services.

Thus, the development of various digital services is prospective today, for example, within the concept of «smart city»: for example, «smart» health care implies the introduction of telemedicine, an effective system of electronic services – from «electronic registry» to getting the test results online, «smart» housing and utilities system implies the control and accounting the data and submitting the information to their users, «smart» «accessible environment» is based on the geographic information systems. But to accelerate the process of digitalization, we need people who will develop technologies. Therefore, universities are also involved in the implementation of digital technologies into the social sphere.

In 2018, at the thematic forum «Community» there were presented projects of digitalization in the most important social sectors such as healthcare, employment and education.

In experts' view, digitalization of the social sphere is a key to the sustainable development and growth of society welfare.

The possibilities of using information systems in federal and regional executive authorities, business and educational structures are actively discussed. Most of them are improving in this work, but they are not related to each other, which may prevent the effective use of these technologies.

The President of Russia emphasized that all interactions between the government and citizens, the government and business should be digitalized.

The «My documents system» and the system of multifunctional centers (MFC) is rapidly developing. The system, which has been deployed for a long time, now works effectively and allows to string on it more and more functions and solve more problems.

Now rapidly developing portal «Work in Russia», which is provided by the Law on employment, and it includes a huge number of transparent and clear communications between an applicant and an employer, because the portal is verified both the Pension Fund and the Tax Service. Single register of the recipient of social services now receives the regional content.

The perspective possibilities of the Unified integrated information system (Ustinovich 2019) are highly appreciated, the necessity of unification of all measures of social support is emphasized.

It is necessary to find ways to improve the organization of medical services for the population through digitalization.

Now there are services provided by the Ministry of Health: the possibility of recording to a doctor, checking the data on the policy and some other possibilities. However, these measures are not enough.

Automated information systems (AIS) in the sphere of child protection are developing. These systems are used as a tool for interdepartmental cooperation and allow to promptly solve various issues, such as AIS «Medical and social expertize».

One of the state information systems is «Federal register of disabled people», which allows to combine in a mobile application all the relevant information for people with disabilities and the institutions that provide them with services. This system allows a citizen to get all the information about disability, recommended and carried out rehabilitation or habilitation activities, public services and payments.

The federal state information system «Federal register of disabled people» allows departments to obtain analytical and statistical data to assist the disabled people in any problem such as providing them with technical means of rehabilitation or providing them with educational services.

4 Conclusions

The effectiveness of a scientific and innovative project, including in the social sphere, is determined by the synergy between the participants, that allows to maximize the planned result, compared with individual, separate work.

Also the direction of effective cooperation is reflected in one of the 17 goals of the Sustainable development of Belarus «Partnership for sustainable development», included into the Agenda for sustainable development for the period up to 2030 (Agenda-2030).

The goal is to strengthen global partnership to promote and achieve ambitious goals by providing knowledge, experience, technologies and financial resources.

Within the goal, a number of indicators are identified, the monitoring of their dynamics determines the degree of effectiveness of the country's sustainable development strategy: the GDP growth rate, the growth rate of goods and services export, the rate of change in the consumer price index.

The development of the social sphere is impossible without the growth in the real sector of the economy. States that are committed to innovation and research uses qualified personnel that is a key resource of digital economies.

At the same time, the undoubted role of science in interaction with the real sector is emphasized by the top leadership of the countries.

The President of the Republic of Belarus Alexander Lukashenko on the II Congress of scientists of the Republic noted the urgent need for cooperation between science, education and business, stressing the leading role of industry: «It is in industry the core of all the innovations. There are no countries with strong science and weak industry. And vice versa < …> industry together with science and the education system should solve main tasks dictated by time: to determine the directions of diversification and modernization of production – from the inspection of new technologies and equipment to participation in forming and commissioning them.

New modern plants should be built on the basis of scientific substantiation. It is necessary to ensure assessment of the product reliability and quality at all stages – from its investigation to production».

At the session of the Council for science and education devoted to the global competitiveness of Russian science, President of the Russian Federation V.V. Putin noted that «interaction of science and business should be a key condition for the implementation of the program «Digital economy», he stressed the need to develop own research infrastructure in the country and emphasized the necessity for the formation of the powerful international research teams in our country, the development of scientific cooperation with other countries and increase the openness of domestic science, expanding cooperation in science with other countries and the formation of powerful international research centers in our country.

That is why the development of higher education adapted to new conditions is relevant both for Russia, Belarus and other countries.

Large-scalc preparations for the digital transformation of universities, which should quickly integrate new and actual knowledge into their educational programs to become real centers of training for the digital economy are underway.

Experts on digital technologies and economics consider that automation will significantly affect the labour market in the next decades.

According to McKinsey Global Institute up to 50% of the world's work processes will be automated by 2036. This will lead to a significant release of staff, reduction of workplaces requiring secondary qualifications and the rising wage gap. At the same time, in Russia, according to the UN and the Federal state statistics service, the number of able-bodied population will decrease in the next two decades. Automation will help to mitigate the negative consequences of this phenomenon. In such circumstances, «digital» personnel is a strategic asset. Its shortage inevitably leads to a slowdown in the growth of both the digital economy and the country's economy as a whole. Thus, the state priority is to provide the necessary number of qualified specialists in digital technologies. And it is necessary to carry out this task through the modern high-quality education system.

A promising factor in the development of the social sphere is formation of conditions for expanding the participation of the non-state sector in providing social services. Therefore, it is important to support the development of social business (Andreeva and Epifanova 2018) in such areas as health care, education, regulation of unemployment and employment, regulation of income of the population, formation of social protection and social assistance to vulnerable categories of the population.

And one of the priorities for support should be the condition of work in the digital environment. Moreover, the digital transformations will contribute to the development of social business itself, which will allow to implement more actively various social missions.

Thus, it can be predicted that in the next years the process of «digitalization» of the social sphere in Russia, the Republic of Belarus and other countries will continue to develop in the direction of expanding the number of industries and the geography of the introduction of new technologies, respectively, the number of digital platforms of social orientation will increase.

References

Andreeva, O.V., Epifanova, T.V.: State Support for Small and Medium-sized Business in the Social Sphere and the Development of Social Business in the Economy of Russia. Modernization of Business Systems of Russian Regions as a Factor of the Growth of Economy: Trends, Challenges, Models and Prospects, Rostov-on-Don, pp. 95–112 (2018)

Babich, A.M., Pavlova, L.H.: State and Municipal Finances. Textbook, 345 p. UNITY, Moscow (2000)

Bregar, L., Puhek, M., Zagmajster, M.: Analiza stanja na področju digitalizacije in e-izobraževanja v visokem šolstvu v Sloveniji. Maribor, DOBA Fakulteta (2017). http://www.doba.si/ftp/dokumenti/fakulteta/gradiva/Bregar_Puhek_2017_IKTvVS.pdf

Dezuanni, M., Foth, M., Mallan, K., Hughes, H.: Digital Participation Through Social Living Labs: Valuing Local Knowledge, Enhancing Engagement. Chandos Publishing, an imprint of Elsevier, Cambridge (2018)

Gartay, J.M.: International experience of digitalization of the economy: Russia and Belarus. In: Working Paper, Modern Eurasian studies, vol. 1, pp. 15–23 (2018)

Keshelava, A.V., Budanov, V.G., Rumyantsev, V.Yu.: Introduction to the Digital Economy, p. 28. Moscow (2017)

Liu, K., Nakata, K., Li, W., Baranauskas, C.: Digitalization, innovation, and transformation. In: Proceedings of the 18-th IFIP WG 8.1 International Conference on Informatics and Semiotics in Organizations, ICISO 2018, Reading, UK, 16–18 July 2018 (2018). https://doi.org/10.1007/978-3-319-94541-5

Speech by the President of Belarus Alexander Lukashenko at the II Congress of Scientists (2017). https://www.belta.by/president/view/vystuplenie-lukashenko-na-ii-sjezde-uchenyh-belarusi-280351-2017. Accessed 12 Nov 2018

Session of the Council for Science and Education (2018). http://www.kremlin.ru/events/president/news/56827. Accessed 24 Mar 2019

Studenikin, N.V.: Influence of Digital Technologies on Social Services: World Experience and Prospects in Russia. News of Tula ЬItate University. Humanities, vol. 1 (2018). https://cyberleninka.ru/article/n/vliyanie-tsifrovyh-tehnologiy-na-sotsialnye-uslugi-mirovoy-opyt-i-perspektivy-v-rossii. Accessed 06 Apr 2019

Pecourt Gracia, J., Rius-Ulldemolins, J.: Digitalization of the Cultural Field and Cultural Intermediaries: a Social Critique of Digital Utopianism La digitalización del campo cultural y los intermediarios culturales: una crítica social del utopismo digital. Revista Española De Investigaciones Sociológicas (REIS) **162**, 73–89 (2018)

Plotnikova, E.: Digitalization of Education in the Leading Universities of Saint Petersburg, 497 p. (2019)

White, A.: Digital Media and Society: Transforming Economics, Politics and Social Practices. Palgrave Macmillan, Basingstoke, Hampshire, New York (2014)

Ustinovich, E.S.: Digitalization of the social sphere in Russia. Soc. Policy Soc. Partn. **2**, 32–36 (2019)

Digital Transformation of the System of Public Finances Management

Natalia G. Vovchenko(✉), Olga B. Ivanova, Elena D. Kostoglodova, Yuliya V. Nerovnya, and Svetlana N. Rykina

Rostov State University of Economics, Rostov-on-Don, Russia
nat.vovchenko@gmail.com, sovet2-1@rsue.ru, ramachka2006@rambler.ru, ulial511@yandex.ru, ya.svetlana-41@ya.ru

Abstract. Financial sector is one of the key drivers of the digital economy in the Russian Federation and in the whole world. Development of financial technologies modernizes the traditional directions of provision of financial and other services in which innovative products and services for final consumers appear. One of serious obstacles on the path of digital transformation of state management in Russia is the problem of ineffective distribution of government authorities and absence of interconnection between the assigned functions and material, personnel, and financial resources that are provided to public authorities.

Based on this, the following tasks are set: considering from theoretical positions the notion of digital transformation, substantiating the modern methodological approaches to management of public finances in the conditions of digitization of economy, and determining the main directions of digital transformation in the sphere of public finances in the context of adoption of a complex of measures for increasing the quality of state management.

The authors come to the conclusion that without systemic transformation of managerial processes and cardinal transformation of the work of the whole government machine it is possible to achieve significant feedback from digitization. Effective and secure functioning of the digital financial space requires implementation of coordinated measures at the level of all its participants and timely and proportional regulation, which will support stability of the financial system, protecting the consumers' rights, and stimulate the development and implementation of digital innovations.

Keywords: Digital transformation · Public finances · Financial technologies

JEL Code: G28 · G32 · G31 · G38 · H70

1 Introduction

Usage of new electronic platforms and digital technologies creates preconditions for transformation of the functions of state management and development of institutional forms, which allow ensuring more flexible and effective interaction between the state

E. G. Popkova and B. S. Sergi (Eds.): ISC 2019, LNNS 87, pp. 940–949, 2020.
https://doi.org/10.1007/978-3-030-29586-8_107

and business for attraction of investments and innovations into strategic "digital" projects (Scientific and analytical report 2018).

Digital transformation is the change on the basis of digitization of the contents of state management, which leads to increase of its quality, i.e.: unjustified state interferences and increase of efficiency and effectiveness of state management.

Digital transformation in state management should not be brought down only to changes during provision of state services. The possibilities of usage of the modern digital technologies for developing the state policy and laws, administration of revenues, management of expenditures and state property, and controlling activities have large potential (Digital future of state management 2019).

The main condition of successful digitization consists in redesigning all processes on the basis of full exclusion of traditional "paper" processes and any personal interactions.

Digital transformation means transition to the data, services, and infrastructure of joint usage, and the main task consists in usage of digital channels for maximum effects for consumers (users, citizens) and increase of effectiveness and efficiency in the public authorities' activities (Digital transformation 2019).

Transition to the path of digitization of the global economy is accompanied by completely new transformation of socio-economic relations in society, which creates completely new opportunities, challenges, and threats. At present, the share of the digital economy in gross domestic product of all countries with developed economy grows and covers all spheres of most countries of the world.

At present, the level of competitiveness of financial organizations is determined by the ability to implement and use the modern digital technologies. This means that ensuring the necessary level of competitiveness of the financial sphere of the Russian economy requires development of the digital segment.

The Russian Federation has the Program of development of the digital economy until 2035, which determines the goals and tasks for provision of Russia's transition to the digital economy. The main task of regulation is creating a favorable environment for development of innovations, which will determine the level of readiness of new products and services for conquering mass markets.

The Passport of the national program "Digital economy of the Russian Federation", which was adopted in autumn of 2018 envisages six federal projects (normative regulation of the digital environment; information infrastructure; personnel for the digital economy, information security; digital technologies, digital state management), which implementation will allow for large increase of internal investments into the digital economy, creation of the modern and safe IT structure, and transfer of government bodies to Russian software (Program of development 2017).

A new stage in development of financial technologies is based on cloud technologies, which opens more opportunities for cardinal changes of the infrastructure and functioning of the whole financial sector of economy. Growth of economy is possible with development of technologies that allow for precise evaluation of the current state of markets and spheres, effective forecasting of their development, and quick reaction to the changes in the situation in national and world markets.

Progressive development of the information environment of society requires refusal from the extensive methods of development of state information systems. At the modern stage, there's a necessity for a shift from the methods of covering the number of directions of analytics to the methods of constructing the unified state digital platform that is based on applying the methods of Big Data analysis – which will expand the horizon of analytics and increase the authenticity of information resources.

Description of the technological ideas shows the principles, goals, and ways of solving the task on transformation and reengineering of technologies of state finances: understanding the technological perspective of development of state finances for the coming five years; the ideology of creation of a technological platform for the purpose of a leap of development of financial technologies (fintech) in the sphere of finances. The description reflects three directions of solving the task – namely, the concept of technological solution, systemic architecture of formation, and plan of achievement of results.

The most problematic tasks include the following:

- imbalance of the system of distribution of government authorities, organizational structure of government bodies, and interaction between them;
- ineffectiveness of budget expenditures for the state machine and execution of functions;
- insufficient level of labor efficiency in the bodies of executive authorities;
- low level of the modern digital competencies and professional qualification of public officers (Digital transformation 2019).

2 Theoretical, Informational, Empirical, and Methodological Bases

The problem of formation and development of the digital economy is important not only from the theoretical point of view but also from the practical point of view at the state level – due to understanding of the decisive role of digital technologies in determining the directions of strategic provision of the country's competitiveness.

Theoretical substantiation of the necessity for digitization of economic processes in Russia is connected to formation of a new paradigm of development of the economy, aimed at provision of increase of involvement of citizens and economic subjects into the digital space, emergence of sustainable digital eco-systems, creation of infrastructure that ensures interaction of subjects in the digital space, etc.

Until now, the notion "digital transformation" has not been assigned a common definition. In a wide sense, digital transformation is treated as changes of all aspects of the society that are connected to application of digital technologies. Digital transformation is considered as the key trend that is peculiar for various spheres and sectors of economy and social sphere and that allows for significant increase of efficiency or expansion of the volume of the organization's operations.

Experts of the Center of Strategic Developments define digital transformation as "deep reorganization, reengineering of business processes with wide application of digital tools as the mechanism of execution of processes, which leads to significant improvement of the characteristics of the processes (reduction of time of their execution, disappearance of the whole groups of sub-processes, increase of output, reduction of resources that are spent for execution of the processes, etc.) and/or emergence of completely new qualities and features (decision making in the automatic regime without human participation)" (Dobrolyubova 2019).

The search for managerial decisions that will allow ensuring achievement of the acceptable growth rates of economy and reduction of administrative load on business requires substantiating the following methodological approaches:

(1) wide implementation of the principles of project management (primarily, for increasing the efficiency of the activities in the sphere of management of public finances), which envisages certain autonomy of the system of project management (project offices, project committees, and project teams) from the traditional linear and functional hierarchy and sets the preconditions for assigning the bodies of project management with the necessary personnel, financial, and other resources;
(2) more consistent implementation of the principles of process management (for reducing the expenditures with simultaneous increase of efficiency of the activities). Wide implementation of comprehensive process management will require decentralization of the management systems and will cause gradual disappearance of borders between the intra-departmental and inter-departmental vertical functional ties. In a lot of cases, an effective process envisages subsidiarity – shifting the operative decisions to the lowest hierarchical level at which they could be made technically;
(3) transition to risk-oriented control and supervision (for reducing the administrative load on business with simultaneous increase of the efficiency of the controlling activities). In the sphere of control and supervision, transition to the risk-oriented methods leads to decentralization of making of a lot of operative decisions that are connected to implementation of these processes. Mutual exchange of information on the results of controlling measures, including information on the determined risks in the financial and budget sphere, including with the usage of the information systems, is one of the most important directions of development of internal state financial control (Maslov 2018);
(4) formation of a platform model (Government as a Platform) in the system of state finances management. This envisages creation of a complex infrastructure for provision of state services and increasing the effectiveness of the system of state management. Development of partnership with companies, non-profit organizations, and citizens within the platform allows reducing the transaction costs and risks and increasing labor efficiency, quality of services, and the level of consumers' satisfaction. The state acquires the functions of creation and management of the eco-system in which all participants of the platform interact. The advantage of the platform approach is that consumer and supplier of product (service) receive a transparent and clear process of interaction, which is projected at the platform and is controlled by it.

Platforms present a clear system of monetization of services for the users and the system of promotion of the user, which leads to benefits of all participants of the platform interaction (Fig. 1) (Digital transformation 2018).

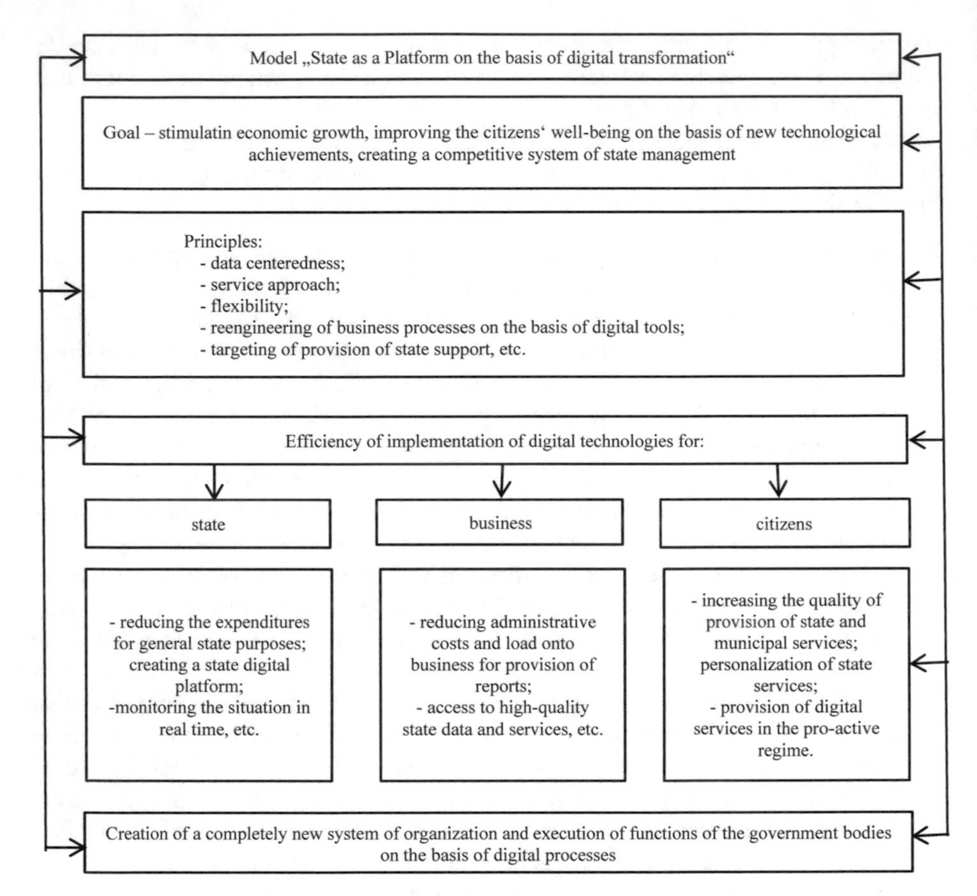

Fig. 1. Conceptual approaches to formation of the model “State as platform” (Petrov et al. 2018).

On the whole, digital platforms allow taking into account the attitude of the regulation’s addressees’ to the developed policy almost in real time and reflect more systemic changes in the process of execution of the state function of this type. (Dobrolyubova 2019);

(5) transition to state management, including by public finances based on Big Data and cloud technologies, which are used at the stages of target setting, development of state policy, decision making, monitoring, and evaluation of results. Big Data is the most important component of the digital eco-system, which value is determined by the fact that companies are able to store large arrays of data and manage and control them with the necessary speed, according to the current business tasks.

Big Data technology eliminates two main drawbacks of the obsolete technologies: absence of flexibility and elasticity and possibilities of expansion of the network.

The purpose of digital transformation of state management is creating digital government that is based on the ideas of client focus, maximization of usefulness of public authorities' usefulness for citizens, and "digitization by default". In the sphere of state management, the principles of "agile management" are developing – they envisage constant usage of the mechanisms of feedback during the whole period of implementing the measures and programs.

3 Results

Formation of the digital economy requires changing the model of managing the public finances and using new financial technologies, infrastructure, methods of Big Data analysis of for applying the modern methods of planning and forecasting of financial indicators, and changing the system of financial control for the purpose of optimizing the usage of financial resources at all levels of the financial system.

Transition to digital technologies allows transforming the information system of development and implementation of the measures of tax policy and policy in the sphere of expenditures. Qualitative improvement of tax administration allowed ensuring significant growth of tax revenues by means of tax bodies' applying new technologies. Digital technologies, including the systems of online payments, reduce expenditures for collection of taxes and create potential for expansion of the tax base (e.g., by improving the process of identification and monitoring of taxpayers and simplifying the execution of tax obligations by taxpayers by means of such means as mobile technologies).

The most important priority in the sphere of public finances management is optimization of budget expenditures, which envisages usage of digital technologies in the part of expansion of the methods of analysis and evaluation of implementing the state programs and projects for the purposes of audit of efficiency and effectiveness of state expenditures. Within such projects it is possible to expand the practice of online research and usage of Big Data. Digital technologies could transform the processes of monitoring and evaluation of the achieved results and ensure the development of the platform solutions in the sphere of result-based management (Ivanova 2018).

Orientation at the indicators of efficiency of budget assets usage and presence of information on such indicators in the budget process is the mandatory attribute of any modern system of state finances management, the most popular variant of implementation of which is program budgeting. Thus, one of the key directions of increasing the effectiveness of budget expenditures in the future period is development of program and targeted budget planning on the basis of state programs. The normative & legal and methodological basis on the issues of formation and implementation of state programs has been formed, and the structure of expenditures of the federal budget conforms to the structure of state programs.

Creation of an open transnational electronic economic space increases the level of significance and changes the structure of state regulation of financial control and supervision in the budget and banking sector of economy and creates the necessity for implementing the changing into the normative & legal provision and ubiquitous practical mastering of the innovative information possibilities (Big Data, Blockchain, etc.) (Burkaltseva 2017).

There's a necessity for provision of interconnections, coordination, and development of effective cooperation during conducting the controlling measures by the subjects of financial control, main custodians of financial assets, and other participants of the economic process at all stages of functioning of financial system of Russia. For these purposes, it is necessary to form the innovative model of financial control (Markova 2018).

Effective financial control is not possible without good interaction between the bodies of financial control and the main custodians of financial assets.

For increasing the level of effectiveness of financial control, it is necessary to provide the all-level exchange of information as to the results that are obtained in the course of controlling and revision measures, which will form the unified base of controlling activities and will allow various bodies of control to minimize the costs for financial control by acknowledging each other's results.

Thus, the evolving digitization has a large potential of transformation of the Russian economy. The eco-system of financial technologies allows forming the effective environment of hi-tech digital platform of state management, which stimulates minimization of human factor, reduction of possible errors, authomatization of collection of statistical, tax, and other reports, ensuring decision making based on analysis of the real situation (Vovchenko 2018).

4 Conclusions and Recommendations

The following is requires for achieving the set goals of digital transformation.

1. Organizing – on the constant basis – the process of digital reengineering of the system, structure, and authorities of the bodies of executive authorities at the federal and regional levels.

For transition to a new system of authorities of the bodies of executive power, which would conform to the tasks of digital transformation, it is necessary to adopt the mechanisms that would ensure connection of the bodies of executive power to the resources for execution, including personnel, financial, and material & technical resources, and to the results of their activities.

For starting the mechanism of regular reconsideration of the structure and contents of government authorities it is expedient to create a digital platform that allows performing operative analysis of data from various state information systems and resources for determining the real load on public officers and their labor efficiency, control over the terms of execution of operations, and transformation of administrative procedures.

2. Implementation and constant conduct of overview of budget expenditures. Overviews of budget expenditures are systemic analysis of the basic (constant) expenditures of the budget, aimed at determining and comparing various variants of saving budget assets and selecting and implementing the most acceptable one. The purpose of overviews of budget expenditures is not optimization of budget expenditures as such but moving the resources that are not used effectively to

solving the top-priority tasks. Without access to authentic and full data on expenditures and results of financing it is impossible to determine and evaluate the opportunities for saving the budget resources. The result of successful overview of expenditures is creation of a data base and quality of the initial information (Bogacheva 2019).

3. Participation of citizens in the budget process. This envisages wider involvement of the institutes of civil society within complex solution of the problems of information support for initiative budgeting. This requires provision of constant actualization of the information resource of the unified portal.
4. Development of the system of long-term (strategic) budget planning. The necessity for achieving the priorities and goals that are set in the documents of strategic planning and qualitative leap in socio-economic development in the conditions of limited budget resources raises the topicality of development and implementation of the system of measures for increasing the effectiveness of activities of public authorities and local administration and modernization of the system of public finances management.
5. Organization and development of unified information platforms, usage of inclusive digital technologies form the conditions for achieving a new quality of state management based on horizontal integration and effective interaction between state government bodies at various levels of executive authorities. Thus, it is possible to speak of transition to a new stage of transformation of the institutes of state management by means of further development of e-government and formation of "digital government", which will create preconditions for solving the tasks of strategic planning on the basis of inter-departmental digital information platforms (Scientific and analytical report 2018).
6. Increase of completeness and quality of financial accounts of the state sector. In the sphere of improvement of financial accounts in the sector of state management a landmark is formation of new accounting and technological model of centralization of the authorities of participants of the budget process on the platforms of the corresponding centers of competencies. Formation of such model aims at reducing the risks of incorrectness of accounting data, increasing the transparency and accountability of expenditures, creating flexible and adapted information technologies, and developing online interaction.
7. One of the main directions of optimizing the system of state management, including by public finances, is centralization of the providing functions of the federal bodies of executive authorities and creation of a register of functions (responsibilities) of the federal bodies of executive authorities. This electronic register will allow for constant monitoring of the functions of the federal bodies and for timely reaction to their increase, which leads to their doubling. The key condition of effective centralization of providing functions is complex automatization of the financial and economic activities of recipients of budget assets on the basis of the modern information technologies.

Thus, within the digital transformation one of the key elements is digitization of state management and creation of state digital platforms, which allow ensuring openness and the level of trust between citizens and the government and creating the service model of public finances management. In its turn, this envisages improvement of the normative and organizational establishment of the order of implementation and financial provision of the measures for digital development. From the point of view of development of the financial system, without improvement of the budget law and change of the approaches to the program and project budgeting it will be difficult to achieve positive results in the sphere of digitization of the process of public finances management. This requires adoption and implementation of a complex of measures for increasing the quality of state management, including in the sphere of public finances, based on creation of a new paradigm of socio-economic development in the conditions of effective implementation of digital technologies.

References

Bogacheva, O.V., Smorodinov, O.V.: Formation of conditions for overviews of expenditures in Russia. Institute of Financial Research. Financ. J. (1), 21–33 (2019)

Burkaltseva, D.D., Babkin, A.V., Vorobyev, Y.N., Kosten, D.G.: Formation of the digital economy in Russia: essence, specific features, technical normalization, and problems of development. Scientific and technical bulletin of St. Petersburg State Polytechnic University. Economic Sciences, no. 3 (2017)

Vovchenko, N.G., Ekimova, K.V., Kostoglodova, E.D.: Development of fintech services in the conditions of digitization of the Russian economy. In: Statistics – Language of the Digital Civilization: Collection of Reports of the International Scientific and Practical Conference "2nd Open Russian Statistical Congress", Rostov-on-Don, 4–6 December 2018, in 2 vol. - vol. 1, pp. 125–130. The Russian Association of Statisticians; Federal State Statistics Service, Rostov State University of Economics, Rostov Regional Branch of Free Economic Community of Russia. AzovPrint Publ., Rostov-on-Don (2018)

Dobrolyubova, E.I., Yuzhakov, V.N., Efremov, A.A., Klochkova, E.N., Talapina, E.V., Startsev, Y.Y.: Digital future of state result-based management. Delo Publishing House, 134 (Scientific reports: state management) (2019)

Ivanova, O.B., Andreeva, O.B.: Digitization of economic processes as a new paradigm of development of the Russian economy. In: Statistics – Language of Digital Civilization: Collection of Reports of the International Scientific and Practical Conference "2nd Open Russian Statistical Congress", Rostov-on-Don, 4–6 December 2018, in 2 vol. - vol. 1, pp. 147–151. The Russian Association of Statisticians; Federal State Statistics Service, Rostov State University of Economics, Rostov Regional Branch of Free Economic Community of Russia. AzovPrint Publ., Rostov-on-Don (2018)

The concept of increasing the effectiveness of budget expenditures in 2019–2024 [adopted by the Decree of the Government of the Russian Federation No. 117-r on 31 January 2019]. http://www.minfin.ru

Markova, E.S.: Problems and perspectives of development of the Russian digital economy at the global market. FES: Finances, Economics, Strategy, no. 1, p. 30 (2018)

Maslov, D.V., Dmitriev, M.E., Ayvazyan, Z.S.: Certain aspects of transformation of state management: processes and quality: analytical overview of the Russian Presidential Academy of National Economy and Public Administration, Moscow, p. 58 (2018)

Petrov, M., Burov, V., Shklyaruk, M., Sharov, A.: State as a platform. Center for Strategic Developments (2018)

Program of development of the digital economy in the Russian Federation until 2035. http://spkurdyumov.ru/uploads/2017/05/strategy.pdf

Formation of the digital economy in Russia: problems, risks, and perspectives: the scientific and analytical report, Moscow, 44 p. (2018)

Dvinskikh, D.Y., Dmitrieva, N.E., Zhulin, A.B., et al.: Digital transformation of state management: myths and reality. In: Dmitrieva, N.E. (ed.) Report for the 20th International Conference on the Issues of Development of Economics and Society, 9–12 April 2019, Moscow, National Research University "Higher School of Economics", Publ. House of Higher School of Economics, p. 43 (2019)

Development of the Program and Project Budgeting in the Conditions of Digitization of the Budget Process

Lyudmila V. Bogoslavtseva(✉), Oksana I. Karepina, Oksana Y. Bogdanova, Aida S. Takmazyan, and Vera V. Terentieva

Rostov State University of Economics, Rostov-on-Don, Russia
b_ludmila@bk.ru, karepindima@mail.ru,
bogdanova1974@mail.ru,
aida.takmazyan@yandex.ru, venver@yandex.ru

Abstract. The purpose of the paper is to determine the directions of development of the program and project tools in the conditions of digital transformation of the budget process. The authors study the accumulated positive experience of using the innovative technologies and tools of program financing. The authors formulate the problems and perspectives of adaptation and development of the program tools to the requirements of digitization. The authors substantiate the application of digital platforms as the key factor of increasing the effectiveness of managerial decisions by bodies of executive power for achievement of the criteria of efficiency of the financed programs. The paper offers the directions of institutional development of program and project budgeting in the conditions of digitization of budget flows at all stages of the budget process.

Keywords: Program and project budgeting · Budget process · Large arrays of data · Digitization of budget flows · Digital platforms

1 Introduction

In the new conditions of digital transformation of financial relations in Russia, modernization of the budget process requires active application of the program and project tools, standards, and the corresponding program provision. Program and project financing envisages interaction of the public authorities, public and legal entities, sectorial economic subjects, and professional community – which allows solving the task of "development of the state digital platforms, ensuring the legal regime and technical tools of functioning of services, and usage of large arrays of data", which are peculiar for state (municipal) programs [1].

Program tools began to be implemented into the budget process since the implementation of the concept of reformation of the budget process in 2004. They originally included the target programs that were aimed at increasing the effectiveness of the budget policy. They allowed solving specific problems of the economic and social character, which changed the forms and methods of budget financing at all stages of the

E. G. Popkova and B. S. Sergi (Eds.): ISC 2019, LNNS 87, pp. 950–959, 2020.
https://doi.org/10.1007/978-3-030-29586-8_108

budget process: formation, discussion, and adoption of the budget project, budget performance, preparation of accounts, and state financial control.

2 Methodology

Since 2010, the development of program and targeted budgeting has been connected to transformation of the functional and departmental expenditures of budgets of the Russian budget system into the program form with assigning the corresponding codes of budget classification, which allow using state programs as a financial tool of program provision of execution of expenditures of budgets.

State program is treated as a document or passport that determines customers and performers, goals and tasks, measures and criteria of efficiency.

State and municipal programs have been actively implemented in the Russian Federation for the purpose of socio-economic development. As of early 2018, there were state programs, grouped according to the following directions: new quality of life; innovative development and modernization of economy; provision of national security; well-balanced regional development; effective state.

Table 1 shows the structure of program expenditures of the federal budget in 2011–2018.

Table 1. The structure of program expenditures of the federal budget in 2011–2018% [2, 3].

Indicator	2011	2012	2013	2014	2015	2016	2017	2018
Expenditures of the federal budget, total	100	100	100	100	100	100	100	100
Program expenditures of the federal budget, total	0.3	2.2	54.9	51.7	66.8	63.4	65.5	81.3
Including for: New quality of life	0.3	0.4	30.1	27.5	41.9	39.4	40.5	50.0
Innovative development and modernization of economy	–	1.3	12.4	12.7	13.4	12.3	11.3	13.2
Effective state	–	–	7.8	7.0	6.8	7.3	9.0	11.8
Well-balanced regional development	–	0.5	4.6	4.5	4.7	4.4	4.7	6.3

Analysis of the data of Table 1 shows quick transfer of expenditures of the federal budget into the program form. While in 2011–2012 the share of program expenditures in the total structure of expenditures of the federal budget constituted 0.3% and 2.2%, expenditures of the program character since 2013 constituted more than 50% of aggregate expenditures of the federal budget, and in 2018 their share exceeded 80%.

The necessity for implementation and practical application of the program and targeted method for modernizing the socio-economic infrastructure and increasing the population's living standards required the development of new forms and methods of digitization of budget flows at all stages of the budget process.

Program provision that is applied by the Ministry of Finance of the Russian Federation is aimed at "increasing budget transparency and monitoring of effectiveness of state projects" and improving the quality of the accounting system – which creates conditions for increasing the efficiency and quality of state financial control [4].

Thus, since 2016, budget appropriations have been reflected not only in view of state programs and sub-programs but also in view of the main events, which allowed evaluating the planned volumes of financing and factual execution for sub-programs and main measures.

Online sources that are to simplify the monitoring of implementation of state programs and sub-programs are also developing. For example, a state integrated information system "Online budget" is being developed; it is used for provision of transparency, openness, and accountability of the activities of government bodies and management bodies of state non-budget funds, local administration, and state and municipal entities. The information system is aimed at provision of the unified information space and application of digital technologies in the sphere of public finances management.

The web-site "Portal of state programs of the Russian Federation" contains the current information on the contents and realization of state programs and sub-programs, including statistical information, accounts on factual and planned target values of the indicators of state programs, list of the main events, and data on their factual implementation and related targeted indicators of state programs and sub-programs.

The Ministry of Finance of the Russian Federation publishes in open access the reports on execution of the state programs as of year-end, which contain explanations for the facts of negative deviations of the factual values of the target indicators from the planned indicators – which positively influences the transparency of state programs and expands the opportunities of their evaluation by the bodies of state financial control [2].

The Federal Treasury of the Russian Federation accumulated large experience in usage of digital technologies at the stage of cash services of execution of the program structure of expenditures of budgets. One cannot but agree with the opinion of the Head of the Federal Treasury, Artykhin, that "Russia's treasure is one of the most innovative – in the aspect of IT – departments under the government" [5].

The Federal Treasury performs the functions on creation, development, and maintenance of the State information system on state and municipal payments, regulates the work of the official web-site of the Unified information system in the sphere of purchases, and successfully implements the software provision of the automatized system "Federal treasury", which ensures treasury services and protection of information data on the budget performance of the budget system of the Russian Federation.

Thus, the information system is a centralized system that ensures acceptance, accounting, and transfer of information between its participants, which are administrators of budget revenues, organizations for acceptance of payments, portals, and multi-functional centers which interaction will be conducted via the system of interdepartmental online interaction. The state information system on state and municipal payments is used for placement and receipt of information on individual and legal entities' performing payments for provision of state and municipal services, which allows them to obtain information about their obligations before the budgets of the budget system of the Russian Federation according to the single-window system.

Thus, this system allows providing documents and information in the electronic form with the usage of the unified system of inter-departmental online interaction.

The Unified information system in the sphere of purchases for the program structure of the budget is especially interesting. As to the volume of accumulation information, the system contains 350 TB—this is Big Data". The annual volume of contracts constitutes RUB 25 trillion [5].

The official web-site of the Unified information system in the sphere of purchases (zakupki.gov.ru) is to provide free access to full and correct information on the contractual system in the sphere of purchases and purchases of goods, works, and services by certain types of legal entities, and for formation, processing, and storing of this information. The order of placement of information on the web-site and its contents are regulated by the Federal Law No. 44-FZ "Regarding the contractual system in the sphere of purchases of goods, works, and services for provision of state and municipal needs" and the Federal Law No. 223-FZ "Regarding the purchases of goods, works, and services by certain types of legal entities" and by the corresponding bylaws.

3 Results

The result of development of treasury technologies within the implementation of the project "Modernization of the treasury system of the Russian Federation" was the system of remote financial document turnover [6, 7].

Implementation of the applied program "Automatized system of the Federal Treasury" into industrial usage allows offering this service to the customers of the Treasury in the form of Internet interaction of the participants of the budget process.

The system of remote financial document turnover is a web-application, which allows the customers of the Federal Treasury to control their payments and financial documents and to have access to current accounts that are formed in the automatized system of the Federal Treasury [8].

However, the procedures of cash services of execution of the program structure of expenditures of budgets of the budget system of the Russian Federation require improvement in the following directions:

- avoiding repeated operations on current and bank accounts;
- quick transfer of budget assets to the recipient and provision of information on the performed operations [9].

Thus, it is necessary to introduce changes into the laws of the Russian Federation in the part of provisions of treasure services and the system of treasury payments.

It seems that integration of the system of treasury payments with the payment systems of the national payment system will allow using the modern services of payment systems; cooperating with the national system of payment cards for transferring social payments; organizing the application of the national payment tools that were issued by credit organizations and participants of the system of treasury payments for performance of budget payments.

Also, as is noted in the economic literature, financial bodies and bodies of state financial control do not always determine the effectiveness of using the budget means. The goal of control at the stage of planning of budget appropriations for state (municipal) programs and at the stage of execution of specific measures is not yet achieved.

For increasing the quality of execution of programs and national projects in the conditions of digitization, together with active development and adoption of state programs, the Federal standards of accounting and accounts of the state sector are developed [10] (Fig. 1).

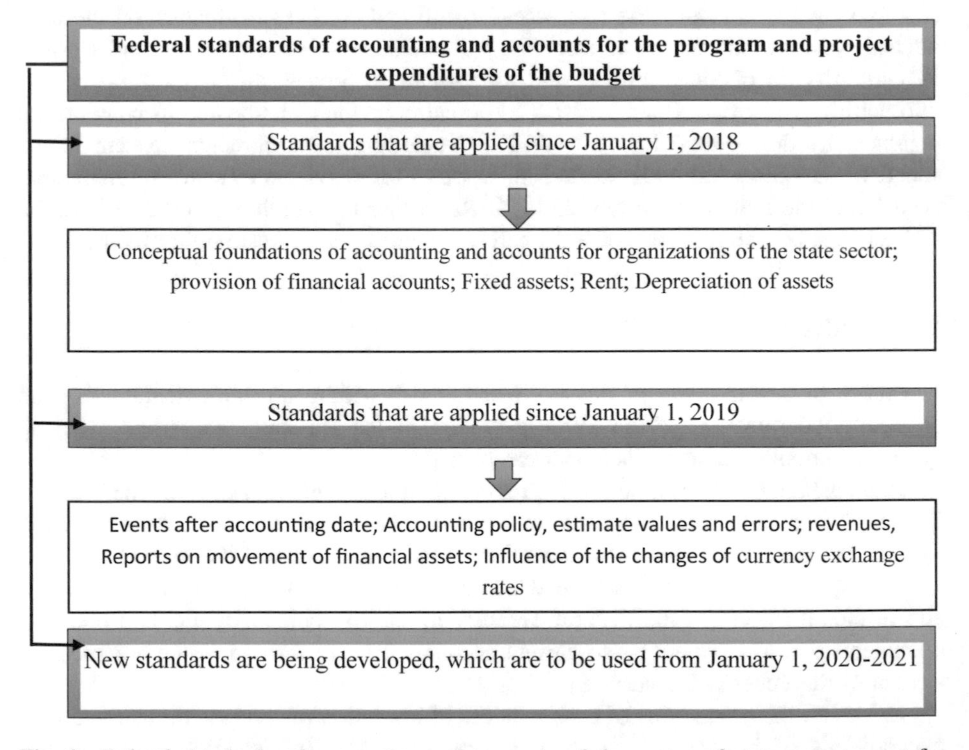

Fig. 1. Federal standards of accounting and accounts of the sector of state management for accounting of program and project expenditures of budgets. (Compiled by the authors.)

Application of standards and formation of the unified requirements, principles, and rules of accounting and provision of accounts in the sector of state management are actualized by the requirements of digitization of the program and project expenditures of budgets.

The modern stage of development of the Federal standards is conducted according to the order of the Ministry of Finances of the Russian Federation dated March, 19, 2019 No. 45n "Regarding the adoption of the program of developing the federal standards of accounting for organizations of the state sector for 2019–2021" (registered in the Ministry of Justice of the Russian Federation on April 12, 2019, No. 54354).

At the modern stage, the Accounts Chamber of the Russian Federation notes such drawback of accounting and accounts as failure to perform expenditures without changing the indicators state programs, which, eventually, distorts the results of inspection. These violations confirm the necessity for development of the program and project tools for the purpose of obtaining the objective information on implementation of state programs.

Control over effectiveness of state programs is performed by the bodies of internal state financial control and bodies of external state financial control. In the process of internal control of effectiveness, the main custodians of the budgets assets and spending units track the quality of the performed work on planning and usage of the budget assets. The role of external control over effectiveness of the program consists in the following:

- on the one hand, increase of society's trust to public authorities via provision of independent and correct information on the results of implementation of the state program;
- on the other hand, the bodies of legislative and executive power receive all necessary information for managing the state program.

Analyzing the organization of internal and external state financial control, it is possible to see their poor interaction in the process of strategic audit and the gap between them, which is caused by drawbacks of the legislative basis and absence of the corresponding digital platforms [3, 11].

Transition to the program principles of organization of the budget process requires increase of strategic audit of development and execution of state programs. For the purpose of increasing the effectiveness and efficiency of control, it is offered to adopt the notion of strategic audit and to emphasize its evaluating character.

Based on consideration of the perspectives of development of state financial control in the conditions of digitization, we offer a proprietary definition of strategic audit as a complex of evaluating measures that include control of effectiveness of planning and execution of documents of strategic planning based on interaction of the bodies of state financial control, executive bodies of public authorities, and spending units of budget assets with the usage of the digital platforms, which allow obtaining objective information for making of managerial decisions by the subjects of public authorities regarding the limited financial resources.

There are no doubts that state programs are documents of strategic planning, which functioning is based on certain principles that are not properly implemented – which is confirmed by the results of strategic audit (control of effectiveness of planning and control of effectiveness of execution of state programs), performed by bodies of external and internal state financial control [12]. With the common goal – control of the program budget expenditures – these bodies work separately and without any coordination, which is caused by drawbacks of the existing legislative basis and absence of digital tools of processing of large arrays of data.

According to the Decree of the President of the Russian Federation dated May 7, 2018, the Government of the Russian Federation is planning global changes with the usage of digital technologies in economy, infrastructure, technologies, and social sphere – i.e., for all program directions and priorities. For achievement of the set goal, 12 national projects were prepared. They include target indicators, volumes of financing, stages, terms, and responsible performers [13].

4 Conclusions

Thus, a new tool of program budgeting, implemented into the budget process since 2018, is national projects, which will require improvement of the methods of state financial control – e.g., development of counter inspections for the purpose of exclusion of repeated actions. [14] We think that quality of such inspections could be ensured only by means of creating the sectorial (industrial) digital platforms for the main subject spheres of the economy (Fig. 2).

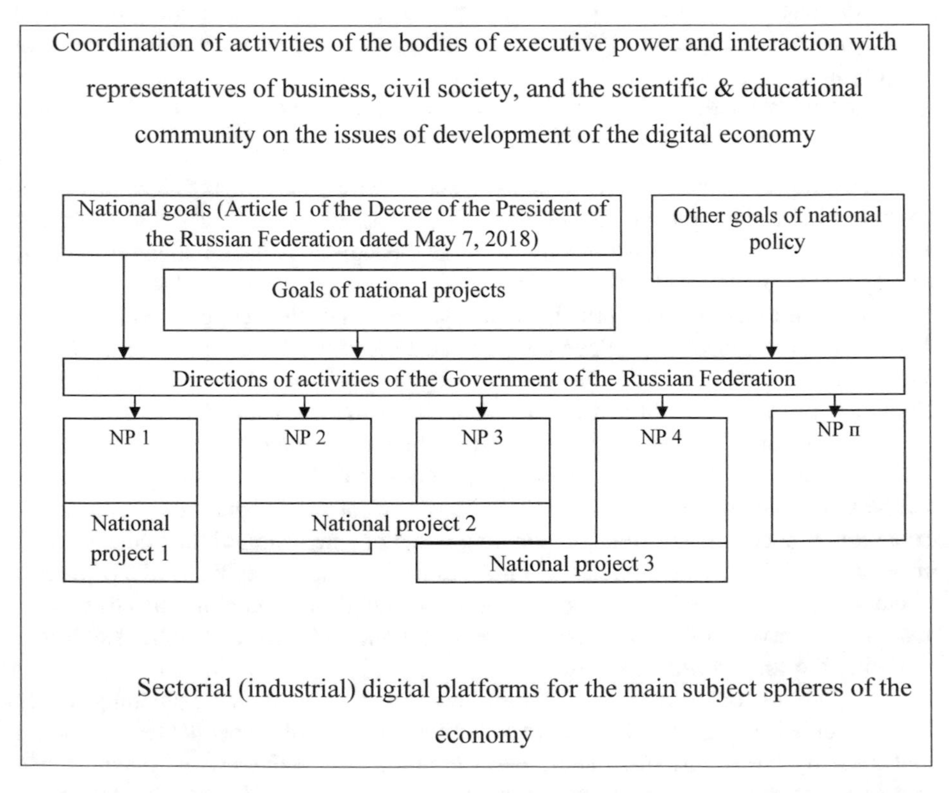

Fig. 2. National projects as an object of creation of sectorial (industrial) digital platforms for the main subject spheres of the economy. (Compiled by the authors.)

It is necessary to pay attention to the necessity of creating the system of regular monitoring of the national project and the operative monitoring and control [15]. Operative monitoring and control should be applied to execution of the national projects on the regular basis at all stages of implementation of the project: from planning to finish. The system of state monitoring will cover the federal and regional levels, and monitoring of the measures will allow tracking dynamics of the project's specific directions. Therefore, digital platforms will provide the public authorities with objective information for making managerial and organizational decisions.

Federal and regional authorities often use the information technologies of project financing, which created the basis for development and application of digital technologies at all stages of the budget process [16, 17]. Unlike the public authorities at the national level. municipal authorities' informatization is at the stage of formation and development:

- solutions in the sphere of information communication at the municipal level are applied;
- automatized information systems and information resources for all top-priority directions of activities of the local administration are created and supported;
- most cities have specialized departments that deal with the issues of informatization;
- there is a direction in activities of developing organizations that is oriented at creation of information systems in the interests of municipalities.

All this took place in rather complex conditions that were caused by the consistent policy of the state in development of information technologies, which was peculiar for constant change of priorities: "e-government, government, and municipality"; electronic digital signature and "trusted third party"; "information society"; provision of state and municipal services in the electronic form and "interdepartmental information interaction in the electronic form".

At the municipal level, there are still problems of implementing informatization and digitization. Thus, more than 90%, or 20,000, of municipal entities of Russia use the existing information products, most of which are poorly adapted to the requirements of the digital economy.

Increase of effectiveness of works on informatization of municipal management in the part of formation and execution of municipal programs and creation and implementation of typical solutions in the interested of local administration could be stimulates by specialized funds and expert centers – e.g., "Fund of development of information technologies of municipalities" and "Expert center of e-government".

Obviously, the work on municipal informatization should not be conducted separately from the development of state information technologies. For development of the digital economy, the municipal authorities require constant interaction with the federal and regional authorities – which stimulates implementation of the principles of openness of local budgets [18].

According to the program "Digital economy in the Russian Federation", digital technologies of public authorities include: large arrays of data, systems of distributed register, new production technologies, wireless technologies, and technologies of virtual and alternate realities. Quick development of digital technologies in the public and legal bodies of local administration will require additional investments and skilled personnel, as well as significant increase of their innovative activity in the following directions:

- normative regulation of the digital environment;
- information infrastructure;
- personnel for the digital economy;
- information security;
- digital technologies;
- digital management at the municipal level.

All the above confirms the necessity for further research of the theoretical and organizational aspects of methodological provision of the processes of planning and performance of state and municipal programs with simultaneous digitization of the corresponding data.

References

1. Decree of the Government of the Russian Federation dated July 28, 2017. No. 1632-r “Regarding adoption of the program ‘Digital economy of the Russian Federation’”. http://www.consultant.ru
2. Masterov, A.I.: Program and targeted budgeting in Russia: problems and methods of solving them. Finances (5), 5–10 (2018)
3. Karepina, O.I.: Development of internal state financial control in Russia. Audit Bull. (2), 54–57 (2018)
4. Ivanova, O.B., Bogoslavtseva, L.V., Karepina, O.I., Kostoglodova, E.D.: Providing the budget transparency and state projects efficiency monitoring in Russia. Eur. Res. Stud. J. **20** (1), 97–104 (2017)
5. Artyukhin, R.E.: Modern treasury – an office that combines functions of a bank, managing company, and payments operator. http://www.roskazna.ru
6. Information of the official web-site of the Federal Treasury. http://www.roskazna.ru/gis/sufd-onlajn/
7. Terentyeva, V.V.: Development of the automatized system of the Federal Treasury. Financ. Res. (4) (2016)
8. Decree of the Government of the Russian Federation dated January 31, 2019, No. 117-R “Regarding adoption of the Concept of increase of effectiveness of budget expenditures in 2019 – 2024”. http://www.consultant.ru/
9. Terentyeva, V.V.: Development of the Russian system of budget payments. In: Materials of the International Conference “Russia and the European Union: Development and Perspectives”, 17–18 November 2016 (2016)
10. Bogoslavtseva, L.V., Takmazyan, A.S.: Implementing the standards of treasury accounting and accounts as a condition of increasing the effectiveness of management of state and municipal finances. Financ. Res. **1**(54), 106–114 (2017)
11. Karepina, O.I.: Development of external state financial control in Russia. Hum. Mod. World **5**(18), 75–81 (2018)
12. Bogoslavtseva, L.V.: Control and audit activities of the bodies of the Federal Treasury: specifics and perspectives. Audit Bull. (3), 39–44 (2018)

13. Samvelovna, T.A., Nikolaevna, S.K., Nikolaevna, R.S.: National projects and their role in solving strategic tasks of social and economic development of the state. In: International Conference "Scientific research of the SCO Countries: Synergy and Integration" Part 3: Participants' reports in English 2019 年4 月9日。中国北京 9 April 2019. Beijing, PRC, pp. 23–28 (2019)
14. Karepina, O.I., Bogoslavtseva, L.V., Bogdanova, O.Y.: Development of state financial control in the context of program and project budgeting. Humanitarian, socio-economic, and social sciences, no. 4 (2019)
15. Ivanova, O.B., Vovchenko, N.G., Kostoglodova, E.D., Bogoslavtseva, L.V., Rukina, S.N., Karepina, O.I.: Financial transparency in budget sector of economy as a necessary condition of clustering. Int. J. Trade Glob. Mark. **10**(2–3), 207–216 (2017)
16. Popova, G.V., Nerovnya, Yu.V., Terentieva, V.V., Shirshov, V.Yu.: Financial policy under the Russian economy stabilization. Eur. Res. Stud. J. **XXI** (2018). Special Issue 1
17. Malakhova, V.V., Samoylova, K.N.: Digitization of the financial sector of the Russian Federation. In: Materials of the 11th International Conference, State and Business, Eco-System of the Digital Economy, St. Petersburg, 24–26 April 2019, vol. 1, pp. 131–134 (2019). http://to-future.ru/konferencii/gosudarstvo-i-biznes/
18. Bogoslavtseva, L.V., Karepina, O.I., Bogdanova, O.Y.: The conceptual approaches of implementing the principle of openness of local budgets as a tool of making of municipal managerial decisions. Eur. Res. Stud. J. (5), 106–112 (2017)

Digitalization of Agro-Industrial Complex as a Basis for Building Organizational-Economic Mechanism of Sustainable Development: Foreign Experience and Perspectives in Russia

Vasilyi U. Boev(✉), Olga D. Ermolenko, Raisa M. Bogdanova, Olga A. Mironova, and Svetlana G. Yaroshenko

Rostov State University of Economics, Rostov-on-Don, Russia
b_v_u@bk.ru, {OlgaErmolenko700, raisa.m1975}@mail.ru, {lady.sensey2010, svet-yaroshenko}@yandex.ru

Abstract. The study examines the problems and perspectives of agro-industrial complex digitalization in Russia and other countries. Implementation of digital technologies in the agro-industrial complex at the current stage of the economic system development is determinant for competitiveness of the industry. Scientific and technological factors are becoming increasingly crucial for development of agricultural industry. The main methodological principles underlying the given study are based on the fundamental research by Russian and foreign experts in the field of agro-industrial complex economics, factors of organizational-economic mechanism formation for the branch's sustainable development based on digital technologies. The study found that currently, the most preferable way for development of agriculture lies in smart and precision farming based on the use of digital technologies, contributing to farming cost reductions and improvement of agricultural profitability.

Keywords: Agro-industrial complex · Digitalization of economy · Digital technologies · Internet of things · Robotization · Organizational-economic mechanism · Sustainable development

JEL Code: O33 · Q000

1 Introduction

The recent years have witnessed major qualitative changes in the development of agricultural industry. This is particularly the case with division of labor, specialization of work, and concentration of production on the basis of cluster structures and agro-industrial integration. Contemporary agriculture is facing the challenge of increase in yield per unit area, which dictates the need for formation of an effective digitalized organizational-economic mechanism providing sustainable development for both agricultural industry and the country's economy as a whole. In this regard, implementation of digital technologies into the agro-industrial complex is currently one of

E. G. Popkova and B. S. Sergi (Eds.): ISC 2019, LNNS 87, pp. 960–968, 2020.
https://doi.org/10.1007/978-3-030-29586-8_109

the key determinants for competitiveness. The use of opportunities and advantages offered by digital economy is the global trend nowadays.

A digitalized comprehensive solution to industrial and social issues is crucial for effective management. Intensification of production, a stronger link between agriculture and other branches of economy presents the farmers with the challenge of environmental protection. The projections for the increase in population on Earth by the year 2050 up to 9.7 bln people have highlighted the need for increase in food production by 70%, which, in its turn, accentuates the issue of food security.

Intensification of production demands more investment from agricultural producers, which inevitably aggravates the issue of economic efficiency, establishment of favorable balance between expenditures on the one hand and output on the other. The contemporary stage of economic development requires a more considerate, complex and systematic approach to all economic, social, technological, and organizational issues. A higher importance is attributed to scientific and technological factors of agricultural development, among which a clear priority is given to development and implementation of digital technologies into agricultural production. In this regard, there is an urgent need to study foreign experience of agro-industrial complex digitalization and its adaptation to Russian conditions.

2 Methodology

Theoretical and methodological base of the study is rooted in the works by Russian and foreign scholars specializing in economics of agro-industrial complex, agrarian reformation, information support for the management of the agricultural economy sectors, sustainability of economic systems' development, theory of economic growth, socio-economic systems' restructuring, provision of the industry's sustainable development based on digital technologies. In justification of the theoretical grounds and reasoning of findings various methodological approaches were used, including the systematic approach in its structural-functional aspect. During the research, as part of the systematic, structural and functional approaches, general scientific and special methods were used: subject-object, logical, comparative and dimensional analysis. The aforementioned methods are applied in various combinations at different stages of the research depending on the goals set and tasks to be solved, which provides fidelity of economic analysis and validity of the findings.

3 Results

Agriculture is a specific rapidly developing production system influenced by various objective factors. A multitude of these factors can be subdivided into five main groups: political, economic, social, natural, and scientific-technological. A combination of these factors forms the system of agriculture and defines the organizational-economic mechanism of the industry's sustainable development. The listed factors are interlinked through systemic interactions, which means that changes in one group lead to changes in another (Ermolenko and Bogdanova 2019).

The increasing significance and influence of scientific-technological factors against the background of the current fourth industrial revolution, as well as development and widespread implementation of digital technologies, is becoming apparent in recent years. Implementation of information technologies, including IoT, is making a revolutionary breakthrough on the global agricultural landscape. The expected result is a higher productivity and resource intensity. However, in developing countries, about 80% of agricultural output is produced by small farmers, who make decisions based on experience and guesses, not on scientifically substantiated guidelines. This approach is not credible according to monitoring and forecasting of certain most important constituents of products' quality. These parameters include: quality of water, condition of soil, air temperature, humidity, irrigation, etc.

Digital information about the weather, soil resources, and condition of crops helps farmers to increase yields. Today, additional smart digital tools are created abroad for development of agricultural industry, conservation of resources, yields, and environmental protection. Digital revolution is changing agriculture by overall implementation of highly automated tractors and combine harvesters equipped with a wide range of sensors collecting data on plant health, crop yields, soil composition and fields' topography. With the use of drones and satellites, databases are formed, which help farmers to run their farms efficiently. The farmers are capable of better forecasting yields, responding to changes and making adjustments quicker to prevent yield losses on the basis of digital technologies. Implementation of data collection technologies makes it possible to develop planting strategies in accordance with the type of soil, which defines the company's level of competitiveness. Agricultural efficiency is achieved by the use of digital technologies for soil analysis, which enables prompter decision-making adapted to certain fields (Ognivtsev 2018).

The level of crop yields is also influenced by the right choice of crop varieties, application of the right dose of fertilizers, determination of the ideal time for taking plant protection measures. At the company Bayer, experts use satellite data for remote crops diagnostics. In the USA, agricultural companies' specialists actively cooperate with different digital organizations. One of such companies is Planetaryresources engaged in space technologies. With the help of geoinformation systems, a database is created, which contains information on the global experience in agricultural technologies over the 30-year period. This database is used for smart connection. Today, Bayer is creating a digital platform capable of providing information on multiple parameters necessary for analyzing successful and efficient farming.

Scholars highlight four key variables influencing the efficiency of agricultural industry. They point out ecological factors (condition of soils, weather conditions, humidity, etc.), as well as pathogens and harmful factors, such as pests and weed vegetation. The third variable is the data on crops and on the way they are influenced by the first two factors. The fourth variable is the system of management: applied plant protection products, and how the farmer manages his/her farm.

The targeted application of fertilizers based on data analysis leads to decrease in the amount of pesticides. Analysis of data on the efficiency of plant protection products together with the data collected from fields enables to estimate the place and amount of the fertilizer's application precisely. The use of analysis based on the data collected with the systematic components of the internet of things makes it possible to increase the efficiency and cost-effectiveness of farmers' activities on the one hand, and to provide high quality products on the other.

In 2016 experts at Bayer tested the system of proportioning combined with the schedule of application time in some farms in Europe. The use of digital technologies enables to formulate individual plant protection recommendations. The mapping smartphone application developed by Bayer provides information on the most suitable proportion of protective products. Later, farmers will learn how to make such maps on their own by scanning a QR-code on the package of the protective product. The software generates the map based on the latest satellite images, data on soils' condition, topographic data, and after this, the obtained information is associated with the corresponding field. The company Bayer has also developed a technology, which keeps track of the crop condition. This system is used to monitor large areas of crops, which allows farmers to take rapid measures and use plant protection and fertilizers for crops loss prevention. Sensing technologies, algorithms and image recognition systems are used by agricultural companies all over the world to improve the quality of produce.

Precise farming improves yields, reduces costs and protects environment. With the help of such systems, it becomes possible to increase efficiency of farmers' work due to better expenditure planning per unit yield. The big data system helps to make decisions based on the information provided by modern sensing technologies and smart software. Using this system, farmers are able to forecast yields, disease threats, etc., and, consequently, make prompter decisions (Ermolenko et al. 2004).

Moreover, an end-to-end platform for internet access has been developed recently for smart farming formation management, which includes efficient use of farming machinery. When devices are connected to the platform, data on supply chains is formed, including third party services. Large scale farms from developed countries have more opportunities for transition to this management model than farmers from developing countries, where the level of agricultural mechanization is still low. Smart and precise farming imply integration of advanced technologies into the existing farming practice in order to increase production efficiency and quality of agricultural produce. Besides, there are additional advantages of digitalization, which improve the life of agricultural workers due to reduction of work intensity (Usenko et al. 2018).

Implementation of digital technologies into every organizational-economic aspect of the farming system presents an opportunity to benefit from the introduction of technologies. Modern technologies and future agricultural technologies are divided into three categories, which will form the basis for the smart farm: autonomous works, drones or UAV, as well as sensors and the internet of things. Replacement of human labor with autonomous and robotized work is a growing tendency in many branches of economy, including agriculture. Robotization of the industry is one of the most promising directions of farming digitalization. The use of robotics developments is more appropriate to fulfill field works, in particular, weeding, planting, and other labor-intensive tasks. Experts use sensing technologies, algorithms and image recognition systems to improve the quality of farming produce. An ideal niche for automatization and implementation of robotics is agriculture, because most of the industry's aspects are labor-intensive and include repetitive and standardized tasks.

Today, farmers begin to use agricultural robots for various tasks: planting, irrigation, harvesting, and sorting yields. Implementation of digital technologies will make it possible to produce more high quality goods at less labor costs, which in its turn will increase profitability and efficiency of farming. (Egorov et al. 2008) The use of autonomous farming machines is increasingly common. One of the earliest machines to be transformed into autonomous use is the tractor. In the early stages, human efforts will be necessary to create borders, develop the best routes, monitor and maintain.

Together with the introduction of video observation, automated vision systems, GPS/GLONASS navigation, IoT connection for remote monitoring and exploitation, the radar and lidar for detection of objects, autonomous machines will become more powerful and self-sufficient. According to some forecasts, concept-based tractors will be able to use “big data”: real-time satellite information on the weather to make the best use of ideal conditions irrespective of man and time of the day automatically.

Incorporating autonomous tractors and smart seeding-machines into the internet of things system will help to reduce seed costs, labor costs, as well as seeding time, and it will increase yields too.

Today, widespread drip irrigation and automatic watering enable farmers to control the time and amount of water their crops get. IoT integration into this system will make this process more precise, autonomous and uninterrupted. The use of the IoT system combined with traditional grain-harvesting and forage-harvesting machines will introduce automation into the harvesting process. (Kuznetsov et al. 2017) These technologies will be especially efficient when reaping fruits and vegetables, such as tomatoes and grapes. Currently, engineers from different fields are developing complex algorithms, which would make it possible to obtain more accurate information on the color, shape and location of the fruit in order to estimate its ripeness.

Today, digitalization of agricultural processes in Western European countries tends to become wider, faster, and more intensive. In Finland, digitalization of agricultural industry is a priority for state policy, even though the industry accounts for only 3% of the country’s GDP. Belgium is an advanced European country, which has also started working with digital technologies in agriculture. The country’s farms implement energy-saving technologies, GPS-developments, smart machines for fertilizers’ and protection application, etc.

As foreign experience of digital technologies implementation shows, many farmers support rational nature management and environmental protection. Bayer Forward Farming is a global information platform of sustainable agriculture.

Today, one of the most important factors defining efficiency of the industry’s development is the formation of organizational-economic mechanism of the industry’s sustainable development based on implementation of digital technologies. Digitalization of the country’s economy as a whole and its agricultural industry in particular requires fundamental changes of economic processes, mechanisms and ways of operating economy, forms and methods of public production management, implementation and development of the digital education system at all levels – from school and university education to professional development and specialists retraining programs, including retraining of farmers (Ermolenko and Bogdanova 2019).

In its essence, the fundamental transformation affects political, social and economic environment, as well as the natural one, which in their turn make up the system of agriculture. All listed above constitutes priority directions in implementing the concept of agro-industrial complex’s sustainable development on the basis of digitalization.

In 2017 the Russian Government adopted the programme "the Digital Economy of the Russian Federation". This programme promotes the use of end-to-end innovative technologies. The aim of the project is digital transformation of the agricultural sector by means of digital technologies implementation for technological breakthrough in the agro-industrial complex and a twofold increase in labour efficiency at agricultural enterprises by 2021. The year 2017 became a turning point, when Russian lawmakers and experts understood the significance of digital technologies for the country's development. It was agreed to begin the project of the agricultural sector's digitalization with implementation of the system of the industrial internet. This system is considered to be one of the most effective tools for achieving a new level of digitalization.

According to calculations by the rating agency Pricewaterhouse Coopers, the effect from IoT implementation into agriculture will be about 2.8 trillion rubles. The advantage of the IoT system in comparison with other promising technologies is economy of scale, which is achieved by means of implementation of this technology, as well as readiness to immediate deployment. In Russian agricultural industry, IoT technology will increase competitiveness and labour efficiency in compliance with the growth of demand for agricultural produce.

Implementation and wide distribution of smart farms, greenhouses and other solutions help to increase yields, reduce fuel consumption for agricultural machinery and water consumption on fields, cut losses during storage and transportation. However, the main consumers of digital technologies in Russia are only large agricultural complexes. The Internet Initiatives Development Fund has produced a roadmap, which describes the plan of innovative technologies implementation into Russian agro-industrial complex in detail. According to some forecasts, the number of machines used in agriculture by 2020 will reach 75 million. In 2018 Ministry of Agriculture set up a working group on developing the programme "Digital Agriculture". The industry's digitalization is based upon a system of tools (Fig. 1).

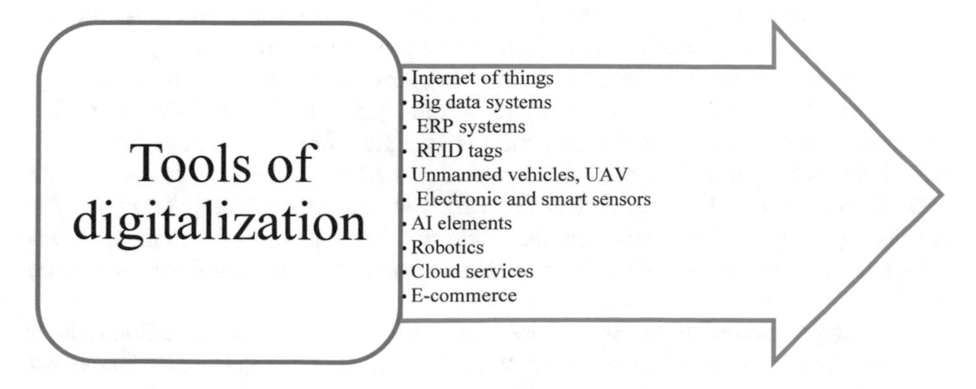

Fig. 1. The system of digitalization tools elements.

The project of agricultural digitalization includes the programme of implementing various decisions at national and regional levels, as well as at the level of agricultural enterprises. The national level involves the use of platforms for end-to-end digital product traceability, organization of electronic trading platforms, centers and programmes of training and retraining of personnel, including analysis of big-data-driven decisions with the tools of distributed ledger and artificial intelligence. The regional level involves implementation of smart sector planning and smart contracts. Development of information systems for comparing the forecast of market needs, widespread implementation of digital agricultural solutions and access to the platform for monitoring lands of agricultural use are all part of the agricultural enterprises level (Boev 2015).

In 2019, digital agriculture is implemented through access to the platform of the seeds traceability system and the livestock production traceability system. Within the period of 2020–2021, it is planned to introduce the systems of monitoring and control over the life cycle of meat products processing and the system of processing plants' end-to-end traceability. By 2021 implementation of these platforms will make it possible to achieve smart planning in all subjects of the Russian Federation for cultivation of the most profitable crops with due regard for transportation to the place of processing and consumption.

However, there is a range of restrictive conditions for implementation of cloud technologies in Russia. Firstly, the country lacks high quality agronomical data, which raises the issue of developing adequate models and their scenario analysis. Besides, there is very little networked technology in Russia, there are no sensors transmitting data from fields to cloud services (Usenko and Udalova 2018).

The problem of technological expansion in Russia is currently connected with a limited understanding of their use by small-scale farms. Most of small and medium-sized agricultural enterprises see the IoT system as only a tool for reduction of production costs. However, technologies of monitoring and control of machines, technologies of precise farming are highly promising and they will become drivers for development of new related markets. These markets can include production of UAVs, drones, autonomous agricultural machines, etc. Sensors, control devices, monitoring software help to increase animals productivity and quality of produce.

According to experts at Pricewaterhouse Coopers, automated systems of feeding, milking and health monitoring could increase milk yields by 30–40%. The expected economic potential of IoT implementation into agriculture in Russia is about 469 billion rubles for the period until 2025. Besides, currently there are problems with network connection of agricultural lands, which is another restricting factor for the development of IoT technology in the country. Development of communications infrastructure requires considerably high investments from communications service providers.

Nowadays, farmers are facing the problem of regional decisions adaptation to their needs and their integration between each other. This situation is connected with the fact that the Russian market lacks complex, bug-free IT-solutions ready for deployment. However, digital technologies implementation helps to cut annual expenditures of agricultural enterprises by over 20%. Another problem of overall digitalization in the country's agro-industrial complex is a low availability of qualified human resources, having skills to make use of technology. To expedite the processes of digitalization, the first electronic educational system "Land of Knowledge" has been launched 2019. The tasks of this system for the period of 2019–2021 include training 55,000 specialists at Russian agricultural enterprises in the skills of digital economy.

4 Conclusions

Thus, current economic conditions make it evident that successful development and effective functioning of the agro-industrial complex require overall implementation and spread of digital technologies. Growth of domestic and external demand for agricultural produce, as well as the need to increase labour efficiency and competitiveness present farmers with new challenges determining the need for the industry's digitalization.

Digital economy in agriculture is the present and the future of the Russian agro-industrial complex. While working out program and regulatory documents aiming at development of the agricultural industry, it is necessary to take into consideration scientific and technological advances and digitalization technologies so as to achieve sustainable development of the agricultural industry and the paradigm of agricultural production growth. Efficiency of agricultural development is achieved by creating an organizational-economic mechanism of sustainable development with the use of digitalization technologies. Digital economy in agriculture will make it possible to develop the model of the industry's growth aimed at both domestic and foreign markets. Growth of the basic products group production aiming at both export demand and import substitution will support demand for produce of the Russian agro-industrial complex. When adopted, these measures will contribute towards sustainable development and growth of production.

References

State Program for the Development of Agriculture and Regulation of Agricultural Commodity Markets in 2013–2020. http://www.mxc.ru/navigation/docfeeder/show/342.htm. Accessed 20 Apr 2019

Regulation of the Government of the Russian Federation No. 1632-r of 28.07.2017 Programme "Digital Economy of the Russian Federation". http://static.government.ru/media/files/9gFM4FHj4PsB79I5v7yLVuPgu4bvR7M0.pdf. Accessed 20 Apr 2019

Boev, V.Yu.: Revisiting the issue of developing the region's innovative potential management. In: The digest: Relevant Issues of Sustainable Development of the Regions of Russia: Proceedings of the International Scientific and Practical Conference, pp. 5–13 (2015)

Egorov, E.A., Serpukhovitina, K.A., Petrov, V.S.: Condition and prospect of scientific support for sustainable development of wine growing. Wine-making and viticulture, vol. 3 (2008)

Ermolenko, V.P., Vasilenko, V.N., Kuznetsov, I.V., et al.: The programme of agro-industrial complex development in Oktyabrsky rayon of Rostov oblast/Monograph. All-Russian Research Institute of Agrarian Economics and Normatives, Rostov-on-Don (2004)

Ermolenko, O.D., Bogdanova, R.M.: Problems and perspectives of wine growing development through state support. In: Development of Russian Economy and Its Security Under Conditions of Modern Challenges and Threats: Proceedings of the International Scientific and Practical Conference, p. 102. OOO AzovPrint, Rostov-on-Don (2019)

Kuznetsov. V.V., Tarasov, A.N., Gaivoronskaya, N.F.: Prediction of parameters for innovative development of the agriculture sector: theory, methodology, practice: monograph/Rostov-on-Don. All-Russian research institute of agrarian economics and normatives, OOO AzovPrint (2017)

Ognivtsev, S.B.: The concept of digital platform for agro-industrial complex. Int. Agric. J. **2** (362) (2018)

Usenko, L.N., Udalova, Z.V., Ermolenko, O.D.: The grape and wine-making subcomplex. In: Altukhov, A.I. (ed.) Food Facility of Russia: Condition and Development Perspectives: Monograpg, pp. 267–297. Federal State Budgetary Scientific Institution All-Russian Horticultural Institute for Breeding, Agrotechnology and Nursery, Non-profit Organization "Horticulture Support and Development Fund", Moscow (2018)

Usenko, L.N., Udalova, Z.V.: Analysis of condition of grape and wine-making subcomplex in agro-industrial complex. Account. Stat. **1**(49) (2018)

The Main Challenges and Threats to the Profession of Jurisprudent in the Conditions of Economy's Digitization

Development of Legal Education and Machine-Readable Law in the Conditions of Economy Digitization

Tatiana V. Epifanova(✉), Natalia G. Vovchenko, Dmitry A. Toporov, and Aleksei N. Pozdnyshov

Rostov State University of Economics, Rostov-on-Don, Russia
profepifanova@gmail.com, nat.vovchenko@gmail.com, toporovda@yandex.ru, nafl978@rambler.ru

Abstract. Purpose: The purpose of the paper is to determine the perspectives of development of legal education and machine-readable law in the conditions of economy digitization.

Design/Methodology/Approach: The authors perform analysis of the Legal-Tech market on the basis of the research that was performed in 2018 – early 2019 by the analytical center NAFI (2019) – the research on the basis of 500 organizations from 42 regions of Russia and the world. The tendencies in development of legal education and machine-readable law in the conditions of economy digitization and perspective directions of development of the Legal-Tech market are determined.

Findings: The article analyzes the issues and problems of development of legal education and machine-readable law in the conditions of economy digitization. The conclusions on the current state of the process of implementation of digital processes and products during training of future specialists in the sphere of jurisprudence and of activities of the acting specialists in law are made. Advantages and drawbacks of the digital technologies in the sphere of legal education and provision of legal service are studied.

Originality/Value: The perspectives of development of legal education and machine-readable law in the conditions of economy digitization on the basis of Legaltech (digital technologies in the sphere of law) and threats to the legal profession are studied. It is determined that the profession of jurisprudent and lawyer will change in the near future, and typical operations will be performed by artificial intelligence; jurisprudents with competencies in the sphere of digital technologies will have much higher professional level – which will increase the competition between jurisprudents and among the educational establishments that deal with training of specialists in the sphere of jurisprudence. It is shown that digital technologies require a serious legal support. Digital law is entering the sphere of civil, criminal, administrative, and financial law and procedural legal spheres.

Keywords: Education · LegalTech · Machine-readable law · Automatization · Digital technologies

JEL Code: I25 · K10 · K19 · K20 · K24 · K40 · K42 · P36

E. G. Popkova and B. S. Sergi (Eds.): ISC 2019, LNNS 87, pp. 971–979, 2020.
https://doi.org/10.1007/978-3-030-29586-8_110

1 Introduction

The world around us is changing quickly, and we face digital transformation, which influences all spheres of life. A lot of work has been performed in the sphere of state management, more services are obtained in the digital form, and a lot of business processes become automatized. For a lot of public officers and jurisprudents, the culture, way of thinking, and skills that allow seeing the opportunities and real application of digital technologies exist separately from the objective reality.

The 2019 Haidar Forum featured the concept "State as a platform". Digital transformation of a state is the challenge that requires the balance between state, business, and civil society. In the new conditions, the profession of jurisprudent requires qualitative modernization. A lot of legal universities, including the faculty of law of Rostov State University of Economics, look for their own model of digital modernization of legal education that would conform to the global standards and digital challenges of our time. One should remember that law is a conservative phenomenon, and, according to the experts' evaluations, the level of presence of digital technologies in the legal sphere is not more than 30% of the market of legal services; on the other hand, digitization enters all spheres of life, correcting the requirements that are set to professional jurisprudents.

The program "Digital economy in the Russian Federation" envisages adoption of the program of training of jurisprudents in the sphere of digital economy by mid-2019, so it is important that this process involve the largest scientific centers, universities, and faculties of law (Government of the Russian Federation 2019). The purpose of the work is to determine the perspectives of development of legal education and machine-readable law in the conditions of economy digitization.

2 Materials and Method

Certain issues of development of legal education and machine-readable law in the conditions of economy digitization are studied in the works Epifanova et al. (2015a), Epifanova et al. (2015b), Epifanova et al. (2016), Epifanova et al. (2017a, b). Epifanova et al. (2018), Epifanova et al. (2017a, b), Popkova et al. (2015), Shatkovskaya and Epifanova (2016), and Vovchenko et al. (2017). The issue of DigiTech is studied in the works Burri (2015), Garcia-Garcia and Gil-Garcia (2018), Pfäffli (2017), and Reyes Olmedo (2017).

In 2018 – early 2019, the analytical center NAFI performed an analysis of the LegalTech market on the basis of 500 organizations from 42 regions of Russia and the world. This allowed determining the following tendencies in development of legal education and machine-readable law in the conditions of economy digitization:

- Active implementation of digital technologies into the modern legal practice and emergence of new legal professions, including legal architect, legal engineer, digital conductors, robot lawyer, and risks management in the sphere of LegalTech;
- growth of demand and development of offer of remote legal services, which are provided much faster;
- oligopolization of the LegalTech market by large legal companies that show the highest results in the sphere of digital modernization.

The future perspectives of development of the LegalTech market for 2020–2030 are connected to implementation of the following investment and innovative projects in development of legal education and machine-readable law in the conditions of economy digitization (Fig. 1).

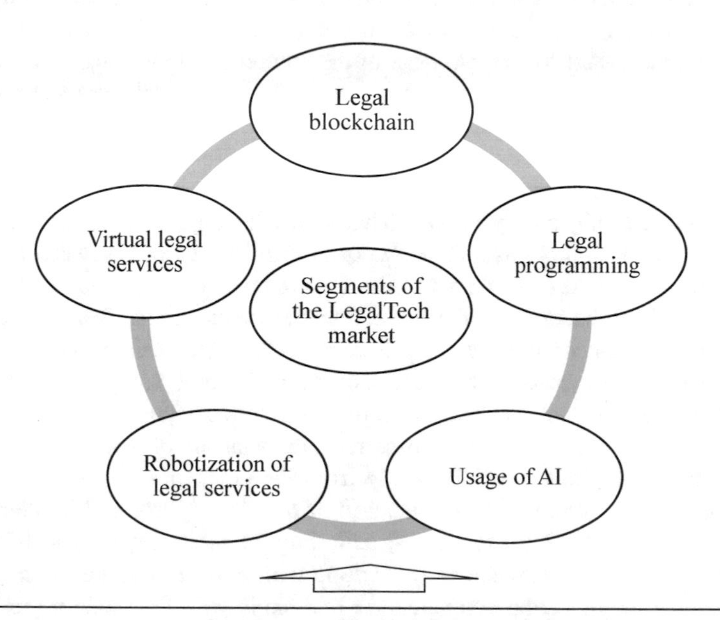

Infrastructure of the LegalTech market:
E-discovery, AI, , LegalResearcg, FinLetigation, Project Management, Notrize, Lowers Search, Trademark, LegalTech Platforms, online-services

Fig. 1. Perspective directions of development of the LegalTech market. Source: compiled by the authors based on NAFI (2019).

Figure 1 shows that the perspective segments (directions, growth vectors) of the LegalTech market for 2020–2030 are legal blockchain (electronic notary services, usage of "smart" legal technologies and technologies of protection of legal data – e.g., cryptotechnologies), legal programming (creation of the technologies of machine-readable legislation – e.g., Scratch), adaptation of artificial intelligence to market needs, robotization of legal services (L2B and L2C platforms), and virtual legal services (e.g., video consultations, translations from crime scenes, presentation of interests in court with the help of the technologies of virtual and alternate reality).

For practical implementation of the above directions, there's a need for new infrastructure of the LegalTech market that includes E-discovery (electronic search for legal services with the help of open data), AI (artificial intelligence), LegalResearch (legal analysis), FinLetigation (financial CRM), Project Management (management of legal projects), Notrize (online notary services), Lowers Search (electronic search for jurisprudent), Trademark (intellectual law), LegalTech Platforms (program provision), and online services for provision of legal services.

3 Results

One of the most discussed topics in the legal community is Legaltech (digital technologies in the sphere of law) and threats to the legal profession. Thus, CEO of Sberbank, German Gref, advised jurisprudents who do not have skills with neural networks to master new technologies or to change the profession. Also, there were discussions at the Gaidar forum that laws of the future are algorithms. In the future, a lot of functions of the state will be automatized – so the normative prescriptions that regulate them have to be transferred to the language of programming. But will it be a law in this case? Law is to be a regulator of public relations, for human relations are regulated by a certain law; algorithm does not regulate but controls and managers – i.e., it does not envisage subjectivity as such. That's why the algorithm-law is not good for regulation of public relations, as it forms superimperative relations between human and AI or between several AI's. At the same time, there are serious and active discussions of the issue of acknowledgement of artificial intelligence an independent subject of law. German Gref's opponent was Mikhail Barshchevsky, who said that "the work of jurisprudents could be replaced by the work of an intellectual platform, but only in the part in which the jurisprudent performs certain formalized functions. The work of a lawyer on complex criminal or civil cases requires analysis of a whole range of circumstances that could be treated differently from the point of view of the law".

Most jurisprudents agreed that digitization of the legal sphere and implementation of artificial intelligence in legal processes, including the judicial sphere, provides large opportunities for jurisprudents and helps to cope with routine tasks faster and better; a threat is the absence of digital competencies and skills with the modern graduates of legal universities. At present, no modern jurisprudent can work without the information and legal systems that allowed increasing the level of legislation and law enforcement and saving a lot of time and efforts for professional jurisprudents. In the past, jurisprudents had to subscribe to a lot of newspapers and journals with the official texts of normative acts, actualize them, systematize them, and form data bases – at present, these functions are performed by the information and legal systems, which allows – in several clicks – receiving full information on a specific issue, forming a selection of the normative and legal acts, court practice, and doctrinal positions (Consultant Plus and GARANT – in Russia; Judicata and Everlaw – in the USA; INFOLEX, PRESTEL, and LEXIS - in the UK, etc.).

Artificial intelligence could become an excellent adviser for a legislator, who performs economic analysis of law – for a lot of legal decisions are made in view of economic effectiveness based on the principles of risk distribution. However, the issue of a data base for AI arises. The court practice cannot be used as such data base: the practice could be unjust and incorrect and different in various regions. However, AI possesses a potential of work with economic data: indicators of GDP, development of certain productions, and information from Federal State Statistics Service. There are perspectives of replacement of judges with robots in the countries where the judicial system is discredited, in countries with high level of trust to the judicial system and the law-enforcement bodies. However, digitization and application of artificial intelligence in the judicial sphere and law enforcement could reduce the level of corruption.

Russia will see the creation of a concept of machine-readable – clear for algorithms – law, which will allow for automatization of certain legal norms. The paragraph on development of such concept in 2019 is present in the plan of the federal project "Normative regulation", which is to be adopted by the government within the work on the national program "Digital economy".

Machine-readable law and automatic law enforcement will change the interrelations of citizen and business with the state. For example, this will take place in the sphere of prevention of car accidents through automatic fines for traffic violations and in the sphere of digitization of the controlling and supervision functions – e.g., when a sensor shows exceeding emissions and sets a fine for the one responsible for the emission. Automatization of law enforcement excludes the human factor and makes responsibility inevitable.

Not only violations could be controlled in this way but also the facts of supply of goods – with the help of scanning of special markings and putting the information into a data base. Tax inspections could be performed automatically for obtaining information on taxpayers' deals and confirmation of the very fact of a certain operation.

The concept is created for using the technologies of AI and Big Data processing in the process of regulation and law enforcement. As of now, adaptation of legislation to the current tasks is very slow and does not show the precision that is required by business. Work with formal legal language will allow reducing labor cost and the number of errors, as well as increasing the speed of implementation of new developments.

Formalized and structured presentation of the legal norms is useful at all stages of the legislative process – this simplifies and accelerates the legal and anti-corruption expertise of the projects of normative acts, allows determining and eliminating the legal gaps and collisions, and increasing the quality of legislation. Artificial intelligence could analyze a large array of court cases of the similar topic and could predict the case's result. Thus, the plaintiff could take an informed decision – whether he should enter the court proceedings and spend money for it.

The LegalTech market volume (unifies technologies in the sphere of legal activities) in the world constitutes USD 16 billion per year. There are no assessments of this market in Russia. However, Russia has a lot of companies that deal with legal technologies: analysis and automatic compilation of documents, remote legal support, search for jurisprudents, automatization of work with agreements and risks, electronic document turnover, creation of information systems, etc. Remote legal support is the most popular direction of their activities.

The moment when LegalTech is widely used in the legal activities is very far – this will require adaptation of not only technologies (creation of working technologies of semantic analysis and special language of cyphering of legal norms) but of the law as well. There are simple repeated processes that could be transferred into the language of algorithms and automatized now – e.g., complaints for fines. The institute of notaries also requires modernization. A lot of processes still require notarized paper documents, and citizens have to put a signature in the physical register (special record book) – all this could be also automatized. As of now, LegalTech's development level does not allow speaking of any significant savings in the sphere. It's improbably that companies will refuse from their jurisprudents or lawyers' services – though personnel who perform simple tasks could be partially replaced by computers.

The advantages of machine-readable legislation are as follows:

(1) Automatization of law enforcement, which means reduction of costs of execution of obligations and risks of further distortion or improper execution of previously coordinated role (in law or agreement).
(2) Automatic determination of contradictions and errors of legislation at the stage of preparation or testing of the normative and legal act – i.e., creation of technologically better legislation.
(3) Quicker and more precise evaluation of the regulating effect of the law (modeling the effect from further changes, together with Big Data).
(4) Emergence of a new large market of program products that solve the legal and related issues, which leads to reduction of the cost of legal services for final consumer.
(5) Quick reduction of the problem of ambiguous treatment of the norms of law and, accordingly, reduction of costs of understanding the norms of law.

Disadvantages (drawbacks) of machine-readable legislation are as follows:

(1) Impossibility of full automatization of legal relations that are connected to the will of people (subjective side of violation of law, strike, and cartel agreement) or legal relations on which a part of objective reality influences (force majeure and circumstances that are directly connected to personality);
(2) Possible failures of automatized systems;
(3) Initial expenditures for purchase, installation, and servicing of automatic systems;
(4) Probability of corrupt usage of technologies;
(5) Social consequences (large reduction of the number of people that are involved in the legal system);
(6) Transfer of legislation into the machine-readable form will require high specification of rules, which might lead to reduction of adaptability of the system, reduction of competition, and other negative consequences.

Let us consider the technological tools and platforms that become more popular in the legal professional sphere:

"FreshDoc. Document constructor" – this service allows for quick creation of legally correct documents (agreements, etc.)

"Autolawyer" – bot that allows forming a complaint to an insurance company for incorrectly calculated coefficient of bonus-malus for third-party only vehicle insurance;

"Patentbot" – full service of automatized registration of a trade mark;

"Flexbby" – organizes effective work with agreements and other documents in the company and automatizes the full cycle of work with legal documents in the company from creation to approval and control over storing of the paper original, creation of the payment schedule, reminder, and control over payments.

"Jeffit" – solves the standards tasks of the lawyer assistant, puts the information on court hearings into the calendar, sends invoices to clients, etc.;

"XSUD" – systematizes and conducts accounting of information on court cases, complaints, inspections, executive proceedings, and agreements; performs automatic downloading on course cases from the portal kad.arbitr.ru; performs setting of a

personal calendar for each jurisprudent with automatic reminders on future events; forms reports on court cases; controls execution of errands; ensures security of access to data.

"Правовед.ру" – bot that answers 85% of questions regarding the Law of protection of consumers' rights.

"Case.pro" – system of automatization of legal processes and optimization of the work in a legal company or legal department; it is synchronized with the data base of courts and integrated with the main postal clients; allows for automatic accounting of time and expenditures for operative billing; flexible system of document management; possibility of controlling deadlines with integration with the Google calendars and Outlook; preparation of reports on performed work with various levels of specification; customer control of doing business and projects.

"Case.look" – the system of search and analysis of court practice of the courts of general jurisdiction and arbitration courts, which allows using more than 30 search filters.

"Case.book" – system for monitoring of court cases and verification of intermediaries; it is integrated and synchronizes with the data base of the courts of the general jurisdiction, arbitration courts, bailiff service, and the federal tax service. The system allows using artificial intelligence for "predicting" the probable result of a certain court case with the precision up to 82% and determining connections between specific persons and companies.

"Form.one" – legal constructor, which allows for independent creation of chat bots for various specific tasks and works. Such chat bot will independently (automatically, without human participation) ask the necessary questions, write down the received answers and data, ask for the copies of the necessary documents, store the received documents, and prepare standards documents (agreements, complaints, etc.) for the client based on the received data and information.

A separate niche is occupied by the technological platforms that allow selecting a specialist for certain tasks, in view of specialization, cost of services, and other parameters. Such projects are created in the USA, the UK, and Russia (Lexoo, Avvo and Legal Space, accordingly). It is possible to speak of uberization of the market of legal services.

Analysis of the implemented developments in the LegalTech sphere shows that all projects in this sphere could be divided into two large groups. The first group includes decisions that conform to the requirements of the professional legal industry – in other words, these are technologies that solve everyday tasks of certain jurisprudents and organizations that provide legal services. The second large group is technological projects that conform to the demands of the final consumers of such legal services.

The first group of LegalTech decisions includes:

(1) Electronic Discovery/E-Discovery
(2) Legal Research (online libraries of legal documents or information data bases)
(3) Law Practice Management Software, with possibilities of time tracking, billing, task management, management of documents, accounts, and reports.

The second group of solutions of LegalTech includes as follows:

(1) Online Legal Services
(2) Lawyer Search or Marketplaces
(3) Litigation Funding
(4) Contract analysis & legal forms and documents creators
(5) Chatbot
(6) Notarization Tools.

At present, the division is rather clear, but technological products that successfully solve the tasks of the legal industry will be adapted to B2C market, excluding an expensive intermediary (jurisprudent).

Obviously, any of the above solutions could help in the activities of private jurisprudents (lawyers) and small legal companies, as well as large legal companies and legal departments.

At the same time, most offers of LegalTech in the Russian market from the first group of solutions have a range of drawbacks, which, in our opinion, restrain ubiquitous implementation of the above technologies into everyday work of jurisprudents. The first main drawback is high cost, which is often set not only as a fixed one-time payment but also as subscription. The second main drawback is complexity of the user interface, which leads to large time delays during implementation, servicing, and usage of the offered products of LegalTech in the sphere of management of a legal company.

4 Conclusion

The profession of jurisprudent and lawyer will change in the near future, and typical operations will be performed by artificial intelligence; jurisprudents with competencies in the sphere of digital technologies will have much higher professional level – which will increase the competition between jurisprudents and among the educational establishments that deal with training of specialists in the sphere of jurisprudence.

Digital technologies require a serious legal support. Digital law enters the sphere of civil, criminal, administrative, and financial law, as well as procedural spheres.

In view of the objectivity of the process of digitization and large potential of implementation of technologies into the profession of jurisprudent, it is necessary to pay attention to the growing value of legal education, which has to ensure the fundamental inter-sectorial training of professional personnel. That's why it is important to perform constant actualization of knowledge and implement professional technological tool and skills of working with them into the educational process.

References

Burri, M.: Public service broadcasting 3.0: Legal design for the digital present. Public Service Broadcasting 3.0: Legal Design for the Digital Present, pp. 1–293 (2015)

Epifanova, T., Albekov, A., Romanova, T., Vovchenko, N.: Study of factors which facilitate increase of effectiveness of university education. Int. J. Educ. Manag. **31**(1), 12–20 (2015a). www.emeraldinsight.com/0951-354X.htm

Epifanova, T.V., Kolesnikov, Y., Usenko, A., Parshina, E., Ostrovskaya, V.: Peculiarities of state regulation of innovational activity of enterprise in the global economy. Contemp. Econ. **10**(4) (2016)

Epifanova, T.V., Romanenko, N.G., Mosienko, T.A., Skvortsova, T.A., Kupchinskiy, A.: Modernization of institutional environment of entrepreneurship in Russia for development of innovation initiative in small business structures. Eur. Res. Stud. J. **8**(3) (2015b)

Epifanova, T.V., Shatkovskaya, T.V., Romanenko, N.G., Mosienko, T.A., Tkachenko, M.A.: Legal provision of clustering in russia as environment for development of innovations. Int. J. Trade Glob. Mark. (2017a). Special Issue "Clusters and Innovational Networks in the Context of Sustainable Development"

Epifanova, T.V., Skvortsova, T.A., Parshina, E.A.: Influence of clustering on innovational development of business structures in region's economy. Int. J. Trade Glob. Mark. (2018). Special Issue "Clusters and Innovational Networks in the Context of Sustainable Development

Epifanova, T.V., Ukraintsev, V.B., Chernenko, O.B., Mishchenko, K.N.: Estimation of threats of the Rostov region economy caused by the collapse of ensuring the object-differentiated approach to rendering the state support: the Russian Federation. Eur. Res. Stud. J. **10**(3), 92–102 (2017b)

Garcia-Garcia, L.M., Gil-Garcia, J.R.: Reconsidering the importance of context for the success of digital government: the case of legal vulnerability and extreme poverty in the provision of migration services at the Southern Mexican border. In: ACM International Conference Proceeding Series, a46 (2018)

Pfäffli, D.: Conference Proceedings of the Weblaw Forum LegalTech 2017—[Tagungsbericht Weblaw Forum LegalTech 2017]. Jusletter IT (2017)

Popkova, E.G., Vovchenko, N.G., Epifanova, T.V., Pogorelenko, N.S.: Did competition help to achieve positive effect of privatization? Revista ESPACIOS (2015). ISSN 0798 1015. http://www.revistaespacios.com/index.html

Reyes Olmedo, P.: Technical-legal management standards for digital legislative information services—[Estándares técnico-jurídicos de gestión para servicios digitales de información legislativa]. Revista Chilena de Derecho y Tecnologia **6**(2), 57–95 (2017)

Shatkovskaya, T.V., Epifanova, T.V.: Correlation of private and public legal interests as theoretical and scientific and practical problem of modern law. J. Adv. Res. Law Econ. **3**(17), 625–643 (2016). https://doi.org/10.14505/jarle.v7.3(17).19

Vovchenko, N.G., Tishenko, E.N., Epifanova, T.V., Gontmacher, M.G.: Electronic currency: the potential risk to national security and method to minimize them. Eur. Res. Stud. J. **10**(1) (2017). Special Issue Dedicated to the International Conference "Russia and EU: Development and Horizons"

NAFI. LegalTech 3.0. The LegalTech market in Russia and the world (2019). https://nafi.ru/projects/predprinimatelstvo/legaltech-3-0-rynok-legaltech-v-rossii-i-v-mire/. Aaccessed 25 Apr 2019

Government of the Russian Federation. Decree dated March 2, 2019 No. 234 "Regarding the system of managing the implementation of the national program "Digital economy of the Russian Federation" (together with "Decree on the system of managing the implementation of the national program "Digital economy of the Russian Federation" (2019). http://www.pravo.gov.ru. Accessed 07 Mar 2019

Digitization of Legal Education: A Popular Direction or Real Necessity?

Larisa I. Poltavtseva[1(✉)] and Alina V. Poltavtseva[2]

[1] Rostov Branch of Russian State University of Justice and Rostov State University of Economics, Rostov-on-Don, Russia
lar_poltav@mail.ru

[2] Rostov State University of Economics, Rostov-on-Don, Russia
alisa55@yandex.ru

Abstract. The purpose of the article is to study the conditions and specific features of the situation in which legal education enters the age of digitization. The authors use the theoretical and empirical materials and methods, including the results of a survey among undergraduates regarding the quality of legal education.

The results of the performed research allow for the conclusion that digitization as a global tendency and strategic line of development of the Russian society influences all spheres of public life, including the sphere of education. Provision of the conditions of growth of the Russian economy sets before the system of education the task of training of competent personnel. Digitization of education does not have a total character and has to be involved into the educational process.

Orientation of the system of education at the digital economy forms new requirements to competencies of specialists in the sphere of jurisprudence and does not contradict nor excludes the goals that are traditional for legal education: usage of knowledge and development of personalities of students. It is expedient to achieve the optimum in integration of the existing approaches to education, as well as methods of training in their traditional and innovative variants. Organization and contents of training should be built in view of opinions of the consumers of educational services (students).

Keywords: Digitization · Legal education · Traditions and innovations of training

JEL Code: I21 - Analysis of education

1 Introduction

The strategic line of development of the Russian society, expressed in the Program "Digital economy of the Russian Federation" [1], influences all spheres of public life, including the sphere of education. Training of skilled jurisprudents is no exception. The article studies the specific features of the situation when legal education enters the age of digitization.

E. G. Popkova and B. S. Sergi (Eds.): ISC 2019, LNNS 87, pp. 980–986, 2020.
https://doi.org/10.1007/978-3-030-29586-8_111

2 Methodology

The methodology of the research is based on the methods of theoretical and methodological analysis of the level of elaboration of the topic and empirical materials and methods, including the results of a survey among undergraduates regarding the quality of legal education.

3 Results

The state and perspectives of development of the Russian legal education is in the focus of researchers' attention. However, this interest grew in the recent decade. The issue of training of jurisprudents was considered at the state level [2] and discussed by the scientific and pedagogical community [3, 4].

The main result of the above processes was stating the crises phenomena in the legal education, which are considered in the context of quantity and quality of training of graduates in law: there a lot of universities that train jurisprudents and a lot of graduates, but there's a lack of skilled specialists in law.

The offered and implemented measures include as follows:

- organizational and legal transformations of the Russian system of legal education, which envisage limitations on training of jurisprudents in non-profile universities and non-profile faculties, reduction of the number of universities that train specialists in the sphere of law, etc.;
- qualitative changes of the training process: increase of the practical aspect of training (increase of practical courses and hours), introduction of the objective criteria of assessment of undergraduates' knowledge (written tests and exams, online testing, introduction of the innovative educational technologies (online courses, usage of multimedia resources, etc.).

Negative assessments of the above measures could be seen in the analyzed literature [5]. However, it is possible to see the state of the modern system of legal education in Russia before digitization and shows the growing need for its improvement.

The sphere of education could be considered in comparison with treatment of the target direction of the Program of digitization and purposes of creation of the Bologna system. They are rather similar and envisage continuity of education, improvement and stabilization of intellectual potential, competitiveness, increase of quality of education, etc. Russia joined the Bologna process in 2003; by now it should have realized and accepted the fact that good purposes in reality did not lead to expected "magic" of unification of the educational process and integrity of the educational space (student mobility, maneuverability of the academic staff, coordination of functioning of Russian and European universities, etc.).

The results of the Russia's transition to the two-level system of education received ambiguous reactions. For example, uncertainty of the status of bachelors in the domestic labor market stimulated the "return" of Russian universities to training by the specialist's program. The transition has not led to a breakthrough in quality of the educational process. There are opinions that originally the main motives of replacement of the traditional Russian system of legal education by a two-level system “bachelor-master” did not seek the goal of increase of the quality of legal education [5].

It should be emphasized that the above analogy concerns only the offered innovation and is to prevent from excessive euphoria in expectation of their implementation. It would be better to orient at the rational and weighted approach in evaluation of obvious advantages and possible risks of digitization of legal education.

Unfortunately, orienting at unsuccessful experience of “popular” enthusiasm regarding the Bologna System and its implementation into the Russian education (which it not always coordinated and perfect), implementation of the idea of digitization could face such danger. The proofs of this (direct and indirect) could be seen even now.

Firstly, it is necessary to note the growing demand for discussion of the topic of digitization at various levels and platforms – which is not bad. Growth of the number of scientific conferences on the current topics is a part of the trend of digitization. One cannot but pay attention to the general tone and enthusiastic statements of certain participants of the discussions. For example, the Parliament hearings in the Council of Federation of the Russian Federation on November 20, 2018 featured the following statements: “Transition of the educational sphere into the digital environment became a global trend”, “Russia must join the global trend in order not to fall behind”, “…Russia has to adapt to the paradigm of so called digital economy”, and “set more progressive vector”, etc. [6]. We do not mention the authors of these statements on purpose.

Secondly, polarization of opinions of discussions’ participants is logical. However, excessive black-and-white thinking in the negative evaluation of innovations is notable [7]. According to the authors of this article, during transfer of educational activities into the digital educational environment the main “threats” are as follows:

- reduction of popularity of the traditional forms of education (refusal from traditional paper books; replacement of interpersonal educational process by a training program; reduction of prestige of a lecturer as a role model);
- reduction of social skills of students, degradation of their personality (digital unawareness, inevitable loss of skills of independent thinking and creative abilities, slow speaking development, loss of writing skills, etc.); dependent behavior (Internet dependence; computer and game dependencies); problems with health.

Such statements contain the opposition to the idea of digitization in the sphere of education. It should be noted that discussions of the perspective of digitization of education also contain other opinions that call for full transfer to remote education (online education).

In order to avoid any guesses and free interpretation of the goals and tasks of digitization let use the original sources. The Program “Digital economy of the Russian Federation” sees the main goals of the direction concerning personnel and education the following: creation of the key conditions for training of personnel in the digital economy; improvement of the system of education that has to provide the digital economy with competent personnel; labor market, which has to conform to the requirements of the digital economy; creation of the system of motivation for mastering of the necessary competencies and participation of personnel in development of the digital economy of Russia.

Regarding personnel and education, the Program orients at achievement of such indicators as the number of graduates of the educational organizations of higher education for the directions of training that are connected with information and telecommunication technologies; number of graduates of higher and secondary vocational education who have competencies in the sphere of information technologies at the average global level; share of population that have digital skills. We see the obvious emphasis on creation of conditions and training of competent personnel for the digital economy who possess competencies in the sphere of information technologies. The program outlines the final results in the form of the share of trained specialists. As for the remote character of education, it clearly states the transition to the remote verification of competencies.

However, formation of the digital educational environment is necessary and very important, and improvement of the system of education is a necessary condition for transition to the digital economy. Advantages and drawbacks of the digital educational environment are widely discussed in literature. Let us consider several publications on this topic.

Thus, Gyulbyakova and Maslovskaya [8] determine the following advantages of online education: online education is open for everyone and access to information and knowledge is unlimited. Online education saves time and money and allows selecting the optimal conditions of training (e.g., at home), which is very good for handicapped people. The teacher could address a large number of students, preserving individual relations with everybody; students actively participate in the training. The drawbacks of remote training are physical absence of teacher; technical problems that are connected to the work of the training systems (malfunctions of communication systems, PC's, terminals, and servers, viruses, hacker attacks, etc.).

The authors state that the near future will see a transition from traditional education to remote education. However, we think that the idea of full transition to remote education requires more attention. In this context, the statement of Churikova might seem too emotional but rather fair: "We must praise God that the competent services of the government have not yet though of remote training of surgeons; but we have to understand that a badly trained jurisprudent in the long-term is no lesser public danger that a badly trained surgeon" [9].

However, the idea of refusal from traditional education via gradual but total transition to remote education has not emerged all by itself. Different works contain the thought on popularization of the ideology of legal utilitarianism, which cannot be ignored. The tendency of treatment of jurisprudence not as a science but as a trade and totality of certain technical skills that are taught to people who want to become lawyers is considered in the article of Churikova [9]. The author comes to the conclusion that the modern European system of professional legal education, which is oriented at training of highly skilled masters of jurisprudence, has a rather utilitarian character.

N.V. Litvak's opinion of the Y. Kuzminov's (rector of the Higher School of Economics) offer to fully replace lecturing by records of online courses, which was supported by the rector of a private European University in St. Petersburg V. Volkov is rather interesting [10]. Thinking that reformation moves according to the negative scenario, the author justly considers that such offer should be confirmed by the results of a sociological survey. Litvak provides an example of the USA – the leader in all technological innovations – which has not refused from "live lectures" – quite on the contrary, attendance of the lectures of the best lecturers is very high, and some scholars charge a large fee for their lectures [11].

The idea of "online education" seems to be very perspective. In order to avoid the seduction of deep modernization of the sphere of education it is necessary to remember a saying "it's easier to rip apart than to build up".

Creation of a digital educational environment is connected to solving the tasks on increase of quality of education, including legal education. The system of continuous information training of legal undergraduates, which is connected to usage of information technologies, is implemented into the educational process [12].

However, is it necessary to refuse from traditional approaches: problem approach; learner-oriented approach; activity approach, etc. Of course no! It is also obvious that each model, including remote training, is interesting and practically necessary. Such though is seen on scientific literature: any approaches to training have both positive and negative characteristics, so the task of modern education consists in their effective combination, and this, in its turn, depends on the goals and created conditions [13]. The issues of conditions of experience and limits of applicability of classical didactics in the digital educational environment are considered [14]. Therefore, it is expedient and necessary to achieve optimum in combination of the well-known traditional and innovative approaches to training.

Thus, using the opinions of the modern users of legal education services on its quality seems to be rational. In the course of the performed survey, sixty second-year students of the faculty of law of Rostov State University of Economics were offered to express their opinion on the quality of teaching in the university and provide their suggestions on increasing it. 38% of the respondents stated that they were fully satisfied with the quality of teaching; 33% think that increase of quality of teaching is connected to the personality of lecturer (the lecturer's interest and his skill to make students interests in studying a specific course); 23% of the respondents stated that the teaching methods could improve the quality of teaching: application of interactive forms of teaching; lectures-discussions; maximum visualization of the educational material. 22% of the respondents noted the necessity for higher involvement of the practical component and visual examples from the jurisprudents' practical activities.

The question "Which form of practical courses stimulates the most effective learning of material?" was answered in the following way: 61% - game forms; 26% - considering real practical situations, including court practice; 21% - group discussions; 18% - dialog between lecturer and students; 16% - creative tasks; 14% - watching movies; 9% - surveys.

It could be possible to see students' orientation at their experience in the interpretation of the obtained results. However, we cannot ignore the need for direct (not remote) communication with lecturer and between the students. At least, the obtained results could be an argument in favor of the offered thesis on expedience of the synthesis of the known and innovative approaches and methods in training.

Neither of the approaches to training can be 100% correct, and each of them has positive and negative characteristics; thus, we want to pay attention to rather high preference of game forms of the lectures (61% of the respondents). As is known, federal state educational standards envisage role (business) games as one of the mandatory active/interactive forms of educational courses, which stimulate formation of the required competencies. According to the results of the survey, game forms are very popular among students. However, a "fly in the ointment" could be the opinion of the professor of the Institute of Psychology of Rostov State Humanitarian University, V. Kudryavtsev. He stated the "delayed" (to the age of thirty) childhood of humankind. Physical childhood passes, and social childhood is extended – and adulthood dost not come. According to the scholar, quick socio-technological boom leads to the situation when people do not live but adapt. "the 'game approach' is popular even in the fundamental science". This leads to the situation when a part of young scholars return to re-invention of 1920's–1930's, but in a simplified form" [15]. Evaluating the results of the survey in the context of the quoted opinion of V. Kudryavtsev, it is possible to agree that modern students are children who need a visual form of presenting information, with practical conformation and "game approach" in mastering of the material. But wouldn't the game hinder the process of their social ageing?

4 Conclusions/Recommendations

1. Digitization as a global tendency and strategic course of development of the Russian society influences all spheres of public life, including the sphere of education. Provision of the conditions of growth of the Russian economy sets before the system of education the task of training of competent personnel.
2. Digitization of education is not total and it should be involved in the educational process. Orientation of the system of education at the digital economy forms new requirements to the competencies of specialists in the sphere of jurisprudence and does not contradict nor excludes the goals that are traditional fro legal education: mastering knowledge and development of the personality of students.
3. The existing approaches to education, as well as methods of teaching in their traditional and innovative variants do not have an unambiguous positive or negative evaluation. It is necessary to achieve optimum in their integration.
4. Organization and contents of training should be built in view of the opinion of consumers of the services of legal education (students).

References

1. Program "Digital economy of the Russian Federation", adopted by the Decree of the Government of the Russian Federation dated July 28, 2017 No. 1632-r. http://static.government.ru/media/files/9gFM4FHj4PsB79I5v7yLVuPgu4bvR7M0.pdf. Accessed 3 May 2019
2. State program of the Russian Federation "Development of education" for 2013–2020. http://new.volsu.ru/upload/medialibrary/809/. Accessed 3 May 2019
3. Bondar, N.S.: Modern landmarks of the Russian legal education: national traditions or multiculturalism illusions? J. Legal Educ. Sci. (1), 7–16 (2013)

4. Ershov, V.V.: Problems and perspectives of development of legal education. J. Sudya (9), 11–18 (2011)
5. Meshcheryakova, O.M.: Russian legal education today: problems and perspectives. Russ. J. Leg. Res. **1**(2), 175–178 (2015)
6. Parliament hearings in the Council of Federation of the Russian Federation, 20 November 2018. https://www.pnp.ru/social/cifrovizaciya-obrazovaniya-potrebuet-izmeneniy-v-zakonodatelstve-schitayut-senatory-i-eksperty.html. Accessed 3 May 2019
7. Kulebyakina, E.: Risks of digitization (2018). https://narasputye.ru/archives/4381. High technologies – a path to degradation. https://narasputye.ru/archives/2495. Accessed 3 May 2019
8. Gyulbyakova, K.N., Maslovskaya, E.A.: Electronic form of training: specific features and perspectives. J. Mod. Probl. Sci. Educ. (4) (2018). http://science-education.ru/ru/article/view?id=27812. Accessed 06 May 2019
9. Churikova, N.N.: Problems of development of legal education in Russia. J. Territ. Sci. (3), 21–25 (2015)
10. Meleshko, V.: "The main trend of the Russian education – digitization". Interview with Y. Kuzminov, rector of the Higher School of Economics, Teacher's Gazette from 23 January 2018 (2018). http://www.ug.ru/article/1029. Accessed 03 May 2019
11. Litvak, N.V.: New reform of domestic higher education: "digitization" and "professorship". J. Sci. Cult. Soc. (2–3) (2018)
12. Kochkarov, R.M.: Legal education in the context of reforms. Bull. High. Sch. Alma Mater (10), 23–27 (2013)
13. Petrochenko, E.N.: The main problems of development of legal education in the Russian Federation. J. Territ. Sci. (2014). https://cyberleninka.ru/journal/n/territoriya-nauki. Accessed 03 May 2019
14. Voskresenskaya, E.V.: Problem of quality of legal education in modern Russia. J. Mod. Issues Sci. Educ. (2) (2017). http://www.science-education.ru/ru/article/view?id=26361. Accessed 03 May 2019
15. Kafidulina, N.N.: Digitization as a trend: growth points for the Russian education. J. Interact. Educ. (1–2), 9 (2018). http://interactiv.su
16. Krashennikova, L.V., Zakharov, K.P.: Development of digital educational environment for implementation of tutor activities. J. Interact. Educ. (4), 6–12 (2019). http://interactiv.su/wp-content/uploads/2018/12/IO_4-interactive-1.pdf. Accessed 03 May 2019
17. Emelyanenko, V.: Why does childhood continues up to thirty and where does this lead. Russian gazette – Federal issue no. 160(7623) (2018). https://rg.ru/2018/07/24/pochemu-segodnia-detstvo-cheloveka-tianetsia-do-30-let-i-k-chemu-eto-vedet.html. Accessed 03 May 2019

The Russian Legal System in the Conditions of Digitization

Irina G. Napalkova(✉) and Petr S. Samygin

Rostov State University of Economics, Rostov-on-Don, Russian Federation
napalkova-ig@mail.ru, samygin78@yandex.ru

Abstract. The article studies the problems of determining the phenomenon of digitization as a modern global trend of social development on the basis of transforming information into the digital form and using the most perfect technologies that transform reality in all its multiple manifestations. The main purpose of the research is to determine the main directions of digitization of the modern law and to study the methodological aspects of digitization's influence on effectiveness of legal regulation and improvement of the Russian legal system.

The authors perform the theoretical and methodological analysis of the processes of digitization of legal reality and determine the main conceptual approaches to understanding the forms of influence of the modern technologies on various spheres of social life that are subject to legal regulation. The authors use the modern methods of cognition of processes that do not have analogs in the previous periods of development of society, which allowed for establishment of a new digital reality, within which law is the main tool of socio-economic development and the sphere of state management. The authors also analyze the process of establishment and development of digital rights of human, provision of public safety, effective mechanism of protection of person, and right of usage of new information technologies.

The authors note that emergence of a new digital paradigm of global development dictates the necessity for forecasting of the future development, determining the dynamics of the changes within the legal field, studying the transformation of the existing norms for provision of digital processes, and modernization of law-making and law enforcement.

The authors pay special attention to preservation of the spiritual and moral foundations of law and its systemic and regulating role in the life of society. In the processes of digitization of the Russian legal system, there's a necessity for accounting of accumulated achievements in the sphere of law and combining past experience and modern achievements for minimizing the possible threats and risks of modern digitization. Such digital system can become a foundation for formation of the most rational strategic decisions at all levels of social life.

Keywords: Legal system · Digitization · Digital rights · Law making · Law enforcement · Information society · Information right

JEL Code: K 10 · K15

E. G. Popkova and B. S. Sergi (Eds.): ISC 2019, LNNS 87, pp. 987–993, 2020.
https://doi.org/10.1007/978-3-030-29586-8_112

1 Introduction

The decisive factor of development of the modern society is formation of a new digital reality, which changes all spheres of social life that influences the mechanism of legal regulation and activities of various legal institutes. The modern sphere of social life is mobile and dynamic; its space expands and its various technologies improve. Under the influence of globalization, the paradigms of thinking changes and institutes of civil society, government bodies, and the system of law transform.

Digitization of social life is determined by society's striving for the most optimal usage of the existing opportunities of the modern technologies and innovative methods. According to the humanitarian goal, scholars actualize the issue of mobilization of multiple means and methods of progressive development of the modern world.

The modern science faces the most important task of consideration – at the theoretical level – of new processes and tendencies in the sphere of law that take place under the influence of digitization and of development of forecasts of further development of law and the system of state management. The law plays the main role in development of these most important spheres of social life. Digital technologies could change the regulating potential of law and ensure its effectiveness and social direction. In the modern conditions there is a need for active research of the influence of digitization on legal reality and its influence on improvement of the mechanism of legal regulation (Vlasova et al. 2012).

2 Methodology

Analysis of digitization in the legal sphere requires determining the methodological basis of the process of cognition. The authors used philosophical, general scientific, and the technical method, which allow determining interdependence of the processes of digitization, causal connections of the changes of law, and new paradigms of development of society. The authors use the potential of new methodological approaches, which allow overcoming the unilateral character of the classical scientific ideas of cognition of reality. The issues of digitization are studied with the usage of the complex and integrative methodology, methodology of comparative law, concrete-historical analysis, and methods of legal modeling.

3 Results

The notion "digitization" is used in the narrow and wide senses. In the narrow sense, digitization is treated as transformation of information into the digital form, which in most cases leads to reduction of costs, appearance of new opportunities, etc. Digitization in the wide sense is treated as the modern global trend of development of society and economy, based on transformation of information into the digital form and leading to increase of effectiveness of economy and improvement of quality of life. Digitization could be considered a trend of effective development if digital transformation covers production, social sphere, and regular life of people, and is accompanied by effective usage of its results, which are accessible to ordinary citizens (Khanina and Chernov 2018).

Digitization of social life led to emergence of previously unknown rights – digital rights. In the conditions of formation of legal state, a special role belongs to provision – at the constitutional level – of digital rights in virtual space, which is specification of universal rights of a human by law, and enabling legislation (Zorkin 2018). The necessity for acknowledgment and protection of digital rights was proclaimed in certain international legal acts. Thus, Okinawa Charter on Global Information Society (Okinawa, July 22, 2000), adopted by representatives of eight leading countries of the world, including Russia, proclaimed the necessity for strengthening of the corresponding policy and normative basis, which stimulate cooperation for optimization of the global goals and fighting misuses that violate the integrity of the network, reduction of the gap in digital technologies, investing into people, and provision of global access and participation in this process (Okinawa Charter on Global Information Society 2000). The text emphasizes that all people must have a right to use the advantages of the global information society: sustainability of the latter is based on democratic values that stimulate the development of human – free exchange of information and knowledge, mutual tolerance, and respect for specific features of other people. The Charter also confirms the states' obligation to coordinate their actions on creation of secure cyber space and security of information systems that are protected from criminals, including transnational organized crime.

The UN General Assembly Resolution dated December 18, 2013, No. 68/167 "Right to Privacy in the Digital Age" states that quick rates of technological development allow people in all regions of the world to use new information and communication technologies and increase the ability of governments, companies, and individuals to track, intercept, or collect information, which could violate human rights (Resolution of the UN General Assembly 2013). It is emphasized that the necessity for provision of public security could justify the selection and protection of certain confidential information, but states should guarantee observation of their international and legal obligations in the sphere of human rights.

The Russian Federation has the effective Federal law dated July 27, 2006 No. 149-FZ "Regarding the information, information technologies, and information protection" and other normative and legal acts, including "Regarding personal data", "Regarding provision of access to information on activities of courts in the Russian Federation", "Regarding protection of children from information that deals damage to their health and development", and other normative and legal acts that regulate the information sphere of social life. In 2017, the Decree of the Government of the Russian Federation No. 1632-r adopted the program "Digital economy of the Russian Federation", which is implemented according to the goals, tasks, directions, volumes, and terms of implementation of the main measures of state policy on creation of the necessary conditions for development of the digital economy, in which the data in the digital form are the key factor of production in all spheres of socio-economic activities, which increases the competitiveness of the country and population's quality of life and ensures economic growth and national sovereignty (Decree of the Government of the Russian Federation 2017).

As of now, there's a necessity for systematization of the modern information law, which could overcome its contradictions and gaps. For this purpose, the Concept of the project of the Information Code of the Russian Federation was developed; it envisages improvement of the constitutional law of citizens for information. The Information Code is to specify the constitutional right of human and citizen for information, regulate general issues of turnover of information in the Russian Federation, establish a range of legal notions that are connected to turnover of information, and establish general requirements to the state information systems, which will be developed and supported by the corresponding bodies of public authorities. The legislator has to determine the main forms of turnover of information, establishing the rights and obligations of its participant and to determine the legal regime of information in the public legal and personal legal spheres and the foundations, forms, and limits of application of the information technologies in the activities of the corresponding subjects of law.

According to T.Y. Khabrieva, "analysis of the processes of digitization allows forecasting the change of the mechanism of law-making and composition of the existing model of social regulation, correction of the limits of known social regulators, and formation of a niche for the program code" (Khabrieva and Chernogor 2018). Large opportunities of digitization are formed in the sphere of private law and civil turnover in the sphere of law-making and law enforcement. It becomes possible to use AI in court proceedings during expertise, obtaining of digital evidences, and organization of the work of a court on the basis of electronic document turnover. The possibilities for electronic court proceedings are being formed. However, according to V.D. Zorkin, the main landmark in this sphere should be the constitutional principles and norms (Zorkin 2018). The processes of digitization should be regulated by the Constitution of the Russian Federation, which has the supreme legal power in the Russian legal system.

A special role in the Russian society belongs to the processes of digitization that are connected to legal regulation of the economic sphere. Analysis of the processes of digitization of the Russian economy allows determining the problems that require special attention. For achieving the positive results, it is necessary to have manual management of the society's socio-economic sphere that would conform to the requirements of digitization as a global trend of effective development of economy. Legal practice requires the synthesis of legal and digital technologies of a large specter of public relations that fall under the sphere of legal regulation, which are determined by the objective conditions of social life (Lazarev 2010). At present, the sphere of legal regulation includes public relations that did not require legal regulation in the past. This sphere includes the relations which participants are anonymous digital subjects that act in the virtual space. These relations require transformation of a lot of legal procedures in the existing law, which are connected primarily to identification of person as a subject of law, realization of human rights in the digital space, financial digital technologies, usage of data bases, definition of the notion of virtual thing, and determination of its creation and other realia of the digital economy (Khlebnikov 2017).

A significant role in digitization of legal reality belongs to the problem of entering the virtual legal space of the processes of implementing multiple functions of the state, activities of municipal bodies, electronic participation of citizens in law making, and conduct of expertise of the projects of the normative and legal acts. The need for digitization of state management with the help of formation of the algorithms of state decisions at the federal and regional levels has grown. The legal space in the conditions of digitization should be evaluated as the basis for emergence of the modern form of interrelations of public authorities and civil society on the moral foundations and preservation of the cultural and historical identity of the peoples of Russia. It is necessary to modernize the system of state control in the sphere of security of personality and society, in fighting corruption, and control over the processes that take place in the cyber space on the basis of usage of all modern information technologies.

Digitization creates a need for increase of the norms in the sphere of civil, administrative, labor, criminal, and other spheres of the Russian law and transformation of the normative complexes in the sphere of information, medicine, and education. In modernization of law, adoption of certain normative prescriptions that determine the possibility of implementation of the most significant interests and needs of the society, their implementation into the existing normative complexes, and re-orientation of the system of law to implementation of the social programs acquires a special importance. Creation of new legal norms requires reconsideration of the processes of their implementation in the law enforcement practice, within which special role belongs to legal qualification, treatment of the norms of law, solving legal collisions, and overcoming legal gaps. Digitization changes the content of law – as new public relation that emerge between the digital virtual entities are formed. These processes stimulate the development of international and legal relations, unification and standardization of law, and formation of intra-state legal standards that conform to the specific features of the national legal system.

Digitization covers the processes of emergence, change, and termination of legal relations that appear in the modern virtual space and implementation of the subjective rights and legal responsibilities of the subjects of legal relations. The processes of digitization have large influence on understanding of the systemic organization of law. Digitization has large influence on the form of law. Digital computer and program formalization of law could stimulate the ordering of the normative and legal acts in the electronic virtual world. The mathematical logic forms the opportunity to determine the contradictions in law and incompleteness of regulation of specific relations (Mikov 2014). In the process of active usage of new technologies of processing of a data base in practical jurisprudence the perspectives of further processes of digital formalization of law appear. Formalization allows presenting the contents with the help of the system of artificial symbols and signs. In the sphere of legal reality formalization is peculiar for establishment of a certain content of a normative and legal act in the verbal and documentary form. According to V.D. Zorkin, one should not neglect the idea of presenting laws as a program code for the purpose of provision of certainty and unambiguity of the contents of the normative and legal acts (Zorkin 2018). Semantically and logically, the contents of the text of a normative and legal act, which expresses a certain idea and thought, is presentation of symbols that could be perceived by a specific subject. The most important sign of law is its formalized character, as the norm of law establishes a certain rule of behavior that is to be applied in typical situations a lot of times and provides it with a quality of stability and sustainability (Maltsev 2011). However, excessively strict formalization of law and its interpretation could lead to dogmatic understanding of law and move away from the true sense of the normative prescription.

As is stated in the scientific literature, “at present, digital law-making is developing on the basis of two approaches: ‘from the model’ (formation of the normative legal acts on the basis of digital models) and ‘from paper’ (transformation of the classical paper law into the rules for usage by a human and machine)” (Averyanov et al. 2018). The formal and legal sources of law are stated in the corresponding electronic versions (“Garant”, “Consultant Plus”, etc.), which are popular among the users, though they are not official sources of publication of legal norms and differ from their official text. The modern “actual version” of the normative and legal acts could become a prototype of the future digital model, which could be created not by the principle of a data base but on the basis of the “knowledge base”.

4 Conclusions

The performed analysis showed that digitization of the current legal reality stimulates the formation of new civilizational phenomena and emergence of new legal consciousness, ideology, and legal culture. New digital reality sets requirements to legal science and practice, connected to improvement of the mechanisms of legal regulation of various spheres of social life on the basis of the modern digital technologies. Further development of the communicative and information technologies ensures a new cycle of development of technological progress and transformation of society at a new phase of development, which is determined by the post-modern age. A large role in the modern conditions of digitization belongs to preservation of the moral foundations of law and its organizing and mobilizing role in the life of society, as law becomes the tool of application of digital technologies and is subject to the influence of the processes of digitization.

The digital space of the modern legal reality should not develop without a system and should not push a person and society to the state of cognitive dissonance; it should not lose the accumulated spiritual and intellectual landmarks. The modern society should use normativity for ensuring regulation of social life in the conditions of digitization on the basis of moral human values. A large role in this process could be belong to digitization and creation of artificial intelligence, if the modern information technologies will be a continuation of the fundamental science, culture, and morality, adding progressive contents to the changing processes of the modern development.

References

Vlasova, N.V., Gracheva, S.A., Meshcheryakova, M.A., et al.: Legal space and human: a monograph, edited by Y.A. Tikhomirov, E.V. Pulyaeva, N.I. Khludeneva, M. (2012)
Maltsev, G.V.: Social foundations of law. Norma - Infra-M, M. (2011)
Lazarev, V.V.: Definition of the sphere of legal regulation. Selected works: in 3 volumes. Vol. 1: Law. Legitimacy. Law enforcement, M. (2010)

Khalin, V.G., Chernova, G.V.: Digitization and its influence on the Russian economy and society: advantages, challenges, threats, and risks. Managerial consulting, No. 10 (2018)
Khlebnikov, P.: Digitization of law as a result of digitization of life. Housing law. No. 9 (2017)
Mikov, A.I.: Presenting ontologies of normative documents with the usage of applied logics. Bulletin of YFU. Series "Technical sciences", No. 6, pp. 60–67 (2014)
Averyanov, M.A., Baranova, O.V., Kochetova, E.Y., Sivakov, R.L.: Digital transformation of the processes of normative regulation: tendencies, approaches, and solutions. Int. J. Open Inf. Technol. **6**(11), 42–49 (2018)
Khabrieva, T.Y., Chernogor, N.N.: Law in the conditions of the digital reality. Journal of Russian Law, No. 1 (2018)
Zorkin, V.D.: Law in the digital world. Reflections at the St. Petersburg International Legal Forum. Russian Newspaper. Capital issue. 30.05.2018. No. 7578 (115). p. 1,4 (2018)
Okinawa Charter on Global Information Society, 22 July 2000. http://www.kremlin.ru/. Accessed 10 Apr 2018
Resolution of the UN General Assembly dated 18 December 2013, No. 68/167 "Right to Privacy in the Digital Age". http://www.un.org/. Accessed 10 Apr 2018
Decree of the Government of the Russian Federation No. 1632-r regarding the adoption of the program "Digital economy of the Russian Federation" (2017). http://www.government.ru/. Accessed 10 Apr 2018

Public Control of the Criminal Sanctions in Information Society International Experience

Gennady Lesnikov[1(✉)], Sergey Ulezko[2], and Alexandra Klochkova[2]

[1] Research Center No. 3 of Research Institute of the Federal Penitentiary Service of Russia, Moscow, Russia
lesnikov07@rambler.ru

[2] Rostov State University of Economics, Rostov-on-Don, Russian Federation
litas2011@yandex.ru, klochkova_alexa86@mail.ru

Abstract. The important part of criminal sanction, which is public control, will be analyzed in this article. The specifics of punishment execution in the form of imprisonment's implementation is connected with providing isolation regimen for convicts alongside with safety of both, prisoners and employees such as custodial supervisor, for example. It is noted that existence of the high latency penitentiary crime possessing and growth of repeated offences lead to the general toughening of confinement conditions. According to the authors, such situation, taking into account privacy of penitentiary facilities, demands effective public control, which in the different countries exercised differently, having regard to national specific.

Keywords: Public control · Correctional system · Recidivism · Criminal sanction · Criminal penalty · Isolation · Prisons · Visitors · Convicts

1 Introduction

In the modern information society developing with rather high extent of communication, ensuring necessary isolation of convicts becomes very problematic. It is caused, first of all, by the fact that the social processes happening at liberty are high-speeded. This fact radically influences the intensity of criminal repression and sentenced imprisonment terms. Long terms of imprisonment become ineffective, incapable to have a positive impact on the convict. The possibility of resocialization of such persons and their subsequent social adaptation in society becomes a serious problem. The purpose of the public organizations, exercising social control of jails, is finding of a certain compromise between objective need of convicts' isolation for suppression of their criminal activity and the corresponding control of the observance of the convicts' rights, providing a possibility of their return to normal life. In other words, public organizations control that the convict was not punished more, than it is defined by court.

In this regard, the purpose of existing research is the comparative analysis between legal support and experience of various organizations, which are engaged in public

E. G. Popkova and B. S. Sergi (Eds.): ISC 2019, LNNS 87, pp. 994–1001, 2020.
https://doi.org/10.1007/978-3-030-29586-8_113

control over penitentiary facilities in the countries of North America, Europe, and the former Soviet Union. Another purpose is to study a specific approach of each country to the solution of corresponding public control issue.

2 Methodology

There are two scientific methods were used in the course of research activity: general (such as analysis and synthesis and method of induction and deduction) and specific (such as historical and legal, comparative and legal, legalistic, formal and logical).

3 Results

Public control of detained persons' rights in European countries is institutionalized and has a long history. Countries, developing in various ways taking into account culture, national peculiarities and interaction of public with law enforcement authorities, accumulated enormous experience in the explored sphere. Some successful solutions of foreign colleagues in the field of public involvement for the purpose to decrease in level of recurrent crime can be introduced in the Russian penal system, which is actively developing public control over human rights in places of detention.

Recidivism is an important efficiency indicator of penal correction system's activity. It is also a source of increased danger as it shows persistent unwillingness of some persons to behave according to the socially accepted norms and their preference of a criminal lifestyle, despite the taken legal measures. Recidivism reveals imperfection of the law-enforcement system as it demonstrates incapability to influence convicts effectively: thus, punishment does not achieve the objectives of correction and re-education of convicts. Recidivists bring the antisocial views and standards of behavior to society, propagandize criminal subculture and popularize "criminal lifestyle".

In Russia from 2012 to 2016, recidivism had grown 16%, in France for the same period it grown 12.67%, on a same time Swedish crime prevention council reports reduction by 0.8%

The Standard Minimum Rules for the Treatment of Prisoners, originally adopted by the First UN Congress on the Prevention of Crime and the Treatment of Offenders in 1955, constitutes: "The purpose and justification of a sentence of imprisonment or a similar measure deprivative of liberty is ultimately to protect society against crime. This end can only be achieved if the period of imprisonment is used to ensure, so far as possible, that upon his return to society the offender is not only willing but able to lead a law-abiding and self-supporting life" [1].

Social control of penitentiary institutions' activity is the main mechanism to achieve this purpose as it allows comparing actual results with the original goals.

Today public control of penitentiary system represents interaction of the associated subjects of control using various social resources (information, labor and so forth) for resocialization of convicts. Public control of health care, education and other social spheres, both in Russia, and in the USA and EU countries is implemented by increase in public authorities' information openness, however access to penitentiary system can be open not for all subjects of control. Public control, on the one hand, has to be exercised, but, on the other hand, it is necessary to maintain the requirement of social isolation and a protection from criminals.

In relation to public control of penitentiary system of foreign practice, there are more often used such concepts as ***civil engagement***, ***public participation***, ***public supervision***, ***public advocacy***, ***formal and informal monitoring***. In the United Nations Office on Drugs and Crime's (UNODC) recommendations on the organization of joint work between governmental and non-governmental bodies in terms of law and order's issues 2014 «Civil society and prison: the invisible bars challenge» it is specified that subjects of control of penitentiary institutions have to:

1. to be independent;
2. to have accurate authority in the legal sphere;
3. to be provided with resources;
4. to work long enough in this sphere and to guarantee stability of their work. The non-state organizations are attracted as addition of official control bodies' work by the mean of holding an interview with prisoners, drawing up complaints answers and assistance in correction of violations.

Now provisions of national preventive mechanisms in the context of the Optional Protocol to the Convention against Torture (OPCAT) 1984 regulate activity of public structures in penal system.

The longest and diverse experience of public control mechanism implementation over penal institutions belongs to Great Britain. The initiative proceeding from the English community leader advancing the ideas of philanthropy, preceded creation of "Prisons' custody societies" – the ancestor of public control in Russian penitentiary sphere.

The history of public participation in work with convicts had begun in England and Wales in 1802. At first, there were patronage societies, created on a basis of private organizations with state support. Each prison had its own society of patronage. In 1860, the Committee of magistrate judges for prisons' supervision was created; it represented interaction between volunteer citizens and special judges (magistrands) who were supervising the activity of jailers. Citizens on a free of charge basis were observing and inspecting the work of penitentiary system. Another responsibility of public was to receive and consider complaints from prisoners and transfer them to the group of judges. Judges upon the results of their consideration, visited prisons and, if necessary, issued warnings to the jailers. Lately in 1898, The House of Commons accepted "Prison Act" which defined the basic principles and mechanisms of the visitors' system – constant social control on prisoner welfare.

Since 2003 observance control of the prisoners' rights is exercised by "Independent Monitoring Boards (England and Wales)" – Councils of independent monitoring of prisons. Councils supervise all prisons of Great Britain. Part 5 of Prison rules 1999 regulates work of Councils. Prison rules defines its main objective as observation of rooms' condition in correctional facilities, organization of management and treatment of prisoners. More than 1600 people from the Public chairpersons, working on a voluntary basis, are a part of Council and at least two of them have to be magistrate judges.

The purpose of visitors' work is to control respect for the principles of honesty and humanity concerning prisoners, discipline and interests of prisons' workers and to be convinced that the convict is not punished more, than it was provided by court. The rights of visitors are regulated by the Prison Act 1972. A subject of visitors' inspection is compliance with the requirements established by Prison Act 1972 to food, work, education, medical support and living conditions of prisoners. Visitors do not represent either prison administration or the interests of prisoners, and act for the benefit of all society. Any candidate for Council of visitors should submit an application, visit appointed prison before being appointed to the member of Council by Secretary of State for Home Affairs on a competitive selective basis. The visitor has to receive the recommendation from other members of Council; however, the Secretary has the right not to approve a visitor, or to approve the one who has no recommendation.

Public visitors are appointed by the order of the Secretary of State for less than three years. Provisions of Rules also establish that the visitor can be exempted from fulfillment of duties in certain cases, such as: unsatisfactory execution of the duties; lack of special preparation; the physical or intellectual disease interfering fulfillment of duties; commission of the crime or other offense not compatible to the status of the member of Council of visitors; conflicts or possible conflicts between Council and member concerning execution of duties or any interest (personal, financial or another).

Visitors undergo special training. They are being explained how to visit prison, what to pay to attention to, what rooms in prison should be visited constantly, how to improve the relations with employees, how to keep records on the audio-and video equipment, which then can serve as proof of violation. In each Council, a training coordinator has to estimate training attainment level of visitors. Every 2–3 year each visitor goes to courses for skilled visitors. There is also a special training for chairpersons of councils, where participant being taught how to interact with chiefs of prisons. Visitors have to be able to communicate and listen appropriately.

Convicts have a right to apply for meeting with a visitor in advance (such application comes to council), as well as to address the visitor directly, when he visits prison. An application should be filled with the prisoner's name, date, place of stay, subject he wants to address. At a meeting with the prisoner, it becomes clear what he had already done to solve his problem. If internal ways of solution of the problem are exhausted, or the convict tried to do something, but could not, then Council of visitors interferes as prison has to solve problems of prisoners. A task of visitors is identification of cases when the administration is not capable to resolve issues independently. Personal address and application are not the only ways to address the Council of visitors. Each prison premises has several delivery boxes for:

- interracial relations (the claims are sorted by special committee under the Chief of prison's chairmanship, visitors also participate in this work),
- appeal to administration of prison,
- a green box for the appeal to Council of visitors.

Standard forms to each of these instances should also be provided to the prisoners. The pink form is confidential.

Annually Council of visitors' reports to the State secretary on technical condition of prison, carried-out work with prisoners, makes the suggestions for improvement of activity of prison. The majority of committee reports addressed to the State secretary in which the condition of separate prisons is analyzed is available to a wide range of the public on the website of a Prison service. Reports could also be found on the public organizations' dealing with problems of the English prison policy or issues of its studying websites. Thus, results of public control become available to all interested sectors of society.

Reports of visitors addressed to the State secretary include such sections as: introduction (general information on establishment); shortcomings of the organization of keeping of prisoners; organization of appointments; problems of the organization of training and labor process; safety issues; medical care; disciplinary practice; questions of training of personnel; miscellaneous. Actually all spheres of activity of correctional facilities are controlled by visitors.

The shortcomings in the listed above directions are described in the visitors' reports, recommendations about their elimination are also provided. For example, a serious shortcoming is the violation of the organization of appointments with prisoners due to various technical reasons, which is direct violation of the prisoners' rights. The committee of visitors has the right to indicate the need to increase the number of health workers or to demand to bring into accord with national standards the equipment of the surgical room. Public visitors have the right to make remarks on the organization of delivery and sending mail of prisoners, to make recommendations for shower installation and arrangements of the medical room or expansion of kitchen space.

It is interesting that public visitors of Great Britain are obliged to investigate and make reports on any human rights violations, which became known to them. They have the right to open a question to the chief of prison or even to address it directly the Secretary. The provision of the law on the state secret under which disclosure of official information without lawful powers and special permission is considered as offense extends to the visitors.

Members of Council of visitors work on a voluntary basis, but the state can compensate such expenses as travelling cost, baby sitter and wage loss.

The important place of work with offenders is taken by the system of the guarantee. It consists of the obligation to "keep the peace" or to be a "good behavior", established at the discretion of the court and cannot exceed one year. Violation of the guarantee allows application of punishment in the form of imprisonment for a period of up to six months. The similar institute works also in the USA.

There is a provision of "assistance to efforts of the convict on resocialization" measures in the Criminal Code of France (Article 132-146). It can be a duty to continue education or professional activity, to live in a certain place, to receive medical treatment, to indemnify, in whole or in part, the loss, even in the absence of the civil suit, not to visit the places of alcoholic drinks sale, not to have weapon, not to visit other convicts, in particular collaborators or accomplices, etc. In France, as well as in Great Britain, imprisonment is an exceptional measure of punishment. In correctional institutions the set of human rights organizations, for example, National Association of visitors of prisons (further – NAPT), carry out their activity.

Powers of members of NAPT are registered and enshrined legislatively in the Code of Criminal Procedure of France: members of NAPT are called to help social employees and the director of penal institution, to give prisoners moral support, to promote their correction, to carry out financial help to the prisoners, released prisoners and their families.

To become the visitor of prisons, it is necessary to get permission of the Ministry of Internal Affairs. Each of members of NAPT is assigned to a certain correctional facility. Before starting their activity, each visitor has to study carefully the Safety rules of establishment to which they are attached. Members of NAPT have exclusive credibility from the chief of institution and social service. For this reason average age of the people, working in NAPT, is 55 years.

Control is exercised by the mean of personal visit of NAPT members in places of detention and through conversations with prisoners. Unlike prison visitors of Great Britain, visitors of prisons in France cannot remain in correctional institution "round the clock". It is recommended to the NAPT members to visit prisons regularly (two-three times a month).

The main feature of public control over observance of prisoners' rights in France is an opportunity for convicts "to replace" their relatives and close by the member to NAPT during imprisonment. It is especially important when the prisoners have no close relatives who could take care of them and protect of their rights and freedoms. In this case such right as permission to bring linen, magazines, books of general education contents for the prisoner, to transfer money into his personal account, to communicate with judges and lawyers have to be authorized to the prison visitor, who has taken a role of "relative". "To replace" the prisoner's relative, the visitor has to get an appointment permission from Social service. At the same time both, the social worker, and the judge or the lawyer can recommend the convict for whom the help from the member of NAPT is necessary. The moral help to prisoners belongs to duties of the prison visitor; they have the right to contact to members of families, both, persons under investigation, and convicts, rendering the necessary help. For strengthening of social connections of prisoners, members of NAPT have to promote in all ways rapprochement of the prisoner with his family. The French visitors of prisons can carry out free of charge function of tutors, for example, when training prisoners for examination. It is also promoted by the French criminal procedure legislation, which allows the visitor of prisons to take out and bring the checked tasks in correctional facility.

In Germany the sum equal to the work remuneration is paid to the studying convicts, it is a state payment called Ausbildungsbeihilfe. Church renders the most significant social help to the former convicts here.

In the Scandinavian countries the network of the parastatal convicts help organizations is developed. In Finland in 1975 the Association for test and the subsequent leaving for the help in the solution of social problems of conditionally released and former prisoners was founded. Activity of the organization is under control of the Ministry of Justice. In Sweden, the "Team of Social Intervention" project is implemented since 2011. It consists of the municipalities' representatives, probations, police, public service of employment, the Agency of social insurance and others. According to results of monitoring of SGI 2014 (Sustainable Governance Indicators), these countries take the leading positions on authorities' control.

Control to observance of the rights of prisoners from the public in places of detention in the European countries is legislatively assigned and has long history of the existence. Today not all developed countries legislatively allocate institute of public control or resort to control from the public in the penitentiary sphere. Therefore, in the USA the isolated institute of public control over observance of the rights of prisoners does not exist. However all programs of rehabilitation of convicts in places of detention rely on continuous participation of public in the course of correction and restoration of prisoners' social communications. Representatives of religious faiths take the most active part in resocialization of prisoners.

For USA the participation of public in behavioral correction of convicts in places of detention, rather than public control on observance of the rights it is typical. As can be seen in this article, countries develop in various ways taking into account culture, national peculiarities and interaction between public and law enforcement agencies. However, the reduction of recurrent crime issue is actual everywhere and Russia not an exception. Therefore, perhaps, it would be effective to introduce some successful developments of foreign colleagues in the field of involvement of public to decrease in level of recurrent crime in the Russian penal system, which is actively developing the system of public control over respect for human rights in places of detention.

References

1. Standard Minimum Rules for the Treatment of Prisoners Adopted by the First United Nations Congress on the Prevention of Crime and the Treatment of Offenders, held at Geneva in 1955, and approved by the Economic and Social Council by its resolutions 663 C (XXIV) of 31 July 1957 and 2076 (LXII) of 13 May 1977. https://www.un.org/ru/documents/decl_conv/conventions/prison.shtml
2. 2013: France, social portrait. http://lexpansion.lexpress.fr/actualite-economique/portrait-de-la-societe-francaise-en-10-chiffres-insolites_1354568.html#xtor = AL-189
3. Danish Penal Code. https://www.forum.yurclub.ru
4. Belgium Penal Code. https://www.ugolovnykodeks.ru/ugolovny-kodeks-belgii
5. French Penal Code. https://www.constitutions.ru
6. Working Paper Series on Prison Reform. https://www.unodc.org/ropan/en/working-paper-series/working-paper-serires.html
7. Ponamarev, S.N., Marukov, A.F., Geranin, V.V.: Prison system of England and modern society: monograph. Ryazan: the academy is right also managements of the Ministry of Justice of the Russian Federation, p. 112 (2002)
8. Tikhomirov, E.V.: Foreign experience of participation of the public in activity of institutions for keeping of minor convicts, pp. 8–10
9. Ponomarev, S.N., Marukov, A.F., Geraninv, B.: Prison system of England and modern society, p. 96
10. National Association of Visitors of Prisons. http://www.prison-visit.org/
11. Pavlov, A.A.: Participation of the public in activity of penal correction system at the present stage: dis…. cand. Vologda, p. 71 (2009)
12. Knaus, A.V.: Public monitoring as form of cooperation of penal correction system and civil society, the Role of penal correction system in prevention of offenses: materials int. sci. conf. Kostanay legal institute of Committee of penal correction system of the Ministry of Justice of Republic of Kazakhstan, Kostanay (2007)

13. Reference Book by the Member of Council Viziterov of Great Britain. https://visit.org/wp-content/docs/spravochnik_visitera.html
14. Korovin, A.S.: Control of public associations of activity of the institutions executing punishments in some foreign states. History of development and the current state of penitentiary science, medicine and practice of execution of punishments: materials of int. conf. Under the editorship of the Dr. in law V.I. Seliverstov M.: Scientific Research Institute FSIN of Russia, Part 1, pp. 280–284 (2009)
15. Sergeyev, V.: Participation of public organizations in reforming of the penitentiary sphere. Crime and punishment, No. 9, pp. 60–61 (2004)
16. Kovalyov, O.G., Sheremetyevo, M.V.: Penal system of the USA: features of the organization and current trends. Penal correction system: right, economy, management, No. 4, pp. 19–22 (2013)
17. Bakhanova, E.V.: Social potential and prospects of public control of a system of execution of punishments: Bulletin of the Nizhny Novgorod University of N.I. Lobachevsky. Series: Social sciences, No. 2(38), pp. 127–132 (2015)

Law Enforcement Problems at Appointment of Administrative Punishment

Veronika V. Kolesnik[1(✉)], Irina V. Kolesnik[1], Natalia V. Fedorenko[2], and Julia V. Fedorenko[2]

[1] Rostov Branch of the Russian State University of Justice, Rostov-on-Don, Russia
nikkipohta@mail.ru
[2] Rostov State University of Economics, Rostov-on-Don, Russia
gr_process38@mail.ru

Abstract. The present article is designed to draw the attention of readers to the existing problems in law-enforcement activity at purpose of administrative punishments. In article the problems of the general and private character which don't have the known decisions so far are analyzed. The attention is focused on an imperfect legislative regulation which, according to authors, is the cornerstone of the most part of problems in law enforcement.

As object of a research purpose of administrative punishments as law-enforcement activity and problems of her implementation acts.

Authors consider problems of interpretation of the concept "administrative punishment", questions of creation of optimum system of the punishments prescribed by the administrative law and also prerequisites of the corruption level of the provisions of the Code of the Russian Federation on Administrative Offences providing the choice of a look and degree of severity of administrative punishments for offenses.

As a result of a research authors formulate conclusions about existence in activity to destination of administrative punishments of essential quantity of unresolved problems, both system, and private character. Besides corruption provisions of the administrative law, according to authors, act as initial prerequisites of key problems of law enforcement.

Keywords: Administrative punishment · Law enforcement problems · System of administrative punishments · Corruptogivity · Dispozitivity

1 Introduction

Activity to destination of administrative punishments, as well as any other law-enforcement activity may contain certain flaws. On the one hand, shortcomings of law enforcement can be caused by an imperfect legislative regulation, with another – the most deformed practice of law enforcement. However much more often problems have complex character and are determined by more significant amount of factors.

Social, economic, personal, political, ideological and many other reasons and conditions can be their cornerstone. And the list of similar problems is quite extensive and doesn't move to calculation as one solved problem leads to a chain of other not

E. G. Popkova and B. S. Sergi (Eds.): ISC 2019, LNNS 87, pp. 1002–1010, 2020.
https://doi.org/10.1007/978-3-030-29586-8_114

resolved questions. For example, with adoption of the resolution of the Plenum of the Supreme Arbitration Court of the Russian Federation of 11.07.14 No. 47[1] many problems connected with administrative prosecution for violations of the "alcoholic" legislation[2] have received the decision, however it, in turn, hasn't led to full elimination of contradictions. And similar examples it is possible to bring in a significant amount.

2 Review of Literature

And it must be kept in mind that problems of law enforcement can't exist in itself in a separation from the theory, from scientific developments and legislative initiatives of various authors. Where theory, there and practice, and vice versa. In essence, the theory is designed to provide practice with new original decisions, and practice the theory – the problems demanding answers. Proceeding from it, within consideration of problems of law enforcement at purpose of administrative punishments also the theoretical aspect is anyway affected.

Such authors as A.B. Agapov, T.Yu. Kourova investigating problems of administrative responsibility, L.L. Popov, N.Yu. Hamaneva raising the general questions of administrative punishments have considerably promoted in a research of problems in this sphere. Besides L.B. Antonova, A.S. Dugenets, A.N. Derygi have essentially affected the level of study of a problem of purpose of administrative punishment of work. Sackcloths and many others.

3 Research Methods

General scientific and chastnonauchny and also special methods of scientific knowledge have acted as the main methods of a research of problems of law enforcement of purpose of administrative punishments. In particular it is possible to allocate such methods as the analysis, synthesis, induction, deduction and also a comparative and legal, historical, legallistic method, a method of the system analysis, etc.

4 Result

Thereby, in our opinion, the system of problems in the explored sphere can be divided into several groups: first, it is the law enforcement problems proceeding directly from the text of the legislation of both administrative, and other look that is defects in the law; secondly, it is the law enforcement problems caused by the wrong practice of application (interpretation) of the law; thirdly, the problems based on a contradiction of the law and practice of its application; fourthly, theoretical problems of purpose of administrative punishments, influencing lawmaking and law enforcement in this sphere, etc.

---

[1] [2].

[2] See about it: [16, Page 117–148; 17, Page 25].

As it is possible to notice, each group of the called problems anyway is connected with the legislation. Laws define how development of practical activities, and can serve as an obstacle for her successful implementation. All this recognizes from the fact that law enforcement and is application of provisions of the law, it can't exist in principle without rule of law. Thereby, the key place in the system of problems of law enforcement at purpose of administrative punishments it is possible to recognize imperfection of the administrative legislation. The last, in turn, generates also contradictions at realization of appropriate authority by public authorities and officials, and discrepancies meanwhile "as it is necessary" (instructions according to a law letter) and that "as actually" (established practices of purpose of administrative punishments), and many other difficulties.

In our opinion, in this case the thesis is fair: "you want to find problems in the legislation address law-enforcement activity and if they need to find the solution, then change the law". At the first approach, this formulation will seem extremely obvious and simple. However, here it is necessary to consider that, in fact, we put the sign of identity between the law and a problem of law enforcement. By and large, it is valid just in that part where it is told about gaps, collisions, deficiency of the law. So far as concerns provisions of the legislation without defects, perhaps, to argue thus it is groundless.

The legislation regulating purpose of administrative punishments, mainly, is submitted by the Russian Federation Code of Administrative Offences (further – the Code of the Russian Federation on Administrative Offences). This law, as we know, represents the two-uniform system of substantive and procedural administrative law as contains also the system of offenses for which there comes administrative responsibility, and procedural provisions defining an order and a form of response to offenses and attraction to the corresponding responsibility. The Code of the Russian Federation on Administrative Offences is also that regulatory legal act which defines not only an order of purpose of administrative punishments, but their types and the purposes and also companion problems. And the last can carry both the general (system), and private character.

Speaking about problems of the general character, it is necessary to pay attention to that part of shortcomings of the Code of the Russian Federation on Administrative Offences which belongs to all types of administrative punishments and to all order of their appointment without exception. The first that gets to the field of research sight, is a legislative formulation of the concept "administrative punishment" which according to Art. 3.1 of the Code of the Russian Federation on Administrative Offences is defined as the measure of responsibility for commission of administrative offense established by the state and is applied for prevention of commission of new offenses by both the offender, and other persons. Some disagreement causes use in definition of the purpose of administrative punishment in the form of prevention of commission of new offenses. In our opinion, the logic of creation of definition isn't absolutely right. The legislator shouldn't have used in definition a mention of the purpose of administrative punishment as, first, similar inclusion breaks sense of a concept into two untied parts (at first it is about a responsibility measure, the word form about the application purpose without the corresponding sheaf is used later), secondly, the instruction as the purpose – prevention of offenses which isn't the only purpose of administrative punishments.

Though according to some authors the purpose of punishment consists there is nothing other, as in prevention of the new acts doing harm to fellow citizens and in deduction of others from similar actions[3].

5 Discussion

The most part of authors adhere to the definition provided by the administrative law. In particular, L.L. Popov[4], N.Yu. Hamaneva[5], A.B. Agapov[6] and others. There are also positions partially other than legislative. So, N.M. Konin noted that it is necessary to understand as administrative punishment "… the administrative measures of responsibility established by the state applied to the guilty legal entities and individuals who have committed these offenses for prevention of commission of new offenses"[7]. As it is possible to notice, this author also left the only purpose – prevention of offenses what we don't agree with. Such representatives of administrative law as M.B. Smolensky and E.V. Drigola speak about a measure of the responsibility applied in the order established by the law to the person who has committed administrative offense[8]. According to us, most we accept such option, considering impossibility of transfer of all is more whole than administrative punishment.

Other problem of the general sense affecting practice of law enforcement at purpose of administrative punishments is the system of administrative punishments, the logician of her construction and optimization. Spoke about need of revision of all set of administrative punishments and problems of its optimization and earlier. For example, A.S. Dugenets specified that "the major problem of the system of administrative punishments existing now in the Russian Federation is the solution of questions of its optimization"[9]. Perhaps, it, really, so. Permission of a question of formation of uniform, complex system of administrative punishments carries the defining value for their application as a set of punishments and a possibility of their choice builds the correct law enforcement. Not for nothing the system of administrative punishments is recognized as internally organized unity consisting of hierarchically ordered set of rather independent types of administrative punishments which set expresses functional mission of administrative punishment in the social environment[10].

Smooth functioning of this system will be possible thanks to a right choice of that or other punishment depending on a situation. And here the hierarchy of punishments for which increase in functionality I died responsibility plays the role it is necessary to give the harmonious and uniform form, to create the accurate system of administrative

[3] [9], Page 105–106.
[4] [4], Page 417.
[5] [5], Page 319.
[6] [3], Page 48.
[7] [12], Page 24.
[8] [15], Page 397.
[9] [11], Page 8–13.
[10] [13], Page 18.

punishments, the general rules of her understanding and application of the types of punishments making her[11].

According to A.S. Mikhlin, the value of system of punishments is very important for ensuring justice of punishment as depending on weight of act punishment will be chosen. In this respect we have a bit different position[12]. Justification of our point of view is built in other plane – in category of problems of private character, that is their distribution is limited to any punishment or a certain administrative offense, etc. For example, problems of purpose of administrative punishment concerning minors[13], problems of purpose of administrative punishment in the form of deprivation of the special right[14], etc.

We will adduce arguments in favor of the fact that not always a possibility of the choice of punishments it is the positive phenomenon. For an example we will take p. 1 Art. 7.27 of the Code of the Russian Federation on Administrative Offences – petty theft in which as the sanction the possibility of the choice or an administrative penalty to the fivefold cost of the stolen property, or administrative detention for a period of up to 15 days, or obligatory works for a period of up to 50 h is specified. Similar running start for decision-making is a source of the raised discretion which allows to manipulate several decisions depending on the subject of law enforcement. First, the law enforcement official, in this case the judge, can appoint even in identical cases for choice one of three types of administrative punishments in spite of the fact that degree of severity of each of them different, secondly, the law enforcement official can appoint concerning each of these types of punishments even in similar affairs various sum (the penalty can vary from 1 thousand rubles to the fivefold cost of the kidnapped person of property), various term (in arrest it can be from one days to 15 days), various duration of punishment (at making decision on appointment as punishment of obligatory works their duration can begin of one hour to fifty).

Similar design very a riskogenic as the choice of the decision is left on a payoff to the law enforcement official. Such provisions are called dispositive, allowing wide limits for a discretion. And the dispozitivity, as we know, can act as the basis for the corruption of risks, that is risks of use of powers, in our example the choice of the decision, for corruption and dangerous acts[15]. And such corruptogenic provisions connected with purpose of administrative punishments in the Code of the Russian Federation on Administrative Offences enough. According to L.B. Antonova "… in the legislation of our country on administrative responsibility establishment of different types of punishments and also the minimum and maximum limits in which these sentences can be imposed is widely used. Theoretically such model is justified, however in real life it attracts a set of negative consequences, beginning from dissatisfaction of a law-abiding part of society and finishing with existence of the environment for

[11] See Foootnote 9.

[12] [14], Page 103.

[13] [10], Page 109–112.

[14] [7], Page 143–145.

[15] See about it, for example: [8], Page 54–89.

corruption manifestations"[16]. Similar dispositive provisions to contain, for example, in Art. 12.7, 14.16, 18.9 of the Code of the Russian Federation on Administrative Offences and in many others. In fact, practically in each article Code of the Russian Federation on Administrative Offences providing administrative responsibility for these or those offenses, sanctions are constructed on such type. In our opinion, similar designs can serve as the base for corruption act. It is simple to be convinced of it having addressed practice of purpose of administrative punishments when for similar acts different sentences are imposed. Certainly, the choice is influenced by a number of other factors (the identity of the violator, frequency, etc.). But if under the same conditions one administrative penalty in the form of the single sum of the stolen property is chosen, and another – administrative detention for a period of 15 days should think about not only of validity of such decision, but of interest of the law enforcement official in this case.

6 Decision

Possibly to reduce quantity of mistakes and level of risks at application of administrative punishments becomes a possible way of use of information technologies which allow to minimize subjectivity and a human factor at decision-making. Similar information technologies densely sat down in a row spheres of activity of the person. The sphere of law did not become in this plan an exception. On the contrary, innovative technologies in a profession of the lawyer allow to facilitate his work, to exempt from routine processes and to give more opportunities to use creative approach in the activity. Requirements of time are that – successful are those who actively master modern IT technologies. The key algorithm which is used at process automation of legal activity consists that a certain hi-tech system, getting into a jungle of branch problems, allows to increase quality of law enforcement, to optimize the legislation, to lower load of subjects of law enforcement, etc. We believe that administrative process, in particular in relation to purpose of administrative punishments, thanks to modern technologies can be quite automated. Here it is not only about such procedures as legal monitoring and creation of electronic codes, but also generation of standard judgments by means of artificial intelligence and creation of the automated control system of judicial practice. The systems of artificial intelligence have extremely wide potential. First, it is connected with the fact that the person still did not learn to use all opportunities not only artificial intelligence and neuronets, but also in principle information technologies and the opportunities. Secondly, traditional practice of law enforcement is exposed against digitalization and technologization of legal process, not to mention the relation to appointment procedure of punishments which frightens off all modern and new. However not everything is lost, a certain progress is planned. In particular, it can be found also in increase in number of information bases, electronic card files, the automated search engines, etc. Despite all this, decisions are still made only by the law

[16] [6], Page 75–76.

enforcement official. From here and defects peculiar to the person. Abuses, excesses by powers, corruption behavior. In our opinion, it is required to develop the systems constructed on algorithms of artificial intelligence which are capable to distinguish signs of administrative offense from introductory information and to build an order of further actions and decisions of the law enforcement official. Now it sounds extremely ephemeral, but certain motions in this direction are already available. Let's give an example.

7 Example

Now in Russia there is a practice of use of technical means at identification of administrative offenses. More it concerns fixing of administrative offenses in the field of traffic. Such trend was widely adopted: not only speeding, but also journey on the forbidding traffic light signal, not granting privilege to the pedestrian, arrival to the stop line, not fastened seat belts, etc. is fixed. But all this turned out to be consequence not of comprehensive implementation of information technologies in activity of law-enforcement bodies, and the shortages of shots. Less frequently now on drags it is possible to see the "living" traffic police officer, in increasing frequency they are replaced by stationary and mobile devices of fixing of administrative offenses. At the same time despite active use of these technical means, a special order of administrative prosecution for traffic offenses at their fixing by the special technical means working in the automatic mode (the protocol on administrative offense is not formed, and the decree on the case of administrative offense is issued without participation of the person against which proceedings on administrative offense (Part 3 of Article 28.6 of the Code of the Russian Federation on Administrative Offences) are initiated, becomes only a fiction. The employee should look through each recorded fact and already on its own behalf to issue the relevant decree. Therefore, the high technological effectiveness comes to an end on a penultimate phase of purpose of administrative punishment. A certain incompleteness of process is observed. It is represented more reasonable, owing to use of technical means of fixing of administrative offenses and appointment procedure of administrative punishment to transfer to the plane of information technologies. Scheme is as follows: data on the administrative offenses fixed in the automatic mode arrive on the uniform center of decision-making which cornerstone the system of artificial intelligence is. This system, using the algorithms put in it and analyzing the arrived data, on the basis of accurate criteria defines a type of administrative offense and issues the decree in the automatic mode. We believe, it can do belongs to administrative offenses not only in the field of traffic, but also to others. For example, fixing by means of radio-controlled devices (quadcopters or other aircraft) offenses (crimes) connected with the illegal cabin of forest plantings, environmental pollution with violation of the rules of improvement of territories, etc. At the same time besides purpose of administrative punishments will probably be carried out by employees.

8 Conclusions

Obviously, it is not the only problems of the general and private character in law-enforcement activity at purpose of administrative punishments. In the existing administrative legislation a set and other unresolved questions. Within this article we have made an attempt to analyse only small part of them. However, in our opinion, the decision them is of particular importance for improvement of quality of law-enforcement activity in this sphere. In particular, the problem of existence of corruption provisions in the administrative law essentially can affect quality of law enforcement at purpose of administrative punishments. In this regard their elimination becomes extremely necessary and relevant. The special place in improvement and optimization of appointment procedure of administrative punishments is occupied by information technologies which in the next years will become irreplaceable means of implementation of legal process. The systems of artificial intelligence are capable to reduce time spent for decision-making on administrative offenses to minimize subjectivity and mistakes at purpose of administrative punishment.

References

1. Russian Federation Code of Administrative Offences of December 30, 2001 No. 195-FZ. Russian newspaper, 31 December 2001
2. The resolution of the Plenum of the Supreme Court of Arbitration of the Russian Federation of July 11, 2014 No. 47 “About some questions of practice of application by arbitration courts of the Federal law “About State Regulation of Production and Turnover of Ethyl Alcohol, Alcoholic and Alcohol-containing Products and about Restriction of Consumption (Drinking) of Alcoholic Products”“. Access from Union of Right Forces ConsultantPlus
3. Agapov, A.B.: Administrative responsibility: textbook. Eksmo, Moscow (2015)
4. Administrative law of Russia: textbook. edition L.L. Popov. Avenue, Moscow (2010)
5. Administrative law of the Russian Federation. Under the editorship of N.Yu. Hamaneva. Lawyer, Moscow (2011)
6. Antonova, L.B.: Problems of purpose of administrative punishments. Bulletin of the Voronezh Institute of the Ministry of Internal Affairs of the Russian Federation, No. 2 (2015)
7. Askerov, M.: Problems of purpose of administrative punishment in the form of deprivation of the special right. Business in the law. Economical and legal magazine, No. 3 (2012)
8. Afanasyev, A.Yu.: Corruption risks of the law of evidence in criminal trial (pre-judicial production): thesis of Candidate of Law Sciences, Nizhny Novgorod (2016)
9. Bekkaria, Ch.: About crimes and punishments, Moscow (1995)
10. Deryga, A.N.: Problems of law enforcement of the material standards of the Code of the Russian Federation on Administrative Offences directed to protection of the rights of minors. The Modern right, No. 9 (2009)
11. Dugenets, A.S.: Optimization of system of administrative punishments. Administrative law and process, No. 3 (2007)
12. Kourova, T.Yu.: Administrative responsibility as one of types of legal responsibility. Modern scientific research and innovations, No. 9–2 (41) (2014)
13. Maximov, I.V.: The system of administrative punishments by the legislation of the Russian Federation: Monograph. SGAP, Saratov (2004)

14. Mikhlin, A.S.: Problems of improvement of system of punishments in the Soviet criminal law. Current problems of criminal law: collection of scientific works. IGP Academy of Sciences of the USSR, Moscow (1988)
15. Smolensk, M.B., Drigola, E.V.: Administrative law: textbook. KNORUS, M. (2010)
16. Yachmenyov, G.G.: The comment to the resolution of the Plenum of the Supreme Arbitration Court of the Russian Federation of 11.07.14, No. 47. Arbitration disputes, No. 4 (2014)
17. Yachmenyov, G.G.: About some controversial issues of qualification of administrative offenses in the field of turnover of alcoholic products. Arbitration disputes, No. 4 (2015)

Digital Rights in Civil Legislation of Russia

Natalia V. Fedorenko[1] and Svetlana E. Hejgetova[1,2(✉)]

[1] Rostov State University of Economics, Rostov-on-Don, Russia
kse2562@mail.ru
[2] Russian Presidential Academy of National Economy and Public Administration (RANEPA), Moscow, Russia

Abstract. The changes in Russian legislation of civil law regulation are analyzed in the present article concerning the introduction of digital technologies in economic turn. The specified features of digital rights as civil law objects are explored.

Keywords: Digital technologies · Digital rights · Digital economy · Single digital environment · Cryptocurrency · Blockchain · Civil law objects · Website · Smart contract · Paycard · Deal · Electronic deal

1 Introduction

The beginning of the 21st century turned with origin of several innovative technologies which made sufficient influence at economic relations and existing business models. Among them re the technologies of Cloud Computing, Big Data, Internet of Things, Augmented Reality and etc.

The particular place among modern information achievements belongs to Blockchain technology, which lies in basis of popular cryptocurrency functioning named Bitcoin. In present Russian Federation the swift economy development of new technological generation takes place so named digital economy.

The government programme "Digital technologies" is worked out and being realized. In aimes of the digital economy development the government of Russian Federation has established in July 2017 The Programme "Digital economy of Russian Federation".

According to the Federal law N149-ФЗ "About information, information technologies and information safety" (art.2) [12], information technologies are the processes, search methods, collecting, storing, processing, introduction, spreading of information and the ways of such processes and methods functioning.

But the modern scientific potential outpaces existing traditional models of civil law regulation that, in its turn, demands deep research and thinking on forming information relations and effective mechanism of legislative regulation working out.

E. G. Popkova and B. S. Sergi (Eds.): ISC 2019, LNNS 87, pp. 1011–1016, 2020.
https://doi.org/10.1007/978-3-030-29586-8_115

2 The Contemporary Digital Technologies in Civil Law Sphere

On the 1st of October 2019 digital rights became a new object of civil rights. The Civil Codex of RF is supplemented by the Federal Law from 18.03.2019 N34-ФЗ with new article 141.1 "Digital right". Under digital rights the legislator understands demands and other rights the consistence and terms of realization of which are in accordance to the rules of information system (for instance, in blockchain).

The electronic civil law turn sphere introduces an opportunity to make contracts and deals in electronic form. The Blockchain technologies are used with the great popularity for former years. There is no legal normative definition of blockchain. In the International Bar Association report the way of Blockchain technology was named: "Blockchain is the accessible for participants cryptographically safe register, storing and tracing data and costs in chronological oder creating safe transaction records. Such records can not be changed or distorted. Each transaction is verified by cryptographically signatures of participants after achieving of decentralized agreement by them and is added to the register as a new block in record chain. The whole cryptochain is in sight for participants that makes all transactions transparent simultaneously preserving personal data closed" [4]. In scientific works Blockchain is defined as "decentralized database ("account book") of all confirmed transactions made in direction of defined active, functionally based on cryptographical algorithms [10, p. 32–60].

As advantages of blockchainsystem we can point at such innovative features like the following:

- an opportunity of fixation form of accordance to defined person true data without any necessity of disturbing the third party specialist;
- an opportunity of direct transfer of information data to other person.

However, the advantages of new technology especially concerning invariability of data in Blockchain pass into confrontation with the main principles of civil law regulation.

The attention in the doctrine is paid to the fact that making contracts through Blockchain is used only in case of full accordance to law demands and absence of misunderstanding between parties. In case of argue based on law the Court may declare the deal invalid. So data placed in the Blockchain system miss law power and must be canceled. But information system is built in such a way that without other party there is no way to use restitution, to return what is gained in the deal [5, p. 24–27]. High level of information invariability guarantees is the advantage of this information system, but at the same time it contradicts to present principles of Civil Codex of RF. The same problems also originate in case of one way denial or breaking the contract (art. 450.1 CC RF). Of special interest are virtual currencies created on Blockchain technology which are not connected with financial system of any state and which can be exchanged for "fiat" money (i.e. supplied only with state trust). In present the approach of various states differs to cryptocurrencies from rejection of any opportunity in its realization to acceptance of them as a legal mean.

Smart-contracts are very popular as a sphere of Blockchain which are a special program code depicting parties agreements and providing automatic execution of their terms. In law science there is an opinion that smart-contract introduction ("clever contracts") may significantly change the present prevailing basis of contract law. For example, participants of civil law relations have agreed to the fact that in some terms like goods receipt marked with RFID-mark to pointed place other party makes payment for it by the means of bitcoin. The following relations were established as "clever contract" ("smart-contract") based on Blockchain. After the system gets data from RFID-mark about goods receipt to pointed place it automatically debits the contractor's account by payment. No additional authorization or confirmation by receiving party is needed. All contract terms are set initially by parties and the computer itself controls their execution.

The task of law nature of such relations and opportunity to consider smart-contracts as civil law contracts is explored in scientific literature. This task is arguable as there is an opinion that the following relations do not satisfy the deal principles but introduce some defined electronic algorithm. As the property trust management can be a potential sphere of smart-contracts as well as heritage relation forms, insurance, the research of body, features and law nature of smart-contracts is of great theoretical and practical interest.

It is marked that almost any kind of contract may be built on Blockchain base and stated in computer code, not in form of classical law text: wedding contracts, crowd-funding relations, establishment contracts and many more. Specialists in IT-sphere think that everything programmed can be a subject of smart-contract. Some representatives of law community support this position. With all these main attention is planned to pay on agreements of shareholders and those referring to title transaction.

It is suggested to use widely the named technologies for making various financial deals. Automatic execution of smart-contract questions necessity to use such forms of payment like letter of credit, collection of payments. Besides, smart-contracts are the alternative for the expense of escrow. However, such use is possible only in terms of established law base, as the use of smart-contracts in commercial turn is connected with high regulative risks which can be minimized only with adequate determination of law regulation regime.

It is suggested to consider smart-contract in Russia as a new kind of civil law contract made in electronic form. Smart-contract in "Digital financial actives" law project is considered as a contract in electronic form, the execution of rights and demands of which is realized in automatic order of digital transactions in spread register of digital transactions in strictly defined by such contract chain and after becoming of determined circumstances defined by such contract. But the following approach to define smart-contract doesn't meet classical presentation of contract relations. Thus, from traditional for Russian civil law understanding position of a contract as a deal, relations, paper such a definition is rather questionable.

There is a contract of electronic mean of payment used in civil turn the subject of which is the providing of pay service. The most widely used electronic mean of payment is the pay card. From literal understanding of p. 19 art.3 of "NPC" law a paycard is the electronic carrier of information. However in law and law literature there is no definition for its law nature as information carrier or electronic information carrier as well as a paycard.

Literal reading of electronic mean of payment legal definition gives an opportunity to resume that technical devices, including electronic infocarriers and infocommunicative technologies are used only with electronic means of payment as they are not the mentioned ones. Nevertheless, paycards exactly and other electronic infocarriers and computer devices and other technologies are the electronic means of payment as they exactly allow to give directions to operator concerning distant transfer of financial means.

However by specialists opinion electronic means of payment are objects of civil rights themselves [1, p. 22–23.]. Concerning the named objects the subjects of civil turn have real and demand rights and the following demands. However in civil right object system in art.128 CC RF the electronic means of payment are not mentioned. We suggest that it is necessary to research the law nature of the following term and determine its place in the system of civil rights.

The contemporary economic turn widely uses the Internet possibilities, featuring website design, for economic and entrepreneur actions. Defining the features of website law regulation it is necessary to consider its private and public rights patterns. Before approbation of the forth part of the Civil Codex of RF there were plenty of opinions on website as the rights object. In scientific research website is considered to be a find of computer program; as a kind of database; as an object of special kind sui generic consisting of various types of information databases [2, p. 110].

Judicial practice considered websites as a single work if the site introduced the result of creative labor for selection or positioning of materials. Website qualification as a kind of complex work was established in adopted changes of the forth part of the CC of RF in new edition of art.1260 p. 2. For safety of validated property master rights in case of information borrowing it is necessary to prove:

- the fact of exclusive right for the object borrowed;
- the fact of object accordance to demands set for objects of author rights.

Thus, from the civil law regulation point of view website is considered as a result of intellectual activity – complex object. In defined terms it can be identified as an object of relative rights – databases.

In terms of the federal project "Information infrastructure" in aims of single digital environment forming it is planned to unify law demands for identification of a citizen. For this purpose the terms of identification and authentication with help of "mobile" and "cloud" electronic signature as well as with use of mobile phone number and driver's license are developed. It is suggested to operate distant identification with use of SSIA - Single system of identification and authentication and biometrical personal data confirmation (picture, voice) in biometrical system. Mechanism of distant identification can be available for individuals for distant providing of banking services, medical services in far and hard gained regions, passing through paid automobile highways.

3 Results

High rates of innovative technologies development often born complex tasks in front of law system.

Development of digital economy in Russian Federation demands for creation of effective law mechanism within innovations, from the one side, will infiltrate into economic activity and, from the other side, all possible risks will be foreseen. As risks one can name impossibility of accurate foreseeing for practical use of digital technologies.

4 Conclusions

Digital rights civil law regulation problems in present are the most complicated demanding detailed research and analysis.

Placing digital rights in the list of civil rights relations objects is an important step in developing law space for adopting new laws about financial actives in Russia (cryptocurrency and tokens) and crowdfunding (investment attraction through electronic platforms).

The absence of effective mechanism for digital rights regulation originates many arguable terms in safety sphere for legal interests of their owners.

It will not be able to transfer all rights into digital technologies but those named in the law.

References

1. Abramova, E.N.: Electronic means of payment as a comprehensive civil rights object, banking law, no 1, pp. 22–32 (2018)
2. Basmanova, E.S.: Website as an object of property rights: Ph.D. thesis (law), Moscow (2010)
3. Belykh, O.A., Bolobonova, M.O.: Some issues of legal regulation of smart contracts in Russia. Legal regulation of economic relations in modern conditions the development of the digital economy: monograph/a.v. Belitskaya, v.s. White, o. Belayeva, etc.; OTV. Ed. V.a. Vajpan, M.a. Yegorov. M.: Justicinform, with 376 (2019)
4. Minina, A.: Blokchejn and its influence on the right, law, 2 May 2018. https://zakon.ru/blog/2018/5/2/blokchejn_i_pravo
5. Nam, K.V.: Legal problems related to the application/blokchejna/Judge, no. 2, pp. 24–27 (2019)
6. Duggal, P.: Blockchain Contracts & Cyberlaw (2015)
7. The ruling of the Supreme Arbitration Court of the Russian Federation dated 22.04.2008 No. 288/8. Bulletin of the Russian Federation, no 7 2008
8. Draft of the federal law No. 419059-7 "Regarding digital assets ". http://asozd2c.duma.gov.ru/addwork/scans.nsf/ID/E426461949B66ACC4325825600217475/$FILE/419059-7_20032018_419059-7.PDF?OpenElement. Accessed 05 May 2019

9. Howlett, R.: The existence of smart contracts outside of legal contracts. https://bitcoinmagazine.com/articles/a-lawyer-s-perspective-can-smart-contracts-exist-outside-the-legal-structure-1468263134/. Accessed 05 May 2019
10. Savelyev, A.I.: Contract law 2.0: smart contracts as the beginning of the end of classical contract law. Bulletin of civil law, no. 3, p. 32–60 (2016)
11. Federal law N 18.03.2019 34-FZ "Regarding amending parts of the first, second and third part 1124 of the Civil Code of the Russian Federation", source published the official Internet-portal of legal information, 18 March 2019. http://www.pravo.gov.ru
12. Federal law dated 27.07.2006 No. 149-FZ (ed. by 18.03.2019) on information, information technology and data protection, collected legislation of RF. 31.07.2006. N 31 (1:00). Church. 3448

Prospects for Further Digitization of Corporate Relations

Tatyana A. Skvortsova[1(✉)], Mikhail M. Skorev[2], Tatyana V. Kulikova[3], Nadezhda V. Nesterova[2], and Mikhail M. Merkulov[1]

[1] Rostov State University of Economics, Rostov-on-Don, Russia
tas24@km.ru, hydrargirum2008@gmail.com

[2] Rostov State Transport University, Rostov-on-Don, Russia
gpip@rgups.ru, nadinalladin@mail.ru

[3] South-Russian Institute of Management Branch of the Presidential Academy of the National Economy and Public Administration, Rostov-on-Don, Russia
tana.72@mail.ru

Abstract. This paper is aimed at studying the prospects for further digitization of corporate relations in Russia. It appears that the introduction of advanced information technologies into corporate practice is a promising direction for the development of corporate relations in the Russian Federation. In this regard, an adequate legal framework must be created, which would allow using information and communication technologies in corporate relations.

In order to achieve the goals set, the authors, firstly, have analyzed the issues of the essence and the content of corporate relations; secondly, consideration has been given to the issues of the use of information technologies during the exercise of right of participants of the corporation to participate in management.

Dialectic and materialistic cognition method, special scientific methods (logical, systematic, functional) and specific scientific methods (technical, documentary analysis) have been used in the course of research.

Based on the research findings, the authors have drawn a conclusion about the need of systematization of Russian legislation in the field of regulation of corporate relations. Furthermore, the authors have shown the need for the introduction of the obligatory use of information and communication technologies by joint stock companies with more than 50 shareholders during the Annual General Meeting of Shareholders.

Research results can be introduced in the form of amendments into the current corporate legislation of the Russian Federation and in the law enforcement practice for the purposes of further digitization of corporate relations in Russia.

Keywords: Corporate relations · Corporate bodies · Joint Stock Companies · General Meeting of Shareholders · Using information and communication technologies

JEL Classification Codes: K 19 · O 17

E. G. Popkova and B. S. Sergi (Eds.): ISC 2019, LNNS 87, pp. 1017–1024, 2020.
https://doi.org/10.1007/978-3-030-29586-8_116

1 Introduction

The development of corporate relations in Russia laid the groundwork for the improvement of legal regulation in this field. The improvement of corporate legislation, being one of significant conditions for the enforcement of property rights protection, must be treated as one of the most important institutional conditions of economic growth.

The legal framework for the regulation of corporate relations is under constant development at present. A relatively recent reforming of civil legislation brought a lot of significant changes (Shevchenko 2014). Thus, the subject-matter of civil regulation which is described in Article 2 of Civil Code of the Russian Federation (hereinafter referred to as the Civil Code of the Russian Federation), has been supplemented by a new large group of social relations – corporate relations. Furthermore, significant changes have been made in Chapter 4 “Legal Entities” of the Civil Code of the Russian Federation, which set new rules for corporations.

As can be seen from the above, modern Russian corporate legislation is at a new stage of development. In this regard, many legal problems appear, one of which consists in finding a balance of interests of all participants of corporate relations, and another consists in setting the legal norms which regulate the relations mentioned above. However, despite the reforming, current legal regulation of corporate relations still needs improvement, so conceptual and practical challenges of the mentioned regulation require further scientific research.

The introduction of digital technologies in corporate relations, which requires amending the legislative regulation, appears to be one of directions for the development of corporate legislation in Russia.

Various legal aspects of corporate relations, their legal regulation was a subject-matter of numerous legal research for a reason.

One of researchers in the field of corporate relations has pointed out that “legal corporate relations is one of pet subjects of the modern civil law” (Babaev 2007). Pokrovskiy stated almost the same opinion at the beginning of the previous century: “Legal entities have been one of pet subjects in civil law literature throughout the entire XIX century” (Pokrovskiy 1998).

According to the first concept, corporate relations are civil-law relations (Lomakin 2004).

The upholders of the second concept do not refer corporate relations to the latter (Gubin et al. 2003).

The view of the property nature of corporate relations is the most traditional one (Shabunova 2004). Some researchers, for example, Bratus refer corporate relations to personal non-proprietary relations (Bratus 1963). However, Erdelevskiy believes that corporate relations are both nonpersonal and non-property relations (Erdelevskiy 1997).

Some researchers refer non-property rights of participants of corporate relations, associated with the participation in management, to a separate group of organizational rights (Molotnikov 2006). Other researchers combine several concepts and treat corporate relations as complex relations – property relations as well as non-property relations (organizational relations) – that are associated with them (Pakhomova 2004).

In our opinion, the most consistent definition of corporate relations is given by Ushnitskiy who believes that corporate relations are a variety of civil-law relations, and equity rights are absolute in nature (Ushnitskiy 2011).

Judging from the review of literature on the research topic, it may be concluded that the level of knowledge of the problem of corporate relations is fairly high. This being said, there is no single approach to the legal nature of corporate relations.

The authors have thrown light on particular aspects of regulation of corporate relations: issues of the legal nature of corporate relations, participation (membership) in corporations, formation of the assets of corporate bodies, corporate relations and others.

The academic literature throws light on the issues associated with the impact of new social relations, resulting from the formation of a digital economy, on the regulatory mechanism (Shatkovskaya et al. 2018). However, the issues of digitization of corporate relations have not been the subject-matter of scientific analysis at all.

2 Methodology

A broad range of methodological approaches has been used during the research study. Dialectical method, systematic, sociological methods, as well as specific scientific methods – comparative legal, historical, structure-functional, normative logical, technical legal, and linguistic methods have been used in the creation of this paper as general scientific cognition methods.

3 Results

The emergence of corporate relations of participation is associated with the emergence of the phenomenon of a corporation – a legal entity. A corporation emerges at that place and at that time when the framework of a common partnership property, the framework of a simple partnership agreement is already unable to meet the requirements of developing civil transactions. The emergence of the corporation as the legal entity which is declared the owner of the property assigned to it, becomes the reason for the creation of legal corporate relations, the cause of occurrence of the legal right of participation in the corporation, to which a proprietary interest in the share in common property of copartners is converted.

At present, corporate relations are included in the subject-matter of civil regulation and are defined as "relations that are associated with the participation in corporate bodies or with management of corporate bodies" (Article 2 of the Civil Code of the Russian Federation).

A key criterion that was chosen in the Civil Code of the Russian Federation as a basis for the identification of corporate relations in the system of other relations that are regulated by civil legislation, requires detailed consideration due to its internal diversification. It should be noted that the need for breakdown of corporate relations into different types has been repeatedly pointed out in legal literature. Lomakin fairly observes that the avoidance of "the elaboration of distinct criteria allowing to single out particular kinds of legal corporate relations… causes the degradation of this concept, elimination of its scientific value" (Lomakin 2004).

Indeed, a rather complex nature of corporate relations which do not fit into the conventional classification of civil-law relations, implies identification of several types within this independent corporate relations group. Given the obvious progressive focus of this approach, according to the fair opinion of Stankevich a major deficiency of the norm under consideration as it reads now consists in the lack of legal definition of designated categories, which requires further formalization and consistent regulation of forms, mechanisms and procedure for the participation in a corporate body, as well as the management of a corporate body (Stankevich 2016).

Based on the interpretation of the latest version of Article 2 of the Civil Code of the Russian Federation, we believe that it is necessary to single out two corporate relations groups: relations of participation in corporate bodies and relations of their management. This being said, in order to prepare draft modifications in civil regulatory legal acts, the content of the mentioned types of corporate relations should be clarified at the legislative level.

Within the framework of this research, we shall particularize the legal regulation of the second corporate relations group – relations of management of corporate bodies – by the norms of the current Russian legislation.

Participants (members) of corporations have the right to participate in management of affairs of the corporation (Paragraph 1 of Article 65.1 of the Civil Code of the Russian Federation). Such right is primarily exercised by means of voting participation in the highest management body of the corporation - the General Meeting of Shareholders. Thus, according to Paragraph 1 of Article 47 of Federal Law No. 208-FZ of 26.12.1995 "Concerning Joint Stock Companies" (hereinafter referred to as the Joint-Stock Companies Act), the company shall be obliged to annually hold the Annual General Meeting of Shareholders.

The requirements as to the procedure for calling, preparing and holding the General Meeting of Shareholders are defined in the Joint-Stock Companies Act. In accordance with the consensus decision of shareholders, the Charter may stipulate the procedure for calling, preparing and holding general meetings, which differs from the procedure stipulated by laws and other legislative and regulatory acts, provided that it does not deprive shareholders of their right to participate in the meeting and to receive information about it (Subparagraph 5 of Paragraph 3 of Article 66.3 of the Civil Code of the Russian Federation).

The Annual General Meeting of Shareholders should address issues of election of the board of directors (supervisory board) of the company, audit commission of the company (if its presence is obligatory in line with the Company Charter), as well as of appointment of an auditor of the company. In addition, the General Meeting should address issues stipulated by Subparagraphs 11 and 11.1 of Paragraph 1 of Article 48 of the mentioned Law; other issues which are attributed to exclusive competence of the General Meeting of Shareholders can be addressed as well; these issues are put on the agenda based on the decision of a body who calls a meeting, or recommendations of shareholders.

It is recommended to stipulate the procedure for calling, preparing and holding the General Meeting of Shareholders in the in-house document of the company (for example, Rules and Regulations for the General Meetings of Shareholders), which is to be approved at the General Meeting (Paragraph 1, Chap. 1, Part "B" of Corporate Governance Code (Approval of the Bank of Russia No. 06-52/2463 of 10.04.2014)).

Preparation of the meeting starts from making a decision to hold it. This decision is made by the board of directors. If its functions are exercised by the General Meeting of Shareholders itself, the meeting should be called and its agenda should be defined by a person or entity specified in the Charter (Paragraph 1 of Article 64, Subparagraph 2, Paragraph 1 of Article 65 of the Joint-Stock Companies Act).

In particular, the form of the meeting should be specified in this decision, too. It should be noted that the Annual General Meeting may not be held via web-conferencing (Paragraph 2 of Article 50 of the Joint-Stock Companies Act). Therefore, the legislator provides for holding the Annual General Meeting of Shareholders only in the form of the presence of shareholders.

In order to notify the attendees of the General Meeting, a notification must be prepared and communicated to the meeting attendees in the prescribed manner (Paragraph 3 of Article 11, Paragraphs 1.1, 1.2 of Article 52 of the Joint-Stock Companies Act). In general, a relevant notification should be delivered to shareholders by registered mail or should be handed to them against acknowledgement. The Charter may provide for other methods of message delivery (of those specified in the law), such as sending an e-mail message (Paragraphs 1.1, 1.2 of Article 52 of the Joint-Stock Companies Act). If the Charter provides for several methods of message delivery, a shareholder may choose one of them by choosing the relevant option in the questionnaire of registered persons in the register of shareholders (Paragraph 3.2 of the Rules and Regulations for the General Meetings of Shareholders (Approval of the Bank of Russia No. 660-P of 16.11.2018)).

Further, the shareholders are entitled to contribute their suggestions containing the information on potential members of governing (supervisory) bodies of the company as well as issues to be put on the agenda. They can be sent by shareholders who own at least 2% of voting shares. They must be received by the Company within 30 days after the end of the current year, if the Charter makes no provision for a later term (Paragraph 1 of Article 53 of the Joint-Stock Companies Act).

The company should develop such procedure of sending suggestions concerning the nomination of potential members of bodies of the Company and putting issues on the agenda of the General Meeting which would be convenient to shareholders (Paragraph 15, Chap. 1, Part "B" of Corporate Governance Code).

As a general rule, the voting at the General Meeting of Shareholders is held with the use of ballot papers. Voting by ballot papers is obligatory for public companies and non-public companies where the number of voting shareholders is at least 50 (Paragraph 1 of Article 60 of the Joint-Stock Companies Act). Other companies may use ballot papers if such use is provided for in their Charter.

In general, the ballot paper is handed against acknowledgement to a person who is included in the list of persons who have the right to participate in the meeting, and entered in the list of meeting attendees (Paragraph 2 of Article 60 of the Joint-Stock Companies Act). Besides, ballot papers must be handed over to the registrar for them to be electronically sent to their nominal holders (Paragraph 3.9 of the Rules and Regulations for the General Meetings of Shareholders).

If, by virtue of law or provisions of the Charter, the obligation has been established to send (deliver) ballot papers before the Meeting, this should be done no later than 20 days before it (Paragraph 2 of Article 60 of the Joint-Stock Companies Act).

Ballot papers are sent (delivered by hand) using the method specified in the Charter, and if it is not stipulated, they should be delivered by registered mail (Paragraph 2 of Article 60 of the Joint-Stock Companies Act). If the Charter (for joint stock companies with more than 500 thousand shareholders) provides for publication of ballot papers, they should be published 20 days before the Meeting (Paragraph 3 of Article 60 of the Joint-Stock Companies Act).

The current Russian Legislation allows holding a meeting which provides a means for using information and communication technologies which "enable the possibility of remote participation in the General Meeting of Shareholders, discuss items on the agenda and make decisions on items put to vote without being actually present at the venue of the General Meeting of Shareholders" (Paragraph 11 of Article 49 of the Joint-Stock Companies Act). In particular, the Rules and Regulations for the General Meetings of Shareholders specify the procedure for the General Meeting of Shareholders, if it is held with an option of filling in the computer-generated form of ballot papers online on the website.

At the same time, it should be pointed out that such procedure for holding a meeting is only possible if it is provided for by the Company Charter. The Joint-Stock Companies Act does not stipulate the obligations of holding a meeting with the use of such information and communication technologies. In our opinion, this procedure must be introduced into corporate practice of joint stock companies with large numbers of shareholders as an obligatory procedure, since it allows minority shareholders to exercise their right of participation in the General Meeting of Shareholders. As long as the use of the abovementioned procedure for holding a meeting is voluntary in nature, majority shareholders will not be interested in making relevant changes in the Company Charter for it to be introduced into corporate practice, since minority shareholders, who are situated far away from the venue of the General Meeting and are forced to bear significant costs to exercise their voting rights during the Annual General Meeting of Shareholders in the form of the presence of shareholders, are more interested in it.

It follows from the foregoing that it is necessary to legislate the obligation of joint stock companies with more than 50 shareholders to hold general meetings of shareholders which provide a means for using information and communication technologies.

With this aim in view, it is recommended to formalize the relevant rule in Article 49 of the Joint-Stock Companies Act.

4 Conclusion/Recommendations

Thus, in accordance with Article 2 of the Civil Code of the Russian Federation, legal corporate relations are a kind of civil-law relations and include relations of participation in corporate bodies and relations of their management. This being said, the legislator has not defined the content of the mentioned groups of relations. We believe that in order to avoid confusion in law enforcement practice in the Civil Code of the Russian Federation, relations that are included in the mentioned groups should be thoroughly regulated.

At present, the Russian Legislation on joint stock companies provides for holding a meeting which provides a means for using information and communication technologies which enable the possibility of remote participation in the General Meeting of Shareholders, discuss items on the agenda and make decisions on items put to vote without being actually present at the venue of the General Meeting of Shareholders

At the same time, such opportunity is only provided when abovementioned provisions are introduced into the Company Charter. We are of the opinion that it is necessary to legislate the obligation to hold general meetings of shareholders which provide a means for using information and communication technologies for joint stock companies with more than 50 shareholders, which is intended to enforce rights of minority shareholders who are situated far away from the venue of the General Meeting, to participate in the discussion of issues on the agenda and in the voting. With this aim in view, it is recommended to formalize a relevant rule in Article 49 of the Joint-Stock Companies Act.

References

Shevchenko, O.M.: Bank of Russia as a regulator of corporate relations. Predprinimatelskoe Pravo **4**, 39–46 (2014)

Civil Code of the Russian Federation (Part One) of 30.11.1994 No. 51-FZ (as amended on 03.08.2018), Collection of Legislative Acts of the Russian Federation, 1994, No. 32, of Article 3301

Babaev, A.B.: Methodologic prerequisites for the study of legal corporate relations. Vestnik Grazhdanskogo Prava Academic periodical **4**(7), 5–22 (2007)

Pokrovskiy, I.A.: Fundamental Problems of Civil Law (Based on Publication of the Year 1917). Statut Publishing House, Moscow (1998)

Lomakin, D.V.: Corporate relations and the subject-matter of civil regulation. Zakonodatelstvo **5**, 58–64 (2004)

Gubin, E.P., Lakhno, P.G. (eds.): Predprinimatelskoe Pravo Rossiyskoy Federatsii. Yurist Publishing House, Moscow (2003)

Shabunova, I.N.: Corporate relations as the subject-matter of civil law. Zhurnal Rossiyskogo Prava **2**, 40–49 (2004)

Erdelevskiy, A.M.: Concerning the protection of non-property rights of shareholders. Khoziaystvo i Pravo **6**, 69–74 (1997)

Molotnikov, A.E.: Responsibilities in Joint Stock Companies. Wolters Kluwer Publishing House, Moscow (2006)

Pakhomova, N.N.: The Elements of Corporate Relations Theory (Legal Aspect). Nalogi i Finansovoe Pravo, Yekaterinburg (2004)

Shatkovskaya, T.V., Yepifanova, T.V., Vovchenko, N.G.: Transformation of the structure of the regulatory mechanism in a digital economy. Problemy Ekonomiki i Yuridicheskoy Praktiki **3**, 142–146 (2018)

Ushnitskiy, R.R.: On civil form of corporate relations. Vestnik Grazhdanskogo Prava **5**, 64–91 (2011)

Stankevich, G.V.: Participation and membership as a prerequisite for corporate relations. Vlast Zakona (1), 30–39 (2016)

Federal Law No. 208-FZ of 26.12.1995 (as amended on 27.12.2018) "Concerning Joint Stock Companies", Collection of Legislative Acts of the Russian Federation, 1996, No. 1, of Article 1

Letter from the Bank of Russia No. 06–52/2463 of 10.04.2014 "On Corporate Governance Code", Bulletin of the Bank of Russia, 18.04.2014, No. 40

Rules and Regulations for the General Meetings of Shareholders (Approval of the Bank of Russia No. 660-P of 16.11.2018), Bulletin of the Bank of Russia, 22.01.2019, No. 3

Bratus, S.N.: Subject and System of Soviet Civil Law. State Publishing House of Legal Literature, Moscow (1963)

Development of Digitization in Contractual Relations

Tatyana A. Skvortsova, Tatyana A. Mosienko, Aelita Yu. Ulezko, Alexander V. Nikolaev, and Andrey A. Arzumanyan

Rostov State University of Economics, Rostov-on-Don, Russia
tas24@km.ru, mosienko-5858@mail.ru, kafedra37@bk.ru, procpravo@yandex.ru, andrew71984@mail.ru

Abstract. The paper is aimed at studying the digitization in contractual relationship in respect of the aspect of legal regulation of smart contracts in Russia. Development of digital technologies in the conclusion of contracts is one of promising directions for the development of contractual relationship in the Russian Federation. Smart contracts came into economic turnover along with blockchain technology and cryptocurrency technology. However, creation of regulatory framework which would be adequate to the existing environment appears to be difficult at present, which can be explained by the complexity of definition of legal nature of smart contracts and cryptocurrency, as well as ambiguousness in terms of approaches to the system of statutory recognition of digital technologies.

In order to achieve the designated goals of research, the authors have analyzed the legal nature of smart contracts, have studied the procedure for the conclusion of civil law contracts and the specifics of the conclusion of smart contracts in the paper.

System method, structured functional method of scientific cognition, method of interpretation of civil law, as well as logical method have been used in writing this paper.

As a result of research, the authors have concluded that smart contract is a specific way of conclusion of a civil law contract. The authors have proposed to add provisions regulating such way of conclusion of contracts to the effective Civil Code of the Russian Federation. In addition, the authors have proposed to regulate the consequences of voidance of concluded smart contracts in a judicial proceeding using the norms of civil legislation.

Research results can be used as recommendations for the introduction of amendments and additions to the Russian civil legislation.

Keywords: Civil law contract · Conclusion of a contract · Digital technologies · Smart contract · Blockchain · Cryptocurrency

JEL Code: K-12 · K-24 · O-30

E. G. Popkova and B. S. Sergi (Eds.): ISC 2019, LNNS 87, pp. 1025–1032, 2020.
https://doi.org/10.1007/978-3-030-29586-8_117

1 Introduction

The vast variety of economic relations in the current context can take the most various form of expression (Epifanova et al. 2017). Nevertheless, no matter what form they are expressed in, contractual framework is still in the core of any of them. Economic activity is primarily based on contractual relationship, according to which the agents which carry out such activity undertake obligations to manufacture or supply particular goods, provide services, perform works, etc. (Rogova 2013).

A contract is one of the most unique legal means within the framework of which the interest of every party can be essentially satisfied only through the satisfaction of the interest of other party. This is what gives rise to the joint interest of the parties in the conclusion of a contract and its proper performance. This is why it is the contract that is based on mutual interest of the parties can guarantee such good organization, good order and stability in economic turnover that cannot be achieved with the use of the most stringent administrative and legal means (Shatkovskaya and Epifanova 2016).

At present, there are global changes areas of society associated with large-scale adoption of digital technologies. One of directions of adoption of digital technologies on the sphere of economic turnover is the use of computer technology in the conclusion and performance of contracts, smart contracts in particular.

At present, issues of the use of blockchain technologies and the use of smart contracts are topical not only in the economy, but also in the world of law. Work is in progress on the introduction of amendments which provide for the definition of the status of digital technologies that are used in financial sector, their concepts (including such as "distributed register technology", "digital letter of credit", "digital mortgage bond", "cryptocurrency", "token", "smart contract") to the legislation of the Russian Federation, judging from the obligatoriness of ruble as the only statutory means of payment in the Russian Federation.

In particular, Draft Federal Law No. 419059-7 "On Digital Financial Assets" (revised, passed by the State Duma of the Federal Assembly of Russian Federation Russia in the first reading on 22.05.2018), which is intended to regulate relationship that arises in case of creation, issue, storage and turnover of digital financial assets, as well as execution of rights and execution of obligations under smart contracts, was drafted and passed in the first reading a year ago.

It should be pointed out that this draft law project was not the only one. This draft law was prepared by the Bank of Russia and the Ministry of Finance of the Russian Federation. During their work they never managed to come to agreement on certain provisions. This is why both of them posted their own draft projects. The Bank of Russia posted its draft project on its website in section "Information Analysis" (subsection "Normative and Other Legal Acts"). The Ministry of Finance of the Russian Federation also posted its draft project on its website in section "Documents" (publication of 25.01.2018) (Bondarchuk 2018).

Nevertheless, eventually the Government of the Russian Federation submitted the draft project for consideration to the State Duma, and the State Duma passed the draft project that was developed by the Ministry of Finance of the Russian Federation, in the first reading.

All of the abovementioned is responsible for the need for scientific analysis of issues of the use of digital technologies in a contractual relationship, including the essence of smart contracts, which are one of the most promising and at the same time contradictory (from the legal viewpoint) directions of the use of such technologies.

The term "smart contract" first appeared in the paper authored by Sabo as early as in 1994, according to whom a self-executed smart contract is a computer based transactional protocol which complies with such contract (Sabo 1994).

Currently, there is no consistent approach to the legal essence of smart contracts in the research literature.

Generally speaking, smart contract can be defined as a programmed contract, conditions of which are prescribed in the program code which is automatically executed by means of a blockchain. In a smart contract that is based on blockchain technology, contract conditions are formulated in a programming language, after which the smart contract is transferred to blockchain (Volos 2018).

Blockchain is a distributed data system (distributed database, book of account), which can be broadly presented as an Excel sheet that is stored in many computers simultaneously and contains certain information. Such information may be, for example, data on whether a person has assets that can symbolize any obligation or affiliation of a tangible object. This information is stored in a blockchain on each device on the network, and the data in the table is identical for all participants (Fedorov 2018).

Smart contracts are based on the computer protocol which sends or receives certain information or changes the data according to rules in effect. The program code which is built-in into the protocol, serves as a basis of this. In such a case, simple rules "if, then" are used: "if" a certain condition will be met, "then" a certain action or transaction will be carried out (Linardatos 2018).

Saveliev distinguishes the following main characteristics of a smart contract:

1. This contract exclusively exists in a digital environment and implies compulsory use of the digital signature that is based on asymmetric encryption technology (enhanced encrypted non-certified digital signature in terms of current legislation).
2. Special form of presentation of conditions. Conditions of such a contract are set forth in one of programming languages, and they are executed through the use of the blockchain database (Saveliev 2016).

From the point of view of law, smart contract is a certain set of smart contracts in a technical sense, which automate certain transactions of fulfillment of obligations by the parties (Saveliev 2017).

According to some authors, currently smart contracts can be integrated in the contract law as one of the ways of fulfillment of obligations (Gromova 2018).

The law-maker is of a different conceptual opinion. By virtue of provisions of Draft Federal Law No. 419059-7 "On Digital Financial Assets" "smart contract is an electronic contract, rights and obligations under which are fulfilled through automatic execution of digital transactions in the distributed register of digital transactions in a strict sequence stipulated by this contract, and upon occurrence of circumstances stipulated in it".

Sinitsyn believes that such statements according to which continuous technical advance and technology boom will be sufficient conditions for the emergence of a new type of contracts are groundless, since these factors can mainly have an impact on the way of conclusion of contracts, but not on its content. For this reason, we cannot agree with the statement that the use of smart contracts will require creation of new types of contractual forms in legislation, since originally it is not about the emergence of a new relationship in the turnover of commodities (if we talk not about the turnover of commodities in an all-sufficient and isolated digital system), but only about the use of digital technologies as records of achievement of digitization and robotization by economic agents in their contractual work when concluding a contract and fulfilling their obligations thereunder (Sinitsyn 2019).

We accept the abovementioned point of view and believe that smart contract should be considered as a specific way of conclusion of a contract. As a consequence, such a way must be regulated at the level of the Civil Code of the Russian Federation, along with other ways of conclusion of a civil law contract.

2 Methodology

The work has been written with the use of conventional and approved methods of research study of civil phenomena. The use of system method and structured functional method has served as a basis of analysis of contractual relations. Consideration of most aspects of the topic under study has been facilitated by the use of methods of interpretation of civil law, as well as logical method.

3 Results

In accordance with Paragraph 1 of Article 420 of the Civil Code of the Russian Federation, the agreement between the two or more parties on the establishment, change or termination of civil rights and obligations shall be recognized as contract.

The existing legislation does not articulate the concept of "conclusion of a contract" as such. Judging from the existing legal regulation in this sphere, conclusion of a contract can be defined as lawful coordinated actions of economic agents aimed at generation of civil law consequences in their property or personal non-property sphere.

In other words, conclusion of a contract is a certain process, which means that most legal norms regulating this process are characterized by organizational (procedural) nature. As for the description of stages that constitute the process of conclusion of a civil law contract, there was no unanimity of opinion on this matter in the science of civil law both in the Soviet period and in our times.

Many researchers believe that the process of conclusion of a contract includes as little as two stages - offer stage and acceptance stage - thus there is no need to distinguish any other stages in it (Zhilinskiy 1998). According to this point of view, actions which constitute the stage of precontractual contacts are merged with the offer stage, and the actions of resolution of precontractual disputes are included in the acceptance stage.

Vitrianskiy has an original point of view on the matter under consideration; he proposed to distinguish the following stages of conclusion of a contract: (1) precontractual contacts of the parties (negotiations); (2) offer; (3) consideration of the offer; (4) acceptance of the offer (Vitrianskiy 2000). In such a case, the optional nature of the first stage is pointed out, as it is used at the discretion of the parties that conclude a contractual relationship; besides, it is pointed out that consideration of the offer as an independent stage of conclusion of a contract usually has legal effect for contracts conclusion of which is mandatory for one of the parties.

It would be fair to assume that consideration of the offer also includes actions of identification of matters in difference in relation to its certain conditions, hence, it is covered by actions of the acceptor of the offer; in other words, this is the acceptance stage.

In the meantime, one should take into account the specifics of conclusion of contracts in a mandatory manner which consists in the fact that such contracts can be concluded in an enforcement procedure - on the basis of a court decision, i.e. actually in the absence of acceptance of a party to which offer was submitted. Hence, one should assume a situation in which the conclusion of a contract is preceded by an offer and its consideration without an acceptance: court decision which obliges a party to conclude the contract can hardly be treated as acceptance.

Therefore, it is reasonable to agree with Vitrianskiy over identification of such stage of conclusion of a contract as consideration of the offer. Judging from the above, the only classification of stages of conclusion of a contract that was set forward by him can be supplemented with a stage of resolution of a dispute over conclusion of a contract in a judicial proceeding and point out the optionality of such stage as acceptance stage (with regard to contracts that are subject to mandatory conclusion).

An important innovation of the Civil Code of the Russian Federation is the establishment of rules of contract negotiations. According to the new Article 434.1 of the Civil Code of the Russian Federation, unless otherwise provided for by any law or contract, individuals and legal entities are free to carry on contract negotiations, they shall bear the costs associated with such negotiations, and that shall not be held liable in case no agreement is reached. When entering into contract negotiations, during contract negotiations, and upon completion of contract negotiations, the parties are obliged to act in good faith, in particular, avoid entering into contract negotiations or continuing such negotiations in case of obvious lack of intent to reach an agreement with other party.

A party which performs or terminates contract negotiations in bad faith, shall be obliged to compensate the incurred losses to the other party. Losses that are subject to compensation by the unconscientious party shall be losses that were incurred by the other party in connection with contract negotiations, as well as in connection with the loss of opportunity to conclude contract with the third party (Spitsyna 2015).

Advancement of the offer is the next stage of conclusion of contracts.

The term “offer” derives from the Latin “offero” - which means a proposal of an individual or legal entity to conclude a civil law contract with an indication of specific conditions - for example the price, description of characteristics, time of delivery of commodities or provision of services.

The concept of offer is directly disclosed in Paragraph 1 of Article 435 of the Civil Code of the Russian Federation, according to which an offer is a proposal which is submitted to one or several particular parties, which is fairly definite and expresses the intent of a party which made a proposal to consider itself as a party which concluded the contract with a consignee who will accept the proposal. Moreover, it is expressly indicated in the Civil Code of the Russian Federation that offer must contain significant conditions of the contract.

According to some researchers, an offer entails the so-called secondary right for the acceptance (Rodionova 2014). The reply of the party (acceptor) to which the offer is submitted, regarding its acceptance shall be recognized as acceptance. Besides, the acceptance must be: (1) performed by the party to which the offer was submitted; (2) complete; (3) irrevocable; (4) submitted within the time limit stipulated in the offer; (5) containing the reply regarding the acceptance of the offer (Krasheninnikov et al. 2010).

If a smart contract is concluded, we should talk about a protocol written in any programming language, functioning in the blockchain and ensuring the self-sufficiency and self-execution of such a contract upon occurrence of circumstances stipulated in it (Saveliev 2016).

Several theses about smart contracts in the contract law are pointed out in the research literature:

1. A smart contract cannot be deleted or amended since it has been loaded into the blockchain. When a smart contract is activated in the blockchain client, the code is locked inside the set of data and is embedded in the blockchain structure.
2. Automated nature of performance: smart contract will be concluded not only in case of coordination of wills, but, by its nature, will be concluded only when the users who came to an agreement will have the opportunity of contemporaneous performance of a smart contract.
3. The preservation and use of most elaborated approaches to smart contracts with regard to the contract law (Rumiantsev 2018).

The abovementioned specifics of the procedure for the conclusion of smart contracts should be taken into consideration by determining the specific character of legal regulation of the conclusion of civil law contracts in a similar way in the Civil Code of the Russian Federation. It appears that Article 449.2 "Specifics of the conclusion of smart contracts" should be added to the Civil Code of the Russian Federation.

As smart contracts are increasingly introduced into the real-world conclusion of contracts, one more problem turns up, which resides in the ensuing of restitutional consequences of voidance of a contract. The existing mechanism of voidance of contracts and application of consequences of their voidness implies the return of the parties to the initial position which existed prior to the settlement of transaction, as the main consequence of its voidness.

If a smart contract is concluded, the next return of the parties to the initial material position on the basis of a court decision with the cancelation of an entry of transaction will be inconsistent with such property of a blockchain as irreversibility of entries. In this regard, Article 167 of the Civil Code of the Russian Federation should provide for special consequences of voidance of concluded smart contracts on the basis of a court decision.

4 Conclusion/Recommendations

Thus, when we talk about smart contracts, we should talk about the use of digital technologies as records of achievement of digitization and robotization by economic agents in their contractual work when concluding a contract and fulfilling their obligations thereunder. In this regard, smart contract should be considered as a specific way of conclusion of a civil law contract which constitutes a protocol written in any programming language, functioning in the blockchain and ensuring the self-sufficiency and self-execution of such a contract upon occurrence of circumstances stipulated in it.

The following action should be taken to regulate this way of conclusion of a civil law contract using the rule of law:

- add Article 449.2 "Specifics of the conclusion of smart contracts" to the Civil Code of the Russian Federation;
- provide for special consequences of voidance of concluded smart contracts on the basis of a court decision in Article 167 of the Civil Code of the Russian Federation.

References

Epifanova, T.V., Shatkovskaya, T.V., Romanenko, N.G., Mosienko, T.A., Tkachenko, M.: Legal provision of clustering in Russia as environment for development of innovations. Int. J. Trade Glob. Markets **10**(2–3), 217–225 (2017)

Rogova, Y.V.: Practical application of the institution of a major change of circumstances in a volatile environment of economic intercourse. Pravo i Ekonomika (1), 19–24 (2013)

Shatkovskaya, T.V., Epifanova, T.V.: Correlation of private and public legal interests as theoretical and scientific and practical problem of modern law. J. Adv. Res. Law Econ. **7**(3), 625–643 (2016)

Bondarchuk, D.: Cryptocurrency, mining and smart contracts will be legalized. What impact will it have on the work of lawyers?, EZH-Yurist (6), 1–2 (2018)

Sabo, N.: Smart contracts (Fourth value revolution) (1994). http://old.computerra.ru/1998/266/194332/. Accessed 29 Sept 1998

Volos, A.A.: Smart contracts and principles of civil law. Rossiyskaya Yustitsiya (12), 5–7 (2018)

Fedorov, D.V.: Tokens, cryptocurrency and smart contracts in domestic draft laws from the perspective of foreign experience. Vestnik Grazhdanskogo Prava (2), 30–74 (2018)

Linardatos, D.: Smart contracts - einige klarstellende Bemerkungen, K&R (2018)

Saveliev, A.I.: Contract Law 2.0: smart contracts as the beginning of the end of conventional contract law. Vestnik Grazhdanskogo Prava (3), 32–60 (2016)

Saveliev, A.I.: Some legal aspects of the use of smart contracts and blockchain technologies in accordance with the Russian law. Zakon (5), 94–117 (2017)

Gromova, E.A.: Smart contracts in Russia: an attempt to determine the legal essence. Pravo i Tsyfrovaya Ekonomika (2), 34–37 (2018)

Sinitsyn, S.A.: A contract: new dimensions of legal regulation and issues of legal consciousness. Zhurnal Rossiyskogo Prava (1), 45–61 (2019)

Zhilinskiy, S.E.: A legal framework of entrepreneurial activities (entrepreneurial law). Norma Publishing House, Moscow (1998)

Vitrianskiy, V.V.: Civil law contract. In: Sukhanov, E.A., et al. (eds.) Civil Law: a Handbook, vol. 2, subvol. 1, pp. 150–202. BEK Publishing House, Moscow (2000)

Spitsyna, T.V.: General provisions about a contract: last minute amendment. Uproshchennaya Sistema Nalogooblozheniya: Bukhgalterskiy Uchet i Nalogooblozheniye (8), 50–60 (2015)

Rodionova, O.M.: Revisiting the civil nature of the offer and acceptance (through the example of conclusion of contracts for procurements, works and services for meeting state and municipal needs). Yurist (5), 38–41 (2014)

Krasheninnikov, P.V., et al.: A contract: Paragraph-to-Paragraph Commentary for Chapters 27, 28 and 29 of the Civil Code of the Russian Federation. Statut Publishing House, Moscow (2010)

Rumiantsev, I.A.: Blockchain and law. In: Rozhkova, M.A., et al. (eds.) Law in the Internet: a Collection of Articles, p. 528. Statut Publishing House, Moscow

Electronic Notarial System as a New Social Institution in a Digital Economy: Quality, Availability, Security

Anna V. Sukhovenko(✉)

Rostov State University of Economics, Rostov-on-Don, Russia
annasuhovenko@gmail.com

Abstract. Purpose: This paper is aimed at substantiating the preferableness of the electronic notarial system and the development of its institutional model in a digital economy.

Design/Methodology/Approach: In the pursuance of this research, the authors are relying on the methodology of neo-institutional economy – the Game Theory. They use it for the expert assessment of the quality (ultimate outcome), availability (expenses) and security (risk) of the two distinguished types of notarial services: notarization of copies of documents and their extracts, notarization of transactions. Efficiency of notarial services is assessed with due account for their different forms using a special formula developed by the authors. It serves as a basis for the determination of the most preferable form of each distinguished type of notarial services (which is characterized by higher efficiency).

Findings: It has been substantiated that electronic form is the most preferred form of each distinguished type of notarial services (characterized by higher efficiency). In order to implement the practice for providing notarial services in electronic form in a digital economy, the authors have developed an institutional model of this process.

Originality/Value: It has been established that a new social institution is being formed in a digital economy – the electronic notarial system, which is a more optimal form of notarial services in terms of their quality, availability and security. The developed institutional model of the electronic notarial system in a digital economy has demonstrated a more complex scheme of its organization compared to the conventional notarial system. At the same time, through the use of modern information and communication technology and the transfer of many functions (for example, search for original documents) to the notary, the process of obtainment of notarial services becomes more simple and convenient for consumers.

Keywords: Electronic notarial system · Social institution · Digital economy · Quality · Availability · Security

JEL codes: D02 · K13 · K32 · L15 · L17 · L86 · O14 · O43

E. G. Popkova and B. S. Sergi (Eds.): ISC 2019, LNNS 87, pp. 1033–1039, 2020.
https://doi.org/10.1007/978-3-030-29586-8_118

1 Introduction

Digital economy creates opportunities and opens prospects for the improvement of the majority of existing business practices on the basis of their technological enhancement. Thus far, considerable experience has been gained in the provision of state services in digital form within the e-Government system, as well as experience in the manufacture and distribution of goods and services in various sectors of economy, including in the settlement of electronic payments on the basis of electronic digital signature.

Electronic notarial system is an advanced and, at the same time, contradictory direction of growth of digital economy. On the one hand, the need for it is fairly high, as the emergence of the possibility of notarization of electronic documents will contribute to the increase in social mobility (including migration) and will give a new impulse to the development of e-Government system, State Procurement System which is based on the online auction due to the removal of the need to submit original documents in paper form as an addition to electronic documents.

On the other hand, notarial activities are characterized by enhanced responsibility. In notarial support of transactions (for example, in case of notarization of testaments and real estate transactions), it is important not only to verify the genuineness of documents but also to make sure of the genuineness of will and expression of will of parties to transactions (lack of coercion), transactional capacity and legal capacity of parties to transactions for the prevention of fraud. The need for social interaction between the notary and his/her consumers is one of the major issues of the electronic notarial system, since fraud risk might increase when electronic form of this interaction is used.

As can be seen from the above, electronic notarial system is a new social institution which emerges in a digital economy and requires scientific and methodological substantiation for the assurance of its high quality, availability and security. This paper is aimed at substantiating the preferableness of the electronic notarial system and the development of its institutional model in a digital economy.

2 Materials and Method

The essence and the specificity of notarial activities in various modern economic and social systems are described in a number of publications (Bratchel 2018, Lavecchia and Stagnaro 2019, Maleshin 2017, Santiago 2017, Yulia et al. 2018). Consideration to individual aspects of the electronic notarial system as an advanced form of notarial services was given in papers of such researchers as (Blibech and Gabillon 2009, Bogoviz et al. 2019, Llopis Benlloch 2018, Maganić 2013, Popkova 2017, Popkova and Sergi 2019, Yüce and Selçuk 2018). A literature review has shown that the institutional nature of the electronic notarial system is poorly enlightened in the existing research papers, thus it requires further study.

In the pursuance of this research, we are relying on the methodology of neo-institutional economy – the Game Theory. We use it to assess the quality (ultimate outcome), availability (expenses) and security (risk) of the two types of notarial services:

1. Notarization of copies of documents and their extracts. In the electronic notarial system, its directions are as follows: (1.1) remote notarization of electronic copies of documents, and (1.2) notarization of electronic copies of documents subject to their originals being submitted;
2. Notarization of transactions. This type of services also includes notarization of testaments, notarization of real estate transactions, notarization of consents, powers of attorney, and other types of notarial acts.
   We perform the assessment using the expert method by means of assigning values to quality indicators (Q, measured in points from 1 to 10, higher is better), availability (A, measured in points from 10 to 1 as a complexity of their obtainment, lower is better) and security (S, measured in fractions from 1) distinguished types of services in case of their provision in the form of the electronic notarial system (en) and in the form of the conventional notarial system (tn). As a result, the indicator of efficiency of notarial services is determined (Ef, measured in points from 1 to 10, higher is better) with the following formula:

$$Ef = (Q/A) * S \tag{1}$$

It serves as a basis for the determination of the most preferable form of each distinguished type of notarial services (which is characterized by higher efficiency).

3 Results

We have used formula (1) to assess efficiency of notarial services of the two distinguished types of notarial services in the form of conventional and electronic notarial systems; its results are presented in Table 1.

Table 1. Assessment of efficiency of notarial services in conventional and electronic forms with the use of the Game Theory methodology

Type of notarial services	Form of notarial services	Assessment of efficiency of notarial services			
		Quality, Q, points 1–10	Availability, points 10–1	Security, S, fractions, from 1	Efficiency, Ef, points 1–10
Notarization of copies of documents (c)	Conventional notarial system (tn)	8	4	1.0	$Ef(tn_c) = 2{,}0$
	Electronic notarial system (en)	10	1	1.0	$Ef(en_c) = 10{,}0$
Notarization of transactions (t)	Conventional notarial system (tn)	10	5	0.9	$Ef(tn_t) = 1{,}8$
	Electronic notarial system (en)	8	1	0.7	$Ef(en_t) = 5{,}6$

Source: Compiled by the authors.

Table 1 shows that we have rated the quality of notarization of copies of documents with the use of the conventional notarial system at 8 points. This is due to the fact that, although notarized paper copies of documents are valid in all agencies and institutions, they should be repeatedly notarized in multiple copies where necessary; thus, the consumer is forced to repeatedly apply for notarial services and submit original documents. We have rated the availability of these services at 4 points, since notarization of documents in conventional form is associated with repeated application for the same notarial act, and, as a result, heavy financial expenses. We have rated the security of these services at 1.0 point (maximum), since fraud is unlikely. This allows us to determine efficiency: Ef(tnc) = (8/4)*1 = 2 points – very low efficiency.

We have rated the quality of notarization of copies of documents with the use of the electronic notarial system at 10 points. This is due to the fact that, although notarized electronic copies of documents are valid in all agencies and institutions; however, their repeated notarization is not required (once notarized, they can further be used times out of number); besides, services can be subdivided into remote notarization of electronic copies of documents and notarization of electronic copies of documents subject to their originals being submitted. We have rated the availability of these services at 1 point, since documents in electronic form can be notarized in a quick and mobile manner, making it possible to reduce expenses of consumers. We have rated the security of these services at 1.0 point (maximum), since fraud is unlikely. This allows us to determine efficiency: Ef(enc) = (10/1)*1 = 10 points – maximum efficiency.

We have rated the quality of notarization of transactions with the use of the conventional notarial system at 10 points. This is due to the fact that face-to-face interaction between the notary and consumers of his services (parties to the transaction) makes it possible to achieve full conformity between will and expression of will. We have rated the availability of these services at 5 points, such as notarization of transactions in conventional form stipulates the presence of all parties to the transaction at a single location at the same time, whereas in-person visit to a notary public is troublesome for certain categories of individuals, when the obtainment of notarial services at home is realizable, but requires additional activities and additional costs. We have rated the security of these services at 0.9 points, such as fraud risk, but it is very low. This allows us to determine efficiency: Ef(tnt) = (10/5)*0.9 = 1.8 points – very low efficiency.

We have rated the quality of notarization of transactions with the use of the electronic notarial system at 8 points. This is due to the fact that remote interaction between the notary and consumers of his services (parties to the transaction) may cause difficulties in mutual understanding and cause the lack of understanding. We have rated the availability of these services at 1 point, since notarization of transactions in electronic form implies remote interaction, which is available, inter alia, to individuals residing in another city, as well as to individuals with disabilities (i.e. complete inclusiveness is achieved). We have rated the security of these services at 0.7 points, as there is moderate fraud risk. This allows us to determine efficiency: Ef(ent) = (8/1)*0.7 = 5.6 points – high efficiency.

The comparison of obtained values of efficiency indicators (Ef) has shown that electronic form is the most preferred form of each distinguished type of notarial services (characterized by higher efficiency). In order to implement the practice for providing notarial services in electronic form in a digital economy, we have developed an institutional model of this process that is shown in Fig. 1.

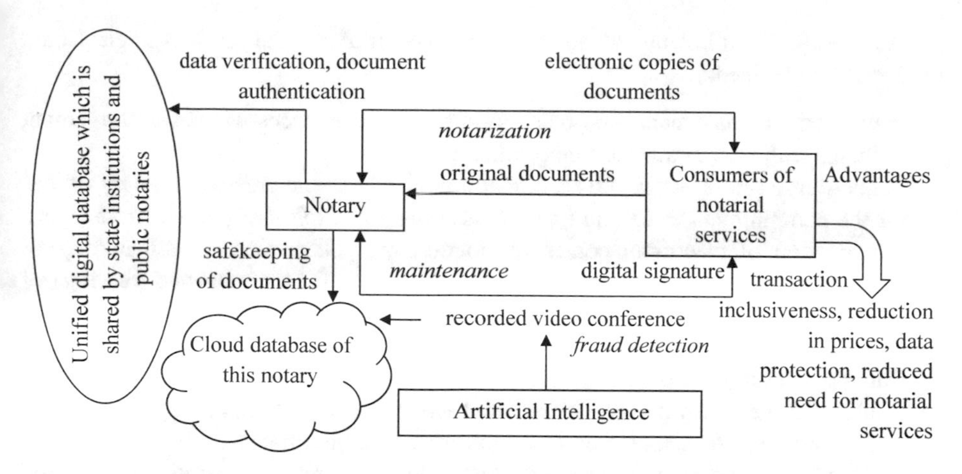

Fig. 1. Institutional model of the electronic notarial system in a digital economy Source: Compiled by the authors.

As is clear from Fig. 1, the institutional model of the electronic notarial system in a digital economy is much more sophisticated as compared to the model of conventional electronic notarial system, since it implies not only the interaction between the notary and consumers of notarial services, but also the use of digital databases in it. In case of remote notarization of copies of documents, the notary verifies data against the unified digital database which is shared by state institutions (for example, Russian State Register, multiservice centers) and public notaries, and verifies the genuineness of documents. If everything is correct and documents are authentic, the notary notarizes their electronic copies and sends them back to consumers.

The notary need not verify documents against the digital database in case of notarization of electronic copies of documents subject to their originals being submitted (for example, if these documents are missing from the database at the initial stage of its creation or due to their foreign origin), and verifies the originals and electronic copies of documents, notarizes them (if they are correct) and sends them back to consumers. Electronic copies of documents are stored in a cloud database of this notary and can be restored in case of their loss (which means repeated notarization of copies of documents will not be required).

In case of electronic notarization of transactions, there is a remote interaction between the notary and consumers by means of a recorded video conference. Artificial Intelligence can be involved in this process in the long view, as it will make it enable detection of fraud (for example, mismatch of identities of parties to the transaction or evidence of coercion to settlement of the transaction). Video record is stored in a cloud database of the notary and can be used in the resolution of further disputes over transaction. Electronic documents that are attached to the transaction, including those that are digitally signed, are stored in a cloud database and can be restored in case of their loss.

As a result, the following advantages of provision of notarial services in electronic form can be obtained:

- Inclusiveness: consumers can obtain certain notarial services whenever convenient to them, without visiting the notary office;
- Reduction in prices: when fully automatic work places are created in notary offices for the committment of certain types of notarial acts exclusively in electronic form (notarization of electronic copies of documents, notarization of authenticity of electronic digital signature on documents), expenses for maintenance (leasing) of premises and for storage of paper copies of documents will be reduced, making it possible to reduce prices for services rendered;
- Data protection: electronic documents and files of transactions are stored in a cloud database of the notary and can be restored whenever necessary, or they can be used in the resolution of disputes over documents and transactions;
- Reduced need for notarial services: notarized electronic documents can be used times out of number; they are less prone to damage or loss.

In addition, it must be noted that in the long view, when it will become possible to use Artificial Intelligence, electronic notarial system can become more secure than conventional notarial system, since it will provide enhanced capabilities in fraud detection - specially trained Artificial Intelligence will most likely detect identity mismatch (for example, using the face recognition technology) or coercion of a party to settlement of the transaction (using the emotion recognition technology), given that a public notary, specializing in legal services, usually has no fraud detection skills.

4 Conclusion

Thus, according to the research findings, a new social institution is being formed in a digital economy – the electronic notarial system, which is a more optimal form of notarial services in terms of their quality, availability and security. The developed institutional model of the electronic notarial system in a digital economy has demonstrated a more complex scheme of its organization compared to the conventional notarial system. At the same time, through the use of modern information and communication technology and the conversion of many notarial acts to digital form, the process of obtainment of notarial services becomes more simple and convenient for consumers.

In summary, it should be pointed out that practical realization of the proposed institutional model of electronic notarial system in a digital economy requires a high level of digital literacy and competence, as well as access to state-of-the-art information and communication technology (for example, for running a video conference over the Internet), digital signing, and, in the long view, using artificial intelligence for fraud detection and prevention). Therefore, in the pursuance of further research, it is recommended to pay attention to the development of measures of state regulation of the digital economy in the interests for the creation of necessary conditions for the institutionalization of the real world application of the electronic notarial system.

References

Blibech, K., Gabillon, A.: A new digital notary system. Lecture Notes in Computer Science (including subseries Lecture Notes in Artificial Intelligence and Lecture Notes in Bioinformatics), vol. 4879, pp. 103–114 (2009)

Bogoviz, A.V., Lobova, S.V., Ragulina, J.V.: Distortions in the theory of costs in the conditions of digital economy. Lect. Notes Netw. Syst. **57**, 1231–1237 (2019)

Bratchel, M.E.: City notaries and the administration of a territory: Lucca, 1430–1501. Pap. Br. Sch. Rome **86**, 183–205 (2018)

Lavecchia, L., Stagnaro, C.: There ain't no such thing as a free deed: the case of Italian notaries. Eur. J. Law Econ. **47**(2), 277–290 (2019)

Llopis Benlloch, J.C.: Notaries and digitalisation of company law. ERA Forum **19**(1), 49–61 (2018)

Maganić, A.: Notaries public in electronic legal transactions | [Javni bilježnik u elektroničkom pravnom prometu]. Zbornik Pravnog Fakulteta u Zagrebu **63**(2), 383–431 (2013)

Maleshin, D.: Chief editor's note on notaries in the BRICS countries. BRICS Law J. **4**(3), 4–5 (2017)

Popkova, E.G.: Economic and Legal Foundations of Modern Russian Society: A New Institutional Theory. Advances in Research on Russian Business and Management. Information Age Publishing, Charlotte (2017)

Popkova, E.G., Sergi, B.S.: Will industry 4.0 and other innovations impact Russia's development? In: Exploring the Future of Russia's Economy and Markets 34–42, pp. 51–68. Emerald Publishing (2019)

Santiago, F.: Implementation of the role of notary through capital market in the ERA of ASEAN economic community. Int. J. Civ. Eng. Technol. **8**(8), 1054–1059 (2017)

Yüce, E., Selçuk, A.A.: Server notaries: a complementary approach to the web PKI trust model. IET Inf. Secur. **12**(5), 34–46 (2018)

Yulia, A., Benny Riyanto, R., Joko Priyono, F.X.: The role of notary public honorary council in the enforcement of the notary code of ethics in Indonesia. IOP Conf. Ser.: Earth Environ. Sci. **175**(1), 012172 (2018)

Technological Prerequisites and Humanitarian Consequences of Ubiquitous Computing and Networking

Anna Guryanova[1(✉)], Elmira Khafiyatullina[2], Marina Petinova[2], Vyacheslav Frolov[1], and Alexander Makhovikov[1]

[1] Samara State University of Economics, Samara, Russia
annaguryanov@yandex.ru, frolov5070@yandex.ru, shentala14@mail.ru
[2] Samara State Technical University, Samara, Russia
dek.fispos2009@yandex.ru, shloss@yandex.ru

Abstract. Ubiquitous computing and networking prerequisites and consequences are analyzed in the article. Special attention is paid to those of them that have the strongest influence on human being. So, technological, ontological and anthropological aspects of modern ubiquitous computing and networking are discussed. The authors accent humanitarian essence of the modern technologies, study their place and role in the lives of modern humans. From the technological point of view 3D printing, Big Data, Cloud Computing and Cognitive Technologies are analyzed. From the ontological point of view the problems of alternative realities' formation and the Internet of Things development are considered. From the anthropological point of view the problems of robotization, Artificial Intelligence, genetic engineering and synthetic biology are the point of the special research interest.

Keywords: Computing · Networking · Human · Technology · Society · Internet of Things · Artificial Intelligence · Virtual reality · Augmented reality · Robotization

JEL Code: Z 10 · Z 13 · Z 19

1 Introduction

Modern society is widely changing under an influence of computing and networking. These areas have got a really ubiquitous character. Many technological and humanitarian transformations are taking place at the moment. The world around us and the human himself, his consciousness, thinking and mentality are radically changing. This happens as a result of the new Internet and computer technologies development. Most of them have become an integral part of our everyday life. Soon they will change greatly the very way of the human existence (Guryanova 2018).

It's even hard today to make correct predictions about the scale and the character of the future transformations. But it's obvious the society will pass significant changes in the nearest future. They'll be caused by various technological innovations having a

E. G. Popkova and B. S. Sergi (Eds.): ISC 2019, LNNS 87, pp. 1040–1047, 2020.
https://doi.org/10.1007/978-3-030-29586-8_119

strong impact on the modern human being. In the present article we'll try to review the most important of them which are relevant to the state of the modern technologies' development.

2 Materials and Methods

Dialectical and information, descriptive and comparative methods, system method and method of classification are used in the present article.

2.1 Dialectical Method

Dialectical method is used to consider various prerequisites and consequences of ubiquitous computing in their cooperation, evolution, interconnection and interdependence.

2.2 Information Method

Information method is used to collect the necessary data proving the findings of the study in the fields of computing and networking, information processing and new cognitive technologies, achievements in genetic engineering and synthetic biology, etc.

2.3 System Method and Method of Classification

System method and method of classification are used to organize various prerequisites and consequences of ubiquitous computing into the three main groups according to their technological, ontological and anthropological orientation.

2.4 Descriptive and Comparative Methods

Descriptive and comparative methods are used to find out the difference between traditional and innovative phenomena of modern life – real, Virtual and Augmented Realities, Artificial Intelligence and human mind, Internet of Things and Industrial Internet of Things, technologies of 2D and 3D printing, etc.

3 Results

Prerequisites and consequences of ubiquitous computing and networking can be traced in different spheres of human activity – scientific and technological progress, social life and economic activity (Guryanova 2019), human psychology, etc. We'll consider those of them which have the strongest influence on the modern society development. The new information technologies are very popular among the modern consumers for the reason of the strongest affecting of their lives. We are talking, first of all, about technological, ontological and anthropological aspects of the modern ubiquitous computing and networking.

3.1 Humanitarian Essence of Modern Technological Innovations

Technological innovations change greatly the way of human life. So, they have a deep humanitarian essence. We are talking about such productive technologies as 3D printing, Big Data, Cloud Computing and Cognitive Technologies.

3.1.1 Humanitarian Potential of 3D Printing

Not very long ago, 3D (three-dimensional) printing technology was only an object of science fiction. But now it's actively used in human life. The 3D printing makes possible to display 3D information and to create 3D physical objects. The main principle of 3D printing is based on the gradual (layer-by-layer) creation of a solid model which is «grown» from a certain material. The 3D printers using is a serious alternative to traditional methods of prototyping and small-scale production.

The advantages of 3D printing over the usual, hand-made methods of models creating are obvious. These are the high speed, the simple character and the low cost. 3D technology eliminates the hand working process and the need of making drawings and calculations on paper. The errors typical for the hand-working become impossible too. The special program allows seeing the model in all its aspects already on the screen. So, the errors can be identified already at the stage of development.

The range of possibilities and applications of 3D printing technology is constantly growing. Today 3D printers can product even organic objects up to donor organs, such as skin tissues and blood vessels, suitable for surgery and transplantation. It's a great humanitarian achievement of nowadays!

3.1.2 Big Data as an Instrument of Effective Information Processing

The number of data and their sources is growing exponentially in the modern world. Therefore, processing technologies have become more and more popular. For example, such an innovative technology preferred by modern consumers as Big Data. Today the flows of data over 100 GB a day are called «Big Data». Big Data is an effective instrument for processing large amounts of information and making individual decisions on this basis. It's very useful for modern humans. The large amounts of data are processed with an aim that the humans can get specific and desired results for their further effective using.

Today, the greater part of the data comes from the three main sources. These are Internet (social networks, forums, blogs, media and other sites), corporate archives of documents, sensor readings and other devices. Amount of the various, fast-changing digital information can't be processed by traditional instruments. Big Data technology allows finding its certain principles and regularities invisible to the humans in normal conditions. It's a real way to optimize all the spheres of human activity – from public administrating to production and telecommunicating. In fact, Big data is a good alternative to the traditional data management systems.

3.1.3 Cloud Computing and Cognitive Technologies in the Human Life

Cloud Computing is a modern information technology conception based on providing ubiquitous and convenient network access on demand to the total amount of configurable computing resources. The last ones can be operatively provided and released with minimal operating costs or calls to the provider. In other words, Cloud Computing is a data processing technology which makes computing resources available to the user on demand as online services.

The other one innovative computing achievement is Cognitive Technology. It serves for information processing in the case of its unstructured and often textual character (the so-called «Unstructured Data»). These technologies don't follow a given algorithm. In contrast, they are able to consider many external factors and to educate themselves using the results of previous calculations and external sources of information (for example, Internet) (Keshelava 2017).

Cloud Computing and Cognitive Technologies will significantly transform the labor costs of the routine office work. They will facilitate much processing of standard documents. Thus, the main part of the wide document flow, as well as any other works connected with information processing, will become extremely automatized. This will make humans free from routine labor operations.

3.2 Ontological Transformations Caused by Ubiquitous Computing

Ubiquitous computing and networking cause great changes in the modern ontology. They are connected with the problem of alternative realities formation, the Internet of Things development, the problem of free access to the Network and digital divides between people and countries.

3.2.1 Virtual and Augmented Realities as a Place of Human Being

Virtual reality (VR) is an important part of the modern human life. It's a special world created by technical resources, which is available to the human through his sensations such as vision, hearing, etc. To make this complex of sensations truer, all actions in the VR take place in a real time. Objects of VR usually act close to the behavior of the similar material objects. So, the user can affect these objects according to the real physical laws (gravity, water characters, pushing of objects, reflection, etc.). However, the users of virtual worlds are often provided more abilities than it's possible in the real life (for example, abilities of flying, creating any objects, etc.). It's usually done for entertainment purposes.

Now VR includes on-line computer games, social networks, forums and many other components. Each of them is a structural point of the virtual environment and at the same time a special link connecting the virtual world and the real one. These two worlds are not only interconnected but also interdependent, as well as a real human and his virtual image in a social network. The process of the real and virtual worlds' unification has already begun and it's impossible to stop it (Keshelava 2017).

However, virtual reality (VR) and augmented reality (AR) are not identical. VR constructs a new artificial world while AR adds some artificial elements to the real world's perception. Today AR is impacting strongly the human life. Besides the picture created by computer doesn't replace the eyes vision, but overlays the objects of the real world. This ability is in demand far beyond the gaming sphere. AR technologies can be widely used in medicine, engineering, construction, design, transport, etc.

3.2.2 Free Access to the Network as a Condition of Human Life

Today there are still many places on the planet where telephone and Internet are absent. This causes many problems for the modern humanity. Today the Internet is a main infrastructure of social communication at local and global levels. According to the World Bank data, the number of Internet users has grown more than three times from 2005 to 2015: in 2005 it was 1 billion and in 2015 – already 3.2 billion of people (World development 2016). In 2017 the every second human on the planet has got an access to the Internet. And in the next 20 years up to 50% operations in the world will become automatic, according to the forecasts of McKinsey Global Institute (Digital Russia 2017).

As we have already mentioned, this process is not ubiquitous. There are digital divides in access and using of the digital technologies. Digital revolution has a small impact on the lives of the great part of the modern people. Only about 15% of them can pay for the Internet access. Mobile phones are the main resources of Internet accessing in developing countries. But only about 80% of people have mobile phones nowadays. Almost 2 billion of people don't have mobile phones, and about 60% of the world's population doesn't have access to the Internet at all (World development 2016).

Experts all over the world try to solve the problem of the global network coverage. But all the previous projects (for example, using satellite phones and devices for receiving Internet signals from satellites) were very expensive for their price. The most progressive modern projects in this field are SpaceX by Elon Musk and OneWeb by Richard Branson. They both want to form a large-scale satellite constellation in a low orbit. This will provide a free access to the Network all over the world.

3.2.3 Internet of Things, Its Place and Role in the Human Life

Internet of Things is a complex conception uniting a wide variety of technologies. It implies an Internet connection and equipment of all devices («things») by sensors. This allows making control, monitoring and management of all the processes they participate in (including automatic mode). The Internet of Things is rapidly developing not only in our everyday life but in the production sphere too. So, two main areas – the Internet of Things (IoT) and the Industrial Internet of Things (IIoT) are co-existing today. In fact, these technologies are similar to each other, but there is also a difference between them. The main task of the IoT is to collect all kinds of data that'll be used in creating models and forecasts. The purpose of the IIoT is to automate production by distance management of resources and capacities with a help of the sensors' data.

In many countries technologies for management of production resources, including interests of their virtual using, are the part of the state programs of the digital economy development. These are, for example, Industrie 4.0 in Germany, Advanced Manufacturing Technology in the USA, Strategic Conception of Production Development in China focusing on quality, innovations and advanced technologies' implementation, Innovate UK in Great Britain, National Digital Economy in Australia, etc.

3.3 Anthropological Consequences of Ubiquitous Computing

Anthropological consequences of ubiquitous computing are really various nowadays. These are the problems of confrontation between robots and humans, Artificial

Intelligence and human mind. Consequences of genetic engineering and synthetic biology and their impact on human life are very actual too.

3.3.1 Social Consequences of Ubiquitous Robotization

Ubiquitous robotization is a reality of nowadays. The newest advances in thc field of robotics are widely used in our everyday life as well as in industrial, medical, military and other spheres. This greatly changes the way of human life.

Modern robots can take care of patients, help us in infrastructure and agriculture and realize many other important services. They take part in automatizing business and education processes (Pecherskaya 2016), in modernizing of the service sector. Over the next 30 years, technologies will probably overcome biological limits of human potential. Wearable devices connected with the Internet will give us a possibility to transmit context-sensitive information related directly to our emotions. Prosthetics united with human brains will return mobility to people with disabilities. Sensors and permanent implants will give us better hearing and vision, deep dive into the virtual and augmented realities (Program 2017). All this seem really good.

But on the other hand, robots are beginning to perform an increasing number of functions and roles traditionally realized by the humans. Thus, ubiquitous robotization can cause a growth of unemployment. It will lead to affecting changes in the labor market (Guryanova 2017), reduction and disappearance of a number of professions in a short time period. In simple terms, humans will be replaced by robots. As a result a greater part of active, well-educated, successfully working people, accustomed to the high quality of life, will be unclaimed. However, experts are sure: if the process of robotization will develop at such a speed it guarantees a personnel deficit in other qualifications. So, everyone who is ready for changes will have enough time to prepare (Keshelava 2017).

3.3.2 Artificial Intelligence Contra the Human Mind

Not very long ago artificial intelligence (AI) was only the theme of science fiction. Today AI technology enters deeply into the human life. And it's an important topic of discussions between the modern scientists. Some of them are sure, if computers become smarter than we are, they will enslave us. For example, Nick Bostrom qualifies AI as one of the main global risks of the modern humanity (Bostrom 2014). But there is also another point of view. Its followers are sure AI is a real tool to reach a high quality of the human life. For example, John Searle thinks computers will never get self-consciousness at all. We can create consciousness similar to the human ones only if we duplicate the real physical and chemical processes taking place in the human brains. In the case of computer intelligence it's impossible (Searle 1992).

In fact, the human brains differ much from the computer. One of the main differences between them is an ability of the humans to learning and improvisation. The humans are able not only to act in accordance with a certain algorithm, but to develop the algorithm itself, to go beyond its boundaries at any time, using, for example, a method of intuition. Strictly speaking, AI technologies have already passed this part of their evolution development by 2017. Today the field of machine learning is rapidly developing. The deep neural networks can assimilate such things that have been only a human prerogative for a long period of time. For example, modern computers can already create artworks…

3.3.3 Genetic Engineering and Synthetic Biology

Genetic engineering is not an invention of our times. In fact, humanity practiced purposeful correction of the genetic code of plants and animals throughout its history. But in the XX century a real breakthrough in the area of genetic engineering has happened. From that very time the scientists have learned to «edit» DNA itself – to «cut» its certain fragments and to «put» them into the right place. Now they are able to correct genetic errors that cause different diseases. They can purposefully create new species of plants and animals and revive the extinct ones. They can also destroy dangerous viruses and bacteria or change their properties in such a way that they don't pose any threat for the humans, etc. And that isn't all. Over the next 30 years, synthetic biology will make engineering organisms that can detect toxins, transform industrial waste into biofuel, and create medicaments through symbiosis with the humans (Program 2017).

But at the same time, synthetic biology poses serious risks because of the opportunity to use its achievements as artificial biological weapons. Invasive synthetic organisms able to destroy natural ecosystems, including the human existence, are also very dangerous. It's obvious that such technologies require a highly responsible application, but their potential is almost limitless. This makes actual the problems of social and ethical character that must be necessarily solved in the modern world. The survival of the humanity itself depends on the success of their solution.

4 Discussions

The new information technologies are very popular among the modern humans for the reason of their strongest influence on their lives. Even technical innovations have a deep humanitarian essence. For example, 3D printers can product donor organs suitable for transplantation. Big Data help humans to process large amounts of information. Cloud Computing and Cognitive Technologies make them free from routine labor operations. These are the great humanitarian achievements of nowadays.

Virtual and Augmented Realities are even an important part of the modern human life. Their influence goes far beyond traditional computer gaming sphere. These three realities seem interconnected and interdependent. So, modern humans spend their life time in both three worlds – real, virtual and augmented. Because of this ontology of human existence changes much.

Ubiquitous robotization is also a reality of nowadays. Modern robots realize many important services. But at the same time they are performing many functions traditionally realized by the humans. Thus, ubiquitous robotization causes a growth of mass unemployment. Are all the people ready for such radical changes? And can they adapt to them in such a short period of time? The authors of the present article have great doubts about these questions. So, they need further discussions in the terms of ethics and humanism.

5 Conclusions

The world around us, the society and the way of human life are radically changing under an influence of ubiquitous computing and networking. Besides, technological innovations have a deep humanitarian essence. For example, such productive technologies as 3D printing, Big Data, cloud computing and cognitive technologies make humans free from routine office operations, help them to process large amounts of information.

Ubiquitous computing and networking cause great changes in the modern ontology. They are connected with dialectic of real, virtual and augmented realities, the Internet of things forming, the free access to the Internet and digital divides between people and countries existence. Anthropological consequences of ubiquitous computing are also various. These are the problems of replacing humans by robots, human mind by artificial intelligence, genetic engineering and synthetic biology development and many others.

It's obvious, the society and the humans will pass significant changes in the nearest future. And the people must be ready for them, for their results and consequences. Today it's time to turn them on the humanitarian way.

References

Bostrom, N.: Superintelligence: Paths, Dangers, Strategies. Oxford University Press, Oxford (2014)

Guryanova, A., Astafeva, N., Filatova, N., Khafiyatullina, E., Guryanov, N.: Philosophical problems of information and communication technology in the process of modern socio-economic development. Adv. Intell. Syst. Comput. **726**, 1033–1040 (2019). https://doi.org/10.1007/978-3-319-90835-9_115

Guryanova, A., Guryanov, N., Frolov, V., Tokmakov, M., Belozerova, O.: Main categories of economics as an object of philosophical analysis. In: Contributions of Economics, Russia and the European Union: Development and Perspectives, pp. 221–228. Springer, Cham (2017). https://doi.org/10.1007/978-3-319-55257-6_30

Guryanova, A., Khafiyatullina, E., Kolibanov, A., Makhovikov, A., Frolov, V.: Philosophical view on human existence in the world of technic and information. Adv. Intell. Syst. Comput. **622**, 97–104 (2018). https://doi.org/10.1007/978-3-319-75383-6_13

Digital Russia: A New Reality. McKinsey & Company, New York (2017). http://www.tadviser.ru/images/c/c2/Digital-Russia-report.pdf

Keshelava, A.V. (ed.): Introduction to Digital Economy. Vniigeosystems, Moscow (2017)

Pecherskaya, E., Averina, L., Kochetckova, N., Chupina, V., Akimova, O.: Methodology of project managers' competency formation in CPE. IJME-Math. Educ. **11**(8), 3066–3075 (2016)

Program of digital economy development in the Russian Federation until 2035 (2017) http://innclub.info/wp-content/uploads/2017/05/strategy.pdf

Searle, J.R.: The Rediscovery of the Mind. MIT Press, Cambridge (1992)

World Development Report 2016: Digital Dividends. Review. World Bank, Washington, D.C. (2016). https://doi.org/10.1596/978-1-4648-0671-1.a. https://openknowledge.worldbank.org/bitstream/handle/10986/23347/21067

Conclusions

It could be concluded that the digital economy is a flexible socio-economic system that might acquire various features, depending on the context and management. The contradiction of the digital economy could and should be overcome with the help of the presented recommendations. However, as the results of the research showed, management of the digital economy should be conducted in various directions – economic, social, and legal, and at various levels – national, regional, and corporate.

The advantage of this volume is that, unlike most publications on the topic of the digital economy in which emphasis is made on breakthrough digital technologies and their adaptation to the current needs of economic activities, here large attention is paid to less elaborated but important social and legal issues of formation and development of the digital economy.

Due to this, the offered strategic approach to state and corporate management of the digital economy is systemic – i.e., it envisages measures of regulation and is aimed at creating advantages not only for entrepreneurship and economic systems of different levels but also for norms of law, social institutes, and each human.

However, despite the significant reduction of uncertainty as to perspectives of development and management of the digital economy for overcoming its contradiction and providing its balance, the obtained results actualized new scientific and practical problems of modern times, which deserve attention of academic and expert societies.

One of the problems is formation of the global competitive environment in the digital economy and unification of all economic systems and economic subjects. A logical substantiation of this problem is treating the digital economy as a mechanism of leveling disproportions in the modern global economy. A serious barrier on the path of solving this problem is imperfection of competition and tendency for monopolization of the markets of the digital economy, which is observed in the modern global economic system. Recently, the digital economy has been increasing imbalance of powers in the global economy and deepening the disproportion of developed and developing countries.

E. G. Popkova and B. S. Sergi (Eds.): ISC 2019, LNNS 87, pp. 1049–1050, 2020.
https://doi.org/10.1007/978-3-030-29586-8

Another problem is managing socialization of new economic agents, which are institutionalized in the conditions of the digital economy: digital machines (e.g., robots that are equipped with AI) and social adaptation of employees, entrepreneurs, and consumers to interaction with new agents. Perspective solutions to the above problems should be developed in further scientific works on the topic of the digital economy.

Author Index

E. G. Popkova and B. S. Sergi (Eds.): ISC 2019, LNNS 87, pp. 1051–1055, 2020.
https://doi.org/10.1007/978-3-030-29586-8

Printed in the United States
By Bookmasters